The Digital Signal Processing Handbook

SECOND EDITION

Digital Signal Processing Fundamentals

EDITOR-IN-CHIEF
Vijay K. Madisetti

CRC Press
Taylor & Francis Group
Boca Raton London New York

CRC Press is an imprint of the
Taylor & Francis Group, an **informa** business

The Electrical Engineering Handbook Series

Series Editor
Richard C. Dorf
University of California, Davis

Titles Included in the Series

The Handbook of Ad Hoc Wireless Networks, Mohammad Ilyas
The Avionics Handbook, Second Edition, Cary R. Spitzer
The Biomedical Engineering Handbook, Third Edition, Joseph D. Bronzino
The Circuits and Filters Handbook, Second Edition, Wai-Kai Chen
The Communications Handbook, Second Edition, Jerry Gibson
The Computer Engineering Handbook, Vojin G. Oklobdzija
The Control Handbook, William S. Levine
The CRC Handbook of Engineering Tables, Richard C. Dorf
The Digital Avionics Handbook, Second Edition Cary R. Spitzer
The Digital Signal Processing Handbook, Second Edition, Vijay K. Madisetti
The Electrical Engineering Handbook, Second Edition, Richard C. Dorf
The Electric Power Engineering Handbook, Second Edition, Leonard L. Grigsby
The Electronics Handbook, Second Edition, Jerry C. Whitaker
The Engineering Handbook, Third Edition, Richard C. Dorf
The Handbook of Formulas and Tables for Signal Processing, Alexander D. Poularikas
The Handbook of Nanoscience, Engineering, and Technology, Second Edition William A. Goddard, III, Donald W. Brenner, Sergey E. Lyshevski, and Gerald J. Iafrate
The Handbook of Optical Communication Networks, Mohammad Ilyas and Hussein T. Mouftah
The Industrial Electronics Handbook, J. David Irwin
The Measurement, Instrumentation, and Sensors Handbook, John G. Webster
The Mechanical Systems Design Handbook, Osita D.I. Nwokah and Yidirim Hurmuzlu
The Mechatronics Handbook, Second Edition, Robert H. Bishop
The Mobile Communications Handbook, Second Edition, Jerry D. Gibson
The Ocean Engineering Handbook, Ferial El-Hawary
The RF and Microwave Handbook, Second Edition, Mike Golio
The Technology Management Handbook, Richard C. Dorf
The Transforms and Applications Handbook, Second Edition, Alexander D. Poularikas
The VLSI Handbook, Second Edition, Wai-Kai Chen

The Digital Signal Processing Handbook, Second Edition

Digital Signal Processing Fundamentals
Video, Speech, and Audio Signal Processing and Associated Standards
Wireless, Networking, Radar, Sensor Array Processing, and Nonlinear Signal Processing

CRC Press
Taylor & Francis Group
6000 Broken Sound Parkway NW, Suite 300
Boca Raton, FL 33487-2742

First issued in paperback 2017

ISBN-13: 978-1-4200-4606-9 (hbk)
ISBN-13: 978-1-138-11374-9 (pbk)

Library of Congress Cataloging-in-Publication Data

Digital signal processing fundamentals / editor, Vijay K. Madisetti.
p. cm.
Includes bibliographical references and index.
ISBN 978-1-4200-4606-9 (alk. paper)
1. Signal processing--Digital techniques. I. Madisetti, V. (Vijay)

TK5102.5.D4485 2009
621.382'2--dc22 2009022327

Visit the Taylor & Francis Web site at
http://www.taylorandfrancis.com

and the CRC Press Web site at
http://www.crcpress.com

Contents

Preface

Digital signal processing (DSP) is concerned with the theoretical and practical aspects of representing information-bearing signals in a digital form and with using computers, special-purpose hardware and software, or similar platforms to extract information, process it, or transform it in useful ways. Areas where DSP has made a significant impact include telecommunications, wireless and mobile communications, multimedia applications, user interfaces, medical technology, digital entertainment, radar and sonar, seismic signal processing, and remote sensing, to name just a few.

Given the widespread use of DSP, a need developed for an authoritative reference, written by the top experts in the world, that would provide information on both theoretical and practical aspects in a manner that was suitable for a broad audience—ranging from professionals in electrical engineering, computer science, and related engineering and scientific professions to managers involved in technical marketing, and to graduate students and scholars in the field. Given the abundance of basic and introductory texts on DSP, it was important to focus on topics that were useful to engineers and scholars without overemphasizing those topics that were already widely accessible. In short, the DSP handbook was created to be relevant to the needs of the engineering community.

A task of this magnitude could only be possible through the cooperation of some of the foremost DSP researchers and practitioners. That collaboration, over 10 years ago, produced the first edition of the successful DSP handbook that contained a comprehensive range of DSP topics presented with a clarity of vision and a depth of coverage to inform, educate, and guide the reader. Indeed, many of the chapters, written by leaders in their field, have guided readers through a unique vision and perception garnered by the authors through years of experience.

The second edition of the DSP handbook consists of volumes on *Digital Signal Processing Fundamentals*; *Video, Speech, and Audio Signal Processing and Associated Standards*; and *Wireless, Networking, Radar, Sensor Array Processing, and Nonlinear Signal Processing* to ensure that each part is dealt with in adequate detail, and that each part is then able to develop its own individual identity and role in terms of its educational mission and audience. I expect each part to be frequently updated with chapters that reflect the changes and new developments in the technology and in the field. The distribution model for the DSP handbook also reflects the increasing need by professionals to access content in electronic form anywhere and at anytime.

Digital Signal Processing Fundamentals, as the name implies, provides a comprehensive coverage of the basic foundations of DSP and includes the following parts: Signals and Systems; Signal Representation and Quantization; Fast Algorithms and Structures; Digital Filtering; Statistical Signal Processing; Adaptive Filtering; Inverse Problems and Signal Reconstruction; and Time–Frequency and Multirate Signal Processing.

I look forward to suggestions on how this handbook can be improved to serve you better.

MATLAB® is a registered trademark of The MathWorks, Inc. For product information, please contact:

The MathWorks, Inc.
3 Apple Hill Drive
Natick, MA 01760-2098 USA
Tel: 508 647 7000
Fax: 508-647-7001
E-mail: info@mathworks.com
Web: www.mathworks.com

Editor

Vijay K. Madisetti is a professor in the School of Electrical and Computer Engineering at the Georgia Institute of Technology in Atlanta. He teaches graduate and undergraduate courses in digital signal processing and computer engineering, and leads a strong research program in digital signal processing, telecommunications, and computer engineering.

Dr. Madisetti received his BTech (Hons) in electronics and electrical communications engineering in 1984 from the Indian Institute of Technology, Kharagpur, India, and his PhD in electrical engineering and computer sciences in 1989 from the University of California at Berkeley.

Hc has authored or edited several books in the areas of digital signal processing, computer engineering, and software systems, and has served extensively as a consultant to industry and the government. He is a fellow of the IEEE and received the 2006 Frederick Emmons Terman Medal from the American Society of Engineering Education for his contributions to electrical engineering.

Contributors

Joseph Arrowood
IvySys Technologies, LLC
Arlington, Virginia

Bruce W. Bomar
Department of Electrical and Computer Engineering
University of Tennessee Space Institute
Tullahoma, Tennessee

C. Sidney Burrus
Department of Electrical and Computer Engineering
Rice University
Houston, Texas

Zhi Ding
Department of Electrical and Computer Engineering
University of California
Davis, California

Petar M. Djurić
Department of Electrical and Computer Engineering
Stony Brook University
Stony Brook, New York

John F. Doherty
Department of Electrical Engineering
The Pennsylvania State University
University Park, Pennsylvania

Scott C. Douglas
Department of Electrical Engineering
Southern Methodist University
Dallas, Texas

Pierre Duhamel
CNRS
Gif sur Yvette, France

Kevin R. Farrell
T-NETIX, Inc.
Englewood, Colorado

Ephraim Feig
Innovations-to-Market
San Diego, California

Georgios B. Giannakis
Department of Electrical and Computer Engineering
University of Minnesota
Minneapolis, Minnesota

Cormac Herley
Microsoft Research
Redmond, Washington

Gabor T. Herman
Department of Computer Science
City University of New York
New York, New York

Alfred Hero
Department of Electrical Engineering and Computer Sciences
University of Michigan
Ann Arbor, Michigan

W. Kenneth Jenkins
Department of Electrical Engineering
The Pennsylvania State University
University Park, Pennsylvania

Thomas Kailath
Department of Electrical Engineering
Stanford University
Stanford, California

Ton Kalker
HP Labs
Palo Alto, California

Lina J. Karam
Department of Electrical, Computer and Energy Engineering
Arizona State University
Tempe, Arizona

Aggelos K. Katsaggelos
Department of Electrical Engineering and Computer Science
Northwestern University
Evanston, Illinois

Steven M. Kay
Department of Electrical, Computer, and Biomedical Engineering
University of Rhode Island
Kingston, Rhode Island

Stephen Kosonocky
Advanced Micro Devices
Fort Collins, Colorado

Jelena Kovačević
Lucent Technologies
Bell Laboratories
Murray Hill, New Jersey

Vic Larson
Science Applications International Corporation
Arlington, Virginia

B.P. Lathi
Department of Electrical Engineering
California State University
Sacramento, California

Vijay K. Madisetti
School of Electrical and Computer Engineering
Georgia Institute of Technology
Atlanta, Georgia

Richard J. Mammone
Department of Electrical and Computer Engineering
Rutgers University
Piscataway, New Jersey

Daniel F. Marshall
Raytheon Company
Lexington, Massachusetts

James H. McClellan
Department of Electrical and Computer Engineering
Georgia Institute of Technology
Atlanta, Georgia

Jerry M. Mendel
Department of Electrical Engineering
University of Southern California
Los Angeles, California

Kambiz Nayebi
Beena Vision Systems Inc.
Roswell, Georgia

Christine Podilchuk
CAIP
Rutgers University
Piscataway, New Jersey

K. Venkatesh Prasad
Ford Motor Company
Detroit, Michigan

Ricardo L. de Queiroz
Engenharia Eletrica
Universidade de Brasilia
Brasília, Brazil

C. Radhakrishnan
Department of Electrical Engineering
The Pennsylvania State University
University Park, Pennsylvania

Ravi P. Ramachandran
Department of Electrical and Computer Engineering
Rowan University
Glassboro, New Jersey

Tami Randolph
Department of Electrical and Computer Engineering
Georgia Institute of Technology
Atlanta, Georgia

Markus Rupp
Mobile Communications Department
Technical University of Vienna
Vienna, Austria

Ali H. Sayed
Department of Electrical Engineering
University of California at Los Angeles
Los Angeles, California

Ivan W. Selesnick
Department of Electrical and Computer Engineering
Polytechnic University
Brooklyn, New York

Mark J.T. Smith
Department of Electrical and Computer Engineering
Purdue University
West Lafayette, Indiana

Iraj Sodagar
PacketVideo
San Diego, California

Clay Stewart
Science Applications International Corporation
Arlington, Virginia

A.C. Surendran
Lucent Technologies
Bell Laboratories
Murray Hill, New Jersey

Charles W. Therrien
Naval Postgraduate School
Monterey, California

Jitendra K. Tugnait
Department of Electrical and Computer Engineering
Auburn University
Auburn, Alabama

Martin Vetterli
École Polytechnique
Lausanne, Switzerland

Douglas B. Williams
Department of Electrical and Computer Engineering
Georgia Institute of Technology
Atlanta, Georgia

Geoffrey A. Williamson
Department of Electrical and Computer Engineering
Illinois Institute of Technology
Chicago, Illinois

Peter Xiao
NeoParadigm Labs. Inc.
San Jose, California

Andrew E. Yagle
Department of Electrical Engineering and Computer Science
University of Michigan
Ann Arbor, Michigan

Jun Zhang
Department of Electrical Engineering and Computer Science
University of Milwaukee
Milwaukee, Wisconsin

Xiaoyu Zhang
CAIP
Rutgers University
Piscataway, New Jersey

I

Signals and Systems

Vijay K. Madisetti
Georgia Institute of Technology

Douglas B. Williams
Georgia Institute of Technology

THE STUDY OF "SIGNALS AND SYSTEMS" has formed a cornerstone for the development of digital signal processing and is crucial for all of the topics discussed in this book. While the reader is assumed to be familiar with the basics of signals and systems, a small portion is reviewed in this section with an emphasis on the transition from continuous time to discrete time. The reader wishing more background may find in it any of the many fine textbooks in this area, for example [1–6].

In Chapter 1, many important Fourier transform concepts in continuous and discrete time are presented. The discrete Fourier transform, which forms the backbone of modern digital signal processing as its most common signal analysis tool, is also described, together with an introduction to the fast Fourier transform algorithms.

In Chapter 2, the author, B.P. Lathi, presents a detailed tutorial of differential and difference equations and their solutions. Because these equations are the most common structures for both implementing and modeling systems, this background is necessary for the understanding of many of the later topics in this book. Of particular interest are a number of solved examples that illustrate the solutions to these formulations.

While most software based on workstations and PCs is executed in single or double precision arithmetic, practical realizations for some high throughput digital signal processing applications must be implemented in fixed point arithmetic. These low cost implementations are still of interest to a wide community in the consumer electronics arena. Chapter 3 describes basic number representations, fixed and floating point errors, roundoff noise, and practical considerations for realizations of digital signal processing applications, with a special emphasis on filtering.

References

1. Jackson, L.B., *Signals, Systems, and Transforms*, Addison-Wesley, Reading, MA, 1991.
2. Kamen, E.W. and Heck, B.S., *Fundamentals of Signals and Systems Using MATLAB*, Prentice-Hall, Upper Saddle River, NJ, 1997.
3. Oppenheim, A.V. and Willsky, A.S., with Nawab, S.H., *Signals and Systems*, 2nd ed., Prentice-Hall, Upper Saddle River, NJ, 1997.
4. Strum, R.D. and Kirk, D.E., *Contemporary Linear Systems Using MATLAB*, PWS Publishing, Boston, MA, 1994.
5. Proakis, J.G. and Manolakis, D.G., *Introduction to Digital Signal Processing*, Macmillan, New York; Collier Macmillan, London, UK, 1988.
6. Oppenheim, A.V. and Schafer, R.W., *Discrete Time Signal Processing*, Prentice-Hall, Englewood Cliffs, NJ, 1989.

1

Fourier Methods for Signal Analysis and Processing

W. Kenneth Jenkins
The Pennsylvania State University

1.1 Introduction

The Fourier transform is a mathematical tool that is used to expand signals into a spectrum of sinusoidal components to facilitate signal representation and the analysis of system performance. In certain applications the Fourier transform is used for spectral analysis, and while in others it is used for spectrum shaping that adjusts the relative contributions of different frequency components in the filtered result. In certain applications the Fourier transform is used for its ability to decompose the input signal into uncorrelated components, so that signal processing can be more effectively implemented on the individual spectral components. Different forms of the Fourier transform, such as the continuous-time (CT) Fourier series, the CT Fourier transform, the discrete-time Fourier transform (DTFT), the discrete

Fourier transform (DFT), and the fast Fourier transform (FFT) are applicable in different circumstances. One goal of this chapter is to clearly define the various Fourier transforms, to discuss their properties, and to illustrate how each form is related to the others in the context of a family tree of Fourier signal processing methods.

Classical Fourier methods such as the Fourier series and the Fourier integral are used for CT signals and systems, i.e., systems in which the signals are defined at all values of t on the continuum $-\infty < t < \infty$. A more recently developed set of discrete Fourier methods, including the DTFT and the DFT, are extensions of basic Fourier concepts for discrete-time (DT) signals and systems. A DT signal is defined only for integer values of n in the range $-\infty < n < \infty$. The class of DT Fourier methods is particularly useful as a basis for digital signal processing (DSP) because it extends the theory of classical Fourier analysis to DT signals and leads to many effective algorithms that can be directly implemented on general computers or special purpose DSP devices.

1.2 Classical Fourier Transform for Continuous-Time Signals

A CT signal $s(t)$ and its Fourier transform $S(j\omega)$ form a transform pair that are related by Equations 1.1a and b for any $s(t)$ for which the integral (Equation 1.1a) converges:

$$S(j\omega) = \int_{-\infty}^{\infty} s(t)e^{-j\omega t}dt \tag{1.1a}$$

$$s(t) = \frac{1}{2\Pi}\int_{-\infty}^{\infty} S(j\omega)e^{j\omega t}d\omega. \tag{1.1b}$$

In most literature Equation 1.1a is simply called the Fourier transform, whereas Equation 1.1b is called the Fourier integral. The relationship $S(j\omega) = F\{s(t)\}$ denotes the Fourier transformation of $s(t)$, where $F\{\cdot\}$ is a symbolic notation for the integral operator, and where ω is the continuous frequency variable expressed in rad/s. A transform pair $s(t) \leftrightarrow S(j\omega)$ represents a one-to-one invertible mapping as long as $s(t)$ satisfies conditions which guarantee that the Fourier integral converges.

In the following discussion the symbol $\delta(t)$ is used to denote a CT impulse function that is defined to be zero for all $t \neq 0$, undefined for $t = 0$, and has unit area when integrated over the range $-\infty < t < \infty$. From Equation 1.1a it is found that $F\{\delta(t - t_0)\} = e^{-j\omega t_0}$ due to the well known sifting property of $\delta(t)$. Similarly, from Equation 1.1b we find that $F^{-1}\{2\pi\delta(\omega - \omega_0)\} = e^{j\omega_0 t}$, so that $\delta(t - t_0) \leftrightarrow e^{-j\omega t_0}$ and $e^{j\omega_0}t \leftrightarrow 2\pi\delta(\omega - \omega_0)$ are Fourier transform pairs. Using these relationships it is easy to establish the Fourier transforms of $\cos(\omega_0 t)$ and $\sin(\omega_0 t)$, as well as many other useful waveforms, many of which are listed in Table 1.1.

The CT Fourier transform is useful in the analysis and design of CT systems, i.e., systems that process CT signals. Fourier analysis is particularly applicable to the design of CT filters which are characterized by Fourier magnitude and phase spectra, i.e., by $|H(j\omega)|$ and $\arg H(j\omega)$, where $H(j\omega)$ is commonly called the frequency response of the filter.

1.2.1 Properties of the Continuous-Time Fourier Transform

The CT Fourier transform has many properties that make it useful for the analysis and design of linear CT systems. Some of the more useful properties are summarized in this section, while a more complete list of the CT Fourier transform properties is given in Table 1.2. Proofs of these properties are found in Oppenheim et al. (1983) and Bracewell (1986). Note that $F\{\cdot\}$ denotes the Fourier transform operation, $F^{-1}\{\cdot\}$ denotes the inverse Fourier transform operation, and "$*$" denotes the linear convolution operation defined as

TABLE 1.1 CT Fourier Transform Pairs

Single	Fourier Transform	Fourier Series Coefficients (If Periodic)
$\sum_{k=-\infty}^{+\infty} a_k e^{\alpha\omega\delta}$	$2\pi \sum_{k=-\infty}^{+\infty} a_k \delta(\omega_k - \omega_0)$	a_k
$e^{j\omega_0 t}$	$2\pi\delta(\omega - \omega_0)$	$a_1 = 1$ $a_k = 0$, otherwise
$\cos \omega_0 t$	$\pi[\delta(\omega - \omega_0) + \delta(\omega + \omega_0)]$	$a_1 = a_{-1} = 1/2$ $a_k = 0$, otherwise
$\sin \omega_0 t$	$\frac{\pi}{j}[\delta(\omega - \omega_0) + \delta(\omega + \omega_0)]$	$a_1 = -a_{-1} = 1/2j$ $a_k = 0$, otherwise
$x(t) = 1$	$2\pi\delta(\omega)$	$a_0 = 1$, $a_k = 0$, $k \neq 0$ (has this Fourier series representation for any choice of $T_0 > 0$)
Periodic square wave $x(t) = \begin{cases} 1, \lvert t\rvert < T_1 \\ 0, T_1 < \lvert t\rvert \leq \frac{T_0}{2} \end{cases}$ and $x(t + T_0) = x(t)$	$\sum_{k=-\infty}^{+\infty} \frac{2 \sin k\omega_0 T_1}{k} \delta(\omega_k \omega_0)$	$\frac{\omega_0 T_1}{\pi} \sin c\left(\frac{k\omega_0 T_1}{\pi}\right) = \frac{\sin k\omega_0 T_1}{k\pi}$
$\sum_{n=-\infty}^{+\infty} \delta(t - nT)$	$\frac{2\pi}{T} \sum_{k=-\infty}^{+\infty} k = -\infty\delta\left(\omega - \frac{2\pi k}{T}\right)$	$a_k = \frac{1}{T}$ for all k
$x(t) = \begin{cases} 1, \lvert t\rvert < T_1 \\ 0, \lvert t\rvert > T_1 \end{cases}$	$2T_1 \sin c\left(\frac{\omega T_1}{\pi}\right) = \frac{2 \sin \omega T_1}{\omega}$	—
$\frac{W}{\pi} \sin c\left(\frac{Wt}{\pi}\right) = \frac{\sin Wt}{\pi t}$	$X(\omega) = \begin{cases} 1, \lvert\omega\rvert < W \\ 0, \lvert\omega\rvert > W \end{cases}$	—
$\delta(t)$	1	—
$u(t)$	$\frac{1}{j\omega} + \pi\delta(\omega)$	—
$\delta(t - t_0)$	$e^{-j\omega r_0}$	—
$e^{-ar}u(t), \mathrm{Re}\{a\} > 0$	$\frac{1}{a + j\omega}$	—
$te^{-at}u(t), \mathrm{Re}\{a\} > 0$	$\frac{1}{(a + j\omega)^2}$	—
$\frac{t^{n-1}}{(n-1)} e^{-at} u(t)$, $\mathrm{Re}\{a\} > 0$	$\frac{1}{(a + j\omega)^n}$	—

Source: Oppenheim, A.V. et al., *Signals and Systems*, Prentice-Hall, Englewood Cliffs, NJ, 1983. With permission.

$$f_1(t) * f_2(t) = \int_{-\infty}^{\infty} f_1(t) f_2(t - \tau)\mathrm{d}\tau.$$

1. Linearity (a and b are complex constants)	$F\{af_1(t) + bf_2(t)\} = aF\{f_1(t)\} + bF\{f_2(t)\}$
2. Time-shifting	$F\{f(t - t_0)\} = e^{-j\omega t_0} F\{f(t)\}$
3. Frequency-shifting	$e^{j\omega_0 t} F^{-1}\{F\{j(\omega - \omega_0)\}$
4. Time-domain convolution	$F\{f_1(t) * f_2(t)\} = F\{f_1(t)\} \cdot F\{f_2(t)\}$
5. Frequency-domain convolution	$F\{f_1(t) \cdot f_2(t)\} = \frac{1}{2\Pi} F\{f_1(t)\} * F\{f_2(t)\}$
6. Time-differentiation	$-j\omega F(j\omega) = F\{\mathrm{d}[f(t)]/\mathrm{d}t\}$
7. Time-integration	$F\left\{\int_{-\infty}^{t} f(\tau)\mathrm{d}\tau\right\} = \frac{1}{j\omega} F(j\omega) + \pi F(0)\delta(\omega)$

TABLE 1.2 Properties of the CT Fourier Transform

Name	If $\mathcal{F} f(t) = F(j\omega)$, then:
Definition	$F(j\omega) = \int_{-\infty}^{\infty} f(t) e^{-j\omega t} dt$
	$f(t) = \frac{1}{2\pi} \int_{-\infty}^{\infty} F(j\omega) e^{j\omega t} d\omega$
Superposition	$\mathcal{F}[af_1(t) + bf_2(t)] = aF_1(j\omega) + bF_2(j\omega)$
Simplification if:	
(a) $f(t)$ is even	$F(j\omega) = 2\int_0^{\infty} f(t) \cos \omega t \, dt$
(b) $f(t)$ is odd	$F(j\omega) = 2j\int_0^{\infty} f(t) \sin \omega t \, dt$
Negative t	$\mathcal{F} f(-t) = F^* (j\omega)$
Scaling:	
(a) Time	$Ff(at) = \frac{1}{\lvert a \rvert} F\left(\frac{j\omega}{a}\right)$
(b) Magnitude	$\mathcal{F} af(t) = aF(j\omega)$
Differentiation	$\mathcal{F}\left[\frac{d^n}{dt^n} f(t)\right] = (j\omega)^n F(j\omega)$
Integration	$\mathcal{F}\left[\int_{-\infty}^{t} f(x) dx\right] = \frac{1}{j\omega} F(j\omega) + \pi F(0)\delta(\omega)$
Time shifting	$\mathcal{F} f(t-a) = F(j\omega) e^{-j\omega a}$
Modulation	$\mathcal{F} f(t) e^{j\omega_0 t} = F[j(\omega - \omega_0)]$
	$\mathcal{F} f(t) \cos \omega_0 t = \frac{1}{2}\{F[j(\omega - \omega_0)] + F[j(\omega + \omega_0)]\}$
	$\mathcal{F} f(t) \sin \omega_0 t = \frac{1}{2} j\{F[j(\omega - \omega_0)] + F[j(\omega + \omega_0)]\}$
Time convolution	$\mathcal{F}^{-1}[F_1(j\omega)F_2(j\omega)] = \int_{-\infty}^{\infty} f_1(\tau) f_2(t-\tau) d\tau$
Frequency convolution	$\mathcal{F}[f_1(t)f_2(t)] = \frac{1}{2\pi}\int_{-\infty}^{\infty} F_1(j\lambda) F_2[j(\omega - \lambda)] d\lambda$

Source: Van Valkinburg, M.E., *Network Analysis*, 3rd ed., Prentice Hall, Englewood Cliffs, NJ, 1974. With permission.

The above properties are particularly useful in CT system analysis and design, especially when the system characteristics are easily specified in the frequency domain, as in linear filtering. Note that properties 1, 6, and 7 are useful for solving differential or integral equations. Property 4 (time-domain convolution) provides the basis for many signal processing algorithms, since many systems can be specified directly by their impulse or frequency response. Property 3 (frequency-shifting) is useful for analyzing the performance of communication systems where different modulation formats are commonly used to shift spectral energy among different frequency bands.

1.2.2 Sampling Models for Continuous- and Discrete-Time Signals

The relationship between the CT and the DT domains is characterized by the operations of sampling and reconstruction. If $s_a(t)$ denotes a signal $s(t)$ that has been uniformly sampled every T seconds, then the mathematical representation of $s_a(t)$ is given by

$$s_a(t) = \sum_{n=-\infty}^{n=\infty} s(t)\delta(t - nT), \tag{1.2a}$$

where $\delta(t)$ is the CT impulse function defined previously. Since the only places where the product $s(t)\delta(t - nT)$ is not identically equal to zero are at the sampling instances, $s(t)$ in Equation 1.2a can be replaced with $s(nT)$ without changing the overall meaning of the expression. Hence, an alternate expression for $s_a(t)$ that is often useful in Fourier analysis is

$$s_a(t) = \sum_{n=-\infty}^{n=\infty} s(nT)\delta(t - nT). \tag{1.2b}$$

The CT sampling model $s_a(t)$ consists of a sequence of CT impulse functions uniformly spaced at intervals of T seconds and weighted by the values of the signal $s(t)$ at the sampling instants, as depicted in Figure 1.1. Note that $s_a(t)$ is not defined at the sampling instants because the CT impulse function itself is not defined at $t = 0$. However, the values of $s(t)$ at the sampling instants are imbedded as "area under the curve" of $s_a(t)$, and as such represent a useful mathematical model of the sampling process. In the DT domain, the sampling model is simply the sequence defined by taking the values of $s(t)$ at the sampling instants, i.e.,

$$s[n] = s(t)\big|_{t=nT}. \tag{1.3}$$

In contrast to $s_a(t)$, which is not defined at the sampling instants, $s[n]$ is well defined at the sampling instants, as illustrated in Figure 1.2. From this discussion it is now clear that $s_a(t)$ and $s[n]$ are different but equivalent models of the sampling process in the CT and DT domains, respectively. They are both useful for signal analysis in their corresponding domains. It will be shown later that their equivalence is established by the fact that they have equal spectra in the Fourier domain, and that the underlying CT signal from which $s_a(t)$ and $s[n]$ are derived can be recovered from either sampling representation provided that a sufficiently high sampling rate is used in the sampling operation.

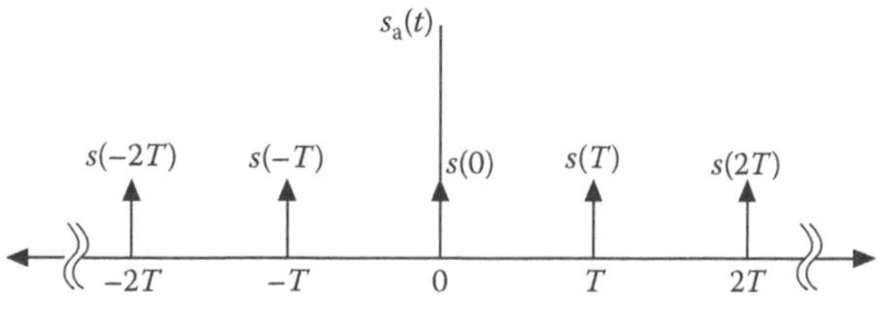

FIGURE 1.1 CT model of a sampled CT signal.

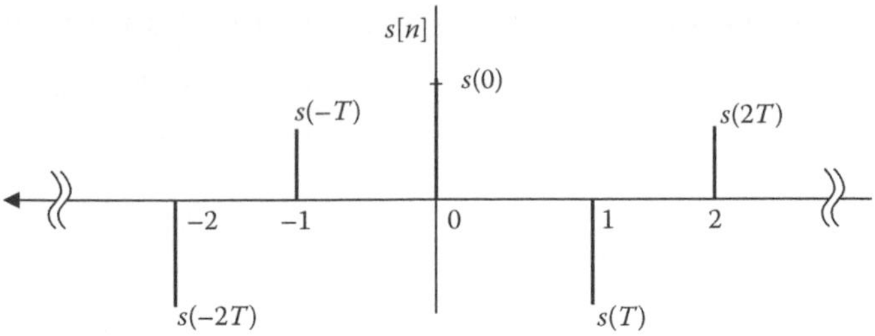

FIGURE 1.2 DT model of a sampled CT signal.

1.2.3 Fourier Spectrum of a Continuous Time Sampled Signal

The operation of uniformly sampling a CT signal $s(t)$ at every T seconds is characterized by Equations 1.2a and b, where $\delta(t)$ is the CT time impulse function defined earlier:

$$s_a(t) = \sum_{n=-\infty}^{\infty} s_a(t)\delta(t-nT) = \sum_{n=-\infty}^{\infty} s_a(nT)\delta(t-nT).$$

Since $s_a(t)$ is a CT signal it is appropriate to apply the CT Fourier transform to obtain an expression for the spectrum of the sampled signal:

$$F\{s_a(t)\} = F\left\{\sum_{n=-\infty}^{\infty} s_a(nT)\delta(t-nT)\right\} = \sum_{n=-\infty}^{\infty} s_a(nT)\left[e^{j\omega T}\right]^{-n}. \tag{1.4}$$

Since the expression on the right-hand side of Equation 1.4 is a function of $e^{j\omega T}$ it is customary to express the transform as $F(e^{j\omega T}) = F\{s_a(t)\}$. If ω is replaced with a normalized frequency $\omega' = \omega/T$, so that $-\pi < \omega' < \pi$, then the right-hand side of Equation 1.4 becomes identical to the DTFT that is defined directly for the sequence $s[n] = s_a(nT)$.

1.2.4 Generalized Complex Fourier Transform

The CT Fourier transform characterized by Equation 1.1 can be generalized by considering the variable $j\omega$ to be the special case of $u = \sigma + j\omega$ with $\sigma = 0$, writing Equation 1.1 in terms of u, and interpreting u as a complex frequency variable. The resulting complex Fourier transform pair is given by Equations 1.5a and b (Bracewell 1986):

$$s(t) = \frac{1}{2\Pi j}\int_{\sigma-j\infty}^{\sigma+j\infty} S(u)e^{jut}\,du \tag{1.5a}$$

$$S(u) = \int_{-\infty}^{\infty} s(t)e^{-jut}\,dt. \tag{1.5b}$$

The set of all values of u for which the integral of Equation 1.5b converges is called the region of convergence, denoted ROC. Since the transform $S(u)$ is defined only for values of u within the ROC, the path of integration in Equation 1.5a must be defined so the entire path lies within the ROC. In some literature this transform pair is called the bilateral Laplace transform because it is the same result obtained by including both the negative and positive portions of the time axis in the classical Laplace transform integral. The complex Fourier transform (bilateral Laplace transform) is not often used in solving practical problems, but its significance lies in the fact that it is the most general form that represents the place where Fourier and Laplace transform concepts merge together. Identifying this connection reinforces the observation that Fourier and Laplace transform concepts share common properties because they result from placing different constraints on the same parent form.

1.3 Fourier Series Representation of Continuous Time Periodic Signals

The classical Fourier series representation of a periodic time domain signal $s(t)$ involves an expansion of $s(t)$ into an infinite series of terms that consist of sinusoidal basis functions, each weighted by a complex constant (Fourier coefficient) that provides the proper contribution of that frequency component to the complete waveform. The conditions under which a periodic signal $s(t)$ can be expanded in a Fourier series are known as the Dirichlet conditions. They require that in each period $s(t)$ has a finite number of discontinuities, a finite number of maxima and minima, and satisfies the absolute convergence criterion of Equation 1.6 (VanValkenburg 1974):

$$\int_{-T/2}^{T/2} |s(t)|\mathrm{d}t < \infty. \tag{1.6}$$

It is assumed throughout the following discussion that the Dirichlet conditions are satisfied by all functions that will be represented by a Fourier series.

1.3.1 Exponential Fourier Series

If $s(t)$ is a CT periodic signal with period T the exponential Fourier series expansion of $s(t)$ is given by

$$s(t) = \sum_{n=-\infty}^{\infty} a_n \mathrm{e}^{jn\omega_0 t}, \tag{1.7a}$$

where $\omega_0 = 2\pi/T$. The a_n's are the complex Fourier coefficients given by

$$a_n = \frac{1}{T}\int_{\frac{-T}{2}}^{\frac{T}{2}} s(t)\mathrm{e}^{-jn\omega_0 t}\mathrm{d}t \quad -\infty < n < \infty. \tag{1.7b}$$

For every value of t where $s(t)$ is continuous the right-hand side of Equation 1.7a converges to $s(t)$. At values of t where $s(t)$ has a finite jump discontinuity, the right-hand side of Equation 1.7a converges to the average of $s(t^-)$ and $s(t^+)$, where $s(t^-) = \lim_{\varepsilon\to 0}(t-\varepsilon)$ and $s(t^+) = \lim_{\varepsilon\to 0}(t+\varepsilon)$.

For example, the Fourier series expansion of the sawtooth waveform illustrated in Figure 1.3 is characterized by $T = 2\pi, \omega_0 = 1, a_0 = 0$, and $a_n = a_{-n} = A\cos(n\pi)/(jn\pi)$ for $n = 1, 2, \ldots$. The coefficients of the exponential Fourier series given by Equation 1.5b can be interpreted as a spectral representation of $s(t)$, since the a_nth coefficient represents the contribution of the $(n\omega_0)$th frequency

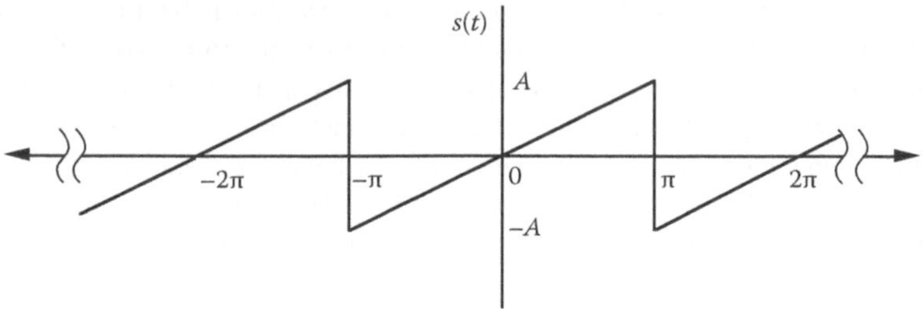

FIGURE 1.3 Periodic CT signal used in Fourier series Example 1.

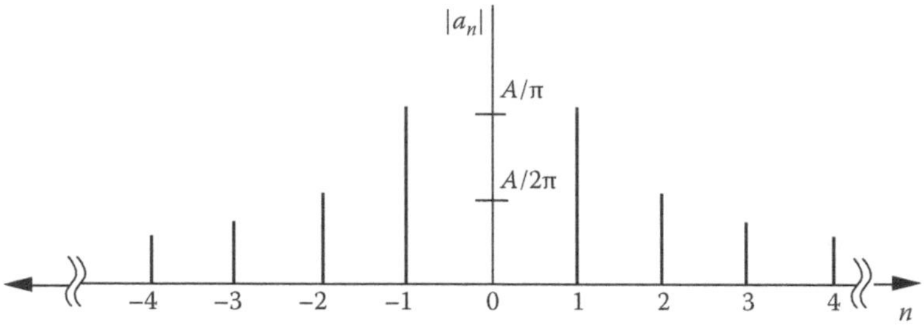

FIGURE 1.4 Magnitude of the Fourier coefficients for Example 1.

component to the complete waveform. Since the a_n's are complex valued, the Fourier domain (spectral) representation has both magnitude and phase spectra. For example, the magnitudes of the a_n's are plotted in Figure 1.4 for the saw tooth waveform of Figure 1.3 (Example 1). The fact that the a_n's constitute a discrete set is consistent with the fact that a periodic signal has a spectrum that contains only integer multiples of the fundamental frequency ω_0. The equation pair given by Equations 1.5a and b can be interpreted as a transform pair that is similar to the CT Fourier transform for periodic signals. This leads to the observation that the classical Fourier series can be interpreted as a special transform that provides a one-to-one invertible mapping between the discrete-spectral domain and the CT domain.

1.3.2 Trigonometric Fourier Series

Although the complex form of the Fourier series expansion is useful for complex periodic signals, the Fourier series can be more easily expressed in terms of real-valued sine and cosine functions for real-valued periodic signals. In the following discussion it is assumed that the signal $s(t)$ is real-valued. When $s(t)$ is periodic and real-valued it is convenient to replace the complex exponential Fourier series with a trigonometric expansion that contains $\sin(\omega_0 t)$ and $\cos(\omega_0 t)$ terms with corresponding real-valued coefficients (VanValkenburg 1974). The trigonometric form of the Fourier series for a real-valued signal $s(t)$ is given by

$$s(t) = \sum_{n=0}^{\infty} b_n \cos(n\omega_0) + \sum_{n=1}^{\infty} c_n \sin(n\omega_0), \tag{1.8a}$$

where $\omega_0 = 2\pi/T$. In Equation 1.8a the b_n's and c_n's are real-valued Fourier coefficients determined by

$$b_0 = \frac{1}{T}\int_{\frac{-T}{2}}^{\frac{T}{2}} s(t)dt$$

$$b_n = \frac{2}{T}\int_{\frac{-T}{2}}^{\frac{T}{2}} s(t)\cos(n\omega_0 t)dt, \quad n = 1, 2, \ldots \tag{1.8b}$$

$$\text{and} \quad c_n = \frac{2}{T}\int_{\frac{-T}{2}}^{\frac{T}{2}} s(t)\sin(n\omega_0 t)dt, \quad n = 1, 2, \ldots.$$

An arbitrary real-valued signal $s(t)$ can be expressed as a sum of even and odd components, $s(t) = s_{\text{even}}(t) + s_{\text{odd}}(t)$, where $s_{\text{even}}(t) = s_{\text{even}}(-t)$ and $s_{\text{odd}}(t) = -s_{\text{odd}}(-t)$, and where $s_{\text{even}}(t) = [s(t) + s(-t)]/2$ and $s_{\text{odd}}(t) = [s(t) - s(-t)]/2$. For the trigonometric Fourier series, it can be shown that $s_{\text{even}}(t)$ is represented by the (even) cosine terms in the infinite series, $s_{\text{odd}}(t)$ is represented by the (odd) sine terms, and b_0 is the DC level of the signal. Therefore, if it can be determined by inspection that a signal has a DC level, or if it is even or odd, then the correct form of the trigonometric series can be chosen to simplify the analysis. For example, it is easily seen that the signal shown in Figure 1.5 (Example 2) is an even signal with a zero DC level, and therefore, can be accurately represented by the cosine series with $b_n = 2A\sin(\pi n/2)/(\pi n/2), n = 1, 2, \ldots$, as shown in Figure 1.6. In contrast note that the sawtooth waveform used in the previous example is an odd signal with zero DC level, so that it can be completely specified by the sine terms of the trigonometric series. This result can be demonstrated by pairing each positive frequency component from the exponential series with its conjugate partner, i.e., $c_n = \sin(n\omega_0 t) = a_n e^{jn\omega_0 t} + a_{-n}e^{-jn\omega_0 t}$, whereby it is found that $c_n = 2A\cos(n\pi)/(n\pi)$ for this example. In general it is found that $a_n = (b_n - jc_n)/2$ for $n = 1, 2, \ldots, a_0 = b_0$, and $a_{-n} = a_n^*$. The trigonometric

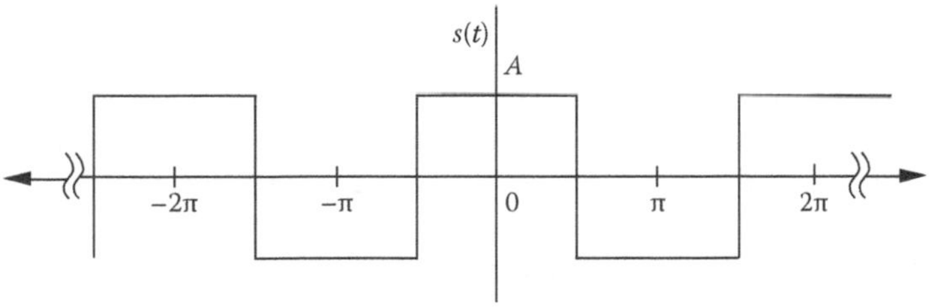

FIGURE 1.5 Periodic CT signal used in Fourier series Example 2.

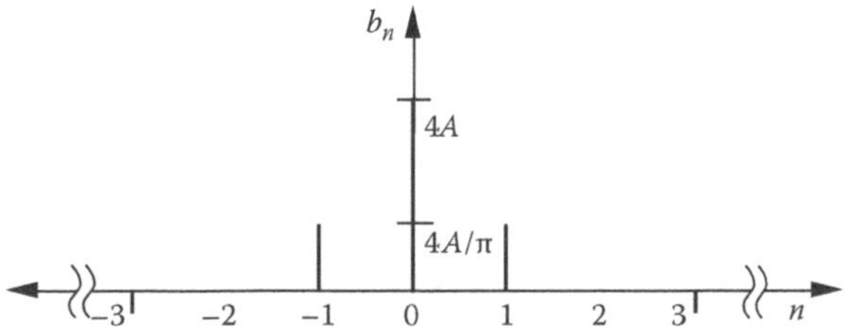

FIGURE 1.6 Fourier coefficients for example of Figure 1.5.

Fourier series is common in the signal processing literature because it replaces complex coefficients with real ones and often results in a simpler and more intuitive interpretation of the results.

1.3.3 Convergence of the Fourier Series

The Fourier series representation of a periodic signal is an approximation that exhibits mean squared convergence to the true signal. If $s(t)$ is a periodic signal of period T, and $s'(t)$ denotes the Fourier series approximation of $s(t)$, then $s(t)$ and $s'(t)$ are equal in the mean square sense if

$$\text{mse} = \int_{\frac{-T}{2}}^{\frac{T}{2}} |s(t) - s\prime(t)|^2 \text{d}t = 0. \tag{1.9}$$

Even with Equation 1.9 is satisfied, mean square error convergence does not guarantee that $s(t) = s'(t)$ at every value of t. In particular, it is known that at values of t where $s(t)$ is discontinuous the Fourier series converges to the average of the limiting values to the left and right of the discontinuity. For example if t_0 is a point of discontinuity, then $s'(t_0) = [s(t_0^-) + s(t_0^+)]/2$, where $s(t_0^-)$ and $s(t_0^+)$ were defined previously (note that at points of continuity, this condition is also satisfied by the very definition of continuity). Since the Dirichlet conditions require that $s(t)$ have at most a finite number of points of discontinuity in one period, the set S_t such that $s(t) \neq s'(t)$ within one period contains a finite number of points, and S_t is a set of measure zero in the formal mathematical sense. Therefore $s(t)$ and its Fourier series expansion $s'(t)$ are equal almost everywhere, and $s(t)$ can be considered identical to $s'(t)$ for analysis in most practical engineering problems.

The condition of convergence almost everywhere is satisfied only in the limit as an infinite number of terms are included in the Fourier series expansion. If the infinite series expansion of the Fourier series is truncated to a finite number of terms, as it must always be in practical applications, then the approximation will exhibit an oscillatory behavior around the discontinuity, known as the Gibbs phenomenon (VanValkenburg 1974). Let $s_N'(t)$ denote a truncated Fourier series approximation of $s(t)$, where only the terms in Equation 1.7a from $n = -N$ to $n = N$ are included if the complex Fourier series representation is used, or where only the terms in Equation 1.8a from $n = 0$ to $n = N$ are included if the trigonometric form of the Fourier series is used. It is well known that in the vicinity of a discontinuity at t_0 the Gibbs phenomenon causes $s_N'(t)$ to be a poor approximation to $s(t)$. The peak magnitude of the Gibbs oscillation is 13% of the size of the jump discontinuity $s(t_0^-) - s(t_0^+)$ regardless of the number of terms used in the approximation. As N increases, the region that contains the oscillation becomes more concentrated in the neighborhood of the discontinuity, until, in the limit as N approaches infinity, the Gibbs oscillation is squeezed into a single point of mismatch at t_0. The Gibbs phenomenon is illustrated in Figure 1.7 where an ideal lowpass frequency response is approximated by impulse response

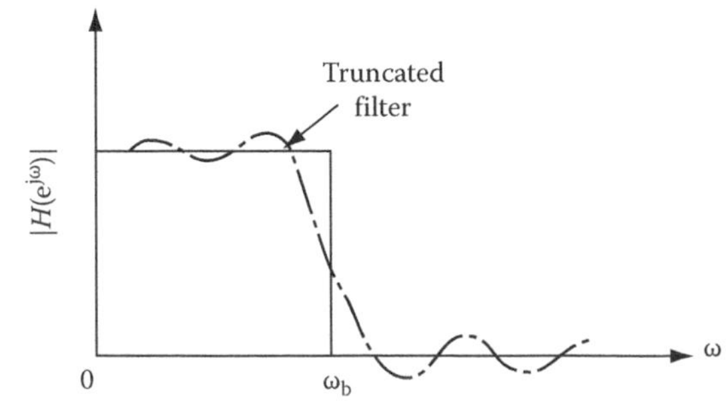

FIGURE 1.7 Gibbs phenomenon in a lowpass digital filter caused by truncating the impulse response to N terms.

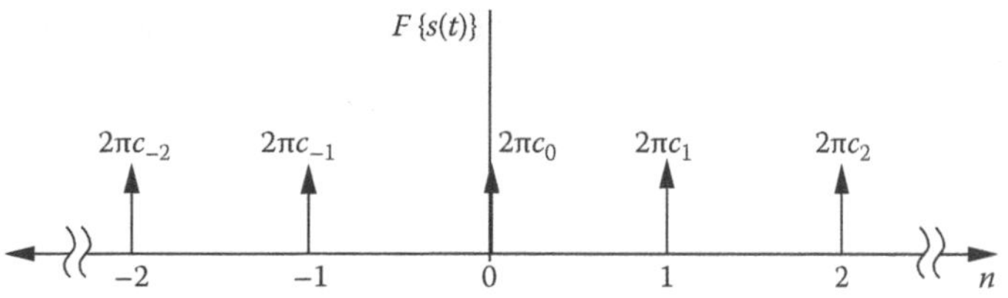

FIGURE 1.8 Spectrum of the Fourier representation of a periodic signal.

function that has been limited to having only N nonzero coefficients, and hence the Fourier series expansion contains only a finite number of terms.

An important property of the Fourier series is that the exponential basis functions $e^{jn\omega_0 t}$ (or $\sin(n\omega_0 t)$ and $\cos(n\omega_0 t)$ for the trigonometric form) for $n = 0, \pm 1, \pm 2, \ldots$ (or $n = 0, 1, 2, \ldots$ for the trigonometric form) constitute an "orthonormal set," i.e., $t_{nk} = 1$ for $n = k$, and $t_{nk} = 0$ for $n \neq k$, where

$$t_{nk} = \frac{1}{T} \int_{\frac{-T}{2}}^{\frac{T}{2}} (e^{-jn\omega_0 t})(e^{jk\omega_0 t})dt.$$

As terms are added to the Fourier series expansion, the orthogonality of the basis functions guarantees that the approximation error decreases in the mean square sense, i.e., that mse_N decreases monotonically as N is increased, where

$$\mathrm{mse}_N = \int_{\frac{-T}{2}}^{\frac{T}{2}} \left|s(t) - s'_N(t)\right|^2 dt.$$

Therefore, when applying Fourier series analysis including more terms always improves the accuracy of the signal representation.

1.3.4 Fourier Transform of Periodic Continuous Time Signals

For a periodic signal $s(t)$ the CT Fourier transform can then be applied to the Fourier series expansion of $s(t)$ to produce a mathematical expression for the "line spectrum" that is characteristic of periodic signals:

$$F\{s(t)\} = F\left\{\sum_{n=-\infty}^{\infty} a_n e^{jn\omega_0 t}\right\} = 2\pi \sum_{n=-\infty}^{\infty} a_n \delta(\omega - \omega_0). \tag{1.10}$$

The spectrum is shown in Figure 1.8. Note the similarity between the spectral representation of Figure 1.8 and the plots of the Fourier coefficients in Figures 1.4 and 1.6, which were heuristically interpreted as a line spectrum. Figures 1.4 and 1.6 are different from Figure 1.8 but they are equivalent representations of the Fourier line spectrum that is characteristic of periodic signals.

1.4 Discrete-Time Fourier Transform

The DTFT is obtained directly in terms of the sequence samples $s[n]$ by taking the relationship obtained in Equation 1.4 to be the definition of the DTFT. Letting $T = 1$ so that the sampling period is removed from the equations and the frequency variable is replaced with a normalized frequency $\omega' = \omega T$, the DTFT pair is defined by Equation 1.11. In order to simplify notation it is not customary to distinguish between ω and ω′, but rather to rely on the context of the discussion to determine whether ω refers to the normalized ($T = 1$) or the un-normalized ($T \neq 1$) frequency variable:

$$S(e^{j\omega'}) = \sum_{n=-\infty}^{\infty} s[n]e^{-j\omega' n} \tag{1.11a}$$

$$s[n] = \frac{1}{2\Pi}\int_{-\Pi}^{\Pi} S(e^{j\omega'})e^{jn\omega'}d\omega'. \tag{1.11b}$$

The spectrum $S(e^{j\omega'})$ is periodic in ω' with period 2π. The fundamental period in the range $-\pi < \omega' \leq \pi$, referred to as the baseband, is the useful frequency range of the DT system because frequency components in this range can be represented unambiguously in sampled form (without aliasing error). In much of the signal processing literature the explicit primed notation is omitted from the frequency variable. However, the explicit primed notation will be used throughout this section because there is a potential for confusion when so many related Fourier concepts are discussed within the same framework.

By comparing Equations 1.4 and 1.11a, and noting that $\omega' = \omega T$, it is seen that

$$F\{s_a(t)\} = \text{DTFT}\{s[n]\},$$

where $s[n] = s_a(t)|_{t=nT}$. This demonstrates that the spectrum of $s_a(t)$ as calculated by the CT Fourier transform is identical to the spectrum of $s[n]$ as calculated by the DTFT. Therefore although $s_a(t)$ and $s[n]$ are quite different sampling models, they are equivalent in the sense that they have the same Fourier domain representation. A list of common DTFT pairs is presented in Table 1.3. Just as the CT Fourier

TABLE 1.3 Some Basic DTFT Pairs

Sequence	Fourier Transform
1. $\delta[n]$	1
2. $\delta[n-n_0]$	$e^{-j\omega n_0}$
3. $1 \ (-\infty < n < \infty)$	$\sum_{k=-\infty}^{\infty} 2\pi\delta(\omega + 2\pi k)$
4. $a^n u[n] \ (\lvert a\rvert < 1)$	$\frac{1}{1 - ae^{-j\omega}}$
5. $u[n]$	$\frac{1}{1 - e^{-j\omega}} + \sum_{k=-\infty}^{\infty} \pi\delta(\omega + 2\pi k)$
6. $(n+1)a^n u[n] \ (\lvert a\rvert < 1)$	$\frac{1}{(1 - ae^{-j\omega})^2}$
7. $\frac{r^n \sin \omega_p(n+1)}{\sin \omega_p} u[n] \ (\lvert r\rvert < 1)$	$\frac{1}{1 - 2r\cos\omega_p e^{-j\omega} + r^2 e^{-j2\omega}}$
8. $\frac{\sin \omega_c n}{\pi n}$	$X(e^{j\omega}) = \begin{cases} 1, & \lvert\omega\rvert < \omega_c, \\ 0, & \omega_c < \lvert\omega\rvert \leq \pi \end{cases}$
9. $x[n] = \begin{cases} 1, & 0 \leq n \leq M \\ 0, & \text{otherwise} \end{cases}$	$\frac{\sin[\omega(M+1)/2]}{\sin(\omega/2)} = e^{-j\omega M/2}$
10. $e^{j\omega_0 n}$	$\sum_{k=-\infty}^{\infty} 2\pi\delta(\omega - \omega_0 + 2\pi k)$
11. $\cos(\omega_0 n + \varphi)$	$\pi \sum_{k=-\infty}^{\infty} [e^{j\varphi}\delta(\omega - \omega_0 + 2\pi k) + e^{j\varphi\delta}(\omega + \omega_0 + 2\pi k)]$

Source: Oppenheim, A.V. and Schafer, R.W., *Discrete-Time Signal Processing*, Prentice-Hall, Englewood Cliffs, NJ, 1989. With permission.

transform is useful in CT signal system analysis and design, the DTFT is equally useful for DT system analysis and design.

In the same way that the CT Fourier transform was found to be a special case of the complex Fourier transform (or bilateral Laplace transform), the DTFT is a special case of the bilateral z-transform with $z = e^{j\omega' t}$. The more general bilateral z-transform is given by

$$S(z) = \sum_{n=-\infty}^{\infty} s[n]z^{-n} \tag{1.12a}$$

$$s[n] = \frac{1}{2\pi j} \oint_C S(z)z^{n-1}dz, \tag{1.12b}$$

where C is a counterclockwise contour of integration which is a closed path completely contained within the region of convergence of $S(z)$. Recall that the DTFT was obtained by taking the CT Fourier transform of the CT sampling model $s_a(t)$. Similarly, the bilateral z-transform results by taking the bilateral Laplace transform of $s_a(t)$. If the lower limit on the summation of Equation 1.12a is taken to be $n = 0$, then Equations 1.12a and b become the one-sided z-transform, which is the DT equivalent of the one-sided Laplace transform for CT signals.

1.4.1 Properties of the Discrete-Time Fourier Transform

Since the DTFT is a close relative of the classical CT Fourier transform, it should come as no surprise that many properties of the DTFT are similar to those of the CT Fourier transform. In fact, for many of the properties presented earlier there is an analogous property for the DTFT. The following list parallels the list that was presented earlier for the CT Fourier transform, to the extent that the same properties exist (a more complete list of DTFT properties is given in Table 1.4). Note that $F\{\cdot\}$ denotes the DTFT

TABLE 1.4 Properties of the DTFT

Sequence $x[n]$ $y[n]$	Fourier $X(e^{j\omega})$ $Y(e^{j\omega})$
1. $ax[n] + by[n]$	$aX(e^{j\omega}) + bY(e^{j\omega})$
2. $x[n - n_d]$ (n_d an integer)	$e^{-j\omega n_d}X(e^{j\omega})$
3. $e^{j\omega_0 n}x[n]$	$X\left[e^{j(\omega-\omega_0)}\right]$
4. $x[-n]$	$X(e^{-j\omega})$ if $x[n]$ real $X^*(e^{j\omega})$
5. $nx[n]$	$j\dfrac{dX(e^{j\omega})}{d\omega}$
6. $x[n] = y[n]$	$X(e^{j\omega})Y(e^{j\omega})$
7. $x[n]y[n]$	$\frac{1}{2\pi}\int_{-x}^{x} X(e^{j\theta})Y\left[e^{j(\omega-\theta)}\right]d\theta$
Parseval's theorem	
8. $\sum_{n=-\infty}^{\infty} \lvert x[n]\rvert^2 = \frac{1}{2\pi}\int_{-x}^{x} \lvert X(e^{j\omega})\rvert^2 d\omega$	
9. $\sum_{n=-\infty}^{\infty} x[n]y^*[n] = \frac{1}{2\pi}\int_{-x}^{x} X(e^{j\omega})Y^*(e^{j\omega})\,d\omega$	

Source: Oppenheim, A.V. and Schafer, R.W., *Discrete-Time Signal Processing*, Prentice-Hall, Englewood Cliffs, NJ, 1989. With permission.

operation, $F^{-1}\{\cdot\}$ denotes the inverse DTFT operation, and "$*$" denotes the DT convolution operation defined as

$$f_1[n] * f_2[n] = \sum_{k=-\infty}^{+\infty} f_1[n] f_2[n-k].$$

1. Linearity (a and b are complex constants)	$\mathrm{DTFT}\{af_1[n] + bf_2[n]\} = a \cdot \mathrm{DTFT}\{f_1[n]\} + b \cdot \mathrm{DTFT}\{f_2[n]\}$
2. Index-shifting	$\mathrm{DTFT}\{f[n-n_0]\} = e^{-j\omega n_0}\mathrm{DTFT}\{f[n]\}$
3. Frequency-shifting	$e^{j\omega_0 n} f[n] = \mathrm{DTFT}^{-1}\{F(j(\omega - \omega_0)\}$
4. Time-domain convolution	$\mathrm{DTFT}\{f_1[n] * f_2[n]\} = F\{f_1[n]\} \cdot F\{f_2[n]\}$
5. Frequency-domain convolution	$\mathrm{DTFT}\{f_1[n] \cdot f_2[n]\} = \frac{1}{2\Pi}\mathrm{DTFT}\{f_1[n]\} * \mathrm{DTFT}\{f_2[n]\}$
6. Frequency-differentiation	$nf[n] = \mathrm{DTFT}^{-1}\{dF(j\omega)/d\omega\}$

Note that the time-differentiation and time-integration properties of the CT Fourier transform do not have analogous counterparts in the DTFT because time domain differentiation and integration are not defined for DT signals. When working with DT systems practitioners must often manipulate difference equations in the frequency domain. For this purpose the properties of linearity and index-shifting are very important. As with the CT Fourier transform time-domain convolution is also important for DT systems because it allows engineers to work with the frequency response of the system in order to achieve proper shaping of the input spectrum, or to achieve frequency selective filtering for noise reduction or signal detection.

1.4.2 Relationship between the CT and DT Spectra

Since DT signals often originate by sampling a CT signal, it is important to develop the relationship between the original spectrum of the CT signal and the spectrum of the DT signal that results. First, the CT Fourier transform is applied to the CT sampling model, and the properties are used to produce the following result:

$$F\{s_a(t)\} = F\left\{s_a(t)\sum_{n=-\infty}^{\infty}\delta(t-nT)\right\} = \frac{1}{2\pi}S_a(j\omega)F\left\{\sum_{n=-\infty}^{\infty}\delta(t-nT)\right\}. \tag{1.13}$$

Since the sampling function (summation of shifted impulses) on the right-hand side of Equation 1.13 is periodic with period T it can be replaced with a CT Fourier series expansion and the frequency-domain convolution property of the CT Fourier transform can be applied to yield two equivalent expressions for the DT spectrum:

$$S(e^{j\omega T}) = \frac{1}{T}\sum_{n=-\infty}^{\infty} S_a(j[\omega - n\omega_s]) \quad \text{or} \quad S(e^{j\omega'}) = \frac{1}{T}\sum_{n=-\infty}^{\infty} S_a(j[\omega' - n2\pi/T]). \tag{1.14}$$

In Equation 1.14 $\omega_s = (2\pi/T)$ is the sampling frequency and $\omega' = \omega T$ is the normalized DT frequency axis expressed in radians. Note that $S(e^{j\omega T}) = S(e^{j\omega'})$ consists of an infinite number of replicas of the CT spectrum $S(j\omega)$, positioned at intervals of $(2\pi/T)$ on the ω-axis (or at intervals of 2π on the ω'-axis), as illustrated in Figure 1.9. Note that if $S(j\omega)$ is band-limited with a bandwidth ω_c, and if T is chosen sufficiently small so that $\omega_s > 2\omega_c$, then the DT spectrum is a copy of $S(j\omega)$ (scaled by $1/T$) in the baseband. The limiting case of $\omega_s = 2\omega_c$ is called the Nyquist sampling frequency. Whenever a CT signal

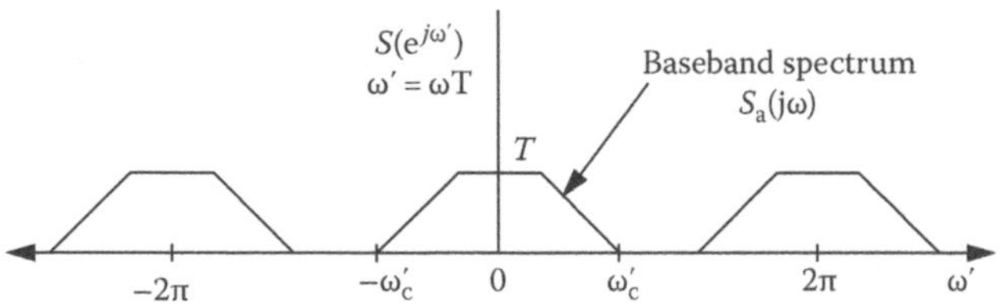

FIGURE 1.9 Relationship between the CT and DT spectra.

is sampled at or above the Nyquist rate, no aliasing distortion occurs (i.e., the baseband spectrum does not overlap with the higher order replicas) and the CT signal can be exactly recovered from its samples by extracting the baseband spectrum of $S(e^{j\omega'})$ with an ideal lowpass filter that recovers the original CT spectrum by removing all spectral replicas outside the baseband and scaling the baseband by a factor of T.

1.5 Discrete Fourier Transform

To obtain the DFT the continuous-frequency domain of the DTFT is sampled at N points uniformly spaced around the unit circle in the z-plane, i.e., at the points $\omega_k = (2\pi k/N), k = 0, 1, \ldots, N-1$. The result is the DFT transform pair defined by Equations 1.15a and b:

$$S[k] = \sum_{n=0}^{N-1} s[n]e^{-j\frac{2\pi kn}{N}}, \quad k = 0, 1, \ldots, N-1 \tag{1.15a}$$

$$s[k] = \frac{1}{N}\sum_{k=0}^{N-1} S[k]e^{j\frac{2\pi kn}{N}}, \quad n = 0, 1, \ldots, N-1, \tag{1.15b}$$

The signal $s[n]$ is either a finite length sequence of length N, or it is a periodic sequence with period N. Regardless of whether $s[n]$ is a finite length or periodic sequence, the DFT treats the N samples of $s[n]$ as though they are one period of a periodic sequence. This is a peculiar feature of the DFT, and one that must be handled properly in signal processing to prevent the introduction of artifacts.

1.5.1 Properties of the DFT

Important properties of the DFT are summarized in Table 1.5. The notation $[k]_N$ denotes k modulo N, and $R_N[n]$ is a rectangular window such that $R_N[n] = 1$ for $n = 0, \ldots, N-1$, and $R_N[n] = 0$ for $n < 0$ and $n \geq N$. The transform relationship given by Equations 1.15a and 1.15b is also valid when $s[n]$ and $S[k]$ are periodic sequences, each of period N. In this case n and k are permitted to range over the complete set of real integers, and $S[k]$ is referred to as the discrete Fourier series (DFS). In some cases the DFS is developed as a distinct transform pair in its own right (Jenkins and Desai 1986). Whether or not the DFT and the DFS are considered identical or distinct is not important in this discussion. The important point to be emphasized here is that the DFT treats $s[n]$ as though it were a single period of a periodic sequence, and all signal processing done with the DFT will inherit the consequences of this assumed periodicity.

Most of the properties listed in Table 1.5 for the DFT are similar to those of the z-transform and the DTFT, although there are important differences. For example, Property 5 (time-shifting property), holds for circular shifts of the finite length sequence $s[n]$, which is consistent with the notion that the DFT treats $s[n]$ as one period of a periodic sequence. Also, the multiplication of two DFTs results in the circular convolution of the corresponding DT sequences, as specified by Property 7. This later property is quite different from the linear convolution property of the DTFT. Circular convolution is simply a linear

TABLE 1.5 Properties of the DFT

Finite-Length Sequence (Length N)	N-Point DFT (Length N)
1. $x[n]$	$X[k]$
2. $x_1[n], x_2[n]$	$X_1[k], X_2[k]$
3. $ax_1[n] + bx_2[n]$	$aX_1[k] + bX_2[k]$
4. $X[n]$	$Nx[(-k)_N]$
5. $x[(n-m)_N]$	$W_N^{km}X[k]$
6. $W_N^{-\ell n}x[n]$	$X[(k-\ell)_N]$
7. $\sum_{m=0}^{N-1} x_1(m)x_2[(n-m)_N]$	$X_1[k]X_2[k]$
8. $x_1[n]x_2[n]$	$\frac{1}{N}\sum_{\ell=0}^{N-1} X_1(\ell)X_2[(k-\ell)_N]$
9. $x^*[n]$	$X^*[(-k)_N]$
10. $x^*[(-n)_N]$	$X^*[k]$
11. $\mathrm{Re}\{x[n]\}$	$X_{ep}[k] = \frac{1}{2}\{X[(k)_N] + X^*[(-k)_N]\}$
12. $j\mathrm{Im}\{x[n]\}$	$X_{op}[k] = \frac{1}{2}\{X[(k)_N] - X^*[(-k)_N]\}$
13. $x_{ep}[n] = \frac{1}{2}\{x[n] + x^*[(-n)_N]\}$	$\mathrm{Re}\{X[k]\}$
14. $x_{op}[n] = \frac{1}{2}\{x[n] - x^*[(-n)_N]\}$	$j\mathrm{Im}\{X[k]\}$
Properties 15–17 apply only when $x[n]$ is real	
15. Symmetry properties	$\begin{cases} X[k] = X^*[(-k)_N] \\ \mathrm{Re}\{X[k]\} = \mathrm{Re}\{X[(-k)_N]\} \\ \mathrm{Im}\{X[k]\} = -\mathrm{Im}\{X[(-k)_N]\} \\ \lvert X[k]\rvert = \lvert X[(-k)_N]\rvert \\ \sphericalangle X[k]\} = -\sphericalangle\{X[(-k)_N]\}. \end{cases}$
16. $x_{ep}[n] = \frac{1}{2}\{x[n] + x[(-n)_N]\}$	$\mathrm{Re}\{X[k]\}$
17. $x_{op}[n] = \frac{1}{2}\{x[n] - x[(-n)_N]\}$	$j\mathrm{Im}\{X[k]\}$

Source: Oppenheim, A.V. and Schafer, R.W., *Discrete-Time Signal Processing*, Prentice-Hall, Englewood Cliffs, NJ, 1989. With permission.

convolution of the periodic extensions of the finite sequences being convolved, where each of the finite sequences of length N defines the structure of one period of the periodic extensions.

For example, suppose it is desired to implement a digital filter with finite impulse response (FIR) $h[n]$. The output in response to $s[n]$ is

$$y[n] = \sum_{k=0}^{N-1} h[k]s[n-k] \tag{1.16}$$

which is obtained by transforming $h[n]$ and $s[n]$ into $H[k]$ and $S[k]$ using the DFT, multiplying the transforms point-wise to obtain $Y[k] = H[k]S[k]$, and then using the inverse DFT to obtain $y[n] = \mathrm{DFT}^{-1}\{Y[k]\}$. If $s[n]$ is a finite sequence of length M, then the results of the circular convolution implemented by the DFT will correspond to the desired linear convolution if and only if the block length of the DFT, N_{DFT}, is chosen sufficiently large so that $N_{\mathrm{DFT}} > N + (M-1)$ and both $h[n]$ and $s[n]$ are padded with zeros to form blocks of length N_{DFT}.

1.5.2 Fast Fourier Transform Algorithms

The DFT is typically implemented in practice with one of the common forms of the FFT algorithm. The FFT is not a Fourier transform in its own right, but rather it is simply a computationally efficient algorithm that reduces the complexity of the computing DFT from Order $\{N^2\}$ to Order $\{N \log_2 N\}$. When N is large, the computational savings provided by the FFT algorithm is so great that the FFT makes real-time DFT analysis practical in many situations which would be entirely impractical without it. There are numerous FFT algorithms, including decimation-in-time (D-I-T) algorithms, decimation-in-frequency (D-I-F) algorithms, bit-reversed algorithms, normally ordered algorithms, mixed-radix algorithms (for block lengths that are not powers-of-2 [PO2]), prime factor algorithms, and Winograd algorithms [Blahut 1985]. The D-I-T and the D-I-F radix-2 FFT algorithms are the most widely used in practice. Detailed discussions of various FFT algorithms can be found in Brigham (1974) and Oppenheim and Schafer (1975).

The FFT is easily understood by examining the simple example of $N = 8$. There are numerous ways to develop the FFT algorithm, all of which deal with a nested decomposition of the summation operator of Equation 1.20a. The development presented here is called an algebraic development of the FFT because it follows straightforward algebraic manipulation. First, each of the summation indices (k, n) in Equation 1.15a is expressed as explicit binary integers, $k = 4k_2 + 2k_1 + k_0$ and $n = 4n_2 + 2n_1 + n_0$, where k_i and n_i are bits that take on the values of either 0 or 1. If these expressions are substituted into Equation 1.20a, all terms in the exponent that contain the factor $N = 8$ can be deleted because $e^{-j2\pi l} = 1$ for any integer l. Upon deleting such terms and re-grouping the remaining terms, the product nk can be expressed in either of two ways:

$$nk = (4k_0)n_2 + (4k_1 + 2k_0)n_1 + (4k_2 + 2k_1 + k_0)n_0 \tag{1.17a}$$

$$nk = (4n_0)k_2 + (4n_1 + 2n_0)k_1 + (4n_2 + 2n_1 + n_0)k_0. \tag{1.17b}$$

Substituting Equation 1.17a into Equation 1.15a leads to the D-I-T FFT, whereas substituting Equation 1.25b leads to the D-I-F FFT. Only the D-I-T FFT is discussed further here. The D-I-F and various related forms are treated in detail in Oppenheim and Schafer (1975).

The D-I-T FFT decomposes into $\log_2 N$ stages of computation, plus a stage of bit reversal,

$$x_1[k_0, n_1, n_0] = \sum_{n_2=0}^{n_2=1} s[n_2, n_1, n_0] W_8^{4k_0 n_2} \quad \text{(stage 1)} \tag{1.18a}$$

$$x_2[k_0, k_1, n_0] = \sum_{n_1=0}^{n_1=1} x_1[k_0, n_1, n_0] W_8^{(4k_1+2k_0)n_1} \quad \text{(stage 2)} \tag{1.18b}$$

$$x_3[k_0, k_1, k_2] = \sum_{n_0=0}^{n_0=1} x_2[k_0, k_1, n_0] W_8^{(4k_2+2k_1+k_0)n_0} \quad \text{(stage 3)} \tag{1.18c}$$

$$s(k_2, k_1, k_0) = x_3(k_0, k_1, k_2) \quad \text{(bit reversal).} \tag{1.18d}$$

In each summation above, one of the n_i's is summed out of the expression, while at the same time a new k_i is introduced. The notation is chosen to reflect this. For example, in stage 3, n_0 is summed out, k_2 is introduced as a new variable, and n_0 is replaced by k_2 in the result. The last operation, called bit reversal, is necessary to correctly locate the frequency samples $X[k]$ in the memory. It is easy to show that if the samples are paired correctly, an in-place computation can be done by a sequence of butterfly operations. The term in-place means that each time a butterfly is to be computed, a pair of data samples is read from memory, and the new data pair produced by the butterfly calculation is written back into the memory locations where the original pair was stored, thereby overwriting the original data. An in-place algorithm is designed so that each data pair is needed for only one butterfly, and so the new results can be immediately stored on top of the old in order to minimize memory requirements.

For example, in stage 3 the $k = 6$ and $k = 7$ samples should be paired, yielding a "butterfly" computation that requires one complex multiply, one complex add, and one subtract:

$$x_3(1,1,0) = x_2(1,1,0) + W_8^3 x_2(1,1,1) \tag{1.19a}$$

$$x_3(1,1,1) = x_2(1,1,0) - W_8^3 x_2(1,1,1) \tag{1.19b}$$

Samples $x_2(6)$ and $x_2(7)$ are read from the memory, the butterfly is executed on the pair, and $x_3(6)$ and $x_3(7)$ are written back to the memory, overwriting the original values of $x_2(6)$ and $x_2(7)$. In general, there are $N/2$ butterflies per stage and $\log_2 N$ stages, so the total number of butterflies is $(N/2)\log_2 N$. Since there is at most one complex multiplication per butterfly, the total number of multiplications is bounded by $(N/2)\log_2 N$ (some of the multiplies involve factors of unity and should not be counted).

Figure 1.10 shows the signal flow graph of the D-I-T FFT for $N = 8$. This algorithm is referred to as an in-place FFT with normally ordered input samples and bit-reversed outputs. Minor variations that include bit-reversed inputs and normally ordered outputs, and non-in-place algorithms with normally ordered inputs and outputs are possible. Also, when N is not a PO2, a mixed-radix algorithm can be used to reduce computation. The mixed-radix FFT is most efficient when N is highly composite, i.e., $N = p_1^{r_1} p_2^{r_2} \cdots p_L^{r_L}$, where the p_i's are small prime numbers and the r_i's are positive integers. It can be shown that the order of complexity of the mixed-radix FFT is Order $\{N[r_1(p_1 - 1) + r_2(p_2 - 1) + \cdots + r_L(p^L - 1)]\}$. Because of the lack of uniformity of structure among stages, this algorithm has not received much attention for hardware implementation. However, the mixed-radix FFT is often used in software applications, especially for processing data recorded in laboratory experiments where it is not convenient to restrict the block lengths to be PO2. Many advanced FFT algorithms, such as higher radix forms, the mixed-radix form, prime-factor algorithm, and the Winograd algorithm are described in Blahut (1985). Algorithms specialized for real-valued data reduce the computational cost by a factor of 2.

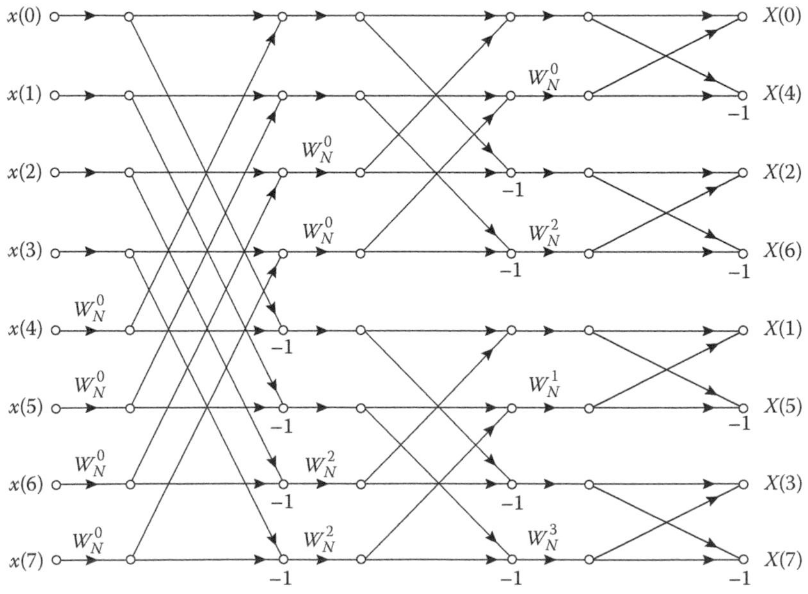

FIGURE 1.10 D-I-T FFT algorithm with normally ordered inputs and bit-reversed outputs.

1.6 Family Tree of Fourier Transforms

Figure 1.11 illustrates the functional relationships among the various forms of CT Fourier transform and DTFT that have been discussed in the previous sections. The family of CT Fourier transforms is shown on the left side of Figure 1.11, whereas the right side of the figure shows the hierarchy of DTFTs. Note that the most general, and consequently the most powerful, Fourier transform is the classical complex Fourier transform (or equivalently, the bilateral Laplace transform). Note also that the complex Fourier transform is identical to the bilateral Laplace transform, and it is at this level that the classical Laplace transform techniques and Fourier transform techniques become identical. Each special member of the CT Fourier family is obtained by impressing certain constraints on the general form, thereby producing special transforms that are simpler and more useful in practical problems where the constraints are met. In Figure 1.11 it is seen that the bilateral z-transform is analogous to the complex Fourier transform, the unilateral z-transform is analogous to the classical (one-sided) Laplace transform, the DTFT is analogous to the classical Fourier (CT) transform, and the DFT is analogous to the classical (CT) Fourier series.

1.6.1 Walsh–Hadamard Transform

The Walsh–Hadamard transform (WHT) is a computationally attractive orthogonal transform that is structurally related to the DFT, and which can be implemented in practical applications without

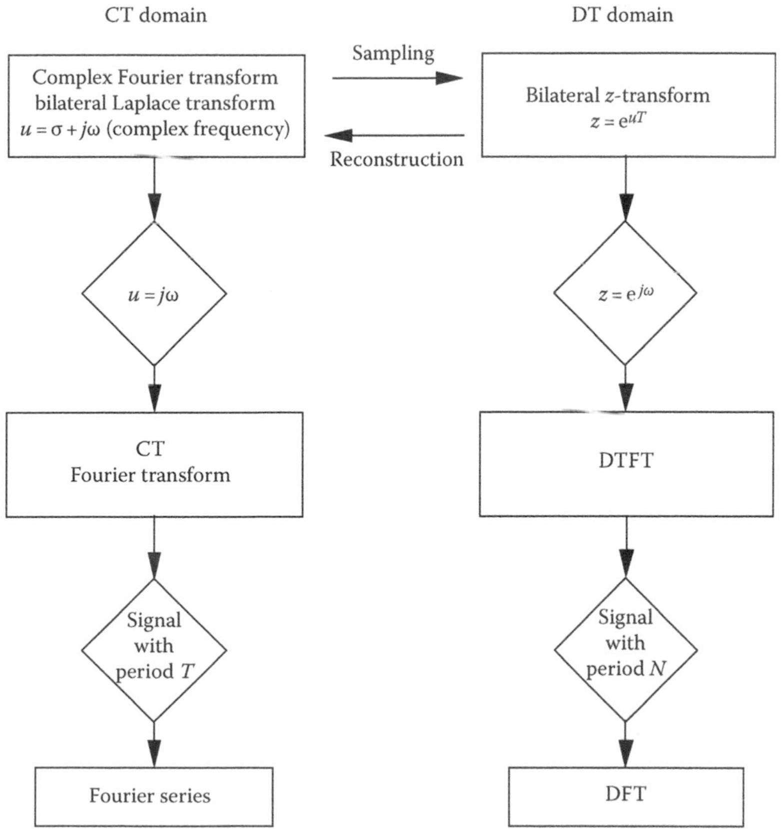

FIGURE 1.11 Functional relationships among various forms of the Fourier transform.

multiplication, and with a computational complexity for addition that is on the same order of complexity as that of an FFT. The t_{mk}th element of the WHT matrix $\mathbf{T}_{\text{WHT}}$ is given by

$$t_{mk} = \frac{1}{\sqrt{N}} \prod_{\ell=0}^{p-1} (-1)^{b_l(m) b_{p-1-\ell}(k)}, \quad m \text{ and } k = 0, \ldots, N-1,$$

where $b_\ell(m)$ is the ℓth order bit in the binary representation of m, and $N = 2^p$. The WHT is defined only when N is a PO2. Note that the columns of $\mathbf{T}_{\text{WHT}}$ form a set of orthogonal basis vectors whose elements are all 1's or -1's, so that the calculation of the matrix-vector product $\mathbf{T}_{\text{WHT}}\mathbf{X}$ can be accomplished with only additions and subtractions. It is well known that $\mathbf{T}_{\text{WHT}}$ of dimension $(N \times N)$, for N a PO2, can be computed recursively according to

$$\mathbf{T}_k = \begin{bmatrix} \mathbf{T}_{k/2} & \mathbf{T}_{k/2} \\ \mathbf{T}_{k/2} & -\mathbf{T}_{k/2} \end{bmatrix} \quad \text{for } K = 4, \ldots, N \text{ (even)} \quad \text{and} \quad \mathbf{T}_2 = \begin{bmatrix} 1 & 1 \\ 1 & -1 \end{bmatrix}.$$

The above relationship provides a convenient way of quickly constructing the Walsh–Hadamard matrix for any arbitrary (even) size N.

Due to structural similarities between the DFT and the WHT matrices, the WHT transform can be implemented using a modified FFT algorithm. The core of any FFT program is a butterfly calculation that is characterized by a pair of coupled equations that have the following form:

$$X_{i+1}(\ell, m) = X_i(\ell, m) + e^{j\theta(\ell,m,k,s)} X_i(k, s)$$
$$X_{i+1}(\ell, m) = X_i(\ell, m) - e^{j\theta(\ell,m,k,s)} X_i(k, s).$$

If the exponential factor in the butterfly calculation is replaced by a "1," so the "modified butterfly" calculation becomes

$$X_{i+1}(\ell, m) = X_i(\ell, m) + X_i(k, s)$$
$$X_{i+1}(\ell, m) = X_i(\ell, m) - X_i(k, s),$$

the modified FFT program will in fact perform a WHT on the input vector. This property not only provides a quick and convenient way to implement the WHT, but is also establishes clearly that in addition to the WHT requiring no multiplication, the number of additions required has order of complexity of $(N/2)\log_2 N$, i.e., the same as the that of the FFT.

The WHT is used in many applications that require signals to be decomposed in real time into a set of orthogonal components. A typical application in which the WHT has been used in this manner is in code division multiple access (CDMA) wireless communication systems. A CDMA system requires spreading of each user's signal spectrum using a PN sequence. In addition to the PN spreading codes, a set of length-64 mutually orthogonal codes, called the Walsh codes, are used to ensure orthogonality among the signals for users received from the same base station. The length $N = 64$ Walsh codes can be thought of as the orthogonal column vectors from a (64×64) Walsh–Hadamard matrix, and the process of demodulation in the receiver can be interpreted as performing a WHT on the complex input signal containing all the modulated user's signals so they can be separated for accurate detection.

1.7 Selected Applications of Fourier Methods

1.7.1 DFT (FFT) Spectral Analysis

An FFT program is often used to perform spectral analysis on signals that are sampled and recorded as part of laboratory experiments, or in certain types of data acquisition systems. There are several issues to be addressed when spectral analysis is performed on (sampled) analog waveforms that are observed over a finite interval of time.

1.7.1.1 Windowing

The FFT treats the block of data as though it were one period of a periodic sequence. If the underlying waveform is not periodic, then harmonic distortion may occur because the periodic waveform created by the FFT may have sharp discontinuities at the boundaries of the blocks. This effect is minimized by removing the mean of the data (it can always be reinserted) and by windowing the data so the ends of the block are smoothly tapered to zero. A good rule of thumb is to taper 10% of the data on each end of the block using either a cosine taper or one of the other common windows (e.g., Hamming, Von Hann, Kaiser windows, etc.). An alternate interpretation of this phenomenon is that the finite length observation has already windowed the true waveform with a rectangular window that has large spectral sidelobes. Hence, applying an additional window results in a more desirable window that minimizes frequency-domain distortion.

1.7.1.2 Zero-Padding

An improved spectral analysis is achieved if the block length of the FFT is increased. This can be done by (1) taking more samples within the observation interval, (2) increasing the length of the observation interval, or (3) augmenting the original data set with zeros. First, it must be understood that the finite observation interval results in a fundamental limit on the spectral resolution, even before the signals are sampled. The CT rectangular window has a $(\sin x)/x$ spectrum, which is convolved with the true spectrum of the analog signal. Therefore, the frequency resolution is limited by the width of the mainlobe in the $(\sin x)/x$ spectrum, which is inversely proportional to the length of the observation interval. Sampling causes a certain degree of aliasing, although this effect can be minimized by using a sufficiently high sampling rate. Therefore, lengthening the observation interval increases the fundamental resolution limit, while taking more samples within the observation interval minimizes aliasing distortion and provides a better definition (more sample points) on the underlying spectrum.

Padding the data with zeros and computing a longer FFT does give more frequency domain points (improved spectral resolution), but it does not improve the fundamental limit, nor does it alter the effects of aliasing error. The resolution limits are established by the observation interval and the sampling rate. No amount of zero padding can improve these basic limits. However, zero padding is a useful tool for providing more spectral definition, i.e., it enables one to get a better look at the (distorted) spectrum that results once the observation and sampling effects have occurred.

1.7.1.3 Leakage and the Picket-Fence Effect

An FFT with block length N can accurately resolve only frequencies $w_k = (2\pi/N)k, k = 0, \ldots, N-1$ that are integer multiples of the fundamental $w_1 = (2\pi/N)$. An analog waveform that is sampled and subjected to spectral analysis may have frequency components between the harmonics. For example, a component at frequency $w_{k+1/2} = (2\pi/N)(k + 1/2)$ will appear scattered throughout the spectrum. The effect is illustrated in Figure 1.12 for a sinusoid that is observed through a rectangular window and then sampled a N points. The "picket-fence effect" means that not all frequencies can be seen by the FFT. Harmonic components are seen accurately, but other components "slip through the picket fence" while their energy is "leaked" into the harmonics. These effects produce artifacts in the spectral domain that must be carefully monitored to assure that an accurate spectrum is obtained from FFT processing.

1.7.2 FIR Digital Filter Design

A common method for designing FIR digital filters is by use of windowing and FFT analysis. In general, window designs can be carried out with the aid of a hand calculator and a table of well-known window

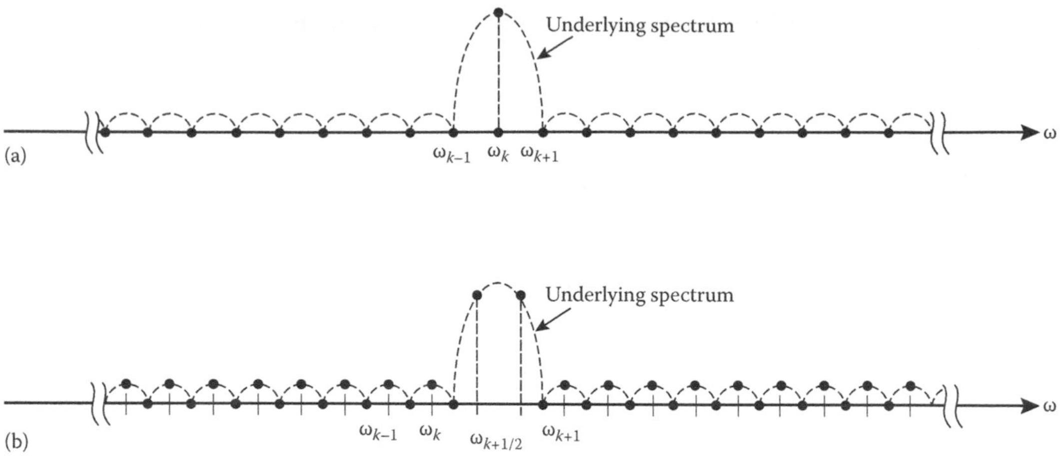

FIGURE 1.12 Illustration of leakage and the picket fence effects. (a) FFT of a windowed sinusoid with frequency $\omega_k = 2\pi k/N$ and (b) leakage for a nonharmonic sinusoidal component.

functions. Let $h[n]$ be the impulse response that corresponds to some desired frequency response, $H(e^{j\omega})$. If $H(e^{j\omega})$ has sharp discontinuities then $h[n]$ will represent an infinite impulse response function. The objective is to time-limit $h[n]$ in such a way as to not distort $H(e^{j\omega})$ any more than necessary. If $h[n]$ is simply truncated, a ripple (Gibbs phenomenon) occurs around the discontinuities in the spectrum, resulting in a distorted filter, as was earlier illustrated in Figure 1.7.

Suppose that $w[n]$ is a window function that time-limits $h[n]$ to create an FIR approximation, $h'[n]$; i.e., $h'[n] = w[n]h[n]$. Then if $W(e^{j\omega})$ is the DTFT of $w[n], h'[n]$ will have a Fourier transform given by $H'(e^{j\omega}) = W(e^{j\omega}) * H(e^{j\omega})$, where $*$ denotes convolution. From this it can be seen that the ripples in $H'(e^{j\omega})$ result from the sidelobes of $W(e^{j\omega})$. Ideally, $W(e^{j\omega})$ should be similar to an impulse so that $H'(e^{j\omega})$ is approximately equal to $H(e^{j\omega})$.

1.7.2.1 Special Case

Let $h[n] = \cos n\omega_0$, for all n. Then $h[n] = w[n] \cos n\omega_0$, and

$$H'(e^{j\omega}) = (1/2)W(e^{j(\omega+\bar{\omega})}) + (1/2)W(e^{j(\omega-\bar{\omega})}) \tag{1.20}$$

as illustrated in Figure 1.13. For this simple class, the center frequency of the passband is controlled by ω_0, and both the shape of the passband and the sidelobe structure are strictly determined by the choice of the window. While this simple class of FIRs does not allow for very flexible designs, it is a simple technique for determining quite useful lowpass, bandpass, and highpass FIR filters.

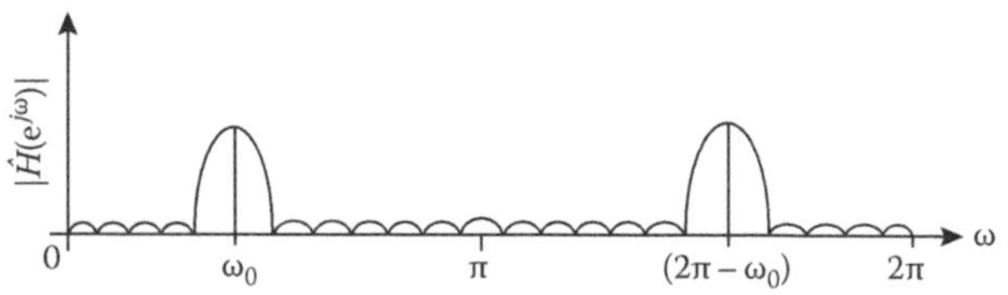

FIGURE 1.13 Design of a simple bandpass FIR filter by windowing.

1.7.2.2 General Case

Specify an ideal frequency response, $H(e^{j\omega})$, and choose samples at selected values of w. Use a long inverse FFT of length N' to find $h'[n]$, an approximation to $h[n]$, where if N is the desired length of the final filter, then $N' \gg N$. Then use a carefully selected window to truncate $h'[n]$ to obtain $h[n]$ by letting $h[n] = w[n]h'[n]$. Finally, use an FFT of length N' to find $H'(e^{j\omega})$. If $H'(e^{j\omega})$ is a satisfactory approximation to $H(e^{j\omega})$, the design is finished. If not, choose a new $H(e^{j\omega})$, or a new $w[n]$ and repeat. Throughout the design procedure it is important to choose $N' = kN$, with k an integer that is typically in the range $[4, \ldots, 10]$. Since this design technique is a trial-and-error procedure, the quality of the result depends to some degree on the skill and experience of the designer.

1.7.3 Fourier Block Processing in Real-Time Filtering Applications

In some practical applications, either the value of M is too large for the memory available, or $s[n]$ may not actually be finite in length, but rather a continual stream of data samples that must be processed by a filter at real time rates. Two well known algorithms are available that partition $s[n]$ into smaller blocks and process the individual blocks with a smaller-length DFT: (1) overlap-save partitioning and (2) overlap-add partitioning. Each of these algorithms is summarized below (Burrus and Parks 1985, Jenkins 2002)

1.7.3.1 Overlap-Save Processing

In this algorithm, N_{DFT} is chosen to be some convenient value with $N_{DFT} > N$. The signal, $s[n]$, is partitioned into blocks which are of length N_{DFT} and which overlap by $N-1$ data points. Hence, the kth block is $s_k[n] = s[n + k(N_{DFT} - N + 1)], n = 0, \ldots, N_{DFT} - 1$. The filter impulse response $h[n]$ is augmented with $N_{DFT} - N$ zeros to produce

$$h_{pad}[n] = \begin{bmatrix} h[n], & n = 0, \ldots, N-1 \\ 0, & n = N, \ldots, N_{DFT} - 1 \end{bmatrix}. \tag{1.21}$$

The DFT is then used to obtain $Y_{pad}[n] = \text{DFT}\{h_{pad}[n]\} \cdot \text{DFT}\{s_k[n]\}$, and $y_{pad}[n] = \text{IDFT}\{Y_{pad}[n]\}$. From the $y_{pad}[n]$ array the values that correctly correspond to the linear convolution are saved; values that are erroneous due to wraparound error caused by the circular convolution of the DFT are discarded. The kth block of the filtered output is obtained by

$$y_k[n] = \begin{bmatrix} y_{pad}[n], & n = 0, \ldots, N-1 \\ 0, & n = N, \ldots, N_{DFT} - 1 \end{bmatrix}. \tag{1.22}$$

For the overlap-save algorithm, each time a block is processed there are $N_{DFT} - N + 1$ points saved and $N-1$ points discarded. Each block moves forward by $N_{DFT} - N + 1$ data points and overlaps the previous block by $N-1$ points.

1.7.3.2 Overlap-Add Processing

This algorithm is similar to the previous one except that the kth input block is defined to be

$$s_k[n] = \begin{bmatrix} s[n], & n = 0, \ldots, L-1 \\ 0, & n = L, \ldots, N_{DFT} - 1 \end{bmatrix}, \tag{1.23}$$

where $L = N_{DFT} - N + 1$. The filter function $h_{pad}[n]$ is augmented with zeros, as before, to create $h_{pad}[n]$, and the DFT processing is executed as before. In each block $y_{pad}[n]$ that is obtained at the output, the first $N-1$ points are erroneous, the last $N-1$ points are erroneous, and the middle $N_{DFT} - 2(N-1)$ points correctly correspond to the linear convolution. However, if the last $N-1$ points from block k are overlapped with the first $N-1$ points from block $k+1$ and added pairwise, correct results corresponding to linear convolution are obtained from these positions, too. Hence, after this addition the number of correct points produced per block is $N_{DFT} - N + 1$, which is the same as that for the overlap-save algorithm. The overlap-add algorithm requires approximately the same amount of computation as the overlap-save algorithm, although the addition of the overlapping portions of blocks is extra. This feature, together with the extra delay of waiting for the next block to be finished before the previous one is complete, has resulted in more popularity for the overlap-save algorithm in practical applications.

Block filtering algorithms make it possible to efficiently filter continual data streams in real time because the FFT algorithm can be used to implement the DFT, thereby minimizing the total computation time and permits reasonably high overall data rates. However, block filtering generates data in bursts, i.e., there is a delay during which no filtered data appears, and then suddenly an entire block is generated. In real-time systems, buffering must be used. The block algorithms are particularly effective for filtering very long sequences of data that are pre-recorded on magnetic tape or disk.

1.7.4 Fourier Domain Adaptive Filtering

A transform domain adaptive filter (TDAF) is a generalization of the well-known least mean square (LMS) adaptive filter in which the input signal is passed through a linear transformation in order to decompose it into a set of orthogonal components and to optimize the adaptive step size for each component and thereby maximize the learning rate of the adaptive filter (Jenkins et al. 1996). The LMS algorithm is an approximation to the steepest descent optimization strategy. For a length N FIR filter with the input expressed as a column vector $\mathbf{x}(n) = [x(n), x(n-1), \ldots, x(n-N+1)]^{\mathrm{T}}$, the filter output $y(n)$ is expressed as

$$y(n) = \mathbf{w}^{\mathrm{T}}(n)\mathbf{x}(n),$$

where

$\mathbf{w}(n) = [w_0(n), w_1(n), \ldots, w_{N-1}(n)]^{\mathrm{T}}$ is the time varying vector of filter coefficients (tap weights) and superscript "T" denotes the vector transpose

The output error is formed as the difference between the filter output and a training signal $d(n)$, i.e. $e(n) = d(n) - y(n)$. Strategies for obtaining an appropriate $d(n)$ vary from one application to another. In many cases the availability of a suitable training signal determines whether an adaptive filtering solution will be successful in a particular application. The ideal cost function is defined by the mean squared error (MSE) criterion, $E\{|e(n)|^2\}$. The LMS algorithm is derived by approximating the ideal cost function by the instantaneous squared error, resulting in $J_{\mathrm{LMS}}(n) = |e(n)|^2$. While the LMS seems to make a rather crude approximation at the very beginning, the approximation results in an unbiased estimator. In many applications the LMS algorithm is quite robust and is able to converge rapidly to a small neighborhood of the Wiener solution.

When a steepest descent optimization strategy is combined with a gradient approximation formed using the LMS cost function $J_{\mathrm{LMS}}(n) = |e(n)|^2$, the conventional LMS adaptive algorithm results

$$\begin{aligned} \mathbf{w}(n+1) &= \mathbf{w}(n) + \mu e(n)\mathbf{x}(n), \\ e(n) &= d(n) - y(n), \end{aligned} \tag{1.24}$$

and

$$y(n) = \mathbf{x}(n)^{\mathrm{T}}\mathbf{w}(n).$$

The convergence behavior of the LMS algorithm, as applied to a direct form FIR filter structure, is controlled by the autocorrelation matrix $\mathbf{R}_x$ of the input process, where

$$\mathbf{R}_x \equiv E[\mathbf{x}^*(n)\mathbf{x}^{\mathrm{T}}(n)]. \tag{1.25}$$

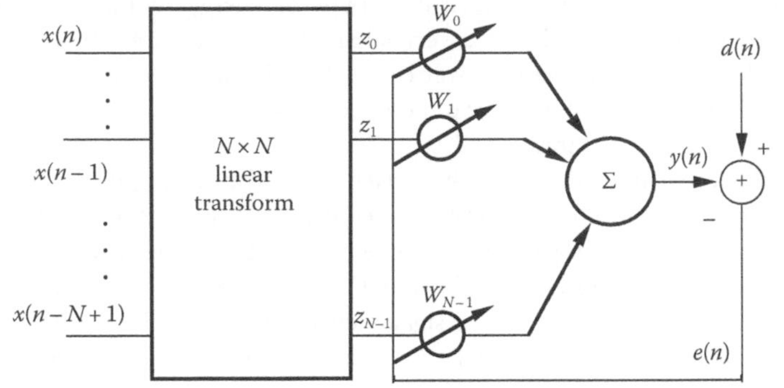

FIGURE 1.14 TDAF structure. (From Jenkins, W. K., Marshall, D. F., Kreidle, J. R., and Murphy, J. J., *IEEE Trans. Circuits Sys.*, 36(4), 474, 1989. With permission.)

The autocorrelation matrix $\mathbf{R}_x$ is usually positive definite, which is one of the conditions necessary to guarantee convergence to the Wiener solution. Another necessary condition for convergence is $0 < m < 1/l_{max}$, where l_{max} is the largest eigenvalue of $\mathbf{R}_x$. It is well established that the convergence of this algorithm is directly related to the eigenvalue spread of $\mathbf{R}_x$. The eigenvalue spread is measured by the condition number of $\mathbf{R}_x$, defined as $k = l_{max}/l_{min}$, where l_{min} is the minimum eigenvalue of $\mathbf{R}_x$. Ideal conditioning occurs when $k = 1$ (white noise); as this ratio increases, slower convergence results. The eigenvalue spread (condition number) depends on the spectral distribution of the input signal, and is related to the maximum and minimum values of the input power spectrum. From this line of reasoning it becomes clear that white noise is the ideal input signal for rapidly training an LMS adaptive filter. The adaptive process is slower and requires more computation for input signals that are colored.

The TDAF structure is shown in Figure 1.14. The input $x(n)$ and the desired signal $d(n)$ are assumed to be zero mean and jointly stationary. The input to the filter is a vector of N current and past input samples, defined in the previous section and denoted as $\mathbf{x}(n)$. This vector is processed by a unitary transform, such as the DFT. Once the filter order N is fixed, the transform is simply an $N \times N$ matrix $\mathbf{T}$, which is in general complex, with orthonormal rows. The transformed outputs form a vector $\mathbf{v}(n)$ which is given by

$$\mathbf{z}(n) = [v_0(n), v_1(n), \ldots, v_{N-1}(n)]^{\mathrm{T}} = \mathbf{T}\mathbf{x}(n).$$

With an adaptive tap vector defined as $\mathbf{w}(n) = [w_0(n), w_1(n), \ldots, w_{N-1}(n)]^{\mathrm{T}}$, the filter output is given by

$$y(n) = \mathbf{w}^{\mathrm{T}}(n)\mathbf{v}(n) = \mathbf{W}^{\mathrm{T}}(n)\mathbf{T}\mathbf{x}(n). \tag{1.26}$$

The instantaneous output error is then formed and used to update the adaptive filter taps using a modified form of the LMS algorithm (Jenkins et al. 1996):

$$\begin{aligned} \mathbf{W}(n+1) &= \mathbf{W}(n) + \mu e(n)\Lambda^{-2}\mathbf{v}^*(n) \\ \Lambda^2 &\equiv \operatorname{diag}\left[\sigma_1^2, \sigma_2^2, \ldots, \sigma_N^2\right], \end{aligned} \tag{1.27}$$

where $\sigma_i^2 = E[|v_i(n)|^2]$.

The power estimates σ_i^2 can be developed on-line by computing an exponentially weighted average of past samples according to

$$\sigma_i^2(n) = \alpha\sigma_i^2(n-1) + |v_i(n)|^2, \quad 0 < a < 1. \tag{1.28}$$

If σ_ι^2 becomes too small due to an insufficient amount of energy in the ith channel, the update mechanism becomes ill-conditioned due to a very large effective step size. In some cases the process will become unstable and register overflow will cause the adaptation to catastrophically fail. So the algorithm given by Equation 1.27 should have the update mechanism disabled for the ith orthogonal channel if σ_i^2 falls below a critical threshold.

The motivation for using the TDAF adaptive system instead of a simpler LMS based system is to achieve rapid convergence of the filters coefficients when the input signal is not white, while maintaining a reasonably low computational complexity requirement. The optimal decorrelating transform is composed of the orthonormal eigenvectors of the input autocorrelation matrix, and is known as the Karhunen–Loéve transform (KLT). The KLT is signal dependent and usually cannot be easily computed in real time. Throughout the literature the DFT, discrete cosine transform (DCT), and WHT have received considerable attention as possible candidates for use in TDAF.

Figure 1.15 shows learning characteristics for computer generated TDAF examples using six different orthogonal transforms to decorrelate the input signal. The examples presented are for system identification experiments, where the desired signal was derived by passing the input through an 8-tap FIR filter that is the "unknown system" to be identified. The filter input was generated by filtering white pseudonoise with a 32-tap linear phase FIR coloring filter to produce an input autocorrelation matrix with a condition number (eigenvalue ratio) of 681. Examples were then produced using the DFT, DCT, WHT, discrete Hartley transform (DHT), and a specially designed computationally efficient PO2 transform. The condition numbers that result from transform processing with each of these transforms are also shown in Figure 1.15. Note that all of the transforms used in this example are able to reduce the input condition number and greatly improve convergence rates, although some transforms are seen to be more effective than others for the coloring chosen for these examples.

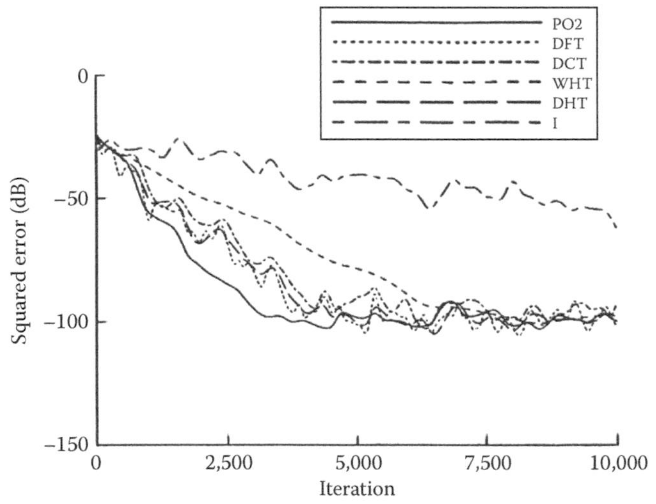

FIGURE 1.15 Comparison of (smoothed) learning curves for five different transforms operating on a colored noise input signal with condition number 681 fault in any of the coefficients. When R redundant coefficients are added as many as R coefficients can fail to adjust without any adverse effect on the filter's ability to achieve the minimum MSE condition. (From Jenkins, W. K., Marshall, D. F., Kreidle, J. R., and Murphy, J. J., *IEEE Trans. Circuits Sys.*, 36(4), 474, 1989. With permission.)

Transform	Effective Input Correlation Matrix Eigenvalue Ratio
Identity (I)	681
DFT	210
DCT	200
WHT	216
DHT	218
PO2 transform	128

1.7.5 Adaptive Fault Tolerance via Fourier Domain Adaptive Filtering

Adaptive systems adjust their parameters to minimize a specified error criterion under normal operating conditions. Fixed errors or Hardware faults would prevent the system to minimize the error criterion, but at the same time the system will adapt the parameters such that the best possible solution is reached. In adaptive fault tolerance the inherent learning ability of the adaptive system is used to compensate for failure of the adaptive coefficients. This mechanism can be used with specially designed structures whose redundant coefficients have the ability to compensate for the adjustment failures of other coefficients [Jenkins et al. 1996].

The FFT-based transform domain fault tolerant adaptive filter (FTAF) is described by the following equations:

$$\begin{aligned} \mathbf{x}[n] &= [\mathbf{x}_{\text{in}}[n], 0\ 0 \cdots 0] \\ \mathbf{x}_{\text{T}}[n] &= \mathbf{Tx}[n] \\ y[n] &= \mathbf{w}_{\text{T}}^{t}[n]\mathbf{x}_{\text{T}}[n] \\ e[n] &= y[n] - d[n], \end{aligned} \tag{1.29}$$

where

- $\mathbf{x}_{\text{in}}[n] = [x[n], x[n-1], \ldots, x[n-N+1]]$ is the vector of the current input and $N-1$ past inputs samples
- $\mathbf{x}[n]$ is $\mathbf{x}_{\text{in}}[n]$ zero-padded with R zeros
- $\mathbf{T}$ is the $M \times M$ DFT matrix where $M = N + R$
- $\mathbf{w}_{\text{T}}[n]$ is the vector of M adaptive coefficients in the transform domain
- $d[n]$ is the desired response
- $e[n]$ is the output error

The FFT-based transform domain FTAF is similar to a standard TDAF except that the input data vector is zero-padded with R zeros before it is multiplied by the transform matrix. Since the input data vector is zero padded the transform domain FTAF maintains a length N impulse response and has R redundant coefficients in the transform domain. When used with the zero padding strategy described above, this structure possesses a property called full fault tolerance, where each redundant coefficient is sufficient to compensate for a single "stuck at" fault condition in any of the coefficients. When R redundant coefficients are added as many as R coefficients can fail without any adverse effect on the filter's ability to achieve the minimum MSE condition.

An example of a transform domain FTAF with one redundant filter tap ($R = 1$) is demonstrated below for the identification of a 64-tap FIR lowpass "unknown" system. The training signal is Gaussian white

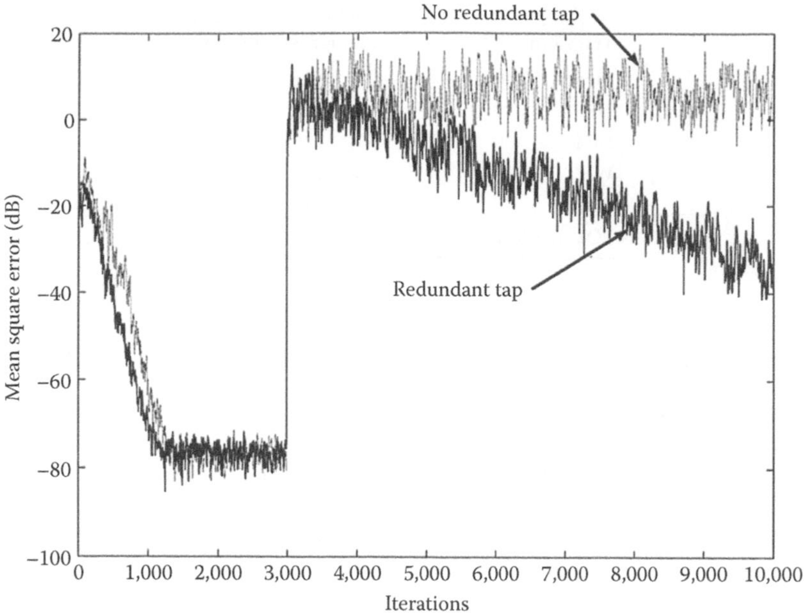

FIGURE 1.16 Learning curve demonstrating post-fault behavior both with and without a redundant tap.

noise with a unit variance and a noise floor of −60 dB. A fixed fault is introduced at iteration 3000 by setting an arbitrary filter coefficient to a random fixed value. Simulated learning curves are shown in Figure 1.16 both demonstrated that the redundant tap allows the filter to re-converge after the occurrence of the fault, although the post-fault convergence rate slowed somewhat due to an increased condition number of the post-fault autocorrelation matrix [Jenkins et al. 1996].

1.8 Summary

Numerous Fourier transform concepts have been presented for both CT and DT signals and systems. Emphasis was placed on illustrating how various forms of the Fourier transform relate to one another, and how they are all derived from more general complex transforms, the complex Fourier (or bilateral Laplace) transform for CT, and the bilateral z-transform for DT. It was shown that many of these transforms have similar properties that are inherited from their parent forms, and that there is a parallel hierarchy among Fourier transform concepts in the CT and DT domains. Both CT and DT sampling models were introduced as a means of representing sampled signals in these two different domains and it was shown that the models are equivalent by virtue of having the same Fourier spectra when transformed into the Fourier domain with the appropriate Fourier transform. It was shown how Fourier analysis properly characterizes the relationship between the spectra of a CT signal and its DT counterpart obtained by sampling, and the classical reconstruction formula was obtained as a result of this analysis. Finally, the DFT, the backbone for much of modern DSP, was obtained from more classical forms of the Fourier transform by simultaneously discretizing the time and frequency domains. The DFT, together with the remarkable computational efficiency provided by the FFT algorithm, has contributed to the resounding success that engineers and scientists have had in applying DSP to many practical scientific problems.

References

Blahut, R. E., *Fast Algorithms for Digital Signal Processing*, Reading, MA: Addison-Wesley Publishing Co., 1985.

Bracewell, R. N., *The Fourier Transform*, 2nd edition, New York: McGraw-Hill, 1986.

Brigham, E. O., *The Fast Fourier Transform*, Englewood Cliffs, NJ: Prentice-Hall, 1974.

Burrus, C. S. and Parks, T. W., *DFT/FFT and Convolution Algorithms*, New York: John Wiley and Sons, 1985.

Jenkins, W. K., Discrete-time signal processing, in *Reference Data for Engineers: Radio, Electronics, Computers, and Communications*, Wendy M. Middleton (editor-in-chief), 9th edition, Carmel, MA: Newnes (Butterworth-Heinemann), 2002, Chapter 28.

Jenkins, W. K. and Desai, M. D., The discrete-frequency Fourier transform, *IEEE Transactions on Circuits and Systems*, CAS-33(7), 732–734, July 1986.

Jenkins, W. K. et al., *Advanced Concepts in Adaptive Signal Processing*, Boston, MA: Kluwer Academic Publishers, 1996.

Oppenheim, A. V. and Schafer, R. W., *Digital Signal Processing*, Englewood Cliffs, NJ: Prentice-Hall, 1975.

Oppenheim, A. V. and Schafer, R. W., *Discrete-Time Signal Processing*, Englewood Cliffs, NJ: Prentice-Hall, 1989.

Oppenheim, A. V., Willsky, A. S., and Young, I.T., *Signals and Systems*, Englewood Cliffs, NJ: Prentice-Hall, 1983.

VanValkenburg, M. E., *Network Analysis*, 3rd edition, Englewood Cliffs, NJ: Prentice-Hall, 1974.

2

Ordinary Linear Differential and Difference Equations

B.P. Lathi
California State University

2.1 Differential Equations

A function containing variables and their derivatives is called a differential expression, and an equation involving differential expressions is called a differential equation. A differential equation is an ordinary differential equation if it contains only one independent variable; it is a partial differential equation if it contains more than one independent variable. We shall deal here only with ordinary differential equations.

In the mathematical texts, the independent variable is generally x, which can be anything such as time, distance, velocity, pressure, and so on. In most of the applications in control systems, the independent variable is time. For this reason we shall use here independent variable t for time, although it can stand for any other variable as well.

The following equation

$$\left(\frac{d^2y}{dt^2}\right)^4 + 3\frac{dy}{dt} + 5y^2(t) = \sin t$$

is an ordinary differential equation of second order because the highest derivative is of the second order. An nth-order differential equation is linear if it is of the form

$$a_n(t)\frac{d^ny}{dt^n} + a_{n-1}(t)\frac{d^{n-1}y}{dt^{n-1}} + \cdots + a_1(t)\frac{dy}{dt} + a_0(t)y(t) = r(t) \tag{2.1}$$

where the coefficients $a_i(t)$ are not functions of $y(t)$. If these coefficients (a_i) are constants, the equation is linear with constant coefficients. Many engineering (as well as nonengineering) systems can be modeled by these equations. Systems modeled by these equations are known as linear time-invariant (LTI) systems. In this chapter we shall deal exclusively with linear differential equations with constant coefficients. Certain other forms of differential equations are dealt with elsewhere in this book.

2.1.1 Role of Auxiliary Conditions in Solution of Differential Equations

We now show that a differential equation does not, in general, have a unique solution unless some additional constraints (or conditions) on the solution are known. This fact should not come as a surprise. A function $y(t)$ has a unique derivative $\mathrm{d}y/\mathrm{d}t$, but for a given derivative $\mathrm{d}y/\mathrm{d}t$ there are infinite possible functions $y(t)$. If we are given $\mathrm{d}y/\mathrm{d}t$, it is impossible to determine $y(t)$ uniquely unless an additional piece of information about $y(t)$ is given. For example, the solution of a differential equation

$$\frac{\mathrm{d}y}{\mathrm{d}t} = 2 \tag{2.2}$$

obtained by integrating both sides of the equation is

$$y(t) = 2t + c \tag{2.3}$$

for any value of c. Equation 2.2 specifies a function whose slope is 2 for all t. Any straight line with a slope of 2 satisfies this equation. Clearly the solution is not unique, but if we place an additional constraint on the solution $y(t)$, then we specify a unique solution.

For example, suppose we require that $y(0) = 5$; then out of all the possible solutions available, only one function has a slope of 2 and an intercept with the vertical axis at 5. By setting $t = 0$ in Equation 2.3 and substituting $y(0) = 5$ in the same equation, we obtain $y(0) = 5 = c$ and

$$y(t) = 2t + 5$$

which is the unique solution satisfying both Equation 2.2 and the constraint $y(0) = 5$.

In conclusion, differentiation is an irreversible operation during which certain information is lost. To reverse this operation, one piece of information about $y(t)$ must be provided to restore the original $y(t)$. Using a similar argument, we can show that, given $\mathrm{d}^2y/\mathrm{d}t^2$, we can determine $y(t)$ uniquely only if two additional pieces of information (constraints) about $y(t)$ are given. In general, to determine $y(t)$ uniquely from its nth derivative, we need n additional pieces of information (constraints) about $y(t)$. These constraints are also called auxiliary conditions. When these conditions are given at $t = 0$, they are called initial conditions.

We discuss here two systematic procedures for solving linear differential equations of the form in Equation 2.1. The first method is the classical method, which is relatively simple, but restricted to a certain class of inputs. The second method (the convolution method) is general and is applicable to all types of inputs. A third method (Laplace transform) is discussed elsewhere in this book. Both the methods discussed here are classified as time-domain methods because with these methods we are able to solve the above equation directly, using t as the independent variable. The method of Laplace transform (also known as the frequency-domain method), on the other hand, requires transformation of variable t into a frequency variable s.

In engineering applications, the form of linear differential equation that occurs most commonly is given by

$$\begin{aligned}\frac{\mathrm{d}^n y}{\mathrm{d}t^n} &+ a_{n-1}\frac{\mathrm{d}^{n-1}y}{\mathrm{d}t^{n-1}} + \cdots + a_1\frac{\mathrm{d}y}{\mathrm{d}t} + a_0 y(t)\\ &= b_m\frac{\mathrm{d}^m f}{\mathrm{d}t^m} + b_{m-1}\frac{\mathrm{d}^{m-1}f}{\mathrm{d}t^{m-1}} + \cdots + b_1\frac{\mathrm{d}f}{\mathrm{d}t} + b_0 f(t)\end{aligned} \tag{2.4a}$$

where all the coefficients a_i and b_i are constants. Using operational notation D to represent d/dt, this equation can be expressed as

$$\begin{aligned}&\left(D^n + a_{n-1}D^{n-1} + \cdots + a_1D + a_0\right)y(t)\\&\quad = \left(b_mD^m + b_{m-1}D^{m-1} + \cdots + b_1D + b_0\right)f(t)\end{aligned} \tag{2.4b}$$

or

$$Q(D)y(t) = P(D)f(t) \tag{2.4c}$$

where the polynomials $Q(D)$ and $P(D)$, respectively, are

$$\begin{aligned}Q(D) &= D^n + a_{n-1}D^{n-1} + \cdots + a_1D + a_0\\P(D) &= b_mD^m + b_{m-1}D^{m-1} + \cdots + b_1D + b_0\end{aligned}$$

Observe that this equation is of the form of Equation 2.1, where $r(t)$ is in the form of a linear combination of $f(t)$ and its derivatives. In this equation, $y(t)$ represents an output variable, and $f(t)$ represents an input variable of an LTI system. Theoretically, the powers m and n in the above equations can take on any value. Practical noise considerations, however, require [1] $m \leq n$.

2.1.2 Classical Solution

When $f(t) \equiv 0$, Equation 2.4 is known as the homogeneous (or complementary) equation. We shall first solve the homogeneous equation. Let the solution of the homogeneous equation be $y_c(t)$, that is,

$$Q(D)y_c(t) = 0$$

or

$$\left(D^n + a_{n-1}D^{n-1} + \cdots + a_1D + a_0\right)y_c(t) = 0$$

We first show that if $y_p(t)$ is the solution of Equation 2.4, then $y_c(t) + y_P(t)$ is also its solution. This follows from the fact that

$$Q(D)y_c(t) = 0$$

If $y_P(t)$ is the solution of Equation 2.4, then

$$Q(D)y_P(t) = P(D)f(t)$$

Addition of these two equations yields

$$Q(D)[y_c(t) + y_P(t)] = P(D)f(t)$$

Thus, $y_c(t) + y_P(t)$ satisfies Equation 2.4 and therefore is the general solution of Equation 2.4. We call $y_c(t)$ the complementary solution and $y_P(t)$ the particular solution. In system analysis parlance, these components are called the natural response and the forced response, respectively.

2.1.2.1 Complementary Solution (the Natural Response)

The complementary solution $y_c(t)$ is the solution of

$$Q(D)y_c(t) = 0 \tag{2.5a}$$

or

$$\left(D^n + a_{n-1}D^{n-1} + \cdots + a_1 D + a_0\right)y_c(t) = 0 \tag{2.5b}$$

A solution to this equation can be found in a systematic and formal way. However, we will take a short cut by using heuristic reasoning. Equation 2.5b shows that a linear combination of $y_c(t)$ and its n successive derivatives is zero, not at some values of t, but for all t. This is possible if and only if $y_c(t)$ and all its n successive derivatives are of the same form. Otherwise their sum can never add to zero for all values of t. We know that only an exponential function $e^{\lambda t}$ has this property. So let us assume that

$$y_c(t) = ce^{\lambda t}$$

is a solution to Equation 2.5b. Now

$$\begin{aligned} Dy_c(t) &= \frac{dy_c}{dt} = c\lambda e^{\lambda t} \\ D^2 y_c(t) &= \frac{d^2 y_c}{dt^2} = c\lambda^2 e^{\lambda t} \\ &\vdots \\ D^n y_c(t) &= \frac{d^n y_c}{dt^n} = c\lambda^n e^{\lambda} t \end{aligned}$$

Substituting these results in Equation 2.5b, we obtain

$$c\left(\lambda^n + a_{n-1}\lambda^{n-1} + \cdots + a_1\lambda + a_0\right)e^{\lambda t} = 0$$

For a nontrivial solution of this equation,

$$\lambda^n + a_{n-1}\lambda^{n-1} + \cdots + a_1\lambda + a_0 = 0 \tag{2.6a}$$

This result means that $ce^{\lambda t}$ is indeed a solution of Equation 2.5 provided that λ satisfies Equation 2.6a. Note that the polynomial in Equation 2.6a is identical to the polynomial $Q(D)$ in Equation 2.5b, with λ replacing D. Therefore, Equation 2.6a can be expressed as

$$Q(\lambda) = 0 \tag{2.6b}$$

When $Q(\lambda)$ is expressed in factorized form, Equation 2.6b can be represented as

$$Q(\lambda) = (\lambda - \lambda_1)(\lambda - \lambda_2)\cdots(\lambda - \lambda_n) = 0 \tag{2.6c}$$

Clearly λ has n solutions: $\lambda_1, \lambda_2, \ldots, \lambda_n$. Consequently, Equation 2.5 has n possible solutions: $c_1 e^{\lambda_1 t}$, $c_2 e^{\lambda_2 t}, \ldots, c_n e^{\lambda_n t}$, with $c_1, c_2, \ldots, c_n$ as arbitrary constants. We can readily show that a general solution is given by the sum of these n solutions,* so that

$$y_c(t) = c_1 e^{\lambda_1 t} + c_2 e^{\lambda_2 t} + \cdots + c_n e^{\lambda_n t} \tag{2.7}$$

where $c_1, c_2, \ldots, c_n$ are arbitrary constants determined by n constraints (the auxiliary conditions) on the solution.

The polynomial $Q(\lambda)$ is known as the characteristic polynomial. The equation

$$Q(\lambda) = 0 \tag{2.8}$$

is called the characteristic or auxiliary equation. From Equation 2.6c, it is clear that $\lambda_1, \lambda_2, \ldots, \lambda_n$ are the roots of the characteristic equation; consequently, they are called the characteristic roots. The terms characteristic values, eigenvalues, and natural frequencies are also used for characteristic roots.† The expotentials $e^{\lambda_i t} (i = 1, 2, \ldots, n)$ in the complementary solution are the characteristic modes (also known as modes or natural modes). There is a characteristic mode for each characteristic root, and the complementary solution is a linear combination of the characteristic modes.

2.1.2.2 Repeated Roots

The solution of Equation 2.5 as given in Equation 2.7 assumes that the characteristic roots $\lambda_1, \lambda_2, \ldots, \lambda_n$ are distinct. If there are repeated roots (same root occurring more than once), the form of the solution is modified slightly. By direct substitution we can show that the solution of the equation

$$(D - \lambda)^2 y_c(t) = 0$$

is given by

$$y_c(t) = (c_1 + c_2 t) e^{\lambda t}$$

In this case the root λ repeats twice. Observe that the characteristic modes in this case are $e^{\lambda t}$ and $te^{\lambda t}$. Continuing this pattern, we can show that for the differential equation

$$(D - \lambda)^r y_c(t) = 0 \tag{2.9}$$

the characteristic modes are $e^{\lambda t}$, $te^{\lambda t}$, $t^2 e^{\lambda t}, \ldots, t^{r-1} e^{\lambda t}$, and the solution is

$$y_c(t) = \left(c_1 + c_2 t + \cdots + c_r t^{r-1}\right) e^{\lambda t} \tag{2.10}$$

---

* To prove this fact, assume that $y_1(t), y_2(t), \ldots, y_n(t)$ are all solutions of Equation 2.5. Then

$$Q(D) y_1(t) = 0$$
$$Q(D) y_2(t) = 0$$
$$\vdots$$
$$Q(D) y_n(t) = 0$$

Multiplying these equations by $c_1, c_2, \ldots, c_n$, respectively, and adding them together yields

$$Q(D)[c_1 y_1(t) + c_2 y_n(t)] = 0$$

This result shows that $c_1 y_1(t) + c_2 y_2(t) + \cdots + c_n y_n(t)$ is also a solution of the homogeneous equation (Equation 2.5).

† The term *eigenvalue* is German for characteristic value.

Consequently, for a characteristic polynomial

$$Q(\lambda) = (\lambda - \lambda_1)^r(\lambda - \lambda_{r+1})\ldots(\lambda - \lambda_n)$$

the characteristic modes are $e^{\lambda_1 t}$, $te^{\lambda_1 t}, \ldots, t^{r-1}\,e^{\lambda t}$, $e^{\lambda_{r+1}t}, \ldots, e^{\lambda_n t}$ and the complementary solution is

$$y_c(t) = \left(c_1 + c_2 t + \cdots + c_r t^{r-1}\right)e^{\lambda t} + c_{r+1}e^{\lambda_{r+1}t} + \cdots + c_n e^{\lambda_n t}$$

2.1.2.3 Particular Solution (the Forced Response): Method of Undetermined Coefficients

The particular solution $y_p(t)$ is the solution of

$$Q(D)y_p(t) = P(D)f(t) \tag{2.11}$$

It is a relatively simple task to determine $y_p(t)$ when the input $f(t)$ is such that it yields only a finite number of independent derivatives. Inputs having the form $e^{\zeta t}$ or t^r fall into this category. For example, $e^{\zeta t}$ has only one independent derivative; the repeated differentiation of $e^{\zeta t}$ yields the same form, that is, $e^{\zeta t}$. Similarly, the repeated differentiation of t^r yields only r independent derivatives. The particular solution to such an input can be expressed as a linear combination of the input and its independent derivatives. Consider, for example, the input $f(t) = at^2 + bt + c$. The successive derivatives of this input are $2at + b$ and $2a$. In this case, the input has only two independent derivatives. Therefore the particular solution can be assumed to be a linear combination of $f(t)$ and its two derivatives. The suitable form for $y_p(t)$ in this case is therefore

$$y_p(t) = \beta_2 t^2 + \beta_1 t + \beta_0$$

The undetermined coefficients β_0, β_1, and β_2 are determined by substituting this expression for $y_p(t)$ in Equation 2.11 and then equating coefficients of similar terms on both sides of the resulting expression.

Although this method can be used only for inputs with a finite number of derivatives, this class of inputs includes a wide variety of the most commonly encountered signals in practice. Table 2.1 shows a variety of such inputs and the form of the particular solution corresponding to each input. We shall demonstrate this procedure with an example.

Note: By definition, $y_p(t)$ cannot have any characteristic mode terms. If any term $p(t)$ shown in the right-hand column for the particular solution is also a characteristic mode, the correct form of the forced response must be modified to $t^i p(t)$, where i is the smallest possible integer that can be used and still can prevent $t^i p(t)$ from having characteristic mode term. For example, when the input is $e^{\zeta t}$, the forced response (right-hand column) has the form $\beta e^{\zeta t}$. But if $e^{\zeta t}$ happens to be a characteristic mode, the correct form of the particular solution is $\beta t e^{\zeta t}$ (see Pair 2). If $te^{\zeta t}$ also happens to be characteristic mode, the correct form of the particular solution is $\beta t^2 e^{\zeta t}$, and so on.

TABLE 2.1 Inputs and Responses for Commonly Encountered Signals

No.	Input $f(t)$	Forced Response
1	$e^{\zeta t}\ \zeta \neq \lambda_i (i = 1, 2, \ldots, n)$	$\beta e^{\zeta t}$
2	$e^{\zeta t}\ \zeta \neq \lambda_i$	$\beta t e^{\zeta t}$
3	k (a constant)	β (a constant)
4	$\cos(\omega t + \theta)$	$(\beta \cos(\omega t + \varphi)$
5	$(t^r + \alpha_{r-1}t^{r-1} + \cdots + \alpha_1 t + \alpha_0)e^{\zeta t}$	$(\beta_r t^r + \beta_{r-1}t^{r-1} + \cdots + \beta_1 t + \beta_0)e^{\zeta t}$

Example 2.1

Solve the differential equation

$$(D^2 + 3D + 2)y(t) = Df(t) \tag{2.12}$$

if the input

$$f(t) = t^2 + 5t + 3$$

and the initial conditions are $y(0^+) = 2$ and $\dot{y}(0^+) = 3$.

The characteristic polynomial is

$$\lambda^2 + 3\lambda + 2 = (\lambda + 1)(\lambda + 2)$$

Therefore the characteristic modes are e^{-t} and e^{-2t}. The complementary solution is a linear combination of these modes, so that

$$y_c(t) = c_1 e^{-t} + c_2 e^{-2t} \quad t \geq 0$$

Here the arbitrary constants c_1 and c_2 must be determined from the given initial conditions.

The particular solution to the input $t^2 + 5t + 3$ is found from Table 2.1 (Pair 5 with $\zeta = 0$) to be

$$y_p(t) = \beta_2 t^2 + \beta_1 t + \beta_0$$

Moreover, $y_p(t)$ satisfies Equation 2.11, that is,

$$(D^2 + 3D + 2)y_p(t) = Df(t) \tag{2.13}$$

Now

$$Dy_p(t) = \frac{d}{dt}\left(\beta_2 t^2 + \beta_1 t + \beta_0\right) = 2\beta_2 t + \beta_1$$

$$D^2 y_p(t) = \frac{d^2}{dt^2}\left(\beta_2 t^2 + \beta_1 t + \beta_0\right) = 2\beta_2$$

and

$$Df(t) = \frac{d}{dt}[t^2 + 5t + 3] = 2t + 5$$

Substituting these results in Equation 2.13 yields

$$2\beta_2 + 3(2\beta_2 t + \beta_1) + 2(\beta_2 t^2 + \beta_1 t + \beta_0) = 2t + 5$$

or

$$2\beta_2 t^2 + (2\beta_1 + 6\beta_2)t + (2\beta_0 + 3\beta_1 + 2\beta_2) = 2t + 5$$

Equating coefficients of similar powers on both sides of this expression yields

$$2\beta_2 = 0$$
$$2\beta_1 + 6\beta_2 = 2$$
$$2\beta_0 + 3\beta_1 + 2\beta_2 = 5$$

Solving these three equations for their unknowns, we obtain $\beta_0 = 1$, $\beta_1 = 1$, and $\beta_2 = 0$. Therefore,

$$y_p(t) = t + 1 \quad t > 0$$

The total solution $y(t)$ is the sum of the complementary and particular solutions. Therefore,

$$\begin{aligned} y(t) &= y_c(t) + y_p(t) \\ &= c_1 e^{-t} + c_2 e^{-2t} + t + 1 \quad t > 0 \end{aligned}$$

so that

$$\dot{y}(t) = -c_1 e^{-t} - 2c_2 e^{-2t} + 1$$

Setting $t = 0$ and substituting the given initial conditions $y(0) = 2$ and $\dot{y}(0) = 3$ in these equations, we have

$$2 = c_1 + c_2 + 1$$
$$3 = -c_1 - 2c_2 + 1$$

The solution to these two simultaneous equations is $c_1 = 4$ and $c_2 = -3$. Therefore,

$$y(t) = 4e^{-t} - 3e^{-2t} + t + 1 \quad t \geq 0$$

2.1.2.4 The Exponential Input $e^{\zeta t}$

The exponential signal is the most important signal in the study of LTI systems. Interestingly, the particular solution for an exponential input signal turns out to be very simple. From Table 2.1 we see that the particular solution for the input $e^{\zeta t}$ has the form $\beta e^{\zeta t}$. We now show that $\beta = Q(\zeta)/P(\zeta)$.* To determine the constant β, we substitute $y_p(t) = \beta e^{\zeta t}$ in Equation 2.11, which gives us

$$Q(D)\left[\beta e^{\zeta t}\right] = P(D)e^{\zeta t} \tag{2.14a}$$

Now observe that

$$\begin{aligned} De^{\zeta t} &= \frac{d}{dt}(e^{\zeta t}) = \zeta e^{\zeta t} \\ D^2 e^{\zeta t} &= \frac{d^2}{dt^2}(e^{\zeta t}) = \zeta^2 e^{\zeta t} \\ &\vdots \\ D^r e^{\zeta t} &= \zeta^r e^{\zeta t} \end{aligned}$$

* This is true only if ζ is not a characteristic root.

Consequently,

$$Q(D)e^{\zeta t} = Q(\zeta)e^{\zeta t} \quad \text{and} \quad P(D)e^{\zeta t} = P(\zeta)e^{\zeta t}$$

Therefore, Equation 2.14a becomes

$$\beta Q(\zeta)e^{\zeta t} = P(\zeta)e^{\zeta t} \tag{2.14b}$$

and

$$\beta = \frac{P(\zeta)}{Q(\zeta)}$$

Thus, for the input $f(t) = e^{\zeta t}$, the particular solution is given by

$$y_p(t) = H(\zeta)e^{\zeta t} \quad t > 0 \tag{2.15a}$$

where

$$H(\zeta) = \frac{P(\zeta)}{Q(\zeta)} \tag{2.15b}$$

This is an interesting and significant result. It states that for an exponential input $e^{\zeta t}$ the particular solution $y_p(t)$ is the same exponential multiplied by $H(\zeta) = P(\zeta)/Q(\zeta)$. The total solution $y(t)$ to an exponential input $e^{\zeta t}$ is then given by

$$y(t) = \sum_{j=1}^{n} c_j e^{\lambda_j t} + H(\zeta)e^{\zeta t}$$

where the arbitrary constants $c_1, c_2, \ldots, c_n$ are determined from auxiliary conditions.

Recall that the exponential signal includes a large variety of signals, such as a constant ($\zeta = 0$), a sinusoid ($\zeta = \pm j\omega$), and an exponentially growing or decaying sinusoid ($\zeta = \sigma \pm j\omega$). Let us consider the forced response for some of these cases.

2.1.2.5 The Constant Input $f(t) = C$

Because $C = Ce^{0t}$, the constant input is a special case of the exponential input $Ce^{\zeta t}$ with $\zeta = 0$. The particular solution to this input is then given by

$$\begin{aligned} y_p(t) &= CH(\zeta)e^{\zeta t} \quad \text{with } \zeta = 0 \\ &= CH(0) \end{aligned} \tag{2.16}$$

2.1.2.6 The Complex Exponential Input $e^{j\omega t}$

Here $\zeta = j\omega$, and

$$y_p(t) = H(j\omega)e^{j\omega t} \tag{2.17}$$

2.1.2.7 The Sinusoidal Input $f(t) = \cos \omega_0 t$

We know that the particular solution for the input $e^{\pm j\omega t}$ is $H(\pm j\omega)e^{\pm j\omega t}$. Since $\cos \omega t = (e^{j\omega t} + e^{-j\omega t})/2$, the particular solution to $\cos \omega t$ is

$$y_p(t) = \frac{1}{2}\left[H(j\omega)e^{j\omega t} + H(-j\omega)e^{-j\omega t}\right]$$

Because the two terms on the right-hand side are conjugates,

$$y_p(t) = \mathrm{Re}\left[H(j\omega)e^{j\omega t}\right]$$

But

$$H(j\omega) = |H(j\omega)|e^{j\angle H(j\omega)}$$

so that

$$\begin{aligned} y_p(t) &= \mathrm{Re}\left\{|H(j\omega)|e^{j[\omega t + \angle H(j\omega)]}\right\} \\ &= |H(j\omega)|\cos[\omega t + \angle H(j\omega)] \end{aligned} \tag{2.18}$$

This result can be generalized for the input $f(t) = \cos(\omega t + \theta)$. The particular solution in this case is

$$y_p(t) = |H(j\omega)|\cos[\omega t + \theta + \angle H(j\omega)] \tag{2.19}$$

Example 2.2

Solve Equation 2.12 for the following inputs:

(a) $10e^{-3t}$ (b) 5 (c) e^{-2t} (d) $10\cos(3t + 30°)$.

The initial conditions are $y(0^+) = 2$, $\dot{y}(0^+) = 3$.

The complementary solution for this case is already found in Example 2.1 as

$$y_c(t) = c_1 e^{-t} + c_2 e^{-2t} \quad t \geq 0$$

For the exponential input $f(t) = e^{\zeta t}$, the particular solution, as found in Equation 2.15 is $H(\zeta)e^{\zeta t}$, where

$$H(\zeta) = \frac{P(\zeta)}{Q(\zeta)} = \frac{\zeta}{\zeta^2 + 3\zeta + 2}$$

(a) For input $f(t) = 10e^{-3t}$, $\zeta = -3$, and

$$\begin{aligned} y_p(t) &= 10H(-3)e^{-3t} \\ &= 10\left[\frac{-3}{(-3)^2 + 3(-3) + 2}\right]e^{-3t} \\ &= -15e^{-3t} \quad t > 0 \end{aligned}$$

The total solution (the sum of the complementary and particular solutions) is

$$y(t) = c_1 e^{-t} + c_2 e^{-2t} - 15e^{-3t} \quad t \geq 0$$

and

$$\dot{y}(t) = -c_1 e^{-t} - 2c_2 e^{-2t} + 45e^{-3t} \quad t \geq 0$$

The initial conditions are $y(0^+) = 2$ and $\dot{y}(0^+) = 3$. Setting $t = 0$ in the above equations and substituting the initial conditions yields

$$c_1 + c_2 - 15 = 2 \quad \text{and} \quad -c_1 - 2c_2 + 45 = 3$$

Solution of these equations yields $c_1 = -8$ and $c_2 = 25$. Therefore,

$$y(t) = -8e^{-t} + 25e^{-2t} - 15e^{-3t} \quad t \geq 0$$

(b) For input $f(t) = 5 = 5e^{0t}$, $\zeta = 0$, and

$$y_p(t) = 5H(0) = 0 \quad t > 0$$

The complete solution is $y(t) = y_c(t) + y_p(t) = c_1 e^{-t} + c_2 e^{-2t}$. We then substitute the initial conditions to determine c_1 and c_2 as explained in (a).

(c) Here $\zeta = -2$, which is also a characteristic root. Hence (see Pair 2, Table 2.1, or the comment at the bottom of the table),

$$y_p(t) = \beta t e^{-2t}$$

To find β, we substitute $y_p(t)$ in Equation 2.11, giving us

$$(D^2 + 3D + 2)y_p(t) = Df(t)$$

or

$$(D^2 + 3D + 2)\left[\beta t e^{-2t}\right] = De^{-2t}$$

But

$$D[\beta t e^{-2t}] = \beta(1 - 2t)e^{-2t}$$
$$D^2[\beta t e^{-2t}] = 4\beta(t - 1)e^{-2t}$$
$$De^{-2t} = -2e^{-2t}$$

Consequently,

$$\beta(4t - 4 + 3 - 6t + 2t)e^{-2t} = -2e^{-2t}$$

or

$$-\beta e^{-2t} = -2e^{-2t}$$

This means that $\beta = 2$, so that

$$y_p(t) = 2te^{-2t}$$

The complete solution is $y(t) = y_c(t) + y_p(t) = c_1e^{-t} + c_2e^{-2t} + 2te^{-2t}$. We then substitute the initial conditions to determine c_1 and c_2 as explained in (a).

(d) For the input $f(t) = 10 \cos (3t + 30°)$, the particular solution (see Equation 2.19) is

$$y_p(t) = 10|H(j3)| \cos[3t + 30° + \angle H(j3)]$$

where

$$H(j3) = \frac{P(j3)}{Q(j3)} = \frac{j3}{(j3)^2 + 3(j3) + 2}$$
$$= \frac{j3}{-7 + j9} = \frac{27 - j21}{130} = 0.263e^{-j37.9°}$$

Therefore,

$$|H(j3)| = 0.263, \quad \angle H(j3) = -37.9°$$

and

$$y_p(t) = 10(0.263) \cos (3t + 30° - 37.9°)$$
$$= 2.63 \cos (3t - 7.9°)$$

The complete solution is $y(t) = y_c(t) + y_p(t) = c_1e^{-t} + c_2e^{-2t} + 2.63 \cos (3t - 7.9°)$. We then substitute the initial conditions to determine c_1 and c_2 as explained in (a).

2.1.3 Method of Convolution

In this method, the input $f(t)$ is expressed as a sum of impulses. The solution is then obtained as a sum of the solutions to all the impulse components. The method exploits the superposition property of the linear differential equations. From the sampling (or sifting) property of the impulse function, we have

$$f(t) = \int_0^t f(x)\delta(t - x)dx \quad t \geq 0 \tag{2.20}$$

The right-hand side expresses $f(t)$ as a sum (integral) of impulse components. Let the solution of Equation 2.4 be $y(t) = h(t)$ when $f(t) = \delta(t)$ and all the initial conditions are zero. Then use of the linearity property yields the solution of Equation 2.4 to input $f(t)$ as

$$y(t) = \int_0^t f(x)h(t - x)dx \tag{2.21}$$

For this solution to be general, we must add a complementary solution. Thus, the general solution is given by

$$y(t) = \sum_{j=1}^{n} c_j e^{\lambda_j t} + \int_0^t f(x)h(t - x)dx \tag{2.22}$$

The first term on the right-hand side consists of a linear combination of natural modes and should be appropriately modified for repeated roots. For the integral on the right-hand side, the lower limit 0 is understood to be 0^- in order to ensure that impulses, if any, in the input $f(t)$ at the origin are accounted for. The integral on the right-hand side of Equation 2.22 is well known in the literature as the convolution integral. The function $h(t)$ appearing in the integral is the solution of Equation 2.4 for the impulsive input $[f(t) = \delta(t)]$. It can be shown that [2]

$$h(t) = P(D)[y_o(t)u(t)] \tag{2.23}$$

where $y_o(t)$ is a linear combination of the characteristic modes subject to initial conditions

$$\begin{aligned} &y_o^{(n-1)}(0) = 1 \\ &y_o(0) = y_o^{(1)}(0) = \cdots = y_o^{(n-2)}(0) = 0 \end{aligned} \tag{2.24}$$

The function $u(t)$ appearing on the right-hand side of Equation 2.23 represents the unit step function, which is unity for $t \geq 0$ and is 0 for $t < 0$.

The right-hand side of Equation 2.23 is a linear combination of the derivatives of $y_o(t)u(t)$. Evaluating these derivatives is clumsy and inconvenient because of the presence of $u(t)$. The derivatives will generate an impulse and its derivatives at the origin [recall that $\frac{d}{dt}u(t) = \delta(t)$]. Fortunately when $m \leq n$ in Equation 2.4, the solution simplifies to

$$h(t) = b_n\delta(t) + [P(D)y_o(t)]u(t) \tag{2.25}$$

Example 2.3

Solve Example 2.2(a) using the method of convolution.

We first determine $h(t)$. The characteristic modes for this case, as found in Example 2.1, are e^{-t} and e^{-2t}. Since $y_o(t)$ is a linear combination of the characteristic modes

$$y_o(t) = K_1e^{-t} + K_2e^{-2t} \quad t \geq 0$$

Therefore,

$$\dot{y}_o(t) = -K_1e^{-t} - 2K_2e^{-2t} \quad t \geq 0$$

The initial conditions according to Equation 2.24 are $\dot{y}_o(0) = 1$ and $y_o(0) = 0$. Setting $t = 0$ in the above equations and using the initial conditions, we obtain

$$K_1 + K_2 = 0 \quad \text{and} \quad -K_1 - 2K_2 = 1$$

Solution of these equations yields $K_1 = 1$ and $K_2 = -1$. Therefore,

$$y_o(t) = e^{-t} - e^{-2t}$$

Also in this case the polynomial $P(D) = D$ is of the first-order, and $b_2 = 0$. Therefore, from Equation 2.25

$$\begin{aligned} h(t) &= [P(D)y_o(t)]u(t) = [Dy_o(t)]u(t) \\ &= \left[\frac{d}{dt}\left(e^{-t} - e^{-2t}\right)\right]u(t) \\ &= (-e^{-t} + 2e^{-2t})u(t) \end{aligned}$$

and

$$\int_0^t f(x)h(t-x)dx = \int_0^t 10e^{-3x}\left[-e^{-(t-x)} + 2e^{-2(t-x)}\right]dx$$
$$= -5e^{-t} + 20e^{-2t} - 15e^{-3t}$$

The total solution is obtained by adding the complementary solution $y_c(t) = c_1e^{-t} + c_2e^{-2t}$ to this component. Therefore,

$$y(t) = c_1e^{-t} + c_2e^{-2t} - 5e^{-t} + 20e^{-2t} - 15e^{-3t}$$

Setting the conditions $y(0^+) = 2$ and $y(0^+) = 3$ in this equation (and its derivative), we obtain $c_1 = -3$, $c_2 = 5$ so that

$$y(t) = -8e^{-t} + 25e^{-2t} - 15e^{-3t} \quad t \geq 0$$

which is identical to the solution found by the classical method.

2.1.3.1 Assessment of the Convolution Method

The convolution method is more laborious compared to the classical method. However, in system analysis, its advantages outweigh the extra work. The classical method has a serious drawback because it yields the total response, which cannot be separated into components arising from the internal conditions and the external input. In the study of systems it is important to be able to express the system response to an input $f(t)$ as an explicit function of $f(t)$. This is not possible in the classical method. Moreover, the classical method is restricted to a certain class of inputs; it cannot be applied to any input.*

If we must solve a particular linear differential equation or find a response of a particular LTI system, the classical method may be the best. In the theoretical study of linear systems, however, it is practically useless. General discussion of differential equations can be found in numerous texts on the subject [1].

2.2 Difference Equations

The development of difference equations is parallel to that of differential equations. We consider here only linear difference equations with constant coefficients. An n th-order difference equation can be expressed in two different forms; the first form uses delay terms such as $y[k-1]$, $y[k-2]$, $f[k-1]$, $f[k-2]$, etc., and the alternative form uses advance terms such as $y[k+1]$, $y[k+2]$, etc. Both forms are useful. We start here with a general nth-order difference equation, using advance operator form.

$$\begin{aligned} y[k+n] + a_{n-1}y[k+n-1] + \cdots + a_1y[k+1] + a_0y[k] \\ = b_mf[k+m] + b_{m-1}f[k+m-1] + \cdots + b_1f[k+1] + b_0f[k] \end{aligned} \tag{2.26}$$

* Another minor problem is that because the classical method yields total response, the auxiliary conditions must be on the total response, which exists only for $t \geq 0^+$. In practice we are most likely to know the conditions at $t = 0^-$ (before the input is applied). Therefore, we need to derive a new set of auxiliary conditions at $t = 0^+$ from the known conditions at $t = 0^-$. The convolution method can handle both kinds of initial conditions. If the conditions are given at $t = 0^-$, we apply these conditions only to $y_c(t)$ because by its definition the convolution integral is 0 at $t = 0^-$.

2.2.1 Causality Condition

The left-hand side of Equation 2.26 consists of values of $y[k]$ at instants $k+n$, $k+n-1$, $k+n-2$, and so on. The right-hand side of Equation 2.26 consists of the input at instants $k+m$, $k+m-1$, $k+m-2$, and so on. For a casual equation, the solution cannot depend on future input values. This show that when the equation is in the advance operator form of Equation 2.26, casuality requires $m \leq n$. For a general casual case, $m = n$, and Equation 2.26 becomes

$$\begin{aligned} &y[k+n] + a_{n-1}y[k+n-1] + \cdots + a_1 y[k+1] + a_0 y[k] \\ &\quad = b_n f[k+n] + b_{n-1} f[k+n-1] + \cdots + b_1 f[k+1] + b_0 f[k] \end{aligned} \quad (2.27a)$$

where some of the coefficients on both sides can be zero. However, the coefficient of $y[k+n]$ is normalized to unity. Equation 2.27a is valid for all values of k. Therefore, the equation is still valid if we replace k by $k-n$ throughout the equation. This yields the alternative form (the delay operator form) of Equation 2.27a

$$\begin{aligned} &y[k] + a_{n-1}y[k-1] + \cdots + a_1 y[k-n+1] + a_0 y[k-n] \\ &\quad = b_n f[k] + b_{n-1} f[k-1] + \cdots + b_1 f[k-n+1] + b_0 f[k-n] \end{aligned} \quad (2.27b)$$

We designate the form of Equation 2.27a the advance operator form, and the form of Equation 2.27b the delay operator form.

2.2.2 Initial Conditions and Iterative Solution

Equation 2.27b can be expressed as

$$\begin{aligned} y[k] = &- a_{n-1}y[k-1] - a_{n-2}y[k-2] - \cdots - a_0 y[k-n] + b_n f[k] \\ &+ b_{n-1} f[k-1] + \cdots + b_0 f[k-n] \end{aligned} \quad (2.27c)$$

This equation shows that $y[k]$, the solution at the k th instant, is computed from $2n+1$ pieces of information. These are the past n values of $y[k]$: $y[k-1], y[k-2], \ldots, y[k-n]$ and the present and past n values of the input: $f[k], f[k-1], f[k-2], \ldots, f[k-n]$. If the input $f[k]$ is known for $k = 0, 1, 2, \ldots$, then the values of $y[k]$ for $k = 0, 1, 2, \ldots$ can be computed from the $2n$ initial conditions $y[-1], y[-2], \ldots, y[-n]$ and $f[-1], f[-2], \ldots, f[-n]$. If the input is causal, that is, if $f[k] = 0$ for $k < 0$, then $f[-1] = f[-2] = \cdots = f[-n] = 0$, and we need only n initial conditions $y[-1], y[-2], \ldots, y[-n]$. This allows us to compute iteratively or recursively the values $y[0], y[1], y[2], y[3], \ldots$, and so on.* For instance, to find $y[0]$ we set $k = 0$ in Equation 2.27c. The left-hand side is $y[0]$, and the right-hand side contains terms $y[-1], y[-2], \ldots, y[-n]$, and the inputs $f[0], f[-1], f[-2], \ldots, f[-n]$.

Therefore, to begin with, we must know the n initial conditions $y[-1], y[-2], \ldots, y[-n]$. Knowing these conditions and the input $f[k]$, we can iteratively find the response $y[0], y[1], y[2], \ldots$, and so on. The following example demonstrates this procedure. This method basically reflects the manner in which a computer would solve a difference equation, given the input and initial conditions.

---

* For this reason Equation 2.27 is called a *recursive difference equation*. However, in Equation 2.27 if $a_0 = a_1 = a_2 = \cdots = a_{n-1} = 0$, then it follows from Equation 2.27c that determination of the present value of $y[k]$ does not require the past values $y[k-1]$, $y\ [k-2]$, etc. For this reason when $a_i = 0$ $(i = 0, 1, \ldots, n-1)$, the difference Equation 2.27 is *nonrecursive*. This classification is important in designing and realizing digital filters. In this discussion, however, this classification is not important. The analysis techniques developed here apply to general recursive and nonrecursive equations. Observe that a nonrecursive equation is a special case of recursive equation with $a_0 = a_1 = \cdots = a_{n-1} = 0$.

Example 2.4

Solve iteratively

$$y[k] - 0.5y[k-1] = f[k] \tag{2.28a}$$

with initial condition $y[-1] = 16$ and the input $f[k] = k^2$ (starting at $k = 0$). This equation can be expressed as

$$y[k] = 0.5y[k-1] + f[k] \tag{2.28b}$$

If we set $k = 0$ in this equation, we obtain

$$\begin{aligned} y[0] &= 0.5y[-1] + f[0] \\ &= 0.5(16) + 0 = 8 \end{aligned}$$

Now, setting $k = 1$ in Equation 2.28b and using the value $y[0] = 8$ (computed in the first step) and $f[1] = (1)^2 = 1$, we obtain

$$y[1] = 0.5(8) + (1)^2 = 5$$

Next, setting $k = 2$ in Equation 2.28b and using the value $y[1] = 5$ (computed in the previous step) and $f[2] = (2)^2$, we obtain

$$y[2] = 0.5(5) + (2)^2 = 6.5$$

Continuing in this way iteratively, we obtain

$$\begin{aligned} y[3] &= 0.5(6.5) + (3)^2 = 12.25 \\ y[4] &= 0.5(12.25) + (4)^2 = 22.125 \end{aligned}$$

and so on.

This iterative solution procedure is available only for difference equations; it cannot be applied to differential equations. Despite the many uses of this method, a closed-form solution of a difference equation is far more useful in the study of system behavior and its dependence on the input and the various system parameters. For this reason we shall develop a systematic procedure to obtain a closed-form solution of Equation 2.27.

2.2.3 Operational Notation

In difference equations it is convenient to use operational notation similar to that used in differential equations for the sake of compactness and convenience. For differential equations, we use the operator D to denote the operation of differentiation. For difference equations, we use the operator E to denote the operation for advancing the sequence by one time interval. Thus,

$$\begin{aligned} Ef[k] &\equiv f[k+1] \\ E^2f[k] &\equiv f[k+2] \\ &\vdots \\ E^nf[k] &\equiv f[k+n] \end{aligned} \tag{2.29}$$

A general n th-order difference Equation 2.27a can be expressed as

$$\begin{aligned}&\left(E^n + a_{n-1}E^{n-1} + \cdots + a_1E + a_0\right)y[k]\\&\quad = \left(b_nE^n + b_{n-1}E^{n-1} + \cdots + b_1E + b_0\right)f[k]\end{aligned} \tag{2.30a}$$

or

$$Q[E]y[k] = P[E]f[k] \tag{2.30b}$$

where $Q[E]$ and $P[E]$ are n th-order polynomial operators, respectively,

$$Q[E] = E^n + a_{n-1}E^{n-1} + \cdots + a_1E + a_0 \tag{2.31a}$$

$$P[E] = b_nE^n + b_{n-1}E^{n-1} + \cdots + b_1E + b_0 \tag{2.31b}$$

2.2.4 Classical Solution

Following the discussion of differential equations, we can show that if $y_p[k]$ is a solution of Equation 2.27 or Equation 2.30, that is,

$$Q[E]y_p[k] = P[E]f[k] \tag{2.32}$$

then $y_p[k] + y_c[k]$ is also a solution of Equation 2.30, where $y_c[k]$ is a solution of the homogeneous equation

$$Q[E]y_c[k] = 0 \tag{2.33}$$

As before, we call $y_p[k]$ the particular solution and $y_c[k]$ the complementary solution.

2.2.4.1 Complementary Solution (the Natural Response)

By definition

$$Q[E]y_c[k] = 0 \tag{2.33a}$$

or

$$(E^n + a_{n-1}E^{n-1} + \cdots + a_1E + a_0)y_c[k] = 0 \tag{2.33b}$$

or

$$y_c[k+n] + a_{n-1}y_c[k+n-1] + \cdots + a_1y_c[k+1] + a_0y_c[k] = 0 \tag{2.33c}$$

We can solve this equation systematically, but even a cursory examination of this equation points to its solution. This equation states that a linear combination of $y_c[k]$ and delayed $y_c[k]$ is zero not for some values of k, but for all k. This is possible if and only if $y_c[k]$ and delayed $y_c[k]$ have the same form. Only an exponential function γ^k has this property as seen from the equation

$$\gamma^{k-m} = \gamma^{-m}\gamma^k$$

This shows that the delayed γ^k is a constant times γ^k. Therefore, the solution of Equation 2.33 must be of the form

$$y_c[k] = c\gamma^k \tag{2.34}$$

To determine c and γ, we substitute this solution in Equation 2.33. From Equation 2.34, we have

$$\begin{aligned} Ey_c[k] &= y_c[k+1] = c\gamma^{k+1} = (c\gamma)\gamma^k \\ E^2y_c[k] &= y_c[k+2] = c\gamma^{k+2} = (c\gamma^2)\gamma^k \\ &\vdots \\ E^ny_c[k] &= y_c[k+n] = c\gamma^{k+n} = (c\gamma^n)\gamma^k \end{aligned} \tag{2.35}$$

Substitution of this in Equation 2.33 yields

$$c\left(\gamma^n + a_{n-1}\gamma^{n-1} + \cdots + a_1\gamma + a_0\right)\gamma^k = 0 \tag{2.36}$$

For a nontrivial solution of this equation

$$\left(\gamma^n + a_{n-1}\gamma^{n-1} + \cdots + a_1\gamma + a_0\right) = 0 \tag{2.37a}$$

or

$$Q[\gamma] = 0 \tag{2.37b}$$

Our solution $c\gamma^k$ (Equation 2.34) is correct, provided that γ satisfies Equation 2.37. Now, $Q[\gamma]$ is an nth-order polynomial and can be expressed in the factorized form (assuming all distinct roots):

$$(\gamma - \gamma_1)(\gamma - \gamma_2)\cdots(\gamma - \gamma_n) = 0 \tag{2.37c}$$

Clearly γ has n solutions $\gamma_1, \gamma_2, \ldots, \gamma_n$ and, therefore, Equation 2.33 also has n solutions $c_1\gamma_1^k, c_2\gamma_2^k, \ldots, c_n\gamma_n^k$. In such a case we have shown that the general solution is a linear combination of the n solutions. Thus,

$$y_c[k] = c_1\gamma_1^k + c_2\gamma_2^k + \cdots + c_n\gamma_n^k \tag{2.38}$$

where $\gamma_1, \gamma_2, \ldots, \gamma_n$ are the roots of Equation 2.37 and $c_1, c_2, \ldots, c_n$ are arbitrary constants determined from n auxiliary conditions. The polynomial $Q[\gamma]$ is called the characteristic polynomial, and

$$Q[\gamma] = 0 \tag{2.39}$$

is the characteristic equation. Moreover, $\gamma_1, \gamma_2, \ldots, \gamma_n$ the roots of the characteristic equation, are called characteristic roots or characteristic values (also eigenvalues). The exponentials $\gamma_i^k (i = 1, 2, \ldots, n)$ are the characteristic modes or natural modes. A characteristic mode corresponds to each characteristic root, and the complementary solution is a linear combination of the characteristic modes of the system.

2.2.4.2 Repeated Roots

For repeated roots, the form of characteristic modes is modified. It can be shown by direct substitution that if a root γ repeats r times (root of multiplicity r), the characteristic modes corresponding to this root are $\gamma^k, k\gamma^k, k^2\gamma^k, \ldots, k^{r-1}\gamma^k$. Thus, if the characteristic equation is

$$Q[\gamma] = (\gamma - \gamma_1)^r(\gamma - \gamma_{r+1})(\gamma - \gamma_{r+2})\cdots(\gamma - \gamma_n) \tag{2.40}$$

the complementary solution is

$$\begin{aligned} y_c[k] &= \left(c_1 + c_2k + c_3k^2 + \cdots + c_rk^{r-1}\right)\gamma_1^k \\ &\quad + c_{r+1}\gamma_{r+1}^k + c_{r+2}\gamma_{r+2}^k + \cdots + c_n\gamma_n^k \end{aligned} \tag{2.41}$$

2.2.4.3 Particular Solution

The particular solution $y_p[k]$ is the solution of

$$Q[E]y_p[k] = p[E]f[k] \tag{2.42}$$

We shall find the particular solution using the method of undetermined coefficients, the same method used for differential equations. Table 2.2 lists the inputs and the corresponding forms of solution with undetermined coefficients. These coefficients can be determined by substituting $y_p[k]$ in Equation 2.42 and equating the coefficients of similar terms.

Note: By definition, $y_p[k]$ cannot have any characteristic mode terms. If any term $p[k]$ shown in the right-hand column for the particular solution should also be a characteristic mode, the correct form of the particular solution must be modified to $k^ip[k]$, where i is the smallest integer that will prevent $k^ip[k]$ from having a characteristic mode term. For example, when the input is r^k, the particular solution in the right-hand column is of the form cr^k. But if r^k happens to be a natural mode, the correct form of the particular solution is βkr^k (see Pair 2).

Example 2.5

Solve

$$(E^2 - 5E + 6)y[k] = (E - 5)f[k] \tag{2.43}$$

if the input $f[k] = (3k + 5)u[k]$ and the auxiliary conditions are $y[0] = 4$, $y[1] = 13$.

The characteristic equation is

$$\gamma^2 - 5\gamma + 6 = (\gamma - 2)(\gamma - 3) = 0$$

TABLE 2.2 Inputs and Forms of Solution

No.	Input $f[k]$	Forced Response $y_p[k]$
1	$r^k \; r \neq \gamma_i \; (i = 1, 2, \ldots, n)$	βr^k
2	$r^k \; r = \gamma_i$	βkr^k
3	$\cos(\Omega k + \theta)$	$\beta \cos(\Omega k + \varphi)$
4	$\left(\sum_{i=0}^{m} a_ik^i\right)r^k$	$\left(\sum_{i=0}^{m} \beta_ik^i\right)r^k$

Therefore, the complementary solution is

$$y_c[k] = c_1(2)^k + c_2(3)^k$$

To find the form of $y_p[k]$ we use Table 2.2, Pair 4 with $r = 1$, $m = 1$. This yields

$$y_p[k] = \beta_1 k + \beta_0$$

Therefore,

$$\begin{aligned} y_p[k+1] &= \beta_1(k+1) + \beta_0 = \beta_1 k + \beta_1 + \beta_0 \\ y_p[k+2] &= \beta_1(k+2) + \beta_0 = \beta_1 k + 2\beta_1 + \beta_0 \end{aligned}$$

Also,

$$f[k] = 3k + 5$$

and

$$f[k+1] = 3(k+1) + 5 = 3k + 8$$

Substitution of the above results in Equation 2.43 yields

$$\begin{aligned} &\beta_1 k + 2\beta_1 + \beta_0 - 5(\beta_1 k + \beta_1 + \beta_0) + 6(\beta_1 k + \beta_0) \\ &\quad = 3k + 8 - 5(3k + 5) \end{aligned}$$

or

$$2\beta_1 k - 3\beta_1 + 2\beta_0 = -12k - 17$$

Comparison of similar terms on two sides yields

$$\left.\begin{aligned} 2\beta_1 &= -12 \\ -3\beta_1 + 2\beta_0 &= -17 \end{aligned}\right\} \Rightarrow \begin{aligned} \beta_1 &= -6 \\ \beta_2 &= -\tfrac{35}{2} \end{aligned}$$

This means

$$y_p[k] = -6k - \frac{35}{2}$$

The total response is

$$\begin{aligned} y[k] &= y_c[k] + y_p[k] \\ &= c_1(2)^k + c_2(3)^k - 6k - \frac{35}{2} \qquad k \geq 0 \end{aligned} \tag{2.44}$$

To determine arbitrary constants c_1 and c_2 we set $k = 0$ and 1 and substitute the auxiliary conditions $y[0] = 4$, $y[1] = 13$, to obtain

$$\left.\begin{aligned} 4 &= c_1 + c_2 - \tfrac{35}{2} \\ 13 &= 2c_1 + 3c_2 - \tfrac{47}{2} \end{aligned}\right\} \Rightarrow \begin{aligned} c_1 &= 28 \\ c_2 &= \tfrac{-13}{2} \end{aligned}$$

Therefore,

$$y_c[k] = 28(2)^k - \frac{13}{2}(3)^k \tag{2.45}$$

and

$$y[k] = \underbrace{28(2)^k - \frac{13}{2}(3)^k}_{y_c[k]} \underbrace{- 6k - \frac{35}{2}}_{y_p[k]} \tag{2.46}$$

2.2.4.4 A Comment on Auxiliary Conditions

This method requires auxiliary conditions $y[0], y[1], \ldots, y[n-1]$, because the total solution is valid only for $k \geq 0$. But if we are given the initial conditions $y[-1], y[-2], \ldots, y[-n]$, we can derive the conditions $y[0], y[1], \ldots, y[n-1]$, using the iterative procedure discussed earlier.

2.2.4.5 Exponential Input

As in the case of differential equations, we can show that for the equation

$$Q[E]y[k] = P[E]f[k] \tag{2.47}$$

the particular solution for the exponential input $f[k] = r^k$ is given by

$$y_p[k] = H[r]r^k \quad r \neq \gamma_i \tag{2.48}$$

where

$$H[r] = \frac{P[r]}{Q[r]} \tag{2.49}$$

The proof follows from the fact that if the input $f[k] = r^k$, then from Table 2.2 (Pair 4), $y_p[k] = \beta r^k$. Therefore,

$$E^i f[k] = f[k+i] = r^{k+i} = r^i r^k \quad \text{and} \quad P[E]f[k] = P[r]r^k$$
$$E^j y_p[k] = \beta r^{k+j} = \beta r^j r^k \quad \text{and} \quad Q[E]y[k] = \beta Q[r]r^k$$

so that Equation 2.47 reduces to

$$\beta Q[r]r^k = P[r]r^k$$

which yields $\beta = P[r]/Q[r] = H[r]$.

This result is valid only if r is not a characteristic root. If r is a characteristic root, the particular solution is βkr^k where β is determined by substituting $y_p[k]$ in Equation 2.47 and equating coefficients of similar terms on the two sides. Observe that the exponential r^k includes a wide variety of signals such as a constant C, a sinusoid $\cos(\Omega k + \theta)$, and an exponentially growing or decaying sinusoid $|\gamma|^k \cos(\Omega k + \theta)$.

2.2.4.6 A Constant Input $f(k) = C$

This is a special case of exponential Cr^k with $r = 1$. Therefore, from Equation 2.48 we have

$$y_p[k] = C\frac{P[1]}{Q[1]}(1)^k = CH[1] \tag{2.50}$$

2.2.4.7 A Sinusoidal Input

The input $e^{j\Omega k}$ is an exponential r^k with $r = e^{j\Omega}$. Hence,

$$y_p[k] = H[e^{j\Omega}]e^{j\Omega k} = \frac{P[e^{j\Omega}]}{Q[e^{j\Omega}]}e^{j\Omega k}$$

Similarly for the input $e^{-j\Omega k}$

$$y_p[k] = H[e^{-j\Omega}]e^{-j\Omega k}$$

Consequently, if the input

$$\begin{aligned} f[k] &= \cos \Omega k = \frac{1}{2}(e^{j\Omega k} + e^{-j\Omega k}) \\ y_p[k] &= \frac{1}{2}\{H[e^{j\Omega}]e^{j\Omega k} + H[e^{-j\Omega}]e^{-j\Omega k}\} \end{aligned}$$

Since the two terms on the right-hand side are conjugates

$$y_p[k] = \text{Re}\{H[e^{j\Omega}]e^{j\Omega k}\}$$

If

$$H[e^{j\Omega}] = |H[e^{j\Omega}]|e^{j\angle H[e^{j\Omega}]}$$

then

$$\begin{aligned} y_p[k] &= \text{Re}\left\{|H[e^{j\Omega}]|e^{j(\Omega k + \angle H[e^{j\Omega}])}\right\} \\ &= |H[e^{j\Omega}]|\cos(\Omega k + \angle H[e^{j\Omega}]) \end{aligned} \tag{2.51}$$

Using a similar argument, we can show that for the input

$$\begin{aligned} f[k] &= \cos(\Omega k + \theta) \\ y_p[k] &= |H[e^{j\Omega}]|\cos(\Omega k + \theta + \angle H[e^{j\Omega}]) \end{aligned} \tag{2.52}$$

Example 2.6

Solve

$$(E^2 - 3E + 2)y[k] = (E + 2)f[k]$$

for $f[k] = (3)^k u[k]$ and the auxiliary conditions $y[0] = 2, y[1] = 1$.

In this case

$$H[r] = \frac{P[r]}{Q[r]} = \frac{r+2}{r^2 - 3r + 2}$$

and the particular solution to input $(3)^k u[k]$ is $H[3](3)^k$, that is,

$$y_p[k] = \frac{3+2}{(3)^2 - 3(3) + 2}(3)^k = \frac{5}{2}(3)^k$$

The characteristic polynomial is $(\gamma^2 - 3\gamma + 2) = (\gamma - 1)(\gamma - 2)$. The characteristic roots are 1 and 2. Hence, the complementary solution is $y_c[k] = c_1 + c_2(2)^k$ and the total solution is

$$y[k] = c_1(1)^k + c_2(2)^k + \frac{5}{2}(3)^k$$

Setting $k = 0$ and 1 in this equation and substituting auxiliary conditions yields

$$2 = c_1 + c_2 + \frac{5}{2} \quad \text{and} \quad 1 = c_1 + 2c_2 + \frac{15}{2}$$

Solution of these two simultaneous equations yields $c_1 = 5.5, c_2 = -5$. Therefore,

$$y[k] = 5.5 - 6(2)^k + \frac{5}{2}(3)^k \quad k \geq 0$$

2.2.5 Method of Convolution

In this method, the input $f[k]$ is expressed as a sum of impulses. The solution is then obtained as a sum of the solutions to all the impulse components. The method exploits the superposition property of the linear difference equations. A discrete-time unit impulse function $\delta[k]$ is defined as

$$\delta[k] = \begin{cases} 1 & k = 0(94) \\ 0 & k \neq 0 \end{cases} \tag{2.53}$$

Hence, an arbitrary signal $f[k]$ can be expressed in terms of impulse and delayed impulse functions as

$$f[k] = f[0]\delta[k] + f[1]\delta[k-1] + f[2]\delta[k-2] + \cdots + f[k]\delta[0] + \cdots \quad k \geq 0 \tag{2.54}$$

The right-hand side expresses $f[k]$ as a sum of impulse components. If $h[k]$ is the solution of Equation 2.30 to the impulse input $f[k] = \delta[k]$, then the solution to input $\delta[k-m]$ is $h[k-m]$. This follows from the fact that because of constant coefficients, Equation 2.30 has time invariance property. Also, because Equation 2.30 is linear, its solution is the sum of the solutions to each of the impulse components of $f[k]$ on the right-hand side of Equation 2.54 Therefore,

$$\begin{aligned} y[k] &= f[0]h[k] + f[1]h[k-1] + f[2]h[k-2] \\ &\quad + \cdots + f[k]h[0] + f[k+1]h[-1] + \cdots \end{aligned}$$

All practical systems with time as the independent variable are causal, that is, $h[k]=0$ for $k<0$. Hence, all the terms on the right-hand side beyond $f[k]h[0]$ are zero. Thus,

$$\begin{aligned} y[k] &= f[0]h[k] + f[1]h[k-1] + f[2]h[k-2] + \cdots + f[k]h[0] \\ &= \sum_{m=0}^{k} f[m]h[k-m] \end{aligned} \tag{2.55}$$

The first term on the right-hand side consists of a linear combination of natural modes and should be appropriately modified for repeated roots. The general solution is obtained by adding a complementary solution to the above solution. Therefore, the general solution is given by

$$y[k] = \sum_{j=1}^{n} c_j \gamma_j^k + \sum_{m=0}^{k} f[m]h[k-m] \tag{2.56}$$

The last sum on the right-hand side is known as the convolution sum of $f[k]$ and $h[k]$.

The function $h[k]$ appearing in Equation 2.56 is the solution of Equation 2.30 for the impulsive input ($f[k]=\delta[k]$) when all initial conditions are zero, that is, $h[-1]=h[-2]=\cdots=h[-n]=0$. It can be shown that [2] $h[k]$ contains an impulse and a linear combination of characteristic modes as

$$h[k] = \frac{b_0}{a_0}\delta[k] + A_1\gamma_1^k + A_2\gamma_2^k + \cdots + A_n\gamma_n^k \tag{2.57}$$

where the unknown constants A_i are determined from n values of $h[k]$ obtained by solving the equation $Q[E]h[k]=P[E]\delta[k]$ iteratively.

Example 2.7

Solve Example 2.5 using convolution method. In other words solve

$$(E^2 - 3E + 2)y[k] = (E+2)f[k]$$

for $f[k]=(3)^k u[k]$ and the auxiliary conditions $y[0]=2, y[1]=1$.

The unit impulse solution $h[k]$ is given by Equation 2.57. In this case $a_0=2$ and $b_0=2$. Therefore,

$$h[k] = \delta[k] + A_1(1)^k + A_2(2)^k \tag{2.58}$$

To determine the two unknown constants A_1 and A_2 in Equation 2.58, we need two values of $h[k]$, for instance $h[0]$ and $h[1]$. These can be determined iteratively by observing that $h[k]$ is the solution of $(E^2-3E+2)h[k]=(E+2)\delta[k]$, that is,

$$h[k+2] - 3h[k+1] + 2h[k] = \delta[k+1] + 2\delta[k] \tag{2.59}$$

subject to initial conditions $h[-1]=h[-2]=0$. We now determine $h[0]$ and $h[1]$ iteratively from Equation 2.59. Setting $k=-2$ in this equation yields

$$h[0] - 3(0) + 2(0) = 0 + 0 \Rightarrow h[0] = 0$$

Next, setting $k = -1$ in Equation 2.59 and using $h[0] = 0$, we obtain

$$h[1] - 3(0) + 2(0) = 1 + 0 \Rightarrow h[1] = 1$$

Setting $k = 0$ and 1 in Equation 2.58 and substituting $h[0] = 0, h[1] = 1$ yields

$$0 = 1 + A_1 + A_2 \quad \text{and} \quad 1 = A_1 + 2A_2$$

Solution of these two equations yields $A_1 = -3$ and $A_2 = 2$. Therefore,

$$h[k] = \delta[k] - 3 + 2(2)^k$$

and from Equation 2.56

$$\begin{aligned} y[k] &= c_1 + c_2(2)^k + \sum_{m=0}^{k} (3)^m \left[\delta[k-m] - 3 + 2(2)^{k-m}\right] \\ &= c_1 + c_2(2)^k + 1.5 - 4(2)^k + 2.5(3)^k \end{aligned}$$

The sums in the above expression are found by using the geometric progression sum formula

$$\sum_{m=0}^{k} r^m = \frac{r^{k+1} - 1}{r - 1} r \neq 1$$

Setting $k = 0$ and 1 and substituting the given auxiliary conditions $y[0] = 2, y[1] = 1$, we obtain

$$2 = c_1 + c_2 + 1.5 - 4 + 2.5 \quad \text{and} \quad 1 = c_1 + 2c_2 + 1.5 - 8 + 7.5$$

Solution of these equations yields $c_1 = 4$ and $c_2 = -2$. Therefore,

$$y[k] = 5.5 - 6(2)^k + 2.5(3)^k$$

which confirms the result obtained by the classical method.

2.2.5.1 Assessment of the Classical Method

The earlier remarks concerning the classical method for solving differential equations also apply to difference equations. General discussion of difference equations can be found in texts on the subject [3].

References

1. Birkhoff, G. and Rota, G.C., *Ordinary Differential Equations*, 3rd edn., John Wiley & Sons, New York, 1978.
2. Lathi, B.P., *Signal Processing and Linear Systems*, Berkeley-Cambridge Press, Carmichael, CA, 1998.
3. Goldberg, S., *Introduction to Difference Equations*, John Wiley & Sons, New York, 1958.

3

Finite Wordlength Effects

Bruce W. Bomar
University of Tennessee Space Institute

3.1 Introduction

Practical digital filters must be implemented with finite precision numbers and arithmetic. As a result, both the filter coefficients and the filter input and output signals are in discrete form. This leads to four types of finite wordlength effects.

Discretization (quantization) of the filter coefficients has the effect of perturbing the location of the filter poles and zeros. As a result, the actual filter response differs slightly from the ideal response. This deterministic frequency response error is referred to as **coefficient quantization error**.

The use of finite precision arithmetic makes it necessary to quantize filter calculations by rounding or truncation. **Roundoff noise** is that error in the filter output that results from rounding or truncating calculations within the filter. As the name implies, this error looks like low-level noise at the filter output.

Quantization of the filter calculations also renders the filter slightly nonlinear. For large signals this nonlinearity is negligible and roundoff noise is the major concern. However, for recursive filters with a zero or constant input, this nonlinearity can cause spurious oscillations called **limit cycles**.

With fixed-point arithmetic it is possible for filter calculations to overflow. The term **overflow oscillation**, sometimes also called **adder overflow limit cycle**, refers to a high-level oscillation that can exist in an otherwise stable filter due to the nonlinearity associated with the overflow of internal filter calculations.

In this chapter, we examine each of these finite wordlength effects. Both fixed-point and floating-point number representations are considered.

3.2 Number Representation

In digital signal processing, $(B+1)$-bit fixed-point numbers are usually represented as two's-complement signed fractions in the format

$$b_0 b_{-1} b_{-2} \ldots b_{-B}$$

The number represented is then

$$X = -b_0 + b_{-1}2^{-1} + b_{-2}2^{-2} + \cdots + b_{-B}2^{-B} \qquad (3.1)$$

where b_0 is the sign bit and the number range is $-1 \leq X < 1$. The advantage of this representation is that the product of two numbers in the range from -1 to 1 is another number in the same range.

Floating-point numbers are represented as

$$X = (-1)^s m 2^c \qquad (3.2)$$

where

s is the sign bit
m is the **mantissa**
c is the characteristic or exponent

To make the representation of a number unique, the mantissa is normalized so that $0.5 \leq m < 1$.

Although floating-point numbers are always represented in the form of Equation 3.2, the way in which this representation is actually stored in a machine may differ. Since $m \geq 0.5$, it is not necessary to store the 2^{-1}-weight bit of m, which is always set. Therefore, in practice numbers are usually stored as

$$X = (-1)^s (0.5 + f) 2^c \qquad (3.3)$$

where f is an unsigned fraction, $0 \leq f < 0.5$.

Most floating-point processors now use the IEEE Standard 754 32-bit floating-point format for storing numbers. According to this standard the exponent is stored as an unsigned integer p where

$$p = c + 126 \qquad (3.4)$$

Therefore, a number is stored as

$$X = (-1)^s (0.5 + f) 2^{p-126} \qquad (3.5)$$

where

s is the sign bit
f is a 23-bit unsigned fraction in the range $0 \leq f < 0.5$
p is an 8-bit unsigned integer in the range $0 \leq p \leq 255$

The total number of bits is $1 + 23 + 8 = 32$. For example, in IEEE format 3/4 is written $(-1)^0(0.5 + 0.25)2^0$ so $s = 0$, $p = 126$, and $f = 0.25$. The value $X = 0$ is a unique case and is represented by all bits zero (i.e., $s = 0$, $f = 0$, and $p = 0$). Although the 2^{-1}-weight mantissa bit is not actually stored, it does exist so the mantissa has 24 bits plus a sign bit.

3.3 Fixed-Point Quantization Errors

In fixed-point arithmetic, a multiply doubles the number of significant bits. For example, the product of the two 5-bit numbers 0.0011 and 0.1001 is the 10-bit number 00.00011011. The extra bit to the left of the decimal point can be discarded without introducing any error. However, the least significant four of the remaining bits must ultimately be discarded by some form of quantization so that the result can be stored to 5 bits for use in other calculations. In the example above this results in 0.0010 (quantization by rounding) or 0.0001 (quantization by truncating). When a sum of products calculation is performed, the quantization can be performed either after each multiply or after all products have been summed with double-length precision.

We will examine three types of fixed-point quantization: rounding, truncation, and magnitude truncation. If X is an exact value, then the rounded value will be denoted $Q_r(X)$, the truncated value $Q_t(X)$, and the magnitude truncated value $Q_{mt}(X)$. If the quantized value has B bits to the right of the decimal point, the quantization step size is

$$\Delta = 2^{-B} \tag{3.6}$$

Since rounding selects the quantized value nearest the unquantized value, it gives a value which is never more than $\pm\Delta/2$ away from the exact value. If we denote the rounding error by $\pm\Delta/2$ away from the exact value. If we denote the rounding error by

$$\varepsilon_r = Q_r(X) - X \tag{3.7}$$

then

$$-\frac{\Delta}{2} \leq \varepsilon_r \frac{\Delta}{2} \tag{3.8}$$

Truncation simply discards the low-order bits, giving a quantized value that is always less than or equal to the exact value so

$$-\Delta < \varepsilon_t \leq 0 \tag{3.9}$$

Magnitude truncation chooses the nearest quantized value that has a magnitude less than or equal to the exact value so

$$-\Delta < \varepsilon_{mt} \leq \Delta \tag{3.10}$$

The error resulting from quantization can be modeled as a random variable uniformly distributed over the appropriate error range. Therefore, calculations with roundoff error can be considered error-free calculations that have been corrupted by additive white noise. The mean of this noise for rounding is

$$m_{\varepsilon_r} = E\{\varepsilon_r\} = \frac{1}{\Delta} \int\limits_{-\Delta/2}^{\Delta/2} \varepsilon_r d\varepsilon_r = 0 \tag{3.11}$$

where $E\{\}$ represents the operation of taking the expected value of a random variable. Similarly, the variance of the noise for rounding is

$$\sigma_{\varepsilon_r}^2 = E\{(\varepsilon_r - m_{\varepsilon_r})^2\} = \frac{1}{\Delta}\int_{-\Delta/2}^{\Delta/2} (\varepsilon_r - m_{\varepsilon_r})^2 d\varepsilon_r = \frac{\Delta^2}{12} \tag{3.12}$$

Likewise, for truncation,

$$\begin{aligned} m_{\varepsilon_t} &= E\{\varepsilon_t\} = -\frac{\Delta}{2} \\ \sigma_{\varepsilon_t}^2 &= E\{(\varepsilon_t - m_{\varepsilon_t})^2\} = \frac{\Delta^2}{12} \end{aligned} \tag{3.13}$$

and, for magnitude truncation,

$$m_{\varepsilon_t} = E\{(\varepsilon_{mt} - m_{\varepsilon_{mt}})^2\} = \frac{\Delta^2}{3} \tag{3.14}$$

3.4 Floating-Point Quantization Errors

With floating-point arithmetic, it is necessary to quantize after both multiplications and additions. The addition quantization arises because, prior to addition, the mantissa of the smaller number in the sum is shifted right until the exponent of both numbers is the same. In general, this gives a sum mantissa that is too long and so must be quantized.

We will assume that quantization in floating-point arithmetic is performed by rounding. Because of the exponent in floating-point arithmetic, it is the relative error that is important. The relative error is defined as

$$\varepsilon_r = \frac{Q_r(X) - X}{X} = \frac{\varepsilon_r}{X} \tag{3.15}$$

Since $X = (-1)^s\, m2^c$, $Q_r(X) = (-1)^s Q_r(m)2^c$ and

$$\varepsilon_r = \frac{Q_r(m) - m}{m} = \frac{\varepsilon}{m} \tag{3.16}$$

If the quantized mantissa has B bits to the right of the decimal point, $|\varepsilon| < \Delta/2$ where, as before, $\Delta = 2^{-B}$. Therefore, since $0.5 \leq m < 1$,

$$|\varepsilon_r| < \Delta \tag{3.17}$$

If we assume that ε is uniformly distributed over the range from $-\Delta/2$ to $\Delta/2$ and m is uniformly distributed over 0.5–1, then

$$m_{\varepsilon_r} = E\left\{\frac{\varepsilon}{m}\right\} = 0$$

and

$$\sigma_{\varepsilon_r}^2 = E\left\{\left(\frac{\varepsilon}{m}\right)^2\right\} = \frac{2}{\Delta}\int_{1/2}^{1}\int_{-\Delta/2}^{\Delta/2}\frac{\varepsilon^2}{m^2}\,d\varepsilon\, dm$$
$$= \frac{\Delta^2}{6} = (0.167)2^{-2B} \tag{3.18}$$

In practice, the distribution of m is not exactly uniform. Actual measurements of roundoff noise in [1] suggested that

$$\sigma_{\varepsilon_r}^2 \approx 0.23\Delta^2 \tag{3.19}$$

while a detailed theoretical and experimental analysis in [2] determined

$$\sigma_{\varepsilon_r}^2 \approx 0.18\Delta^2 \tag{3.20}$$

From Equation 3.15, we can represent a quantized floating-point value in terms of the unquantized value and the random variable ε_r using

$$Q_r(X) = X(1 + \varepsilon_r) \tag{3.21}$$

Therefore, the finite-precision product X_1X_2 and the sum $X_1 + X_2$ can be written as

$$fl(X_1X_2) = X_1X_2(1 + \varepsilon_r) \tag{3.22}$$

and

$$fl(X_1 + X_2) = (X_1 + X_2)(1 + \varepsilon_r) \tag{3.23}$$

where ε_r is zero-mean with the variance of Equation 3.20.

3.5 Roundoff Noise

To determine the roundoff noise at the output of a digital filter, we will assume that the noise due to a quantization is stationary, white, and uncorrelated with the filter input, output, and internal variables. This assumption is good if the filter input changes from sample to sample in a sufficiently complex manner. It is not valid for zero or constant inputs for which the effects of rounding are analyzed from a limit-cycle perspective.

To satisfy the assumption of a sufficiently complex input, roundoff noise in digital filters is often calculated for the case of a zero-mean white noise filter input signal $x(n)$ of variance σ_x^2. This simplifies calculation of the output roundoff noise because expected values of the form $E\{x(n)x(n-k)\}$ are zero for $k \neq 0$ and give σ_x^2 when $k = 0$. This approach to analysis has been found to give estimates of the output roundoff noise that are close to the noise actually observed for other input signals.

Another assumption that will be made in calculating roundoff noise is that the product of two quantization errors is zero. To justify this assumption, consider the case of a 16-bit fixed-point processor. In this case, a quantization error is of the order 2^{-15}, while the product of two quantization errors is of the order 2^{-30}, which is negligible by comparison.

If a linear system with impulse response $g(n)$ is excited by white noise with mean m_x and variance σ_x^2, the output is noise of mean [3, pp. 788–790]

$$m_y = m_x \sum_{n=-\infty}^{\infty} g(n) \tag{3.24}$$

and variance

$$\sigma_y^2 = \sigma_x^2 \sum_{n=-\infty}^{\infty} g^2(n) \tag{3.25}$$

Therefore, if $g(n)$ is the impulse response from the point where a roundoff takes place to the filter output, the contribution of that roundoff to the variance (mean-square value) of the output roundoff noise is given by Equation 3.25 with σ_x^2 replaced with the variance of the roundoff. If there is more than one source of roundoff error in the filter, it is assumed that the errors are uncorrelated so the output noise variance is simply the sum of the contributions from each source.

3.5.1 Roundoff Noise in FIR Filters

The simplest case to analyze is a finite impulse response (FIR) filter realized via the convolution summation

$$y(n) = \sum_{k=0}^{N-1} h(k)x(n-k) \tag{3.26}$$

When fixed-point arithmetic is used and quantization is performed after each multiply, the result of the N multiplies is N-times the quantization noise of a single multiply. For example, rounding after each multiply gives, from Equations 3.6 and 3.12, an output noise variance of

$$\sigma_o^2 = N\frac{2^{-2B}}{12} \tag{3.27}$$

Virtually all digital signal processor integrated circuits contain one or more double-length accumulator registers which permit the sum-of-products in Equation 3.26 to be accumulated without quantization. In this case only a single quantization is necessary following the summation and

$$\sigma_o^2 = \frac{2^{-2B}}{12} \tag{3.28}$$

For the floating-point roundoff noise case we will consider Equation 3.26 for $N=4$ and then generalize the result to other values of N. The finite-precision output can be written as the exact output plus an error term $e(n)$. Thus,

$$\begin{aligned} y(n) + e(n) = (\{[h(0)x(n)[1+\varepsilon_1(n)] \\ + h(1)x(n-1)[1+\varepsilon_2(n)]][1+\varepsilon_3(n)] \\ + h(2)x(n-2)[1+\varepsilon_4(n)]\}\{1+\varepsilon_5(n)\} \\ + h(3)x(n-3)[1+\varepsilon_6(n)])[1+\varepsilon_7(n)] \end{aligned} \tag{3.29}$$

In Equation 3.29, $\varepsilon_1(n)$ represents the error in the first product, $\varepsilon_2(n)$ the error in the second product, $\varepsilon_3(n)$ the error in the first addition, etc. Notice that it has been assumed that the products are summed in the order implied by the summation of Equation 3.26.

Expanding Equation 3.29, ignoring products of error terms, and recognizing $y(n)$ gives

$$\begin{aligned} e(n) = {} & h(0)x(n)[\varepsilon_1(n) + \varepsilon_3(n) + \varepsilon_5(n) + \varepsilon_7(n)] \\ & + h(1)x(n-1)[\varepsilon_2(n) + \varepsilon_3(n) + \varepsilon_5(n) + \varepsilon_7(n)] \\ & + h(2)x(n-2)[\varepsilon_4(n) + \varepsilon_5(n) + \varepsilon_7(n)] \\ & + h(3)x(n-3)[\varepsilon_6(n) + \varepsilon_7(n)] \end{aligned} \tag{3.30}$$

Assuming that the input is white noise of variance σ_x^2 so that $E\{x(n)x(n-k)\}$ is zero for $k \neq 0$, and assuming that the errors are uncorrelated,

$$E\{e^2(n)\} = \left[4h^2(0) + 4h^2(1) + 3h^2(2) + 2h^2(3)\right]\sigma_x^2\sigma_{\varepsilon_r}^2 \tag{3.31}$$

In general, for any N,

$$\sigma_o^2 = E\{e^2(n)\} = \left[Nh^2(0) + \sum_{k=1}^{N-1}(N+1-k)h^2(k)\right]\sigma_x^2\sigma_{\varepsilon_r}^2 \tag{3.32}$$

Notice that if the order of summation of the product terms in the convolution summation is changed, then the order in which the $h(k)$'s appear in Equation 3.32 changes. If the order is changed so that the $h(k)$ with smallest magnitude is first, followed by the next smallest, etc., then the roundoff noise variance is minimized. However, performing the convolution summation in nonsequential order greatly complicates data indexing and so may not be worth the reduction obtained in roundoff noise.

3.5.2 Roundoff Noise in Fixed-Point IIR Filters

To determine the roundoff noise of a fixed-point infinite impulse response (IIR) filter realization, consider a causal first-order filter with impulse response

$$h(n) = a^n u(n) \tag{3.33}$$

realized by the difference equation

$$y(n) = ay(n-1) + x(n) \tag{3.34}$$

Due to roundoff error, the output actually obtained is

$$\hat{y}(n) = Q\{ay(n-1) + x(n)\} = ay(n-1) + x(n) + e(n) \tag{3.35}$$

where $e(n)$ is a random roundoff noise sequence. Since $e(n)$ is injected at the same point as the input, it propagates through a system with impulse response $h(n)$. Therefore, for fixed-point arithmetic with rounding, the output roundoff noise variance from Equations 3.6, 3.12, 3.25, and 3.33 is

$$\sigma_o^2 = \frac{\Delta^2}{12}\sum_{n=-\infty}^{\infty} h^2(n) = \frac{\Delta^2}{12}\sum_{n=0}^{\infty} a^{2n} = \frac{2^{-2B}}{12}\,\frac{1}{1-a^2} \tag{3.36}$$

With fixed-point arithmetic there is the possibility of overflow following addition. To avoid overflow it is necessary to restrict the input signal amplitude. This can be accomplished by either placing a scaling multiplier at the filter input or by simply limiting the maximum input signal amplitude. Consider the case of the first-order filter of Equation 3.34. The transfer function of this filter is

$$H(e^{j\omega}) = \frac{Y(e^{j\omega})}{X(e^{j\omega})} = \frac{1}{e^{j\omega} - a} \tag{3.37}$$

so

$$\left|H(e^{j\omega})\right|^2 = \frac{1}{1 + a^2 - 2a\cos(\omega)} \tag{3.38}$$

and

$$\left|H(e^{j\omega})\right|_{\max} = \frac{1}{1 - |a|} \tag{3.39}$$

The peak gain of the filter is $1/(1 - |a|)$ so limiting input signal amplitudes to $|x(n)| \leq 1 - |a|$ will make overflows unlikely.

An expression for the output roundoff noise-to-signal ratio can easily be obtained for the case where the filter input is white noise, uniformly distributed over the interval from $-(1 - |a|)$ to $(1 - |a|)$ [4,5]. In this case,

$$\sigma_x^2 = \frac{1}{2(1 - |a|)} \int\limits_{-(1-|a|)}^{1-|a|} x^2 \mathrm{d}x = \frac{1}{3}(1 - |a|)^2 \tag{3.40}$$

so, from Equation 3.25,

$$\sigma_y^2 = \frac{1}{3} \frac{(1 - |a|)^2}{1 - a^2} \tag{3.41}$$

Combining Equations 3.36 and 3.41 then gives

$$\frac{\sigma_o^2}{\sigma_y^2} = \left(\frac{2^{-2B}}{12} \frac{1}{1 - a^2}\right)\left(3 \frac{1 - a^2}{(1 - |a|)^2}\right) = \frac{2^{-2B}}{12} \frac{3}{(1 - |a|)^2} \tag{3.42}$$

Notice that the noise-to-signal ratio increases without bound as $|a| \to 1$.

Similar results can be obtained for the case of the causal second-order filter realized by the difference equation

$$y(n) = 2r\cos(\theta)y(n - 1) - r^2 y(n - 2) + x(n) \tag{3.43}$$

This filter has complex-conjugate poles at $re^{\pm j\theta}$ and impulse response

$$h(n) = \frac{1}{\sin(\theta)} r^n \sin[(n + 1)\theta]u(n) \tag{3.44}$$

Due to roundoff error, the output actually obtained is

$$\hat{y}(n) = 2r\cos(\theta)y(n-1) - r^2y(n-2) + x(n) + e(n) \tag{3.45}$$

There are two noise sources contributing to $e(n)$ if quantization is performed after each multiply, and there is one noise source if quantization is performed after summation. Since

$$\sum_{n=-\infty}^{\infty} h^2(n) = \frac{1+r^2}{1-r^2}\frac{1}{(1+r^2)^2 - 4r^2\cos^2(\theta)} \tag{3.46}$$

the output roundoff noise is

$$\sigma_o^2 = \nu\frac{2^{-2B}}{12}\frac{1+r^2}{1-r^2}\frac{1}{(1+r^2)^2 - 4r^2\cos^2(\theta)} \tag{3.47}$$

where $\nu = 1$ for quantization after summation, and $\nu = 2$ for quantization after each multiply.

To obtain an output noise-to-signal ratio we note that

$$H(e^{j\omega}) = \frac{1}{1 - 2r\cos(\theta)e^{-j\omega} + r^2e^{-j2\omega}} \tag{3.48}$$

and, using the approach of [6],

$$\left|H(e^{j\omega})\right|^2_{\max} = \frac{1}{4r^2\left\{\left[\mathrm{sat}\left(\frac{1+r^2}{2r}\cos(\theta)\right) - \frac{1+r^2}{2r}\cos(\theta)\right]^2 + \left[\frac{1-r^2}{2r}\sin(\theta)\right]^2\right\}} \tag{3.49}$$

where

$$\mathrm{sat}(\mu) = \begin{cases} 1 & \mu > 1 \\ \mu & -1 \le \mu \le 1 \\ -1 & \mu < -1 \end{cases} \tag{3.50}$$

Following the same approach as for the first-order case then gives

$$\frac{\sigma_o^2}{\sigma_y^2} = \nu\frac{2^{-2B}}{12}\frac{1+r^2}{1-r^2}\frac{3}{(1+r^2)^2 - 4r^2\cos^2(\theta)} \times \frac{1}{4r^2\left\{\left[\mathrm{sat}\left(\frac{1+r^2}{2r}\cos(\theta)\right) - \frac{1+r^2}{2r}\cos(\theta)\right]^2 + \left[\frac{1-r^2}{2r}\sin(\theta)\right]^2\right\}} \tag{3.51}$$

Figure 3.1 is a contour plot showing the noise-to-signal ratio of Equation 3.51 for $\nu = 1$ in units of the noise variance of a single quantization, $2^{-2B}/12$. The plot is symmetrical about $\theta = 90°$, so only the range from 0° to 90° is shown. Notice that as $r \to 1$, the roundoff noise increases without bound. Also notice that the noise increases as $\theta \to 0°$.

It is possible to design state-space filter realizations that minimize fixed-point roundoff noise [7–10]. Depending on the transfer function being realized, these structures may provide a roundoff noise level that is orders-of-magnitude lower than for a nonoptimal realization. The price paid for this reduction in roundoff noise is an increase in the number of computations required to implement the filter. For an

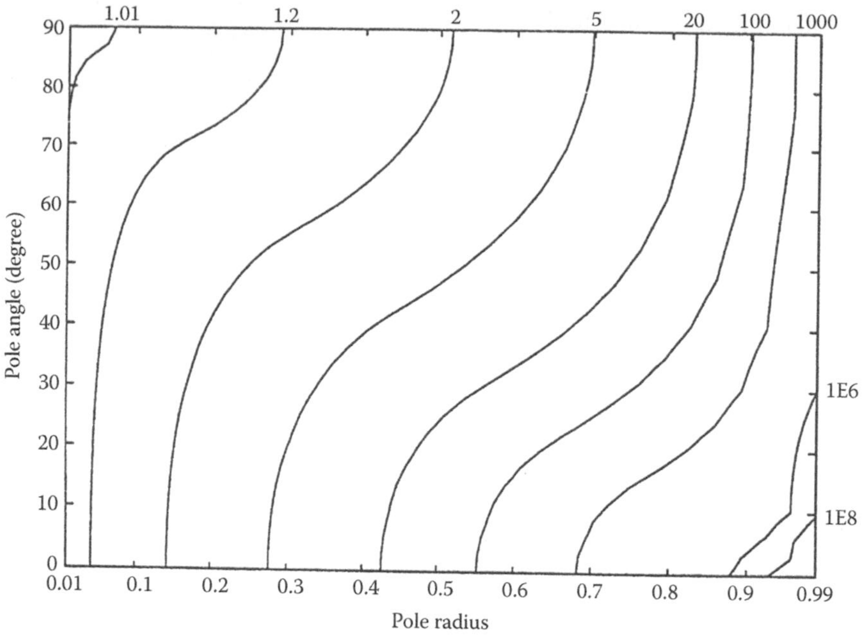

FIGURE 3.1 Normalized fixed-point roundoff noise variance.

Nth-order filter the increase is from roughly $2N$ multiplies for a direct form realization to roughly $(N+1)^2$ for an optimal realization. However, if the filter is realized by the parallel or cascade connection of first- and second-order optimal subfilters, the increase is only to about $4N$ multiplies. Furthermore, near-optimal realizations exist that increase the number of multiplies to only about $3N$ [10].

3.5.3 Roundoff Noise in Floating-Point IIR Filters

For floating-point arithmetic it is first necessary to determine the injected noise variance of each quantization. For the first-order filter this is done by writing the computed output as

$$y(n) + e(n) = [ay(n-1)(1+\varepsilon_1(n)) + x(n)](1+\varepsilon_2(n)) \tag{3.52}$$

where

$\varepsilon_1(n)$ represents the error due to the multiplication
$\varepsilon_2(n)$ represents the error due to the addition

Neglecting the product of errors, Equation 3.52 becomes

$$y(n) + e(n) \approx ay(n-1) + x(n) + ay(n-1)\varepsilon_1(n) + ay(n-1)\varepsilon_2(n) + x(n)\varepsilon_2(n) \tag{3.53}$$

Comparing Equations 3.34 and 3.53, it is clear that

$$e(n) = ay(n-1)\varepsilon_1(n) + ay(n-1)\varepsilon_2(n) + x(n)\varepsilon_2(n) \tag{3.54}$$

Taking the expected value of $e^2(n)$ to obtain the injected noise variance then gives

$$\begin{aligned} E\{e^2(n)\} &= a^2E\{y^2(n-1)\}E\{\varepsilon_1^2(n)\} + a^2E\{y^2(n-1)\}E\{\varepsilon_2^2(n)\} \\ &\quad + E\{x^2(n)\}E\{\varepsilon_2^2(n)\} + E\{x(n)y(n-1)\}E\{\varepsilon_2^2(n)\} \end{aligned} \tag{3.55}$$

To carry this further it is necessary to know something about the input. If we assume the input is zero-mean white noise with variance σ_x^2, then $E\{x^2(n)\} = \sigma_x^2$ and the input is uncorrelated with past values of the output so $E\{x(n)y(n-1)\} = 0$ giving

$$E\{e^2(n)\} = 2a^2\sigma_y^2\sigma_{\varepsilon_r}^2 + \sigma_x^2\sigma_{\varepsilon_r}^2 \tag{3.56}$$

and

$$\sigma_o^2 = \left(2a^2\sigma_y^2\sigma_{\varepsilon_r}^2 + \sigma_x^2\sigma_{\varepsilon_r}^2\right)\sum_{n=-\infty}^{\infty} h^2(n) = \frac{2a^2\sigma_y^2 + \sigma_x^2}{1-a^2}\sigma_{\varepsilon_r}^2 \tag{3.57}$$

However,

$$\sigma_y^2 = \sigma_x^2 \sum_{n=-\infty}^{\infty} h^2(n) = \frac{\sigma_x^2}{1-a^2} \tag{3.58}$$

so

$$\sigma_o^2 = \frac{1+a^2}{(1-a^2)^2}\sigma_{\varepsilon_r}^2\sigma_x^2 = \frac{1+a^2}{1-a^2}\sigma_{\varepsilon_r}^2\sigma_y^2 \tag{3.59}$$

and the output roundoff noise-to-signal ratio is

$$\frac{\sigma_o^2}{\sigma_y^2} = \frac{1+a^2}{1-a^2}\sigma_{\varepsilon_r}^2 \tag{3.60}$$

Similar results can be obtained for the second-order filter of Equation 3.43 by writing

$$y(n) + e(n) = \left(\left[2r\cos(\theta)y(n-1)(1+\varepsilon_1(n)) - r^2y(n-2)(1+\varepsilon_2(n))\right] \times [1+\varepsilon_3(n)] + x(n)\right)(1+\varepsilon_4(n)) \tag{3.61}$$

Expanding with the same assumptions as before gives

$$\begin{aligned} e(n) \approx\ & 2r\cos(\theta)y(n-1)[\varepsilon_1(n) + \varepsilon_3(n) + \varepsilon_4(n)] \\ & - r^2y(n-2)[\varepsilon_2(n) + \varepsilon_3(n) + \varepsilon_4(n)] + x(n)\varepsilon_4(n) \end{aligned} \tag{3.62}$$

and

$$\begin{aligned} E\{e^2(n)\} = {} & 4r^2\cos^2(\theta)\sigma_y^2 3\sigma_{\varepsilon_r}^2 + r^2\sigma_y^2 3\sigma_{\varepsilon_r}^2 \\ & + \sigma_x^2\sigma_{\varepsilon_r}^2 - 8r^3\cos(\theta)\sigma_{\varepsilon_r}^2 E\{y(n-1)y(n-2)\} \end{aligned} \tag{3.63}$$

However,

$$\begin{aligned} E\{y(n-1)y(n-2)\} &= E\{[2r\cos(\theta)y(n-2) - r^2y(n-3) + x(n-1)]y(n-2)\} \\ &= 2r\cos(\theta)E\{y^2(n-2)\} - r^2E\{y(n-2)y(n-3)\} \\ &= 2r\cos(\theta)E\{y^2(n-2)\} - r^2E\{y(n-1)y(n-2)\} \\ &= \frac{2r\cos(\theta)}{1+r^2}\sigma_y^2 \end{aligned} \tag{3.64}$$

so

$$E\{e^2(n)\} = \sigma_{\varepsilon_r}^2\sigma_x^2 + \left[3r^2 + 12r^2\cos^2(\theta) - \frac{16r^4\cos^2(\theta)}{1+r^2}\right]\sigma_{\varepsilon_r}^2\sigma_y^2 \tag{3.65}$$

and

$$\sigma_o^2 = E\{e^2(n)\}\sum_{n=-\infty}^{\infty} h^2(n)\xi\left[\sigma_{\varepsilon_r}^2\sigma_x^2 + \left[3r^4 + 12r^2\cos^2(\theta) - \frac{16r^4\cos^2(\theta)}{1+r^2}\right]\sigma_{\varepsilon_r}^2\sigma_y^2\right] \tag{3.66}$$

where from Equation 3.46,

$$\xi = \sum_{n=-\infty}^{\infty} h^2(n) = \frac{1+r^2}{1-r^2}\frac{1}{(1+r^2)^2 - 4r^2\cos^2(\theta)} \tag{3.67}$$

Since $\sigma_y^2 = \xi\sigma_x^2$, the output roundoff noise-to-signal ratio is then

$$\frac{\sigma_o^2}{\sigma_y^2} = \xi\left[1 + \xi\left[3r^2 + 12r^2\cos^2(\theta) - \frac{16r^4\cos^2(\theta)}{1+r^2}\right]\right]\sigma_{\varepsilon_r}^2 \tag{3.68}$$

Figure 3.2 is a contour plot showing the noise-to-signal ratio of Equation 3.68 in units of the noise variance of a single quantization $\sigma_{\varepsilon_r}^2$. The plot is symmetrical about $\theta = 90°$, so only the range from 0° to 90° is shown. Notice the similarity of this plot to that of Figure 3.1 for the fixed-point case. It has been observed that filter structures generally have very similar fixed-point and floating-point roundoff characteristics [2]. Therefore, the techniques of [7–10], which were developed for the fixed-point case,

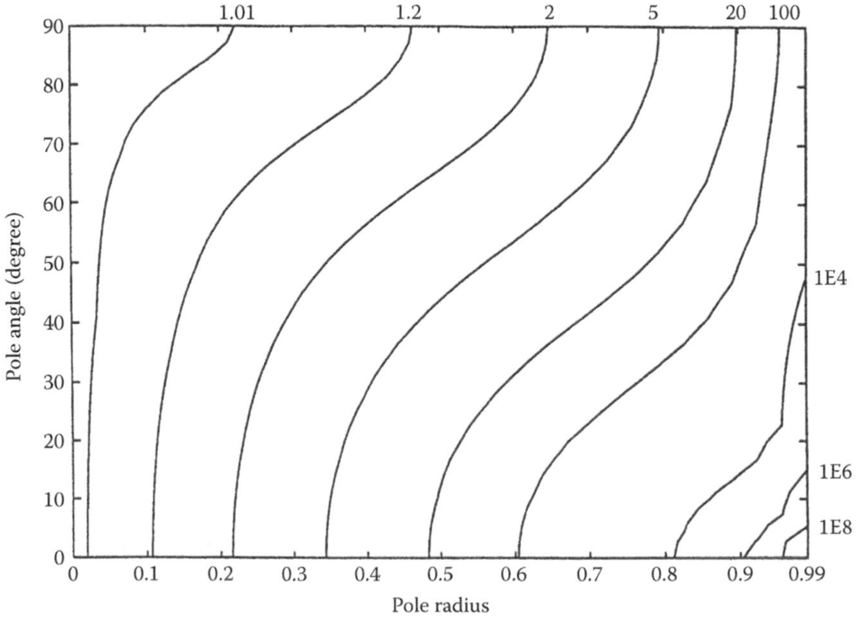

FIGURE 3.2 Normalized floating-point roundoff noise variance.

can also be used to design low-noise floating-point filter realizations. Furthermore, since it is not necessary to scale the floating-point realization, the low-noise realizations need not require significantly more computation than the direct form realization.

3.6 Limit Cycles

A limit cycle, sometimes referred to as a **multiplier roundoff limit cycle**, is a low-level oscillation that can exist in an otherwise stable filter as a result of the nonlinearity associated with rounding (or truncating) internal filter calculations [11]. Limit cycles require recursion to exist and do not occur in nonrecursive FIR filters.

As an example of a limit cycle, consider the second-order filter realized by

$$y(n) = Q_r\left\{\frac{7}{8}y(n-1) - \frac{5}{8}y(n-2) + x(n)\right\} \tag{3.69}$$

where $Q_r\{\}$ represents quantization by rounding. This is stable filter with poles at $0.4375 \pm j0.6585$. Consider the implementation of this filter with 4-bit (3-bit and a sign bit) two's complement fixed-point arithmetic, zero initial conditions ($y(-1) = y(-2) = 0$), and an input sequence $x(n) = \frac{3}{8}\delta(n)$, where $\delta(n)$ is the unit impulse or unit sample. The following sequence is obtained:

$$\begin{aligned}
y(0) &= Q_r\left\{\frac{3}{8}\right\} = \frac{3}{8}\\
y(1) &= Q_r\left\{\frac{21}{64}\right\} = \frac{3}{8}\\
y(2) &= Q_r\left\{\frac{3}{32}\right\} = \frac{1}{8}\\
y(3) &= Q_r\left\{-\frac{1}{8}\right\} = -\frac{1}{8}\\
y(4) &= Q_r\left\{-\frac{3}{16}\right\} = -\frac{1}{8}\\
y(5) &= Q_r\left\{-\frac{1}{32}\right\} = 0\\
y(6) &= Q_r\left\{\frac{5}{64}\right\} = \frac{1}{8}\\
y(7) &= Q_r\left\{\frac{7}{64}\right\} = \frac{1}{8}\\
y(8) &= Q_r\left\{\frac{1}{32}\right\} = 0\\
y(9) &= Q_r\left\{-\frac{5}{64}\right\} = -\frac{1}{8}\\
y(10) &= Q_r\left\{-\frac{7}{64}\right\} = -\frac{1}{8}\\
y(11) &= Q_r\left\{-\frac{1}{32}\right\} = 0\\
y(12) &= Q_r\left\{\frac{5}{64}\right\} = \frac{1}{8}
\end{aligned} \tag{3.70}$$

Notice that while the input is zero except for the first sample, the output oscillates with amplitude 1/8 and period 6.

Limit cycles are primarily of concern in fixed-point recursive filters. As long as floating-point filters are realized as the parallel or cascade connection of first- and second-order subfilters, limit cycles will generally not be a problem since limit cycles are practically not observable in first- and second-order systems implemented with 32-bit floating-point arithmetic [12]. It has been shown that such systems must have an extremely small margin of stability for limit cycles to exist at anything other than underflow levels, which are at an amplitude of less than 10^{-38} [12].

There are at least three ways of dealing with limit cycles when fixed-point arithmetic is used. One is to determine a bound on the maximum limit cycle amplitude, expressed as an integral number of quantization steps [13]. It is then possible to choose a wordlength that makes the limit cycle amplitude acceptably low. Alternately, limit cycles can be prevented by randomly rounding calculations up or down [14]. However, this approach is complicated to implement. The third approach is to properly choose the filter realization structure and then quantize the filter calculations using magnitude truncation [15,16]. This approach has the disadvantage of producing more roundoff noise than truncation or rounding (see Equations 3.12 through 3.14).

3.7 Overflow Oscillations

With fixed-point arithmetic it is possible for filter calculations to overflow. This happens when two numbers of the same sign add to give a value having magnitude greater than one. Since numbers with magnitude greater than one are not representable, the result overflows. For example, the two's complement numbers 0.101 (5/8) and 0.100 (4/8) add to give 1.001 which is the two's complement representation of −7/8.

The overflow characteristic of two's complement arithmetic can be represented as $R\{\}$ where

$$R\{X\} = \begin{cases} X-2 & X \geq 1 \\ X & -1 \leq X < 1 \\ X+2 & X < -1 \end{cases} \tag{3.71}$$

For the example just considered, $R\{9/8\} = -7/8$.

An overflow oscillation, sometimes also referred to as an adder overflow limit cycle, is a high-level oscillation that can exist in an otherwise stable fixed-point filter due to the gross nonlinearity associated with the overflow of internal filter calculations [17]. Like limit cycles, overflow oscillations require recursion to exist and do not occur in nonrecursive FIR filters. Overflow oscillations also do not occur with floating-point arithmetic due to the virtual impossibility of overflow.

As an example of an overflow oscillation, once again consider the filter of Equation 3.69 with 4-bit fixed-point two's complement arithmetic and with the two's complement overflow characteristic of Equation 3.71:

$$y(n) = Q_r\left\{R\left[\frac{7}{8}y(n-1) - \frac{5}{8}y(n-2) + x(n)\right]\right\} \tag{3.72}$$

In this case, we apply the input

$$\begin{aligned} x(n) &= -\frac{3}{4}\delta(n) - \frac{5}{8}\delta(n-1) \\ &= \left\{-\frac{3}{4}, -\frac{5}{8}, 0, 0, \ldots\right\} \end{aligned} \tag{3.73}$$

giving the output sequence

$$\begin{aligned} y(0) &= Q_r\left\{R\left[-\frac{3}{4}\right]\right\} = Q_r\left\{-\frac{3}{4}\right\} = -\frac{3}{4} \\ y(1) &= Q_r\left\{R\left[-\frac{41}{32}\right]\right\} = Q_r\left\{\frac{23}{32}\right\} = \frac{3}{4} \\ y(2) &= Q_r\left\{R\left[\frac{9}{8}\right]\right\} = Q_r\left\{-\frac{7}{8}\right\} = -\frac{7}{8} \\ y(3) &= Q_r\left\{R\left[-\frac{79}{64}\right]\right\} = Q_r\left\{\frac{49}{64}\right\} = \frac{3}{4} \\ y(4) &= Q_r\left\{R\left[\frac{77}{64}\right]\right\} = Q_r\left\{-\frac{51}{64}\right\} = -\frac{3}{4} \\ y(5) &= Q_r\left\{R\left[-\frac{9}{8}\right]\right\} = Q_r\left\{\frac{7}{8}\right\} = \frac{7}{8} \\ y(6) &= Q_r\left\{R\left[\frac{79}{64}\right]\right\} = Q_r\left\{-\frac{49}{64}\right\} = -\frac{3}{4} \\ y(7) &= Q_r\left\{R\left[-\frac{77}{64}\right]\right\} = Q_r\left\{\frac{51}{64}\right\} = \frac{3}{4} \\ y(8) &= Q_r\left\{R\left[\frac{9}{8}\right]\right\} = Q_r\left\{-\frac{7}{8}\right\} = -\frac{7}{8} \end{aligned} \tag{3.74}$$

This is a large-scale oscillation with nearly full-scale amplitude.

There are several ways to prevent overflow oscillations in fixed-point filter realizations. The most obvious is to scale the filter calculations so as to render overflow impossible. However, this may unacceptably restrict the filter dynamic range. Another method is to force completed sums-of-products to saturate at ± 1, rather than overflowing [18,19]. It is important to saturate only the completed sum, since intermediate overflows in two's complement arithmetic do not affect the accuracy of the final result. Most fixed-point digital signal processors provide for automatic saturation of completed sums if their saturation arithmetic feature is enabled. Yet another way to avoid overflow oscillations is to use a filter structure for which any internal filter transient is guaranteed to decay to zero [20]. Such structures are desirable anyway, since they tend to have low roundoff noise and be insensitive to coefficient quantization [21].

3.8 Coefficient Quantization Error

Each filter structure has its own finite, generally nonuniform grids of realizable pole and zero locations when the filter coefficients are quantized to a finite wordlength. In general the pole and zero locations desired in filter do not correspond exactly to the realizable locations. The error in filter performance (usually measured in terms of a frequency response error) resulting from the placement of the poles and zeros at the nonideal but realizable locations is referred to as coefficient quantization error.

Consider the second-order filter with complex-conjugate poles

$$\begin{aligned} \lambda &= re^{\pm j\theta} \\ &= \lambda_r \pm j\lambda_i \\ &= r\cos(\theta) \pm jr\sin(\theta) \end{aligned} \tag{3.75}$$

and transfer function

$$H(z) = \frac{1}{1 - 2r\cos(\theta)z^{-1} + r^2z^{-2}} \tag{3.76}$$

realized by the difference equation

$$y(n) = 2r\cos(\theta)y(n-1) - r^2y(n-2) + x(n) \tag{3.77}$$

Figure 3.3 from [5] shows that quantizing the difference equation coefficients results in a nonuniform grid of realizable pole locations in the z-plane. The grid is defined by the intersection of vertical lines corresponding to quantization of $2\lambda_r$ and concentric circles corresponding to quantization of $-r^2$. The sparseness of realizable pole locations near $Z = \pm 1$ will result in a large coefficient quantization error for poles in this region.

Figure 3.4 gives an alternative structure to Equation 3.77 for realizing the transfer function of Equation 3.76. Notice that quantizing the coefficients of this structure corresponds to quantizing λ_r and λ_i. As shown in Figure 3.5 from [5], this results in a uniform grid of realizable pole locations. Therefore, large coefficient quantization errors are avoided for all pole locations.

It is well established that filter structures with low roundoff noise tend to be robust to coefficient quantization, and visa versa [22–24]. For this reason, the uniform grid structure of Figure 3.4 is also popular because of its low roundoff noise. Likewise, the low-noise realizations of [7–10] can be expected to be relatively insensitive to coefficient quantization, and digital wave filters and lattice filters that are derived from low-sensitivity analog structures tend to have not only low coefficient sensitivity, but also low roundoff noise [25,26].

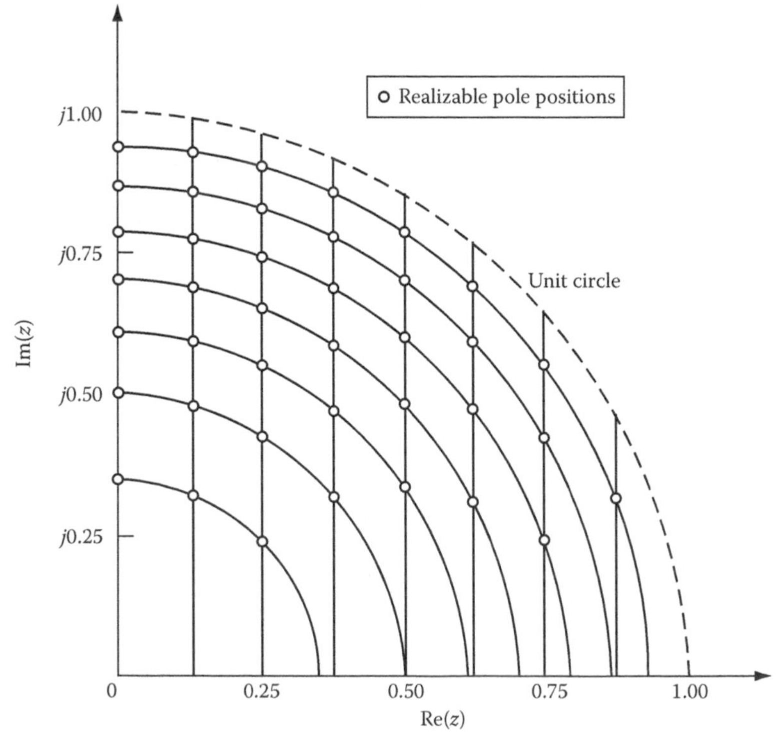

FIGURE 3.3 Realizable pole locations for the difference equation of Equation 3.76.

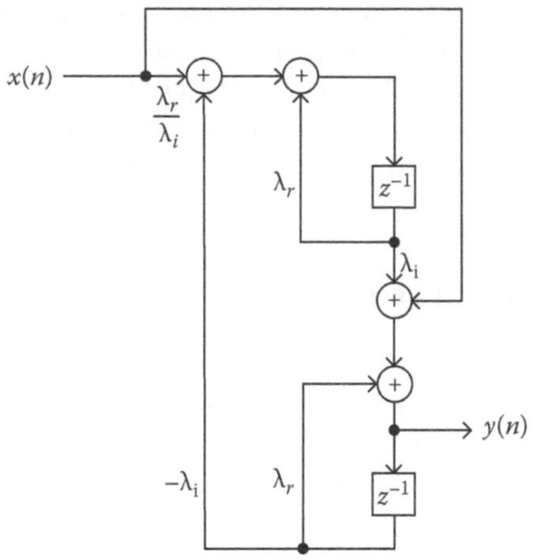

FIGURE 3.4 Alternate realization structure

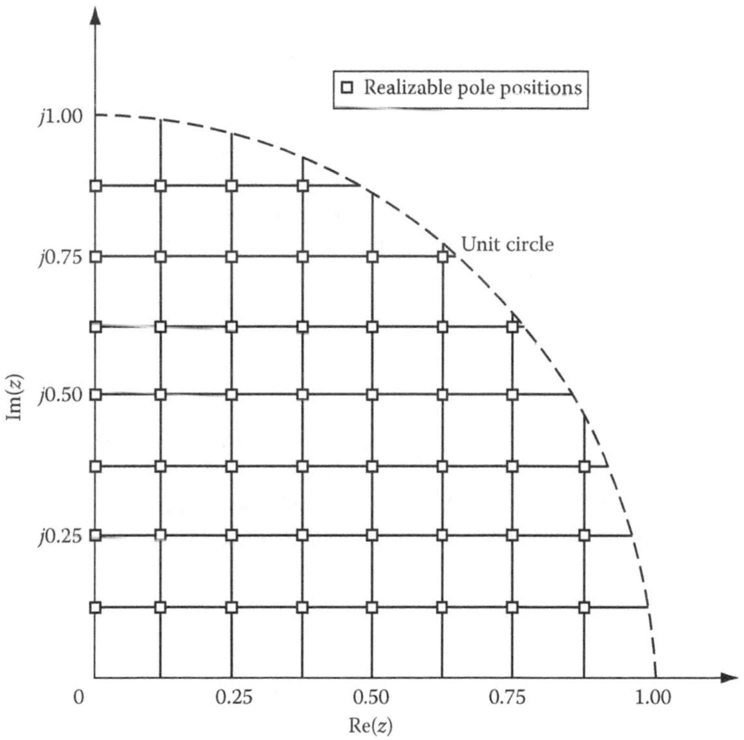

FIGURE 3.5 Realizable pole locations for the alternate realization structure.

It is well known that in a high-order polynomial with clustered roots, the root location is a very sensitive function of the polynomial coefficients. Therefore, filter poles and zeros can be much more accurately controlled if higher order filters are realized by breaking them up into the parallel or cascade connection of first- and second-order subfilters. One exception to this rule is the case of linear-phase FIR filters in which the symmetry of the polynomial coefficients and the spacing of the filter zeros around the unit circle usually permits an acceptable direct realization using the convolution summation.

Given a filter structure it is necessary to assign the ideal pole and zero locations to the realizable locations. This is generally done by simply rounding or truncating the filter coefficients to the available number of bits, or by assigning the ideal pole and zero locations to the nearest realizable locations. A more complicated alternative is to consider the original filter design problem as a problem in discrete optimization, and choose the realizable pole and zero locations that give the best approximation to the desired filter response [27–30].

3.9 Realization Considerations

Linear-phase FIR digital filters can generally be implemented with acceptable coefficient quantization sensitivity using the direct convolution sum method. When implemented in this way on a digital signal processor, fixed-point arithmetic is not only acceptable but may actually be preferable to floating-point arithmetic. Virtually all fixed-point digital signal processors accumulate a sum of products in a double-length accumulator. This means that only a single quantization is necessary to compute an output. Floating-point arithmetic, on the other hand, requires a quantization after every multiply and after every add in the convolution summation. With 32-bit floating-point arithmetic these quantizations introduce a small enough error to be insignificant for many applications.

When realizing IIR filters, either a parallel or cascade connection of first- and second-order subfilters is almost always preferable to a high-order direct-form realization. With the availability of very low-cost floating-point digital signal processors, like the Texas Instruments TMS320C32, it is highly recommended that floating-point arithmetic be used for IIR filters. Floating-point arithmetic simultaneously eliminates most concerns regarding scaling, limit cycles, and overflow oscillations. Regardless of the arithmetic employed, a low roundoff noise structure should be used for the second-order sections. Good choices are given in [2] and [10]. Recall that realizations with low fixed-point roundoff noise also have low floating-point roundoff noise. The use of a low roundoff noise structure for the second-order sections also tends to give a realization with low coefficient quantization sensitivity. First-order sections are not as critical in determining the roundoff noise and coefficient sensitivity of a realization, and so can generally be implemented with a simple direct form structure.

References

1. Weinstein, C. and Oppenheim, A.V., A comparison of roundoff noise in floating-point and fixed-point digital filter realizations, *Proc. IEEE*, 57, 1181–1183, June 1969.
2. Smith, L.M., Bomar, B.W., Joseph, R.D., and Yang, G.C., Floating-point roundoff noise analysis of second-order state-space digital filter structures, *IEEE Trans. Circuits Syst. II*, 39, 90–98, Feb. 1992.
3. Proakis, G.J. and Manolakis, D.J., *Introduction to Digital Signal Processing*, New York: Macmillan, 1988.
4. Oppenheim, A.V. and Schafer, R.W., *Digital Signal Processing*, Englewood Cliffs, NJ: Prentice-Hall, 1975.
5. Oppenheim, A.V. and Weinstein, C.J., Effects of finite register length in digital filtering and the fast Fourier transform, *Proc. IEEE*, 60, 957–976, Aug. 1972.
6. Bomar, B.W. and Joseph, R.D., Calculation of L_∞ norms for scaling second-order state-space digital filter sections, *IEEE Trans. Circuits Syst.*, CAS-34, 983–984, Aug. 1987.
7. Mullis, C.T. and Roberts, R.A., Synthesis of minimum roundoff noise fixed-point digital filters, *IEEE Trans. Circuits Syst.*, CAS-23, 551–562, Sept. 1976.
8. Jackson, L.B., Lindgren, A.G., and Kim, Y., Optimal synthesis of second-order state-space structures for digital filters, *IEEE Trans. Circuits Syst.*, CAS-26, 149–153, Mar. 1979.
9. Barnes, C.W., On the design of optimal state-space realizations of second-order digital filters, *IEEE Trans. Circuits Syst.*, CAS-31, 602–608, July 1984.

10. Bomar, B.W., New second-order state-space structures for realizing low roundoff noise digital filters, *IEEE Trans. Acoust. Speech Signal Process.*, ASSP-33, 106–110, Feb. 1985.
11. Parker, S.R. and Hess, S.F., Limit-cycle oscillations in digital filters, *IEEE Trans. Circuit Theory*, CT-18, 687–697, Nov. 1971.
12. Bauer, P.H., Limit cycle bounds for floating-point implementations of second-order recursive digital filters, *IEEE Trans. Circuits Syst. II*, 40, 493–501, Aug. 1993.
13. Green, B.D. and Turner, L.E., New limit cycle bounds for digital filters, *IEEE Trans. Circuits Syst.*, 35, 365–374, Apr. 1988.
14. Buttner, M., A novel approach to eliminate limit cycles in digital filters with a minimum increase in the quantization noise, in *Proceedings of the 1976 IEEE International Symposium on Circuits and Systems*, Munich, Germany, Apr. 1976, pp. 291–294.
15. Diniz, P.S.R. and Antoniou, A., More economical state-space digital filter structures which are free of constant-input limit cycles, *IEEE Trans. Acoust. Speech Signal Process.*, ASSP-34, 807–815, Aug. 1986.
16. Bomar, B.W., Low-roundoff-noise limit-cycle-free implementation of recursive transfer functions on a fixed-point digital signal processor, *IEEE Trans. Ind. Electron.*, 41, 70–78, Feb. 1994.
17. Ebert, P.M., Mazo, J.E., and Taylor, M.G., Overflow oscillations in digital filters, *Bell Syst. Tech. J.*, 48, 2999–3020, Nov. 1969.
18. Willson, A.N., Jr., Limit cycles due to adder overflow in digital filters, *IEEE Trans. Circuit Theory*, CT-19, 342–346, July 1972.
19. Ritzerfield, J.H.F., A condition for the overflow stability of second-order digital filters that is satisfied by all scaled state-space structures using saturation, *IEEE Trans. Circuits Syst.*, 36, 1049–1057, Aug. 1989.
20. Mills, W.T., Mullis, C.T., and Roberts, R.A., Digital filter realizations without overflow oscillations, *IEEE Trans. Acoust. Speech Signal Process.*, ASSP-26, 334–338, Aug. 1978.
21. Bomar, B.W., On the design of second-order state-space digital filter sections, *IEEE Trans. Circuits Syst.*, 36, 542–552, Apr. 1989.
22. Jackson, L.B., Roundoff noise bounds derived from coefficient sensitivities for digital filters, *IEEE Trans. Circuits Syst.*, CAS-23, 481–485, Aug. 1976.
23. Rao, D.B.V., Analysis of coefficient quantization errors in state-space digital filters, *IEEE Trans. Acoust. Speech Signal Process.*, ASSP-34, 131–139, Feb. 1986.
24. Thiele, L., On the sensitivity of linear state-space systems, *IEEE Trans. Circuits Syst.*, CAS-33, 502–510, May 1986.
25. Antoniou, A., *Digital Filters: Analysis and Design*, New York: McGraw-Hill, 1979.
26. Lim, Y.C., On the synthesis of IIR digital filters derived from single channel AR lattice network, *IEEE Trans. Acoust. Speech Signal Process.*, ASSP-32, 741–749, Aug. 1984.
27. Avenhaus, E., On the design of digital filters with coefficients of limited wordlength, *IEEE Trans. Audio Electroacoust.*, AU-20, 206–212, Aug. 1972.
28. Suk, M. and Mitra, S.K., Computer-aided design of digital filters with finite wordlengths, *IEEE Trans. Audio Electroacoust.*, AU-20, 356–363, Dec. 1972.
29. Charalambous, C. and Best, M.J., Optimization of recursive digital filters with finite wordlengths, *IEEE Trans. Acoust. Speech Signal Process.*, ASSP-22, 424–431, Dec. 1979.
30. Lim, Y.C., Design of discrete-coefficient-value linear-phase FIR filters with optimum normalized peak ripple magnitude, *IEEE Trans. Circuits Syst.*, 37, 1480–1486, Dec. 1990.

II

Signal Representation and Quantization

Jelena Kovačević
Lucent Technologies, Bell Laboratories

Christine Podilchuk
Rutgers University

SAMPLING THEOREMS CAN BE TRACED to the original paper by Whittaker in 1915 on interpolation. He proved the exactness of a method for interpolating between the samples from a function. Nyquist then presented the sampling theory for sampled telephone signals in 1928 establishing for the first time the term "Nyquist frequency." Shannon in 1948 and Kotel'nikov in 1933 wrote additional treatises on this topic [1–4].

Extensions from one-dimensional to multidimensional sampling can be traced to papers by Bracewell in 1956, and to Miyakawa in 1959. Multidimensional Fourier analysis, however, can be traced back to papers by Germain and Navier in the early eighteenth and nineteenth centuries [5–7].

In this section, Chapter 4 presents a thorough discussion of the techniques that are currently used and their underlying theory.

Of related interest is structure of the conversion process from the analog domain to the digital domain, and Chapter 5 presents a thorough survey of the various architectures for analog-to-digital conversion.

Finally, the process of quantization of discrete samples is discussed in Chapter 6. This discussion considers the accuracy issues arising due to quantization, in addition to other related topics.

References

1. Whittaker, E. T., On the functions which are represented by the expansions of the interpolation theory, *Proc. R. Soc. Edinburgh* 35:181–194, 1915.
2. Nyquist, H., Certain topics in telegraph transmission theory, *Trans. AIEE* 47:617–644, 1928.
3. Shannon, C. E., A mathematical theory of communication, *Bell Syst. Tech. J.* 27:379–423, 1948.
4. Sullivan, W. et al., *The Early Years of Radio Astronomy*, Cambridge University Press, Cambridge, UK, 1984.
5. Bracewell, R. N., Two-dimensional aerial smoothing in radio astronomy, *Aust. J. Phys.* 9:197–314, 1956.
6. Miyakawa, K., Sampling theory of stationary stochastic variables in multidimensional space, *J. Inst. Electron. Commn. (Jpn.)* 4(2):421–427, 1959.
7. Bracewell, R. N., *Two-Dimensional Imaging*, Prentice-Hall, Englewood Cliffs, NJ, 1995.

4

On Multidimensional Sampling

Ton Kalker
HP Labs

This chapter gives an overview of the most relevant facts of sampling theory, paying particular attention to the multidimensional aspect of the problem. It is shown that sampling theory formulated in a multidimensional setting provides insight to the supposedly simpler situation of one-dimensional sampling.

4.1 Introduction

The signals we encounter in the physical reality around us almost invariably have a continuous domain of definition. We like to model a speech signal as continuous function of amplitudes, where the domain of definition is a (finite) length interval of real numbers. A video signal is most naturally viewed as continuous function of luminance (chrominance) values, where the domain of definition is some volume in space-time.

In modern electronic systems we deal with many (in essence) continuous signals in a digital fashion. This means that we do not deal with these signals directly, but only with sampled versions of it: we only retain the values of these signals at a discrete set of points. Moreover, due to the inherently finite precision arithmetic capabilities of digital systems, we only record an approximated (quantized) value at every point of the sampling set. If we define sampling as the process of restricting a signal to a discrete set, explicitly without quantization of the sampled values, we can describe the contribution of this chapter as a study of the relation between continuous signals and their sampled versions.

Many textbooks start this topic by only considering sampling in the one-dimensional case. Digressions into the multidimensional case are usually made in later and more advanced sections. In this chapter we will start from the outset with the multidimensional case. It will be argued that this is the most natural setting, and that this approach will even lead to greater understanding of the one-dimensional case.

I will assume that not every reader is familiar with the concept of a lattice. As lattices are the most basic kind of sets onto which to sample signals, this chapter will start with a crash course on lattices in Section 4.2. After this the real work starts in Section 4.3 with an overview of the sampling theory for continuous functions. The central theme of this section is the intimate relationship between sampling and the discrete space-time Fourier transform (DSFT). In Section 4.4 we consider simultaneous sampling in both spatial and frequency domain. The central theme in this section is the relationship with the discrete Fourier transform (DFT). We continue with a digression on cascaded sampling (Section 4.5), and with some useful results on changing variables (Section 4.6). We end with an application of sampling theory to HDTV-to-SDTV conversion. The proofs (or hints to it) of the stated result can be found in the appendix.

We end this introduction with some conventions. We will refer to a signal as a function, defined on some appropriate domain. As all of our functions are in principle multidimensional, we will lighten the burden of notation by suppressing the multidimensional character of variables involved wherever possible. In particular we will use $f(x)$ to denote a function $f(x_1, \ldots, x_n)$ on some continuous domain (say $\mathbb{R}^n$). Similarly we will use $f(k)$ to denote a function $f(k_1, \ldots, k_n)$ on some discrete domain (say $\mathbb{Z}^n$). By abuse of terminology we will refer to a function defined on a continuous domain as a continuous function and to a function on discrete domain as discrete function.

4.2 Lattices

Although sampling of a function can in principle be done with respect to any set of points (nonuniform sampling), the most common form of sampling is done with respect to sets of points which have a certain algebraic structure and are known as lattices. They are the object of study in this section.

4.2.1 Definition

Formally, the definition of a lattice is given as

Definition 4.1: A (sub)lattice $\mathcal{L}$ of $\mathbb{C}^n$ ($\mathbb{R}^n$, $\mathbb{Z}^n$) is a set of points satisfying that

1. There is a shortest nonzero element.
2. If $\lambda_1, \lambda_2 \in \mathcal{L}$, then $a\lambda_1 + b\lambda_2 \in \mathcal{L}$ for all integers a and b.
3. $\mathcal{L}$ contains n linearly independent elements.

This definition may seem to make lattices rather abstract objects, but they can be made more tangible by representing them by generating matrices. Namely, one can show that every lattice $\mathcal{L}$ contains a set of linearly independent points $\{\lambda_1, \ldots, \lambda_n\}$ such that every other point $\lambda \in \mathcal{L}$ is an integer linear combination $\sum_{i=1}^{n} a_i\lambda_i$. Arranging such a set in a matrix $L = [\lambda_1, \ldots, \lambda_n]$ yields a generating matrix L of $\mathcal{L}$. It has the property that every $\lambda \in \mathcal{L}$ can be written as $\lambda = Lk$, where $k \in \mathbb{Z}^n$ is an integer vector. At this point it is important to note that there is no such thing as the generating matrix L of a lattice $\mathcal{L}$. Defining a unimodular matrix U as an integer matrix with $|\det(U)| = 1$, every other generating matrix is of the form LU, and every such matrix is a generating matrix. However, this also shows that the determinant of a generating matrix is determined up to a sign.

Definition 4.2: Let $\mathcal{L}$ be a lattice and let L be a generating matrix of $\mathcal{L}$. Then the determinant of $\mathcal{L}$ is defined by

$$\det(\mathcal{L}) = |\det(L)|.$$

In case the dimension is 1 ($n = 1$), every lattice is given as all the integer multiples of a single scalar. This scalar is unique up to a sign, and by convention one usually defines the positive scalar as the sampling period T (for time):

$$\mathcal{L}_T = \{nT : n \in \mathbb{Z}\} \subset \mathbb{C}, \mathbb{R}, \mathbb{Z}. \tag{4.1}$$

In case the dimension is 2 ($n = 2$) it is no longer possible to single out a natural candidate as the generating matrix for a lattice. As an example consider the lattice $\mathcal{L}$ generated by the matrix (see also Figure 4.1):

$$L_1 = \begin{bmatrix} \sqrt{3} & \sqrt{3} \\ -1 & 1 \end{bmatrix}.$$

There is no reason to consider the matrix L_1 as the generating matrix of the lattice $\mathcal{L}$, and in fact the matrix

$$L_2 = \begin{bmatrix} \sqrt{3} & 2\sqrt{3} \\ 1 & 0 \end{bmatrix}.$$

is just as valid a generating matrix as L_1.

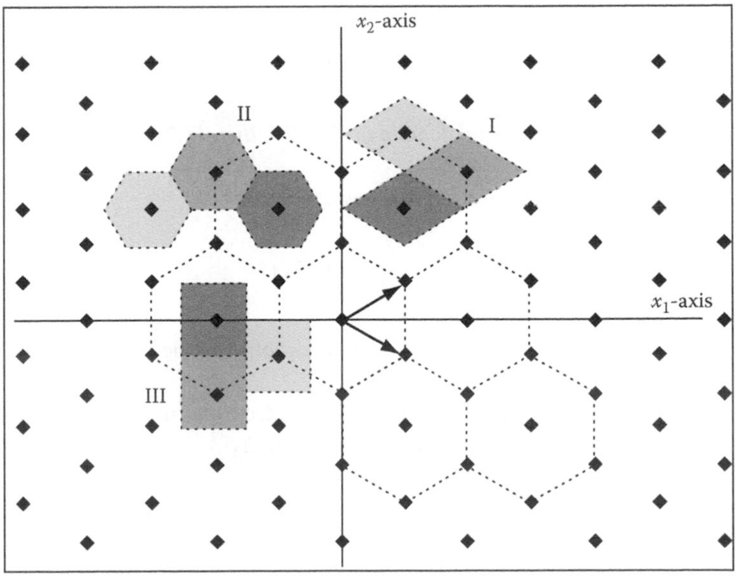

FIGURE 4.1 A hexagonal lattice in the continuous plane.

4.2.2 Fundamental Domains and Cosets

Each lattice $\mathcal{L}$ can be used to partition its embedding space into so-called fundamental domains. The importance of the concept of fundamental domains lies in their ability to define $\mathcal{L}$-periodic functions, i.e., functions $f(x)$ for which $f(x) = f(x + \lambda)$ for every $\lambda \in \mathcal{L}$. Knowing a $\mathcal{L}$-periodic function $f(x)$ on a fundamental domain is sufficient to know the complete function. Periodic functions will emerge naturally when we come to speak about sampling of continuous functions.

Let $\mathcal{L} \subset \mathcal{D}$ be a lattice, where $\mathcal{D}$ is either a lattice $\mathcal{M} \subset \mathbb{R}^n$ or the space $\mathbb{R}^n$ itself. Let L be a generating matrix of $\mathcal{L}$, and let P be an arbitrary subset of $\mathcal{D}$. With every $p \in P$ we can associate a translated version or coset $p + \mathcal{L}$ of $\mathcal{L}$. The set of cosets is referred to as the coset group of $\mathcal{L}$ with respect to $\mathcal{D}$ and is denoted by the expression $\mathcal{D}/\mathcal{L}$. A fundamental domain is defined as a subset $P \subset \mathcal{D}$ which intersects every coset in exactly one point.

Definition 4.3: The set P is called a fundamental domain of the lattice $\mathcal{L}$ in $\mathcal{D}$ if and only if

1. $p \neq q$ implies $p + \mathcal{L} \neq q + \mathcal{L}$, and
2. $\bigcup_{p \in P} p + \mathcal{L} = \mathcal{D}$.

A fundamental domain is not a uniquely defined object. For example, the shaded areas in Figure 4.1 show three possibilities for the choice of a fundamental domain. Although the shapes may differ, their volume is defined by the lattice $\mathcal{L}$.

THEOREM 4.1

Let P be a fundamental domain of the lattice $\mathcal{L}$ in D, and assume that P is measurable, i.e., that its volume is defined.

1. *If $\mathcal{D} = \mathbb{R}^n$, then the volume of P is given by*

$$\text{vol}(P) = \det(\mathcal{L}).$$

2. *If $\mathcal{D} = \mathcal{M}$, and if Q is a fundamental domain of $\mathcal{L}$ in $\mathbb{R}^n$, then $Q \cap \mathcal{M}$ is a fundamental domain of $\mathcal{L}$ in $\mathcal{M}$.*
3. *If $\mathcal{D} = \mathcal{M}$, then the number of points in P is given by*

$$\#(P) = \det(\mathcal{L})/\det(\mathcal{M}).$$

This number is referred to as the index of $\mathcal{L}$ in $\mathcal{M}$, and is denoted by the symbol $\iota(\mathcal{L}, \mathcal{M})$.

As a consequence of assertion 1 of this theorem, all the shaded areas in Figure 4.1, being fundamental domains of the same hexagonal lattice, have a volume equal to $2\sqrt{3}$.

4.2.3 Reciprocal Lattices

For any lattice $\mathcal{L}$ there exists a reciprocal lattice $\mathcal{L}^*$ as defined below. Reciprocal lattices appear in the theory of Fourier transforms of sampled continuous functions (see Section 4.3).

Definition 4.4: Let $\mathcal{L}$ be a lattice. Its reciprocal lattice $\mathcal{L}^*$ is defined by

$$\mathcal{L}^* = \{\lambda^* : \langle \lambda^*, \lambda \rangle \in \mathbb{Z} \forall \lambda \in \mathcal{L}\},$$

where $\langle \lambda^*, \lambda \rangle$ denotes the usual inner product $\sum_i \lambda_i{}^* \lambda_i$.

This notion of reciprocal lattice is made more tangible by the observation that the reciprocal lattice of $[L]$ is the lattice $[L^{-t}]$, where $[M]$ denotes the lattice generated by a matrix M. In particular $\det(\mathcal{M}^*) = \det(\mathcal{M})^{-1}$. For example, the reciprocal lattice of the lattice of Figure 4.1 is generated by the matrix

$$\frac{1}{2\sqrt{3}} \begin{bmatrix} 1 & 1 \\ -\sqrt{3} & \sqrt{3} \end{bmatrix}.$$

This lattice is very similar to the original lattice: it differs by a rotation by $\pi/2$, and a scaling factor of $1/2\sqrt{3}$. In particular, the volume of a fundamental domain of $\mathcal{L}^*$ is equal to $1/2\sqrt{3}$.

An important property of reciprocal lattices is that subset inclusions are reversed. To be precise, the inclusion $\mathcal{M} \subset \mathcal{L}$ holds if and only if $\mathcal{L}^* \subset \mathcal{M}^*$. Using some elementary math it follows that the coset groups $\mathcal{L}/\mathcal{M}$ and $\mathcal{M}^*/\mathcal{L}^*$ have the same number of elements.

4.3 Sampling of Continuous Functions

In this section we will give the main results on the theory of sampled continuous functions. It will be shown that there is a strong relationship between sampling in the spatial domain and periodizing in the frequency domain. In order to state this result this section starts with a short overview of multidimensional Fourier transforms. This allows us to formulate the main result (Theorem 4.3), which states very informally that sampling in the spatial domain is equivalent to periodizing in the frequency domain.

4.3.1 The Continuous Space-Time Fourier Transform

Let $f(x)$ be a nice* function defined on the continuous domain $\mathbb{R}^n$. Let its continuous space-time Fourier transform (CSFT)† $F(\nu)$ be defined by

$$F(\nu) = F(f)(\nu) = \int_{\mathbb{R}^n} e^{-2\pi i \langle x, \nu \rangle} f(x) \mathrm{d}x \tag{4.2}$$

with inverse transform given by

$$F(\nu) = F^{-1}(f)(x) = \int_{\mathbb{R}^n} e^{2\pi i \langle x, \nu \rangle} F(\nu) \mathrm{d}\nu. \tag{4.3}$$

* Nice means in this context that all sums, integrals, Fourier transforms, etc. involving the function exist and are finite.

† Contrary to the conventional wisdom, we choose to exclude the factor 2π from the frequency term $\omega = 2\pi\nu$. This has the advantage that the Fourier transform is orthogonal, without any need for normalizing factors.

Forgetting many technicalities, the CSFT has the following basic properties:

- The CSFT is an isometry, i.e., it preserves inner products:

$$\langle f, g\rangle = \langle \mathcal{F}(f), \mathcal{F}(g)\rangle.$$

- The CSFT of the point-wise multiplication of two functions is the convolution of the two separate CSFTs:

$$\mathcal{F}(f \cdot g) = \mathcal{F}(f) * \mathcal{F}(g).$$

A special class of functions* is the class of lattice combs (Figure 4.2 illustrates the lattice comb of the quincunx lattice generated by the matrix $\begin{bmatrix} 1 & -1 \\ 1 & 1 \end{bmatrix}$). If $\mathcal{L}$ is a lattice, the lattice comb $\coprod_{\mathcal{L}}$ is a set of δ functions with support on $\mathcal{L}$ and is formally defined by

$$\coprod_{\mathcal{L}}(x) = \sum_{\lambda \in \mathcal{L}} \delta_\lambda(x). \quad (4.4)$$

The following theorem states the most important facts about lattice combs.

THEOREM 4.2

With notations as above, we have the following properties:

$$\coprod_{\mathcal{L}}(x) = \frac{1}{\det(\mathcal{L})} \sum_{\lambda^* \in \mathcal{L}^*} e^{-2\pi i\langle x, \lambda^*\rangle} \quad (4.5)$$

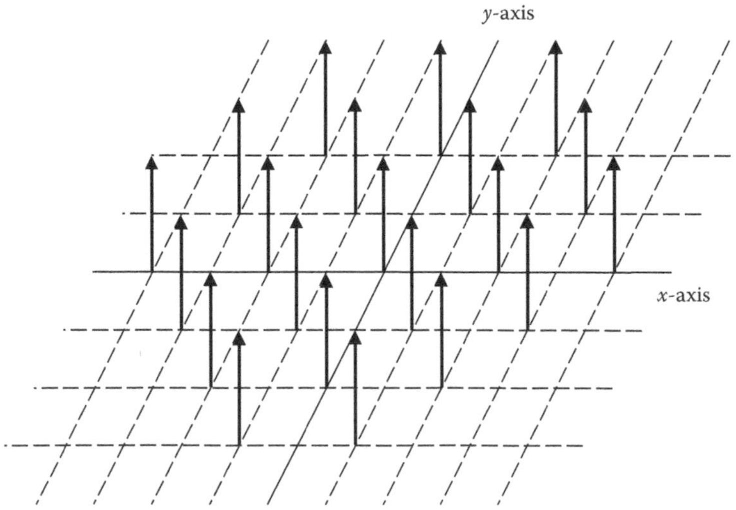

FIGURE 4.2 Lattice comb for the quincunx lattice.

* Actually distributions.

$$\mathcal{F}\left(\coprod_{\mathcal{L}}\right)(\nu) = \sum_{\lambda \in L} e^{-2\pi i \langle \lambda, \nu \rangle}$$
$$= \det(\mathcal{L}^*) \coprod_{\mathcal{L}^*}(\nu). \quad (4.6)$$

The last equation says that the CSFT of a lattice comb is the lattice comb of the reciprocal lattice, up to a constant.

4.3.2 The Discrete Space-Time Fourier Transform

The CSFT is a functional on continuous functions. We also need a similar functional on (multidimensional) sequences. This functional will be the DSFT. In this section, we will only state the definition. The properties of this functional and its relation to the CSFT will be highlighted in the next section. So let $\mathcal{L}$ be a lattice and let P^* be a fundamental domain of the reciprocal lattice $\mathcal{L}^*$. Let $\tilde{f}(x) = \Sigma_{\mathcal{L}}(f)(x)$ be the sampled version of f, and let $\tilde{F}(\nu) = \Pi_{\mathcal{L}^*}(F)(\nu)$ be the periodized version of $F(\nu)$. Then we define the forward and backward DSFT by

$$\tilde{\mathcal{F}}(\tilde{f})(\nu) = \sum_{x \in \mathcal{L}} e^{-2\pi i \langle x, \nu \rangle} \tilde{f}(x), \quad (4.7)$$

and

$$\tilde{\mathcal{F}}^{-1}(\tilde{F})(\nu) = \det(\mathcal{L}) \int_{P^*} e^{2\pi i \langle x, \nu \rangle} \tilde{F}(\nu) d\nu, \quad (4.8)$$

respectively.

Note that the function $\tilde{F}(\tilde{f})(\nu)$ is a $\mathcal{L}^*$-periodic function. This implies that the formula for the inverse DSFT is independent of the choice of the fundamental domain P^*.

4.3.3 Sampling and Periodizing

One of the most important issues in the sampling of functions concerns the relationship between the CSFT of the original function and the DSFT of a sampled version. In this section, we will state the main theorem (Theorem 4.3) of sampling theory.

Before continuing we need two definitions. If $f(x)$ is a function and $\mathcal{L} \subset \mathbb{R}^n$ is a lattice, sampling $f(x)$ on $\mathcal{L}$ is defined by

$$\Sigma_{\mathcal{L}}(f)(x) = \begin{cases} f(x) & \text{if } x \in \mathcal{L} \\ 0 & \text{if } x \notin \mathcal{L}\,. \end{cases} \quad (4.9)$$

The above definition has to be read carefully: sampling a function $f(x)$ on a lattice means that we modify $f(x)$ by putting all its values outside of the lattice to 0. It does not mean that we forget how the lattice is embedded in the continuous domain. For example, when we sample a one-dimensional continuous function $f(x)$ on the set of even numbers, the downsampled function $f_s(k)$ is not defined by $f_s(k) = f(2k)$, but by $f_s(k) = f(k)$ when k is even, and 0 otherwise.

Closely related to the sampling operator is the periodizing operator $\Pi_{\mathcal{L}}$, which modifies a function $f(x)$ such that it becomes $\mathcal{L}$-periodic. This operator is defined by

$$\Pi_{\mathcal{L}}(f)(x) = \det(\mathcal{L}) \sum_{\lambda \in \mathcal{L}} f(x - \lambda). \quad (4.10)$$

Clearly $\Pi_{\mathcal{L}}(f)(x)$ is $\mathcal{L}$-periodic, i.e., $\Pi_{\mathcal{L}}(f)(x) = \Pi_{\mathcal{L}}(f)(x - \lambda)$ for all $\lambda \in \mathcal{L}$. With these tools at our disposal we are now in a position to formulate the main theorem of sampling theory.

THEOREM 4.3

With definitions and notations as above, consider the following diagram:

$$\begin{array}{ccc} f & \xrightarrow{\mathcal{F}} & F \\ \downarrow\Sigma_{\mathcal{L}} & & \downarrow\Pi_{\mathcal{L}^*} \\ \tilde{f} & \xrightarrow{\tilde{\mathcal{F}}} & \tilde{F} \end{array}$$

The following assertions hold:

1. *The above diagram commutes,* i.e., whichever way we take to go from top left to bottom right, the result is the same. Informally this can be formulated as saying that first sampling and taking the DSFT is the same as first taking the CSFT and then periodizing.*
2. $\sqrt{\det(\mathcal{L})}\tilde{\mathcal{F}}$ *(and, therefore,* $\sqrt{\det(\mathcal{L}^*)}\tilde{\mathcal{F}}^{-1}$ *) is an isometry with respect to the inner products*

$$\langle \tilde{f}, \tilde{g} \rangle_{\mathcal{L}} = \sum_{\lambda \in \mathcal{L}} \tilde{f}^{\dagger}(\lambda)\tilde{g}(\lambda)$$

and

$$\langle \tilde{F}, \tilde{G} \rangle_{P^*} = \int_{P^*} \tilde{F}^{\dagger}(\nu)\tilde{G}(\nu)d\nu,$$

respectively.

Proof 4.1 The proof relies heavily on the property of lattice combs and can be found in the appendix.

This theorem has many important consequences, the best known of which is the Shannon sampling theorem. This theorem says that a function can be retrieved from a sampled version if the support of its CSFT is contained within a fundamental domain of the reciprocal lattice. Given the above theorem this result is immediate: we only need to verify that a function $F(\nu)$ can be retrieved from $\Pi_{\mathcal{L}^*}(F)$ by restriction to a fundamental domain when $F(\nu)$ has sufficiently restricted support.

THEOREM 4.4 (Shannon)

Let $\mathcal{L}$ be a lattice, and let $f(x)$ be a continuous function with CSFT $F(\nu)$. Let $\tilde{f} = \Sigma_{\mathcal{L}}(f)$. The function $f(x)$ can be retrieved from $\tilde{f}(\lambda)$ if and only if the support of $F(\nu)$ is contained in some fundamental domain P^* of the reciprocal lattice $\mathcal{L}^*$. In that case we can retrieve $f(x)$ from $\tilde{f}(\lambda)$ with the formula

$$f(x) = \sum_{\lambda \in \mathcal{L}} \tilde{f}(\lambda)\text{Int}(x - \lambda),$$

* Commuting diagrams are a common mathematical tool to describe that certain sequences of function applications are equivalent.

where

$$\mathrm{Int}(x) = \det(\mathcal{L}) \int\limits_{P^*} e^{2\pi i \langle x,\nu\rangle} d\nu.$$

Proof 4.2 We only need to prove the interpolation formula.

$$\begin{aligned} f(x) &= \int\limits_{P^*} e^{2\pi i\langle x,\nu\rangle} F(\nu) d\nu \\ &= \det(\mathcal{L}) \sum_{\lambda\in\mathcal{L}} f(\lambda) \int\limits_{P^*} e^{2\pi i\langle x-\lambda,\nu\rangle} d\nu \\ &= \sum_{\lambda\in\mathcal{L}} f(\lambda)\mathrm{Int}(x-\lambda). \end{aligned} \tag{4.11}$$

We end this section with an example showing all the aspects of Theorem 4.3.

Example 4.1

Let $\mathcal{L} \subset \mathbb{Z}^2$ be the quincunx sampling lattice generated by the matrix $L = \frac{1}{2}\begin{bmatrix} 1 & -1 \\ 1 & 1 \end{bmatrix}$. Let

$$f(x_1, x_2) = \sin c(x_1 - x_2) \sin c(x_1 + x_2).$$

A simple computation shows that CSFT $F(\nu_1, \nu_2)$ (of $f(x_1, x_2)$ is given by

$$F(\nu_1, \nu_2) = \frac{1}{2}\chi_\Lambda(\nu_1, \nu_2),$$

where Λ is the set $\Lambda = \{(\nu_1, \nu_2) : |\nu_1| + |\nu_2| \leq 1\}$. Observing that $\mathcal{L}^*$ is generated by $\begin{bmatrix} 1 & -1 \\ 1 & 1 \end{bmatrix}$, we find that the periodized function $\Pi_{\mathcal{L}^*}(F)$ is constant with value 1.

Sampling $f(x)$ on the quincunx lattice yields the function $\tilde{f}(\lambda)$:

$$\tilde{f}(\lambda_1, \lambda_2) = \begin{cases} 1 & \text{if } (\lambda_1, \lambda_2) = (0,0) \\ 0 & \text{if } (\lambda_1, \lambda_2) \neq (0,0) \end{cases}.$$

It is now trivial to check that $\tilde{\mathcal{F}}(\tilde{f}) = \tilde{F}$, as predicted by Theorem 4.3. Moreover, as

$$\|\tilde{f}\|_2^2 = \sum_{\lambda\in\mathcal{L}} \delta_0(\lambda)^2 = 1$$

and

$$\|\tilde{F}\|_2^2 = \int\limits_{\Lambda} d\nu = 1/2,$$

it follows that $\|\tilde{\mathcal{F}}\|$ and $\|\tilde{f}\|$ differ by a factor of $\sqrt{2} = \sqrt{\det(\mathcal{L}^*)}\sqrt{2} = \sqrt{\det(\mathcal{L}^*)}$, again as predicted by Theorem 4.3.

4.4 From Infinite Sequences to Finite Sequences

In the previous section we considered sampling in the spatial domain and saw that this was equivalent to periodizing in the frequency domain. One obvious question now arises: what happens if we sample the DSFT of a (spatially) sampled function? In this section we will answer this question and show that sampling in both spatial and frequency domains simultaneously is closely related to properties of the DFT.

4.4.1 The Discrete Fourier Transform

The DFT is a frequency transform on finite sequences. In a multidimensional context the DFT is best defined by assuming two lattices $\mathcal{L}$ and $\mathcal{M}$, $\mathcal{M} \subset \mathcal{L} \subset \mathbb{R}^n$. Let P be a fundamental domain of $\mathcal{L}$ in $\mathcal{M}$, and let P^* be a fundamental domain of $\mathcal{M}^*$ in $\mathcal{L}^*$ (recall that lattice inclusions invert when going over to the reciprocal domain [Section 4.2]). Note that both P and P^* have the same number points, viz. $\#(P) = \#(P^*) = \iota(\mathcal{L}^*, \mathcal{M}^*) = \iota(\mathcal{M}, \mathcal{L})$. Let $\hat{f}(p)$, $p \in P$ be a finite sequence over P. The DFT $\hat{\mathcal{F}}$ is now defined as functional which maps sequences $\hat{f}$ to sequences $\hat{F}$ over P^*. The formal definitions of $\hat{\mathcal{F}}$ and $\hat{\mathcal{F}}^{-1}$ are as follows.

Definition 4.5:

$$\hat{\mathcal{F}}(\hat{f})(p^*) = \frac{1}{\det(\mathcal{M})} \sum_{p \in P} e^{-2\pi i \langle p, p^* \rangle} \hat{f}(p) \tag{4.12}$$

$$\hat{\mathcal{F}}^{-1}(\hat{F})(p) = \frac{1}{\det(\mathcal{L}^*)} \sum_{p^* \in P^*} e^{2\pi i \langle p, p^* \rangle} \hat{F}(p^*). \tag{4.13}$$

It is obvious that the conventional one-dimensional DFT is a special case of the more general multidimensional DFT defined above. The next example makes this more explicit.

Example 4.2

Let $\mathcal{M} \subset \mathcal{L} \subset \mathbb{R}$ be defined by $\mathcal{M} = \mathbb{Z}$ for some positive integer p, and let $\mathcal{L} = \frac{1}{p}\mathbb{Z}$. One easily checks that the set P and P^* can be chosen as $\{0/p, \ldots, (p-1)/p\}$ and $\{0, \ldots, p-1\}$, respectively. If x_n and X_m are the values of $\hat{f}$ on $n/p \in P$ and of $\hat{F}$ on $m \in P^*$, respectively, then the functionals $\hat{\mathcal{F}}$ and $\hat{\mathcal{F}}^{-1}$ are defined in the (x_n, X_m) domain as

$$X_m = \sum_{n=0}^{p-1} e^{-\frac{2\pi i n m}{p}} x_n, \tag{4.14}$$

$$x_n = \frac{1}{p} \sum_{m=0}^{p-1} e^{\frac{2\pi i n m}{p}} X_m. \tag{4.15}$$

This is, of course, nothing else but the usual definition of the one-dimensional DFT on finite sequences of length p.

The following example shows the general DFT at work in a two-dimensional setting.

Example 4.3

(Example 4.1 continued) Continuing Example 4.1, we choose the lattice $\mathcal{M} = \mathbb{Z}^2$ as the periodizing lattice. We can then choose

$$P = \{p_0, p_1\} = \left\{(0, 0), \left(\frac{1}{2}, \frac{1}{2}\right)\right\}$$

and

$$P^* = \{p_0^*, p_1^*\} = \{(0, 0), (1, 0)\}.$$

The functional $\hat{\mathcal{F}}$ is then given by

$$\begin{aligned} X_0 &= x_0 e^{-2\pi i \langle p_0, p_0^* \rangle} + x_1 e^{-2\pi i \langle p_1, p_0^* \rangle} \\ &= x_0 + x_1 \\ X_1 &= x_0 e^{-2\pi i \langle p_0, p_1^* \rangle} + x_1 e^{-2\pi i \langle p_1, p_1^* \rangle} \\ &= x_0 - x_1, \end{aligned}$$

and the functional $\hat{\mathcal{F}}^{-1}$ by

$$\begin{aligned} x_0 &= \frac{1}{2}\left(X_0 e^{-2\pi i \langle p_0, p_0^* \rangle} + X_1 e^{-2\pi i \langle p_0, p_1^* \rangle}\right) \\ &= \frac{1}{2}(X_0 + X_1) \\ x_1 &= \frac{1}{2}\left(X_0 e^{-2\pi i \langle p_1, p_0^* \rangle} + X_1 e^{-2\pi i \langle p_1, p_1^* \rangle}\right) \\ &= \frac{1}{2}(X_0 - X_1). \end{aligned}$$

4.4.2 Combined Spatial and Frequency Sampling

We start with setting up the context of the problem. So let $f(x)$ be a nice continuous function on $\mathbb{R}^n$ and let $\mathcal{M}$ and $\mathcal{L}$ be two lattices such that $\mathcal{M} \subset \mathcal{L} \subset \mathbb{R}^n$. Sampling $f(x)$ on $\mathcal{L}$ and periodizing on $\mathcal{M}$ we construct a function $\hat{f}(x)$ that has support on $\mathcal{L}$ and is $\mathcal{M}$-periodic. In formula:

$$\hat{f}(x) = \begin{cases} \det(M) \sum_{\mu \in M} f(x - \mu) & \text{if } x \in \mathcal{L} \\ 0 & \text{if } x \notin \mathcal{L} \end{cases}$$

A similar definition can be given for the function $\hat{F}(\nu)$, which is obtained from the CSFT $F(\nu)$ of $f(x)$ by periodizing on $\mathcal{L}^*$ and sampling on $\mathcal{M}^*$.

One easily verifies that $\hat{f}(x)$ is completely specified by its values on a (finite) fundamental domain P of $\mathcal{M}$ in $\mathcal{L}$. Similarly $\hat{D}(\nu)$ is completely specified by its values on a fundamental domain P^* of $\mathcal{L}^*$ in $\mathcal{M}^*$. Now we are in a position to extend the commutative diagram of Theorem 4.3.

THEOREM 4.5

With notations and definitions as above, consider the following extensions of the diagram of Theorem 4.3:

$$
\begin{array}{ccc}
f & \stackrel{\mathcal{F}}{\longrightarrow} & F \\
\downarrow \Sigma_{\mathcal{L}} & & \downarrow \Pi_{\mathcal{L}^*} \\
\tilde{f} & \stackrel{\tilde{\mathcal{F}}}{\longrightarrow} & \tilde{\mathcal{F}} \\
\downarrow \Pi_{\mathcal{M}} & & \downarrow \Sigma_{\mathcal{M}^*} \\
\hat{f} & \stackrel{\hat{\mathcal{F}}}{\longrightarrow} & \hat{\mathcal{F}}
\end{array}
$$

The following assertions hold:

1. *The above diagram commutes*
2. *The functionals* $\sqrt{\det(\mathcal{L})}\sqrt{\det(\mathcal{M})}\hat{\mathcal{F}}$ and $\sqrt{\det(\mathcal{L}^*)}\sqrt{\det(\mathcal{M}^*)}\hat{\mathcal{F}}^{-1}$ *are isometries with respect to the inner products*

$$\langle \hat{f}, \hat{g} \rangle_P = \sum_{p \in P} \hat{f}^\dagger(p)\hat{g}(p)$$

and

$$\langle \hat{F}, \hat{G} \rangle_{P^*} = \sum_{p^* \in P^*} \hat{F}^\dagger(p^*)\hat{G}(p^*).$$

Proof 4.3 See appendix.

The theorem above says that sampling the Fourier transform of a sampled function amounts to periodizing that sampled version. In this process only a finite number of data points in both the spatial and the frequency domain are sufficient to specify the resulting functions. Moreover, the CSFT can be pushed down to a DFT to provide for a one-to-one orthogonal correspondence between the two domains.

We close this section with two examples.

Example 4.4

(Example 4.2 continued) The formulas for the DFT obtained in Example 4.2 are not orthonormal. According to Theorem 4.5 above we have to multiply the forward transform with $\sqrt{\det(\mathcal{L})\det(\mathcal{M})} = \frac{1}{\sqrt{p}}$ and the backward transform with the inverse of this number to obtain orthonormal versions of the DFT. This result in the following well-known formulas for the orthonormal one-dimensional DFT.

$$X_m = \frac{1}{\sqrt{p}} \sum_{n=0}^{p-1} e^{-\frac{2\pi inm}{p}} x_n, \tag{4.16}$$

$$x_n = \frac{1}{\sqrt{p}} \sum_{m=0}^{p-1} e^{\frac{2\pi inm}{p}} X_m. \tag{4.17}$$

Example 4.5

(Example 4.3 continued) With $\mathcal{L}$, $\mathcal{M}$, $f(x)$, P and P^* as in Example 4.3, we find that the periodized sampled function $\hat{f}$ is represented by the pair $(1, 0)$, and that the periodized sampled CSFT $\hat{F}$ of F is represented by the pair $(1, 1)$. Using the formulas for the DFT of Example 4.3 is now easy to verify that $\hat{\mathcal{F}}(\{1, 0\}) = \{1, 1\}$ and $\hat{\mathcal{F}}^{-1}(\{1, 1\}) = \{1, 0\}$, as predicted by Theorem 4.5.

4.5 Lattice Chains

In the previous section we considered the sampling of continuous functions. In this section we will consider the sampling of discrete functions. The necessity of studying this topic comes from the fact that very often the sampling of a continuous function $f(x)$ is done in steps: $f(x)$ is first sampled to a fine grid $\mathcal{L}_1$, and subsequently sampled to a coarser grid $\mathcal{L}_2$, $\mathcal{L}_2 \subset \mathcal{L}_1$. Letting $\tilde{f}^{(i)} = \Sigma_{\mathcal{L}_i}(f)$ and letting $\tilde{F}^{(i)}$ be the corresponding DFST, a natural question is whether we can obtain $\tilde{F}^{(2)}$ directly from $\tilde{F}^{(1)}$, without having to go back to CSFT of $f(x)$. This question is addressed in the following theorem and answered affirmatively.

THEOREM 4.6

With notation as above, and letting P^* be a fundamental domain of $\mathcal{L}_1^*$ in $\mathcal{L}_2^*$, we have the following result:

$$\tilde{F}^{(2)}(\nu) = \frac{1}{\#(P^*)} \sum_{p^* \in P^*} \tilde{F}^{(1)}(\nu - p^*).$$

Proof 4.4 See appendix.

The above result has a natural interpretation. The function $\tilde{F}^{(1)}$ is by construction $\mathcal{L}_1^*$-periodic. The function $\tilde{F}^{(2)}$ has more symmetries as it is $\mathcal{L}_2^*$-periodic. The above theorem can be phrased as saying that $\tilde{F}^{(2)}$ is obtained from $\tilde{F}^{(1)}$ by periodizing (and thereby enlarging the set of symmetries) and averaging (dividing by $\#(P^*)$). The following example shows an application of Theorem 4.6 in the one-dimensional case.

Example 4.6

Let $f(x) = \text{sinc}(x/2)$. Let $\mathcal{L}_1 = \mathbb{Z}$ be the lattice of integers and let $\mathcal{L}_2 = 2\mathbb{Z}$ be the lattice of even integers. Let as before $\tilde{F}^{(i)}(x)$ denote the sampled versions of $f(x)$. Then one easily computes that

$$\tilde{F}^{(1)}(\nu) = 2 \sum_{\lambda^* \in Z} \chi_{[-1/4;1/4]}(\nu - \lambda^*),$$

$$\tilde{F}^{(2)}(\nu) = 1,$$

where χ_A denotes the characteristic function of a set A.

Using Theorem 4.6 above we can also compute $\tilde{F}^{(2)}(\nu)$ directly from $\tilde{F}^{(1)}(\nu)$. We proceed as follows. Computing the reciprocal lattices we find $\mathcal{L}_1^* = \mathbb{Z}$ and $\mathcal{L}_2^* = \frac{1}{2}\mathbb{Z}$. We find two shifted versions of $\mathcal{L}_1^*$ within $\mathcal{L}_2^*$, viz. $\mathcal{L}_1^*$ and $\frac{1}{2} + \mathcal{L}_1^*$. Picking an arbitrary point in each coset, say 0 and $\frac{1}{2}$ respectively, we find

$$\tilde{F}^{(2)}(\nu) = \frac{1}{2}\left(\tilde{F}^{(1)}(\nu) + \tilde{F}^{(1)}\left(\nu - \frac{1}{2}\right)\right)$$
$$= 1.$$

4.6 Change of Variables

Consider the case of a one-dimensional continuous function $f(x)$. It is not always the case that $f(x)$ has a nice form, suitable for direct mathematical treatment. In such a situation a change of variables can sometimes help out. If A is an invertible linear transformation on $\mathbb{R}^n$, it might be more convenient to work with the variable $y = Ax$. Substituting $x = A^{-1}y$ we formally define the change of variable functional $f(x) \to f^A(x)$ by

$$f^A(x) = f(A^{-1}x).$$

A similar approach can be used for discrete functions. Instead of using a linear transform A on some continuous domain, we need in this case an isomorphism $A: \mathcal{L}_1 \to \mathcal{L}_2$ between two lattices $\mathcal{L}_1$ and $\mathcal{L}_2$. If $\tilde{F}(k)$ is a discrete function on $\mathcal{L}_1$, a change of variables by A yields a discrete function on $\mathcal{L}_2$ defined by

$$\tilde{f}^A(k) = \tilde{f}(A^{-1}k).$$

A typical example for a change of variables on discrete functions is the following. Let the lattice $\mathcal{L}_1 = 2\mathbb{Z}$, let $\mathcal{L}_2 = \mathbb{Z}$ and define $A: \mathcal{L}_1 \to \mathcal{L}_2$ by $2k \to k$. Given a function $f(x)$ on $\mathbb{R}$, downsampling it to $\mathcal{L}_1$ and changing variables with A, yield a discrete function $\tilde{f}(k)$ on $\mathbb{Z}$ defined by $\tilde{f}(k) = f(2k)$. In many textbooks this function $\tilde{f}(k)$ is referred to as the downsampled version of $f(x)$, but our analysis shows that it is better to view the discrete function $\tilde{f}(k)$ as the result of two consecutive operations: downsampling and change of variables.

The following two theorems address the question of how the CSFT and DSFT behave under a change of variables for the continuous and discrete case, respectively.

THEOREM 4.7

Let A be an invertible linear transform on $\mathbb{R}^n$, and let $f(x)$ be a function on $\mathbb{R}^n$. Then the CSFT of $f^A(x)$ is given by

$$\mathcal{F}(f^A) = |\det(A)| \mathcal{F}(f)^{A^{-t}}.$$

Proof 4.5 See appendix.

THEOREM 4.8

Let $A: \mathcal{L}_1 \to \mathcal{L}_2$ be an isomorphism of lattices, and let $\tilde{f}(k)$ be a function on $\mathcal{L}_1$. Then the DSFT of $\tilde{f}^A(k)$ is given by

$$\tilde{\mathcal{F}}(\tilde{f}^A) = \tilde{\mathcal{F}}(\tilde{f})^{A^{-t}}.$$

Proof 4.6 See appendix.

Note that in the assertion of Theorem 4.7 a factor $|\det(A)|$ is present, which is lacking in the assertion of Theorem 4.8. The last theorem of this section addresses the situation in which a function is extended by zero-padding to a larger domain.

THEOREM 4.9

Let $\mathcal{L}$, $\mathcal{L} \subset \mathcal{D}$ be a lattice, where $\mathcal{D}$ is either a lattice $\mathcal{M}$ or the ambient space $\mathbb{R}^n$. Let $\tilde{f}(\lambda)$ be a function on $\mathcal{L}$. Define the $\mathcal{D}$-extension $\tilde{f}_D$ of $\tilde{f}$ by

$$\tilde{f}_D(x) = \begin{cases} \tilde{f}(x) & \text{if } x \in L \\ 0 & \text{otherwise.} \end{cases}$$

Define $\Phi(\nu)$ by

$$\Phi(\nu) = \begin{cases} \mathcal{F}(\tilde{f}_D)(\nu) & \text{if } \mathcal{D} = \mathbb{R}^n \\ \tilde{\mathcal{F}}(\tilde{f}_D)(\nu) & \text{if } \mathcal{D} = \mathcal{M}, \end{cases}$$

i.e., $\Phi(\nu)$ is the appropriate Fourier transform of $\tilde{f}_D$. Then the equality $\Phi(\nu) = \tilde{\mathcal{F}}(\tilde{f})(\nu)$ holds.

Informally, the above theorem says that the Fourier transform of an extended function is equal to the Fourier transform of the function itself, i.e., extending a function does not change the Fourier transform. We will now apply the three theorems above in two examples.

Example 4.7

Let $A: \mathbb{Z}^n \to \mathbb{R}^n$ be a nonsingular linear mapping, and let $\mathcal{L} = [A]$ be the lattice generated by A. Let $f(x)$ be a continuous function on $\mathbb{R}^n$, and let $g = f^{A^{-1}}$. Define a discrete function $\tilde{g}(m)$ on $\mathbb{Z}^n$ by the rule*

$$\tilde{g}(m) = f(Am).$$

The question is how the Fourier transforms of $f(x)$ and $\tilde{g}(k)$ are related. To answer this question we define $\tilde{f}(\lambda)$ to be the sampled version $\Sigma_{\mathcal{L}}(f)(\lambda)$ of $f(x)$. The following commutative diagram results:

$$\begin{array}{ccc} (\mathbb{R}^n, g) & \xleftarrow{A^{-1}} & (R^n, f) \\ \downarrow \Sigma_{\mathbb{Z}^n} & & \downarrow \Sigma_{\mathcal{L}} \\ (\mathbb{Z}^n, \tilde{g}) & \xleftarrow{A^{-1}} & (\mathcal{L}, \tilde{f}) \end{array}$$

Tracing the diagram from top right to bottom right to bottom left we find

$$\begin{aligned} \tilde{\mathcal{F}}(\tilde{g})(\nu) &= \left[\tilde{\mathcal{F}}(\tilde{f})\right]^{A^t}(\nu) \\ &= \det(\mathcal{L}^*) \sum_{\lambda^* \in \mathcal{L}^*} \left(\mathcal{F}(f)^{A^t}(\nu - \lambda^*)\right) \\ &= \frac{1}{\det(A)} \sum_{\lambda^* \in \mathcal{L}^*} F(f)(A^{-t}\nu - \lambda^*), \end{aligned}$$

* This is a common situation when we have to sample a continuous function (on points of the form An) and store it in some rectangular storage space (with addresses n).

where we have used Theorem 4.8 and Theorem 4.3 in the first and second steps, respectively. Of course we should find the same result tracing the diagram from top right to top left to bottom left:

$$\begin{aligned}\tilde{\mathcal{F}}(\tilde{g})(\nu) &= \sum_{k\in\mathbb{Z}^n} \mathcal{F}(g)(\nu - k)\\ &= \sum_{k\in\mathbb{Z}^n} \mathcal{F}(f^{A^{-1}})(\nu - k)\\ &= \frac{1}{\det(A)} \sum_{k\in\mathbb{Z}^n} \mathcal{F}(f)^{A^t}(\nu - k)\\ &= \frac{1}{\det(A)} \sum_{k\in\mathbb{Z}^n} \mathcal{F}(f)^{A^t}(\nu - k)\\ &= \frac{1}{\det(A)} \sum_{\lambda\in\mathcal{L}^*} \mathcal{F}(f)(A^{-t}\nu - \lambda^*),\end{aligned}$$

where we have first applied Theorem 4.3, followed by an application of Theorem 4.8. As one sees, both calculations end up with the same result.

Example 4.8

Let $\mathcal{L}_1$ and $\mathcal{L}_2$ be two lattices. Let $A: \mathcal{L}_1 \to \mathcal{L}_2$ be a nonsingular linear mapping, and let $\tilde{f}$ be a function on $\mathcal{L}_1$. Let $\mathcal{L}_3$ be the lattice generated by A, $\mathcal{L}_3 = [A] \subset \mathcal{L}_2$. Define $\tilde{g}$ on $\mathcal{L}_2$ by

$$\tilde{g}(\lambda_2) = \begin{cases} \tilde{f}(\lambda_1) & \text{if } \lambda_2 = A\lambda_1 \\ 0 & \text{otherwise.} \end{cases}$$

The question is to find an expression for the DSFT of $\tilde{g}$. To this end we define $\tilde{h}$ on $\mathcal{L}_3$ by $\tilde{h} = \tilde{f}^A$. The following diagram results:

$$(\mathcal{L}_1, \tilde{f}) \xrightarrow{A} (\mathcal{L}_3, \tilde{h}) \xrightarrow{\text{extension}} (\mathcal{L}_2, \tilde{g})$$

For the DSFT of $\tilde{g}$ we find

$$\begin{aligned}\tilde{\mathcal{F}}(\tilde{g})(\nu) &= \tilde{\mathcal{F}}(\tilde{h})(\nu)\\ &= \tilde{\mathcal{F}}(\tilde{f}^A)(\nu)\\ &= \tilde{\mathcal{F}}(\tilde{f})^{A^{-t}}(\nu)\\ &= \tilde{\mathcal{F}}(\tilde{f})(A^t\nu),\end{aligned}$$

where we have used Theorem 4.9 and Theorem 4.8 in the first and second step, respectively.

4.7 An Extended Example: HDTV-to-SDTV Conversion

This section will introduce an application of sampling theory as it occurs in the problem of interlaced high definition television (HDTV) to interlaced standard definition television (SDTV) conversion. This problem exists because an HDTV broadcast can at present only be viewed by a minority of people. Most people can only view SDTV broadcast. As broadcasters like their programs to be viewed by as many customers as possible, they are interested in (preferably inexpensive) schemes which can convert HDTV in SDTV. In this section we present an approach to this conversion problem as has been suggested in [1].

In order to keep the notational burden low, our television signal will be one-dimensional. This leaves us with a spatial axis, referred to as the y-axis (y for vertical), and a time axis, referred to as the t-axis.

An interlaced television signal is constructed by sampling a continuous luminance signal with at times kT, but only even lines for even k and only the odd lines for odd k. Choosing T to be 1 in some unit of time, and recalling that we assume one-dimensional images, we may model an interlaced HDTV signal as a luminance signal sampled at the quincunx lattice $\mathcal{L}_2$ by the matrix

$$\begin{bmatrix} 1 & -1 \\ 1 & 1 \end{bmatrix}.$$

In order to prevent alias distortion, i.e., in order to prevent that frequencies overlap after sampling, the continuous luminance signal has to be sufficiently band limited. An often-used pass band region is given by the diamond in Figure 4.3c.

An SDTV interlaced signal has half the vertical resolution of the HDTV signal, but the same temporal resolution, and we may model this as the sampling of the continuous luminance signal on the skew quincunx lattice $\mathcal{L}_1$ generated by the matrix

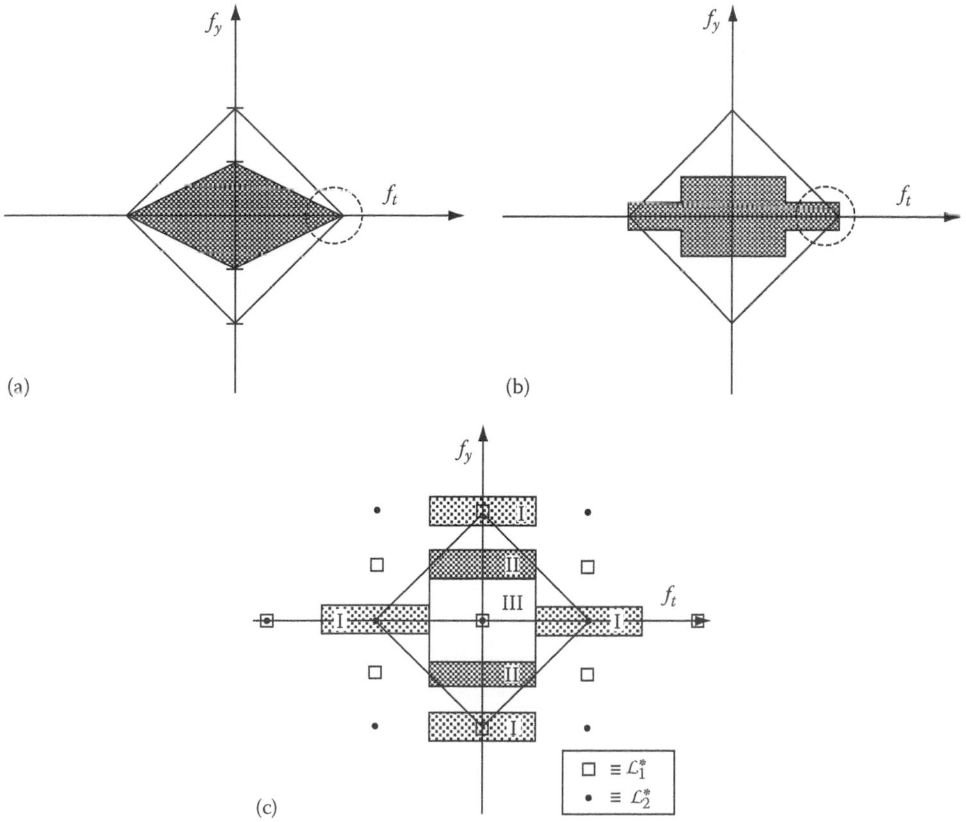

FIGURE 4.3 HDTV-to-SDTV conversion in the frequency domain.

$$\begin{bmatrix} 1 & -1 \\ 2 & 2 \end{bmatrix}.$$

Note that the lattice $\mathcal{L}_1$ is not a sublattice of the $\mathcal{L}_2$. This has the consequence that the extraction of an SDTV signal from an HDTV signal is not simply a question of subsampling the HDTV signal; interpolation is needed to compute the values of the luminance signal at the missing points. In the frequency domain this is equivalent to restricting the pass band region of the HDTV signal to a smaller pass band region, such that no alias occurs when the interpolated signal is sampled to the SDTV lattice.

Figure 4.3a gives a possible solution. The SDTV pass band region is chosen as the skew diamond region within the HDTV pass band (the outer diamond). This solution has several disadvantages. One disadvantage is the fact that the realization of this diamond pass band region can only be realized by non-separable filters, and, therefore, that it is expensive. A second disadvantage is the temporal attenuation at maximum temporal frequency, which may introduce visible artifacts for moving video.

As argued in [1], the best compromise between vertical resolution and temporal attenuation at maximum temporal frequency is given by a pass band of the form as given in Figure 4.3b. This pass band can even be realized cheaply.

Following [1] we note that the temporal information at maximum frequency (region I on the f_t-axis in Figure 4.3c) is repeated at maximal vertical frequency (region I on the f_y-axis in Figure 4.3c). This is simply a consequence of the fact that the DSFT of the HDTV signal is $\mathcal{L}_2^*$-periodic. We can retain this information by using an appropriately chosen vertical high pass filter. In a practical implementation this implies that (after temporal low-pass filtering) we extract from the HDTV signal a base-band signal using a vertical low-pass filter (the rectangle III in Figure 4.3c) and a temporal band using a vertical high-pass filter. The temporal band is now modulated to position II in Figure 4.3c by multiplying the sample at position $(2k, t)$ with $(-1)^k$.

The base band and the temporal band are now merged and sampled to the SDTV lattice. Due to this last sampling operation, region II is repeated at its original position I in frequency space: this follows immediately from computing the reciprocal SDTV quincunx lattice.

This proves (as first shown in [1]) that a high quality HDTV-to-SDTV conversion can be achieved using only separable filters.

4.8 Conclusions

We have presented the basic facts of multidimensional sampling theory. Particular attention has been paid to the interaction of the different kinds of Fourier transforms, the sampling operator, and the periodizing operator. Every basic result is accompanied by one or more examples. An application of the theory to a format conversion problem has been presented.

Appendix

4.A.1 Proof of Theorem 4.3

Proof 4.7 We first observe that

$$\Sigma_{\mathcal{L}}(f) = f \cdot \coprod_{\mathcal{L}},$$

$$\prod_{\mathcal{L}}(F) = F * \coprod_{\mathcal{L}^*}.$$

It follows immediately that $\mathcal{F}[\Sigma_{\mathcal{L}}(f)] = \coprod_{\mathcal{L}^*}[\mathcal{F}(f)]$. To prove the first assertion of this theorem, it suffices to verify that $\tilde{\mathcal{F}}(\tilde{f}) = \tilde{\mathcal{F}}$:

$$\begin{aligned}
\tilde{F}(\nu) &= \mathcal{F}(f \cdot \coprod_{\mathcal{L}})(\nu) \\
&= \int_{\mathbb{R}^n} \sum_{\lambda \in \mathcal{L}} e^{-2\pi i \langle x,\nu \rangle} f(x) \delta_\lambda(x) dx \\
&= \sum_{\lambda \in \mathcal{L}} e^{-2\pi i \langle \lambda,\nu \rangle} f(\lambda) \\
&= \tilde{\mathcal{F}}(\tilde{f}).
\end{aligned}$$

The second assertion of the theorem, viz. the isometry property of the DSFT, follows from

$$\begin{aligned}
\langle \tilde{F}, \tilde{G} \rangle_{P^*} &= \frac{1}{\det(\mathcal{L})^2} \int_{P^*} \left\langle \coprod_{\mathcal{L}^*} * F, \coprod_{\mathcal{L}^*} * G \right\rangle_{P^*} \\
&= \frac{1}{\det(\mathcal{L})^2} \int_{P^*} \left(\sum_{\lambda_1^* \in \mathcal{L}^*} F(\nu - \lambda_1^*) \right) \left(\sum_{\lambda_1^* \in \mathcal{L}^*} G(\nu - \lambda_2^*) \right) d\nu \\
&= \frac{1}{\det(\mathcal{L})^2} \int_{\mathbb{R}^n} F(\nu) \left(\sum_{\lambda^* \in \mathcal{L}^*} G(\nu - \lambda^*) \right) d\nu \\
&= \frac{1}{\det(\mathcal{L})^2} \langle F, \tilde{G} \rangle \\
&= \frac{1}{\det(\mathcal{L})} \langle f, \tilde{g} \rangle \\
&= \frac{1}{\det(\mathcal{L})} \langle f, \tilde{g} \rangle \mathcal{L}.
\end{aligned}$$

4.A.2 Proof of Theorem 4.5

Proof 4.8 Similar to the proof of Theorem 4.3, to prove the first assertion it suffices to show that $\tilde{\mathcal{F}}(\hat{f}) = \hat{F}$:

$$\begin{aligned}
\tilde{\mathcal{F}}(\hat{f})(\nu) &= \sum_{\lambda \in \mathcal{L}} e^{-2\pi i \langle \lambda,\nu \rangle} \hat{f}(\lambda) \\
&= \left(\sum_{\mu \in \mathcal{M}} e^{-2\pi i \langle \mu,\nu \rangle} \right) \left(\sum_{p \in P} e^{-2\pi i \langle p,\nu \rangle} \hat{f}(p) \right) \\
&= \frac{1}{\det(\mathcal{M})} \coprod_{\mathcal{L}^*} \cdot \left(\sum_{p \in P} e^{-2\pi i \langle p,\nu \rangle} \hat{f}(p) \right) \\
&= \coprod_{\mathcal{L}^*} \cdot \tilde{\mathcal{F}}(\hat{f})(\nu).
\end{aligned}$$

The isometry property of the DFT follows from

$$
\begin{aligned}
\langle \hat{f}, \hat{g} \rangle_P &= \sum_{p \in P} \hat{f}^{\dagger}(p)\hat{g}(p) \\
&= \det(\mathcal{M})^2 \sum_{p \in P} \left(\sum_{\mu_1 \in \mathcal{M}} \tilde{f}^{\dagger}(p - \mu_1) \right) \left(\sum_{\mu_2 \in \mathcal{M}} \tilde{g}(p - \mu_2) \right) \\
&= \det(\mathcal{M})^2 \sum_{\lambda \in L} \tilde{f}^{\dagger}(\lambda) \left(\sum_{\mu \in \mathcal{M}} \tilde{g}(\lambda - \mu) \right) \\
&= \det(\mathcal{M}) \langle \tilde{f}, \hat{g} \rangle_L \\
&= \det(\mathcal{M})^2 \left\langle f, \coprod_{\mathcal{L}} \cdot (\coprod_{\mathcal{M}} * g \right\rangle \\
&= \frac{\det(\mathcal{M})}{\det(\mathcal{L})} \left\langle F, \coprod_{L^*} * (\coprod_{M^*} \cdot G \right\rangle \\
&= \frac{\det(\mathcal{M})}{\det(\mathcal{L})} \left\langle F, \coprod_{M^*} \cdot (\coprod_{L^*} * G \right\rangle \\
&= \det(\mathcal{M})\det(\mathcal{L}) \langle \hat{F}, \hat{G} \rangle_{P^*}
\end{aligned}
$$

The last step in this derivation follows from reversing the other steps, replacing the spatial functions f and g by their frequency domain counterparts F and G.

4.A.3 Proof of Theorem 4.6

Proof 4.9

$$
\begin{aligned}
\tilde{F}^{(2)}(\nu) &= \frac{1}{\det(L_2)} \sum_{\lambda_2^* \in L_2^*} F(\nu - \lambda_2^*) \\
&= \frac{1}{\det(L_2)} \sum_{p^* \in P^*} \sum_{\lambda_1^* \in L_1^*} F(\nu - p^* - \lambda_1^*) \\
&= \frac{\det(L_1)}{\det(L_2)} \sum_{p^* \in P^*} \tilde{F}^{(1)}(\nu - p^*) \\
&= \frac{1}{\iota(L_2, L_1)} \sum_{p^* \in P^*} \tilde{F}^{(1)}(\nu - p^*) \\
&= \frac{1}{\#(P^*)} \sum_{p^* \in P^*} \tilde{F}^{(1)}(\nu - p^*).
\end{aligned}
$$

4.A.4 Proof of Theorem 4.7

Proof 4.10

$$
\begin{aligned}
F(f^A)(\nu) &= \int_{R^n} e^{-2\pi i \langle x,\nu \rangle} f^A(x) dx \\
&= \int_{R^n} e^{-2\pi i \langle x,\nu \rangle} f(A^{-1}x) dx \\
&= |\det(A)| \int_{\mathbb{R}^n} e^{-2\pi i \langle Ay,\nu \rangle} f(y) dy \\
&= |\det(A)| \int_{\mathbb{R}^n} e^{-2\pi i \langle y,A^t\nu \rangle} f(y) dy \\
& |\det(A)| F(A^t \nu) \\
& |\det(A)| F^{A^{-t}}(\nu).
\end{aligned}
$$

4.A.5 Proof of Theorem 4.8

Proof 4.11

$$
\begin{aligned}
\tilde{F}(\tilde{f}^A)(\nu) &= \sum_{\lambda_2 \in L_2} e^{-2\pi i (\lambda_2,\nu)} \tilde{f}^A(\lambda_2) \\
&= \sum_{\lambda_2 \subset L_2} e^{-2\pi i (\lambda_2,\nu)} \tilde{f}(A^{-1}\lambda_2) \\
&= \sum_{\lambda_1 \in L_1} e^{-2\pi i (A\lambda_1,\nu)} \tilde{f}(\lambda_1) \\
&= \sum_{\lambda_1 \in L_1} e^{-2\pi i (\lambda_1,A^t\nu)} \tilde{f}(\lambda_1) \\
&= \tilde{\mathcal{F}}(\tilde{f})^{A^{-t}}(\nu).
\end{aligned}
$$

Glossary of Symbols and Expressions

$\mathbb{Z}^n$	n-dimensional integer space
$\mathbb{R}^n$	n-dimensional real space
$\mathbb{C}^n$	n-dimensional complex space
CSFT	continuous space-time Fourier transform
DSFT	discrete space-time Fourier transform
DFT	discrete Fourier transform
$\mathcal{L}, \mathcal{M}$	sampling lattice
λ, μ	elements of lattice $\mathcal{L}, \mathcal{M}$
λ^*, μ^*	element of reciprocal lattice $\mathcal{L}^*, \mathcal{M}^*$

$[L]$	lattice generated by matrix L
$\#(A)$	number of points of set A
$\mathrm{vol}(A)$	volume (measure) of set A
$\det(\mathcal{L})$	determinant of lattice $\mathcal{L}$
$\iota(\mathcal{M}, \mathcal{L})$	index of lattice $\mathcal{M}$ with respect to lattice $\mathcal{L}$
$\mathcal{L}/\mathcal{M}$	coset group of lattice $\mathcal{M}$ with respect to lattice $\mathcal{L}$
$\mathcal{L}^*$	reciprocal lattice of $\mathcal{L}$
$\amalg_{\mathcal{L}}$	lattice comb
P	fundamental domain
$\Vert\alpha\Vert_2$	L_2-norm of α
α^t	Hermitian transpose of α
$\langle\alpha, \beta\rangle\mathcal{N}$	inner products of α and β with respects to $\mathcal{N}$-norm
$\alpha^\dagger$	complex conjugate of α
$\alpha \cdot \beta$	point-wise multiplication
$\alpha * \beta$	convolution
$f^A(x)$	change of variables $f(A^{-1}x)$
X_A	characteristic function of set A
$\mathcal{F}$	continuous space-time Fourier transform
$\tilde{\mathcal{F}}$	discrete space-time Fourier transform
$\hat{\mathcal{F}}$	discrete Fourier transform
$\Sigma_{\mathcal{L}}$	sampling operator
$\Pi_{\mathcal{L}}$	periodizing operator

$$\sin c(x)\begin{cases} \sin(\pi x)/\pi x & \text{if } x \neq 0 \\ 1 & \text{if } x = 0 \end{cases}$$

References

1. Albani, L., Mian, G., and Rizzi, A., A new intra-frame solution for HDTV-to-SDTV down-conversion, in *HDTV-1995 International Workshop and the Evolution of Television*, 1995.
2. Cassels, J., *An Introduction to the Geometry of Numbers*. Springer-Verlag, Berlin, Germany, 1971.
3. Hungerford, T., *Algebra, Graduate Texts in Mathematics*, Vol. 73. Springer-Verlag, New York, 1974.
4. Dudgeon, D.E. and Mersereau, R.M., *Multidimensional Digital Signal Processing*. Signal Processing Series, Prentice-Hall, Englewood Cliffs, NJ, 1984.
5. Dubois, E., The sampling and reconstruction of time-varying imagery with application in video systems, *Proc. IEEE*, 73: 502–522, April 1985.
6. Viscito, E. and Allebach, J., The analysis and design of multidimensional FIR perfect reconstruction filter banks for arbitrary sampling lattices, *IEEE Trans. Circuits Syst.*, 38: 29–42, January 1991.
7. Chen, T. and Vaidyanathan, P., Recent developments in multidimensional multirate systems, *IEEE Trans. Circuits Syst. Video Technol.*, 3: 116–137, April 1993.
8. Vetterli, M. and Kovačević, J., *Wavelets and Subband Coding*. Signal Processing Series, Prentice-Hall, Englewood Cliffs, NJ, 1995.
9. Jerri, A., The Shannon sampling theorem—its various extensions and applications: A tutorial review, *Proc. IEEE*, 1565–1596, November 1977.

5

Analog-to-Digital Conversion Architectures

Stephen Kosonocky
Advanced Micro Devices

Peter Xiao
NeoParadigm Labs. Inc.

5.1 Introduction

Digital signal processing (DSP) methods fundamentally require that signals are quantized at discrete time instances and represented as a sequence of words consisting of 1's and 0's. In nature, signals are usually nonquantized and continuously varied with time. Natural signals such as air pressure waves as a result of speech are converted by a transducer to a proportional analog electrical signal. Consequently, it is necessary to perform a conversion of the analog electrical signal to a digital representation or vice versa if an analog output is desired. The number of quantization levels used to represent the analog signal and the rate at which it is sampled is a function of the desired accuracy, bandwidth that is required, and the cost of the system. Figure 5.1 shows the basic elements of a DSP system. The analog signal is first converted to a discrete time signal by a sample-and-hold (S/H) circuit. The output of the S/H is then applied to an analog-to-digital (A/D) converter circuit where the sampled analog signal is converted to a digitally coded signal. The digital signal is then applied to the DSP system where the desired DSP algorithm is performed. Depending on the application, the output of the DSP system can be used directly in digital form or converted back to an analog signal by a digital-to-analog (D/A) converter. A digital filtering application may produce an analog signal as its output, whereas a speech recognition system may pass the digital output of the DSP system to a computer system for further processing. This section will describe basic converter terminology and a sample of common architectures for both conventional Nyquist rate converters and oversampled delta–sigma converters.

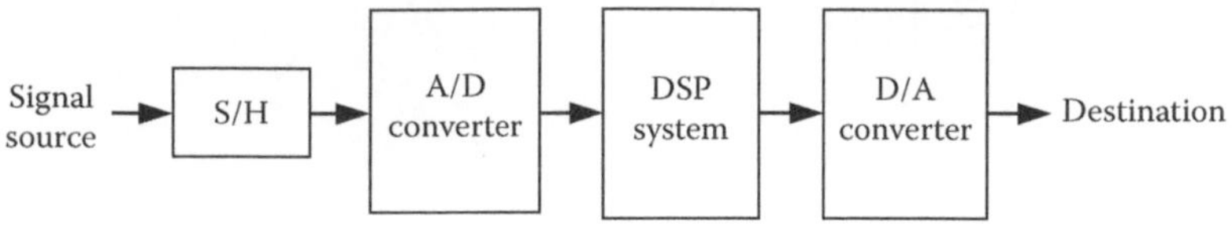

FIGURE 5.1 DSP system.

5.2 Fundamentals of A/D and D/A Conversions

The analog signal can be given as either a voltage signal or current signal, depending on the signal source. Figure 5.2 shows the ideal transfer characteristics for a 3-bit A/D conversion. The output of the converter is an n-bit digital code given as

$$D = \frac{A_{\text{sig}}}{\text{FS}} = \frac{b_n}{2^n} + \frac{b_{n-1}}{2^{n-1}} + \cdots + \frac{b_1}{2^1}, \tag{5.1}$$

where

A_{sig} is the analog signal
FS is the analog full scale level
b_n is a digital value of either 0 or 1

As shown in the figure, each digital code represents a quantized analog level. The width of the quantized region is one least-significant bit (LSB) and the ideal response line passes through the center of each quantized region. The converse D/A operation can be represented as viewing the digital code in Figure 5.2 as the input and the analog signal as the output. An n-bit D/A converter transfer equation is given as

$$A_{\text{sig}} = \text{FS}\left(\frac{b_n}{2^n} + \frac{b_{n-1}}{2^{n-1}} + \cdots + \frac{b_1}{2^1}\right), \tag{5.2}$$

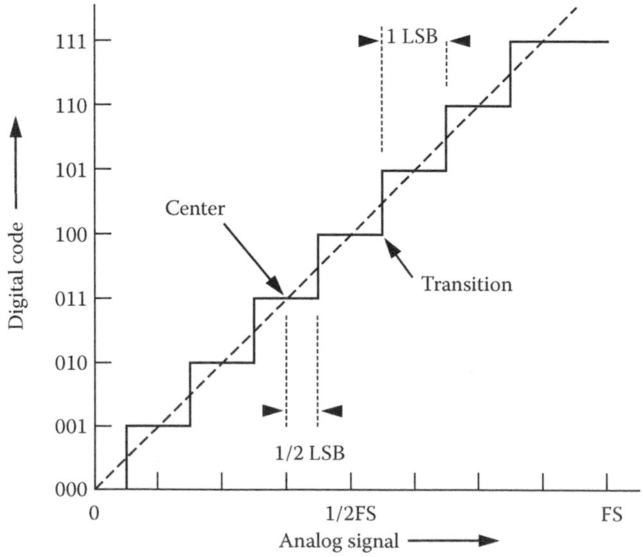

FIGURE 5.2 Ideal transfer characteristics for an A/D converter.

where

A_{sig} is the analog output signal
FS is the analog full scale level
b_n is a binary coefficient

The resolution of a converter is defined as the smallest distinct change that can be resolved (produced) at an analog input (output) for an A/D (D/A) converter. This can be expressed as

$$\Delta A_{sig} = \frac{FS}{2^N} \tag{5.3}$$

where ΔA_{sig} is the smallest reproducible analog signal for an N-bit converter with full scale analog signal of FS.

The accuracy of a converter, often referred to also as relative accuracy, is the worst-case error between the actual and the ideal converter output after gain and offset errors are removed [1]. This can be quantified as the number of equivalent bits of resolution or as a fraction of an LSB.

The conversion rate specifies the rate at which a digital code (analog signal) can be accurately converted into an analog signal (digital code). Accuracy is often expressed as a function of conversion rate and the two are closely linked. The conversion rate is often an underlying factor in choosing the converter architecture. The speed and accuracy of analog components are a limiting factor. Sensitive analog operations can either be done in parallel, at the expense of accuracy, or cyclicly reused to allow high accuracy with lower conversion speeds.

5.2.1 Nonideal A/D and D/A Converters

Actual A/D and D/A converters exhibit deviations from the ideal characteristics shown in Figure 5.2. Integration of a complete converter on a single monolithic circuit or as a macro within a very large scale integration (VLSI) DSP system presents formidable design challenges. Converter architectures and design trade-offs are most often dictated by the fabrication process and available device types. Device parameters such as voltage threshold, physical dimensions, etc. vary across a semiconductor die. These variations can manifest themselves into errors. The following terms are used to describe converter nonideal behavior:

1. Offset error, described in Figure 5.3, is a d.c. error between the actual response with the ideal response. This can usually be removed by trimming techniques.
2. Gain error is defined as an error in the slope of the transfer characteristic shown in Figure 5.4, which can also usually be removed by trimming techniques.
3. Integral nonlinearity is the measure of worst-case deviation from an ideal line drawn between the full scale analog signal and zero. This is shown in Figure 5.5 as a monotonic nonlinearity.
4. Differential nonlinearity is the measure of nonuniform step sizes between adjacent steps in a converter. This is usually specified as a fraction of an LSB.
5. Monotonicity in a converter specifies that the output will increase with an increasing input. Certain converter architectures can guarantee monotonicity for a specified number of bits of resolution. A nonmonotonic transfer characteristic is detailed in Figure 5.6.
6. Settling time for D/A converters refers to the time taken from a change of the digital code to the point at which the analog output settles within some tolerance around the final value.
7. Glitches can occur during changes in the output at major transitions, i.e., at 1 MSB, 1/2 MSB, or 1/4 MSB. During large changes, switching time delays between internal signal paths can cause a spike in the output.

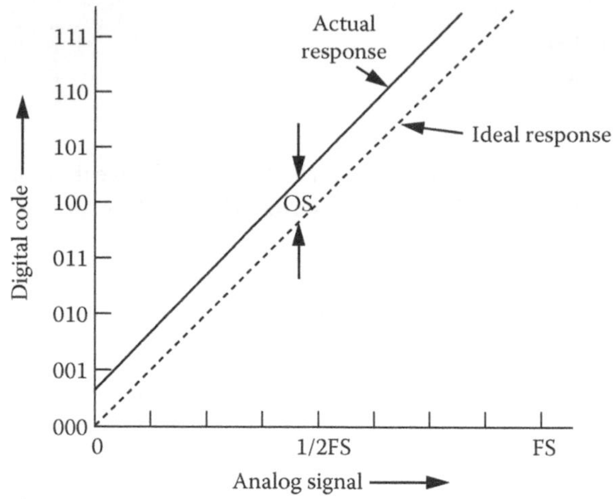

FIGURE 5.3 Offset error.

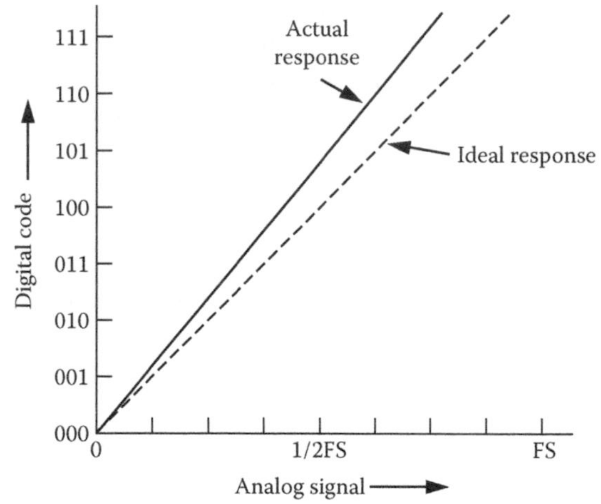

FIGURE 5.4 Gain error.

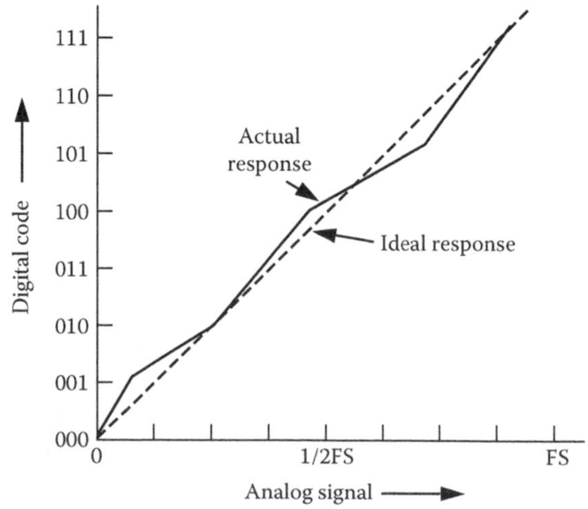

FIGURE 5.5 Monotonic nonlinearity.

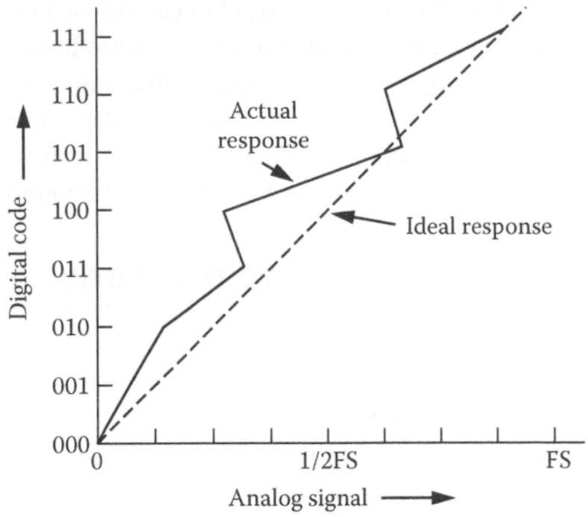

FIGURE 5.6 Nonmonotonic nonlinearity.

The choice of converter architecture can greatly affect the relative weight of each of these errors. Data converters are often designed for low cost implementation in standard digital processes, i.e., digital CMOS, which often do not have well-controlled resistors or capacitors. Absolute values of these devices can vary by as much as ±20% under typical process tolerances. Post-fabrication trimming techniques can be used to compensate for process variations, but at the expense of added cost and complexity to the manufacturing process. As will be shown, various architectural techniques can be used to allow high-speed or highly accurate data conversion with such variations of process parameters.

5.3 Digital-to-Analog Converter Architecture

The D/A converter, also known as a DAC, decodes a digital word into a discrete analog level. Depending on the application, this can be either a voltage or current. Figure 5.7 shows a high-level block diagram of a D/A converter. A binary word is latched and decoded and drives a set of switches that control a scaling network. A basic analog scaling network can be based on voltage scaling, current scaling, or charge scaling [1,2]. The scaling network scales the appropriate analog level from the analog reference circuit and applies it to the output driver. A simple serial string of identical resistors between a reference voltage

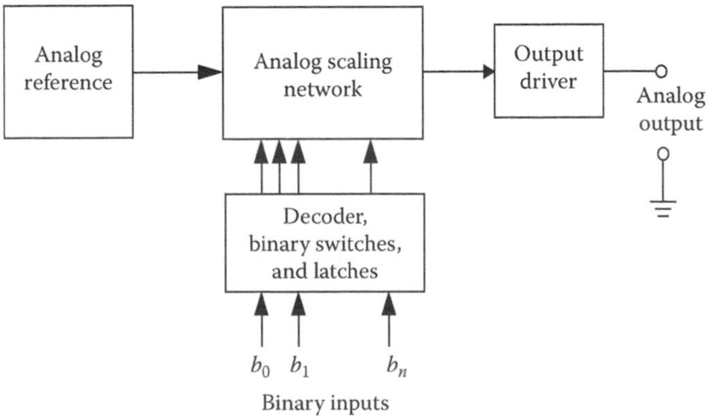

FIGURE 5.7 Basic D/A converter block diagram.

and ground can be used as a voltage scaling network. Switches can be used to tap voltages off the resistors and apply them to the output driver. Current scaling approaches are based on switched scaled current sources. Charge scaling is achieved by applying a reference voltage to a capacitor divider using scaled capacitors where the total capacitance value is determined by the digital code [1]. Choice of the architecture depends on the available components in the target technology, conversion rate, and resolution. Detailed description of these trade-offs and designs can be found in the references [1–5].

5.4 Analog-to-Digital Converter Architectures

The A/D converter, also known as an ADC, encodes an analog signal into a digital word. Conventional converters work by sampling the time varying analog signal at a sufficient rate to fully resolve the highest frequency components. According to the sampling theorem, the minimum sampling rate is twice the frequency of the highest frequency contained in the signal source. The sampling rate requirement thus becomes the major deterministic factor in choosing a proper converter architecture. Certain architectures exploit parallelism to achieve high-speed operation on the order of 100's of MHz, and others which can be used for high-accuracy 16-bit resolution for signals with maximum frequencies on the order of tens of kilohertz.

5.4.1 Flash A/D

The flash A/D, also known as a parallel A/D, is the highest speed architecture for A/D conversion since maximum parallelism is used. Figure 5.8 shows a block diagram of a 3-bit flash A/D converter. A flash

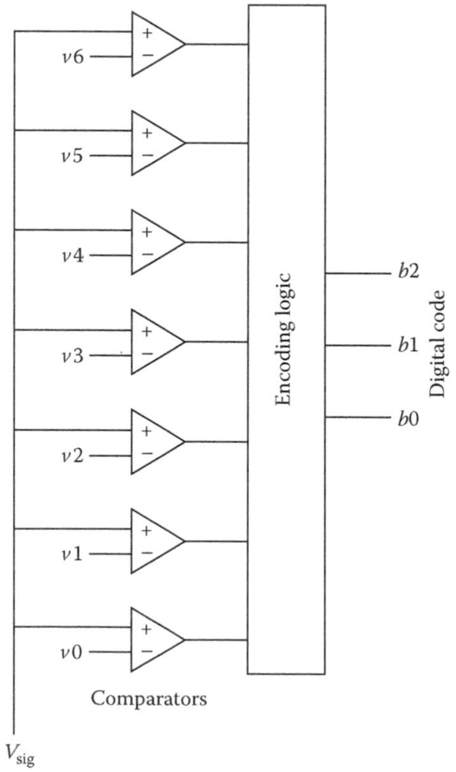

FIGURE 5.8 A 3-bit flash A/D converter.

converter requires $2^n - 1$ analog comparators, $2^n - 1$ reference voltages, and a digital encoder. The reference voltages are required to be evenly spaced between 0.5 LSB above the most negative signal and 1.5 LSB below the most positive signal and spaced 1 LSB apart. Each reference voltage is applied to the negative input of a comparator and the analog signal voltage is applied simultaneously to all the comparators. A thermometer code results at the output of the comparators which is converted to a digital word by encoding logic. The speed of the converter is limited by the time delay through a comparator and the encoding logic. This speed is gained at the expense of accuracy, which is limited by the ability to generate evenly spaced reference voltages and the precision of the comparators. Each analog comparator must be precisely matched in order to achieve acceptable performance at a given resolution. For these reasons, flash A/D converters are typically used only for very high-speed low-resolution applications.

5.4.2 Successive Approximation A/D Converter

A successive approximation A/D converter is formed creating a feedback loop around a D/A converter. Figure 5.9 shows a block diagram for an 8-bit successive approximation A/D. The operation of the converter works by initializing the successive approximation register (SAR) to a value where all bits are set to 0 except the MSB which is set to 1. This represents the mid-level value. The analog signal is applied to an S/H circuit, and on the first clock cycle the DAC converts the digital code stored in the SAR into an analog signal. The comparator is used to determine whether the analog signal is greater or less than the mid level, and control logic determines whether to leave the MSB set to 1 or to change it back to 0. The process is repeated on the next clock cycle, but instead the next MSB is tested. For an n-bit converter n clock cycles are required to fully quantize each S/H signal. The speed of the successive approximation converter is largely limited by the speed of the DAC and the time delay through the comparator. This type of converter is widely used for medium speed and medium accuracy applications. The resolution is limited by the DAC converter and the comparator.

5.4.3 Pipelined A/D Converter

A pipelined A/D converter achieves high-speed conversion and high accuracy at the expense of latency in the conversion process. A pipelined A/D converter block diagram is shown in Figure 5.10. The conversion process is broken into multiple stages where, at each stage, a partial conversion is done and the converted bits are shifted down the pipeline in digital registers. Figure 5.11 shows the detail of a single pipeline stage. The analog signal is applied to a S/H circuit and the output is applied to an n-bit flash

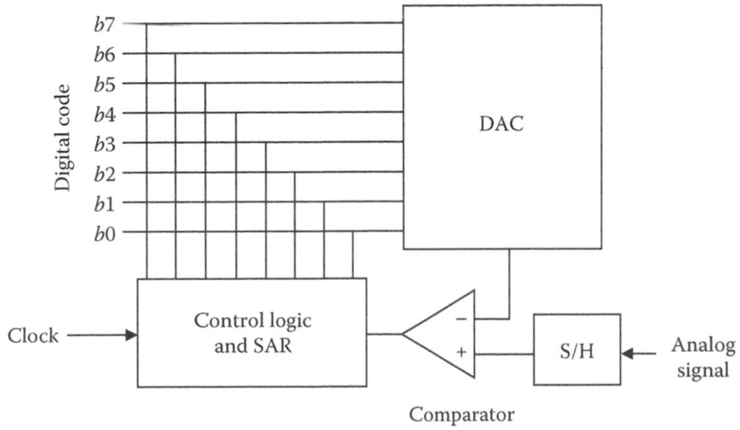

FIGURE 5.9 An 8-bit successive approximation A/D converter.

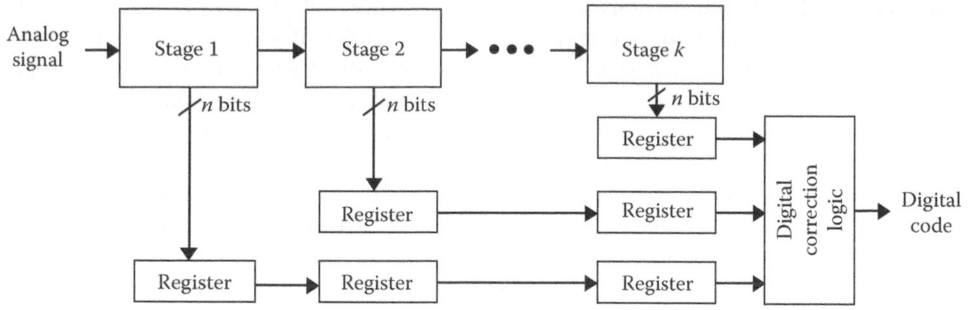

FIGURE 5.10 Pipelined A/D converter.

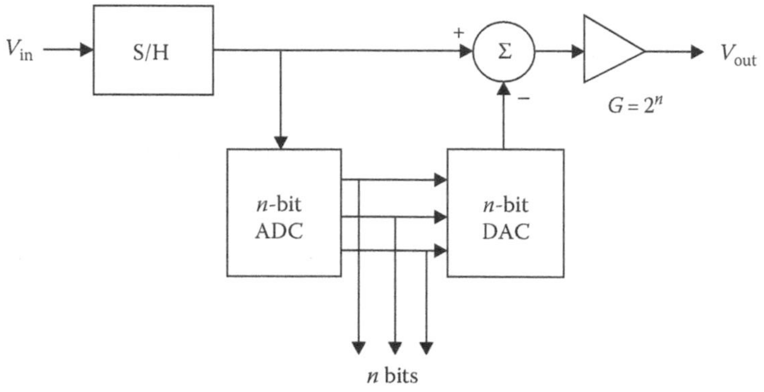

FIGURE 5.11 Diagram of single pipelined A/D converter stage.

ADC where n is less then the total desired resolution. The outputs of the ADC are connected directly to a DAC, and the output of the DAC is subtracted from the original analog signal stored in the S/H to produce a residual signal. The residual signal is then amplified by 2^n so that it will vary within the entire full scale range of the next stage and is transferred on the next clock cycle. At this point, the first stage begins conversion on the next analog sample. The maximum conversion rate is determined by the time delay through a single stage. Pipelining allows high-resolution conversion without the need for many comparators. An 8-bit converter can be ideally constructed with $k = 4$ stages with $n = 2$ bits of resolution per stage, requiring only 12 total comparators. This can be contrasted with an 8-bit flash converter requiring 255 comparators. Each pipeline stage adds an additional cycle of latency before the final code is converted. Pipelined converters also accommodate digital correction schemes for errors generated in the analog circuitry. Digital correction can be achieved by using higher resolution ADC and DAC circuits in each stage than required so that errors in the preceding stage can be detected and corrected digitally [5]. Auto calibration can also be achieved by adding additional stages after the required stages to convert errors in the DAC values and storing these digitally to be added to the final result [6].

5.4.4 Cyclic A/D Converter

Cyclic A/D converters, also known as algorithmic converters, trade off conversion speed for high accuracy without the need for calibration or device trimming. Figure 5.12 shows a block diagram of a cyclic A/D converter [5]. Here the same analog components are cyclicly reused for conversion of each bit for each analog sample. The conversion process works by initially sampling the input signal by setting

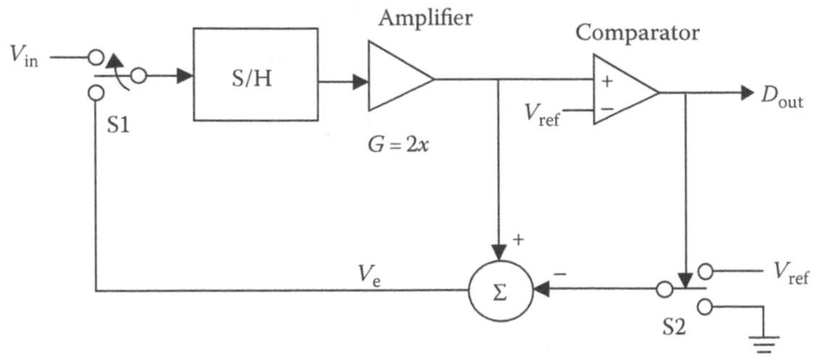

FIGURE 5.12 Block diagram of a cyclic A/D converter.

switch S1 appropriately. The sampled signal is then amplified by a factor of two and applied to a comparator where it is compared to a reference level, V_{ref}. If the voltage exceeds the reference level, a bit value of 1 is produced and the reference voltage is subtracted from the amplified signal by control of switch S2 to produce the residual voltage V_e. If the amplified signal is less than the reference voltage, V_{ref}, the comparator outputs a 0, and V_e represents the unchanged amplified signal. On the remaining cycles for the sample, switch S1 changes so that the residual voltage V_e is applied to the S/H circuit. The cycle is repeated for each remaining bit. Operation on the conversion process produces a serial stream of digital bit values from output of the comparator. An n-bit converter requires n conversion cycles for each sampled signal.

5.5 Delta–Sigma Oversampling Converter

The oversampling delta–sigma A/D converter was first proposed 30 years ago [7], while it only became popular after the maturity of the VLSI digital technology. With the advancement of semiconductor technology, an increasing portion of signal processing tasks have been shifted from the usual analog domain to digital domain. For digital systems to interact with analog signal sources, such as voice, data, and video, the role of A/D interface is essential. In voice data processing and communication, an accurate digital form is often desired to represent the voice. Due to the large demand of these systems, the cost must be kept at a minimum. All these requirements call upon a need to implement monolithic high-resolution A/D interfaces in economical semiconductor technology. However, with the increasing complexity of integration and a trend of reducing supply voltage, the accuracy of device components and analog signal dynamic range deteriorate. It becomes more difficult to realize high resolution conversions by conventional Nyquist rate converter architecture.

Compared to Nyquist rate converters, the oversampling converters use coarse analog components at the front end and employ more DSP in the later stages. High-resolution conversions are achieved by trading off speed and DSP complexity, both of which can be easily realized in modern VLSI technology.

The oversampling A/D converter and Nyquist rate converter are compared in Figure 5.13. A nonoversampled A/D converter has an anti-aliasing lowpass filter in the front. The anti-aliasing filter attenuates high-frequency components buried in the analog input and prevents them from being aliased into the signal frequency band. Because the converter is sampled at the Nyquist rate, which is twice the input signal bandwidth, the anti-aliasing filter's transition band must be very narrow and its stopband must have enough suppression of the out-of-band noise. This requirement makes the filter very complex and adds to the complexity that a nonoversampled A/D already has.

In comparison, an oversampled delta–sigma A/D converter, as shown in Figure 5.13b, is sampled at a higher rate than the input Nyquist rate. A simple first-order lowpass filter is sufficient to attenuate the noise components at the sampling frequency region to avoid the noise aliasing. This is because only

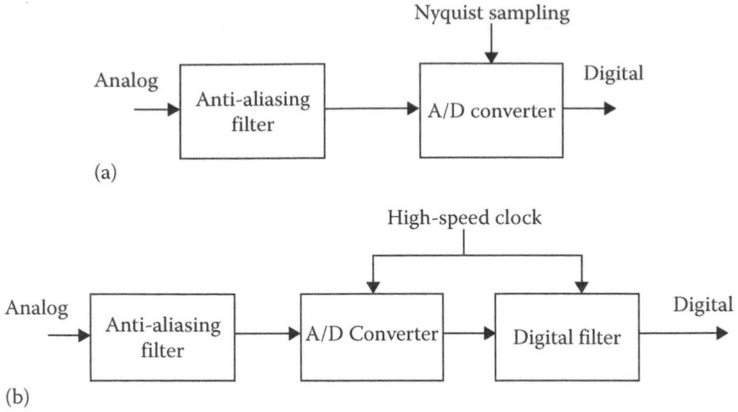

FIGURE 5.13 (a) Nonoversampled A/D converter and (b) oversampled A/D converter.

the noise components close to the sampling frequency can be aliased back into the signal band. This arrangement simplifies the design and implementation of the filter. The complexity of the A/D itself is much simpler than the nonoversampled A/D converters, as we will see later. The only extra complexity in the oversampled A/D converters is that more DSP is required after the A/D conversion. But this becomes less and less an issue with the advancement of the VLSI technology. In the following sections, we will explain the conversion principle and various architectures of the oversampling delta–sigma converter.

5.5.1 Delta–Sigma A/D Converter Architecture

5.5.1.1 Delta–Sigma Oversampling A/D Converter Principle

The structure of a first-order delta–sigma converter is shown in Figure 5.14. The input signal is sampled at a frequency $f_s (T = 1/f_s)$. A feedback signal from a 1-bit D/A converter is subtracted from the input and the residue signal is accumulated by an integrator. The output of the integrator is quantized to generate a 1-bit digital stream. This digital output sets the sign of the feedback. If the digital output is, it feeds back a large negative signal to subtract from the input signal. The net effect of the feedback loop is to keep the output of the integrator small so that the output digits always track the amplitudes of the input signal.

The resolution of an A/D converter is determined by the quantization noise generated in the process. Even though a delta-sigma converter only has an 1-bit quantizer, much higher resolution is achieved by

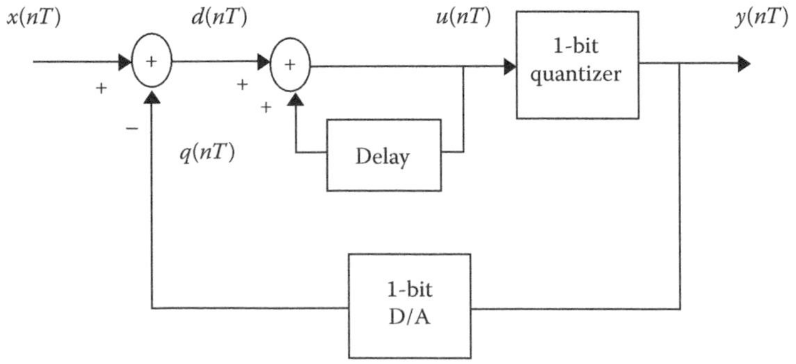

FIGURE 5.14 The modulator of a first-order delta–sigma converter: T is the sampling period and n is the index.

employing the noise shaping mechanism to move the noise out of the signal band and later blocking it using a lowpass digital filter.

Quantization is a nonlinear process and the feedback mechanism makes the noise highly dependent on the input signal spectrum. Rigorous treatment of this noise component in a delta-sigma converter can be found in the literature [8]. Useful information can still be obtained by linearizing the quantization process. The noise component is approximated by white additive noise uniformly distributed up to half of the sampling frequency. This approximation is valid because over a long period of time, the input to the quantizer will spread over a large number of values and appear to be quasi-random, so the noise introduced is quasi-random as well. Similar to a nonoversampled A/D converter, the rms value of the noise is $e_{rms}^2 = \frac{\Delta^2}{12}$, where Δ is the quantization step. When the quantizer is sampled at f_s, the noise power is sampled into a frequency band: $0 \leq f < f_s/2$ and its spectral density is

$$Q(f) = \sqrt{2} \cdot e_{rms}, \tag{5.4}$$

where f is normalized to $f_{\_s}$.

The delta–sigma converter can be generalized as shown in Figure 5.15. The forward path is modeled by transfer function $B(z)$ plus the noise, and the feedback path can be modeled by $C(z)$. The system output and input transfer function is governed by

$$Y(z) = \frac{B(z) \cdot X(z) + Q}{1 + B(z) \cdot C(z)}. \tag{5.5}$$

To achieve high-resolution A/D conversion, the system needs to convert the input signal within a specified frequency bandwidth and minimize the noise component in that band. One method is to pass the signal component and block the noise component. This can be expressed as

$$Y(z) = X(z) + H_{ns}(z) \cdot Q, \tag{5.6}$$

where the input $X(z)$ passes through the system, but the quantization noise is modified by a noise-shaping function $H_{ns}(z)$.

Comparing Equation 5.5 to Equation 5.6, to achieve the noise-shaping effect, the system in Figure 5.15 needs to have the following property:

$$C(z) = 1 - \frac{1}{B(z)} \tag{5.7}$$

$$B(z) = \frac{1}{H_{ns}(z)}.$$

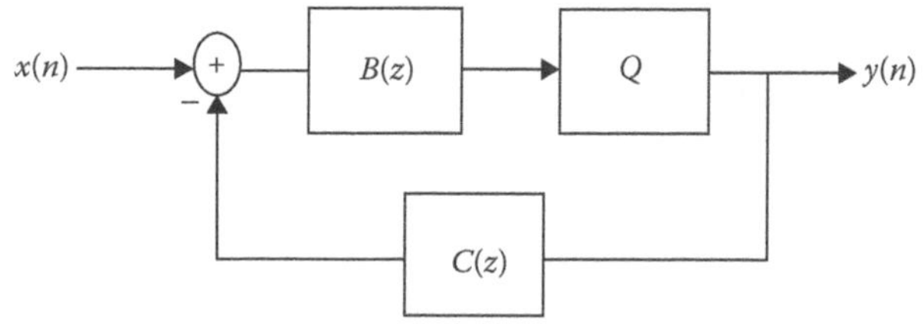

FIGURE 5.15 General feedback system.

Now, we can see the delta–sigma A/D converter shown in Figure 5.14 as a noise-shaping data converter. The transfer function of the integrator in the forward pass is $\frac{1}{1-z^{-1}}$; the D/A converter in the feedback path is equivalent to a delay element and its transfer function is z^{-1}. They satisfy the relation required by a noise-shaping converter in Equation 5.7. Therefore, its noise-shaping function $H_{ns}(z)$ is

$$H_{ns}(z) = \frac{1}{B(z)} = 1 - z^{-1}, \tag{5.8}$$

which is a highpass filtering function. The amplitude of its response is

$$|H_{ns}(z)| = |1 - z^{-1}| = 2\sin(\pi f), \tag{5.9}$$

where f is the normalized frequency with respect to f_s. This function is plotted in Figure 5.16. As shown in the figure, the noise is evenly distributed across the frequency, before applying the noise shaping function. The noise power in the signal band is the area of a region highlighted by the grey color underneath the flat line. After applying the noise-shaping function, the noise in the signal band is suppressed to a much lower level and the total noise power left (dark gray region) is much smaller than the original noise power. The high-frequency noise portion will be filtered by the digital filter. Therefore, the signal-to-noise ratio (SNR) of the converter is greatly enhanced.

Quantitatively, the noise power left in the signal band is the integration of its spectrum up to signal bandwidth f_b as

$$N^2 = \int_0^{f_b/f_s} \left(|H_{ns}(z)|^2 Q^2\right) df = \frac{2\Delta^2}{3f_s} \int_0^{f_b/f_s} [\sin(\pi f)]^2 df, \tag{5.10}$$

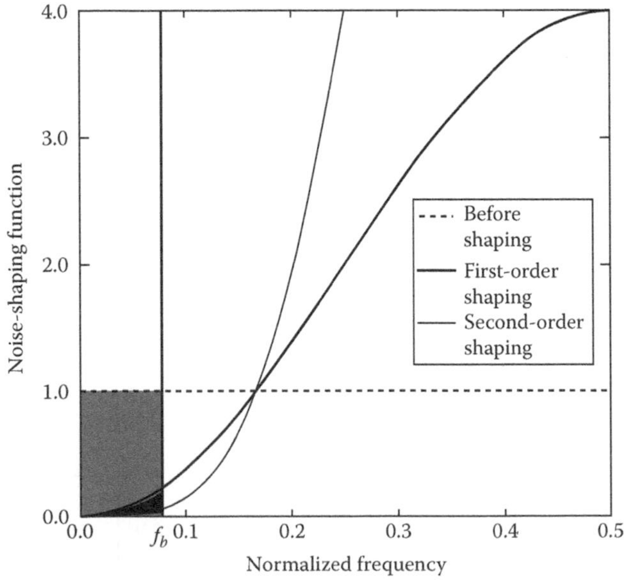

FIGURE 5.16 Plot of noise-shaping effect of the delta–sigma modulator comparing the noise power left within the baseband f_h. The noise (cross-hatched region) of a first-order modulator is much less than the noise before shaping (shaded region). Noise from the second-order shaping is even less.

where Q^2 is substituted for the noise spectral density in Equation 5.4. In a delta–sigma converter, the signal bandwidth is significantly lower than the sampling frequency. The resulting integration is

$$N_q^2 = \frac{2\pi^2\Delta^2}{9}\left(\frac{f_b}{f_s}\right)^3. \quad (5.11)$$

For a sine wave input, the maximum signal amplitude is $\frac{\Delta}{2}$ and its average power is $\frac{\Delta^2}{8}$. This gives a peak SNR as

$$\frac{S^2}{N^2} = \frac{9}{16\pi^2}\left(\frac{f_s}{f_b}\right)^3. \quad (5.12)$$

We can see that the peak SNR is only a function of the frequency ratio $\frac{f_s}{f_b}$. The faster the converter is sampled, the higher the resolution can be achieved. The expression in Equation 5.12 can be transformed into

$$\mathrm{SNR} = 10\log_{10}\frac{S^2}{N^2} = 20\log_{10}\left(\frac{3}{\sqrt{2}\pi}\right) + 9\log_2 M[\mathrm{dB}], \quad (5.13)$$

where M is an important parameter called the oversampling ratio, defined as the ratio of the sampling frequency over the Nyquist sampling frequency $2f_b$. From this expression, we can see that we can get 9 dB of increase in SNR for every doubling of the sampling frequency. This corresponds to 1.5 bits. For example, if $M = 128$, we have 11.5 bits more resolution than sampling at the Nyquist rate. This method allows a high resolution A/D conversion by using only a 1-bit quantizer.

We can see that higher resolution is achieved by trading off the input signal bandwidth. In order to get 1.5 more bits, the bandwidth has to be cut by a half in this structure. To have a more favorable resolution and bandwidth trade-off, we can go to higher order delta–sigma converters.

5.5.1.2 Higher Order Single-Stage Converters

In the first-order delta–sigma converter, the noise-shaping function is $H_{ns}(z) = 1 - z^{-1}$. Higher order converters can allow the noise-shaping function go up to Lth power, given as

$$H_{ns}(z) = \left(1 - z^{-1}\right)^L, \quad (5.14)$$

where L is an integer >1. Thus, the magnitude of this noise-shaping function is

$$|H_{ns}(z)| = \left|\left(1 - z^{-1}\right)^L\right| = [2\sin(\pi f)]^L. \quad (5.15)$$

This function is also plotted in Figure 5.16 for $L = 2$. As seen in the figure, more noise from the signal band is blocked than with the first-order function. Integrating Equation 5.14 over the signal band allows calculation of the SNR of an Lth order delta–sigma converter as

$$\frac{S^2}{N^2} = \frac{3(2L+1)}{2^{2L+2}\cdot\pi^{2L}}\cdot\left(\frac{f_s}{f_b}\right)^{2L+1}, \quad (5.16)$$

which is equivalent to

$$\mathrm{SNR} = 20\log_{10} = \left(\frac{\sqrt{3(2L+1)/2}}{\pi^L}\right) + 3(2L+1)\log_2 M[\mathrm{dB}], \quad (5.17)$$

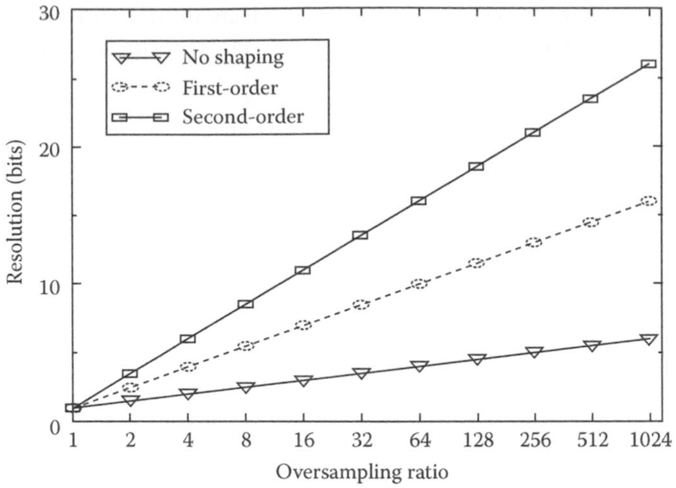

FIGURE 5.17 A plot of the resolution vs. oversampling ratio for different types of delta–sigma converters and Nyquist sampling converter.

where M is the oversampling ratio. For every doubling of the sampling frequency, the SNR is increased by $3(2L+1)$ dB, i.e., $L+0.5$ bits more resolution. For example, $L=2$ adds 2.5 bits and $L=3$ adds 3.5 bits of resolution. Therefore, compared to the first-order system, by employing a higher order delta–sigma converter architecture, the same resolution can be achieved with a lower sampling frequency, or a higher input bandwidth can be allowed at the same resolution with the same sampling frequency. Figure 5.17 shows a plot of Equation 5.17 comparing resolution vs. oversampling ratio for different order delta–sigma converters.

A second-order delta–sigma converter can be realized as shown in Figure 5.18 with two integrators. Higher order converters can be similarly constructed. However, when the order of the converter is greater than two, special care must be taken to insure the converter stability [9]. More zeroes are introduced in the transfer function of the forward path to suppress the signal swing after the integrators.

Other methods can be used to improve the resolution of the delta–sigma converter. A first-order and a second-order converter can be cascaded to achieve the same performance as a third-order converter, but with better stability over the frequency range [10]. A multi-bit quantizer can also be used to replace the 1-bit quantizer in the architecture presented here [11]. This improves the resolution at the same sampling speed. Interested readers are referred to reference articles.

In an oversampling converter, the digital decimation filter is also an integral part. Only after the decimation filter is the resolution of the converter realized. The design of decimation filters are discussed in other sections of this book and can also be found in the reference article by Candy [12].

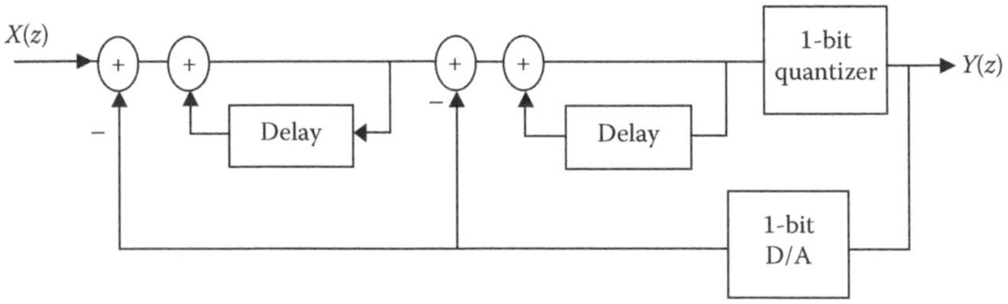

FIGURE 5.18 Block diagram of a second-order delta–sigma modulator.

References

1. Grebene, A.B., *Bipolar and MOS Analog Integrated Circuit Design*, John Wiley & Sons, New York, 1984.
2. Sheingold, D.H. (ed.), *Analog-Digital Conversion Handbook*, Prentice-Hall, Englewood Cliffs, NJ, 1986.
3. Toumazou, C., Lidgey F.J., and Haigh, D.G. (eds.), *Analogue IC Design: The Current-Mode Approach*, Peter Peregrinus Ltd., London, 1990.
4. Gray, P.R., Hodges, D.A., and Broderson, R.W. (eds.), *Analog MOS Integrated Circuits*, IEEE Press, New York, 1980.
5. Gray, P.R., Wooley, B.A., and Broderson, R.W. (eds.), *Analog MOS Integrated Circuits, II*, IEEE Press, New York, 1989.
6. Lee, S.H. and Song, B.S., Digital-domain calibration of multistep analog-to-digital converters, *IEEE J. Solid-State Circuits*, 27(12): 1679–1688, Dec. 1992.
7. Inose, H. and Yasuda, Y., A unity bit coding method by negative feedback, *Proc. IEEE*, 51: 1524–1535, Nov. 1963.
8. Gray, R.M., Oversampled sigma-delta modulation, *IEEE Trans. Commun.*, 35: 481–489, May 1987.
9. Chao, K.C.-H., Nadeem, S., Lee, W.L., and Sodini, C.G., A higher order topology for interpolative modulators for oversampled A/D converters, *IEEE Trans. Circuits Syst.*, CAS-37: 309–318, Mar. 1990.
10. Matsuya, Y., Uchimura, K., Iwata, A., Kobayashi, T., Ishikawa, M., and Yoshitoma, T., A 16-bit oversampling A-to-D conversion technology using triple-integration noise shaping, *IEEE J. Solid-State Circuits*, SC-22: 921–929, Dec. 1987.
11. Larson, L.E., Cataltepe, T., and Temes, G.C., Multibit oversampled $\Sigma - \Delta$ A/D converter with digital error correction, *Electron. Lett.*, 24: 1051–1052, Aug. 1988.
12. Candy, J.C., Decimation for sigma delta modulation, *IEEE Trans. Commun.*, COM-24: 72–76, Jan. 1986.

6

Quantization of Discrete Time Signals

Ravi P. Ramachandran
Rowan University

6.1 Introduction

Signals are usually classified into four categories. A continuous time signal $x(t)$ has the field of real numbers $\mathbf{R}$ as its domain in that t can assume any real value. If the range of $x(t)$ (values that $x(t)$ can assume) is also $\mathbf{R}$, then $x(t)$ is said to be a continuous time, continuous amplitude signal. If the range of $x(t)$ is the set of integers $\mathbf{Z}$, then $x(t)$ is said to be a continuous time, discrete amplitude signal. In contrast, a discrete time signal $x(n)$ has $\mathbf{Z}$ as its domain. A discrete time, continuous amplitude signal has $\mathbf{R}$ as its range. A discrete time, discrete amplitude signal has $\mathbf{Z}$ as its range. Here, the focus is on discrete time signals. Quantization is the process of approximating any discrete time, continuous amplitude signal into one of a finite set of discrete time, continuous amplitude signals based on a particular distortion or distance measure. This approximation is merely signal compression in that an infinite set of possible signals is converted into a finite set. The next step of encoding maps the finite set of discrete time, continuous amplitude signals into a finite set of discrete time, discrete amplitude signals.

A signal $x(n)$ is quantized one block at a time in that p (almost always consecutive) samples are taken as a vector $\mathbf{x}$ and approximated by a vector $\mathbf{y}$. The signal or data vectors $\mathbf{x}$ of dimension p (derived from $x(n)$) are in the vector space $\mathbf{R}^p$ over the field of real numbers $\mathbf{R}$. Vector quantization is achieved by mapping the infinite number of vectors in $\mathbf{R}^p$ to a finite set of vectors in $\mathbf{R}^p$. There is an inherent compression of the data vectors. This finite set of vectors in $\mathbf{R}^p$ is encoded into another finite set of vectors in a vector space of dimension q over a finite field (a field consisting of a finite set of numbers). For communication applications, the finite field is the binary field (0,1). Therefore, the original vector $\mathbf{x}$ is converted or compressed into a bit stream either for transmission over a channel or for storage purposes. This compression is necessary due to channel bandwidth or storage capacity constraints in a system.

The purpose of this chapter is to describe the basic definition and properties of vector quantization, introduce the practical aspects of design and implementation, and relate important issues. Note that two excellent review articles [1,2] give much insight into the subject. The outline of the chapter is as follows. The basic concepts are elaborated on in Section 6.2. Design algorithms for scalar and vector quantizers are described in Section 6.3. A design example is also provided. The practical issues are discussed in Section 6.4. The multistage and split manifestations of vector quantizers are described in Section 6.5. In Section 6.6, two applications of vector quantization in speech processing are discussed.

6.2 Basic Definitions and Concepts

In this section, we elaborate on the definitions of a vector and scalar quantizer, discuss some commonly used distance measures, and examine the optimality criteria for quantizer design.

6.2.1 Quantizer and Encoder Definitions

A quantizer, Q, is mathematically defined as a mapping [3] $Q: \mathbf{R}^p \rightarrow C$. This means that the p-dimensional vectors in the vector space $\mathbf{R}^p$ are mapped into a finite collection C of vectors that are also in $\mathbf{R}^p$. This collection C is called the codebook and the number of vectors in the codebook, N, is known as the codebook size. The entries of the codebook are known as codewords or codevectors. If $p = 1$, we have a scalar quantizer (SQ). If $p > 1$, we have a vector quantizer (VQ).

A quantizer is completely specified by p, C and a set of disjoint regions in $\mathbf{R}^p$ which dictate the actual mapping. Suppose C has N entries $\mathbf{y}_1, \mathbf{y}_2, \ldots, \mathbf{y}_N$. For each codevector, $\mathbf{y}_i$, there exists a region, $\boldsymbol{R}_i$, such that any input vector $\mathbf{x} \in \boldsymbol{R}_i$ gets mapped or quantized to $\mathbf{y}_i$. The region $\boldsymbol{R}_i$ is called a Voronoi region [3,4] and is defined to be the set of all $\mathbf{x} \in \mathbf{R}^p$ that are quantized to $\mathbf{y}_i$. The properties of Voronoi regions are as follows:

1. Voronoi regions are convex subsets of $\mathbf{R}^p$.
2. $\bigcup_{i=1}^{N} \boldsymbol{R}_i = \mathbf{R}^p$.
3. $R_i \cap R_j$ is the null set for $i \neq j$.

It is seen that the quantizer mapping is nonlinear and many to one and hence noninvertible.

Encoding the codevectors y_i is important for communications. The encoder, E, is mathematically defined as a mapping $E: C \rightarrow C_{\mathrm{B}}$. Every vector $\mathbf{y}_i \in C$ is mapped into a vector $\mathbf{t}_i \in C_{\mathrm{B}}$ where $\mathbf{t}_i$ belongs to a vector space of dimension $q = [\log_2 N]$ over the binary field (0, 1). The encoder mapping is one to one and invertible. The size of C_{B} is also N. As a simple example, suppose C contains four vectors of dimension p, namely ($\mathbf{y}_1$, $\mathbf{y}_2$, $\mathbf{y}_3$, $\mathbf{y}_4$). The corresponding mapped vectors in C_{B} are $\mathbf{t}_1 = [0\ 0]$, $\mathbf{t}_2 = [0\ 1]$, $\mathbf{t}_3 = [1\ 0]$, and $\mathbf{t}_4 = [1\ 1]$. The decoder D described by $D: C_{\mathrm{B}} \rightarrow C$ performs the inverse operation of the encoder.

A block diagram of quantization and encoding for communications applications is shown in Figure 6.1. Given that the final aim is to transmit and reproduce $\mathbf{x}$, the two sources of error are due to quantization and Cchannel. The quantization error is $\mathbf{x} - \mathbf{y}_i$ and is heavily dealt with in this chapter. The channel introduces errors that transform $\mathbf{t}_i$ into $\mathbf{t}_j$ thereby reproducing $\mathbf{y}_j$ instead of $\mathbf{y}_i$ after decoding. Channel errors are ignored for the purposes of this chapter.

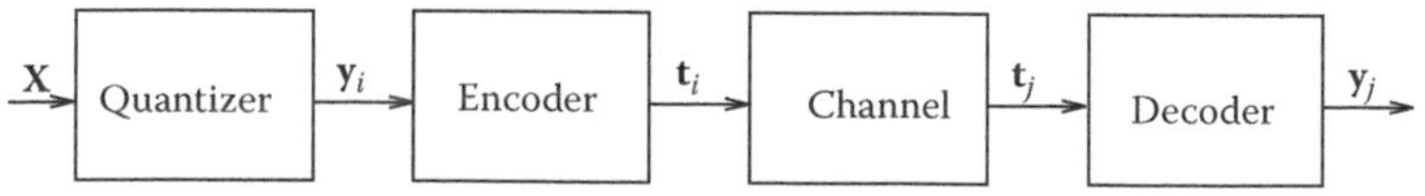

FIGURE 6.1 Block diagram of quantization and encoding for communication systems.

6.2.2 Distortion Measure

A distortion or distance measure between two vectors $\mathbf{x} = [x_1\, x_2\, x_3 \cdots x_p]^T \in \mathbf{R}^p$ and $\mathbf{y} = [y_1\, y_2\, y_3 \cdots y_p]^T \in \mathbf{R}^p$ where the superscript T denotes transposition is symbolically given by $d(\mathbf{x}, \mathbf{y})$. Most distortion measures satisfy three properties given by

1. Positivity: $d(\mathbf{x}, \mathbf{y})$ is a real number greater than or equal to zero with equality if and only if $\mathbf{x} = \mathbf{y}$
2. Symmetry: $d(\mathbf{x}, \mathbf{y}) = d(\mathbf{y}, \mathbf{x})$
3. Triangle inequality: $d(\mathbf{x}, \mathbf{z}) \leq d(\mathbf{x}, \mathbf{y}) + d(\mathbf{y}, \mathbf{z})$

To qualify as a valid measure for quantizer design, only the property of positivity needs to be satisfied. The choice of a distance measure is dictated by the specific application and computational considerations. We continue by giving some examples of distortion measures.

Example 6.1: The L_r Distance

The L_r distance is given by

$$d(\mathbf{x}, \mathbf{y}) = \sum_{i=1}^{p} |x_i - y_i|^r \tag{6.1}$$

This is a computationally simple measure to evaluate. The three properties of positivity, symmetry, and the triangle inequality are satisfied. When $r = 2$, the squared Euclidean distance emerges and is very often used in quantizer design. When $r = 1$, we get the absolute distance. If $r = \infty$, it can be shown that [2]

$$\lim_{r \to \infty} d(\mathbf{x}, \mathbf{y})^{1/r} = \max_i |x_i - y_i| \tag{6.2}$$

This is the maximum absolute distance taken over all vector components.

Example 6.2: The Weighted L_2 Distance

The weighted L_2 distance is given by

$$d(\mathbf{x}, \mathbf{y}) = (\mathbf{x} - \mathbf{y})^T \mathbf{W} (\mathbf{x} - \mathbf{y}) \tag{6.3}$$

where $\mathbf{W}$ is the matrix of weights. For positivity, $\mathbf{W}$ must be positive-definite. If $\mathbf{W}$ is a constant matrix, the three properties of positivity, symmetry, and the triangle inequality are satisfied. In some applications, $\mathbf{W}$ is a function of $\mathbf{x}$. In such cases, only the positivity of $d(\mathbf{x}, \mathbf{y})$ is guaranteed to hold. As a particular case, if $\mathbf{W}$ is the inverse of the covariance matrix of $\mathbf{x}$, we get the Mahalanobis distance [2]. Other examples of weighting matrices will be given when we discuss the applications of quantization.

6.2.3 Optimality Criteria

There are two necessary conditions for a quantizer to be optimal [2,3]. As before, the codebook C has N entries $\mathbf{y}_1, \mathbf{y}_2, \ldots, \mathbf{y}_N$ and each codevector $\mathbf{y}_i$ is associated with a Voronoi region R_i. The first condition known as the nearest neighbor rule states that a quantizer maps any input vector $\mathbf{x}$ to the codevector closest to it. Mathematically speaking, $\mathbf{x}$ is mapped to $\mathbf{y}_i$ if and only if $d(\mathbf{x}, \mathbf{y}_i) \leq d(\mathbf{x}, \mathbf{y}_j) \forall j \neq i$. This enables us to more precisely define a Voronoi region as

$$R_i = \{\mathbf{x} \in \mathbf{R}^p : d(\mathbf{x}, \mathbf{y}_i) \leq d(\mathbf{x}, \mathbf{y}_j) \forall j \neq i\} \tag{6.4}$$

The second condition specifies the calculation of the codevector $\mathbf{y}_i$ given a Voronoi region R_i. The codevector $\mathbf{y}_i$ is computed to minimize the average distortion in R_i which is denoted by D_i where

$$D_i = E\left[d(\mathbf{x}, \mathbf{y}_i) \mid \mathbf{x} \in R_i\right] \tag{6.5}$$

6.3 Design Algorithms

Quantizer design algorithms are formulated to find the codewords and the Voronoi regions so as to minimize the overall average distortion D given by

$$D = E\left[d(\mathbf{x}, \mathbf{y})\right] \tag{6.6}$$

If the probability density $p(\mathbf{x})$ of the data $\mathbf{x}$ is known, the average distortion is [2,3]

$$D = \int d(\mathbf{x}, \mathbf{y}) p(\mathbf{x}) d\mathbf{x} \tag{6.7}$$

$$= \sum_{i=1}^{N} \int_{R_i} d(\mathbf{x}, \mathbf{y}_i) p(\mathbf{x}) d\mathbf{x} \tag{6.8}$$

Note that the nearest neighbor rule has been used to get the final expression for D. If the probability density is not known, an empirical estimate is obtained by computing many sampled data vectors. This is called training data, or a training set, and is denoted by $T = \{\mathbf{x}_1, \mathbf{x}_2, \mathbf{x}_3, \ldots, \mathbf{x}_M\}$ where M is the number of vectors in the training set. In this case, the average distortion is

$$D = \frac{1}{M} \sum_{k=1}^{M} d(\mathbf{x}_k, \mathbf{y}) \tag{6.9}$$

$$= \frac{1}{M} \sum_{i=1}^{N} \sum_{\mathbf{x}_k \in R_i} d(\mathbf{x}_k, \mathbf{y}_i) \tag{6.10}$$

Again, the nearest neighbor rule has been used to get the final expression for D.

6.3.1 Lloyd–Max Quantizers

The Lloyd–Max method is used to design SQs and assumes that the probability density of the scalar data $p(x)$ is known [5,6]. Let the codewords be denoted by $y_1, y_2, \ldots, y_N$. For each codeword y_i, the Voronoi region is a continuous interval $R_i = (v_i, v_{i+1}]$. Note that $v_1 = -\infty$ and $v_{N+1} = \infty$. The average distortion is

$$D = \sum_{i=1}^{N} \int_{v_i}^{v_{i+1}} d(x, y_i) p(x) dx \tag{6.11}$$

Setting the partial derivatives of D with respect to v_i and y_i to zero gives the optimal Voronoi regions and codewords.

In the particular case when $d(x, y_i) = (x - y_i)^2$, it can be shown that [5] the optimal solution is

$$v_i = \frac{y_i + y_{i+1}}{2} \tag{6.12}$$

for $2 \leq i \leq N$ and

$$y_i = \frac{\int_{v_i}^{v_{i+1}} xp(x)dx}{\int_{v_i}^{v_{i+1}} p(x)dx} \tag{6.13}$$

for $1 \leq i \leq N$. The overall iterative algorithm is

1. Start with an initial codebook and compute the resulting average distortion.
2. Solve for v_i.
3. Solve for y_i.
4. Compute the resulting average distortion.
5. If the average distortion decreases by a small amount that is less than a given threshold, the design terminates. Otherwise, go back to Step 2.

The extension of the Lloyd–Max algorithm for designing VQs has been considered [7]. One practical difficulty is whether the multidimensional probability density function (pdf) $p(\mathbf{x})$ is known or must be estimated. Even if this is circumvented, finding the multidimensional shape of the convex Voronoi regions is extremely difficult and practically impossible for dimensions >5 [7]. Therefore, the Lloyd–Max approach cannot be extended to multidimensions and methods have been configured to design a VQ from training data. We will now elaborate on one such algorithm.

6.3.2 Linde–Buzo–Gray Algorithm

The input to the Linde–Buzo–Gray (LBG) algorithm [7] is a training set $T = \{\mathbf{x}_1, \mathbf{x}_2, \mathbf{x}_3, \ldots, \mathbf{x}_M\} \in \mathbf{R}^p$ having M vectors, a distance measure $d(\mathbf{x}, \mathbf{y})$, and the desired size of the codebook N. From these inputs, the codewords $\mathbf{y}_i$ are iteratively calculated. The probability density $p(\mathbf{x})$ is not explicitly considered and the training set serves as an empirical estimate of $p(\mathbf{x})$. The Voronoi regions are now expressed as

$$R_i = \{\mathbf{x}_k \in T : d(\mathbf{x}_k, \mathbf{y}_i) \leq d(\mathbf{x}_k, \mathbf{y}_j) \forall j \neq i\} \tag{6.14}$$

Once the vectors in R_i are known, the corresponding codevector y_i is found to minimize the average distortion in R_i as given by

$$D_i = \frac{1}{M_i} \sum_{\mathbf{x}_k \in R_i} d(\mathbf{x}_k, \mathbf{y}_i) \tag{6.15}$$

where M_i is the number of vectors in R_i. In terms of D_i, the overall average distortion D is

$$D = \sum_{i=1}^{N} \frac{M_i}{M} D_i \tag{6.16}$$

Explicit expressions for $\mathbf{y}_i$ depend on $d(\mathbf{x}, \mathbf{y}_i)$ and two examples are given. For the L_1 distance,

$$\mathbf{y}_i = \text{median}[\mathbf{x}_k \in \boldsymbol{R}_i] \tag{6.17}$$

For the weighted $\boldsymbol{L}_2$ distance in which the matrix of weights $\mathbf{W}$ is constant,

$$\mathbf{y}_i = \frac{1}{\boldsymbol{M}_i} \sum_{\mathbf{x}_k \in R_i} \mathbf{x}_k \tag{6.18}$$

which is merely the average of the training vectors in $\boldsymbol{R}_i$. The overall methodology to get a codebook of size N is

1. Start with an initial codebook and compute the resulting average distortion.
2. Find $\boldsymbol{R}_i$.
3. Solve for $\mathbf{y}_i$.
4. Compute the resulting average distortion.
5. If the average distortion decreases by a small amount that is less than a given threshold, the design terminates. Otherwise, go back to Step 2.

If N is a power of 2 (necessary for coding), a growing algorithm starting with a codebook of size 1 is formulated as follows:

1. Find codebook of size 1.
2. Find initial codebook of double the size by doing a binary split of each codevector. For a binary split, one codevector is split into two by small perturbations.
3. Invoke the methodology presented earlier of iteratively finding the Voronoi regions and codevectors to get the optimal codebook.
4. If the codebook of the desired size is obtained, the design stops. Otherwise, go back to Step 2 in which the codebook size is doubled.

Note that with the growing algorithm, a locally optimal codebook is obtained. Also, SQ design can also be performed.

Here, we present a numerical example in which $p=2$, $\boldsymbol{M}=4$, $\boldsymbol{N}=2$, $T=\{\mathbf{x}_1=[0\quad 0], \mathbf{x}_2=[0\quad 1], \mathbf{x}_3=[1\quad 0], \mathbf{x}_4=[1\quad 1]\}$, and $d(\mathbf{x},\mathbf{y})=(\mathbf{x}-\mathbf{y})^{\mathrm{T}}(\mathbf{x}-\mathbf{y})$. The codebook of size 1 is $\mathbf{y}_1=[0.5\quad 0.5]$. We will invoke the LBG algorithm twice, each time using a different binary split. For the first run,

1. Binary split: $\mathbf{y}_1=[0.51\quad 0.5]$ and $\mathbf{y}_2=[0.49\quad 0.5]$
2. Iteration 1:
   a. $\boldsymbol{R}_1=\{\mathbf{x}_3, \mathbf{x}_4\}$ and $\boldsymbol{R}_2=\{\mathbf{x}_1, \mathbf{x}_2\}$
   b. $\mathbf{y}_1=[1\quad 0.5]$ and $\mathbf{y}_2=[0\quad 0.5]$
   c. Average distortion: $D=0.25[(0.5)^2+(0.5)^2+(0.5)^2+(0.5)^2]=0.25$
3. Iteration 2:
   a. $\boldsymbol{R}_1=\{\mathbf{x}_3, \mathbf{x}_4\}$ and $\boldsymbol{R}_2=\{\mathbf{x}_1, \mathbf{x}_2\}$
   b. $\mathbf{y}_1=[1\quad 0.5]$ and $\mathbf{y}_2=[0\quad 0.5]$
   c. Average distortion: $D=0.25[(0.5)^2+(0.5)^2+(0.5)^2+(0.5)^2]=0.25$
4. No change in average distortion, the design terminates

For the second run,

1. Binary split: $\mathbf{y}_1=[0.5\quad 0.51]$ and $\mathbf{y}_2=[0.5\quad 0.49]$
2. Iteration 1:
   a. $\boldsymbol{R}_1=\{\mathbf{x}_2, \mathbf{x}_4\}$ and $\boldsymbol{R}_2=\{\mathbf{x}_1, \mathbf{x}_3\}$
   b. $\mathbf{y}_1=[0.5\quad 1]$ and $\mathbf{y}_2=[0.5\quad 0]$
   c. Average distortion: $D=0.25[(0.5)^2+(0.5)^2+(0.5)^2+(0.5)^2]=0.25$
3. Iteration 2:
   a. $\boldsymbol{R}_1=\{\mathbf{x}_2, \mathbf{x}_4\}$ and $\boldsymbol{R}_2=\{\mathbf{x}_1, \mathbf{x}_3\}$
   b. $\mathbf{y}_1=[0.5\quad 1]$ and $\mathbf{y}_2=[0.5\quad 0]$
   c. Average distortion: $D=0.25[(0.5)^2+(0.5)^2+(0.5)^2+(0.5)^2]=0.25$
4. No change in average distortion, the design terminates

The two codebooks are equally good locally optimal solutions that yield the same average distortion. The initial condition as determined by the binary split influences the final solution.

6.4 Practical Issues

When using quantizers in a real environment, there are many practical issues that must be considered to make the operation feasible. First we enumerate the practical issues and then discuss them in more detail. Note that the issues listed below are interrelated.

1. Parameter set
2. Distortion measure
3. Dimension
4. Codebook storage
5. Search complexity
6. Quantizer type
7. Robustness to different inputs
8. Gathering of training data

A parameter set and distortion measure are jointly configured to represent and compress information in a meaningful manner that is highly relevant to the particular application. This concept is best illustrated with an example. Consider linear predictive (LP) analysis [8] of speech that is performed by the autocorrelation method. The resulting minimum phase nonrecursive filter

$$A(z) = 1 - \sum_{k=1}^{p} a_k z^{-k} \tag{6.19}$$

removes the near-sample redundancies in the speech. The filter $1/A(z)$ describes the spectral envelope of the speech. The information regarding the spectral envelope as contained in the LP filter coefficients a_k must be compressed (quantized) and coded for transmission. This is done in predictive speech coders [9]. There are other parameter sets that have a one-to-one correspondence to the set a_k. An equivalent parameter set that can be interpreted in terms of the spectral envelope is desired. The line spectral frequencies (LSFs) [10,11] have been found to be the most useful.

The distortion measure is significant for meaningful quantization of the information and must be mathematically tractable. Continuing the above example, the LSFs must be quantized such that the spectral distortion (SD) between the spectral envelopes they represent is minimized. Mathematical tractability implies that the computation involved for (1) finding the codevectors given the Voronoi regions (as part of the design procedure) and (2) quantizing an input vector with the least distortion given a codebook is small. The L_1, L_2, and weighted L_2 distortions are mathematically feasible. For quantizing LSFs, the L_2 and weighted L_2 distortions are often used [12–14]. More details on LSF quantization will be provided in a forthcoming section on applications. At this point, a general description is provided just to illustrate the issues of selecting a parameter set and a distortion measure.

The issues of dimension, codebook storage, and search complexity are all related to computational considerations. A higher dimension leads to an increase in the memory requirement for storing the codebook and in the number of arithmetic operations for quantizing a vector given a codebook (search complexity). The dimension is also very important in capturing the essence of the information to be quantized. For example, if speech is sampled at 8 kHz, the spectral envelope consists of 3–4 formants (vocal tract resonances) which must be adequately captured. By using LSFs, a dimension of 10–12 suffices for capturing the formant information. Although a higher dimension leads to a better description of the fine details of the spectral envelope, this detail is not crucial for speech coders. Moreover, this higher dimension imposes more of a computational burden. The codebook storage requirement depends on the codebook size N. Obviously, a smaller value of N imposes less of a memory requirement. Also for coding, the number of bits to be transmitted should be minimized, thereby diminishing the memory requirement. The search complexity is directly related to the codebook size and dimension. However, it is also influenced by the type of distortion measure.

The type of quantizer (scalar or vector) is dictated by computational considerations and the robustness issue (discussed later). Consider the case when a total of 12 bits are used for quantization, the dimension is 6, and the L_2 distance measure is utilized. For a VQ, there is one codebook consisting of $2^{12} = 4{,}096$ codevectors each having 6 components. A total of $4{,}096 \times 6 = 24{,}576$ numbers need to be stored. Computing the L_2 distance between an input vector and one codevector requires 6 multiplications and 11 additions. Therefore, searching the entire codebook requires $6 \times 4{,}096 = 24{,}576$ multiplications and $11 \times 4{,}096 = 45{,}056$ additions. For an SQ, there are 6 codebooks, one for each dimension. Each codebook requires 2 bits or $2^2 = 4$ codewords. The overall codebook size is $4 \times 6 = 24$. Hence, a total of 24 numbers needs to be stored. Consider the first component of an input vector. Four multiplications and four additions are required to find the best codeword. Hence, for all 6 components, 24 multiplications and 24 additions are needed to complete the search. The storage and search complexity are always much less for an SQ.

The quantizer type is also closely related to the robustness issue. A quantizer is said to be robust to different test input vectors if it can maintain the same performance for a large variety of inputs. The performance of a quantizer is measured as the average distortion resulting from the quantization of a set of test inputs. A VQ takes advantage of the multidimensional probability density of the data as empirically estimated by the training set. An SQ does not consider the correlations among the vector components as a separate design is performed for each component based on the probability density of that component. For test data having a similar density to the training data, a VQ will outperform an SQ given the same overall codebook size. However, for test data having a density that is different from that of the training data, an SQ will outperform a VQ given the same overall codebook size. This is because an SQ can accomplish a better coverage of a multidimensional space. Consider the example in Figure 6.2. The vector space is of two dimensions ($p = 2$). The component x_1 lies in the range 0 to x_1(max) and x_2 lies between 0 and x_2(max). The multidimensional pdf $p(x_1, x_2)$ is shown as the region ABCD in Figure 6.2. The training data will represent this pdf and can be used to design a vector and SQ of the same overall codebook size. The VQ will perform better for test data vectors in the region ABCD. Due to the individual ranges of the values of x_1 and x_2, the SQ will cover the larger space OKLM. Therefore, the SQ will perform better for test data vectors in OKLM but outside ABCD. An SQ is more robust in that it performs better for data with a density different from that of the training set. However, a VQ is preferable if the test data is known to have a density that resembles that of the training set.

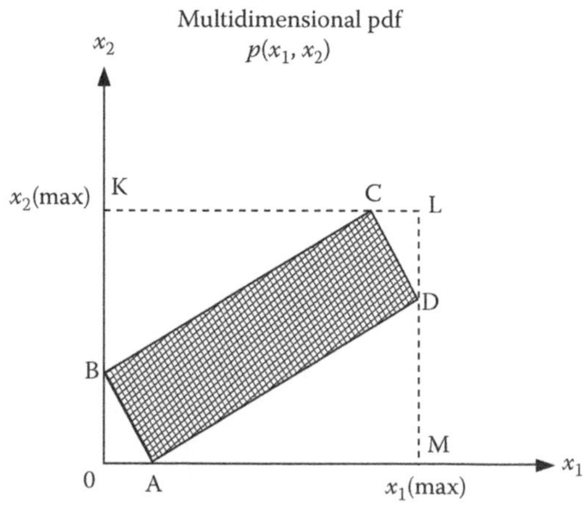

FIGURE 6.2 Example of a multidimensional probability density for explanation of the robustness issue.

In practice, the true multidimensional pdf of the data is not known as the data may emanate from many different conditions. For example, LSFs are obtained from speech material derived from many environmental conditions (like different telephones and noise backgrounds). Although getting a training set that is representative of all possible conditions gives the best estimate of the multidimensional pdf, it is impossible to configure such a set in practice. A versatile training set contributes to the robustness of the VQ but increases the time needed to accomplish the design.

6.5 Specific Manifestations

Thus far, we have considered the implementation of a VQ as being a one-step quantization of x. This is known as full VQ and is definitely the optimal way to do quantization. However, in applications such as LSF coding, quantizers between 25 and 30 bits are used. This leads to a prohibitive codebook size and search complexity. Two suboptimal approaches are now described that use multiple codebooks to alleviate the memory and search complexity requirements.

6.5.1 Multistage VQ

In multistage VQ consisting of R stages [3], there are R quantizers, $\mathbf{Q}_1, \mathbf{Q}_2, \ldots, \mathbf{Q}_R$. The corresponding codebooks are denoted as $C_1, C_2, \ldots, C_R$. The sizes of these codebooks are $N_1, N_2, \ldots, N_R$. The overall codebook size is $N = N_1 + N_2 + \cdots + N_R$. The entries of the ith codebook C_i are $\mathbf{y}_1^{(i)}, \mathbf{y}_2^{(i)}, \ldots, \mathbf{y}_{N_i}^{(i)}$. Figure 6.3 shows a block diagram of the entire system.

The procedure for multistage VQ is as follows. The input $\mathbf{x}$ is first quantized by $\mathbf{Q}_1$ to $\mathbf{y}_k^{(1)}$. The quantization error is $\mathbf{e}_1 = \mathbf{x} - \mathbf{y}_k^{(1)}$, which is in turn quantized by $\mathbf{Q}_2$ to $\mathbf{y}_k^{(2)}$. The quantization error at the second stage is $\mathbf{e}_2 = \mathbf{e}_1 - \mathbf{y}_k^{(2)}$. This error is quantized at the third stage. The process repeats and at the Rth stage, $\mathbf{e}_{R-1}$ is quantized by $\mathbf{Q}_R$ to $\mathbf{y}_k^{(R)}$ such that the quantization error is $\mathbf{e}_R$. The original vector $\mathbf{x}$ is quantized to $\mathbf{y} = \mathbf{y}_k^{(1)} + \mathbf{y}_k^{(2)} + \cdots + \mathbf{y}_k^{(R)}$. The overall quantization error is $\mathbf{x} - \mathbf{y} = \mathbf{e}_R$.

The reduction in the memory requirement and search complexity is best illustrated by a simple example. A full VQ of 30 bits will have one codebook of 2^{30} codevectors (cannot be used in practice). An equivalent multistage VQ of $R = 3$ stages will have three 10-bit codebooks C_1, C_2, and C_3. The total number of codevectors to be stored is 3×2^{10}, which is practically feasible. It follows that the search complexity is also drastically reduced over that of a full VQ.

The simplest way to train a multistage VQ is to perform sequential training of the codebooks. We start with a training set $T = \{\mathbf{x}_1, \mathbf{x}_2, \mathbf{x}_3, \ldots, \mathbf{x}_M\} \in \mathbf{R}^p$ to get C_1. The entire set T is quantized by Q_1 to get a training set for the next stage. The codebook C_2 is designed from this new training set. This procedure is repeated so that all the R codebooks are designed. A joint design procedure for multistage VQ has been recently developed in [15] but is outside the scope of this chapter.

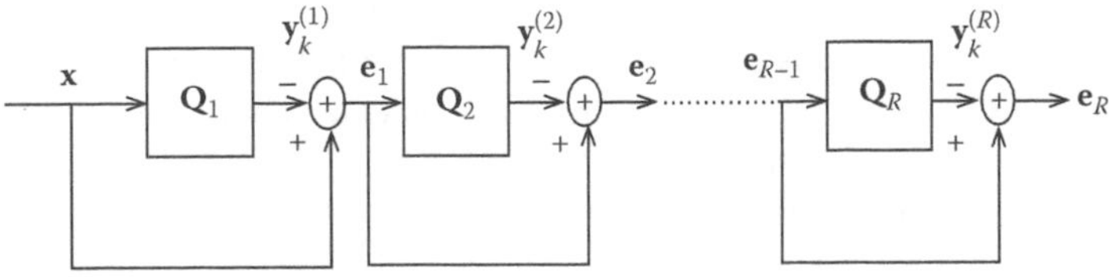

FIGURE 6.3 Multistage vector quantization.

6.5.2 Split VQ

In split VQ [3], $\mathbf{x} = [x_1\ x_2\ x_3 \cdots x_p]^T \in \mathbf{R}^p$ is split or partitioned into R subvectors of smaller dimension as $\mathbf{x} = [\mathbf{x}^{(1)}\ \mathbf{x}^{(2)}\ \mathbf{x}^{(3)} \cdots \mathbf{x}^{(R)}]^T$. The ith subvector $\mathbf{x}^{(i)}$ has dimension d_i. Therefore, $p = d_1 + d_2 + \cdots + d_R$. Specifically,

$$\mathbf{x}^{(1)} = [x_1\ x_2\ \cdots\ x_{d_1}]^T \tag{6.20}$$

$$\mathbf{x}^{(2)} = [x_{d_1+1}\ x_{d_1+2}\ \cdots\ x_{d_1+d_2}]^T \tag{6.21}$$

$$\mathbf{x}^{(3)} = [x_{d_1+d_2+1}\ x_{d_1+d_2+2}\ \cdots\ x_{d_1+d_2+d_3}]^T \tag{6.22}$$

and so forth.

There are R quantizers, one for each subvector. The subvectors $\mathbf{x}^{(i)}$ are individually quantized to $\mathbf{y}_k^{(i)}$ so that the full vector $\mathbf{x}$ is quantized to $\mathbf{y} = \left[\mathbf{y}_k^{(1)}\ \mathbf{y}_k^{(2)}\ \mathbf{y}_k^{(3)} \cdots \mathbf{y}_k^{(R)}\right]^T \in \mathbf{R}^p$. The quantizers are designed using the appropriate subvectors in the training set T. The extreme case of a split VQ is when $R = p$. Then, $d_1 = d_2 = \cdots = d_p = 1$ and we get an SQ.

The reduction in the memory requirement and search complexity is again illustrated by a similar example as for multistage VQ. Suppose the dimension $p = 10$. A full VQ of 30 bits will have one codebook of 2^{30} codevectors. An equivalent split VQ of $R = 3$ splits uses subvectors of dimensions $d_1 = 3$, $d_2 = 3$, and $d_3 = 4$. For each subvector, there will be a 10-bit codebook having 2^{10} codevectors.

Finally, note that split VQ is feasible if the distortion measure is separable in that

$$d(\mathbf{x}, \mathbf{y}) = \sum_{i=1}^{R} d\left(\mathbf{x}^{(i)}, \mathbf{y}_k^{(i)}\right) \tag{6.23}$$

This property is true for the L_r distance and for the weighted L_2 distance if the matrix of weights $\mathbf{W}$ is diagonal.

6.6 Applications

In this chapter, two applications of quantization are discussed. One is in the area of speech coding and the other is in speaker identification. Both are based on LP analysis of speech [8] as performed by the autocorrelation method. As mentioned earlier, the predictor coefficients, a_k, describe a minimum phase nonrecursive LP filter $A(z)$ as given by Equation 6.19. We recall that the filter $1/A(z)$ describes the spectral envelope of the speech, which in turn gives information about the formants.

6.6.1 Predictive Speech Coding

In predictive speech coders, the predictor coefficients (or a transformation thereof) must be quantized. The main aim is to preserve the spectral envelope as described by $1/A(z)$ and, in particular, preserve the formants. The coefficients a_k are transformed into an LSF vector $\mathbf{f}$. The LSFs are more clearly related to the spectral envelope in that (1) the spectral sensitivity is local to a change in a particular frequency and (2) the closeness of two adjacent LSFs indicates a formant.

Ideally, LSFs should be quantized to minimize the SD given by

$$\mathrm{SD} = \sqrt{\frac{1}{B}\int_R \left[10\log\left(\left|A_q(e^{j2\pi f})\right|^2 / \left|A(e^{j2\pi f})\right|^2\right)\right]^2 df} \tag{6.24}$$

where

$A(\cdot)$ refers to the original LP filter
$A_q(\cdot)$ refers to the quantized LP filter
B is the bandwidth of interest
R is the frequency range of interest

The SD is not a mathematically tractable measure and is also not separable if split VQ is to be used. A weighted L_2 measure is used in which $\mathbf{W}$ is diagonal and the ith diagonal element is $w(i)$ is given by [14]:

$$w(i) = \frac{1}{f_i - f_{i-1}} + \frac{1}{f_{i+1} - f_i} \tag{6.25}$$

where

$\mathbf{f} = [f_1\, f_2\, f_3 \cdots f_p]^T \in \mathbf{R}^p$
f_0 is taken to be zero
f_{p+1} is taken to be the highest digital frequency (π or 0.5 if normalized)

Regarding this distance measure, note the following:

1. The LSFs are ordered ($f_{i+1} > f_i$) if and only if the LP filter $A(z)$ is minimum phase. This guarantees that $w(i) > 0$.
2. The weight $w(i)$ is high if two adjacent LSFs are close to each other. Therefore, more weight is given to regions in the spectrum having formants.
3. The weights are dependent on the input vector $\mathbf{f}$. This makes the computation of the codevectors using the LBG algorithm different from the case when the weights are constant. However, for finding the codevector given a Voronoi region, the average of the training vectors in the region is taken so that the ordering property is preserved.
4. Mathematical tractability and separability of the distance measure are obvious.

A quantizer can be designed from a training set of LSFs using the weighted L_2 distance. Consider LSFs obtained from speech that is lowpass filtered to 3400 Hz and sampled at 8 kHz. If there are additional highpass or bandpass filtering effects, some of the LSFs tend to migrate [16]. Therefore, a VQ trained solely on one filtering condition will not be robust to test data derived from other filtering conditions [16]. The solution in [16] to robustize a VQ is to configure a training set consisting of two main components. First, LSFs from different filtering conditions are gathered to provide a reasonable empirical estimate of the multidimensional pdf. Second, a uniformly distributed set of vectors provides for coverage of the multidimensional space (similar to what is accomplished by an SQ). Finally, multistage or split LSF quantizers are used for practical feasibility [13,15,16].

6.6.2 Speaker Identification

Speaker recognition is the task of identifying a speaker by his or her voice. Systems performing speaker recognition operate in different modes. A closed set mode is the situation of identifying a particular speaker as one in a finite set of reference speakers [17]. In an open set system, a speaker is either identified as belonging to a finite set or is deemed not to be a member of the set [17]. For speaker verification, the claim of a speaker to be one in a finite set is either accepted or rejected [18]. Speaker recognition can either be done as a text-dependent or text-independent task. The difference is that in the former case, the speaker is constrained as to what must be said, while in the latter case no constraints are imposed. In this chapter, we focus on the closed set, text-independent mode. The overall system will have three components, namely, (1) LP analysis for parameterizing the spectral envelope, (2) feature extraction for ensuring speaker discrimination, and (3) classifier for making a decision. The input to the system will be a speech signal. The output will be a decision regarding the identity of the speaker.

After LP analysis of speech is carried out, the LP predictor coefficients, a_k, are converted into the LP cepstrum. The cepstrum is a popular feature as it provides for good speaker discrimination. Also, the cepstrum lends itself to the L_2 or weighted L_2 distance that is simple and yet reflective of the log SD between two LP filters [19]. To achieve good speaker discrimination, the formants must be captured. Hence, a dimension of 12 is usually used.

The cepstrum is used to develop a VQ classifier [20] as shown in Figure 6.4. For each speaker enrolled in the system, a training set is established from utterances spoken by that speaker. From the training set, a VQ codebook is designed that serves as a speaker model. The VQ codebook represents a portion of the multidimensional space that is characteristic of the feature or cepstral vectors for a particular speaker. Good discrimination is achieved if the codebooks show little or no overlap as illustrated in Figure 6.5 for

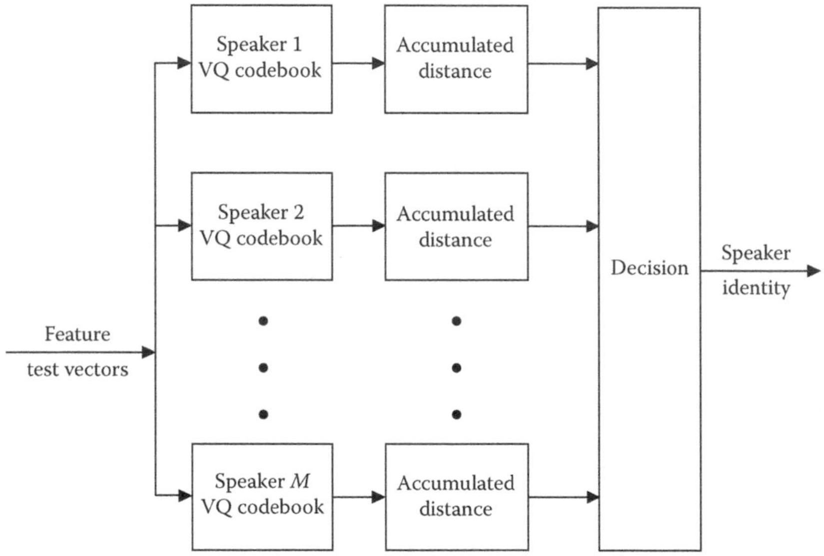

FIGURE 6.4 A VQ-based classifier for speaker identification.

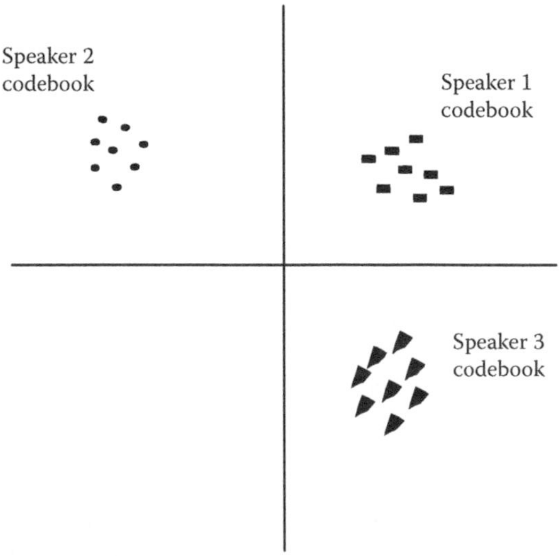

FIGURE 6.5 VQ codebooks for three speakers.

the case of three speakers. Usually, a small codebook size of 64 or 128 codevectors is sufficient [21]. Even if there are 50 speakers enrolled, the memory requirement is feasible for real-time applications. An SQ is of no use because the correlations among the vector components are crucial for speaker discrimination. For the same reason, multistage or split VQ is also of no use. Moreover, full VQ can easily be used given the relatively smaller codebook size as compared to coding.

Given a random speech utterance, the testing procedure for identifying a speaker is as follows (see Figure 6.4). First, the S test feature (cepstrum) vectors are computed. Consider the first vector. It is quantized by the codebook for speaker 1 and the resulting minimum L_2 or weighted L_2 distance is recorded. This quantization is done for all S vectors and the resulting minimum distances are accumulated (added up) to get an overall score for speaker 1. In this manner, an overall score is computed for all the speakers. The identified speaker is the one with the least overall score. Note that with the small codebook sizes, the search complexity is practically feasible. In fact, the overall score for the different speakers can be obtained in parallel. The performance measure for a speaker identification system is the identification success rate, which is the number of test utterances for which the speaker is identified correctly divided by the total number of test utterances.

The robustness issue is of great significance and emerges when the cepstral vectors derived from certain test speech material have not been considered in the training phase. This phenomenon of a full VQ not being robust to a variety of test inputs has been mentioned earlier and has been encountered in our discussion on LSF coding. The use of different training and testing conditions degrades performance since the components of the cepstrum vectors (such as LSFs) tend to migrate. Unlike LSF coding, appending the training set with a uniformly distributed set of vectors to accomplish coverage of a large space will not work as there will be much overlap among the codebooks of different speakers. The focus of the research is to develop more robust features that show little variation as the speech material changes [22,23].

6.7 Summary

This chapter has presented a tutorial description of quantization. Starting from the basic definition and properties of vector and scalar quantization, design algorithms are described. Many practical aspects of design and implementation (such as distortion measure, memory, search complexity, and robustness) are discussed. These practical aspects are interrelated. Two important applications of vector quantization in speech processing are discussed in which these practical aspects play an important role.

References

1. Gray, R.M., Vector quantization, *IEEE Acoust. Speech Signal Process.*, 1, 4–29, Apr. 1984.
2. Makhoul, J., Roucos, S., and Gish, H., Vector quantization in speech coding, *Proc. IEEE*, 73: 1551–1588, Nov. 1985.
3. Gersho, A. and Gray, R.M., *Vector Quantization and Signal Compression*, Kluwer Academic Publishers, Norwell, MA, 1991.
4. Gersho, A., Asymptotically optimal block quantization, *IEEE Trans. Inf. Theory*, IT-25: 373–380, July 1979.
5. Jayant, N.S. and Noll, P., *Digital Coding of Waveforms, Principles and Applications to Speech and Video*, Prentice-Hall, Englewood Cliffs, NJ, 1984.
6. Max, J., Quantizing for minimum distortion, *IEEE Trans. Inf. Theory*, IT-6(2): 7–12, Mar. 1960.
7. Linde, Y., Buzo, A., and Gray, R.M., An algorithm for vector quantizer design, *IEEE Trans. Commun.*, COM-28: 84–95, Jan. 1980.
8. Rabiner, L.R. and Schafer, R.W., *Digital Processing of Speech Signals*, Prentice-Hall, Englewood Cliffs, NJ, 1978.
9. Atal, B.S., Predictive coding of speech at low bit rates, *IEEE Trans. Commun.*, COM-30: 600–614, Apr. 1982.

10. Itakura, F., Line spectrum representation of linear predictor coefficients of speech signals, *J. Acoust. Soc. Am.*, 57: S35(A), 1975.
11. Wakita, H., Linear prediction voice synthesizers: Line spectrum pairs (LSP) is the newest of several techniques, *Speech Technol.*, 17–22, Fall 1981.
12. Soong, F.K. and Juang, B.H., Line spectrum pair (LSP) and speech data compression, *IEEE International Conference on Acoustics, Speech and Signal Processing*, San Diego, CA, Mar. 1984, pp. 1.10.1–1.10.4.
13. Paliwal, K.K. and Atal, B.S., Efficient vector quantization of LPC parameters at 24 bits/frame, *IEEE Trans. Speech Audio Process.*, 1: 3–14, Jan. 1993.
14. Laroia, R., Phamdo, N., and Farvardin, N., Robust and efficient quantization of speech LSP parameters using structured vector quantizers, *IEEE International Conference on Acoustics, Speech and Signal Processing*, Toronto, Canada, May 1991, pp. 641–644.
15. LeBlanc, W.P., Cuperman, V., Bhattacharya, B., and Mahmoud, S.A., Efficient search and design procedures for robust multi-stage VQ of LPC parameters for 4 kb/s speech coding, *IEEE Trans. Speech Audio Process.*, 1: 373–385, Oct. 1993.
16. Ramachandran, R.P., Sondhi, M.M., Seshadri, N., and Atal, B.S., A two codebook format for robust quantization of line spectral frequencies, *IEEE Trans. Speech Audio Process.*, 3: 157–168, May 1995.
17. Doddington, G.R., Speaker recognition—identifying people by their voices, *Proc. IEEE*, 73: 1651–1664, Nov. 1985.
18. Furui, S., Cepstral analysis technique for automatic speaker verification, *IEEE Trans. Acoust. Speech Signal Process.*, ASSP-29: 254–272, Apr. 1981.
19. Rabiner, L.R. and Juang, B.-H., *Fundamentals of Speech Recognition*, Prentice-Hall, Englewood Cliffs, NJ, 1993.
20. Rosenberg, A.E. and Soong, F.K., Evaluation of a vector quantization talker recognition system in text independent and text dependent modes, *Comput. Speech Lang.*, 22: 143–157, 1987.
21. Farrell, K.R., Mammone, R.J., and Assaleh, K.T., Speaker recognition using neural networks versus conventional classifiers, *IEEE Trans. Speech Audio Process.*, 2: 194–205, Jan. 1994.
22. Assaleh, K.T. and Mammone, R.J., New LP-derived features for speaker identification, *IEEE Trans. Speech Audio Process.*, 2: 630–638, Oct. 1994.
23. Zilovic, M.S., Ramachandran, R.P., and Mammone, R.J., Speaker identification based on the use of robust cepstral features derived from pole-zero transfer functions, *IEEE Trans. Speech Audio Process.*, 6(3): 260–267, May 1998.

III

Fast Algorithms and Structures

Pierre Duhamel
CNRS

THE FIELD OF DIGITAL SIGNAL PROCESSING grew rapidly and achieved its current prominence primarily through the discovery of efficient algorithms for computing various transforms (mainly the Fourier transforms) in the 1970s. In addition to fast Fourier transforms, discrete cosine transforms have also gained importance owing to their performance being very close to the statistically optimum Karhunen Loeve transform.

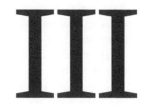

Transforms, convolutions, and matrix-vector operations form the basic tools utilized by the signal processing community, and this section reviews and presents the state of art in these areas of increasing importance.

Chapter 7 presents a thorough discussion of this important transform. Chapter 8 presents an excellent survey of filtering and convolution techniques.

One approach to understanding the time and space complexities of signal processing algorithms is through the use of quantitative complexity theory, and Feig's Chapter 9 applies quantitative measures to the computation of transforms. Finally, Chapter 10 presents a comprehensive discussion of matrix computations in signal processing.

7

Fast Fourier Transforms: A Tutorial Review and State of the Art*

Pierre Duhamel
CNRS

Martin Vetterli
École Polytechnique

The publication of the Cooley–Tukey fast Fourier transform (CTFFT) algorithm in 1965 has opened a new area in digital signal processing by reducing the order of complexity of some crucial computational tasks such as Fourier transform and convolution from N^2 to $N \log_2 N$, where N is the problem size. The

* Reprinted from *Signal Processing*, 19, 259–299, 1990 with kind permission from Elsevier Science-NL, Sara Burger-Hartstraat 25, 1055 KV Amsterdam, the Netherlands.

development of the major algorithms (CTFFT, split-radix fast Fourier transform [SRFFT], prime factor algorithm [PFA], and Winograd fast Fourier transform [FFT]) is reviewed. Then, an attempt is made to indicate the state of the art on the subject, showing the standing of research, open problems, and implementations.

7.1 Introduction

Linear filtering and Fourier transforms are among the most fundamental operations in digital signal processing. However, their wide use makes their computational requirements a heavy burden in most applications. Direct computation of both convolution and discrete Fourier transform (DFT) requires on the order of N^2 operations where N is the filter length or the transform size. The breakthrough of the CTFFT comes from the fact that it brings the complexity down to an order of $N \log_2 N$ operations. Because of the convolution property of the DFT, this result applies to the convolution as well. Therefore, FFT algorithms have played a key role in the widespread use of digital signal processing in a variety of applications such as telecommunications, medical electronics, seismic processing, radar or radio astronomy to name but a few.

Among the numerous further developments that followed Cooley and Tukey's original contribution, the FFT introduced in 1976 by Winograd [54] stands out for achieving a new theoretical reduction in the order of the multiplicative complexity. Interestingly, the Winograd algorithm uses convolutions to compute DFTs, an approach which is just the converse of the conventional method of computing convolutions by means of DFTs. What might look like a paradox at first sight actually shows the deep interrelationship that exists between convolutions and Fourier transforms.

Recently, the Cooley–Tukey type algorithms have emerged again, not only because implementations of the Winograd algorithm have been disappointing, but also due to some recent developments leading to the so-called split-radix algorithm [27]. Attractive features of this algorithm are both its low arithmetic complexity and its relatively simple structure.

Both the introduction of digital signal processors and the availability of large scale integration has influenced algorithm design. While in the 1960s and early 1970s, multiplication counts alone were taken into account, it is now understood that the number of addition and memory accesses in software and the communication costs in hardware are at least as important.

The purpose of this chapter is first to look back at 20 years of developments since the Cooley–Tukey paper. Among the abundance of literature (a bibliography of more than 2500 titles has been published [33]), we will try to highlight only the key ideas. Then, we will attempt to describe the state of the art on the subject. It seems to be an appropriate time to do so, since on the one hand, the algorithms have now reached a certain maturity, and on the other hand, theoretical results on complexity allow us to evaluate how far we are from optimum solutions. Furthermore, on some issues, open questions will be indicated.

Let us point out that in this chapter we shall concentrate strictly on the computation of the DFT, and not discuss applications. However, the tools that will be developed may be useful in other cases. For example, the polynomial products explained in Section 7.5.1 can immediately be applied to the derivation of fast running FIR algorithms [73,81].

The chapter is organized as follows.

Section 7.2 presents the history of the ideas on FFTs, from Gauss to the split-radix algorithm.

Section 7.3 shows the basic technique that underlies all algorithms, namely the divide and conquer approach, showing that it always improves the performance of a Fourier transform algorithm.

Section 7.4 considers Fourier transforms with twiddle factors, that is, the classic Cooley–Tukey type schemes and the split-radix algorithm. These twiddle factors are unavoidable when the transform length is composite with non-coprime factors. When the factors are coprime, the divide and conquer scheme can be made such that twiddle factors do not appear.

This is the basis of Section 7.5, which then presents Rader's algorithm for Fourier transforms of prime lengths, and Winograd's method for computing convolutions. With these results established, Section 7.5 proceeds to describe both the PFA and the Winograd Fourier transform algorithm (WFTA).

Section 7.6 presents a comprehensive and critical survey of the body of algorithms introduced thus far, then shows the theoretical limits of the complexity of Fourier transforms, thus indicating the gaps that are left between theory and practical algorithms.

Structural issues of various FFT algorithms are discussed in Section 7.7.

Section 7.8 treats some other cases of interest, like transforms on special sequences (real or symmetric) and related transforms, while Section 7.9 is specifically devoted to the treatment of multidimensional transforms.

Finally, Section 7.10 outlines some of the important issues of implementations. Considerations on software for general purpose computers, digital signal processors, and vector processors are made. Then, hardware implementations are addressed. Some of the open questions when implementing FFT algorithms are indicated.

The presentation we have chosen here is constructive, with the aim of motivating the "tricks" that are used. Sometimes, a shorter but "plug-in" like presentation could have been chosen, but we avoided it because we desired to insist on the mechanisms underlying all these algorithms. We have also chosen to avoid the use of some mathematical tools, such as tensor products (that are very useful when deriving some of the FFT algorithms) in order to be more widely readable.

Note that concerning arithmetic complexities, all sections will refer to synthetic tables giving the computational complexities of the various algorithms for which software is available. In a few cases, slightly better figures can be obtained, and this will be indicated.

For more convenience, the references are separated between books and papers, the latter being further classified corresponding to subject matters (one-dimensional [1-D] FFT algorithms, related ones, multidimensional transforms and implementations).

7.2 A Historical Perspective

The development of the FFT will be surveyed below because, on the one hand, its history abounds in interesting events, and on the other hand, the important steps correspond to parts of algorithms that will be detailed later.

A first subsection describes the pre-Cooley–Tukey area, recalling that algorithms can get lost by lack of use, or, more precisely, when they come too early to be of immediate practical use. The developments following the Cooley–Tukey algorithm are then described up to the most recent solutions. Another subsection is concerned with the steps that lead to the WFTA and to the PFA, and finally, an attempt is made to briefly describe the current state of the art.

7.2.1 From Gauss to the CTFFT

While the publication of a fast algorithm for the DFT by Cooley and Tukey [25] in 1965 is certainly a turning point in the literature on the subject, the divide and conquer approach itself dates back to Gauss as noted in a well-documented analysis by Heideman et al. [34]. Nevertheless, Gauss's work on FFTs in the early nineteenth century (around 1805) remained largely unnoticed because it was only published in Latin and this after his death.

Gauss used the divide and conquer approach in the same way as Cooley and Tukey have published it later in order to evaluate trigonometric series, but his work predates even Fourier's work on harmonic analysis (1807)! Note that his algorithm is quite general, since it is explained for transforms on sequences with lengths equal to any composite integer.

During the nineteenth century, efficient methods for evaluating Fourier series appeared independently at least three times [33], but were restricted on lengths and number of resulting points. In 1903, Runge derived an algorithm for lengths equal to powers of 2 which was generalized to powers of 3 as well and used in the 1940s. Runge's work was thus quite well known, but nevertheless disappeared after the war.

Another important result useful in the most recent FFT algorithms is another type of divide and conquer approach, where the initial problem of length $N_1 \cdot N_2$ is divided into subproblems of lengths N_1 and N_2 without any additional operations, N_1 and N_2 being coprime.

This result dates back to the work of Good [32] who obtained this result by simple index mappings. Nevertheless, the full implication of this result will only appear later, when efficient methods will be derived for the evaluation of small, prime length DFTs. This mapping itself can be seen as an application of the Chinese remainder theorem (CRT), which dates back to 100 years AD! [10–18].

Then, in 1965, appeared a brief article by Cooley and Tukey, entitled "An algorithm for the machine calculation of complex Fourier series" [25], which reduces the order of the number of operations from N^2 to $N \log_2(N)$ for a length $N = 2^n$ DFT.

This turned out to be a milestone in the literature on fast transforms, and was credited [14,15] with the tremendous increase of interest in digital signal processor (DSP) beginning in the 1970s. The algorithm is suited for DFTs on any composite length, and is thus of the type that Gauss had derived almost 150 years before. Note that all algorithms published in-between were more restrictive on the transform length [34].

Looking back at this brief history, one may wonder why all previous algorithms had disappeared or remained unnoticed, whereas the Cooley–Tukey algorithm had such a tremendous success. A possible explanation is that the growing interest in the theoretical aspects of digital signal processing was motivated by technical improvements in semiconductor technology. And, of course, this was not a one-way street.

The availability of reasonable computing power produced a situation where such an algorithm would suddenly allow numerous new applications. Considering this history, one may wonder how many other algorithms or ideas are just sleeping in some notebook or obscure publication.

The two types of divide and conquer approaches cited above produced two main classes of algorithms. For the sake of clarity, we will now skip the chronological order and consider the evolution of each class separately.

7.2.2 Development of the Twiddle Factor FFT

When the initial DFT is divided into sublengths which are not coprime, the divide and conquer approach as proposed by Cooley and Tukey leads to auxiliary complex multiplications, initially named twiddle factors, which cannot be avoided in this case.

While Cooley–Tukey's algorithm is suited for any composite length, and explained in [25] in a general form, the authors gave an example with $N = 2^n$, thus deriving what is now called a radix-2 decimation in time (DIT) algorithm (the input sequence is divided into decimated subsequences having different phases). Later, it was often falsely assumed that the initial CTFFT was a DIT radix-2 algorithm only.

A number of subsequent papers presented refinements of the original algorithm, with the aim of increasing its usefulness.

The following refinements were concerned:

- With the structure of the algorithm: it was emphasized that a dual approach leads to "decimation in frequency" (DIF) algorithms.
- With the efficiency of the algorithm, measured in terms of arithmetic operations: Bergland showed that higher radices, for example radix-8, could be more efficient [21].
- With the extension of the applicability of the algorithm: Bergland [60], again, showed that the FFT could be specialized to real input data, and Singleton gave a mixed radix FFT suitable for arbitrary composite lengths.

While these contributions all improved the initial algorithm in some sense (fewer operations and/or easier implementations), actually no new idea was suggested.

Interestingly, in these very early papers, all the concerns guiding the recent work were already here: arithmetic complexity, but also different structures and even real-data algorithms.

In 1968, Yavne [58] presented a little-known paper that sets a record: his algorithm requires the least known number of multiplications, as well as additions for length-2^n FFTs, and this both for real and complex input data. Note that this record still holds, at least for practical algorithms. The same number of operations was obtained later on by other (simpler) algorithms, but due to Yavne's cryptic style, few researchers were able to use his ideas at the time of publication.

Since twiddle factors lead to most computations in classical FFTs, Rader and Brenner [44], perhaps motivated by the appearance of the Winograd Fourier transform which possesses the same characteristic, proposed an algorithm that replaces all complex multiplications by either real or imaginary ones, thus substantially reducing the number of multiplications required by the algorithm. This reduction in the number of multiplications was obtained at the cost of an increase in the number of additions, and a greater sensitivity to roundoff noise. Hence, further developments of these "real factor" FFTs appeared in [24,42], reducing these problems. Bruun [22] also proposed an original scheme particularly suited for real data. Note that these various schemes only work for radix-2 approaches.

It took more than 15 years to see again algorithms for length-2^n FFTs that take as few operations as Yavne's algorithm. In 1984, four papers appeared or were submitted almost simultaneously [27,40,46,51] and presented so-called "split-radix" algorithms. The basic idea is simply to use a different radix for the even part of the transform (radix-2) and for the odd part (radix-4). The resulting algorithms have a relatively simple structure and are well adapted to real and symmetric data while achieving the minimum known number of operations for FFTs on power of 2 lengths.

7.2.3 FFTs without Twiddle Factors

While the divide and conquer approach used in the Cooley–Tukey algorithm can be understood as a "false" mono- to multidimensional mapping (this will be detailed later), Good's mapping, which can be used when the factors of the transform lengths are coprime, is a true mono- to multidimensional mapping, thus having the advantage of not producing any twiddle factor.

Its drawback, at first sight, is that it requires efficiently computable DFTs on lengths that are coprime: For example, a DFT of length 240 will be decomposed as $240 = 16 \cdot 3 \cdot 5$, and a DFT of length 1008 will be decomposed in a number of DFTs of lengths 16, 9, and 7. This method thus requires a set of (relatively) small-length DFTs that seemed at first difficult to compute in less than N_i^2 operations. In 1968, however, Rader [43] showed how to map a DFT of length N, N prime, into a circular convolution of length $N - 1$. However, the whole material to establish the new algorithms was not ready yet, and it took Winograd's work on complexity theory, in particular on the number of multiplications required for computing polynomial products or convolutions [55] in order to use Good's and Rader's results efficiently.

All these results were considered as curiosities when they were first published, but their combination, first done by Winograd and then by Kolba and Parks [39] raised a lot of interest in that class of algorithms. Their overall organization is as follows.

After mapping the DFT into a true multidimensional DFT by Good's method and using the fast convolution schemes in order to evaluate the prime length DFTs, a first algorithm makes use of the intimate structure of these convolution schemes to obtain a nesting of the various multiplications. This algorithm is known as the Winograd Fourier transform algorithm [54], an algorithm requiring the least known number of multiplications among practical algorithms for moderate lengths DFTs. If the nesting is not used, and the multidimensional DFT is performed by the row–column method, the resulting algorithm is known as the prime factor algorithm [39], which, while using more multiplications, has less additions and a better structure than the WFTA.

From the above explanations, one can see that these two algorithms, introduced in 1976 and 1977, respectively, require more mathematics to be understood [19]. This is why it took some effort to translate the theoretical results, especially concerning the WFTA, into actual computer code.

It is even our opinion that what will remain mostly of the WFTA are the theoretical results, since although a beautiful result in complexity theory, the WFTA did not meet its expectations once implemented, thus leading to a more critical evaluation of what "complexity" meant in the context of real life computers [41,108,109].

The result of this new look at complexity was an evaluation of the number of additions and data transfers as well (and no longer only of multiplications). Furthermore, it turned out recently that the theoretical knowledge brought by these approaches could give a new understanding of FFTs with twiddle factors as well.

7.2.4 Multidimensional DFTs

Due to the large amount of computations they require, the multidimensional DFTs as such (with common factors in the different dimensions, which was not the case in the multidimensional translation of a mono-dimensional problem by PFA) were also carefully considered.

The two most interesting approaches are certainly the vector radix FFT (a direct approach to the multidimensional problem in a Cooley–Tukey mood) proposed in 1975 by Rivard [91] and the polynomial transform solution of Nussbaumer and Quandalle [87,88] in 1978.

Both algorithms substantially reduce the complexity over traditional row-column computational schemes.

7.2.5 State of the Art

From a theoretical point of view, the complexity issue of the DFT has reached a certain maturity. Note that Gauss, in his time, did not even count the number of operations necessary in his algorithm. In particular, Winograd's work on DFTs whose lengths have coprime factors both sets lower bounds (on the number of multiplications) and gives algorithms to achieve these [35,55], although they are not always practical ones. Similar work was done for length-2^n DFTs, showing the linear multiplicative complexity of the algorithm [28,35,105] but also the lack of practical algorithms achieving this minimum (due to the tremendous increase in the number of additions [35]).

Considering implementations, the situation is of course more involved since many more parameters have to be taken into account than just the number of operations.

Nevertheless, it seems that both the radix-4 and the split-radix algorithm are quite popular for lengths which are powers of 2, while the PFA, thanks to its better structure and easier implementation, wins over the WFTA for lengths having coprime factors.

Recently, however, new questions have come up because in software on the one hand, new processors may require different solutions (vector processors, signal processors), and on the other hand, the advent of VLSI for hardware implementations sets new constraints (desire for simple structures, high cost of multiplications vs. additions).

7.3 Motivation (or Why Dividing Is Also Conquering)

This section is devoted to the method that underlies all fast algorithms for DFT, that is, the "divide and conquer" approach.

The DFT is basically a matrix-vector product. Calling $(x_0, x_1, \ldots, x_{N-1})^T$ the vector of the input samples,

$$(X_0, X_1, \ldots, X_{N-1})^T$$

the vector of transform values, and W_N the primitive N th root of unity ($W_N = e^{-j2\pi/N}$), the DFT can be written as

$$\begin{bmatrix} X_0(1) \\ X_1(2) \\ X_2(3) \\ \vdots \\ X_{N-1} \end{bmatrix} = \begin{bmatrix} 1 & 1 & 1 & 1 & \cdots & 1 \\ 1 & W_N & W_N^2 & W_N^3 & \cdots & W_N^{N-1} \\ 1 & W_N^2 & W_N^4 & W_N^6 & \cdots & W_N^{2(N-1)} \\ \vdots & \vdots & \vdots & \vdots & \vdots & \vdots \\ 1 & W_N^{N-1} & W_N^{2(N-1)} & \cdots & \cdots & W_N^{(N-1)(N-1)} \end{bmatrix} \begin{bmatrix} x_0 \\ x_1 \\ x_2 \\ x_3 \\ \vdots \\ x_{N-1} \end{bmatrix}. \tag{7.1}$$

The direct evaluation of the matrix-vector product in Equation 7.1 requires of the order of N^2 complex multiplications and additions (we assume here that all signals are complex for simplicity).

The idea of the "divide and conquer" approach is to map the original problem into several subproblems in such a way that the following inequality is satisfied:

$$\sum \text{cost(subproblems)} + \text{cost(mapping)} < \text{cost(original problem)}. \tag{7.2}$$

But the real power of the method is that, often, the division can be applied recursively to the subproblems as well, thus leading to a reduction of the order of complexity.

Specifically, let us have a careful look at the DFT transform in Equation 7.3 and its relationship with the z-transform of the sequence $\{x_n\}$ as given in Equation 7.4.

$$X_k = \sum_{i=0}^{N-1} x_i W_N^{ik}, \quad k = 0, \ldots, N-1, \tag{7.3}$$

$$x(z) = \sum_{i=0}^{N-1} x_i z^{-i}. \tag{7.4}$$

$\{X_k\}$ and $\{x_i\}$ form a transform pair, and it is easily seen that X_k is the evaluation of $X(z)$ at point $z = W_N^{-k}$:

$$X_k = X(z)_{z=W_N^{-k}}. \tag{7.5}$$

Furthermore, due to the sampled nature of $\{x_n\}$, $\{X_k\}$ is periodic, and vice versa: since $\{X_k\}$ is sampled, $\{x_n\}$ must also be periodic.

From a physical point of view, this means that both sequences $\{x_n\}$ and $\{X_k\}$ are repeated indefinitely with period N. This has a number of consequences as far as fast algorithms are concerned.

All fast algorithms are based on a divide and conquer strategy; we have seen this in Section 7.2. But how shall we divide the problem (with the purpose of conquering it)?

The most natural way is, of course, to consider subsets of the initial sequence, take the DFT of these subsequences, and reconstruct the DFT of the initial sequence from these intermediate results.

Let $\boldsymbol{I}_l$, $l = 0, \ldots, r-1$ be the partition of $\{0, 1, \ldots, N-1\}$ defining the r different subsets of the input sequence. Equation 7.4 can now be rewritten as

$$X(z) = \sum_{i=0}^{N-1} x_i z^{-i} = \sum_{l=0}^{r-1} \sum_{i \in \boldsymbol{I}_l} x_i z^{-i}, \tag{7.6}$$

and, normalizing the powers of z with respect to some x_{0l} in each subset $\boldsymbol{I}_l$,

$$X(z) = \sum_{l=0}^{r-1} z^{-i_{0l}} \sum_{i \in \boldsymbol{I}_l} x_i z^{-i+i_{0l}}. \tag{7.7}$$

From the considerations above, we want the replacement of z by W_N^{-k} in the innermost sum of Equation 7.7 to define an element of the DFT of $\{x_i | i \in \boldsymbol{I}_l\}$. Of course, this will be possible only if the subset $\{x_i | i \in \boldsymbol{I}_l\}$, possibly permuted, has been chosen in such a way that it has the same kind of periodicity as the initial sequence. In what follows, we show that the three main classes of FFT algorithms can all be casted into the form given by Equation 7.7.

- In some cases, the second sum will also involve elements having the same periodicity, hence will define DFTs as well. This corresponds to the case of Good's mapping: all the subsets $\boldsymbol{I}_l$, have the same number of elements $m = N/r$ and $(m, r) = 1$.
- If this is not the case, Equation 7.7 will define one step of an FFT with twiddle factors: when the subsets $^{\boldsymbol{I}_l}$ all have the same number of elements, Equation 7.7 defines one step of a radix-r FFT.
- If $r = 3$, one of the subsets having $N/2$ elements, and the other ones having $N/4$ elements, Equation 7.7 is the basis of a split-radix algorithm.

Furthermore, it is already possible to show from Equation 7.7 that the divide and conquer approach will always improve the efficiency of the computation.

To make this evaluation easier, let us suppose that all subsets $\boldsymbol{I}_l$ have the same number of elements, say N_1. If $N = N_1 \cdot N_2, r = N_2$, each of the innermost sums of Equation 7.7 can be computed with N_1^2 multiplications, which gives a total of $N_2 N_1^2$, when taking into account the requirement that the sum over $i \in \boldsymbol{I}_l$ defines a DFT. The outer sum will need $r = N_2$ multiplications per output point, that is, $N_2 \cdot N$ for the whole sum.

Hence, the total number of multiplications needed to compute Equation 7.7 is

$$N_2 \cdot N + N_2 \cdot N_1^2 = N_1 \cdot N_2(N_1 + N_2) < N_1^2 \cdot N_2^2 \quad \text{if } N_1, N_2 > 2, \tag{7.8}$$

which shows clearly that the divide and conquer approach, as given in Equation 7.7, has reduced the number of multiplications needed to compute the DFT.

Of course, when taking into account that, even if the outermost sum of Equation 7.7 is not already in the form of a DFT, it can be rearranged into a DFT plus some so-called twiddle-factors, this mapping is always even more favorable than is shown by Equation 7.8, especially for small N_1, N_2 (e.g., the length-2 DFT is simply a sum and difference).

Obviously, if N is highly composite, the division can be applied again to the subproblems, which results in a number of operations generally several orders of magnitude better than the direct matrix-vector product.

The important point in Equation 7.2 is that two costs appear explicitly in the divide and conquer scheme: the cost of the mapping (which can be zero when looking at the number of operations only) and the cost of the subproblems. Thus, different types of divide and conquer methods attempt to find various balancing schemes between the mapping and the subproblem costs. In the radix-2 algorithm, for example, the subproblems end up being quite trivial (only sum and differences), while the mapping requires twiddle factors that lead to a large number of multiplications. On the contrary, in the PFA, the mapping requires no arithmetic operation (only permutations), while the small DFTs that appear as subproblems will lead to substantial costs since their lengths are coprime.

7.4 FFTs with Twiddle Factors

The divide and conquer approach reintroduced by Cooley and Tukey [25] can be used for any composite length N but has the specificity of always introducing twiddle factors. It turns out that when the factors of N are not coprime (e.g., if $N=2^n$), these twiddle factors cannot be avoided at all. This section will be devoted to the different algorithms in that class.

The difference between the various algorithms will consist in the fact that more or fewer of these twiddle factors will turn out to be trivial multiplications, such as $1, -1, j$, and $-j$.

7.4.1 The Cooley–Tukey Mapping

Let us assume that the length of the transform is composite: $N=N_1 \cdot N_2$.

As we have seen in Section 7.3, we want to partition $\{x_i \mid i=0,\ldots,N-1\}$ into different subsets $\{x_i \mid i \in I_l\}$ in such a way that the periodicities of the involved subsequences are compatible with the periodicity of the input sequence, on the one hand, and allow to define DFTs of reduced lengths on the other hand.

Hence, it is natural to consider decimated versions of the initial sequence:

$$I_{n_1} = \{n_2N_1 + n_1\}, \quad n_1 = 0,\ldots,N_1-1, \quad n_2 = 0,\ldots,N_2-1, \tag{7.9}$$

which, introduced in Equation 7.6, gives

$$X(z) = \sum_{n_1=0}^{N_1-1}\sum_{n_2=0}^{N_2-1} x_{n_2N_1+n_1} z^{-(n_2N_1+n_1)}, \tag{7.10}$$

and, after normalizing with respect to the first element of each subset,

$$\begin{aligned} X(z) &= \sum_{n_1=0}^{N_1-1} z^{-n_1} \sum_{n_2=0}^{N_2-1} x_{n_2N_1+n_1} z^{-n_2N_1}, \\ X_k &= X(z)\big|_{z=W_N^{-k}} \\ &= \sum_{n_1=0}^{N_1-1} W_N^{n_1k} \sum_{n_2=0}^{N_2-1} x_{n_2N_1+n_1} W_N^{n_2N_1k}. \end{aligned} \tag{7.11}$$

Using the fact that

$$W_N^{iN_1} = e^{-j2\pi N_1 i/N} = e^{-j2\pi/N_2} = W_{N_2}^i, \tag{7.12}$$

Equation 7.11 can be rewritten as

$$X_k = \sum_{n_1=0}^{N_1-1} W_N^{n_1k} \sum_{n_2=0}^{N_2-1} x_{n_2N_1+n_1} W_{N_2}^{n_2k}. \tag{7.13}$$

Equation 7.13 is now nearly in its final form, since the right-hand sum corresponds to N_1 DFTs of length N_2, which allows the reduction of arithmetic complexity to be achieved by reiterating the process. Nevertheless, the structure of the CTFFT is not fully given yet.

Call $Y_{n_1,k}$ the kth output of the n_1th such DFT:

$$Y_{n_1,k} = \sum_{n_2=0}^{N_2-1} x_{n_2N_1+n_1} W_{N_2}^{n_2k}. \tag{7.14}$$

Note that in $Y_{n_1,k}$ can be taken modulo N_2, because

$$W_{N_2}^{k} = W_{N_2}^{N_2+k'} = W_{N_2}^{N_2} \cdot W_{N_2}^{k'} = W_{N_2}^{k'}. \tag{7.15}$$

With this notation, X_k becomes

$$X_k = \sum_{n_1=0}^{N_1-1} Y_{n_1,k} W_N^{n_1k}. \tag{7.16}$$

At this point, we can notice that all the X_k for k's being congruent modulo N_2 are obtained from the same group of N_1 outputs of $Y_{n_1,k}$. Thus, we express k as

$$k = k_1N_2 + k_2, \quad k_1 = 0, \ldots, N_1 - 1, \quad k_2 = 0, \ldots, N_2 - 1. \tag{7.17}$$

Obviously, $Y_{n_1,k}$ is equal to Y_{n_1,k_2} since k can be taken modulo N_2 in this case (see Equations 7.12 and 7.15). Thus, we rewrite Equation 7.16 as

$$X_{k_1N_2+k_2} = \sum_{n_1=0}^{N_1-1} Y_{n_1,k_2} W_N^{n_1(k_1N_2+k_2)}, \tag{7.18}$$

which can be reduced, using Equation 7.12, to

$$X_{k_1N_2+k_2} = \sum_{n_1=0}^{N_1-1} Y_{n_1,k_2} W_N^{n_1k_2} W_{N_1}^{n_1k_1}. \tag{7.19}$$

Calling Y'_{n_1,k_2} the result of the first multiplication (by the twiddle factors) in Equation 7.19, we get

$$Y'_{n_1,k_2} = Y_{n_1,k_2} W_N^{n_1k_2}. \tag{7.20}$$

We see that the values of $X_{k_1N_2+k_2}$ are obtained from N_2 DFTs of length N_1 applied on Y'_{n_1,k_2}:

$$X_{k_1N_2+k_2} = \sum_{n_1=0}^{N_1-1} Y'_{n_1,k_2} W_{N_1}^{n_1k_1}. \tag{7.21}$$

We recapitulate the important steps that led to Equation 7.21. First, we evaluated N_1 DFTs of length N_2 in Equation 7.14. Then, N multiplications by the twiddle factors were performed in Equation 7.20. Finally, N_2 DFTs of length N_1 led to the final result (Equation 7.21).

A way of looking at the change of variables performed in Equations 7.9 and 7.17 is to say that the 1-D vector x_i has been mapped into a 2-D vector x_{n_1,n_2} having N_1 lines and N_2 columns. The computation of the DFT is then divided into N_1 DFTs on the lines of the vector x_{n_1,n_2}, a point by point multiplication with the twiddle factors and finally N_2 DFTs on the columns of the preceding result.

Until recently, this was the usual presentation of FFT algorithms, by the so-called "index mappings" [4,23]. In fact, Equations 7.9 and 7.17, taken together, are often referred to as the "Cooley–Tukey mapping" or "common factor mapping." However, the problem with the 2-D interpretation is that it does not include all algorithms (like the split-radix algorithm that will be seen later). Thus, while this interpretation helps the understanding of some of the algorithms, it hinders the comprehension of others. In our presentation, we tried to enhance the role of the periodicities of the problem, which result from the initial choice of the subsets.

Nevertheless, we illustrate pictorially a length-15 DFT using the 2-D view with $N_1=3$ and $N_2=5$ (see Figure 7.1), together with the Cooley–Tukey mapping in Figure 7.2, to allow a precise comparison

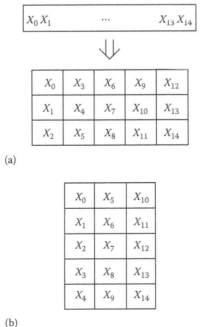

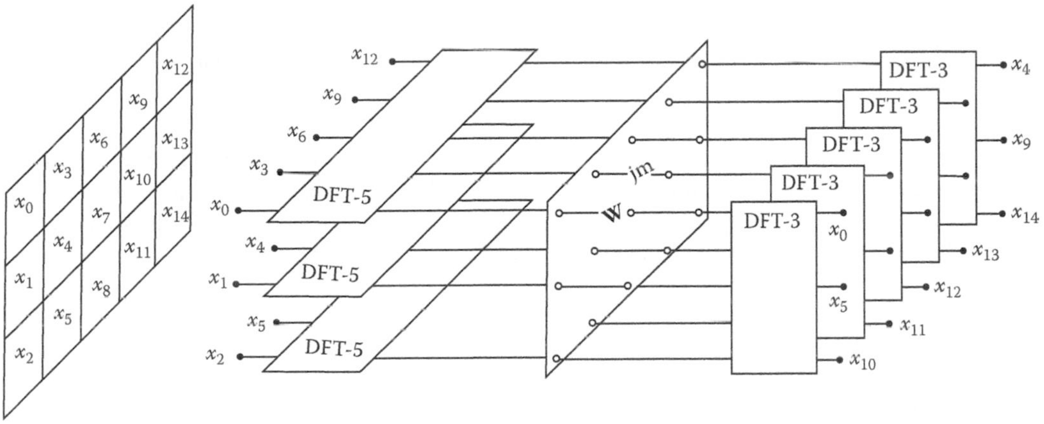

FIGURE 7.1 2-D view of the length-15 CTFFT.

X_0 X_1	⋯	X_{13} X_{14}

⇓

X_0	X_3	X_6	X_9	X_{12}
X_1	X_4	X_7	X_{10}	X_{13}
X_2	X_5	X_8	X_{11}	X_{14}

(a)

X_0	X_5	X_{10}
X_1	X_6	X_{11}
X_2	X_7	X_{12}
X_3	X_8	X_{13}
X_4	X_9	X_{14}

(b)

FIGURE 7.2 Cooley–Tukey mapping: (a) $N_1=3$, $N_2=5$ and (b) $N_1=5$, $N_2=3$.

with Good's mapping that leads to the other class of FFTs: the FFTs without twiddle factors. Note that for the case where N_1 and N_2 are coprime, the Good's mapping will be more efficient as shown in the next section, and thus this example is for illustration and comparison purpose only. Because of the twiddle factors in Equation 7.20, one cannot interchange the order of DFTs once the input mapping has been chosen. Thus, in Figure 7.2a, one has to begin with the DFTs on the rows of the matrix. Choosing $N_1 = 5$ and $N_2 = 3$ would lead to the matrix of Figure 7.2b, which is obviously different from just transposing the matrix of Figure 7.2a. This shows again that the mapping does not lead to a true 2-D transform (in that case, the order of row and column would not have any importance).

7.4.2 Radix-2 and Radix-4 Algorithms

The algorithms suited for lengths equal to powers of 2 (or 4) are quite popular since sequences of such lengths are frequent in signal processing (they make full use of the addressing capabilities of computers or DSP systems).

We assume first that $N = 2^n$. Choosing $N_1 = 2$ and $N_2 = 2^{n-1} = N/2$ in Equations 7.9 and 7.10 divides the input sequence into the sequence of even- and odd-numbered samples, which is the reason why this approach is called "decimation in time". Both sequences are decimated versions, with different phases, of the original sequence. Following Equation 7.17, the output consists of $N/2$ blocks of 2 values. Actually, in this simple case, it is easy to rewrite Equations 7.14 and 7.21 exhaustively:

$$x_{K_2} = \sum_{n_2=0}^{N/2-1} x_{2n_2} W_{N/2}^{n_2k_2} + W_N^{k_2} \sum_{n_2=0}^{N/2-1} x_{2n_2+1} W_{N/2}^{n_2k_2}, \tag{7.22a}$$

$$X_{N/2+k_2} = \sum_{n_2=0}^{N/2-1} x_{2n_2} W_{N/2}^{n_2k_2} - W_N^{k_2} \sum_{n_2=0}^{N/2-1} x_{2n_2+1} W_{N/2}^{n_2k_2}. \tag{7.22b}$$

Thus, X_m and $X_{N/2+m}$ are obtained by 2-point DFTs on the outputs of the length-$N/2$ DFTs of the even- and odd-numbered sequences, one of which is weighted by twiddle factors. The structure made by a sum and difference followed (or preceded) by a twiddle factor is generally called a "butterfly." The DIT radix-2 algorithm is schematically shown in Figure 7.3.

Its implementation can now be done in several different ways. The most natural one is to reorder the input data such that the samples of which the DFT has to be taken lie in subsequent locations. This results in the bit-reversed input, in-order output DIT algorithm. Another possibility is to selectively compute the DFTs over the input sequence (taking only the even- and odd-numbered samples), and perform an in-place computation. The output will now be in bit-reversed order. Other implementation schemes can lead to constant permutations between the stages (constant geometry algorithm [15]).

If we reverse the role of N_1 and N_2, we get the DIF version of the algorithm. Inserting $N_1 = N/2$ and $N_2 = 2$ into Equation 7.9, Equation 7.10 leads to (again from Equations 7.14 and 7.21)

$$X_{2k_1} = \sum_{n_1=0}^{N/2-1} W_{N/2}^{n_1k_1} \left(x_{n_1} + x_{N/2+n_1}\right), \tag{7.23a}$$

$$X_{2k_1+1} = \sum_{n_1=0}^{N/2-1} W_{N/2}^{n_1k_1} W_N^{n_1} \left(x_{n_1} - x_{N/2+n_1}\right). \tag{7.23b}$$

This first step of a DIF algorithm is represented in Figure 7.5a, while a schematic representation of the full DIF algorithm is given in Figure 7.4. The duality between division in time and division in frequency is obvious, since one can be obtained from the other by interchanging the role of $\{x_i\}$ and $\{X_k\}$.

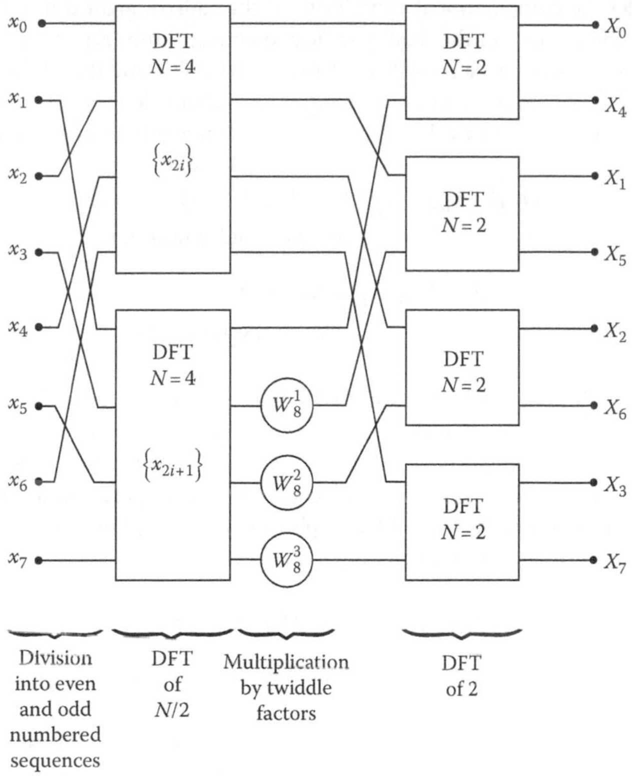

FIGURE 7.3 DIT radix-2 FFT.

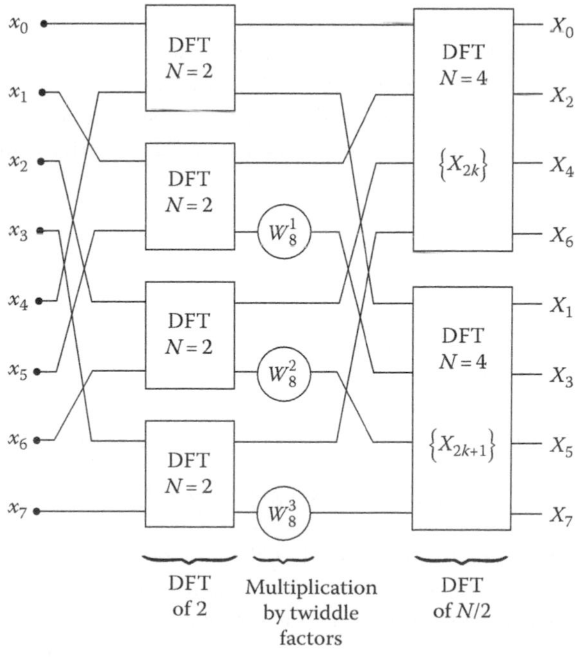

FIGURE 7.4 DIF radix-2 FFT.

Let us now consider the computational complexity of the radix-2 algorithm (which is the same for the DIF and DIT version because of the duality indicated above). From Equation 7.22 or 7.23, one sees that a DFT of length N has been replaced by two DFTs of length $N/2$, and this at the cost of $N/2$ complex multiplications as well as N complex additions. Iterating the scheme $\log_2 N - 1$ times in order to obtain trivial transforms (of length 2) leads to the following order of magnitude of the number of operations:

$$O_M[\text{DFT}_{\text{radix-2}}] \approx N/2(\log_2 N - 1) \quad \text{complex multiplications,} \tag{7.24a}$$

$$O_A[\text{DFT}_{\text{radix-2}}] \approx N(\log_2 N - 1) \quad \text{complex additions.} \tag{7.24b}$$

A closer look at the twiddle factors will enable us to still reduce these numbers. For comparison purposes, we will count the number of real operations that are required, provided that the multiplication of a complex number x by W_N^i is done using three real multiplications and three real additions [12]. Furthermore, if i is a multiple of $N/4$, no arithmetic operation is required, and only two real multiplications and additions are required if i is an odd multiple of $N/8$. Taking into account these simplifications results in the following total number of operations [12]:

$$M[\text{DFT}_{\text{radix-2}}] = 3N/2\log_2 N - 5N + 8, \tag{7.25a}$$

$$A[\text{DFT}_{\text{radix-2}}] = 7N/2\log_2 N - 5N + 8. \tag{7.25b}$$

Nevertheless, it should be noticed that these numbers are obtained by the implementation of four different butterflies (one general plus three special cases), which reduces the regularity of the programs. An evaluation of the number of real operations for other number of special butterflies is given in [4], together with the number of operations obtained with the usual 4-mult, 2-adds complex multiplication algorithm.

Another case of interest appears when N is a power of 4. Taking $N_1 = 4$ and $N_2 = N/4$, Equation 7.13 reduces the length-N DFT into 4 DFTs of length $N/4$, about $3N/4$ multiplications by twiddle factors, and $N/4$ DFTs of length 4. The interest of this case lies in the fact that the length-4 DFTs do not cost any multiplication (only 16 real additions). Since there are $\log_4 N - 1$ stages and the first set of twiddle factors (corresponding to $n_1 = 0$ in Equation 7.20) is trivial, the number of complex multiplications is about

$$O_M[\text{DFT}_{\text{radix-4}}] \approx 3N/4(\log_4 N - 1). \tag{7.26}$$

Comparing Equation 7.26 to Equation 7.24a shows that the number of multiplications can be reduced with this radix-4 approach by about a factor of 3/4. Actually, a detailed operation count using the simplifications indicated above gives the following result [12]:

$$M[\text{DFT}_{\text{radix-4}}] = 9N/8\log_2 N - 43N/12 + 16/3, \tag{7.27a}$$

$$A[\text{DFT}_{\text{radix-4}}] = 25N/8\log_2 N - 43N/12 + 16/3. \tag{7.27b}$$

Nevertheless, these operation counts are obtained at the cost of using six different butterflies in the programming of the FFT. Slight additional gains can be obtained when going to even higher radices (like 8 or 16) and using the best possible algorithms for the small DFTs. Since programs with a regular structure are generally more compact, one often uses recursively the same decomposition at each stage, thus leading to full radix-2 or radix-4 programs, but when the length is not a power of the radix (e.g., 128 for a radix-4 algorithm), one can use smaller radices towards the end of the decomposition. A length-256 DFT could use two stages of radix-8 decomposition, and finish with one stage of radix-4. This approach is called the "mixed-radix" approach [45] and achieves low arithmetic complexity while allowing flexible transform length (e.g., not restricted to powers of 2), at the cost of a more involved implementation.

7.4.3 Split-Radix Algorithm

As already noted in Section 7.2, the lowest known number of both multiplications and additions for length-2^n algorithms was obtained as early as 1968 and was again achieved recently by new algorithms. Their power was to show explicitly that the improvement over fixed- or mixed-radix algorithms can be obtained by using a radix-2 and a radix-4 simultaneously on different parts of the transform. This allowed the emergence of new compact and computationally efficient programs to compute the length-2^n DFT.

Below, we will try to motivate (*a posteriori*!) the split-radix approach and give the derivation of the algorithm as well as its computational complexity.

When looking at the DIF radix-2 algorithm given in Equation 7.23, one notices immediately that the even indexed outputs X_{2k_1} are obtained without any further multiplicative cost from the DFT of a length-$N/2$ sequence, which is not so well-done in the radix-4 algorithm for example, since relative to that length-$N/2$ sequence, the radix-4 behaves like a radix-2 algorithm. This lacks logical sense because it is well known that the radix-4 is better than the radix-2 approach.

From that observation, one can derive a first rule: the even samples of a DIF decomposition X_{2k} should be computed separately from the other ones, with the same algorithm (recursively) as the DFT of the original sequence (see [53] for more details).

However, as far as the odd indexed outputs X_{2k+1} are concerned, no general simple rule can be established, except that a radix-4 will be more efficient than a radix-2, since it allows computation of the samples through two $N/4$ DFTs instead of a single $N/2$ DFT for a radix-2, and this at the same multiplicative cost, which will allow the cost of the recursions to grow more slowly. Tests showed that computing the odd indexed output through radices higher than 4 was inefficient.

The first recursion of the corresponding "split-radix" algorithm (the radix is split in two parts) is obtained by modifying Equation 7.23 accordingly: X

$$X_{2k_1} = \sum_{n_1=0}^{N/2-1} W_{N/2}^{n_1k_1}\left(x_{n_1} + x_{N/2+n_1}\right), \tag{7.28a}$$

$$X_{4k_1+1} = \sum_{n_1=0}^{N/4-1} W_{N/4}^{n_1k_1} W_N^{n_1}\left[\left(x_{n_1} - x_{N/2+n_1}\right) + j\left(x_{n_1+N/4} - x_{n_1+3N/4}\right)\right], \tag{7.28b}$$

$$X_{4k_1+3} = \sum_{n_1=0}^{N/4-1} W_{N/4}^{n_1k_1} W_N^{3n}\left[\left(x_{n_1} + x_{N/2+n_1}\right) - j\left(x_{n_1+N/4} - x_{n_1+3N/4}\right)\right]. \tag{7.28c}$$

The above approach is a DIF SRFFT, and is compared in Figure 7.5 with the radix-2 and radix-4 algorithms. The corresponding DIT version, being dual, considers separately the subsets $\{x_{2i}\}$, $\{x_{4i+1}\}$, and $\{x_{4i+3}\}$ of the initial sequence.

Taking $\boldsymbol{I}_0 = \{2i\}$, $\boldsymbol{I}_1 = \{4i+1\}$, and $\boldsymbol{I}_2 = \{4i+3\}$ and normalizing with respect to the first element of the set in Equation 7.7 leads to

$$X_k = \sum_{I_0} x_{2i} W_N^{k(2i)} + W_N^k \sum_{I_1} x_{4i+1} W_N^{k(4i+1)-k} + W_N^{3k} \sum_{I_2} x_{4i+3} W_N^{k(4i+3)-3k}, \tag{7.29}$$

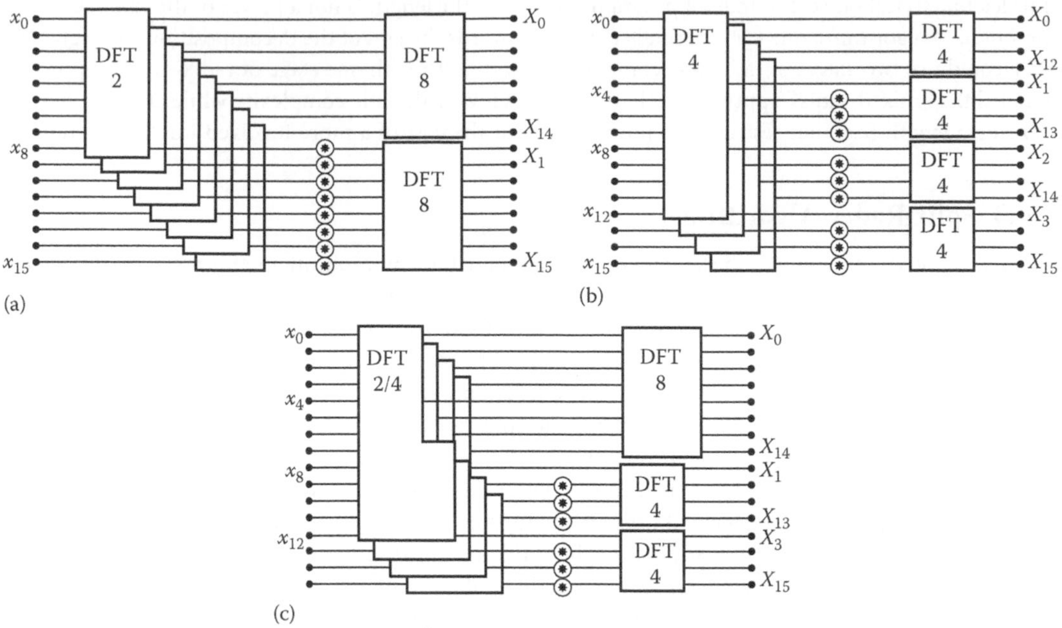

FIGURE 7.5 Comparison of various DIF algorithms for the length-16 DFT: (a) radix-2, (b) radix-4, and (c) split-radix.

which can be explicitly decomposed in order to make the redundancy between the computation of X_k, $X_{k+N/4}$, $X_{k+N/2}$, and $X_{k+3N/4}$ more apparent:

$$X_k = \sum_{i=0}^{N/2-1} x_{2i} W_{N/2}^{ik} + W_N^k \sum_{i=0}^{N/4-1} x_{4i+1} W_{N/4}^{ik} + W_N^{3k} \sum_{i=0}^{N/4-1} x_{4i+3} W_{N/4}^{ik}, \tag{7.30a}$$

$$X_{k-n/4} = \sum_{i=0}^{N/2-1} x_{2i} W_{N/2}^{ik} + jW_N^k \sum_{i=0}^{N/4-1} x_{4i+1} W_{N/4}^{ik} - jW_N^{3k} \sum_{i=0}^{N/4-1} x_{4i+3} W_{N/4}^{ik}, \tag{7.30b}$$

$$X_{k+N/2} = \sum_{i=0}^{N/2-1} x_{2i} W_{N/2}^{ik} - W_N^k \sum_{i=0}^{N/4-1} x_{4i+1} W_{N/4}^{ik} - W_N^{3k} \sum_{i=0}^{N/4-1} x_{4i+3} W_{N/4}^{ik}, \tag{7.30c}$$

$$X_{K+3N/4} = \sum_{i=0}^{N/2-1} x_{2i} W_{N/2}^{ik} - jW_N^k \sum_{i=0}^{N/4-1} x_{4i+1} W_{N/4}^{ik} + jW_N^{3k} \sum_{i=0}^{N/4-1} x_{4i+3} W_{N/4}^{ik}. \tag{7.30d}$$

The resulting algorithms have the minimum known number of operations (multiplications plus additions) as well as the minimum number of multiplications among practical algorithms for lengths which are powers of 2. The number of operations can be checked as being equal to

$$M\left[\mathrm{DFT}_{\text{split-radix}}\right] = N \log_2 N - 3N + 4 \tag{7.31a}$$

$$A\left[\mathrm{DFT}_{\text{split-radix}}\right] = 3N \log_2 N - 3N + 4 \tag{7.31b}$$

These numbers of operations can be obtained with only four different building blocks (with a complexity slightly lower than the one of a radix-4 butterfly), and are compared with the other algorithms in Tables 7.1 and 7.2.

TABLE 7.1 Number of Nontrivial Real Multiplication for Various FFTs on Complex Data

N	Radix-2	Radix-4	SRFFT	PFA	Winograd
16	24	20	20		
30				100	68
32	88		68		
60				200	136
64	264	208	196		
120				460	276
128	712		516		
240				1,100	632
256	1,800	1,392	1,284		
504				2,524	1,572
512	4,360		3,076		
1,008				5,804	3,548
1,024	10,248	7,856	7,172		
2,048	23,560		16,388		
2,520				17,660	9,492

TABLE 7.2 Number of Real Additions for Various FFTs on Complex Data

N	Radix-2	Radix-4	SRFFT	PFA	Winograd
16	152	148	148		
30				384	384
32	408		388		
60				888	888
64	1,032	976	964		
120				2,076	2,076
128	2,504		2,308		
240				4,812	5,016
256	5,896	5,488	5,380		
504				13,388	14,540
512	13,566		12,292		
1,008				29,548	34,668
1,024	30,728	28,336	27,652		
2,048	38,616		61,444		
2,520				84,076	99,628

Of course, due to the asymmetry in the decomposition, the structure of the algorithm is slightly more involved than for fixed-radix algorithms. Nevertheless, the resulting programs remain fairly simple [113] and can be highly optimized. Furthermore, this approach is well suited for applying FFTs on real data. It allows an in-place, butterfly style implementation to be performed [65,77].

The power of this algorithm comes from the fact that it provides the lowest known number of operations for computing length-2^n FFTs, while being implemented with compact programs. We shall see later that there are some arguments tending to show that it is actually the best possible compromise.

Note that the number of multiplications in Equation 7.31a is equal to the one obtained with the so-called "real-factor" algorithms [24,44]. In that approach, a linear combination of the data, using additions only, is made such that all twiddle factors are either pure real or pure imaginary. Thus, a multiplication of a complex number by a twiddle factor requires only two real multiplications. However, the real factor algorithms are quite costly in terms of additions, and are numerically ill-conditioned (division by small constants).

7.4.4 Remarks on FFTs with Twiddle Factors

The Cooley–Tukey mapping in Equations 7.9 and 7.17 is generally applicable, and actually the only possible mapping when the factors on N are not coprime. While we have paid particular attention to the case $N = 2^n$, similar algorithms exist for $N = p^m$ (p an arbitrary prime). However, one of the elegances of the length-2^n algorithms comes from the fact that the small DFTs (lengths 2 and 4) are multiplication-free, a fact that does not hold for other radices like 3 or 5, for instance. Note, however, that it is possible, for radix-3, either to completely remove the multiplication inside the butterfly by a change of base [26], at the cost of a few multiplications and additions, or to merge it with the twiddle factor [49] in the case where the implementation is based on the 4-mult 2-add complex multiplication scheme. It was also recently shown that, as soon as a radix p^2 algorithm was more efficient than a radix-p algorithm, a split-radix p/p^2 was more efficient than both of them [53]. However, unlike the 2^n case, efficient implementations for these p^n split-radix algorithms have not yet been reported. More efficient mixed-radix algorithms also remain to be found (initial results are given in [40]).

7.5 FFTs Based on Costless Mono- to Multidimensional Mapping

The divide and conquer strategy, as explained in Section 7.3, has few requirements for feasibility: N needs only to be composite, and the whole DFT is computed from DFTs on a number of points which is a factor of N (this is required for the redundancy in the computation of Equation 7.11 to be apparent). This requirement allows the expression of the innermost sum of Equation 7.11 as a DFT, provided that the subsets $\boldsymbol{I}_1$, have been chosen in such a way that x_i, $i \in \boldsymbol{I}_1$, is periodic. But, when N factors into relatively prime factors, say $N = N_1 \cdot N_2$, $(N_1, N_2) = 1$, a very simple property will allow a stronger requirement to be fulfilled.

Starting from any point of the sequence x_i, you can take as a first subset with compatible periodicity either $\{x_{i+N_1 \cdot n_2} | n_2 = 1, \ldots, N_2 - 1\}$ or, equivalently $\{x_{i+N_2 \cdot n_1} | n_1 = 1, \ldots, N_1 - 1\}$, and both subsets only have one common point x_i (by compatible, it is meant that the periodicity of the subsets divides the periodicity of the set). This allows a rearrangement of the input (periodic) vector into a matrix with a periodicity in both dimensions (rows and columns), both periodicities being compatible with the initial one (see Figure 7.6).

7.5.1 Basic Tools

FFTs without twiddle factors are all based on the same mapping, which is explained in Section 7.5.1.1. This mapping turns the original transform into sets of small DFTs, the lengths of which are coprime. It is therefore necessary to find efficient ways of computing these short-length DFTs. Section 7.5.1.2 explains how to turn them into cyclic convolutions for which efficient algorithms are described in Section 7.5.1.3.

7.5.1.1 The Mapping of Good

Performing the selection of subsets described in the introduction of Section 7.5 for any index i is equivalent to writing i as

$$i = \langle n_1 \cdot N_2 + n_2 \cdot N_1 \rangle_N, \quad n_1 = 1, \ldots, N_1 - 1, \quad n_2 = 1, \ldots, N_2 - 1, \; N = N_1 N_2, \tag{7.32}$$

0	1	2	3	4	5	6	7	8	9	10	11	12	13	14

⇓

0	3	6	9	12
5	8	11	14	2
10	13	1	4	7

(a)

0	6	12	3	9
10	1	7	13	4
5	11	2	8	14

(b)

FIGURE 7.6 Prime factor mapping for $N = 15$. (a) Good's mapping and (b) CRT mapping.

and, since N_1 and N_2 are coprime, this mapping is easily seen to be one to one [32]. (It is obvious from the right-hand side of Equation 7.32 that all congruences modulo N_1 are obtained for a given congruence modulo N_2, and vice versa.)

This mapping is another arrangement of the "CRT" mapping, which can be explained as follows on index k.

The CRT states that if we know the residue of some number k modulo two relatively prime numbers N_1 and N_2, it is possible to reconstruct $\langle k \rangle_{N_1 N_2}$ as follows:

Let $\langle k \rangle_{N_1} = k_1$ and $\langle k \rangle_{N_2} = k_2$. Then the value of k mod $N(N = N_1 \cdot N_2)$ can be found by

$$k = \langle N_1 t_1 k_2 + N_2 t_2 k_1 \rangle_N, \tag{7.33}$$

t_1 being the multiplicative inverse of N_1 mod N_2, that is $\langle t_1, N_1 \rangle_{N_2} = 1$, and t_2 the multiplicative inverse of N_2 mod N_1 (these inverses always exist, since N_1 and N_2 are coprime: $(N_1, N_2) = 1$).

Taking into account these two mappings in the definition of the DFT equation (Equation 7.3) leads to

$$X_{N_1 t_1 k_2 + N_2 t_2 k_1} = \sum_{n_1=0}^{N_1-1} \sum_{n_2=0}^{N_2-1} x_{n_1 N_2 + n_2 N_1} W_N^{(n_1 N_2 + N_1 n_2)(N_1 t_1 k_2 + N_2 t_2 k_1)}, \tag{7.34}$$

but

$$W_N^{N_2} = W_{N_1} \tag{7.35}$$

and

$$W_{N_1}^{N_2 t_2} = W_{N_1}^{\langle N_2 t_2 \rangle_{N_1}} = W_{N_1}, \tag{7.36}$$

which implies

$$X_{N_1 t_1 k_2 + N_2 t_2 k_1} = \sum_{n_1=0}^{N_1-1} \sum_{n_2=0}^{N_2-1} x_{n_1 N_2 + n_2 N_1} W_{N_1}^{n_1 k_2} W_{N_2}^{n_2 k_2}, \tag{7.37}$$

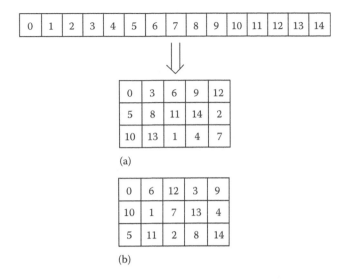

which, with

$$x'_{n_1,n_2} = x_{n_1N_2+n_2N_1}$$

and

$$X'_{k_1,k_2} = X_{N_1t_1k_2+N_2t_2k_1},$$

leads to a formulation of the initial DFT into a true bidimensional transform:

$$X'_{k_1k_2} = \sum_{n_1=0}^{N_1-1}\sum_{n_2=0}^{N_2-1} x'_{n_1n_2} W_{N_1}^{n_1k_1} W_{N_2}^{n_2k_2} \tag{7.38}$$

An illustration of the prime factor mapping is given in Figure 7.6a for the length $N = 15 = 3 \cdot 5$, and Figure 7.6b provides the CRT mapping. Note that these mappings, which were provided for a factorization of N into two coprime numbers, easily generalizes to more factors, and that reversing the roles of N_1, and N_2 results in a transposition of the matrices of Figure 7.6.

7.5.1.2 DFT Computation as a Convolution

With the aid of Good's mapping, the DFT computation is now reduced to that of a multidimensional DFT, with the characteristic that the lengths along each dimension are coprime. Furthermore, supposing that these lengths are small is quite reasonable, since Good's mapping can provide a full multidimensional factorization when N is highly composite.

The question is now to find the best way of computing this multidimensional DFT and these small-length DFTs. A first step in that direction was obtained by Rader [43], who showed that a DFT of prime length could be obtained as the result of a cyclic convolution: Let us rewrite Equation 7.1 for a prime length $N = 5$:

$$\begin{bmatrix} X_0 \\ X_1 \\ X_2 \\ X_3 \\ X_4 \end{bmatrix} = \begin{bmatrix} 1 & 1 & 1 & 1 & 1 \\ 1 & W_5^1 & W_5^2 & W_5^3 & W_5^4 \\ 1 & W_5^2 & W_5^4 & W_5^1 & W_5^3 \\ 1 & W_5^3 & W_5^1 & W_5^4 & W_5^2 \\ 1 & W_5^4 & W_5^3 & W_5^2 & W_5^1 \end{bmatrix} \begin{bmatrix} x_0 \\ x_1 \\ x_2 \\ x_3 \\ x_4 \end{bmatrix}. \tag{7.39}$$

Obviously, removing the first column and first row of the matrix will not change the problem, since they do not involve any multiplication. Furthermore, careful examination of the remaining part of the matrix shows that each column and each row involves every possible power of W_5, which is the first condition to be met for this part of the DFT to become a cyclic convolution. Let us now permute the last two rows and last two columns of the reduced matrix:

$$\begin{bmatrix} X'_1 \\ X'_2 \\ X'_4 \\ X'_3 \end{bmatrix} = \begin{bmatrix} W_5^1 & W_5^2 & W_5^4 & W_5^3 \\ W_5^2 & W_5^4 & W_5^3 & W_5^1 \\ W_5^4 & W_5^3 & W_5^1 & W_5^2 \\ W_5^3 & W_5^1 & W_5^2 & W_5^4 \end{bmatrix} \begin{bmatrix} x_1 \\ x_2 \\ x_4 \\ x_3 \end{bmatrix}. \tag{7.40}$$

Equation 7.40 is then a cyclic correlation (or a convolution with the reversed sequence).

It turns out that this a general result.

It is well known in number theory that the set of numbers lower than a prime p admits some primitive elements g such that the successive powers of g modulo p generate all the elements of the set. In the example above, $p=5$ and $g=2$, and we observe that

$$g^0 = 1, \quad g^1 = 2, \quad g^2 = 4, \quad \text{and} \quad g^3 = 8 = 3 \pmod 5.$$

The above result (Equation 7.40) is only the writing of the DFT in terms of the successive powers of W_p^g:

$$X'_k = \sum_{i=1}^{p-1} x_i W_p^{ik}, \quad k = 1, \ldots, p-1, \tag{7.41}$$

$$\begin{aligned} \langle ik \rangle_p &= \left\langle \langle i \rangle_p \cdot \langle k \rangle_p \right\rangle_p = \left\langle \langle g^{u_i} \rangle_p \cdot \langle g_p^{v_k} \rangle \right\rangle_p, \\ X'_{g^{v_i}} &= \sum_{u_i=0}^{p-2} x_{g^{u_i}} \cdot \left(W_p^g\right)^{u_i+v_i}, \quad v_i = 0, \ldots, p-2, \end{aligned} \tag{7.42}$$

and the length-p DFT turns out to be a length $(p-1)$ cyclic correlation:

$$\left\{X'_g\right\} = \{x_g\} * \left\{W_p^g\right\}. \tag{7.43}$$

7.5.1.3 Computation of the Cyclic Convolution

Of course Equation 7.42 has changed the problem, but it is not solved yet. And in fact, Rader's result was considered as a curiosity up to the moment when Winograd [55] obtained some new results on the computation of cyclic convolution.

And, again, this was obtained by application of the CRT. In fact, the CRT, as explained in Equation 7.33, Equation 7.34 can be rewritten in the polynomial domain: if we know the residues of some polynomial $K(z)$ modulo two mutually prime polynomials

$$\begin{aligned} \langle K(z) \rangle_{P_1(z)} &= K_1(z), \\ \langle K(z) \rangle_{P_2(z)} &= K_2(z), \end{aligned} \quad (P_1(z), P_2(z)) = 1, \tag{7.44}$$

we shall be able to obtain

$$K(z) \bmod P_1(z) \cdot P_2(z) = P(z)$$

by a procedure similar to that of Equation 7.33.

This fact will be used twice in order to obtain Winograd's method of computing cyclic convolutions:

A first application of the CRT is the breaking of the cyclic convolution into a set of polynomial products. For more convenience, let us first state Equation 7.43 in polynomial notation:

$$X'(z) = x'(z) \cdot w(z) \bmod (z^{p-1} - 1). \tag{7.45}$$

Now, since $p-1$ is not prime (it is at least even), $z^{p-1} - 1$ can be factorized at least as

$$z^{p-1} - 1 = (z^{(p-1)/2} + 1)(z^{(p-1)/2} - 1), \tag{7.46}$$

and possibly further, depending on the value of p. These polynomial factors are known and named cyclotomic polynomials $\varphi_q(z)$. They provide the full factorization of any $z^N - 1$:

$$z^N - 1 = \prod_{q|N} \varphi_q(z). \tag{7.47}$$

A useful property of these cyclotomic polynomials is that the roots of $\varphi_q(z)$ are all the q th primitive roots of unity, hence degree $\{\varphi_q(z)\} = \varphi(q)$, which is by definition the number of integers lower than q and coprime with it. Namely, if $w_q = e^{-j2\pi/q}$, the roots of $\varphi_q(z)$ are $\{W_q^r | (r,q) = 1\}$.

As an example, for $p = 5, z^{p-1} - 1 = z^4 - 1$,

$$\begin{aligned} z^4 - 1 &= \varphi_1(z) \cdot \varphi_2(z) \cdot \varphi_4(z) \\ &= (z-1)(z+1)(z^2+1). \end{aligned}$$

The first use of the CRT to compute the cyclic convolution (Equation 7.45) is then as follows:

1. Compute

$$\left.\begin{aligned} x'_q(z) &= x'(z) \bmod \varphi_q(z), \\ w'_q(z) &= w(z) \bmod \varphi_q(z) \end{aligned}\right. \quad q|p-1$$

2. Then obtain

$$X'_q(z) = x'_q(z) \cdot w'_q(z) \bmod \varphi_q(z)$$

3. Reconstruct $X'(z) \bmod z^{p-1} - 1$ from the polynomials $X'_q(z)$ using the CRT

Let us apply this procedure to our simple example:

$$\begin{aligned} x'(z) &= x_1 + x_2 z + x_4 z^2 + x_3 z^3, \\ w(z) &= W_5^1 + W_5^2 z + W_5^4 z^2 + W_5^3 z^3. \end{aligned}$$

Step 1:

$$\begin{aligned} w_4(z) &= w(z) \bmod \varphi_4(z) \\ &= \left(W_5^1 - W_5^4\right) + \left(W_5^2 - W_5^3\right) z, \\ w_2(z) &= w(z) \bmod \varphi_2(z) \\ &= \left(W_5^1 + W_5^4 - W_5^2 - W_5^3\right), \\ w_1(z) &= w(z) \bmod \varphi_1(z) \\ &= \left(W_5^1 + W_5^4 + W_5^2 + W_5^3\right) [= -1], \\ x'_4(z) &= (x_1 - x_4) + (x_2 - x_3) z, \\ x'_2(z) &= (x_1 + x_4 - x_2 - x_3), \\ x'_1(z) &= (x_1 + x_4 + x_2 + x_3). \end{aligned}$$

Step 2:

$$\begin{aligned} X'_4(z) &= x'_4(z) \cdot w_4(z) \bmod \varphi_4(z), \\ X'_2(z) &= x'_2(z) \cdot w_2(z) \bmod \varphi_2(z), \\ X'_1(z) &= x'_1(z) \cdot w_1(z) \bmod \varphi_1(z). \end{aligned}$$

Step 3:

$$X'(z) = \left[X_1'(z)(1+z)/2 + X_2'(z)(1-z)/2\right] \\ \times (1+z^2)/2 + X_4'(z)(1-z^2)/2.$$

Note that all the coefficients of $W_q(z)$ are either real or purely imaginary. This is a general property due to the symmetries of the successive powers of W_p.

The only missing tool needed to complete the procedure now is the algorithm to compute the polynomial products modulo the cyclotomic factors. Of course, a straightforward polynomial product followed by a reduction modulo $\varphi_q(z)$ would be applicable, but a much more efficient algorithm can be obtained by a second application of the CRT in the field of polynomials.

It is already well-known that knowing the values of an N th degree polynomial at $N+1$ different points can provide the value of the same polynomial anywhere else by Lagrange interpolation. The CRT provides an analogous way of obtaining its coefficients.

Let us first recall the equation to be solved:

$$X_q'(z) = x_q'(z) \cdot w_q(z) \bmod \varphi_q(z), \tag{7.48}$$

with

$$\deg \varphi_q(z) = \varphi(q).$$

Since $\varphi_q(z)$ is irreducible, the CRT cannot be used directly. Instead, we choose to evaluate the product $X_q''(z) = x_q'(z) \cdot w_q(z)$ modulo an auxiliary polynomial $A(z)$ of degree greater than the degree of the product. This auxiliary polynomial will be chosen to be fully factorizable. The CRT hence applies, providing

$$X_q''(z) = x_q'(z) \cdot w_q(z),$$

since the mod $A(z)$ is totally artificial, and the reduction modulo $\varphi_q(z)$ will be performed afterwards.

The procedure is then as follows:

Let us evaluate both $x_q'(z)$ and $w_q(z)$ modulo a number of different monomials of the form

$$(z - a_i), \quad i = 1, \ldots, 2\varphi(q) - 1.$$

Then compute

$$X_q''(a_i) = x_q'(a_i) w_q(a_i), \quad i = 1, \ldots, 2\varphi(q) - 1. \tag{7.49}$$

The CRT then provides a way of obtaining

$$X_q''(z) \bmod A(z), \tag{7.50}$$

with

$$A(z) = \prod_{i=1}^{2\varphi(q)-1} (z - a_i),$$

which is equal to $X''_q(z)$ itself, since

$$\deg X''_q(z) = 2\varphi(q) - 2. \tag{7.51}$$

Reduction of $X''_q(z)$ mod $\varphi_z(z)$ will then provide the desired result.

In practical cases, the points $\{a_i\}$ will be chosen in such a way that the evaluation of $w'_q(a_i)$ involves only additions (i.e., $a_i = 0, \pm 1, \ldots$).

This limits the degree of the polynomials whose products can be computed by this method. Other suboptimal methods exist [12], but are nevertheless based on the same kind of approach (the "dot products" (Equation 7.49) become polynomial products of lower degree, but the overall structure remains identical).

All this seems fairly complicated, but results in extremely efficient algorithms that have a low number of operations. The full derivation of our example ($p = 5$) then provides the following algorithm:

5 point DFT:

$$
\begin{aligned}
& u = 2\pi/5 \\
& t_1 = x_1 + x_4, t_2 = x_2 + x_3 \quad \text{(reduction modulo } z^2 - 1\text{)}, \\
& t_3 = x_1 - x_4, t_4 = x_3 - x_2 \quad \text{(reduction modulo } z^2 + 1\text{)}, \\
& t_5 = t_1 + t_2 \quad \text{(reduction modulo } z - 1\text{)}, \\
& t_6 = t_1 - t_2 \quad \text{(reduction modulo } z + 1\text{)}, \\
& m_1 = [(\cos u + \cos 2u)/2]t_5 \quad \left(X'_1(z) = x'_1(z) \cdot w_1(z) \bmod \varphi_1(z)\right), \\
& m_2 = [(\cos u - \cos 2u)/2]t_6 \quad \left(X'_2(z) = x'_2(z) \cdot w_2(z) \bmod \varphi_2(z)\right), \\
& \qquad \text{polynomial product modulo } z^2 + 1 \quad \left(X'_4(z) = x'_4(z) \cdot w_4(z) \bmod \varphi_u(z)\right), \\
& m_3 = -j(\sin u)(t_3 + t_4), \\
& m_4 = -j(\sin u + \sin 2u)t_4, \\
& m_5 = j(\sin u - \sin 2u)t_3, \\
& s_1 = m_3 - m_4, \\
& s_2 = m_3 + m_5 \\
& \quad \text{(reconstruction following Step 3, the 1/2 terms} \\
& \quad \text{have been included into the polynomial products)}, \\
& s_3 = x_0 + m_1, \\
& s_4 = s_3 + m_2, \\
& s_5 = s_3 - m_2, \\
& X_0 = x_0 + t_5, \\
& X_1 = s_4 + s_1, \\
& X_2 = s_5 + s_2, \\
& X_3 = s_5 - s_2, \\
& X_4 = s_4 - s_1.
\end{aligned}
$$

When applied to complex data, this algorithm requires 10 real multiplications and 34 real additions vs. 48 real multiplications and 88 real additions for a straightforward algorithm (matrix-vector product).

In matrix form, and slightly changed, this algorithm may be written as follows:

$$\left(X'_0, X'_1, \ldots, X'_4\right)^{\mathrm{T}} = C \cdot D \cdot B \cdot (x_0, x_1, \ldots, x_4)^{\mathrm{T}}, \tag{7.52}$$

with

$$C = \begin{bmatrix} 1 & 0 & 0 & 0 & 0 & 0 \\ 1 & 1 & 1 & 1 & -1 & 0 \\ 1 & 1 & -1 & 1 & 0 & 1 \\ 1 & 1 & -1 & -1 & 0 & -1 \\ 1 & 1 & 1 & -1 & 1 & 0 \end{bmatrix},$$

$$D = \mathrm{diag}[1, (\cos u + \cos 2u)/2 - 1, (\cos u - \cos 2u)/2, -j \sin u, -j(\sin u + \sin 2u), j(\sin u - \sin 2u)],$$

$$B = \begin{bmatrix} 1 & 1 & 1 & 1 & 1 \\ 0 & 1 & 1 & 1 & 1 \\ 0 & 1 & -1 & -1 & 1 \\ 0 & 1 & -1 & 1 & -1 \\ 0 & 0 & -1 & 1 & 0 \\ 0 & 1 & 0 & 0 & 1 \end{bmatrix}.$$

By construction, D is a diagonal matrix, where all multiplications are grouped, while C and B only involve additions (they correspond to the reductions and reconstructions in the applications of the CRT).

It is easily seen that this structure is a general property of the short-length DFTs based on CRT: all multiplications are "nested" at the center of the algorithms. By construction, also, D has dimension M_p, which is the number of multiplications required for computing the DFT, some of them being trivial (at least one, needed for the computation of X_0). In fact, using such a formulation, we have $M_p \geq p$. This notation looks awkward, at first glance (why include trivial multiplications in the total number?), but Section 7.5.3 will show that it is necessary in order to evaluate the number of multiplications in the Winograd FFT.

It can also be proven that the methods explained in this section are essentially the only ways of obtaining FFTs with the minimum number of multiplications. In fact, this gives the optimum structure, mathematically speaking. These methods always provide a number of multiplications lower than twice the length of the DFT:

$$M_{N_1} < 2N_1.$$

This shows the linear complexity of the DFT in this case.

7.5.2 Prime Factor Algorithms

Let us now come back to the initial problem of this section: the computation of the bidimensional transform given in Equation 7.38 [95]. Rearranging the data in matrix form, of size $N_1 N_2$, and F_1 (resp. F_2) denoting the Fourier matrix of size N_1 (resp. N_2) results in the following notation, often used in the context of image processing:

$$X = F_1 x F_2^{\mathrm{T}}. \tag{7.53}$$

Performing the FFT algorithm separately along each dimension results in the so-called PFA.

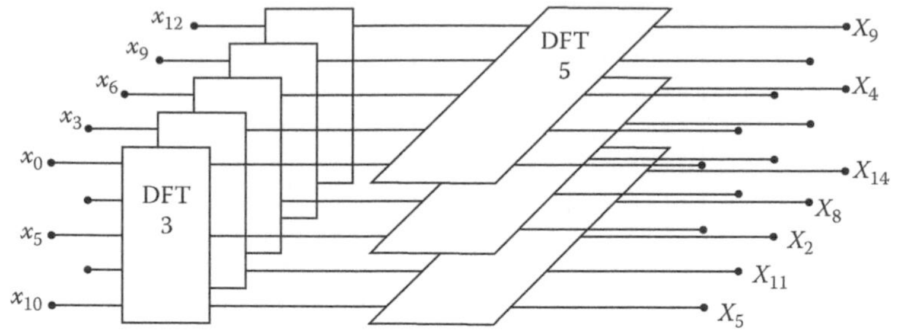

FIGURE 7.7 Schematic view of PFA for $N = 15$.

To summarize, PFA makes use of Good's mapping (Section 7.5.1.1) to convert the length $N_1 \cdot N_2$ 1-D DFT into a size $N_1 \times N_2$ 2-D DFT, and then computes this 2-D DFT in a row–column fashion, using the most efficient algorithms along each dimension.

Of course, this applies recursively to more than two factors, the constraints being that they must be mutually coprime. Nevertheless, this constraint implies the availability of a whole set of efficient small DFTs ($N_i = 2, 3, 4, 5, 7, 8$, and 16 is already sufficient to provide a dense set of feasible lengths).

A graphical display of PFA for length $N = 15$ is given in Figure 7.7. Since there are N_2 applications of length N_1 FFT and N_1, applications of length N_2 FFTs, the computational costs are as follows:

$$\begin{aligned} M_{N_1N_2} &= N_1M_2 + N_2M_1, \\ A_{N_1N_2} &= N_1A_2 + N_2A_1, \end{aligned} \tag{7.54}$$

or, equivalently, the number of operations to be performed per output point is the sum of the individual number of operations in each short algorithm: let m_N and a_N be these reduced numbers

$$\begin{aligned} m_{N_1N_2N_3N_4} &= m_{N_1} + m_{N_2} + m_{N_3} + m_{N_4}, \\ a_{N_1N_2N_3N_4} &= a_{N_1} + a_{N_2} + a_{N_3} + a_{N_4}. \end{aligned} \tag{7.55}$$

An evaluation of these figures is provided in Tables 7.1 and 7.2.

7.5.3 Winograd's Fourier Transform Algorithm

Winograd's FFT [56] makes full use of all the tools explained in Section 7.5.1.

Good's mapping is used to convert the length $N_1 \cdot N_2$ 1-D DFT into a length $N_1 \times N_2$ 2-D DFT, and the intimate structure of the small-length algorithms is used to nest all the multiplications at the center of the overall algorithm as follows.

Reporting Equation 7.52 into Equation 7.53 results in

$$X = C_1D_1B_1xB_2^{\mathrm{T}}D_2C_2^{\mathrm{T}}. \tag{7.56}$$

Since C and B do not involve any multiplication, the matrix $(B_1xB_2^{\mathrm{T}})$ is obtained by only adding properly chosen input elements. The resulting matrix now has to be multiplied on the left and on the right by diagonal matrices D_1 and D_2, of respective dimensions M_1 and M_2. Let M_1' and M_2' be the numbers of trivial multiplications involved.

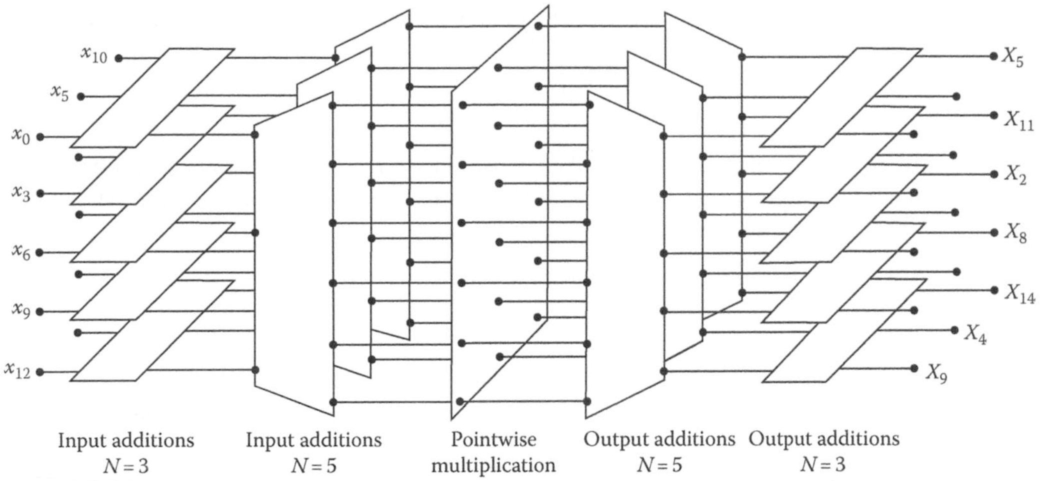

FIGURE 7.8 Schematic view of WFTA for $N = 15$.

Premultiplying by the diagonal matrix D_1 multiplies each row by some constant, while postmultiplying does it for each column. Merging both multiplications leads to a total number of

$$M_{N_1N_2} = M_{N_1} \cdot M_{N_2} \tag{7.57}$$

out of which $M'_{N_1} \cdot M'_{N_2}$ are trivial.

Pre- and postmultiplying by C_1 and C_2^{T} will then complete the algorithm.

A graphical display of WFTA for length $N = 15$ is given in Figure 7.8, which clearly shows that this algorithm cannot be performed in place.

The number of additions is more intricate to obtain.

Let us consider the pictorial representation of Equation 7.56 as given in Figure 7.8.

Let C_1 involve A_1^1 additions (output additions) and B_1 involve A_2^1 additions (input additions). (Which means that there exists an algorithm for multiplying C_1 by some vector involving A_1^1 additions. This is different from the number of ± 1's in the matrix—see the $p = 5$ example.)

Under these conditions, obtaining xB_2 will cost $A_2^2 \cdot N_1$ additions, $B_1(xB_2^{\mathrm{T}})$ will cost $A_1^2 \cdot M_2$ additions, $C_1(D_1B_1xB_2^{\mathrm{T}})$ will cost $A_1^1 \cdot M_2$ additions and $(C_1D_1B_1xB_2^{\mathrm{T}})C_2$ will cost $A_2^1 \cdot N_1$ additions, which gives a total of

$$A_{N_1N_2} = N_1A_2 + M_2A_1. \tag{7.58}$$

This formula is not symmetric in N_1 and N_2. Hence, it is possible to interchange N_1 and N_2, which does not change the number of multiplications. This is used to minimize the number of additions.

Since $M_2 \geq N_2$, it is clear that WFTA will always require at least as many additions as PFA, while it will always need fewer multiplications, as long as optimum short length DFTs are used. The demonstration is as follows.

Let

$$\begin{aligned}
M_1 &= N_1 + \varepsilon_1, \quad M_2 = N_2 + \varepsilon_2, \\
M_{\mathrm{PFA}} &= N_1M_2 + N_2M_1 \\
&= 2N_1N_2 + N_1\varepsilon_2 + N_2\varepsilon_1, \\
M_{\mathrm{WFTA}} &= M_1 \cdot M_2 \\
&= N_1N_2 + \varepsilon_1\varepsilon_2 + N_1\varepsilon_2 + N_2\varepsilon_1.
\end{aligned}$$

Since ε_1 and ε_2 are strictly smaller than N_1 and N_2 in optimum short-length DFTs, we have, as a result,

$$M_{\text{WFTA}} < M_{\text{PFA}}.$$

Note that this result is not true if suboptimal short-length FFTs are used. The numbers of operations to be performed per output point (to be compared with Equation 7.55) are as follows in the WFTA:

$$m_{N_1N_2} = m_{N_1} \cdot M_{N_2}, \; a_{N_1N_2} = a_{N_2} + m_{N_2}a_{N_1}. \tag{7.59}$$

These numbers are given in Tables 7.1 and 7.2.

Note that the number of additions in the WFTA was reduced later by Nussbaumer [12] with a scheme called "split nesting," leading to the algorithm with the least known number of operations (multiplications + additions).

7.5.4 Other Members of This Class

PFA and WFTA are seen to be both described by the following equation [38]:

$$X = C_1 D_1 B_1 x B_2^{\text{T}} D_2 C_2^{\text{T}}. \tag{7.60}$$

Each of them is obtained by different ordering of the matrix products.

- The PFA multiplies $(C_1\ D_1\ B_1)x$ first, and then the result is postmultiplied by $(B_2^{\text{T}} D_2 C_2^{\text{T}})$.
- The WFTA starts with $B_1 x B_2^{\text{T}}$, then $(D_1 \times D_2)$, then C_1 and finally C_2^{T}.

Nevertheless, these are not the only ways of obtaining X: C and B can be factorized as two matrices each, to fully describe the way the algorithms are implemented. Taking this fact into account allows a great number of different algorithms to be obtained. Johnson and Burrus [38] systematically investigated this whole class of algorithms, obtaining interesting results, such as

- Some WFTA-type algorithms, with reduced number of additions
- Algorithms with lower number of multiplications than both PFA and WFTA in the case where the short-length algorithms are not optimum

7.5.5 Remarks on FFTs without Twiddle Factors

It is easily seen that members of this class of algorithms differ fundamentally from FFTs with twiddle factors.

Both classes of algorithms are based on a divide and conquer strategy, but the mapping used to eliminate the twiddle factors introduced strong constraints on the type of lengths that were possible with Good's mapping.

Due to those constraints, the elaboration of efficient FFTs based on Good's mapping required considerable work on the structure of the short FFTs. This resulted in a better understanding of the mathematical structure of the problem, and a better idea of what was feasible and what was not.

This new understanding has been applied to the study of FFTs with twiddle factors. In this study, issues, such as optimality, distance (in cost) of the practical algorithms from the best possible ones, and the structural properties of the algorithms, have been prominent in the recent evolution of the field of algorithms.

7.6 State of the Art

FFT algorithms have now reached a great maturity, at least in the 1-D case, and it is now possible to make strong statements about what eventual improvements are feasible and what are not.

In fact, lower bounds on the number of multiplications necessary to compute a DFT of given length can be obtained by using the techniques described in Section 7.5.1.

7.6.1 Multiplicative Complexity

Let us first consider the FFTs with lengths that are powers of two.

Winograd [57] was first able to obtain a lower bound on the number of complex multiplications necessary to compute length 2^n DFTs. This work was then refined in [28], which provided realizable lower bounds, with the following multiplicative complexity:

$$\mu_c[\mathrm{DFT}2^n] = 2^{n+1} - 2n^2 + 4n - 8. \tag{7.61}$$

This means that there will never exist any algorithm computing a length 2^n DFT with a lower number of nontrivial complex multiplications than the one in Equation 7.61.

Furthermore, since the demonstration is constructive [28], this optimum algorithm is known. Unfortunately, it is of no practical use for lengths greater than 64 (it involves much too many additions).

The lower part of Figure 7.9 shows the variation of this lower bound and of the number of complex multiplications required by some practical algorithms (radix-2, radix-4, and SRFT). It is clearly seen that SRFFT follows this lower bound up to $N=64$, and is fairly close for $N=128$. Divergence is quite fast afterwards.

It is also possible to obtain a realizable lower bound on the number of real multiplications [35,36]:

$$\mu_r[\mathrm{DFT}2^n] = 2^{n+2} - 2n^2 - 2n + 4. \tag{7.62}$$

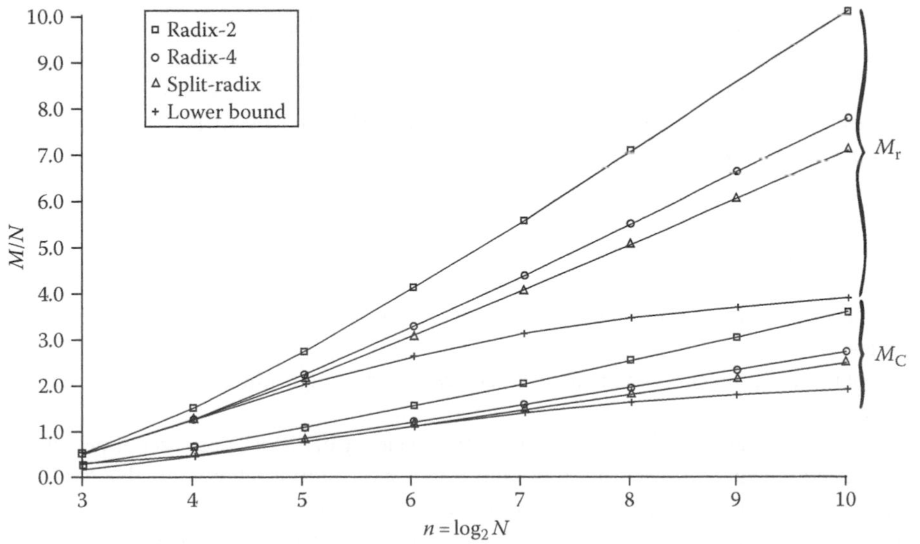

FIGURE 7.9 Number of nontrivial real or complex multiplications per output point.

The variation of this bound, together with that of the number of real multiplications required by some practical algorithms is provided on the upper part of Figure 7.9. Once again, this realizable lower bound is of no practical use above a certain limit. But, this time, the limit is much lower: SRFFT, together with radix-4, meets the lower bound on the number of real multiplications up to $N = 16$, which is also the last point where one can use an optimal polynomial product algorithm (modulo $u^2 + 1$) which is still practical. ($N = 32$ would require an optimal product modulo $u^4 + 1$ that requires a large number of additions.)

It was also shown [31,76] that all of the three following algorithms: optimum algorithm minimizing complex multiplications, optimum algorithm minimizing real multiplications and SRFFT, had exactly the same structure. They performed the decomposition into polynomial products exactly in the same manner, and they differ only in the way the polynomial products are computed.

Another interesting remark is as follows: the same number of multiplications as in SRFFT could also be obtained by so-called "real factor radix-2 FFTs" [24,42,44] (which were, on another respect, somewhat numerically ill-conditioned and needed about 20% more additions). They were obtained by making use of some computational trick to replace the complex twiddle factors by purely real or purely imaginary ones. Now, the question is: Is it possible to do the same kind of thing with radix-4, or even SRFFT? Such a result would provide algorithms with still fewer operations. The knowledge of the lower bound tells us that it is impossible because, for some points (e.g., $N = 16$) this would produce an algorithm with better performance than the lower bound. The challenge of eventually improving SRFFT is now as follows.

Comparison of SRFFT with $\mu_c[\text{DFT } 2^n]$ tells us that no algorithm using complex multiplications will be able to improve significantly SRFFT for lengths less than 512. Furthermore, the trick allowing real factor algorithms to be obtained cannot be applied to radices greater than 2 (or at least not in the same manner).

The above discussion thus shows that there remain very few approaches (yet unknown) that could eventually improve the best known length 2^n FFT.

And what is the situation for FFTs based on Good's mapping?

Realizable lower bounds are not so easily obtained. For a given length $N = \prod N_i$, they involve a fairly complicated number theoretic function [8], and simple analytical expressions cannot be obtained. Nevertheless, programs can be written to compute $\mu_r\{\text{DFTN}_N\}$, and are given in [36]. Table 7.3 provides numerical values for a number of lengths of interest.

Careful examination of Table 7.3 provides a number of interesting conclusions.

First, one can see that, for comparable lengths (since SRFFT and WFTA cannot exist for the same lengths), a classification depending on the efficiency is as follows: WFTA always requires the lowest number of multiplications, followed by PFA, and followed by SRFFT, all fixed- or mixed-radix FFTs being next. Nevertheless, none of these algorithms attains the lower bound, except for very small lengths.

Another remark is that the number of multiplications required by WFTA is always smaller than the lower bound for the corresponding length that is a power of 2. This means, on the one hand, that transform lengths for which Good's mapping can be applied are well suited for a reduction in the number of multiplications, and on the other hand, that they are very efficiently computed by WFTA, from this point of view.

And this states the problem of the relative efficiencies of these algorithms: How close are they to their respective lower bound?

The last column of Table 7.3 shows that the relative efficiency of SRFFT decreases almost linearly with the length (it requires about twice the minimum number of multiplications for $N = 2048$), while the relative efficiency of WFTA remains almost constant for all the lengths of interest (it would not be the same result for much greater N). Lower bounds for Winograd-type lengths are also seen to be smaller than for the corresponding power of 2 lengths.

TABLE 7.3 Practical Algorithms vs. Lower Bounds (Number of Nontrivial Real Multiplications for FFTs on Real Data)

N	SRFFT	WFTA	Lower Bound	SRFT (Lower Bound)	WFTA (Lower Bound)
16	20		20	1	
30		68	56		1.21
32	68		64	1.06	
60		136	112		1.21
64	196		168	1.16	
120		276	240		1.15
128	516		396	1.3	
240		632	548		1.15
256	1,284		876	1.47	
504		1,572	1,320		1.19
512	3,076		1,864	1.64	
1,008		3,548	2,844		1.25
1,024	7,172		3,872	1.85	
2,048	16,388		7,876	2.08	
2,520		9,492	7,440		1.27

All these considerations result in the following conclusion: lengths for which Good's mapping is applicable allow a greater reduction of the number of multiplications (which is due directly to the mathematical structure of the problem). And, furthermore, they allow a greater relative efficiency of the actual algorithms vs. the lower bounds (and this is due indirectly to the mathematical structure).

7.6.2 Additive Complexity

Nevertheless, the situation is not the same as regards the number of additions.

Most of the work on optimality was concerned with the number of multiplications. Concerning the number of additions, one can distinguish between additions due to the complex multiplications and the ones due to the butterflies. For the case $N=2^n$, it was shown in [106,110] that the latter number, which is achieved in actual algorithms, is also the optimum. Differences between the various algorithms is thus only due to varying numbers of complex multiplications. As a conclusion, one can see that the only way to decrease the number of additions is to decrease the number of true complex multiplications (which is close to the lower bound).

Figure 7.10 gives the variation of the total number of operations (multiplications plus additions) for these algorithms, showing that SRFFT has the lowest operation count. Furthermore, its more regular structure results in faster implementations.

Note that all the numbers given here concern the initial versions of SRFFT, PFA, and WFTA, for which FORTRAN programs are available. It is nevertheless possible to improve the number of additions in WFTA by using the so-called split-nesting technique [12] (which is used in Figure 7.10), and the number of multiplications of PFA by using small-length FFTs with scaled output [12], resulting in an overall scaled DFT.

As a conclusion, one can realize that we now have practical algorithms (mainly WFTA and SRFFT) that follow the mathematical structure of the problem of computing the DFT with the minimum number of multiplications, as well as a knowledge of their degree of suboptimality.

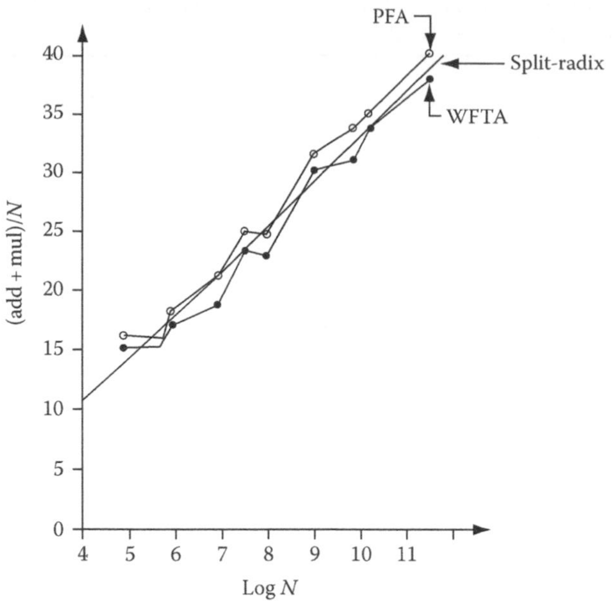

FIGURE 7.10 Total number of operations per output point for different algorithms.

7.7 Structural Considerations

This section is devoted to some points that are important in the comparison of different FFT algorithms, namely easy obtention of inverse FFT, in-place computation, regularity of the algorithm, quantization noise, and parallelization, all of which are related to the structure of the algorithms.

7.7.1 Inverse FFT

FFTs are often used regardless of their "frequency" interpretation for computing FIR filtering in blocks, which achieves a reduction in arithmetic complexity compared to the direct algorithm. In that case, the forward FFT has to be followed, after pointwise multiplication of the result, by an inverse FFT. It is of course possible to rewrite a program along the same lines as the forward one, or to reorder the outputs of a forward FFT. A simpler way of computing an inverse FFT by using a forward FFT program is given (or reminded) in [99], where it is shown that, if CALL FFT (XR, Xl, N) computes a forward FFT of the sequence $\{XR(i) + jXI(i)|i = 0, \ldots, N-1\}$, CALL FFT(XI, XR, N) will compute an inverse FFT of the same sequence, whatever the algorithm is. Thus, all FFT algorithms on complex data are equivalent in that sense.

7.7.2 In-Place Computation

Another point in the comparison of algorithms is the memory requirement: most algorithms (CTFFT, SRFFT, and PFA) allow in-place computation (no auxiliary storage of size depending on N is necessary), while WFTA does not. And this may be a drawback for WFTA when applied to rather large sequences.

CTFFT and SRFFT also allow rather compact programs [4,113], the size of which is independent of the length of the FFT to be computed.

On the contrary, PFA and WFTA will require longer and longer programs when the upper limit on the possible lengths is increased: an 8-module program ($n = 2, 4, 8, 16, 3, 5, 7$, and 9) allows obtaining a rather dense set of lengths up to $N = 5040$ only. Longer transforms can only be obtained either by the use of rather "exotic" modules that can be found in [37], or by some kind of mixture between CTFFT (or SRFFT) and PFA.

7.7.3 Regularity and Parallelism

Regularity has been discussed for nearly all algorithms when they were described. Let us recall here that CTFFT is very regular (based on repetitive use of a few modules) and SRFFT follows (repetitive use of very few modules in a slightly more involved manner). Then, PFA requires repetitive use (more intricate than CTFFT) of more modules, and finally WFTA requires some combining of parts of these modules, which means that, even if it has some regularity, this regularity is more hidden.

Let us point out also that the regularity of an algorithm cannot really be seen from its flowgraph. The equations describing the algorithm, as given in Equation 7.13 or 7.38, do not fully define the implementations, which is partially done in the flowgraph. The reordering of the nodes of a flowgraph may provide a more regular one. (The classical radix-2 and radix-4 CTFFT can be reordered into a constant geometry algorithm. See also [30] for SRFFT.)

Parallelization of CTFFT and SRFFT is fairly easy, since the small modules are applied on sets of data that are separable and contiguous, while it is slightly more difficult with PFA, where the data required by each module are not in contiguous locations.

Finally, let us point out that mathematical tools such as tensor products can be used to work on the structure of the FFT algorithms [50,101], since the structure of the algorithm reflects the mathematical structure of the underlying problem.

7.7.4 Quantization Noise

Roundoff noise generated by finite precision operations inside the FFT algorithm is also of importance. Of course, fixed point implementations of CTFFT for lengths 2^n were studied first, and it was shown that the error-to-signal ratio of the FFT process increases as $\sqrt{N}$ (which means 1/2 bit per stage) [117]. SRFFT and radix-4 algorithms were also reported to generate less roundoff than radix-2 [102].

Although the WFTA requires fewer multiplications than the CTFFT (hence has less noise sources), it was soon recognized that proper scaling was difficult to include in the algorithm, and that the resulting noise-to-signal ratio was higher. It is usually thought that two more bits are necessary for representing data in the WFTA to give an error of the same order as CTFFT (at least for practical lengths). A floating point analysis of PFA is provided in [104].

7.8 Particular Cases and Related Transforms

The previous sections have been devoted exclusively to the computation of the matrix-vector product involving the Fourier matrix. In particular, no assumption has been made on the input or output vector. In the following subsections, restrictions will be put on these vectors, showing how the previously described algorithms can be applied when the input is, for example, real-valued, or when only a part of the output is desired. Then, transforms closely related to the DFT will be discussed as well.

7.8.1 DFT Algorithms for Real Data

Very often in applications, the vector to be transformed is made up of real data. The transformed vector then has an Hermitian symmetry, that is,

$$X_{N-k} = X_k^*, \tag{7.63}$$

as can be seen from the definition of the DFT. Thus, X_0 is real, and when N is even, $X_{N/2}$ is real as well. That is, the N input values map to 2 real and $N/2 - 1$ complex conjugate values when N is even, or 1 real and $(N-1)/2$ complex conjugate values when N is odd (which leaves the number of free variables unchanged).

This redundancy in both input and output vectors can be exploited in the FFT algorithms in order to reduce the complexity and storage by a factor of 2. That the complexity should be half can be shown by the following argument. If one takes a real DFT of the real and imaginary parts of a complex vector separately, then $2N$ additions are sufficient in order to obtain the result of the complex DFT [3]. Therefore, the goal is to obtain a real DFT that uses half as many multiplications and less than half as many additions. If one could do better, then it would improve the complex FFT as well by the above construction.

For example, take the DIF SRFFT algorithm (Equation 7.28). First, X_{2k} requires a half-length DFT on real data, and thus the algorithm can be reiterated. Then, because of the Hermitian symmetry property (Equation 7.63),

$$X_{4k+1} = X^{\star}_{4(N/4-k-1)+3}, \tag{7.64}$$

and therefore Equation 7.28c is redundant and only one DFT of size $N/4$ on complex data needs to be evaluated for Equation 7.28b. Counting operations, this algorithm requires exactly half as many multiplications and slightly less than half as many additions as its complex counterpart, or [30]

$$M(\mathrm{R}-\mathrm{DFT}(2^m)) = 2^{n-1}(n-3)+2, \tag{7.65}$$

$$A(\mathrm{R}-\mathrm{DFT}(2^m)) = 2^{n-1}(3n-5)+4. \tag{7.66}$$

Thus, the goal for the real DFT stated earlier has been achieved. Similar algorithms have been developed for radix-2 and radix-4 FFTs as well. Note that even if DIF algorithms are more easily explained, it turns out that DIT ones have a better structure when applied to real data [29,65,77].

In the PFA case, one has to evaluate a multidimensional DFT on real input. Because the PFA is a row–column algorithm, data become Hermitian after the first 1-D FFTs, hence an accounting has to be made of the real and conjugate parts so as to divide the complexity by 2 [77]. Finally, in the WFTA case, the input addition matrix and the diagonal matrix are real, and the output addition matrix has complex conjugate rows, showing again the saving of 50% when the input is real. Note, however, that these algorithms generally have a more involved structure than their complex counterparts (especially in the PFA and WFTA cases). Some algorithms have been developed which are inherently "real," like the real-factor FFTs [22,44] or the FFCT algorithm [51], and do not require substantial changes for real input.

A closely related question is how to transform (or actually back transform) data that possess Hermitian symmetry. An actual algorithm is best derived by using the transposition principle: since the Fourier transform is unitary, its inverse is equal to its Hermitian transpose, and the required algorithm can be obtained simply by transposing the flowgraph of the forward transform (or by transposing the matrix factorization of the algorithm). Simple graph theoretic arguments show that both the multiplicative and additive complexities are exactly conserved.

Assume next that the input is real and that only the real (or imaginary) part of the output is desired. This corresponds to what has been called a cosine (or sine) DFT, and obviously, a cosine and a sine DFT on a real vector can be taken altogether at the cost of a single real DFT. When only a cosine DFT has to be computed, it turns out that algorithms can be derived so that only half the complexity of a real DFT (i.e., the quarter of a complex DFT) is required [30,52], and the same holds for the sine DFT as well [52]. Note that the above two cases correspond to DFTs on real and symmetric (or antisymmetric) vectors.

7.8.2 DFT Pruning

In practice, it may happen that only a small number of the DFT outputs are necessary, or that only a few inputs are different from zero. Typical cases appear in spectral analysis, interpolation, and fast convolution applications. Then, computing a full FFT algorithm can be wasteful, and advantage should be taken of the inputs and outputs that can be discarded.

We will not discuss "approximate" methods which are based on filtering and sampling rate changes [2] but only consider "exact" methods. One such algorithm is due to Goertzel [68] which is based on the complex resonator idea. It is very efficient if only a few outputs of the FFT are required. A direct approach to the problem consists in pruning the flowgraph of the complete FFT so as to disregard redundant paths (corresponding to zero inputs or unwanted outputs). As an inspection of a flowgraph quickly shows, the achievable gains are not spectacular, mainly because of the fact that data communication is not local (since all arithmetic improvements in the FFT over the DFT are achieved through data shuffling).

More complex methods are therefore necessary in order to achieve the gains one would expect. Such methods lead to an order of $N \log_2 K$ operations, where N is the transform size and K the number of active inputs or outputs [48]. Reference [78] also provides a method combining Goertzel's method with shorter FFT algorithms. Note that the problems of input and output pruning are dual, and that algorithm for one problem can be applied to the other by transposition.

7.8.3 Related Transforms

Two transforms which are intimately related to the DFT are the discrete Hartley transform (DHT) [61,62] and the discrete cosine transform (DCT) [1,59]. The former has been proposed as an alternative for the real DFT and the latter is widely used in image processing.

The DHT is defined by

$$X_k = \sum_{n=0}^{N-1} x_n(\cos(2\pi nk/N) + \sin(2\pi nk/N)) \tag{7.67}$$

and is self-inverse, provided that X_0 is further weighted by $1/\sqrt{2}$. Initial claims for the DHT were

- Improved arithmetic efficiency. This was soon recognized to be false, when compared to the real DFT. The structures of both programs are very similar and their arithmetic complexities are equivalent (DHTs actually require slightly more additions than real-valued FFTs).
- Self-inverse property. It has been explained above that the inverse real DFT on Hermitian data has exactly the same complexity as the real DFT (by transposition). If the transposed algorithm is not available, it can be found in [65] how to compute the inverse of a real DFT with a real DFT with only a minor increase in additive complexity.

Therefore, there is no computational gain in using a DHT, and only a minor structural gain if an inverse real DFT cannot be used.

The DCT, on the other hand, has found numerous applications in image and video processing. This has led to the proposal of several fast algorithms for its computation [51,64,70,72]. The DCT is defined by

$$X_k = \sum_{n=0}^{N-1} x_n \cos(2\pi(2k+1)n/4N). \tag{7.68}$$

A scale factor of $1/\sqrt{2}$ for X_0 has been left out in Equation 7.68, mainly because the above transform appears as a subproblem in a length-$4N$ real DFT [51]. From this, the multiplicative complexity of the DCT can be related to that of the real DFT as [69]

$$\mu(\mathrm{DCT}(N)) = (\mu(\mathrm{real}-\mathrm{DFT}(4N)) - \mu(\mathrm{real}-\mathrm{DFT}(2N)))/2. \tag{7.69}$$

Practical algorithms for the DCT depend, as expected, on the transform length.

- N odd: The DCT can be mapped through permutations and sign changes only into a same length real DFT [69].
- N even: The DCT can be mapped into a same length real DFT plus $N/2$ rotations [51]. This is not the optimal algorithm [69,100] but, however, a very practical one.

Other sinusoidal transforms [71], like the discrete sine transform, can be mapped into DCTs as well, with permutations and sign changes only. The main point of this paragraph is that DHTs, DCTs, and other related sinusoidal transforms can be mapped into DFTs, and therefore one can resort to the vast and mature body of knowledge that exists for DFTs. It is worth noting that so far, for all sinusoidal transforms that have been considered, a mapping into a DFT has always produced an algorithm that is at least as efficient as any direct factorization. And if an improvement is ever achieved with a direct factorization, then it could be used to improve the DFT as well. This is the main reason why establishing equivalences between computational problems is fruitful, since it allows improvement of the whole class when any member can be improved.

Figure 7.11 shows the various ways the different transforms are related: starting from any transform with the best-known number of operations, you may obtain by following the appropriate arrows the corresponding transform for which the minimum number of operations will be obtained as well.

1. a	Complex DFT 2^n	2 real DFT's 2^n
		$+2^{n+1} - 4$ additions
b	Real DFT 2^n	1 real DFT $2^{n-1} + 1$ complex DFT 2^{n-2}
		$+(3.2^{n-2} - 4)$ multiplications $+ (2^n + 3.2^{n-2} - n)$ additions
2. a	Real DFT 2^n	1 real DFT $2^{n-1} + 2$ DCT's 2^{n-2}
		$+3.2^{n-1} - 2$ additions
b	DCT 2^n	1 real DFT 2^n
		$+(3.2^{n-1} - 2)$ multiplications $+ (3.2^{n-1} - 3)$ additions
3. a	Complex DFT 2^n	1 odd DFT $2^{n-1} + 1$ complex DFT 2^{n-1}
		$+2^{n+1}$ additions
b	Odd DFT 2^{n-1}	2 complex DFT's 2^{n-2}
		$+2(3.2^{n-2} - 4)$ multiplications $+ (2^n + 3.2^{n-1} - 8)$ additions
4. a	Real DFT 2^n	1 DHT 2^n
		-2 additions
b	DHT 2^n	1 real DFT 2^n
		$+2$ additions
5.	Complex DFT $2^n \times 2^n$	3.2^{n-1} odd DFT $2^{n-1} + 1$ complex DFT $2^{n-1} \times 2^{n-1}$
		$+n \cdot 2^n$ additions
6. a	Real DFT 2^n	1 real symmetric DFT $2^n + 1$ real antisymmetric DFT 2^n
		$+(6n + 10) \cdot 4^{n-1}$ additions
b	Real symm DFT 2^n	1 real symmetric DFT $2^{n-1} + 1$ inverse real DFT
		$+3(2^{n-3} - 1) + 1$ multiplications $+ (3n-4) \cdot 2^{n-3} + 1$ additions

FIGURE 7.11 (a) Consistency of the split-radix-based algorithms. Path showing the connections between the various transforms.

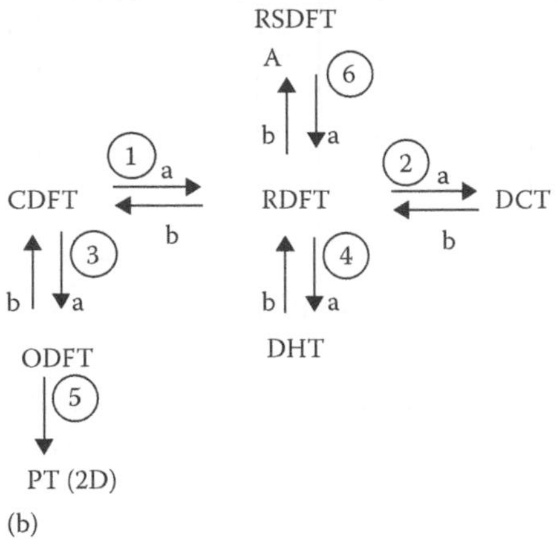

FIGURE 7.11 (continued) (b) Consistency of the split-radix-based algorithms. Weighting of each connection in terms of real operations.

7.9 Multidimensional Transforms

We have already seen in Sections 7.4 and 7.5 that both types of divide and conquer strategies resulted in a multidimensional transform with some particularities. In the case of the Cooley–Tukey mapping, some "twiddle factors" operations had to be performed between the treatment of both dimensions, while in the Good's mapping, the resulting array had dimensions that were coprime.

Here, we shall concentrate on true 2-D FFTs with the same size along each dimension (generalization to more dimensions is usually straightforward).

Another characteristic of the 2-D case is the large memory size required to store the data. It is therefore important to work in-place. As a consequence, in-place programs performing FFTs on real data are also more important in the 2-D case, due to this memory size problem. Furthermore, the required memory is often so large that the data are stored in mass memory and brought into core memory when required, by rows or columns. Hence, an important parameter when evaluating 2-D FFT algorithms is the amount of memory calls required for performing the algorithm.

The 2-D DFT to be computed is defined as follows:

$$X_{k,r} = \sum_{i=0}^{N-1} \sum_{j=0}^{N-1} x_{i,j} W_N^{ik+jr}, \quad k,r = 0,\ldots,N-1. \tag{7.70}$$

The methods for computing this transform are distributed in four classes: row-column algorithms, vector-radix (VR) algorithms, nested algorithms, and polynomial transform algorithms. Among them, only the VR and the polynomial transform were specifically designed for the 2-D case. We shall only give the basic principles underlying these algorithms and refer to the literature for more details.

7.9.1 Row–Column Algorithms

Since the DFT is separable in each dimension, the 2-D transform given in Equation 7.70 can be performed in two steps, as was explained for the PFA:

- First compute N FFTs on the columns of the data
- Then compute N FFTs on the rows of the intermediate result

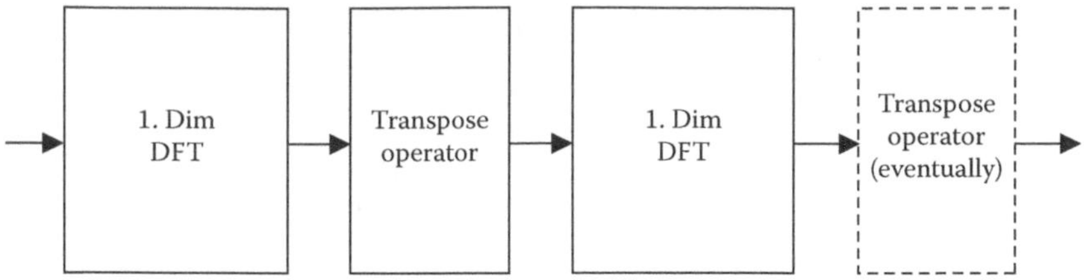

FIGURE 7.12 Row–column implementation of the 2-D FFT.

Nevertheless, when considering 2-D transforms, one should not forget that the size of the data becomes huge quickly: a length 1024×1024 DFT requires 10^6 words of storage, and the matrix is therefore stored in mass memory. But, in that case, accessing a single data is not more costly than reading the whole block in which it is stored. An important parameter is then the number of memory accesses required for computing the 2-D FFT.

This is why the row–column FFT is often performed as shown in Figure 7.12, by performing a matrix transposition between the FFTs on the columns and the FFTs on the rows, in order to allow an access to the data by blocks. Row–column algorithms are very easily implemented and only require efficient 1-D FFTs, as described before, together with a matrix transposition algorithm (for which an efficient algorithm [84] was proposed). Note, however, that the access problem tends to be reduced with the availability of huge core memories.

7.9.2 Vector-Radix Algorithms

A computationally more efficient way of performing the 2-D FFT is a direct approach to the multidimensional problem: the VR algorithm [85,91,92].

They can easily be understood through an example: the radix-2 DIT VRFFT.

This algorithm is based on the following decomposition:

$$X_{k,r} = \sum_{i=0}^{N/2-1} \sum_{j=0}^{N/2-1} x_{2i,2j} W_{N/2}^{ik+jr} + W_N^k \sum_{i=0}^{N/2-1} \sum_{j=0}^{N/2-1} x_{2i+1,2j} W_{N/2}^{ik+jr}$$
$$+ W_N^r \sum_{i=0}^{N/2-1} \sum_{j=0}^{N/2-1} x_{2i,2j+1} W_{N/2}^{ik+jr} + W_N^{k+r} \sum_{i=0}^{N/2-1} \sum_{j=0}^{N/2-1} x_{2i+1,2j+1} W_{N/2}^{ik+jr}, \quad (7.71)$$

and the redundancy in the computation of $X_{k,r}$, $X_{k+N/2,r}$, $X_{k,r+N/2}$, and $X_{k+N/2,r+N/2}$ leads to simplifications which allow reduction of the arithmetic complexity.

This is the same approach as was used in the CTFFTs, the decomposition being applied to both indices altogether.

Of course, higher radix decompositions or split-radix decompositions are also feasible [86], the main difference being that the vector-radix SRFFT, as derived in [86], although being more efficient than the one in [90], is not the algorithm with the lowest arithmetic complexity in that class: For the 2-D case, the best algorithm is not only a mixture of radices 2 and 4.

Figure 7.13 shows what kinds of decompositions are performed in the various algorithms. Due to the fact that the VR algorithms are true generalizations of the Cooley–Tukey approach, it is easy to realize that they will be obtained by repetitive use of small blocks of the same type (the "butterflies," by extension). Figure 7.14 provides the basic butterfly for a vector radix-2 FFT, as derived by Equation 7.71. It should be clear, also, from Figure 7.13 that the complexity of these butterflies increases very quickly with the radix: a radix-2 butterfly involves 4 inputs (it is a 2×2 DFT followed by some "twiddle factors"), while VR4 and VSR butterflies involve 16 inputs.

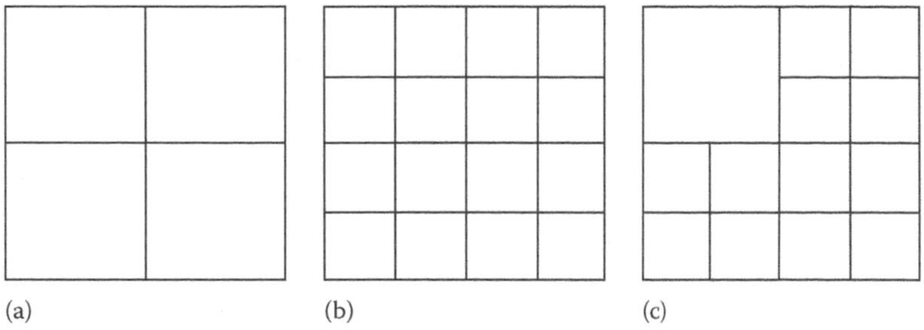

FIGURE 7.13 Decomposition performed in various vector-radix algorithms: (a) VR2, (b) VR4, and (c) VSR.

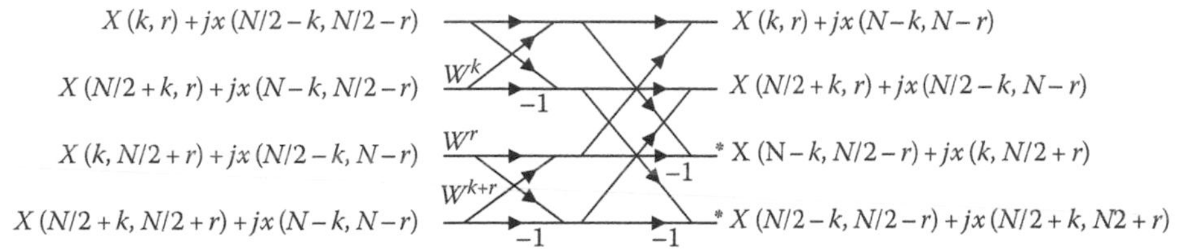

FIGURE 7.14 General VR2 butterfly.

Note also that the only VR algorithms that have seriously been considered all apply to lengths that are powers of 2, although other radices are of course feasible.

The number of read/write cycles of the whole set of data needed to perform the various FFTs of this class, compared to the row–column algorithm, can be found in [86].

7.9.3 Nested Algorithms

They are based on the remark that the nesting property used in Winograd's algorithm, as explained in Section 7.5.3, is not bound to the fact that the lengths are coprime (this requirement was only needed for Good's mapping). Hence, if the length of the DFT allows the corresponding 1-D DFT to be of a nested type (product of mutually prime factors), it is possible to nest further the multiplications, so that the overall 2-D algorithm is also nested.

The number of multiplications thus obtained are very low (see Table 7.4), but the main problem deals with memory requirements: WFTA is not performed in-place, and since all multiplications are nested,

TABLE 7.4 Number of Nontrivial Real Multiplications per Output Point for Various 2-D FFTs on Real Data

$N \times N$ (WFTA)	$N \times N$ (Others)	R.C.	VR2	VR4	VSR	WFTA	PT
	2×2	0	0		0		0
	4×4	0	0	0	0		0
	8×8	0.5	0.375		0.375		0.375
	16×16	1.25	1.25	0.844	0.844		0.844
30×30	32×32	2.125	2.062		1.43	1.435	1.336
	64×64	3.0625	3.094	2.109	2.02		1.834
120×120	128×128	4.031	4.172		2.655	1.4375	2.333
240×240	256×256	5.015	5.273	3.48	3.28	1.82	2.833
504×504	512×512	6.008	6.386		3.92	2.47	3.33
1008×1008	1024×1024	7.004	7.506	4.878	4.56	3.12	3.83

it requires the availability of a number of memory locations equal to the number of multiplications involved in the algorithms. For a length 1008 × 1008 FFT, this amounts to about 6×10^6 locations. This restricts the practical usefulness of these algorithms to small- or medium-length DFTs.t

7.9.4 Polynomial Transform

Polynomial transforms were first proposed by Nussbaumer [74] for the computation of 2-D cyclic convolutions. They can be seen as a generalization of Fourier transforms in the field of polynomials. Working in the field of polynomials resulted in a simplification of the multiplications by the root of unity, which was changed from a complex multiplication to a vector reordering. This powerful approach was applied in [87,88] to the computation of 2-D DFTs as follows.

Let us consider the case where $N = 2^n$, which is the most common case. The 2-D DFT of Equation 7.70 can be represented by the following three polynomial equations:

$$X_i(z) = \sum_{j=0}^{N-1} x_{i,j} \cdot z^j, \tag{7.72a}$$

$$\bar{X}_k(z) = \sum_{i=0}^{N-1} X_i(z) W_N^{ik} \bmod (z^N - 1), \tag{7.72b}$$

$$X_{k,r} = X_{k,r} \bar{X}_k(z) \bmod \left(z - W_N^r\right). \tag{7.72c}$$

This set of equations can be interpreted as follows: Equation 7.72a writes each row of the data as a polynomial, Equation 7.72b computes explicitly the DFTs on the columns, while Equation 7.72c computes the DFTs on the rows as a polynomial reduction (it is merely the equivalent of Equation 7.5). Note that the modulo operation in Equation 7.72b is not necessary (no polynomial involved has a degree greater than N), but it will allow a divide and conquer strategy on Equation 7.72c.

In fact, since $(z^N - 1) = (z^{N/2} - 1)(z^{N/2} + 1)$, the set of two Equations 7.72b and 7.72c can be separated into two cases, depending on the parity of r:

$$\bar{X}_k^1(z) = \sum_{i=0}^{N-1} X_i(z) W_N^{ik} \bmod (z^{N/2} - 1), \tag{7.73a}$$

$$X_{k,2r} = \bar{X}_k^1(z) \bmod \left(z - W_N^{2r}\right), \tag{7.73b}$$

$$\bar{X}_k^2(z) = \sum_{i=0}^{N-1} X_i(z) W_N^{ik} \bmod (z^{N/2} + 1), \tag{7.74a}$$

$$X_{k,2r+1} = \bar{X}_k^2(z) \bmod \left(z - W_N^{2r+1}\right). \tag{7.74b}$$

Equation 7.73 is still of the same type as the initial one, hence the same procedure as the one being derived will apply. Let us now concentrate on Equation 7.74 which is now recognized to be the key aspect of the problem.

Since $(2r + 1, N) = 1$, the permutation $(2r + 1) \cdot k (\bmod N)$ maps all values of k, and replacing k with $(2r + 1) \cdot k$ in Equation 7.73a will merely result in a reordering of the outputs:

$$\bar{X}_{k(2r+1)}^2(z) = \sum_{i=0}^{N-1} X_i(z) W_N^{(2r+1)ik} \bmod (z^{N/2} + 1), \tag{7.75a}$$

$$X_{k(2r+1),2r+1} = \bar{X}_{k(2r+1)}^2(z) \bmod \left(z - W_N^{2r+1}\right), \tag{7.75b}$$

and, since $z = W_N^{2r+1}$ in Equation 7.75b, we can replace W_N^{2r+1} by z in Equation 7.75a:

$$\bar{X}^2_{k(2r+1)}(z) = \sum_{i=0}^{N-1} X_i(z) z^{ik} \bmod (Z^{N/2} + 1), \tag{7.76}$$

which is exactly a polynomial transform, as defined in [74]. This polynomial transform can be computed using an FFT-type algorithm, without multiplications, and with only $N^2/2\log_2 N$ additions.

$X_{k,2r+1}$ will now be obtained by application of Equation 7.75b. $\bar{X}^2(z)$ being computed mod $(z^{N/2} + 1)$ is of degree $N/2 - 1$. For each k, Equation 7.75b will then correspond to the reduction of one polynomial modulo the odd powers of W_N. From Equation 7.5, this is seen to be the computation of the odd outputs of a length N DFT, which is sometimes called an odd DFT.

The terms $X_{k,2r+1}$ are seen to be obtained by one reduction mod $(z^{N/2} + 1)$ (Equation 7.74), one polynomial transform of N terms mod $Z^{N/2} + 1$ (Equation 7.76), and N odd DFTs. This procedure is then iterated on the terms $X_{2k+1,2r}$, by using exactly the same algorithm, the role of k and r being interchanged. $X_{2k,2r}$ is exactly a length $N/2 \times N/2$ DFT, on which the same algorithm is recursively applied.

In the first version of the polynomial transform computation of the 2-D FFT, the odd DFT was computed by a real-factor algorithm, resulting in an excess in the number of additions required.

As seen in Tables 7.4 and 7.5, where the number of multiplications and additions for the various 2-D FFT algorithms are given, the polynomial transform approach results in the algorithm requiring the lowest arithmetic complexity, when counting multiplications and additions altogether. The addition counts given in Table 7.5 are updates of the previous ones, assuming that the odd DFTs are computed by a split-radix algorithm.

Note that the same kind of performance was obtained by Auslander et al. [82,83] with a similar approach which, while more sophisticated, gave a better insight on the mathematical structure of this problem. Polynomial transforms were also applied to the computation of 2-D DCT [52,79].

7.9.5 Discussion

A number of conclusions can be stated by considering Tables 7.4 and 7.5, keeping the principles of the various methods in mind.

VR2 is more complicated to implement than row–column algorithms, and requires more operations for lengths greater than equal to 32. Therefore, it should not be considered. Note that this result holds only because efficient and compact 1-D FFTs, such as SRFFT, have been developed.

The row–column algorithm is the one allowing the easiest implementation, while having a reasonable arithmetic complexity. Furthermore, it is easily parallelized, and simplifications can be found for the reorderings (bit reversal and matrix transposition [66]), allowing one of them to be free in nearly any

TABLE 7.5 Number of Real Additions per Output Point for Various 2-D FFTs on Real Data

$N \times N$	$N \times N$ (Others)	R.C.	VR2	VR4	VSR	WFTA	PT
	2×2	2	2		2		2
	4×4	3.25	3.25	3.25	3.25		3.25
	8×8	5.56	5.43		5.43		5.43
	16×16	8.26	8.14	7.86	7.86		7.86
	32×32	11.13	11.06		10.43	12.98	10.34
	64×64	14.06	14.09	13.11	13.02		12.83
	128×128	17.03	17.17		15.65	17.48	15.33
	256×256	20.01	20.27	18.48	17.67	22.79	17.83
	512×512	23.00	23.38		20.92	34.42	20.33
	1024×1024	26.00	26.5	23.88	23.56	45.30	22.83

kind of implementation. WFTA has a huge number of additions (twice the number required for the other algorithms for $N = 1024$), requires huge memory, has a difficult implementation, but requires the least multiplications. Nevertheless, we think that, in today's implementations, this advantage will in general not outweigh its drawbacks.

VSR is difficult to implement, and will certainly seldom defeat VR4, except in very special cases (huge memory available and N very large).

VR4 is a good compromise between structural and arithmetic complexity. When row–column algorithms are not fast enough, we think it is the next choice to be considered.

Polynomial transforms have the greatest possibilities: lowest arithmetic complexity, possibility of in-place computation, but very little work was done on the best way of implementing them. It was even reported to be slower than VR2 [103]. Nevertheless, it is our belief that looking for efficient implementations of polynomial transform based FFTs is worth the trouble. The precise understanding of the link between VR algorithms and polynomial transforms may be a useful guide for this work.

7.10 Implementation Issues

It is by now well recognized that there is a strong interaction between the algorithm and its implementation. For example, regularity, as discussed before, will only pay off if it is closely matched by the target architecture. This is the reason why we will discuss in the sequel different types of implementations. Note that very often, the difference in computational complexity between algorithms is not large enough to differentiate between the efficiency of the algorithm and the quality of the implementation.

7.10.1 General Purpose Computers

FFT algorithms are built by repetitive use of basic building blocks. Hence, any improvement (even small) in these building blocks will pay in the overall performance. In the Cooley–Tukey or the split-radix case, the building blocks are small and thus easily optimizable, and the effect of improvements will be relatively more important than in the PFA/WFTA case where the blocks are larger.

When monitoring the amount of time spent in various elementary ftoating point operations, it is interesting to note that more time is spent in load/store operations than in actual arithmetic computations [30,107,109] (this is due to the fact that memory access times are comparable to ALU cycle times on current machines). Therefore, the locality of the algorithm is of paramount importance. This is why the PFA and WFTA do not meet the performance expected from their computational complexity only.

On another side, this drawback of PFA is compensated by the fact that only a few coefficients have to be stored. On the contrary, classical FFTs must store a large table of sine and cosine values, calculate them as needed, or update them with resulting roundoff errors.

Note that special automatic code generation techniques have been developed in order to produce efficient code for often used programs like the FFT. They are based on a "de-looping" technique that produces loop free code from a given piece of code [107]. While this can produce unreasonably large code for large transforms, it can be applied successfully to sub-transforms as well.

7.10.2 Digital Signal Processors

DSPs strongly favor multiply/accumulate based algorithms. Unfortunately, this is not matched by any of the fast FFT algorithms (where sums of products have been changed to fewer but less regular computations). Nevertheless, DSPs now take into account some of the FFT requirements, like modulo counters and bit-reversed addressing. If the modulo counter is general, it will help the implementation of all FFT algorithms, but it is often restricted to the CTFFT/SRFFT case only (modulo a power of 2) for which efficient timings are provided on nearly all available machines by manufacturers, at least for small to medium lengths.

7.10.3 Vector Processor and Multiprocessor

Implementations of Fourier transforms on vectorized computers must deal with two interconnected problems [93]. First, the vector (the size of data that can be processed at the maximal rate) has to be full as often as possible. Then, the loading of the vector should be made from data available inside the cache memory (as in general purpose computers) in order to save time. The usual hardware design parameters will, in general, favor length-2^m FFT implementations. For example, a radix-4 FFT was reported to be efficiently realized on a commercial vector processor [93].

In the multiprocessor case, the performance will be dependent on the number and power of the processing nodes but also strongly on the available interconnection network. Because the FFT algorithms are deterministic, the resource allocation problem can be solved off-line. Typical configurations include arithmetic units specialized for butterfly operations [98], arrays with attached shuffle networks, and pipelines of arithmetic units with intermediate storage and reordering [17]. Obviously, these schemes will often favor classical Cooley–Tukey algorithms because of their high regularity. However, SRFFT or PFA implementations have not been reported yet, but could be promising in high-speed applications.

7.10.4 VLSI

The discussion of partially dedicated multi-processors leads naturally to fully dedicated hardware structures like the ones that can be realized in very large scale integration (VLSI) [9,11]. As a measure of efficiency, both chip area (A) and time (T) between two successive DFT computations (setup times are neglected since only throughput is of interest) are of importance. Asymptotic lower bounds for the product $A \cdot T^2$ have been reported for the FFT [116] and lead to

$$\Omega_{AT^2}(\mathrm{DFT}(N)) = N^2 \log^2(N), \tag{7.77}$$

that is, no circuit will achieve a better behavior than Equation 7.77 for large N. Interestingly, this lower bound is achieved by several algorithms, notably the algorithms based on shuffle-exchange networks and the ones based on square grids [96,114]. The trouble with these optimal schemes is that they outperform more traditional ones, like the cascade connection with variable delay [98] (which is asymptotically suboptimal), only for extremely large N's and are therefore not relevant in practice [96].

Dedicated chips for the FFT computation are therefore often based on some traditional algorithm which is then efficiently mapped into a layout. Examples include chips for image processing with small size DCTs [115] as well as wafer scale integration for larger transforms. Note that the cost is dominated both by the number of multiplications (which outweigh additions in VLSI) and the cost of communication. While the former figure is available from traditional complexity theory, the latter one is not yet well studied and depends strongly on the structure of the algorithm as discussed in Section 7.7. Also, dedicated arithmetic units suited for the FFT problem have been devised, like the butterfly unit [98] or the CORDIC unit [94,97] and contribute substantially to the quality of the overall design. But, similarly to the software case, the realization of an efficient VLSI implementation is still more an art than a mere technique.

7.11 Conclusion

The purpose of this chapter has been threefold: a tutorial presentation of classic and recent results, a review of the state of the art, and a statement of open problems and directions.

After a brief history of the FFT development, we have shown by simple arguments, that the fundamental technique used in all FFT algorithms, namely the divide and conquer approach, will always improve the computational efficiency.

Then, a tutorial presentation of all known FFT algorithms has been made. A simple notation, showing how various algorithms perform various divisions of the input into periodic subsets, was used as the basis for a unified presentation of CTFFT, SRFFT, PFA, and Winograd FFT algorithms. From this chapter, it is clear that Cooley–Tukey and split-radix algorithms are instances of one family of FFT algorithms, namely FFTs with twiddle factors.

The other family is based on a divide and conquer scheme (Good's mapping) which is costless (computationally speaking). The necessary tools for computing the short-length FFTs which then appear were derived constructively and led to the discussion of the PFA and of the WFTA.

These practical algorithms were then compared to the best possible ones, leading to an evaluation of their suboptimality. Structural considerations and special cases were addressed next. In particular, it was shown that recently proposed alternative transforms like the Hartley transform do not show any advantage when compared to real-valued FFTs.

Special attention was then paid to multidimensional transforms, where several open problems remain. Finally, implementation issues were outlined, indicating that most computational structures implicitly favor classical algorithms. Therefore, there is room for improvements if one is able to develop architectures that match more recent and powerful algorithms.

Acknowledgments

The authors would like to thank Professor M. Kunt for inviting them to write this chapter, as well as for his patience. Professor C. S. Burrus, Dr. J. Cooley, Dr. M. T. Heideman, and Professor H. J. Nussbaumer are also thanked for fruitful interactions on the subject of this chapter. We are indebted to J. S. White, J. C. Bic, and P. Gole for their careful reading of the manuscript.

References

Books

1. Ahmed, N. and Rao, K.R., *Orthogonal Transforms for Digital Signal Processing*, Springer, Berlin, Germany, 1975.
2. Blahut, R.E., *Fast Algorithms for Digital Signal Processing*, Addison-Wesley, Reading, MA, 1986.
3. Brigham, E.O., *The Fast Fourier Transform*, Prentice-Hall, Englewood Cliffs, NJ, 1974.
4. Burrus, C.S. and Parks, T.W., *DFT/FFT and Convolution Algorithms*, John Wiley & Sons, New York, 1985.
5. Burrus, C.S., Efficient Fourier transform and convolution algorithms, in: J.S. Lim and A.V. Oppenheim (Eds.), *Advanced Topics in Digital Signal Processing*, Prentice-Hall, Englewood Cliffs, NJ, 1988.
6. Digital Signal Processing Committee (Ed.), *Selected Papers in Digital Signal Processing*, Vol. II, IEEE Press, New York, 1975.
7. Digital Signal Processing Committee (Ed.), *Programs for Digital Signal Processing*, IEEE Press, New York, 1979.
8. Heideman, M.T., *Multiplicative Complexity, Convolution and the DFT*, Springer, Berlin, Germany, 1988.
9. Kung, S.Y., Whitehouse, H.J., and Kailath, T. (Eds.), *VLSI and Modern Signal Processing*, Prentice-Hall, Englewood Cliffs, NJ, 1985.
10. McClellan, J.H. and Rader, C.M., *Number Theory in Digital Signal Processing*, Prentice-Hall, Englewood Cliffs, NJ, 1979.
11. Mead, C. and Conway, L., *Introduction to VLSI*, Addison-Wesley, Reading, MA, 1980.
12. Nussbaumer, H.J., *Fast Fourier Transform and Convolution Algorithms*, Springer, Berlin, Germany, 1982.
13. Oppenheim, A.V. (Ed.), *Papers on Digital Signal Processing*, MIT Press, Cambridge, MA, 1969.
14. Oppenheim, A.V. and Schafer, R.W., *Digital Signal Processing*, Prentice-Hall, Englewood Cliffs, NJ, 1975.

15. Rabiner, L.R. and Rader, C.M. (Eds.), *Digital Signal Processing*, IEEE Press, New York, 1972.
16. Rabiner, L.R. and Gold, B., *Theory and Application of Digital Signal Processing*, Prentice-Hall, Englewood Cliffs, NJ, 1975.
17. Schwartzlander, E.E., *VLSI Signal Processing Systems*, Kluwer Academic Publishers, Dordrecht, the Netherlands, 1986.
18. Soderstrand, M.A., Jenkins, W.K., Jullien, G.A., and Taylor, F.J. (Eds.), *Residue Number System Arithmetic: Modern Applications in Digital Signal Processing*, IEEE Press, New York, 1986.
19. Winograd, S., *Arithmetic Complexity of Computations*, SIAM CBMS-NSF Series, No. 33, SIAM, Philadelphia, PA, 1980.

1-D FFT Algorithms

20. Agarwal, R.C. and Burrus, C.S., Fast one-dimensional digital convolution by multi-dimensional techniques, *IEEE Trans. Acoust. Speech Signal Process.*, ASSP-22(1): 1–10, February 1974.
21. Bergland, G.D., A fast Fourier transform algorithm using base 8 iterations, *Math. Comp.*, 22(2): 275–279, April 1968 (reprinted in [13]).
22. Bruun, G., z-Transform DFT filters and FFTs, *IEEE Trans. Acoust. Speech Signal Process.*, ASSP-26 (1): 56–63, February 1978.
23. Burrus, C.S., Index mappings for multidimensional formulation of the DFT and convolution, *IEEE Trans. Acoust. Speech Signal Process.*, ASSP-25(3): 239–242, June 1977.
24. Cho, K.M. and Temes, G.C., Real-factor FFT algorithms, *Proceedings of the IEEE International Conference on Acoustics, Speech and Signal Processing*, Tulsa, OK, April 1978, pp. 634–637.
25. Cooley, J.W. and Tukey, J.W., An algorithm for the machine calculation of complex Fourier series, *Math. Comp.*, 19: 297–301, April 1965.
26. Dubois, P. and Venetsanopoulos, A.N., A new algorithm for the radix-3 FFT, *IEEE Trans. Acoust. Speech Signal Process.*, ASSP-26: 222–225, June 1978.
27. Duhamel, P. and Hollmann, H., Split-radix FFT algorithm, *Electron. Lett.*, 20(1): 14–16, January 5, 1984.
28. Duhamel, P. and Hollmann, H., Existence of a 2^n FFT algorithm with a number of multiplications lower than 2^{n+1}, *Electron. Lett.*, 20(17): 690–692, August 1984.
29. Duhamel, P., Un algorithme de transformation de Fourier rapide à double base, *Annales des Telecommunications*, 40(9–10): 481–494, September 1985.
30. Duhamel, P., Implementation of "split-radix" FFT algorithms for complex, real and real-symmetric data, *IEEE Trans. Acoust. Speech Signal Process.*, ASSP-34(2): 285–295, April 1986.
31. Duhamel, P., Algorithmes de transformés discrètes rapides pour convolution cyclique et de convolution cyclique pour transformés rapides, Thèse de doctorat d'état, Université Paris XI, Paris, September 1986.
32. Good, I.J., The interaction algorithm and practical Fourier analysis, *J. R. Stat. Soc. Ser. B*, B-20: 361–372, 1958; B-22, 372–375, 1960.
33. Heideman, M.T. and Burrus, C.S., A bibliography of fast transform and convolution algorithms II, Technical Report No. 8402, Rice University, Houston, TX, February 24, 1984.
34. Heideman, M.T., Johnson, D.H., and Burrus, C.S., Gauss and the history of the FFT, *IEEE Acoust. Speech Signal Process.*, 1(4): 14–21, October 1984.
35. Heideman, M.T. and Burrus, C.S., On the number of multiplications necessary to compute a length-2^n DFT, *IEEE Trans. Acoust. Speech Signal Process.*, ASSP-34(1): 91–95, February 1986.
36. Heideman, M.T., Application of multiplicative complexity theory to convolution and the discrete Fourier transform, PhD Thesis, Department of Electrical and Computer Engineering, Rice University, Houston, TX, April 1986.
37. Johnson, H.W. and Burrus, C.S., Large DFT modules: 11, 13, 17, 19, and 25, Technical Report No. 8105, Department of Electrical and Computer Engineering, Rice University, Houston, TX, December 1981.

38. Johnson, H.W. and Burrus, C.S., The design of optimal DFT algorithms using dynamic programming, *IEEE Trans. Acoust. Speech Signal Process.*, ASSP-31(2): 378–387, 1983.
39. Kolba, D.P. and Parks, T.W., A prime factor algorithm using high-speed convolution, *IEEE Trans. Acoust. Speech Signal Process.*, ASSP-25: 281–294, August 1977.
40. Martens, J.B., Recursive cyclotomic factorization—A new algorithm for calculating the discrete Fourier transform, *IEEE Trans. Acoust. Speech Signal Process.*, ASSP-32(4): 750–761, August 1984.
41. Nussbaumer, H.J., Efficient algorithms for signal processing, *Second European Signal Processing Conference*, EUSIPC0-83, Erlangen, Germany, September 1983.
42. Preuss, R.D., Very fast computation of the radix-2 discrete Fourier transform, *IEEE Trans. Acoust. Speech Signal Process.*, ASSP-30: 595–607, August 1982.
43. Rader, C.M., Discrete Fourier transforms when the number of data samples is prime, *Proc. IEEE*, 56: 1107–1008, 1968.
44. Rader, C.M. and Brenner, N.M., A new principle for fast Fourier transformation, *IEEE Trans. Acoust. Speech Signal Process.*, ASSP-24: 264–265, June 1976.
45. Singleton, R., An algorithm for computing the mixed radix fast Fourier transform, *IEEE Trans. Audio Electroacoust.*, AU-17: 93–103, June 1969 (reprinted in [13]).
46. Stasinski, R., Asymmetric fast Fourier transform for real and complex data, *IEEE Trans. Acoust. Speech Signal Process.*, unpublished manuscript.
47. Stasinski, R., Easy generation of small-N discrete Fourier transform algorithms, *IEE Proc.*, Part G, 133(3): 133–139, June 1986.
48. Stasinski, R., FFT pruning. A new approach, *Proc. Eusipco 86*, 1986, pp. 267–270.
49. Suzuki, Y., Sone, T., and Kido, K., A new FFT algorithm of radix 3, 6, and 12, *IEEE Trans. Acoust. Speech Signal Process.*, ASSP-34(2): 380–383, April 1986.
50. Temperton, C., Self-sorting mixed-radix fast Fourier transforms, *J. Comput. Phys.*, 52(1): 1–23, October 1983.
51. Vetterli, M. and Nussbaumer, H.J., Simple FFT and DCT algorithms with reduced number of operations, *Signal Process.*, 6(4): 267–278, August 1984.
52. Vetterli, M. and Nussbaumer, H.J., Algorithmes de transformé de Fourier et cosinus mono et bi-dimensionnels, *Annales des Télécommunications*, Tome 40(9–10): 466–476, September–October 1985.
53. Vetterli, M. and Duhamel, P., Split-radix algorithms for length-p^m DFTs, *IEEE Trans. Acoust. Speech Signal Process.*, ASSP-37(1): 57–64, January 1989.
54. Winograd, S., On computing the discrete Fourier transform, *Proc. Nat. Acad. Sci. U.S.A.*, 73: 1005–1006, April 1976.
55. Winograd, S., Some bilinear forms whose multiplicative complexity depends on the field of constants, *Math. Syst. Theory*, 10(2): 169–180, 1977 (reprinted in [10]).
56. Winograd, S., On computing the DFT, *Math. Comp.*, 32(1): 175–199, January 1978 (reprinted in [10]).
57. Winograd, S., On the multiplicative complexity of the discrete Fourier transform, *Adv. Math.*, 32(2): 83–117, May 1979.
58. Yavne, R., An economical method for calculating the discrete Fourier transform, *AFIPS Proceedings, Fall Joint Computer Conference*, Washington D.C., 1968, Vol. 33, pp. 115–125.

Related Algorithms

59. Ahmed, N., Natarajan, T., and Rao, K.R., Discrete cosine transform, *IEEE Trans. Comput.*, C-23: 88–93, January 1974.
60. Bergland, G.D., A radix-eight fast Fourier transform subroutine for real-valued series, *IEEE Trans. Audio Electroacoust.*, 17(1): 138–144, June 1969.
61. Bracewell, R.N., Discrete Hartley transform, *J. Opt. Soc. Am.*, 73(12): 1832–1835, December 1983.
62. Bracewell, R.N., The fast Hartley transform, *Proc. IEEE*, 22(8): 1010–1018, August 1984.

63. Burrus, C.S., Unscrambling for fast DFT algorithms, *IEEE Trans. Acoust. Speech Signal Process.*, ASSP-36(7): 1086–1087, July 1988.
64. Chen, W.-H., Smith, C.H., and Fralick, S.C., A fast computational algorithm for the discrete cosine transform, *IEEE Trans. Commn.*, COM-25: 1004–1009, September 1977.
65. Duhamel, P. and Vetterli, M., Improved Fourier and Hartley transform algorithms. Application to cyclic convolution of real data, *IEEE Trans. Acoust. Speech Signal Process.*, ASSP-35(6): 818–824, June 1987.
66. Duhamel, P. and Prado, J., A connection between bit-reverse and matrix transpose. Hardware and software consequences, *Proceedings of the IEEE Acoustics, Speech and Signal Process*ing, New York, 1988, pp. 1403–1406.
67. Evans, D.M., An improved digit reversal permutation algorithm for the fast Fourier and Hartley transforms, *IEEE Trans. Acoust. Speech Signal Process.*, ASSP-35(8): 1120–1125, August 1987.
68. Goertzel, G., An algorithm for the evaluation of finite Fourier series, *Am. Math. Mon.*, 65(1): 34–35, January 1958.
69. Heideman, M.T., Computation of an odd-length DCT from a real-valued DFT of the same length, *IEEE Trans. Acoust. Speech Signal Process.*, 40(1): 54–61, January 1992.
70. Hou, H.S., A fast recursive algorithm for computing the discrete Fourier transform, *IEEE Trans. Acoust. Speech Signal Process.*, ASSP-35(10): 1455–1461, October 1987.
71. Jain, A.K., A sinusoidal family of unitary transforms, *IEEE Trans. PAMI*, 1(4): 356–365, October 1979.
72. Lee, B.G., A new algorithm to compute the discrete cosine transform, *IEEE Trans. Acoust. Speech Signal Process.*, ASSP-32: 1243–1245, December 1984.
73. Mou, Z.J. and Duhamel, P., Fast FIR filtering: Algorithms and implementations, *Signal Process.*, 13(4): 377–384, December 1987.
74. Nussbaumer, H.J., Digital filtering using polynomial transforms, *Electron. Lett.*, 13(13): 386–386, June 1977.
75. Polge, R.J., Bhaganan, B.K., and Carswell, J.M., Fast computational algorithms for bit-reversal, *IEEE Trans. Comput.*, 23(1): 1–9, January 1974.
76. Duhamel, P., Algorithms meeting the lower bounds on the multiplicative complexity of length-2^n DFTs and their connection with practical algorithms, *IEEE Trans. Acoust. Speech Signal Process.*, ASSP-38: 1504–1511, September 1990.
77. Sorensen, H.V., Jones, D.L., Heideman, M.T., and Burrus, C.S., Real-valued fast Fourier transform algorithms, *IEEE Trans. Acoust. Speech Signal Process.*, ASSP-35(6): 849–863, June 1987.
78. Sorensen, H.V., Burrus, C.S., and Jones, D.L., A new efficient algorithm for computing a few DFT points, *Proceedings of the IEEE International Symposium on Circuits and Systems*, Espoo, Finland, June 1988, pp. 1915–1918.
79. Vetterli, M., Fast 2-D discrete cosine transform, *Proceedings of the IEEE International Conference on Acoustics, Speech, and Signal Processing*, Tampa, FL, March 1985, pp. 1538–1541.
80. Vetterli, M., Analysis, synthesis and computational complexity of digital filter banks, PhD Thesis, Ecole Polytechnique Federale de Lausanne, Switzerland, April 1986.
81. Vetterli, M., Running FIR and IIR filtering using multirate filter banks, *IEEE Trans. Acoust. Speech Signal Process.*, ASSP-36(5): 730–738, May 1988.

Multidimensional Transforms

82. Auslander, L., Feig, E., and Winograd, S., New algorithms for the multidimensional Fourier transform, *IEEE Trans. Acoust. Speech Signal Process.*, ASSP-31(2): 338–403, April 1983.
83. Auslander, L., Feig, E., and Winograd, S., Abelian semisimple algebras and algorithms for the discrete Fourier transform, *Adv. Appl. Math.*, 5: 31–55, 1984.
84. Eklundh, J.O., A fast computer method for matrix transposing, *IEEE Trans. Comput.*, 21(7): 801–803, July 1972 (reprinted in [6]).

85. Mersereau, R.M. and Speake, T.C., A unified treatment of Cooley-Tukey algorithms for the evaluation of the multidimensional DFT, *IEEE Trans. Acoust. Speech Signal Process.*, ASSP-22(5): 320–325, October 1981.
86. Mou, Z.J. and Duhamel, P., In-place butterfly-style FFT of 2-D real sequences, *IEEE Trans. Acoust. Speech Signal Process.*, ASSP-36(10): 1642–1650, October 1988.
87. Nussbaumer, H.J. and Quandalle, P., Computation of convolutions and discrete Fourier transforms by polynomial transforms, *IBM J. Res. Develop.*, 22: 134–144, 1978.
88. Nussbaumer, H.J. and Quandalle, P., Fast computation of discrete Fourier transforms using polynomial transforms, *IEEE Trans. Acoust. Speech Signal Process.*, ASSP-27: 169–181, 1979.
89. Pease, M.C., An adaptation of the fast Fourier transform for parallel processing, *J. Assoc. Comput. Mach.*, 15(2): 252–264, April 1968.
90. Pei, S.C. and Wu, J.L., Split-vector radix 2-D fast Fourier transform, *IEEE Trans. Circuits Syst.*, 34 (1): 978–980, August 1987.
91. Rivard, G.E., Algorithm for direct fast Fourier transform of bivariant functions, *1975 Annual Meeting of the Optical Society of America*, Boston, MA, October 1975.
92. Rivard, G.E., Direct fast Fourier transform of bivariant functions, *IEEE Trans. Acoust. Speech Signal Process.*, 25(3): 250–252, June 1977.

Implementations

93. Agarwal, R.C. and Cooley, J.W., Fourier transform and convolution subroutines for the IBM 3090 Vector Facility, *IBM J. Res. Dev.*, 30(2): 145–162, March 1986.
94. Ahmed, H., Delosme, J.M., and Morf, M., Highly concurrent computing structures for matrix arithmetic and signal processing, *IEEE Trans. Comput.*, 15(1): 65–82, January 1982.
95. Burrus, C.S. and Eschenbacher, P.W., An in-place, in-order prime factor FFT algorithm, *IEEE Trans. Acoust. Speech Signal Process.*, ASSP-29(4): 806–817, August 1981.
96. Card, H.C., VLSI computations: From physics to algorithms, *Integration*, 5: 247–273, 1987.
97. Despain, A.M., Fourier transform computers using CORDIC iterations, *IEEE Trans. Comput.*, 23 (10): 993–1001, October 1974.
98. Despain, A.M., Very fast Fourier transform algorithms hardware for implementation, *IEEE Trans. Comput.*, 28(5): 333–341, May 1979.
99. Duhamel, P., Piron, B., and Etcheto, J.M., On computing the inverse DFT, *IEEE Trans. Acoust. Speech Signal Process.*, ASSP-36(2): 285–286, February 1988.
100. Duhamel, P. and H'mida, H., New 2^n DCT algorithms suitable for VLSI implementation, *Proceedings of the IEEE International Conference on Acoustics, Speech, and Signal Processing*, Dallas, TX, April 1987, pp. 1805–1809.
101. Johnson, J., Johnson, R., Rodriguez, D., and Tolimieri, R., A methodology for designing, modifying, and implementing Fourier transform algorithms on various architectures, preliminary draft, *Circuits Syst. Signal Process.*, 9(4): 449–500, December 1990.
102. Elterich, A. and Stammler, W., Error analysis and resulting structural improvements for fixed point FFT's, *Proceedings of the IEEE International Conference on Acoustics, Speech, and Signal Processing*, New York, April 11–14, 1988, Vol. 3, pp. 1419–1422.
103. Lhomme, B., Morgenstern, J., and Quandalle, P., Implantation de transformés de Fourier de dimension 2^n, *Techniques et Science Informatiques*, 4(2): 324–328, 1985.
104. Manson, D.C. and Liu, B., Floating point roundoff error in the prime factor FFT, *IEEE Trans. Acoust. Speech Signal Process.*, 29(4): 877–882, August 1981.
105. Mescheder, B., On the number of active *-operations needed to compute the DFT, *Acta Informatica*, 13: 383–408, May 1980.
106. Morgenstern, J., The linear complexity of computation, *Assoc. Comput. Mach.*, 22(2): 184–194, April 1975.

107. Morris, L.R., Automatic generation of time efficient digital signal processing software, *IEEE Trans. Acoust. Speech Signal Process.*, ASSP-25: 74–78, February 1977.
108. Morris, L.R., A comparative study of time efficient FFT and WFTA programs for general purpose computers, *IEEE Trans. Acoust. Speech Signal Process.*, ASSP-26: 141–150, April 1978.
109. Nawab H. and McClellan, J.H., Bounds on the minimum number of data transfers in WFTA and FFT programs, *IEEE Trans. Acoust. Speech Signal Process.*, ASSP-27: 394–398, August 1979.
110. Pan, V.Y., The additive and logical complexities of linear and bilinear arithmetic algorithms, *J. Algorithms*, 4(1): 1–34, March 1983.
111. Rothweiler, J.H., Implementation of the in-order prime factor transform for variable sizes, *IEEE Trans. Acoust. Speech Signal Process.*, ASSP-30(1): 105–107, February 1982.
112. Silverman, H.F., An introduction to programming the Winograd Fourier transform algorithm, *IEEE Trans. Acoust. Speech Signal Process.*, ASSP-25(2): 152–165, April 1977, with corrections in: *IEEE Trans. Acoust Speech Signal Process.*, ASSP-26(3): 268, June 1978, and in ASSP-26(5): 482, October 1978.
113. Sorensen, H.V., Heideman, M.T., and Burrus, C.S., On computing the split-radix FFT, *IEEE Trans. Acoust. Speech Signal Process.*, ASSP-34(1): 152–156, February 1986.
114. Thompson, C.D., Fourier transforms in VLSI, *IEEE Trans. Comput.*, 32(11): 1047–1057, November 1983.
115. Vetterli, M. and Ligtenberg, A., A discrete Fourier-cosine transform chip, *IEEE J. Selected Areas Commn.*, Special Issue on VLSI in Telecommunications, SAC-4(1): 49–61, January 1986.
116. Vuillemin, J., A combinatorial limit to the computing power of VLSI circuits, *Proceedings of the 21st Annual Symposium on Foundations of Computer Science*, IEEE Computer Society, Syracuse, NY, October 13–15, 1980, pp. 294–300.
117. Welch, P.D., A fixed-point fast Fourier transform error analysis, *IEEE Trans. Audio Electroacoust.*, 15(2): 70–73, June 1969 (reprinted in [13] and [15]).

Software

FORTRAN (or DSP) code can be found in the following references:

[7] contains a set of classical FFT algorithms.
[111] contains a prime factor FFT program.
[4] contains a set of classical programs and considerations on program optimization, as well as TMS 32010 code.
[113] contains a compact split-radix Fortran program.
[29] contains a speed-optimized SRFFT.
[77] contains a set of real-valued FFTs with twiddle factors.
[65] contains a split-radix real-valued FFT, as well as a Hartley transform program.
[112] as well as [7] contain a Winograd Fourier transform Fortran program.
[66], [67], and [75] contain improved bit-reversal algorithms.

8

Fast Convolution and Filtering

Ivan W. Selesnick
Polytechnic University

C. Sidney Burrus
Rice University

8.1 Introduction

One of the first applications of the Cooley–Tukey fast Fourier transform (FFT) algorithm was to implement convolution faster than the usual direct method [13,25,30]. Finite impulse response (FIR) digital filters and convolution are defined by

$$y(n) = \sum_{k=0}^{L-1} h(k)x(n-k), \tag{8.1}$$

where, for an FIR filter,

$x(n)$ is a length-N sequence of numbers considered to be the input signal
$h(n)$ is a length-L sequence of numbers considered to be the filter coefficients
$y(n)$ is the filtered output

Examination of this equation shows that the output signal $y(n)$ must be a length-$(N + L - 1)$ sequence of numbers, and the direct calculation of this output requires NL multiplications and approximately NL additions (actually, $(N - 1)(L - 1)$). If the signal and filter length are both length-N, we say the arithmetic complexity is of order N^2, $O(N^2)$. Our goal is to calculate this convolution or filtering faster than directly implementing Equation 8.1. The most common way to achieve "fast convolution" is to section or block the signal and use the FFT on these blocks to take advantage of the efficiency of the FFT. Clearly, one disadvantage of this technique is an inherent delay of one block length.

Indeed, this approach is so common as to be almost synonymous with fast convolution. The problem is to implement ongoing, noncyclic convolution with the finite-length, cyclic convolution that the FFT gives. An answer was quickly found in a clever organization of piecing together blocks of data using what is now called the overlap-add method and the overlap-save method. These two methods convolve length-L blocks using one length-L FFT, L complex multiplications, and one length-L inverse FFT [22].

Later this was generalized to arbitrary length blocks or sections to give block convolution and block recursion [5]. By allowing the block lengths to be even shorter than one word (bits and bytes!) we come up with an interesting implementation called distributed arithmetic that requires no explicit multiplications [7,34].

Another approach for improving the efficiency of convolution and recursion uses fast algorithms other than the traditional FFT. One possibility is to use a transform based on number-theoretic roots of unity rather than the usual complex roots of unity [17]. This gives rise to number-theoretic transforms that require no multiplications and no trigonometric functions. Still another method applies Winograd's fast algorithms directly to convolution rather than through the Fourier transform. Finally, we remark that some filters $h(n)$ require fewer arithmetic operations because of their structure.

8.2 Overlap-Add and Overlap-Save Methods for Fast Convolution

If one implements convolution by use of the FFT, then it is cyclic convolution that is obtained. In order to use the FFT, zeros are appended to the signal or filter sequence until they are both the same length. If the FFT of the signal $x(n)$ is term-by-term multiplied by the FFT of the filter $h(n)$, the result is the FFT of the output $y(n)$. However, the length of $y(n)$ obtained by an inverse FFT is the same as the length of the input. Because the DFT or FFT is a periodic transform, the convolution implemented using this FFT approach is cyclic convolution, which means the output of Equation 8.1 is wrapped or aliased. The tail of $y(n)$ is added to it head—but that is not usually what is wanted for filtering or normal convolution and correlation. This aliasing, the effects of cyclic convolution, can be overcome by appending zeros to both $x(n)$ and $h(n)$ until their lengths are $N + L - 1$ and by then using the FFT. The part of the output that is aliased is zero and the result of the cyclic convolution is exactly the same as noncyclic convolution. The cost is taking the FFT of lengthened sequences—sequences for which about half the numbers are zero. Now that we can do noncyclic convolution with the FFT, how do we account for the effects of sectioning the input and output into blocks?

8.2.1 Overlap-Add

Because convolution is linear, the output of a long sequence can be calculated by simply summing the outputs of each block of the input. What is complicated is that the output blocks are longer than the input. This is dealt with by overlapping the tail of the output from the previous block with the beginning of the output from the present block. In other words, if the block length is N and it is greater than the filter length L, the output from the second block will overlap the tail of the output from the first block and they will simply be added. Hence the name "overlap-add." Figure 8.1 illustrates why the overlap-add method works, for $N = 10$ and $L = 5$.

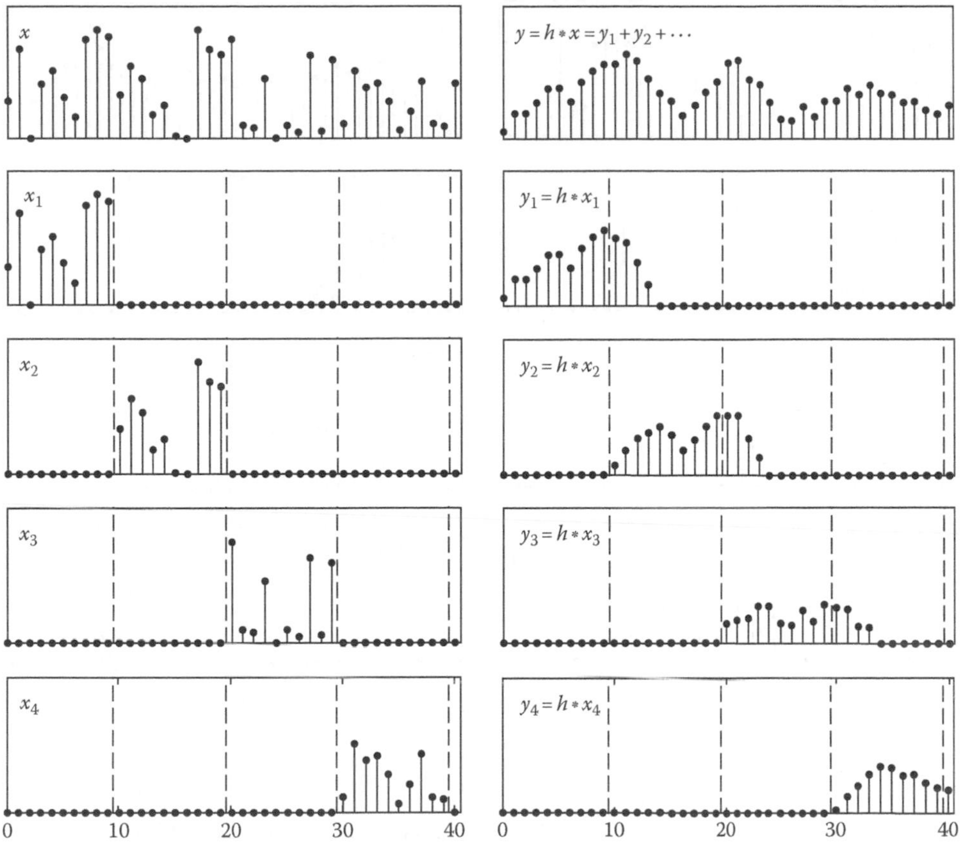

FIGURE 8.1 Overlap-add algorithm. The sequence $y(n)$ is the result of convolving $x(n)$ with an FIR filter $h(n)$ of length 5. In this example, $h(n) = 0.2$ for $n = 0, \ldots, 4$. The block length is 10, the overlap is 4. As illustrated in the figure, $x(n) = x_1(n) + x_2(n) + \cdots$ and $y(n) = y_1(n) + y_2(n) + \cdots$, where $y_i(n)$ is the result of convolving $x_i(n)$ with the filter $h(n)$.

Combining the overlap-add organization with use of the FFT yields a very efficient algorithm for calculating convolution that is faster than direct calculation for lengths above 20–50. This crossover point depends on the computer being used and the overhead needed by use of the FFTs.

8.2.2 Overlap-Save

A slightly different organization of the above approach is also often used for high-speed convolution. Rather than sectioning the input and then calculating the output from overlapped outputs from these individual input blocks, we will section the output and then use whatever part of the input contributes to that output block. In other words, to calculate the values in a particular output block, a section of length $N + L - 1$ from the input will be needed. The strategy is to save the part of the first input block that contributes to the second output block and use it in that calculation. It turns out that exactly the same amount of arithmetic and storage are used by these two approaches. Because it is the input that is now overlapped and, therefore, must be saved, this second approach is called overlap-save.

This method has also been called overlap-discard in [12] because, rather than adding the overlapping output blocks, the overlapping portion of the output blocks are discarded. As illustrated in Figure 8.2, both the head and the tail of the output blocks are discarded. It may appear in Figure 8.2 that an FFT

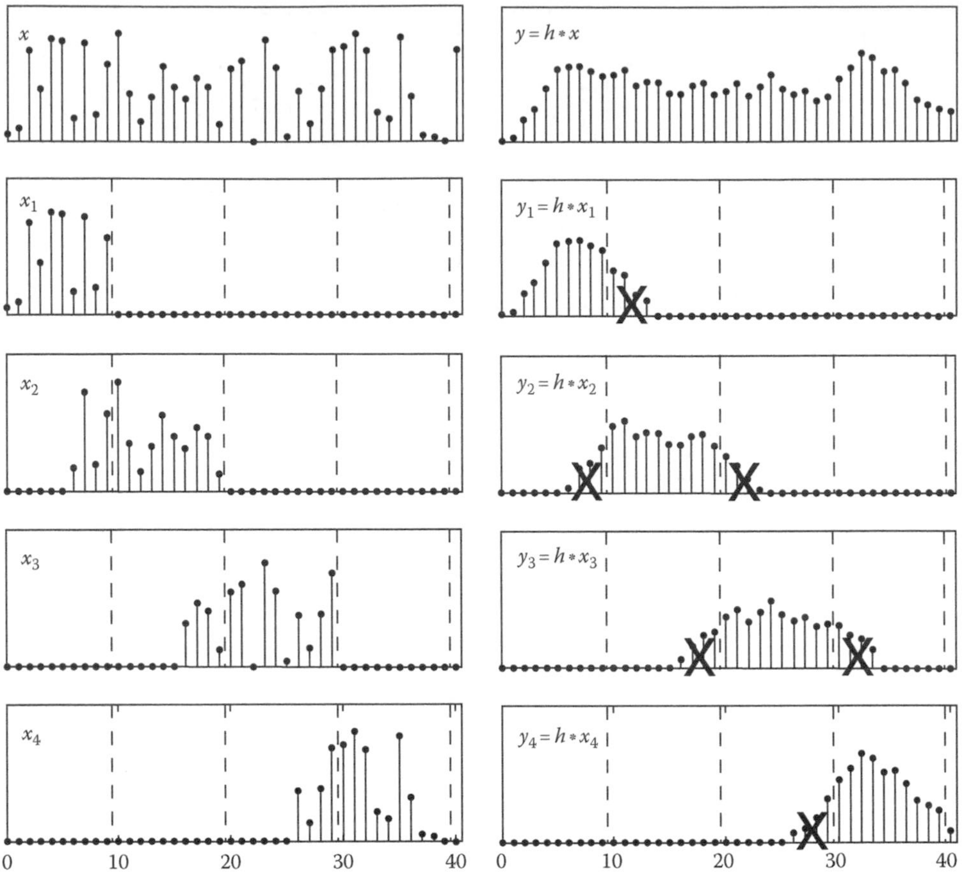

FIGURE 8.2 Overlap-save algorithm. The sequence $y(n)$ is the result of convolving $x(n)$ with an FIR filter $h(n)$ of length 5. In this example, $h(n) = 0.2$ for $n = 0, \ldots, 4$. The block length is 10, the overlap is 4. As illustrated in the figure, the sequence $y(n)$ is obtained, block by block, from the appropriate block of $y_i(n)$, where $y_i(n)$ is the result of convolving $x_i(n)$ with the filter $h(n)$.

of length 18 is needed. However, with the use of the FFT (to get cyclic convolution), the head and the tail overlap, so the FFT length is 14. (In practice, block lengths are generally chosen so that the FFT length $N + L - 1$ is a power of 2.)

8.2.3 Use of the Overlap Methods

Because the efficiency of the FFT is $O[N \log(N)]$, the efficiency of the overlap methods for convolution increases with length. To use the FFT for convolution will require one length-N forward FFT, N complex multiplications, and one length-N inverse FFT. The FFT of the filter is done once and stored rather than done repeatedly for each block. For short lengths, direct convolution will be more efficient. The exact length of filter where the efficiency crossover occurs depends on the computer and software being used.

If it is determined that the FFT is potentially faster than direct convolution, the next question is what block length to use. Here, there is a compromise between the improved efficiency of long FFTs and the fact you are processing a lot of appended zeros that contribute nothing to the output. An empirical plot of multiplication (and, perhaps, additions) per output point vs. block length will have a minimum that may be several times the filter length. This is an important parameter that should be optimized for each implementation. Remember that this increased block length may improve efficiency but it adds a delay and requires memory for storage.

8.3 Block Convolution

The operation of an FIR filter is described by a finite convolution as

$$y(n) = \sum_{k=0}^{L-1} h(k)x(n-k), \tag{8.2}$$

where

$x(n)$ is casual
$h(n)$ is causal and of length L
the time index n goes from zero to infinity or some large value

With a change of index variables this becomes

$$y(n) = \sum_{k=0}^{n} h(n-k)x(k), \tag{8.3}$$

which can be expressed as a matrix operation by

$$\begin{bmatrix} y_0 \\ y_1 \\ y_2 \\ \vdots \end{bmatrix} = \begin{bmatrix} h_0 & 0 & 0 & \cdots & 0 \\ h_1 & h_0 & 0 & & \\ h_2 & h_1 & h_0 & & \\ \vdots & & & & \vdots \end{bmatrix} \begin{bmatrix} x_0 \\ x_1 \\ x_2 \\ \vdots \end{bmatrix}. \tag{8.4}$$

The H matrix of impulse response values is partitioned into N square submatrices and the X and Y vectors are partitioned into length-N blocks or sections. This is illustrated for $N=3$ by

$$H_0 = \begin{bmatrix} h_0 & 0 & 0 \\ h_1 & h_0 & 0 \\ h_2 & h_1 & h_0 \end{bmatrix}, \quad H_1 = \begin{bmatrix} h_3 & h_2 & h_1 \\ h_4 & h_3 & h_2 \\ h_5 & h_4 & h_3 \end{bmatrix}, \quad \text{etc.} \tag{8.5}$$

$$\underline{x}_0 = \begin{bmatrix} x_0 \\ x_1 \\ x_2 \end{bmatrix}, \quad \underline{x}_1 = \begin{bmatrix} x_3 \\ x_4 \\ x_5 \end{bmatrix}, \quad \underline{y}_0 = \begin{bmatrix} y_0 \\ y_1 \\ y_2 \end{bmatrix}, \quad \text{etc.} \tag{8.6}$$

Substituting these definitions into Equation 8.4 gives

$$\begin{bmatrix} \underline{y}_0 \\ \underline{y}_1 \\ \underline{y}_2 \\ \vdots \end{bmatrix} = \begin{bmatrix} H_0 & 0 & 0 & \cdots & 0 \\ H_1 & H_0 & 0 & & \\ H_2 & H_1 & H_0 & & \\ \vdots & & & & \vdots \end{bmatrix} \begin{bmatrix} \underline{x}_0 \\ \underline{x}_1 \\ \underline{x}_2 \\ \vdots \end{bmatrix} \tag{8.7}$$

The general expression for the nth output block is

$$\underline{y}_n = \sum_{k=0}^{n} H_{n-k}\, \underline{x}_k, \tag{8.8}$$

which is a vector or block convolution. Since the matrix-vector multiplication within the block convolution is itself a convolution, Equation 8.9 is a sort of convolution of convolutions and the finite length matrix-vector multiplication can be carried out using the FFT or other fast convolution methods.

The equation for one output block can be written as the product

$$\underline{y}_2 = [H_2 \quad H_1 \quad H_0] \begin{bmatrix} \underline{x}_0 \\ \underline{x}_1 \\ \underline{x}_2 \end{bmatrix} \tag{8.9}$$

and the effects of one input block can be written

$$\begin{bmatrix} H_0 \\ H_1 \\ H_2 \end{bmatrix} \underline{x}_1 = \begin{bmatrix} \underline{y}_0 \\ \underline{y}_1 \\ \underline{y}_2 \end{bmatrix}. \tag{8.10}$$

These are generalized statements of overlap-add [11,30]. The block length can be longer, shorter, or equal to the filter length.

8.3.1 Block Recursion

Although less well known, infinite impulse response (IIR) filters can be implemented with block processing [5,6]. The block form of an IIR filter is developed in much the same way as the block convolution implementation of the FIR filter. The general constant coefficient difference equation which describes an IIR filter with recursive coefficients a_l, convolution coefficients b_k, input signal $x(n)$, and output signal $y(n)$ is given by

$$y(n) = \sum_{l=1}^{N-1} a_l y_{n-l} + \sum_{k=0}^{M-1} b_k x_{n-k} \tag{8.11}$$

using both functional notation and subscripts, depending on which is easier and clearer. The impulse response $h(n)$ is

$$h(n) = \sum_{l=1}^{N-1} a_l h(n-l) + \sum_{k=0}^{M-1} b_k \delta(n-k), \tag{8.12}$$

which, for $N = 4$, can be written in matrix operator form

$$\begin{bmatrix} 1 & 0 & 0 & \cdots & 0 \\ a_1 & 1 & 0 & & \\ a_2 & a_1 & 1 & & \\ a_3 & a_2 & a_1 & & \\ 0 & a_3 & a_2 & & \\ \vdots & & & & \vdots \end{bmatrix} \begin{bmatrix} h_0 \\ h_1 \\ h_2 \\ h_3 \\ h_4 \\ \vdots \end{bmatrix} = \begin{bmatrix} b_0 \\ b_1 \\ b_2 \\ b_3 \\ 0 \\ \vdots \end{bmatrix}.$$

In terms of smaller submatrices and blocks, this becomes

$$\begin{bmatrix} A_0 & 0 & 0 & \cdots & 0 \\ A_1 & A_0 & 0 & & \\ 0 & A_1 & A_0 & & \\ \vdots & & & & \vdots \end{bmatrix} \begin{bmatrix} \underline{h}_0 \\ \underline{h}_1 \\ \underline{h}_2 \\ \vdots \end{bmatrix} = \begin{bmatrix} \underline{b}_0 \\ \underline{b}_1 \\ 0 \\ \vdots \end{bmatrix} \tag{8.13}$$

for blocks of dimension two. From this formulation, a block recursive equation can be written that will generate the impulse response block by block:

$$A_0\underline{h}_n + A_1\underline{h}_{n-1} = 0 \quad \text{for } n \geq 2 \tag{8.14}$$

$$h_n = -A_0^{-1}A_1\underline{h}_{n-1} = K\underline{h}_{n-1} \quad \text{for } n \geq 2 \tag{8.15}$$

with

$$h_1 = -A_0^{-1}A_1A_0^{-1}\underline{b}_0 + A_0^{-1}\underline{b}_1. \tag{8.16}$$

Next, we develop the recursive formulation for a general input as described by the scalar difference equation (Equation 8.12) and in matrix operator form by

$$\begin{bmatrix} 1 & 0 & 0 & \cdots & 0 \\ a_1 & 1 & 0 & & \\ a_2 & a_1 & 1 & & \\ a_3 & a_2 & a_1 & & \\ 0 & a_3 & a_2 & & \\ \vdots & & & & \vdots \end{bmatrix} \begin{bmatrix} y_0 \\ y_1 \\ y_2 \\ y_3 \\ y_4 \\ \vdots \end{bmatrix} = \begin{bmatrix} b_0 & 0 & 0 & \cdots & 0 \\ b_1 & b_0 & 0 & & \\ b_2 & b_1 & b_0 & & \\ 0 & b_2 & b_1 & & \\ 0 & 0 & b_2 & & \\ \vdots & & & & \vdots \end{bmatrix} \begin{bmatrix} x_0 \\ x_1 \\ x_2 \\ x_3 \\ x_4 \\ \vdots \end{bmatrix}, \tag{8.17}$$

which, after substituting the definitions of the submatrices and assuming the block length is larger than the order of the numerator or denominator, becomes

$$\begin{bmatrix} A_0 & 0 & 0 & \cdots & 0 \\ A_1 & A_0 & 0 & & \\ 0 & A_1 & A_0 & & \\ \vdots & & & & \vdots \end{bmatrix} \begin{bmatrix} \underline{y}_0 \\ \underline{y}_1 \\ \underline{y}_2 \\ \vdots \end{bmatrix} = \begin{bmatrix} B_0 & 0 & 0 & \cdots & 0 \\ B_1 & B_0 & 0 & & \\ 0 & B_1 & B_0 & & \\ \vdots & & & & \vdots \end{bmatrix} \begin{bmatrix} \underline{x}_0 \\ \underline{x}_1 \\ \underline{x}_2 \\ \vdots \end{bmatrix}. \tag{8.18}$$

From the partitioned rows of Equation 8.19, one can write the block recursive relation as

$$A_0\underline{y}_{n+1} + A_1\underline{y}_n = B_0\underline{x}_{n+1} + B_1\underline{x}_n. \tag{8.19}$$

Solving for $\underline{y}_{n+1}$ gives

$$\underline{y}_{n+1} = -A_0^{-1}A_1\underline{y}_n + A_0^{-1}B_0\underline{x}_{n+1} + A_0^{-1}B_1\underline{x}_n \tag{8.20}$$

$$y_{n+1} = K\underline{y}_n + H_0\underline{x}_{n+1} + \tilde{H}_1\underline{x}_n, \tag{8.21}$$

which is a first-order vector difference equation [5,6]. This is the fundamental block recursive algorithm that implements the original scalar difference equation in Equation 8.12. It has several important characteristics:

1. The block recursive formulation is similar to a state variable equation but the states are blocks or sections of the output [6].
2. If the block length were shorter than the denominator, the vector difference equation would be higher than first order. There would be a nonzero A_2. If the block length were shorter than the numerator, there would be a nonzero B_2 and a higher order block convolution operation. If the

block length were one, the order of the vector equation would be the same as the scalar equation. They would be the same equation.

3. The actual arithmetic that goes into the calculation of the output is partly recursive and partly convolution. The longer the block, the more the output is calculated by convolution, and the more arithmetic is required.
4. There are several ways of using the FFT in the calculation of the various matrix products in Equation 8.20. Each has some arithmetic advantage for various forms and orders of the original equation. It is also possible to implement some of the operations using rectangular transforms, number theoretic transforms (NTTs), distributed arithmetic, or other efficient convolution algorithms [6,36].

8.4 Short- and Medium-Length Convolutions

For the cyclic convolution of short- ($n \leq 10$) and medium-length sequences ($n \leq 100$), special algorithms are available. For short lengths, algorithms that require the minimum number of multiplications possible have been developed by Winograd [8,17,35]. However, for longer lengths, Winograd's algorithms, based on his theory of multiplicative complexity, require a large number of additions and become cumbersome to implement. Nesting algorithms, such as the Agarwal–Cooley and split-nesting algorithms, are methods that combine short convolutions. By nesting Winograd's short convolution algorithms, efficient medium-length convolution algorithms can thereby be obtained.

In the following section, we give a matrix description of these algorithms and of the Toom–Cook algorithm. Descriptions based on polynomials can be found in [4,8,19,21,24]. The presentation that follows relies upon the notions of similarity transformations, companion matrices, and Kronecker products. With them, the algorithms are described in a manner that brings out their structure and differences. It is found that when companion matrices are used to describe cyclic convolution, the algorithms block-diagonalize the cyclic shift matrix.

8.4.1 Toom–Cook Method

A basic technique in fast algorithms for convolution is interpolation: two polynomials are evaluated at some common points, these values are multiplied, and by computing the polynomial interpolating these products, the product of the two original polynomials is determined [4,19,21,31]. This interpolation method is often called the Toom–Cook method and can be described by a bilinear form. Let $n = 2$,

$$\begin{aligned} X(s) &= x_0 + x_1 s + x_2 s^2 \\ H(s) &= h_0 + h_1 s + h_2 s^2 \\ Y(s) &= y_0 + y_1 s + y_2 s^2 + y_3 s^3 + y_4 s^4. \end{aligned}$$

The linear convolution of x and h can be represented by a matrix-vector product $y = Hx$,

$$\begin{bmatrix} y_0 \\ y_1 \\ y_2 \\ y_3 \\ y_4 \end{bmatrix} = \begin{bmatrix} h_0 & & \\ h_1 & h_0 & \\ h_2 & h_1 & h_0 \\ & h_2 & h_1 \\ & & h_2 \end{bmatrix} \begin{bmatrix} x_0 \\ x_1 \\ x_2 \end{bmatrix}$$

or as a polynomial product $Y(s) = H(s)X(s)$. In the former case, the linear convolution matrix can be written as $h_0 H_0 + h_1 H_1 + h_2 H_2$ where the meaning of H_k is clear. In the later case, one obtains the expression

$$y = C\{Ah * Ax\}, \tag{8.22}$$

where * denotes point-by-point multiplication. The terms Ah and Ax are the values of $H(s)$ and $X(s)$ at some points $i_1, \ldots, i_{2n-1}(n=2)$. The point-by-point multiplication gives the values $Y(i_1), \ldots, Y(i_{2n-1})$. The operation of C obtains the coefficients of $Y(s)$ from its values at the point $i_1, \ldots, i_{2n-1}$. Equation 8.22 is a bilinear form and it implies that

$$H_k = C\ \mathrm{diag}(Ae_k)A,$$

where e_k is the kth standard basis vector. (Ae_k is the kth column of A). However, A and C do not need to be Vandermonde matrices as suggested above. As long as A and C are matrices such that $H_k = C\ \mathrm{diag}(Ae_k)A$, then the linear convolution of x and h is given by the bilinear form $y = C\{Ah * Ax\}$. More generally, as long as A, B, and C are matrices satisfying $H_k = C\ \mathrm{diag}(Be_k)A$, then $y = C\{Bh * Ax\}$ computes the linear convolution of h and x. For convenience, if $C\{Bh * Ax\}$ computes the n point linear convolution of h and x (both h and x are n point sequences), then we say "(A, B, C) describes a bilinear form for n point linear convolution."

Example 8.1

(A, A, C) describes a two-point linear convolution where

$$A = \begin{bmatrix} 1 & 0 \\ 1 & 1 \\ 0 & 1 \end{bmatrix} \quad \text{and} \quad C = \begin{bmatrix} 1 & 0 & 0 \\ 0 & 1 & 0 \\ -1 & -1 & 1 \end{bmatrix}. \qquad (8.23)$$

8.4.2 Cyclic Convolution

The cyclic convolution of x and h can be represented by a matrix-vector product

$$\begin{bmatrix} y_0 \\ y_1 \\ y_2 \end{bmatrix} = \begin{bmatrix} h_0 & h_2 & h_1 \\ h_1 & h_0 & h_2 \\ h_2 & h_1 & h_0 \end{bmatrix} \begin{bmatrix} x_0 \\ x_1 \\ x_2 \end{bmatrix}$$

or as the remainder of a polynomial product after division by $s^n - 1$, denoted by $Y(s) = \langle H(s)X(s) \rangle_{s^n-1}$. In the former case, the cyclic convolution matrix can be written as $h_0 I + h_1 S_2 + h_2 S_2^2$ where S_n is the cyclic shift matrix,

$$S_n = \begin{bmatrix} & & & 1 \\ 1 & & & \\ & \ddots & & \\ & & 1 & \end{bmatrix}.$$

It will be useful to make a more general statement.

The companion matrix of a monic polynomial, $M(s) = m_0 + m_1 s + \cdots + m_{n-1}s^{n-1} + s^n$ is given by

$$C_M = \begin{bmatrix} & & & -m_0 \\ 1 & & & -m_1 \\ & \ddots & & \vdots \\ & & 1 & -m_{n-1} \end{bmatrix}.$$

Its usefulness in the following discussion comes from the following relation, which permits a matrix formulation of convolution:

$$Y(s) = \langle H(s)X(s)\rangle_{M(s)} \quad \Leftrightarrow \quad y = \left(\sum_{k=0}^{n-1} h_k C_M^k\right)x, \tag{8.24}$$

where

x, h, and y are the vectors of coefficients
C_M is the companion matrix of $M(s)$

In Equation 8.24, y is the convolution of x and h with respect to $M(s)$. In the case of cyclic convolution, $M(s) = s^n - 1$ and C_{s^n-1} is the cyclic shift matrix, S_n.

Similarity transformations can be used to interpret the action of some convolution algorithms. If $C_M = T^{-1}QT$ for some matrix $T(C_M$ and Q are similar, denoted $C_M \sim Q)$, then Equation 8.24 becomes

$$y = T^{-1}\left(\sum_{k=0}^{n-1} h_k Q^k\right)Tx.$$

That is, by employing the similarity transformation given by T in this way, the action of S_n^k is replaced by that of Q^k. Many cyclic convolution algorithms can be understood, in part, by understanding the manipulations made to S_n and the resulting new matrix Q. If the transformation T is to be useful, it must satisfy two requirements: (1) Tx must be simple to compute and (2) Q must have some advantageous structure. For example, by the convolution property of the DFT, the DFT matrix F diagonalizes S_n and, therefore, it diagonalizes every circulant matrix. In this case, Tx can be computed by an FFT and the structure of Q is the simplest possible: a diagonal.

8.4.3 Winograd Short Convolution Algorithm

The Winograd algorithm [35] can be described using the notation above. Suppose $M(s)$ can be factored as $M(s) = M_1(s)\, M_2(s)$ where $M_1(s)$ and $M_2(s)$ have no common roots, then $C_M \sim (C_{M_1} \oplus C_{M_2})$ where $\oplus$ denotes the matrix direct sum. Using this similarity and recalling in Equation 8.24, the original convolution can be decomposed into two disjoint convolutions. This is a statement of the Chinese remainder theorem (CRT) for polynomials expressed in matrix notation. In the case of cyclic convolution, $s^n - 1$ can be written as the product of cyclotomic polynomials—polynomials whose coefficients are small integers. Denoting the dth cyclotomic polynomial by $\Phi_d(s)$, one has $s^n - 1 = \prod_{d|n}\Phi_d(s)$. Therefore, S_n can be transformed to a block diagonal matrix,

$$S_n \sim \begin{bmatrix} C_{\Phi_1} & & & \\ & C_{\Phi_d} & & \\ & & \ddots & \\ & & & C_{\Phi_n} \end{bmatrix} = \left(\bigoplus_{d|n} C_{\Phi_d}\right). \tag{8.25}$$

The symbol $\oplus$ denotes the matrix direct sum (diagonal concatenation). Each matrix on the diagonal is the companion matrix of a cyclotomic polynomial.

Example 8.2

$$
\begin{aligned}
s^{15} - 1 &= \Phi_1(s)\Phi_3(s)\Phi_5(s)\Phi_{15}(s)\\
&= (s-1)(s^2+s+1)(s^4+s^3+s^2+s+1)(s^8-s^7+s^5-s^4+s^3-s+1)
\end{aligned}
$$

$$
S_{15} = T^{-1}\begin{bmatrix}
1 & & & & & & & & & & & & & & \\
 & & -1 & & & & & & & & & & & & \\
 & 1 & -1 & & & & & & & & & & & & \\
 & & & & & & -1 & & & & & & & & \\
 & & & 1 & & & -1 & & & & & & & & \\
 & & & & 1 & & -1 & & & & & & & & \\
 & & & & & 1 & -1 & & & & & & & & \\
 & & & & & & & & & & & & & & -1 \\
 & & & & & & & 1 & & & & & & & 1 \\
 & & & & & & & & 1 & & & & & & \\
 & & & & & & & & & 1 & & & & & -1 \\
 & & & & & & & & & & 1 & & & & 1 \\
 & & & & & & & & & & & 1 & & & -1 \\
 & & & & & & & & & & & & 1 & & \\
 & & & & & & & & & & & & & 1 & 1
\end{bmatrix} T. \tag{8.26}
$$

Each block represents a convolution with respect to a cyclotomic polynomial, or a "cyclotomic convolution." When n has several prime divisors the similarity transformation T becomes quite complicated. However, when n is a prime power, the transformation is very structured, as described in [29].

As in the previous section, we can write a bilinear form for cyclotomic convolution. Let d be any positive integer and let $X(s)$ and $H(s)$ be polynomials of degree $\varphi(d) - 1$ where $\varphi(\cdot)$ is the Euler totient function. If A, B, and C are matrices satisfying $(C_{\Phi_d})^k = C\,\mathrm{diag}(Be_k)A$ for $0 \le k \le \varphi(d) - 1$, then the coefficients of $Y(s) = \langle X(s)H(s)\rangle_{\Phi_d(s)}$ are given by $y = C\{Bh * Ax\}$. As above, for such A, B, and C, we say "(A, B, C) describes a bilinear form for $\Phi_d(s)$ convolution."

But since $\langle X(s)H(s)\rangle_{\Phi_d(s)}$ can be found by computing the product of $X(s)$ and $H(s)$ and reducing the result, a cyclotomic convolution algorithm can always be derived by following a linear convolution algorithm by the appropriate reduction operation: If G is the appropriate reduction matrix and if (A, B, C) describes a bilinear form for a $\varphi(d)$ point linear convolution, then (A, B, GC) describes a bilinear form for $\Phi_d(s)$ convolution. That is, $y = GC\{Bh * Ax\}$ computes the coefficients of $\langle X(s)H(s)\rangle_{\Phi_d(s)}$.

Example 8.3

A bilinear form for $\Phi_3(s)$ convolution is described by (A, A, GC) where A and C are given in Equation 8.23 and G is given by

$$
G = \begin{bmatrix} 1 & 0 & -1 \\ 0 & 1 & -1 \end{bmatrix}.
$$

The Winograd short cyclic convolution algorithm decomposes the convolution into smaller (cyclotomic) ones, and can be described as follows. If (A_d, B_d, C_d) describes a bilinear form for $\Phi_d(s)$ convolution, then a bilinear form for cyclic convolution is provided by

$$
A = \left(\oplus_{d|n} A_d\right)T, \quad B = \left(\oplus_{d|n} B_d\right)T, \quad \text{and} \quad C = T^{-1}\left(\oplus_{d|n} C_d\right).
$$

The matrix T decomposes the problem into disjoint parts, and T^{-1} recombines the results.

8.4.4 Agarwal–Cooley Algorithm

The Agarwal–Cooley [3] algorithm uses a similarity of another form. Namely, when $n = n_1 n_2$, and $(n_1, n_2) = 1$

$$S_n = P^t (S_{n_1} \otimes S_{n_2}) P, \tag{8.27}$$

where

$\otimes$ denotes the Kronecker product

P is a permutation matrix

The permutation is $k \longrightarrow \langle k \rangle_{n_1} + n_1 \langle k \rangle_{n_2}$. This converts a one-dimensional cyclic convolution of length n into a two-dimensional one of length n_1 along one dimension and length n_2 along the second. Then an n_1-point and an n_2-point cyclic convolution algorithm can be combined to obtain an n-point algorithm.

8.4.5 Split-Nesting Algorithm

The split-nesting algorithm [21] combines the structures of the Winograd and Agarwal–Cooley methods, so that S_n is transformed to a block diagonal matrix as in Equation 8.25:

$$S_n \sim \bigoplus_{d|n} \Psi(d). \tag{8.28}$$

Here $\Psi(d) = \otimes_{p|d, p \in P} C_{\Phi_{H_d(p)}}$, where $H_d(p)$ is the highest power of p dividing d and P is the set of primes. An example clarifies this decomposition.

Example 8.4

$$S_{45} = P^t R^{-1} \begin{bmatrix} 1 & & & & & \\ & C_{\Phi_3} & & & & \\ & & C_{\Phi_9} & & & \\ & & & C_{\Phi_5} & & \\ & & & & C_{\Phi_3} \otimes C_{\Phi_5} & \\ & & & & & C_{\Phi_9} \otimes C_{\Phi_5} \end{bmatrix} RP, \tag{8.29}$$

where

P is the same permutation matrix of Equation 8.27

R is a matrix described in [29]

In the split-nesting algorithm, each matrix along the diagonal represents a multidimensional cyclotomic convolution rather than a one-dimensional one. To obtain a bilinear form for the split-nesting method, bilinear forms for one-dimensional convolutions can be combined to obtain bilinear forms for multi-dimensional cyclotomic convolution. This is readily explained by an example.

Example 8.5

A 45-point circular convolution algorithm:

$$y = P^t R^{-1} C\{BRPh * ARPx\}, \tag{8.30}$$

where

$$A = \oplus A_3 \oplus A_9 \oplus A_5 \oplus (A_3 \otimes A_5) \oplus (A_9 \otimes A_5)$$
$$B = 1 \oplus B_3 \oplus B_9 \oplus B_5 \oplus (B_3 \otimes B_5) \oplus (B_9 \otimes B_5)$$
$$C = 1 \oplus C_3 \oplus C_9 \oplus C_5 \oplus (C_3 \otimes C_5) \oplus (C_9 \otimes C_5)$$

and where $\left(A_{p^i}, B_{p^i}, C_{p^i}\right)$ describes a bilinear form for $\Phi_{p^i}(s)$ convolution.

Split-nesting (1) requires a simpler similarity transformation than the Winograd algorithm and (2) decomposes cyclic convolution into several disjoint multidimensional convolutions. For these reasons, for medium lengths, split-nesting can be more efficient than the Winograd convolution algorithm, even though it does not achieve the minimum number of multiplications. An explicit matrix description of the similarity transformation is provided in [29].

8.5 Multirate Methods for Running Convolution

While fast FIR filtering, based on block processing and the FFT, is computationally efficient, for real-time processing it has three drawbacks: (1) a delay is incurred; (2) the multiply-accumulate (MAC) structure of the convolutional sum, a command for which DSPs are optimized, is lost; and (3) extra memory and communication (data transfer) time is needed. For real-time applications, this has motivated the development of alternative methods for convolution that partially retain the FIR filtering structure [18,33]. In the z-domain, the running convolution of x and h is described by a polynomial product

$$Y(z) = H(z)X(z), \tag{8.31}$$

where
$X(z)$ and $Y(z)$ are of infinite degree
$H(z)$ is of finite degree

Let us write the polynomials as follows:

$$X(z) = X_0(z^2) + z^{-1}X_1(z^2) \tag{8.32}$$

$$Y(z) = Y_0(z^2) + z^{-1}Y_1(z^2) \tag{8.33}$$

$$H(z) = H_0(z^2) + z^{-1}H_1(z^2), \tag{8.34}$$

where

$$X_0(z) = \sum_{i=0}^{\infty} x_{2i}z^{-i}, \quad X_1(z) = \sum_{i=0}^{\infty} x_{2i+1}z^{-i}$$

and Y_0, Y_1, H_0, and H_1 are similarly defined. (These are known as polyphase components, although that is not important here.)

The polynomial product (Equation 8.31) can then be written as

$$Y_0(z^2) + z^{-1}Y_1(z^2) = \left(H_0(z^2) + z^{-1}H_1(z^2)\right)\left(X_0(z^2) + z^{-1}X_1(z^2)\right) \tag{8.35}$$

or in matrix form as

$$\begin{bmatrix} Y_0 \\ Y_1 \end{bmatrix} = \begin{bmatrix} H_0 & z^{-2}H_1 \\ H_1 & H_0 \end{bmatrix} \begin{bmatrix} X_0 \\ X_1 \end{bmatrix}, \tag{8.36}$$

where $Y_0 = Y_0(z^2)$, etc.

The general form of Equation 8.34 is given by

$$X(z) = \sum_{k=0}^{N-1} z^{-1} X_k(z^N),$$

where

$$X_k(z) = \sum_i x_{Ni+k} z^{-i}$$

and similarly for H and Y. For clarity, $N = 2$ is used in this exposition.

Note that the right-hand side of Equation 8.35 is a product of two polynomials of degree N, where the coefficients are themselves polynomials, either of finite degree (H_i), or of infinite degree (X_i). Accordingly, the Toom–Cook algorithm described previously can be employed, in which case the sums and products become polynomial sums and products. The essential key is that the polynomial products are themselves equivalent to FIR filtering, with shorter filters.

A Toom–Cook algorithm for carrying out Equation 8.35 is given by

$$\begin{bmatrix} Y_0 \\ Y_1 \end{bmatrix} = C\left\{ A \begin{bmatrix} H_0 \\ H_1 \end{bmatrix} * A \begin{bmatrix} X_0 \\ X_1 \end{bmatrix} \right\},$$

where

$$A = \begin{bmatrix} 1 & 0 \\ 1 & 1 \\ 0 & 1 \end{bmatrix} \quad \text{and} \quad C = \begin{bmatrix} 1 & 0 & z^{-2} \\ -1 & 1 & -1 \end{bmatrix}.$$

This Toom–Cook algorithm yields the multirate filter bank structure shown in Figure 8.3. The outputs of the two downsamplers, on the left side of the structure shown in the figure, are $X_0(z)$ and $X_1(z)$. The outputs of the two upsamplers, on the right side of the structure, are $Y_0(z^2)$ and $Y_1(z^2)$. Note that the three filters H_0, $H_0 + H_1$, and H_1 operate at half the sampling rate. The right-most operation shown in Figure 8.3 is not an arithmetic addition—it is a merging of the two sequences, $Y_0(z^2)$ and $z^{-1}Y_1(z^2)$, by

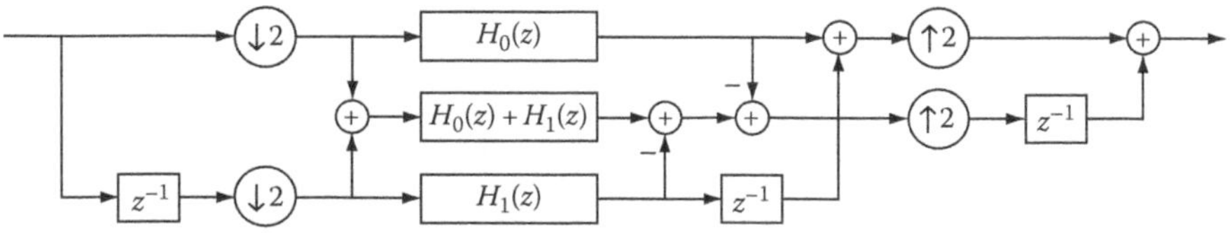

FIGURE 8.3 Filter structure based on a two-point convolution algorithm. Let H_0 be the even coefficients of a filter H, let H_1 be the odd coefficients. The structure implements the filter H using three half-length filters, each running at half rate of H.

TABLE 8.1 Computation of Running Convolution

Method	Subsampling	Delay	Multiplications/Points
1 32-point FIR filter	1	0	32
3 16-point FIR filters	2	1	24
9 8-point FIR filters	4	3	18
27 4-point FIR filters	8	7	13.5
8.1 2-point FIR filters	16	15	10.125
243 1-point multiplications	32	31	7.59

Source: Vetterli, M., *IEEE Trans. Acoust. Speech Signal Process.*, 36(5), 730, May 1988.
Note: Based on repeated application of two-point convolution structure in Figure 8.3.

interleaving. The arithmetic overhead is one "input" addition and three "output" additions per two samples; that is a total of two additions per sample.

If the original filter $H(z)$ is of length L and operates at the rate f_s, then the structure in Figure 8.3 is an implementation of $H(z)$ that employs three filters of length $L/2$, each operating at the rate $\frac{1}{2}f_s$.

The convolutional sum for $H(z)$, when implemented directly, requires L multiplications per output point and $L-1$ additions per output point. Per output point, the structure in Figure 8.3 requires $\frac{3}{4}L$ multiplications and $2+\frac{3}{2}(L/2-1)=\frac{3}{4}L+\frac{1}{2}$ additions.

The decomposition can be repeatedly applied to each of the three filters; however, the benefit diminishes for small L, and quantization errors may accumulate. Table 8.1 gives the number of multiplications needed to implement a length 32 FIR filter, using various levels of decomposition.

Other short linear convolution algorithms can be obtained from existing ones by a technique known as transposition. The transposed form of a short convolution algorithm has the same arithmetic complexity, but in a different arrangement. It was observed in [18] that the transposed forms generally have more input additions and fewer output additions. Consequently, the transposed forms should be more robust to quantization noise.

Various short-length convolution algorithms that are appropriate for this approach are provided in [18]. Also addressed is the issue of when to stop successive decompositions—and the problem of finding the best way to combine small-length filters, depending on various criteria. In particular, it is noted that DSPs generally perform an MAC operation in a single clock cycle, in which case a MAC should be considered a single operation. It appears that this approach is amenable to (1) efficient multiprocessor implementations due to their inherent parallelism and (2) efficient VLSI realization, since the implementation requires only local communication, instead of global exchange of data as in the case of FFT-based algorithms.

In [33], the following is noted. The mapping of long convolutions into small, subsampled convolutions is attractive in hardware (VLSI), software (signal processors), and multiprocessor implementations since the basic building blocks remain convolutions which can be computed efficiently once small enough.

8.6 Convolution in Subbands

Maximally decimated perfect reconstruction filter banks have been used for a variety of applications where processing in subbands is advantageous. Such filter banks can be regarded as generalizations of the short-time Fourier transform, and it turns out that the convolution theorem can be extended to them [23,32]. In other words, the convolution of two signals can be found by directly convolving the subband signals and combining the results. In [23], both uniform and nonuniform decimation ratios are considered for orthonormal and biorthonormal filter banks. In [32], the results of [23] are generalized.

The advantage of this method is that the subband signals can be quantized based on the signal variance in each subband and other perceptual considerations, as in traditional subband coding. Instead of quantizing $x(n)$ and then convolving with $g(n)$, the subbands $x_k(n)$ and $g_k(n)$ are quantized, and the results are added. When quantizing in the subbands, the subband energy distribution can be exploited and bits can be allocated to subbands accordingly. For a fixed bit rate, this approach increases the accuracy of the overall convolution—that is, this approach offers a coding gain.

In [23] an optimal bit allocation formula and the optimized coding gain is derived for orthogonal filter banks. The contribution to coding gain comes partly from the nonuniformity of the signal spectrum and partly from the nonuniformity of the filter spectrum. When the filter impulse response is taken to be the unit impulse $\delta(n)$, the formulas for the bit allocation and coding gain reduce to those for traditional subband and transform coding.

The efficiency that is gained from subband convolution comes from the ability to use a fewer number of bits to achieve a given level of accuracy. In addition, in [23], low sensitivity filter structures are derived from the subband convolution theorem and examined.

8.7 Distributed Arithmetic

Rather than grouping the individual scalar data values in a discrete-time signal into blocks, the scalar values can be partitioned into groups of bits. Because multiplication of integers, multiplication of polynomials, and discrete-time convolution are the same operations, the bit-level description of multiplication can be mixed with the convolution of the signal processing. The resulting structure is called distributed arithmetic [7,34].

8.7.1 Multiplication Is Convolution

To simplify the presentation, we will assume the data and coefficients to be positive integers with simple binary coding and the problem of carrying will be omitted. Assume the product of two B-bit words is desired

$$y = ax, \tag{8.37}$$

where

$$a = \sum_{i=0}^{B-1} a_i 2^i \quad \text{and} \quad x = \sum_{i=0}^{B-1} a_j 2^j \tag{8.38}$$

with $a_i, x_j \in \{0, 1\}$. This gives

$$y = \sum_i a_i 2^i \sum_j x_j 2^j, \tag{8.39}$$

which, with a change of variables $k = i + j$ becomes

$$y = \sum_k \sum_i a_i x_{k-i} 2^k. \tag{8.40}$$

Using the binary description of y as

$$y = \sum_k y_k 2^k, \tag{8.41}$$

we have for the binary coefficients

$$y_k = \sum_i a_i x_{k-i} \tag{8.42}$$

as a convolution of the binary coefficients for a and x. We see that multiplying two numbers is the same as convolving their coefficient representation any base. Multiplication is convolution.

8.7.2 Convolution Is Two Dimensional

Consider the following convolution of number strings (FIR filtering)

$$y(n) = \sum_\ell a(\ell)x(n-\ell). \tag{8.43}$$

Using the binary representation of the coefficients and data, we have

$$y(n) = \sum_\ell \sum_i a_i(\ell)2^i \sum_j x_j(n-\ell)2^j \tag{8.44}$$

$$y(n) = \sum_\ell \sum_i \sum_i a_i(\ell)x_j(n-\ell)2^{i+j}, \tag{8.45}$$

which after changing variables, $k=i+j$ becomes

$$y(n) = \sum_k \sum_l \sum_i a_i(l)x_{k-i}(n-l)2^k. \tag{8.46}$$

A one-dimensional convolution of numbers is a two-dimensional convolution of the binary (or other base) representations of the numbers.

8.7.3 Distributed Arithmetic by Table Lookup

The usual way that distributed arithmetic convolution is calculated does the arithmetic in a special concentrated algorithm or piece of hardware. We are now going to reorder the very general description in Equation 8.46 to allow some of the operations to be precomputed and stored in a lookup table. The arithmetic will then be distributed with the convolution itself.

If Equation 8.46 is summed over the index i, we have

$$y(n) = \sum_j \sum_\ell a(\ell)x_j(n-\ell)2^j. \tag{8.47}$$

Each sum of ℓ convolves the word string $a(n)$ with the bit string $x_j(n)$ to produce a partial product which is then shifted and added by the sum over j to give $y(n)$.

If Equation 8.47 is summed over ℓ to form a table which can be addressed by the binary numbers $x_j(n)$, we have

$$y(n) = \sum_j f\left(x_j(n), x_j(n-1), \ldots\right)2^j, \tag{8.48}$$

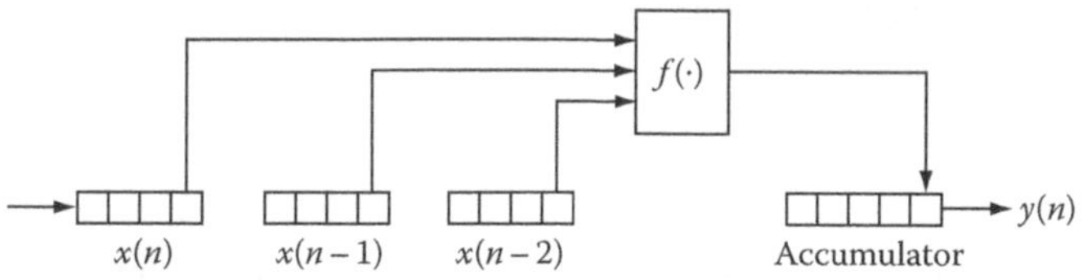

FIGURE 8.4 Distributed arithmetic by table lookup. In this example, a sequence $x(n)$ is filtered with a length 3 FIR filter. The wordlength for $x(n)$ is 4 bits. The function $f(.)$ is a function of three binary variables, and can be implemented by table lookup. The bits of $x(n)$ are shifted, bit by bit, through the input registers. Accordingly, the bits of $y(n)$ are shifted through the accumulator—after 4-bit shifts, a new output $y(n)$ becomes available.

where

$$f\left(x_j(n), x_j(n-1), \ldots\right) = \sum_{\ell} a(\ell)x_j(n-\ell). \tag{8.49}$$

The numbers $a(i)$ are the coefficients of the filter, which as usual is assumed to be fixed. Consider a filter of length L. This function $f(\,)$ is a function of L binary variables and, therefore, takes on 2^L possible values. The function is determined by the filter, $a(i)$. For example, if $L = 3$, the table (function values) would contain eight values:

$$0, a(0), a(1), a(2), (a(0) + a(1)), (a(1) + a(2)), (a(0) + a(2)), (a(0) + a(1) + a(2)) \tag{8.50}$$

and if the words were stored as B bits, they would require $2^L\,B$ bits of memory.

There are extensions and modifications of this basic idea to allow a very flexible trade of memory for logic. The idea is to precompute as much as possible, store it in a table, and fetch it when needed. The two extremes of this are on one hand to compute all possible outputs and simply fetch them using the input as an address. The other extreme is the usual system which simply stores the coefficients and computes what is needed as needed.

This table lookup is illustrated in Figure 8.4 where the blocks represent 4-bit words, where the least significant bit of each of the four most recent data words form the address for the table lookup from memory. After 4-bit shifts and accumulates, the output word $y(n)$ is available, using no multiplications.

Distributed arithmetic with table lookup can be used with FIR and IIR filters and can be arranged in direct, transpose, cascade, parallel, etc. structures. It can be organized for serial or parallel calculations or for combinations of the two. Because most microprocessors or DSP chips do not have appropriate instructions or architectures for distributed arithmetic, it is best suited for special purpose VLSI design and in those cases, it can be extremely fast.

An alternative realization of these ideas can be developed using a form of periodically time varying system that is oversampled [10].

8.8 Fast Convolution by Number Theoretic Transforms

If one performs all calculations in a finite field or ring of integers rather than the usual infinite field of real or complex numbers, a very efficient type of Fourier transform can be formulated that requires no floating point operations—it supports exact convolution with finite precision arithmetic [1,2,17,26]. This is particularly interesting because a digital computer is a finite machine and arithmetic over finite systems fits it perfectly. In the following, all arithmetic operations are performed modulo for some integer M, called the modulus. A bit of number theory can be found in [17,20,28].

8.8.1 Number Theoretic Transforms

Here we look at the conditions placed on a general linear transform in order for it to support cyclic convolution. The form of a linear transformation of a length-N sequence of number is given by

$$X(k) = \sum_{n=0}^{N-1} t(n, k)x(n) \bmod M \tag{8.51}$$

for $k = 0, 1, \ldots, (N-1)$. The definition of cyclic convolution of two sequences in Z_M is given by

$$y(n) = \sum_{m=0}^{N-1} x(m)h(n-m) \bmod M \tag{8.52}$$

for $n = 0, 1, \ldots, (N-1)$ where all indices are evaluated modulo N. We would like to find the properties of the transformation such that it will support cyclic convolution. This means that if $X(k)$, $H(k)$, and $Y(k)$ are the transforms of $x(n)$, $h(n)$, and $y(n)$ respectively, then

$$Y(k) = X(k)H(k). \tag{8.53}$$

The conditions are derived by taking the transform defined in Equation 8.1 of both sides of Equation 8.52 which gives the form for our general linear transform (Equation 8.51) as

$$X(k) = \sum_{n=0}^{N-1} \alpha^{nk} x(n), \tag{8.54}$$

where α is a root of order N, which means that N is the smallest integer such that $\alpha^N = 1$.

THEOREM 8.1

The transform (Equation 8.11*) supports cyclic convolution if and only if α is a root of order N and N^{-1} mod M is defined.*

This is discussed in [1,2]. This transform supports N-point cyclic convolution only if a particular relationship between the modulus M and the data length N is satisfied. The following theorem describes that relationship.

THEOREM 8.2

The transform (Equation 8.11*) supports N-point cyclic convolution if and only if*

$$N | O(M), \tag{8.55}$$

where

$$O(M) = \gcd\{p_1 - 1, p_2 - 1, \ldots, p_l - 1\} \tag{8.56}$$

and the prime factorization of M is

$$M = p_1^{r_1} p_2^{r_2} \cdots p_l^{r_l}. \tag{8.57}$$

Equivalently, N must divide $p_i - 1$ for every prime p_i dividing M. This theorem is a more useful form of Theorem 8.1. Notice that $N_{\max} = O(M)$.

One needs to find appropriate N, M, and α such that

- N should be appropriate for a fast algorithm and handle the desired sequence lengths.
- M should allow the desired dynamic range of the signals and should allow simple modular arithmetic.
- α should allow a simple multiplication for $\alpha^{nk}x(n)$.

We see that if M is even, it has a factor of 2 and, therefore, $O(M) = N_{\max} = 1$ which implies M should be odd. If M is prime the $O(M) = M - 1$ which is as large as could be expected in a field of M integers. For $M = 2^k - 1$, let k be a composite $k = pq$ where p is prime. Then $2^p - 1$ divides $2^{pq} - 1$ and the maximum possible length of the transform will be governed by the length possible for $2^p - 1$. Therefore, only the prime k need be considered interesting. Numbers of this form are known as Mersenne numbers and have been used by Rader [26]. For Mersenne number transforms, it can be shown that transforms of length at least $2p$ exist and the corresponding $\alpha = -2$. Mersenne number transforms are not of as much interest because $2p$ is not highly composite and, therefore, we do not have FFT-type algorithms.

For $M = 2^k + 1$ and k odd, 3 divides $2^k + 1$ and the maximum possible transform length is 2. Thus, we consider only even k. Let $k = s2^t$, where s is an odd integer. Then 2^{2^t} divides $2^{s2^t} + 1$ and the length of the possible transform will be governed by the length possible for $2^{2^t} + 1$. Therefore, integers of the form $M = 2^{2^t} + 1$ are of interest. These numbers are known as Fermat numbers [26]. Fermat numbers are prime for $0 \leq t \leq 4$ and are composite for all $t \geq 5$.

Since Fermat numbers up to F_4 are prime, $O(F_t) = 2^b$ where $b = 2^t$ and $t \leq 4$, we can have a Fermat number transform for any length $N = 2^m$ where $m \leq b$. For these Fermat primes the integer $\alpha = 3$ is of order $N = 2^b$ allowing the largest possible transform length. The integer $\alpha = 2$ is of order $N = 2b = 2^{t+1}$. Then all multiplications by powers of α are bit shifts—which is particularly attractive because in Equation 8.51, the data values are multiplied by powers of α.

Table 8.2 gives possible parameters for various Fermat number moduli. This table gives values of N for the two most important values of α which are 2 and $\sqrt{2}$. The second column gives the approximate number of bits in the number representation. The third column gives the Fermat number modulus, the fourth is the maximum convolution length for $\alpha = 2$, the fifth is the maximum length for $\alpha = \sqrt{2}$, the sixth is the maximum length for any α, and the seventh is the α for that maximum length. Remember that the first two rows have a Fermat number modulus which is prime and the second two rows have a composite Fermat number as modulus. Note the differences.

The NTT itself seems to be very difficult to interpret or use directly. It seems to be useful only as a means for high-speed convolution where it has remarkable characteristics. The books, articles, and presentations that discuss NTT and related topics are [4,17,21]. A recent book discusses NTT in a signal processing context [14].

TABLE 8.2 Fermat Number Moduli

t	B	$M = F_t$	N_2	$N_{\sqrt{2}}$	$N_{\max}$	α for $N_{\max}$
3	8	$2^8 + 1$	16	32	256	3
4	16	$2^{16} + 1$	32	64	65,536	3
5	32	$2^{32} + 1$	64	128	128	$\sqrt{2}$
6	64	$2^{64} + 1$	128	256	256	$\sqrt{2}$

8.9 Polynomial-Based Methods

The use of polynomials in representing elements of a digital sequence and in representing the convolution operation has led to the development of a family of algorithms based on the fast polynomial transform [4,16,21]. These algorithms are especially useful for two-dimensional convolution. The CRT for polynomials, which is central to Winograd's short convolution algorithm, is also conveniently described in polynomial notation. An interesting approach combines the use of the polynomial-based methods with the number theoretic approach to convolution (NTTs), wherein the elements of a sequence are taken to lie in a finite field [9,15]. In [15] the CRT is extended to the case of a ring of polynomials with coefficients from a finite ring of integers. It removes the limitations on both word length and sequence length of NNTs and serves as a link between the two methods (CRT and NNT). The new result so obtained, which specializes to both the NNTs and the CRT for polynomials, has been called the AICE-CRT (the American-Indian-Chinese extension of the CRT). A complex version has also been derived.

8.10 Special Low-Multiply Filter Structures

In the use of convolution for digital filtering, the convolution operation can be simplified, if the filter $h(n)$ is chosen appropriately.

Some filter structures are especially simple to implement. Some examples are

- A simple implementation of the recursive running sum is based on the factorization $\sum_{k=0}^{L-1} z^k = (z^L + 1)/(z - 1)$.
- If the transfer function $H(z)$ of the filter possesses a root at $z = -1$ of multiplicity K, the factor $(z+1)/2$ can be extracted from the transfer function. The factor $(z+1)/2$ can be implemented very simply.
- This idea is extended in prefiltering and IFIR filtering techniques—a filter is implemented as a cascade of two filters: one with a crude response that is simple to implement, another that makes up for it, but requires the usual implementation complexity. The overall response satisfies specifications and can be implemented with reduced complexity.
- The maximally flat symmetric FIR filter can be implemented without multiplications using the De Casteljau algorithm [27].

In summary, a filter can often be designed so that the convolution operation can be performed with less computational complexity and/or at a faster rate. Much work has focused on methods that take into account implementation complexity during the approximation phase of the filter design process (see Chapter 11).

References

1. Agarwal, R.C. and Burrus, C.S., Fast convolution using Fermat number transforms with applications to digital filtering, *IEEE Trans. Acoust. Speech Signal Process.*, ASSP-22(2): 87–97, April 1974. Reprinted in [17].
2. Agarwal, R.C. and Burrus, C.S., Number theoretic transforms to implement fast digital convolution, *Proc. IEEE*, 63(4): 550–560, April 1975. (Also in IEEE Press DSP Reprints II, 1979).
3. Agarwal, R.C. and Cooley, J.W., New algorithms for digital convolution, *IEEE Trans. Acoust. Speech Signal Process.*, 25(5): 392–410, October 1977.
4. Blahut, R.E., *Fast Algorithms for Digital Signal Processing*, Addison-Wesley, Reading, MA, 1985.
5. Burrus, C.S., Block implementation of digital filters, *IEEE Trans. Circuit Theory*, CT-18(6): 697–701, November 1971.

6. Burrus, C.S., Block realization of digital filters, *IEEE Trans. Audio Electroacoust.*, AU-20(4): 230–235, October 1972.
7. Burrus, C.S., Digital filter structures described by distributed arithmetic, *IEEE Trans. Circuits Syst.*, CAS-24(12): 674–680, December 1977.
8. Burrus, C.S., Efficient Fourier transform and convolution algorithms, in Jae S. Lim and Alan V. Oppenheim (Eds.), *Advanced Topics in Signal Processing*, Prentice-Hall, Englewood Cliffs, NJ, 1988.
9. Garg, H.K., Ko, C.C., Lin, K.Y., and Liu, H., On algorithms for digital signal processing of sequences, *Circuits Syst. Signal Process.*, 15(4): 437–452, 1996.
10. Ghanekar, S.P., Tantaratana, S., and Franks, L.E., A class of high-precision multiplier-free FIR filter realizations with periodically time-varying coefficients, *IEEE Trans. Signal Process.*, 43(4): 822–830, 1995.
11. Gold, B. and Rader, C.M., *Digital Processing of Signals*, McGraw-Hill, New York, 1969.
12. Harris, F.J., Time domain signal processing with the DFT, in D.F. Elliot (Ed.), *Handbook of Digital Signal Processing*, Academic Press, New York, 1987, ch. 8, pp. 633–699.
13. Helms, H.D., Fast Fourier transform method of computing difference equations and simulating filters, *IEEE Trans. Audio Electroacoust.*, AU-15: 85–90, June 1967.
14. Krishna, H., Krishna, B., Lin, K.-Y, and Sun, J.-D., *Computational Number Theory and Digital Signal Processing*, CRC Press, Boca Raton, FL, 1994.
15. Lin, K.Y., Krishna, H., and Krishna, B., Rings, fields the Chinese remainder theorem and an American-Indian-Chinese extension—Part I: Theory. *IEEE Trans. Circuits Syst. II*, 41(10): 641–655, 1994.
16. Loh, A.M. and Siu, W.-C., Improved fast polynomial transform algorithm for cyclic convolutions, *Circuits Syst. Signal Process.*, 14(5): 603–614, 1995.
17. McClellan, J.H. and Rader, C.M., *Number Theory in Digital Signal Processing*, Prentice-Hall, Englewood Cliffs, NJ, 1979.
18. Mou, Z.-J. and Duhamel, P., Short-length FIR filters and their use in fast nonrecursive filtering, *IEEE Trans. Signal Process.*, 39(6): 1322–1332, June 1991.
19. Myers, D.G., *Digital Signal Processing: Efficient Convolution and Fourier Transform Techniques*, Prentice-Hall, Englewood Cliffs, NJ, 1990.
20. Niven, I. and Zuckerman, H.S., *An Introduction to the Theory of Numbers*, 4th ed., John Wiley & Sons, New York, 1980.
21. Nussbaumer, H.J., *Fast Fourier Transform and Convolution Algorithms*, Springer-Verlag, New York, 1982.
22. Oppenheim, A.V. and Schafer, R.W., *Discrete-Time Signal Processing*, Prentice-Hall, Englewood Cliffs, NJ, 1989.
23. Phoong, S.-M. and Vaidyanathan, P.P., One- and two-level filter-bank convolvers, *IEEE Trans. Signal Process.*, 43(1): 116–133, January 1995.
24. Proakis, J.G., Rader, C.M., Ling, F., and Nikias, C.L., *Advanced Digital Signal Processing*, Macmillan, New York, 1992.
25. Rabiner, L.R. and Gold, B., *Theory and Application of Digital Signal Processing*, Prentice-Hall, Englewood Cliffs, NJ, 1975.
26. Rader, C.M., Discrete convolution via Mersenne transforms, *IEEE Trans. Comput.*, 21(12): 1269–1273, December 1972.
27. Samadi, S., Cooklev, T., Nishihara, A., and Fujii, N., Multiplierless structure for maximally flat linear phase FIR filters, *Electron. Lett.*, 29(2): 184–185, January 21, 1993.
28. Schroeder, M.R., *Number Theory in Science and Communication*, 2nd ed., Springer-Verlag, Berlin, Germany, 1986.
29. Selesnick, I.W. and Burrus, C.S., Automatic generation of prime length FFT programs, *IEEE Trans. Signal Process.*, 44(1): 14–24, January 1996.

30. Stockham, T.G., High speed convolution and correlation, in *AFIPS Conference Proceedings, 1966 Spring Joint Computer Conference*, Vol. 28, 1966, pp. 229–233.
31. Tolimieri, R., An, M., and Lu, C., *Algorithms for Discrete Fourier Transform and Convolution*, Springer-Verlag, New York, 1989.
32. Vaidyanathan, P.P, Orthonormal and biorthonormal filter banks as convolvers, and convolutional coding gain, *IEEE Trans. Signal Process.*, 41(6): 2110–2129, June 1993.
33. Vetterli, M., Running FIR and IIR filtering using multirate filter banks, *IEEE Trans. Acoust. Speech Signal Process.*, 36(5): 730–738, May 1988.
34. White, S.A., Applications of distributed arithmetic to digital signal processing, *IEEE Acoust. Speech Signal Process. Mag.*, 6(3): 4–19, July 1989.
35. Winograd, S., *Arithmetic Complexity of Computations*, SIAM, Philadelphia, PA, 1980.
36. Zalcstein, Y., A note on fast cyclic convolution, *IEEE Trans. Comput.*, 20: 665–666, June 1971.

9

Complexity Theory of Transforms in Signal Processing

Ephraim Feig
Innovations-to-Market

9.1 Introduction

Complexity theory of computation attempts to determine how "inherently" difficult are certain tasks. For example, how inherently complex is the task of computing an inner product of two vectors of length N? Certainly one can compute the inner product $\sum_{j=1}^{N} x_j y_j$ by computing the N products $x_j y_j$ and then summing them. But can one compute this inner product with fewer than N multiplications? The answer is no, but the proof of this assertion is no trivial matter. One first abstracts and defines the notions of the algorithm and its components (such as addition and multiplication); then a theorem is proven that any algorithm for computing a bilinear form which uses K multiplications can be transformed to a quadratic algorithm (some algorithm of a very special form, which uses no divisions, and whose multiplications only compute quadratic forms) which uses at most K multiplications [21]; and finally a proof by induction on the length N of the summands in the inner product is made to obtain the lower bound result [7,14,22,25]. We will not present the details here; we just want to let the reader know that the process for even proving what seems to be an intuitive result is quite complex.

Consider next the more complex task of computing the product of an N point vector by an $M \times N$ matrix. This corresponds to the task of computing M separate inner products of N-point vectors. It is tempting to jump to the conclusion that this task requires MN multiplications. But we should not jump to fast conclusions. First, the M inner products are separate, but not independent (the term is used loosely, and not in any linear algebra sense). After all, the second factor in the M inner products is always the same. It turns out [7,22,25] that, indeed, our intuition this time is correct again. And the proof is really not much more difficult than the proof for the complexity result for inner products. In fact, once the general machinery is built, the proof is a slight extension of the previous case. So far intuition proved accurate.

In complexity theory one learns early on to be skeptical of intuitions. An early surprising result in complexity theory—and to date still one of its most remarkable—contradicts the intuitive guess that computing the product of two 2×2 matrices requires 8 multiplications. Remarkably, Strassen [20] has shown that it can be done with 7 multiplications. His algorithm is very nonintuitive; I am not aware of any good algebraic explanation for it except for the assertion that the mathematical identities which define the algorithm indeed are valid. It can also be shown [16] that 7 is the minimum number of multiplications required for the task.

The consequences of Strassen's algorithm for general matrix multiplication tasks are profound. The task of computing the product of two 4×4 matrices with real entries can be viewed as a task of computing two 2×2 matrices whose entries are themselves 2×2 matrices. Each of the 7 multiplications in Strassen's algorithm now become matrix multiplications requiring 7 real multiplications plus a bunch of additions; and each addition in Strassen's algorithm becomes an addition of 2×2 matrices, which can be done with 4 real additions. This process of obtaining algorithms for large problems, which are built up of smaller ones in a structures manner, is called the "nesting" procedure [25]. It is a very powerful tool in both complexity theory and algorithm design. It is a special form of recursion.

The set of $N \times N$ matrices form a noncommutative algebra. A branch of complexity theory called "multiplicative complexity theory" is quite well established for certain relatively few algebras, and wide open for the rest. In this theory complexity is measured by the number of "essential multiplications." Given an algebra over a field F, an algorithm is a sequence of arithmetic operations in the algebra. A multiplication is called essential if neither factor is an element in F. If one of the factors in a multiplication is an element in F, the operation is called a scaling.

Consider an algebra of dimension N over a field F, with basis $b_1, \ldots, b_N$. An algorithm for computing the product of two elements $\sum_{j=1}^{N} f_j b_j$ and $\sum_{j=1}^{N} g_j b_j$ with $f_j, g_j \in F$ is called bilinear, if every multiplication in the algorithm is of the form $L_1(f_1, \ldots, f_N) * L_2(g_1, \ldots, g_N)$, where L_1 and L_2 are linear forms and * is the product in the algebra, and it uses no divisions. Because none of the arithmetic operations in bilinear algorithms rely on the commutative nature of the underlying field, these algorithms can be used to build recursively via the nesting process algorithms for noncommutative algebras of increasingly large dimensions, which are built from the smaller algebras via the tensor product. For example, the algebra of 4×4 matrices (over some field F; I will stop adding this necessary assumption, as it will be obvious from content) is isomorphic to the tensor product of the algebra of 2×2 matrices with itself. Likewise, the algebra of 16×16 matrices is isomorphic to the tensor product of the algebra of 4×4 matrices with itself. And this proceeds to higher and higher dimensions.

Suppose we have a bilinear algorithm for computing the product in an algebra T_1 of dimension D, which uses M multiplications and A additions (including subtractions) and S scalings. The algebra $T_2 = T_1 \otimes T_1$ has dimension D^2. By the nesting procedure we can obtain an algorithm for computing the product in T_2 which uses M multiplications of elements in T_1, A additions of elements in T_1, and S scalings of elements in T_1. Each multiplication in T_1 requires M multiplications, A additions, and S scalings; each addition in T_1 requires D additions; and each scaling in T_1 requires D scalings. Hence, the total computational requirements for this new algorithm is M^2 multiplications, $A(M+D)$ additions, and $S(M+D)$ scalings. If the nesting procedure is continued to yield an algorithm for the product in the D^4 dimensional algebra $T_4 = T_2 \otimes T_2$, then its computational requirements would be M^4 multiplications, $A(M+D)(M^2+D^2)$ additions, and $S(M+D)(M^2+D^2)$ scalings. One more iteration would yield an algorithm for the D^8 dimensional algebra $T_8 = T_4 \otimes T_4$, which uses M^8 multiplications, $A(M+D)(M^2+D^2)(M^4+D^4)$ additions, and $S(M+D)(M^2+D^2)(M^4+D^4)$ scalings. The general pattern should be apparent by now. We see that the growth of the number of operations (i.e., the high order term) is governed by M and not by A or S. A major goal of complexity theory is the understanding of computational requirements as problem sizes increase, and nesting is the natural way of building algorithms for larger and larger problems. We see one reason why counting multiplications (as opposed to all arithmetic operations) became so important in complexity theory. (Historically, in the early days multiplications were indeed much more expensive than additions.)

Algebras of polynomials are important in signal processing; filtering can be viewed as polynomial multiplications. The product of two polynomials of degrees d_1 and d_2 can be computed with $d_1 + d_2 - 1$ multiplications. Furthermore, it is rather easy to prove (a straightforward dimension argument) that this is the minimal number of multiplications necessary for this computation. Algorithms which compute these products with these numbers of multiplications (so-called optimal algorithms) are obtained using Lagrange interpolation techniques. For even moderate values of d_j, they use inordinately many additions and scalings. Indeed, they use $(d_1 + d_2 - 3)(d_1 + d_2 - 2)$ additions, and a half as many scalings. So these algorithms are not very practical, but they are of theoretical interest. Also of interest is the asymptotic complexity of polynomial products. They can be computed by embedding them in cyclic convolutions of sizes at most twice as long. Using FFT techniques, these can be achieved with order D log D arithmetic operations, where D is the maximum of the degrees. With optimal algorithms, while the number of (essential) multiplications is linear, the total number of operations is quadratic. If nesting is used, then the asymptotic behavior of the number of multiplications is also quadratic.

Convolution algebras are derived from algebras of polynomials. Given a polynomial $P(u)$ of degree D, one can define an algebra of dimension D whose entries are all polynomials of degree less than D, with addition defined in the standard way, and multiplication is modulo $P(u)$. Such algebras are called convolution algebras. For polynomials $P(u) = u^D - 1$, the algebras are cyclic convolutions of dimension D. For polynomials $P(u) = u^D + 1$, these algebras are called signed-cyclic convolutions. The product of two polynomials modulo $P(u)$ can be obtained from the product of the two polynomials without any extra essential multiplications. Hence, if the degree of $P(u)$ is D, then the product modulo $P(u)$ can be done with $2D - 1$ multiplications. But can it be done with fewer multiplications?

Whereas complexity theory has huge gaps in almost all areas, it has triumphed in convolution algebras. The minimum number of multiplications required to compute a product in an algebra is called the multiplicative complexity of the algebra. The multiplicative complexity of convolution algebras (over infinite fields) is completely determined [22]. If $P(u)$ factors (over the base field; the role of the field will be discussed in greater detail soon) to a product of k irreducible polynomials, then the multiplicative complexity of the algebra is $2D - k$. So if $P(u)$ is irreducible, then the answer to the question in the previous paragraph is no. Otherwise, it is yes.

The above complexity result for convolution algebras is a sharp bound. It is a lower bound in that every algorithm for computing the product in the algebra requires at least $2D - k$ multiplications, where k is the number of factors of the defining polynomial $P(u)$. It is also an upper bound, in that there are algorithms which actually achieve it. Let us factor $P(u) = \Pi P_j(u)$ into a product of irreducible polynomials (here we see the role of the field; more about this soon). Then the convolution algebra modulo $P(u)$ is isomorphic to a direct sum of algebras modulo $P_j(u)$; the isomorphism is via the Chinese remainder theorem. The multiplicative complexity of the direct summands is $2d_j - 1$, where d_j are the degrees of $P_j(u)$; these are sharp bounds. The algorithm for the algebra modulo $P(u)$ is derived from these smaller algorithms; because of the isomorphism, putting them all together requires no extra multiplications. The proof that this is a lower bound, first given by Winograd [23], is quite complicated.

The above result is an example of a "direct sum theorem." If an algebra is decomposable to a direct sum of subalgebras, then clearly the multiplicative complexity of the algebra is less than or equal to the sum of the multiplicative complexities of the summands. In some (relatively rare) circumstances equality can be shown. The example of convolution algebras is such a case. The results for convolution algebras are very strong. Winograd has shown that every minimal algorithm for computing products in a convolution algebra is bilinear and is a direct sum algorithm. The latter means that the algorithm actually computes a minimal algorithm for each direct summand and then combines these results without any extra essential multiplications to yield the product in the algebra itself.

Things get interesting when we start considering algebras which are tensor products of convolution algebras (these are called multidimensional convolution algebras). A simple example already is enlightening. Consider the algebra C of polynomial multiplications modulo $u^2 + 1$ over the rationals Q; this algebra is called the Gaussian rationals. The polynomial $u^2 + 1$ is irreducible over Q (the algebra is a field), so by the previous result, its multiplicative complexity is 3. The nesting procedure would yield an algorithm the product in $C \otimes C$ which uses 9 multiplications. But it can in fact be computed with 6 multiplications. The reason is due to an old theorem, probably due to Kroeneker (though I cannot find the original proof); the reference I like best is Adrian Albert's book [1]. The theorem asserts that the tensor product of fields is isomorphic to a direct sum of fields, and the proof of the theorem is actually a construction of this isomorphism. For our example, the theorem yields that the tensor product $C \otimes C$ is isomorphic to a direct sum of two copies of C. The product in $C \otimes C$ can, therefore, be computed by computing separately the product in each of the two direct summands, each with 3 multiplications, and the final result can be obtained without any more essential multiplications. The explicit isomorphism was presented to the complexity theory community by Winograd [22]. Since the example is sufficiently simple to work out, and the results so fundamental to much of our later discussions, we will present it here explicitly.

Consider A, the polynomial ring modulo $u^2 + 1$ over the Q. This is a field of dimension 2 over Q, and it has the matrix representation (called its regular representation) given by

$$\rho(a + bu) = \begin{pmatrix} a & -b \\ b & a \end{pmatrix}. \tag{9.1}$$

While for all $b \neq 0$, the matrix above is not diagonalizable over Q, the field (algebra) is diagonalizable over the complexes. Namely,

$$\begin{pmatrix} 1 & i \\ 1 & -i \end{pmatrix}\begin{pmatrix} a & -b \\ b & a \end{pmatrix}\begin{pmatrix} 1 & i \\ 1 & -i \end{pmatrix}^{-1} = \begin{pmatrix} a+ib & 0 \\ 0 & a-ib \end{pmatrix}. \tag{9.2}$$

The elements 1 and i of A correspond (in the regular representation) in the tensor algebra $A \otimes A$ to the matrices

$$\rho(1) = \begin{pmatrix} 1 & 0 \\ 0 & 1 \end{pmatrix} \tag{9.3}$$

and

$$\rho(i) = \begin{pmatrix} 0 & -1 \\ 1 & 0 \end{pmatrix}, \tag{9.4}$$

respectively. Hence, the 4×4 matrix

$$R = \begin{pmatrix} \rho(1) & \rho(i) \\ \rho(1) & \rho(-i) \end{pmatrix} \tag{9.5}$$

diagonalizes the algebra $A \otimes A$. Explicitly, we can compute

$$\begin{pmatrix} 1 & 0 & 0 & -1(6) \\ 0 & 1 & 1 & 0(7) \\ 1 & 0 & 0 & 1(8) \\ 0 & 1 & -1 & 0 \end{pmatrix}\begin{pmatrix} x_0 & -x_1 & -x_2 & -x_3(9) \\ x_1 & x_0 & -x_3 & x_2(10) \\ x_2 & -x_3 & x_0 & -x_1(11) \\ x_3 & x_2 & x_1 & x_0 \end{pmatrix}$$
$$\begin{pmatrix} 1 & 0 & 0 & -1(12) \\ 0 & 1 & 1 & 0(13) \\ 1 & 0 & 0 & 1(14) \\ 0 & 1 & -1 & 0 \end{pmatrix} = \begin{pmatrix} y_0 & -y_1 & 0 & 0(15) \\ y_1 & y_0 & 0 & 0(16) \\ 0 & 0 & y_2 & -y_2(17) \\ 0 & 0 & y_3 & y_3 \end{pmatrix}, \tag{9.6}$$

where $y_0 = x_0 - x_3$, $y_1 = x_1 + x_2$, $y_2 = x_0 + x_3$, and $y_3 = x_1 - x_2$. A simple way to derive this is by setting X_0 to be the top left 2×2 minor of the matrix with x_j entries in the above equation, X_1 to be its bottom left 2×2 minor, and observing that

$$R\begin{pmatrix} X_0 & -X_1 \\ X_1 & X_0 \end{pmatrix}R^{-1} = \begin{pmatrix} \rho(1)X_0 + \rho(i)X_1 & \\ & \rho(0)X_0 - \rho(i)X_1 \end{pmatrix}. \tag{9.7}$$

The algorithmic implications are straightforward. The product in $A \otimes A$ can be computed with fewer multiplications than the nesting process would yield. Straightforward extensions of the above construction yield recipes for obtaining minimal algorithms for products in algebras which are tensor products of convolution algebras. The example also highlights the role of the base field. The complexity of A as an algebra over Q is 3; the complexity of A as an algebra over the complexes is 2, as over the complexes this algebra diagonalizes.

Historically, multiplicative complexity theory generalized in two ways (and in various combinations of the two). The first addressed the question: What happens when one of the factors in the product is not an arbitrary element but a fixed element not in the basefield? The second addressed: What is the complexity of semi-direct systems—those in which several products are to be computed, and one factor is arbitrary but fixed, while the others are arbitrary? Computing an arbitrary product in an n-dimensional algebra can be thought of (via the regular representation) as computing a product of a matrix $A(X)$ times a vector Y, where the entries in the matrix $A(X)$ are linear combinations of n indeterminates $x_1, \ldots, x_n$ and y is a vector of n indeterminates $y_1, \ldots, y_n$. When one factor is a fixed element in an extension field, the entries in $A(X)$ are now entries in some extension field of the basefield which may have algebraic relations. For example, consider

$$G = \begin{pmatrix} \gamma(1, 8) & -\gamma(3, 8) \\ \gamma(3, 8) & \gamma(1, 8) \end{pmatrix}, \tag{9.8}$$

where $\gamma(m, n) = \cos(2\pi m/n)$. The complex numbers $\gamma(1, 8)$ and $\gamma(3, 8)$ are linearly independent over Q, but they satisfy the algebraic relation $\gamma(1,8)/\gamma(3,8) = \sqrt{2}$. This algebraic relation gives a relation of the two numbers to the rationals, namely $\gamma(1, 8)^2/\gamma(3, 8)^2 = 2$. Now this is not a linear relation; linear independence over Q has complexity ramifications. But this algebraic relation also has algorithmic ramifications. The linear independence implies that the multiplicative complexity of multiplying an arbitrary vector by G is 3. But because of the algebraic relation, it is not true (as is the case for quadratic extensions by indeterminates) that all minimal algorithms for this product are quadratic. A non-quadratic minimal algorithm is given via the factorization

$$G = \begin{pmatrix} \gamma(1,8) & 0 \\ 0 & \gamma(1,8) \end{pmatrix}\begin{pmatrix} 1 & 1-\sqrt{2} \\ \sqrt{2}-1 & 1 \end{pmatrix}. \tag{9.9}$$

As for computing the product of G and k distinct vectors, theory has it that the multiplicative complexity is $3k$ [3]. In other words, a direct sum theorem holds for this case. This result, and its generalization, due to Auslander and Winograd [3], is very deep; its proof is very complicated. But it yields great rewards.

The multiplicative complexity of all DFTs and DCTs are established using this result. The key to obtaining multiplicative complexity results for DFTs and DCTs is to find the appropriate block diagonalizations that transform these linear operators to such direct sums, and then to invoke this fundamental theorem. We will next cite this theorem, and then describe explicitly how we apply it to DFTs and DCTs.

FUNDAMENTAL THEOREM (Auslander–Winograd):

Let P_j be polynomials of degrees d_j, respectively, over a field φ. Let F_j denote polynomials of degree $d_j - 1$ with complex coefficients (i.e., they are complex numbers). For nonnegative integers k_j, let $T(k_j, F_j, P_j)$ denote the task of computing k_j products of arbitrary polynomials by F_j modulo P_j. Let $\sum_j T\left(k_j, F_j, P_j\right)$ denote the task of simultaneously computing all of these products. If the coefficients span a vector space of dimension $\sum_j d_j$ over φ, then the multiplicative complexity of $\sum_j T\left(k_j, F_j, P_j\right)$ is $\sum_j k_j\left(2d_j - 1\right)$. In other words, if the dimension assumption holds, then so does the direct sum theorem for this case.

Multiplicative complexity results for DFTs and DCTs assert that their computation is linear in the size of the input. The measure is number of nonrational multiplications. More specifically, in all cases (arbitrary input sizes, arbitrary dimensions), the number of nonrational multiplications necessary for computing these transforms is always less than twice the size of the input. The exact numbers are interesting, but more important is the algebraic structure of the transforms which lead to these numbers. This is what will be emphasized in the remainder of this chapter. Some special cases will be discussed in greater detail; general results will be reviewed rather briefly.

The following notation will be convenient. If A, B are matrices with real entries, and R, S are invertible rational matrices such that $A = RBS$, then we will say that A is rationally equivalent (or more plainly, equivalent) to B and write $A \approx B$. The multiplicative complexity of A is the same as that of B.

9.2 One-Dimensional DFTs

We will build up the theory for the DFT in stages. The one-dimensional DFT on input size N is a linear operator whose matrix is given by $F_N = (w^{jk})$, where $w = e^{2\pi i/N}$, and j, k index the rows and columns of the matrix, respectively. The first row and first column of F_N have all entries equal to 1, so the multiplicative complexity of F_N are the same as that of its "core" C_N, its minor comprising its last $N - 1$ rows and $N - 1$ columns. The first results were for one-dimensional DFTs on input sizes which are prime [24]. For p a prime integer, the set of integers between 0 and $p - 1$ form a cyclic group under multiplication modulo p. It was shown by Rader [19] that there exist permutations of the rows and columns of the core C_N that bring it to the cyclic convolution $w^{g^{j+k}}$, where g is any generator of the cyclic group described above. Using the decomposition for cyclic convolutions described above, we decompose the core to a direct sum of convolutions modulo the irreducible factors of $u^{p-1} - 1$. This decomposition into cyclotomic polynomials is well known [18]. There are $\tau(p - 1)$ irreducible factors, where $\tau(n)$ is the number of positive divisors of the positive integer n. One direct summand is the 1×1 matrix corresponding to the factor $u - 1$, and its entry is -1 (in particular, rational). Also, the coefficients of the other polynomials comprising the direct summands are all linearly independent over Q, hence the fundamental theorem (in its weakest form) applies. It yields that the multiplicative complexity of F_p for p a prime is $2p - \tau(p - 1) - 3$.

Next is the case for $N = p^k$ where p is an odd prime and the integer k is greater than 1. The group of units comprising those integers between 0 and $p - 1$ which are relatively prime to p, and under multiplication modulo p, is of order $p^k - p^{k-1}$. A Rader-like permutation [24] brings the sub-core, whose rows and columns are indexed by the entries in this group of units, to a cyclic convolution. The group of units, when multiplied by p, forms an orbit of order $p^{k-1} - p^{k-2}$ (p elements in the group of units map to the same element in the orbit), and the Rader-like permutations induces a permutation on the orbit, which yields cyclic convolutions of the sizes of the orbit. This proceeds until the final orbit of size $p - 1$. These cyclic convolutions are decomposed via the Chinese remainder theorem, and (after much cancellation and rearrangement) it can be shown that the core C_N in this case reduces to k direct summands, each of which is a semi-direct sum of $j(p - 1)(p^{k-j} - p^{k-j-1})$ dimensional convolutions modulo irreducible polynomials, $j = 1, 2, \ldots, k$. Also, the dimension of the coefficients of the polynomials is precisely $\sum_{j=1}^{k}(p-1)(p^{k-j}-p^{k-j-1})$. These are precisely the conditions sufficient to invoke the fundamental theorem. This algebraic decomposition yields minimal algorithms. When one adds all these up, the numerical result is that the multiplicative complexity for the DFT on p^k points where p is an odd prime and k a positive integer, is $2p^k - k - 2 - \frac{k^2+k}{2}\tau(p-1)$.

The case of the one-dimensional DFT on $N=2^n$ points is most familiar. In this case,

$$F_N = P_N \begin{pmatrix} F_{N/2} & \\ & G_{N/2} \end{pmatrix} R_N, \tag{9.10}$$

where

P_N is the permutation matrix which rearranges the output to even entries followed by odd entries
R_N is a rational matrix for computing the so-called "butterfly additions"
$G_{N/2} = D_{N/2}F_{N/2}$, where $D_{N/2}$ is a diagonal matrix whose entries are the so-called "twiddle factors"

This leads to the classical divide-and-conquer algorithm called the FFT. For our purposes, $G_{N/2}$ is equivalent to a direct sum of two polynomial products modulo u^{2^j}, $j=0,\ldots,n-3$. It is routine to proceed inductively, and then show that the hypothesis of the fundamental theorem are satisfied. Without details, the final result is that the complexity of the DFT on $N=2^n$ points is $2^{n+1}-n^2-n-2$. Again, the complexity is below $2N$.

For the general one-dimensional DFT case, we start with the equivalence $F_{mn} \approx F_m \otimes F_n$, whenever m and n are relatively prime, and where $\otimes$ denotes the tensor product. If m and n are of the forms p^k for some prime p and positive integer k, then from above, both F_m and F_n are equivalent to direct sums of polynomial products modulo irreducible polynomials. Applying the theorem of Kroeneker/Albert, which states that the tensor product of algebraic extension fields is isomorphic to a direct sum of fields, we have that F_{mn} is, therefore, equivalent to a direct sum of polynomial products modulo irreducible polynomials. When one follows the construction suggested by the theorem and counts the dimensionality of the coefficients, one can show that this direct sum system satisfies the hypothesis of the fundamental theorem. This argument extends to the general one-dimensional case of F_N where $N = \Pi_j p_j^{k_j}$ with p_j distinct primes.

9.3 Multidimensional DFTs

The k-dimensional DFT on $N_1,\ldots,N_k$ points is equivalent to the tensor product $F_{N_1} \otimes \cdots \otimes F_{N_k}$. Directly from the theorem of Kroeneker/Albert, this is equivalent to a direct sum of polynomial products modulo irreducible polynomials. It can be shown that this system satisfies the hypothesis of the fundamental theorem so that complexity results can be directly invoked for the general multidimensional DFT. Details can be found in [6]. More interesting than the general case are some special cases with unique properties.

The k-dimensional DFT on $p,\ldots,p$ points, where p is an odd prime, is quite remarkable. The core of this transform is a cyclic convolution modulo $u^{p^k-1}-1$. The core of the matrix corresponding to $F_p \otimes \cdots \otimes F_p$, which is the entire matrix minus its first row and column, can be brought into this large cyclic convolution by a permutation derived from a generator of the group of units of the field with p^k elements. The details are in [4]. Even more remarkably, this large cyclic convolution is equivalent to a direct sum of $p+1$ copies of the same cyclic convolution obtainable from the core of the one-dimensional DFT on p points. In other words, the k-dimensional DFT on $p,\ldots,p$ points, where p is an odd prime, is equivalent to a direct sum of $p+1$ copies of the one-dimensional DFT on p points. In particular, its multiplicative complexity is $(p+1)[2p-\tau(p-1)-3]$.

Another particularly interesting case is the k-dimensional DFT on $N,\ldots,N$ points, where $N=2^k$. This transform is equivalent to the k-fold tensor product $F_N \otimes \cdots \otimes F_N$, and we have seen above the recursive decomposition of F_N to a direct sum of $F_{N/2}$ and $G_{N/2}$. The semi-simple Abelian construction [5,9] yields that $F_{N/2} \otimes G_{N/2}$ is equivalent to $N/2$ copies of $G_{N/2}$, and likewise that $F_{N/2} \otimes G_{N/2}$ is equivalent to $N/2$ copies of $G_{N/2}$. Hence, F_N and F_N is equivalent to $3N/2$ copies of $G_{N/2}$ plus $F_{N/2} \otimes F_{N/2}$. This leads recursively to a complete decomposition of the two-dimensional DFT to a direct sum of polynomial products modulo irreducible polynomials (of the form $u^{2^m} + 1$ in this case). The extensions to arbitrary dimensions are quite detailed but straightforward.

9.4 One-Dimensional DCTs

As in the case of DFTs, DCTs are also all equivalent to direct sums of polynomial multiplications modulo irreducible polynomials and satisfy the hypothesis of the fundamental theorem. In fact, some instances are easier to handle. A fast way to see the structure of the DCT is by relating it to the DFT. Let C_N denote the one-dimensional DCT on N points; recall we defined F_N to be the one-dimensional DFT on N points.

It can be shown [15] that F_{4N} is equivalent to a direct sum of two copies of C_N plus one copy of F_{2N}. This is sufficient to yield complexity results for all one-dimensional DCTs. But for some special cases, direct derivations are more revealing. For example, when $N = 2^k$, C_N is equivalent to a direct sum of polynomial products modulo $u^{2^j} + 1$, for $j = 1, \ldots, k - 1$. This is a much simpler form than the corresponding one for the DFT on 2^k points. It is then straightforward to check that this direct sum system satisfies the hypothesis of the fundamental theorem, and then that the multiplicative complexity of C_{2^k} is $2^{k+1} - n - 2$.

Another (not so) special case is when N is an odd integer. Then C_N is equivalent to F_N, from which complexity results follow directly. Another useful result is that, as in the case of the DFT, C_{pq} is equivalent to $C_p \otimes C_q$ where p and q are relatively prime [26]. We can then use the theorem of Kroeneker/Albert [11] to build direct sum structures for DCTs of composites given direct sums of the various components.

9.5 Multidimensional DCTs

Here too, once the one-dimensional DCT structures are known, their extensions to multidimensions via tensor products, utilizing the theorem of Kroeneker/Albert, is straightforward. This leads to the appropriate direct sum structures, proving that the coefficients satisfy the hypothesis of the fundamental theorem does require some careful applications of elementary number theory. This is done in [11].

A most interesting special case is multidimensional DCT on input sizes which are powers of 2 in each dimension. If the input is k dimensional with size $2^{j_1} \times \cdots \times 2^{j_k}$, and $j_1 \leq j_i$, $i = 2, \ldots, k$, then the multidimensional DCT is equivalent to $2^{j_2} \times \cdots \times 2^{j_k}$ copies of the one-dimensional DCT on 2^{j_1} points [12]. This is a much more straightforward result than the corresponding one for multidimensional DFTs.

9.6 Nonstandard Models and Problems

DCTs have become popular because of their role in compression. In such roles, the DCT is usually followed by quantization. Therefore, in such applications, one need not actually compute the DCT but a scaled version of it, and then absorb the scaling into the quantization step. For the one-dimensional case this means that one can replace the computation of a product by C with a product by a matrix DC, where D is diagonal. It turns out [2,10] that for propitious choices of D, the computation of the product by DC is easier than that by C. The question naturally arises: What is the minimum number of steps required to compute a product of the form DC, where D can be any diagonal matrix? Our ability to answer such a question is very limited. All we can say today is that if we can compute a scaled DCT on N points with m multiplications, then certainly we can compute a DCT on N multiplications with $m + N$ points. Since we know the complexity of DCTs, this gives a lower bound on the complexity of scaled DCTs. For example, the one-dimensional DCT on 8 points (the most popular applied case) requires 12 multiplications. (The reader may see the number 11 in the literature; this is for the case of the "unnormalized DCT" in which the *DC* component is scaled. The unnormalized DCT is not orthogonal.) Suppose a scaled DCT on 8 points can be done with m multiplications. Then $8 + m \geq 12$, or $m \geq 4$. An algorithm for the scaled DCT on 8 points which uses 5 multiplications is known [2,10]. It is an open question whether one can actually do it in 4 multiplications or not. Similarly, the two-dimensional DCT on 8×8 points can be done with 54 multiplications [10,13], and theory says that at least 24 are needed [12]. The gap is very wide, and I know of stronger results as of this writing.

Machines whose primitive operations are fused multiply-accumulate are becoming very popular, especially in the higher end workstation arena. Here a single cycle can yield a result of the form $ab + c$ for arbitrary floating point numbers a, b, and c; we call such an operation a "mutiply/add." Lower bounds are obviously bounded below by lower bounds for number of multiplications and also for lower bounds on number of additions. The latter is a wide open subject. A simple yet instructive example involves multiplications of a 4×4 Hadamard matrix. It is well known that, in general, multiplication by an $N \times N$ Hadamard matrix, where N is a power of 2, can be done with $N\log_2 N$ additions. Recently it was shown [8] that the 4×4 case can be done with 7 multiply/add operations [8]. This result has not been extended, and it may in fact be rather hard to extend except in most trivial (and uninteresting) ways.

Upper bounds of DFTs have been obtained. It was shown in [17] that a complex DFT on $N = 2^k$ points can be done with $\frac{8}{3}Nk - \frac{16}{9}N + 2 - \frac{2}{9}(-1)^k$ real multiply/adds. For real input, an upper bound of $\frac{4}{3}Nk - \frac{17}{9}N + 3 - \frac{2}{9}(-1)^k$ real multiply/adds was given. These were later improved slightly using the results of the Hadamard transform computation. Similar multidimensional results were also obtained.

In the past several years new, more powerful, processors have been introduced. Sun and HP have incorporated new vector instructions. Intel has introduced its aggressive Intel's MMX architecture. And new multimedia signal processors from Philips, Samsung, and Chromatic are pushing similar designs even more aggressively. These will lead to new models of computation. Astounding (though probably not surprising) upper bounds will be announced; lower bounds are sure to continue to baffle.

References

1. Albert, A., *Structure of Algebras*, AMS Colloqium Publications, Vol. 21, New York, 1939.
2. Arai, Y., Agui, T., and Nakajima, M., A fast DCT-SQ scheme for images, *Trans. IEICE*, E-71(11): 1095–1097, Nov. 1988.
3. Auslander, L. and Winograd, S., The multiplicative complexity of certain semilinear systems defined by polynomials, *Adv. Appl. Math.*, 1(3): 257–299, 1980.
4. Auslander, L., Feig, E., and Winograd, S., New algorithms for the multidimensional discrete Fourier transform, *IEEE Trans. Acoust. Speech Signal Process.*, ASSP-31(2): 388–403, Apr. 1983.
5. Auslander, L., Feig, E., and Winograd, S., Abelian semi-simple algebras and algorithms for the discrete Fourier transform, *Adv. Appl. Math.*, 5: 31–55, Mar. 1984.
6. Auslander, L., Feig, E., and Winograd, S., The multiplicative complexity of the discrete Fourier transform, *Adv. Appl. Math.*, 5: 87–109, Mar. 1984.
7. Brocket, R.W. and Dobkin, D., On the optimal evaluation of a set of bilinear forms, *Linear Algebra Appl.*, 19(3): 207–235, 1978.
8. Coppersmith, D., Feig, E., and Linzer, E., Hadamard transforms on multiply/add architectures, *IEEE Trans. Signal Process.*, 46(4): 969–970, Apr. 1994.
9. Feig, E., New algorithms for the 2-dimensional discrete Fourier transform, IBM RC 8897 (No. 39031), June 1981.
10. Feig, E., A fast scaled DCT algorithm, *Proceedings of the SPIE-SPSE*, Santa Clara, CA, Feb. 11–16, 1990.
11. Feig, E. and Linzer, E., The multiplicative complexity of discrete cosine transforms, *Adv. Appl. Math.*, 13: 494–503, 1992.
12. Feig, E. and Winograd, S., On the multiplicative complexity of discrete cosine transforms, *IEEE Trans. Inf. Theory*, 38(4): 1387–1391, July 1992.

13. Feig, E. and Winograd, S., Fast algorithms for the discrete cosine transform, *IEEE Trans. Signal Process.*, 40(9): 2174–2193, Sept. 1992.
14. Fiduccia, C.M. and Zalcstein, Y., Algebras having linear multiplicative complexities, *J. ACM*, 24(2): 311–331, 1977.
15. Heideman, M.T., *Multiplicative Complexity, Convolution, and the DFT*, Springer-Verlag, New York, 1988.
16. Hopcroft, J. and Kerr, L., On minimizing the number of multiplications necessary for matrix multiplication, *SIAM J. Appl. Math.*, 20: 30–36, 1971.
17. Linzer, E. and Feig, E., Modified FFTs for fused multiply-add architectures, *Math. Comput.*, 60(201): 347–361, Jan. 1993.
18. Niven, I. and Zuckerman, H.S., *An Introduction to the Theory of Numbers*, John Wiley & Sons, New York, 1980.
19. Rader, C.M., Discrete Fourier transforms when the number of data samples is prime, *Proc. IEEE*, 56(6): 1107–1108, June 1968.
20. Strassen, V., Gaussian elimination is not optimal, *Numer. Math.*, 13: 354–356, 1969.
21. Strassen, V., Vermeidung con divisionen, *J. Reine Angew. Math.*, 264: 184–202, 1973.
22. Winograd, S., On the number of multiplications necessary to compute certain functions, *Commn. Pure Appl. Math.*, 23: 165–179, 1970.
23. Winograd, S., Some bilinear forms whose multiplicative complexity depends on the field of constants, *Math. Syst. Theory*, 10(2): 169–180, 1977.
24. Winograd, S., On the multiplicative complexity of the discrete Fourier transform, *Adv. Math.*, 32(2): 83–117, May, 1979.
25. Winograd, S., Arithmetic complexity of computations, *CBMS-NSF Regional Conference Series in Applied Mathematics*, Vol. 33, SIAM, Philadelphia, PA, 1980.
26. Yang, P.P.N. and Narasimha, M.J., Prime factor decomposition of the discrete cosine transform and its hardware realization, *Proceedings of the IEEE International Conference on Acoustics, Speech and Signal Processing*, 1985.

10

Fast Matrix Computations

Andrew E. Yagle
University of Michigan

10.1 Introduction

This chapter presents two major approaches to fast matrix multiplication. We restrict our attention to matrix multiplication, excluding matrix addition and matrix inversion, since matrix addition admits no fast algorithm structure (save for the obvious parallelization), and matrix inversion (i.e., solution of large linear systems of equations) is generally performed by iterative algorithms that require repeated matrix-matrix or matrix-vector multiplications. Hence, matrix multiplication is the real problem of interest.

We present two major approaches to fast matrix multiplication. The first is the divide-and-conquer strategy made possible by Strassen's [1] remarkable reformulation of noncommutative 2×2 matrix multiplication. We also present the APA (arbitrary precision approximation) algorithms, which improve on Strassen's result at the price of approximation, and a recent result that reformulates matrix multiplication as convolution and applies number theoretic transforms (NTTs). The second approach is to use a wavelet basis to sparsify the representation of Calderon–Zygmund operators as matrices. Since electromagnetic Green's functions are Calderon–Zygmund operators, this has proven to be useful in solving integral equations in electromagnetics. The sparsified matrix representation is used in an iterative algorithm to solve the linear system of equations associated with the integral equations, greatly reducing the computation. We also present some new insights that make the wavelet-induced sparsification seem less mysterious.

10.2 Divide-and-Conquer Fast Matrix Multiplication

10.2.1 Strassen Algorithm

It is not obvious that there should be any way to perform matrix multiplication other than using the definition of matrix multiplication, for which multiplying two $N \times N$ matrices requires N^3

multiplications and additions (N for each of the N^2, elements of the resulting matrix). However, in 1969, Strassen [1] made the remarkable observation that the product of two 2×2 matrices

$$\begin{bmatrix} a_{1,1} & a_{1,2} \\ a_{2,1} & a_{2,2} \end{bmatrix} \begin{bmatrix} b_{1,1} & b_{1,2} \\ b_{2,1} & b_{2,2} \end{bmatrix} = \begin{bmatrix} c_{1,1} & c_{1,2} \\ c_{2,1} & c_{2,2} \end{bmatrix} \tag{10.1}$$

may be computed using only seven multiplications (fewer than the obvious eight), as

$$\begin{aligned} m_1 &= (a_{1,2} - a_{2,2})(b_{2,1} + b_{2,2}); \quad m_3 = (a_{1,1} - a_{2,1})(b_{1,1} + b_{1,2}); \\ m_2 &= (a_{1,1} + a_{2,2})(b_{1,1} + b_{2,2}); \\ m_4 &= (a_{1,1} + a_{1,2})b_{2,2}; \quad m_7 = (a_{2,1} + a_{2,2})b_{1,1}; \\ m_5 &= a_{1,1}(b_{1,2} - b_{2,2}); \quad m_6 = a_{2,2}(b_{2,1} - b_{1,1}); \\ c_{1,1} &= m_1 + m_2 - m_4 + m_6; \quad c_{1,2} = m_4 + m_5; \\ c_{2,2} &= m_2 - m_3 + m_5 - m_7; \quad c_{2,1} = m_6 + m_7 \end{aligned} \quad . \tag{10.2}$$

A vital feature of Equation 10.2 is that it is noncommutative, i.e., it does not depend on the commutative property of multiplication. This can be seen easily by noting that each of the m_i are the product of a linear combination of the elements of A by a linear combination of the elements of B, in that order, so that it is never necessary to use, say $a_{2,2}b_{2,1} = b_{2,1}a_{2,2}$. We note there exist commutative algorithms for 2×2 matrix multiplication that require even fewer operations, but they are of little practical use.

The significance of noncommutativity is that the noncommutative algorithm (Equation 10.2) may be applied as is to block matrices. That is, if the $a_{i,j}$, $b_{i,j}$, and $c_{i,j}$ in Equations 10.1 and 10.2 are replaced by block matrices, Equation 10.2 is still true. Since matrix multiplication can be subdivided into block submatrix operations (i.e., Equation 10.1 is still true if $a_{i,j}$, $b_{i,j}$, and $c_{i,\,j}$ are replaced by block matrices), this immediately leads to a divide-and-conquer fast algorithm.

10.2.2 Divide-and-Conquer

To see this, consider the $2^n \times 2^n$ matrix multiplication $AB = C$, where A, B, and C are all $2^n \times 2^n$ matrices. Using the usual definition, this requires $(2^n)^3 = 8^n$ multiplications and additions. But if A, B, and C are subdivided into $2^{n-1} \times 2^{n-1}$ blocks $a_{i,j}$, $b_{i,j}$, and $c_{i,j}$, then $AB = C$ becomes Equation 10.1, which can be implemented with Equation 10.2 since Equation 10.2 does not require the products of subblocks of A and B to commute. Thus the $2^n \times 2^n$ matrix multiplication $AB = C$ can actually be implemented using only seven matrix multiplications of $2^{n-1} \times 2^{n-1}$ subblocks of A and B. And these subblock multiplications can in turn be broken down by using Equation 10.2 to implement them as well. The end result is that the $2^n \times 2^n$ matrix multiplication $AB = C$ can be implemented using only 7^n multiplications, instead of 8^n.

The computational savings grow as the matrix size increases. For $n = 5$ (32×32 matrices) the savings is about 50%. For $n = 12$ (4096×4096 matrices) the savings is about 80%. The savings as a fraction can be made arbitrarily close to unity by taking sufficiently large matrices. Another way of looking at this is to note that $N \times N$ matrix multiplication requires $O(N^{\log_2 7}) = O(N^{2.807}) < N^3$ multiplications using Strassen.

Of course we are not limited to subdividing into $2 \times 2 = 4$ subblocks. Fast noncommutative algorithms for 3×3 matrix multiplication requiring only $23 < 3^3 = 27$ multiplications were found by exhaustive search in [2,3]; 23 is now known to be optimal. Repeatedly subdividing $AB = C$ into $3 \times 3 = 9$ subblocks computes a $3^n \times 3^n$ matrix multiplication in $23^n < 27^n$ multiplications; $N \times N$ matrix multiplication requires $O(N^{\log_3 23}) = O(N^{2.854})$ multiplications, so this is not quite as good as using Equation 10.2. A fast noncommutative algorithm for 5×5 matrix multiplication requiring only $102 < 5^3 = 125$ multiplications was found in [4]; this also seems to be optimal. Using this algorithm, $N \times N$ matrix multiplication requires $O(N^{\log_5 102}) = O(N^{2.874})$ multiplications, so this is even worse. Of course, the idea is to write $N = 2^a 3^b 5^c$ for some a, b, c and subdivide into $2 \times 2 = 4$ subblocks a times, then subdivide into $3 \times 3 = 9$ subblocks b times, etc. The total number of multiplications is then $7^a 23^b 102^c < 8^a 27^b 125^c = N^3$.

Note that we have not mentioned additions. Readers familiar with nesting fast convolution algorithms will know why; now we review why reducing multiplications is much more important than reducing additions when nesting algorithms. The reason is that at each nesting stage (reversing the divide-and-conquer to build up algorithms for multiplying large matrices from Equation 10.2), each scalar addition is replaced by a matrix addition (which requires N^2 additions for $N \times N$ matrices), and each scalar multiplication is replaced by a matrix multiplication (which requires N^3 multiplications and additions for $N \times N$ matrices). Although we are reducing N^3 to about $N^{2.8}$, it is clear that each multiplication will produce more multiplications and additions as we nest than each addition. So reducing the number of multiplications from eight to seven in Equation 10.2 is well worth the extra additions incurred. In fact, the number of additions is also $O(N^{2.807})$.

The design of these base algorithms has been based on the theory of bilinear and trilinear forms. The review paper [5] and book [6] of Pan are good introductions to this theory. We note that reducing the exponent of N in $N \times N$ matrix multiplication is an area of active research. This exponent has been reduced to below 2.5; a known lower bound is 2. However, the resulting algorithms are too complicated to be useful.

10.2.3 Arbitrary Precision Approximation Algorithms

APA algorithms are noncommutative algorithms for 2×2 and 3×3 matrix multiplication that require even fewer multiplications than the Strassen-type algorithms, but at the price of requiring longer wordlengths. Proposed by Bini [7], the APA algorithm for multiplying two 2×2 matrices is this:

$$
\begin{aligned}
p_1 &= (a_{2,1} + \varepsilon a_{1,2})(b_{2,1} + \varepsilon b_{1,2});\\
p_2 &= (-a_{2,1} + \varepsilon a_{1,1})(b_{1,1} + \varepsilon b_{1,2});\\
p_3 &= (a_{2,2} - \varepsilon a_{1,2})(b_{2,1} + \varepsilon b_{2,2});\\
p_4 &= a_{2,1}(b_{1,1} - b_{2,1});\\
p_5 &= (a_{2,1} + a_{2,2})b_{2,1};\\
c_{1,1} &= (p_1 + p_2 + p_4)/\varepsilon - \varepsilon(a_{1,1} + a_{1,2})b_{1,2};\\
c_{2,1} &= p_4 + p_5;\\
c_{2,2} &= (p_1 + p_3 - p_5)/\varepsilon - \varepsilon a_{1,2}(b_{1,2} - b_{2,2}).
\end{aligned}
\tag{10.3}
$$

If we now let $\varepsilon \to 0$, the second terms in Equation 10.3 become negligible next to the first terms, and so they need not be computed. Hence, three of the four elements of $C = AB$ may be computed using only five multiplications. $c_{1,2}$ may be computed using a sixth multiplication, so that, in fact, two 2×2 matrices may be multiplied to arbitrary accuracy using only six multiplications. The APA 3×3 matrix multiplication algorithm requires 21 multiplications. Note that APA algorithms improve on the exact Strassen-type algorithms ($6 < 7$, $21 < 23$).

The APA algorithms are often described as being numerically unstable, due to roundoff error as $\varepsilon \to 0$. We believe that an electrical engineering perspective on these algorithms puts them in a light different from that of the mathematical perspective. In fixed point implementation, the computation $AB = C$ can be scaled to operations on integers, and the p_i can be bounded. Then it is easy to set ε a sufficiently small (negative) power of two to ensure that the second terms in Equation 10.3 do not overlap the first terms, provided that the wordlength is long enough. Thus, the reputation for instability is undeserved. However, the requirement of large wordlengths to be multiplied seems also to have escaped notice; this may be a more serious problem in some architectures.

The divide-and-conquer and resulting nesting of APA algorithms work the same way as for the Strassen-type algorithms. $N \times N$ matrix multiplication using Equation 10.3 requires $O\left(N^{\log_2(6)}\right) = O(N^{2.585})$ multiplications, which improves on the $O(N^{2.807})$ multiplications using Equation 10.2. But the wordlengths are longer.

A design methodology for fast matrix multiplication algorithms by grouping terms has been proposed in a series of papers by Pan (see [5,6]). While this has proven quite fruitful, the methodology of grouping terms becomes somewhat ad hoc.

10.2.4 Number Theoretic Transform Based Algorithms

An approach similar in flavor to the APA algorithms, but more flexible, has been taken recently in [8]. First, matrix multiplication is reformulated as a linear convolution, which can be implemented as the multiplication of two polynomials using the z-transform. Second, the variable z is scaled, producing a scaled convolution, which is then made cyclic. This aliases some quantities, but they are separated by a power of the scaling factor. Third, the scaled convolution is computed using pseudo-NTTs. Finally, the various components of the product matrix are read off of the convolution, using the fact that the elements of the product matrix are bounded. This can be done without error if the scaling factor is sufficiently large.

This approach yields algorithms that require the same number of multiplications or fewer as APA for 2×2 and 3×3 matrices. The multiplicands are again sums of scaled matrix elements as in APA. However, the design methodology is quite simple and straightforward, and the reason why the fast algorithm exists is now clear, unlike the APA algorithms. Also, the integer computations inherent in this formulation make possible the engineering insights into APA noted above.

We reformulate the product of two $N \times N$ matrices as the linear convolution of a sequence of length N^2 and a sparse sequence of length $N^3 - N + 1$. This results in a sequence of length $N^3 + N^2 - N$, from which elements of the product matrix may be obtained. For convenience, we write the linear convolution as the product of two polynomials. This result (of [8]) seems to be new, although a similar result is briefly noted in [3] (p. 197). Define

$a_{i,j} = a_{i+jN}; \quad b_{i,j} = b_{N-1-i+jN}; \quad 0 \le i, j \le N-1$

$$\left(\sum_{i=0}^{N-1}\sum_{j=0}^{N-1} a_{i+jN}x^{i+jN}\right)\left(\sum_{i=0}^{N-1}\sum_{j=0}^{N-1} b_{N-1-i+jN}x^{N(N-1-i+jN)}\right) = \sum_{i=0}^{N^3+N^2-N-1} c_i x^i; \tag{10.4}$$

$$c_{i,j} = c_{N^2-N+i+jN^2}, \quad 0 \le i, j \le N-1.$$

Note that coefficients of all three polynomials are read off of the matrices A, B, and C column-by-column (each column of B is reversed), and the result is noncommutative. For example, the 2×2 matrix multiplication (Equation 10.1) becomes

$$\begin{aligned}&(a_{1,1} + a_{2,1}x + a_{1,2}x^2 + a_{2,2}x^3)(b_{2,1} + b_{1,1}x^2 + b_{2,2}x^4 + b_{1,2}x^6)\\&= * + {}^*x + c_{1,1}x^2 + c_{2,1}x^3 + {}^*x^4 + {}^*x^5 + c_{1,2}x^6 + c_{2,2}x^7 + {}^*x^8 + {}^*x^9,\end{aligned} \tag{10.5}$$

where * denotes an irrelevant quantity. In Equation 10.5, substitute $x = sz$ and take the result $mod(z^6 - 1)$. This gives

$$\begin{aligned}&(a_{1,1} + a_{2,1}sz + a_{1,2}s^2z^2 + a_{2,2}s^3z^3)((b_{2,1} + b_{1,2}s^6) + b_{1,1}s^2z^2 + b_{2,2}s^4z^4)\\&= (* + c_{1,2}s^6) + ({}^*s + c_{2,2}s^7)z + (c_{1,1}s^2 + {}^*s^8)z^2\\&\quad + (c_{2,1}s^3 + {}^*s^9)z^3 + {}^*z^4 + {}^*z^5; \quad mod(z^6 - 1).\end{aligned} \tag{10.6}$$

If $|c_{i,j}|, |*| < s^6$, then the * and $c_{i,j}$ may be separated without error, since both are known to be integers. If s is a power of 2, $c_{0,1}$ may be obtained by discarding the $6\log_2 s$ least significant bits in the binary representation of $* + c_{0,1}s^6$. The polynomial multiplication $mod(z^6 - 1)$ can be computed using NTTs [9] using 6 multiplications. Hence, 2×2 matrix multiplication requires 6 multiplications. Similarly, 3×3 matrices may be multiplied using 21 multiplications. Note these are the same numbers required by the APA algorithms, quantities multiplied are again sums of scaled matrix elements, and results are again sums in which one quantity is partitioned from another quantity which is of no interest.

However, this approach is more flexible than the APA approach (see [8]). As an extreme case, setting $z = 1$ in Equation 10.5 computes a 2×2 matrix multiplication using ONE (very long wordlength) multiplication! For example, using $s = 100$

$$\begin{bmatrix} 2 & 4 \\ 3 & 5 \end{bmatrix}\begin{bmatrix} 9 & 8 \\ 7 & 6 \end{bmatrix} = \begin{bmatrix} 46 & 40 \\ 62 & 54 \end{bmatrix} \tag{10.7}$$

becomes the single scalar multiplication

$$(5, 040, 302)(8, 000, 600, 090, 007) = 40, 325, 440, 634, 862, 462, 114. \tag{10.8}$$

This is useful in optical computing architectures for multiplying large numbers.

10.3 Wavelet-Based Matrix Sparsification

10.3.1 Overview

A common application of solving large linear systems of equations is the solution of integral equations arising in, say, electromagnetics. The integral equation is transformed into a linear system of equations using Galerkin's method, so that entries in the matrix and vectors of knowns and unknowns are coefficients of basis functions used to represent the continuous functions in the integral equation. Intelligent selection of the basis functions results in a sparse (mostly zero entries) system matrix. The sparse linear system of unknowns is then usually solved using an iterative algorithm, which is where the sparseness becomes an advantage (iterative algorithms require repeated multiplication of the system matrix by the current approximation to the vector of unknowns).

Recently, wavelets have been recognized as a good choice of basis function for a wide variety of applications, especially in electromagnetics. This is true because in electromagnetics the kernel of the integral equation is a two-dimensional (2-D) or three-dimensional (3-D) Green's function for the wave equation, and these are Calderon–Zygmund operators. Using wavelets as basis functions makes the matrix representation of the kernel drop off rapidly away from the main diagonal, more rapidly than discretization of the integral equation would produce.

Here we quickly review the wavelet transform as a representation of continuous functions and show how it sparsifies Calderon–Zygmund integral operators. We also provide some insight into why this happens and present some alternatives that make the sparsification less mysterious. We present our results in terms of continuous (integral) operators, rather than discrete matrices, since this is the proper presentation for applications, and also since similar results can be obtained for the explicitly discrete case.

10.3.2 Wavelet Transform

We will not attempt to present even an overview of the rich subject of wavelets. The reader is urged to consult the many papers and textbooks (e.g., [10]) now being published on the subject. Instead, we restrict our attention to aspects of wavelets essential to sparsification of matrix operator representations. The wavelet transform of an L^2 function $f(x)$ is defined as

$$f_i(n) = 2^{i/2} \int_{-\infty}^{\infty} f(x)\psi(2^i x - n)\mathrm{d}x, \quad f(x) = \sum_i \sum_n f_i(n)\psi(2^i x - n)2^{i/2}, \tag{10.9}$$

where $\{\psi(2^i x - n), i, n \in Z\}$ is a complete orthonormal basis for L^2. That is, L^2 (the space of square-integrable functions) is spanned by dilations (scaling) and translations of a wavelet basis function $\psi(x)$. Constructing this $\psi(x)$ is nontrivial, but has been done extensively in the literature.

Since the summations must be truncated to finite intervals in practice, we define the wavelet scaling function $\varphi(x)$ whose translations on a given scale span the space spanned by the wavelet basis function $\psi(x)$ at all translations and at scales coarser than the given scale. Then we can write

$$\begin{aligned} f(x) &= 2^{I/2} \sum_n c_I(n)\varphi(2^I x - n) + \sum_{i=I}^{\infty} \sum_n f_i(n)\psi(2^i x - n)2^{i/2} \\ c_I(n) &= 2^{I/2} \int_{-\infty}^{\infty} f(x)\varphi(2^I x - n)\mathrm{d}x. \end{aligned} \tag{10.10}$$

So the projection $c_I(n)$ of $f(x)$ on the scaling function $\varphi(x)$ at scale I replaces the projections $f_i(n)$ on the basis function $\psi(x)$ on scales coarser (smaller) than I. The scaling function $\varphi(x)$ is orthogonal to its translations but (unlike the basis function $\psi(x)$) is not orthogonal between scales. Truncating the summation at the upper end approximates $f(x)$ at the resolution defined by the finest (largest) scale i; this is somewhat analogous to truncating Fourier series expansions and neglecting high-frequency components.

We also define the 2-D wavelet transform of $f(x, y)$ as

$$\begin{aligned} f_{i,j}(m,n) &= 2^{i/2}2^{j/2} \int_{-\infty}^{\infty}\int_{-\infty}^{\infty} f(x,y)\psi(2^i x - m)\psi(2^j y - n)\mathrm{d}x\,\mathrm{d}y \\ f(x,y) &= \sum_{i,j,m,n} f_{i,j}(m,n)\psi(2^i x - m)\psi(2^j y - n)2^{i/2}2^{i/2}. \end{aligned} \tag{10.11}$$

However, it is more convenient to use the 2-D counterpart of Equation 10.10, which is

$$
\begin{aligned}
c_1(m,n) &= 2^I \int_{-\infty}^{\infty}\int_{-\infty}^{\infty} f(x,y)\varphi(2^I x - m)\varphi(2^I y - n)\mathrm{d}x\,\mathrm{d}y \\
f_i^1(m,n) &= 2^i \int_{-\infty}^{\infty}\int_{-\infty}^{\infty} f(x,y)\varphi(2^i x - m)\psi(2^i y - n)\mathrm{d}x\,\mathrm{d}y \\
f_i^2(m,n) &= 2^i \int_{-\infty}^{\infty}\int_{-\infty}^{\infty} f(x,y)\psi(2^i x - m)\varphi(2^i y - n)\mathrm{d}x\,\mathrm{d}y \\
f_i^3(m,n) &= 2^i \int_{-\infty}^{\infty}\int_{-\infty}^{\infty} f(x,y)\psi(2^i x - m)\psi(2^i y - n)\mathrm{d}x\,\mathrm{d}y \\
c_1(m,n) &= 2^I \int_{-\infty}^{\infty}\int_{-\infty}^{\infty} f(x,y)\varphi(2^I x - m)\varphi(2^I y - n)\mathrm{d}x\,\mathrm{d}y \\
f_i^1(m,n) &= 2^i \int_{-\infty}^{\infty}\int_{-\infty}^{\infty} f(x,y)\varphi(2^i x - m)\psi(2^i y - n)\mathrm{d}x\,\mathrm{d}y \qquad (10.12) \\
f_i^2(m,n) &= 2^i \int_{-\infty}^{\infty}\int_{-\infty}^{\infty} f(x,y)\psi(2^i x - m)\varphi(2^i y - n)\mathrm{d}x\,\mathrm{d}y \\
f_i^3(m,n) &= 2^i \int_{-\infty}^{\infty}\int_{\infty}^{\infty} f(x,y)\psi(2^i x - m)\psi(2^i y - n)\mathrm{d}x\,\mathrm{d}y \\
f(x,y) &= \sum_{m,n} c_I(m,n)\varphi(2^I x - m)\varphi(2^I y - n)2^I \\
&\quad + \sum_{i=I}^{\infty}\sum_{m,n} f_i^1(m,n)\varphi(2^i x - m)\psi(2^i y - n)2^i \\
&\quad + \sum_{i=I}^{\infty}\sum_{m,n} f_i^2(m,n)\psi(2^i x - m)\varphi(2^i y - n)2^i \\
&\quad + \sum_{i=I}^{\infty}\sum_{m,n} f_i^3(m,n)\psi(2^i x - m)\psi(2^i y - n)2^i.
\end{aligned}
$$

Once again the projection $c_I(m,n)$ on the scaling function at scale I replaces all projections on the basis functions on scales coarser than M.

Some examples of wavelet scaling and basis functions:

Scaling	Pulse	B-Spline	Sinc	Softsinc	Daubechies
Wavelet	Haar	Battle–Lemarie	Paley–Littlewood	Meyer	Daubechies

An important property of the wavelet basis function $\psi(x)$ is that its first k moments can be made zero, for any integer k [10]:

$$\int_{-\infty}^{\infty} x^i \psi(x) dx = 0, \quad i = 0, \ldots, k \tag{10.13}$$

10.3.3 Wavelet Representations of Integral Operators

We wish to use wavelets to sparsify the L^2 integral operator $K(x, y)$ in

$$g(x) = \int_{-\infty}^{\infty} K(x, y) f(y) dy. \tag{10.14}$$

A common situation: Equation 10.14 is an integral equation with known kernel $K(x, y)$ and known $g(x)$ in which the goal is to compute an unknown function $f(y)$. Often the kernel $K(x, y)$ is the Green's function (spatial impulse response) relating observed wave field or signal $g(x)$ to unknown source field or signal $f(y)$.

For example, the Green's function for Laplace's equation in free space is

$$G(r) = -\frac{1}{2\pi} \log r \ \text{(2-D)}, \quad \frac{1}{4\pi r} \ \text{(3-D)}, \tag{10.15}$$

where r is the distance separating the points of source and observation. Now consider a line source in an infinite 2-D homogeneous medium, with observations made along the same line. The observed field strength $g(x)$ at position x is

$$g(x) = -\frac{1}{2\pi} \int_{-\infty}^{\infty} \log |x - y| f(y) dy, \tag{10.16}$$

where $f(y)$ is the source strength at position y.

Using Galerkin's method, we expand $f(y)$ and $g(x)$ as in Equation 10.9 and $K(x, y)$ as in Equation 10.11. Using the orthogonality of the basis functions yields

$$\sum_j \sum_n K_{i,j}(m, n) f_j(n) = g_i(m). \tag{10.17}$$

Expanding $f(y)$ and $g(x)$ as in Equation 10.10 and $K(x, y)$ as in Equation 10.12 leads to another system of equations which is difficult notationally to write out in general, but can clearly be done in individual applications. We note here that the entries in the system matrix in this latter case can be rapidly generated using the fast wavelet algorithm of Mallat (see [10]).

The point of using wavelets is as follows. $K(x, y)$ is a Calderon–Zygmund operator if

$$\left| \frac{\partial^k}{\partial x^k} K(x, y) \right| + \left| \frac{\partial^k}{\partial y^k} K(x, y) \right| \leq \frac{C_k}{|x - y|^{k+1}} \tag{10.18}$$

for some $k \geq 1$. Note in particular that the Green's functions in Equation 10.15 are Calderon–Zygmund operators. Then the representation in Equation 10.12 of $K(x, y)$ has the property [11]

$$|f_i^1(m,n)| + |f_i^2(m,n)| + |f_i^3(m,n)| \leq \frac{C_k}{1+|m-n|^{k+1}}, \quad |m-n| > 2k \tag{10.19}$$

if the wavelet basis function $\psi(x)$ has its first k moments zero (Equation 10.13).

This means that using wavelets satisfying Equation 10.13 sparsifies the matrix representation of the kernel $K(x, y)$. For example, a direct discretization of the 3-D Green's function in Equation 10.15 decays as $1/|m-n|$ as one moves away from the main diagonal $m=n$ in its matrix representation. However, using wavelets, we can attain the much faster decay rate $1/(1+|m-n|^{k+1})$ far away from the main diagonal. By neglecting matrix entries less than some threshold (typically 1% of the largest entry) a sparse and mostly banded matrix is obtained. This greatly speeds up the following matrix computations:

1. Multiplication by the matrix for solving the forward problem of computing the response to a given excitation (as in Equation 10.16).
2. Fast solution of the linear system of equations for solving the inverse problem of reconstructing the source from a measured response (solving Equation 10.16 as an integral equation). This is typically performed using an iterative algorithm such as conjugate gradient method. Sparsification is essential for convergence in a reasonable time.

A typical sparsified matrix from an electromagnetics application is shown in Figure 6 of [12]. Battle–Lemarie wavelet basis functions were used to sparsify the Galerkin method matrix in an integral equation for planar dielectric millimeter-wave waveguides and a 1% threshold applied (see [12] for details). Note that the matrix is not only sparse but (mostly) banded.

10.3.4 Heuristic Interpretation of Wavelet Sparsification

Why does this sparsification happen? Considerable insight can be gained using Equation 10.13. Let $\hat{\psi}(\omega)$ be the Fourier transform of the wavelet basis function $\psi(x)$. Since the first k moments of $\psi(x)$ are zero by Equation 10.13 we can expand $\hat{\psi}(\omega)$ in a power series around $\omega = 0$:

$$\hat{\psi}(\omega) \approx \omega^k, \quad |\omega| \ll 1. \tag{10.20}$$

This shows that for small $|\omega|$ taking the wavelet transform of $f(x)$ is roughly equivalent to taking the kth derivative of $f(x)$. This can be confirmed that many wavelet basis functions bear a striking resemblance to the impulse responses of regularized differentiators. Since $K(x, y)$ is assumed a Calderon–Zygmund operator, its kth derivatives in x and y drop off as $1/|x-y|^{k+1}$. Thus, it is not surprising that the wavelet transform of $K(x, y)$, which is roughly taking kth derivatives, should drop off as $1/|m-n|^{k+1}$. Of course there is more to it, but this is why it happens.

It is not surprising that $K(x, y)$ can be sparsified by taking advantage of its derivatives being small. To see a more direct way of accomplishing this, apply integration by parts to Equation 10.14 and take the partial derivative with respect to x. This gives

$$\frac{\mathrm{d}g(x)}{\mathrm{d}x} = -\int_{-\infty}^{\infty} \left(\frac{\partial}{\partial x}\frac{\partial}{\partial y}K(x,y)\right)\left(\int_{-\infty}^{y} f(y')\mathrm{d}y'\right)\mathrm{d}y, \tag{10.21}$$

which will likely sparsify a smooth $K(x, y)$. Of course, higher derivatives can be used until a condition like Equation 10.18 is reached. The operations of integrating $f(y)$ and $\frac{\partial^k g}{\partial x^k}$ (to get $g(x)$) k times can be accomplished using $nk \ll n^2$ additions, so considerable savings can result. This is different from using wavelets, but in the same spirit.

References

1. Strassen, V., Gaussian elimination is not optimal, *Numerische Math.*, 13: 354–356, 1969.
2. Landerman, J.D., A noncommutative algorithm for multiplying 3×3 matrices using 23 multiplications, *Bull. Am. Math. Soc.*, 82: 127–128, 1976.
3. Johnson, R.W. and McLoughlin, A.M., Noncommutative bilinear algorithms for 3×3 matrix multiplication, *SIAM J. Comput.*, 15: 595–603, 1976.
4. Makarov, O.M., A noncommutative algorithm for multiplying 5×5 matrices using 102 multiplications, *Inf. Proc. Lett.*, 23: 115–117, 1986.
5. Pan, V., How can we speed up matrix multiplication? *SIAM Rev.*, 26(3): 393–415, 1984.
6. Pan, V., *How Can We Multiply Matrices Faster?* Springer-Verlag, New York, 1984.
7. Bini, D., Capovani, M., Lotti, G., and Romani, F., $O(n^{2.7799})$ complexity for matrix multiplication, *Inf. Proc. Lett.*, 8: 234–235, 1979.
8. Yagle, A.E., Fast algorithms for matrix multiplication using pseudo number theoretic transforms, *IEEE Trans. Signal Process.*, 43: 71–76, 1995.
9. Nussbaumer, H.J., *Fast Fourier Transforms and Convolution Algorithms*, Springer-Verlag, Berlin, Germany, 1982.
10. Daubechies, I., *Ten Lectures on Wavelets*, SIAM, Philadelphia, PA, 1992.
11. Beylkin, G., Coifman, R., and Rokhlin, V., Fast wavelet transforms and numerical algorithms I, *Commn. Pure Appl. Math.*, 44: 141–183, 1991.
12. Sabetfakhri, K. and Katehi, L.P.B., Analysis of integrated millimeter wave and submillimeter wave waveguides using orthonormal wavelet expansions, *IEEE Trans. Microw. Theory Technol.*, 42: 2412–2422, 1994.

IV

Digital Filtering

Lina J. Karam
Arizona State University

James H. McClellan
Georgia Institute of Technology

DIGITAL FILTERING IS ONE OF THE MOST IMPORTANT FUNCTIONS in digital signal processing, and this single chapter not only provides a thorough coverage of conventional topics such as FIR (finite-duration impulse response) and IIR (infinite-duration impulse response) filtering, it also presents material on design methods and new research directions that have not been widely available in the open literature.

Karam and McClellan present an "Introduction to Digital Filtering," followed by Karam's "Steps in Filter Design." A comprehensive coverage of FIR and IIR classical filter design follows in "FIR Design Methods" by Selesnick, Burrus, Karam, and McClellan, and "IIR Design Methods" by Karam, Selesnick, and Burrus, respectively. Unique to this chapter is the special discussion of "Software Tools" for filtering by McClellan.

The various topics covered in this section are integrated together very well, ensuring a coherent and authoritative coverage of the filtering area.

11

Digital Filtering

Lina J. Karam
Arizona State University

James H. McClellan
Georgia Institute of Technology

Ivan W. Selesnick
Polytechnic University

C. Sidney Burrus
Rice University

11.1 Introduction

Digital filters are widely used in processing digital signals of many diverse applications, including speech processing and data communications, image and video processing, sonar, radar, seismic and oil exploration, and consumer electronics. One class of digital filters, the linear shift-invariant (LSI) type, are the most frequently used because they are simple to analyze, design, and implement. This chapter treats the LSI case only; other filter types, such as adaptive filters, require quite different design methodologies.

An LSI digital filter can be uniquely identified in the time/space domain by its impulse response $h(n)$ (where n is an integer index). Alternatively, the LSI digital filter can be uniquely characterized in the frequency domain by its frequency response $H(\omega)$ (where ω is a real-valued frequency variable in radians), which is also the discrete-time Fourier transform (DTFT) of the sequence $h(n)$. LSI digital filters are of two main types: finite-duration impulse response (FIR) filters for which the impulse response $h(n)$ is nonzero for only a finite number of samples, and infinite-duration impulse response (IIR) filters for which $h(n)$ has an infinite number of nonzero samples. In the FIR case, the samples of the sequence $h(n)$ are commonly referred to as the filter coefficients; for the IIR case, the filter coefficients include feedback terms in a difference equation.

Digital filter design has been extensively addressed within the last 25 years. The design and realization of digital filters involve a blend of theory, applications, and technologies. For most applications, it is desirable to design frequency-selective filters which alter or pass unchanged different frequency components. In this case, the desired design specifications are given in the frequency domain by specifying a desired frequency response $D(f)$. Note that $D(f)$ is, in general, complex valued, consisting of a desired magnitude response $|D(f)|$ and a desired phase response $\angle D(f)$. One of the most important problems is the design of a highly frequency-selective filter with sharp cutoff edges (short transition bands). However, ideal sharp edges correspond mathematically to discontinuities and cannot be realized in practice. Therefore, the filter design problem consists in finding an implementable filter whose order is low and whose frequency response $H(f)$ best approximates the specified ideal magnitude and phase responses which are given as the desired design specifications or constraints.

The design of digital filters is typically done by performing the following steps:

1. Convert the desired design constraints into precise specifications of the desired magnitude and phase responses, designed filter type (FIR or IIR), filter order, error tolerance, or criteria
2. Approximate the design specifications (of Step 1) by finding the implementable FIR or IIR filter such that the obtained filter frequency response best meets the design specs according to a mathematical error criterion
3. Realize the filter using the digital technology most suitable for the considered application

While Step 2 is performed using mathematical optimization and approximation methods, Step 1 is highly dependent on the application and the detail provided by the user. Step 3 depends on the technology or software used to build the filter.

Nowadays, the optimization needed in Step 2 is usually done with computer software that implements sophisticated numerical optimization routines. In addition, these design packages usually have a convenient graphical user interface to aid in the conversion of specs needed in Step 1. With such software, a filter design can be carried out quickly so that many designs can be tried in the process of getting the best filter. Since most filter design techniques involve the trade-off among competing parameters, the software can also incorporate design rules that allow the user to predict the order needed for certain specs without actually designing the filter, for example.

This chapter is organized as follows. Section 11.2 provides a discussion of Steps 1 and 3, including creating the design specifications, selecting the filter type and order, specifying the error tolerances and criteria, and realizing the designed filter. Step 2 is treated in Sections 11.3 and 11.4. Section 11.3 describes the classical FIR and IIR design methods. Section 11.4 presents nonclassic and more recently developed design methods with added efficiency and/or flexibility. Finally, Section 11.5 gives examples of some of the currently available software design tools and describes the characteristics that a user can expect from such tools.

11.2 Steps in Filter Design

Lina J. Karam

The general filter design problem can be briefly stated as follows. Given some ideal frequency response, $D(\omega)$, find a realizable IIR or FIR digital filter whose frequency response, $H(\omega)$, approximates $D(\omega)$. The realizable filter is found by optimizing some measure of the filter's performance, for example, minimizing the filter order (IIR) or the filter length (FIR), or minimizing the width of the transition bands, or reducing the passband error and/or stopband error. Setting up the specifications for the general filter design problem will define these parameters and show which trade-offs are possible.

11.2.1 Creating the Design Specifications

Since the frequency response of a digital filter is always periodic in the frequency variable ω with a period of 2π, the design specifications need only be specified for one period; usually, over the frequency region $[-\pi, \pi]$. Furthermore, when the frequency response is conjugate-symmetric (i.e., $D^*(\omega) = D(-\omega)$), then it is sufficient to specify the response only on the positive frequency interval $[0, \pi]$. The conjugate-symmetric case is the most common, because it corresponds to filters with real coefficients.

The simplest case is that of an ideal lowpass digital filter with zero phase, whose frequency response can be expressed as

$$D(\omega) = \begin{cases} 1, & |\omega| < \omega_c \\ 0, & \omega_c < |\omega| < \pi, \end{cases} \tag{11.1}$$

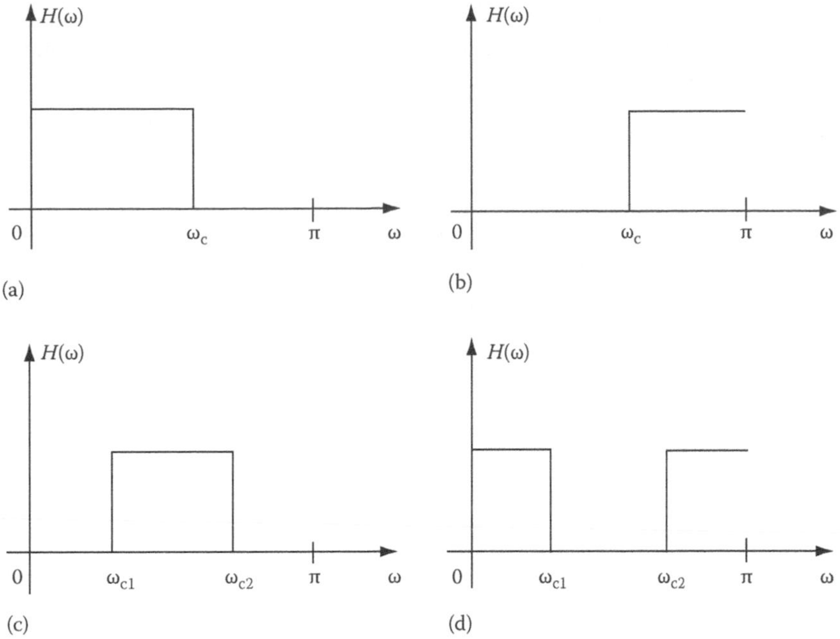

FIGURE 11.1 Common ideal digital filter types: (a) ideal lowpass, (b) ideal highpass, (c) ideal bandpass, and (d) ideal bandstop.

where ω_c is the cutoff frequency corresponding to the location of a sharp cutoff edge, as shown in Figure 11.1a. In this case, the frequency response, $D(\omega)$, is real-valued and, therefore, corresponds also to the magnitude response of the filter (since the phase is zero). Ideal frequency responses of other commonly used frequency-selective filters are shown in Figure 11.1.

These ideal filters have frequency responses with sharp cutoff edges (discontinuities) and cannot be implemented directly. They must be approximated with a realizable system—the sharp cutoff edges need to be replaced with transition bands in which the designed frequency response would change smoothly in going from one band to the other. So, design templates need to be provided where the sharp cutoff edges are replaced with nonzero width transition bands located around the ideal cutoff edges. A typical design template for a lowpass filter is shown in Figure 11.2, where

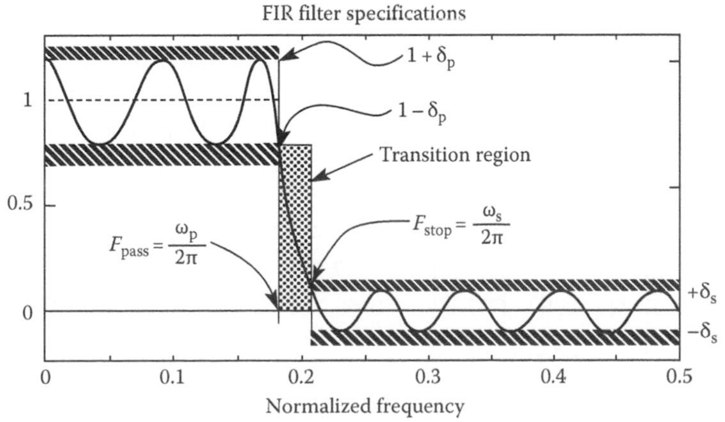

FIGURE 11.2 Design template for a lowpass filter.

- ω_p is the passband cutoff frequency.
- ω_s is the stopband cutoff frequency. The cutoff frequency ω_c is usually taken to be midway between the passband and stopband cutoff frequencies.
- The open interval (ω_p, ω_s) is the transition band of width $\Delta\omega_t = \omega_s - \omega_p$. In the common design methods, no design specifications are given in the transition bands which are therefore commonly known as "don't care bands." However, it is usually desirable to have the frequency response change smoothly (i.e., no fluctuations or overshoots) in the transition bands; this requirement might not be satisfied by a design method that places no design constraints on the frequency response in the transition bands.
- δ_p is known as the passband ripple and is the maximum allowable error in the passband.
- δ_s is known as the stopband ripple and is the maximum allowable error in the stopband.

The objective of filter design then is to find a realizable FIR or IIR filter whose frequency response $H(\omega)$ approximates the specified design constraints given by the design template. Ideally, the filter design process would make each of the following parameters as small as possible: δ_p, δ_s, $\Delta\omega_t$, IIR filter order (number of poles of $H(z)$ which is a rational function), or FIR filter length (number of zeros of $H(z)$ which is a finite polynomial). Practically, the filter design process minimizes one of these parameters while holding the others fixed.

Traditionally, many of the filters designed in practice are specified in terms of constraints on the magnitude response and no constraints on the phase response other than those imposed implicitly by stability and/or causality requirements (e.g., poles inside unit circle [U.C.] in the complex Z-plane for IIR, and linear-phase for FIR [1]). More recently, design methods that include phase design specifications have been presented [2–5]. In this latter case, two design templates must be provided, one for the magnitude response and another for the (passband) phase response. An ideal phase response is most likely a constant slope phase function:

$$\angle D(\omega) = -M\omega.$$

The parameter M is equivalent to the desired delay of the filter (in samples). An error template for the phase would be a tolerance about the desired phase, for example, δ_ϕ would denote the maximum allowable phase ripple, so that we require

$$|\angle H(\omega) - \angle D(\omega)| < \delta_\phi.$$

11.2.2 Specs Derived from Analog Filtering

Often, the desired design specifications are not given directly in the digital domain. Instead, an equivalent analog filtering operation is desired but is to be performed using an embedded digital filter. Figure 11.3 shows a standard system for processing continuous-time (continuous-space) signals using a digital filter.

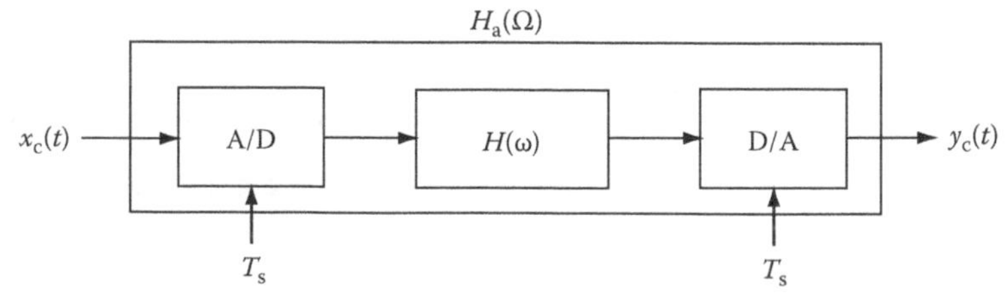

FIGURE 11.3 Standard system for processing analog signals using a digital (discrete-time) filter.

The analog input signal is first transformed into a digital signal through an analog-to-digital (A/D) conversion operation; then, filtering is carried out using a digital filter; finally, the filtered digital output is converted back to the analog domain using a digital-to-analog (D/A) converter. For this system, if the sampling period T_s of the A/D and D/A converters is chosen appropriately to avoid aliasing of the input spectrum, the overall system (consisting of the A/D converter, the digital filter, and the D/A converter) behaves as an equivalent analog filter. In this case, the frequency response $H_a(\Omega)$ of the equivalent analog filter is related to the frequency response $H(\omega)$ of the digital filter through a simple linear scaling relation between the digital frequency ω and the analog frequency Ω. This linear scaling relation is given by

$$\omega = \Omega T_s \tag{11.2}$$

leading to the following expression of the analog $H_a(\Omega)$ in terms of the digital $H(\omega)$:

$$H_a(\Omega) = \begin{cases} H(\Omega T_s), & |\Omega| < \frac{\pi}{T_s} \\ 0, & |\Omega| \geq \frac{\pi}{T_s}. \end{cases} \tag{11.3}$$

Equivalently, $H(\omega)$ can also be expressed in terms of $H_a(\Omega)$ as follows:

$$H(\omega) = H_a(\omega/T_s), \quad |\omega| < \pi. \tag{11.4}$$

A typical filter design problem corresponding to this system is to design the digital filter such that the overall equivalent analog filter best approximates some ideal analog specifications. So, if we are given the desired analog specifications of the overall analog system, these can be turned into specifications for the desired digital filter by using Equation 11.4. Then, a digital filter $H(\omega)$ can be designed to approximate the derived desired digital specifications. Finally, the resulting analog frequency response of the overall system can be found using Equation 11.3, for example, to compare with the ideal analog response.

11.2.3 Specifying an Error Measure

An error measure is needed to assess how much the designed filter $H(\omega)$ deviates from the desired filter $D(\omega)$. Defining the pointwise error $E(\omega)$ as

$$E(\omega) = [D(\omega) - H(\omega)], \tag{11.5}$$

we must reduce $E(\omega)$ to a scalar error measure (also called an error norm). With a correctly chosen norm, there are many possible optimization algorithms that will compute the best filter parameters to minimize the chosen error norm. The following error norms are the most commonly used in filter design:

- Mean squared error (MSE) or L_2 norm

$$E_2 = \left[\frac{1}{2\pi}\int_B |E(\omega)|^2 d\omega\right]^{1/2} \tag{11.6}$$

- L_p norm which is a generalization of the L_2 norm and where p is a nonzero integer

$$E_p = \left[\frac{1}{2\pi}\int_B |E(\omega)|^p d\omega\right]^{1/p} \tag{11.7}$$

- Chebyshev or L_∞ norm

$$E_\infty = \max_{\omega \in B} |E(\omega)| \tag{11.8}$$

The Chebyshev error norm limits the worst case deviation from the ideal specifications.

In the above definitions, $|\cdot|$ denotes the complex error magnitude and B is the frequency region of interest over which the error norm is to be minimized. The frequency subset $B \subset [-\pi, \pi)$ is taken to be the union of the desired passbands and stopbands.

A more selective control of the approximation accuracy can be achieved by introducing a weighting function $W(\omega)$ in Equation 11.5 as follows:

$$E(\omega) = W(\omega)[D(\omega) - H(\omega)]. \tag{11.9}$$

The weighting function $W(\omega)$ must be a real, strictly positive and continuous function on B. It can force a better match over selected regions or frequency points relative to other regions in B. Alternatively, note that Equation 11.5 reduces to Equation 11.9 if we replace $D(\omega)$ with $W(\omega)D(\omega)$ and $H(\omega)$ with $W(\omega)H(\omega)$.

11.2.4 Selecting the Filter Type and Order

As mentioned in Section 11.1, there are two main types of filters, namely FIR and IIR. These differ in their characteristics and in the way they are designed. Since the design algorithm depends strongly on the choice of IIR vs. FIR filter, the designer should make this decision as early as possible. Although the desired frequency response specifications can be approximated with either type of filter, deciding which of the two filter types to use depends on many factors including the implementation hardware, as well as the magnitude and phase characteristics of the resulting filter. To aid in this decision, the main characteristics of FIR and IIR filters are discussed below.

11.2.4.1 FIR Characteristics

1. The impulse response $h(n)$ has a finite length, i.e., $h(n)$ is nonzero only for a finite range of indices n. For a general N-length FIR system, $h(n) \neq 0$ only for $N_1 \leq n \leq N_2 = (N_1 + N - 1)$. When $N_1 \geq 0$, the filter is also causal.
2. The FIR frequency response $H(\omega)$ is a finite-degree polynomial in $e^{j\omega}$ of the form

$$H(\omega) = \sum_{n=N_1}^{N_2} h(n)(e^{j\omega})^{-n}, \tag{11.10}$$

   where N_1 and N_2 are (negative or positive) integers corresponding to the indices of the first and last samples of $h(n)$, respectively. The N impulse response samples are the free parameters of the design procedure. This form is general enough to represent noncausal filters such as zero-phase filters.
3. Designing an FIR filter consists in finding the polynomial $H(\omega)$ that best approximates the design specifications. This is done by computing the "optimal" (relative to some criteria) impulse response samples $\{h(n)\}_{n=N_1}^{N_2}$, which correspond to the unknown coefficients of the polynomial $H(\omega)$. The impulse response length N is usually fixed, but it could also be considered as a free parameter to be optimized. Procedures for designing FIR filters are given in Sections 11.3.1 and 11.4.1.
4. The filter transfer function, denoted by $H(z)$, is the z-transform of $h(n)$ and is useful for studying the stability of the system. For FIR filters, $H(z)$ is a finite-degree polynomial in the complex variable z and is given by

$$H(z) = H(e^{j\omega})|_{e^{j\omega}=z} = \sum_{n=N_1}^{N_2} h(n)z^{-n}. \tag{11.11}$$

It follows that the function $H(z)$ has no poles except possibly at 0 or ∞, i.e., it cannot be infinite for any point z with $0 < |z| < \infty$. It has only zeros (points z at which $H(z) = 0$). Therefore, an FIR filter is always stable.

5. FIR filters allow the design of causal linear-phase systems which are very important and widely used in practice. In fact, in many signal processing applications, such as speech and image processing, it is desirable to pass some portion of the signal frequency band with minimal distortion. For that purpose, linear-phase systems are particularly desirable since the effect of the linear-phase is a pure time delay. For a more detailed discussion of linear-phase systems, the reader is referred to [1].
6. Because the impulse response is of finite length, FIR filters are realized using the convolution operation [1] which can be implemented directly in the time/space domain, or in terms of the FFT in the frequency domain. More details about the implementation will be given in Section 11.2.6.
7. Since FIR filters have no feedback loops, they are relatively insensitive to roundoff noise. Noise due to coefficient quantization can be a problem for very long filters, but can be mitigated by avoiding the direct-form structures, and using special structures such as the cascade form for implementation.
8. FIR filters with very long impulse responses ($N \approx 500$) might be required to meet certain design specifications, for example, high accuracy and/or short transition bands. Longer filters lead to an increased complexity for both design and implementation. They require significant computing time to optimize all the parameters $h(n)$, and also many operations per second in the actual filter implementation.
9. The trade-off among the filter design parameters has been determined empirically for some types of FIR designs. The following simple (approximate) formula shows the relationship among the ripples, bandedges, and filter length (N) for one method, the Parks–McClellan (PM) algorithm (Section 11.3.1.3):

$$(N-1)\Delta\omega \approx \frac{-20\log_{10}\sqrt{\delta_p\delta_s} - 13}{2.324},$$

where $\Delta\omega = \omega_s - \omega_p$ is the transition width. This formula allows the designer to predict the value of N that will be needed to satisfy specs given for $\{\omega_p, \omega_s, \delta_p, \delta_s\}$. Other design formulas are given in Section 11.3.1.

11.2.4.2 IIR Characteristics

1. The impulse response $h(n)$ has an infinite number of nonzero samples (infinite length). As an example, for a general IIR filter, $h(n) \neq 0$ only for $N_o \leq n \leq \infty$, where N_o is a nonnegative integer (commonly, N_o is taken to be 0; in this case, the filter is said to be causal).
2. The frequency response $H(\omega)$ is a rational function, i.e., a ratio of two finite-degree polynomials in $e^{j\omega}$ of the form

$$H(\omega) = \frac{B(\omega)}{A(\omega)} = e^{-j\omega N_o}\frac{\sum_{k=0}^{M} b_k e^{-j\omega k}}{\sum_{k=0}^{N} a_k e^{-j\omega k}}, \tag{11.12}$$

where N_o is an integer constant. The order of an IIR filter is equal to N, which is the degree of the denominator in Equation 11.12; usually the degree of the numerator M is no greater than N. The order N also determines the number of previous output samples that need to be stored and then fed back to compute the current output sample. Therefore, IIR systems are also known as feedback systems. The filter coefficients $\{b_n\}$ and $\{a_n\}$ in Equation 11.12 correspond to the unknown (free) parameters of the design.

3. Designing an IIR filter amounts to finding the rational function $H(\omega)$ that best approximates the design specifications. In the frequency domain, this is done by computing the "optimal" (relative to some criteria) coefficients $\{b_n\}$ and $\{a_n\}$ in Equation 11.12 for the rational function $H(\omega)$. The filter order N is usually fixed, but can also be considered as a free parameter to be optimized. Procedures for designing IIR filters are given in Sections 11.3.2 and 11.4.2.
4. As mentioned previously in section on pages 11–16, the filter transfer function, denoted by $H(z)$, is the z-transform of $h(n)$ and is useful for studying the stability of the system. In the context of LSI filters, stability implies that a bounded input to the filter will always result in a bounded output. For IIR filters, $H(z)$ is a rational function in the complex variable z and is given by

$$H(z) = H(e^{j\omega})|_{e^{j\omega}=z} = z^{-N_o} \frac{\sum_{k=0}^{M} b_k z^{-k}}{\sum_{k=0}^{N} a_k z^{-k}}. \tag{11.13}$$

The roots of the denominator polynomial are poles of the function $H(z)$, i.e., $H(z)$ is infinite at points z with $0 \leq |z| < \infty$. Stability then requires that no poles lie on the U.C. ($|z| = 1$) in the z-plane. Causality and stability require that the poles lie inside the U.C. in the z-plane. So, it is possible to obtain a resulting IIR filter that is unstable. Also, coefficient quantization noise might severely affect the response of the filter and its stability by disturbing the poles locations and by driving some of the poles closer to or onto the U.C.

5. It is not possible to design causal linear-phase IIR filters. The resulting IIR causal realizable filters must have a nonlinear phase response. Forward-backward filtering can be used as an implementation to approximate a zero-phase response [1].
6. Because the impulse response is infinitely long, convolution can no longer be used to implement the IIR filters. Instead, IIR filters are efficiently implemented using feedback difference equations as described in Section 11.2.6.
7. The noise characteristics of an IIR filter can be a major consideration when doing an implementation, especially in fixed-point arithmetic. Coefficient quantization degrades the actual filter response from that designed by high-precision software. More critical is roundoff noise sensitivity which can be amplified by the feedback loops in the filter.
8. Compared to FIR filters, IIR filters can achieve the desired design specifications with a relatively low order (as few as 4–6 poles). So, fewer unknown parameters need to be computed and stored, which might lead to a lower design and implementation complexity. However, the phase response of IIR filters is never linear, which leads to the use of all-pass filters to compensate the group delay, and thus raises the order of the filter and the complexity of the design process.
9. IIR filters are commonly designed by using closed-form design formulas corresponding to classical filter types. While for FIR filters the length-estimating formulas are only approximate, the order-estimating formulas for IIR filters are exact since they are derived from the mathematical properties of the classical prototypes. These formulas are very useful to obtain the IIR filter order needed to satisfy the desired design specifications.

11.2.5 Designing the Filter

After the designed filter type (FIR or IIR) is specified, a suitable design procedure can be selected depending on the chosen filter type. Popular design procedures are based on computing the unknown filter parameters by optimizing one of the error criteria indicated in Section 11.2.3.

For FIR filters, the two main classical methods are the windowing method [1] and the PM (Remez) algorithm [6]. The windowing method minimizes the MSE when a rectangular window (corresponding to pure truncation of the ideal impulse response) is used at the expense of possible large overshoots near the bandedges and large ripples in the resulting frequency response. It is suboptimal when other general windows are used. However, the edge overshoot, transition width, and ripple height can be controlled by using different types of windows as described in section on pages 11–12. The PM (Remez) algorithm minimizes the Chebyshev (L_∞) error norm resulting in optimal equiripple designs. However, the original PM algorithm is restricted to the design of linear-phase filters with a symmetric magnitude response. An extension of this algorithm that allows the design of optimal FIR filters with arbitrary magnitude and phase specifications has been presented by Karam and McClellan in [2,3]. Linear-programming-based [4,7] and Constrained least-square [8] optimization methods also have been presented to allow the inclusion of additional important design constraints. These and other FIR design procedures are described in Sections 11.3.2 and 11.4.1.

While the design of FIR filters is typically performed directly in the digital domain, IIR filters are commonly designed by transforming the digital design specifications into analog design specifications and performing the filter design in the analog domain. The resulting analog filter is then transformed into a digital filter using a suitable transformation. One important classical IIR design method is the bilinear transformation method. Digital-only IIR design methods have also been presented. A description of IIR design procedures is given in Sections 11.3.2 and 11.4.2.

11.2.6 Realizing the Designed Filter

Realizing the designed digital filter corresponds to computing the output of the filter in response to any given input. For LSI filters, this is simplified by the fact that the input and output signals are related through a simple convolution operation in the time/space domain. If $x(n)$ is the input, $y(n)$ the corresponding output, and $h(n)$ the impulse response of the LSI filter, then this relation is given by

$$y(n) = h(n) * x(n) = \sum_{k=N_1}^{N_2} h(k)x(n-k), \qquad (11.14)$$

where N_1 and N_2 are the indices of the first and last nonzero samples of $h(n)$. In the frequency (Fourier transform) domain, the convolution relation (Equation 11.14) corresponds to a multiplication of the respective Fourier transforms:

$$Y(\omega) = H(\omega)X(\omega), \qquad (11.15)$$

where $X(\omega)$, $H(\omega)$, and $Y(\omega)$ are the DTFT of $x(n)$, $h(n)$, and $y(n)$, respectively. The variable ω in Equation 11.15 is continuous and, therefore, Equation 11.15 cannot be implemented in practice. An implementable version of Equation 11.15 is obtained by using the discrete Fourier transform (DFT), which is a sampled version of the DTFT and which consists of samples of the DTFT evaluated at the points $\omega = (2\pi k/N_{DFT})$, $k = 0, \ldots, (N-1)$. N_{DFT} is the size of the DFT and corresponds to the number of sample points within the period 2π. It is a known fact that the time/space digital signal can be exactly recovered from its DFT if N_{DFT} is chosen to be greater than or equal to the length of the time/space signal. Using the DFT, Equation 11.15 becomes

$$Y(k) = H(k)X(k), \quad k = 0, \ldots, N_{DFT} \qquad (11.16)$$

where $N_{DFT} \geq \max$ [length of $x(n)$ + length of $h(n) - 1$] in order to perform the pointwise multiplication. The DFT can be computed very efficiently using the fast Fourier transform (FFT) algorithm.

11.2.6.1 Realizing FIR Filters

For FIR filters, the impulse response has a finite length and, therefore, N_1 and N_2 in Equation 11.14 are finite. Also, in this case, a finite-size DFT is sufficient to exactly represent $h(n)$ $[N_{DFT} \geq (N_2 - N_1 + 1)]$. Consequently, for finite-length input signals $x(n)$, Equation 11.14 or 11.16 can be directly used to realize the designed FIR filter in software or hardware. Commonly, the FIR filter coefficients $h(n)$ (or the DFT values if Equation 11.16 is used) are quantized to the precision of the processor or chip, stored, and used as in Equation 11.14 to realize the designed FIR filter. While for Equation 11.14 the storage can be fixed to the size of $h(n)$ and is independent of the input, the size of the DFTs in Equation 11.16 and, therefore, the needed storage vary with the size of the input signal. To overcome this problem and to handle the processing of large-size signals, block-based convolution (also known as sectioned or high-speed convolution) is used where the input signal is divided into blocks (sections) of fixed equal size; then, the convolution of each input block with $h(n)$ is computed using Equation 11.16 with $X(k)$ being, in this case, the DFT of the considered block; the computed block convolutions are finally properly combined to lead the final output $y(n)$. Two popular ways of performing block convolutions are [1,9] (1) overlap-add and (2) overlap-save.

11.2.6.2 Realizing IIR Filters

For IIR filters, the impulse response has infinite length and, therefore, the summation in Equation 11.14 involves an infinite number of terms (N_1 and/or N_2 infinite). This makes Equation 11.14 not suitable for realizing IIR filters. Similarly, the direct realization of Equation 11.16 would require computing the infinite-length DFT $H(k)$, which is not possible. These problems are overcome by using feedback difference equations to realize the designed IIR filters. In fact, using Equation 11.15 with $H(\omega)$ replaced by Equation 11.12, we get

$$Y(\omega) = e^{-j\omega N_o} \frac{\sum_{k=0}^{M} b_k e^{-j\omega k}}{\sum_{k=0}^{N} a_k e^{-j\omega k}} X(\omega). \tag{11.17}$$

For simplicity and without loss of generality, assume $N_o = 0$; we can rewrite Equation 11.17 as

$$\sum_{k=0}^{N} a_k e^{-j\omega k} Y(\omega) = \sum_{k=0}^{M} b_k e^{-j\omega k} X(\omega). \tag{11.18}$$

Taking the inverse DTFT of both sides of Equation 11.18 and noting that multiplication by $e^{-j\omega k}$ corresponds to a shift by k in the time/space domain, we obtain the input–output relation of the system in the time/space domain:

$$\sum_{k=0}^{N} a_k y(n-k) = \sum_{n=0}^{M} b_k x(n-k). \tag{11.19}$$

The difference equation (Equation 11.19) can be rearranged leading a recursive (feedback) input–output relation. For instance, in order to compute the right-sided output sequence $y(n)$, for $n \geq n_o$ (n_o integer constant), Equation 11.19 can be rewritten as

$$y(n) = \sum_{n=0}^{M} \frac{b_k}{a_o} x(n-k) - \sum_{k=1}^{N} \frac{a_k}{a_o} y(n-k), \tag{11.20}$$

where a_o is commonly taken to be 1, without loss of generality, since it can be integrated into the parameters b_k and a_k. Realizing Equation 11.20 requires that N initial output values, $y(n_o - 1), \ldots, y(n_o - N)$, be specified. For LSI filters, initial rest conditions are required: if the input $x(n) = 0$ for $n < n_o$, then set $y(n) = 0$ for $n < n_o$.

For a left-sided output sequence, Equation 11.19 can be rearranged as follows:

$$\underbrace{y(n-N)}_{m} = \sum_{k=0}^{M} \frac{b_k}{a_N} x(n-k) - \sum_{k=0}^{N-1} \frac{a_k}{a_N} y(n-k). \tag{11.21}$$

So, Equation 11.21 can be used to compute $y(m)$, $m \leq n_o$, by setting $n = m + N$ and specifying the N initial values $y(n_o + 1), \ldots, y(n_o + N)$.

The feedback difference Equations 11.20 and 11.21 are simple to implement in software or hardware. The MATLAB®* software command **y = filter(b, a, x)** implements Equation 11.20. In hardware, typical DSP chips implement low-order filters ($N = 1$ or $N = 2$); the low-order filters can be combined together (in cascade and/or parallel) to produce the desired higher order filters (see Section 11.5). To implement the filter in hardware, the difference equations (or, equivalently, the rational frequency response) are represented by structures, which are flow graphs describing the algorithm, that is to be implemented, in terms of basic building blocks [1, Chapter 6]. The basic building blocks include adders, multipliers, branch points, and delay elements.

11.2.6.3 Quantization: Finite Wordlength Effect

In the design step, the filter coefficients are usually computed with a very high precision. In practice, these coefficients can be implemented with finite wordlength only. Since the design algorithm yields coefficients computed to the highest precision available (e.g., double-precision floating-point), the filter coefficients must be quantized to the internal format of the DSP. In addition, fixed-point chips are widely used since they generally provide higher processing speed at lower cost than do the floating point systems. In the case of a fixed-point DSP, this quantization also requires scaling of the coefficients to a predetermined maximum value. The quantization and/or truncation of the coefficients will generally cause the frequency response of the implemented filter to deviate from the designed filter frequency response. The deviation from the desired specifications will depend on the chosen filter type and on the structure used to implement the filter. For IIR filters, the quantization of the coefficients might turn a stable filter into an unstable one. Other effects are due to the fact that arithmetic operations performed on finite wordlength numbers generally result in numbers with larger wordlengths, which then need to be quantized or truncated to the allowable precision. Therefore, it is important to specify the required minimum wordlength that can be tolerated. As indicated in Section 11.5, very few design algorithms perform the optimization of quantized coefficients. Studies of the different wordlength effects have resulted in "rules of thumb" for the design and realization of a system such that the desired properties can be achieved with reduced errors and expense. A detailed study of the wordlength effects and the characterization of the resulting errors can be found in Sections 6.7 through 6.10 of [1] and Sections 7.5 through 7.7 of [9].

11.3 Classical Filter Design Methods

The methods described in this section are magnitude-only approximation methods, i.e., the desired phase response is assumed to be constant or linear and is not included in the design. These classical methods mainly design frequency-selective filters with real-valued coefficients $h(n)$.

Methods for the design of filters with general specifications [2,4,10] have been developed more recently and are presented in Section 11.4.

* MATLAB is a registered trademark of the Mathworks, Inc.

11.3.1 FIR Design Methods

Ivan W. Selesnick, C. Sidney Burrus, Lina J. Karam, and James H. McClellan

The classical FIR design methods are mainly concerned with the design of linear-phase FIR filters with real-valued coefficients $h(n)$. These filters are of four possible types [1,11]. The properties of the four types of linear-phase filters are summarized in Table 11.1 and illustrated in Figure 11.4.

TABLE 11.1 Summary of the Four Types of Linear-Phase FIR Filters

	Odd Length (N)	Even Length (N)
Even Symmetry	Type I	Type II
$h(\alpha + n) = h(\alpha - n)$	$\sum_{k=0}^{\frac{1}{2}(N-1)} a(k)\cos(\omega k)$	$\sum_{k=1}^{\frac{1}{2}N} b(k)\cos\left[\omega\left(k - \frac{1}{2}\right)\right]$
$\alpha = \frac{N-1}{2}$	$a(0) = h\left(\frac{N-1}{2}\right)$	zero at $\omega = \pi$
$\beta = 0$	$a(k) = 2h\left(\frac{N-1}{2} - k\right)$	$b(k) = 2h\left(\frac{N}{2} - k\right)$
		$\cos\left(\frac{1}{2}\omega\right) \sum_{k=0}^{\frac{1}{2}N-1} b(k)\cos(\omega k)$
Odd Symmetry	Type III	Type IV
$h(\alpha + n) = -h(\alpha - n)$	$\sum_{k=1}^{\frac{1}{2}(N-1)} c(k)\sin(\omega k)$	$\sum_{k=1}^{\frac{1}{2}} d(k)\sin\left[\omega\left(k - \frac{1}{2}\right)\right]$
$\alpha = \frac{N-1}{2}$	zeros at $\omega = 0, \pi$	zero at $\omega = 0$
$\beta = \frac{\pi}{2}$	$c(k) = 2h\left(\frac{N-1}{2} - k\right)$	$d(k) = 2h\left(\frac{N}{2} - k\right)$
	$h\left(\frac{N-1}{2}\right) = 0$	
	$\sin(\omega) \sum_{k=0}^{\alpha-1} \check{c}(k)\cos(\omega k)$	$\sin\frac{1}{2}(\omega) \sum_{k=0}^{\frac{1}{2}N-1} \check{c}(k)\cos(\omega k)$

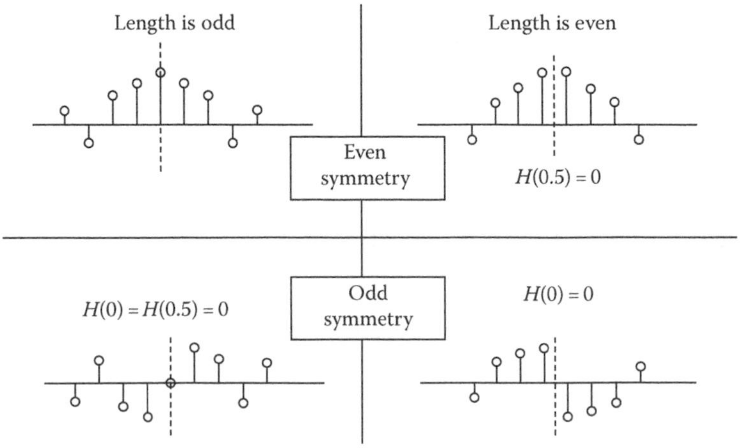

FIGURE 11.4 Examples of impulse responses corresponding to the four types of linear-phase filters. $H(f)$ is the corresponding frequency response, where the normalized frequency variable $f = \omega/2\pi$.

11.3.1.1 Design by Windowing

The Fourier relationship between the impulse response and $H(\omega)$ suggests that $h(n)$ can be obtained via

$$h(n) = \frac{1}{2\pi}\int_{-\pi}^{\pi} D(\omega)e^{j\omega n}d\omega, \tag{11.22}$$

where $D(\omega)$ is the desired frequency response. However, these Fourier series coefficients are usually infinitely supported. The windowing technique proposes that the infinitely supported Fourier series be truncated and multiplied by an appropriate function (a "window") to obtain an FIR filter. For the design of odd length symmetric filters, it is appropriate that $D(\omega)$ be a real-valued even function—then $h(n)$ is real and $h(n) = h(-n)$. A casual filter is obtained by then shifting $h(n)$.

11.3.1.1.1 Steps in Window Filter Design

1. Create the ideal impulse response, using the inverse DTFT to obtain $h_d[n]$:

$$h_d[n] = \frac{1}{2\pi}\int_{-\pi}^{\pi} D(\omega)e^{j\omega n}d\omega,$$

   where $D(\omega)$ is the ideal frequency response. For example, $D(\omega)$ might be the ideal LPF.
2. Note: If the length of the window is N, then the "ideal" frequency response must contain a linear phase term. For example, the ideal LPF would be specified as

$$D(\omega) = \begin{cases} 1 \cdot e^{-j\omega(N-1)/2} & -\omega_c \leq \omega \leq +\omega_c \\ 0 & \omega_c \leq |\omega| < \pi. \end{cases}$$

   This allows both even-length and odd-length filters to be designed.
3. Create the FIR filter coefficients by multiplying by the window:

$$h[n] = w[n] \cdot h_d[n] \quad n = 0, 1, \ldots, N-1.$$

4. In the frequency domain, this windowing operation results in a convolution of the ideal frequency response with the Fourier transform of the window, $W(\omega)$:

$$H(\omega) = \frac{1}{2\pi}\int_{-\pi}^{\pi} D(\theta)W(\omega-\theta)d\theta.$$

   Note that this convolution is periodic with period 2π.
5. Transition Width: The result is that the ideal frequency response is smeared by the convolution, so the actual frequency response has a smooth roll-off from the passband to the stopband.
6. Passband and Stopband Deviations: In addition, all windows have sidelobes in their Fourier transforms, so the convolution gives rise to ripples in the frequency response of the FIR filter.

Examples of commonly used windows and their transforms are shown in Figure 11.5. Windowed filter design examples are shown in Figure 11.6.

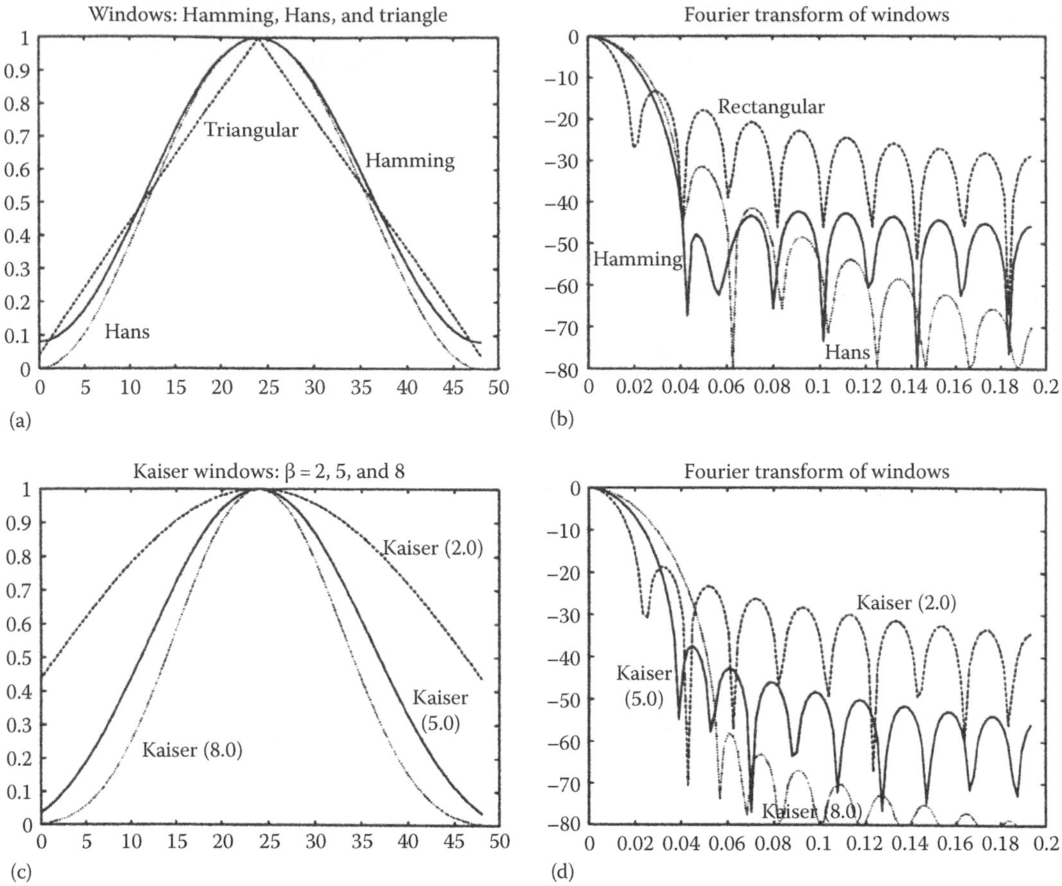

FIGURE 11.5 Common windows and their Fourier transforms. The window length is $N = 49$. (a) Index (a), (b) normalized frequency (sampling frequency = 1), (c) Index (a), and (d) normalized frequency (sampling frequency = 1).

11.3.1.1.2 Window Selection

Let $D(\omega)$ be the response of an ideal lowpass filter with cutoff frequency ω_c, illustrated in Figure 11.7. The Fourier series of $D(\omega)$ are samples of the **sinc** function:

$$\text{sinc}(n) = \begin{cases} \dfrac{\omega_c}{\pi} \dfrac{\sin(\omega_c n)}{\omega_c n} & n \neq 0 \\ \dfrac{\omega_c}{\pi} & n = 0. \end{cases} \tag{11.23}$$

Simple truncation of the sinc function samples is generally not found to be acceptable because the frequency responses of filters so obtained have large errors near the cutoff frequency. Moreover, as the filter length is increased, the size of this error does not diminish to zero (although the square error does). This is known as Gibbs phenomenon. Figure 11.8 illustrates a filter obtained by truncating the sinc function.

To overcome this problem, the windowing technique obtains $h(n)$ by multiplying the sinc function by a "window" that is tapered near its endpoints:

$$h(n) = w(n) \cdot \text{sinc}(n). \tag{11.24}$$

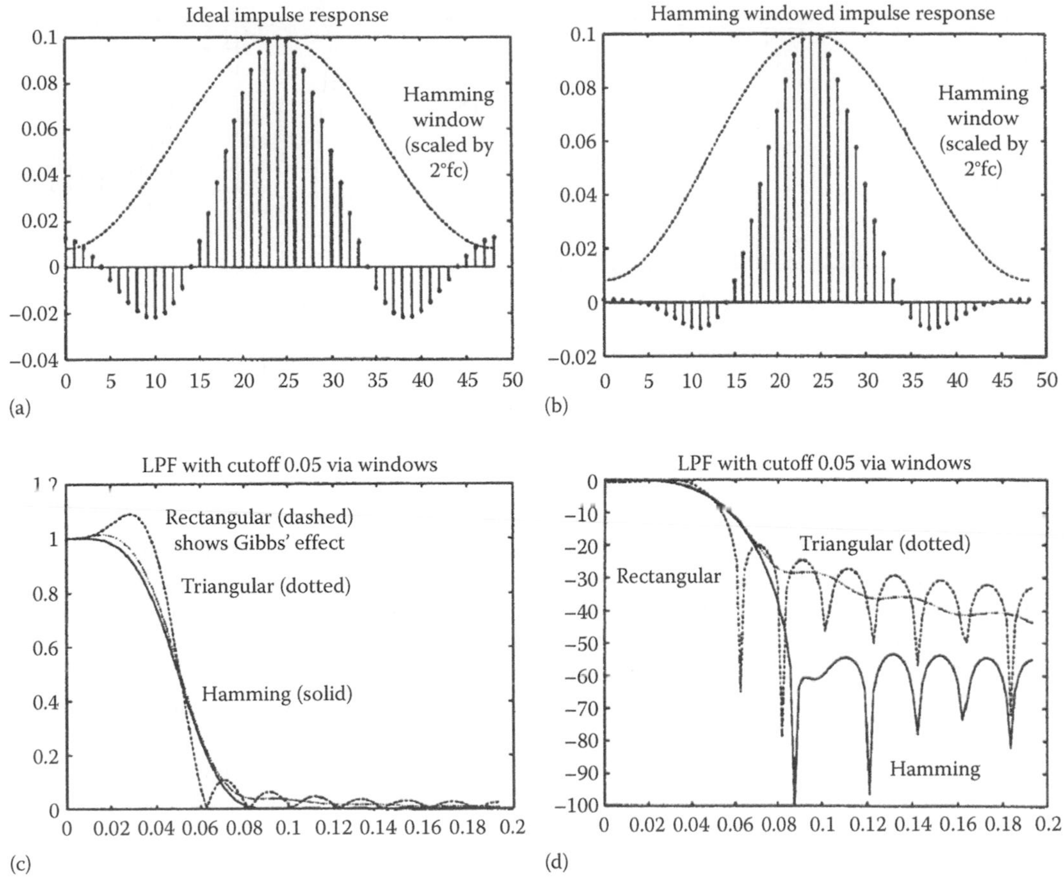

FIGURE 11.6 Examples of windowed filter design. The window length is $N = 49$. (a) Index (a), (b) index (a), (c) normalized frequency (sampling frequency = 1), and (d) normalized frequency (sampling frequency = 1).

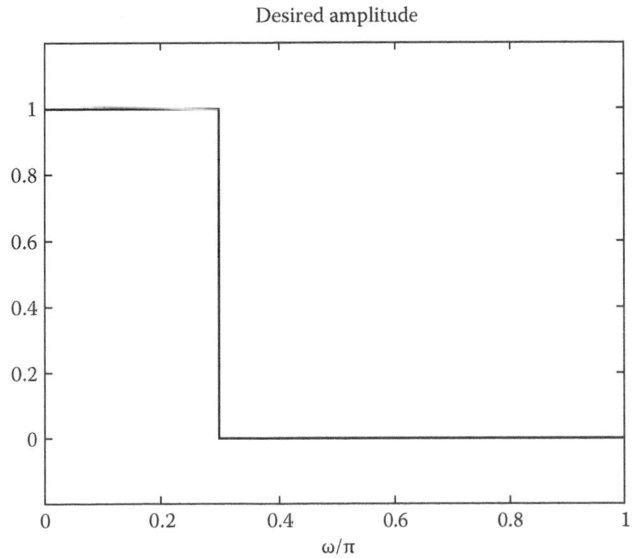

FIGURE 11.7 Ideal lowpass filter, $\omega_c = 0.3\pi$.

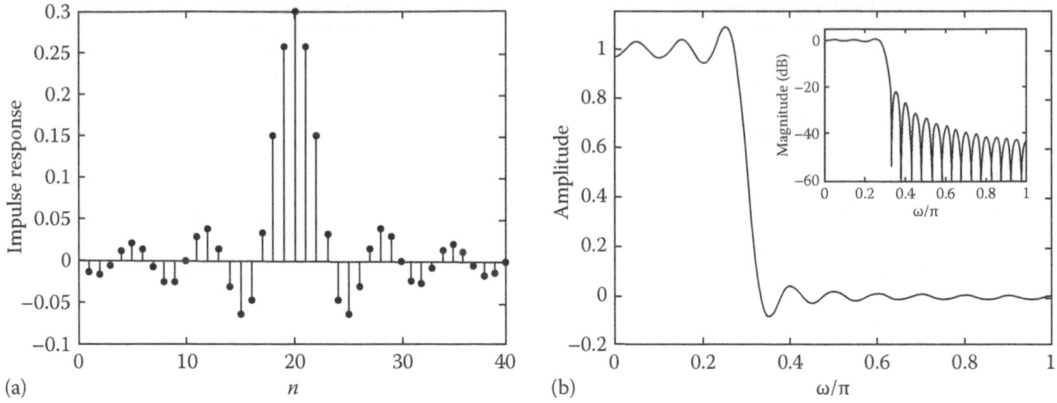

FIGURE 11.8 Lowpass filter obtained by sinc function truncation, $\omega_c = 0.3\pi$.

The generalized cosine windows and the Bartlett (triangular) window are examples of well-known windows. A useful window function has a frequency response that has a narrow mainlobe, a small relative peak sidelobe height, and good sidelobe Roughly, the width of the mainlobe affects the width of the transition band of $H(\omega)$, while the relative height of the sidelobes affects the size of the ripples in $H(\omega)$. These cannot be made arbitrarily good at the same time. There is a trade-off between mainlobe width and relative sidelobe height. Some windows, such as the Kaiser window [12], provide a parameter that can be varied to control this trade-off.

One approach to window design computes the window sequence that has most of its energy in a given frequency band, say $[-B, B]$. Specifically, the problem is formulated as follows. Find $w(n)$ of specified finite support that maximizes

$$\lambda = \frac{\int_{-B}^{B} |W(\omega)|^2 d\omega}{\int_{-\pi}^{\pi} |W(\omega)|^2 d\omega}, \tag{11.25}$$

where $W(\omega)$ is the Fourier transform of $w(n)$. The solution is a particular discrete prolate spheroidal (DPS) sequence [13], The solution to this problem was traditionally found by finding the largest eigenvector* of a matrix whose entries are samples of the sinc function [13]. However, that eigenvalue problem is numerically ill conditioned—the eigenvalues cluster around to 0 and 1. Recently, an alternative eigenvalue problem has become more widely known, that has exactly the same eigenvectors as the first eigenvalue problem (but different eigenvalues), and is numerically well conditioned [14–16]. The well-conditioned eigenvalue problem is described by $\mathbf{A}\mathbf{v} = \theta\mathbf{v}$ where $\mathbf{A}$ is tridiagonal and has the following form:

$$\mathbf{A}_{i,j} = \begin{cases} \frac{1}{2} i(N-i) & j = i-1 \\ \left(\frac{N-1}{2} - i\right)^2 \cos B & j = i \\ \frac{1}{2}(i+1)(N-1-i) & j = i+1 \\ 0 & |j-i| > 1 \end{cases} \tag{11.26}$$

* The eigenvector with the largest eigenvalue.

for $i,j = 0,\ldots,N-1$. Again, the eigenvector with the largest eigenvalue is the sought solution. The advantage of **A** in Equation 11.26 over the first eigenvalue problem is twofold: (1) the eigenvalues of **A** in Equation 11.26 are well spread (so that the computation of its eigenvectors is numerically well conditioned) and (2) the matrix **A** in Equation 11.26 is tridiagonal, facilitating the computation of the largest eigenvector via the power method.

By varying the bandwidth, B, a family of DPS windows is obtained. By design, these windows are optimal in the sense of energy concentration. They have good mainlobe width and relative peak sidelobe height characteristics. However, it turns out that the sidelobe roll-off of the DPS windows is relatively poor, as noted in [16].

The Kaiser [12] and Saramäki [17,18] windows were originally developed in order to avoid the numerically ill-conditioning of the first matrix eigenvalue problem described above. They approximate the prolate spheroidal sequence, and do not require the solution to an eigenvalue problem. Kaiser's approximation to the prolate spheroidal window [12] is given by

$$w(n) = \frac{I_0\left(\beta\sqrt{1-(n-M)^2/M^2}\right)}{I_0(\beta)} \quad \text{for } n = 0,1,\ldots,N-1, \tag{11.27}$$

where

$M = \frac{1}{2}(N-1)$

β is an adjustable parameter

$I_0(x)$ is the modified zeroth-order Bessel function of the first kind

The window in Equation 11.27 is known as the Kaiser window of length N. For an odd-length window, the midpoint M is an integer. The parameter β controls the trade-off between the mainlobe width and the peak sidelobe level—it should be chosen to lie between 0 and 10 for useful windows. High values of β produce filters having high stopband attenuation, but wide transition widths. The relationship between β and the ripple height in the stopband (or passband) is illustrated in Figure 11.9 and is given by

$$\beta = \begin{cases} 0 & \text{ATT} < 21 \\ 0.5842(\text{ATT}-21)^{0.4} + 0.07886(\text{ATT}-21) & 21 \le \text{ATT} \le 50 \\ 0.1102(\text{ATT}-8.7) & 50 < \text{ATT}, \end{cases} \tag{11.28}$$

where $\text{ATT} = -20\log_{10}\delta_s$ is the ripple height in decibel.

For lowpass FIR filter design, the following design formula helps the designer to estimate the Kaiser window length N in terms of the desired maximum passband and stopband error δ,* and transition width $\Delta F = (\omega_p - \omega_s)/2\pi$:

$$N \approx \frac{-20\log_{10}(\delta) - 7.95}{14.357\Delta F} + 1. \tag{11.29}$$

Examples of filter designs using the Kaiser window are shown in Figure 11.10.

A second approach to window design minimizes the relative peak sidelobe height. The solution is the Dolph–Chebyshev window [17,19], all the sidelobes of which have equal height. Saramäki has described a family of transitional windows that combine the optimality properties of the DPS window and the

* For Kaiser window designs, $\delta = \delta_p = \delta_s$.

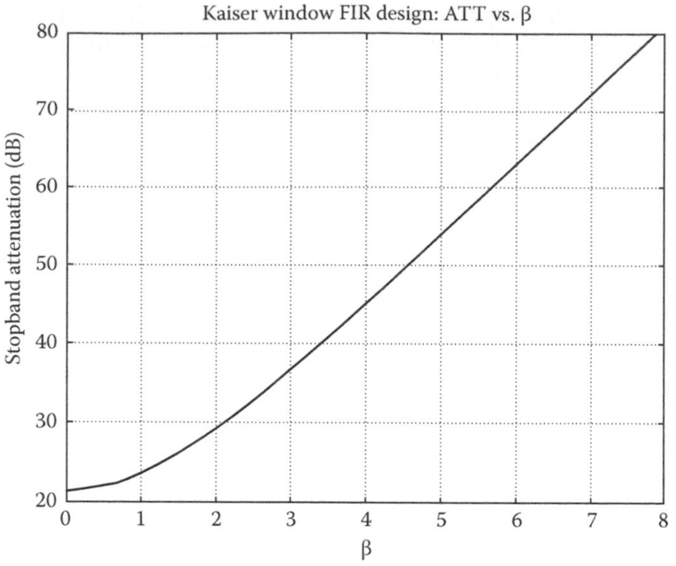

FIGURE 11.9 Kaiser window: stopband attenuation vs. β.

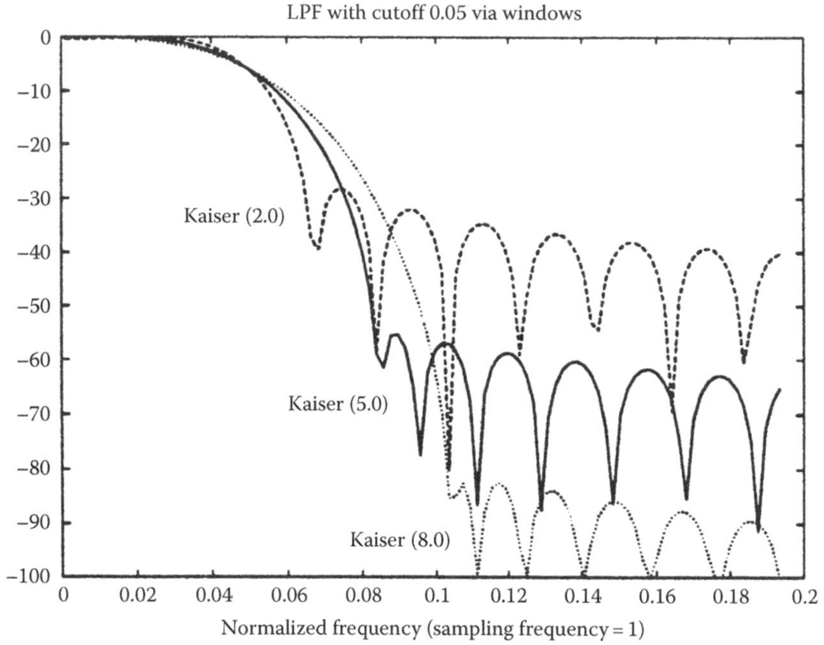

FIGURE 11.10 Frequency responses (log scale) of filters designed using the Kaiser window with selected values for the parameter β. Note the trade-off between mainlobe width and sidelobes height.

Dolph–Chebyshev window. He has found that the transitional window yields better results than both the DPS window and the Dolph–Chebyshev window, in terms of attenuation vs. transition width [17].

An extensive list and analysis of windows is given in [19]. In addition, the use of nonsymmetric windows for the design of fractional delay filters has been discussed in [20,21].

11.3.1.1.3 Remarks

- The technique is conceptually and computationally simple.
- Using the window method, it is not possible to weight the passband and stopband differently.
- The ripple sizes in each band will be approximately the same. But requirements are often more strict in the stopband.
- It is difficult to specify the bandedges and maximum ripple size precisely.
- The technique is not suitable for arbitrary desired responses.
- The use of windows for filter design is generally considered suboptimal because they do not solve a clear optimization problem, but see [22].

11.3.1.2 Optimal Square Error Design

The formulation is as follows. Given a filter length N, a desired amplitude function $D(\omega)$, and a nonnegative function $W(\omega)$, find the symmetric filter that minimizes the weighted integral square error (or "L_2 error"), defined by

$$\|E(\omega)\|_2 = \left(\frac{1}{\pi}\int_0^{\pi} W(\omega)[A(\omega) - D(\omega)]^2 \mathrm{d}\omega\right)^{\frac{1}{2}}. \tag{11.30}$$

For simplicity, symmetric odd-length filters* will be discussed here, in which case $A(\omega)$ can be written as

$$A(\omega) = \frac{1}{\sqrt{2}}a(0) + \sum_{n=1}^{M} a(n)\cos n\omega, \tag{11.31}$$

where $N = 2M + 1$ and where the impulse response coefficients $h(n)$ are related to the cosine coefficients $a(n)$ by

$$h(n) = \begin{cases} \frac{1}{2}a(M-n) & \text{for } 0 \le n \le M-1 \\ \frac{1}{\sqrt{2}}a(0) & \text{for } n = M \\ \frac{1}{2}a(n-M) & \text{for } M+1 \le n \le N-1 \\ 0 & \text{otherwise.} \end{cases} \tag{11.32}$$

The nonstandard choice of $\frac{1}{\sqrt{2}}$ here simplifies the notation below.

The coefficients $\mathbf{a} = [a(0), \ldots, a(M)]^t$ are found by solving the linear system

$$\mathbf{Ra} = \mathbf{c}, \tag{11.33}$$

* To treat the four linear phase types together, see Equations 11.51 through 11.55 in the sequel. Then, $\|E(\omega)\|_2$ becomes $\left\{\frac{1}{\pi}\int_0^{\pi} \bar{W}(\omega)[A(\omega) - \bar{D}(\omega)]^2 \mathrm{d}\omega\right\}^{\frac{1}{2}}$, where $\bar{W}(\omega) = W(\omega)Q^2(\omega)$ and $\bar{D}(\omega) = D(\omega)/Q^2(\omega)$ $A(\omega)$ is as in Equation 11.31.

where the elements of the vector **c** are given by

$$c_0 = \frac{\sqrt{2}}{\pi} \int_0^{\pi} W(\omega) D(\omega) d\omega \tag{11.34}$$

$$c_k = \frac{2}{\pi} \int_0^{\pi} W(\omega) D(\omega) \cos k\omega \, d\omega, \tag{11.35}$$

and the elements of the matrix **R** are given by

$$R_{0,0} = \frac{1}{\pi} \int_0^{\pi} W(\omega) d\omega \tag{11.36}$$

$$R_{0,k} = R_{k,0} = \frac{\sqrt{2}}{\pi} \int_0^{\pi} W(\omega) \cos k\omega \, d\omega \tag{11.37}$$

$$R_{k,l} = R_{l,k} = \frac{2}{\pi} \int_0^{\pi} W(\omega) \cos k\omega \cos l\omega \, d\omega \tag{11.38}$$

for $l, k = 1, \ldots, M$. Often it is desirable that the coefficients satisfy some linear constraints, say $\mathbf{Ga} = \mathbf{b}$. Then the solution, found with the use of Lagrange multipliers, is given by the linear system

$$\begin{bmatrix} \mathbf{R} & \mathbf{G}^t \\ \mathbf{G} & 0 \end{bmatrix} \begin{bmatrix} \mathbf{a} \\ \mu \end{bmatrix} = \begin{bmatrix} \mathbf{c} \\ \mathbf{b} \end{bmatrix}, \tag{11.39}$$

the solution of which is easily verified to be given by

$$\mu = (\mathbf{G}\mathbf{R}^{-1}\mathbf{G}^t)^{-1}(\mathbf{G}\mathbf{R}^{-1}\mathbf{c} - \mathbf{b}) \quad \mathbf{a} = \mathbf{R}^{-1}(\mathbf{c} - \mathbf{G}^t \mu), \tag{11.40}$$

where μ are the Lagrange multipliers.

In the unweighted case ($W(\omega) = 1$) the solution is given by a simpler system:

$$\begin{bmatrix} \mathbf{I}_{M+1} & \mathbf{G}^t \\ \mathbf{G} & 0 \end{bmatrix} \begin{bmatrix} \mathbf{a} \\ \mu \end{bmatrix} = \begin{bmatrix} \mathbf{c} \\ \mathbf{b} \end{bmatrix}. \tag{11.41}$$

In Equation 11.41, $\mathbf{I}_{M+1}$ is the $(M+1)$ by $(M+1)$ identity matrix. It is interesting to note that in the unweighted case, the least square filter minimizes a worst case pointwise error in the time domain over a set of bounded energy input signals [23].

In the unweighted case with no constraint, the solution becomes $\mathbf{a} = \mathbf{c}$. This is equivalent to truncation of the Fourier series coefficients (the "rectangular window" method). This simple solution is due to the orthogonality of the basis functions $\{\frac{1}{\sqrt{2}}, \cos \omega, \cos 2\omega, \ldots\}$ when $W(\omega) = 1$. In general, whenever the basis functions are orthogonal, then the solution takes this simple form.

11.3.1.2.1 Discrete Squares Error

When $D(\omega)$ is simple, the integrals above can be found analytically. Otherwise, entries of **R** and **b** can be found numerically.

Define a dense uniform grid of frequencies over $[0, \pi)$ as $\omega_i = i\pi/L$ for $i = 0, \ldots, L-1$ and for some large L (say $L \approx 10M$). Let $\mathbf{d}$ be the vector given by $\mathbf{d}_i = D(\omega_i)$ and $\mathbf{C}$ be the L by $M+1$ matrix of cosine terms: $\mathbf{C}_{i,0} = \frac{1}{\sqrt{2}}$, $\mathbf{C}_{i,k} = \cos k\omega_i$ for $k = 1, \ldots, M$. ($\mathbf{C}$ has many more rows than columns.) Let $\mathbf{W}$ be the diagonal weighting matrix diag $[W(\omega_i)]$. Then

$$\mathbf{R} \approx \frac{2}{L\pi}\mathbf{C}^t\mathbf{W}\mathbf{C} \quad \mathbf{c} \approx \frac{2}{L\pi}\mathbf{C}^t\mathbf{W}\mathbf{d}. \tag{11.42}$$

Using these numerical approximations for $\mathbf{R}$ and $\mathbf{c}$ is equivalent to minimizing the discrete squares error,

$$\sum_{i=0}^{L-1} W(\omega_i)(D(\omega_i) - A(\omega_i))^2 \tag{11.43}$$

that approximates the integral square error. In this way, an FIR filter can be obtained easily, whose response approximates an arbitrary $D(\omega)$ with an arbitrary $W(\omega)$. This makes the least squares error approach very useful. It should be noted that the minimization of Equation 11.43 is most naturally formulated as the least squares solution to an over-determined linear system of equations, an approach described in [11]. The solution is the same, however.

11.3.1.2.2 Transition Regions

As an example, the least squares design of a length $N = 2M+1$ symmetric lowpass filter according to the desired response and weight functions

$$D(\omega) = \begin{cases} 1 & \omega \in [0, \omega_p] \\ 0 & \omega \in [\omega_s, \pi] \end{cases} \qquad W(\omega) = \begin{cases} K_p & \omega \in [0, \omega_p] \\ 0 & \omega \in [\omega_p, \omega_s] \\ K_s & \omega \in [\omega_s, \pi] \end{cases} \tag{11.44}$$

is developed. For this $D(\omega)$ and $W(\omega)$, the vector $\mathbf{c}$ in Equation 11.33 is given by

$$c_k = \frac{2K_p \sin(k\omega_p)}{k\pi} \qquad 1 \le k \le M \tag{11.45}$$

and the matrix $\mathbf{R}$ is given by

$$\mathbf{R} = \mathbf{T}[\text{toeplitz}(\mathbf{p}, \mathbf{p}) + \text{hankel}(\mathbf{p}, \mathbf{q})]\mathbf{T}, \tag{11.46}$$

where the matrix $\mathbf{T}$ is the identity matrix everywhere except for $T_{0,0}$, which is $\frac{1}{\sqrt{2}}$. The vectors $\mathbf{p}$ and $\mathbf{q}$ are given by

$$p_0 = \frac{K_p\omega_p + K_s(\pi - \omega_s)}{\pi} \tag{11.47}$$

$$p_k = \frac{K_p \sin(k\omega_p) - K_s \sin(k\omega_s)}{k\pi} \qquad 1 \le k \le M \tag{11.48}$$

$$q_k = \frac{K_p \sin((k+M)\omega_p) - K_s \sin((k+M)\omega_s)}{(k+M)\pi} \qquad 0 \le k \le M. \tag{11.49}$$

The matrix toeplitz $(\mathbf{p}, \mathbf{p})$ is a symmetric matrix with constant diagonals, the first row and column of which is $\mathbf{p}$. The matrix hankel $(\mathbf{p}, \mathbf{q})$ is a symmetric matrix with constant anti-diagonals, the first column of which is $\mathbf{p}$, the last row of which is $\mathbf{q}$. The structure of the matrix $\mathbf{R}$ makes possible the efficient solution of $\mathbf{Ra} = \mathbf{b}$ [24].

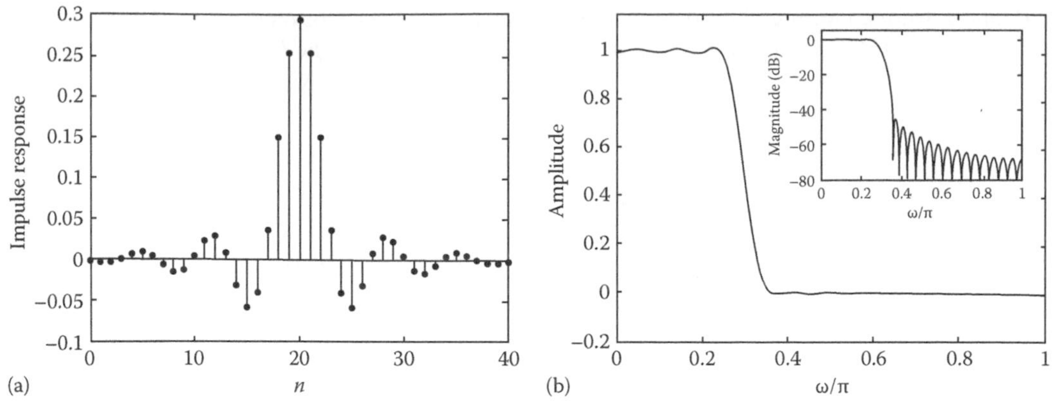

FIGURE 11.11 Weighted least squares example: $N = 41$, $\omega_p = 0.25\pi$, $\omega_s = 0.35\pi$, and $K = 4$.

Because the error is weighted by zero in the transition band $[\omega_p, \omega_s]$, the Gibbs phenomenon is eliminated: the peak error diminishes to zero as the filter length is increased. Figure 11.11 illustrates an example.

11.3.1.2.3 Other Least Squares Approaches

Another approach modifies the discontinuous ideal lowpass response of Figure 11.7 so that a fractional order spline is used to continuously connect the passband and stopband [25]. In this case, with uniform error weighting, (1) a simple closed-form expression for the least squares error solution is available, and (2) Gibbs phenomenon is eliminated. The use of spline transition regions also facilitates the design of multiband filters by combining various lowpass filters [26]. In that case, a least squares error multiband filter can be obtained via closed-form expressions, where the transition region widths can be independently specified.

Similar expressions can be derived for the even length filter and the odd symmetric filters. It should also be noted that the least squares error approach is directly applicable to the design of nonsymmetric FIR filters, complex-valued FIR filters, and two-dimensional (2-D) FIR filters.

In addition, another approach to filter design according to a square error criterion produces filters known as eigenfilters [27]. This approach gives the filter coefficients as the largest eigenvalue of a matrix that is readily constructed.

11.3.1.2.4 Remarks

- Optimal with respect to square error criterion
- Simple, non-iterative method
- Analytic solutions sometimes possible, otherwise solution is obtained via solution to linear system of equations
- Allows the use of a frequency-dependent weighting function
- Suitable for arbitrary $D(\omega)$ and $W(\omega)$
- Easy to include arbitrary linear constraints
- Does not allow direct control of maximum ripple size

11.3.1.3 Equiripple Optimal Chebyshev Filter Design

The minimization of the Chebyshev norm is useful because it permits the user to explicitly specify band-edges and relative error sizes in each band. Furthermore, the designed equiripple FIR filters have the smallest transition width among all FIR filters with the same deviation.

Linear phase FIR filters that minimize a Chebyshev error criterion can be obtained with the Remez exchange algorithm [28,29] or by linear programming techniques [30]. Both these methods are iterative numerical procedures and are applicable to arbitrary desired frequency response amplitudes.

11.3.1.3.1 Remez Exchange (Parks–McClellan)

Parks and McClellan proposed the use of the Remez algorithm for FIR filter design and made programs available [6,29,31]. Many texts describe the PM algorithm in detail [1,11].

11.3.1.3.2 Problem Formulation

Given a filter length, N, a desired (real-valued) amplitude function, $D(\omega)$, and a nonnegative weighting function, $W(\omega)$, find the symmetric (or antisymmetric) filter that minimizes the weighted Chebyshev error, defined by

$$||E(\omega)||_\infty = \max_{\omega \in B} |W(\omega)(A(\omega) - D(\omega))|, \tag{11.50}$$

where B is a closed subset of $[0, \pi]$. Both $D(\omega)$ and $W(\omega)$ should be continuous over B. The solution to this problem is called the best weighted Chebyshev approximation to $D(\omega)$ over B.

To treat each of the four linear phase cases together, note that in each case, the amplitude $A(\omega)$ can be written as [32]

$$A(\omega) = Q(\omega)P(\omega), \tag{11.51}$$

where $P(\omega)$ is a cosine polynomial (Table 11.1). By expressing $A(\omega)$ in this way, the weighted error function in each of the four cases can be written as

$$E(\omega) = W(\omega)[A(\omega) - D(\omega)] \tag{11.52}$$

$$= W(\omega)Q(\omega)\left[P(\omega) - \frac{D(\omega)}{Q(\omega)}\right]. \tag{11.53}$$

Therefore, an equivalent problem is the minimization of

$$||E(\omega)||_\infty = \max_{\omega \in B} |\bar{W}(\omega)[P(\omega) - \bar{D}(\omega)]|, \tag{11.54}$$

where

$$\bar{W}(\omega) = W(\omega)Q(\omega), \quad \bar{D}(\omega) = \frac{D(\omega)}{Q(\omega)}, \quad P(\omega) = \sum_{k=0}^{r-1} a(k)\cos k\omega, \tag{11.55}$$

and $\bar{B} = B -$ [endpoints where $Q(\omega) = 0$].

The Remez exchange algorithm, for computing the best Chebyshev solution, uses the alternation theorem. This theorem characterizes the best Chebyshev solution.

11.3.1.3.3 Alternation Theorem

If $P(\omega)$ is given by Equation 11.55, then a necessary and sufficient condition that $P(\omega)$ be the unique minimizer of Equation 11.54 is that there exist in $\bar{B}$ at least $r + 1$ extremal points $\omega_1, \ldots, \omega_{r+1}$ (in order: $\omega_1 < \omega_2 < \cdots < \omega_{r+1}$), such that

$$E(\omega_i) = c \cdot (-1)^i ||E(\omega)||_\infty \quad \text{for } i = 1, \ldots, r+1, \tag{11.56}$$

where c is either 1 or -1.

The alternation theorem states that $|E(\omega)|$ attains its maximum value at a minimum of $r+1$ points, and that the weighted error function alternates sign on at least $r+1$ of those points. Consequently, the weighted error functions of best Chebyshev solutions exhibit an equiripple behavior.

For lowpass filter design via the PM algorithm, the functions $D(\omega)$ and $W(\omega)$ in Equation 11.44 are usually used. For lowpass filters so obtained, the deviations δ_p and δ_s satisfy the relation $\delta_p/\delta_s = K_s/K_p$. For example, consider the design of a real symmetric lowpass filter of length $N=41$. Then $Q(\omega)=1$ and $r=(N+1)/2=21$. With the desired amplitude and weight function, Equation 11.44, with $K=4$ and $\omega_p=0.25\pi$, $\omega_s=0.35\pi$, the best Chebyshev solution and its weighted error function are illustrated in Figure 11.12. The maximum errors in the passband and stopband are $\delta_p=0.0178$ and $\delta_s=0.0714$, respectively. The circular marks in Figure 11.12c indicate the extremal points of the alternation theorem.

To elaborate on the alternation theorem, consider the design of a length 21 lowpass filter and a length 41 bandpass filter. Several optimal Chebyshev filters are illustrated in Figures 11.13 through 11.16. It can be verified by inspection that each of the filters illustrated in Figures 11.13 through 11.16 is Chebyshev optimal, by verifying that the alternation theorem is satisfied. In each case, a set of $r+1$ extremal points, which satisfies the necessary and sufficient conditions of the alternation theorem, is indicated by circular marks in Figures 11.13 through 11.16.

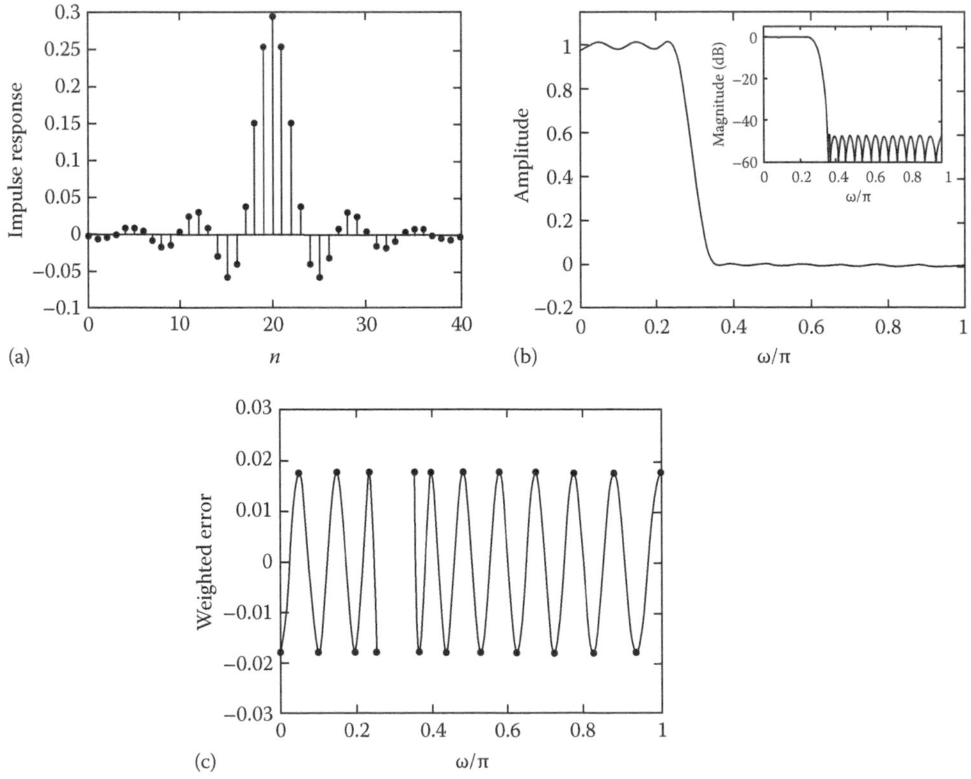

FIGURE 11.12 Equiripple lowpass filter obtained via the PM algorithm: $N=41$, $\omega_p=0.25\pi$, $\omega_s=0.35\pi$, and $\delta_p/\delta_s=4$.

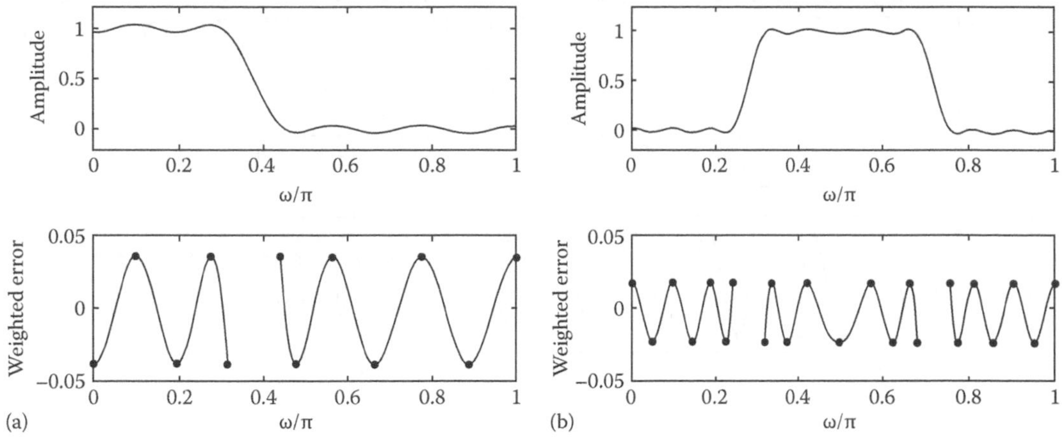

FIGURE 11.13 PM example. (a) Lowpass: $N = 21$, $\omega_p = 0.3161\pi$, and $\omega_s = 0.4444\pi$. (b) Bandpass: $N = 41$, $\omega_1 = 0.2415\pi$, $\omega_2 = 0.3189\pi$, $\omega_3 = 0.6811\pi$, and $\omega_4 = 0.7585\pi$.

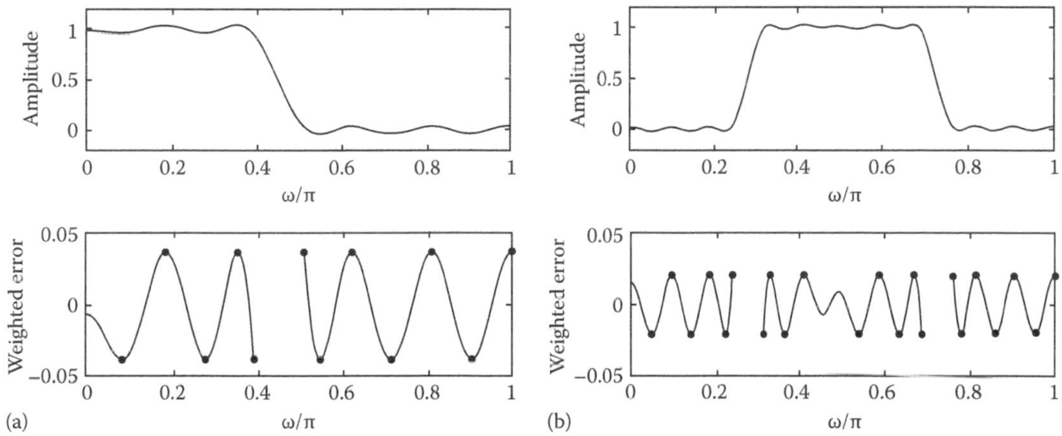

FIGURE 11.14 PM example. (a) Lowpass: $N = 21$, $\omega_p = 0.3889\pi$, and $\omega_s = 0.5082\pi$. (b) Bandpass: $N = 41$, $\omega_1 = 0.2378\pi$, $\omega_2 = 0.3132\pi$, $\omega_3 = 0.6870\pi$, and $\omega_4 = 0.7621\pi$.

Several remarks regarding the weighted error function of a best Chebyshev solution are worth noting.

1. $E(\omega)$ may have local minima and maxima in $\bar{B}$ at which $|E(\omega)|$ does not attain its maximum value. See Figure 11.14.
2. $|E(\omega)|$ may attain its maximum value at more than $r + 1$ points in $\bar{B}$. See Figure 11.15.
3. If there exists in $\bar{B}$ s ordered points $\omega_1, \ldots, \omega_s$, with $s > r + 1$, at which $|E(\omega_i)| = \|E(\omega)\|_\infty$ (i.e., there are more than $r + 1$ extremal points), then it is possible that $E(\omega_i) = E(\omega_{i+1})$ for some i. See Figure 11.16. This is rare and, for lowpass filter design, impossible.

Figure 11.14 illustrates two filters that possess "scaled-extra ripples" (ripples of non-maximal size [30]). Figure 11.15 illustrates two maximal ripple filters. Maximal ripple filters are a subset of optimal Chebyshev filters that occur for special values of ω_p, ω_s, etc. (The first algorithms for equiripple filter

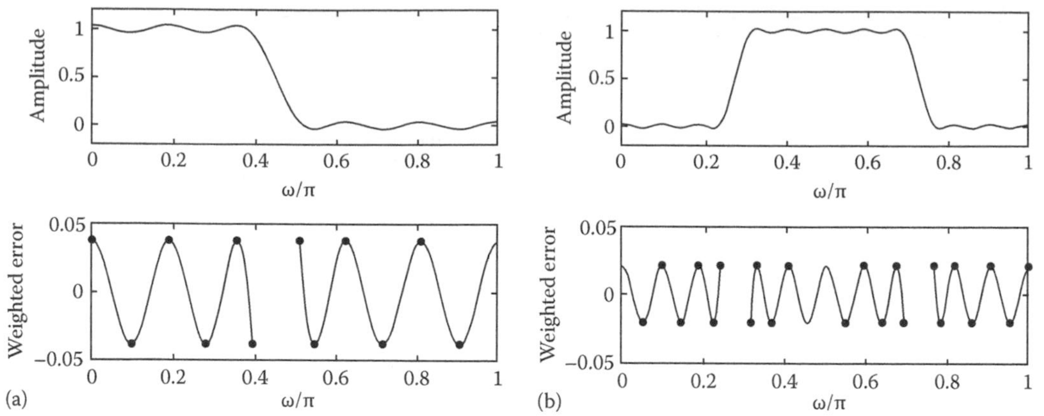

FIGURE 11.15 PM example. Lowpass: $N=21$, $\omega_p=0.3919\pi$, and $\omega_s=0.5103\pi$. Bandpass: $N=41$ $\omega_1=0.2370\pi$, $\omega_2=0.3115\pi$, $\omega_3=0.6885\pi$, and $\omega_4=0.7630\pi$.

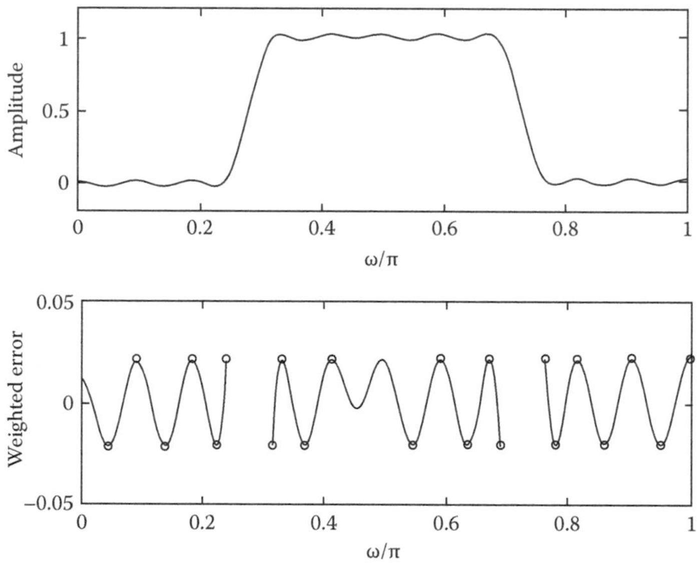

FIGURE 11.16 PM example: $N=41$, $\omega_1=0.2374\pi$, $\omega_2=0.3126\pi$, $\omega_3=0.6876\pi$, and $\omega_4=0.7624\pi$.

design produced only maximal ripple filters [33,34]). Figure 11.16 illustrates a filter that possesses two scaled-extra ripples and one extra ripple of maximal size. These extra ripples have no bearing on the alternation theorem. The set of $r+1$ points, indicated in Figure 11.16, is a set that satisfies the alternation theorem; therefore, the filter is optimal in the Chebyshev sense.

11.3.1.3.4 Remez Algorithm

To understand the Remez exchange algorithm, first note that Equation 11.56 can be written as

$$\sum_{k=0}^{r-1} a(k)\cos k\omega_i - \frac{(-1)^i\delta}{\bar{W}(\omega_i)} = \bar{D}(\omega_i) \quad \text{for } i=1,\ldots,r+1, \tag{11.57}$$

where δ represents $\|E(\omega)\|_\infty$, and consider the following. If the set of extremal points in the alternation theorem were known in advance, then the solution could be found by solving the system of Equation 11.57.

The system in Equation 11.57 represents an interpolation problem, which in matrix form becomes

$$\begin{bmatrix} 1 & \cos\omega_1 & \cdots & \cos(r-1)\omega_1 & 1/\bar{W}(\omega_1)(37) \\ 1 & \cos\omega_2 & \cdots & \cos(r-1)\omega_2 & -1/\bar{W}(\omega_2)(38) \\ \vdots & & & & \vdots(39) \\ & & & & (40) \\ 1 & \cos\omega_{r+1} & \cdots & \cos(r-1)\omega_{r+1} & (-1)^r/\bar{W}(\omega_{r+1})(41) \end{bmatrix} \begin{bmatrix} a(0)(42) \\ a(1)(43) \\ \vdots(44) \\ a(r-1)(45) \\ \delta \end{bmatrix} = \begin{bmatrix} \bar{D}(\omega_1)(46) \\ \bar{D}(\omega_2)(47) \\ \vdots(48) \\ (49) \\ \bar{D}(\omega_{r+1}) \end{bmatrix} \quad (11.58)$$

to which there is a unique solution. Therefore, the problem becomes one of finding the correct set of points over which to solve the interpolation problem in Equation 11.57.

The Remez exchange algorithm proceeds by iteratively

1. Solving the interpolation problem in Equation 11.58 over a specified set of $r+1$ points (a reference set)
2. Updating the reference set (by an exchange procedure)

The initial reference set can be taken to be $r+1$ points uniformly spaced over $\bar{B}$. Convergence is achieved when $\|E(\omega)\|_\infty - |\delta| < \varepsilon$, where ε is a small number (such as 10^{-6}) indicating the numerical accuracy desired.

During the interpolation step, the solution to Equation 11.58 is facilitated by the use of a closed-form solution for δ and interpolation formulas [29].

After the interpolation step is performed, the reference set is updated as follows. The weighted error function is computed, and a new reference set $\omega_1, \ldots, \omega_{r+1}$ is found such that (1) the current weighted error function $E(\omega)$ alternates sign on the new reference set, (2) $|E(\omega_i)| \geq |\delta|$ for each point ω_i of the new reference set, and (3) $|E(\omega_i)| > |\delta|$ for at least one point ω_i of the new reference set. Generally, the new reference set is found by taking the set of local minima and maxima of $E(\omega)$ that exceed the current value of δ, and taking a subset of this set that satisfies the alternation property. Figure 11.17 illustrates the operation of the PM algorithm.

11.3.1.3.5 Design Rules for Lowpass Filters

While the PM algorithm is applicable for the approximation of arbitrary responses $D(\omega)$, the lowpass case has received particular attention [12,35–37]. In the design of lowpass filters via the PM algorithm, there are five parameters of interest: the filter length N, the passband and stopband edges ω_p and ω_s, and the maximum error in the passband and stopband δ_p and δ_s. Their values are not independent—any four determines the fifth. Formulas for predicting the required filter length for a given set of specifications make this clear. Kaiser developed the following approximate relation for estimating the equiripple FIR filter length for meeting the specifications:

$$N \approx \frac{-20\log_{10}\left(\sqrt{\delta_p\delta_s}\right) - 13}{14.6\Delta F} + 1, \quad (11.59)$$

where $\Delta F = (\omega_s - \omega_p)/(2\pi)$. Defining the filter attenuation ATT to be $-20\log_{10}(\sqrt{\delta_p\delta_s})$, and comparing Equation 11.29 with Equation 11.59, it can be seen that the optimal Chebyshev design results in filters with about 5 dB more attenuation than the windowed designed filters when the same specs are used for the other design parameters (N and ΔF). Figure 11.18 compares window-based designs with Chebyshev (PM)-based designs.

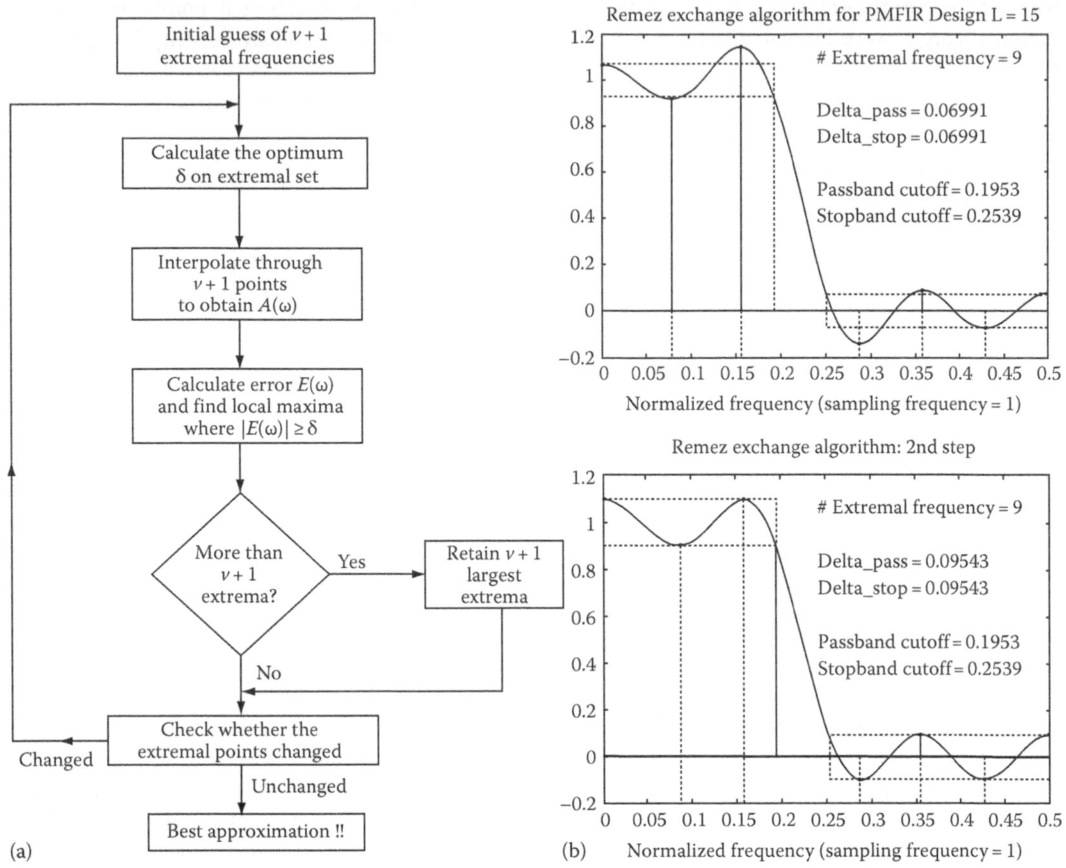

FIGURE 11.17 Operation of the PM algorithm: (a) block diagram and (b) exchange steps. Extremal points constituting the current extremal set are shown as solid circles; extremal points selected to form the new extremal set are shown as solid squares.

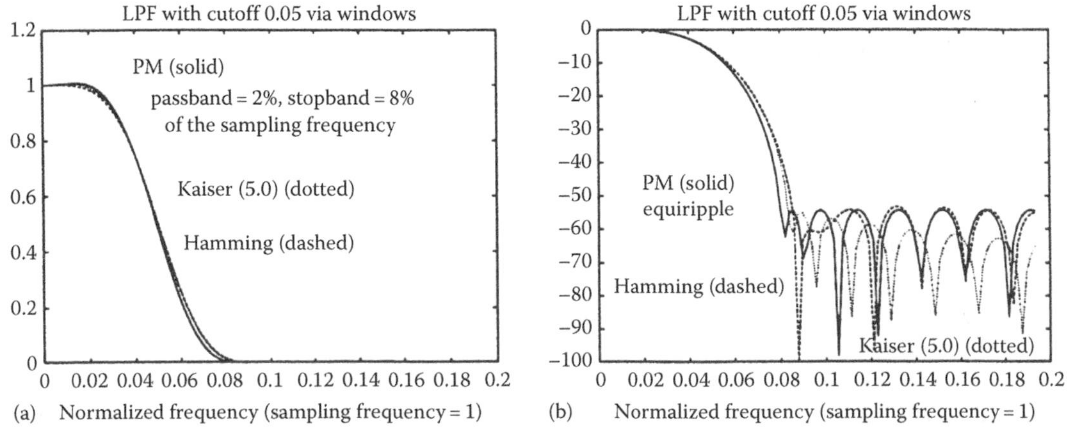

FIGURE 11.18 Comparison of window designs with optimal Chebyshev (PM) designs. The window length is $N = 49$: (a) frequency response of designed filter using linear scale and (b) frequency response of designed filter using log (decibel) scale.

Herrmann et al. gave a somewhat more accurate design formula for the optimal Chebyshev FIR filter design [37]:

$$N \approx \frac{D_\infty(\delta_p, \delta_s) - f(\delta_p, \delta_s)(\Delta F)^2}{\Delta F} + 1, \tag{11.60}$$

where

$$\begin{aligned} D_\infty(\delta_p, \delta_s) &= 0.005309\left(\log_{10}^2\delta_p + 0.07114\log_{10}\delta_p - 0.4761\right)\log_{10}\delta_s \\ &\quad - \left(0.00266\log_{10}^2\delta_p + 0.5941\log_{10}\delta_p + 0.4278\right), \\ f(\delta_p, \delta_s) &= 11.01217 + 0.51244(\log_{10}\delta_p - \log_{10}\delta_s). \end{aligned} \tag{11.61}$$

These formulas assume that $\delta_s < \delta_p$. If otherwise, then interchange δ_p and δ_s. Equation 11.60 is the one used in the MATLAB implementation (**remezord()** function) as part of the MATLAB signal processing toolbox.

To use the PM algorithm for lowpass filter design, the user specifies $N, \omega_p, \omega_s, \delta_p/\delta_s$. The PM algorithm can be modified so that the user specifies other parameter sets [38]. For example, with one modification, the user specifies $N, \omega_p, \delta_p, \delta_s$; or similarly, $N, \omega_s, \delta_p, \delta_s$. With a second modification, the user specifies $N, \omega_p, \omega_s, \delta_p$; or similarly, $N, \omega_p, \omega_s, \delta_s$.

Note that Equation 11.59 states that the filter length N and the transition width ΔF are inversely proportional. This is in contrast to the relation for maximally flat symmetric filters. For equiripple filters with fixed δ_p and δ_s, ΔF diminishes like $1/N$; while for maximally flat filters, ΔF diminishes like $1/\sqrt{N}$.

11.3.1.3.6 Remarks

- Optimal with respect to Chebyshev norm
- Explicit control of bandedges and relative ripple sizes
- Efficient algorithm, always converges
- Allows the use of a frequency-dependent weighting function
- Suitable for arbitrary $D(\omega)$ and $W(\omega)$
- Does not allow arbitrary linear constraints

11.3.1.3.7 Summary of Optimal Chebyshev Linear Phase FIR Filter Design

1. The desired frequency response can be written as

$$D(\omega) = A(\omega)e^{-j(\alpha\omega+\beta)},$$

   where $\alpha = (N-1)/2$ always, and $\beta = 0$ for filters with even symmetry. Since $A(\omega)$ is a real-valued function, the Chebyshev approximation is applied to $A(\omega)$ and the linear phase comes for free. However, the delay will be proportional to the designed filter length.
2. The mathematical theory of Chebyshev approximation is applied. In this type of optimization, the maximum value of the error is minimized, as opposed to the error energy as in least squares. Minimizing the maximum error is consistent with the desire to keep the passband and stopband deviations as small as possible. (Recall that least squares suffers from the Gibbs effect). However, minimization of the maximum error does not permit the use of derivatives to find the optimal solution.
3. The alternation theorem gives the necessary and sufficient conditions for the optimum in terms of equal-height ripples in the (weighted) error function.

4. The Remez exchange algorithm will compute the optimal approximation by searching for the locations of the peaks in the error function. This algorithm is iterative.
5. The inputs to the algorithm are the filter length N, the locations of the passband and stopband cutoff frequencies ω_p and ω_s, and a weight function to weight the error in the passband and stopband differently.
6. The Chebyshev approximation problem can also be reformulated as a linear program. This is useful if additional linear design constraints need to be included.
7. Transition width is minimized among all FIR filters with the same deviations.
8. Passband and stopband deviations: The response is equiripple, it does not fall off away from the transition region. Compared to the Kaiser window design, the optimal Chebyshev FIR design gives about 5 dB more attenuation (where attenuation is given by $-20 \log_{10}\delta$ and δ is the stopband or passband error) for the same specs on all other filter design parameters.

11.3.1.3.7.1 Linear Programming Often it is desirable that an FIR filter be designed to minimize the Chebyshev error subject to linear constraints that the PM algorithm does not allow. An example described by Rabiner and Gold includes time domain constraints—in that example [30], the oscillatory behavior of the step response of a lowpass filter is included in the design formulation.

Another example comes from a communication application [39]—given $h_1(n)$, design $h_2(n)$ so that $h(n) = (h_1 * h_2)(n)$ is an Mth band filter (i.e., $h(Mn) = 0$ for all $n \neq 0$ and $M \neq 0$). Such constraints are linear in $h_1(n)$. (In the special case that $h_1(n) = \delta(n)$, $h_2(n)$ is itself an Mth band filter, and is often used for interpolation.)

Linear programming formulations of approximation problems (and optimization problems in general) are very attractive because well-developed algorithms exist (namely the simplex algorithm and more recently, interior point methods) for solving such problems. Although linear programming requires significantly more computation than the methods described above, for many problems it is a very rapid and viable technique [7]. Furthermore, this approach is very flexible—it allows arbitrary linear equality and inequality constraints.

The problem of minimizing the weighted Chebyshev error $W(\omega)[A(\omega) - D(\omega)]$ where $A(\omega)$ is given by $Q(\omega)\sum_{k=0}^{r-1} a(k) \cos k\omega$ can be formulated as a linear program as follows:

$$\text{minimize } \delta \tag{11.62}$$

subject to

$$A(\omega) - \frac{\delta}{W(\omega)} \leq D(\omega) \tag{11.63}$$

$$-A(\omega) - \frac{\delta}{W(\omega)} \leq -D(\omega). \tag{11.64}$$

The variables are $a(0), \ldots, a(r-1)$ and δ. The cost function and the constraints are linear functions of the variables, hence the formulation is that of a linear program.

11.3.1.3.7.2 Remarks

- Optimal with respect to chosen criteria
- Easy to include arbitrary linear constraints
- Criteria limited to linear programming formulation
- High computational cost

11.3.2 IIR Design Methods

Lina J. Karam, Ivan W. Selesnick, and C. Sidney Burrus

The objective in IIR filter design is to find a rational function $H(\omega)$ (as in Equation 11.12) that approximates the ideal specifications according to some design criteria.

The approximation of an arbitrary specified frequency response is more difficult for IIR filters than is so for FIR filters. This is due to the nonlinear dependence of $H(\omega)$ on the filter coefficients in the IIR case. However, for the ideal lowpass response, there exist analytic techniques to directly obtain IIR filters. These techniques are based on converting analog filters into IIR digital filters. One such popular IIR design method is the bilinear transformation method [1,11]. Other types of frequency-selective filters (shown in Figure 11.1) can be obtained from the designed lowpass prototype using additional frequency transformations [1, Chapter 7].

Direct "discrete-time" iterative IIR design methods have also been proposed (see Section 11.4.2). While these methods can be used to approximate general magnitude responses (i.e., not restricted to the design of the standard frequency-selective filters), they are iterative and slower than the traditional "continuous-time/space" based approaches that make use of simple and efficient closed-form design formulas.

11.3.2.1 Bilinear Transformation Method

The traditional IIR design approaches reduce the "discrete-time/space" (digital) filter design problem into a "continuous-time/space" (analog) filter design problem, which can be solved using well-developed and relatively simple design procedures based on closed-form design formulas. Then, a transformation is used to map the designed analog filter into a digital filter meeting the desired specifications.

Let $H(z)$ denote the transfer function of a digital filter (i.e., $H(z)$ is the Z-transform of the filter impulse response $h(n)$) and let $H_a(s)$ denote the transfer function of an analog filter (i.e., $H_a(s)$ is the Laplace transform of the continuous-time filter impulse response $h(t)$). The bilinear transformation is a mapping between the complex variables s and z and is given by

$$s = K\left(\frac{1 - z^{-1}}{1 + z^{-1}}\right), \tag{11.65}$$

where K is a design parameter. Replacing s by Equation 11.65 in $H_a(s)$, the analog filter with transfer function $H_a(s)$ can be converted into a digital filter whose transfer function is equal to

$$H(z) = H_a(s)\big|_{s=K\left(\frac{1-z^{-1}}{1+z^{-1}}\right)}. \tag{11.66}$$

Alternatively, the mapping can be used to convert a digital filter into an analog filter by expressing z in function of s.

Note that the analog frequency variable Ω corresponds to the imaginary part of s (i.e., $s = \sigma + j\Omega$), while the digital frequency variable ω (in radians) corresponds to the angle (phase) of z (i.e., $z = re^{j\omega}$). The bilinear transformation (Equation 11.65) was constructed such that it satisfies the following important properties:

1. The left-half plane (LHP) of the s-plane maps into the inside of the U.C. in the z-plane. As a result, a stable and causal analog filter will always result in a stable and causal digital filter.
2. The $j\Omega$ axis (imaginary axis) in the s-plane maps into the U.C. in the z-plane (i.e., $z = e^{j\omega}$). This results in a direct relationship between the continuous-time frequency Ω and the discrete-time frequency ω. Replacing z by $e^{j\omega}$ (U.C.) in Equation 11.65, we obtain the following relation:

$$\Omega = K\tan(\omega/2) \tag{11.67}$$

or, equivalently,

$$\omega = 2\arctan(\Omega/K). \tag{11.68}$$

The design parameter K can be used to map one specific frequency point in the analog domain to a selected frequency point in the digital domain, and to control the location of the designed filter cutoff frequency. Equations 11.67 and 11.68 are nonlinear, resulting in a warping of the frequency axis as the filter frequency response is transformed from one domain to another. This follows from the fact that the bilinear transformation maps (via Equations 11.67 or 11.68) the entire $j\Omega$ axis, i.e., $-\infty \leq \Omega \leq \infty$, onto one period $-\pi \leq \omega \leq \pi$ (which corresponds to one revolution of the U.C. in the z-plane).

The bilinear transformation design procedure can be summarized as follows:

1. Transform the digital frequency domain specifications to the analog domain using Equation 11.67. The frequency domain specs are given typically in terms of magnitude response specs as shown in Figure 11.2. After the transformation, the digital magnitude response specs are converted into specs on the analog magnitude response.
2. Design a stable and causal analog filter with transfer function $H_a(s)$ such that $|H_a(s=j\Omega)|$ approximates the derived analog specs. This is typically done by using one of the classical frequency-selective analog filters whose magnitude responses are given in terms of closed-form formulas; the parameters in the closed-form formulas (e.g., needed analog filter order and analog cutoff frequency) can then be computed to meet the desired analog specs. Typical analog prototypes include Butterworth, Chebyshev, and elliptic filters; the characteristics of these filters are discussed in section on pages 11–32. The closed-form formulas give only the magnitude response $|H_a(j\Omega)|$ of the analog filter and, therefore, do not uniquely specify the complete frequency response (or corresponding transfer function) which also should include a phase response. From all the filters having magnitude response $|H_a(j\Omega)|$, we need to select the filter that is stable and, if needed, causal. Using the fact that the computed magnitude-squared response $|H_a(j\Omega)|^2 = |H_a(s)|^2$, for $s=j\Omega$, and that $|H_a(s)|^2 = H_a(s)H_a^*(-s^*)$, where s^* denotes the complex conjugate of s, the system function $H_a(s)$ of the desired stable and causal filter is obtained by selecting the poles of $|H_a(j\Omega)|^2$ lying in the LHP of the s-plane [11].
3. Obtain the transfer function $H(z)$ for the digital filter by applying the bilinear transformation (Equation 11.65) to $H_a(s)$. The design parameter K can be fixed or chosen to map one analog frequency point Ω (e.g., the passband or stopband cutoff) into a desired digital frequency point ω.
4. The frequency response $H(\omega)$ of the resulting stable digital filter can be obtained from the transfer function $H(z)$ by replacing z by $e^{j\omega}$, i.e.,

$$H(\omega) = H(z)\big|_{z=e^{j\omega}}. \tag{11.69}$$

11.3.2.2 Classical IIR Filter Types

The four standard classical analog filter types are known as (1) Butterworth, (2) Chebyshev I, (3) Chebyshev II, and (4) elliptic [1,11]. The characteristics of these analog filters are described briefly below.

Digital versions of these filters are obtained via the bilinear transformation [1,11], and examples are illustrated in Figure 11.19.

11.3.2.2.1 Butterworth

The magnitude-squared function of an Nth order Butterworth lowpass filter is given by

$$|H_a(\phi\Omega)|^2 = \frac{1}{1 + (\Omega/\Omega_c)^{2N}}, \tag{11.70}$$

where Ω_c is the cutoff frequency.

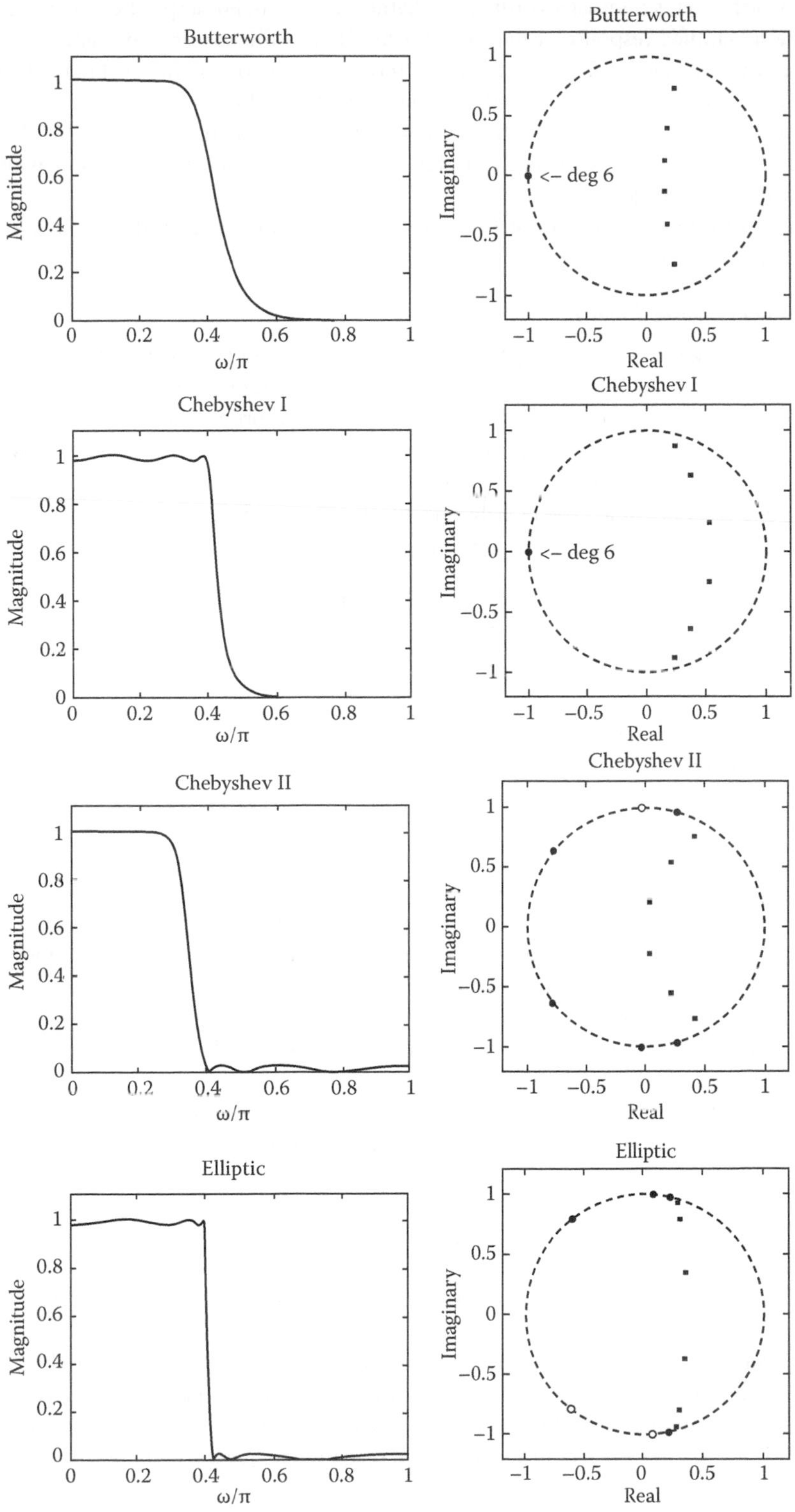

FIGURE 11.19 Classical IIR digital filters.

The Butterworth filter is optimal according to a flatness criterion. For a specified filter order and cutoff frequency, the magnitude response of the Butterworth filter is the solution that attains the maximum number of derivatives equal to 0 at $\Omega = 0$ and ∞ ($\omega = 0$ and π for the digital filter). This magnitude response is maximally flat in the passband (i.e., the first $(2N - 1)$ derivatives of $|H_a(j\Omega)|^2$ are zero at $\Omega = 0$), and it decreases monotonically in the passband and stopband. Note that $|H_a(\Omega = 0)| = 1$ and $|H_a(\Omega = \Omega_c)| = 1/\sqrt{2}$, for all N. Also, as the filter order N increases, the transition width decreases, yielding a sharper cutoff edge.

The Butterworth filter has the poorest frequency selectivity compared to the Chebyshev and elliptic filters, but it is the simplest to design.

11.3.2.2.2 Chebyshev: Types I and II

If the filter specs are given in terms of passband and stopband ripples (as shown in Figure 11.2), then these specs are exceeded for a Butterworth filter because of the monotonic behavior of the magnitude response. The specs can be met more efficiently with a lower order filter if the error is distributed uniformly over the passband or the stopband or (best) both. This can be accomplished by choosing an approximating filter with an equiripple behavior.

The magnitude response of a Type I Chebyshev filter is equiripple in the passband and monotonic in the stopband. The magnitude-squared response is given by

$$|H_a(\phi\Omega)|^2 = \frac{1}{1 + \varepsilon^2 T_N^2(\Omega/\Omega_c)}, \tag{11.71}$$

where

$T_N(x)$ is the Nth degree Chebyshev polynomial in x
ε is a parameter specified by the allowable passband ripple
Ω_c is the filter cutoff frequency
N is the filter order

The Type I Chebyshev filter is optimal according to a Chebyshev criterion in the passband and a flatness criterion in the stopband. For a specified filter order and passband edge, the magnitude response of this filter attains the minimum Chebyshev error in the passband and the maximum number of vanishing derivatives at $\Omega = \infty$ ($\omega = \pi$ for the digital filter).

Note that $|H_a(j\Omega)|^2$ ripples between 1 and $1/(1 + \varepsilon^2)$ in the passband ($0 \le |\Omega| \le \Omega_c$) since $0 \le T_N^2(x) \le 1$ for $0 \le x \le 1$. For $x > 1$, $T_N^2(x)$ increases monotonically; so, $|H_a(j\Omega)|^2$ decreases monotonically in the stopband ($\Omega > \Omega_c$).

From Equation 11.71, three parameters are required to specify the filter: ε, Ω_c, and N. In a typical design, ε is specified by the allowable passband ripple δ_p by solving

$$\frac{1}{1 + \varepsilon^2} = (1 - \delta_p)^2. \tag{11.72}$$

Ω_c is specified by the desired passband cutoff frequency, and N is then chosen so that the stopband specs are met.

A similar treatment can be made for Chebyshev II filters (also called inverse Chebyshev). The Type II Chebyshev filter has a magnitude response that is monotonic in the passband and equiripple in the stopband. It can be obtained from the Type I Chebyshev filter by replacing $\varepsilon^2 T_N^2(\Omega/\Omega_c)$ in Equation 11.73 by $\left[\varepsilon^2 T_N^2(\Omega_c/\Omega)\right]^{-1}$, resulting in the following magnitude-squared function:

$$|H_a(\phi\Omega)|^2 = \frac{1}{1 + \left[\varepsilon^2 T_N^2(\Omega_c/\Omega)\right]^{-1}}. \tag{11.73}$$

For the Chebyshev II filter, the parameter ε is determined by the allowable stopband ripple δ_s as follows:

$$\frac{\varepsilon^2}{1+\varepsilon^2} = (1-\delta_s)^2. \tag{11.74}$$

The order N is determined so that the passband specs are met.

The Chebyshev filter is so called because the Chebyshev polynomials are used in the formula.

11.3.2.2.3 Elliptic

The magnitude response of an elliptic filter is equiripple in both the passband and stopband. It is optimal according to a weighted Chebyshev criterion. For a specified filter order and bandedges, the magnitude response of the elliptic filter attains the minimum weighted Chebyshev error. In addition, for a given order N, the transition width is minimized among all filters with the same passband and stopband deviations.

The magnitude-squared response of an elliptic filter is given by

$$|H_a(\phi\Omega)|^2 = \frac{1}{1+\varepsilon^2 E_N^2(\Omega)}, \tag{11.75}$$

where $E_N(\Omega)$ is a Jacobian elliptic function [11]. Elliptic filters are so called because elliptic functions are used in the formula.

11.3.2.2.4 Remarks

Note that, for these four filter types, the approximation is in the magnitude and no phase approximation is achieved. Also note that each of these filter types has a symmetric FIR counterpart.

The four types of IIR filters shown in Figure 11.19 are usually obtained from analog prototypes via the bilinear transformation (BLT), as described in section on pages 11–30. The analog filter $H(s)$ is designed to approximate the ideal lowpass filter over the imaginary axis. The BLT maps the imaginary axis to the U.C. $|z|=1$, and is given by the change of variables, $s = K\left(\frac{z-1}{z+1}\right)$. This mapping preserves the optimality of the four classical filter types. Another method for obtaining IIR digital filters from analog prototypes is the impulse-invariant method [11]. In this method, the impulse response of a digital filter is obtained by sampling the continuous-time/space impulse response of the analog prototype. However, the impulse invariance method usually results in aliasing distortion and is appropriate only for bandlimited filters. For this reason, the bilinear transformation method is usually preferred.

Note that, for the four analog prototypes described above, the numerator degree of the designed digital IIR filter equals the denominator degree.* For the design of digital IIR filters with unequal numerator and denominator degree, analytic techniques are available only for special cases (see Section 11.4.2). For other cases, iterative numerical methods are required.

Highpass, bandpass, and band-reject filters can also be obtained from analog prototypes (or from the digital versions) by appropriate frequency transformations [11]. Those transformations are generally useful only when the IIR filter has equal degree numerator and denominator, which is the case for the digital versions of the classical analog prototypes.

A fifth IIR filter for which closed-form expressions are readily available is the all-pole filter that possesses a maximally flat group delay at $\omega=0$. In this case, no magnitude approximation is achieved. It should be noted that this filter is not obtained directly from the analog equivalent, the Bessel filter (the BLT does not preserve the maximally flat group delay characteristic). Instead, it can be derived directly in the digital domain [40]. For a specified filter order and DC group delay, the group delay of

* Possibly, however, a single pole is located at $z=0$, in which case their degrees differ by one.

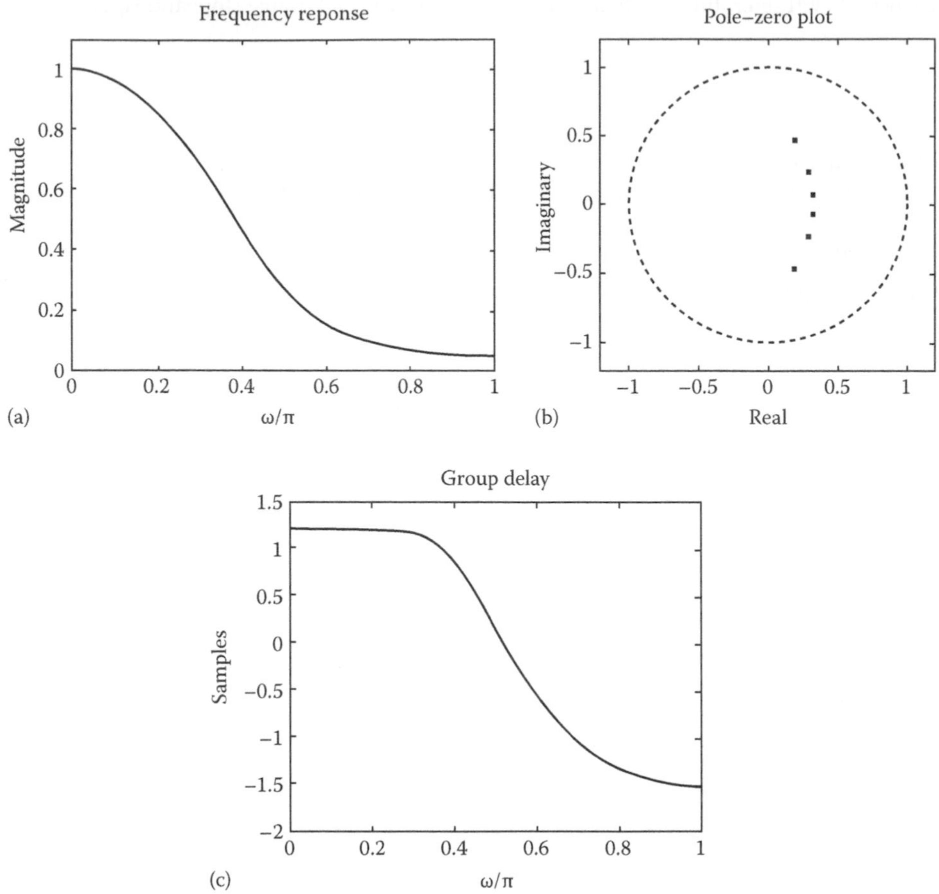

FIGURE 11.20 Maximally flat delay IIR filter: $N = 6$ and $\tau = 1.2$.

this filter attains the maximal number of vanishing derivatives at $\omega = 0$. The particularly simple formula for $H(z)$ is

$$H(z) = \frac{\sum_{k=0}^{N} a_k}{\sum_{k=0}^{N} a_k z^{-k}} \quad \text{where } a_k = (-1)^k \binom{N}{k} \frac{(2\tau)_k}{(2\tau + N + 1)_k}, \tag{11.76}$$

where

τ is the DC group delay

the Pochhammer symbol $(x)_k$ denotes the rising factorial: $(x)(x+1)(x+2) \cdots (x+k-1)$.

An example is shown in Figure 11.20, where it is evident that the magnitude response makes a poor lowpass filter. However, such a filter (1) can be cascaded with a symmetric FIR filter that improves the magnitude without affecting its phase linearity [41], and (2) is useful for fractional delay allpass filters as described in Section 11.4.2.2.

11.3.2.3 Comments and Generalizations

The design of IIR digital filters by transformation of classical analog prototypes is attractive because formulas exist for these filters. Unfortunately, digital filters so obtained necessarily possess an equal number of poles and zeros away from the origin. For some specifications, it is desired that the numerator and denominator degrees not be restricted to be equal.

Several authors have addressed the design and the advantages of IIR filters with unequal numerator and denominator degrees [42–48]. In [46,49], Saramäki finds that the classical elliptic and Chebyshev filter types are seldom the best choice. In [42], Jackson improves the Martinez–Parks algorithm and notes that, for equiripple filters, the use of just two poles "is often the most attractive compromise between computational complexity and other performance measures of interest."

Generally, the design of recursive digital filters having unequal denominator and numerator degrees requires the use of iterative numerical methods. However, for some special cases, formulas are available. For example, a digital generalization of the classical Butterworth filter can be obtained with the formulas given in [50]. Figure 11.21 illustrates an example. It is evident from the figure, that some zeros of the filter contribute to the shaping of the passband. The zeros at $z = -1$ produce a flat behavior at $\omega = \pi$, while the remaining zeros, together with the poles, produce a flat behavior at $\omega = 0$. The specified cutoff frequency determines the way in which the zeros are split between the $z = -1$ and the passband.

To illustrate the effect of various numerator and denominator degrees, examine a set of filters for which (1) the sum of the numerator degree and the denominator degree is constant, say 20, and (2) the cutoff frequency is constant, say $\omega_c = 0.6\pi$. By varying the number of poles from 0 to 10 in steps of 2 (so that the number of zeros is decreased from 20 to 10 in steps of 2), the filters shown in Figure 11.22 are obtained.

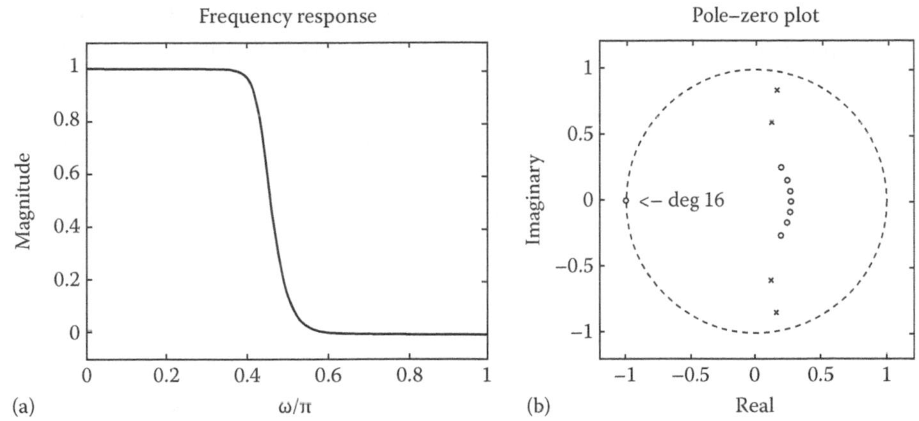

FIGURE 11.21 Generalized Butterworth filter.

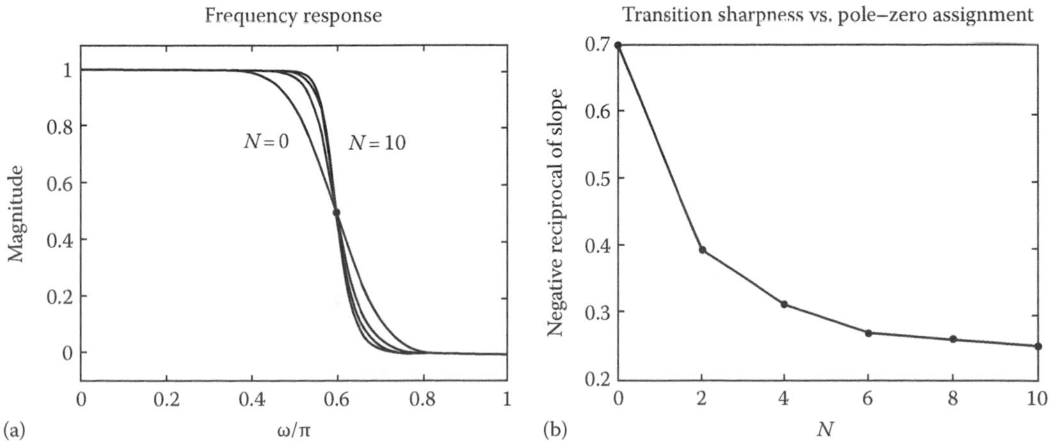

FIGURE 11.22 The filters for which the cutoff frequency is $\omega_o = 0.6\,\pi$, and for which the sum of the number of poles and the number of zeros is 20. N denotes the number of poles.

Figure 11.22 also shows the negative reciprocal of the slope of the magnitude response at the cutoff frequency—this indicates the width of the transition band. Notice that, for this example, as the number of poles and zeros become more equal, the transition becomes sharper. It is interesting to note that the improvement is greatest when the number of poles is increased from 0 to 2. When implementation issues are taken into consideration, the filters with two or four poles appear to attain a good trade-off between performance and implementation complexity.

11.4 Other Developments in Digital Filter Design

11.4.1 FIR Filter Design

Ivan W. Selesnick, C. Sidney Burrus, Lina J. Karam, and James H. McClellan

11.4.1.1 Maximally Flat Real Symmetric FIR Filters

By requiring the derivatives of the amplitude function $A(\omega)$ to satisfy derivative constraints at $\omega = 0$ and $\omega = \pi$, a lowpass filter is obtained having a very flat monotone response, see Figure 11.23. The resulting design is very simple, efficient implementations of such filters exist [51,52], and the filters have been found to be useful when used together [53] or in conjunction with other filters [54]. Such filters preserve the input signal around $\omega = 0$ very well, and achieve very high attenuation in the stopband.

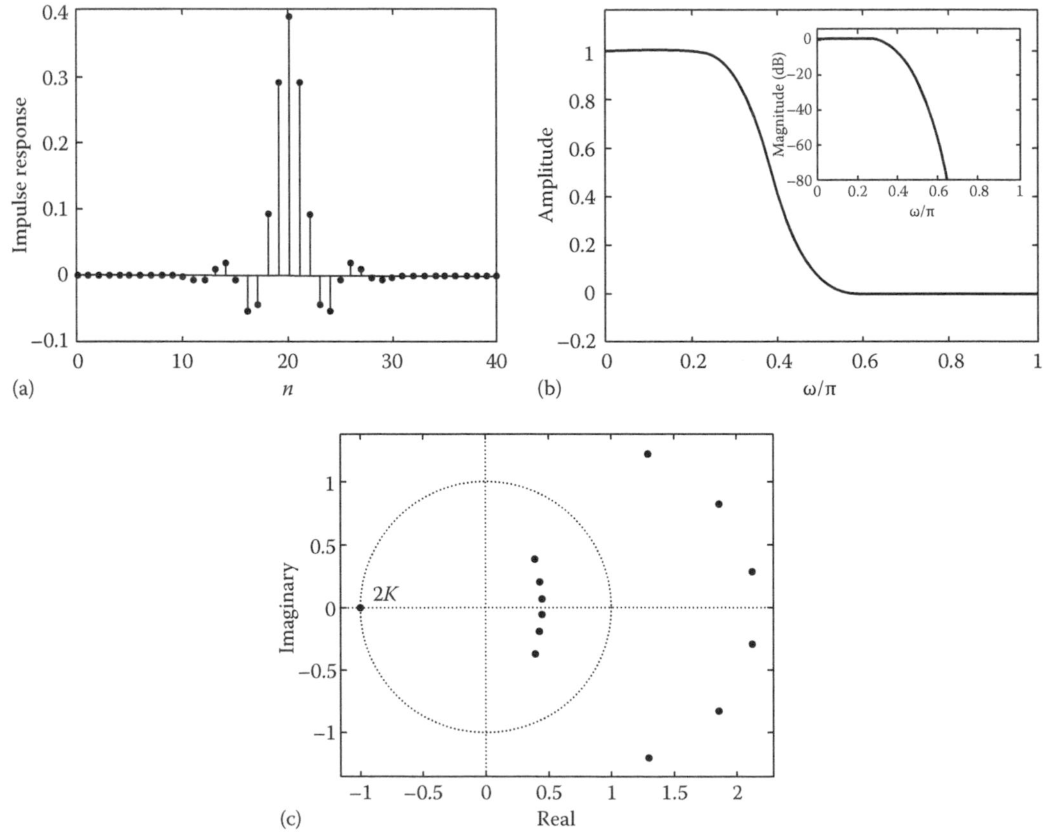

FIGURE 11.23 Maximally flat filter: $N = 41$ and $K = 14$.

The transition between the passband and stopband is wide, however. This design problem was introduced by Herrmann [55] and is formulated as follows.

Given $N = 2M + 1$ and K $(1 \leq K \leq M)$, find a symmetric filter of length N such that the amplitude response, given by

$$A(\omega) = h(M) + 2\sum_{n=1}^{M} h(M-n)\cos n\omega \tag{11.77}$$

satisfies the following constraints:

1. $A(\omega = 0) = 1$
2. $\frac{\partial^{2i}}{\partial^{2i}\omega} A(\omega = 0) = 0 \quad \text{for } i = 1, 2, \ldots, M-K$
3. $\frac{\partial^{2i}}{\partial^{2i}\omega} A(\omega = \pi) = 0 \quad \text{for } i = 0, 1, \ldots, K-1$

The odd indexed derivatives of $A(\omega)$ are automatically zero at $\omega = 0$, so they do not need to be specified. The solution has the property that $A^{(i)}(\omega = 0) = 0$ for $i = 1, \ldots, 2(M-K)+1$ and $A^{(i)}(\omega = \pi) = 0$ for $i = 1, \ldots, 2K-1$. These equations are linear in the unknown filter coefficients; however, they are ill-conditioned. Fortunately, the solution can be written in closed form in several ways [55,56].

It is convenient to use the transformation $x = \frac{1}{2}(1 - \cos\omega)$, then the solution can be written [55] as

$$A(x) = (1-x)^K \sum_{n=0}^{M-K} d(n)x^n, \tag{11.78}$$

where

$$d(n) = \binom{K-1+n}{n} = \frac{(K-1+n)!}{(K-1)!n!}. \tag{11.79}$$

The transfer function has $2K$ zeros at $z = -1$, and these are the only stopband zeros. The zeros not lying at $z = -1$ can be found by computing the roots of $\sum_{n=0}^{M-K} d(n)x^n$ and mapping them back to the z domain via $z = 1 - 2x \pm \sqrt{(2x-1) - 1}$. This equation is understood by writing $\cos\omega$ as $\frac{1}{2}(e^{j\omega} + e^{-j\omega})$ and, in turn, as $\frac{1}{2}(z + \frac{1}{z})$.

For the special case $2K = M + 1$, the polynomial $A(x)$ in Equation 11.78 has become famous for its role in Daubechies' construction of compactly supported orthogonal wavelets [57].

Given a desired cutoff frequency and transition width, design formulas have been found [55,58] that give approximate values for N and K. In particular, Kaiser reported that the filter length is approximately inversely proportional to the square of the transition width: $M \approx (\frac{\pi}{\omega_b - \omega_a})^2$ where ω_b is that frequency at which $A(\omega) = 0.05$ and ω_a is that frequency at which $A(\omega) = 0.95$. Accordingly, halving the width of the transition band requires increasing the filter length by roughly a factor of 4.

Because the filter has $2K$ zeros at $z = -1$ the number of multiplications can be reduced by extracting the factor $(\frac{1+z^{-1}}{2})^{2K}$ as is indicated in Equation 11.78. (This factor can be implemented without multiplications.) The large dynamic range of $d(n)$ can be avoided by using the structure suggested by Vaidyanathan [52] that uses the observation $d(n) = \frac{K+n-1}{n} d(n-1)$. A multiplierless implementation based on the De Casteljau algorithm is described in [51].

The formulas above permit only an approximate specification of the cutoff frequency—the only parameters the user controls is N and K. For $N = 21$, Figure 11.24 illustrates the filters obtained by letting $K = 5$ and $K = 6$. Call them $h_1(n)$ and $h_2(n)$. To obtain a maximally flat symmetric filter having a half-magnitude frequency* ω_o between those of h_1 and h_2, a weighted average of h_1 and h_2 can be

* The half-magnitude frequency ω_o is that frequency such that $A(\omega_o) = \frac{1}{2}$.

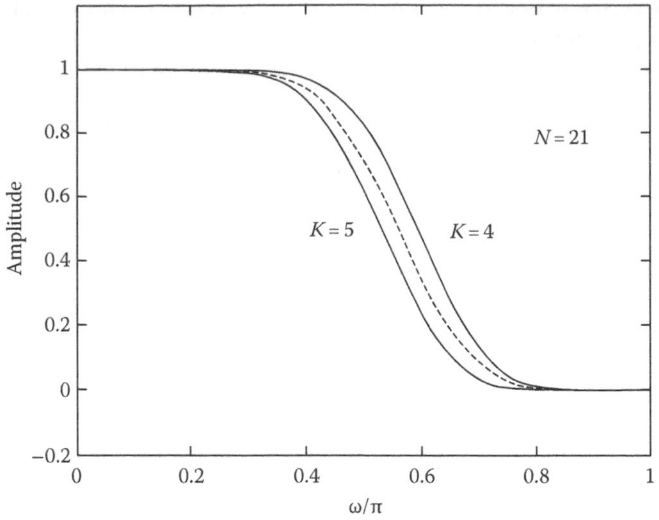

FIGURE 11.24 Three maximally flat filters: $N = 21$.

used [59,60]. The desired filter is $h(n) = c \cdot h_1(n) + (1 - c) \cdot h_2(n)$ where $c = [0.5 - H_2(\omega_o)]/[H_1(\omega_o) - H_2(\omega_o)]$. For $\omega_o = 0.56\pi$, the response of the new filter $h(n)$ is shown as a dashed line in Figure 11.24.

Remarks

1. Extremely good at $\omega = 0$ and $\omega = \pi$
2. Simple design
3. Efficient implementations
4. Smooth impulse response
5. Wide transition

11.4.1.2 Affine Filter Structure

It is frequently useful to employ the structure shown in Figure 11.25, the transfer function of which is

$$H(z) = H_1(z)H_2(z) + H_3(z). \tag{11.80}$$

In many cases, $H_2(z)$ and $H_3(z)$ are already known or determined, and it is desired that $H_1(z)$ be designed so that the overall transfer function approximates a desired transfer function $D(z)$ according to some chosen criteria.

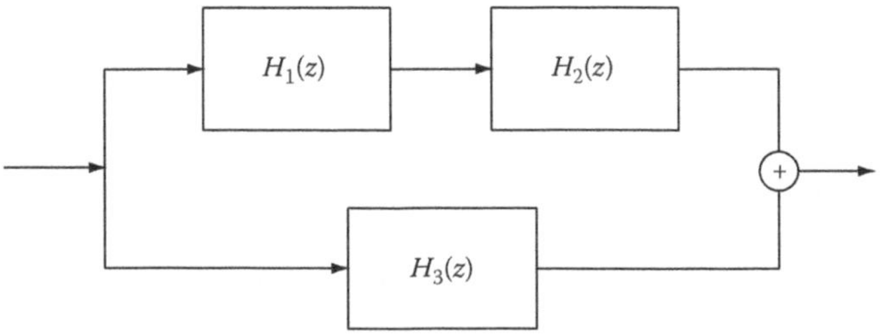

FIGURE 11.25 Affine filter structure.

Note that (1) if h_1, h_2, and h_3 are each symmetric, (2) if $h_1 * h_2$ has the same type of symmetry as h_3, and (3) if $h_1 * h_2$ and h_3 are of the same length, then the filter Equation 11.80 is itself symmetric. In this case, designing $H_1(z)$ by minimizing either the weighted square error or the weighted Chebyshev error is particularly straightforward. An equivalent problem is obtained as follows, having a modified desired function and a modified weighting function.

Let the amplitudes of the filters be $A_1(\omega)$, $A_2(\omega)$, and $A_3(\omega)$, where $A_1(\omega) = Q(\omega)P(\omega)$ and $P(\omega)$ is a cosine polynomial as in Table 11.1. Then $A(\omega) = Q(\omega)P(\omega)A_2(\omega) + A_3(\omega)$. First consider the design via the Chebyshev norm:

$$\|E(\omega)\|_\infty = \max_\omega |\bar{W}(\omega)[P(\omega) - \bar{D}(\omega)]| \tag{11.81}$$

$$\bar{W}(\omega) = W(\omega)Q(\omega)A_2(\omega) \quad \bar{D}(\omega) = \frac{D(\omega) - A_3(\omega)}{Q(\omega)A_2(\omega)}. \tag{11.82}$$

The minimization of Equation 11.81 can be accomplished by the PM algorithm or by linear programming if it is required that additional linear constraints be satisfied.

For the least squares error,

$$\|E(\omega)\|_2 = \left\{ \frac{1}{\pi} \int_0^\pi \bar{W}(\omega)[P(\omega) - \bar{D}(\omega)]^2 d\omega \right\}^{\frac{1}{2}}, \tag{11.83}$$

where

$$\bar{W}(\omega) = W(\omega)[Q(\omega)A_2(\omega)]^2 \quad \text{and} \quad \bar{D}(\omega) = \frac{D(\omega) - A_3(\omega)}{(Q(\omega)A_2(\omega))^2}. \tag{11.84}$$

The minimization of Equation 11.83 can be accomplished by solving the linear system Equation 11.33, or Equation 11.34 if it is required that additional linear constraints be satisfied.

In some design problems, the form of Equation 11.83 is useful because it describes a parameterization (or constraint) where $H_1(z)$ represents the available degrees of freedom [61–63].

11.4.1.2.1 Prefilters

In addition, the design of filters having low implementation complexity often employs the structure in Figure 11.25. One strategy is to choose transfer functions $H_2(z)$, $H_3(z)$, having very low implementation complexity—such filters may have crude frequency responses, but they can often be implemented without multipliers and few additions. $H_1(z)$ is then designed so that the overall transfer function meets the specified requirements.

This approach, introduced in [64], is often called "prefiltering," especially when $H_3(z) = 0$. In this case, $H_2(z)$ is the prefilter. Prefilters are filters having (1) very low implementation complexity, but (2) imperfect frequency responses. In this case, $H_1(z)$ is sometimes called an equalizer. In [64], it is shown that this approach provides benefits in (1) reduced computational complexity, (2) reduced sensitivity to coefficient quantization, and (3) reduced roundoff noise. For narrowband filters, this approach gives a particularly good reduction in implementation complexity.

One class of prefilters [64,65] is obtained by combining recursive running sum (RRS) building blocks.* The RRS filter is simple to implement and has all its zeros equally spaced on the U.C. (except at $z = 1$). Other prefilters are obtained from cyclotomic polynomials (CPs) [66]—all the roots of which lie on the

* Based on the factorization $\sum_{k=0}^{L-1} z^k = (z^L - 1)/(z - 1)$, the RRS filter is a recursive implementation of the running sum.

U.C. Because all the coefficients are simple small integers (the first 105 CPs have coefficients in $\{-1, 0, 1\}$), CPs can be implemented as filters without requiring multipliers. In [67], it is shown that the problem of designing prefilters from CPs can be formulated as an optimization problem with linear objective functions by applying the logarithm to the transfer function of the CP prefilter. The design problem is then solved in [67] by mixed integer linear programming.

11.4.1.2.2 IFIR Filters

Another useful structure has the transfer function $H_1(z^M)H_2(z)$[54]. The impulse response of $H_1(z^M)$ is sparse, so arithmetic complexity is reduced. A time-domain interpretation emerges by considering the convolution of $h_1(Mn)$ and $h_2(n) \cdot h_2(n)$ fills in, or interpolates, the gaps in $h_1(Mn)$. This structure is particularly well suited for efficient implementations of narrow band lowpass filters. For other frequency responses, the generalization is masking (see, e.g., [17]).

11.4.1.3 Nonsymmetric or Nonlinear Phase FIR Filter Design

Although the requirement that an FIR filter be real and symmetric simplifies the filter approximation problem, it is sometimes more restrictive than is desirable. The following scenarios motivate the consideration of nonsymmetric and/or nonlinear phase FIR filters:

1. In some cases, phase linearity is of little importance and it is more important that the delay be low. Recall that the group delay of a symmetric filter is half its filter order. This delay is higher than necessary. In other cases, exactly linear phase is not required, but some degree of phase linearity is desired. It is then desirable to sacrifice exactly linear phase in exchange for delay reduction and/or delay control. The desired constant delay can be specified by explicitly including the phase or desired group delay as part of the design specifications as indicated in the following subsection on optimal design of FIR filters. The resulting designed nearly linear-phase filter has a conjugate symmetric frequency response and a real-valued, nonsymmetric, impulse response (see Section 11.4.1.4.5).
2. Sometimes it is required that $H(\omega)$ approximate a desired nonsymmetric or nonlinear phase frequency response $D(\omega)$.* Examples include equalizer design [68], fractional delay filter design [21], and seismic migration filter design [2].

In each case, the additional degrees of freedom that are made available by giving up symmetry or phase linearity can be used to improve the phase and/or magnitude response.

Approaches to the design of nonsymmetric and/or nonlinear phase FIR filters fall roughly into at least three categories:

1. General complex approximation (see Section 11.4.1.4). Given an arbitrary desired frequency response $D(\omega)$, the best Chebyshev, or least square, approximation is found. For the Chebyshev criterion, the approximation is significantly more difficult in the general complex case than in the real symmetric case. Recently, several algorithms have been presented for designing general filters in the Chebyshev sense [2–5,69,141,143].
2. Design of minimum-phase filters by spectral factorization of square magnitude approximation [70]. This is a very effective technique, and it can be used in conjunction with the maximally flat, least square, and Chebyshev criterion.
3. The simultaneous approximation of magnitude and group delay. There is little theory to facilitate the solution to this nonlinear problem, but see [71–75,142] and Section 11.4.1.6.

* Note that the frequency response of a filter can be symmetric with a *nonlinear* phase (e.g., seismic migration filters designed in Section 11.4.1.4.5.3).

11.4.1.4 Optimal Design of FIR Filters with Arbitrary Magnitude and Phase

As indicated before, the alternation theorem [76] is at the basis of the PM (second Remez exchange) algorithm described in Section 11.3.1. Karam and McClellan recently extended the alternation theorem from the real-only to the general complex case [2]. As a result, they derived an efficient multiple-exchange algorithm [3,10] for the design of optimal FIR filters with arbitrary magnitude and phase specifications approximated in the Chebyshev sense. Both causal and noncausal filters with complex or real-valued impulse responses can be designed. In addition, the Karam–McClellan algorithm exactly reduces to the classic PM (second Remez exchange) algorithm when real-only or imaginary-only filters are designed and is, therefore, a true generalization of the classic Remez algorithm to the complex case. A version of the Karam–McClellan algorithm (**cremez**) is currently available as part of the Signal Processing Toolbox in MATLAB (Version 5).

11.4.1.4.1 Problem Formulation

The complex FIR filter design problem may be stated as follows. Let $D(\omega)$ be the desired magnitude and phase of the filter frequency response defined on a compact frequency subset $B \subset [-\pi, \pi)$. $D(\omega)$ is to be approximated by an FIR filter having a frequency response $H(\omega)$ and an impulse response h_n, $n = N_1, \ldots, N_2$, of length $N = N_2 - N_1 + 1$. The filter design problem consists in finding the filter coefficients $\{h_n\}$ that will minimize the Chebyshev error norm:

$$\|E(\omega)\| = \max_{\omega \in B} \{|D(\omega) - H(\omega)|\}, \tag{11.85}$$

where

$$H(\omega) = \sum_{n=N_1}^{N_2} h_n e^{-j\omega n}. \tag{11.86}$$

The error norm (Equation 11.85) can include a real, strictly positive, and continuous weighting function $W(\omega)$ on B by simply replacing $D(\omega)$ with $W(\omega)D(\omega)$ and $H(\omega)$ with $W(\omega)H(\omega)$.

Note that this formulation will handle both causal filters ($N_1 \geq 0$) and noncausal filters ($N_1 < 0$). Although some authors [77] have reported an ill-conditioned behavior when using Equation 11.86, the error (Equation 11.85) can be rewritten so that the problem is well posed by removing a linear phase term due to N_1. This new problem, with a guaranteed unique optimal solution, results by rewriting $D(\omega)$ and $H(\omega)$ with respect to a linear phase term as

$$D(\omega) = e^{-j\frac{N_1+N_2}{2}\omega} A(\omega) \tag{11.87}$$

and

$$H(\omega) = e^{-j\frac{N_1+N_2}{2}\omega} H_{nc}(\omega). \tag{11.88}$$

The linear phase $e^{-j\frac{N_1+N_2}{2}\omega}$ does not affect the magnitude of the error (Equation 11.85); so the design problem works with the following equivalent expression for the error magnitude:

$$|E(\omega)| = |A(\omega) - H_{nc}(\omega)|. \tag{11.89}$$

The function $H_{nc}(\omega)$ can be expressed as a linear combination of real basis functions satisfying the Haar condition [2,78]:

$$H_{nc}(\omega) = \begin{cases} \sum_{k=0}^{(N-1)/2} \alpha_k \cos k\omega + \sum_{k=1}^{(N-1)/2} \beta_k \sin k\omega, & N \text{ odd} \\ \sum_{k=0}^{(N-2)/2} \left[\alpha_k \cos\left(k + \frac{1}{2}\right)\omega + \beta_k \sin\left(k + \frac{1}{2}\right)\omega \right], & N \text{ even.} \end{cases} \quad (11.90)$$

The Haar condition [76,79], which is satisfied by the cos() and sin() basis functions, guarantees that the optimal solution is unique and that the set of extremal points of the optimal error function, $E_o(\omega)$, consists of at least $n+1$ points, where n is the number of approximating basis functions.

The parameters $\{\alpha_k, \beta_k\}$ in Equation 11.90 are the complex coefficients that need to be determined such that $H_{nc}(\omega)$ best approximates $A(\omega)$. The filter coefficients $\{h_n\}$ can be very easily obtained from $\{\alpha_k, \beta_k\}$ [78]. Usually, the number of approximating basis functions in Equation 11.90 is $n = N$, but this number is reduced by half when $A(\omega)$ is symmetric (all $\{\beta_k\}$ are equal to 0), or antisymmetric (all $\{\alpha_k\}$ are equal to 0).

11.4.1.4.2 Design Algorithm

A main strategy in Chebyshev approximation is to work on sparse finite subsets, B_s, of the desired frequency set B and relate the optimal error on B_s to the optimal error on B. The norm of the optimal error on B_s will always be a lower bound to the error norm on B [79]. If $\|E_s\|$ denotes the optimal error norm on the sparse set B_s, and $\|E_o\|$ the optimal error norm on B, the design problem on B is solved by finding the subset B_s on which $\|E_s\|$ is maximal and equal to its upper bound $\|E_o\|$. This could be done by iteratively constructing new subsets B_s with monotonically increasing error norms $\|E_s\|$. For that purpose, two main issues must be addressed in developing the approximation algorithm:

1. Finding an efficient way to compute the best approximation $H_s(\omega)$ on a given subset B_s of r points ($r \geq n+1$).
2. Devising a simple strategy to construct a new subset B_s where the optimal error norm $\|E_s\|$ is guaranteed to increase.

While in the real case it is sufficient to consider subsets containing $r = n+1$ points, the minimal subset size r is not known *a priori* in the complex case. The fundamental theorem of complex Chebyshev approximation tells us that r can take any value between $n+1$ and $2n+1$. It is desirable, whenever possible, to keep the size of the subsets, B_s, small since the computational complexity increases with the size of B_s. The case where $r = n+1$ points is important because, in that case, it was shown [2] that the best approximation on a subset of $n+1$ points can be simply computed by solving a linear system of equations. So, the first issue is directly resolved.

In addition, by exploiting the alternation property* of the complex optimal error on B_s efficient multi-point exchange rules can be derived and the second issue is easily resolved. These exchange rules were derived in [2,78] resulting in the very efficient complex Remez algorithm which iteratively constructs best approximations on subsets of $n+1$ points with monotonically increasing error norms $\|E_s\|$.

The complex Remez algorithm terminates when finding the set B_s having the largest error norm ($\|E_s\| = |\delta|$) among all subsets consisting of exactly $n+1$ points. This complex Remez multiple-exchange algorithm converges to the optimal Chebyshev solution on B when the optimal error $E_o(\omega)$ satisfies an alternating property [78]. Otherwise, the computed solution is optimal over a reduced set $B' \subset B$. In this latter case, the maximal error norm $|\delta|$ over the sets of $n+1$ points is strictly less than, but usually very

* *Alternation* in the complex case corresponds to a phase shift of π when going from one extremal point to the next in sequence.

close to, the upper bound $\|E_o\|$. To compute the optimum over B, subsets consisting of more than $n+1$ ($r > n+1$) need to be considered. Such sets are constructed by the second stage of the new algorithm presented in [3,10], starting with the solution generated by the initial complex Remez stage.

When $r > n+1$, both issues mentioned above are much harder to resolve. In particular, a simple and efficient point-exchange strategy, where the size of B_s is kept minimal and constant, does not seem possible when $r > n+1$. The approach in [3,10] is to use a second ascent stage for constructing a sequence of best approximations on subsets of r points ($r > n+1$) with monotonically increasing error norms (ascent strategy). The algorithm starts with the best approximation on subsets of $n+1$ points (minimum possible size) using the very efficient complex Remez algorithm [2] and then continues constructing the sequence of best approximations with increasing error norms on subsets of B_s more than $n+1$ points by means of a second stage. Since the continuous domain B is represented by a dense set of discrete points, the proposed design algorithm must yield an approximation of maximum norm in a finite number of iterations since there is a finite number of distinct subsets B_s containing $r(n+1 \leq r \leq 2n+1)$ points in the discrete set B.

A detailed block diagram of the design algorithm is shown in Figure 11.26. The two stages of the new algorithm have the same basic ascent structure. They both consist of the two main steps shown in Figure 11.26, and they only differ in the way these steps are implemented.

A detailed block diagram of the complex Remez stage (Stage 1) is also shown in Figure 11.27. Note that when $D(\omega)$ is real-valued, δ will also be real and, therefore, the real phase-rotated error $E_r(\omega)$ is equal to $\pm E(\omega)$. In this case, the presented algorithm reduces to the PM algorithm as modified by McCallig [80] for approximating general real-valued frequency responses in the Chebyshev sense. Moreover, for many problems, the resulting initial approximation computed by the complex Remez method is the optimal Chebyshev solution and, thus, the second stage of the algorithm does not need to execute. Even when the resulting initial solution is not optimal, it has been observed that the computed deviation $|\delta|$ is very close to the optimal error norm $\|E_o\|$ (its upper bound).

As indicated above, the second stage is invoked only when the complex Remez stage (Stage 1) results in a subset optimal solution. In this case, the initial set B_s of Stage 2 is formed by taking the set of all local maxima of the error corresponding to the final solution computed by Stage 1. The resulting $B_s \subset B$ would then contain r points, where $n+1 < r \leq 2n+1$. The best approximation on the constructed subset, B_s, is computed by means of a generalized descent method [10,78] suitably adapted for minimizing the nondifferentiable Chebyshev error norm. The total number of ascent iterations is independent of the method used for computing the best solution $H_s(\omega)$ on B_s. Then, the new sets, B_s, are constructed by locating and adding the new local maxima of the error on B to the current subset, B_s, and by removing from B_s those points where the error magnitude is relatively small. So, the size of the constructed subsets varies up and down. The algorithm terminates when all the extremal points of $E(\omega)$ are in B_s.

It should be noted that each iteration of Stage 2 includes descent iterations, which we will refer to as descent steps.* An observation in relation to the complexity of the two stages of the algorithm is in order. The initial complex Remez stage is extremely efficient and does not produce any significant overhead. However, one iteration of the second stage includes several descent steps, each one having higher computational complexity than the initial complex Remez stage. For convenience, the term major iterations will be used to refer to the iterations of the second stage. From the discussion above, it follows that the initial complex Remez stage is comparable to one step in a major iteration and can thus be regarded as an initialization step in the first major iteration.

An interesting analogy of the proposed two-stage algorithm with the first and second algorithms of Remez can be made. It should be noted that both Remez algorithms can be used for solving real 1-D Chebyshev approximation problems satisfying the Haar condition. The two real Remez algorithms involve the solution of a sequence of discrete problems [81]: at each iteration, a finite discrete subset, B_s, is defined and the best Chebyshev approximation is computed on B_s. In the second algorithm of

* The simplex method of linear programming could also be used for the descent steps.

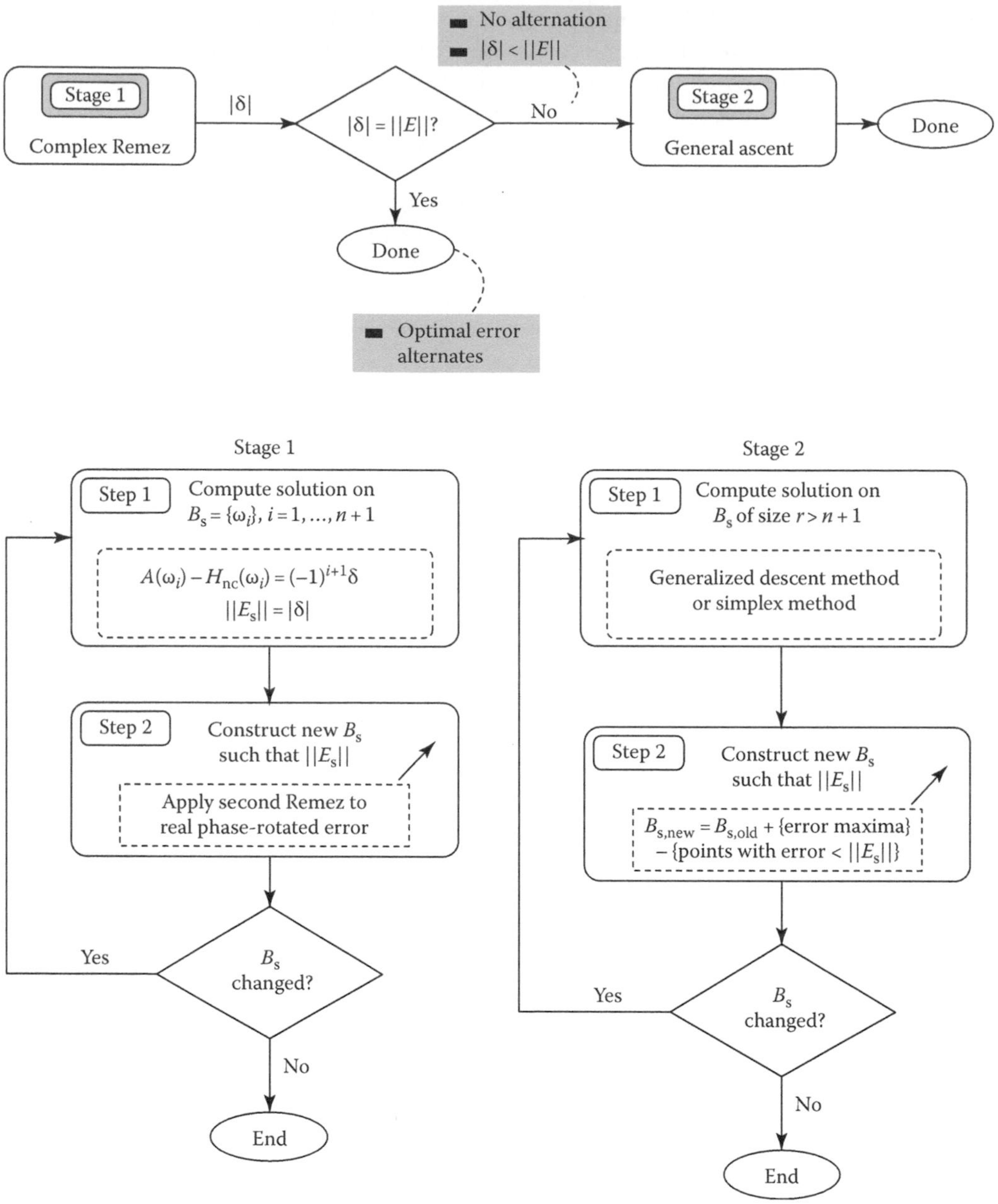

FIGURE 11.26 Block diagram of the Karam–McClellan design algorithm. $|\delta|$ is the maximal optimal deviation on the sets B_s consisting of $n+1$ points in B. $||E||$ is the Chebyshev error norm on B.

Remez, the successive subsets B_s contain exactly $n+1$ points: an initial subset of $n+1$ points is replaced by $n+1$ local maxima of the current real error function. In the first algorithm of Remez, the initial point set contains at least $n+1$ points, and these points are supplemented at each iteration by the global maximum of the current approximation error. As shown in [2], the complex Remez stage (Stage 1) of the new proposed algorithm is a generalization of the second Remez algorithm to the complex case and reduces to it when real-valued or pure imaginary functions are approximated. On the other hand, the second stage of the proposed algorithm can be compared to the first Remez algorithm in that the size of

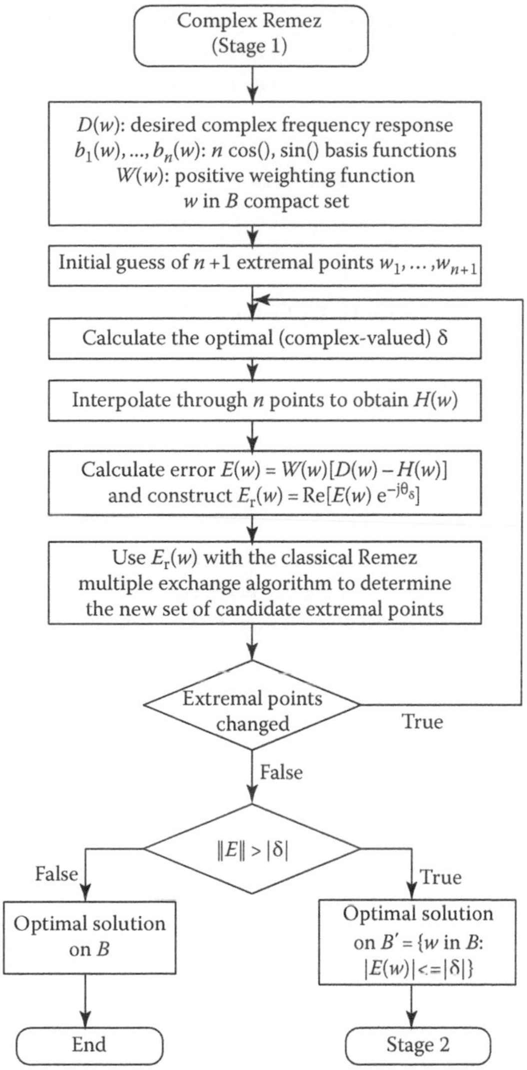

FIGURE 11.27 Block diagram of the complex Remez (Stage 1) algorithm.

the constructed subsets B_s is variable and is greater than $n+1$, except at the initial iteration. A main difference between the second stage and the first Remez algorithm is that the second stage is based on a multiple-exchange strategy while the first algorithm of Remez is a single-exchange method.

11.4.1.4.3 Descent Steps

In what follows, we describe the generalized descent method and the simplex method which can be used in Step 1 of Stage 2 to compute the optimal Chebyshev solution on the discrete set of points B_s. The descent method presented in this section is based on the work of Demjanov–Malozemov [82,83] and Wolfe [84], and is suitably adapted for minimizing the nondifferentiable Chebyshev error norm.

Let $D(\omega)$ be the function that is to be approximated on B_s, and let $H_{s,0}(\omega)$ be an initial approximation given by the basis coefficient vector

$$\mathbf{c}_0 = [c_{01}, c_{02}, \ldots, c_{0n}]^T \tag{11.91}$$

whose elements are the n (complex or real) coefficients associated with the cos() and/or sin() basis functions $\{\phi_i\}_{i=0}^{n}$. The superscript T in Equation 11.91 refers to the transpose operation. The descent method iteratively generates a sequence $\{\mathbf{c}_k\}$ of basis coefficient vectors, $\{\mathbf{d}_k\}$ of perturbation vectors, and $\{t_k\}$ of positive scalars such that

$$\mathbf{c}_{k+1} = \mathbf{c}_k + t_k\mathbf{d}_k \tag{11.92}$$

and

$$\|E_{s,k+1}(\omega)\| \leq \|E_{s,k}(\omega)\| \quad \text{for } \omega \in B_s, \tag{11.93}$$

where $E_{s,k}(\omega)$ is the approximation error

$$E_{s,k}(\omega) = D(\omega) - H_{s,k}(\omega) = D(\omega) - \sum_{i=1}^{n} c_{ki}\phi_i(\omega) \tag{11.94}$$

and k is the iteration number. The perturbation vectors $\{\mathbf{d}_k\}$ correspond to descent directions and $\{t_k\}$ must be chosen so that $\|E_k(\omega)\|$ would significantly decrease at the next iteration. Once $\mathbf{d}_k$ is chosen, a line search method could be used to find the optimal t_k for a maximum decrease of $\|E_{s,k}(\omega)\|$ along the direction $\mathbf{d}_k$. Alternatively, a more efficient procedure for finding the best t_k was presented in [83, pp. 109–112]. Standard gradient techniques cannot be used in this case for generating the directions $\{\mathbf{d}_k\}$ since the Chebyshev error norm is a nondifferentiable function of the coefficient vector $\mathbf{c}$.

With r denoting the number of points in B_s, the Chebyshev approximation problem can be reformulated as the minimization of the function

$$\phi(\mathbf{c}) = \max_{i\in(1,\ldots,r)} e_i(\mathbf{c}), \tag{11.95}$$

where

$$e_i(\mathbf{c}) = \left|D(\omega_i) - \Phi_i^{\mathrm{T}}\mathbf{c}\right|^2 \tag{11.96}$$

and

$$\Phi_i = [\phi_1(\omega_i), \phi_2(\omega_i), \ldots, \phi_n(\omega_i)]^{\mathrm{T}}. \tag{11.97}$$

Each $e_i(\mathbf{c})$ is a convex differentiable function with a complex gradient vector g_i given by

$$g_i = \frac{\partial e_i(\mathbf{c})}{\partial \mathbf{c}} = -2\bar{\Phi}_i E_i, \tag{11.98}$$

where $\bar{\Phi}_i$ is the complex conjugate of Φ_i, and $E_i = D(\omega_i) - \Phi_i^{\mathrm{T}}\mathbf{c}$. Note that g_i is a vector in the n-dimensional complex space Z_n which is isomorphic to the $2n$-dimensional real Euclidean space R_{2n}. A point $z = (z_1, \ldots, z_n) \in Z_n$, with complex coordinates $z_j = \alpha_j + j\beta_j$, corresponds to the point $z = (\alpha_1, \ldots, \alpha_n, \beta_1, \ldots, \beta_n) \in R_{2n}$. In what follows, g_i refers to the real vector in R_{2n}.

For a given coefficient vector $\mathbf{c}$, consider the set of extremal indices $I_e(\mathbf{c})$ defined as

$$I_e(\mathbf{c}) = \{i \in (1, \ldots, r) : e_i(\mathbf{c}) = \varphi(\mathbf{c})\}. \tag{11.99}$$

In other words, $I_e(\mathbf{c})$ contains every index i (corresponding to the ith point ω_i in B_s) for which $E(\omega)$ attains its maximum on B_s. Letting

$$G(\mathbf{c}) = \{g_i : i \in I_e(\mathbf{c})\}, \tag{11.100}$$

consider the convex hull $G_c(\mathbf{c})$ of $G(\mathbf{c}) \cdot G_c(\mathbf{c})$ is a polyhedron in R_{2n} and there is a unique point $g_{\min} \in G_c(\mathbf{c})$ having minimum Euclidean norm [85]. The following gradient characterization results for $\varphi(\mathbf{c})$ [82,85]:

$$\nabla\varphi(\mathbf{c}) = g_{\min} \tag{11.101}$$

and $-g_{\min}$ is the direction of steepest descent at $\mathbf{c}$. Note that $\nabla\varphi(\mathbf{c})$ depends only on the set of extremal points represented by $I_e(\mathbf{c})$. So, the problem of finding the steepest descent direction reduces to the problem of finding the point of smallest norm in the convex hull of a given finite point set. An algorithm especially designed for that calculation has been presented by Wolfe [84]. The filter coefficient vector $\mathbf{c}_o$ minimizes $\varphi(\mathbf{c})$, and therefore the approximation error norm $\|E_s\|$, if and only if

$$\nabla\varphi(\mathbf{c}_o) = 0 \tag{11.102}$$

or, equivalently (see Equation 11.98),

$$0 \in G_c(\mathbf{c}_o). \tag{11.103}$$

Using Equation 11.98, it can be shown that the optimality condition (Equation 11.103) reduces to the Kolmogoroff optimality criterion for Chebyshev approximation [86, p. 21].

While a direct generalization of the steepest descent method does not in general lead to convergence [82,85], successive approximation and conjugate subgradient methods based on Equation 11.101 have been developed for minimizing nondifferentiable functions [83,85,87]. The descent method presented in this section is based on the techniques presented in [83,84]. It is suitably adapted for solving the Chebyshev approximation problem, which was reformulated as Equations 11.95 through 11.97, and, consequently, for solving the filter design problem. Before describing the steps of the proposed descent method, some new definitions are needed. Define

$$I_{e,\varepsilon}(\mathbf{c}) = \{i \in (1,\ldots,r) : \varphi(\mathbf{c}) - e_i(\mathbf{c}) \leq \varepsilon\}, \quad \varepsilon \geq 0 \tag{11.104}$$

and

$$G_\varepsilon(\mathbf{c}) = \{g_i : i \in I_{e,\varepsilon}(\mathbf{c})\}. \tag{11.105}$$

Also, let $G_{c,\varepsilon}(\mathbf{c})$ denote the convex hull of $G_\varepsilon(\mathbf{c})$ and $g_{\min,\varepsilon}$ the point in $G_{c,\varepsilon}(\mathbf{c})$ nearest to the origin. Clearly, $I_{e,0}(\mathbf{c})$, $I_e(\mathbf{c})$, $G_0(\mathbf{c}) = G(\mathbf{c})$, $G_{c,0}(\mathbf{c}) = G_c(\mathbf{c})$, and $g_{\min,0} = g_{\min}$.

The basic steps of the descent algorithm can now be summarized as follows:

1. **Set initial parameters.** Fix two parameters $\varepsilon_0 > 0$ and $\rho_0 > 0$, and take an initial approximation $\mathbf{c}_0$ on the desired set B_s, i.e., $\phi_{s,0}(x) = \sum_{i=1}^{n} c_{0i}\phi_i(x)$. Suggested values for ε_0 and ρ_0 are $\varepsilon_0 = 0.012$ and $\rho_0 = 1.0$. Since the passage from $\mathbf{c}_k$ to $\mathbf{c}_{k+1}$ $(k = 0, 1, \ldots)$ is effected the same way, suppose that the kth approximation $\mathbf{c}_k$ is already computed.
2. **Set current approximation and accuracy.** Set $\mathbf{c} = \mathbf{c}_k$, $\varepsilon = \varepsilon_0/2^k$, and $\rho = \rho_0/2^k$.
3. **Compute the ε-gradient, $g_{\min,\varepsilon}$.** Find the point $g_{\min,\varepsilon}$ of $G_{c,\varepsilon}(\mathbf{c})$ nearest to the origin using the technique by Wolfe [84].

4. **Check accuracy of current approximation.** If $||g_{\min,\varepsilon}|| \leq \rho$, go to Step 8.
5. **Compute the ε-steepest descent direction $\mathbf{d}_k$:**

$$\mathbf{d}_k = -\frac{g_{\min,\varepsilon}}{||g_{\min,\varepsilon}||}. \tag{11.106}$$

6. **Determine the best step size t_k.** Consider the ray

$$\mathbf{c}(t) = \mathbf{c} + t\mathbf{d}_k \tag{11.107}$$

and determine $t_k \geq 0$ such that

$$\varphi[\mathbf{c}(t_k)] = \min_{t \geq 0} \varphi[\mathbf{c}(t)]. \tag{11.108}$$

7. **Refine approximation accuracy.** Set $\mathbf{c} = \mathbf{c}(t_k)$ and repeat from Step 3.
8. **Compute generalized gradient, $g_{\min}$.** The technique by Wolfe [84] is used to find the point $g_{\min}$ of $G_c(\mathbf{c}_k)$ nearest to the origin (see also [83, Appendix IV]).
9. **Check stopping criteria.** If $g_{\min} \equiv$, then $\mathbf{c}$ is the vector of the coefficients of the best approximation $H_s(\omega)$ of the function $D(\omega)$ on $B_s = \{\omega_i : i = 1, \ldots, r\}$ and the algorithm terminates.
10. **Update approximation and repeat with higher accuracy.** The approximation $\mathbf{c}_{k+1}$ is now given by

$$\mathbf{c}_{k+1} = \mathbf{c}. \tag{11.109}$$

Return to Step 2.
This successive approximation descent method is guaranteed to converge, as shown in [83].

11.4.1.4.4 Descent via the Simplex Method

Other general optimization techniques (e.g., the simplex method of linear programming [4,88]) can also be used instead of the descent method in the second stage of the proposed algorithm. The advantage of the linear-programming method over the generalized descent method is that additional linear constraints can be incorporated into the design problem.

Using the real rotation theorem [11, p. 122],

$$|z| = \max_{-\pi \leq \theta < \pi} \text{Re}\{ze^{j\theta}\}, \quad \text{where } z \text{ complex}, \tag{11.110}$$

the complex filter design problem on the frequency set B_s can be restated as the following linear approximation problem: find the optimal length-N impulse response $\mathbf{h}^* = [h_{N_1} \cdots h_{N_2}]^*$ such that

$$\delta(\mathbf{h}^*) = \min_{\mathbf{h}} \delta(\mathbf{h}), \tag{11.111}$$

where

$$\delta(\mathbf{h}) = \max_{\omega \in B_s} \max_{-\pi \leq \theta < \pi} \left[\text{Re}\left\{E(\omega)e^{j\theta}\right\}\right]$$

$$E(\omega) = D(\omega) - H(\omega) \quad \text{(refer to Equation 11.86).}$$

This problem can, in turn, be formulated as a linear program by defining

$$\mathbf{u} = \left[h^r_{N_1}, \ldots, h^r_{N_2}, h^i_{N_1}, \ldots, h^i_{N_1}, \delta\right]$$
$$\mathbf{k} = [0, \ldots, 0, 0, \ldots, 0, 1]$$

where $h^r_n = \text{Re}\{h_n\}$, $h^i_n = \text{Re}\{h_n\}$, and $\delta = \mathbf{uk}^{\text{T}}$. The resulting linear program becomes

$$\min_{\text{u}} \mathbf{uk}^{\text{T}} \tag{11.112}$$

subject to $\text{Re}\{E(\omega)e^{j\theta}\} \leq \delta$, for all $\omega \in B_s$ and $\theta \in [-\pi, \pi]$. Alternatively, the dual linear program can be formulated and solved [4,88].

11.4.1.4.5 Design Examples

In the following design examples, the filter specifications are given in terms of the normalized frequency $f = \omega/2\pi$.

11.4.1.4.5.1 Low Delay Filters with Nearly Linear Phase In many signal processing applications, linear-phase systems are particularly desirable because the effect of exact linear-phase is a perfect delay. While exactly linear-phase causal filters exhibit a constant group delay, the delay they introduce is proportional to the filter length N and is always equal to $(N-1)/2$. This delay may be unacceptably large, especially when using filters having a high degree of selectivity (sharp cutoff edges). Furthermore, in real-time applications (e.g., real-time speech and video processing), selective filters are required to have a constant group delay that is as small as possible. Minimum-phase FIR filters cause less delay, but introduce phase distortion which may have a severe effect on the shape of the processed signal. Chen and Parks [89] observed that the desired group delay which gives the minimum error deviation can be smaller than that of an exactly linear-phase filter of the same length.

Complex approximation can be used to design filters that have less delay than the exactly linear-phase filter of the same length, and which have approximately a constant group delay in the passband. The resulting complex filters are called "nearly linear-phase" and are obtained by defining the desired linear-phase frequency response to be

$$D(\omega) = \begin{cases} e^{-j\tau_i\omega}, & \omega \in i\text{th passband} \\ 0, & \omega \in \text{stopbands} \end{cases} \tag{11.113}$$

where τ_i is the desired group delay in the ith passband. Since the phase term is explicitly included in the approximation problem, the desired delay is fixed (τ_i) and is not determined by the FIR filter length N. Moreover, increasing N does not increase the group delay, but potentially leads to a better approximation of the desired constant delay. The definition of $D(\omega)$ should be conjugate symmetric. Then the frequency response of the optimal Chebyshev filter approximating Equation 11.113 will also be conjugate symmetric [79, p. 27] and, therefore, the approximating filter coefficients will be real-valued.

Figure 11.28 shows the properties of a reduced delay, length-32, FIR filter designed with the following desired specifications:

$$\text{Desired: } D(f) = \begin{cases} e^{-j(2\pi f)12.5} & \text{if } 0 \leq |f| \leq 0.06 \\ 0 & \text{if } 0.12 \leq |f| \leq 0.5 \end{cases}$$

$$\text{Weight: } W(f) = \begin{cases} 1 & \text{if } |f| \leq 0.06 \\ 10 & \text{if } 0.12 \leq |f| \leq 0.5 \end{cases}$$

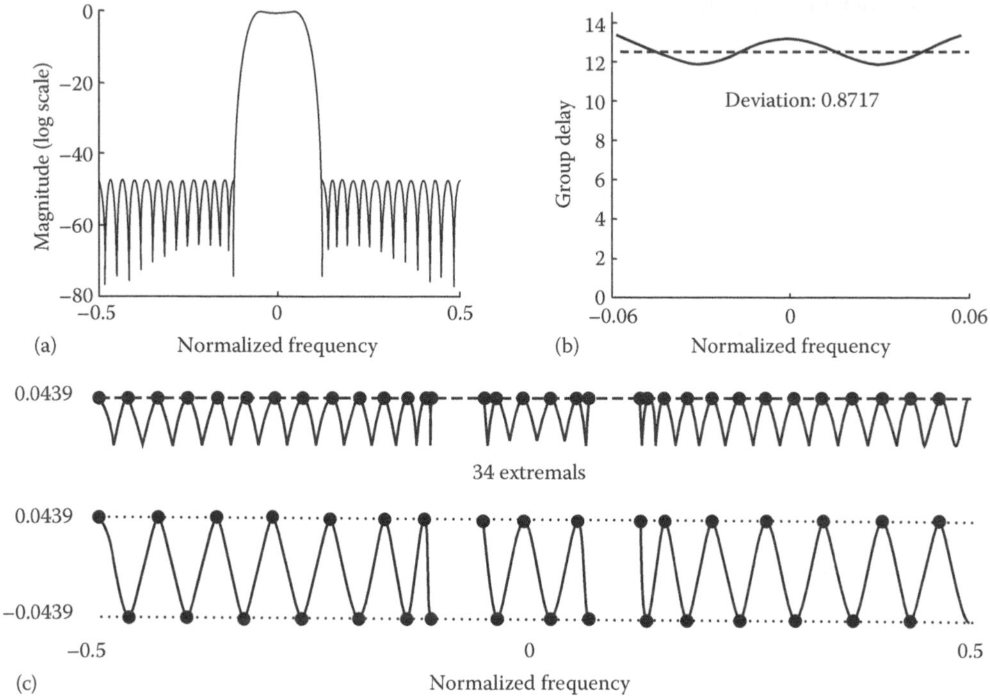

FIGURE 11.28 Example 1—nearly linear-phase filter, $N = 32$: (a) FIR filter magnitude response in decibel, (b) FIR filter (solid) and desired filter (dashed) passband group delays, and (c) magnitude of the weighted error (top) and real phase-rotated weighted error (bottom).

The term "reduced delay" refers to the fact that the desired group delay ($\tau = 12.5$) is set to be smaller than $(N-1)/2 = 15.5$, which is the delay of an exactly linear-phase filter. The complex Remez stage of the multiple-exchange algorithm converges to the optimal solution 11 in exchange steps. The resulting optimal filter has a Chebyshev error norm $\|E_{opt}\| = |\delta| = 0.0439$. The FIR filter's group delay (Figure 11.28b) corresponds to the nearly linear-phase characteristic in the passband. Note that the optimal error (Figure 11.28c) assumes its maximum value at $N+2$ extremal points with an alternating phase shift of π. This alternation can be clearly seen from the plot of the real phase-rotated error $E_r(f)$ in Figure 11.28c.

11.4.1.4.5.2 Real-Valued or Exactly Linear-Phase Filters The real-valued filter design problem corresponds to the case where the function $A(\omega)$, given by Equation 11.87, reduces to a real-valued function. In this case, the initial complex Remez stage (Stage 1 described above) always converges to the unique optimal solution of the desired function $D(\omega)$ on the specified frequency bands B.

Figure 11.29 shows the characteristics of the exactly linear-phase filter corresponding to the nearly linear-phase filter shown in Figure 11.28; i.e., the same design specifications were used except that the delay is set to be equal to $(N-1)/2 = 15.5$ in the passband, because $N = 32$ in this example. The complex Remez stage converged to the optimal solution in eight exchange steps. The resulting optimal filter has an optimal error norm $\|E_{opt}\| = |\delta| = 0.04956$ which is larger than the corresponding reduced delay, nearly linear-phase filter of Figure 11.28, which indicates that a nearly linear-phase filter not only can have less group delay but also a reduced Chebyshev error norm. Note that although the phase specifications were explicitly included as part of the approximation problem, the exactly linear-phase optimal solution was obtained as expected.

11.4.1.4.5.3 Seismic Migration Filters The objective of seismic migration is to define the boundaries of the earth layers [90, 91]. For this purpose, downward propagating waves are initiated by acoustic sources

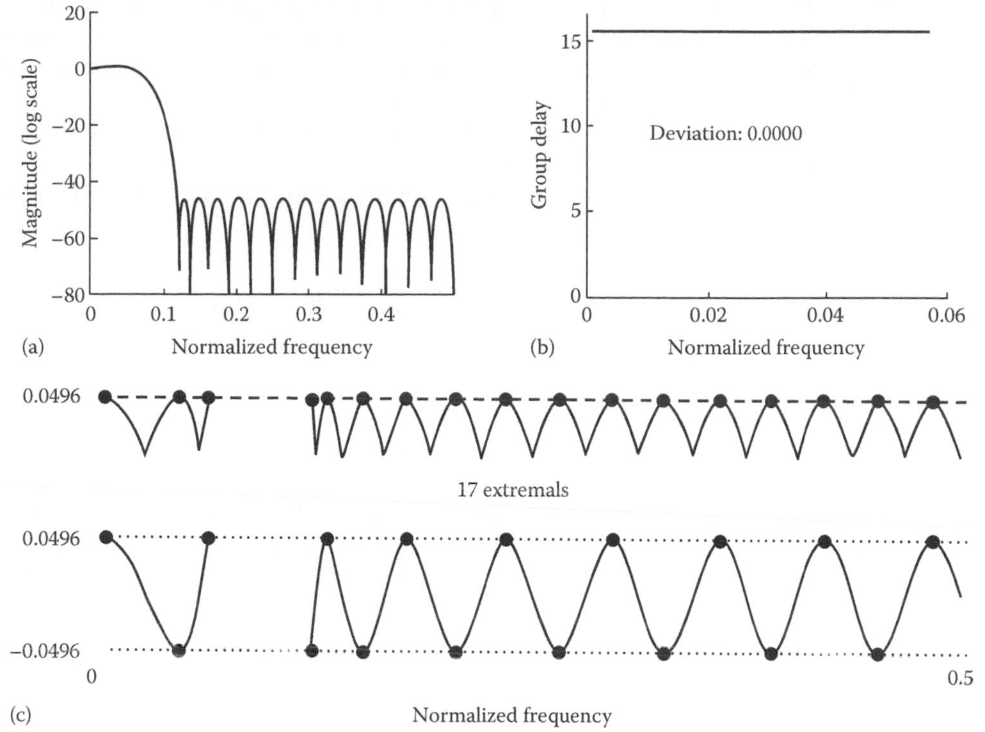

FIGURE 11.29 Example 2—exactly linear-phase lowpass filter, $N = 32$: (a) FIR filter magnitude response in decibel, (b) FIR filter (solid) and desired filter (dashed) passband group delays, (c) magnitude of the weighted error (top) and real phase-rotated weighted error (bottom).

at the earth. Then the migration procedure starts with the wave field measured at the earth's surface and computes the wave field values at all desired depths. This extrapolation operation is performed using a digital space-time filter whose frequency response approximates

$$D(k_x, \omega) = D_A\left(\frac{k_x}{\Delta x}, \frac{\omega}{\Delta t}\right) = \exp\left\{ j\frac{\Delta z}{\Delta x}\sqrt{\frac{\Delta x^2}{\Delta t^2}\frac{\omega^2}{v^2} - k_x^2} \right\}, \tag{11.114}$$

where

k_x is the spatial frequency
ω is the temporal frequency

The extrapolation in depth is usually done for a fixed frequency ω_o and a fixed velocity v_o. The process is repeated for other frequency and velocity values using frequency- and velocity-dependent migration filters. Therefore, for a fixed ratio ω_o/v_o, $D(k_x, \omega_o) = D(k_x)$ is a 1-D migration filter with a cutoff frequency equal to $\alpha = (\Delta x/\Delta t)(\omega_o/v_o)$. The objective of the migration filter design problem is to approximate the ideal frequency response $D(k_x)$. Since $D(k_x)$ is a complex-valued even-symmetric function, it can be approximated by an N-length FIR digital filter whose frequency response is given by [92]

$$H(k_x) = h_o + 2\sum_{n=1}^{(N+1)/2} h_n \cos nk_x, \tag{11.115}$$

where the filter coefficients h_n are complex-valued. Note that, even if a symmetry constraint is not imposed on the FIR filter, the resulting optimal filter will be even-symmetric when an odd-length filter is used to approximate $D(k_x)$. This property follows directly from Chebyshev approximation theory [79, p. 27]. The approximation of $D(k_x)$ needs mostly to be accurate in the region $|k_x| < |\alpha|$ (passband) which corresponds to the wavenumbers (k_x) for which waves are propagating [90,93]. The evanescent region $|k_x| > |\alpha|$ (stopband) will contain little or no energy.

Figure 11.30 displays the properties of the optimal, length-31, seismic migration filter that approximates the following specifications:

$$\text{Desired: } D(f) = \begin{cases} e^{j\sqrt{(2)^2-(2\pi f)^2}} & \text{if } |f| \leq 2\sin(75^\circ)/2\pi \\ 0 & \text{if } 2/2\pi \leq |f| \leq 0.5 \end{cases}$$

$$\text{Weight: } W(f) = \begin{cases} 500 & \text{if } |f| \leq 2\sin(75^\circ)/2\pi \\ 1 & \text{if } 2/2\pi \leq |f| \leq 0.5. \end{cases}$$

For this design example, the starting impulse response index is $N_1 = -15$. The large passband weighting is used to force an almost perfect match in the passband. In fact, for migration filters, the approximation need only be accurate in the passband as long as the stopband magnitude deviation is not larger than

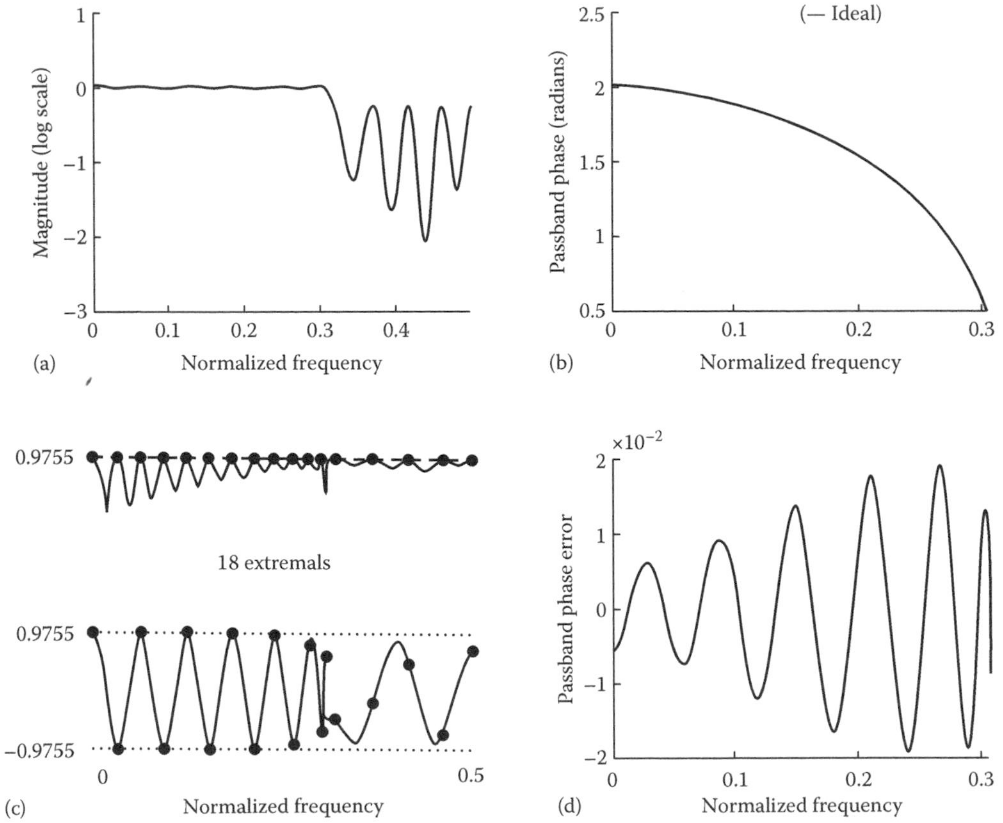

FIGURE 11.30 Example 3—seismic migration filter, $N = 31$: (a) FIR filter magnitude response in decibel, (b) FIR filter (solid) and desired filter (dashed) passband phase responses, (c) magnitude of the weighted error (top) and real phase-rotated weighted error (bottom), and (d) phase error in passband.

unity [93, p. 361]. The optimal solution was obtained in five major iterations. Since $D(f)$ is symmetric with respect to $f = 0$, the resulting optimal filter is also symmetric with an optimal error $\|E_o\| = 0.9755$. The lower bound $|\delta|$, which is computed by the complex Remez stage, is 0.9704.

11.4.1.5 Design of Minimum-Phase FIR Filters

A minimum-phase FIR lowpass filter whose magnitude response is optimal in the Chebyshev sense is most conveniently designed by first designing a symmetric FIR filter [63,70]. By modifying an equiripple symmetric filter, a new filter can be obtained, the amplitude of which is nonnegative. That filter can then be spectrally factored to obtain a filter whose magnitude is equiripple. For example, see Figure 11.31.

The top row in Figure 11.31 illustrates a filter $h(n)$ obtained using the PM program. Let δ_1 and δ_2 denote the deviations from 1 and 0 in the passband and stopband, respectively. By adding δ_2 to $h(15)$, and then scaling $h(n)$ appropriately, the filter illustrated in the second row of Figure 11.31 is obtained. The amplitude response of that filter is nonnegative and the zeros of that filter lying on the U.C. are double zeros. That being the case, that filter can be spectrally factored to obtain the filter shown in the third row of Figure 11.31. The new nonsymmetric filter has a smaller delay, although its phase is nonlinear. Note that the frequency response magnitude of the nonsymmetric filter is the square root of that of the filter from which it was obtained. Denote the deviations from 1 and 0 in the passband and stopband of the nonsymmetric filter by δ_p and δ_s. Then, the following relationship holds [63]:

$$\delta_1 = \frac{4\delta_p}{2 + 2\delta_p^2 - \delta_s^2} \tag{11.116}$$

$$\delta_2 = \frac{\delta_s^2}{2 + 2\delta_p^2 - \delta_s^2}. \tag{11.117}$$

Given specifications for δ_p and δ_s, Equations 11.116 and 11.117 give the appropriate values to guide the design of the prototype symmetric FIR filter.

This method can also be used for the Chebyshev design of minimum-phase bandpass filters, as long as the stopband error in each stopband is equally weighted. If this is not the case, then the Remez exchange algorithm, used in the PM algorithm, can be modified so that it produces symmetric filters whose amplitude functions are nonnegative. The appropriate modification is simple: the interpolation equations in the Remez algorithm of the form $A(\omega_i) = -\delta$ are to be replaced by interpolation equations of the form $A(\omega_i) = 0$. The resulting symmetric FIR filters can then be spectrally factored. This modification makes possible the design of equiripple minimum-phase FIR bandpass filters where the stopband ripple in each band does not have to be the same. If a least squares error criterion is used, then symmetric filters with non-negative amplitudes can be obtained by using a constrained least squares method.

11.4.1.6 Delay Variation of Maximally Flat FIR Filters

Consider the problem of giving up exactly linear-phase for approximately linear-phase in return for a smaller delay. This problem was also considered in section on pages 11–43 By subjecting the frequency response magnitude and the group delay (individually) to differing numbers of flatness constraints, a family of nonsymmetric lowpass maximally flat FIR filters is obtained [94]. This approach is appropriate when

1. Exactly linear phase is not required
2. Some degree of phase linearity is desired
3. A maximally flat frequency response is desired

The resulting filters can be made to have approximately linear phase in the passband and a smaller group delay at $\omega = 0$, in comparison to a symmetric filter of equal length.

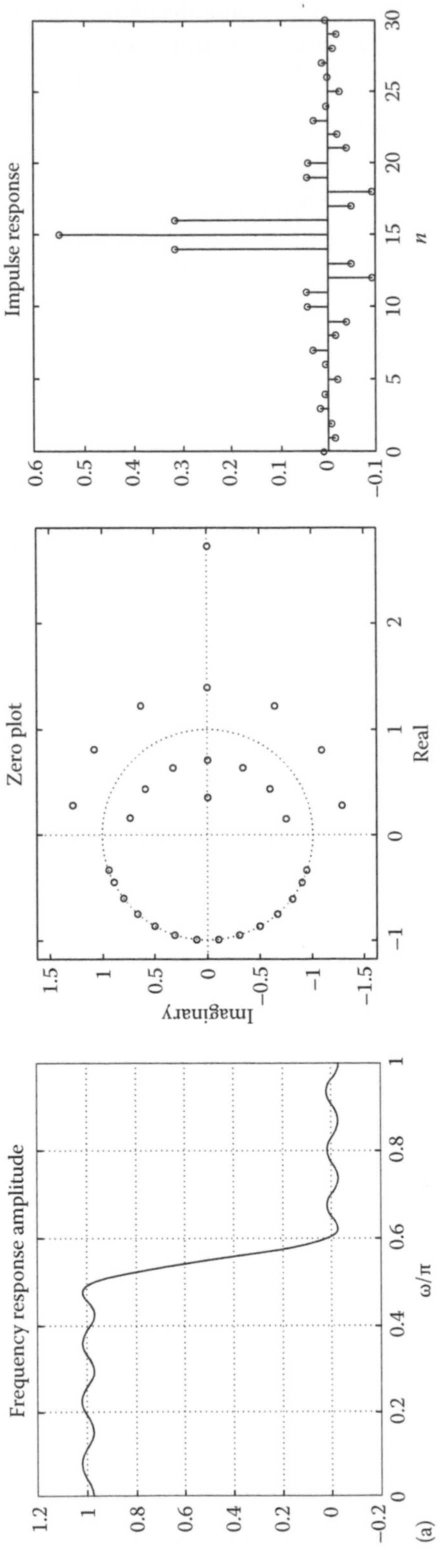
Frequency response amplitude
ω/π
(a)
Zero plot
Real
Imaginary
Impulse response
n
Frequency response amplitude
ω/π
(b)
Zero plot
Real
Imaginary
Impulse response
n

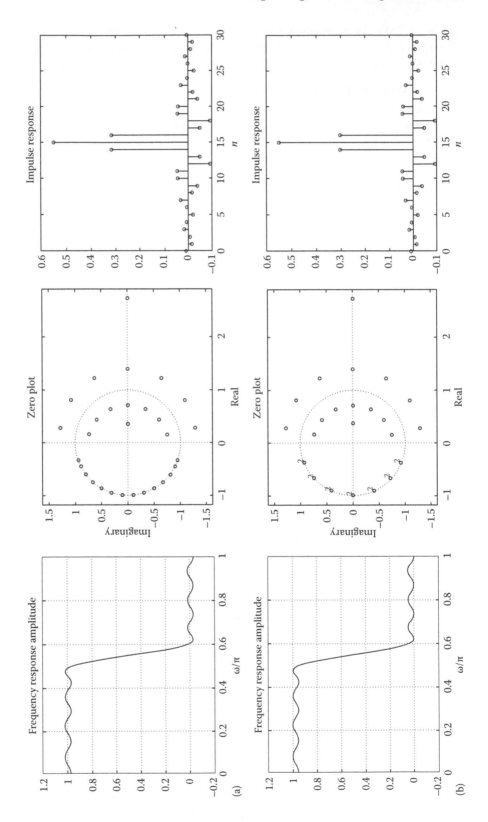

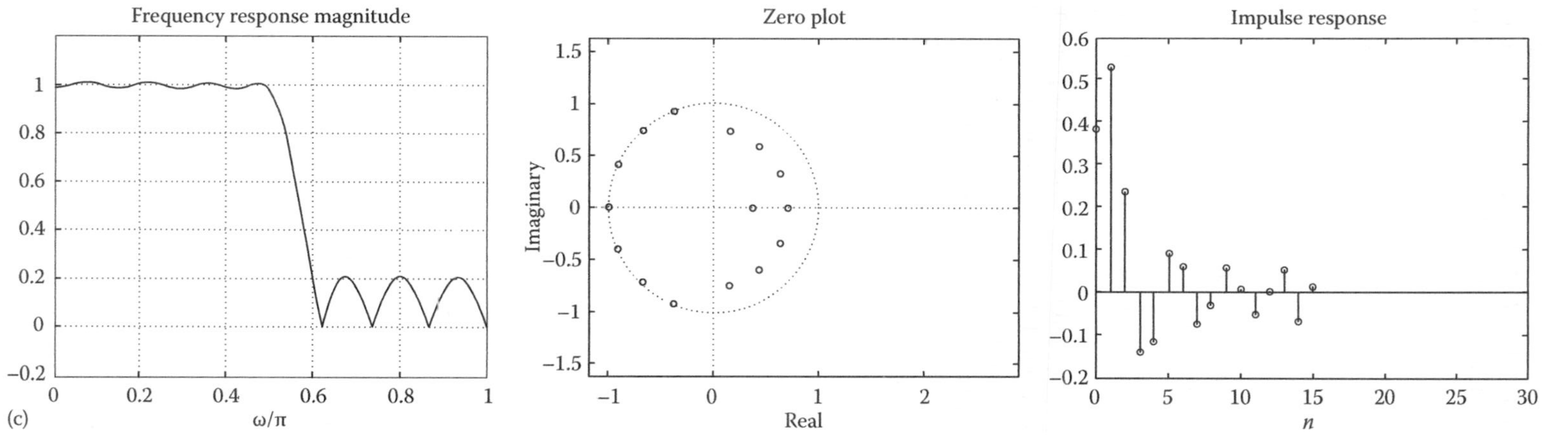

FIGURE 11.31 Design of a minimum-phase FIR filter whose magnitude response is optimal in the Chebyshev sense.

11.4.1.6.1 Problem Formulation

Let $F(\omega)$ denote the square magnitude response: $F(\omega) = |H(\omega)|^2$, let $G(\omega)$ denote the group delay: $G(\omega) = -\frac{\partial}{\partial\omega}\angle H(\omega)$. Given the flatness parameters K, L, M, (with $K > 0$, $M \geq 0$, $L \leq M$), find N filter coefficients $h(0), \ldots, h(N-1)$ such that

1. $N = K + L + M + 1$
2. $F(0) = 1$
3. $H(z)$ has a root at $z = -1$ of order K
4. $F^{(2i)}(0) = 0$ for $i = 1, \ldots, M$
5. $G^{(2i)}(0) = 0$ for $i = 1, \ldots, L$

The odd indexed derivatives of $F(\omega)$ and $G(\omega)$ are automatically zero at $\omega = 0$, so they do not need to be specified. Linear-phase filters and minimum-phase filters result from the special cases $L = M$ and $L = 0$, respectively.

This problem gives rise to nonlinear equations. Consequently, the existence of multiple solutions should not be surprising and, indeed, that is true here. It is informative to construct a table indicating the number of solutions as a function of K, L, and M. It turns out that the number of solutions is independent of K. The number of solutions as a function of L and M is indicated in Table 11.2 for the first few L and M. Many solutions have complex coefficients or possess frequency response magnitudes that are unacceptable between 0 and π. For this reason, it is useful to tabulate the number of real solutions possessing monotonic responses, as is done in Table 11.3. From Table 11.3, two distinct regions emerge. Define two regions in the (L, M) plane. Define region I as all pairs (L, M) for which

TABLE 11.2 Total Number of Solutions

				L				
	0	1	2	3	4	5	6	7
0	1							
1	2	3						
2	4	4	5					
3	8	6	6	7				
4	16	8	8	8	9			
5	32	16	10	10	10	11		
6	64	26	12	12	12	12	13	
7	128	48	24	14	14	14	14	15

TABLE 11.3 Number of Real Monotonic Solutions, Not Counting Time-Reversals

				L				
M	0	1	2	3	4	5	6	7
0	1							
1	1	1						
2	1	1	1					
3	2	1	1	1				
4	2	1	1	1	1			
5	4	2	1	1	1	1		
6	4	2	1	1	1	1	1	
7	8	4	2	1	1	1	1	1

TABLE 11.4 Regions I and II

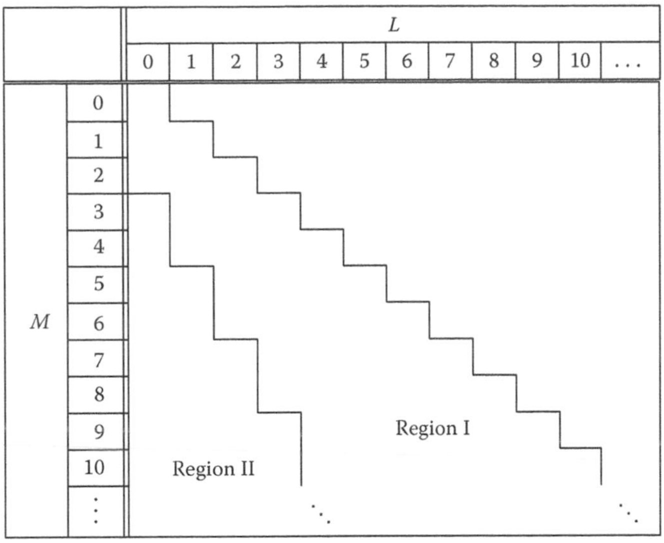

$$\left\lfloor \frac{M-1}{2} \right\rfloor \leq L \leq M.$$

Define region II as all pairs (L, M) for which

$$0 \leq L \leq \left\lfloor \frac{M-1}{2} \right\rfloor - 1.$$

See Table 11.4. It turns out that for (L, M) in region I, all the variables in the problem formulation, except $G(0)$, are linearly related and can be eliminated, yielding a polynomial in $G(0)$; the details are given in [94]. For region II, no similarly simple technique is yet available (except for $L = 0$).

11.4.1.6.2 Design Examples

Figures 11.32 and 11.33 illustrate four different FIR filters of length 13 for which $K + L + M = 12$. Each of these filters has 6 zeros at $z = -1$ ($K = 6$) and 6 zeros contributing to the flatness of the passband at $z = 1$ ($L + M = 6$). The four filters shown were obtained using the four values $L = 0, 1, 2, 3$.

When $L = 3$ and $M = 3$, the symmetric filter shown in Figure 11.32 is obtained. This filter is most easily obtained using formulas for maximally flat symmetric filters [55]. When $L = 0$, $M = 6$, the minimum-phase filter shown in Figure 11.33 is obtained. This filter is most easily obtained by spectrally factoring a length 25 maximally flat symmetric filter. The other two filters shown ($L = 2$, $M = 4$, and $L = 1$, $M = 5$) cannot be obtained using the formulas of Herrmann. They provide a compromise solution.

Observe that for the filters shown, the way in which the passband zeros are split between the interior of the U.C. and its exterior is given by the values L and M.

It may be observed that the cutoff frequencies of the four filters in Figure 11.32 are unequal. This is to be expected because the cutoff frequency (denoted ω_o) was not included in the problem formulation above. In the problem formulation, both the cutoff frequency and the DC group delay can be only indirectly controlled by specifying K, L, and M.

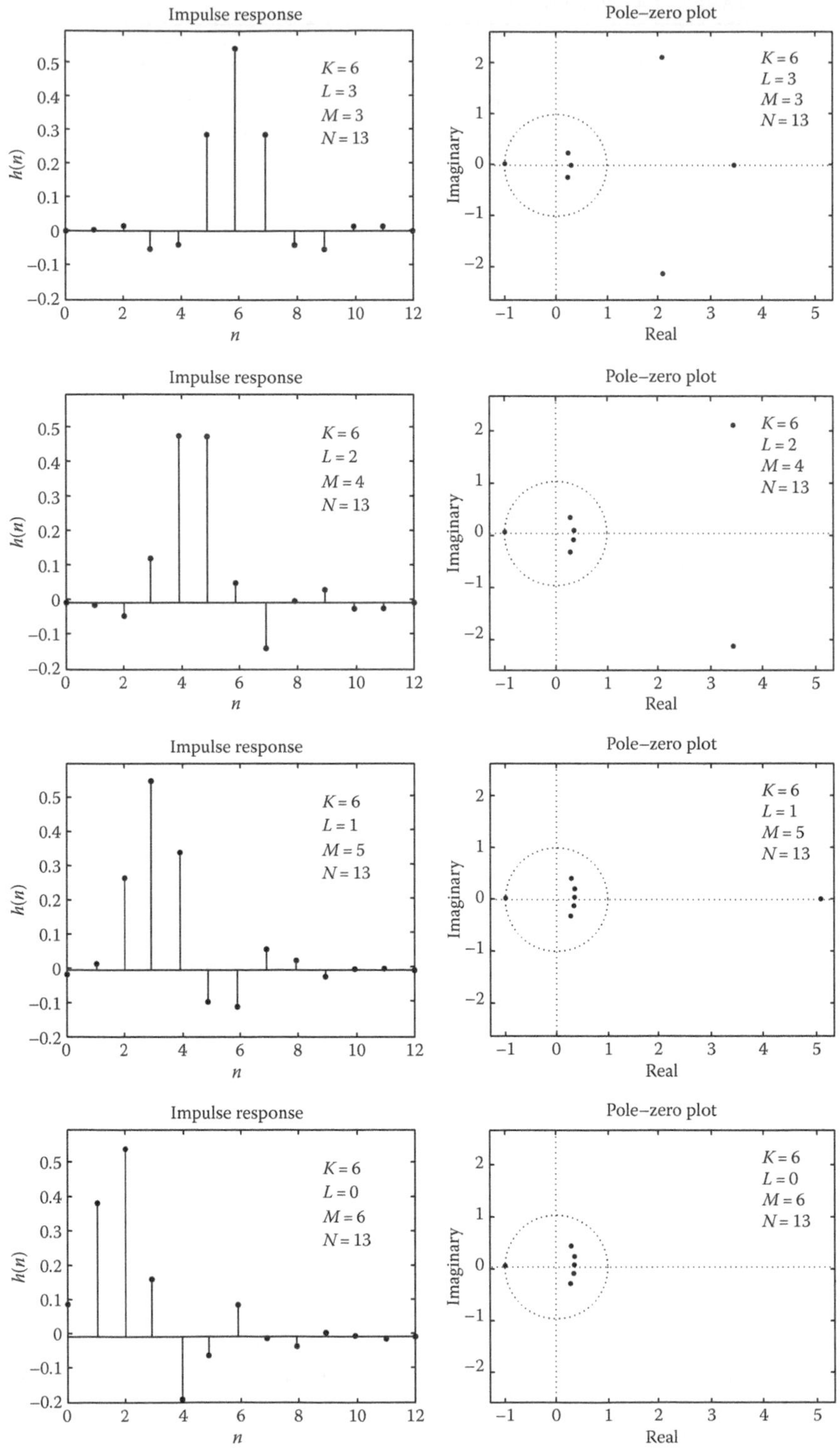

FIGURE 11.32 A selection of nonlinear-phase maximally flat filters of length (for which $K + L + M = 12$). For each filter shown, the zero at $z = -1$ is of multiplicity 6.

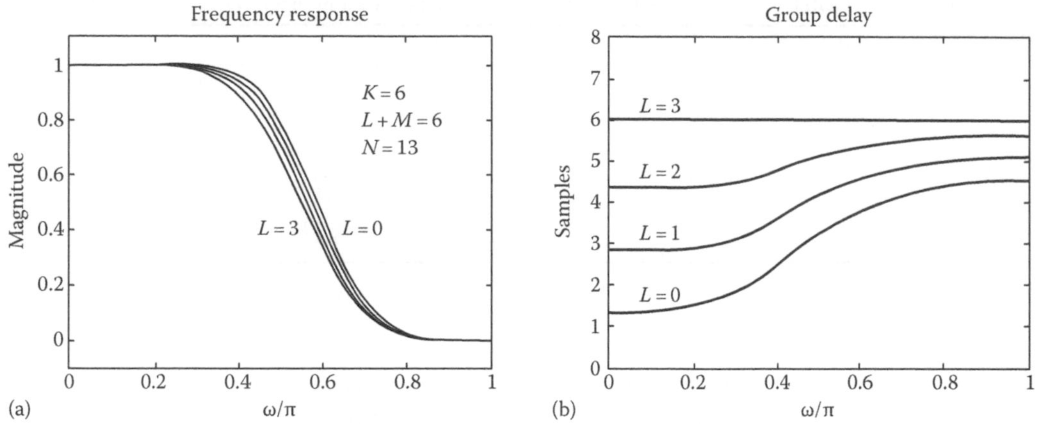

FIGURE 11.33 The magnitude responses and group delays of the filters shown in Figure 11.32.

11.4.1.6.3 Continuously Tuning ω_o and G(0)

To understand the relationship between ω_o, $G(0)$, and K, L, M, it is useful to consider ω_o and $G(0)$ as coordinates in a plane. Then each solution can be indicated by a point in the ω_o–$G(0)$ plane. For $N = 13$, those region I filters that are real and possess monotonic responses appear as the vertices in Figure 11.34.

To obtain filters of length 13 for which $(\omega_o, G(0))$ lie within one of the sectors, two degrees of flatness must be given up. (Then $K + L + M + 3 = N$, in contrast to item 1 in the problem formulation above.)

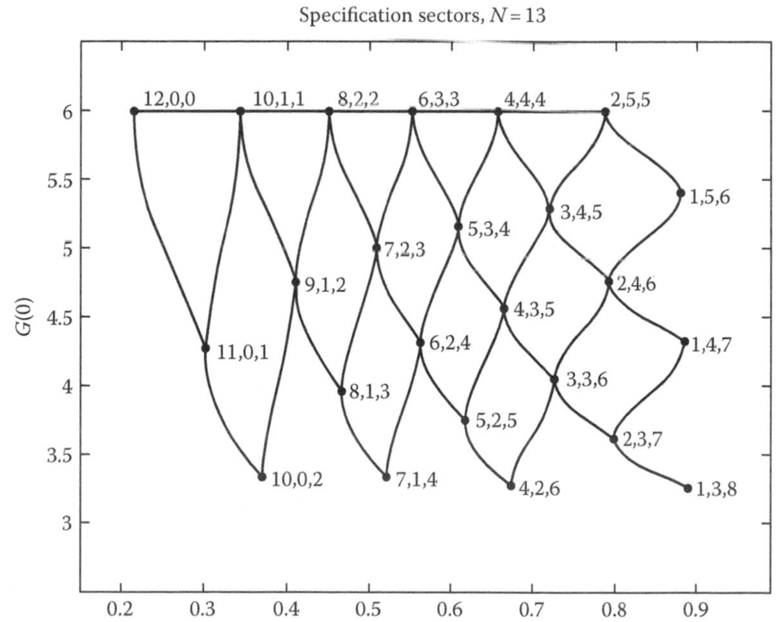

FIGURE 11.34 Specification sectors in the $\omega_o - G(0)$ plane for length 13 filters in region I. The vertices are points at which $K + L + M + 1 = 13$. The three integers by each vertex are the flatness parameters (K, L, M).

TABLE 11.5 Flatness Parameters for the Filters Shown in Figure 11.35

N	ω_o/π	$G(0)$	K	L	M
		3.5	3	2	5
		4	3	2	5
		4.5	4	2	4
13	0.636	5	3	3	4
		5.5	3	3	4
		6	4	3	3

In this way arbitrary (noninteger) DC group delays and cutoff frequencies can be achieved exactly. This is ideally suited for applications requiring fractional delay lowpass filters.

The flatness parameters of a point in the $\omega_o - G(0)$ plane are the (component-wise) minimum of the flatness parameters of the vertices of the sector in which the point lies [94].

11.4.1.6.4 Reducing the Delay

To design a set of filters of length 13 for which $\omega_o = 0.636\,\pi$ and for which $G(0)$ is varied from 3.5 to 6 in increments of 0.5, Figure 11.34 is used to determine the appropriate flatness parameters—they are tabulated in Table 11.5. The resulting responses are shown in Figure 11.35. It can be seen that the delay can be reduced while maintaining relatively constant group delay around $\omega = 0$, with no magnitude response degradation.

11.4.1.7 Combining Criteria in FIR Filter Design

Ivan W. Selesnick and C. Sidney Burrus

11.4.1.7.1 Savitzky–Golay Filters

The Savitzky–Golay filters are one example where two of the above described criteria are combined. The two criteria that are combined in the Savitzky–Golay filter are (1) maximally flat behavior (section on pages 11–38) and (2) least squares error (section on pages 11–18). Interestingly, the Savitzky–Golay

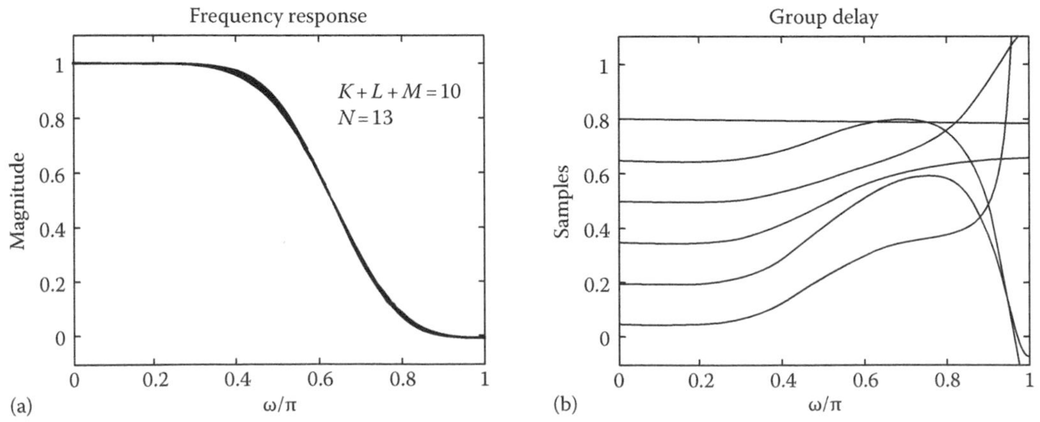

FIGURE 11.35 Length 13 filters obtained by giving up two degrees of flatness and by specifying that the cutoff frequency be 0.636π, and that the specified DC group delay be varied from 3.5 to 6.

filters illustrate an equivalence between digital lowpass filtering and the smoothing of noisy data by polynomials [63,95,96]. As a consequence of this equivalence, Savitzky–Golay filters can be obtained by two different derivations. Both derivations assume that a sequence $x(n)$ is available, where $x(n)$ is composed of an unknown sequence of interest $s(n)$, corrupted by an additive zero-mean white noise sequence $r(n)$: $x(n) = s(n) + r(n)$. The problem is the estimation of $s(n)$ from $x(n)$ in a way that minimizes the distortion suffered by $s(n)$. Two approaches yield the Savitzky–Golay filters: (1) polynomial smoothing and (2) moment preserving maximal noise reduction.

11.4.1.7.2 Polynomial Smoothing

Suppose a set of $N = 2M + 1$ contiguous samples of $x(n)$, centered around n_0, can be well approximated by a degree L polynomial in the least squares sense. Then an estimate of $s(n_0)$ is given by $p(n_0)$ where $p(n)$ is the degree L polynomial that minimizes

$$\sum_{k=-M}^{M} (p(n_o + k) - x(n_o + k))^2. \tag{11.118}$$

It turns out that the estimate of $s(n_0)$ provided by $p(n_0)$ can be written as

$$p(n_0) = (h * x)(n_0), \tag{11.119}$$

where $h(n)$ is the Savitzky–Golay filter of length $N = 2M + 1$ and smoothing parameter L. Therefore, the smoothing of noisy data by polynomials is equivalent to lowpass FIR filtering. Assuming L is odd, with $L = 2K + 1$, $h(n)$ can be written [63] as

$$h(n) = \begin{cases} C_K \frac{1}{n} q_{2K+1}(n) & n = \pm 1, \ldots, \pm M \\ C_K q_{2K+1}(0) & n = 0, \end{cases} \tag{11.120}$$

where

$$C_K = (-1)^K \frac{(2K+1)!}{(K!)^2} \prod_{k=-K}^{K} \frac{1}{2M + 2k + 1} \tag{11.121}$$

and the polynomials q_l are generated via the recurrence

$$q_0(n) = 1 \quad q_1(n) = n \tag{11.122}$$

$$q_{l+1}(n) = \frac{2l+1}{l+1} n q_l(n) - \frac{l(2M+1+l)(2M+1-l)}{4(l+1)} q_{l-1}(n), \tag{11.123}$$

$q_l(n)$ denotes the derivative of $q_l(n)$.

The impulse response (shifted so that it is casual) and frequency response amplitude of a length 41, $L = 13$, Savitzky–Golay filter is shown in Figure 11.36. As is evident from the figure, Savitzky–Golay filters have poor stopband attenuation—however, they are optimal according to the criteria by which they are designed.

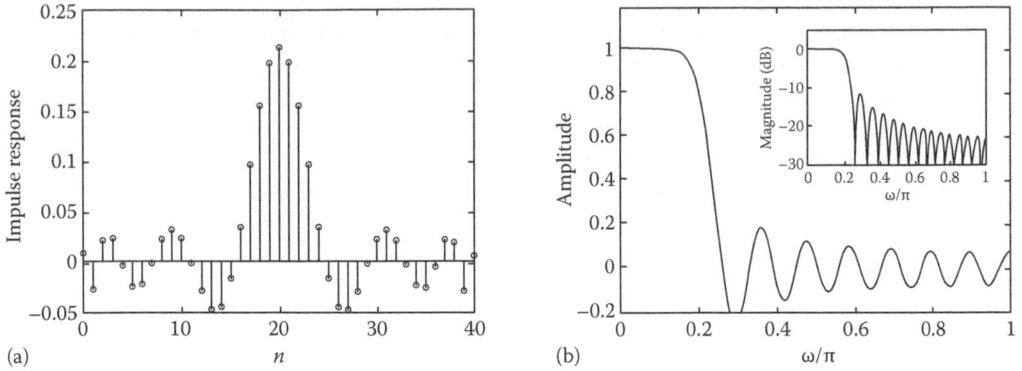

FIGURE 11.36 Savitzky–Golay filter, $N=41$, $L=13$, and $K=6$: (a) impulse response and (b) magnitude response.

11.4.1.7.3 Moment Preserving Maximal Noise Reduction

Consider again the problem of estimating $s(n)$ from $x(n)$ via FIR filtering.

$$y(n) = (h_1 * x)(n) \tag{11.124}$$

$$= (h_1 * s)(n) + (h_1 * r)(n) \tag{11.125}$$

$$= y_1(n) + e_r(n), \tag{11.126}$$

where $y_1(n) = (h_1 * s)(n)$ and $e_r(n) = (h_1 * r)(n)$. Consider designing $h_1(n)$ by minimizing the variance of $e_r(n)$, $\sigma^2(n) = E\left[e_r^2(n)\right]$. Because $\sigma^2(n)$ is proportional to $\|h_1\|_2^2 = \sum_{n=-M}^{M} h_1^2(n)$, the filter minimizing $\sigma^2(n)$ is the zero filter, $h_1(n) \equiv 0$. However, the zero filter also eliminates $s(n)$. A more useful approach requires that $h_1(n)$ preserve the moments of $s(n)$ up to a specified order L. Define the l th moment:

$$m_l[s] = \sum_{n=-M}^{M} n^l s(n). \tag{11.127}$$

The requirement that $m_l[y_1] = m_l[s]$ for $l = 0, \ldots, L$, is equivalent to the requirement that $m_0[h_1] = 1$ and $m_l[h_1] = 0$ for $l = 1, \ldots, L$. The filter $h_1(n)$ is then obtained by the problem formulation

$$\text{minimize} \|h_1\|_2^2 \tag{11.128}$$

subject to

$$m_0[h_1] = 1 \tag{11.129}$$

$$m_l[h_1] = 0 \quad \text{for } l = 1, \ldots, L. \tag{11.130}$$

As shown in [63,96], the solution $h_1(n)$ is the Savitzky–Golay filter (Equation 11.120).

It should be noted that the problem formulated in Equations 11.128 through 11.130 is equivalent to the least squares approach, as described in section on pages 11–40: minimize Equation 11.30 with $D(\omega) = 0$, $W(\omega) = 1$ subject to the constraints

$$A(\omega = 0) = 1 \tag{11.131}$$

$$A^{(i)}(\omega = 0) = 0 \quad \text{for } i = 1, \ldots, L. \tag{11.132}$$

(These derivative constraints can be expressed as $\mathbf{Ga} = \mathbf{b}$). As such, the solution to Equation 11.41 is the Savitzky–Golay filter (Equation 11.120)—however, with the constraints in Equations 11.131 and 11.132, the resulting linear system (Equation 11.41) is numerically ill-conditioned. Fortunately, the explicit solution (Equation 11.120) eliminates the need to solve ill-conditioned equations.

11.4.1.7.4 Structure for Symmetric FIR Filter Having Flat Passband

Define the transfer function $G(z) = z^{-M} - H(z)$, where $H(z) = \sum_{n=0}^{2M+1} h(n)z^{-n}$ and $h(n)$ is the length $N = 2M + 1$ Savitzky–Golay filter in Equation 11.120, shifted so that it is casual, as in Figure 11.36. The filter $G(z)$ is a highpass filter that satisfies derivative constraints at $\omega = 0$. It follows that $G(z)$ possesses a zero at $z = 1$ of order $2K + 2$, and so can be expressed as $G(z) = (-1)^{K+1}(\frac{1-z^{-1}}{2})^{2K+2}H_1(z)$. Accordingly,* the transfer function of a symmetric filter of length $N = 2M + 1$, satisfying Equations 11.131 and 11.132, can be written as

$$H(z) = z^{-M} - (-1)^{K+1}\left(\frac{1 - z^{-1}}{2}\right)^{2K+2} H_1(z), \tag{11.133}$$

where $H_1(z)$ is a symmetric filter of length $N - 2K - 2 = 2(M - K) - 1$. The amplitude response of $H(z)$ is

$$A(\omega) = 1 - \left(\frac{1 - \cos\omega}{2}\right)^{K+1} A_1(\omega), \tag{11.134}$$

where $A_1(\omega)$ is the amplitude response of $H_1(z)$. Equation 11.133 structurally imposes the desired derivative constraints in Equations 11.131 and 11.132 with $L = 2K + 1$, and reduces the implementation complexity by extracting the multiplierless factor $(\frac{1-z^{-1}}{2})^{2K+2}$. In addition, this structure possesses good passband sensitivity properties with respect to coefficient quantization [97].

Equation 11.133 is a special case of the affine form 11.80. Accordingly, as discussed in section on pages 11–40, $h_1(n)$ in Equation 11.133 could be obtained by minimizing Equation 11.83, with suitably defined $\bar{D}(\omega)$ and $\bar{W}(\omega)$. Although this is unnecessary for the design of Savitzky–Golay filters, it is useful for the design of other symmetric filters for which $A(\omega)$ is flat at $\omega = 0$, for example, the design of such filters in the least squares sense with various $W(\omega)$ and $D(\omega)$, or the design of such filters according to the Chebyshev norm.

Remarks

- Solution to two optimal smoothing techniques: (1) polynomial smoothing and (2) moment preserving maximal noise reduction
- Explicit formulas for solution
- Excellent at $\omega = 0$
- Polynomial assumption for $s(n)$
- Poor stopband attenuation

11.4.1.7.5 Flat Passband, Chebyshev Stopband

The use of a filter having a very flat passband is desirable because it minimizes the distortion of low frequency signals. However, in the removal of high frequency noise from a low frequency signal by lowpass filtering, it is often desirable that the stopband attenuation be greater than that offered by a Savitzky–Golay filter. One approach [98] minimizes the weighted Chebyshev error, subject to the derivative constraints in Equations 11.131 and 11.132 imposed at $\omega = 0$. As discussed above, the form of Equation 11.133 facilitates the design and implementation of such filters. To describe this approach

* Note that $-1\left(\frac{1-z^{-1}}{2}\right)^2\Big|_{z=e^{j\omega}} = e^{-j\omega}\left(\frac{1-\cos\omega}{2}\right) - 1\left(\frac{1-z^{-1}}{2}\right)^2\left(\frac{1-\cos\omega}{2}\right)$.

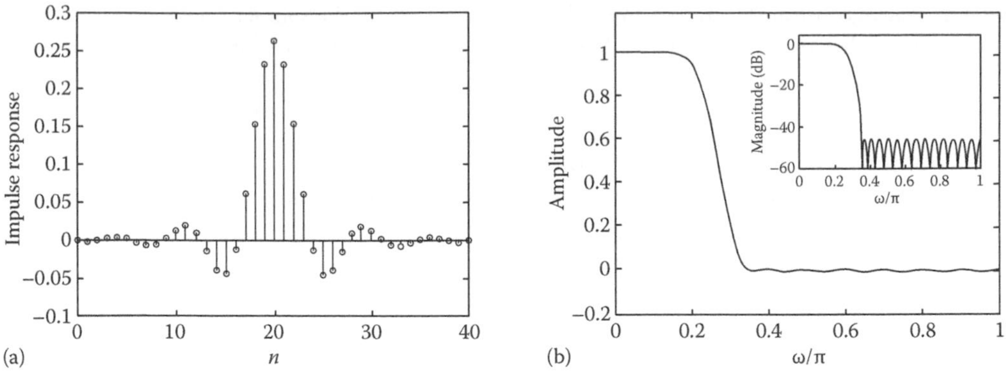

FIGURE 11.37 Lowpass FIR filter designed via the minimization of stopband Chebyshev error subject to derivative constraints at $\omega = 0$. (a) Impulse response and (b) magnitude response.

[97], let the desired amplitude and weight function be as in Equation 11.44. For the form of Equation 11.133, $A_2(\omega)$ and $A_3(\omega)$ in section on pages 11–40 are given by $A_2(\omega) = -\left(\frac{1-\cos\omega}{2}\right)^K$ and $A_3(\omega) = 1$. $H_1(z)$ can then be designed by minimizing Equation 11.81 via the PM algorithm. Passband monotonicity, which is sometimes desired, can be ensured by setting $K_p = 0$ in Equation 11.44 [99]. Then the passband is shaped by the derivative constraints at $\omega = 0$ that are structurally imposed by Equation 11.133.

Figure 11.37 illustrates a length 41 symmetric filter, whose passband is monotonic. The filter shown was obtained with $K = 6$ and

$$D(\omega) = 0 \quad \omega \in [\omega_s, \pi] \quad W(\omega) = \begin{cases} 0 & \omega \in [0, \omega_s] \\ 1 & \omega \in [\omega_s, \pi] \end{cases}, \tag{11.135}$$

where $\omega_s = 0.3387\pi$. Because $W(\omega)$ is positive only in the stopband, ω_p is not part of the problem formulation.

11.4.1.7.6 Bandpass Filters

To design bandpass filters having very flat passbands, one specifies a passband frequency, ω_p, where one wishes to impose flatness constraints. The appropriate form is $H(z) = z^{-(N-1)/2} + H_1(z)H_2(z)$ with

$$H_2(z) = \left(\frac{1 - 2(\cos\omega_p)z^{-1} + z^{-2}}{4}\right)^K, \tag{11.136}$$

where

N is odd

$H_1(z)$ is a filter whose impulse response is symmetric and of length $N - 2K$

The overall frequency response amplitude $A(\omega)$ is given by

$$A(\omega) = 1 + (-1)^K \left(\frac{\cos\omega_p - \cos\omega}{2}\right)^K A_1(\omega). \tag{11.137}$$

As above, $H_1(z)$ can be found via the PM algorithm. Monotonicity of the passband on either side of ω_p can be ensured by weighting the passband by 0, and by taking K to be even. The filter of length 41

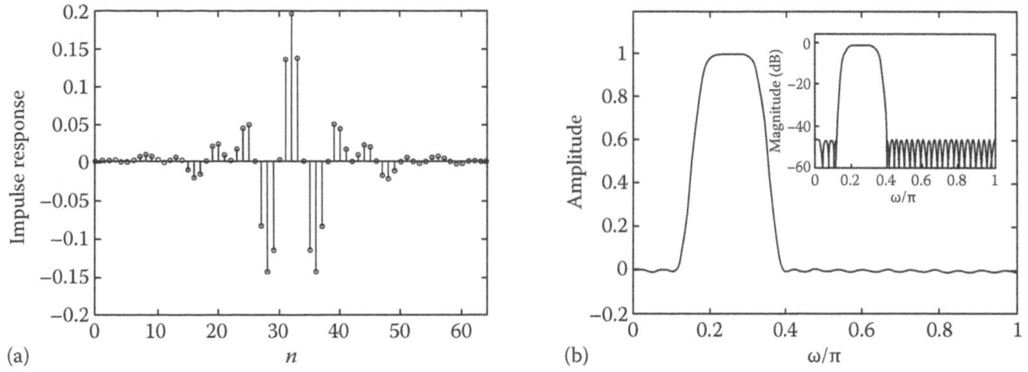

FIGURE 11.38 Bandpass FIR filter designed via the minimization of stopband Chebyshev error subject to derivative constraints at $\omega = 0.25\pi$. (a) Impulse response and (b) magnitude response.

illustrated in Figure 11.38 was obtained by minimizing the Chebyshev error with $\omega_p = 0.25\pi$, $K = 8$, and

$$D(\omega) = 0 \quad W(\omega) = \begin{cases} 1 & \omega \in [0, \omega_1] \\ 0 & \omega \in [\omega_1, \omega_2], \\ 1 & \omega \in [\omega_2, \pi] \end{cases} \tag{11.138}$$

where $\omega_1 = 0.1104\pi$ and $\omega_2 = 0.3889\pi$.

11.4.1.7.7 Constrained Least Square

The constrained least square approach to filter design provides a compromise between the square error and Chebyshev criteria. This approach produces least square error and best Chebyshev filters as special cases, and is motivated by an observation made by Adams [100]. Least square filter design is based on the assumption that the size of the peak error can be ignored. Likewise, filter design according to the Chebyshev norm assumes the integral square error is irrelevant. In practice, however, both of these criteria are often important. Furthermore, the peak error of a least square filter can be reduced with only a slight increase in the square error. Similarly, the square error of an equiripple filter can be reduced with only a slight increase in the Chebyshev error [8,100]. In Adams' terminology, both equiripple filters and least square filters are inefficient.

11.4.1.7.8 Problem Formulation

Suppose the following are given: the filter length N, the desired response $D(\omega)$, a lower bound function $L(\omega)$, and an upper bound function $U(\omega)$, where $D(\omega)$, $L(\omega)$, and $U(\omega)$ satisfy

1. $L(\omega) \leq D(\omega)$
2. $U(\omega) \geq D(\omega)$
3. $U(\omega) > L(\omega)$

Find the filter of length N that minimizes

$$\|E\|_2^2 = \frac{1}{\pi} \int_0^{\pi} W(\omega)(A(\omega) - D(\omega))^2 d\omega \tag{11.139}$$

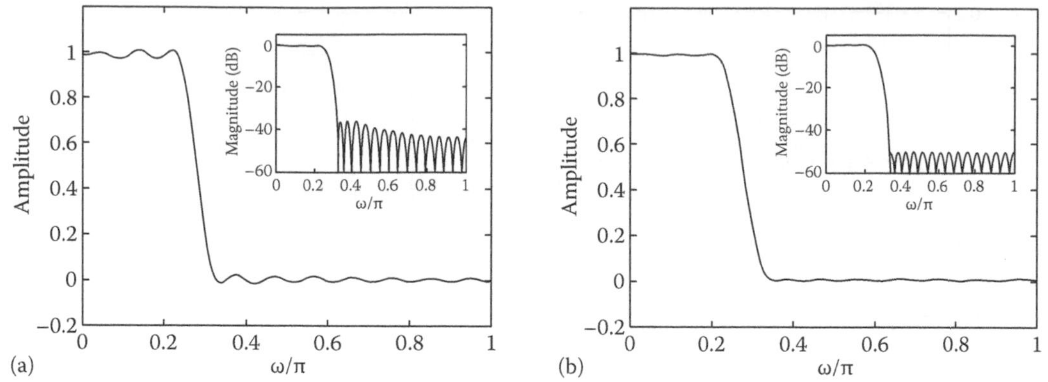

FIGURE 11.39 Lowpass filter design via bound-constrained least squares. (a) $\delta = 0.0178$ (−35 dB) and (b) $\delta = 0.0032$ (−50 dB).

such that (1) the local maxima of $A(\omega)$ do not exceed $U(\omega)$ and (2) the local minima of $A(\omega)$ do not fall below $L(\omega)$.

11.4.1.7.9 Design Examples

Figure 11.39 illustrates two length 41 filters obtained by minimizing Equation 11.139, subject to the bound constraints, where

$$D(\omega) = \begin{cases} 1 & \omega \in [0, \omega_c](64) \\ 0 & \omega \in (\omega_c, \pi] \end{cases} \quad (11.140)$$

$$W(\omega) = \begin{cases} 1 & \omega \in [0, \omega_c](66) \\ 20 & \omega \in (\omega_c, \pi] \end{cases} \quad (11.141)$$

$$L(\omega) = \begin{cases} 1 - \delta_p & \omega \in [0, \omega_c](68) \\ -\delta_s & \omega \in (\omega_c, \pi] \end{cases} \quad (11.142)$$

$$U(\omega) = \begin{cases} 1 + \delta_p & \omega \in [0, \omega_c](70) \\ \delta_s & \omega \in (\omega_c, \pi] \end{cases} \quad (11.143)$$

and where $\omega_c = 0.3\pi$. For the filter on the left of the figure, $\delta_p = \delta_s = 0.0178 = 10^{-35/20}$; for the filter on the right of the figure, $\delta_p = \delta_s = 0.0032 = 10^{-50/20}$. The extremal points of $A(\omega)$ lie within the upper and lower bound functions. Note that the filter on the right is an equiripple filter—it could have been obtained with the PM algorithm, given the appropriate parameter values.

This approach is not a quadratic program (QP) because the domain of the constraints are not explicit. Two observations regarding this formulation and example should be noted:

1. For a fixed length, the maximum ripple size can be made arbitrarily small. When the specified values δ_p and δ_s are small enough, the solution is an equiripple filter. As the constraints are made more strict, the transition width of the solution becomes wider. The width of the transition automatically increases as appropriate.
2. As the example illustrates, it is not necessary to use a "don't care" band, for example, it is not necessary to exclude from the square error a region around the discontinuity of the ideal low-pass filter. The problem formulation, however, does not preclude the use of a zero-weighted transition band.

11.4.1.7.10 Quadratic Programming Approach

Some lowpass filter specifications require that $A(\omega)$ lie within $U(\omega)$ and $L(\omega)$ for all $\omega \in [0, \omega_p] \cup [\omega_s, \pi]$ for given bandedges ω_p and ω_s. While the approach described above ensures that the local maxima and minima of $A(\omega)$ lie below $U(\omega)$ and above $L(\omega)$, respectively, it does not ensure that this is true at the given bandedges ω_p and ω_s. This is because ω_p and ω_s are not generally extremal points of $A(\omega)$. The approach described above can be modified so that bandedge constraints are satisfied; however, it should be recognized that in this case, a QP formulation is possible.

Adams formulates the constrained least square filter design problem as a QP and describes algorithms for solving the relevant QP in [100,101]. The design of a lowpass filter, for example, can be formulated as a QP as follows.

11.4.1.7.10.1 QP Formulation Suppose the following are given: the filter length, N, the bandedges, ω_p and ω_s, and maximum allowable deviations, δ_p and δ_s. Find the filter that minimizes the square error:

$$\|E\|_2^2 = \frac{1}{\pi} \int_0^{\pi} W(\omega)[A(\omega) - D(\omega)]^2 d\omega \tag{11.144}$$

such that

$$L(\omega) \le A(\omega) \le U(\omega) \quad \omega \in [0, \omega_p] \cup [\omega_s, \pi], \tag{11.145}$$

where

$$D(\omega) = \begin{cases} 1 & \omega \in [0, \omega_p] \\ 0 & \omega \in [\omega_s, \pi] \end{cases} \tag{11.146}$$

$$W(\omega) = \begin{cases} K_p & \omega \in [0, \omega_p] \\ 0 & \omega \in [\omega_p, \omega_s] \\ K_s & \omega \in [\omega_s, \pi] \end{cases} \tag{11.147}$$

$$L(\omega) = \begin{cases} 1 - \delta_p & \omega \in [0, \omega_p] \\ -\delta_s & \omega \in [\omega_s, \pi] \end{cases} \tag{11.148}$$

$$U(\omega) = \begin{cases} 1 + \delta_p & \omega \in [0, \omega_p] \\ \delta_s & \omega \in [\omega_s, \pi] \end{cases}. \tag{11.149}$$

This is a QP because the constraints are linear inequality constraints and the cost function is a quadratic function of the variables. The QP formulation is useful because it is very general and flexible. For example, it can be used for arbitrary $D(\omega)$, $W(\omega)$, and arbitrary constraint functions.

Note, however, that for a fixed filter length and a fixed δ_p and δ_s (each less than 0.5), it is not possible to obtain an arbitrarily narrow transition band. Therefore, if the bandedges ω_p and ω_s are taken to be too close together, then the QP has no solution. Similarly, for a fixed ω_p and ω_s, if δ_p and δ_s are taken too small, then there is again no solution.

Remarks

- Compromise between square error and Chebyshev criterion.
- Two options: formulation without bandedge constraints or as a QP.
- QP allows (requires) bandedge constraints, but may have no solution.
- Formulation without bandedge constraints can satisfy arbitrarily strict bound constraints.
- QP is well formulated for arbitrary $D(\omega)$ and $W(\omega)$.
- QP is well formulated for the inclusion of arbitrary linear constraints.

11.4.2 IIR Filter Design

Ivan W. Selesnick and C. Sidney Burrus

11.4.2.1 Numerical Methods for Magnitude-Only IIR Design

Numerical methods for magnitude only approximation for IIR filters generally proceed by constructing a noncausal symmetric IIR filter whose amplitude response is nonnegative. Equivalently, a rational function is found, the numerator and denominator of which are both symmetric polynomials of odd degree, with two properties: (1) all zeros lying on the U.C. $|z| = 1$ have even multiplicity and (2) no poles lie on the U.C. A spectral factorization then yields a stable casual digital filter.

The differential correction algorithm for Chebyshev approximation by rational functions, and variations thereof, have been applied to IIR filter design [102–106]. This algorithm is guaranteed to converge to an optimal solution, and is suitable for arbitrary desired magnitude responses. However, (1) it does not utilize the characterization theorem (see [28] for a characterization theorem for rational Chebyshev approximation), and (2) it proceeds by solving a sequence of (semi-infinite) linear programs. Therefore, it can be slow and computationally intensive.

A Remez algorithm for rational Chebyshev approximation [28] is applicable to IIR filter design, but it is not guaranteed to converge. Deczky's numerical optimization program [107] is also applicable to this problem, as are other optimization methods. It should be noted that general optimization methods can be used for IIR filter design according to a variety of criteria, but the following aspects make it a challenge: (1) initialization, (2) local optimal (nonglobal) solutions, and (3) ensuring the filter's stability.

11.4.2.2 Allpass (Phase-Only) IIR Filter Design

An allpass filter is a filter with a frequency response $H(\omega)$ for which $|H(\omega)| = 1$ for all frequencies ω. The only FIR allpass filter is the trivial delay $h(n) = \delta(n - k)$. IIR allpass filters, on the other hand, must have a transfer function of the form

$$H(z) = \frac{z^N P(z^{-1})}{P(z)} \tag{11.150}$$

where $P(z)$ is a degree N polynomial in z. The problem is the design of the polynomial $P(z)$ so that the phase, or group delay, of $H(z)$ approximates a desired function. The form in Equation 11.150 structurally imposes the allpass property of $H(z)$.

The design of digital allpass filters has received much attention, for (1) low complexity structures with low roundoff noise behavior are available for allpass filters [108,109] and (2) they are useful components in a variety of applications. Indeed, while the traditional application of allpass filters is phase equalization [68,107], their uses in fractional delay design [21], multirate filtering, filterbanks, notch filtering, recursive phase splitters, and other applications have also been described [63,110]. Of particular recent interest has been the design of frequency selective filters realizable as a parallel combination of two allpasses:

$$H(z) = \frac{1}{2}[A_1(z) + A_2(z)]. \tag{11.151}$$

It is interesting to note that digital filters, obtained from the classical analog (Butterworth, Chebyshev, and elliptic) prototypes via the bilinear transformation, can be realized as allpass sums [109,111,112]. As allpass sums, such filters can be realized with low complexity structures that are robust to finite precision effects [109]. More importantly, the allpass sum is a generalization of the classical transfer functions that is honored with a number of benefits. Certainly, examples have been given where the utility of allpass sums is well illustrated [113,114]. Specifically, when some degree of phase linearity is desired, nonclassical filters of the form in Equation 11.151 can be designed that achieve superior results with respect to implementation complexity, delay, and phase linearity.

The desired degree of phase linearity can, in fact, be structurally incorporated. If one of the allpass branches in an allpass sum contains only delay elements, then the allpass sum exhibits approximately linear phase in the passbands [115,116]. The frequency selectivity is then obtained by appropriately designing the remaining allpass branch. Interestingly, by varying the number of delay elements used and the degrees of $A_1(z)$ and $A_2(z)$, the phase linearity can be affected. Simultaneous approximation of the phase and magnitude is a difficult problem in general, so the ability to structurally incorporate this aspect of the approximation problem is most useful.

While general procedures for allpass design [117–122] are applicable to the design of frequency selective allpass sums, several publications have addressed, in addition to the general problem, the details specific to allpass sums [63,123–125]. Of particular interest are the recently described iterative Remez-like exchange algorithms for the design of allpass filters and allpass sums according to the Chebyshev criterion [113,114,126,127].

A simple procedure for obtaining a fractional delay allpass filter uses the maximally flat delay all-pole filter (Equation 11.76). By using the denominator of that IIR filter for $P(z)$ in Equation 11.150, a fractional delay filter is obtained [21]. The group delay of the allpass filter is $2\tau + N$ where τ is that of the all-pole filter used and N is the filter order.

11.4.2.3 Magnitude and Phase Approximation

The optimal frequency domain design of an IIR filter where both the magnitude and the phase are specified, is more difficult than the approximation of one alone. One of the difficulties lies in the choice of the phase function. If the chosen phase function is inconsistent with a stable filter, then the best approximation according to a chosen norm may be unstable. In that case, additional stability constraints must be made explicit. Nevertheless, several numerical methods have been described for the approximation of both magnitude and phase. Let $D(e^{j\omega})$ denote the complex valued desired frequency response.

The minimization of the weighed integral square error

$$\int_0^{\pi} W(\omega)\left|\frac{B(e^{j\omega})}{A(e^{j\omega})} - D(e^{j\omega})\right|^2 d\omega \tag{11.152}$$

is a nonlinear optimization problem. If a good initial solution is known, and if the phase of $D(e^{j\omega})$ is chosen appropriately, then Newton's method, or other optimization algorithms, can be successfully used [107,128]. A modified minimization problem, that comes from the observation that $B/A \approx D \rightarrow B \approx DA$ is the minimization of the weighted equation error [11]:

$$\int_0^{\pi} W(\omega)\left|B(e^{j\omega}) - D(e^{j\omega})A(e^{j\omega})\right|^2 d\omega \tag{11.153}$$

which is linear in the filter coefficients. There is a family of iterative methods [129] based on iteratively minimizing the weighted equation error, or a variation thereof, with a weighting function that is appropriately modified from one iteration to the next.

The minimization of the complex Chebyshev error has also been addressed by several authors. The Ellacott–Williams algorithm for complex Chebyshev approximation by rational functions, and variations thereof, have been applied to this problem [130]. This algorithm calls for the solution to a sequence of complex polynomial Chebyshev problems, and is guaranteed to converge to a local minimum.

11.4.2.3.1 Structure-Based Methods

Several approaches to the problem of magnitude and phase approximation, or magnitude and group delay approximation, use a combination of filters. There are at least three such approaches.

1. One approach cascades (1) a magnitude optimal IIR filters and (2) an allpass filter [107]. The allpass filter is designed to equalize the phase.
2. A second approach cascades (1) a phase optimal IIR filter and (2) a symmetric FIR filter [41]. The FIR filter is designed to equalize the magnitude.
3. A third approach employs a parallel combination of allpass filters. Their phases can be designed so that their combined frequency response is selective and has approximately linear phase [113].

11.4.2.4 Time-Domain Approximation

Another approach is based on knowledge of the time domain behavior of the filter sought. Prony's method [11] obtains filter coefficients of an IIR filter that has specified impulse response values $h(0), \ldots, h(K-1)$, where K is the total number of degrees of freedom in the filter coefficients. To obtain an IIR filter whose impulse response approximates desired values $d(0), \ldots, d(L-1)$, where $L > K$, an equation error approach can be minimized, as above, by solving a linear system. The true square error, a nonlinear function of the coefficients, can be minimized by iterative methods [131]. As above, initialization, local-minima, and stability can make this problem difficult.

A more general problem is the requirement that the filter approximately reproduce other input-output data. In those cases, where the sought filter is given only by input-output data, the problem is the identification of the system. The problem of designing an IIR filter that reproduces observed input-output data is an important modeling problem in system and control theory, some methods for which can be used for filter design [129].

11.4.2.5 Model Order Reduction

Model order reduction (MOR) techniques, developed largely in the control theory literature, are generally noniterative linear algebraic techniques. Given a transfer function, these techniques produce a second transfer function of specified (lower) degree that approximates the given transfer function. Suppose input–output data of an unknown system is available. One two-step modeling approach proceeds by first constructing a high order model that well reproduces the observed input–output data and, second, obtains a lower order model by reducing the order of the high-order model. Two common methods for MOR are (1) balanced model truncation [132] and (2) optimal Hankel norm MOR [133]. These methods, developed for both continuous and discrete time, produce stable models for which the numerator and denominator degrees are equal.

MOR has been applied to filter design in [134–137]. One approach [134] begins with a high-order FIR filter (obtained by any technique), and uses MOR to obtain a lower order IIR filter, that approximates the FIR filter. As noted above, the phase of the FIR filter used can be important. MOR techniques can yield different results when applied to minimum, maximum, and linear phase FIR filters [134].

11.5 Software Tools

James H. McClellan

Over the past 30 years, many design algorithms have been introduced for optimizing the characteristics of frequency-selective digital filters. Most of these algorithms now rely on numerical optimization, especially when the number of filter coefficients is large. Many sophisticated computer optimization methods have been programmed and distributed for widespread use in the DSP engineering community. Since it is challenging to learn the details of every one of these methods and to understand subtleties of various methods, a designer must now rely on software packages that contain a subset of the available methods. With the proliferation of DSP boards for PCs, the manufacturers have been eager to place design tools in the hands of their users so that the complete design process can be accomplished with one piece of software. This software includes the filter design and optimization, followed by a filter implementation stage. The steps in the design process include

1. Filter specification via a graphical user interface.
2. Filter design via numerical optimization algorithms. This includes the order estimation stage where the filter specifications are used to compute a predicted filter length (FIR) or number of poles (IIR).
3. Coefficient formatting for the DSP board. Since the design algorithm yields coefficients computed to the highest precision available (e.g., double-precision floating-point), the filter coefficients must be quantized to the internal format of the DSP. In the extreme case of a fixed-point DSP, this quantization also requires scaling of the coefficients to a predetermined maximum value.
4. Optimization of the quantized coefficients. Very few design algorithms perform this step. Given the type of arithmetic in the DSP and the structure for the filter, search algorithms can be programmed to find the best filter; however, it is easier to use some "rules of thumb" that are based on approximations.
5. Downloading the coefficients. If the DSP board is attached to a host computer, then the filter coefficients must be loaded to the DSP and the filtering program started.

11.5.1 Filter Design: Graphical User Interface

Operating systems and application programs based on windowing systems have interface building tools that provide an easy way to unify many algorithms under one view. This view concentrates on the filter specifications, so the designer can set up the problem once and then try many different approaches. If the view is a graphical rendition of the tolerance scheme, then the designer can also see the difference between the actual frequency response and the template. Buttons or menu choices can be given for all the different algorithms and parameters available.

With such a graphical user interface (GUI), the human is placed in the filter design loop. It has always been necessary for the human to be in the loop because filter design is the art of trading off many competing objectives. The filter design programs will optimize a mathematical criterion such as minimum L_p error, but that result might not exactly meet all the expectations of the designer. For example, trades between the length of an FIR implementation and the order of an IIR implementation can only be done by designing the individual filters and then comparing the order vs. length in a proposed implementation.

One implementation of the GUI approach to filter design can be found in a recent version of the MATLAB software.* The screen shot in Figure 11.40 shows the GUI window presented by sptool, which is the graphical tool for various signal processing operations, including filter design, in MATLAB version 5.0. In this case, the filter being designed is a length-23 FIR filter optimized for minimum Chebyshev error via the PM method for FIR design. The filter order was estimated from the ripples and bandedges, but in this case N is too small. The simultaneous graphical view of both the specifications and the actual frequency response makes it clear that the designed filter does meet the desired specifications.

In the MATLAB GUI, the user interface contains two types of controls: display modes and filter design specifications. The display mode buttons are located across the top of the window and are self-explanatory. The filter design specification fields and menus are at the left side of the window. Figure 11.41 shows these in more detail. Previously, we listed the different parameters needed to define the filter specifications: bandedges, ripple heights, etc. In the GUI, we see that each of these has an entry. The available design methods come from the pop-up menu that is presently set to "elliptic" in Figure 11.41.

* The screen shots were made with permission of the Mathworks, Inc.

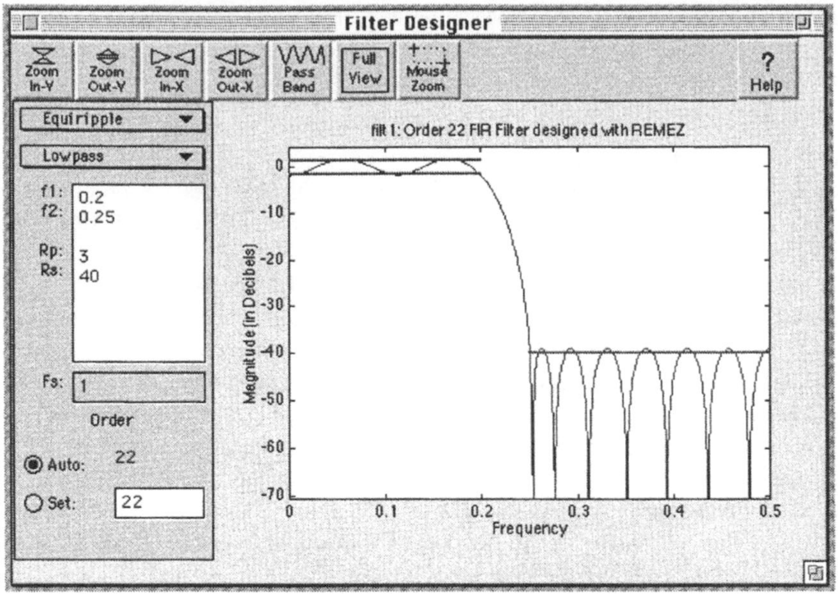

FIGURE 11.40 Screen shot from the MATLAB filter design tool called sptool. The equiripple filter was designed by the MATLAB function remez.

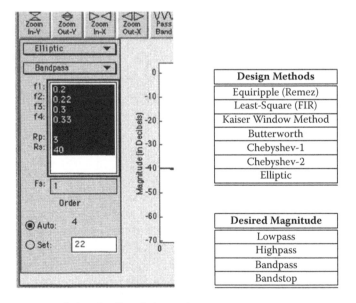

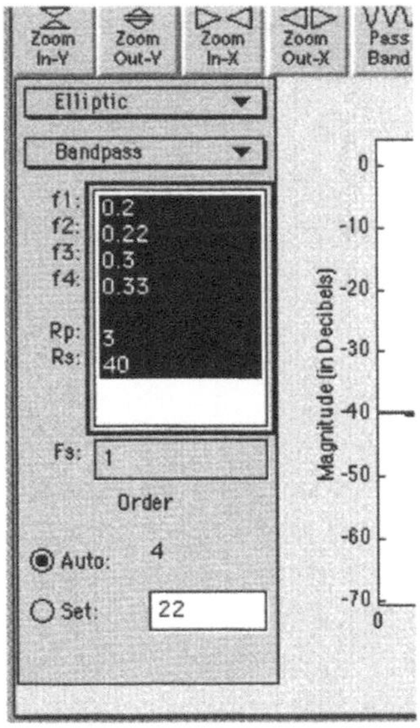

Design Methods
Equiripple (Remez)
Least-Square (FIR)
Kaiser Window Method
Butterworth
Chebyshev-1
Chebyshev-2
Elliptic

Desired Magnitude
Lowpass
Highpass
Bandpass
Bandstop

FIGURE 11.41 Pop-up menu choices for filter design options.

The design method must be chosen from the list given in Figure 11.41. The shape of the desired magnitude response must also be chosen from four types; in Figure 11.41, the type is set to "Bandpass," but the other choices are given in the list "Desired Magnitude." This elliptic bandpass filter is shown in Figure 11.44.

11.5.1.1 Bandedges and Ripples

An open box is provided so the user can enter numerical values for the parameters that define the boundaries of the tolerance scheme. In the bandpass case, four bandedges are needed, as well as the desired ripple heights for the passband and the two stopbands. The bandedges are denoted by f_1, f_2, f_3, and f_4 in Figure 11.41; the ripple heights (in decibel) by R_p and R_s. A value of $R_s = 40$ dB is taken to mean 40 dB of attenuation in both stopbands, i.e., $|\delta_s| \leq 0.01$. For the elliptic filter design, the ripples cannot be different in the two stopbands. The passband specification is the difference between the positive-going ripples at 1 and the negative-going ripples at $1 - \delta_p$:

$$R_p = -20\log_{10}(1 - \delta_p).$$

In the FIR case, the specification for R_p can be confusing because it is the total ripple which is the difference between the positive-going ripples at $1 + \delta_p$ and the negative-going ripples at $1 - \delta_p$:

$$R_p = 20\log_{10}(1 + \delta_p) - 20\log_{10}(1 - \delta_p).$$

In Figure 11.42, the value 3 dB is the same as $\delta_p \approx 0.171$. As the expanded view of the passband in Figure 11.42 shows, the ripples are not expected to be symmetric on a logarithmic scale. This expanded view for the FIR filter from Figure 11.40 was obtained by pressing the Pass Band button at the top.

11.5.1.2 Graphical Manipulation of the Specification Template

With the graphical view of the filter specifications, it is possible to use a pointing device such as a mouse to "grab" the specifications and move them around. This has the advantage that the relative placement of bandedges can be visualized while the movement is taking place. In the MATLAB GUI, the filter is quickly redesigned every time the mouse is released, so the user also gets immediate feedback on how close the filter approximation can be to the new specification. Order estimation is also done instantaneously, so the designer can develop some intuition concerning trade-offs such as transition width vs. filter order.

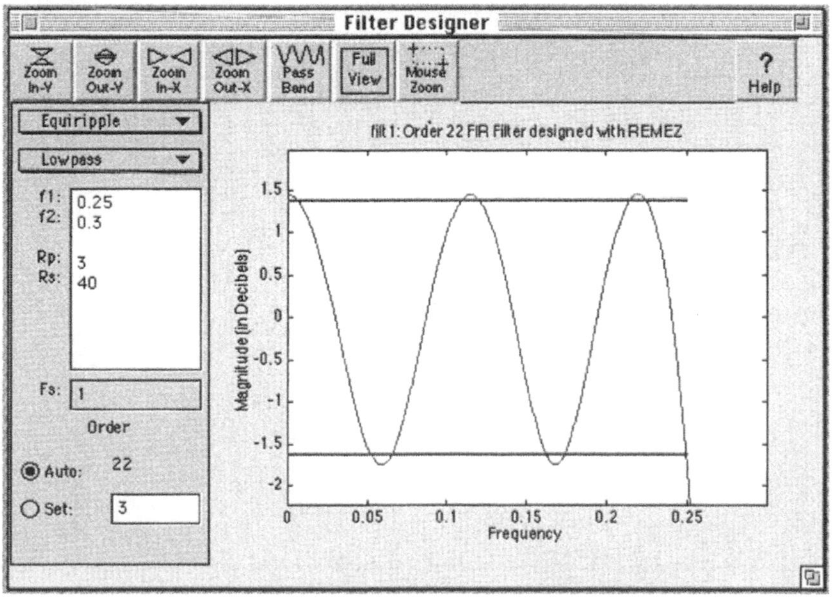

FIGURE 11.42 Expanded view of the passband of the lowpass filter from Figure 11.40.

11.5.1.3 Frequency Scaling

The field for F_s is useful when the filter specifications come from the "analog world", and are expressed in hertz with the sampling frequency given separately. Then the sampling frequency can be specified, and the horizontal axis is labeled and scaled in terms of F_s. Since the design is only carried out for $0 \leq \omega \leq \pi$, the highest frequency on the horizontal axis will be $F_s/2$. When $F_s = 1$, we say that the frequency is normalized and the numbers on the horizontal axis can be interpreted as a percentage

11.5.1.4 Automatic Order Estimation

Perhaps the most important feature of a software filter design package is its use of design rules. Since the design problem is always trying to trade off among the parameters of the specification, it is useful to be able to predict what the result will be without actually carrying out the design. A typical design formula involves the bandedges, the desired ripples and the filter order. For example, a simple approximate formula [12,37] for FIR filters designed by the Remez exchange method is

$$N(\omega_s - \omega_p) = \frac{-20\log_{10}\sqrt{\delta_p\delta_s} - 13}{2.324}. \tag{11.154}$$

Most often the desired filter is specified by $\{\omega_p, \omega_s, \delta_p, \delta_s\}$, so the design formula can be used to predict the filter order. Since most algorithms must work with a fixed number of parameters (determined by N) in doing optimization, this step is necessary before an iterative numerical optimization can be done.

The MATLAB GUI allows the user to turn on this order-estimating feature, so that an estimate of the filter order is calculated automatically whenever the filter specifications change. In the case of the FIR filters, the order-estimating formulae are only approximate—being derived from an empirical study of the parameters taken over many different designs. In some cases, the length N obtained is not large enough, and when the filter is designed it will fail to meet the desired specifications (see Figure 11.40). On the other hand, the Kaiser window design in Figure 11.43 does meet the specifications, even though its length (47) was also estimated from an approximate formula [12] similar to Equation 11.154.

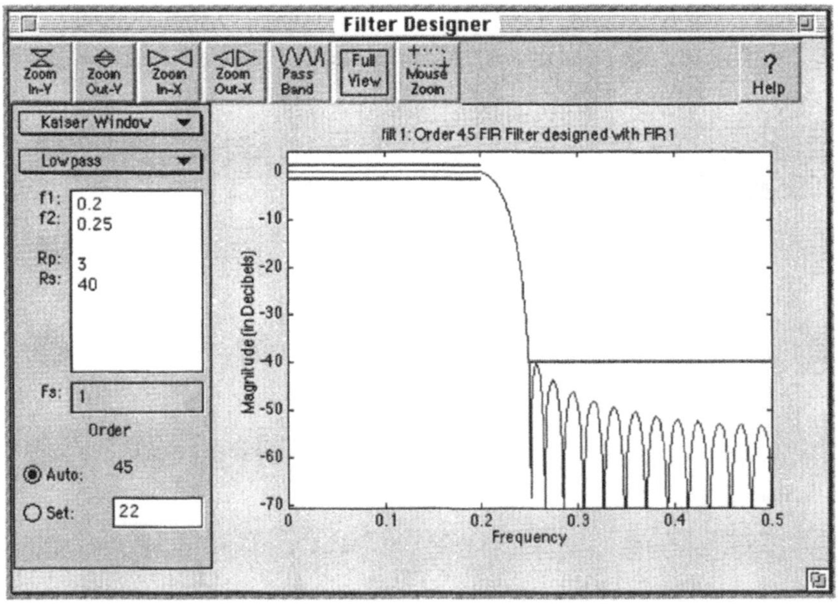

FIGURE 11.43 Length-47 FIR filter designed by the Kaiser window method. The order was estimated to be 46, and in this case the filter does meet the desired specifications.

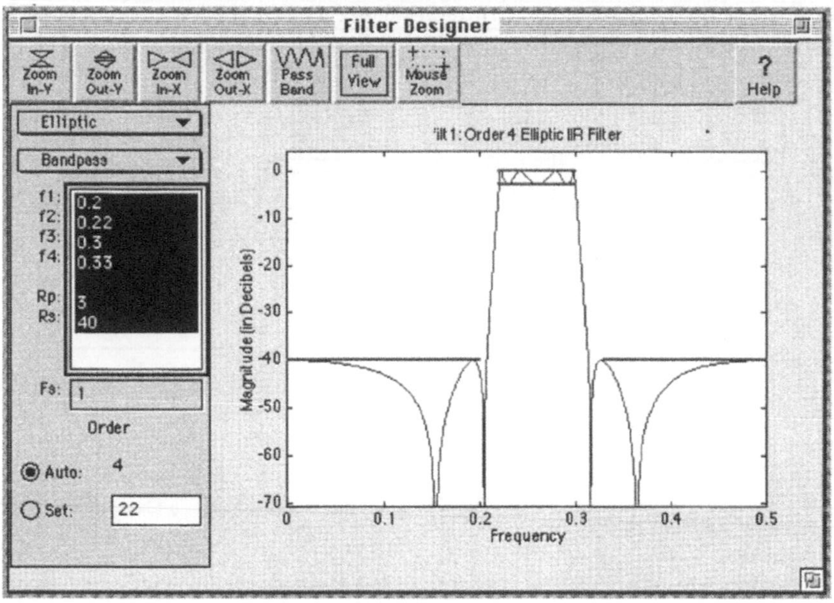

FIGURE 11.44 Eight-pole elliptic bandpass filter. The order was calculated to be 4, but the filter exceeds the desired specifications by quite a bit.

For the IIR case, however, the formulas are exact because they are derived from the mathematical properties of the Chebyshev polynomials or elliptic functions that define the classical filter types. Typically, the bandedges and the bilinear transformation define several simultaneous nonlinear equations that must be satisfied, but these can be solved in succession to get an order N that is guaranteed to work. The filter in Figure 11.44 shows the case where the order estimate was used for the bandpass design and the filter meets the specifications; but in Figure 11.45 the filter order was set to 3, which gave a sixth-order bandpass that fails to meet the specifications because its transition regions are too wide.

11.5.2 Filter Implementation

Another type of filter design tool ties in the filter's implementation with the design. Many DSP board vendors offer software products that perform filter design and then download the filter information to a DSP to process the data stream. Representative of this type of design is the DFDP-4/plus software* shown in the screen shots of Figures 11.46 through 11.51.

Similar to the MATLAB software, DFDP-4 can do the specification and design of the filter coefficients. In fact, it possesses an even wider range of filter design methods that includes filter banks and other special structures. It can design FIR filters based on the window method and the PM algorithm (an example is shown in Figure 11.46). For the IIR problem, the classical filter types (Butterworth, Chebyshev, and elliptic) are provided; Figure 11.47 shows an elliptic bandpass filter. In addition to the standard lowpass, highpass, and bandpass filter shapes, DFDP-4 can also handle the multiband case as well as filters with an arbitrary desired magnitude (as in Figure 11.51). When designing IIR filters, the phase response presents a difficulty because it is not linear or close to linear. The screen shot in

* DFDP is a trademark of Atlanta Signal Processors, Inc. The screen shots were made with permission of Atlanta Signal Processors, Inc.

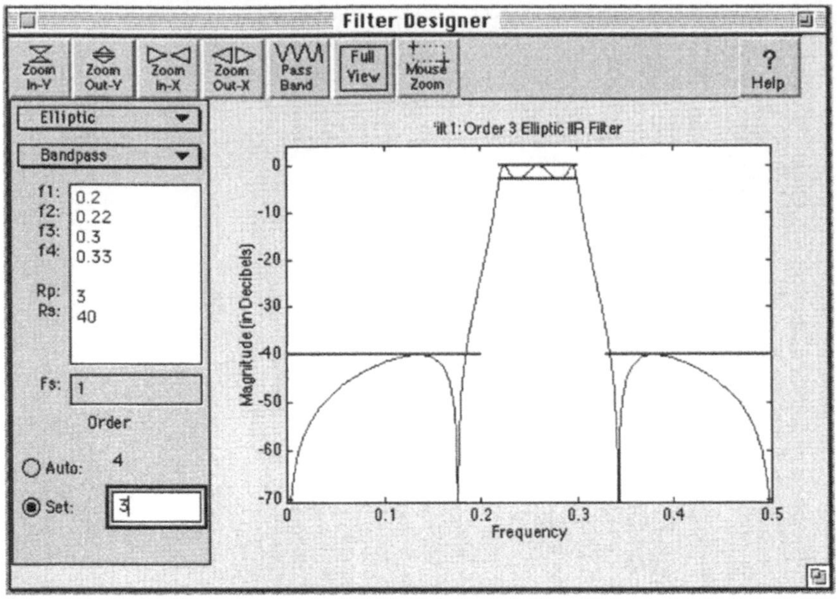

FIGURE 11.45 Six-pole elliptic bandpass filter. The order was set at 3, which is too small to meet the desired specifications.

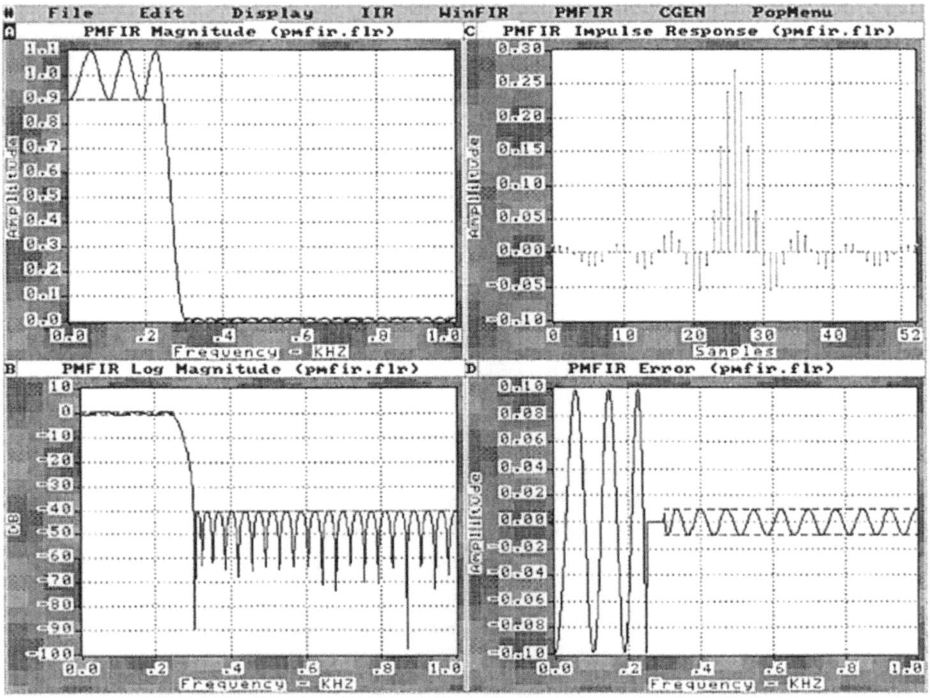

FIGURE 11.46 Length-57 FIR filter designed by the PM method, using the ASPI DFDP-4/plus software.

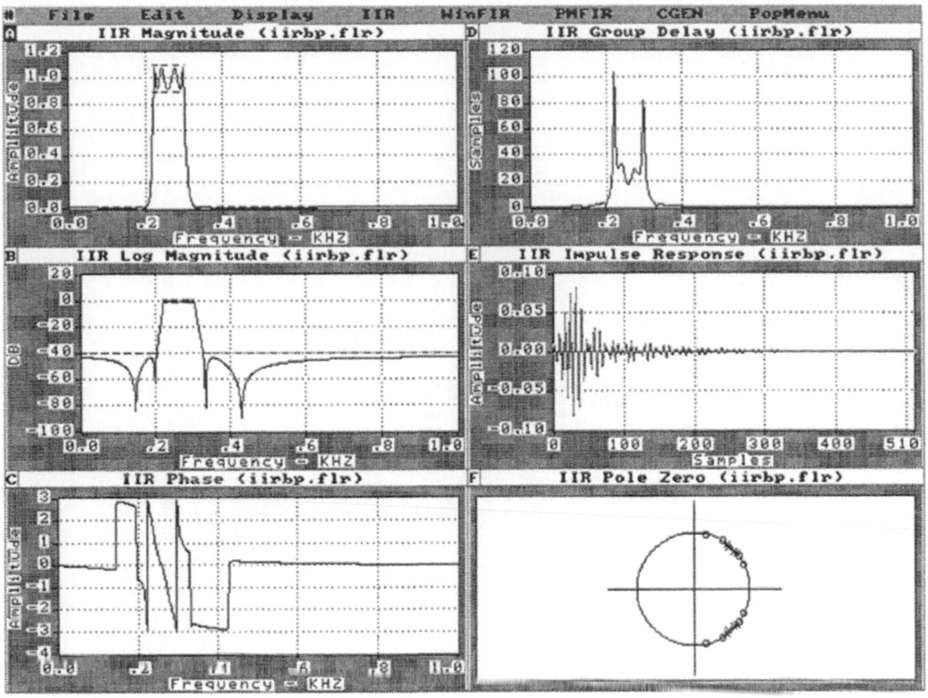

FIGURE 11.47 Eighth-order IIR bandpass elliptic filter designed using DFDP-4.

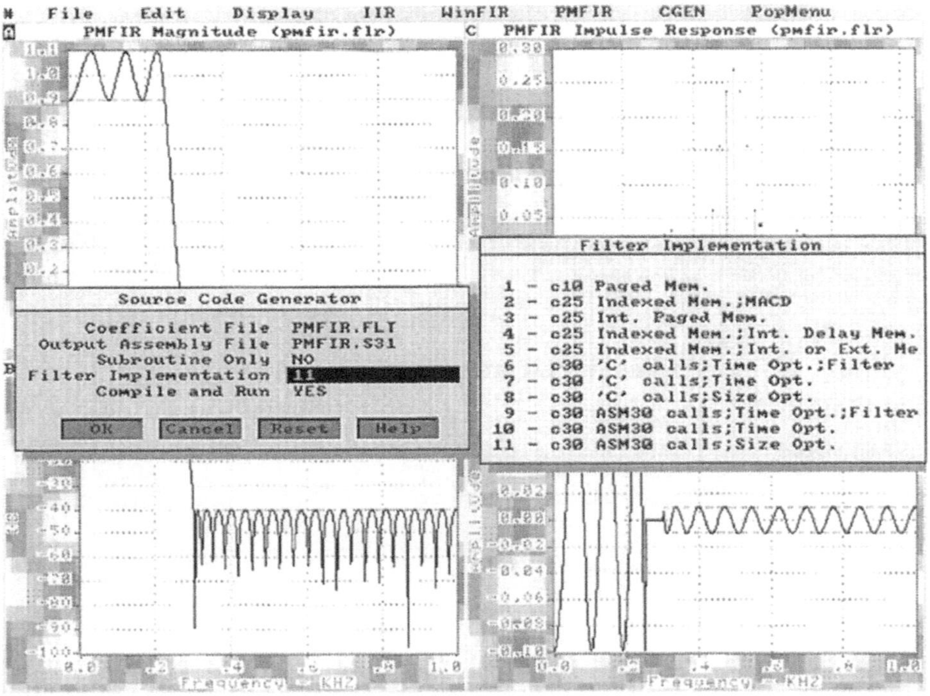

FIGURE 11.48 Code generation for an FIR filter using DFDP-4.

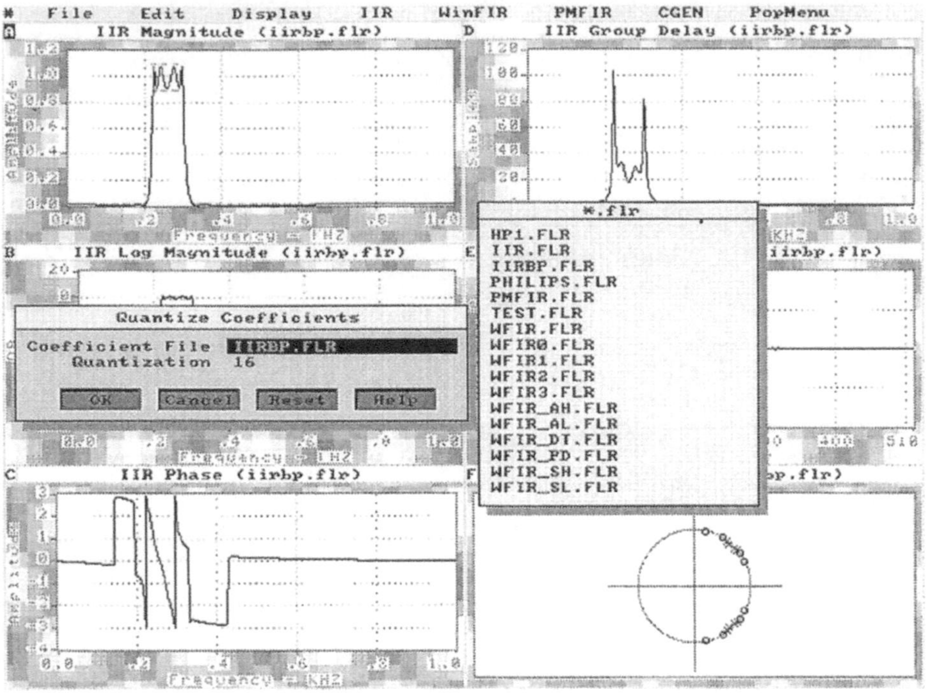

FIGURE 11.49 Eighth-order IIR bandpass elliptic filter with quantized coefficients.

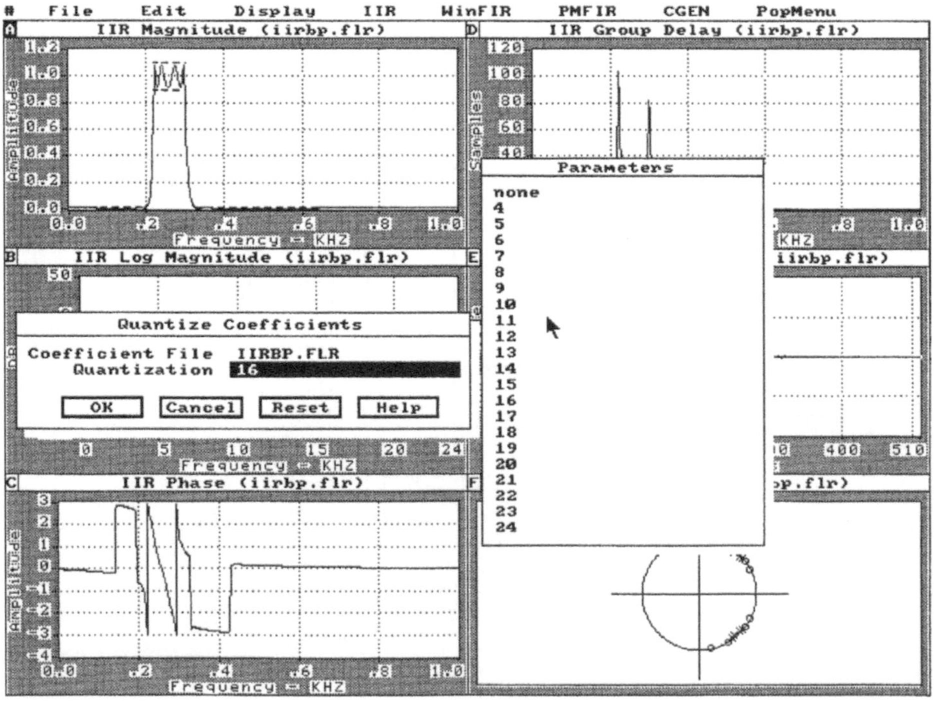

FIGURE 11.50 Eighth-order IIR bandpass elliptic filter, saving 16-bit coefficients.

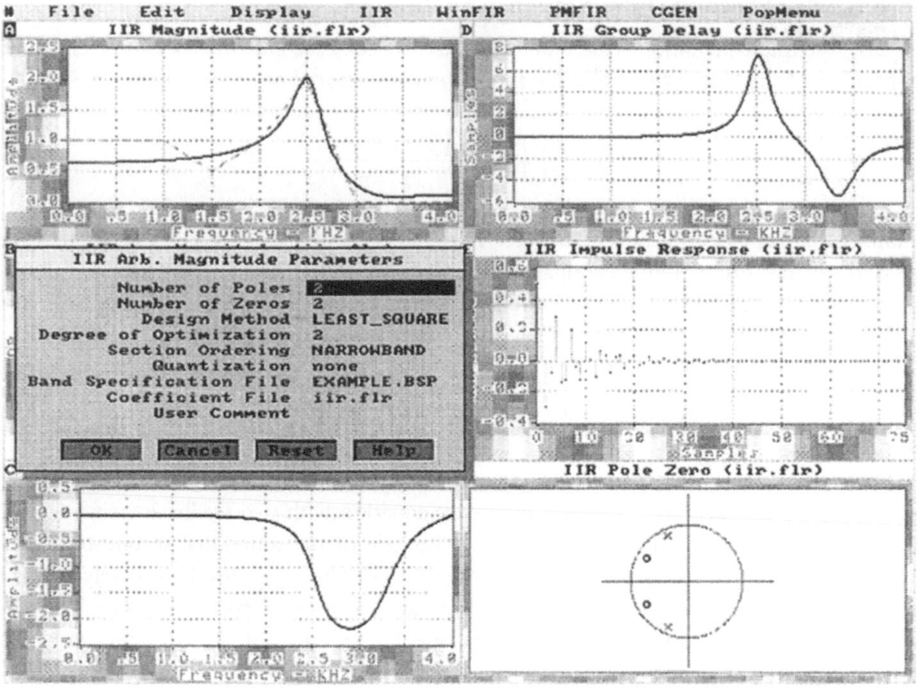

FIGURE 11.51 Arbitrary magnitude IIR filter.

Figure 11.47 shows the phase response in the lower left-hand panel and the group delay in the upper right-hand. The wide variation in the group delay, which is the derivative of the phase, indicates that the phase is far from linear. DFDP-4 provides an algorithm to optimize the group delay, which is a useful feature to compensate the phase response of an elliptic filter by using several allpass sections to flatten the group delay.

In DFDP-4, the filter design stage is specified by entering the bandedges and the desired ripples in dialog boxes until all the parameters are filled in for that type of design. Conflicts among the specifications can be resolved at this point before the design algorithm is invoked. For some designs such as the arbitrary magnitude design, the specification can involve many parameters to properly define the desired magnitude.

The filter design stage is followed by an implementation stage in which DFDP-4 produces the appropriate filter coefficients for either a fixed-point or floating-point implementation, targeted to a specific DSP microprocessor. The filter coefficients can be quantized over a range from 4 to 24 bits, as shown in Figure 11.50. The filter's frequency response would then be checked after quantization to compare with the designed filter and the original specifications. In the FIR case, coefficient quantization is the primary step needed prior to generating code for the DSP microprocessor, since the preferred implementation on a DSP is direct form. Internal wordlength scaling is also needed if a fixed-point implementation is being done. Once the wordlength is chosen, DFDP-4 will generate the entire assembly language program needed for the TMS-320 processor used on the boards supported by ASPI. As shown in Figure 11.48, there are a variety of supported processors, and even within a given processor family, the user can choose options such as "time optimization," "size optimization," etc. In Figure 11.48, the choice of "11" dictates a filter implementation on a TMS 320-C30, with ASM30 assembly language calls, and size optimization. The filter coefficients are taken from the file called PMFIR.FLT, and the assembly code is written to the file PMFIR.S31.

11.5.2.1 Cascade of Second-Order Sections

In the IIR case, the implementation is often done with a cascade of second-order sections. The numerator and denominator of the transfer function $H(z)$ must first be factored as

$$H(z) = \frac{B(z)}{A(z)} = \frac{G\prod_{i=1}^{M}(1 - z_i z^{-1})}{\prod_{i=1}^{N}(1 - p_i z^{-1})}, \tag{11.155}$$

where p_i and z_i are the poles and zeros of the filter. In the screen shot of Figure 11.47 we see that the poles and zeros of the eighth-order elliptic bandpass filter are displayed to the user. The second-order sections are obtained by grouping together two poles and two zeros to create each second-order section; conjugate pairs must be kept together if the filter coefficients are going to be real:

$$H(z) = \frac{B(z)}{A(z)} = \prod_{k=1}^{N/2} \frac{\beta_{0k} + \beta_{1k}z^{-1} + \beta_{2k}z^{-2}}{1 + \alpha_{1k}z^{-1} + \alpha_{2k}z^{-2}}. \tag{11.156}$$

Each second-order factor defines a recursive difference equation with two feedback terms: α_{1k} and α_{2k}. The product of all the sections is implemented as a cascade of the individual second-order feedback filters. This implementation has the advantage that the overall filter response is relatively insensitive to coefficient quantization and roundoff noise when compared to a direct form structure. Therefore, the cascaded second-order sections provide a robust implementation, especially for IIR filters with poles very close to the U.C.

Clearly, there are many different ways to pair the poles and zeros when defining the second-order sections. Furthermore, there are many different orderings for the cascade, and each one will produce different noise gains through the filter. Sections with a pole pair close to the U.C. will be extremely narrowband with a very high gain at one frequency. The rules of thumb originally developed by Jackson [138] give good orderings depending on the nature of the input signal—wideband vs. narrowband. This choice can be seen in Figure 11.51 where the section ordering slot is set to NARROWBAND.

11.5.2.2 Scaling for Fixed-Point

A second consideration when ordering the second-order sections is the problem of scaling to avoid overflow. This issue only arises when the IIR filter is targeted to a fixed-point DSP microprocessor. Since the gain of individual sections may vary widely, the fixed-point data might overflow beyond the maximum value allowed by the wordlength. To combat this problem, multipliers (or shifters that multiply by a power of 2) can be inserted in-between the cascaded sections to guard against overflow. However, dividing by two will shift bits off the lower end of the fixed-point word, thereby introducing more roundoff noise. The value of the scaling factor can be approximated via a worst-case analysis that prevents overflow entirely, or a mean square method that reduces the likelihood of overflow depending on the input signal characteristics.

Proper treatment of the scaling problem requires that it be solved in conjunction with the ordering of sections for minimal roundoff noise. Similar "rules of thumb" can be employed to get a good (if not optimal) implementation that simultaneously addresses ordering, pole–zero pairing, and scaling [138]. The theoretical problem of optimizing the implementation for word length and noise performance is rarely done because it is such a difficult problem, and not one for which an efficient solution has been found. Thus, most software tools rely on approximations to perform the implementation and code-generation steps quickly.

Once the transfer function is factored into second-order sections, the code-generation phase creates the assembly language program that will actually execute in the DSP and downloads it to the DSP board. Coefficient quantization is done as part of the assembly code generation. With the program loaded into the DSP, tests on real-time data streams can be conducted.

11.5.2.3 Comments and Summary

The two design tools presented here are representative of the capabilities that one should expect in a state of the art filter design package. There are many software design products available and most of them have similar characteristics, but may be more powerful in some respects, for example, more design algorithm choices, different DSP microprocessor support, alternative display options, etc. A user can choose a design tool with these criteria in mind, confident that the GUI will make it relatively easy to use the powerful mathematical design algorithms without learning the idiosyncrasies of each method. The uniform view of the GUI as managing the filter specifications should simplify the design process, while allowing the best possible filters to be designed through trial and comparison.

One limiting aspect of the GUI filter design tool is that it can easily do magnitude approximation, but only for the standard cases of bandpass and multiband filters. It is easy to envision, however, that the GUI could support graphical user entry of the specifications by having the user draw the desired magnitude. Then other magnitude shapes could be supported, as in DFDP-4. Another extension would be to provide a graphical input for the desired phase response, or group delay, in addition to the magnitude specification. Although a great majority of filter designs are done for the bandpass case, there has been a recent surge of interest in having the flexibility to do simultaneous magnitude and phase approximation. With the development of better general magnitude and phase design methods, the filter design packages now offer this capability.

References

1. Oppenheim, A.V. and Schafer, R.W. *Discrete-Time Signal Processing*, Prentice-Hall, Englewood Cliffs, NJ, 1989.
2. Karam, L.J. and McClellan, J.H. Complex Chebyshev approximation for FIR filter design, *IEEE Trans. Circuits Sys. II*, 42, 207–216, Mar. 1995.
3. Karam, L.J. and McClellan, J.H. Design of optimal digital FIR filters with arbitrary magnitude and phase responses, *Proceedings of the IEEE International Symposium on Circuits and Systems*, Atlanta, GA, May 1996, Vol. 2, pp. 385–388.
4. Burnside, D. and Parks, T.W. Optimal design of FIR filters with the complex Chebyshev error criteria, *IEEE Trans. Signal Process.*, 43, 605–616, Mar. 1995.
5. Preuss, K. On the design of FIR filters by complex Chebyshev approximation, *IEEE Trans. Acoust. Speech Signal Process.*, 37, 702–712, May 1989.
6. Parks, T.W. and McClellan, J.H. Chebyshev approximation for nonrecursive digital filters with linear phase, *IEEE Trans. Circuit Theory*, CT-19, 189–194, Mar. 1972.
7. Steiglitz, K., Parks, T.W., and Kaiser, J.F. METEOR: A constraint-based FIR filter design program, *IEEE Trans. Signal Process.*, 40, 1901–1909, Aug. 1992.
8. Selesnick, I.W., Lang, M., and Burrus, C.S. Constrained least square design of FIR filters without specified transition bands, *IEEE Trans. Signal Process.*, 44, 1879–1892, Aug. 1996.
9. Proakis, J.G. and Manolakis, D.G. *Digital Signal Processing: Principles, Algorithms, and Applications*, Prentice-Hall, Englewood Cliffs, NJ, 1996.
10. Karam, L.J. and McClellan, J.H. Design of optimal digital FIR filters with arbitrary magnitude and phase responses, in *Circuits and Systems, ISCAS'96, Connecting the World, 1996 IEEE International Symposium*, 2, 385–388, May 1996.
11. Parks, T.W. and Burrus, C.S. *Digital Filter Design*, John Wiley & Sons, New York, 1987.
12. Kaiser, J.F. Nonrecursive digital filter design using the I_0–sinh window function, *Proceedings of the IEEE International Symposium on Circuits and Systems (ISCAS)*, San Francisco, CA, Apr. 1974, pp. 20–23.
13. Slepian, D. Prolate spheroidal wave functions, Fourier analysis and uncertainty, *Bell Syst. Tech. J.*, 57, 1371–1430, May–June 1978.

14. Gruenbacher, D.M. and Hummels, D.R. A simple algorithm for generating discrete prolate spheroidal sequences, *IEEE Trans. Signal Process.*, 42, 3276–3278, Nov. 1994.
15. Percival, D.B. and Walden, A.T. *Spectral Analysis for Physical Applications: Multitaper and Conventional Univariate Techniques*, Cambridge University Press, Cambridge, U.K., 1993.
16. Verma, T., Bilbao, S., and Meng, T.H.Y. The digital prolate spheroidal window, *Proceedings of the IEEE International Conference on Acoustics, Speech, and Signal Processing (ICASSP)*, Atlanta, GA, May 7–10, 1996, Vol. 3, pp. 1351–1354.
17. Saramäki, T. Finite impulse response filter design, in *Handbook for Digital Signal Processing*, Mitra, S.K. and Kaiser, J.F. (Eds.), John Wiley & Sons, New York, 1993, Chapter 4, pp. 155–277.
18. Saramäki, T. Adjustable windows for the design of FIR filters—A tutorial, *Proceedings of the 6th Mediterranean Electrotechnical Conference*, Ljubljana, Yugoslavia, May 22–24, 1991, pp. 28–33.
19. Elliot, D.F. *Handbook of Digital Signal Processing*, Academic Press, New York, 1987.
20. Cain, G.D., Yardim, A., and Henry, P. Offset windowing for FIR fractional-sample delay, *Proceedings of the IEEE International Conference on Acoustics, Speech, and Signal Processing (ICASSP)*, Detroit, MI, May 9–12, 1995, pp. 1276–1279.
21. Laakso, T.I., Välimäki, V., Karjalainen, M., and Laine, U.K. Splitting the unit delay, *IEEE Signal Process. Mag.*, 13, 30–60, Jan. 1996.
22. Gopinath, R.A. Thoughts on least square-error optimal windows, *IEEE Trans. Signal Process.*, 44, 984–987, Apr. 1996.
23. Weisburn, E.A., Parks, T.W., and Shenoy, R.G. Error criteria for filter design, *Proceedings of the IEEE International Conference on Acoustics, Speech, and Signal Processing (ICASSP)*, Adelaide, Australia, April 19–22, 1994, Vol. 3, pp. 565–568.
24. Merchant, G.A. and Parks, T.W. Efficient solution of a Toeplitz-plus-Hankel coefficient matrix system of equations, *IEEE Trans. Acoust. Speech Signal Process.*, 30, 40–44, Feb. 1982.
25. Burrus, C.S., Soewito, A.W., and Gopinath, R.A. Least squared error FIR filter design with transition bands, *IEEE Trans. Signal Process.*, 40, 1327–1340, June 1992.
26. Burrus, C.S. Multiband least squares FIR filter design, *IEEE Trans. Signal Process.*, 43, 412–421, Feb. 1995.
27. Vaidyanathan, P.P. and Nguyen, T.Q. Eigenfilters: A new approach to least-squares FIR filter design and applications including nyquist filters, *IEEE Trans. Circuits Syst.*, 34, 11–23, Jan. 1987.
28. Powel, M.J.D. *Approximation Theory and Methods*, Cambridge University Press, New York, 1981.
29. Rabiner, L.R., McClellan, J.H., and Parks, T.W. FIR digital filter design techniques using weighted Chebyshev approximation, *Proc. IEEE*, 63, 595–610, Apr. 1975.
30. Rabiner, L.R. and Gold, B. *Theory and Application of Digital Signal Processing*, Prentice-Hall, Englewood Cliffs, NJ, 1975.
31. McClellan, J.H., Parks, T.W., and Rabiner, L.R. A computer program for designing optimum FIR linear phase digital filters, *IEEE Trans. Audio Electroacoust.*, 21, 506–526, Dec. 1973.
32. McClellan, J.H. On the design of one-dimensional and two-dimensional fir digital filters, PhD thesis, Rice University, Houston, TX, Apr. 1973.
33. Herrmann, O. Design of nonrecursive filters with linear phase, *Electron. Lett.*, 6, 328–329, May 28, 1970.
34. Hofstetter, E., Oppenheim, A., and Siegel, J. A new technique for the design of nonrecursive digital filters, *Proceedings of Fifth Annual Princeton Conference on Information Sciences and Systems*, Princeton, NJ, Oct. 1971, pp. 64–72.
35. Parks, T.W. and McClellan, J.H. On the transition region width of finite impulse-response digital filters, *IEEE Trans. Audio Electroacoust.*, 21, 1–4, Feb. 1973.
36. Rabiner, L.R. Approximate design relationships for lowpass FIR digital filters, *IEEE Trans. Audio Electroacoust.*, 21, 456–460, Oct. 1973.
37. Herrmann, O., Rabiner, L.R., and Chan, D.S.K. Practical design rules for optimum finite impulse response lowpass digital filters, *Bell Syst. Tech. J.*, 52, 769–799, 1973.

38. Selesnick, I.W. and Burrus, C.S. Exchange algorithms that complement the Parks-McClellan algorithm for linear phase FIR filter design, *IEEE Trans. Circuits Syst. II*, 44(2), 137–143, Feb. 1997.
39. de Saint-Martin, F.M. and Siohan, P. Design of optimal linear-phase transmitter and receiver filters for digital systems, *Proceedings of IEEE International Symposium Circuits and Systems (ISCAS)*, Seattle, WA, Apr. 30–May 3, 1995, Vol. 2, pp. 885–888.
40. Thiran, J.P. Recursive digital filters with maximally flat group delay, *IEEE Trans. Circuit Theory*, 18, 659–664, Nov. 1971.
41. Saramäki, T. and Neuvo, Y. Digital filters with equiripple magnitude and group delay, *IEEE Trans. Acoust. Speech Signal Process.*, 32, 1194–1200, Dec. 1984.
42. Jackson, L.B. An improved Martinez/Parks algorithm for IIR design with unequal numbers of poles and zeros, *IEEE Trans. Signal Process.*, 42, 1234–1238, May 1994.
43. Liang, J. and Figueiredo, R.J.P.D. An efficient iterative algorithm for designing optimal recursive digital filters, *IEEE Trans. Acoust. Speech Signal Process.*, 31, 1110–1120, Oct. 1983.
44. Martinez, H.G. and Parks, T.W. Design of recursive digital filters with optimum magnitude and attenuation poles on the unit circle, *IEEE Trans. Acoust. Speech Signal Process.*, 26, 150–156, Apr. 1978.
45. Saramäki, T. Design of optimum wideband recursive digital filters, *Proceedings of IEEE International Symposium on Circuits and Systems (ISCAS)*, Rome, Italy, May 10–12, 1982, pp. 503–506.
46. Saramäki, T. Design of digital filters with maximally flat passband and equiripple stopband magnitude, *Int. J. Circuit Theory Appl.*, 13, 269–286, Apr. 1985.
47. Unbehauen, R. On the design of recursive digital low-pass filters with maximally flat pass-band and Chebyshev stop-band attenuation, *Proceedings of IEEE International Symposium on Circuits and Systems (ISCAS)*, Chicago, IL, 1981, pp. 528–531.
48. Zhang, X. and Iwakura, H. Design of IIR digital filters based on eigenvalue problem, *IEEE Trans. Signal Process.*, 44, 1325–1333, June 1996.
49. Saramäki, T. Design of optimum recursive digital filters with zeros on the unit circle, *IEEE Trans. Acoust. Speech Signal Process.*, 31, 450–458, Apr. 1983.
50. Selesnick, I.W. and Burrus, C.S. Generalized digital Butterworth filter design, *Proceedings of IEEE International Conference on Acoustics, Speech, and Signal Processing (ICASSP)*, Atlanta, GA, May 7–10, 1996, pp. 1367–1370.
51. Samadi, S., Cooklev, T., Nishihara, A., and Fujii, N. Multiplierless structure for maximally flat linear phase FIR filters, *Electron. Lett.*, 29, 184–185, Jan. 21, 1993.
52. Vaidyanathan, P.P. On maximally-flat linear-phase FIR filters, *IEEE Trans. Circuits Syst.*, 31, 830–832, Sept. 1984.
53. Vaidyanathan, P.P. Efficient and multiplierless design of FIR filters with very sharp cutoff via maximally flat building blocks, *IEEE Trans. Circuits Syst.*, 32, 236–244, Mar. 1985.
54. Neuvo, Y., Dong, C.-Y., and Mitra, S.K. Interpolated finite impulse response filters, *IEEE Trans. Acoust. Speech Signal Process.*, 32, 563–570, June 1984.
55. Herrmann, O. On the approximation problem in nonrecursive digital filter design, *IEEE Trans. Circuit Theory*, 18, 411–413, May 1971.
56. Rajagopal, L.R. and Roy, S.C.D. Design of maximally-flat FIR filters using the Bernstein polynomial, *IEEE Trans. Circuits Syst.*, 34, 1587–1590, Dec. 1987.
57. Daubechies, I. *Ten Lectures On Wavelets*, SIAM, Philadelphia, PA, 1992.
58. Kaiser, J.F. Design subroutine (MXFLAT) for symmetric FIR low pass digital filters with maximally-flat pass and stop bands, in *Programs for Digital Signal Processing*, I.A.S. Digital Signal Processing Committee (Ed.), IEEE Press, New York, 1979, Chapter 5.3, pp. 5.3-1–5.3-6.
59. Jinaga, B.C. and Roy, S.C.D. Coefficients of maximally flat low and high pass nonrecursive digital filters with specified cutoff frequency, *Signal Process.*, 9, 121–124, Sept. 1985.
60. Thajchayapong, P., Puangpool, M., and Banjongjit, S. Maximally flat FIR filter with prescribed cutoff frequency, *Electron. Lett.*, 16, 514–515, June 19, 1980.

61. Rabenstein, R. Design of FIR digital filters with flatness constraints for the error function, *Circuits Syst. Signal Process.*, 13(1), 77–97, 1993.
62. Schüssler, H.W. and Steffen, P. An approach for designing systems with prescribed behavior at distinct frequencies regarding additional constraints, *Proceedings of IEEE International Conference on Acoustics, Speech, and Signal Processing (ICASSP)*, Tampa, FL, Apr. 1985, Vol. 10, pp. 61–64.
63. Schüssler, H.W. and Steffen, P. Some advanced topics in filter design, in *Advanced Topics in Signal Processing*, Lim, J.S. and Oppenheim, A.V. (Eds.), Prentice-Hall, Englewood Cliffs, NJ, 1988, Chapter 8, pp. 416–491.
64. Adams, J.W. and Willson, A.N., Jr., A new approach to FIR digital filter with fewer multipliers and reduced sensitivity, *IEEE Trans. Circuits Syst.*, 30, 277–283, May 1983.
65. Adams, J.W. and Willson, A.N., Jr., Some efficient prefilter structures, *IEEE Trans. Circuits Syst.*, 31, 260–266, Mar. 1984.
66. Hartnett, R.J. and Boudreaux-Bartels, G.F. On the use of cyclotomic polynomials prefilters for efficient FIR filter design, *IEEE Trans. Signal Process.*, 41, 1766–1779, May 1993.
67. Oh, W.J. and Lee, Y.H. Design of efficient FIR filters with cyclotomic polynomial prefilters using mixed integer linear programming, *Proceedings of IEEE International Conference on Acoustics, Speech, and Signal Processing (ICASSP)*, Atlanta, GA, May 1996, pp. 1287–1290.
68. Lang, M. Optimal weighted phase equalization according to the l_∞-norm, *Signal Process.*, 27, 87–98, Apr. 1992.
69. Leeb, F. and Henk, T. Simultaneous amplitude and phase approximation for FIR filters, *Int. J. Circuit Theory Appl.*, 17, 363–374, July 1989.
70. Herrmann, O. and Schüssler, H.W. Design of nonrecursive filters with minimum phase, *Electron. Lett.*, 6, 329–330, May 28, 1970.
71. Baher, H. FIR digital filters with simultaneous conditions on amplitude and delay, *Electron. Lett.*, 18, 296–297, Apr. 1, 1982.
72. Calvagno, G., Cortelazzo, G.M., and Mian, G.A. A technique for multiple criterion approximation of FIR filters in magnitude and group delay, *IEEE Trans. Signal Process.*, 43, 393–400, Feb. 1995.
73. Rhodes, J.D. and Fahmy, M.I.F. Digital filters with maximally flat amplitude and delay characteristics, *Int. J. Circuit Theory Appl.*, 2, 3–11, Mar. 1974.
74. Sullivan, J.L. and Adams, J.W. A new nonlinear optimization algorithm for asymmetric FIR digital filters, *Proceedings of IEEE International Symposium on Circuits and Systems (ISCAS)*, London, U.K., May 30–June 2, 1994, Vol. 2, pp. 541–544.
75. Scanlan, S.O. and Baher, H. Filters with maximally flat amplitude and controlled delay responses, *IEEE Trans. Circuits and Systems*, 23, 270–278, May 1976.
76. Rice, J.R. *The Approximation of Functions*, Addison-Wesley, Reading, MA, 1969.
77. Alkhairy, A.S., Christian, K.S., and Lim, J.S. Design and characterization of optimal FIR filters with arbitrary phase, *IEEE Trans. Signal Process.*, 41, 559–572, Feb. 1993.
78. Karam, L.J. Design of complex digital FIR filters in the Chebyshev sense, PhD thesis, Georgia Institute of Technology, Atlanta, GA, Mar. 1995.
79. Meinardus, G. *Approximation of Functions: Theory and Numerical Methods*, Springer-Verlag, New York, 1967.
80. McCallig, M.T. Design of digital FIR filters with complex conjugate pulse responses, *IEEE Trans. Circuit Syst.*, CAS-25, 1103–1105, Dec. 1978.
81. Cheney, E.W. *Introduction to Approximation Theory*, McGraw-Hill, New York, 1966.
82. Demjanov, V.F. Algorithms for some minimax problems, *J. Comput. Syst. Sci.*, 2, 342–380, 1968.
83. Demjanov, V.F and Malozemov, V.N. *Introduction to Minimax*, John Wiley & Sons, New York, 1974.
84. Wolfe, P. Finding the nearest point in a polytope, *Math. Programming*, 11, 128–149, 1976.
85. Wolfe, P. A method of conjugate subgradients for minimizing nondifferentiable functions, *Math. Programming Study*, 3, 145–173, 1975.

86. Lorentz, G.G. *Approximation of Functions*, Holt, Rinehart and Winston, New York, 1966.
87. Feuer, A. Minimizing well-behaved functions, *Proceedings of 12th Annual Allerton Conference on Circuit and System Theory*, Allerton, IL, Oct. 1974, pp. 15–34.
88. Watson, G.A. The calculation of best restricted approximations, *SIAM J. Numerical Anal.*, 11, 693–699, Sept. 1974.
89. Chen, X. and Parks, T.W. Design of FIR filters in the complex domain, *IEEE Trans. Acoust. Speech Signal Process.*, ASSP-35, 144–153, Feb. 1987.
90. Harris, D.B. Design and implementaion of rational 2-D digital filters, PhD thesis, Massachusetts Institute of Technology, Cambridge, MA, Nov. 1979.
91. Claerbout, J. *Fundamentals of Geophysical Data Processing*, McGraw-Hill, New York, 1976.
92. Hale, D. 3-D depth migration via McClellan transformations, *Geophysics*, 56, 1778–1785, Nov. 1991.
93. Dudgeon, D.E. and Mersereau, R.M. *Multidimensional Digital Signal Processing*, Prentice-Hall, Englewood Cliffs, NJ, 1984.
94. Selesnick, I.W. New techniques for digital filter design, PhD thesis, Rice University, Houston, TX, 1996.
95. Orfanidis, S.J. *Introduction to Signal Processing*, Prentice-Hall, Englewood Cliffs, NJ, 1996.
96. Steffen, P. On digital smoothing filters: A brief review of closed form solutions and two new filter approaches, *Circuits Syst. Signal Process.*, 5(2), 187–210, 1986.
97. Vaidyanathan, P.P. Optimal design of linear-phase FIR digital filters with very flat passbands and equiripple stopbands, *IEEE Trans. Circuits Syst.*, 32, 904–916, Sept. 1985.
98. Kaiser, J.F. and Steiglitz, K. Design of FIR filters with flatness constraints, *Proceedings of IEEE International Conference on Acoustics, Speech, and Signal Processing (ICASSP)*, Boston, MA, 1983, Vol. 8, pp. 197–200.
99. Selesnick, I.W. and Burrus, C.S. Exchange algorithms for the design of linear phase FIR filters and differentiators having flat monotonic passbands and equiripple stopbands, *IEEE Trans. Circuits Syst. II*, 43, 671–675, Sept. 1996.
100. Adams, J.W. FIR digital filters with least squares stop bands subject to peak-gain constraints, *IEEE Trans. Circuits Syst.*, 39, 376–388, Apr. 1991.
101. Adams, J.W., Sullivan, J.L., Hashemi, R., Ghadimi, R., Franklin, J., and Tucker, B. New approaches to constrained optimization of digital filters, *Proceedings of IEEE International Symposium on Circuits and Systems (ISCAS)*, Chicago, IL, May 1993, Vol. 1, pp. 80–83.
102. Barrodale, I., Powell, M.J.D., and Roberts, F.D.K. The differential correction algorithm for rational L_∞-approximation, *SIAM J. Numerical Anal.*, 9, 493–504, Sept. 1972.
103. Crosara, S. and Mian, G.A. A note on the design of IIR filters by the differential-correction algorithm, *IEEE Trans. Circuits Syst.*, 30, 898–903, Dec. 1983.
104. Dudgeon, D.E. Recursive filter design using differential correction, *IEEE Trans. Acoust. Speech Signal Process.*, 22, 443–448, Dec. 1974.
105. Kaufman, E.H., Jr., Leeming, D.J., and Taylor, G.D. A combined Remes-differential correction algorithm for rational approximation, *Math. Comput.*, 32, 233–242, Jan. 1978.
106. Rabiner, L.R., Graham, N.Y., and Helms, H.D. Linear programming design of IIR digital filters with arbitrary magnitude function, *IEEE Trans. Acoust. Speech Signal Process.*, 22, 117–123, Apr. 1974.
107. Deczky, A.G. Synthesis of recursive digital filters using the minimum p-error criterion, *IEEE Trans. Audio Electroacoust.*, 20, 257–263, Oct. 1972.
108. Renfors, M. and Zigouris, E. Signal processor implementation of digital all-pass filters, *IEEE Trans. Acoust. Speech Signal Process.*, 36, 714–729, May 1988.
109. Vaidyanathan, P.P., Mitra, S.K., and Neuvo, Y. A new approach to the realization of low-sensitivity IIR digital filters, *IEEE Trans. Acoust. Speech Signal Process.*, 34, 350–361, Apr. 1986.
110. Regalia, P.A., Mitra, S.K., and Vaidyanathan, P.P. The digital all-pass filter: A versatile signal processing building block, *Proc. IEEE*, 76, 19–37, Jan. 1988.

111. Vaidyanathan, P.P., Regalia, P.A., and Mitra, S.K. Design of doubly-complementary IIR digital filters using a single complex allpass filter, with multirate applications, *IEEE Trans. Circuits Syst.*, 34, 378–389, Apr. 1987.
112. Vaidyanathan, P.P. *Multirate Systems and Filter Banks*, Prentice-Hall, Englewood Cliffs, NJ, 1993.
113. Gerken, M., Schüßler, H.W., and Steffen, P. On the design of digital filters consisting of a parallel connection of allpass sections and delay elements, *Archiv für Electronik und Übertragungstechnik (AEÜ)*, 49, 1–11, Jan. 1995.
114. Jaworski, B. and Saramäki, T. Linear phase IIR filters composed of two parallel allpass sections, *Proceedings of IEEE International Symposium on Circuits and Systems (ISCAS)*, London, U.K., May 30–June 2, 1994, Vol. 2, pp. 537–540.
115. Kim, C.W. and Ansari, R. Approximately linear phase IIR filters using allpass sections, *Proceedings of IEEE International Symposium on Circuits and Systems (ISCAS)*, San Jose, CA, May 5–7, 1986, pp. 661–664.
116. Renfors, M. and Saramäki, T. A class of approximately linear phase digital filters composed of allpass subfilters, *Proceedings of IEEE International Symposium on Circuits and Systems (ISCAS)*, San Jose, CA, May 5–7, 1986, pp. 678–681.
117. Chen, C.-K. and Lee, J.-H. Design of digital all-pass filters using a weighted least squares approach, *IEEE Trans. Circuits Syst. II*, 41, 346–351, May 1994.
118. Kidambi, S.S. Weighted least-squares design of recursive allpass filters, *IEEE Trans. Signal Process.*, 44, 1553–1556, June 1996.
119. Lang, M. and Laakso, T. Simple and robust method for the design of allpass filters using least-squares phase error criterion, *IEEE Trans. Circuits Syst. II*, 41, 40–48, Jan. 1994.
120. Nguyen, T.Q., Laakso, T.I., and Koilpillai, R.D. Eigenfilter approach for the design of allpass filters approximating a given phase response, *IEEE Trans. Signal Process.*, 42, 2257–2263, Sept. 1994.
121. Pei, S.-C. and Shyu, J.-J. Eigenfilter design of 1-D and 2-D IIR digital all-pass filters, *IEEE Trans. Signal Process.*, 42, 966–968, Apr. 1994.
122. Schüßler, H.W. and Steffan, P. On the design of allpasses with prescribed group delay, *Proceedings of IEEE International Conference on Acoustics, Speech, and Signal Processing (ICASSP)*, Albuquerque, NM, Apr. 3–6, 1990, pp. 1313–1316.
123. Anderson, M.S. and Lawson, S.S. Direct design of approximately linear phase (ALP) 2-D IIR digital filters, *Electron. Lett.*, 29, 804–805, Apr. 29, 1993.
124. Ansari, R. and Liu, B. A class of low-noise computationally efficient recursive digital filters with applications to sampling rate alterations, *IEEE Trans. Acoust. Speech Signal Process.*, 33, 90–97, Feb. 1985.
125. Saramäki, T. On the design of digital filters as a sum of two all-pass filters, *IEEE Trans. Circuits Syst.*, 32, 1191–1193, Nov. 1985.
126. Lang, M. Allpass filter design and applications, *Proceedings of IEEE International Conference on Acoustics, Speech, and Signal Processing (ICASSP)*, Detroit, MI, May 9–12, 1995, pp. 1264–1267.
127. Schüssler, H.W. and Weith, J. On the design of recursive Hilbert-transformers, *Proceedings of IEEE International Conference on Acoustics, Speech, and Signal Processing (ICASSP)*, Dallas, TX, Apr. 6–9, 1987, pp. 876–879.
128. Steiglitz, K. Computer-aided design of recursive digital filters, *IEEE Trans. Audio Electroacoust.*, 18, 123–129, 1970.
129. Shaw, A.K. Optimal design of digital IIR filters by model-fitting frequency response data, *IEEE Trans. Circuits Syst. II*, 42, 702–710, Nov. 1995.
130. Chen, X. and Parks, T.W. Design of IIR filters in the complex domain, *Proceedings of IEEE International Conference on Acoustics, Speech, and Signal Processing (ICASSP)*, New York, Apr. 11–14, 1988, Vol. 3, pp. 1443–1446.
131. Therrian, C.W. and Velasco, C.H. An iterative Prony method for ARMA signal modeling, *IEEE Trans. Signal Process.*, 43, 358–361, Jan. 1995.

132. Pernebo, L. and Silverman, L.M. Model reduction via balanced state space representations, *IEEE Trans. Autom. Control*, 27, 382–387, Apr. 1982.
133. Glover, K. All optimal Hankel-norm approximations of linear multivariable systems and their l^{∞}-error bounds, *Int. J. Control*, 39(6), 1115–1193, 1984.
134. Beliczynski, B., Kale, I., and Cain, G.D. Approximation of FIR by IIR digital filters: An algorithm based on balanced model reduction, *IEEE Trans. Signal Process.*, 40, 532–542, Mar. 1992.
135. Chen, B.-S., Peng, S.-C., and Chiou, B.-W. IIR filter design via optimal Hankel-norm approximation, *IEE Proc., Part G*, 139, 586–590, Oct. 1992.
136. Rudko, M. A note on the approximation of FIR by IIR digital filters: An algorithm based on balanced model reduction, *IEEE Trans. Signal Process.*, 43, 314–316, Jan. 1995.
137. Tufan, E. and Tavsanoglu, V. Design of two-channel IIR PRQMF banks based on the approximation of FIR filters, *Electron. Lett.*, 32, 641–642, Mar. 28, 1996.
138. Jackson, L.B. *Digital Filters and Signal Processing (3rd ed.) with MATLAB Exercises*, Kluwer Academic Publishers, Amsterdam, the Netherlands, 1996.
139. Committee, I.D. Ed., *Selected Papers in Digital Signal Processing, II*, IEEE Press, New York, 1976.
140. Rabiner, L.R. and Rader, C.M. Eds., *Digital Signal Processing*, IEEE Press, New York, 1972.
141. Potchinkov, A. and Reemtsen, R., The design of FIR filters in the complex plane by convex optimization, *Signal Process.*, 46, 127–146, 1995.
142. Potchinkov, A. and Reemtsen, R., The simultaneous approximation of magnitude and phase by FIR digital filters, I and II, *Int. J. Circuit Theory Appl.*, 25, 167–197, 1997.
143. Lang, M.C., Design of nonlinear phase FIR digital filters using quadratic programming, in *Proceedings of IEEE International Conference on Acoustics, Speech, and Signal Processing*, Munich, Germany, Apr. 1997, Vol. 3, pp. 2169–2172.

V

Statistical Signal Processing

Georgios B. Giannakis
University of Minnesota

Statistical signal processing deals with random signals, their acquisition, their properties, their transformation by system operators, and their characterization in the time and frequency domains. The goal is to extract pertinent information about the underlying mechanisms that generate them or transform them. The area is grounded in the theories of signals and systems, random variables and stochastic processes, detection and estimation, and mathematical statistics. Random signals are temporal or spatial and can be derived from man-made (e.g., binary communication signals) or natural (e.g., thermal noise in a sensory array) sources. They can be continuous or discrete in their amplitude or index, but no exact expression describes their evolution. Signals are often described statistically when the engineer has incomplete knowledge about their description or origin. In these cases, statistical descriptors are used to characterize one's degree of knowledge (or ignorance) about the randomness. Especially interesting are those signals (e.g., stationary and ergodic) that can be described using deterministic quantities computable from finite data records. Applications of statistical signal processing algorithms to random signals are omnipresent in science and engineering in such areas as speech, seismic, imaging, sonar, radar, sensor arrays, communications, controls, manufacturing, atmospheric sciences, econometrics, and medicine, just to name a few. This section deals with the fundamentals of statistical signal processing, including some interesting topics that deviate from traditional assumptions. The focus is on discrete index random signals (i.e., time series) with possibly continuous-valued amplitudes. The reason is twofold: measurements are often made in discrete fashion (e.g., monthly temperature data) and continuously recorded signals (e.g., speech data) are often sampled for parsimonious representation and efficient processing by computers.

Chapter 12 reviews definitions, characterization, and estimation problems entailing random signals. The important notions outlined are stationarity, independence, ergodicity, and Gaussianity. The basic operations involve correlations, spectral densities, and linear time-invariant transformations. Stationarity reflects invariance of a signal's statistical description with index shifts. Absence (or presence) of relationships among samples of a signal at different points is conveyed by the notion of (in)dependence, which provides information about the signal's dynamical behavior and memory as it evolves in time or space. Ergodicity allows computation of statistical descriptors from finite data records. In increasing order of computational complexity, descriptors include the mean (or average) value of the signal, the autocorrelation, and higher than second-order correlations which reflect relations among two or more signal samples. Complete statistical characterization of random signals is provided by probability density and distribution functions. Gaussianity describes probabilistically a particular distribution of signal values which is characterized completely by its first- and second-order statistics. It is often encountered in practice because, thanks to the central limit theorem, averaging a sufficient number of random signal values (an operation often performed by, e.g., narrowband filtering) yields outputs which are (at least approximately) distributed according to the Gaussian probability law. Frequency-domain statistical descriptors inherit all the merits of deterministic Fourier transforms and can be computed efficiently using the fast Fourier transform. The standard tool here is the power spectral density which describes how average power (or signal variance) is distributed across frequencies; but polyspectral densities are also important for capturing distributions of higher order signal moments across frequencies. Random input signals passing through linear systems yield random outputs. Input–output auto- and cross-correlations and spectra characterize not only the random signals themselves but also the transformation induced by the underlying system.

Many random signals as well as systems with random inputs and outputs possess finite degrees of freedom and can thus be modeled using finite parameters. Depending on *a priori* knowledge, one estimates parameters from a given data record, treating them either as random or deterministic. Various approaches become available by adopting different figures of merit (estimation criteria). Those outlined in this chapter include the maximum likelihood, minimum variance, and least-squares criteria for deterministic parameters. Random parameters are estimated using the maximum *a posteriori* and Bayes criteria. Unbiasedness, consistency, and efficiency are important properties of estimators which, together with performance bounds and computational complexity, guide the engineer to select the proper criterion and estimation algorithm.

While estimation algorithms seek values in the continuum of a parameter set, the need arises often in signal processing to classify parameters or waveforms as one or another of prespecified classes. Decision making with two classes is sought frequently in practice, including as a special case the simpler problem of detecting the presence or absence of an information-bearing signal observed in noise. Such signal detection and classification problems along with the associated theory and practice of hypotheses testing are the subject of Chapter 13. The resulting strategies are designed to minimize the average number of decision errors. Additional performance measures include receiver operating characteristics, signal-to-noise ratios, probabilities of detection (or correct classification), false alarm (or misclassification) rates, and likelihood ratios. Both temporal and spatiotemporal signals are considered, focusing on linear single- and multivariate Gaussian models. Trade-offs include complexity versus optimality, off-line vs. real-time processing, and separate vs. simultaneous detection and estimation for signal models containing unknown parameters.

Parametric and nonparametric methods are described in Chapter 14 for the basic problem of spectral estimation. Estimates of the power spectral density have been used over the last century and continue to be of interest in numerous applications involving retrieval of hidden periodicities, signal modeling, and time series analysis problems. Starting with the periodogram (normalized square magnitude of the data Fourier transform), its modifications with smoothing windows, and moving on to the more recent minimum variance and multiple window approaches, the nonparametric methods described here constitute the first step used to characterize the spectral content of stationary stochastic signals. Factors dictating the designer's choice include computational complexity, bias-variance, and resolution trade-offs. For data adequately described by a parametric model, such as the auto-regressive (AR), moving-average (MA), or ARMA model, spectral analysis reduces to estimating the model parameters. Such a data reduction step achieved by modeling offers parsimony and increases resolution and accuracy, provided that the model and its order (number of parameters) fit well the available time series. Processes containing harmonic tones (frequencies) have line spectra, and the task of estimating frequencies appears in diverse applications in science and engineering. The methods presented here include both the traditional periodogram as well as modern subspace approaches such as the MUSIC and its derivatives.

Estimation from discrete-time observations is the theme of Chapter 15. The unifying viewpoint treats both parameter and waveform (or signal) estimation from the perspective of minimizing the averaged square error between observations and input–output or state variable signal models. Starting from the traditional linear least-squares formulation, the exposition includes weighted and recursive forms, their properties, and optimality conditions for estimating deterministic parameters as well as their minimum mean-square error and maximum *a posteriori* counterparts for estimating random parameters. Waveform estimation, on the other hand, includes not only input–output signals but also state space vectors in linear and nonlinear state variable models. Prediction, smoothing, and the celebrated Kalman filtering problems are outlined in this framework and relationships are highlighted with the Wiener filtering formulation. Nonlinear least-squares and iterative minimization schemes are discussed for problems where the desired parameters are nonlinearly related with the data. Nonlinear equations can often be linearized, and the extended Kalman filter is described briefly for estimating nonlinear state variable models. Minimizing the mean-square error criterion leads to the basic orthogonality principle which appears in both parameter and waveform estimation problems. Generally speaking, the mean-square error criterion possesses rather universal optimality when the underlying models are linear and the random data involved are Gaussian distributed.

Before accessing applicability and optimality of estimation algorithms in real-life applications, models need to be checked for linearity, and the random signals involved need to tested for Gaussianity and stationarity. Performance bounds and parameter confidence intervals must also be derived in order to evaluate the fit of the model. Finally, diagnostic tools for model falsification are needed to validate that the chosen model represents faithfully the underlying physical system. These important issues are discussed in Chapter 16. Stationarity, Gaussianity, and linearity tests are presented in a hypothesis-testing framework relying upon second-order and higher order statistics of the data. Tests are also described for estimating the number of parameters (or degrees of freedom) necessary for parsimonious modeling. Model validation is accomplished by checking for whiteness and independence of the error processes formed by subtracting model data from measured data. Tests may declare signal or noise data as non-Gaussian and/or nonstationary. The non-Gaussian models outlined here include the generalized Gaussian, Middleton's class, and the stable noise distribution models.

As for nonstationary signals and time-varying systems, detection and estimation tasks become more challenging and solutions are not possible in the most general case. However, structured nonstationarities such as those entailing periodic and almost periodic variations in their statistical descriptors are tractable. The resulting random signals are called (almost) cyclostationary and their analysis is the theme of Chapter 17. The exposition starts with motivation and background material including links between cyclostationary signals and multivariate stationary processes, time-frequency representations, and multirate operators. Examples of cyclostationary signals and cyclostationarity-inducing operations are also described along with applications to signal processing and communication problems with emphasis on signal separation and channel equalization.

Modern theoretical directions in the field appear toward non-Gaussian, nonstationary, and nonlinear signal models. Advanced statistical signal processing tools (algorithms, software, and hardware) are of interest in current applications such as manufacturing, biomedicine, multimedia services, and wireless communications. Scientists and engineers will continue to search and exploit determinism in signals that they create or encounter, and find it convenient to model, as random.

12

Overview of Statistical Signal Processing

Charles W. Therrien
Naval Postgraduate School

12.1 Discrete Random Signals

Many or most signals of interest in the real world cannot be written as an explicit mathematical formula. These real signals representing speech, noise, music, data, etc., are often described by a probabilistic model and statistical methods are used for their analysis. While the associated physical phenomena are often continuous, the signals are usually sampled and processed digitally. This leads to the concept of a discrete random signal or sequence.

12.1.1 Random Signals and Sequences

The following can be used as a working definition of a discrete random signal.

Definition 12.1: *A discrete random signal is an indexed sequence* $x[n]$ *such that, for any choice of the index or independent variable, say* $n = n_o$, $x[n_o]$ *is a random variable.*

If the index n represents time, as is usually the case, any realization of the random sequence may be referred to as a "time series." The index could represent another quantity, however, such as the position in a uniform linear array.

The underlying model that represents the random sequence is known as a random process or a stochastic process. Figure 12.1 shows some examples of discrete random signals. The noise signal of Figure 12.1a can take on any real value, while the binary data sequence of Figure 12.1b (in the absence of noise) can take on only two discrete values ($+1$ and -1). The examples in Figure 12.1c and d are interesting because, while they satisfy the definition of a random signal, their evolution (in time) is

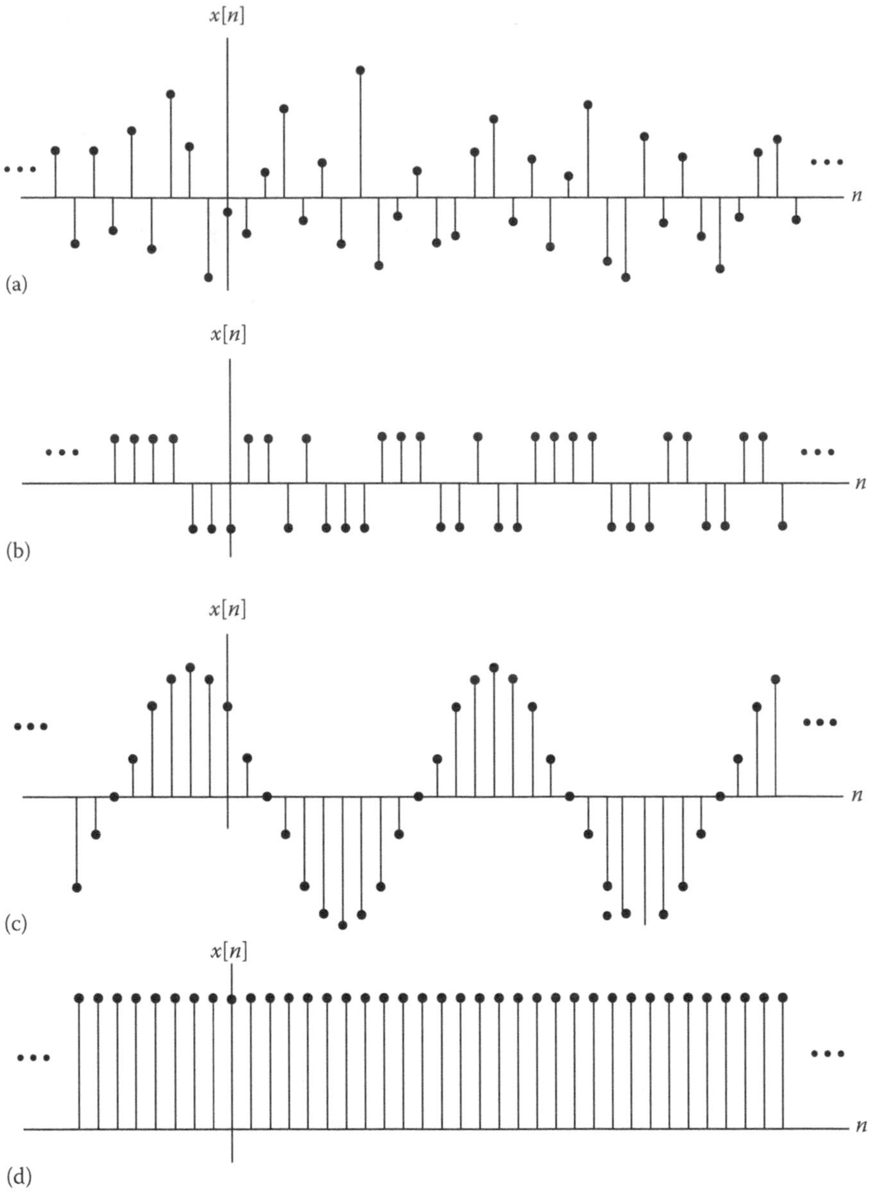

FIGURE 12.1 Examples of discrete random signals: (a) sampled noise, (b) binary data, (c) random sinusoid, and (d) constant random voltage. (From Therrien, C.W., *Discrete Random Signals and Statistical Signal Processing*, Prentice Hall, Inc., Upper Saddle River, NJ, 1992. With permission.)

known forever once a few values of the process are observed. In the case of the sinusoid, its amplitude and/or phase may be random variables, but its future values can be determined from any two consecutive values of the signal. In the case of a constant voltage, its value is a random variable, but any one sample of the signal specifies the signal for all time. Such random signals are called *predictable* and form a set of processes distinct from those such as in Figure 12.1a and b, which are said to be *regular*. Predictable random processes can be predicted perfectly (i.e., with zero error) from a linear combination of past values of the process.

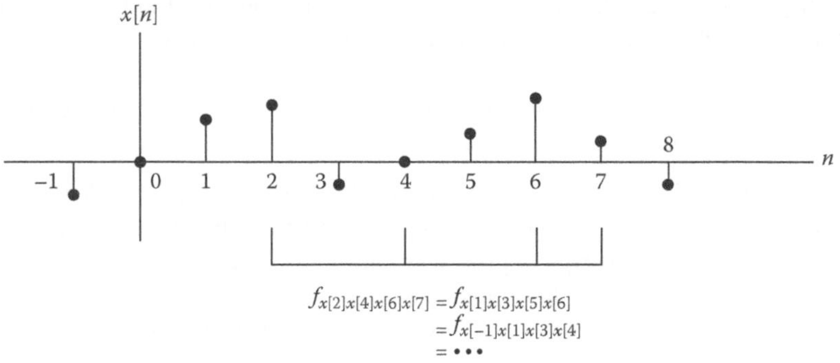

FIGURE 12.2 Stationary random process. Any set of samples with the same spacing has the same probability density function. (From Therrien, C.W., *Discrete Random Signals and Statistical Signal Processing*, Prentice Hall, Inc., Upper Saddle River, NJ, 1992. With permission.)

The fundamental statistical characterization of a random process is through the joint probability distribution or joint density function of its samples. For purposes of this chapter, it is sufficient to work with the *density* function, using impulses to formally represent any discrete probability values.* To characterize the signal completely, it must be possible to form the joint density of any set of samples of the process, as shown in Figure 12.2. If this density function is independent of where the samples are taken in the process as long as the spacing is the same, then the process is said to be stationary in the strict sense (see Figure 12.2). A formal definition follows.

Definition 12.2: *A random process is stationary in the strict sense if and only if*

$$f_{x[n_0],x[n_1],\ldots,x[n_L]} = f_{x[n_0+k],x[n_1+k],\ldots,x[n_L+k]} \tag{12.1}$$

for all choices of the n_i, and all values of the integers k and L.

Some related ideas are the concepts of periodicity and cyclostationarity for random processes:

Definition 12.3: *A random process is periodic if there exists an integer P such that*

$$f_{x[n_0],x[n_1],\ldots,x[n_L]} = f_{x[n_0+k_0P],x[n_1+k_1P],\ldots,x[n_L+k_LP]} \tag{12.2}$$

for all choices of the n_i, for any set of integers k_i, and for any value of L. If Equation 12.2 holds only for equal values of the integers $k_0 = k_1 = \cdots = k_L = k$, then the process is said to be cyclostationary in the discrete-time sense.

---

* For example, the probability density for a sample of the binary random signal of Figure 21.1b taking on values of ±1 would be written as $f_{x[n]}(x) = P\delta_c(x-1) + (1-P)\delta_c(x+1)$, where P is the probability of a positive value (+1) and $\delta_c(x)$ is the "continuous" impulse function defined by its action on any continuous function $g(x)$: $g(x) = \int_{-\infty}^{\infty} g(s)\delta_c(x-s)ds$. The subscript c is added to distinguish it from the discrete impulse or unit sample function defined by $\delta[n] = 1$ for $n = 0$ and zero otherwise.

Periodic random processes usually have an explicit dependence on a sinusoid or complex exponential (term of the form $e^{j\omega n}$). This need not be true for cyclostationary processes.

There are three main cases that occur in signal processing where a complete statistical characterization of the random signal is possible. These are as follows:

1. When the samples of the signal are *independent*. In that case, the joint density for any set of samples can be written as a product of the density functions for the individual samples. If the samples have mean zero, this type of process is known as a *strictly white* process.
2. When the conditional density for the samples $f_{x[n]|x[n-1],x[n-2],\ldots}$ depends only on the previous sample $x[n-1]$ (or on the previous p samples). This type of process is known as a Markov process (or a pth-order Markov process).
3. When the samples of the process are jointly Gaussian. This is called a *Gaussian* random process and occurs frequently in real life, for example, when the random sequence is a sampled version of noise (see [1] for a more complete discussion).

In a great many cases, however, there is incomplete knowledge of the statistical distribution of the signals; nevertheless, a very useful analysis can still be carried out using only certain statistical moments of the signal.

12.1.1.1 Moments of Random Processes

For a real-valued sequence the first- and second-order moments are denoted by

$$M_x^{(1)}[n] \stackrel{\text{def}}{=} E\{x[n]\} \tag{12.3}$$

and

$$M_x^{(2)}[n; l] \stackrel{\text{def}}{=} E\{x[n]x[n+l]\} \tag{12.4}$$

where $E\{\cdot\}$ denotes expectation. Notice that the first moment $M_x^{(1)}$ in general depends on the time n and that the second moment $M_x^{(2)}$ expresses the correlation between a point in the random process at time n and another point at time $n' = n + l$. (Note that l may be positive or negative.)

In most modern electrical engineering treatments, the second moment is replaced by the autocorrelation function, defined as

$$R_x[n; l] \stackrel{\text{def}}{=} E\{x[n]x[n-l]\} \tag{12.5}$$

so that $R_x[n; l] = M_x^{(2)}[n; -l]$. The notation $[n; l]$ for the arguments of the autocorrelation function, while not entirely standard, is useful in that it focuses on a particular time instant n and a point located at position l relative to the first point (see Figure 12.3). The variable l is known as the *lag*. Moreover, certain general properties of random processes are reflected in the autocorrelation function using the definition of Equation 12.5 [2]:

1. For a stationary random process, $R_x[n; l]$ is independent of n. (It depends only on the lag l.)
2. For a cyclostationary random process, $R_x[n; l]$ is periodic in n (but not in l).
3. For a periodic random process, $R_x[n; l]$ is periodic in both n and l.

These properties can usually be exploited to advantage in signal procession algorithms.

Higher-order moments, say of orders 3 and 4, are defined in a way analogous to Equations 12.3 and 12.4:

$$M_x^{(3)}[n; l_1, l_2] \stackrel{\text{def}}{=} E\{x[n]x[n+l_1]x[n+l_2]\} \tag{12.6}$$

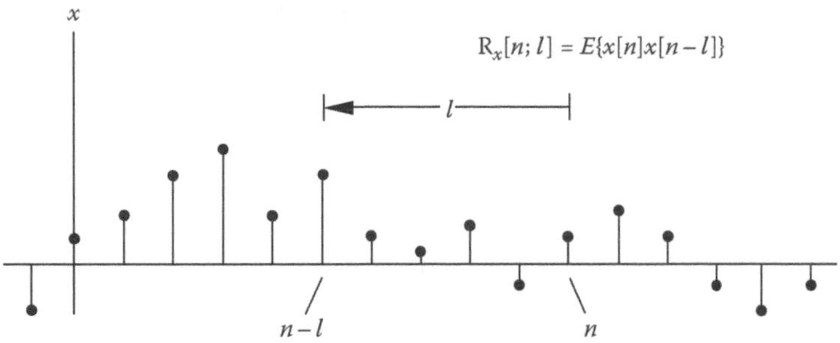

FIGURE 12.3 Illustration of correlation for a random process. (From Therrien, C.W., *Discrete Random Signals and Statistical Signal Processing*, Prentice Hall, Inc., Upper Saddle River, NJ, 1992. With permission.)

$$M_x^{(4)}[n; l_1, l_2, l_3] \stackrel{\text{def}}{=} E\{x[n]x[n+l_1]x[n+l_2]x[n+l_3]\} \tag{12.7}$$

More general moments can be represented by expressions such as

$$E\{x^{p_0}[n]x^{p_1}[n+l_1]\cdots x^{p_L}[n+l_L]\}$$

for various selections of the powers p_i, lags l_i, and number of terms $L+1$. Moments are usually not known a priori but must be estimated from data. In the case of a stationary random process, it is useful if the moment computed from the signal average defined as

$$\langle x^{p_0}[n]x^{p_1}[n+l_1]\cdots x^{p_L}[n+l_L]\rangle \stackrel{\text{def}}{=} \lim_{N\to\infty}\frac{1}{2N+1}\sum_{n=-N}^{N} x^{p_0}[n]x^{p_1}[n+l_1]\cdots x^{p_L}[n+l_L] \tag{12.8}$$

satisfies the property

$$\langle x^{p_0}[n]x^{p_1}[n+l_1]\cdots x^{p_L}[n+l_L]\rangle \doteq E\{x^{p_0}[n]x^{p_1}[n+l_1]\cdots x^{p_L}[n+l_L]\} \tag{12.9}$$

where the notation "$\doteq$" means that the event

$$\langle x^{p_0}[n]x^{p_1}[n+l_1]\cdots x^{p_L}[n+l_L]\rangle = E\{x^{p_0}[n]x^{p_1}[n+l_1]\cdots x^{p_L}[n+l_L]\}$$

has probability 1. If Equation 12.9 is satisfied for all L; all choices of the spacings $l_1, l_2, \ldots, l_L$; and all choices of the powers $p_0, p_1, \ldots, p_L$, then the process is said to be strictly ergodic. A random process that satisfies only the condition

$$\langle x[n]\rangle \doteq E\{x[n]\} \tag{12.10}$$

is said to be "ergodic in the mean" while one that satisfies

$$\langle x[n]x[n+l]\rangle \doteq E\{x[n]x[n+l]\} \tag{12.11}$$

is said to be "ergodic in correlation." These last two conditions are sufficient for many applications.

Ergodicity implies that statistical moments can be estimated from a single realization of a random process, which is sometimes all that is available in a practical situation. A noise process such as that depicted in Figure 12.1a is typically an ergodic process while the battery voltage depicted in Figure 12.1d is not. (Averaging Figure 12.1d in time will produce only the value of the signal in the given realization, not the mean of the distribution from which the random signal was drawn.)

12.1.1.2 Complex Random Signals

In some signal processing applications, the signals are complex-valued. Such signals have a real and imaginary part and can be written as

$$x[n] = x_r[n] + jx_i[n] \tag{12.12}$$

where x_r and x_i are two real-valued sequences. Strictly speaking, complex-valued random processes must be characterized by joint probability density functions or joint moments between the two real-valued components. In many cases, however, certain symmetries arise in the statistics that allow for a simplified description using the signal and its complex conjugate. For example, the autocorrelation function for a complex random process is defined as

$$R_x[n;\, l] \overset{\text{def}}{=} E\{x[n]x^*[n-l]\} \tag{12.13}$$

It can be seen, by substituting Equation 12.12 in Equation 12.13 and expanding, that the sums of products are present in the expectation and individual terms such as $E\{x_r[n]x_r[n-l]\}$ or $E\{x_r[n]x_i[n-l]\}$ are not represented. In order to find these terms and thus completely characterize the second moments of the complex random signal, it is necessary to know the additional complex quantity

$$R'_x[n;\, l] \overset{\text{def}}{=} E\{x[n]x[n-l]\} \tag{12.14}$$

which is defined *without* the conjugate. $R'_x[n;\, l]$ is known as the complementary autocorrelation function, the pseudo-autocorrelation function, or the *relation* function [3,4]. With this additional information, the individual moments can be computed from expressions such as $E\{x_r[n]x_r[n-l]\} = \frac{1}{2}\mathrm{Re}\left(\mathrm{R}_x[n;\, l] + \mathrm{R}'_x[n;\, l]\right)$ or $E\{x_r[n]x_i[n-l]\} = -\frac{1}{2}\mathrm{Im}\left(\mathrm{R}_x[n;\, l] - \mathrm{R}'_x[n;\, l]\right)$.

A special case occurs when $R'_x[n;\, l]$ is identically zero. In this case, the random process is said to be *circular* [5] (or *proper* in the context of complex Gaussian random processes [3,6]) and thus the individual correlation terms can be derived from $R_x[n;\, l]$ alone.

An alternate definition of circularity is that the second-order statistics of $x[n]$ are invariant to a phase shift ($e^{j\theta}x[n]$ for any θ) [4]. Stationary random processes always exhibit circularity; however, processes that are nonstationary may or may not be circular. Traditional analyses of complex random processes have either ignored the issue of circularity or assumed that $E\{x[n]x[n-l]\}$ is zero. In cases where $R'_x[n;\, l]$ is not truly zero, however, the performance of signal processing algorithms can be enhanced by acknowledging this lack of circularity and including it in the signal model. Further discussion for the need to acknowledge circularity (or the lack thereof) in certain applications such as digital communications can be found in the literature (e.g., [3,7,8]).

The sections to follow focus on the case where the random processes are in fact stationary and develop the methods that are commonly applied to such signals. Since stationary random signals are also circular, any further discussion of circularity can be deferred to Section 12.3 on random vectors.

12.1.2 Characterization of Stationary Random Signals

12.1.2.1 Moments and Cumulants

It follows from Definition 12.1 that the moments of a stationary random process are independent of the time index n. Thus, the mean is a constant and can be defined by

$$m_x \overset{\text{def}}{=} E\{x[n]\} \tag{12.15}$$

The autocorrelation function depends on only the time difference or lag l between the two signal samples and can now be defined as

$$R_x[l] \overset{\text{def}}{=} E\{x[n]x^*[n-l]\} \tag{12.16}$$

The autocovariance function is likewise defined as

$$C_x[l] \overset{\text{def}}{=} E\{(x[n]-m_x)(x^*[n-l]-m_x^*)\} \tag{12.17}$$

and satisfies the relation

$$R_x[l] = C_x[l] + |m_x|^2 \tag{12.18}$$

If a random signal is not strictly stationary, but its mean is constant and its autocorrelation function depends only on l (not n), then the process is called *wide-sense* stationary. Most often when the term "stationary" is used without further qualification, the term is intended to mean "wide-sense stationary."* The specific values $R_x[0] = E\{|x[n]|^2\}$ and $C_x[0] = E\{|x[n] - m_x|^2\}$ represent the power and the variance of the signal, respectively.

An example of a seemingly trivial but fundamental autocorrelation function is that of a white noise process. A white noise process is any process having mean zero and uncorrelated samples; that is, $R_x[l] = 0$ for $l \neq 0$. A white noise process thus has correlation and covariance functions of the form

$$R_x[l] = C_x[l] = \sigma_o^2\delta[l] \tag{12.19}$$

where
- $\delta[l]$ is the unit sample function (discrete-time impulse)
- σ_o^2 is the variance of any sample of the process

Any sequence of zero-mean independently-distributed random variables forms a white noise process. For example, a binary-valued sequence formed by assigning +1 and −1 to the flips of a coin is white noise. In electrical engineering applications, however, the noise may be Gaussian or follow some other distribution. The term "white" applies in all of these cases as long as Equation 12.19 is satisfied.

The assumption of stationarity implies circularity of the random process (see Section 12.1.1). Therefore, all necessary second-moment statistics can be derived from Equations 12.16 and 12.15 or Equations 12.17 and 12.15. In particular, if the signal is stationary and written as in Equation 12.12, then the autocorrelation functions for the real and imaginary parts of the signal are equal and are given by

$$R_{x_r}[l] = R_{x_i}[l] = 1/2\,\mathrm{Re}(R_x[l]) \tag{12.20}$$

while the cross-correlation functions between the real and imaginary parts (see Equation 12.28 for definition of cross-correlation) must satisfy

$$R_{x_r x_i}[l] = -R_{x_i x_r}[l] = -1/2\,\mathrm{Im}(R_x[l]) \tag{12.21}$$

In defining autocorrelation and autocovariance for real-valued random processes, the complex conjugate in Equations 12.16 and 12.17 can be safely ignored. The foregoing discussion should serve

* The abbreviation wss is also used frequently in the literature.

to emphasize, however, that for complex random processes, the conjugate is essential. In fact if the conjugate is dropped from the second term in Equation 12.16, then $E\{x[n]x[n-l]\}$ is identically zero for all values of l due to the circularity property of stationary random processes.

The autocorrelation (or autocovariance) function has two defining properties:

1. Conjugate symmetric:

$$R_x[l] = R_x^*[-l] \tag{12.22}$$

2. Positive semidefinite:

$$\sum_{n_1=-\infty}^{\infty} \sum_{n_0=-\infty}^{\infty} a^*[n_1]R_x[n_1 - n_0]a[n_0] \geq 0 \quad \text{for } any \text{ sequence } a[n] \tag{12.23}$$

These properties follow easily from the definitions [1]. The second property can be shown to imply that

$$R_x[0] \geq |R_x[l]| \quad l \neq 0$$

Note, however, that this is a derived property and not a fundamental defining property for the correlation function, that is, it is a necessary but not a sufficient condition.

Higher-order moments and cumulants are sometimes used in modern signal processing as well. The third- and fourth-order moments for a stationary random process are usually written as

$$M_x^{(3)}[l_1, l_2] = E\{x^*[n]x[n+l_1]x[n+l_2]\} \tag{12.24}$$

$$M_x^{(4)}[l_1, l_2, l_3] = E\{x^*[n]x^*[n+l_1]x[n+l_2]x[n+l_3]\} \tag{12.25}$$

while for a zero-mean random process the third- and fourth-order cumulants are given by

$$C_x^{(3)}[l_1, l_2] = E\{x^*[n]x[n+l_1]x[n+l_2]\} \tag{12.26}$$

$$\begin{aligned} C_x^{(4)}[l_1, l_2, l_3] = {} & E\{x^*[n]x^*[n+l_1]x[n+l_2]x[n+l_3]\} \\ & - C_x^{(2)}[l_2]C_x^{(2)}[l_3 - l_1] - C_x^{(2)}[l_3]C_x^{(2)}[l_2 - l_1] \\ & \text{(complex random process)} \end{aligned} \tag{12.27a}$$

$$\begin{aligned} C_x^{(4)}[l_1, l_2, l_3] = {} & E\{x[n]x[n+l_1]x[n+l_2]x[n+l_3]\} - C_x^{(2)}[l_1]C_x^{(2)}[l_3 - l_2] \\ & - C_x^{(2)}[l_2]C_x^{(2)}[l_3 - l_1] - C_x^{(2)}[l_3]C_x^{(2)}[l_2 - l_1] \\ & \text{(real random process)} \end{aligned} \tag{12.27b}$$

where $C_x^{(2)}[l] = E\{x^*[n]x[n+l]\}$ is the second-order cumulant, identical (in this zero-mean case) to the covariance function. It should be noted that unlike the second-order moments, the definition of these statistics for a complex random process is not standard, so alternate definitions to Equations 12.24 through 12.27 with different placement of the complex conjugate may be encountered.

For most analyses, cumulants are preferred to moments because the cumulants of order 3 and higher for a Gaussian process are identically zero. Thus, signal processing methods based on higher-order cumulants have the advantage of being "blind" to any form of Gaussian noise.

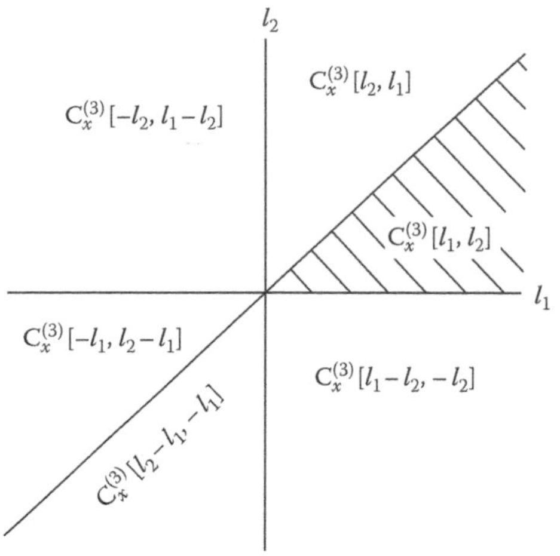

FIGURE 12.4 Regions of symmetry for the third-order cumulant of real-valued signals. (From Therrien, C.W., *Discrete Random Signals and Statistical Signal Processing*, Prentice Hall, Inc., Upper Saddle River, NJ, 1992. With permission.)

For real-valued signals, these higher-order cumulants have many regions of symmetry. The symmetry regions for the third-order cumulant are shown in Figure 12.4. Symmetry regions for the third-order cumulant of complex signals consist only of the half planes defined by $C_x^{(3)}[l_1, l_2] = C_x^{(3)}[l_2, l_1]$.

Cross-moments between two or more random signals are also of utility. For two jointly stationary* random signals x and y the cross-correlation and cross-covariance functions are defined by

$$R_{xy}[l] = E\{x[n]y^*[n-l]\} \tag{12.28}$$

and

$$C_{xy}[l] = E\{(x[n] - m_x)(y[n-l] - m_y)^*\} \tag{12.29}$$

and satisfy the relation

$$R_{xy}[l] = C_{xy}[l] + m_x m_y^* \tag{12.30}$$

These cross-moment functions have no particular properties except that $R_{xy}[l] = R_{yx}^*[-l]$ and $C_{xy}[l] = C_{yx}^*[-l]$. Higher-order cross-moments and cumulants can be defined in an analogous way to Equations 12.24 through 12.27 and are also encountered in some applications.

12.1.2.2 Frequency and Transform Domain Characterization

Random signals can be characterized in the frequency domain as well as in the signal domain. The power spectral density function is defined by the Fourier transform of the autocorrelation function

$$S_x(e^{j\omega}) = \sum_{l=-\infty}^{\infty} R_x[l]e^{-j\omega l} \tag{12.31}$$

* Two signals are said to be *jointly stationary* (in the wide sense), if each of the signals is itself wide-sense stationary, and the cross-correlation is a function of only the time difference, or lag, l.

with inverse transform

$$R_x[l] = \frac{1}{2\pi} \int_{-\pi}^{\pi} S_x(e^{j\omega}) e^{j\omega l} d\omega \tag{12.32}$$

The name "power spectral density" comes from the fact that

$$\text{average power} = E\{|x[n]|^2\} = R_x[0] = \frac{1}{2\pi} \int_{-\pi}^{\pi} S_x(e^{j\omega}) d\omega$$

which follows directly from Equations 12.16 and 12.32. Since the power spectral density may contain both continuous and discrete components (see Figure 12.5), its general form is

$$S_x(e^{j\omega}) = S'_x(e^{j\omega}) + \sum_i 2\pi \mathbf{P}_i \delta_c(e^{j\omega} - e^{j\omega_i}) \tag{12.33}$$

where $S'_x(e^{j\omega})$ represents the continuous part of the spectrum while the sum of weighted impulses represents the discrete part or "lines" in the spectrum. Impulses or lines arise from periodic or almost periodic random signals such as those of Figure 12.1c and d.

The two defining properties for the autocorrelation function (Equations 12.22 and 12.23) are manifested as two corresponding properties of the power spectral density function, namely,

1. $S_x(e^{j\omega})$ is real.
2. $S_x(e^{j\omega})$ is nonnegative: $S_x(e^{j\omega}) \geq 0$.

In addition, for real-valued random signals, $S_x(e^{j\omega})$ is an *even* function of frequency.

The white noise process, introduced on page 7, has a power spectral density function that is a constant $S_x(e^{j\omega}) = \sigma_o^2$. The term "white" refers to the fact that the spectrum, like that of ideal white light, is flat and represents all frequencies in equal proportions.

The multidimensional Fourier transforms of the cumulants are also of considerable importance and are referred to generically as cumulant spectra, higher-order spectra, or polyspectra. For the third- and fourth-order cumulants, these higher-order spectra are called the bispectrum and trispectrum, respectively, and are defined by

$$B_x(\omega_1, \omega_2) = \sum_{l_1=-\infty}^{\infty} \sum_{l_2=-\infty}^{\infty} C_x^{(3)}[l_1, l_2] e^{-j(\omega_1 l_1 + \omega_2 l_2)} \tag{12.34}$$

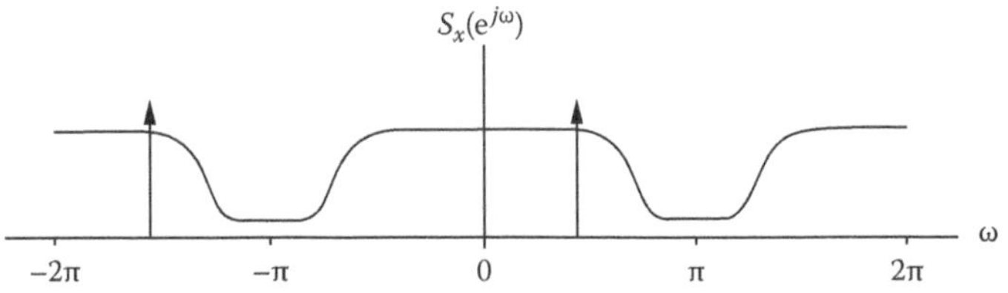

FIGURE 12.5 Typical power density spectrum for a complex random process showing continuous and discrete components. (From Therrien, C.W., *Discrete Random Signals and Statistical Signal Processing*, Prentice Hall, Inc., Upper Saddle River, NJ, 1992. With permission.)

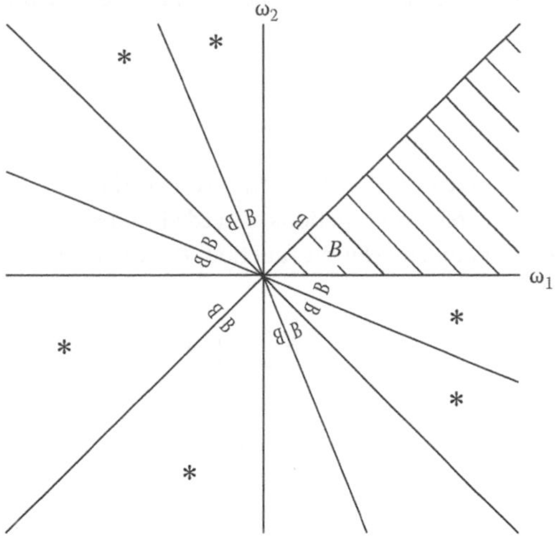

FIGURE 12.6 Regions of symmetry for the bispectrum of a real-valued signal. (From Therrien, C.W., *Discrete Random Signals and Statistical Signal Processing*, Prentice Hall, Inc., Upper Saddle River, NJ, 1992. With permission.)

and

$$T_x(\omega_1, \omega_2, \omega_3) = \sum_{l_1=-\infty}^{\infty} \sum_{l_2=-\infty}^{\infty} \sum_{l_3=-\infty}^{\infty} C_x^{(4)}[l_1, l_2, l_3] e^{-j(\omega_1 l_1 + \omega_2 l_2 + \omega_3 l_3)} \tag{12.35}$$

These quantities have many regions of symmetry. The regions of symmetry of the bispectrum of a real-valued signal are shown in Figure 12.6. For a complex signal there is only symmetry between half planes.

Higher-order processes whose cumulants are proportional to the unit sample function and whose higher-order spectra are therefore constant are sometimes called higher-order white noise processes. For a "strictly white" process (see page 4), the cumulants of all orders are impulses and thus the polyspectra of all orders are constant functions of frequency.

Cross-power spectral density functions are also defined as Fourier transforms of the corresponding cross-correlation functions, for example,

$$S_{xy}(e^{j\omega}) = \sum_{l=-\infty}^{\infty} R_{xy}[l] e^{-j\omega l} \tag{12.36}$$

Since the cross-correlation function has no particular properties, the cross-power spectral density function will also have no distinctive properties; it is complex-valued in general. The cross-spectral density evaluated at a particular point in frequency can be interpreted as a measure of the correlation that exists between components of the two processes at the chosen frequency. The normalized cross-spectrum

$$\Gamma_{xy}(e^{j\omega}) \stackrel{\text{def}}{=} \frac{S_{xy}(e^{j\omega})}{\sqrt{S_x(e^{j\omega})}\sqrt{S_y(e^{j\omega})}} \tag{12.37}$$

is called the coherence function and its squared magnitude

$$\left|\Gamma_{xy}(e^{j\omega})\right|^2 = \frac{\left|S_{xy}(e^{j\omega})\right|^2}{S_x(e^{j\omega}) S_y(e^{j\omega})} \tag{12.38}$$

is called the magnitude-squared coherence (MSC). The MSC is often used instead of $|S_{xy}(e^{j\omega})|$ and has the convenient property

$$0 \leq \left|\Gamma_{xy}(e^{j\omega})\right|^2 \leq 1 \tag{12.39}$$

Random signals can also be characterized in the z (transform) domain. In particular, the z-transform of the autocorrelation and cross-correlation functions is needed in many analyses such as in the design of filters for random signals. For the autocorrelation function, the quantity

$$S_x(z) = \sum_{l=-\infty}^{\infty} R_x[l]z^{-l} \tag{12.40}$$

is known as the *complex spectral density function*. It has the basic symmetry property

$$S_x(z) = S_x^*(1/z^*) \tag{12.41}$$

and is real and nonnegative on the unit circle. For real-valued random processes, Equation 12.41 can be expressed as

$$S_x(z) = S_x(z^{-1})$$

but expressing the property in this way sometimes hides the function's true features. For a rational* complex spectral density function, Equation 12.41 implies that for any root of the numerator or denominator, say at location z_o, there is a corresponding root at the conjugate reciprocal position, $1/z_o^*$. This also implies that zeros on the unit circle occur in even multiplicities. (Poles are not allowed to occur on the unit circle.) In addition, since a real-valued random process has real coefficients in the polynomials that define $S_x(z)$, the complex roots of such processes occur in conjugate pairs. Therefore, for real-valued processes, poles or zeros, not on the real axis, occur in groups of four:

$$z_o, \quad 1/z_o, \quad z_o^*, \quad \text{and} \quad 1/z_o^*$$

The autocorrelation function can be obtained from the inverse transform

$$R_x[l] = \frac{1}{2\pi j}\oint_C S_x(z)z^{l-1}dz \tag{12.42}$$

which involves a contour integral in the region of convergence of the transform [1]. Because of the symmetry, the region of convergence is always an annular region of the form

$$a < |z| < \frac{1}{a}$$

and the integral can be evaluated using the method of residues [9].

The complex cross-spectral density function $S_{xy}(z)$ and its inverse are defined by equations analogous to Equations 12.40 and 12.42. Again, because the cross-correlation function has no special properties, none are imparted to the complex cross-spectral density function.

* The term *rational* is used to describe functions of the form $S_x(z) = N(z)/D(z)$, where $N(z)$ and $D(z)$ are polynomials.

A simple but useful real autocorrelation function has the exponential form*

$$R_x[l] = \sigma^2 \rho^{|l|} \tag{12.43}$$

This function and its corresponding power spectral density

$$S_x(e^{j\omega}) = \frac{\sigma^2(1-\rho^2)}{1+\rho^2 - 2\rho\cos\omega} \tag{12.44}$$

are illustrated in Figure 12.7. The corresponding complex spectral density function can be expressed as

$$S_x(z) = \frac{\sigma^2(1-\rho^2)}{-\rho z + (1+\rho^2) - \rho z^{-1}} \tag{12.45}$$

This function has one pair of real axis poles at $z=\rho$ and $z=1/\rho$ and the region of convergence lies between the two poles.

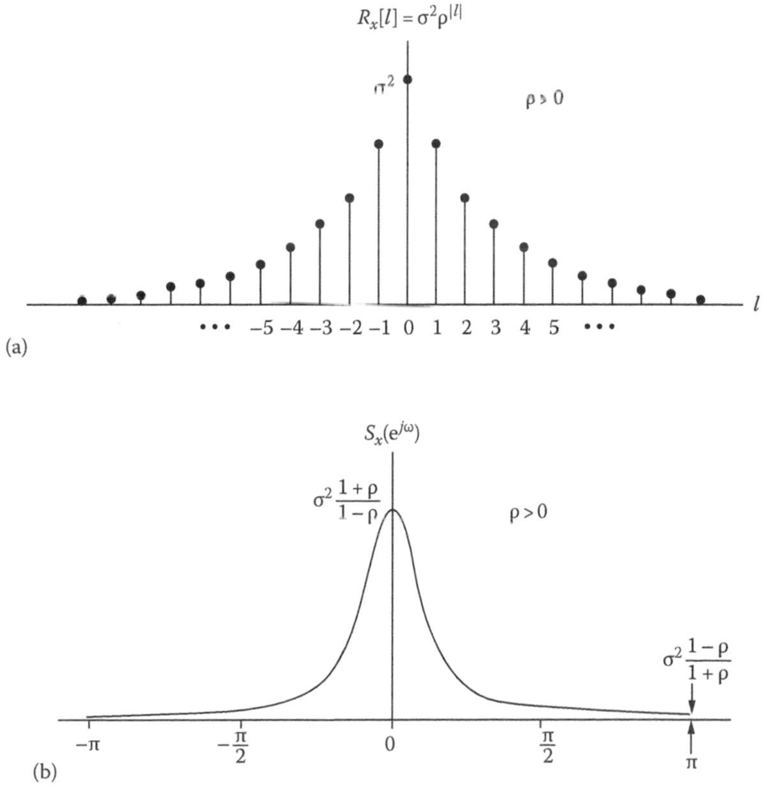

FIGURE 12.7 Real exponential autocorrelation function and corresponding power spectral density ($\rho > 0$): (a) autocorrelation function and (b) power spectral density function. (From Therrien, C.W., *Discrete Random Signals and Statistical Signal Processing*, Prentice Hall, Inc., Upper Saddle River, NJ, 1992. With permission.)

* A complex version of this autocorrelation function can be found in [1].

12.2 Linear Transformations

A linear shift-invariant system can be represented in the signal domain by its impulse response sequence $h[n]$. If a random process $x[n]$ is applied to the linear system, the output $y[n]$ is given by the convolution

$$y[n] = \sum_{k=-\infty}^{\infty} h[k]x[n-k] \tag{12.46}$$

If $x[n]$ is stationary, then $y[n]$ will also be stationary [1]. Taking expectations on both sides of the equation yields

$$E\{y[n]\} = \sum_{k=-\infty}^{\infty} h[k]E\{x[n-k]\}$$

or

$$m_y = m_x \cdot \sum_{k=-\infty}^{\infty} h[k] \tag{12.47}$$

The output autocorrelation function can be computed by the following steps. Multiplying Equation 12.46 on both sides by $y^*[n-l]$ and taking the expectation yields

$$E\{y[n]y^*[n-l]\} = \sum_{k=-\infty}^{\infty} h[k]E\{x[n-k]y^*[n-l]\}$$

or

$$R_y[l] = \sum_{k=-\infty}^{\infty} h[k]R_{xy}[l-k]$$

which will be written as

$$R_y[l] = h[l] * R_{xy}[l] \tag{12.48}$$

using "*" to denote convolution of the sequences. Multiplying Equation 12.46 by $x^*[n-l]$ and performing similar steps yields

$$R_{yx}[l] = h[l] * R_x[l] \tag{12.49}$$

Conjugating terms and noting that $R_{xy}[l] = R^*_{yx}[-l]$ and $R_x[l] = R^*_x[-l]$ permits Equation 12.49 to be written as

$$R_{xy}[l] = h^*[-l] * R_x[l] \tag{12.50}$$

Combining Equations 12.48 and 12.50 then yields

$$R_y[l] = h[l] * h^*[-l] * R_x[l] \tag{12.51}$$

TABLE 12.1 Linear Transformation Relations

System Defined by $y[n] = h[n] * x[n]$		
$R_{yx}[l] = h[l] * R_x[l]$	$S_{yx}(e^{j\omega}) = H(e^{j\omega})S_x(e^{j\omega})$	$S_{yx}(z) = H(z)S_x(z)$
$R_{xy}[l] = h^*[-l] * R_x[l]$	$S_{xy}(e^{j\omega}) = H^*(e^{j\omega})S_x(e^{j\omega})$	$S_{xy}(z) = H^*(1/z^*)S_x(z)$
$R_y[l] = h[l] * R_{xy}[l]$	$S_y(e^{j\omega}) = H(e^{j\omega})S_{xy}(e^{j\omega})$	$S_y(z) = H(z)S_{xy}(z)$
$R_y[l] = h[l] * h^*[-l] * R_x[l]$	$S_y(e^{j\omega}) = \|H(e^{j\omega})\|^2 S_x(e^{j\omega})$	$S_y(z) = H(z)H^*(1/z^*)S_x(z)$

Source: Therrien, C.W., *Discrete Random Signals and Statistical Signal Processing*, Prentice Hall, Inc., Upper Saddle River, NJ, 1992. With permission.
Note: For real $h[n]$, $H^*(1/z^*) = H(z^{-1})$.

Equation 12.51 shows that the output autocorrelation function is obtained as a double convolution of the input autocorrelation function with the impulse response and the reversed conjugated impulse response. It can easily be shown that the autocovariance and cross-covariance functions also satisfy the relations in Equations 12.48 through 12.51.

By using the Fourier and z-transform relations and the last four equations it is easy to derive expressions for the results of a linear transformation in the frequency and transform domains. The complete set of relations is listed in Table 12.1; those for the output process are the ones most frequently used and appear in the last row of the table.

As an example of the use of linear transformations, consider the simple first-order causal system described by the difference equation

$$y[n] = \rho y[n-1] + x[n]$$

with real parameter ρ. The system has an impulse response given by $h[n] = \rho^n u[n]$ where $u[n]$ is the unit step function, and a transfer function given by [17]

$$H(z) = \frac{1}{1 - \rho z^{-1}}$$

If the input is a white noise process with $S_x(z) = \sigma_o^2$, and all signals are real, then the output complex spectral density function is (see Table 12.1)

$$S_y(z) = H(z)H(z^{-1})S_x(z) = \frac{\sigma_o^2}{(1 - \rho z^{-1})(1 - \rho z)}$$

This is identical in form to Equation 12.45 with $\sigma_o^2 = \sigma^2(1 - \rho^2)$. It follows that the autocorrelation function and power spectral density function of the output also have the forms as in Equations 12.43 and 12.44. (This could be shown directly by applying the other relations in the table.) Thus, a process with exponential autocorrelation function can be obtained by driving a first-order filter with white noise.

The higher-order moments and cumulants of the output of a linear system can also be computed from the corresponding input quantities, although the formulas are more complicated. For the third- and fourth-order cumulants the formulas are

$$C_y^{(3)}[l_1, l_2] = \sum_{k_0=-\infty}^{\infty} \sum_{k_1=-\infty}^{\infty} \sum_{k_2=-\infty}^{\infty} C_x^{(3)}[l_1 - k_1 + k_0, l_2 - k_2 + k_0]h[k_2]h[k_1]h^*[k_0] \tag{12.52}$$

and

$$C_y^{(4)}[l_1, l_2, l_3] = \sum_{k_0=-\infty}^{\infty} \sum_{k_1=-\infty}^{\infty} \sum_{k_2=-\infty}^{\infty} \sum_{k_3=-\infty}^{\infty} C_x^{(4)}[l_1 - k_1 + k_0, l_2 - k_2 + k_0, l_3 - k_3 + k_0] \cdot h[k_3]h[k_2]h^*[k_1]h^*[k_0] \qquad (12.53)$$

These formulas can be interpreted as a sequence of convolutions with the filter impulse response in various directions (see [1]).

The corresponding frequency domain expressions are relatively simpler since they contain only products of terms. The expressions for the bispectrum and trispectrum are

$$B_y(\omega_1, \omega_2) = H^*[e^{j(\omega_1+\omega_2)}]H(e^{j\omega_1})H(e^{j\omega_2})B_x(\omega_1, \omega_2) \qquad (12.54)$$

and

$$T_y(\omega_1, \omega_2, \omega_3) = H^*[e^{j(\omega_1+\omega_2+\omega_3)}]H^*(e^{-j\omega^{(1)}})H(e^{j\omega_2})H(e^{j\omega_3})T_x(\omega_1, \omega_2, \omega_3) \qquad (12.55)$$

Unlike the power spectral density function, these higher-order spectra are affected by the phase of the linear system. For example, the phase of the output bispectrum is given by

$$\angle B_y(\omega_1, \omega_2) = -\angle H[e^{j(\omega_1+\omega_2)}] + \angle H(e^{j\omega_1}) + \angle H(e^{j\omega_2}) + \angle B_x(\omega_1, \omega_2)$$

Using higher-order statistics it is possible to identify both the magnitude and phase of a linear system, while with second-order statistics it is possible to identify only the magnitude.

12.3 Representation of Signals as Random Vectors

12.3.1 Statistical Description of Random Vectors

It is often useful to define a random vector $\boldsymbol{x}$ consisting of N consecutive values of a random signal as shown in Figure 12.8. The joint density function of these N values is referred to as the probability density function of the random vector and is written as $f_x(\mathbf{x})$. Consider the case of a real-valued signal first. If $\mathbf{x}^o$ denotes a particular value of the random vector

$$\mathbf{x}^o = \begin{bmatrix} x_0^o \\ x_1^o \\ \vdots \\ x_{N-1}^o \end{bmatrix}$$

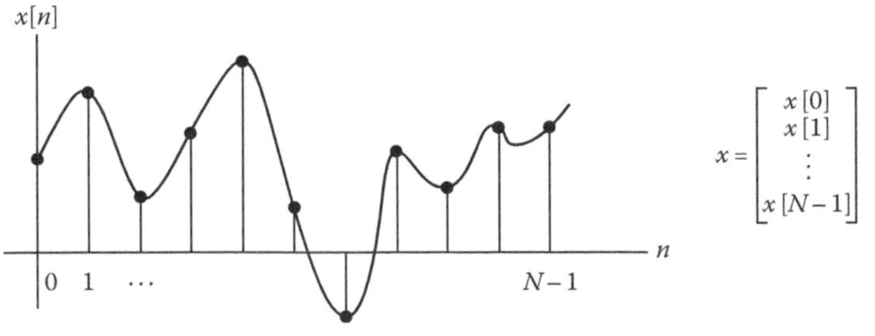

FIGURE 12.8 Representation of a random sequence as a random vector. (From Therrien, C.W., *Discrete Random Signals and Statistical Signal Processing*, Prentice Hall, Inc., Upper Saddle River, NJ, 1992. With permission.)

and if small increments Δx_i are taken in each of the components, the expression

$$f_x(\mathbf{x}^o)\Delta x_0 \Delta x_1 \cdots \Delta x_{N-1}$$

represents the probability that the signal (i.e., the random vector $\boldsymbol{x}$) lies in a small region of the vector space described by

$$x_0^o < x[0] \leq x_0^o + \Delta x_0, \ldots, x_{N-1}^o < x[N-1] \leq x_{N-1}^o + \Delta x_{N-1} \tag{12.56}$$

For a complex-valued random signal, $\boldsymbol{x}$ has complex components and $f_x(\mathbf{x})$ represents the joint density between the $2N$ real and imaginary parts of the components of $\boldsymbol{x}$. Conditional and joint densities for random vectors are defined in a corresponding way [1] and have interpretations that are analogous to those for scalar random variables.

12.3.2 Moments

The first- and second-moment properties for random vectors are considerably important and are represented as follows. The mean vector is defined by

$$\mathbf{m}_x \overset{\text{def}}{=} E\{\boldsymbol{x}\} = \begin{bmatrix} m_0 \\ m_1 \\ \vdots \\ m_{N-1} \end{bmatrix} \tag{12.57}$$

where $m_i = E\{x[i]\}$ for $i = 0, 1, \ldots, N-1$. In the case of a stationary signal, all of the m_i have the same value (frequently zero).

The correlation matrix* is defined by

$$\mathbf{R}_x \overset{\text{def}}{=} E\{\mathbf{x}\mathbf{x}^{*T}\} \tag{12.58}$$

Note that this expression represents an outer product of vectors, not an inner product, so the result is an $N \times N$ square matrix with the element in row i and column j given by $E\{x[i]x^*[j]\}$. For a stationary random process, $E\{x[i]x^*[j]\}$ is equal to $R_x[i-j]$, so the matrix has the form

$$\mathbf{R}_x = \begin{bmatrix} R_x[0] & R_x[-1] & \cdots & R_x[-N+1] \\ R_x[1] & R_x[0] & \ddots & \vdots \\ \vdots & \ddots & \ddots & R_x[-1] \\ R_x[N-1] & \cdots & R_x[1] & R_x[0] \end{bmatrix}$$

The correlation matrix is Hermitian symmetric ($\mathbf{R}_x = \mathbf{R}_x^{*T}$) and Toeplitz (all elements on each diagonal are equal). The Hermitian symmetry property follows from the basic definition, Equation 12.58, and is true for all correlation matrices; the Toeplitz property occurs only for correlation matrices of stationary random processes.

The covariance matrix is defined as

$$\mathbf{C}_x = E\{(\boldsymbol{x} - \mathbf{m}_x)(\boldsymbol{x} - \mathbf{m}_x)^{*T}\} \tag{12.59}$$

* Sometimes called the *autocorrelation* matrix.

TABLE 12.2 Relations for the Complex Correlation Matrix of a Circular (Proper) Random Vector

Complex correlation matrix	$\mathbf{R}_x = E\{\boldsymbol{x}\boldsymbol{x}^{*\mathrm{T}}\} = 2\mathbf{R}_x^{\mathrm{E}} + j2\mathbf{R}_x^{\mathrm{o}}$
Correlation matrices for components	$\mathbf{R}_x^{\mathrm{E}} = E\{\boldsymbol{x}_{\mathrm{r}}\boldsymbol{x}_{\mathrm{r}}^{\mathrm{T}}\} = E\{\boldsymbol{x}_{\mathrm{i}}\boldsymbol{x}_{\mathrm{i}}^{\mathrm{T}}\}$
	$\mathbf{R}_x^{\mathrm{o}} = -E\{\boldsymbol{x}_{\mathrm{r}}\boldsymbol{x}_{\mathrm{i}}^{\mathrm{T}}\} = E\{\boldsymbol{x}_{\mathrm{i}}\boldsymbol{x}_{\mathrm{r}}^{\mathrm{T}}\}$

and satisfies the relation

$$\mathbf{R}_x = \mathbf{C}_x + \mathbf{m}_x\mathbf{m}_x^{*\mathrm{T}} \tag{12.60}$$

The covariance matrix is thus the correlation matrix of the vector with the mean removed.

For nonstationary random processes, the mean vector and correlation matrix may not be sufficient to describe the complete second-order statistics of a complex random vector [3,6,10]. In general, the relation matrix* defined by

$$\mathbf{R}'_x \stackrel{\mathrm{def}}{=} E\{\mathbf{x}\mathbf{x}^{\mathrm{T}}\} \tag{12.61}$$

is also needed. If the random vector is derived from a random process exhibiting circularity, however, then $\mathbf{R}'_x$ is identically zero and the random vector $\boldsymbol{x}$ is likewise said to be circular (or proper). In this case, the correlation and cross-correlation matrices for the real and imaginary parts of the complex random vector $\boldsymbol{x}$ are related to the real and imaginary parts of the correlation matrix as shown in Table 12.2.

The correlation and covariance matrices of any random vector are positive semidefinite, that is,

$$\mathbf{a}^{*\mathrm{T}}\mathbf{R}_x\mathbf{a} \geq 0$$

(and $\mathbf{a}^{*\mathrm{T}}\mathbf{C}_x\mathbf{a} \geq 0$) for any vector $\mathbf{a}$. The correlation matrix for a *regular* random process is in fact strictly positive definite ($>$ rather than $\geq$), while that for a predictable random process is just positive semidefinite, if the size is sufficiently large.

Cross-correlation and cross-covariance matrices for two random signals or two random vectors $\boldsymbol{x}$ and $\boldsymbol{y}$ can also be defined as

$$\mathbf{R}_{xy} = E\{\boldsymbol{x}\boldsymbol{y}^{*\mathrm{T}}\} \tag{12.62}$$

and

$$\mathbf{C}_{xy} = E\{(\boldsymbol{x} - \mathbf{m}_x)(\boldsymbol{y} - \mathbf{m}_y)^{*\mathrm{T}}\} \tag{12.63}$$

These matrices have no particular properties and are not even square if $\boldsymbol{x}$ and $\boldsymbol{y}$ have different sizes. They exhibit a Toeplitz-like structure, however (all terms on the same diagonal are equal), if the two random processes are jointly stationary.

12.3.3 Linear Transformation of Random Vectors

When a vector $\boldsymbol{y}$ is defined by a linear transformation

$$\boldsymbol{y} = \mathbf{A}\boldsymbol{x} \tag{12.64}$$

* Also called the complementary correlation matrix or pseudo-correlation matrix.

the mean of $\boldsymbol{y}$ is given by $E\{\boldsymbol{y}\} = \mathbf{A}E\{\boldsymbol{x}\}$ or

$$\mathbf{m}_y = \mathbf{A}\mathbf{m}_x \tag{12.65}$$

while the correlation matrix is given by $E\{\boldsymbol{y}\boldsymbol{y}^{*\mathrm{T}}\} = \mathbf{A}E\{\boldsymbol{x}\boldsymbol{x}^{*\mathrm{T}}\}\mathbf{A}^{*\mathrm{T}}$ or

$$\mathbf{R}_y = \mathbf{A}\mathbf{R}_x\mathbf{A}^{*\mathrm{T}} \tag{12.66}$$

From these last two equations and Equation 12.60, it can be shown that the covariance matrix transforms in a similar manner, that is,

$$\mathbf{C}_y = \mathbf{A}\mathbf{C}_x\mathbf{A}^{*\mathrm{T}} \tag{12.67}$$

Transformations that result in random vectors with uncorrelated components are of special interest. Strictly speaking, the term "uncorrelated" applies to the covariance matrix. That is, if a random vector has uncorrelated components, its covariance matrix is diagonal. It is common practice, however, to assume that the mean is zero and discuss the methods using the correlation matrix. If the mean is nonzero, then the components are said to be *orthogonal* rather than uncorrelated.

Since correlation matrices are Hermitian symmetric and positive semidefinite, their eigenvalues are nonnegative and eigenvectors are orthogonal (see, e.g., [11,12]). Any correlation matrix can therefore be factored as

$$\mathbf{R}_x = \mathbf{E}\boldsymbol{\Lambda}\mathbf{E}^{*\mathrm{T}} \tag{12.68}$$

where

$\mathbf{E}$ is a unitary matrix ($\mathbf{E}^{*\mathrm{T}}\mathbf{E} = \mathbf{I}$) whose columns are the eigenvectors
$\boldsymbol{\Lambda}$ is a diagonal matrix whose elements are the eigenvalues

Since the inverse of a unitary matrix is its Hermitian transpose, the last equation can be rewritten as

$$\boldsymbol{\Lambda} = \mathbf{E}^{*\mathrm{T}}\mathbf{R}_x\mathbf{E}$$

Comparing this with Equation 12.66 shows that if $\boldsymbol{y}$ is defined by

$$\boldsymbol{y} = \mathbf{E}^{*\mathrm{T}}\boldsymbol{x} \tag{12.69}$$

then $\mathbf{R}_y$ will be equal to $\boldsymbol{\Lambda}$, a diagonal matrix. Since $\mathbf{R}_y$ is diagonal, the components of $\boldsymbol{y}$ are uncorrelated ($E\{y_i y_j^*\} = 0,\ i \neq j$). Thus, one way to produce a vector with uncorrelated components is to apply the eigenvector transformation (Equation 12.69).

Another way to produce a vector with uncorrelated components involves triangular decomposition of the correlation matrix. Matrices that satisfy certain conditions of their principal minors [13] can be factored into a product of a lower triangular and an upper triangular matrix. (This is called "*LU*" decomposition). Correlation matrices always satisfy the needed conditions, and since they are Hermitian symmetric, they can be written as a unique product

$$\mathbf{R}_x = \mathbf{L}\mathbf{D}\mathbf{L}^{*\mathrm{T}} \tag{12.70}$$

where

L is a lower triangular matrix with ones on the diagonal

D is a diagonal matrix

The product $\mathbf{DL}^{*T}$ is the upper triangular matrix "*U*" in the *LU* decomposition.

Equation 12.70 can be rewritten as

$$\mathbf{D} = \mathbf{L}^{-1}\mathbf{R}_x(\mathbf{L}^{-1})^{*T} \tag{12.71}$$

where it can be shown that $\mathbf{L}^{-1}$ is of the same form as **L** (i.e., lower triangular with ones on the diagonal). From Equations 12.71 and 12.66, it can be recognized that **D** is the correlation matrix for a random vector $\boldsymbol{y}$ defined by

$$\boldsymbol{y} = \mathbf{L}^{-1}\boldsymbol{x} \tag{12.72}$$

Since **D** is a diagonal matrix, the components of $\boldsymbol{y}$ are seen to be uncorrelated.

The two transformations Equations 12.69 and 12.72 correspond to two fundamentally different ways of decorrelating a signal. The eigenvector transformation represents the signal in terms of an orthogonal set of basis functions (the eigenvectors) and has important geometric interpretations (see Section 12.3.4 and [1, Chapter 2]). It is also the basis for modern subspace methods of spectrum analysis and array processing. The transformation defined by the triangular decomposition has the advantage that it can be implemented by a causal linear filter. Thus, it has important practical applications. It is the transformation that naturally arises in the very important area of signal processing known as linear predictive filtering [1].

12.3.4 Gaussian Density Function

One of the cases mentioned in Section 12.1.1 in which a complete statistical description of a random process is possible is the Gaussian case. The form of the probability density function is slightly different in the real and the complex cases.

12.3.4.1 Real Gaussian Density

When a random signal is Gaussian, the density function for the random vector $\boldsymbol{x}$ representing that signal is specified in terms of just the mean vector and covariance matrix. For a real random signal this density function has the form

$$f_x(\mathbf{x}) = \frac{1}{(2\pi)^{\frac{N}{2}}|\mathbf{C}_x|^{\frac{1}{2}}} e^{-\frac{1}{2}(\mathbf{x}-\mathbf{m}_x)^T\mathbf{C}_x^{-1}(\mathbf{x}-\mathbf{m}_x)} \quad \text{(real random vector)} \tag{12.73}$$

where N is the dimension of $\boldsymbol{x}$.

The contours of the density function defined by

$$f_x(\mathbf{x}) = \text{constant} \tag{12.74}$$

are ellipsoids centered about the mean vector as shown in Figure 12.9 for a dimension $N = 2$. These are known as concentration ellipsoids (because they represent regions where the data is concentrated) and are useful in representing the signal from a geometric point of view. The orientation and eccentricity of the ellipsoid depend on the correlation between the components of the random vector.

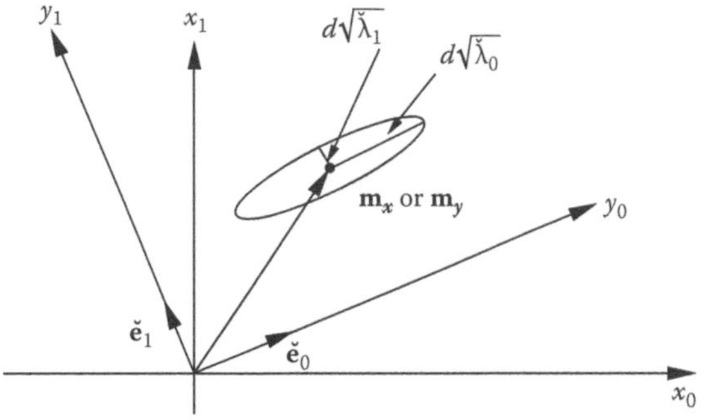

FIGURE 12.9 Typical contour of a Gaussian density function. (From Therrien, C.W., *Discrete Random Signals and Statistical Signal Processing*, Prentice Hall, Inc., Upper Saddle River, NJ, 1992. With permission.)

To prove that the Gaussian density contours are ellipsoids, observe that the Equation 12.74, which defines the contours, implies that the quadratic form in the exponent of Equation 12.73 satisfies the condition

$$(\mathbf{x}-\mathbf{m}_x)^{*\mathrm{T}}\mathbf{C}_x^{-1}(\mathbf{x}-\mathbf{m}_x)=d^2 \tag{12.75}$$

where d is a positive real constant.* By using the eigenvector decomposition as in Equation 12.68, this quadratic form can be rewritten as

$$\begin{aligned}(\mathbf{x}-\mathbf{m}_x)^{*\mathrm{T}}\mathbf{C}_x^{-1}(\mathbf{x}-\mathbf{m}_x)&=(\mathbf{x}-\mathbf{m}_x)^{*\mathrm{T}}\breve{\mathbf{E}}\breve{\mathbf{\Lambda}}^{-1}\breve{\mathbf{E}}^{*\mathrm{T}}(\mathbf{x}-\mathbf{m}_x)\\&=(y-\mathbf{m}_y)^{*\mathrm{T}}\breve{\mathbf{\Lambda}}^{-1}(y-\mathbf{m}_y)=d^2\end{aligned} \tag{12.76}$$

where $y=\breve{\mathbf{E}}^{*T}x$ and "hats" have been added to the variables to indicate that they pertain to the covariance matrix rather than the correlation matrix. Since $\breve{\mathbf{\Lambda}}^{-1}$ is diagonal, this last expression can be written in expanded form as

$$\frac{|y_0-\mathbf{m}_{y0}|^2}{\breve{\lambda}_0}+\frac{|y_1-\mathbf{m}_{y1}|^2}{\breve{\lambda}_1}+\cdots+\frac{|y_{N-1}-\mathbf{m}_{yN-1}|^2}{\breve{\lambda}_{N-1}}=d^2 \tag{12.77}$$

which is the equation of an N-dimensional ellipsoid with center at $\mathbf{m}_y$. The transformation $y=\breve{\mathbf{E}}^{*\mathrm{T}}x$ represents a rotation of the coordinate system to one aligned with the eigenvectors, which are parallel to the axes of the ellipsoid. The sizes of the axes are proportional to the square roots of the eigenvalues.

12.3.4.2 Complex Gaussian Density

For a complex random vector, the probability density function is really a joint density function for the real and imaginary parts of the vector. If this joint density is expressed in terms of the vector and its conjugate, it can be written (with some abuse of notation) as a product [6]

$$f'(\mathbf{x},\mathbf{x}^*)=f_x(\mathbf{x})\cdot f(\mathbf{x}^*|\mathbf{x}) \tag{12.78}$$

* The parameter d is known as the *Mahalanobis distance* between the random vector $\mathbf{x}$ and the mean $\mathbf{m}_x$.

The first term on the right is given by

$$f_x(\mathbf{x}) = \frac{1}{\pi^N |\mathbf{C}_x|} \exp\left[-(\mathbf{x} - \mathbf{m}_x)^{*\mathrm{T}} \mathbf{C}_x^{-1} (\mathbf{x} - \mathbf{m}_x)\right] \quad \text{(complex random vector)} \tag{12.79}$$

and is known as the complex Gaussian density function [1,14]. It involves only the mean and covariance matrix and is the form most commonly found in the literature. This form is strictly correct, however, only when the zero-mean relation function

$$\mathbf{C}_x' = E\left\{(\boldsymbol{x} - \mathbf{m}_x)(\boldsymbol{x} - \mathbf{m}_x)^{\mathrm{T}}\right\}$$

is zero, that is, when the random vector satisfies circularity. The abuse of notation occurs in part because $f_x(\mathbf{x})$ is not a true analytic function of a complex random variable unless it is written as a function of both $\mathbf{x}$ and $\mathbf{x}^*$.

The second term in Equation 12.78 makes the expression for the Gaussian density function completely general and must be included when the random vector does not satisfy circularity. This term has the form [6]

$$f(\mathbf{x}^*|\mathbf{x}) = \frac{1}{|\mathbf{I} - \mathbf{G}^{*\mathrm{T}}\mathbf{G}|} \exp\{-\mathbf{x}^{*\mathrm{T}}\mathbf{G}^{*\mathrm{T}}\mathbf{P}^{-1}\mathbf{G}\mathbf{x} + \Re[\mathbf{x}^{\mathrm{T}}\mathbf{G}^{\mathrm{T}}(\mathbf{P}^{-1})^*\mathbf{x}]\}$$

where

$$\mathbf{G} = \mathbf{C}_x'^{*\mathrm{T}}\mathbf{C}_x^{-1}$$

and

$$\mathbf{P} = \mathbf{C}_x^* - \mathbf{C}_x'^{*\mathrm{T}}\mathbf{C}_x^{-1}\mathbf{C}_x'$$

When $\mathbf{C}_x' = \mathbf{0}$, $f(\mathbf{x}^*|\mathbf{x})$ becomes equal to one so that $f'(\mathbf{x}, \mathbf{x}^*) = f_x(\mathbf{x})$. The term $f(\mathbf{x}^*|\mathbf{x})$ can be thought of as a conditional density function for $\boldsymbol{x}^*$ given $\boldsymbol{x}$ (although again it is not a true analytic function). The parameter $\mathbf{G}$ can be interpreted as a matrix of coefficients for estimating $\boldsymbol{x}^*$ from $\boldsymbol{x}$ while $\mathbf{P}$ is the corresponding error covariance matrix. Estimation is discussed further below.

12.4 Fundamentals of Estimation

Problems of statistical estimation deal with deriving values for quantities that cannot be observed or measured directly from quantities that *can* be observed. These problems may deal with finding the parameters for a model of a signal from direct measurements of that signal or for deducing one signal from another (say a clean transmitted signal from a received noisy and distorted one). The former type of problem is known as parameter estimation while the latter is called random variable estimation. Although the two types of problems have much in common, it is convenient to consider them separately here.

12.4.1 Estimation of Parameters

12.4.1.1 Maximum Likelihood Estimation

The problem of parameter estimation begins with having observations of a random variable. The random variable is described by a probability density function of some known form, but having one or more parameters whose values are unknown. The problem is to determine values for the parameter or parameters from observations of the random variable.

In many cases, the observed random variable and the parameters are vector quantities. The density function is thus denoted by

$$f_{x;\boldsymbol{\theta}}(\mathbf{x};\boldsymbol{\theta}) \tag{12.80}$$

where

$\mathbf{x}$ represents the observations
$\boldsymbol{\theta}$ is the parameter (or set of parameters) to be determined

The value of $\boldsymbol{\theta}$ determined by the estimation procedure is called an "estimate" of the parameter and is denoted by $\hat{\boldsymbol{\theta}}$. (Estimates of parameters are conventionally denoted by putting a "hat" over the symbol.)

A particularly important estimate is the maximum likelihood estimate. For a given fixed set of observations, say $\mathbf{x}=\mathbf{x}^o$, the maximum likelihood estimate of the parameter is the value of $\boldsymbol{\theta}$ that maximizes $f_{x;\boldsymbol{\theta}}(\mathbf{x}^o;\boldsymbol{\theta})$. This estimate of $\boldsymbol{\theta}$ will be denoted by $\hat{\boldsymbol{\theta}}_{ml}$.

To see why this may be a good estimate and why it is called "maximum likelihood," consider the following simple example involving a scalar parameter. It is desired to estimate the mean m of a Gaussian density function

$$f_{x;m}(\mathrm{x};m) = \frac{1}{\sqrt{2\pi}}\mathrm{e}^{-\frac{(\mathrm{x}-m)^2}{2}}$$

using a single observation x^o. (The variance is assumed known and equal to one.) The density function is depicted in Figure 12.10 for several proposed values of the mean. It is obvious that if the true value of the mean were any of m_1, m_2, or m_4, the given observation x^o would be unlikely. The choice m_3, however, which places the peak of the density function at the location x^o, makes the given observation most likely (i.e., it maximizes the quantity $f_{x;m}(\mathrm{x}^o;m)$) and is therefore the maximum likelihood estimate.

When $f_{x;\boldsymbol{\theta}}(\mathbf{x};\boldsymbol{\theta})$ of Equation 12.80 is viewed as a function of $\boldsymbol{\theta}$ rather than as a function of $\mathbf{x}$, it is known as the *likelihood function*. Thus, the maximum likelihood estimate of a parameter maximizes the likelihood function. If the likelihood function is continuous, and the maximum does not occur at a boundary, then the maximum likelihood estimate for a real scalar parameter θ can be found through either of the necessary conditions

$$\left.\frac{\partial f_{x;\theta}(\boldsymbol{x};\theta)}{\partial\theta}\right|_{\theta=\hat{\theta}_{ml}} = 0 \tag{12.81a}$$

$$\left.\frac{\partial \ln f_{x;\theta}(\boldsymbol{x};\theta)}{\partial\theta}\right|_{\theta=\hat{\theta}_{ml}} = 0 \tag{12.81b}$$

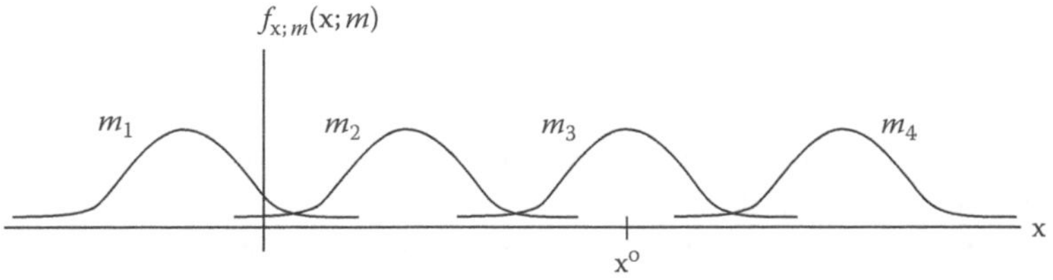

FIGURE 12.10 Maximum likelihood estimation of the mean. (From Therrien, C.W., *Discrete Random Signals and Statistical Signal Processing*, Prentice Hall, Inc., Upper Saddle River, NJ, 1992. With permission.)

These equations are sometimes referred to as the likelihood equation and the log likelihood equation, respectively. Equation 12.81b follows from Equation 12.81a because the logarithm is a strict monotonically increasing function; this form is convenient in many problems where the likelihood function involves an exponential. For vector parameters $\boldsymbol{\theta}$, the corresponding likelihood and log likelihood equations involve setting the gradient or vector of derivatives with respect to all components of the parameter to zero. (See [1] for procedures involving both real and complex vector parameters.)

12.4.1.2 Properties of Estimates

The maximum likelihood estimate for a parameter $\boldsymbol{\theta}$ can be written as

$$\hat{\boldsymbol{\theta}}_{ml}(\boldsymbol{x}) = \text{argmax} f_{\boldsymbol{x};\boldsymbol{\theta}}(\boldsymbol{x};\, \boldsymbol{\theta}) \tag{12.82}$$

from which it is clear that the estimate is a function of the observations. Moreover, since the observation vector $\boldsymbol{x}$ is a random vector, the estimate is itself a random variable and has a mean, covariance, density function, and so on. Not all estimates are maximum likelihood estimates. For example, the estimate for the mean of a Gaussian random signal, given by Equation 12.89, is a maximum likelihood estimate, while the following estimate for the variance (where $\hat{m}_x$ is given by Equation 12.89) is not:

$$\hat{\sigma}_x^2 = \frac{1}{N}\sum_{n=0}^{N-1}(x[n] - \hat{m}_x)^2$$

All estimates are functions of the observations; however, it is useful to denote a general estimate by

$$\hat{\boldsymbol{\theta}}_N = \hat{\boldsymbol{\theta}}_N(\boldsymbol{x})$$

where the subscript N denotes the number of observations (dimension of $\boldsymbol{x}$), and examine its statistical properties.

Among the properties of estimates that are most useful are the following:

1. An estimate $\hat{\boldsymbol{\theta}}_N$ is unbiased if

$$E\{\hat{\boldsymbol{\theta}}_N\} = \boldsymbol{\theta}$$

Otherwise the estimate is biased with bias $\mathbf{b}(\boldsymbol{\theta}) = E\{\hat{\boldsymbol{\theta}}_N\} - \boldsymbol{\theta}$. An estimate is asymptotically unbiased if

$$\lim_{N\to\infty} E\{\hat{\boldsymbol{\theta}}_N\} = \boldsymbol{\theta}$$

2. An estimate $\hat{\boldsymbol{\theta}}_N$ is consistent if

$$\lim_{N\to\infty} \Pr[|\hat{\boldsymbol{\theta}}_N - \boldsymbol{\theta}| < \varepsilon] = 1$$

for any arbitrarily small number ε. The sequence of estimates $\{\hat{\boldsymbol{\theta}}_N\}$ is then said to converge in probability to the parameter $\boldsymbol{\theta}$.

3. An estimate $\hat{\boldsymbol{\theta}}$ is said to be efficient with respect to another estimate $\hat{\boldsymbol{\theta}}'$, if the difference of their covariance matrices $\mathbf{C}_{\hat{\boldsymbol{\theta}}'} - \mathbf{C}_{\hat{\boldsymbol{\theta}}}$ is positive definite (written $\mathbf{C}_{\hat{\boldsymbol{\theta}}'} > \mathbf{C}_{\hat{\boldsymbol{\theta}}}$). This implies that the variance of every component of $\hat{\boldsymbol{\theta}}$ must be smaller than the variance of the corresponding component of $\hat{\boldsymbol{\theta}}'$. If $\hat{\boldsymbol{\theta}}_N$ is unbiased and efficient with respect to $\hat{\boldsymbol{\theta}}_{N-1}$ for all N then $\hat{\boldsymbol{\theta}}_N$ is a consistent estimate.

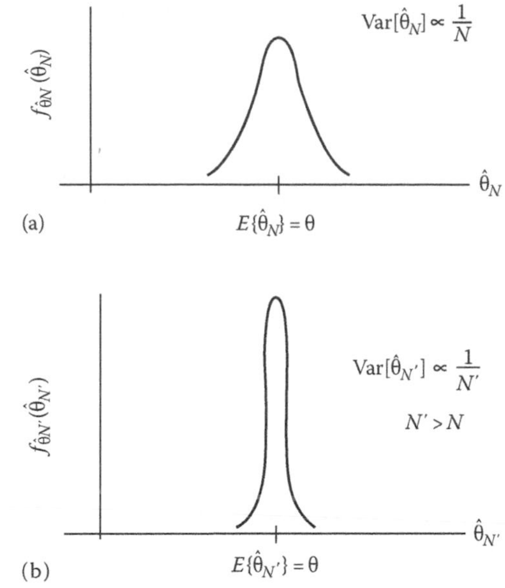

FIGURE 12.11 Density function for an unbiased estimate whose variance decreases with N: (a) density function of the estimate $\hat{\theta}_N$ and (b) density function of the estimate $\hat{\theta}_{N'}$ with $N' > N$. (From Therrien, C.W., *Discrete Random Signals and Statistical Signal Processing*, Prentice Hall, Inc., Upper Saddle River, NJ, 1992. With permission.)

The last statement needs a little more explanation which can best be given for the case of a scalar estimate. For a scalar estimate property, (3) is a statement about its variance. The Tchebycheff inequality (see, e.g., [15]) states that

$$\Pr[|\hat{\theta}_N - \theta| \geq \varepsilon] \leq \frac{\mathrm{Var}[\hat{\theta}_N]}{\varepsilon^2}$$

Thus, if the variance of $\hat{\theta}_N$ decreases with N, the probability that $|\hat{\theta}_N - \theta| \geq \varepsilon$ approaches zero as $N \rightarrow \infty$. In other words, the probability that $|\hat{\theta}_N - \theta| < \varepsilon$ approaches one. This last property is illustrated in Figure 12.11.

The variance of any unbiased estimate can be bounded with a powerful result known as the Cramér–Rao inequality. For the case of a scalar parameter, the Cramér–Rao bound has the form

$$\mathrm{Var}[\hat{\theta}] \geq \frac{1}{E\left\{\left(\frac{\partial \ln f_{x;\theta}(x;\,\theta)}{\partial\theta}\right)^2\right\}} = \frac{1}{-E\left\{\frac{\partial^2 \ln f_{x;\theta}(x;\,\theta)}{\partial\theta^2}\right\}} \tag{12.83}$$

where equality occurs if and only if

$$\hat{\theta}(x) - \theta = K(\theta)\cdot\frac{\partial \ln f_{x;\theta}(x;\,\theta)}{\partial\theta}$$

The two alternate expressions on the right-hand side are valid as long as the partial derivatives exist and are absolutely integrable.

The general form of the Cramér–Rao bound for vector parameters is usually written as

$$\mathbf{C}_{\hat{\theta}} \geq \mathbf{J}^{-1} \tag{12.84}$$

meaning that the difference matrix $\mathbf{C}_{\hat{\boldsymbol{\theta}}} - \mathbf{J}^{-1}$ is positive semidefinite. The bounding matrix on the right-hand side of Equation 12.84 is the inverse of the Fisher information matrix defined by

$$\mathbf{J}(\boldsymbol{\theta}) = E\{\boldsymbol{s}(\boldsymbol{x};\, \boldsymbol{\theta})\boldsymbol{s}^{\mathrm{T}}(\boldsymbol{x};\, \boldsymbol{\theta})\} \tag{12.85}$$

where $\boldsymbol{s}(\boldsymbol{x};\, \boldsymbol{\theta})$ is a vector whose ith component is the derivative of $\ln f_{\boldsymbol{x};\boldsymbol{\theta}}(\boldsymbol{x};\, \boldsymbol{\theta})$ with respect to $\hat{\theta}_i$, the ith component of $\boldsymbol{\theta}$. Equation 12.84 implies that the variance of $\hat{\theta}_i$ is bounded by

$$\mathrm{Var}\left[\hat{\theta}_i\right] \geq j_{ii}^{(-1)} \tag{12.86}$$

where $j_{ii}^{(-1)}$ is the ith diagonal element of the inverse Fisher information matrix. The bound, Equation 12.84, is satisfied with equality if and only if the estimate satisfies an equation of the form

$$\hat{\boldsymbol{\theta}}(\boldsymbol{x}) - \boldsymbol{\theta} = \mathbf{K}(\boldsymbol{\theta})\boldsymbol{s}(\boldsymbol{x};\, \boldsymbol{\theta}) \tag{12.87}$$

In this case, $\mathbf{K}$ is uniquely defined by

$$\mathbf{K}(\boldsymbol{\theta}) = \mathbf{J}^{-1}(\boldsymbol{\theta}) \tag{12.88}$$

(see [1]). An estimate satisfying the bound with equality is known as a minimum-variance estimate. It can be shown that if an unbiased minimum variance estimate exists and the maximum likelihood estimate does not occur at a boundary, then the maximum likelihood estimate is that minimum-variance estimate.

An interpretation of the Cramér–Rao bound in terms of concentration ellipsoids is given in Figure 12.12. If the *deviation* in the estimate is defined as

$$\boldsymbol{\delta}(\boldsymbol{x};\, \boldsymbol{\theta}) \stackrel{\text{def}}{=} \hat{\boldsymbol{\theta}}(\boldsymbol{x}) - \boldsymbol{\theta}$$

then the bias of the estimate $\mathbf{b}(\boldsymbol{\theta})$ is the mean deviation (i.e., its expected value). The concentration ellipse for the deviation, with covariance $\mathbf{C}_{\hat{\boldsymbol{\theta}}}$, is shown in the figure. The minimum-deviation covariance of the Cramér–Rao bound is represented by the smaller ellipse with covariance $\mathbf{J}^{-1}$. Geometrically the bound states that the $\mathbf{J}^{-1}$ ellipsoid lies entirely within the $\mathbf{C}_{\hat{\boldsymbol{\theta}}}$ ellipsoid. In the best case (when $\hat{\boldsymbol{\theta}}$ is the maximum likelihood estimate), the two ellipsoids coincide.

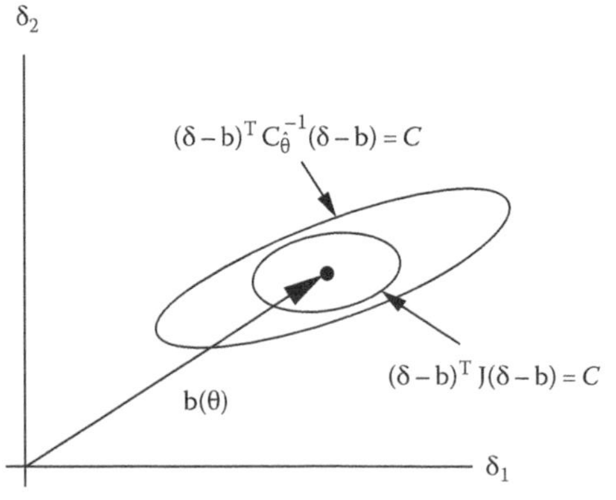

FIGURE 12.12 Concentration ellipses for the deviation of the estimate of a vector parameter; geometric interpretation of the Cramér–Rao bound. (From Therrien, C.W., *Discrete Random Signals and Statistical Signal Processing*, Prentice Hall, Inc., Upper Saddle River, NJ, 1992. With permission.)

12.4.1.3 Estimates for Moments of Discrete Random Signals

Some of the most important parameters for random signals are their mean, autocorrelation (or autocovariance) functions, and perhaps higher-order statistics. Under conditions of stationarity and ergodicity, these parameters can be estimated from a given realization of the signal (see Section 12.1.1). Some common forms of these estimates and some of their statistical properties are cited here.

Given N time samples of a random signal, an estimate for the mean can be formed as

$$\hat{m}_x = \frac{1}{N}\sum_{n=0}^{N} x[n] \tag{12.89}$$

This estimate, known as the sample mean, is unbiased and efficient and therefore a consistent estimate. An expression for the variance of the estimate is not difficult to derive in terms of the autocovariance function for the process (see [1]).

The autocorrelation function is usually estimated by one of the two formulas

$$\hat{R}_x[l] = \frac{1}{N-l}\sum_{n=0}^{N-1-l} x[n+l]x^*[n], \quad 0 \le l < N \tag{12.90}$$

or

$$\hat{R}_x[l] = \frac{1}{N}\sum_{n=0}^{N-1-l} x[n+l]x^*[n], \quad 0 \le l < N \tag{12.91}$$

Values of the autocorrelation for negative lags are computed via the relation $\hat{R}_x[-l] = \hat{R}_x^*[l]$. Equation 12.90 generates an unbiased estimate, while Equation 12.91 is only asymptotically unbiased. Explicit expressions for the variance of these estimates can be derived in the Gaussian case [1]; these expressions have been found to be good approximations for non-Gaussian cases as well when $l \ll N$. Both estimates are found to be efficient and therefore consistent, but Equation 12.90 has one serious defect: it cannot be guaranteed to generate a positive semidefinite sequence. Therefore, Equation 12.91 is usually the preferred estimate for the autocorrelation function.

Estimates for the cross-correlation function of two sequences x and y can be generated by substituting $y^*[n]$ for $x^*[n]$ in either of the above two equations. The cross-correlation estimates have similar statistical properties. Since the cross-correlation function is not symmetric, however, explicit values for negative lags have to be computed by interchanging x and y in these equations and using the relation $\hat{R}_{xy}[-l] = \hat{R}_{yx}^*[l]$.

Estimates for higher-order moments and cumulants can be obtained by an analogous procedure. For example, after removal of the mean, an estimate for the third-order cumulant is given by

$$\hat{C}_x^{(3)}[l_1, l_2] = \frac{1}{N}\sum_{n=n_I}^{n_F} x^*[n]x[n+l_1]x[n+l_2] \qquad 0 \le l_2 \le l_1 \le N-1 \tag{12.92}$$

where

$$n_I = \max(0, -l_1, -l_2) \quad \text{and} \quad n_F = \min(N-1, N-1-l_1, N-1-l_2)$$

Estimates of the correlation matrix can be found by using Equation 12.91 and generating a corresponding Toeplitz correlation matrix. It is not advisable to use Equation 12.90 in this procedure because of the possibility of generating an estimate for the correlation matrix which is not positive semidefinite. In many applications, however, estimates for the correlation matrix are generated directly from products of suitably arranged matrices of the data. These methods, such as the autocorrelation method, the covariance method, and the modified covariance method, can be found in many references dealing with signal modeling and statistical signal processing (e.g., [1]).

12.4.2 Estimation of Random Variables

The problems of filtering, prediction, and many others can be treated in the general context of random variable estimation. A typical problem is illustrated in Figure 12.13, where the signal, treated as random variable y, is to be estimated from related observations $x_1, x_2, \ldots, x_N$. The form of the estimate is

$$\hat{y} = \hat{y}(\boldsymbol{x}) = \phi(x_1, x_2, \ldots, x_N) \tag{12.93}$$

where the function ϕ is nonlinear in general.

A framework for this problem is provided by the procedure known as Bayes estimation. Here one seeks to minimize the risk defined by

$$\mathcal{R} = E\{C(y, \hat{y})\} \tag{12.94}$$

where C is a function known as the *cost* of the estimate which depends on both y and $\hat{y}$.

Two special cases for the cost function are shown in Figure 12.14. In both cases, the cost depends on the difference $y - \hat{y}$, otherwise known as the error. In the case depicted by Figure 12.14a, the cost is a quadratic function and the risk becomes equal to the mean-square error; while in case Figure 12.14b, the cost of any error that is not less than an arbitrarily small amount ε is uniform and equal to 1. It can be shown that the optimal estimate in both of these cases depends only on the conditional density function $f_{y|\boldsymbol{x}}$ and is given by

$$\hat{y}_{\text{ms}}(\boldsymbol{x}) = \int_{-\infty}^{\infty} y f_{y|\boldsymbol{x}}(y|\boldsymbol{x})\text{d}y \tag{12.95}$$

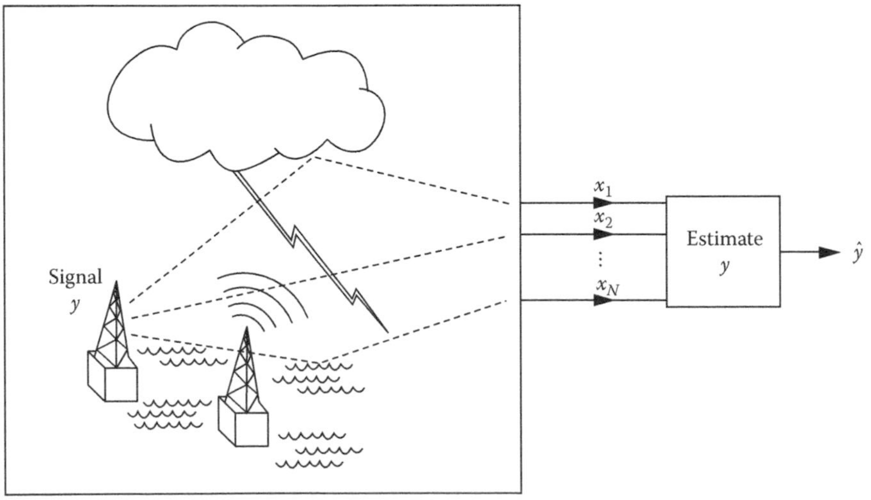

FIGURE 12.13 Estimation of a random variable y from related observations $x_1, x_2, \ldots, x_N$. (From Therrien, C.W., *Discrete Random Signals and Statistical Signal Processing*, Prentice Hall, Inc., Upper Saddle River, NJ, 1992. With permission.)

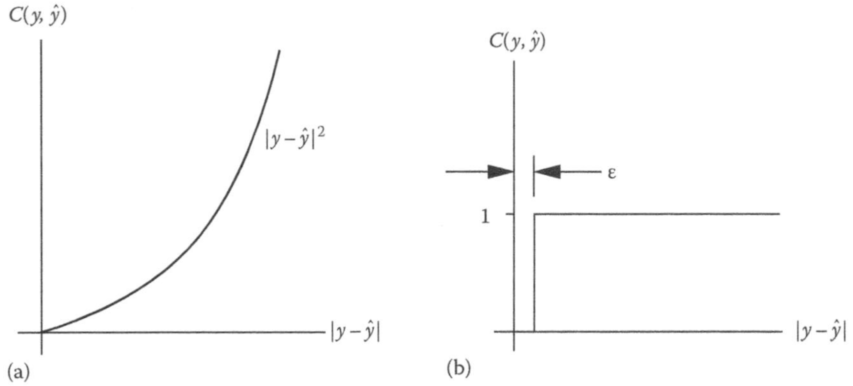

FIGURE 12.14 Cost functions for Bayes estimation: (a) mean-square and (b) uniform. (From Therrien, C.W., *Discrete Random Signals and Statistical Signal Processing*, Prentice Hall, Inc., Upper Saddle River, NJ, 1992. With permission.)

for the mean-square case and

$$\hat{y}_{\text{MAP}}(\boldsymbol{x}) = \operatorname{argmax} \quad f_{y|\boldsymbol{x}}(y|\boldsymbol{x}) \tag{12.96}$$

for the uniform cost function. Notice that Equation 12.95 is the mean of the conditional density, while Equation 12.96 is the maximum of the conditional density function. The latter is called the maximum *a posteriori* (MAP) estimate because it maximizes the posterior density (i.e., the density after knowing the observations).

In general, the mean-square and the MAP estimates are different nonlinear functions of the observations. The choice of estimates depends on the particular problem at hand. In the case of zero-mean jointly Gaussian random variables, however, both estimates become identical and are linear functions of the observations. This is particularly useful for signal processing applications since the estimates can be implemented with linear filters. Signal processing considerations thus motivate finding an estimate that is linear, regardless of the distribution of the random variables. Such a linear estimate can easily be found if the mean-square error criterion is used. Further, the resulting estimate depends only upon second-moment statistics of the signals. This topic, which is taken up below, is of fundamental importance to most areas of modern statistical signal processing.

12.4.3 Linear Mean-Square Estimation

The problem to be addressed here is an estimation problem of the type depicted in Figure 12.13 where the estimate is restricted to be of the form

$$\hat{y} = \mathbf{a}^{*\mathrm{T}}\boldsymbol{x} \tag{12.97}$$

where

$\boldsymbol{x}$ is the vector of observations $x_1, x_2, \ldots, x_N$

$\mathbf{a} = [a_1\, a_2 \cdots a_N]^{\mathrm{T}}$ is a vector of weighting coefficients chosen to minimize the mean-square error

$$E\{|y - \hat{y}|^2\} \tag{12.98}$$

If the mean of $\boldsymbol{x}$ and y are not zero, the mean-square error can be further reduced by using an estimate of the form

$$\hat{y} = \mathbf{a}^{*T}\boldsymbol{x} + b$$

where it can be shown that the constant b should be chosen as $b = m_y - \mathbf{a}^{*T}\mathbf{m}_x$. Since this estimate is not "linear" in the strict mathematical sense, however, it is traditional to consider only the form represented by Equation 12.97. To obtain the best estimate in practice, one should remove the mean of the variables before considering the estimation problem. This is equivalent to using covariances instead of correlations in the solution that appears below and then adding m_y back on to the resulting linear estimate.

The solution to the linear mean-square estimation problem is based on a key idea called the orthogonality principle, which can be stated as follows:

THEOREM 1.1 (Orthogonality):

Let $\varepsilon = y - \hat{y}$ be the error in estimation. Then the weighting vector **a** *minimizes the mean-square error $\sigma_\varepsilon^2 = E\{|y - \hat{y}|^2\}$ if* **a** *is chosen such that $E\{x_i\varepsilon^*\} = 0$, $i = 1, 2, \ldots, N$. Further, the minimum mean-square error is then given by $\sigma_\varepsilon^2 = E\{y\varepsilon^*\}$.*

The condition $E\{x_i\varepsilon^*\} = 0$ is known as orthogonality of the random variables ε and x_i. While the proof of this theorem is not difficult, a geometric argument is perhaps more illuminating.

Let us consider the case for $N = 2$. If the random variables x_1, x_2, y, and e are thought of as "vectors" in an abstract vector space,* then the linear form of the estimate

$$\hat{y} = a_1^* x_1 + a_2^* x_2$$

implies that $\hat{y}$ lies in a plane defined by x_1 and x_2 (see Figure 12.15). The random variable represented by the "vector" y is in general not restricted to this plane and the error ε is the difference between the two vectors, extending from the tip of $\hat{y}$ to the tip of y. From the illustration, it is clear that the magnitude of the error is minimized when ε is made orthogonal to the plane. This is equivalent to requiring that ε be orthogonal to both x_1 and x_2, which is what the orthogonality principle states.

In order to define orthogonality in an abstract vector space, a suitable inner product between vectors is needed. It can be shown that the operation $E\{x_i\varepsilon^*\}$ satisfies the necessary conditions to serve as this definition of inner product. With this form of inner product, it is also clear from Figure 12.15 that the magnitude of the squared error $\sigma_\varepsilon^2 = E\{|\varepsilon|^2\}$ is equal to the inner product $E\{y\varepsilon^*\}$.

The orthogonality principle can be used to easily derive the equations needed to find the linear mean-square estimate. The theorem requires that

$$E\{\boldsymbol{x}\varepsilon^*\} = E\{\boldsymbol{x}(y - \mathbf{a}^{*T}\boldsymbol{x})^*\} = E\{\boldsymbol{x}(y^* - \boldsymbol{x}^{*T}\mathbf{a})\} = 0 \tag{12.99}$$

* Hopefully readers will not be confused by the use of the word "vector" in a new context here. The term "vector" does *not* refer to the array of observations $\boldsymbol{x} = [x_1\, x_2 \cdots x_N]^T$. Rather the term is used to refer to objects in some abstract space, which in this case are random variables. Mathematically, a *vector space* $\mathcal{V}$ is a set of elements $u, v, \ldots$ such that if $u \in \mathcal{V}$ and $v \in \mathcal{V}$ then there is a unique element $u + v \in \mathcal{V}$ called the *sum*. Further if c is an element from an associated field such as the field of real or complex numbers, then the *scalar product* $c \cdot u$, with certain associative and distributive properties, is also an element of $\mathcal{V}$ (see, e.g., [16]). Random variables with the associated field of complex numbers satisfy these conditions.

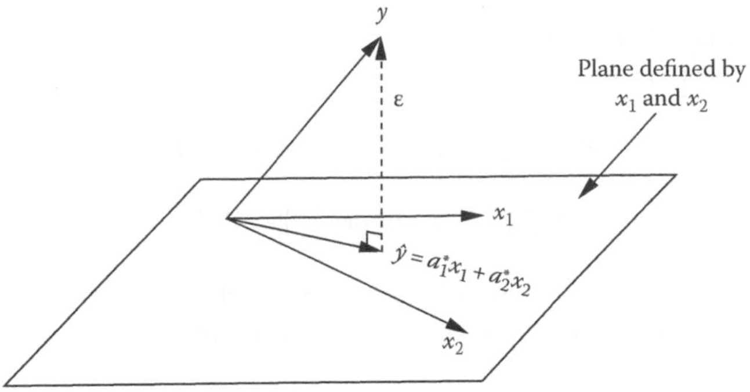

FIGURE 12.15 Vector space interpretation of linear mean-square estimation. (From Therrien, C.W., *Discrete Random Signals and Statistical Signal Processing*, Prentice Hall, Inc., Upper Saddle River, NJ, 1992. With permission.)

Taking the expectation then yields

$$\mathbf{R}_x \mathbf{a} = \mathbf{r}_{x_y} \tag{12.100}$$

where

$$\mathbf{R}_x = E\{\boldsymbol{x}\boldsymbol{x}^{*T}\} \tag{12.101}$$

is the correlation matrix of the observations and

$$\mathbf{r}_{x_y} = E\{\boldsymbol{x}y^*\} \tag{12.102}$$

is the cross-correlation between the $\boldsymbol{x}$ and the random variable y to be estimated. The coefficients $\mathbf{a}$ to produce the optimal estimate are thus obtained by solving the linear equations (Equation 12.100).

The minimum mean-square error follows directly from the second part of the theorem. That is,

$$\sigma_\varepsilon^2 = E\{y\varepsilon^*\} = E\{y(y^* - \boldsymbol{x}^{*T}\mathbf{a})\} = \sigma_y^2 - \mathbf{r}_{x_y}^{*T}\mathbf{a} \tag{12.103}$$

Note, as anticipated, the calculations in Equations 12.100 and 12.103 involve only second-moment statistics.

12.4.3.1 Example

This section concludes with a simple example. Consider the estimation of a stationary random signal in additive noise. The observations are given by

$$x[n] = s[n] + \eta[n]$$

where
- s is the desired signal
- η is the additive noise

If the signal and noise are uncorrelated and the signal and/or noise has mean zero, then the autocorrelation function for the observations is given by

$$R_x[l] = R_s[l] + R_\eta[l] \tag{12.104}$$

where

R_s is the signal autocorrelation function
R_η is the noise autocorrelation function

The cross-correlation function for the signal and the observation sequence is given by

$$R_{sx}[l] = E\{s[n]x^*[n-l]\} = R_s[l] \tag{12.105}$$

Suppose the optimal estimator is to be implemented as a linear filter with finite length impulse response $h[0], h[1], \ldots h[N-1]$. Then the estimate is generated by the expression

$$\hat{s}[n] = h[0]x[n] + h[1]x[n-1] + \cdots + h[N-1]x[n-N+1]$$

which is of the form Equation 12.97. If the mean-square error $E\{|\varepsilon[n]|^2\}$ is to be minimized, where $\varepsilon[n] = s[n] - \hat{s}[n]$, then this is a problem in linear mean-square estimation.

The correlation matrix and cross-correlation vector needed in Equation 12.100 are easily formulated using the expressions for the corresponding autocorrelation functions. To illustrate, for the case $N=3$ these equations take the form

$$\begin{bmatrix} R_x[0] & R_x[-1] & R_x[-2] \\ R_x[1] & R_x[0] & R_x[-1] \\ R_x[2] & R_x[1] & R_x[0] \end{bmatrix} \begin{bmatrix} h[0] \\ h[1] \\ h[2] \end{bmatrix} = \begin{bmatrix} R_{sx}[0] \\ R_{sx}[1] \\ R_{sx}[2] \end{bmatrix} \tag{12.106}$$

where R_x and R_{sx} are computed from Equations 12.104 and 12.105. Equation 12.103 for the mean-square error becomes

$$\sigma_\varepsilon^2 = R_s[0] - \sum_{l=0}^{N-1} h^*[l]R_{sx}[l] \tag{12.107}$$

Equation 12.106, which is a special form of Equation 12.100 for optimal filtering problems, is known as the Wiener–Hopf equation. It appears in various forms for many important problems including those of signal modeling and linear prediction.

Bibliography

1. Therrien, C.W. *Discrete Random Signals and Statistical Signal Processing.* Prentice Hall, Inc., Upper Saddle River, NJ, 1992.
2. Therrien, C.W. Some considerations for statistical characterization of non-stationary random processes. In *Proceedings of the 36th Asilomar Conference on Signals, Systems, and Computers,* Pacific Grove, CA, November 2002, pp. 1554–1558.
3. Neeser, F.D. and Massey, J.L. Proper complex random processes with applications to information theory. *IEEE Transactions on Information Theory,* 39(4): 1293–1302, July 1993.
4. Picinbono, B. and Bondon, P. Second-order statistics of complex signals. *IEEE Transactions on Signal Processing,* 45(2): 411–420, February 1997.
5. Picinbono, B. On circularity. *IEEE Transactions on Signal Processing,* 42(12): 3473–3482, December 1994.
6. Picinbono, B. Second-order complex random vectors and normal distributions. *IEEE Transactions on Signal Processing,* 44(10): 2637–2640, October 1996.

7. Picinbono, B. and Chevalier, P. Widely linear estimation with complex data. *IEEE Transactions on Signal Processing*, 43(8): 2030–2033, August 1995.
8. Schreier, P.J. and Scharf, L.L. Second-order analysis of improper complex random vectors and processes. *IEEE Transactions on Signal Processing*, 51(3): 714–725, March 2003.
9. Churchill, R.V. and Brown, J.W. *Complex Variables and Applications*, 4th edition. McGraw-Hill Book Company, New York, 1984.
10. Adriaan van den Bos, A. The multivariate complex normal distribution—a generalization. *IEEE Transactions on Information Theory*, 41(2): 537–539, March 1995.
11. Wilkinson, J.H. *The Algebraic Eigenvalue Problem*. Oxford University Press, New York, 1965.
12. Strang, G. *Linear Algebra and Its Applications*, 4th edition. Brooks-Cole, Pacific Grove, CA, 2005.
13. Golub, G.H. and Van Loan, C.F. *Matrix Computations*, 3rd edition. The Johns Hopkins University Press, Baltimore, MD, 1996.
14. Miller, K.S. *Complex Stochastic Processes*. Addison-Wesley, Reading, MA, 1974.
15. Papoulis, A. *Probability, Random Variables, and Stochastic Processes*, 3rd edition. McGraw-Hill, New York, 1991.
16. Birkhoff, G. and Mac Lane, S. *A Survey of Modern Algebra*, 4th edition. MacMillan, New York, 1977.
17. Oppenheim, A.V. and Schafer R.W., *Discrete Time Signal Processing*, Prentice Hall, Englewood Cliffs, 1989.

13

Signal Detection and Classification

Alfred Hero
University of Michigan

13.1 Introduction

Detection and classification arise in signal processing problems whenever a decision is to be made among a finite number of hypotheses concerning an observed waveform. Signal detection algorithms decide whether the waveform consists of "noise alone" or "signal masked by noise." Signal classification algorithms decide whether a detected signal belongs to one or another of the prespecified classes of signals. The objective of signal detection and classification theory is to specify systematic strategies for designing algorithms which minimize the average number of decision errors. This theory is grounded in the mathematical discipline of statistical decision theory where detection and classification are respectively called binary and M-ary hypothesis testing [1,2]. However, signal processing engineers must also contend with the exceedingly large size of signal processing datasets, the absence of reliable and tractible signal models, the associated requirement of fast algorithms, and the requirement for real time imbedding of unsupervised algorithms into specialized software or hardware. While ad hoc statistical detection algorithms were implemented by engineers before 1950, the systematic development of signal detection theory was first undertaken by radar and radio engineers in the early 1950s [3,4].

This chapter provides a brief and limited overview of some of the theory and practice of signal detection and classification. The focus will be on the Gaussian observation model. For more details and examples see the last section of this chapter and the cited references.

13.2 Signal Detection

Assume that for some physical measurement a sensor produces an output waveform $x = \{x(t) : t \in [0, T]\}$ over a time interval $[0, T]$. Assume that the waveform may have been produced by ambient noise alone or by an impinging signal of known form plus the noise. These two possibilities are called the null hypothesis H and the alternative hypothesis K, respectively, and are commonly written in the compact notation:

$$\begin{aligned} H&: x = \text{noise alone} \\ K&: x = \text{signal} + \text{noise}. \end{aligned}$$

The hypotheses H and K are called simple hypotheses when the statistical distributions of x under H and K involve no unknown parameters such as signal amplitude, signal phase, or noise power. When the statistical distribution of x under a hypothesis depends on unknown (nuisance) parameters, the hypothesis is called a composite hypothesis.

To decide between the null and alternative hypotheses, one might apply a high threshold to the sensor output x and make a decision that the signal is present if and only if the threshold is exceeded at some time within $[0, T]$. The engineer is then faced with practical question of where to set the threshold so as to ensure that the number of decision errors is small. There are two types of errors possible: the error of missing the signal (decide H under K [signal is present]) and the error of false alarm (decide K under H [no signal is present]). There is always a compromise between choosing a high threshold to make the average number of false alarms small versus choosing a low threshold to make the average number of misses small. To quantify this compromise it becomes necessary to specify the statistical distribution of x under each of the hypotheses H and K.

13.2.1 ROC Curve

Let the aforementioned threshold be denoted γ. Define the K decision region $\mathcal{R}_K = \{x : x(t) > \gamma, \text{ for some} t \in [0, T]\}$. This region is also called the critical region and simply specifies the conditions on x for which the detector declares the signal to be present. Since the detectors make mutually exclusive binary decisions the critical region completely specifies the operation of the detector. The probabilities of false alarm and miss are functions of γ given by $P_{FA} = P(\mathcal{R}_K \mid H)$ and $P_M = 1 - P(\mathcal{R}_K \mid K)$ where $P(A \mid H)$ and $P(A \mid K)$ denote the probabilities of arbitrary event A under hypothesis H and hypothesis K, respectively. The probability of correct detection $P_D = P(\mathcal{R}_K \mid K)$ is commonly called the power of the detector and P_{FA} is called the *level* of the detector.

The plot of the pair $P_{FA} = P_{FA}(\gamma)$ and $P_D = P_D(\gamma)$ over the range of thresholds $-\infty < \gamma < \infty$ produces a curve called the receiver-operating characteristic (ROC) which completely describes the error rate of the detector as a function of γ (Figure 13.1). Good detectors have ROC curves which have desirable properties such as concavity (negative curvature), monotone increase in P_D as P_{FA} increases, high slope of P_D at the point $(P_{FA}, P_D) = (0, 0)$, etc. [5]. For the energy detection example shown in Figure 13.1 it is evident that regardless of the actual energy σ^2 an increase in the rate of correct detections P_D can be bought only at the expense of increasing the rate of false alarms P_{FA}. Simply stated, the job of the signal processing engineer is to find ways to test between K and H, which push the ROC curve toward the upper left corner of Figure 13.1 where P_D is high for low P_{FA}: this is the regime of P_D and P_{FA} where reliable signal detection can occur.

13.2.2 Detector Design Strategies

When the signal waveform and the noise statistics are fully known, the hypotheses are simple and an optimal detector exists which has a ROC curve that upper bounds the ROC of any other detector, i.e., it has the highest possible power P_D for any fixed level P_{FA}. This optimal detector is called the most powerful (MP) test and is specified by the ubiquitous likelihood ratio test (LRT) described below. In the

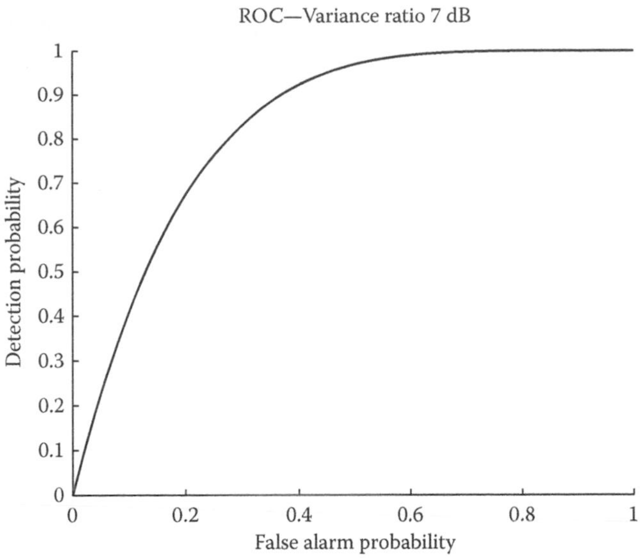

FIGURE 13.1 The ROC curve describes the trade-off between maximizing the power P_D and minimizing the probability of false alarm P_{FA} of a test between two hypotheses H and K. Shown is the ROC curve of the LRT (energy detector) which tests $H: x =$ complex random variable with variance $\sigma^2 = 1$ versus $K: x =$ complex random variable with variance $\sigma^2 = 5$ (7 dB variance ratio).

more common case where the signal and/or noise are described by unknown parameters, at least one hypothesis is composite and a detector has different ROC curves for different values of the parameters (see Figure 13.2). Unfortunately, there seldom exists a uniformly MP detector whose ROC curves remain upper bounds for the entire range of unknown parameters. Therefore, for composite hypotheses, other design strategies must generally be adopted to ensure reliable detection performance. There are a wide range of different strategies available including: Bayesian detection [5] and hypothesis testing [6], min–max hypothesis testing [2], constant false alarm rate (CFAR) detection [7] and similar, unbiased hypothesis testing [1], invariant hypothesis testing [8,9], sequential detection [10], simultaneous detection and estimation [11], and nonparametric detection [12]. Detailed discussion of these strategies is outside of the scope of this chapter. However, all of these strategies have a common link: their application produces one form or another of the LRT.

13.2.3 Likelihood Ratio Test

Here, we introduce an unknown parameter θ to simplify the upcoming discussion on composite hypothesis testing. Define the probability density of the measurement x as $f(x \mid \theta)$ where θ belongs to a parameter space Θ. It is assumed that $f(x \mid \theta)$ is a known function of x and θ. We can now state the detection problem as the problem of testing between

$$H: x \sim f(x \mid \theta), \theta \in \Theta_H \tag{13.1}$$

$$K: x \sim f(x \mid \theta), \theta \in \Theta_K, \tag{13.2}$$

where Θ_H and Θ_K are nonempty sets which partition the parameter space into two regions. Note it is essential that Θ_H and Θ_K be disjoint ($\Theta_H \cap \Theta_K = \emptyset$) so as to remove any ambiguity on the decisions, and exhaustive ($\Theta_H \cup \Theta_K = \Theta$) to ensure that all states of nature in Θ are accounted for. Let a detector be

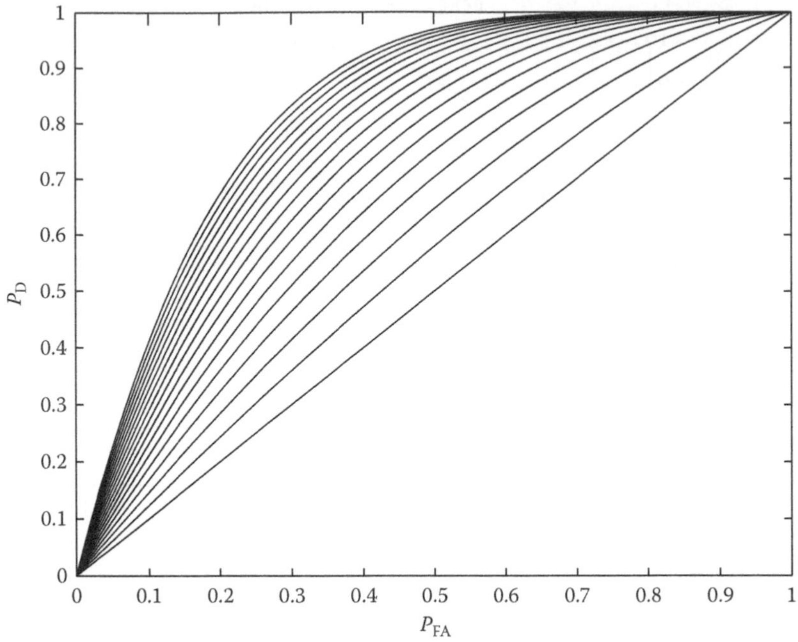

FIGURE 13.2 Eight members of the family of ROC curves for the LRT (energy detector) which tests $H: x =$ complex random variable with variance $\sigma^2 = 1$ versus composite K: $x =$ complex random variable with variance $\sigma^2 > 1$. ROC curves shown are indexed over a range [0 dB, 21 dB] of variance ratios in equal 3 dB increments. ROC curves approach a step function as variance ratio increases.

specified by a critical region $\mathcal{R}_K$. Then for any pair of parameters $\theta_H \in \Theta_H$ and $\theta_K \in \Theta_K$, the level and power of the detector can be computed by integrating the probability density $f(x\,|\,\theta)$ over $\mathcal{R}_K$:

$$P_{FA} = \int_{x \in \mathcal{R}_K} f(x\,|\,\theta_H)\mathrm{d}x \tag{13.3}$$

and

$$P_D = \int_{x \in \mathcal{R}_K} f(x\,|\,\theta_K)\mathrm{d}x. \tag{13.4}$$

The hypotheses in Equations 13.1 and 13.2 are simple when $\Theta = \{\theta_H, \theta_K\}$ consists of only two values and $\Theta_H = \{\theta_H\}$ and $\Theta_K = \{\theta_K\}$ are point sets. For simple hypotheses, the Neyman–Pearson lemma [1] states that there exists a MP test which maximizes P_D subject to the constraint that $P_{FA} \leq \alpha$, where α is a prespecified maximum level of false alarm. This test has the form of a threshold test known as the likelihood ratio test:

$$L(x) \stackrel{\text{def}}{=} \frac{f(x\,|\,\theta_K)}{f(x\,|\,\theta_H)} \underset{H}{\overset{K}{\gtrless}} \eta, \tag{13.5}$$

where η is a threshold which is determined by the constraint $P_{FA} = \alpha$:

$$\int_{\eta}^{\infty} g(l \mid \theta_H) dl = \alpha. \tag{13.6}$$

Here $g(l \mid \theta)$ is the probability density function of the likelihood ratio statistic $L(x)$. It must also be mentioned that if the density $g(l \mid \theta_H)$ contains delta functions, a simple randomization [1] of the LRT may be required to meet the false alarm constraint (Equation 13.6).

The test statistic $L(x)$ is a measure of the strength of the evidence provided by x such that the probability density $f(x \mid \theta_K)$ produces x as opposed to the probability density $f(x \mid \theta_H)$. Similarly, the threshold η represents the detector designer's prior level of "reasonable doubt" about the sufficiency of the evidence—only above a level η is the evidence sufficient for rejecting H.

When θ takes on more than two values at least one of the hypotheses (Equation 13.1 or 13.2) is composite and the Neyman–Pearson lemma no longer applies. A popular but ad hoc alternative which enjoys some asymptotic optimality properties is to implement the generalized likelihood ratio test (GLRT):

$$L_g(x) \stackrel{\text{def}}{=} \frac{\max_{\theta_K \in \Theta_K} f(x \mid \theta_K)}{\max_{\theta_H \in \Theta_H} f(x \mid \theta_H)} \underset{H}{\overset{K}{\gtrless}} \eta \tag{13.7}$$

where, if possible, the threshold η is set to attain a specified level of P_{FA}. The GLRT can be interpreted as a LRT which is based on the most likely values of the unknown parameters θ_H and θ_K, i.e., the values which maximize the likelihood functions $f(x \mid \theta_H)$ and $f(x \mid \theta_K)$, respectively (see next section).

13.3 Signal Classification

When, based on a noisy observed waveform x, one must decide among a number of possible signal waveforms $s_1, \ldots, s_p$, $p > 1$, we have a p-ary signal classification problem. Denoting $f(x \mid \theta_i)$ the density function of x when signal s_i is present, the classification problem can be stated as the problem of testing between the p hypotheses:

$$\begin{aligned} H_1 \quad &: \quad x \sim f(x \mid \theta_1),\ \theta_1 \in \Theta_1 \\ &\vdots \qquad\qquad \vdots \\ H_p \quad &: \quad x \sim f(x \mid \theta_p),\ \theta_p \in \Theta_p \end{aligned}$$

where Θ_i is a space of unknowns which parameterize the signal s_i. As before, it is essential that the hypotheses be disjoint, which ensures that $\{f(x \mid \theta_i)\}_{i=1}^{p}$ are distinct functions of x for all $\theta_i \in \Theta_i$, $i = 1, \ldots, p$, and that they be exhaustive, which ensures that the true density of x is included in one of the hypotheses. Similar to the case of detection, a classifier is specified by a partition of the space of observations x into p disjoint decision regions $\mathcal{R}_{H_1}, \ldots, \mathcal{R}_{H_p}$. Only $p - 1$ of these decision regions are needed to specify the operation of the classifier. The performance of a signal classifier is characterized by its set of p misclassification probabilities $P_{M_1} = 1 - P(x \in \mathcal{R}_{H_1} \mid H_1), \ldots, P_{M_p} = P(x \in \mathcal{R}_{H_p} \mid H_p)$. Unlike in the case of detection, even for simple hypotheses, where $\Theta_i = \theta_i$ consists of a single point, $i = 1, \ldots, p$, optimal p-ary classifiers that uniformly minimize all P_{M_i} s do not exist for $p > 2$. However, classifiers can be designed to minimize other weaker criteria such as average misclassification probability $\frac{1}{p}\sum_{i=1}^{p} P_{M_i}$ [5], worst-case misclassification probability $\max_i P_{M_i}$ [2], Bayes posterior misclassification probability [13], and others.

The maximum likelihood (ML) classifier is a popular classification technique which is closely related to ML parameter estimation. This classifier is specified by the rule:

$$\text{decide } H_j \text{ if and only if } \max_{\theta_j \in \Theta_j} f(x \mid \theta_j) \geq \max_k \max_{\theta_k \in \Theta_k} f(x \mid \theta_k), \quad j = 1, \ldots, p. \tag{13.8}$$

When the signal waveforms and noise statistics subsumed by the hypotheses $H_1, \ldots, H_p$ are fully known, the ML classifier takes the simpler form:

$$\text{decide } H_j \text{ if and only if } f_j(x) \geq \max_k f_k(x), \quad j = 1, \ldots, p,$$

where f_k denotes the known density function of x when the kth signal is present. For this simple case, it can be shown that the ML classifier is an optimal decision rule which minimizes the total misclassification error probability, as measured by the average $\frac{1}{p}\sum_{i=1}^{p} P_{M_i}$. In some cases, a weighted average $\frac{1}{p}\sum_{i=1}^{p} \beta_i P_{M_i}$ is a more appropriate measure of total misclassification error, e.g., when β_i is the prior probability of H_i, $i = 1, \ldots, p$, $\sum_{i=1}^{p} \beta_i = 1$. For this case, the optimal classifier is given by the maximum *a posteriori* decision rule [5,13]:

$$\text{decide } H_j \text{ if and only if } f_j(x)\beta_j \geq \max_k f_k(x)\beta_k, \quad j = 1, \ldots, p.$$

13.4 Linear Multivariate Gaussian Model

Assume that $\mathbf{X}$ is an $m \times n$ matrix of complex-valued Gaussian random variables which obeys the following linear model [9,14]:

$$\mathbf{X} = \mathbf{ASB} + \mathbf{W}, \tag{13.9}$$

where

- $\mathbf{A}$, $\mathbf{S}$, and $\mathbf{B}$ are rectangular $m \times q$, $q \times p$, and $p \times n$ complex matrices
- $\mathbf{W}$ is an $m \times n$ matrix whose n columns are i.i.d. zero-mean circular complex Gaussian vectors each with positive definite covariance matrix $\mathbf{R}_w$

We will assume that $n \geq m$. This model is very general and, as will be seen in subsequent sections, covers many signal processing applications.

A few comments about random matrices are now in order. If $\mathbf{Z}$ is an $m \times n$ random matrix, the mean, $E[\mathbf{Z}]$, of $\mathbf{Z}$ is defined as the $m \times n$ matrix of means of the elements of $\mathbf{Z}$, and the covariance matrix is defined as the $mn \times mn$ covariance matrix of the $mn \times 1$ vector, vec[$\mathbf{Z}$], formed by stacking columns of $\mathbf{Z}$. When the columns of $\mathbf{Z}$ are uncorrelated and each have the same $m \times m$ covariance matrix $\mathbf{R}$, the covariance of $\mathbf{Z}$ is block diagonal:

$$\text{Cov}[\mathbf{Z}] = \mathbf{R} \otimes \mathbf{I}_n, \tag{13.10}$$

where $\mathbf{I}_n$ is the $n \times n$ identity matrix. For $p \times q$ matrix $\mathbf{C}$ and $r \times s$ matrix $\mathbf{D}$, the notation $\mathbf{C} \otimes \mathbf{D}$ denotes the Kronecker product which is the following $pr \times qs$ matrix:

$$\mathbf{C} \otimes \mathbf{D} = \begin{bmatrix} \mathbf{C}\, d_{11} & \mathbf{C}\, d_{12} & \dots & \mathbf{C}\, d_{1s} \\ \mathbf{C}\, d_{21} & \mathbf{C}\, d_{22} & \dots & \mathbf{C}\, d_{2s} \\ \vdots & \vdots & \vdots & \vdots \\ \mathbf{C}\, d_{r1} & \mathbf{C}\, d_{r2} & \dots & \mathbf{C}\, d_{rs} \end{bmatrix}. \tag{13.11}$$

The density function of **X** has the form [14]

$$f(\mathbf{X};\theta) = \frac{1}{\pi^{mn}\,|\,\mathbf{R}_w|^n} \exp\left(-\mathrm{tr}\left\{[\mathbf{X}-\mathbf{ASB}]\,[\mathbf{X}-\mathbf{ASB}]^H\mathbf{R}_w^{-1}\right\}\right), \tag{13.12}$$

where

$|\mathbf{C}|$ is the determinant

$\mathrm{tr}\{\mathbf{D}\}$ is the trace of square matrices **C** and **D**

For convenience, we will use the shorthand notation

$$\mathbf{X} \sim \mathcal{N}_{mn}(\mathbf{ASB}, \mathbf{R}_w \otimes \mathbf{I}_n),$$

which is to be read as **X** is distributed as an $m \times n$ complex Gaussian random matrix with mean **ASB**, and covariance $\mathbf{R}_w \otimes \mathbf{I}_n$.

In the examples presented in the next section, several distributions associated with the complex Gaussian distribution will be seen to govern the various test statistics. The complex noncentral chi-square distribution with p degrees of freedom and vector of noncentrality parameters $(\rho, \underline{d})$ plays a very important role here. This is defined as the distribution of the random variable $\chi^2(\rho, \underline{d}) \overset{\text{def}}{=} \sum_{i=1}^{p} d_i\,|\,z_i|^2 + \rho$ where the z_is are independent univariate complex Gaussian random variables with zero mean and unit variance and where ρ is scalar and $\underline{d}$ is a (row) vector of positive scalars. The complex noncentral chi-square distribution is closely related to the real noncentral chi-square distribution with $2p$ degrees of freedom and noncentrality parameters $(\rho, \mathrm{diag}([\underline{d}, \underline{d}])$ defined in [9]. The case of $\rho = 0$ and $\underline{d} = [1, \dots, 1]$ corresponds to the standard (central) complex chi-square distribution. For derivations and details on this and other related distributions see [14].

13.5 Temporal Signals in Gaussian Noise

Consider the time-sampled superposed signal model

$$x(t_i) = \sum_{j=1}^{p} s_j b_j(t_i) + w(t_i),\; i = 1, \dots, n,$$

where we interpret t_i as time; but it could also be space or other domain. The temporal signal waveforms $\underline{b}_j = [b_j(t_1), \dots, b_j(t_n)]^{\mathrm{T}}, j = 1, \dots, p$ are assumed to be linearly independent where $p \le n$. The scalar s_j is a time-independent complex gain applied to the jth signal waveform. The noise $w(t)$ is complex Gaussian with zero mean and correlation function $r_w(t,\tau) = E[w(t)w^*(\tau)]$. By concatenating the samples into a column vector $\underline{x} = [x(t_1), \dots, x(t_n)]^{\mathrm{T}}$, the above model is equivalent to

$$\underline{x} = \mathbf{B}\underline{s} + \underline{w}, \tag{13.13}$$

where $\mathbf{B} = [\underline{b}_1, \dots, \underline{b}_p]$ and $\underline{s} = [s_1, \dots, s_p]^{\mathrm{T}}$. Therefore, the density function (Equation 13.12) applies to the transpose $\underline{x}^{\mathrm{T}}$ with $\mathbf{R}_w = \mathrm{Cov}(\underline{w})$, $m = q = 1$, and $\mathbf{A} = 1$.

13.5.1 Signal Detection: Known Gains

For known gain factors s_i, known signal waveforms $\underline{b}_i$, and known noise covariance $\mathbf{R}_w$, the LRT (Equation 13.5) is the MP signal detector for deciding between the simple hypotheses $H : \underline{x} \sim \mathcal{N}_n(0, \mathbf{R}_w)$ versus $K : \underline{x} \sim \mathcal{N}_n(\mathbf{B}\underline{s}, \mathbf{R}_w)$. The LRT has the form

$$L(x) = \exp\left(-2 * \text{Re}\left\{\underline{x}^{\text{H}}\mathbf{R}_w^{-1}\mathbf{B}\underline{s}\right\} + \underline{s}^{\text{H}}\mathbf{B}^{\text{H}}\mathbf{R}_w^{-1}\mathbf{B}\underline{s}\right) \underset{H}{\overset{K}{\gtrless}} \eta. \tag{13.14}$$

This test is equivalent to a linear detector with critical region $\mathcal{R}_K = \{x: T(x) > \gamma\}$ where

$$T(x) = \text{Re}\left\{\underline{x}^{\text{H}}\mathbf{R}_w^{-1}\underline{s}_c\right\}$$

and $\underline{s}_c = \mathbf{B}\underline{s} = \sum_{j=1}^{p} s_j \underline{b}_j$ is the observed compound signal component.

Under both hypotheses H and K, the test statistic T is Gaussian distributed with common variance but different means. It is easily shown that the ROC curve is monotonically increasing in the detectability index $\rho = \underline{s}_c^{\text{H}}\mathbf{R}_w^{-1}\underline{s}_c$. It is interesting to note that when the noise is white, $\mathbf{R}_w = \sigma^2\mathbf{I}_n$ and the ROC curve depends on the form of the signals only through the signal-to-noise ratio $\rho = \frac{\|\underline{s}_c\|^2}{\sigma^2}$. In this special case, the linear detector can be written in the form of a correlator detector:

$$T(x) = \text{Re}\left\{\sum_{i=1}^{n} s_c^*(t_i)x(t_i)\right\} \underset{H}{\overset{K}{\gtrless}} \gamma,$$

where $s_c(t) = \sum_{j=1}^{p} s_j b_j(t)$. When the sampling times t_i are equispaced, e.g., $t_i = i$, the correlator takes the form of a matched filter:

$$T(x) = \text{Re}\left\{\sum_{i=1}^{n} h(n-i)x(i)\right\} \underset{H}{\overset{K}{\gtrless}} \gamma,$$

where $h(i) = s_c^*(-i)$. Block diagrams for the correlator and the matched filter implementations of the LRT are shown in Figures 13.3 and 13.4.

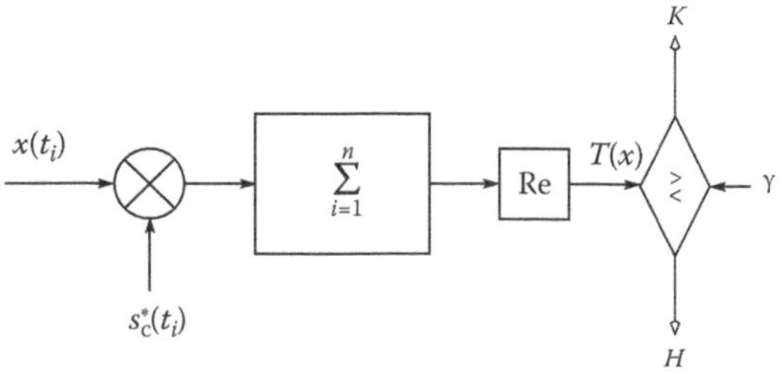

FIGURE 13.3 The correlator implementation of the MP LRT for signal component $s_c(t_i)$ in additive Gaussian white noise. For nonwhite noise, a prewhitening transformation must be performed on $x(t_i)$ and $s_c(t_i)$ prior to implementation of correlator detector.

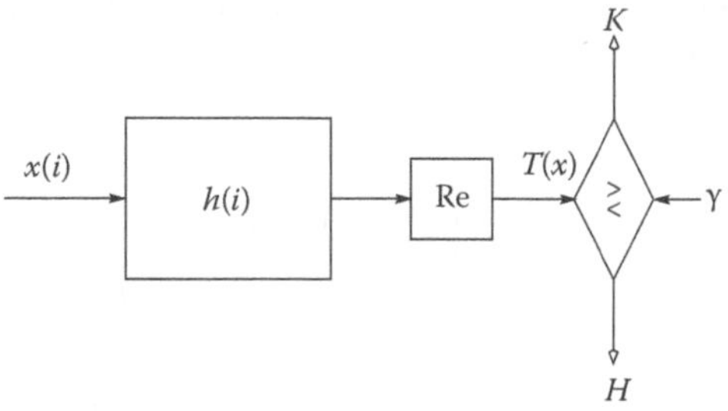

FIGURE 13.4 The matched filter implementation of the MP LRT for signal component $s_c(i)$ in additive Gaussian white noise. Matched filter impulse response is $h(i) = s_c^*(-i)$. For nonwhite noise, a prewhitening transformation must be performed on $x(i)$ and $s_c(i)$ prior to implementation of matched filter detector.

13.5.2 Signal Detection: Unknown Gains

When the gains s_j are unknown, the alternative hypothesis K is composite, the critical region $\mathcal{R}_K$ depends on the true gains for $p > 1$, and no MP test for $H: \underline{x} \sim \mathcal{N}_n(0, \mathbf{R}_w)$ versus $K: \underline{x} \sim \mathcal{N}_n(\mathbf{B}\underline{s}, \mathbf{R}_w)$ exists. However, the GLRT (Equation 13.7) can easily be derived by maximizing the likelihood ratio for known gains (Equation 13.14) over $\underline{s}$. Recalling from least-squares theory that $\min_{\underline{s}} (\underline{x} - \mathbf{B}\underline{s})^{\mathrm{H}}\mathbf{R}_w^{-1}(\underline{x} - \mathbf{B}\underline{s}) = \underline{x}^{\mathrm{H}}\mathbf{R}_w^{-1}\underline{x} - \underline{x}^{\mathrm{H}}\mathbf{R}_w^{-1}\mathbf{B}\left[\mathbf{B}^{\mathrm{H}}\mathbf{R}_w^{-1}\mathbf{B}\right]^{-1}\mathbf{B}^{\mathrm{H}}\mathbf{R}_w^{-1}\underline{x}$, the GLRT can be shown to take the form

$$T_g(x) = \underline{x}^{\mathrm{H}}\mathbf{R}_w^{-1}\mathbf{B}\left[\mathbf{B}^{\mathrm{H}}\mathbf{R}_w^{-1}\mathbf{B}\right]^{-1}\mathbf{B}^{\mathrm{H}}\mathbf{R}_w^{-1}\underline{x} \underset{H}{\overset{K}{\gtrless}} \gamma.$$

A more intuitive form for the GLRT can be obtained by expressing T_g in terms of the prewhitened observations $\tilde{\underline{x}} = \mathbf{R}_w^{-\frac{1}{2}}\underline{x}$ and prewhitened signal waveform matrix $\tilde{\mathbf{B}} = \mathbf{R}_w^{-\frac{1}{2}}\mathbf{B}$, where $\mathbf{R}_w^{-\frac{1}{2}}$ is the right Cholesky factor of $\mathbf{R}_w^{-1}$:

$$T_g(x) = \|\tilde{\mathbf{B}}\left[\tilde{\mathbf{B}}^{\mathrm{H}}\tilde{\mathbf{B}}\right]^{-1}\tilde{\mathbf{B}}^{\mathrm{H}}\tilde{\underline{x}}\|^2. \tag{13.15}$$

$\tilde{\mathbf{B}}[\tilde{\mathbf{B}}^{\mathrm{H}}\tilde{\mathbf{B}}]^{-1}\tilde{\mathbf{B}}^{\mathrm{H}}$ is the idempotent $n \times n$ matrix which projects onto column space of the prewhitened signal waveform matrix $\tilde{\mathbf{B}}$ (whitened signal subspace). Thus, the GLRT decides that some linear combination of the signal waveforms $\underline{b}_1, \ldots, \underline{b}_p$ is present only if the energy of the component of x lying in the whitened signal subspace is sufficiently large.

Under the null hypothesis, the test statistic T_g is distributed as a complex central chi-square random variable with p degrees of freedom, while under the alternative hypothesis T_g is a noncentral chi-square with noncentrality parameter vector $\left(\underline{s}^{\mathrm{H}}\mathbf{B}^{\mathrm{H}}\mathbf{R}_w^{-1}\mathbf{B}\underline{s}, 1\right)$. The ROC curve is indexed by the number of signals p and the noncentrality parameter but is not expressible in the closed form for $p > 1$.

13.5.3 Signal Detection: Random Gains

In some cases, a random Gaussian model for the gains may be more appropriate than the unknown gain model considered above. When the p-dimensional gain vector $\underline{s}$ is multivariate normal with zero mean and $p \times p$ covariance matrix $\mathbf{R}_s$, the compound signal component $\underline{s}_c = \mathbf{B}_{\underline{s}}$ is an n-dimensional random Gaussian vector with zero mean and rank p covariance matrix $\mathbf{B}\mathbf{R}_s\mathbf{B}^{\mathrm{H}}$. A standard assumption is that the gains and the additive noise are statistically independent. The detection problem can then be stated as testing the two simple hypotheses $H: \underline{x} \sim \mathcal{N}_n(0, \mathbf{R}_w)$ versus $K: \underline{x} \sim \mathcal{N}_n(0, \mathbf{B}\mathbf{R}_s\mathbf{B}^{\mathrm{H}} + \mathbf{R}_w)$. It can be shown that the MP LRT has the form

$$T(x) = \sum_{i=1}^{p} \left(\frac{\lambda_i}{1 + \lambda_i} \right) \left| \underline{v}_i^* \mathbf{R}_w^{-\frac{1}{2}} \underline{x} \right|^2 \begin{matrix} K \\ > \\ < \\ H \end{matrix} \gamma, \tag{13.16}$$

where $\{\lambda_i\}_{i=1}^p$ are the nonzero eigenvalues of the matrix $\mathbf{R}_w^{-\frac{1}{2}} \mathbf{B}\mathbf{R}_s\mathbf{B}^{\mathrm{H}} \mathbf{R}_w^{-\frac{\mathrm{H}}{2}}$ and $\{\underline{v}_i\}_{i=1}^p$ are the associated eigenvectors. Under H, the test statistic $T(x)$ is distributed as complex noncentral chi-square with p degrees of freedom and noncentrality parameter vector $(0, \underline{d}_H)$ where $\underline{d}_H = [\lambda_1/(1 + \lambda_1), \ldots, \lambda_p/(1 + \lambda_p)]$. Under the alternative hypothesis, T is also distributed as noncentral complex chi-square, however with noncentrality vector $(0, \underline{d}_K)$ where $\underline{d}_K$ are the nonzero eigenvalues of $\mathbf{B}\mathbf{R}_s\mathbf{B}^{\mathrm{H}}$. The ROC is not available in closed form for $p > 1$.

13.5.4 Signal Detection: Single Signal

We obtain a unification of the GLRT for unknown gain and the LRT for random gain in the case of a single impinging signal waveform: $\mathbf{B} = \underline{b}_1, p = 1$. In this case, the test statistic T_g in Equation 13.15 and T in Equation 13.16 reduce to the identical form and we get the same detector structure

$$\frac{\left| \underline{x}^{\mathrm{H}} \mathbf{R}_w^{-1} \underline{b}_1 \right|^2}{\underline{b}_1^{\mathrm{H}} \mathbf{R}_w^{-1} \underline{b}_1} \begin{matrix} K \\ > \\ < \\ H \end{matrix} \eta.$$

This establishes that the GLRT is uniformly MP over all values of the gain parameter s_1 for $p = 1$. Note that even though the form of the unknown parameter GLRT and the random parameter LRT are identical for this case, their ROC curves and their thresholds γ will be different since the underlying observation models are not the same. When the noise is white, the test simply compares the magnitude squared of the complex correlator output $\sum_{i=1}^{n} b_1^*(t_i)x(t_i)$ to a threshold γ.

13.6 Spatiotemporal Signals

Consider the general spatiotemporal model

$$\underline{x}(t_i) = \sum_{j=1}^{q} \underline{a}_j \sum_{k=1}^{p} s_{jk} b_k(t_i) + \underline{w}(t_i), \quad i = 1, \ldots, n.$$

This model applies to a wide range of applications in narrowband array processing and has been thoroughly studied in the context of signal detection in [14]. The m-element vector $\underline{x}(t_i)$ is a snapshot at time t_i of the m-element array response to p impinging signals arriving from q different directions. The vector $\underline{a}_j$ is a known steering vector which is the complex response of the array to signal energy arriving from the jth direction. From this direction, the array receives the superposition $\sum_{k=1}^{p} s_{jk}\underline{b}_k$ of p known as time-varying signal waveforms $\underline{b}_k = [b_k(t_1), \ldots, b_k(t_n)]^{\mathrm{T}}, k = 1, \ldots, p$. The presence of the superposition accounts for both direct and multipath arrivals and allows for more signal sources than directions of arrivals when $p > q$. The complex Gaussian noise vectors $\underline{w}(t_i)$ are spatially correlated with spatial covariance $\mathrm{Cov}[\underline{w}(t_i)] = \mathbf{R}_w$, but are temporally uncorrelated $\mathrm{Cov}[\underline{w}(t_i), \underline{w}(t_j)] = 0$, $i \neq j$.

By arranging the n column vectors $\{\underline{x}(t_i)\}_{i=1}^{n}$ in an $m \times n$ matrix $\mathbf{X}$, we obtain the equivalent matrix model

$$\mathbf{X} = \mathbf{ASB}^{\mathrm{H}} + \mathbf{W},$$

where

$\mathbf{S} = (s_{ij})$ is a $q \times p$ matrix whose rows are vectors of signal gain factors for each different direction of arrival

$\mathbf{A} = [\underline{a}_1, \ldots, \underline{a}_q]$ is an $m \times q$ matrix whose columns are steering vectors for different directions of arrival

$\mathbf{B} = [\underline{b}_1, \ldots, \underline{b}_p]^{\mathrm{T}}$ is a $p \times n$ matrix whose rows are different signal waveforms

To avoid singular detection, it is assumed that $\mathbf{A}$ is of rank q, $q \leq m$, and that $\mathbf{B}$ is of rank p, $p \leq n$. We consider only a few applications of this model here. For many others see [14].

13.6.1 Detection: Known Gains and Known Spatial Covariance

First we assume that the gain matrix $\mathbf{S}$ and the spatial covariance $\mathbf{R}_w$ are known. This case is only relevant when one knows the direct path and multipath geometry of the propagation medium ($\mathbf{S}$), the spatial distribution of the ambient (possibly coherent) noise ($\mathbf{R}_w$), the q directions of the impinging superposed signals ($\mathbf{A}$), and the p signal waveforms ($\mathbf{B}$). Here, the detection problem is stated in terms of the simple hypotheses $H: \mathbf{X} \sim \mathcal{N}_{nm}(0, \mathbf{R}_w \otimes \mathbf{I}_n)$ versus $K: \mathbf{X} \sim \mathcal{N}_{nm}(\mathbf{ASB}, \mathbf{R}_w \otimes \mathbf{I}_n)$. For this case, the LRT (Equation 13.5) is the MP test and, using Equation 13.12, has the form

$$T(x) = \mathrm{Re}\left(\mathrm{tr}\left\{\mathbf{A}^{\mathrm{H}}\mathbf{R}_w^{-1}\mathbf{X}\mathbf{B}^{\mathrm{H}}\mathbf{S}^{\mathrm{H}}\right\}\right) \underset{H}{\overset{K}{\gtrless}} \gamma.$$

Since the test statistic is Gaussian under H and K, the ROC curve is of similar form to the ROC for detection of temporal signals with known gains.

Identifying the quantities $\tilde{\mathbf{X}} = \mathbf{R}_w^{-\frac{1}{2}}\mathbf{X}$ and $\tilde{\mathbf{A}} = \mathbf{R}_w^{-\frac{1}{2}}\mathbf{A}$ as the spatially whitened measurement matrix and spatially whitened array response matrix, respectively, the test statistic T can be interpreted as a multivariate spatiotemporal correlator detector. In particular, when there is only one signal impinging on the array from a single direction, then $p = q = 1$, $\tilde{\mathbf{A}} = \underline{\tilde{a}}$ a column vector, $\mathbf{B} = \underline{b}^{\mathrm{T}}$ a row vector, $\mathbf{S} = s$ a complex scalar, and the test statistic becomes

$$\begin{aligned} T(x) &= \mathrm{Re}\left\{\underline{\tilde{a}}^{\mathrm{H}} \cdot_s \tilde{\mathbf{X}} \cdot_t \underline{b}^* \, s^*\right\} \\ &= \mathrm{Re}\left\{ s^* \sum_{j=1}^{m} \tilde{a}_j^* \sum_{i=1}^{n} b^*(t_i)\tilde{x}_j(t_i) \right\}, \end{aligned}$$

where the multiplication notation $\cdot_s$ and $\cdot_t$ are used to simply emphasize the respective matrix multiplication operations (correlation) which occur over the spatial domain and the time domain. It can be shown that the ROC curve monotonically increases in the detectability index $\rho = n\underline{a}^{\mathrm{H}}\mathbf{R}_w^{-1}\underline{a} \cdot \|s\underline{b}\|^2$.

13.6.2 Detection: Unknown Gains and Unknown Spatial Covariance

By assuming the gain matrix $\mathbf{S}$ and $\mathbf{R}_w$ to be unknown, the detection problem becomes one of testing for noise alone against noise plus p coherent signal waveforms, where the waveforms lie in the subspace formed by all linear combinations of the rows of **B** but are otherwise unknown. This gives a composite null and alternative hypothesis for which the GLRT can be derived by maximizing the known gain likelihood ratio over the gain matrix **S**. The result is the GLRT [14]:

$$T_g(x) = \frac{|\mathbf{A}^H \hat{\hat{\mathbf{R}}}_K^{-1} \mathbf{A}|}{|\mathbf{A}^H \hat{\mathbf{R}}_H^{-1} \mathbf{A}|} \underset{H}{\overset{K}{\gtrless}} \gamma,$$

where

$|\cdot|$ denotes the determinant

$\hat{\mathbf{R}}_H = \frac{1}{n}\mathbf{X}\mathbf{X}^H$ is a sample estimate of the spatial covariance matrix using all of the snapshots

$\hat{\hat{\mathbf{R}}}_K = \frac{1}{n}\mathbf{X}\left[\mathbf{I}_n - \mathbf{B}^H[\mathbf{B}\mathbf{B}^H]^{-1}\mathbf{B}\right]\mathbf{X}^H$ is the sample estimate using only those components of the snapshots lying outside of the row space of the signal waveform matrix **B**

To gain insight into the test statistic T_g, consider the asymptotic convergence of T_g as the number of snapshots n goes to infinity. By the strong law, $\hat{\hat{\mathbf{R}}}_K$ converges to the covariance matrix of $\mathbf{X}[\mathbf{I}_n - \mathbf{B}^H[\mathbf{B}\mathbf{B}^H]^{-1}\mathbf{B}]$. Since $\mathbf{I}_n - \mathbf{B}^H[\mathbf{B}\mathbf{B}^H]^{-1}\mathbf{B}$ annihilates the signal component **ASB**, this covariance is the same quantity **R**, $\mathbf{R} \leq \mathbf{R}_w$, under both H and K. On the other hand, $\hat{\mathbf{R}}_H$ converges to $\mathbf{R}_w$ under H, while it converges to $\mathbf{R}_w + \mathbf{A}\mathbf{S}\mathbf{B}\mathbf{B}^H\mathbf{S}^H\mathbf{A}^H$ under K. Hence, when strong signals are present, T_g tends to take on very large values near the quantity $(|\mathbf{A}^H\mathbf{R}^{-1}\mathbf{A}|)/(|\mathbf{A}^H[\mathbf{R}_w + \mathbf{A}\mathbf{S}\mathbf{B}\mathbf{B}^H\mathbf{S}^H\mathbf{A}^H]^{-1}\mathbf{A}^H|) \gg 1$.

The distribution of T_g under H (K) can be derived in terms of the distribution of a sum of central (noncentral) complex β random variables. See [14] for discussion of performance and algorithms for data recursive computation of T_g. Generalizations of this GLRT exist which incorporate nonzero mean [14,15].

13.7 Signal Classification

Typical classification problems arising in signal processing are classifying an individual signal waveform out of a set of possible linearly independent waveforms, classifying the presence of a particular set of signals as opposed to other sets of signals, classifying among specific linear combinations of signals, and classifying the number of signals present. The problem of classification of the number of signals, also known as the order selection problem, is treated in the Section 16.3 of this book. While the spatiotemporal model could be treated in analogous fashion, for concreteness we focus on the case of the Gaussian temporal signal model (Equation 13.13).

13.7.1 Classifying Individual Signals

Here, it is of interest to decide which one of the p-scaled signal waveforms $s_1\underline{b}_1, \ldots, s_p\underline{b}_p$ is present in the observations $\underline{x} = [x(t_1), \ldots, x(t_n)]^T$. Denote by H_k the hypothesis that $\underline{x} = s_k\underline{b}_k + \underline{w}$. Signal classification can then be stated as the problem of testing between the following simple hypotheses:

$$\begin{aligned} H_1 &: \quad x = s_1\underline{b}_1 + \underline{w} \\ &\vdots \qquad\quad \vdots \\ H_p &: \quad x = s_p\underline{b}_p + \underline{w}. \end{aligned}$$

For known gain factors s_k, known signal waveforms $\underline{b}_k$, and known noise covariance $\mathbf{R}_w$, these hypotheses are simple, the density function $f(x \mid s_k, \underline{b}_k) = \mathcal{N}_n(s_k\underline{b}_k, \mathbf{R}_w)$ under H_k involves no unknown parameters and the ML classifier (Equation 13.8) reduces to the decision rule

$$\text{decide } H_j \text{ if and only if } j = \operatorname{argmin}_{k=1,\ldots,p} (\underline{x} - s_k\underline{b}_k)^H \mathbf{R}_w^{-1} (\underline{x} - s_k\underline{b}_k). \tag{13.17}$$

Thus, the classifier chooses the most likely signal as that signal $s_j \underline{b}_j$ which has minimum normalized distance from the observed waveform $\underline{x}$. The classifier can also be interpreted as a minimum distance classifier, which chooses the signal that minimizes the Euclidean distance $\|\tilde{\underline{x}} - s_k \tilde{\underline{b}}_k\|$ between the prewhitened signal $\tilde{\underline{b}}_k = \mathbf{R}_w^{-\frac{1}{2}} \underline{b}_k$ and the prewhitened measurement $\tilde{\underline{x}} = \mathbf{R}_w^{-\frac{1}{2}} \underline{x}$.

Written in the minimum normalized distance form, the ML classifier appears to involve nonlinear statistics. However, an obvious simplification of Equation 13.17 reveals that the ML classifier actually only requires computing linear functions of $\underline{x}$:

$$\text{decide } H_j \text{ if and only if } j = \text{argmax}_{k=1,\ldots,p} \left\{ \text{Re}\left(\underline{x}^{\mathrm{H}} \mathbf{R}_w^{-1} \underline{b}_k s_k\right) - \frac{1}{2} \mid s_k|^2 \underline{b}_k^{\mathrm{H}} \mathbf{R}_w^{-1} \underline{b}_k \right\}.$$

Note that this linear reduction only occurs when the covariances $\mathbf{R}_w$ are identical under each H_k, $k = 1, \ldots$, p. In this case, the ML classifier can be implemented using prewhitening filters followed by a bank of correlators or matched filters, an offset adjustment, and a maximum selector (Figure 13.5).

An additional simplification occurs when the noise is white, $\mathbf{R}_w = \mathbf{I}_n$, and all signal energies $\mid s_k|^2 \|\underline{b}_k^{\mathrm{H}}\|^2$ are identical: the classifier chooses the most likely signal as that signal $b_j(t_i)s_j$ which is maximally correlated with the measurement x:

$$\text{decide } H_j \text{ if and only if } j = \text{argmax}_{k=1,\ldots,p} \left\{ \text{Re}\left(s_k \sum_{i=1}^{n} b_k^*(t_i) x(t_i) \right) \right\}.$$

The decision regions $\mathcal{R}_{H_k} = \{x\text{: decide } H_k\}$ induced by Equation 13.17 are piecewise linear regions, known as Voronoi cells $\mathcal{V}_k$, centered at each of the prewhitened signals $s_k \tilde{\underline{b}}_k$. The misclassification error probabilities $P_{\mathrm{M}_k} = 1 - P(x \in \mathcal{R}_{H_k} \mid H_k) = 1 - \int_{x \in \mathcal{V}_k} f(x \mid H_k) \mathrm{d}x$ must generally be computed by integrating complex multivariate Gaussian densities $f(x \mid H_k) = \mathcal{N}_n(s_k \underline{b}_k, \mathbf{R}_w)$ over these regions. In the case of orthogonal signals $\underline{b}_i \mathbf{R}_w^{-1} \underline{b}_j = 0$, $i \neq j$, this integration reduces to a single integral of a univariate $\mathcal{N}_1(\rho_k, \rho_k)$ density function times the product of $p - 1$ univariate $\mathcal{N}_1(0, \rho_i)$ cumulative distribution

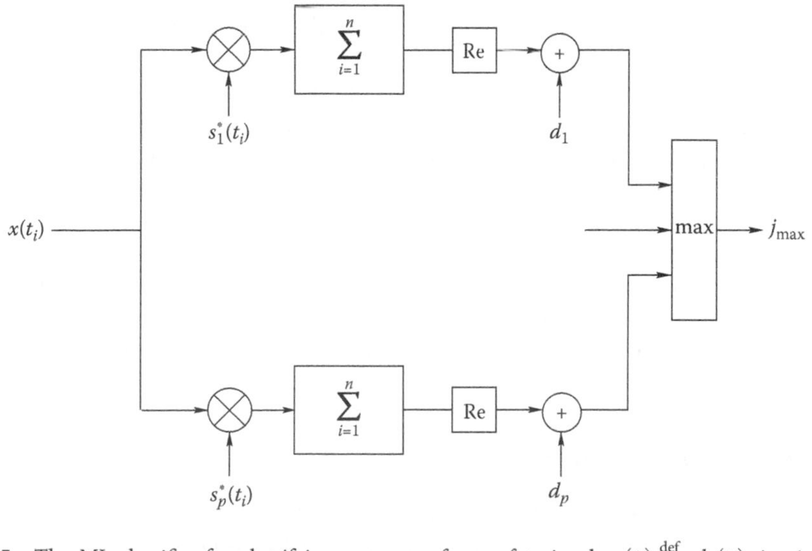

FIGURE 13.5 The ML classifier for classifying presence of one of p signals $s_j(t_i) \stackrel{\text{def}}{=} s_j \underline{b}_j(t_i)$, $j = 1, \ldots, p$, under additive Gaussian white noise. $d_j = -\frac{1}{2}|s_j|^2 \|\underline{b}_j\|^2$ and $j_{\max}$ is index of correlator output which is maximum. For nonwhite noise, a prewhitening transformation must be performed on $x(t_i)$ and the $b_j(t_i)$s prior to implementation of ML classifier.

functions, $i = 1, \ldots, p, i \neq k$, where $\rho_k = \underline{b}_k^H \mathbf{R}_w^{-1} \underline{b}_k$. Even for this case, no general closed-form expressions for P_{M_k} are available. However, analytical lower bounds on P_{M_k} and on average misclassification probability $\frac{1}{p}\sum_{k=1}^{p} P_{M_k}$ can be used to qualitatively assess classifier performance [13].

13.7.2 Classifying Presence of Multiple Signals

We conclude by treating the problem where the signal component of the observation is the linear combination of one of J hypothesized subsets $\mathcal{S}_k, k = 1, \ldots, J$, of the signal waveforms $\underline{b}_1, \ldots, \underline{b}_p$. Assume that the subset $\mathcal{S}_k$ contains p_k signals and that the $\mathcal{S}_k, k = 1, \ldots, J$, are disjoint, i.e., they do not contain any signals in common. Define the $n \times p_k$ matrix $\mathbf{B}_k$ whose columns are formed from the subset $\mathcal{S}_k$. We can now state the classification problem as testing between the J composite hypotheses

$$
\begin{array}{lll}
H_1 & : & \underline{x} = \mathbf{B}_1 \underline{s}_1 + \underline{w}, \quad \underline{s}_1 \in \mathbb{C}^{p_1} \\
\vdots & \vdots & \\
H_J & : & \underline{x} = \mathbf{B}_J \underline{s}_J + \underline{w}, \quad \underline{s}_J \in \mathbb{C}^{p_J}
\end{array}
$$

where $\underline{s}_k$ is a column vector of p_k unknown complex gains.

The density function under H_k, $f(x \mid \underline{s}_k, \mathbf{B}_k) = \mathcal{N}_n(\mathbf{B}_k\underline{s}_k, \mathbf{R}_w)$, is a function of unknown parameters $\underline{s}_k$ and therefore the ML classifier (Equation 13.8) involves finding the largest among MLs $\max_{\underline{s}_k} f(x \mid \underline{s}_k, \mathbf{B}_k)$, $k = 1, \ldots, J$. This yields the following form for the ML classifier:

$$
\text{decide } H_j \text{ if and only if } j = \operatorname{argmin}_{k=1,\ldots,J} (\underline{x} - \mathbf{B}_k \underline{s}_k)^H \mathbf{R}_w^{-1} (\underline{x} - \mathbf{B}_k \underline{s}_k),
$$

where $\underline{s}_k = \left[\mathbf{B}_k^H \mathbf{R}_w^{-1} \mathbf{B}_k\right]^{-1} \mathbf{B}_k^H \mathbf{R}_w^{-1} \underline{x}$ is the ML gain vector estimate. The decision regions are once again piecewise linear but with Voronoi cells having centers at the least-squares estimates of the hypothesized signal components $\mathbf{B}_k \underline{s}_k$, $k = 1, \ldots, J$.

Similar to the case of noncomposite hypotheses considered in the previous subsection, a simplification of Equation 13.18 is possible:

$$
\text{decide } H_j \text{ if and only if } j = \operatorname{argmax}_{k=1,\ldots,J} \underline{x}^H \mathbf{R}_w^{-1} \mathbf{B}_k \left(\mathbf{B}_k^H \mathbf{R}_w^{-1} \mathbf{B}_k\right)^{-1} \mathbf{B}_k^H \mathbf{R}_w^{-1} \underline{x}
$$

Defining the prewhitened versions $\underline{x} = \mathbf{R}_w^{-\frac{1}{2}} \underline{x}$ and $\tilde{\mathbf{B}}_k = \mathbf{R}_w^{-\frac{1}{2}} \mathbf{B}_k$ of the observations and the kth signal matrix, the ML classifier is seen to decide that the linear combination of the p_j signals in H_j is present when the length $\|\tilde{\mathbf{B}}_j [\tilde{\mathbf{B}}_j^H \tilde{\mathbf{B}}_j]^{-1} \tilde{\mathbf{B}}_j^H]\underline{x}\|$ of the projection of $\underline{x}$ onto the jth signal space (colspan$\{\tilde{\mathbf{B}}_j\}$) is greatest. This classifer can be implemented as a bank of p adaptive matched filters each matched to one of the least-squares estimates $\tilde{\mathbf{B}}_k \underline{s}_k$, $k = 1, \ldots, p$, of the prewhitened signal component. Under any H_i, the quantities $\underline{x}^H \mathbf{R}_w^{-1} \mathbf{B}_k [\mathbf{B}_k^H \mathbf{R}_w^{-1} \mathbf{B}_k]^{-1} \mathbf{R}_w^{-1} \underline{x}$, $k = 1, \ldots, J$, are distributed as complex noncentral chi-square with p_k degrees of freedom. For the special case of orthogonal prewhitened signals $\underline{b}_i \mathbf{R}_w^{-1} \underline{b}_j = 0$, $i \neq j$, these variables are also statistically independent and P_{M_i} can be computed as a one-dimensional integral of a univariate noncentral chi-square density times the product of $J - 1$ univariate noncentral chi-square cumulative distribution functions.

13.8 Additional Reading

There are now many classic books that treat signal detection theory, including [5,7,8,16,17]. There are many more that are relevant to signal detection, e.g., books that treat pattern recognition and machine learning [18–20], multiuser detection [21], nonparametric inference [22], and robust statistics [23]. It is of course not possible to give a comprehensive list here. Let it suffice to cite a few of this author's favorite recent books on detection theory. The classic text by Van Trees [5] has been recently updated [24] and it includes many additional applications and recent developments, including signal detection for arrays. Another recent book is Levy's textbook [25] which provides a comprehensive treatment of signal detection with a chapter on Markov chain applications. The textbook [26] by Kay offers an excellent and accessible treatment of detection theory oriented toward signal processing. Finally, many signal detection problems, including the ones outlined in this chapter, can be put into the framework of statistical inference in linear multivariate analysis. The book by Anderson [27] is the seminal reference text in this area.

References

1. E. L. Lehmann, *Testing Statistical Hypotheses*, Wiley, New York, 1959.
2. T. S. Ferguson, *Mathematical Statistics—A Decision Theoretic Approach*, Academic Press, Orlando, FL, 1967.
3. D. Middleton, *An Introduction to Statistical Communication Theory*, Peninsula Publishing Co, Los Altos, CA (reprint of 1960 McGraw-Hill edition), 1987.
4. W. B. Davenport, W. L. Root, *An Introduction to the Theory of Random Signals and Noise*, IEEE Press, New York (reprint of 1958 McGraw-Hill edition), 1987.
5. H. L. Van-Trees, *Detection, Estimation, and Modulation Theory: Part I*, Wiley, New York, 1968.
6. D. Blackwell, M. A. Girshik, *Theory of Games and Statistical Decisions*, Wiley, New York, 1954.
7. C. Helstrom, *Elements of Signal Detection and Estimation*, Prentice-Hall, Englewood Cliffs, NJ, 1995.
8. L. L. Scharf, *Statistical Signal Processing: Detection, Estimation, and Time Series Analysis*, Addison-Wesley, Reading, MA, 1991.
9. R. J. Muirhead, *Aspects of Multivariate Statistical Theory*, Wiley, New York, 1982.
10. D. Siegmund, *Sequential Analysis: Tests and Confidence Intervals*, Springer-Verlag, New York, 1985.
11. B. Baygun, A. O. Hero, Optimal simultaneous detection and estimation under a false alarm constraint, *IEEE Trans. Inf. Theory*, 41(3): 688–703, 1995.
12. S. A. Kassam, J. B. Thomas, *Nonparametric Detection—Theory and Applications*, Dowden, Hutchinson and Ross, Stroudburg, PA, 1980.
13. K. Fukunaga, *Statistical Pattern Recognition*, 2nd ed., Academic Press, San Diego, CA, 1990.
14. E. J. Kelly, K. M. Forsythe, Adaptive detection and parameter estimation for multidimensional signal models, Technical Report No. 848, M.I.T. Lincoln Laboratory, April 1989.
15. T. Kariya, B. K. Sinha, *Robustness of Statistical Tests*, Academic Press, San Diego, CA, 1989.
16. H. V. Poor, *An Introduction to Signal Detection and Estimation*, Springer-Verlag, New York, 1988.
17. A. D. Whalen, *Detection of Signals in Noise*, 2nd ed., Academic Press, Orlando, FL, 1995.
18. C. M. Bishop, *Pattern Recognition and Machine Learning*, Springer, New York, 2006.
19. C. M. Bishop, *Information Theory, Inference and Learning Algorithms*, Cambridge University Press, Cambridge, UK, 2003.
20. T. Hastie, R. Tibshirani, J. H. Friedman, *The Elements of Statistical Learning: Data Mining, Inference, and Prediction*, Springer, New York, 2001.
21. S. Verdu, *Multiuser Detection*, Cambridge University Press, Cambridge, UK, 1998.
22. M. Hollander, D. A. Wolfe, *Nonparametric Statistical Methods*, 2nd ed., Wiley, New York, 1991.
23. P. J. Huber, *Robust Statistics*, Wiley, New York, 1981.
24. H. L. VanTrees, *Detection, Estimation, and Modulation Theory, Optimum Array Processing*, John Wiley & Sons, New York, 2002.
25. B. C. Levy, *Principles of Signal Detection and Parameter Estimation*, Springer, New York, 2008.
26. S. M. Kay, *Fundamentals of Statistical Signal Processing, Volume 2: Detection Theory*, Prentice-Hall, Englewood-Cliffs, NJ, 1998.
27. T. W. Anderson, *An Introduction to Multivariate Statistical Analysis*, Wiley, New York, 2003.

14

Spectrum Estimation and Modeling

Petar M. Djurić
Stony Brook University

Steven M. Kay
University of Rhode Island

14.1 Introduction

The main objective of spectrum estimation is the determination of the power spectral density (PSD) of a random process. The PSD is a function that plays a fundamental role in the analysis of stationary random processes in which it quantifies the distribution of total power as a function of frequency. The estimation of the PSD is based on a set of observed data samples from the process. A necessary assumption is that the random process is at least wide-sense stationary, that is, its first- and second-order statistics do not change with time. The estimated PSD provides information about the structure of the random process which can then be used for refined modeling, prediction, or filtering of the observed process.

Spectrum estimation has a long history with beginnings in ancient times [20]. The first significant discoveries that laid the grounds for later developments, however, were made in the early years of the eighteenth century. They include one of the most important advances in the history of mathematics, Fourier's theory. According to this theory, an arbitrary function can be represented by an infinite summation of sine and cosine functions. Later came the Sturm–Liouville spectral theory of differential equations, which was followed by the spectral representations in quantum and classical physics developed by John von Neuman and Norbert Wiener, respectively. The statistical theory of spectrum estimation started practically in 1949 when Tukey introduced a numerical method for computation of spectra from empirical data. A very important milestone for further development of the field was the reinvention of the fast Fourier transform (FFT) in 1965, which is an efficient algorithm for computation of the discrete Fourier transform (DFT). Shortly thereafter came the work of John Burg, who proposed a fundamentally new approach to spectrum estimation based on the principle of maximum entropy. In the past three decades, his work was followed up by many researchers who have developed numerous new spectrum estimation procedures and applied them to various physical processes from diverse scientific fields. Today, spectrum estimation is a vital scientific discipline which plays a major role in many applied sciences such as radar, speech processing, underwater acoustics, biomedical signal processing, sonar, seismology, vibration analysis, control theory, and econometrics.

14.2 Important Notions and Definitions

14.2.1 Random Processes

The objects of interest of spectrum estimation are random processes. They represent time fluctuations of a certain quantity which cannot be fully described by deterministic functions. The voltage waveform of a speech signal, the bit stream of zeros and ones of a communication message, or the daily variations of the stock market index are examples of random processes. Formally, a random process is defined as a collection of random variables indexed by time. (The family of random variables may also be indexed by a different variable, for example, space, but here we will consider only random time processes.) The index set is infinite and may be continuous or discrete. If the index set is continuous, the random process is known as a continuous-time random process, and if the set is discrete, it is known as a discrete-time random process. The speech waveform is an example of a continuous random process and the sequence of zeros and ones of a communication message, a discrete one. We shall focus only on discrete-time processes where the index set is the set of integers.

A random process can be viewed as a collection of a possibly infinite number of functions, also called realizations. We shall denote the collection of realizations by $\{\tilde{x}[n]\}$ and an observed realization of it by $\{x[n]\}$. For a fixed n, $\{\tilde{x}[n]\}$ represents a random variable, also denoted as $\tilde{x}[n]$, and $x[n]$ is the nth sample of the realization $\{x[n]\}$. If the samples $x[n]$ are real, the random process is real, and if they are complex, the random process is complex. In the discussion to follow, we assume that $\{\tilde{x}[n]\}$ is a complex random process.

The random process $\{\tilde{x}[n]\}$ is fully described if for any set of time indices $n_1, n_2, \ldots, n_m$, the joint probability density function of $\tilde{x}[n_1], \tilde{x}[n_2], \ldots,$ and $\tilde{x}[n_m]$ is given. If the statistical properties of the process do not change with time, the random process is called stationary. This is always the case if for any choice of random variables $\tilde{x}[n_1], \tilde{x}[n_2], \ldots,$ and $\tilde{x}[n_m]$, their joint probability density function is identical to the joint probability density function of the random variables $\tilde{x}[n_1+k], \tilde{x}[n_2+k], \ldots,$ and $\tilde{x}[n_m+k]$ for any k. Then we call the random process strictly stationary. For example, if the samples of the random process are independent and identically distributed random variables, it is straightforward to show that the process is strictly stationary. Strict stationarity, however, is a very severe requirement and is relaxed by introducing the concept of wide-sense stationarity. A random process is wide-sense stationary if the following two conditions are met:

$$E(\tilde{x}[n]) = \mu \tag{14.1}$$

and

$$\begin{aligned} r[n, n+k] &= E(\tilde{x}^*[n]\tilde{x}[n+k]) \\ &= r[k] \end{aligned} \tag{14.2}$$

where

$E(\cdot)$ is the expectation operator

$\tilde{x}^*[n]$ is the complex conjugate of $\tilde{x}[n]$

$r[k]$ is the autocorrelation function of the process

Thus, if the process is wide-sense stationary, its mean value μ is constant over time, and the autocorrelation function depends only on the lag k between the random variables. For example, if we consider the random process

$$\tilde{x}[n] = a\cos(2\pi f_0 n + \tilde{\theta}), \tag{14.3}$$

where the amplitude a and the frequency f_0 are constants, and the phase $\tilde{\theta}$ is a random variable that is uniformly distributed over the interval $(-\pi, \pi)$, one can show that

$$E(\tilde{x}[n]) = 0 \tag{14.4}$$

and

$$\begin{aligned} r[n, n+k] &= E(\tilde{x}^*[n]\tilde{x}[n+k]) \\ &= \frac{a^2}{2}\cos(2\pi f_0 k). \end{aligned} \tag{14.5}$$

Thus, Equation 14.3 represents a wide-sense stationary random process.

14.2.2 Spectra of Deterministic Signals

Before we define the concept of spectrum of a random process, it will be useful to review the analogous concept for deterministic signals, which are signals whose future values can be exactly determined without any uncertainty. Besides their description in the time domain, the deterministic signals have a very useful representation in terms of superposition of sinusoids with various frequencies, which is given by the discrete-time Fourier transform (DTFT). If the observed signal is $\{g[n]\}$ and it is not periodic, its DTFT is the complex-valued function $G(f)$ defined by

$$G(f) = \sum_{n=-\infty}^{\infty} g[n]e^{-j2\pi fn}, \tag{14.6}$$

where
- $j = \sqrt{-1}$
- f is the normalized frequency, $0 \leq f < 1$
- $e^{j2\pi fn}$ is the complex exponential given by

$$e^{j2\pi fn} = \cos(2\pi fn) + j\sin(2\pi fn). \tag{14.7}$$

The sum in Equation 14.6 converges uniformly to a continuous function of the frequency f if

$$\sum_{n=-\infty}^{\infty} |g[n]| < \infty. \tag{14.8}$$

The signal $\{g[n]\}$ can be determined from $G(f)$ by the inverse DTFT defined by

$$g[n] = \int_0^1 G(f)e^{j2\pi fn}\,df, \tag{14.9}$$

which means that the signal $\{g[n]\}$ can be represented in terms of complex exponentials whose frequencies span the continuous interval $[0, 1)$.

The complex function $G(f)$ can be alternatively expressed as

$$G(f) = |G(f)|e^{j\phi(f)}, \tag{14.10}$$

where

$|G(f)|$ is called the amplitude spectrum of $\{g[n]\}$

$\phi(f)$ the phase spectrum of $\{g[n]\}$.

For example, if the signal $\{g[n]\}$ is given by

$$g[n] = \begin{cases} 1, & n = 1 \\ 0, & n \neq 1 \end{cases} \tag{14.11}$$

then

$$G(f) = e^{-j2\pi f} \tag{14.12}$$

and the amplitude and phase spectra are

$$\begin{aligned} |G(f)| &= 1, & 0 \leq f < 1 \\ \phi(f) &= -2\pi f, & 0 \leq f < 1. \end{aligned} \tag{14.13}$$

The total energy of the signal is given by

$$\mathcal{E} = \sum_{n=-\infty}^{\infty} |g[n]|^2 \tag{14.14}$$

and according to Parseval's theorem, it can also be obtained from the amplitude spectrum of the signal, that is,

$$\sum_{n=-\infty}^{\infty} |g[n]|^2 = \int_0^1 |G(f)|^2 df. \tag{14.15}$$

From Equation 14.15, we deduce that $|G(f)|^2\, df$ is the contribution to the total energy of the signal from the frequency band $(f, f + df)$. Therefore, we say that $|G(f)|^2$ represents the energy density spectrum of the signal $\{g[n]\}$.

When $\{g[n]\}$ is periodic with period N, that is,

$$g(n) = g(n + N) \tag{14.16}$$

for all n, and where N is the period of $\{g[n]\}$, we use the DFT to express $\{g[n]\}$ in the frequency domain, that is,

$$G(f_k) = \sum_{n=0}^{N-1} g[n]e^{-j2\pi f_k n}, \quad f_k = \frac{k}{N}, \quad k \in \{0, 1, \ldots, N-1\}. \tag{14.17}$$

Note that the frequency here takes values from a discrete set. The inverse DFT is defined by

$$g[n] = \frac{1}{N}\sum_{k=0}^{N-1} G(f_k)e^{j2\pi f_k n}, \quad f_k = \frac{k}{N}. \tag{14.18}$$

Now Parseval's relation becomes

$$\sum_{n=0}^{N-1} |g[n]|^2 = \frac{1}{N}\sum_{k=0}^{N-1} |G(f_k)|^2, \quad f_k = \frac{k}{N}, \tag{14.19}$$

where the two sides are the total energy of the signal in one period. If we define the total power of the discrete-time signal by

$$P = \frac{1}{N}\sum_{n=0}^{N-1} |g[n]|^2, \tag{14.20}$$

then from Equation 14.19

$$P = \frac{1}{N^2}\sum_{k=0}^{N-1} |G(f_k)|^2, \quad f_k = \frac{k}{N}. \tag{14.21}$$

Thus, $|G(f_k)|^2/N^2$ is the contribution to the total power from the term with frequency f_k, and so it represents the power spectrum "density" of $\{g(n)\}$ at that frequency. For example, if the periodic signal in one period is defined by

$$g[n] = \begin{cases} 1, & n = 0 \\ 0, & n = 1, 2, \ldots, N-1, \end{cases} \tag{14.22}$$

its PSD $P(f_k)$ is

$$P(f_k) = \frac{1}{N^2}, \quad f_k = \frac{k}{N}, \quad k \in \{0, 1, \ldots, N-1\}. \tag{14.23}$$

Again, note that the PSD is defined for a discrete set of frequencies.

In summary, the spectra of deterministic aperiodic signals are energy densities defined on the continuous set of frequencies $C_f = [0, 1)$. On the other hand, the spectra of periodic signals are power densities defined on the discrete set of frequencies $D_f = \{0, 1/N, 2/N, \ldots, (N-1)/N\}$, where N is the period of the signal.

14.2.3 Spectra of Random Processes

Suppose that we observe one realization of the random process $\{\tilde{x}[n]\}$, $\{x[n]\}$. From the definition of the DTFT and the assumption of wide-sense stationarity of $\{\tilde{x}[n]\}$, it is obvious that we cannot use the DTFT to obtain $X(f)$ from $\{x[n]\}$ because Equation 14.8 does not hold when we replace $g[n]$ by $x[n]$. And indeed, if $\{x[n]\}$ is a realization of a wide-sense stationary process, its energy is infinite. Its power, however, is finite as was the case with the periodic signals. So if we observe $\{x[n]\}$ from $-N$ to N, $\{x[n]\}_{-N}^{N}$, and assume that outside this interval the samples $x[n]$ are equal to zero, we can find its DTFT, $X_N(f)$ from

$$X_N(f) = \sum_{n=-N}^{N} x[n]e^{-j2\pi fn}. \quad (14.24)$$

Then according to Equation 14.15, $|X_N(f)|^2 df$ represents the energy of the truncated realization that is contributed by the components whose frequencies are between f and $f + df$. The power due to these components is given by

$$\frac{|X_N(f)|^2 df}{2N+1} \quad (14.25)$$

and $|X_N(f)|^2/(2N+1)$ can be interpreted as power density. If we let $N \to \infty$, under suitable conditions [17],

$$\lim_{N\to\infty} \frac{|X_N(f)|^2}{2N+1} \quad (14.26)$$

is finite for all f, and this is then the PSD of $\{x[n]\}$. We would prefer to find, however, the PSD of $\{\tilde{x}[n]\}$, which we define as

$$P(f) = \lim_{N\to\infty} E\left\{\frac{|\tilde{X}_N(f)|^2}{2N+1}\right\}, \quad (14.27)$$

where $\tilde{X}_N(f)$ is the DTFT of $\{\tilde{x}[n]\}_{-N}^{N}$. Clearly, $P(f)df$ is interpreted as the average contribution to the total power from the components of $\{\tilde{x}[n]\}$ whose frequencies are between f and $f + df$.

There is a very important relationship between the PSD of a wide-sense stationary random process and its autocorrelation function. By Wold's theorem, which is the analog of Wiener–Khintchine theorem for continuous-time random processes, the PSD in Equation 14.27 is the DTFT of the autocorrelation function of the process [17], that is,

$$P(f) = \sum_{k=-\infty}^{\infty} r[k]e^{-j2\pi fk}, \quad (14.28)$$

where $r[k]$ is defined by Equation 14.2.

For all practical purposes, there are three different types of $P(f)$ [17]. If $P(f)$ is an absolutely continuous function of f, the random process has a purely continuous spectrum. If $P(f)$ is identically equal to zero for all f except for some frequencies $f = f_k$, $k = 1, 2, \ldots$, where it is infinite, the random process has a line spectrum. In this case, a useful representation of the spectrum is given by the Dirac δ-functions,

$$P(f) = \sum_k P_k \delta(f - f_k), \quad (14.29)$$

where P_k is the power associated with the kth line component. Finally, the spectrum of a random process may be mixed, if it is a combination of a continuous and line spectra. Then $P(f)$ is a superposition of a continuous function of the frequency f and δ-functions.

14.3 The Problem of Power Spectrum Estimation

The problem of power spectrum estimation can be stated as follows: Given a set of N samples $\{x[0], x[1], \ldots, x[N-1]\}$ of a realization of the random process $\{\tilde{x}[n]\}$, denoted also by $\{x[n]\}_0^{N-1}$, estimate the PSD of the random process, $P(f)$. Obviously this task amounts to estimation of a function and is distinct from the typical problem in elementary statistics where the goal is to estimate a finite set of parameters.

Spectrum estimation methods can be classified into two categories: nonparametric and parametric [14,24]. The nonparametric approaches do not assume any specific parametric model for the PSD. They are based solely on the estimate of the autocorrelation sequence of the random process from the observed data. For the parametric approaches on the other hand, we first postulate a model for the process of interest, where the model is described by a small number of parameters. Based on the model, the PSD of the process can be expressed in terms of the model parameters. Then the PSD estimate is obtained by substituting the estimated parameters of the model in the expression for the PSD. For example, if a random process $\{\tilde{x}[n]\}$ can be modeled by

$$\tilde{x}[n] = a\tilde{x}[n-1] + \tilde{w}[n], \tag{14.30}$$

where a is an unknown parameter and $\{\tilde{w}[n]\}$ is a zero-mean wide-sense stationary random process whose random variables are uncorrelated and with the same variance σ^2, it can be shown that the PSD of $\{\tilde{x}[n]\}$ is

$$P(f) = \frac{\sigma^2}{|1 + ae^{-j2\pi f}|^2}. \tag{14.31}$$

Thus, to find $P(f)$ it is sufficient to estimate a and σ^2.

The performance of a PSD estimator is evaluated by several measures of goodness. One is the bias of the estimator defined by

$$b(f) = E[\hat{P}(f)] - P(f), \tag{14.32}$$

where $\hat{P}(f)$ and $P(f)$ are the estimated and true PSD, respectively. If the bias $b(f)$ is identically equal to zero for all f, the estimator is said to be unbiased, which means that on average it yields the true PSD. Among the unbiased estimators, we search for the one that has minimal variability. The variability is measured by the variance of the estimator

$$v(f) = E([\hat{P}(f) - E[P(f)]]^2). \tag{14.33}$$

A measure that combines the bias and the variance is the relative mean-square error given by [17]

$$\vartheta(f) = \frac{v(f) + b(f)^2}{P(f)}. \tag{14.34}$$

The variability of a PSD estimator is also measured by the normalized variance [9]

$$\psi(f) = \frac{v(f)}{E^2[\hat{P}(f)]}. \tag{14.35}$$

Finally, another important metric for comparison is the resolution of the PSD estimators. It corresponds to the ability of the estimator to provide the fine details of the PSD of the random process. For example if the PSD of the random process has two peaks at frequencies f_1 and f_2, then the resolution of the estimator would be measured by the minimum separation of f_1 and f_2 for which the estimator still reproduces two peaks at f_1 and f_2.

14.4 Nonparametric Spectrum Estimation

When the method for PSD estimation is not based on any assumptions about the generation of the observed samples other than wide-sense stationarity, then it is termed a nonparametric estimator. According to Equation 14.28, $P(f)$ can be obtained by first estimating the autocorrelation sequence from the observed samples $x[0], x[1], \ldots, x[N-1]$, and then applying the DTFT to these estimates. One estimator of the autocorrelation is given by

$$\hat{r}[k] = \frac{1}{N}\sum_{n=0}^{N-1-k} x^*[n]x[n+k], \quad 0 \le k \le N-1. \tag{14.36}$$

The estimates of $\hat{r}[k]$ for $-N < k < 0$ are obtained from the identity

$$\hat{r}[-k] = \hat{r}^*[k] \tag{14.37}$$

and those for $|k| \ge N$ are set equal to zero. This estimator, although biased, has been preferred over others. An important reason for favoring it is that it always yields nonnegative estimates of the PSD, which is not the case with the unbiased estimator.

Many nonparametric estimators rely on using Equation 14.36 and then transform the obtained autocorrelation sequence to estimate the PSD. Other nonparametric methods, however, operate directly on the observed data.

14.4.1 Periodogram

The periodogram was introduced by Schuster in 1898 when he was searching for hidden periodicities while studying sunspot data [22]. To find the periodogram of the data $\{x[n]\}_0^{N-1}$, first we determine the autocorrelation sequence $r[k]$ for $-(N-1) \le k \le N-1$ and then take the DTFT, that is,

$$\hat{P}_{\mathrm{PER}}(f) = \sum_{k=-N+1}^{N-1} \hat{r}[k]e^{-j2\pi fk}. \tag{14.38}$$

It is more convenient to write the periodogram directly in terms of the observed samples $x[n]$. It is then defined as

$$\hat{P}_{\mathrm{PER}}(f) = \frac{1}{N}\left|\sum_{n=0}^{N-1} x[n]e^{-j2\pi fn}\right|^2. \tag{14.39}$$

Thus, the periodogram is proportional to the squared magnitude of the DTFT of the observed data. In practice, the periodogram is calculated by applying the FFT, which computes it at a discrete set of frequencies $D_{\mathrm{f}} = \{f_k : f_k = k/N, k = 0, 1, 2, \ldots, (N-1)\}$. The periodogram is then expressed by

$$\hat{P}_{\mathrm{PER}}(f_k) = \frac{1}{N}\left|\sum_{n=0}^{N-1} x[n]e^{-j2\pi kn/N}\right|^2, \quad f_k \in D_{\mathrm{f}}. \tag{14.40}$$

To allow for finer frequency spacing in the computed periodogram, we define a zero-padded sequence according to

$$x'[n] = \begin{cases} x[n], & n = 0, 1, \ldots, N-1 \\ 0, & n = N, N+1, \ldots, N'. \end{cases} \quad (14.41)$$

Then we specify the new set of frequencies $D'_f = \{f_k : f_k = k/N', k \in \{0, 1, 2, \ldots, (N'-1)\}\}$, and obtain

$$\hat{P}_{PER}(f_k) = \frac{1}{N}\left|\sum_{n=0}^{N-1} x[n]e^{-j2\pi kn/N'}\right|^2, \quad f_k \in D'_f. \quad (14.42)$$

A general property of good estimators is that they yield better estimates when the number of observed data samples increases. Theoretically, if the number of data samples tends to infinity, the estimates should converge to the true values of the estimated parameters. So, in the case of a PSD estimator, as we get more and more data samples, it is desirable that the estimated PSD tends to the true value of the PSD. In other words, if for finite number of data samples the estimator is biased, the bias should tend to zero as $N \to \infty$ as should the variance of the estimate. If this is indeed the case, the estimator is called consistent. Although the periodogram is asymptotically unbiased, it can be shown that it is not a consistent estimator. For example, if $\{\tilde{x}[n]\}$ is real zero-mean white Gaussian noise, which is a process whose random variables are independent, Gaussian, and identically distributed with variance σ^2, the variance of $\hat{P}_{PER}(f)$ is equal to σ^4 regardless of the length N of the observed data sequence [14]. The performance of the periodogram does not improve as N gets larger because as N increases, so does the number of parameters that are estimated, $P(f_0), P(f_1), \ldots, P(f_{N-1})$. In general, for the variance of the periodogram for frequencies not near 0 or $\pm 1/2$ (0 or 1), we can write [14]

$$\text{var}(\hat{P}_{PER}) \approx P^2(f), \quad (14.43)$$

where $P(f)$ is the true PSD.

Interesting insight can be gained if one writes the periodogram as follows:

$$\begin{aligned} \hat{P}_{PER}(f) &= \frac{1}{N}\left|\sum_{n=0}^{N-1} x[n]e^{-j2\pi fn}\right|^2 \\ &= \frac{1}{N}\left|\sum_{n=-\infty}^{\infty} x[n]w_R[n]e^{-j2\pi fn}\right|^2, \end{aligned} \quad (14.44)$$

where $w_R[n]$ is a rectangular window defined by

$$w_R[n] = \begin{cases} 1, & n \in \{0, 1, \ldots, N-1\} \\ 0, & \text{otherwise.} \end{cases} \quad (14.45)$$

Thus, we can regard the finite data record used for estimating the PSD as being obtained by multiplying the whole realization of the random process by a rectangular window. Then it is not difficult to show that the expected value of the periodogram is given by [9]

$$E\{\hat{P}_{PER}(f)\} = \frac{1}{N}\int_0^1 |W_R(f-\xi)|^2 P(\xi)d\xi, \quad (14.46)$$

where $W_R(f)$ is the DTFT of the rectangular window. Hence, the mean value of the periodogram is a smeared version of the true PSD. Since the implementation of the periodogram as defined in Equation 14.44 implies the use of a rectangular window, a question arises as to whether we could use a window of different shape to reduce the variance of the periodogram. The answer is yes, and indeed many windows have been proposed which weight the data samples in the middle of the observed segment more than those toward the ends of the segment. Some frequently used alternatives to the rectangular window are the windows of Bartlett, Hanning, Hamming, and Blackman. The magnitude of the DTFT of a window provides two important characteristics about it. One is the width of the window's mainlobe and the other is the strength of its sidelobes. A narrow mainlobe allows for a better resolution, and low sidelobes improve the smoothing of the estimated spectrum. Unfortunately, the narrower its mainlobe, the higher the sidelobes, which is a typical trade-off in spectrum estimation. It turns out that the rectangular window allows for the best resolution but has the largest sidelobes.

14.4.2 Bartlett Method

One approach to reduce the variance of the periodogram is to subdivide the observed data record into K nonoverlapping segments, find the periodogram of each segment, and finally evaluate the average of the so-obtained periodograms. This spectrum estimator, also known as the Bartlett's estimator, has variance that is smaller than the variance of the periodogram.

Suppose that the number of data samples N is equal to KL, where K is the number of segments and L is their length. If the ith segment is denoted by $\{x_i[n]\}_0^{L-1}, i = 1, 2, \ldots, K$, where

$$x_i[n] = x[n + (i - 1)L], \quad n \in \{0, 1, \ldots, L - 1\} \tag{14.47}$$

and its periodogram by

$$\hat{P}_{\text{PER}}^{(i)}(f) = \frac{1}{L}\left|\sum_{n=0}^{L-1} x_i[n]e^{-j2\pi fn}\right|^2, \tag{14.48}$$

then the Bartlett spectrum estimator is

$$\hat{P}_{\text{B}}(f) = \frac{1}{K}\sum_{i=1}^{K}\hat{P}_{\text{PER}}^{(i)}(f). \tag{14.49}$$

This estimator is consistent and its variance compared to the variance of the periodogram is reduced by a factor of K. This reduction, however, is paid by a decrease in resolution. The Bartlett estimator has a resolution K times less than that of the periodogram. Thus, this estimator allows for a straightforward trading of resolution for variance.

14.4.3 Welch Method

The Welch method is another estimator that exploits the periodogram. It is based on the same idea as Bartlett's approach of splitting the data into segments and finding the average of their periodograms. The difference is that the segments are overlapped, where the overlaps are usually 50% or 75%, and the data within a segment are windowed. Let the length of the segments be L, the ith segment be denoted again by $\{x_i[n]\}_0^{L-1}$, and the offset of successive sequences be D samples. Then

$$N = L + D(K - 1), \tag{14.50}$$

where
 N is the total number of observed samples
 K the total number of sequences.

Note that if there is no overlap, $K = N/L$, and if there is 50% overlap, $K = 2N/L - 1$. The ith sequence is defined by

$$x_i[n] = x[n + (i - 1)D], \quad n \in \{0, 1, \ldots, L - 1\}, \tag{14.51}$$

where $i = 1, 2, \ldots, K$, and its periodogram by

$$\hat{P}_{\mathrm{M}}^{(i)}(f) = \frac{1}{L}\left|\sum_{n=0}^{L-1} w[n]x_i[n]e^{-j2\pi fn}\right|^2. \tag{14.52}$$

Here, $\hat{P}_{\mathrm{M}}^{(i)}(f)$ is the modified periodogram of the data because the samples $x[n]$ are weighted by a nonrectangular window $w[n]$. The Welch spectrum estimate is then given by

$$\hat{P}_{\mathrm{B}}(f) = \frac{1}{K}\sum_{i=1}^{K} \hat{P}_{\mathrm{M}}^{(i)}(f). \tag{14.53}$$

By permitting an overlap of sequences, we can form more segments than in the case of Bartlett's method. Also, if we keep the same number of segments, the overlap allows for longer segments. The increased number of segments reduces the variance of the estimator, and the longer segments improve its resolution. Thus, with the Welch method, we can trade reduction in variance for improvement in resolution in many more ways than with the Bartlett method. It can be shown that if the overlap is 50%, the variance of the Welch estimator is approximately 9/16 of the variance of the Bartlett estimator [9].

14.4.4 Blackman–Tukey Method

The periodogram can be expressed in terms of the estimated autocorrelation lags as

$$\hat{P}_{\mathrm{PER}}(f) = \sum_{k=-(N-1)}^{N-1} \hat{r}[k]e^{-j2\pi fk}, \tag{14.54}$$

where

$$\hat{r}[k] = \begin{cases} \frac{1}{N}\sum_{n=0}^{N-1-k} x^*[n]x[n+k], & k = 0, 1, \ldots, N-1 \\ \hat{r}^*[-k], & k = -(N-1), -(N-2), \ldots, -1. \end{cases} \tag{14.55}$$

From Equations 14.54 and 14.55, we see that the estimated autocorrelation lags are given the same weight in the periodogram regardless of the difference of their variances. From Equation 14.55, however, it is obvious that the autocorrelations with smaller lags will be estimated more accurately than the ones with lags close to N because of the different number of terms that are used in the summation. For example, $\hat{r}[N-1]$ has only the term $x^*[0]x[n-1]$ compared to the N terms used in the computation of $\hat{r}[0]$. Therefore, the large variance of the periodogram can be ascribed to the large weight given to the poor autocorrelation estimates used in its evaluation.

Blackman and Tukey proposed to weight the autocorrelation sequence so that the autocorrelations with higher lags are weighted less [3]. Their estimator is given by

$$\hat{P}_{BT}(f) = \sum_{k=-(N-1)}^{N-1} w[k]\hat{r}[k]e^{-j2\pi fk}, \tag{14.56}$$

where the window $w[k]$ is a real, nonnegative, symmetric, and nonincreasing sequence with $|k|$, that is,

$$\begin{aligned} &1.\quad 0 \le w[k] \le w[0] = 1,\\ &2.\quad w[-k] = w[k], \text{ and}\\ &3.\quad w[k] = 0, \quad M < |k|, M \le N-1. \end{aligned} \tag{14.57}$$

Note that the symmetry property of $w[k]$ ensures that the spectrum is real.

The Blackman–Tukey estimator can be expressed in the frequency domain by the convolution

$$\hat{P}_{BT}(f) = \int_0^1 W(f-\xi)\hat{P}_{PER}(\xi)d\xi. \tag{14.58}$$

From Equation 14.58, we deduce that the window's DTFT should satisfy

$$W(f) \ge 0, \quad f \in (0,1) \tag{14.59}$$

so that the spectrum is guaranteed to be a nonnegative function, that is,

$$\hat{P}_{BT}(f) \ge 0, \quad 0 \le f < 1. \tag{14.60}$$

The bias, the variance, and the resolution of the Blackman–Tukey method depend on the applied window. For example, if the window is triangular (Bartlett),

$$w_B[k] = \begin{cases} \frac{M-|k|}{M}, & |\mathrm{k}\,| \le \mathrm{M} \\ 0, & \text{otherwise} \end{cases} \tag{14.61}$$

and if $N \gg M \gg 1$, the variance of the Blackman–Tukey estimator is [14]

$$\mathrm{var}(\hat{P}_{BT}) \approx \frac{2M}{3N}P^2(f), \tag{14.62}$$

where $P(f)$ is the true spectrum of the process. Compared to Equation 14.43, it is clear that the variance of this estimator may be significantly smaller than the variance of the periodogram. However, as M decreases, so does the resolution of the Blackman–Tukey estimator.

14.4.5 Minimum Variance Spectrum Estimator

The periodogram (Equation 14.44) can also be written as

$$\begin{aligned} \hat{P}_{PER}(f) &= \frac{1}{N}\left|\mathbf{e}^H(f)\mathbf{x}\right|^2 \\ &= N\left|\mathbf{h}^H(f)\mathbf{x}\right|^2, \end{aligned} \tag{14.63}$$

where $\mathbf{e}(f)$ is an $N \times 1$ vector defined by

$$\mathbf{e}(f) = \left[1 e^{j2\pi f} \; e^{j4\pi f} \; \ldots \; e^{j2(N-1)\pi f}\right]^{\mathrm{T}} \tag{14.64}$$

and $\mathbf{h}(f) = \mathbf{e}(f)/N$ with superscript H denoting complex conjugate transpose. We could interpret $\mathbf{h}(f)$ as a filter's finite impulse response (FIR). It is easy to show that $\mathbf{h}(f)$ is a bandpass filter centered at f with a bandwidth of approximately $1/N$. Then starting with Equation 14.63, we can prove that the value of the periodogram at frequency f can be obtained by squaring the magnitude of the filter output at $N-1$. Such filters exist for all the frequencies where the periodogram is evaluated, and they all have the same bandwidth. Thus, the periodogram may be viewed as a bank of FIR filters with equal bandwidths.

Capon proposed a spectrum estimator for processing large seismic arrays which, like the periodogram, can be interpreted as a bank of filters [5]. The width of these filters, however, is data dependent and optimized to minimize their response to components outside the band of interest. If the impulse response of the filter centered at f_0 is $\mathbf{h}(f_0)$, then it is desired to minimize

$$\rho = \int_0^1 |H(f)|^2 P(f) \mathrm{d}f \tag{14.65}$$

subject to the constraint

$$H(f_0) = 1, \tag{14.66}$$

where $H(f)$ is the DTFT of $\mathbf{h}(f_0)$. This is a constrained minimization problem, and the solution provides the optimal impulse response. When the solutions are used to determine the PSD of the observed data, we obtain the minimum variance (MV) spectrum estimator

$$\hat{P}_{\mathrm{MV}}(f) = \frac{N}{\mathbf{e}^{\mathrm{H}}(f)\hat{\mathbf{R}}^{-1}\mathbf{e}(f)}, \tag{14.67}$$

where $\hat{\mathbf{R}}^{-1}$ is the inverse matrix of the $N \times N$ estimated autocorrelation matrix $\hat{\mathbf{R}}$ defined by

$$\hat{\mathbf{R}} = \begin{bmatrix} \hat{r}[0] & \hat{r}[-1] & \hat{r}[-2] & \cdots & \hat{r}[-N+1] \\ \hat{r}[1] & \hat{r}[0] & \hat{r}[-1] & \cdots & \hat{r}[-N+2] \\ \vdots & \vdots & \vdots & \vdots & \vdots \\ \hat{r}[N-1] & \hat{r}[N-2] & \hat{r}[N-3] & \cdots & \hat{r}[0] \end{bmatrix}. \tag{14.68}$$

The length of the FIR filter does not have to be N, especially if we want to avoid the use of the unreliable estimates of $r[k]$. If the length of the filter's response is $p < N$, then the vector $\mathbf{e}(f)$, the autocorrelation matrix $\hat{\mathbf{R}}$, and the spectrum estimate $\hat{P}_{\mathrm{MV}}(f)$ are defined by Equations 14.64, 14.68, and 14.67, respectively, with N replaced by p [14].

The MV estimator has better resolution than the periodogram and the Blackman–Tukey estimator. The resolution and the variance of the MV estimator depend on the choice of the filter length p. If p is large, the bandwidth of the filter is small, which allows for better resolution. A larger p, however, requires more autocorrelation lags in the autocorrelation matrix $\hat{\mathbf{R}}$, which increases the variance of the estimated spectrum. Again, we have a trade-off between resolution and variance.

14.4.6 Multiwindow Spectrum Estimator

Many efforts have been made to improve the performance of the periodogram by multiplying the data with a nonrectangular window. The introduction of such windows has been more or less ad hoc, although they have been constructed to have narrow mainlobes and low sidelobes. By contrast, Thomson has proposed a spectrum estimation method that also involves the use of windows but is derived from fundamental principles. The method is based on the approximate solution of a Fredholm equation using an eigenexpansion [25]. The method amounts to applying multiple windows to the data, where the windows are discrete prolate spheroidal (Slepian) sequences. These sequences are orthogonal and their Fourier transforms have the maximum energy concentration in a given bandwidth W.

The multiwindow (MW) spectrum estimator is given by [25]

$$\hat{P}_{\mathrm{MW}}(f) = \frac{1}{m}\sum_{i=0}^{m-1}\hat{P}_i(f), \tag{14.69}$$

where the $\hat{P}_i(f)$ is the ith eigenspectrum defined by

$$\hat{P}_i(f) = \frac{1}{\lambda_i}\left|\sum_{n=0}^{N-1} x[n]w_i[n]e^{-j2\pi fn}\right|^2, \tag{14.70}$$

where

$w_i[n]$ is the ith Slepian sequence
λ_i the ith Slepian eigenvalue
W the analysis bandwidth.

The steps for obtaining $\hat{P}_{\mathrm{MW}}(f)$ are [26] the following:

1. Selection of the analysis bandwidth W whose typical values are between $1.5/N$ and $20/N$. The number of windows m depends on the selected W, and is given by $\lfloor 2NW \rfloor$, where $\lfloor x \rfloor$ denotes the largest integer less than or equal to x. The spectrum estimator has a resolution equal to W.
2. Evaluation of the m eigenspectra according to Equation 14.70, where the Slepian sequences and eigenvalues satisfy

$$\mathbf{C}\mathbf{w}_i = \lambda_i \mathbf{w}_i, \tag{14.71}$$

   with the elements of the matrix $\mathbf{C}$ being given by

$$c_{mn} = \frac{\sin(2\pi W(m-n))}{\pi(m-n)}, \quad m, n = 1, 2, \ldots, N. \tag{14.72}$$

   In the evaluation of the eigenspectra, only the Slepian sequences that correspond to the m largest eigenvalues of $\mathbf{C}$ are used.
3. Computation of the average spectrum according to Equation 14.69.

If the spectrum is mixed, that is, the observed data contain harmonics, the MW method uses a likelihood ratio test to determine if harmonics are present. If the test shows that there is a harmonic around the frequency f_0, the spectrum is reshaped by adding an impulse at f_0 followed by correction of the "local" spectrum for the inclusion of the impulse. For details, see [10,25,26].

The MW method is consistent, and its variance for fixed W tends to zero as $1/N$ when $N \to \infty$. The variance, however, as well as the bias and the resolution depend on the bandwidth W.

14.5 Parametric Spectrum Estimation

A philosophically different approach to spectrum estimation of a random process is the parametric one, which is based on the assumption that the process can be described by a parametric model. Based on the model, the spectrum of the process can then be expressed in terms of the parameters of the model. The approach thus consists of three steps: (1) selection of an appropriate parametric model (usually based on a priori knowledge about the process), (2) estimation of the model parameters, and (3) computation of the spectrum using the so-obtained parameters. In the literature, the parametric spectrum estimation methods are known as high-resolution methods because they can achieve better resolution than the nonparametric methods.

The most frequently used models in the literature are the autoregressive (AR), the moving average (MA), the autoregressive moving average (ARMA), and the sum of harmonics (complex sinusoids) embedded in noise. With the AR model, we assume that the observed data have been generated by a system whose input–output difference equation is given by

$$x[n] = -\sum_{k=1}^{p} a_k x[n-k] + e[n], \tag{14.73}$$

where

$x[n]$ is the observed output of the system
$e[n]$ is the unobserved input of the system
a_k's are its coefficients.

The input $e[n]$ is a zero-mean white noise process with unknown variance σ^2, and p is the order of the system. This model is usually abbreviated as AR(p). The MA model is given by

$$x[n] = \sum_{k=0}^{q} b_k e[n-k], \tag{14.74}$$

where

b_k's denote the MA parameters
$e[n]$ is a zero-mean white noise process with unknown variance σ^2
q is the order of the model.

The first MA coefficient b_0 is set usually to be $b_0 = 1$, and the model is denoted by MA(q). The ARMA model combines the AR and MA models and is described by

$$x[n] = -\sum_{k=1}^{p} a_k x[n-k] + \sum_{k=0}^{q} b_k e[n-k]. \tag{14.75}$$

Since the AR and MA orders are p and q, respectively, the model in Equation 14.75 is referred to as ARMA (p, q). Finally, the model of complex sinusoids in noise is

$$x[n] = \sum_{i=1}^{m} A_i e^{j2\pi f_i n} + e[n], \quad n = 0, 1, \ldots, N-1, \tag{14.76}$$

where

m is the number of complex sinusoids
A_i and f_i are the complex amplitude and frequency of the ith complex sinusoid, respectively
$e[n]$ is a sample of a noise process, which is not necessarily white.

Frequently, we assume that the samples $e[n]$ are generated by a certain parametric probability distribution whose parameters are unknown, or $e[n]$ itself is modeled as an AR, MA, or ARMA process.

14.5.1 Spectrum Estimation Based on Autoregressive Models

When the model of $x[n]$ is AR(p), the PSD of the process is given by

$$P_{\mathrm{AR}}(f) = \frac{\sigma^2}{\left|1 + \sum_{k=1}^{p} a_k e^{-j2\pi fk}\right|^2}. \quad (14.77)$$

Thus, to find $P_{\mathrm{AR}}(f)$ we need the estimates of the AR coefficients a_k and the noise variance σ^2.

If we multiply the two sides of Equation 14.73 by $x^*[n-k]$, $k \geq 0$, and take their expectations, we obtain

$$E(x[n]x^*[n-k]) = -\sum_{l=1}^{p} a_l E(x[n-l]x^*[n-k]) + E(e[n]x^*[n-k]) \quad (14.78)$$

or

$$r[k] = \begin{cases} -\sum_{l=1}^{p} a_l r[k-l], & k > 0 \\ -\sum_{l=1}^{p} a_l r[k-l] + \sigma^2, & k = 0. \end{cases} \quad (14.79)$$

The expressions in Equation 14.79 are known as the Yule–Walker equations. To estimate the p unknown AR coefficients from Equation 14.79, we need at least p equations as well as the estimates of the appropriate autocorrelations. The set of equations that requires the estimation of the minimum number of correlation lags is

$$\hat{\mathbf{R}}\mathbf{a} = -\hat{\mathbf{r}}, \quad (14.80)$$

where $\hat{\mathbf{R}}$ is the $p \times p$ matrix:

$$\hat{\mathbf{R}} = \begin{bmatrix} \hat{r}[0] & \hat{r}[-1] & \hat{r}[-2] & \cdots & \hat{r}[-p+1] \\ \hat{r}[1] & \hat{r}[0] & \hat{r}[-1] & \cdots & \hat{r}[-p+2] \\ \vdots & \vdots & \vdots & \vdots & \vdots \\ \hat{r}[p-1] & \hat{r}[p-2] & \hat{r}[p-3] & \cdots & \hat{r}[0] \end{bmatrix} \quad (14.81)$$

and

$$\hat{\mathbf{r}} = [\hat{r}[1]\hat{r}[2] \cdots \hat{r}[p]]^{\mathrm{T}}. \quad (14.82)$$

The parameters $\mathbf{a}$ are estimated by

$$\hat{\mathbf{a}} = -\hat{\mathbf{R}}^{-1}\hat{\mathbf{r}} \quad (14.83)$$

and the noise variance is found from

$$\hat{\sigma}^2 = \hat{r}[0] + \sum_{k=1}^{p} \hat{a}_k \hat{r}_k^*[k]. \tag{14.84}$$

The PSD estimate is obtained when $\hat{\mathbf{a}}$ and $\hat{\sigma}^2$ are substituted in Equation 14.77. This approach for estimating the AR parameters is known in the literature as the autocorrelation method.

Many other AR estimation procedures have been proposed including the maximum likelihood method, the covariance method, and the Burg method [14]. Burg's work in the late 1960s has a special place in the history of spectrum estimation because it kindled the interest in this field. Burg showed that the AR model provides an extrapolation of a known autocorrelation sequence $r[k]$, $|k| \leq p$, for $|k|$ beyond p so that the spectrum corresponding to the extrapolated sequence is the flattest of all spectra consistent with the $2p+1$ known autocorrelations [4].

An important issue in finding the AR PSD is the order of the assumed AR model. There exist several model-order selection procedures, but the most widely used are the Information Criterion A, also known as Akaike information criterion (AIC), due to Akaike [1] and the Information Criterion B, also known as Bayesian information criterion (BIC), also known as the minimum description length (MDL) principle, of Rissanen [18] and Schwarz [23]. According to the AIC criterion, the best model is the one that minimizes the function AIC(k) over k defined by

$$\text{AIC}(k) = N \log \hat{\sigma}_k^2 + 2k, \tag{14.85}$$

where

k is the model order

$\hat{\sigma}_k^2$ is the estimated noise variance of that model.

Similarly, the MDL criterion chooses the order which minimizes the function MDL(k) defined by

$$\text{MDL}(k) = N \log \hat{\sigma}_k^2 + k \log N, \tag{14.86}$$

where N is the number of observed data samples. It is important to emphasize that the MDL rule can be derived if, as a criterion for model selection, we use the maximum *a posteriori* principle. It has been found that the AIC is an inconsistent criterion, whereas the MDL rule is consistent. Consistency here means that the probability of choosing the correct model order tends to one as $N \to \infty$.

The AR-based spectrum estimation methods show very good performance if the processes are narrowband and have sharp peaks in their spectra. Also, many good results have been reported when they are applied to short data records.

14.5.2 Spectrum Estimation Based on Moving Average Models

The PSD of a moving average process is given by

$$P_{\text{MA}}(f) = \sigma^2 \left| 1 + \sum_{k=1}^{q} b_k e^{-j2\pi fk} \right|^2. \tag{14.87}$$

It is not difficult to show that the $r[k]$ s for $|k| > q$ of an MA(q) process are identically equal to zero, and that Equation 14.87 can be expressed also as

$$P_{\text{MA}}(f) = \sum_{k=-q}^{q} r[k] e^{-j2\pi fk}. \tag{14.88}$$

Thus, to find $\hat{P}_{MA}(f)$ it would be sufficient to estimate the autocorrelations $r[k]$ and use the found estimates in Equation 14.88. Obviously, this estimate would be identical to $\hat{P}_{BT}(f)$ when the applied window is rectangular and of length $2q + 1$.

A different approach is to find the estimates of the unknown MA coefficients and σ^2 and use them in Equation 14.87. The equations of the MA coefficients are nonlinear, which makes their estimation difficult. Durbin has proposed an approximate procedure that is based on a high-order AR approximation of the MA process. First the data are modeled by an AR model of order L, where $L \gg q$. Its coefficients are estimated from Equation 14.83 and $\hat{\sigma}^2$ according to Equation 14.84. Then the sequence 1, $\hat{a}_1, \hat{a}_2, \ldots, \hat{a}_L$, is fitted with an AR($q$) model, whose parameters are also estimated using the autocorrelation method. The estimated coefficients $\hat{b}_1, \hat{b}_2, \ldots, \hat{b}_q$ are subsequently substituted in Equation 14.87 together with $\hat{\sigma}^2$.

Good results with MA models are obtained when the PSD of the process is characterized by broad peaks and sharp nulls. The MA models should not be used for processes with narrowband features.

14.5.3 Spectrum Estimation Based on Autoregressive Moving Average Models

The PSD of a process that is represented by the ARMA model is given by

$$P_{ARMA}(f) = \sigma^2 \frac{\left|1 + \sum_{k=1}^{q} b_k e^{-j2\pi f k}\right|^2}{\left|1 + \sum_{k=1}^{p} a_k e^{-j2\pi f k}\right|^2}. \tag{14.89}$$

The ML estimates of the ARMA coefficients are difficult to obtain, so we usually resort to methods that yield suboptimal estimates. For example, we can first estimate the AR coefficients based on the following equation:

$$\begin{bmatrix} \hat{r}[q] & \hat{r}[q-1] & \cdots & \hat{r}[q-p+1] \\ \hat{r}[q+1] & \hat{r}[q] & \cdots & \hat{r}[q-p+2] \\ \vdots & \vdots & \vdots & \vdots \\ \hat{r}[M-1] & \hat{r}[M-2] & \cdots & \hat{r}[M-p] \end{bmatrix} \begin{bmatrix} a_1 \\ a_2 \\ \vdots \\ a_p \end{bmatrix} + \begin{bmatrix} \varepsilon_{q+1} \\ \varepsilon_{q+2} \\ \vdots \\ \varepsilon_M \end{bmatrix} = - \begin{bmatrix} \hat{r}[q+1] \\ \hat{r}[q+2] \\ \vdots \\ \hat{r}[M] \end{bmatrix} \tag{14.90}$$

or

$$\hat{\mathrm{R}}a + \varepsilon = -\hat{\mathrm{r}}, \tag{14.91}$$

where the vector ε models the errors in the Yule–Walker equations due to the estimation errors of the autocorrelation lags, and $M \geq p + q$. From Equation 14.91, we can find the least-squares estimates of $\mathbf{a}$ by

$$\hat{\mathbf{a}} = -\left(\hat{\mathbf{R}}^{\mathrm{H}}\hat{\mathbf{R}}\right)^{-1}\hat{\mathbf{R}}^{\mathrm{H}}\hat{\mathbf{r}}. \tag{14.92}$$

This procedure is known as the least-squares-modified Yule–Walker equation method. Once the AR coefficients are estimated, we can filter the observed data

$$y[n] = x[n] + \sum_{k=1}^{p} \hat{a}_k x[n-k] \tag{14.93}$$

and obtain a sequence that is approximately modeled by an MA(q) model. From the data $y[n]$ we can estimate the MA PSD by Equation 14.88 and obtain the PSD estimate of the data $x[n]$:

$$\hat{P}_{\text{ARMA}}(f) = \frac{\hat{P}_{\text{MA}}(f)}{\left|1 + \sum_{k=1}^{p} \hat{a}_k e^{-j2\pi fk}\right|^2} \tag{14.94}$$

or estimate the parameters $b_1, b_2, \ldots, b_q$ and σ^2 by Durbin's method, for example, and then use

$$\hat{P}_{\text{ARMA}}(f) = \hat{\sigma}^2 \frac{\left|1 + \sum_{k=1}^{q} \hat{b}_k e^{-j2\pi fk}\right|^2}{\left|1 + \sum_{k=1}^{p} \hat{a}_k e^{-j2\pi fk}\right|^2}. \tag{14.95}$$

The ARMA model has an advantage over the AR and MA models because it can better fit spectra with nulls and peaks. Its disadvantage is that it is more difficult to estimate its parameters than the parameters of the AR and MA models.

14.5.4 Pisarenko Harmonic Decomposition Method

Let the observed data represent m complex sinusoids in noise, that is,

$$x[n] = \sum_{i=1}^{m} A_i e^{j2\pi f_i n} + e[n], \quad n = 0, 1, \ldots, N-1, \tag{14.96}$$

where

f_i is the frequency of the ith complex sinusoid
A_i is the complex amplitude of the ith sinusoid

$$A_i = |A_i| e^{j\phi_i}, \tag{14.97}$$

ϕ_i being a random phase of the ith complex sinusoid
$e[n]$ is a sample of a zero-mean white noise process.

The PSD of the process is a sum of the continuous spectrum of the noise and a set of impulses with area $|A_i|^2$ at the frequencies f_i, or

$$P(f) = \sum_{i=1}^{m} |A_i|^2 \delta(f - f_i) + P_e(f), \tag{14.98}$$

where $P_e(f)$ is the PSD of the noise process.

Pisarenko studied the model in Equation 14.96 and found that the frequencies of the sinusoids can be obtained from the eigenvector corresponding to the smallest eigenvalue of the autocorrelation matrix. His method, known as Pisarenko harmonic decomposition (PHD), led to important insights and stimulated further work which resulted in many new procedures known today as "signal and noise subspace" methods.

When the noise $\{\tilde{e}[n]\}$ is zero-mean white with variance σ^2, the autocorrelation of $\{\tilde{x}[n]\}$ can be written as

$$r[k] = \sum_{i=1}^{m} |A_i|^2 e^{j2\pi f_i k} + \sigma^2 \delta[k] \tag{14.99}$$

or the autocorrelation matrix can be represented by

$$\mathbf{R} = \sum_{i=1}^{m} |A_i|^2 \mathbf{e}_i \mathbf{e}_i^{\mathrm{H}} + \sigma^2 \mathbf{I}, \tag{14.100}$$

where

$$\mathbf{e}_i = [1 e^{j2\pi f_i} \ e^{j4\pi f_i} \ e^{j2\pi(N-1)f_i}]^{\mathrm{T}} \tag{14.101}$$

and $\mathbf{I}$ is the identity matrix. It is seen that the autocorrelation matrix $\mathbf{R}$ is composed of the sum of signal and noise autocorrelation matrices:

$$\mathbf{R} = \mathbf{R}_s + \sigma^2 \mathbf{I}, \tag{14.102}$$

where

$$\mathbf{R}_s = \mathbf{E}\mathbf{P}\mathbf{E}^{\mathrm{H}} \tag{14.103}$$

for

$$\mathbf{E} = [\mathbf{e}_1 \ \mathbf{e}_2 \ \cdots \ \mathbf{e}_m] \tag{14.104}$$

and $\mathbf{P}$ is a diagonal matrix:

$$\mathbf{P} = \mathrm{diag}\{|A_1|^2, |A_2|^2, \ldots, |A_m|^2\}. \tag{14.105}$$

If the matrix $\mathbf{R}_s$ is $M \times M$, where $M \geq m$, its rank will be equal to the number of complex sinusoids m. Another important representation of the autocorrelation matrix $\mathbf{R}$ is via its eigenvalues and eigenvectors, that is,

$$\mathbf{R} = \sum_{i=1}^{m} (\lambda_i + \sigma^2) \mathbf{v}_i \mathbf{v}_i^{\mathrm{H}} + \sum_{i=m+1}^{M} \sigma^2 \mathbf{v}_i \mathbf{v}_i^{\mathrm{H}}, \tag{14.106}$$

where the λ_is, $i = 1, 2, \ldots, m$, are the nonzero eigenvalues of $\mathbf{R}_s$. Let the eigenvalues of $\mathbf{R}$ be arranged in decreasing order so that $\lambda_1 \geq \lambda_2 \geq \cdots \geq \lambda_M$, and let $\mathbf{v}_i$ be the eigenvector corresponding to λ_i. The space spanned by the eigenvectors $\mathbf{v}_i$, $i = 1, 2, \ldots, m$, is called the signal subspace, and the space spanned by $\mathbf{v}_i$, $i = m+1, m+2, \ldots, M$, the noise subspace. Since the set of eigenvectors are orthonormal, that is,

$$\mathbf{v}_i^{\mathrm{H}} \mathbf{v}_l = \begin{cases} 1, & i = l \\ 0, & i \neq l \end{cases} \tag{14.107}$$

the two subspaces are orthogonal. In other words if $\mathbf{s}$ is in the signal subspace, and $\mathbf{z}$ is in the noise subspace, then $\mathbf{s}^{\mathrm{H}}\mathbf{z} = 0$.

Now suppose that the matrix $\mathbf{R}$ is $(m+1)\times(m+1)$. Pisarenko observed that the noise variance corresponds to the smallest eigenvalue of $\mathbf{R}$ and that the frequencies of the complex sinusoids can be estimated by using the orthogonality of the signal and noise subspaces, that is,

$$\mathbf{e}_i^{\mathrm{H}} \mathbf{v}_{m+1} = 0, \quad i = 1, 2, \ldots, m. \tag{14.108}$$

We can estimate the f_i s by forming the pseudospectrum

$$\hat{P}_{\mathrm{PHD}}(f) = \frac{1}{|\mathbf{e}^{\mathrm{H}}(f)\mathbf{v}_{m+1}|^2}, \tag{14.109}$$

which should theoretically be infinite at the frequencies f_i. In practice, however, the pseudospectrum does not exhibit peaks exactly at these frequencies because **R** is not known and, instead, is estimated from finite data records.

The PSD estimate in Equation 14.109 does not include information about the power of the noise and the complex sinusoids. The powers, however, can easily be obtained by using Equation 14.98. First note that $P_e(f) = \sigma^2$ and $\hat{\sigma}^2 = \lambda_{m+1}$. Second, the frequencies f_i are determined from the pseudospectrum Equation 14.109, so it remains to find the powers of the complex sinusoids $P_i = |A_i|^2$. This can readily be accomplished by using the set of m linear equations:

$$\begin{bmatrix} |\hat{\mathbf{e}}_1^{\mathrm{H}}\mathbf{v}_1|^2 & |\hat{\mathbf{e}}_2^{\mathrm{H}}\mathbf{v}_1|^2 & \cdots & |\hat{\mathbf{e}}_m^{\mathrm{H}}\mathbf{v}_1|^2 \\ |\hat{\mathbf{e}}_1^{\mathrm{H}}\mathbf{v}_2|^2 & |\hat{\mathbf{e}}_2^{\mathrm{H}}\mathbf{v}_2|^2 & \cdots & |\hat{\mathbf{e}}_m^{\mathrm{H}}\mathbf{v}_2|^2 \\ \vdots & \vdots & \vdots & \vdots \\ |\hat{\mathbf{e}}_1^{\mathrm{H}}\mathbf{v}_m|^2 & |\hat{\mathbf{e}}_2^{\mathrm{H}}\mathbf{v}_m|^2 & \cdots & |\hat{\mathbf{e}}_m^{\mathrm{H}}\mathbf{v}_m|^2 \end{bmatrix} \begin{bmatrix} P_1 \\ P_2 \\ \vdots \\ P_m \end{bmatrix} = \begin{bmatrix} \lambda_1 - \hat{\sigma}^2 \\ \lambda_2 - \hat{\sigma}^2 \\ \vdots \\ \lambda_m - \hat{\sigma}^2 \end{bmatrix}, \tag{14.110}$$

where

$$\hat{\mathbf{e}}_i = [1 e^{j2\pi\hat{f}_i}\ e^{j4\pi\hat{f}_i}\ \cdots\ e^{j2\pi(N-1)\hat{f}_i}]^{\mathrm{T}}. \tag{14.111}$$

In summary, Pisarenko's method consists of four steps:

1. Estimate the $(m+1) \times (m+1)$ autocorrelation matrix **R** (provided it is known that the number of complex sinusoids is m)
2. Evaluate the minimum eigenvalue λ_{m+1} and the eigenvectors of $\hat{\mathbf{R}}$
3. Set the white noise power to $\sigma^2 = \lambda_{m+1}$, estimate the frequencies of the complex sinusoids from the peak locations of $\hat{P}_{\mathrm{PHD}}(f)$ in Equation 14.109, and compute their powers from Equation 14.110
4. Substitute the estimated parameters in Equation 14.98

Pisarenko's method is not used frequently in practice because its performance is much poorer than the performance of some other signal and noise subspace-based methods developed later.

14.5.5 Multiple Signal Classification

A procedure very similar to Pisarenko's is the MUltiple SIgnal Classification (MUSIC) method, which was proposed in the late 1970s by Schmidt [21]. Suppose again that the process $\{\tilde{x}[n]\}$ is described by m complex sinusoids in white noise. If we form an $M \times M$ autocorrelation matrix **R**, find its eigenvalues and eigenvectors and rank them as before, then as mentioned in the previous subsection, its m eigenvectors corresponding to the m largest eigenvalues span the signal subspace. Then, the remaining eigenvectors span the noise subspace. According to MUSIC, we estimate the noise variance from the $M - m$ smallest eigenvalues of $\hat{\mathbf{R}}$

$$\hat{\sigma}^2 = \frac{1}{M-m} \sum_{i=m+1}^{M} \lambda_i \tag{14.112}$$

and the frequencies from the peak locations of the pseudospectrum

$$\hat{P}_{\mathrm{MU}}(f) = \frac{1}{\sum_{i=m+1}^{M} \left| \mathbf{e}(f)^{\mathrm{H}} \mathbf{v}_i \right|^2}. \tag{14.113}$$

It should be noted that there are other ways of estimating the f_is. Finally the powers of the complex sinusoids are determined from Equation 14.110, and all the estimated parameters are substituted in Equation 14.98.

MUSIC has better performance than Pisarenko's method because of the introduced averaging via the extra noise eigenvectors. The averaging reduces the statistical fluctuations present in Pisarenko's pseudospectrum, which arise due to the errors in estimating the autocorrelation matrix. These fluctuations can further be reduced by applying the Eigenvector method [12], which is a modification of MUSIC and whose pseudospectrum is given by

$$\hat{P}_{\mathrm{EV}}(f) = \frac{1}{\sum_{i=m+1}^{M} \left| \frac{1}{\lambda_i} \mathbf{e}(f)^{\mathrm{H}} \mathbf{v}_i \right|^2}. \tag{14.114}$$

Pisarenko's method, MUSIC, and its variants exploit the noise subspace to estimate the unknown parameters of the random process. There are, however, approaches that estimate the unknown parameters from vectors that lie in the signal subspace. The main idea there is to form a reduced rank autocorrelation matrix which is an estimate of the signal autocorrelation matrix. Since this estimate is formed from the m principal eigenvectors and eigenvalues, the methods based on them are called principal component spectrum estimation methods [9,14]. Once the signal autocorrelation matrix is obtained, the frequencies of the complex sinusoids are found, followed by estimation of the remaining unknown parameters of the model.

14.6 Further Developments

Spectrum estimation continues to attract the attention of many researchers. The answers to many interesting questions are still unknown, and many problems still need better solutions. The field of spectrum estimation is constantly enriched with new theoretical findings and a wide range of results obtained from examinations of various physical processes. In addition, new concepts are being introduced that provide tools for improved processing of the observed signals and allow for a better understanding. Many new developments are driven by the need to solve specific problems that arise in applications, such as in sonar and communications.

For example, one of these advances is the introduction of canonical autoregressive decomposition [16]. The decomposition is a parametric approach for the estimation of mixed spectra where the continuous part of the spectrum is modeled by an AR model. In [13], it is shown how to obtain maximum likelihood frequency estimates for sinusoids in white Gaussian noise by using the mean likelihood estimator, which is implemented by the concept of importance sampling. Another development is related to Bayesian spectrum estimation. Jaynes has introduced it in [11] and some interesting results for spectra of harmonics in white Gaussian noise have been reported in [8]. A Bayesian spectrum estimate is based on

$$\hat{P}_{\mathrm{BA}}(f) = \int_{\Theta} P(f, \theta) f(\theta | \{x[n]\}_0^{N-1}) \mathrm{d}\theta, \tag{14.115}$$

where

$P(f, \theta)$ is the theoretical parametric spectrum
θ denotes the parameters of the process
Θ is the parameter space
$f(\theta|\{x[n]\}_0^{N-1})$ is the *a posteriori* probability density function of the process parameters.

Therefore, the Bayesian spectrum estimate is defined as the expected value of the theoretical spectrum over the joint posterior density function of the model parameters. Typically, the closed form solutions of the integral in Equation 14.115 cannot be obtained, and one has to rely on Monte Carlo-based solutions that include use of Markov chain Monte Carlo sampling or the population Monte Carlo method [6,19].

The processes that we have addressed here are wide-sense stationary. The stationarity assumption, however, is often a mathematical abstraction and only an approximation in practice. Many physical processes are actually nonstationary and their spectra change with time. In biomedicine, speech analysis, and sonar, for example, it is typical to observe signals whose power during some time intervals is concentrated at high frequencies and, shortly thereafter, at low or middle frequencies. In such cases, it is desirable to describe the PSD of the process at every instant of time, which is possible if we assume that the spectrum of the process changes smoothly over time. Such a description requires a combination of the time- and frequency-domain concepts of signal processing into a single framework [7]. So there is an important distinction between the PSD estimation methods discussed here and the time–frequency representation approaches. The former provide the PSD of the process for all times, whereas the latter yield the local PSDs at every instant of time. This area of research is well developed but still far from complete. Although many theories have been proposed and developed, including evolutionary spectra [17], the Wigner–Wille method [15], and the kernel choice approach [2], time-varying spectrum analysis has remained a challenging and fascinating area of research.

References

1. Akaike, H., A new look at the statistical model identification, *IEEE Trans. Autom. Control*, AC-19: 716–723, 1974.
2. Amin, M.G., Time-frequency spectrum analysis and estimation for nonstationary random processes, in *Time-Frequency Signal Analysis*, B. Boashash (Ed.), Longman Cheshire, Melbourne, Australia, 1992, pp. 208–232.
3. Blackman, R.B. and Tukey, J.W., *The Measurement of Power Spectra from the Point of View of Communications Engineering*, Dover Publications, New York, 1958.
4. Burg, J.P., Maximum entropy spectral analysis, Ph.D. dissertation, Stanford University, Stanford, CA, 1975.
5. Capon, J., High-resolution frequency-wavenumber spectrum analysis, *Proc. IEEE*, 57: 1408–1418, 1969.
6. Cappé, O., Guillin, A., and Robert, C.P., Population Monte Carlo, *J. Comput. Graphical Stat.*, 13: 907–929, 2004.
7. Cohen, L., *Time-Frequency Analysis*, Prentice Hall, Englewood Cliffs, NJ, 1995.
8. Djurić, P.M. and Li, H.-T., Bayesian spectrum estimation of harmonic signals, *Signal Process. Lett.*, 2: 213–215, 1995.
9. Hayes, M.S., *Statistical Digital Signal Processing and Modeling*, John Wiley & Sons, New York, 1996.
10. Haykin, S., *Advances in Spectrum Analysis and Array Processing*, Prentice Hall, Englewood Cliffs, NJ, 1991.

11. Jaynes, E.T., Bayesian spectrum and chirp analysis, in *Maximum Entropy and Bayesian Spectral Analysis and Estimation Problems*, C.R. Smith and G.J. Erickson (Eds.), D. Reidel, Dordrecht, the Netherlands, 1987, pp. 1–37.
12. Johnson, D.H. and DeGraaf, S.R., Improving the resolution of bearing in passive sonar arrays by eigenvalue analysis, *IEEE Trans. Acoust. Speech Signal Process.*, ASSP-30: 638–647, 1982.
13. Kay, S. and Saha, S., Mean likelihood frequency estimation, *IEEE Trans. Signal Process.*, SP-48: 1937–1946, 2000.
14. Kay, S.M., *Modern Spectral Estimation*, Prentice Hall, Englewood Cliffs, NJ, 1988.
15. Martin, W. and Flandrin, P., Wigner-Ville spectral analysis of nonstationary processes, *IEEE Trans. Acoust. Speech Signal Process.*, 33: 1461–1470, 1985.
16. Nagesha, V. and Kay, S.M., Spectral analysis based on the canonical autoregressive decomposition, *IEEE Trans. Signal Process.*, SP-44: 1719–1733, 1996.
17. Priestley, M.B., *Spectral Analysis and Time Series*, Academic Press, New York, 1981.
18. Rissanen, J., Modeling by shortest data description, *Automatica*, 14: 465–471, 1978.
19. Robert, C.P., *The Bayesian Choice*, Springer, New York, 2007.
20. Robinson, E.A., A historical perspective of spectrum estimation, *Proc. IEEE*, 70: 885–907, 1982.
21. Schmidt, R., Multiple emitter location and signal parameter estimation, *Proceedings of the RADC Spectrum Estimation Workshop*, Rome, NY, 1979, pp. 243–258.
22. Schuster, A., On the investigation on hidden periodicities with application to a supposed 26-day period of meteorological phenomena, *Terrestrial Magnetism*, 3: 13–41, 1898.
23. Schwarz, G., Estimating the dimension of the model, *Ann. Stat.*, 6: 461–464, 1978.
24. Stoica, P. and Moses, R., *Spectral Analysis of Signals*, Prentice Hall, Upper Saddle River, NJ, 2005.
25. Thomson, D.J., Spectrum estimation and harmonic analysis, *Proc. IEEE*, 70: 1055–1096, 1982.
26. Thomson, D.J., Quadratic-inverse spectrum estimates: Applications to paleoclimatology, *Philos. Trans. R. Soc. London A*, 332: 539–597, 1990.

15

Estimation Theory and Algorithms: From Gauss to Wiener to Kalman

Jerry M. Mendel
University of Southern California

15.1 Introduction

Estimation is one of four modeling problems. The other three are representation (how something should be modeled), measurement (which physical quantities should be measured and how they should be measured), and validation (demonstrating confidence in the model). Estimation, which fits in between the problems of measurement and validation, deals with the determination of those physical quantities that cannot be measured from those that can be measured. We shall cover a wide range of estimation techniques including weighted least squares, best linear unbiased, maximum-likelihood, mean-squared, and maximum-*a posteriori*. These techniques are for parameter or state estimation or a combination of the two, as applied to either linear or nonlinear models.

The discrete-time viewpoint is emphasized in this chapter because (1) much real data is collected in a digitized manner, so it is in a form ready to be processed by discrete-time estimation algorithms and (2) the mathematics associated with discrete-time estimation theory is simpler than with continuous-time estimation theory. We view (discrete-time) estimation theory as the extension of classical signal processing to the design of discrete-time (digital) filters that process uncertain data in an optimal manner. Estimation theory can, therefore, be viewed as a natural adjunct to digital signal processing theory. Mendel [12] is the primary reference for all the material in this chapter.

Estimation algorithms process data and, as such, must be implemented on a digital computer. Our computation philosophy is, whenever possible, leave it to the experts. Many of our chapter's algorithms can be used with MATLAB® and appropriate toolboxes (MATLAB is a registered trademark of The MathWorks, Inc.). See [12] for specific connections between MATLAB and toolbox M-files and the algorithms of this chapter.

The main model that we shall direct our attention to is linear in the unknown parameters, namely

$$\mathbf{Z}(k) = \mathbf{H}(k)\theta + \mathbf{V}(k). \tag{15.1}$$

In this model, which we refer to as a "generic linear model," $\mathbf{Z}(k) = \text{col}(z(k), z(k-1), \ldots, z(k-N+1))$, which is $N \times 1$, is called the measurement vector. Its elements are $z(j) = \mathbf{h}'(j)\theta + \nu(j)$; θ, which is $n \times 1$, is called the parameter vector, and contains the unknown deterministic or random parameters that will be estimated using one or more of this chapter's techniques; $\mathbf{H}(k)$, which is $N \times n$, is called the observation matrix; and, $\mathbf{V}(k)$, which is $N \times 1$, is called the measurement noise vector. By convention, the argument "k" of $\mathbf{Z}(k)$, $\mathbf{H}(k)$, and $\mathbf{V}(k)$ denotes the fact that the last measurement used to construct Equation 15.1 is the kth.

Examples of problems that can be cast into the form of the generic linear model are: identifying the impulse response coefficients in the convolutional summation model for a linear time-invariant system from noisy output measurements, identifying the coefficients of a linear time-invariant finite-difference equation model for a dynamical system from noisy output measurements, function approximation, state estimation, estimating parameters of a nonlinear model using a linearized version of that model, deconvolution, and identifying the coefficients in a discretized Volterra series representation of a nonlinear system.

The following estimation notation is used throughout this chapter: $\hat{\theta}(k)$ denotes an estimate of θ and $\tilde{\theta}(k)$ denotes the error in estimation, i.e., $\tilde{\theta}(k) = \theta - \hat{\theta}(k)$. The generic linear model is the starting point for the derivation of many classical parameter estimation techniques, and the estimation model for $\mathbf{Z}(k)$ is $\hat{\mathbf{Z}}(k) = \mathbf{H}(k)\hat{\theta}(k)$. In the rest of this chapter we develop specific structures for $\hat{\theta}(k)$. These structures are referred to as estimators. Estimates are obtained whenever data are processed by an estimator.

15.2 Least-Squares Estimation

The method of least squares dates back to Karl Gauss around 1795 and is the cornerstone for most estimation theory. The weighted least-squares estimator (WLSE), $\hat{\theta}_{\text{WLS}}(k)$, is obtained by minimizing the objective function $J[\hat{\theta}(k)] = \tilde{\mathbf{Z}}'(k)\mathbf{W}(k)\tilde{\mathbf{Z}}(k)$, where (using Equation 15.1) $\tilde{\mathbf{Z}}(k) = \mathbf{Z}(k) - \hat{\mathbf{Z}}(k) = \mathbf{H}(k)\tilde{\theta}(k) + \mathbf{V}(k)$, and weighting matrix $\mathbf{W}(k)$ must be symmetric and positive definite. This weighting matrix can be used to weight recent measurements more (or less) heavily than past measurements. If $\mathbf{W}(k) = c\mathbf{I}$, so that all measurements are weighted the same, then weighted least-squares reduces to least squares, in which case, we obtain $\hat{\theta}_{\text{LS}}(k)$. Setting $dJ[\hat{\theta}(k)]/d\hat{\theta}(k) = 0$, we find that

$$\hat{\theta}_{\text{WLS}}(k) = [\mathbf{H}'(k)\mathbf{W}(k)\mathbf{H}(k)]^{-1}\mathbf{H}'(k)\mathbf{W}(k)\mathbf{Z}(k) \tag{15.2}$$

and, consequently,

$$\hat{\theta}_{\text{LS}}(k) = [\mathbf{H}'(k)\mathbf{H}(k)]^{-1}\mathbf{H}'(k)\mathbf{Z}(k). \tag{15.3}$$

Note, also, that $J[\hat{\theta}_{\text{WLS}}(k)] = \mathbf{Z}'(k)\mathbf{W}(k)\mathbf{Z}(k) - \hat{\theta}'_{\text{WLS}}(k)\mathbf{H}'(k)\mathbf{W}(k)\mathbf{H}(k)\hat{\theta}_{\text{WLS}}(k)$.

Matrix $\mathbf{H}'(k)\mathbf{W}(k)\mathbf{H}(k)$ must be nonsingular for its inverse in Equation 15.2 to exist. This is true if $\mathbf{W}(k)$ is positive definite, as assumed, and $\mathbf{H}(k)$ is of maximum rank. We know that $\hat{\theta}_{WLS}(k)$ minimizes $J[\hat{\theta}_{WLS}(k)]$ because $d^2J[\hat{\theta}(k)]/d\hat{\theta}^2(k) = 2\mathbf{H}'(k)\mathbf{W}(k)\mathbf{H}(k) > 0$, since $\mathbf{H}'(k)\mathbf{W}(k)\mathbf{H}(k)$ is invertible. Estimator $\hat{\theta}_{WLS}(k)$ processes the measurements $\mathbf{Z}(k)$ linearly; hence, it is referred to as a linear estimator. In practice, we do not compute $\hat{\theta}_{WLS}(k)$ using Equation 15.2, because computing the inverse of $\mathbf{H}'(k)\mathbf{W}(k)\mathbf{H}(k)$ is fraught with numerical difficulties. Instead, the so-called normal equations $[\mathbf{H}'(k)\mathbf{W}(k)\mathbf{H}(k)]\hat{\theta}_{WLS}(k) = \mathbf{H}'(k)\mathbf{W}(k)\mathbf{Z}(k)$ are solved using stable algorithms from numerical linear algebra (e.g., [3] indicating that one approach to solving the normal equations is to convert the original least squares problem into an equivalent, easy-to-solve problem using orthogonal transformations such as Householder or Givens transformations). Note, also, that Equations 15.2 and 15.3 apply to the estimation of either deterministic or random parameters, because nowhere in the derivation of $\hat{\theta}_{WLS}(k)$ did we have to assume that θ was or was not random. Finally, note that WLSEs may not be invariant under changes of scale. One way to circumvent this difficulty is to use normalized data.

Least-squares estimates can also be computed using the singular-value decomposition (SVD) of matrix $\mathbf{H}(k)$. This computation is valid for both the overdetermined ($N < n$) and underdetermined ($N > n$) situations and for the situation when $\mathbf{H}(k)$ may or may not be of full rank. The SVD of $K \times M$ matrix $\mathbf{A}$ is

$$\mathbf{U}'\mathbf{A}\mathbf{V} = \begin{bmatrix} \Sigma & 0 \\ 0 & 0 \end{bmatrix}, \tag{15.4}$$

where

$\mathbf{U}$ and $\mathbf{V}$ are unitary matrices

$\Sigma = \text{diag}(\sigma_1, \sigma_2, \ldots, \sigma_r), \sigma_1 \geq \sigma_2 \geq \cdots \geq \sigma_r > 0$, where the σ_i's are the singular values of $\mathbf{A}$ and r is the rank of $\mathbf{A}$

Let the SVD of $\mathbf{H}(k)$ be given by Equation 15.4. Even if $\mathbf{H}(k)$ is not of maximum rank, then

$$\ddot{\theta}_{LS}(k) = \mathbf{V}\begin{bmatrix} \Sigma^{-1} & 0 \\ 0 & 0 \end{bmatrix}\mathbf{U}'\mathbf{Z}(k), \tag{15.5}$$

where

$\Sigma^{-1} = \text{diag}\left(\sigma_1^{-1}\sigma_2^{-1}, \ldots, \sigma_r^{-1}\right)$

r is the rank of $\mathbf{H}(k)$

Additionally, in the overdetermined case,

$$\hat{\theta}_{LS}(k) = \sum_{i=1}^{r} \frac{\mathbf{v}_i(k)}{\sigma_i^2(k)}\mathbf{v}_i'(k)\mathbf{H}'(k)\mathbf{Z}(k). \tag{15.6}$$

Similar formulas exist for computing $\hat{\theta}_{WLS}(k)$.

Equations 15.2 and 15.3 are batch equations, because they process all of the measurements at one time. These formulas can be made recursive in time by using simple vector and matrix partitioning techniques. The information form of the recursive WLSE is

$$\hat{\theta}_{WLS}(k+1) = \hat{\theta}_{WLS}(k) + \mathbf{K}_w(k+1)[z(k+1) - \mathbf{h}'(k+1)\hat{\theta}_{WLS}(k)], \tag{15.7}$$

$$\mathbf{K}_w(k+1) = \mathbf{P}(k+1)\mathbf{h}(k+1)w(k+1), \tag{15.8}$$

$$\mathbf{P}^{-1}(k+1) = \mathbf{P}^{-1}(k) + \mathbf{h}(k+1)w(k+1)\mathbf{h}'(k+1). \tag{15.9}$$

Equations 15.8 and 15.9 require the inversion of $n \times n$ matrix $\mathbf{P}$. If n is large, then this will be a costly computation. Applying a matrix inversion lemma to Equation 15.9, one obtains the following alternative covariance form of the recursive WLSE (Equation 15.7), and

$$\mathbf{K}_\mathbf{w}(k+1) = \mathbf{P}(k)\mathbf{h}(k+1)\left[\mathbf{h}'(k+1)\mathbf{P}(k)\mathbf{h}(k+1) + \frac{1}{w(k+1)}\right]^{-1}, \tag{15.10}$$

$$\mathbf{P}(k+1) = [\mathbf{I} - \mathbf{K}_\mathbf{w}(k+1)\mathbf{h}'(k+1)]\mathbf{P}(k). \tag{15.11}$$

Equations 15.7 through 15.9 or Equations 15.7, 15.10, and 15.11 are initialized by $\hat{\theta}_{\text{WLS}}(n)$ and $\mathbf{P}^{-1}(n)$, where $\mathbf{P}(n) = [\mathbf{H}'(n)\mathbf{W}(n)\mathbf{H}(n)]^{-1}$, and are used for $k = n, n+1, \ldots, N-1$.

Equation 15.7 can be expressed as

$$\hat{\theta}_{\text{WLS}}(k+1) = [\mathbf{I} - \mathbf{K}_\mathbf{w}(k+1)\mathbf{h}'(k+1)]\hat{\theta}_{\text{WLS}}(k) + \mathbf{K}_\mathbf{w}(k+1)z(k+1), \tag{15.12}$$

which demonstrates that the recursive WLSE is a time-varying digital filter that is excited by random inputs (i.e., the measurements), one whose plant matrix $[\mathbf{I} - \mathbf{K}_\mathbf{w}(k+1)\mathbf{h}'(k+1)]$ may itself be random because $\mathbf{K}_\mathbf{w}(k+1)$ and $\mathbf{h}(k+1)$ may be random, depending upon the specific application. The random natures of these matrices make the analysis of this filter exceedingly difficult.

Two recursions are present in the recursive WLSEs. The first is the vector recursion for $\hat{\theta}_{\text{WLS}}$ given by Equation 15.7. Clearly, $\hat{\theta}_{\text{WLS}}(k+1)$ cannot be computed from this expression until measurement $z(k+1)$ is available. The second is the matrix recursion for either $\mathbf{P}^{-1}$ given by Equation 15.9 or $\mathbf{P}$ given by Equation 15.11. Observe that values for these matrices can be precomputed before measurements are made. A digital computer implementation of Equations 15.7 through 15.9 is $\mathbf{P}^{-1}(k+1) \rightarrow \mathbf{P}(k+1) \rightarrow \mathbf{K}_\mathbf{w}(k+1) \rightarrow \hat{\theta}_{\text{WLS}}(k+1)$, whereas for Equations 15.7, 15.10, and 15.11, it is $\mathbf{P}(k) \rightarrow \mathbf{K}_\mathbf{w}(k+1) \rightarrow \hat{\theta}_{\text{WLS}}(k+1) \rightarrow \mathbf{P}(k+1)$. Finally, the recursive WLSEs can even be used for $k = 0, 1, \ldots, N-1$. Often $z(0) = 0$, or there is no measurement made at $k = 0$, so that we can set $z(0) = 0$. In this case we can set $w(0) = 0$, and the recursive WLSEs can be initialized by setting $\hat{\theta}_{\text{WLS}}(0) = 0$ and $\mathbf{P}(0)$ to a diagonal matrix of very large numbers. This is very commonly done in practice. Fast fixed-order recursive least-squares algorithms that are based on the Givens rotation [3] and can be implemented using systolic arrays are described in [5] and the references therein.

15.3 Properties of Estimators

How do we know whether or not the results obtained from the WLSE, or for that matter any estimator, are good? To answer this question, we must make use of the fact that all estimators represent transformations of random data; hence, $\hat{\theta}(k)$ is itself random, so that its properties must be studied from a statistical viewpoint. This fact and its consequences, which seem so obvious to us today, are due to the eminent statistician R.A. Fischer.

It is common to distinguish between small-sample and large-sample properties of estimators. The term "sample" refers to the number of measurements used to obtain $\hat{\theta}$, i.e., the dimension of $\mathbf{Z}$. The phrase "small sample" means any number of measurements (e.g., 1, 2, 100, 10^4, or even an infinite number), whereas the phrase "large sample" means "an infinite number of measurements." Large-sample properties are also referred to as asymptotic properties. If an estimator possesses as small-sample property, it also possesses the associated large-sample property; but the converse is not always true. Although large sample means an infinite number of measurements, estimators begin to enjoy large-sample properties for much fewer than an infinite number of measurements. How few usually depends on the dimension of θ, n, the memory of the estimators, and in general on the underlying, albeit unknown, probability density function.

A thorough study into $\hat{\theta}$ would mean determining its probability density function $p(\hat{\theta})$. Usually, it is too difficult to obtain $p(\hat{\theta})$ for most estimators (unless $\hat{\theta}$ is multivariate Gaussian); thus, it is customary to emphasize the first- and second-order statistics of $\hat{\theta}$ (or its associated error $\tilde{\theta} = \theta - \hat{\theta}$), the mean, and the covariance.

Small-sample properties of an estimator are unbiasedness and efficiency. An estimator is unbiased if its mean value is tracking the unknown parameter at every value of time, i.e., the mean value of the estimation error is zero at every value of time. Dispersion about the mean is measured by error variance. Efficiency is related to how small the error variance will be. Associated with efficiency is the very famous Cramér–Rao inequality (Fisher information matrix, in the case of a vector of parameters) which places a lower bound on the error variance, a bound that does not depend on a particular estimator.

Large-sample properties of an estimator are asymptotic unbiasedness, consistency, asymptotic normality, and asymptotic efficiency. Asymptotic unbiasedness and efficiency are limiting forms of their small sample counterparts, unbiasedness and efficiency. The importance of an estimator being asymptotically normal (Gaussian) is that its entire probabilistic description is then known, and it can be entirely characterized just by its asymptotic first- and second-order statistics. Consistency is a form of convergence of $\hat{\theta}(k)$ to θ; it is synonymous with convergence in probability. One of the reasons for the importance of consistency in estimation theory is that any continuous function of a consistent estimator is itself a consistent estimator, i.e., "consistency carries over." It is also possible to examine other types of stochastic convergence for estimators, such as mean-squared convergence and convergence with probability 1. A general carryover property does not exist for these two types of convergence; it must be established case-by-case (e.g., [11]).

Generally speaking, it is very difficult to establish small sample or large sample properties for least-squares estimators, except in the very special case when $\mathbf{H}(k)$ and $\mathbf{V}(k)$ are statistically independent. While this condition is satisfied in the application of identifying an impulse response, it is violated in the important application of identifying the coefficients in a finite difference equation, as well as in many other important engineering applications. Many large sample properties of LSEs are determined by establishing that the LSE is equivalent to another estimator for which it is known that the large sample property holds true. We pursue this below.

Least-squares estimators require no assumptions about the statistical nature of the generic model. Consequently, the formula for the WLSE is easy to derive. The price paid for not making assumptions about the statistical nature of the generic linear model is great difficulty in establishing small or large sample properties of the resulting estimator.

15.4 Best Linear Unbiased Estimation

Our second estimator is both unbiased and efficient by design, and is a linear function of measurements $\mathbf{Z}(k)$. It is called a best linear unbiased estimator (BLUE), $\hat{\theta}_{\mathrm{BLU}}(k)$. As in the derivation of the WLSE, we begin with our generic linear model; but, now we make two assumptions about this model, namely: (1) $\mathbf{H}(k)$ must be deterministic and (2) $\mathbf{V}(k)$ must be zero mean with positive definite known covariance matrix $\mathbf{R}(k)$. The derivation of the BLUE is more complicated than the derivation of the WLSE because of the design constraints; however, its performance analysis is much easier because we build good performance into its design.

We begin by assuming the following linear structure for $\hat{\theta}_{\mathrm{BLU}}(k)$, $\hat{\theta}_{\mathrm{BLU}}(k) = \mathbf{F}(k)\mathbf{Z}(k)$. Matrix $\mathbf{F}(k)$ is designed such that (1) $\hat{\theta}_{\mathrm{BLU}}(k)$ is an unbiased estimator of θ and (2) the error variance for each of the n parameters is minimized. In this way, $\hat{\theta}_{\mathrm{BLU}}(k)$ will be unbiased and efficient (within the class of linear estimators) by design. The resulting BLUE estimator is

$$\hat{\theta}_{\mathrm{BLU}}(k) = [\mathbf{H}'(k)\mathbf{R}^{-1}(k)\mathbf{H}(k)]\mathbf{H}'(k)\mathbf{R}^{-1}(k)\mathbf{Z}(k). \tag{15.13}$$

A very remarkable connection exists between the BLUE and WLSE, namely, the BLUE of θ is the special case of the WLSE of θ when $\mathbf{W}(k) = \mathbf{R}^{-1}(k)$. Consequently, all results obtained in our section above for

$\hat{\theta}_{WLS}(k)$ can be applied to $\hat{\theta}_{BLU}(k)$ by setting $\mathbf{W}(k) = \mathbf{R}^{-1}(k)$. Matrix $\mathbf{R}^{-1}(k)$ weights the contributions of precise measurements heavily and deemphasizes the contributions of imprecise measurements. The best linear unbiased estimation design technique has led to a weighting matrix that is quite sensible.

If $\mathbf{H}(k)$ is deterministic and $\mathbf{R}(k) = \sigma_\nu^2\mathbf{I}$, then $\hat{\theta}_{BLU}(k) = \hat{\theta}_{LS}(k)$. This result, known as the Gauss–Markov theorem, is important because we have connected two seemingly different estimators, one of which, $\hat{\theta}_{BLU}(k)$, has the properties of unbiasedness and minimum variance by design; hence, in this case $\hat{\theta}_{LS}(k)$ inherits these properties.

In a recursive WLSE, matrix $\mathbf{P}(k)$ has no special meaning. In a recursive BLUE (which is obtained by substituting $\mathbf{W}(k) = \mathbf{R}^{-1}(k)$ into Equations 15.7 through 15.9, or Equations 15.7, 15.10, and 15.11), matrix $\mathbf{P}(k)$ is the covariance matrix for the error between θ and $\hat{\theta}_{BLU}(k)$, i.e., $\mathbf{P}(k) = [\mathbf{H}'(k)\mathbf{R}^{-1}(k)\mathbf{H}(k)]^{-1} = \text{Cov}[\tilde{\theta}_{BLU}(k)]$. Hence, every time $\mathbf{P}(k)$ is calculated in the recursive BLUE, we obtain a quantitative measure of how well we are estimating θ.

Recall that we stated that WLSEs may change in numerical value under changes in scale. BLUEs are invariant under changes in scale. This is accomplished automatically by setting $\mathbf{W}(k) = \mathbf{R}^{-1}(k)$ in the WLSE.

The fact that $\mathbf{H}(k)$ must be deterministic severely limits the applicability of BLUEs in engineering applications.

15.5 Maximum-Likelihood Estimation

Probability is associated with a forward experiment in which the probability model, $p(\mathbf{Z}(k)|\theta)$, is specified, including values for the parameters, θ, in that model (e.g., mean and variance in a Gaussian density function), and data (i.e., realizations) are generated using this model. Likelihood, $l(\theta|\mathbf{Z}(k))$, is proportional to probability. In likelihood, the data is given as well as the nature of the probability model; but the parameters of the probability model are not specified. They must be determined from the given data. Likelihood is, therefore, associated with an inverse experiment.

The maximum-likelihood method is based on the relatively simple idea that different (statistical) populations generate different samples and that any given sample (i.e., set of data) is more likely to have come from some populations than from others.

In order to determine the maximum-likelihood estimate (MLE) of deterministic $\theta, \hat{\theta}_{ML}$, we need to determine a formula for the likelihood function and then maximize that function. Because likelihood is proportional to probability, we need to know the entire joint probability density function of the measurements in order to determine a formula for the likelihood function. This, of course, is much more information about $\mathbf{Z}(k)$ than was required in the derivation of the BLUE. In fact, it is the most information that we can ever expect to know about the measurements. The price we pay for knowing so much information about $\mathbf{Z}(k)$ is complexity in maximizing the likelihood function. Generally, mathematical programming must be used in order to determine $\hat{\theta}_{ML}$.

Maximum-likelihood estimates are very popular and widely used because they enjoy very good large sample properties. They are consistent, asymptotically Gaussian with mean θ and covariance matrix $\frac{1}{N}\mathbf{J}^{-1}$, in which $\mathbf{J}$ is the Fisher information matrix, and are asymptotically efficient. Functions of maximum-likelihood estimates are themselves maximum-likelihood estimates, i.e., if $\mathbf{g}(\theta)$ is a vector function mapping θ into an interval in r-dimensional Euclidean space, then $\mathbf{g}(\hat{\theta}_{ML})$ is a MLE of $\mathbf{g}(\theta)$. This "invariance" property is usually not enjoyed by WLSEs or BLUEs.

In one special case it is very easy to compute $\hat{\theta}_{ML}$, i.e., for our generic linear model in which $\mathbf{H}(k)$ is deterministic and $\mathbf{V}(k)$ is Gaussian. In this case $\hat{\theta}_{ML} = \hat{\theta}_{BLU}$. These estimators are unbiased, because $\hat{\theta}_{BLU}$ is unbiased; efficient (within the class of linear estimators), because $\hat{\theta}_{BLU}$ is efficient; consistent, because $\hat{\theta}_{ML}$ is consistent; and Gaussian, because they depend linearly on $\mathbf{Z}(k)$, which is Gaussian. If, in addition, $\mathbf{R}(k) = \sigma_\nu^2\mathbf{I}$, then $\hat{\theta}_{ML}(k) = \hat{\theta}_{BLU}(k) = \hat{\theta}_{LS}(k)$, and these estimators are unbiased, efficient (within the class of linear estimators), consistent, and Gaussian.

The method of maximum-likelihood is limited to deterministic parameters. In the case of random parameters, we can still use the WLSE or the BLUE, or, if additional information is available, we can use either a mean-squared or maximum-*a posteriori* estimator, as described below. The former does not use statistical information about the random parameters, whereas the latter does.

15.6 Mean-Squared Estimation of Random Parameters

Given measurements $\mathbf{z}(1), \mathbf{z}(2), \ldots, \mathbf{z}(k)$, the mean-squared estimator (MSE) of random θ, $\hat{\theta}_{MS}(k) = \phi[\mathbf{z}(i), i = 1, 2, \ldots, k]$, minimizes the mean-squared error $J[\tilde{\theta}_{MS}(k)] = \mathbf{E}\{\tilde{\theta}'_{MS}(k)\tilde{\theta}_{MS}(k)\}$, where $\tilde{\theta}_{MS}(k) = \theta - \hat{\theta}_{MS}(k)$. The function $\phi[\mathbf{z}(i), i=1,2,\ldots,k]$ may be nonlinear or linear. Its exact structure is determined by minimizing $J[\tilde{\theta}_{MS}(k)]$.

The solution to this mean-squared estimation problem, which is known as the fundamental theorem of estimation theory, is

$$\hat{\theta}_{MS}(k) = \mathbf{E}\{\theta|\mathbf{Z}(k)\}. \tag{15.14}$$

As it stands, Equation 15.14 is not terribly useful for computing $\hat{\theta}_{MS}(k)$. In general, we must first compute $p[\theta|\mathbf{Z}(k)]$ and then perform the requisite number of integrations of $\theta p[\theta|\mathbf{Z}(k)]$ to obtain $\hat{\theta}_{MS}(k)$. It is useful to separate this computation into two major cases: (1) θ and $\mathbf{Z}(k)$ are jointly Gaussian—the Gaussian case, and (2) θ and $\mathbf{Z}(k)$ are not jointly Gaussian—the non-Gaussian case.

When θ and $\mathbf{Z}(k)$ are jointly Gaussian, the estimator that minimizes the mean-squared error is

$$\hat{\theta}_{MS}(k) = \mathbf{m}_\theta + \mathbf{P}_{\theta z}(k)\mathbf{P}_z^{-1}(k)[\mathbf{Z}(k) - \mathbf{m}_z(k)], \tag{15.15}$$

where

$\mathbf{m}_\theta$ is the mean of θ
$\mathbf{m}_z(k)$ is the mean of $\mathbf{Z}(k)$
$\mathbf{P}_z(k)$ is the covariance matrix of $\mathbf{Z}(k)$
$\mathbf{P}_{\theta z}(k)$ is the cross-covariance between θ and $\mathbf{Z}(k)$

Of course, to compute $\hat{\theta}_{MS}(k)$ using Equation 15.15, we must somehow know all of these statistics, and we must be sure that θ and $\mathbf{Z}(k)$ are jointly Gaussian. For the generic linear model, $\mathbf{Z}(k) = \mathbf{H}(k)\theta + \mathbf{V}(k)$, in which $\mathbf{H}(k)$ is deterministic, $\mathbf{V}(k)$ is Gaussian noise with known invertible covariance matrix $\mathbf{R}(k)$, θ is Gaussian with mean $\mathbf{m}_\theta$ and covariance matrix $\mathbf{P}_\theta$, and θ and $\mathbf{V}(k)$ are statistically independent, then θ and $\mathbf{Z}(k)$ are jointly Gaussian, and Equation 15.15 becomes

$$\hat{\theta}_{MS}(k) = \mathbf{m}_\theta + \mathbf{P}_\theta\mathbf{H}'(k)[\mathbf{H}(k)\mathbf{P}_\theta\mathbf{H}'(k) + \mathbf{R}(k)]^{-1}[\mathbf{Z}(k) - \mathbf{H}(k)\mathbf{m}_\theta], \tag{15.16}$$

where error-covariance matrix $\mathbf{P}_{MS}(k)$, which is associated with $\hat{\theta}_{MS}(k)$, is

$$\begin{aligned}\mathbf{P}_{MS}(k) &= \mathbf{P}_\theta - \mathbf{P}_\theta\mathbf{H}'(k)[\mathbf{H}(k)\mathbf{P}_\theta\mathbf{H}'(k) + \mathbf{R}(k)]^{-1}\mathbf{H}(k)\mathbf{P}_\theta \\ &= \left[\mathbf{P}_\theta^{-1} + \mathbf{H}'(k)\mathbf{R}^{-1}(k)\mathbf{H}(k)\right]^{-1}.\end{aligned} \tag{15.17}$$

Using Equation 15.17 in Equation 15.16, $\hat{\theta}_{MS}(k)$ can be reexpressed as

$$\hat{\theta}_{MS}(k) = \mathbf{m}_\theta + \mathbf{P}_{MS}(k)\mathbf{H}'(k)\mathbf{R}^{-1}(k)[\mathbf{Z}(k) - \mathbf{H}(k)\mathbf{m}_\theta]. \tag{15.18}$$

Suppose θ and $\mathbf{Z}(k)$ are not jointly Gaussian and that we know $\mathbf{m}_\theta$, $\mathbf{m}_z(k)$, $\mathbf{P}_z(k)$, and $\mathbf{P}_{\theta z}(k)$. In this case, the estimator that is constrained to be an affine transformation of $\mathbf{Z}(k)$ and that minimizes the mean-squared error is also given by Equation 15.15.

We now know the answer to the following important question: When is the linear (affine) mean-squared estimator the same as the mean-squared estimator? The answer is when θ and $\mathbf{Z}(k)$ are jointly

Gaussian. If θ and $\mathbf{Z}(k)$ are not jointly Gaussian, then $\hat{\theta}_{MS}(k) = \mathbf{E}\{\theta|\mathbf{Z}(k)\}$, which, in general, is a nonlinear function of measurements $\mathbf{Z}(k)$, i.e., it is a nonlinear estimator.

Associated with mean-squared estimation theory is the orthogonality principle: Suppose $\mathbf{f}[\mathbf{Z}(k)]$ is any function of the data $\mathbf{Z}(k)$; then the error in the mean-squared estimator is orthogonal to $\mathbf{f}[\mathbf{Z}(k)]$ in the sense that $\mathbf{E}\{[\theta - \hat{\theta}_{MS}(k)]\mathbf{f}'[\mathbf{Z}(k)]\} = 0$. A frequently encountered special case of this occurs when $\mathbf{f}[\mathbf{Z}(k)] = \hat{\theta}_{MS}(k)$, in which case $\mathbf{E}\{\tilde{\theta}_{MS}(k)\tilde{\theta}_{MS}(k)\} = 0$.

When θ and $\mathbf{Z}(k)$ are jointly Gaussian, $\hat{\theta}_{MS}(k)$ in Equation 15.15 has the following properties: (1) it is unbiased; (2) each of its components has the smallest error variance; (3) it is a "linear" (affine) estimator; (4) it is unique; and (5) both $\hat{\theta}_{MS}(k)$ and $\tilde{\theta}_{MS}(k)$ are multivariate Gaussian, which means that these quantities are completely characterized by their first- and second-order statistics. Tremendous simplifications occur when θ and $\mathbf{Z}(k)$ are jointly Gaussian!

Many of the results presented in this section are applicable to objective functions other than the mean-squared objective function. See the supplementary material at the end of Lesson 13 in [12] for discussions on a wide number of objective functions that lead to $\mathbf{E}\{\theta|\mathbf{Z}(k)\}$ as the optimal estimator of θ, as well as discussions on a full-blown nonlinear estimator of θ.

There is a connection between the BLUE and the MSE. The connection requires a slightly different BLUE, one that incorporates the *a priori* statistical information about random θ. To do this, we treat $\mathbf{m}_\theta$ as an additional measurement that is augmented to $\mathbf{Z}(k)$. The additional measurement equation is obtained by adding and subtracting θ in the identity $\mathbf{m}_\theta = \mathbf{m}_\theta$, i.e., $\mathbf{m}_\theta = \theta + (\mathbf{m}_\theta - \theta)$. Quantity $(\mathbf{m}_\theta - \theta)$ is now treated as zero-mean measurement noise with covariance $\mathbf{P}_\theta$. The augmented linear model is

$$\left[\frac{\mathbf{Z}(k)}{\mathbf{m}_\theta}\right] = \left[\frac{\mathbf{H}(k)}{\mathbf{I}}\right]\theta + \left[\frac{\mathbf{V}(k)}{\mathbf{m}_\theta - \theta}\right]. \tag{15.19}$$

Let the BLUE estimator for this augmented model be denoted $\hat{\theta}^a_{BLU}(k)$. Then it is always true that $\hat{\theta}_{MS}(k) = \hat{\theta}^a_{BLU}(k)$. Note that the weighted least-squares objective function that is associated with $\hat{\theta}^a_{BLU}(k)$ is $J_a[\hat{\theta}^a(k)] = [\mathbf{m}_\theta - \hat{\theta}^a(k)]'\mathbf{P}_\theta^{-1}[\mathbf{m}_\theta - \hat{\theta}^a(k)] + \tilde{\mathbf{Z}}'(k)\mathbf{R}^{-1}(k)\tilde{\mathbf{Z}}(k)$.

15.7 Maximum A Posteriori Estimation of Random Parameters

Maximum *a posteriori* (MAP) estimation is also known as Bayesian estimation. Recall Bayes's rule: $p[\theta|\mathbf{Z}(k)] = p[\mathbf{Z}(k)|\theta]p(\theta)/p[\mathbf{Z}(k)]$ in which density function $p[\theta|\mathbf{Z}(k)]$ is known as the *a posteriori* (or posterior) conditional density function, and $p(\theta)$ is the prior density function for θ. Observe that $p[\theta|\mathbf{Z}(k)]$ is related to likelihood function $l\{\theta|\mathbf{Z}(k)\}$, because $l\{\theta|\mathbf{Z}(k)\} \propto p[\mathbf{Z}(k)|\theta]$. Additionally, because $p[\mathbf{Z}(k)]$ does not depend on θ, $p[\theta|\mathbf{Z}(k)] \propto p[\mathbf{Z}(k)|\theta]p(\theta)$. In MAP estimation, values of θ are found that maximize $p[\mathbf{Z}(k)|\theta]p(\theta)$. Obtaining a MAP estimate involves specifying both $p[\mathbf{Z}(k)|\theta]$ and $p(\theta)$ and finding the value of θ that maximizes $p[\theta|\mathbf{Z}(k)]$. It is the knowledge of the *a priori* probability model for θ, $p(\theta)$, that distinguishes the problem formulation for MAP estimation from MS estimation.

If $\theta_1, \theta_2, \ldots, \theta_n$ are uniformly distributed, then $p[\theta|\mathbf{Z}(k)] \propto p[\mathbf{Z}(k)|\theta]$, and the MAP estimator of θ equals the ML estimator of θ. Generally, MAP estimates are quite different from ML estimates. For example, the invariance property of MLEs usually does not carry over to MAP estimates. One reason for this can be seen from the formula $p[\theta|\mathbf{Z}(k)] \propto p[\mathbf{Z}(k)|\theta]p(\theta)$. Suppose, for example, that $\phi = \mathbf{g}(\theta)$ and we want to determine $\hat{\phi}_{MAP}$ by first computing $\hat{\theta}_{MAP}$. Because $p(\theta)$ depends on the Jacobian matrix of $\mathbf{g}^{-1}(\phi)$, $\hat{\phi}_{MAP} \neq \mathbf{g}(\hat{\theta}_{MAP})$. Usually $\hat{\theta}_{MAP}$ and $\hat{\theta}_{ML}(k)$ are asymptotically identical to one another since in the large sample case the knowledge of the observations tends to swamp the knowledge of the prior distribution [10].

Generally speaking, optimization must be used to compute $\hat{\theta}_{MAP}(k)$. In the special but important case, when $\mathbf{Z}(k)$ and θ are jointly Gaussian, then $\hat{\theta}_{MAP}(k) = \hat{\theta}_{MS}(k)$. This result is true regardless of the nature of the model relating θ to $\mathbf{Z}(k)$. Of course, in order to use it, we must first establish that $\mathbf{Z}(k)$ and θ are jointly Gaussian. Except for the generic linear model, this is very difficult to do.

When $\mathbf{H}(k)$ is deterministic, $\mathbf{V}(k)$ is white Gaussian noise with known covariance matrix $\mathbf{R}(k)$, and θ is multivariate Gaussian with known mean $\mathbf{m}_\theta$ and covariance $\mathbf{P}_\theta$, $\hat{\theta}_{MAP}(k) = \hat{\theta}^a_{BLU}(k)$; hence, for the generic linear Gaussian model, MS, MAP, and BLUE estimates of θ are all the same, i.e., $\hat{\theta}_{MS}(k) = \hat{\theta}^a_{BLU}(k) = \hat{\theta}_{MAP}(k)$.

15.8 The Basic State-Variable Model

In the rest of this chapter we shall describe a variety of mean-squared state estimators for a linear, (possibly) time-varying, discrete-time, dynamical system, which we refer to as the basic state-variable model. This system is characterized by $n \times 1$ state vector $\mathbf{x}(k)$ and $m \times 1$ measurement vector $\mathbf{z}(k)$, and is

$$\mathbf{x}(k+1) = \Phi(k+1,k)\mathbf{x}(k) + \Gamma(k+1,k)\mathbf{w}(k) + \Psi(k+1,k)\mathbf{u}(k) \tag{15.20}$$

$$\mathbf{z}(k+1) = \mathbf{H}(k+1)\mathbf{x}(k+1) + \mathbf{v}(k+1), \tag{15.21}$$

where $k = 0, 1, \ldots$. In this model $\mathbf{w}(k)$ and $\mathbf{v}(k)$ are $p \times 1$ and $m \times 1$ mutually uncorrelated (possibly nonstationary) jointly Gaussian white noise sequences, i.e., $\mathbf{E}\{\mathbf{w}(i)\mathbf{w}'(j)\} = \mathbf{Q}(i)\delta_{ij}$, $\mathbf{E}\{\mathbf{v}(i)\mathbf{v}'(j)\} = \mathbf{R}(i)\delta_{ij}$ and $\mathbf{E}\{\mathbf{w}(i)\mathbf{v}'(j)\} = \mathbf{S} = 0$, for all i and j. Covariance matrix $\mathbf{Q}(i)$ is positive semi-definite and $\mathbf{R}(i)$ is positive definite (so that $\mathbf{R}^{-1}(i)$ exists). Additionally, $\mathbf{u}(k)$ is an $l \times 1$ vector of known system inputs, and initial state vector $\mathbf{x}(0)$ is multivariate Gaussian, with mean $\mathbf{m}_\mathbf{x}(0)$ and covariance $\mathbf{P}_\mathbf{x}(0)$, and $\mathbf{x}(0)$ is not correlated with $\mathbf{w}(k)$ and $\mathbf{v}(k)$. The dimensions of matrices Φ, Γ, Ψ, $\mathbf{H}$, $\mathbf{Q}$, and $\mathbf{R}$ are $n \times n$, $n \times p$, $n \times l$, $m \times n$, $p \times p$, and $m \times m$, respectively. The double arguments in matrices Φ, Γ, and Ψ may not always be necessary, in which case we replace $(k+1, k)$ by k.

Disturbance $\mathbf{w}(k)$ is often used to model disturbance forces acting on the system, errors in modeling the system, or errors due to actuators in the translation of the known input, $\mathbf{u}(k)$, into physical signals. Vector $\mathbf{v}(k)$ is often used to model errors in measurements made by sensing instruments, or unavoidable disturbances that act directly on the sensors.

Not all systems are described by this basic model. In general, $\mathbf{w}(k)$ and $\mathbf{v}(k)$ may be correlated, some measurements may be made so accurate that, for all practical purposes, they are "perfect" (i.e., no measurement noise is associated with them), and either $\mathbf{w}(k)$ or $\mathbf{v}(k)$, or both, may be nonzero mean or colored noise processes. How to handle these situations is described in Lesson 22 of [12].

When $\mathbf{x}(0)$ and $\{\mathbf{w}(k), k = 0, 1, \ldots\}$ are jointly Gaussian, then $\{\mathbf{x}(k), k = 0, 1, \ldots\}$ is a Gauss–Markov sequence. Note that if $\mathbf{x}(0)$ and $\mathbf{w}(k)$ are individually Gaussian and statistically independent, they will be jointly Gaussian. Consequently, the mean and covariance of the state vector completely characterize it. Let $\mathbf{m}_\mathbf{x}(k)$ denote the mean of $\mathbf{x}(k)$. For our basic state-variable model, $\mathbf{m}_\mathbf{x}(k)$ can be computed from the vector recursive equation

$$\mathbf{m}_\mathbf{x}(k+1) = \Phi(k+1,k)\mathbf{m}_\mathbf{x}(k) + \Psi(k+1,k)\mathbf{u}(k), \tag{15.22}$$

where $k = 0, 1, \ldots$, and $\mathbf{m}_\mathbf{x}(0)$ initializes Equation 15.22. Let $\mathbf{P}_\mathbf{x}(k)$ denote the covariance matrix of $\mathbf{x}(k)$. For our basic state-variable model, $\mathbf{P}_\mathbf{x}(k)$ can be computed from the matrix recursive equation

$$\mathbf{P}_\mathbf{x}(k+1) = \Phi(k+1,k)\mathbf{P}_\mathbf{x}(k)\Phi'(k+1,k) + \Gamma(k+1,k)\mathbf{Q}(k)\Gamma'(k+1,k), \tag{15.23}$$

where $k = 0, 1, \ldots$, and $\mathbf{P}_\mathbf{x}(0)$ initializes Equation 15.23. Equations 15.22 and 15.23 are easily programmed for a digital computer.

For our basic state-variable model, when $\mathbf{x}(0)$, $\mathbf{w}(k)$, and $\mathbf{v}(k)$ are jointly Gaussian, then $\{\mathbf{z}(k), k = 1, 2, \ldots\}$ is Gaussian, and

$$\mathbf{m}_\mathbf{z}(k+1) = \mathbf{H}(k+1)\mathbf{m}_\mathbf{x}(k+1) \tag{15.24}$$

and

$$\mathbf{P_z}(k+1) = \mathbf{H}(k+1)\mathbf{P_x}(k+1)\mathbf{H}'(k+1) + \mathbf{R}(k+1), \tag{15.25}$$

where $\mathbf{m_x}(k+1)$ and $\mathbf{P_x}(k+1)$ are computed from Equations 15.22 and 15.23, respectively.

For our basic state-variable model to be stationary, it must be time-invariant, and the probability density functions of $\mathbf{w}(k)$ and $\mathbf{v}(k)$ must be the same for all values of time. Because $\mathbf{w}(k)$ and $\mathbf{v}(k)$ are zero-mean and Gaussian, this means that $\mathbf{Q}(k)$ must equal the constant matrix $\mathbf{Q}$ and $\mathbf{R}(k)$ must equal the constant matrix $\mathbf{R}$. Additionally, either $\mathbf{x}(0) =$ or $\Phi(k, 0)\mathbf{x}(0) \approx 0$ when $k > k_0$; in both cases $\mathbf{x}(k)$ will be in its steady-state regime, so stationarity is possible.

If the basic state-variable model is time-invariant and stationary and if Φ is associated with an asymptotically stable system (i.e., one whose poles all lie within the unit circle), then [1] matrix $\mathbf{P_x}(k)$ reaches a limiting (steady-state) solution $\bar{\mathbf{P}}_\mathbf{x}$ and $\bar{\mathbf{P}}_\mathbf{x}$ is the solution of the following steady-state version of Equation 15.23: $\bar{\mathbf{P}}_\mathbf{x} = \Phi\bar{\mathbf{P}}_\mathbf{x}\Phi' + \Gamma\mathbf{Q}\Gamma'$. This equation is called a discrete-time Lyapunov equation.

15.9 State Estimation for the Basic State-Variable Model

Prediction, filtering, and smoothing are three types of mean-squared state estimation that have been developed since 1959. A predicted estimate of a state vector $\mathbf{x}(k)$ uses measurements which occur earlier than t_k and a model to make the transition from the last time point, say t_j, at which a measurement is available, to t_k. The success of prediction depends on the quality of the model. In state estimation we use the state equation model. Without a model, prediction is dubious at best.

A recursive mean-squared state filter is called a Kalman filter, because it was developed by Kalman around 1959 [9]. Although it was originally developed within a community of control theorists, and is regarded as the most widely used result of so-called "modern control theory," it is no longer viewed as a control theory result. It is a result within estimation theory; consequently, we now prefer to view it as a signal processing result. A filtered estimate of state vector $\mathbf{x}(k)$ uses all of the measurements up to and including the one made at time t_k.

A smoothed estimate of state vector $\mathbf{x}(k)$ not only uses measurements which occur earlier than t_k plus the one at t_k, but also uses measurements to the right of t_k. Consequently, smoothing can never be carried out in real time, because we have to collect "future" measurements before we can compute a smoothed estimate. If we don't look too far into the future, then smoothing can be performed subject to a delay of LT seconds, where T is our data sampling time and L is a fixed positive integer that describes how many sample points to the right of t_k are to be used in smoothing.

Depending upon how many future measurements are used and how they are used, it is possible to create three types of smoother: (1) the fixed-interval smoother, $\hat{\mathbf{x}}(k|N)$, $k = 0, 1, \ldots, N-1$, where N is a fixed positive integer; (2) the fixed-point smoother, $\hat{\mathbf{x}}(k|j)$, $j = k+1,\ k+2, \ldots,$ where k is a fixed positive integer; and (3) the fixed-lag smoother, $\hat{\mathbf{x}}(k|k+L)$, $k = 0, 1, \ldots,$ where L is a fixed positive integer.

15.9.1 Prediction

A single-stage predicted estimate of $\mathbf{x}(k)$ is denoted $\hat{\mathbf{x}}(k|k-1)$. It is the mean-squared estimate of $\mathbf{x}(k)$ that uses all the measurements up to and including the one made at time t_{k-1}; hence, a single-stage predicted estimate looks exactly one time point into the future. This estimate is needed by the Kalman filter. From the fundamental theorem of estimation theory, we know that $\hat{\mathbf{x}}(k|k-1) = \mathbf{E}\{\mathbf{x}(k)|\mathbf{Z}(k-1)\}$ where $\mathbf{Z}(k-1) = \text{col}(\mathbf{z}(1), \mathbf{z}(2), \ldots, \mathbf{z}(k-1))$, from which it follows that

$$\hat{\mathbf{x}}(k|k-1) = \Phi(k, k-1)\hat{\mathbf{x}}(k-1|k-1) + \Psi(k, k-1)\mathbf{u}(k-1), \tag{15.26}$$

where $k = 1, 2, \ldots$. Observe that $\hat{\mathbf{x}}(k|k-1)$ depends on the filtered estimate $\hat{\mathbf{x}}(k-1|k-1)$ of the preceding state vector $\mathbf{x}(k-1)$. Therefore, Equation 15.26 cannot be used until we provide the Kalman filter.

Let $\mathbf{P}(k|k-1)$ denote the error-covariance matrix that is associated with $\hat{\mathbf{x}}(k|k-1)$, i.e.,

$$\mathbf{P}(k|k-1) = \mathbf{E}\{[\tilde{\mathbf{x}}(k|k-1) - \mathbf{m}_{\tilde{\mathbf{x}}}(k|k-1)][\tilde{\mathbf{x}}(k|k-1) - \mathbf{m}_{\tilde{\mathbf{x}}}(k|k-1)]'\},$$

where $\tilde{\mathbf{x}}(k|k-1) = \mathbf{x}(k) - \hat{\mathbf{x}}(k|k-1)$. Additionally, let $\mathbf{P}(k-1|k-1)$ denote the error-covariance matrix that is associated with $\hat{\mathbf{x}}(k-1|k-1)$, i.e.,

$$\mathbf{P}(k-1|k-1) = \mathbf{E}\{[\tilde{\mathbf{x}}(k-1|k-1) - \mathbf{m}_{\tilde{\mathbf{x}}}(k-1|k-1)][\tilde{\mathbf{x}}(k-1|k-1) - \mathbf{m}_{\tilde{\mathbf{x}}}(k-1|k-1)]'\},$$

where $\tilde{\mathbf{x}}(k-1|k-1) = \mathbf{x}(k-1) - \hat{\mathbf{x}}(k-1|k-1)$. Then

$$\mathbf{P}(k|k-1) = \boldsymbol{\Phi}(k, k-1)\mathbf{P}(k-1|k-1)\boldsymbol{\Phi}'(k, k-1) + \boldsymbol{\Gamma}(k, k-1)\mathbf{Q}(k-1)\boldsymbol{\Gamma}'(k, k-1), \tag{15.27}$$

where $k = 1, 2, \ldots$.

Observe, from Equations 15.26 and 15.27, that $\hat{\mathbf{x}}(0|0)$ and $\mathbf{P}(0|0)$ initialize the single-stage predictor and its error covariance, where $\hat{\mathbf{x}}(0|0) = \mathbf{m}_{\mathbf{x}}(0)$ and $\mathbf{P}(0|0) = \mathbf{P}(0)$.

A more general state predictor is possible, one that looks further than just one step. See Lesson 16 of [12] for its details.

The single-stage predicted estimate of $\mathbf{z}(k+1)$, $\hat{\mathbf{z}}(k+1|k)$, is given by $\hat{\mathbf{z}}(k+1|k) = \mathbf{H}(k+1)\hat{\mathbf{x}}(k+1|k)$. The error between $\mathbf{z}(k+1)$ and $\hat{\mathbf{z}}(k+1|k)$ is $\tilde{\mathbf{z}}(k+1|k)$; $\tilde{\mathbf{z}}(k+1|k)$ is called the innovations process (or prediction error process, or measurement residual process), and this process plays a very important role in mean-squared filtering and smoothing. The following representations of the innovations process $\tilde{\mathbf{z}}(k+1|k)$ are equivalent:

$$\begin{aligned}\tilde{\mathbf{z}}(k+1|k) &= \mathbf{z}(k+1) - \hat{\mathbf{z}}(k+1|k) = \mathbf{z}(k+1) - \mathbf{H}(k+1)\hat{\mathbf{x}}(k+1|k) \\ &= \mathbf{H}(k+1)\tilde{\mathbf{x}}(k+1|k) + \mathbf{v}(k+1).\end{aligned} \tag{15.28}$$

The innovations is a zero-mean Gaussian white noise sequence, with

$$\mathbf{E}\{\tilde{\mathbf{z}}(k+1|k)\tilde{\mathbf{z}}'(k+1|k)\} = \mathbf{H}(k+1)\mathbf{P}(k+1|k)\mathbf{H}'(k+1) + \mathbf{R}(k+1). \tag{15.29}$$

The paper by Kailath [7] gives an excellent historical perspective of estimation theory and includes a very good historical account of the innovations process.

15.9.2 Filtering (Kalman Filter)

The Kalman filter (KF) and its later extensions to nonlinear problems represent the most widely applied by-product of modern control theory. We begin by presenting the KF, which is the mean-squared filtered estimator of $\mathbf{x}(k+1)$, $\hat{\mathbf{x}}(k+1|k+1)$, in predictor-corrector format:

$$\hat{\mathbf{x}}(k+1|k+1) = \hat{\mathbf{x}}(k+1|k) + \mathbf{K}(k+1)\tilde{\mathbf{z}}(k+1|k) \tag{15.30}$$

for $k = 0, 1, \ldots$, where $\hat{\mathbf{x}}(0|0) = \mathbf{m}_{\mathbf{x}}(0)$ and $\tilde{\mathbf{z}}(k+1|k)$ is the innovations sequence in Equation 15.28 (use the second equality to implement the KF). Kalman gain matrix $\mathbf{K}(k+1)$ is $n \times m$, and is specified by the set of relations:

$$\mathbf{K}(k+1) = \mathbf{P}(k+1|k)\mathbf{H}'(k+1)[\mathbf{H}(k+1)\mathbf{P}(k+1|k)\mathbf{H}'(k+1) + \mathbf{R}(k+1)]^{-1}, \tag{15.31}$$

$$\mathbf{P}(k+1|k) = \boldsymbol{\Phi}(k+1,k)\mathbf{P}(k|k)\boldsymbol{\Phi}'(k+1,k) + \boldsymbol{\Gamma}(k+1,k)\mathbf{Q}(k)\boldsymbol{\Gamma}'(k+1,k), \tag{15.32}$$

and

$$\mathbf{P}(k+1|k+1) = [\mathbf{I} - \mathbf{K}(k+1)\mathbf{H}(k+1)]\mathbf{P}(k+1|k) \tag{15.33}$$

for $k = 0, 1, \ldots,$ where $\mathbf{I}$ is the $n \times n$ identity matrix, and $\mathbf{P}(0|0) = \mathbf{P}_{\mathbf{x}}(0)$.

The KF involves feedback and contains within its structure a model of the plant. The feedback nature of the KF manifests itself in two different ways: in the calculation of $\hat{\mathbf{x}}(k+1|k+1)$ and also in the calculation of the matrix of gains, $\mathbf{K}(k+1)$. Observe, also from Equations 15.26 and 15.32, that the predictor equations, which compute $\hat{\mathbf{x}}(k+1|k)$ and $\mathbf{P}(k+1|k)$, use information only from the state equation, whereas the corrector equations, which compute $\mathbf{K}(k+1)$, $\hat{\mathbf{x}}(k+1|k+1)$, and $\mathbf{P}(k+1|k+1)$, use information only from the measurement equation. Once the gain is computed, then Equation 15.30 represents a time-varying recursive digital filter. This is seen more clearly when Equations 15.26 and 15.28 are substituted into Equation 15.30. The resulting equation can be rewritten as

$$\begin{aligned}\hat{\mathbf{x}}(k+1|k+1) = {} & [\mathbf{I} - \mathbf{K}(k+1)\mathbf{H}(k+1)]\boldsymbol{\Phi}(k+1,k)\hat{\mathbf{x}}(k|k) + \mathbf{K}(k+1)\mathbf{z}(k+1) \\ & + [\mathbf{I} - \mathbf{K}(k+1)\mathbf{H}(k+1)]\boldsymbol{\Psi}(k+1,k)\mathbf{u}(k)\end{aligned} \tag{15.34}$$

for $k = 0, 1, \ldots$. This is a state equation for state vector $\hat{\mathbf{x}}$, whose time-varying plant matrix is $[\mathbf{I} - \mathbf{K}(k+1)\mathbf{H}(k+1)]\boldsymbol{\Phi}(k+1,k)$. Equation 15.34 is time-varying even if our basic state-variable model is time-invariant and stationary, because gain matrix $\mathbf{K}(k+1)$ is still time-varying in that case. It is possible, however, for $\mathbf{K}(k+1)$ to reach a limiting value (i.e., steady-state value, $\bar{\mathbf{K}}$), in which case Equation 15.34 reduces to a recursive constant coefficient filter. Equation 15.34 is in recursive filter form, in that it relates the filtered estimate of $\mathbf{x}(k+1)$, $\hat{\mathbf{x}}(k+1|k+1)$, to the filtered estimate of $\mathbf{x}(k)$, $\hat{\mathbf{x}}(k|k)$. Using substitutions similar to those in the derivation of Equation 15.34, we can also obtain the following recursive predictor form of the KF:

$$\begin{aligned}\hat{\mathbf{x}}(k+1|k) = {} & \boldsymbol{\Phi}(k+1,k)[\mathbf{I} - \mathbf{K}(k)\mathbf{H}(k)]\hat{\mathbf{x}}(k|k-1) \\ & + \boldsymbol{\Phi}(k+1,k)\mathbf{K}(k)\mathbf{z}(k) + \boldsymbol{\Psi}(k+1,k)\mathbf{u}(k).\end{aligned} \tag{15.35}$$

Observe that in Equation 15.35 the predicted estimate of $\mathbf{x}(k+1)$, $\hat{\mathbf{x}}(k+1|k)$, is related to the predicted estimate of $\mathbf{x}(k)$, $\hat{\mathbf{x}}(k|k-1)$, and that the time-varying plant matrix in Equation 15.35 is different from the time-varying plant matrix in Equation 15.34.

Embedded within the recursive KF is another set of recursive Equations 15.31 through 15.33. Because $\mathbf{P}(0|0)$ initializes these calculations, these equations must be ordered as follows: $\mathbf{P}(k|k) \to \mathbf{P}(k+1|k) \to \mathbf{K}(k+1) \to \mathbf{P}(k+1|k+1) \to$, etc. By combining these equations, it is possible to get a matrix equation for $\mathbf{P}(k+1|k)$ as a function of $\mathbf{P}(k|k-1)$ or a similar equation for $\mathbf{P}(k+1|k+1)$ as a function of $\mathbf{P}(k|k)$. These equations are nonlinear and are known as matrix Riccati equations.

A measure of recursive predictor performance is provided by matrix $\mathbf{P}(k+1|k)$, and a measure of recursive filter performance is provided by matrix $\mathbf{P}(k+1|k+1)$. These covariances can be calculated prior to any processing of real data, using Equations 15.31 through 15.33. These calculations are often referred to as a performance analysis, and $\mathbf{P}(k+1|k+1) \neq \mathbf{P}(k+1|k)$. It is indeed interesting that the KF utilizes a measure of its mean-squared error during its real-time operation.

Because of the equivalence between mean-squared, BLUE, and WLS filtered estimates of our state vector $\mathbf{x}(k)$ in the Gaussian case, we must realize that the KF equations are just a recursive solution to a system of normal equations. Other implementations of the KF that solve the normal equations using stable algorithms from numerical linear algebra (see, e.g., [2]) and involve orthogonal transformations have better numerical properties than Equations 15.30 through 15.33 (see, e.g., [4]).

A recursive BLUE of a random parameter vector θ can be obtained from the KF equations by setting $\mathbf{x}(k) = \theta, \Phi(k+1,k) = \mathbf{I}, \Gamma(k+1,k) = 0, \Psi(k+1,k) = 0$, and $\mathbf{Q}(k) = 0$. Under these conditions we see that $\mathbf{w}(k) = 0$ for all k, and $\mathbf{x}(k+1) = \mathbf{x}(k)$, which means, of course, that $\mathbf{x}(k)$ is a vector of constants, θ. The KF equations reduce to $\hat{\theta}(k+1|k+1) = \hat{\theta}(k|k) + \mathbf{K}(k+1)[\mathbf{z}(k+1) - \mathbf{H}(k+1)\hat{\theta}(k|k)], \mathbf{P}(k+1|k) = \mathbf{P}(k|k), \mathbf{K}(k+1) = \mathbf{P}(k|k)\mathbf{H}'(k+1)[\mathbf{H}(k+1)\mathbf{P}(k|k)\mathbf{H}'(k+1) + \mathbf{R}(k+1)]^{-1}$, and $\mathbf{P}(k+1|k+1) = [\mathbf{I} - \mathbf{K}(k+1)\mathbf{H}(k+1)]\mathbf{P}(k|k)$. Note that it is no longer necessary to distinguish between filtered and predicted quantities, because $\hat{\theta}(k+1|k) = \hat{\theta}(k|k)$ and $\mathbf{P}(k+1|k) = \mathbf{P}(k|k)$; hence, the notation $\hat{\theta}(k|k)$ can be simplified to $\hat{\theta}(k)$, for example, which is consistent with our earlier notation for the estimate of a vector of constant parameters.

A divergence phenomenon may occur when either the process noise or measurement noise or both are too small. In these cases the Kalman filter may lock onto wrong values for the state, but believes them to the true values; i.e., it "learns" the wrong state too well. A number of different remedies have been proposed for controlling divergence effects, including: (1) adding fictitious process noise, (2) finite-memory filtering, and (3) fading memory filtering. Fading memory filtering seems to be the most successful and popular way to control divergence effects. See [6] or [12] for discussions about these remedies.

For time-invariant and stationary systems, if $\lim_{k\to\infty}\mathbf{P}(k+1|k) = \mathbf{P}_p$ exists, then $\lim_{k\to\infty}\mathbf{K}(k) = \bar{\mathbf{K}}$ and the Kalman filter becomes a constant coefficient filter. Because $\mathbf{P}(k+1|k)$ and $\mathbf{P}(k|k)$ are intimately related, then if $\mathbf{P}_p$ exists, $\lim_{k\to\infty}\mathbf{P}(k|k) = \mathbf{P}_f$ also exists. If the basic state-variable model is time-invariant, stationary, and asymptotically stable, then (a) for any nonnegative symmetric initial condition $\mathbf{P}(0|-1)$, we have $\lim_{k\to\infty}\mathbf{P}(k+1|k) = \mathbf{P}_p$ with $\mathbf{P}_p$ independent of $\mathbf{P}(0|-1)$ and satisfying the following steady-state algebraic matrix Riccati equation,

$$\mathbf{P}_p = \Phi\mathbf{P}_p[\mathbf{I} - \mathbf{H}'(\mathbf{H}\mathbf{P}_p\mathbf{H}' + \mathbf{R})^{-1}\mathbf{H}\mathbf{P}_p]\Phi' + \Gamma\mathbf{Q}\Gamma' \tag{15.36}$$

and (b) the eigenvalues of the steady-state KF, $\lambda[\Phi - \bar{\mathbf{K}}\mathbf{H}\Phi]$, all lie within the unit circle, so that the filter is asymptotically stable, i.e., $|\lambda[\Phi - \bar{\mathbf{K}}\mathbf{H}\Phi]| < 1$. If the basic state-variable model is time-invariant and stationary, but is not necessarily asymptotically stable (e.g., it may have a pole on the unit circle), the points (a) and (b) still hold as long as the basic state-variable model is completely stabilizable and detectable (e.g., [8]). To design a steady-state KF: (1) given $(\Phi, \Gamma, \Psi, \mathbf{H}, \mathbf{Q}, \mathbf{R})$, compute $\mathbf{P}_p$, the positive definite solution of Equation 15.36; (2) compute $\bar{\mathbf{K}}$, as $\bar{\mathbf{K}} = \mathbf{P}_p\mathbf{H}'(\mathbf{H}\mathbf{P}_p\mathbf{H}' + \mathbf{R})^{-1}$; and (3) use $\bar{\mathbf{K}}$ in

$$\begin{aligned}\hat{\mathbf{x}}(k+1|k+1) &= \Phi\hat{\mathbf{x}}(k|k) + \Psi\mathbf{u}(k) + \bar{\mathbf{K}}\tilde{\mathbf{z}}(k+1|k)\\ &= (\mathbf{I} - \bar{\mathbf{K}}\mathbf{H})\Phi\hat{\mathbf{x}}(k|k) + \bar{\mathbf{K}}\mathbf{z}(k+1) + (\mathbf{I} - \bar{\mathbf{K}}\mathbf{H})\Psi\mathbf{u}(k).\end{aligned} \tag{15.37}$$

Equation 15.37 is a steady-state filter state equation. The main advantage of the steady-state filter is a drastic reduction in online computations.

15.9.3 Smoothing

Although there are three types of smoothers, the most useful one for digital signal processing is the fixed-interval smoother, hence, we only discuss it here. The fixed-interval smoother is $\hat{\mathbf{x}}(k|N)$, $k = 0, 1, \ldots, N-1$, where N is a fixed positive integer. The situation here is as follows: with an experiment completed, we have measurements available over the fixed interval $1 \le k \le N$. For each time point within this interval we wish to obtain the optimal estimate of the state vector $\mathbf{x}(k)$, which is based on all the available measurement data $\{\mathbf{z}(j), j = 1, 2, \ldots, N\}$. Fixed-interval smoothing is very useful in signal processing situations, where the processing is done after all the data are collected. It cannot be carried out online during an experiment like filtering can. Because all the available data are used, we cannot hope to do better (by other forms of smoothing) than by fixed-interval smoothing.

A mean-squared fixed-interval smoothed estimate of $\mathbf{x}(k), \hat{\mathbf{x}}(k|N)$, is

$$\hat{\mathbf{x}}(k|N) = \hat{\mathbf{x}}(k|k-1) + \mathbf{P}(k|k-1)\mathbf{r}(k|N), \tag{15.38}$$

where $k = N-1, N-2, \ldots, 1$, and $n \times 1$ vector $\mathbf{r}$ satisfies the backward-recursive equation

$$\mathbf{r}(j|N) = \Phi'_p(j+1, j)\mathbf{r}(j+1|N) + \mathbf{H}'(j)[\mathbf{H}(j)\mathbf{P}(j|j-1)\mathbf{H}'(j) + \mathbf{R}(j)]^{-1}\tilde{\mathbf{z}}(j|j-1), \tag{15.39}$$

where $\Phi_p(k+1, k) = \Phi(k+1, k)[\mathbf{I} - \mathbf{K}(k)\mathbf{H}(k)]$ and $j = N, N-1, \ldots, 1$, and $\mathbf{r}(N+1|N) = 0$. The smoothing error-covariance matrix, $\mathbf{P}(k|N)$, is

$$\mathbf{P}(k|N) = \mathbf{P}(k|k-1) - \mathbf{P}(k|k-1)\mathbf{S}(k|N)\mathbf{P}(k|k-1), \tag{15.40}$$

where $k = N-1, N-2, \ldots, 1$, and $n \times n$ matrix $\mathbf{S}(j|N)$, which is the covariance matrix of $\mathbf{r}(j|N)$, satisfies the backward-recursive equation

$$\begin{aligned}\mathbf{S}(j|N) = {} & \Phi'_p(j+1, j)\mathbf{S}(j+1|N)\Phi_p(j+1, j) \\ & + \mathbf{H}'(j)[\mathbf{H}(j)\mathbf{P}(j|j-1)\mathbf{H}'(j) + \mathbf{R}(j)]^{-1}\mathbf{H}(j),\end{aligned} \tag{15.41}$$

where $j = N, N-1, \ldots, 1$, and $\mathbf{S}(N+1|N) = 0$. Observe that fixed-interval smoothing involves a forward pass over the data, using a KF, and then a backward pass over the innovations, using Equation 15.39. The smoothing error-covariance matrix, $\mathbf{P}(k|N)$, can be precomputed; but, it is not used during the computation of $\hat{\mathbf{x}}(k|N)$. This is quite different than the active use of the filtering error-covariance matrix in the KF.

An important application for fixed-interval smoothing is deconvolution. Consider the single-input single-output system:

$$z(k) = \sum_{i=1}^{k} \mu(i)h(k-i) + \nu(k), \quad k = 1, 2, \ldots, N, \tag{15.42}$$

where

$\mu(j)$ is the system's input, which is assumed to be white, and not necessarily Gaussian
$h(j)$ is the system's impulse response

Deconvolution is the signal-processing procedure for removing the effects of $h(j)$ and $\nu(j)$ from the measurements so that we are left with an estimate of $\mu(j)$. In order to obtain a fixed-interval smoothed estimate of $\mu(j)$, we must first convert Equation 15.42 into an equivalent state-variable model. The single-channel state-variable model $\mathbf{x}(k+1) = \Phi\mathbf{x}(k) + \gamma\mu(k)$ and $z(k) = \mathbf{h}'\mathbf{x}(k) + \nu(k)$ is equivalent to Equation 15.42 when $\mathbf{x}(0) = 0$, $\mu(0) = 0$, $h(0) = 0$, and $h(l) = \mathbf{h}'\Phi^{l-i}\gamma(l = 1, 2, \ldots)$. A two-pass fixed-interval smoother for $\mu(k)$ is $\hat{\mu}(k|N) = q(k)\gamma'\mathbf{r}(k+1|N)$ where $k = N-1, N-2, \ldots, 1$. The smoothing error variance, $\sigma^2_\mu(k|N)$, is $\sigma^2_\mu(k|N) = q(k) - q(k)\gamma'\,\mathbf{S}(k+1|N)\gamma q(k)$. In these formulas $\mathbf{r}(k|N)$ are computed using Equations 15.39 and 15.41, respectively, and $\mathbf{E}\{\mu^2(k)\} = q(k)$.

15.10 Digital Wiener Filtering

The steady-state KF is a recursive digital filter with filter coefficients equal to $h_f(j)$, $j = 0, 1, \ldots$. Quite often $h_f(j) \approx 0$ for $j \geq J$, so that the transfer function of this filter, $H_f(z)$, can be truncated, i.e., $H_f(z) \approx h_f(0) + h_f(1)z^{-1} + \cdots + h_f(J)z^{-J}$. The truncated steady-state, KF can then be implemented as a finite-impulse response (FIR) digital filter. There is, however, a more direct way for designing a FIR minimum mean-squared error filter, i.e., a digital Wiener filter (WF).

Consider the scalar measurement case, in which measurement $z(k)$ is to be processed by a digital filter $F(z)$, whose coefficients, $f(0), f(1), \ldots, f(\eta)$, are obtained by minimizing the mean-squared error $I(\mathbf{f}) = \mathbf{E}\{[d(k) - y(k)]^2\} = \mathbf{E}\{e^2(k)\}$, where $y(k) = f(k) * z(k) = \sum_{i=0}^{n} f(i)z(k-i)$ and $d(k)$ is a desired filter output signal. Using calculus, it is straightforward to show that the filter coefficients that minimize $I(\mathbf{f})$ satisfy the following discrete-time Wiener–Hopf equations:

$$\sum_{i=0}^{\eta} f(i)\phi_{zz}(i-j) = \phi_{zd}(j), \quad j = 0, 1, \ldots, \eta, \tag{15.43}$$

where

$\phi_{zd}(i) = \mathbf{E}\{d(k)z(k-i)\}$
$\phi_{zz}(i-m) = \mathbf{E}\{z(k-i)z(k-m)\}$

Observe that Equation 15.43 are a system of normal equations and can be solved in many different ways, including the Levinson algorithm. The minimum mean-squared error, $I*(\mathbf{f})$, in general, approaches a nonzero limiting value which is often reached for modest values of filter length η.

To relate this FIR WF to the truncated steady-state KF, we must first assume a signal-plus-noise model for $z(k)$, because a KF uses a system model, i.e., $z(k) = s(k) + v(k) = h(k) * w(k) + v(k)$, where $h(k)$ is the IR of a linear time-invariant system and, as in our basic state-variable model, $w(k)$ and $v(k)$ are mutually uncorrelated (stationary) white noise sequences with variances q and r, respectively. We must also specify an explicit form for "desired signal" $d(k)$. We shall require that $d(k) = s(k) = h(k) * w(k)$, which means that we want the output of the FIR digital WF to be as close as possible to signal $s(k)$. The resulting Wiener–Hopf equations are

$$\sum_{i=0}^{\eta} f(i)\left[\frac{q}{r}\phi_{hh}(j-i) + \delta(j-i)\right] = \frac{q}{r}\phi_{hh}(j), \quad j = 0, 1, \ldots, \eta, \tag{15.44}$$

where $\phi_{hh}(i) = \sum_{l=0}^{\infty} h(l)h(l+i)$. The truncated steady-state KF is a FIR digital WF. For a detailed comparison of Kalman and Wiener filters, see Lesson 19 of [12].

To obtain a digital Wiener deconvolution filter, we assume that filter $F(z)$ is an infinite impulse response (IIR) filter, with coefficients $\{f(j), j = 0, \pm 1, \pm 2, \ldots\}$; $d(k) = \mu(k)$ where $\mu(k)$ is a white noise sequence and $\mu(k)$ and $v(k)$ are stationary and uncorrelated. In this case, Equation 15.43 becomes

$$\sum_{i=-\infty}^{\infty} f(i)\phi_{zz}(i-j) = \phi_{z\mu}(j) = qh(-j), \quad j = 0, \pm 1, \pm 2, \ldots. \tag{15.45}$$

This system of equations cannot be solved as a linear system of equations, because there are a doubly infinite number of them. Instead, we take the discrete-time Fourier transform of Equation 15.45, i.e., $F(\omega)\Phi_{zz}(\omega) = qH*(\omega)$, but, from Equation 15.42, $\Phi_{zz}(\omega) = q|H(\omega)|^2 + r$; hence,

$$F(\omega) = \frac{qH*(\omega)}{q|H(\omega)|^2 + r}. \tag{15.46}$$

The inverse Fourier transform of Equation 15.46, or spectral factorization, gives $\{f(j), j = 0, \pm 1, \pm 2, \ldots\}$.

15.11 Linear Prediction in DSP and Kalman Filtering

A well-studied problem in digital signal processing (e.g., [5]), is the linear prediction problem, in which the structure of the predictor is fixed ahead of time to be a linear transformation of the data. The "forward" linear prediction problem is to predict a future value of stationary discrete-time random sequence $\{y(k),\ k = 1, 2, \ldots\}$ using a set of past samples of the sequence. Let $\hat{y}(k)$ denote the predicted value of $y(k)$ that uses M past measurements, i.e.,

$$\hat{y}(k) = \sum_{i=1}^{M} a_{M,i}y(k-i). \tag{15.47}$$

The forward prediction error filter (PEF) coefficients, $a_{M,1}, \ldots, a_{M,M}$, are chosen so that either the mean-squared or least-squared forward prediction error (FPE), $f_M(k)$, is minimized, where $f_M(k) = y(k) - \hat{y}(k)$. Note that in this filter design problem the length of the filter, M, is treated as a design variable, which is why the PEF coefficients are augmented by M. Note, also, that the PEF coefficients do not depend on t_k; i.e., the PEF is a constant coefficient predictor, whereas our mean-squared state-predictor and filter are time-varying digital filters.

Predictor $\hat{y}(k)$ uses a finite window of past measurements: $y(k-1), y(k-2), \ldots, y(k-M)$. This window of measurements is different for different values of t_k. This use of measurements is quite different than our use of the measurements in state prediction, filtering, and smoothing. The latter are based on an expanding memory, whereas the former is based on a fixed memory.

Digital signal-processing specialists have invented a related type of linear prediction named backward linear prediction in which the objective is to predict a past value of a stationary discrete-time random sequence using a set of future values of the sequence. Of course, backward linear prediction is not prediction at all; it is smoothing. But the term backward linear prediction is firmly entrenched in the DSP literature. Both forward and backward PEFs have a filter architecture associated with them that is known as a tapped delay line. Remarkably, when the two filter design problems are considered simultaneously, their solutions can be shown to be coupled, and the resulting architecture is called a lattice. The lattice filter is doubly recursive in both time, k, and filter order, M. The tapped delay line is only recursive in time. Changing its filter length leads to a completely new set of filter coefficients. Adding another stage to the lattice filter does not affect the earlier filter coefficients. Consequently, the lattice filter is a very powerful architecture. No such lattice architecture is known for mean-squared state estimators.

In a second approach to the design of the FPE coefficients, the constraint that the FPE coefficients are constant is transformed into the state equations:

$$a_{M,1}(k+1) = a_{M,1}(k), a_{M,2}(k+1) = a_{M,2}(k), \ldots, a_{M,M}(k+1) = a_{M,M}(k).$$

Equation 15.47 then plays the role of the observation equation in our basic state-variable model, and is one in which the observation matrix is time-varying. The resulting mean-squared error design is then referred to as the Kalman filter solution for the PEF coefficients. Of course, we saw above that this solution is a very special case of the KF, the BLUE. In yet a third approach, the PEF coefficients are modeled as

$$\begin{aligned} a_{M,1}(k+1) &= a_{M,1}(k) + w_1(k), a_{M,2}(k+1) \\ &= a_{M,2}(k) + w_2(k), \ldots, a_{M,M}(k+1) = a_{M,M}(k) + w_M(k), \end{aligned}$$

where $w_i(k)$ are white noises with variances q_i. Equation 15.47 again plays the role of the measurement equation in our basic state-variable model and is one in which the observation matrix is time-varying. The resulting mean-squared error design is now a full-blown KF.

15.12 Iterated Least Squares

Iterated least squares (ILS) is a procedure for estimating parameters in a nonlinear model. Because it can be viewed as the basis for the extended KF, which is described in the next section, we describe ILS briefly here. To keep things simple, we describe ILS for the scalar parameter model $z(k) = f(\theta, k) + \nu(k)$ where $k = 1, 2, \ldots, N$. ILS is basically a four-step procedure:

1. Linearize $f(\theta, k)$ about a nominal value of θ, θ^*. Doing this, we obtain the perturbation measurement equation

$$\delta z(k) = F_\theta(k;\theta^*)\delta\theta + \nu(k), \quad k = 1, 2, \ldots, N \tag{15.48}$$

where $\delta z(k) = z(k) - z^*(k) = z(k) - f(\theta^*, k)$, $\delta\theta = \theta - \theta^*$, and $F_\theta(k;\theta^*) = \partial f(\theta, k)/\partial\theta|_{\theta=\theta^*}$.

2. Concatenate Equation 15.48 for the N values of k and compute $\hat{\delta\theta}_{\text{WLS}}(N)$ using Equation 15.2.
3. Solve the equation $\hat{\delta\theta}_{\text{WLS}}(N) = \hat{\theta}_{\text{WLS}}(N) - \theta^*$ for $\hat{\theta}_{\text{WLS}}(N)$, i.e., $\hat{\theta}_{\text{WLS}}(N) = \theta^* + \hat{\delta\theta}_{\text{WLS}}(N)$.
4. Replace θ^* with $\hat{\theta}_{\text{WLS}}(N)$ and return to Step 1.

Iterate through these steps until convergence occurs. Let $\hat{\theta}^i_{\text{WLS}}(N)$ and $\hat{\theta}^{i+1}_{\text{WLS}}(N)$ denote estimates of θ obtained at iterations i and $i+1$, respectively. Convergence of the ILS method occurs when $|\hat{\theta}^{i+1}_{\text{WLS}}(N) - \hat{\theta}^i_{\text{WLS}}(N)| < \varepsilon$ where ε is a prespecified small positive number.

Observe from this four-step procedure that ILS uses the estimate obtained from the linearized model to generate the nominal value of θ about which the nonlinear model is relinearized. Additionally, in each complete cycle of this procedure, we use both the nonlinear and linearized models. The nonlinear model is used to compute $z^*(k)$ and subsequently $\delta z(k)$. The notions of relinearizing about a filter output and using both the nonlinear and linearized models are also at the very heart of the extended KF.

15.13 Extended Kalman Filter

Many real-world systems are continuous-time in nature and are also nonlinear. The extended Kalman filter (EKF) is the heuristic, but very widely used, application of the KF to estimation of the state vector for the following nonlinear dynamical system:

$$\dot{\mathbf{x}}(t) = \mathbf{f}[\mathbf{x}(t), \mathbf{u}(t), i] + \mathbf{G}(t)\mathbf{w}(t) \tag{15.49}$$

$$\mathbf{z}(t) = \mathbf{h}[\mathbf{x}(t), \mathbf{u}(t), t] + \mathbf{v}(t) \quad t = t_i, \quad i = 1, 2, \ldots. \tag{15.50}$$

In this model measurement, Equation 15.50 is treated as a discrete-time equation, whereas state Equation 15.49 is treated as a continuous-time equation; $\dot{x}(t)$ is short for $d\mathbf{x}(t)/dt$; both $\mathbf{f}$ and $\mathbf{h}$ are continuous and continuously differentiable with respect to all elements of $\mathbf{x}$ and $\mathbf{u}$; $\mathbf{w}(t)$ is a zero-mean continuous-time white noise process, with $\mathbf{E}\{\mathbf{w}(t)\mathbf{w}'(\tau)\} = \mathbf{Q}(t)\delta(t-\tau)$; $\mathbf{v}(t_i)$ is a discrete-time zero-mean white noise sequence, with $\mathbf{E}\{\mathbf{v}(t_i)\mathbf{v}'(t_j)\} = \mathbf{R}(t_i)\delta_{ij}$; and $\mathbf{w}(t)$ and $\mathbf{v}(t_i)$ are mutually uncorrelated at all $t = t_i$, i.e., $\mathbf{E}\{\mathbf{w}(t)\mathbf{v}'(t_i)\} = 0$ for $t = t_i$, $i = 1, 2, \ldots$.

In order to apply the KF to Equations 15.49 and 15.50, we must linearize and discretize these equations. Linearization is done about a nominal input $\mathbf{u}^*(t)$ and nominal trajectory $\mathbf{x}^*(t)$, whose choices we discuss below. If we are given a nominal input $\mathbf{u}^*(t)$, then $\mathbf{x}^*(t)$ satisfies the nonlinear differential equation:

$$\dot{\mathbf{x}}^*(t) = \mathbf{f}[\mathbf{x}^*(t), \mathbf{u}^*(t), t] \tag{15.51}$$

and associated with $\mathbf{x}^*(t)$ and $\mathbf{u}^*(t)$ is the following nominal measurement, $\mathbf{z}^*(t)$, where

$$\mathbf{z}^*(t) = \mathbf{h}[\mathbf{x}^*(t), \mathbf{u}^*(t), t] \quad t = t_i, \quad i = 1, 2, \ldots \tag{15.52}$$

Equations 15.51 and 15.52 are referred to as the nominal system model. Letting $\delta\mathbf{x}(t) = \mathbf{x}(t) - \mathbf{x}^*(t)$, $\delta\mathbf{u}(t) = \mathbf{u}(t) - \mathbf{u}^*(t)$, and $\delta\mathbf{z}(t) = \mathbf{z}(t) - \mathbf{z}^*(t)$, we have the following linear perturbation state-variable model:

$$\delta\dot{\mathbf{x}}(t) = \mathbf{F_x}[\mathbf{x}^*(t), \mathbf{u}^*(t), t]\delta\mathbf{x}(t) + \mathbf{F_u}[\mathbf{x}^*(t), \mathbf{u}^*(t), t]\delta\mathbf{u}(t) + \mathbf{G}(t)\mathbf{w}(t) \tag{15.53}$$

$$\delta\mathbf{z}(t) = \mathbf{H_x}[\mathbf{x}^*(t), \mathbf{u}^*(t), t]\delta\mathbf{x}(t) + \mathbf{H_u}[\mathbf{x}^*(t), \mathbf{u}^*(t), t]\delta\mathbf{u}(t) + \mathbf{v}(t),$$

$$\delta\mathbf{z}(t) = \mathbf{H_x}[\mathbf{x}^*(t), \mathbf{u}^*(t), t]\delta\mathbf{x}(t) + \mathbf{H_u}[\mathbf{x}^*(t), \mathbf{u}^*(t), t]\delta\mathbf{u}(t) + \mathbf{v}(t), \quad t = t_i, \quad i = 1, 2, \ldots, \tag{15.54}$$

where $\mathbf{F_x}[\mathbf{x}^*(t), \mathbf{u}^*(t), t]$, for example, is the following time-varying Jacobian matrix:

$$\mathbf{F_x}[\mathbf{x}^*(t), \mathbf{u}^*(t), t] = \begin{pmatrix} \partial f_1/\partial x_1^* & \cdots & \partial f_1/\partial x_n^* \\ \vdots & \ddots & \vdots \\ \partial f_n/\partial x_1^* & \cdots & \partial f_n/\partial x_n^* \end{pmatrix} \tag{15.55}$$

in which $\partial f_i/\partial x_j^* = \partial f_i[\mathbf{x}(t), \mathbf{u}(t), t]/\partial x_j(t)|_{\mathbf{x}(t)=\mathbf{x}^*(t),\, \mathbf{u}(t)=\mathbf{u}^*(t)}$.

Starting with Equations 15.53 and 15.54, we obtain the following discretized perturbation state variable model:

$$\delta\mathbf{x}(k+1) = \Phi(k+1, k;^*)\delta\mathbf{x}(k) + \Psi(k+1, k;^*)\delta\mathbf{u}(k) + \mathbf{w}_\mathrm{d}(k) \tag{15.56}$$

$$\delta\mathbf{z}(k+1) = \mathbf{H_x}(k+1;^*)\delta\mathbf{x}(k+1) + \mathbf{H_u}(k+1;^*)\delta\mathbf{u}(k+1) + \mathbf{v}(k+1), \tag{15.57}$$

where the notation $\Phi(k+1, k;^*)$, for example, denotes the fact that this matrix depends on $\mathbf{x}^*(t)$ and $\mathbf{u}^*(t)$. In Equation 15.56, $\Phi(k+1, k;^*) = \Phi(t_{k+1}, t_k;^*)$, where

$$\dot{\Phi}(t, \tau;^*) = \mathbf{F_x}[\mathbf{x}^*(t), \mathbf{u}^*(t), t]\Phi(t, \tau;^*), \quad \Phi(t, t;^*) = \mathbf{I}. \tag{15.58}$$

Additionally,

$$\Psi(k+1, k;^*) = \int_{t_k}^{t_{k+1}} \Phi(t_{k+1}, \tau;^*)\mathbf{F_u}[\mathbf{x}^*(\tau), \mathbf{u}^*(\tau), \tau]d\tau \tag{15.59}$$

and $\mathbf{w}_\mathrm{d}(k)$ is a zero-mean noise sequence that is statistically equivalent to $\int_{t_k}^{t_{k+1}} \Phi(t_{k+1}, \tau)\mathbf{G}(\tau)\mathbf{w}(\tau)d\tau$; hence, its covariance matrix, $\mathbf{Q}_\mathrm{d}(k+1, k)$, is

$$\mathbf{E}\{\mathbf{w}_\mathrm{d}(k)\mathbf{w}'_\mathrm{d}(k)\} = \mathbf{Q}_\mathrm{d}(k+1, k) = \int_{t_k}^{t_{k+1}} \Phi(t_{k+1}, \tau)\mathbf{G}(\tau)\mathbf{Q}(\tau)\mathbf{G}'(\tau)\Phi'(t_{k+1}, \tau)d\tau. \tag{15.60}$$

Great simplifications of the calculations in Equations 15.58 through 15.60 occur if $\mathbf{F}(t)$, $\mathbf{B}(t)$, $\mathbf{G}(t)$, and $\mathbf{Q}(t)$ are approximately constant during the time interval $t \in [t_k, t_{k+1}]$, i.e., if $\mathbf{F}(t) \approx \mathbf{F}_k$, $\mathbf{B}(t) \approx \mathbf{B}_k$, $\mathbf{G}(t) \approx \mathbf{G}_k$, and $\mathbf{Q}(t) \approx \mathbf{Q}_k$ for $t \in [t_k, t_{k+1}]$. In this case, $\Phi(k+1, k) = e^{\mathbf{F}_k T}$, $\Psi(k+1, k) \approx \mathbf{B}_k T = \Psi(k)$, and $\mathbf{Q}_\mathrm{d}(k+1, k) \approx \mathbf{G}_k\mathbf{Q}_k\mathbf{G}'_k T = \mathbf{Q}_\mathrm{d}(k)$ where $T = t_{k+1} - t_k$.

Suppose $\mathbf{x}^*(t)$ is given *a priori*; then we can compute predicted, filtered, or smoothed estimates of $\delta\mathbf{x}(k)$ by applying all of our previously derived state estimators to the discretized perturbation state-variable model in Equations 15.56 and 15.57. We can precompute $\mathbf{x}^*(t)$ by solving the nominal differential equation (Equation 15.51). The KF associated with using a precomputed $\mathbf{x}^*(t)$ is known as a relinearized KF. A relinearized KF usually gives poor results, because it relies on an open-loop strategy for choosing $\mathbf{x}^*(t)$. When $\mathbf{x}^*(t)$ is precomputed, there is no way of forcing $\mathbf{x}^*(t)$ to remain close to $\mathbf{x}(t)$, and this must be done or else the perturbation state-variable model is invalid.

The relinearized KF is based only on the discretized perturbation state-variable model. It does not use the nonlinear nature of the original system in an active manner. The EKF relinearizes the nonlinear system about each new estimate as it becomes available, i.e., at $k = 0$, the system is linearized about $\hat{\mathbf{x}}(0|0)$. Once $\mathbf{z}(1)$ is processed by the EKF so that $\hat{\mathbf{x}}(1|1)$ is obtained, the system is linearized about $\hat{\mathbf{x}}(1|1)$. By "linearize about $\hat{\mathbf{x}}(1|1)$," we mean $\hat{\mathbf{x}}(1|1)$ is used to calculate all the quantities needed to make the transition from $\hat{\mathbf{x}}(1|1)$ to $\hat{\mathbf{x}}(2|1)$ and subsequently $\hat{\mathbf{x}}(2|2)$. The purpose of relinearizing about the filter's output is to use a better reference trajectory for $\mathbf{x}^*(t)$. Doing this, $\delta\mathbf{x} = \mathbf{x} - \hat{\mathbf{x}}$ will be held as small as possible, so that our linearization assumptions are less likely to be violated than in the case of the relinearized KF.

The EKF is available only in predictor–corrector format [6]. Its prediction equation is obtained by integrating the nominal differential equation for $\mathbf{x}^*(t)$ from t_k to t_{k+1}. Its correction equation is obtained by applying the KF to the discretized perturbation state-variable model. The equations for the EKF are

$$\hat{\mathbf{x}}(k+1|k) = \hat{\mathbf{x}}(k|k) + \int_{t_k}^{t_{k+1}} \mathbf{f}[\hat{\mathbf{x}}(t|t_k), \mathbf{u}^*(t), t]\mathrm{d}t, \tag{15.61}$$

which must be evaluated by numerical integration formulas that are initialized by $\mathbf{f}[\hat{\mathbf{x}}(t_k|t_k), \mathbf{u}^*(t_k), t_k]$,

$$\begin{aligned}\hat{\mathbf{x}}(k+1|k+1) = {} & \hat{\mathbf{x}}(k+1|k) + \mathbf{K}(k+1;^*) \\ & \{\mathbf{z}(k+1) - \mathbf{h}[\hat{\mathbf{x}}(k+1|k), \mathbf{u}^*(k+1), k+1] \\ & - \mathbf{H}_{\mathbf{u}}(k+1;^*)\delta\mathbf{u}(k+1)\}\end{aligned} \tag{15.62}$$

$$\begin{aligned}\mathbf{K}(k+1;^*) = {} & \mathbf{P}(k+1|k;^*)\mathbf{H}'_{\mathbf{x}}(k+1;^*) \\ & [\mathbf{H}_{\mathbf{x}}(k+1;^*)\mathbf{P}(k+1|k;^*)\mathbf{H}'_{\mathbf{x}}(k+1;^*) + \mathbf{R}(k+1)]^{-1}\end{aligned} \tag{15.63}$$

$$\mathbf{P}(k+1|k;^*) = \mathbf{\Phi}(k+1, k;^*)\mathbf{P}(k|k;^*)\mathbf{\Phi}'(k+1, k;^*) + \mathbf{Q}_{\mathrm{d}}(k+1, k;^*) \tag{15.64}$$

$$\mathbf{P}(k+1|k+1;^*) = [\mathbf{I} - \mathbf{K}(k+1;^*)\mathbf{H}_{\mathbf{x}}(k+1;^*)]\mathbf{P}(k+1|k;^*). \tag{15.65}$$

In these equations, $\mathbf{K}(k+1;^*)$, $\mathbf{P}(k+1|k;^*)$, and $\mathbf{P}(k+1|k+1;^*)$ depend on the nominal $\mathbf{x}^*(t)$ that results from prediction, $\hat{\mathbf{x}}(k+1|k)$. For a complete flowchart of the EKF, see Figure 24.2 in [12].

The EKF is very widely used; however, it does not provide an optimal estimate of $\mathbf{x}(k)$. The optimal mean-squared estimate of $\mathbf{x}(k)$ is still $\mathbf{E}\{\mathbf{x}(k)|\mathbf{Z}(k)\}$, regardless of the linear or nonlinear nature of the system's model. The EKF is a first-order approximation of $\mathbf{E}\{\mathbf{x}(k)|\mathbf{Z}(k)\}$ that sometimes works quite well, but cannot be guaranteed to always work well. No convergence results are known for the EKF; hence, the EKF must be viewed as an ad hoc filter. Alternatives to the EKF, which are based on nonlinear filtering, are quite complicated and are rarely used.

The EKF is designed to work well as long as $\delta\mathbf{x}(k)$ is "small." The iterated EKF [6] is designed to keep $\delta\mathbf{x}(k)$ as small as possible. The iterated EKF differs from the EKF in that it iterates the correction equation L times until $\|\hat{\mathbf{x}}_L(k+1|k+1) - \hat{\mathbf{x}}_{L-1}(k+1|k+1)\| \leq \varepsilon$. Corrector 1 computes $\mathbf{K}(k+1;^*)$, $\mathbf{P}(k+1|k;^*)$, and $\mathbf{P}(k+1|k+1;^*)$ using $\mathbf{x}^* = \hat{\mathbf{x}}(k+1|k)$; corrector 2 computes these quantities using $\mathbf{x}^* = \hat{\mathbf{x}}_1(k+1|k+1)$; corrector 3 computes these quantities using $\mathbf{x}^* = \hat{\mathbf{x}}_2(k+1|k+1)$; etc. Often, just adding one additional corrector (i.e., $L = 2$) leads to substantially better results for $\hat{\mathbf{x}}(k+1|k+1)$ than are obtained using the EKF.

Acknowledgment

The author gratefully acknowledges Prentice-Hall for extending permission to include summaries of materials that appeared originally in *Lessons in Estimation Theory for Signal Processing, Communications, and Control* [12].

Further Information

Recent articles about estimation theory appear in many journals, including the following engineering journals: *AIAA J.*, *Automatica*, *IEEE Transactions on Aerospace and Electronic Systems*, *IEEE Transactions on Automatic Control*, *IEEE Transactions on Information Theory*, *IEEE Transactions on Signal Processing*, *International Journal of Adaptive Control and Signal Processing*, and *International Journal of Control* and *Signal Processing*. Nonengineering journals that also publish articles about estimation theory include *Annals of the Institute of Statistical Mathematics*, *Annals of Mathematical Statistics*, *Annals of Statistics*, *Bulletin of the International Statistical Institute*, and *Sankhya*.

Some engineering conferences that continue to have sessions devoted to aspects of estimation theory include American Automatic Control Conference, IEEE Conference on Decision and Control, IEEE International Conference on Acoustics, Speech and Signal Processing, IFAC International Congress, and some IFAC Workshops.

MATLAB toolboxes that implement some of the algorithms described in this chapter are Control Systems, Optimization, and System Identification. See [12], at the end of each lesson, for descriptions of which M-files in these toolboxes are appropriate. Additionally, [12] lists six estimation algorithm M-files that do not appear in any MathWorks toolboxes or in MATLAB. They are **rwlse**, a recursive least-squares algorithm; **kf**, a recursive KF; **kp**, a recursive Kalman predictor; **sof**, a recursive suboptimal filter in which the gain matrix must be prespecified; **sop**, a recursive suboptimal predictor in which the gain matrix must be prespecified; and **fis**, a fixed-interval smoother.

References

1. Anderson, B.D.O. and Moore, J.B., *Optimal Filtering*, Prentice-Hall, Englewood Cliffs, NJ, 1979.
2. Bierman, G.J., *Factorization Methods for Discrete Sequential Estimation*, Academic Press, New York, 1977.
3. Golub, G.H. and Van Loan, C.F., *Matrix Computations*, 2nd ed., Johns Hopkins University Press, Baltimore, MD, 1989.
4. Grewal, M.S. and Andrews, A.P., *Kalman Filtering: Theory and Practice*, Prentice-Hall, Englewood Cliffs, NJ, 1993.
5. Haykin, S., *Adaptive Filter Theory*, 2nd ed., Prentice-Hall, Englewood Cliffs, NJ, 1991.
6. Jazwinski, A.H., *Stochastic Processes and Filtering Theory*, Academic Press, New York, 1970.
7. Kailath, T.K., A view of three decades of filtering theory, *IEEE Trans. Info. Theory*, IT-20: 146–181, 1974.
8. Kailath, T.K., *Linear Systems*, Prentice-Hall, Englewood Cliffs, NJ, 1980.
9. Kalman, R.E., A new approach to linear filtering and prediction problems, *Trans. ASME J. Basic Eng. Ser. D*, 82: 35–46, 1960.
10. Kashyap, R.L. and Rao, A.R., *Dynamic Stochastic Models from Empirical Data*, Academic Press, New York, 1976.
11. Ljung, L., *System Identification: Theory for the User*, Prentice-Hall, Englewood Cliffs, NJ, 1987.
12. Mendel, J.M., *Lessons in Estimation Theory for Signal Processing, Communications, and Control*, Prentice-Hall PTR, Englewood Cliffs, NJ, 1995.

16

Validation, Testing, and Noise Modeling

Jitendra K. Tugnait
Auburn University

16.1 Introduction

Linear parametric models of stationary random processes, whether signal or noise, have been found to be useful in a wide variety of signal processing tasks such as signal detection, estimation, filtering, and classification, and in a wide variety of applications such as digital communications, automatic control, radar and sonar, and other engineering disciplines and sciences. A general representation of a linear discrete-time stationary signal $x(t)$ is given by

$$x(t) = \sum_{i=0}^{\infty} h(i)\varepsilon(t-i), \tag{16.1}$$

where

$\{\varepsilon(t)\}$ is a zero-mean, i.i.d. (independent and identically distributed) random sequence with finite variance

$\{h(i),\ i \geq 0\}$ is the impulse response of the linear system such that $\sum_{i=-\infty}^{\infty} h^2(i) < \infty$

Much effort has been expended on developing approaches to linear model fitting given a single measurement record of the signal (or noisy signal) [1,2]. Parsimonious parametric models such as AR (autoregressive), MA (moving average), ARMA or state-space, as opposed to impulse response modeling, have been popular together with the assumption of Gaussianity of the data.

Define

$$H(q) = \sum_{i=0}^{\infty} h(i)q^{-i}, \tag{16.2}$$

where q^{-1} is the backward shift operator (i.e., $q^{-1}x(t) = x(t-1)$, etc.). If q is replaced with the complex variable z, then $H(z)$ is the Z-transform of $\{h(i)\}$, i.e., it is the system transfer function. Using Equation 16.2, Equation 16.1 may be rewritten as

$$x(t) = H(q)\varepsilon(t). \tag{16.3}$$

Fitting linear models to the measurement record requires estimation of $H(q)$, or equivalently of $\{h(i)\}$ (without observing $\{\varepsilon(t)\}$). Typically $H(q)$ is parameterized by a finite number of parameters, say by the parameter vector $\theta^{(M)}$ of dimension M. For instance, an AR model representation of order M means that

$$H_{\mathrm{AR}}(q;\theta^{(M)}) = \frac{1}{1+\sum_{i=1}^{M} a_i q^{-i}}, \quad \theta^{(M)} = (a_1, a_2, \ldots, a_M)^{\mathrm{T}}. \tag{16.4}$$

This reduces the number of estimated parameters from a "large" number to M.

In this section several aspects of fitting models such as Equation 16.1 through 16.3 to the given measurement record are considered. These aspects are (see also Figure 16.1)

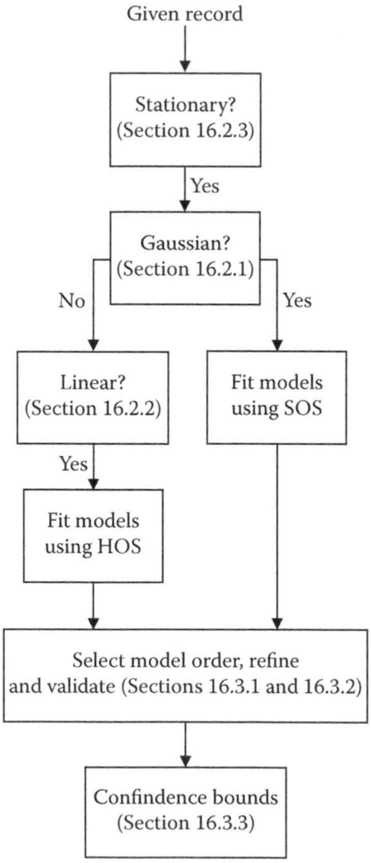

FIGURE 16.1 Section outline (SOS, second-order statistics and HOS, higher order statistics).

- Is the model of the type (Equation 16.1) appropriate to the given record? This requires testing for linearity and stationarity of the data.
- Linear Gaussian models have long been dominant both for signals as well as for noise processes. Assumption of Gaussianity allows implementation of statistically efficient parameter estimators such as maximum likelihood estimators. A Gaussian process is completely characterized by its second-order statistics (autocorrelation function or, equivalently, its power spectral density). Since the power spectrum of $\{x(t)\}$ of Equation 16.1 is given by

$$S_{xx}(\omega) = \sigma_\varepsilon^2 |H(e^{j\omega})|^2, \quad \sigma_\varepsilon^2 = E\{\varepsilon^2(t)\}. \tag{16.5}$$

- One cannot determine the phase of $H(e^{j\omega})$ independent of $|H(e^{j\omega})|$. Determination of the true phase characteristic is crucial in several applications such as blind equalization of digital communications channels. Use of higher order statistics allows one to uniquely identify non-minimum-phase parametric models. Higher order cumulants of Gaussian processes vanish, hence, if the data are stationary Gaussian, a minimum-phase (or maximum-phase) model is the "best" that one can estimate. Therefore, another aspect considered in this section is testing for non-Gaussianity of the given record.
- If the data are Gaussian, one may fit models based solely upon the second-order statistics of the data—else use of higher order statistics in addition to or in lieu of the second-order statistics is indicated, particularly if the phase of the linear system is crucial. In either case, one typically fits a model $H(q;\theta^{(M)})$ by estimating the M unknown parameters through optimization of some cost function. In practice (the model order), M is unknown and its choice has a significant impact on the quality of the fitted model. In this section another aspect of the model-fitting problem considered is that of order selection.
- Having fitted a model $H(q;\theta^{(M)})$, one would also like to know how good are the estimated parameters? Typically this is expressed in terms of error bounds or confidence intervals on the fitted parameters and on the corresponding model transfer function.
- Having fitted a model, a final step is that of model falsification. Is the fitted model an appropriate representation of the underlying system? This is referred to variously as model validation, model verification, or model diagnostics.
- Finally, various models of univariate noise pdf (probability density function) are discussed to complete the discussion of model fitting.

16.2 Gaussianity, Linearity, and Stationarity Tests

Given a zero-mean, stationary random sequence $\{x(t)\}$, its third-order cumulant function $C_{xxx}(i,k)$ is given by [12]

$$C_{xxx}(i,k) := E\{x(t+i)x(t+k)x(t)\}. \tag{16.6}$$

Its bispectrum $B_{xxx}(\omega_1, \omega_2)$ is defined as [12]

$$B_{xxx}(\omega_1, \omega_2) = \sum_{i=-\infty}^{\infty} \sum_{k=-\infty}^{\infty} C_{xxx}(i,k) e^{-j(\omega_1 i + \omega_2 k)}. \tag{16.7}$$

Similarly, its fourth-order cumulant function $C_{xxxx}(i,k,l)$ is given by [12]

$$\begin{aligned} C_{xxxx}(i,k,l){:} = {} & E\{x(t)x(t+i)x(t+k)x(t+l)\} \\ & - E\{x(t)x(t+i)\}E\{x(t+k)x(t+l)\} \\ & - E\{x(t)x(t+i)\}E\{x(t+l)x(t+i)\} \\ & - E\{x(t)x(t+l)\}E\{x(t+k)x(t+i)\}. \end{aligned} \tag{16.8}$$

Its trispectrum is defined as [12]

$$T_{xxxx}(\omega_1,\omega_2,\omega_3){:} = \sum_{i=-\infty}^{\infty}\sum_{k=-\infty}^{\infty}\sum_{l=-\infty}^{\infty} C_{xxxx}(i,k,l)\mathrm{e}^{-j(\omega_1 i+\omega_2 k+\omega_3 l)}. \tag{16.9}$$

If $\{x(t)\}$ obeys Equation 16.1, then [12]

$$B_{xxx}(\omega_1,\omega_2) = \gamma_{3\varepsilon}H(\mathrm{e}^{j\omega_1})H(\mathrm{e}^{j\omega_2})H^*\left[\mathrm{e}^{j(\omega_1+\omega_2)}\right] \tag{16.10}$$

and

$$T_{xxxx}(\omega_1,\omega_2,\omega_3) = \gamma_{4\varepsilon}H(\mathrm{e}^{j\omega_1})H(\mathrm{e}^{j\omega_2})H(\mathrm{e}^{j\omega_3})H^*\left[\mathrm{e}^{j(\omega_1+\omega_2+\omega_3)}\right], \tag{16.11}$$

where

$$\gamma_{3\varepsilon} = C_{\varepsilon\varepsilon\varepsilon}(0,0,0) \quad \text{and} \quad \gamma_{4\varepsilon} = C_{\varepsilon\varepsilon\varepsilon\varepsilon}(0,0,0,0). \tag{16.12}$$

For Gaussian processes, $B_{xxx}(\omega_1, \omega_2) \equiv 0$ and $T_{xxxx}(\omega_1,\omega_2,\omega_3) \equiv 0$; equivalently, $C_{xxx}(i,k) \equiv 0$ and $C_{xxxx}(i,k,l) \equiv 0$. This forms a basis for testing Gaussianity of a given measurement record. When $\{x(t)\}$ is linear (i.e., it obeys Equation 16.1), then using Equations 16.5 and 16.10,

$$\frac{|B_{xxx}(\omega_1,\omega_2)|^2}{S_{xx}(\omega_1)S_{xx}(\omega_1)S_{xx}(\omega_1+\omega_2)} = \frac{\gamma_{3\varepsilon}}{\sigma_\varepsilon^6} = \text{constant} \quad \forall\omega_1,\omega_2, \tag{16.13}$$

and using Equations 16.5 and 16.11,

$$\frac{|T_{xxxx}(\omega_1,\omega_2,\omega_3)|^2}{S_{xx}(\omega_1)S_{xx}(\omega_1)S_{xx}(\omega_3)S_{xx}(\omega_1+\omega_2+\omega_3)} = \frac{\gamma_{4\varepsilon}}{\sigma_\varepsilon^8} = \text{constant} \quad \forall\omega_1,\omega_2,\omega_3. \tag{16.14}$$

The above two relations form a basis for testing linearity of a given measurement record. How the tests are implemented depends upon the statistics of the estimators of the higher order cumulant spectra as well as that of the power spectra of the given record.

16.2.1 Gaussianity Tests

Suppose that the given zero-mean measurement record is of length N denoted by $\{x(t),\ t=1,2,\ldots,N\}$. Suppose that the given sample sequence of length N is divided into K nonoverlapping segments each of size N_B samples so that $N=KN_\mathrm{B}$. Let $X^{(i)}(\omega)$ denote the discrete Fourier transform of the ith block $\{x[t+(i-1)N_\mathrm{B}],\ 1\le t\le N_\mathrm{B}\}$ $(i=1,2,\ldots,K)$ given by

$$X^{(i)}(\omega_m) = \sum_{l=0}^{N_\mathrm{B}-1} x[l+1+(i-1)N_\mathrm{B}]\mathrm{e}^{-j\omega_m l}, \tag{16.15}$$

where

$$\omega_m = \frac{2\pi}{N_B} m, \quad m = 0, 1, \ldots, N_B - 1. \tag{16.16}$$

Denote the estimate of the bispectrum $B_{xxx}(\omega_m, \omega_n)$ at bifrequency $\left(\omega_m = \frac{2\pi}{N_B} m, \omega_n = \frac{2\pi}{N_B} n\right)$ as $\hat{B}_{xxx}(m, n)$, given by averaging over K blocks

$$\hat{B}_{xxx}(m, n) = \frac{1}{K} \sum_{i=1}^{K} \left\{ \frac{1}{N_B} X^{(i)}(\omega_m) X^{(i)}(\omega_n) [X^{(i)}(\omega_m + \omega_n)]^* \right\}, \tag{16.17}$$

where X^* denotes the complex conjugate of X. A principal domain of $\hat{B}_{xxx}(m, n)$ is the triangular grid

$$D = \left\{ (m, n) | 0 \le m \le \frac{N_B}{2}, 0 \le n \le m, \; 2m + n \le N_B \right\}. \tag{16.18}$$

Values of $\hat{B}_{xxx}(m, n)$ outside D can be inferred from that in D.

Select a coarse frequency grid $(\overline{m}, \overline{n})$ in the principal domain D as follows. Let d denote the distance between two adjacent coarse frequency pairs such that $d = 2r + 1$ with r a positive integer. Set $n_0 = 2 + r$ and $\overline{n} = n_0, n_0 + d, \ldots, n_0 + (L_{\overline{n}} - 1)d$ where $L_{\bar{n}} = \left\lfloor \frac{\lfloor \frac{N_B}{3} \rfloor - 1}{d} \right\rfloor$. For a given $\overline{n}$, set $m_{0,\bar{n}} = \left\lfloor \frac{N_B - \bar{n}}{2} \right\rfloor - r$, $\bar{m} = \bar{m}_{\bar{n}} = m_{0,\bar{n}}, m_{0,\bar{n}} - d, \ldots, m_{0,\bar{n}} - (L_{\bar{m},\bar{n}} - 1)d$ where $L_{\bar{m},\bar{n}} = \left\lfloor \frac{m_{0,\bar{n}} - (\bar{n} + r + 1)}{d} \right\rfloor + 1$. Let P denote the number of points on the coarse frequency grid as defined above so that $P = \sum_{\bar{n}=1}^{L_{\bar{n}}} L_{\bar{m},\bar{n}}$. Suppose that $(\overline{m}, \overline{n})$ is a coarse point, then select a fine grid $(\overline{m}, n_{\overline{n}k})$ and $(m_{\overline{m}i}, n_{\overline{n}k})$ consisting of

$$m_{\bar{m}i} = \bar{m} + i, \quad |i| \le r, \quad n_{\bar{n}k} = \bar{n} + k, \quad |k| \le r, \tag{16.19}$$

for some integer $r > 0$ such that $(2r + 1)^2 > P$; see also Figure 16.2. Order the $L(= (2r + 1)^2)$ estimates $\hat{B}_{xxx}(m_{\overline{m}i}, n_{\overline{n}k})$ on the fine grid around the bifrequency pair $(\overline{m}, \overline{n})$ into an L-vector, which after relabeling, may be denoted as ν_{ml}, $l = 1, 2, \ldots, L$, $m = 1, 2, \ldots, P$, where m indexes the coarse grid and l indexes the fine grid. Define P-vectors

$$\Psi_i = (\nu_{1i}, \nu_{2i}, \ldots, \nu_{Pi})^T \quad (i = 1, 2, \ldots, L). \tag{16.20}$$

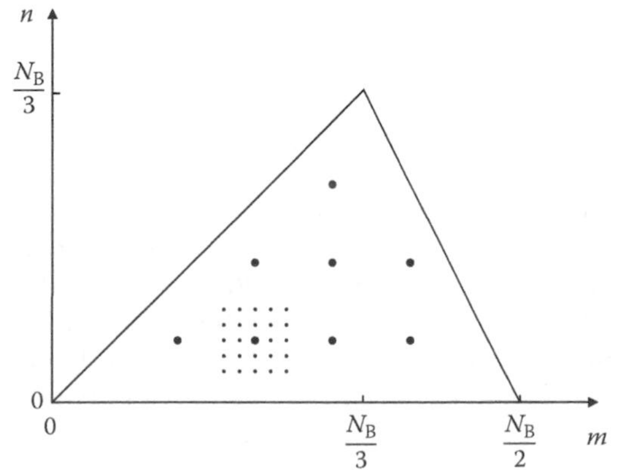

FIGURE 16.2 Coarse and fine grids in the principal domain.

Consider the estimates

$$\overline{M} = \frac{1}{L}\sum_{i=1}^{L}\overline{\Psi}_i \quad \text{and} \quad \overline{\Sigma} = \frac{1}{L}\sum_{i=1}^{L}(\overline{\Psi}_i - \overline{M})(\overline{\Psi}_i - \overline{M})^{\mathrm{H}}. \tag{16.21}$$

Define

$$F_{\mathrm{G}} = \frac{2(L-P)}{2P}\overline{M}^{\mathrm{H}}\overline{\Sigma}^{-1}\overline{M}. \tag{16.22}$$

If $\{x(t)\}$ is Gaussian, then F_{G} is distributed as a central F (Fisher) with $(2P, 2(L-P))$ degrees of freedom. A statistical test for testing Gaussianity of $\{x(t)\}$ is to declare it to be a non-Gaussian sequence if $F_{\mathrm{G}} > T_\alpha$ where T_α is selected to achieve a fixed probability of false alarm $\alpha(=\Pr\{F_{\mathrm{G}} > T_\alpha\}$ with F_{G} distributed as a central F with $(2P, 2(L-P))$ degrees of freedom). If $F_{\mathrm{G}} \leq T_\alpha$, then either $\{x(t)\}$ is Gaussian or it has zero bispectrum.

The above test is patterned after [3]. It treats the bispectral estimates on the "fine" bifrequency grid as a "data set" from a multivariable Gaussian distribution with unknown covariance matrix. Hinich [4] has simplified the test of [3] by using the known asymptotic expression for the covariance matrix involved, and his test is based upon χ^2 distributions. Notice that $F_{\mathrm{G}} \leq T_\alpha$ does not necessarily imply that $\{x(t)\}$ is Gaussian; it may result from that fact that $\{x(t)\}$ is non-Gaussian with zero bispectrum. Therefore, a next logical step would be to test for vanishing trispectrum of the record. This has been done in [14] using the approach of [4]; extensions of [3] are too complicated. Computationally simpler tests using "integrated polyspectrum" of the data have been proposed in [6]. The integrated polyspectrum (bispectrum or trispectrum) is computed as cross-power spectrum and it is zero for Gaussian processes. Alternatively, one may test if $C_{xxx}(i, k) \equiv 0$ and $C_{xxxx}(i, k, l) \equiv 0$. This has been done in [8].

Other tests that do not rely on higher order cumulant spectra of the record may be found in [13].

16.2.2 Linearity Tests

Denote the estimate of the power spectral density $S_{xx}(\omega_m)$ of $\{x(t)\}$ at frequency $\omega_m = \frac{2\pi}{N_{\mathrm{B}}}m$ as $\hat{S}_{xx}(m)$ given by

$$\hat{S}_{xx}(m) = \frac{1}{K}\sum_{i=1}^{K}\left\{\frac{1}{N_{\mathrm{B}}}X^{(i)}(\omega_m)\left[X^{(i)}(\omega_m)\right]^*\right\}. \tag{16.23}$$

Consider

$$\hat{\gamma}_x(m,n) = \frac{|\hat{B}_{xxx}(m,n)|^2}{\hat{S}_{xx}(m)\hat{S}_{xx}(n)\hat{S}_{xx}(m+n)}. \tag{16.24}$$

It turns out that $\hat{\gamma}_x(m,n)$ is a consistent estimator of the left side of Equation 16.13, and it is asymptotically distributed as a Gaussian random variable, independent at distinct bifrequencies in the interior of D. These properties have been used by Subba Rao and Gabr [3] to design a test of linearity.

Construct a coarse grid and a fine grid of bifrequencies in D as before. Order the L estimates $\hat{\gamma}_x(m_{\overline{m}i}, n_{\overline{n}k})$ on the fine grid around the bifrequency pair $(\overline{m}, \overline{n})$ into an L -vector, which after relabeling, may be denoted as β_{ml}, $l = 1, 2, \ldots, L$, $m = 1, 2, \ldots, P$, where m indexes the coarse grid and l indexes the fine grid. Define P-vectors

$$\Psi_i = (\beta_{1i}, \beta_{2i}, \ldots, \beta_{Pi})^{\mathrm{T}}, \quad (i = 1, 2, \ldots, L). \tag{16.25}$$

Consider the estimates

$$M = \frac{1}{L}\sum_{i=1}^{L}\Psi_i \quad \text{and} \quad \sum = \frac{1}{L}\sum_{i=1}^{L}(\Psi_i - M)(\Psi_i - M)^{\mathrm{T}}. \tag{16.26}$$

Define a $(P-1)\times P$ matrix B whose ij th element B_{ij} is given by $B_{ij}=1$ if $i=j;=-1$ if $j=i+1;=0$ otherwise. Define

$$F_{\mathrm{L}} = \frac{L-P+1}{P-1}(BM)^{\mathrm{T}}\left(B\sum \mathbf{B}^{\mathrm{T}}\right)^{-1}BM. \tag{16.27}$$

If $\{x(t)\}$ is linear, then F_{L} is distributed as a central F with $(P-1,\ L-P+1)$ degrees of freedom. A statistical test for testing linearity of $\{x(t)\}$ is to declare it to be a nonlinear sequence if $F_{\mathrm{L}} > T_\alpha$ where T_α is selected to achieve a fixed probability of false alarm $\alpha(=\Pr\{F_{\mathrm{L}} > T_\alpha\}$ with F_{L} distributed as a central F with $(P-1,\ L-P+1)$ degrees of freedom). If $F_{\mathrm{L}} \leq T_\alpha$, then either $\{x(t)\}$ is linear or it has zero bispectrum.

The above test is patterned after [3]. Hinich [4] has "simplified" the test of [3]. Notice that $F_{\mathrm{L}} \leq T_\alpha$ does not necessarily imply that $\{x(t)\}$ is nonlinear; it may result from that fact that $\{x(t)\}$ is non-Gaussian with zero bispectrum. Therefore, a next logical step would be to test if Equation 16.14 holds true. This has been done in [14] using the approach of [4]; extensions of [3] are too complicated. The approaches of [3] and [4] will fail if the data are noisy. A modification to [3] is presented in [7] when additive Gaussian noise is present. Finally, other tests that do not rely on higher order cumulant spectra of the record may be found in [13].

16.2.3 Stationarity Tests

Various methods exist for testing whether a given measurement record may be regarded as a sample sequence of a stationary random sequence. A crude yet effective way to test for stationarity is to divide the record into several (at least two) nonoverlapping segments and then test for equivalency (or compatibility) of certain statistical properties (mean, mean-square value, power spectrum, etc.) computed from these segments. More sophisticated tests that do not require *a priori* segmentation of the record are also available.

Consider a record of length N divided into two nonoverlapping segments each of length $N/2$. Let $KN_{\mathrm{B}} = N/2$ and use the estimators such as Equation 16.23 to obtain the estimator $\hat{S}_{xx}^{(l)}(m)$ of the power spectrum $S_{xx}^{(l)}(\omega_m)$ of the l–th segment $(l=1,\ 2)$, where ω_m is given by Equation 16.16. Consider the test statistic

$$Y = \frac{2}{N_{\mathrm{B}}-2}\sqrt{\frac{K}{2}}\sum_{m=1}^{\frac{N_{\mathrm{B}}}{2}-1}\left[\ln \hat{S}_{xx}^{(1)}(m) - \ln \hat{S}_{xx}^{(2)}(m)\right]. \tag{16.28}$$

Then, asymptotically Y is distributed as zero-mean, unit variance Gaussian if $\{x(t)\}$ is stationary. Therefore, if $|Y| > T_\alpha$, then $\{x(t)\}$ is declared to be nonstationary where the threshold T_α is chosen to achieve a false-alarm probability of $\alpha(=\Pr\{|Y| > T_\alpha\}$ with Y distributed as zero-mean, unit variance Gaussian). If $|Y| \leq T_\alpha$, then $\{x(t)\}$ is declared to be stationary. Notice that similar tests based upon higher order cumulant spectra can also be devised.

The above test is patterned after [10]. More sophisticated tests involving two model comparisons as above but without prior segmentation of the record are available in [11] and references therein. A test utilizing evolutionary power spectrum may be found in [9].

16.3 Order Selection, Model Validation, and Confidence Intervals

As noted earlier, one typically fits a model $H(q; \theta^{(M)})$ to the given data by estimating the M unknown parameters through optimization of some cost function. A fundamental difficulty here is the choice of M. There are two basic philosophical approaches to this problem: one consists of an iterative process of model fitting and diagnostic checking (model validation), and the other utilizes a more "objective" approach of optimizing a cost w.r.t. M (in addition to $\theta^{(M)}$).

16.3.1 Order Selection

Let $f_{\theta^{(M)}}(\mathbf{X})$ denote the pdf of $\mathbf{X} = [x(1), x(2), \ldots, x(N)]^T$ parameterized by the parameter vector $\theta^{(M)}$ of dimension M. A popular approach to model order selection in the context of linear Gaussian models is to compute the Akaike information criterion (AIC)

$$\mathrm{AIC}(M) = -2 \ln f_{\hat{\theta}^{(M)}}(\mathbf{X}) + 2M, \tag{16.29}$$

where $\hat{\theta}^{(M)}$ maximizes $f_{\theta^{(M)}}(\mathbf{X})$ given the measurement record $\mathbf{X}$. Let $\overline{M}$ denote an upper bound on the true model order. Then the minimum AIC estimate, the selected model order, is given by the minimizer of AIC(M) over $M = 1, 2, \ldots, \overline{M}$. Clearly one needs to solve the problem of maximization of $\ln f_{\theta^{(M)}}(X)$ w.r.t. $\theta^{(M)}$ for each value of $M = 1, 2, \ldots, \overline{M}$. The second term on the right side of Equation 16.29 penalizes overparametrization.

Rissanen's minimum description length (MDL) criterion is given by

$$\mathrm{MDL}(M) = -2 \ln f_{\hat{\theta}^{(M)}}(X) + M \ln N. \tag{16.30}$$

It is known that if $\{x(t)\}$ is a Gaussian AR model, then AIC is an inconsistent estimator of the model order whereas MDL is consistent, i.e., MDL picks the correct model order with probability one as the data length tends to infinity, whereas there is a nonzero probability that AIC will not. Several other variations of these criteria exist [15].

Although the derivation of these order selection criteria is based upon Gaussian distribution, they have frequently been used for non-Gaussian processes with success provided attention is confined to the use of second-order statistics of the data. They may fail if one fits models using higher order statistics.

16.3.2 Model Validation

Model validation involves testing to see if the fitted model is an appropriate representation of the underlying (true) system. It involves devising appropriate statistical tools to test the validity of the assumptions made in obtaining the fitted model. It is also known as model falsification, model verification, or diagnostic checking. It can also be used as a tool for model order selection. It is an essential part of any model fitting methodology.

Suppose that $\{x(t)\}$ obeys Equation 16.1. Suppose that the fitted model corresponding to the estimated parameter $\hat{\theta}^{(M)}$ is $H(q; \hat{\theta}^{(M)})$. Assuming that the true model $H(q)$ is invertible, in the ideal case one should get $\varepsilon(t) = H^{-1}(q)x(t)$ where $\{\varepsilon(t)\}$ is zero-mean, i.i.d. (or at least white when using second-order statistics). Hence, if the fitted model $H(q; \hat{\theta}^{(M)})$ is a valid description of the underlying true system, one expects $\varepsilon'(t) = H^{-1}(q; \hat{\theta}^{(M)})x(t)$ to be zero-mean, i.i.d. One of the diagnostic checks then is to test for whiteness or independence of the inverse filtered data (or the residuals or linear innovations, in case second-order statistics are used). If the fitted model is unable to "adequately" capture the underlying true system, one expects $\{\varepsilon'(t)\}$ to deviate from i.i.d. distribution. This is one of the most widely used and useful diagnostic checks for model validation.

A test for second-order whiteness of $\{\varepsilon'(t)\}$ is as follows [15]. Construct the estimates of the covariance function as

$$\hat{r}_{\varepsilon}(\tau) = N^{-1}\sum_{t=1}^{N-\tau} \varepsilon'(t+\tau)\varepsilon'(t) \quad (\tau \geq 0). \tag{16.31}$$

Consider the test statistic

$$R = \frac{N}{\hat{r}_{\varepsilon}^{2}(0)} \sum_{i=1}^{m} \hat{r}_{\varepsilon}^{2}(i), \tag{16.32}$$

where m is some *a priori* choice of the maximum lag for whiteness testing. If $\{\varepsilon'(t)\}$ is zero-mean white, then R is distributed as $\chi^2(m)$ (χ^2 with m degrees of freedom). A statistical test for testing whiteness of $\{\varepsilon'(t)\}$ is to declare it to be a nonwhite sequence (hence invalidate the model) if $R > T_\alpha$ where T_α is selected to achieve a fixed probability of false alarm $\alpha(=\Pr\{R > T_\alpha\}$ with R distributed as $\chi^2(m)$). If $R \leq T_\alpha$, then $\{\varepsilon'(t)\}$ is second-order white, hence the model is validated.

The above procedure only tests for second-order whiteness. In order to test for higher order whiteness, one needs to examine either the higher order cumulant functions or the higher order cumulant spectra (or the integrated polyspectra) of the inverse-filtered data. A statistical test using bispectrum is available in [5]. It is particularly useful if the model fitting is carried out using higher order statistics. If $\{\varepsilon'(t)\}$ is third-order white, then its bispectrum is a constant for all bifrequencies. Let $\hat{B}_{\varepsilon'\varepsilon'\varepsilon'}(m,n)$ denote the estimate of the bispectrum $B_{\varepsilon'\varepsilon'\varepsilon'}(\omega_m, \omega_n)$ mimicking Equation 16.17. Construct a coarse grid and a fine grid of bifrequencies in D as before. Order the L estimates $\hat{B}_{\varepsilon'\varepsilon'\varepsilon'}(m_{\overline{m}i}, n_{\overline{n}k})$ on the fine grid around the bifrequency pair $(\overline{m}, \overline{n})$ into an L-vector, which after relabeling may be denoted as μ_{ml}, $l = 1, 2, \ldots, L$, $m = 1, 2, \ldots, P$, where m indexes the coarse grid and l indexes the fine grid. Define P-vectors

$$\tilde{\Psi}_i = (\mu_{1i}, \mu_{2i}, \ldots, \mu_{Pi})^{\mathrm{T}}, \quad (i = 1, 2, \ldots, L). \tag{16.33}$$

Consider the estimates

$$\tilde{M} = \frac{1}{L}\sum_{i=1}^{L} \tilde{\Psi}_i \quad \text{and} \quad \tilde{\Sigma} = \frac{1}{L}\sum_{i=1}^{L} (\tilde{\Psi}_i - \tilde{M})(\tilde{\Psi}_i - \tilde{M})^{\mathrm{H}}. \tag{16.34}$$

Define a $(P-1) \times P$ matrix B whose ij th element B_{ij} is given by $B_{ij} = 1$ if $i = j; = -1$ if $j = i + 1; = 0$ otherwise. Define

$$F_{\mathrm{W}} = \frac{2(L-P+1)}{2P-2}(B\tilde{M})^{\mathrm{H}}(B\tilde{\Sigma}B^{\mathrm{T}})^{-1}B\tilde{M}. \tag{16.35}$$

If $\{\varepsilon'(t)\}$ is third-order white, then F_{W} is distributed as a central F with $(2P-2, 2(L-P+1))$ degrees of freedom. A statistical test for testing third-order whiteness of $\{\varepsilon'(t)\}$ is to declare it to be a nonwhite sequence if $F_{\mathrm{W}} > T_\alpha$ where T_α is selected to achieve a fixed probability of false alarm α $(=\Pr\{F_{\mathrm{W}} > T_\alpha\}$ with F_{W} distributed as a central F with $(2P-2, 2(L-P+1))$ degrees of freedom). If $F_{\mathrm{W}} \leq T_\alpha$, then either $\{\varepsilon'(t)\}$ is third-order white or it has zero bispectrum.

The above model validation test can be used for model order selection. Fix an upper bound on the model orders. For every admissible model order, fit a linear model and test its validity. From among the validated models, select the "smallest" order as the correct order. It is easy to see that this procedure will work only so long as the various candidate orders are nested. Further details may be found in [5] and [15].

16.3.3 Confidence Intervals

Having settled upon a model order estimate M, let $\hat{\theta}_N^{(M)}$ be the parameter estimator obtained by minimizing a cost function $V_N[\theta^{(M)}]$, given a record of length N, such that $V_\infty\,(\theta) := \lim_{N\to\infty} V_N(\theta)$ exists. For instance, using the notation of the section on order selection, one may take $V_N[\theta^{(M)}] = -N^{-1} \ln f_{\theta^{(M)}}(X)$. How reliable are these estimates? An assessment of this is provided by confidence intervals.

Under some general technical conditions, it usually follows that asymptotically (i.e., for large N), $\sqrt{N}\left[\hat{\theta}_N^{(M)} - \theta_0\right]$ is distributed as a Gaussian random vector with zero-mean and covariance matrix P where θ_0 denotes the true value of $\theta^{(M)}$. A general expression for P is given by [15]

$$P = \left[V''_\infty(\theta_0)\right]^{-1} P_\infty \left[V''_\infty(\theta_0)\right]^{-1}, \tag{16.36}$$

where

$$P_\infty = \lim_{N\to\infty} E\left\{N V_N'^{T}(\theta_0) V_N'(\theta_0)\right\} \tag{16.37}$$

and V' (a row vector) and V'' (a square matrix) denote the gradient and the Hessian, respectively, of V.

The above result can be used to evaluate the reliability of the parameter estimator. It follows from the above results that

$$\eta_N = N\left[\hat{\theta}_N^{(M)} - \theta_0\right]^{\mathrm{T}} P^{-1} \left[\hat{\theta}_N^{(M)} - \theta_0\right] \tag{16.38}$$

is asymptotically $\chi^2(M)$. Define $\chi^2_\alpha(M)$ via $\Pr\{y > \chi^2_\alpha(M)\} = \alpha$ where y is distributed as $\chi^2(M)$. For instance, $\chi^2_{0.05} = 9.49$ so that $\Pr\{\eta_N > 9.49\} = 0.05$. The ellipsoid $\eta_N \le \chi^2_\alpha(M)$ then defines the 95% confidence ellipsoid for the estimate $\hat{\theta}_N^{(M)}$. It implies that θ_0 will lie with probability 0.95 in this ellipsoid around $\hat{\theta}_N^{(M)}$.

In practice obtaining expression for P is not easy; it requires knowledge of θ_0. Typically, one replaces θ_0 with $\hat{\theta}_N^{(M)}$. If a closed-form expression for P is not available, it may be approximated by a sample average [16].

16.4 Noise Modeling

As for signal models, Gaussian modeling of noise processes has long been dominant. Typically the central limit theorem is invoked to justify this assumption; thermal noise is indeed Gaussian. Another reason is analytical tractability when the Gaussian assumption is made. Nevertheless, non-Gaussian noise occurs often in practice. For instance, underwater acoustic noise, low-frequency atmospheric noise, radar clutter noise, and urban and man-made radio-frequency noise all are highly non-Gaussian [17]. All these types of noise are impulsive in character, i.e., the noise produces large-magnitude observations more often than predicted by a Gaussian model. This fact has led to development of several models of univariate non-Gaussian noise pdf, all of which have their tails decay at rates lower than the rate of decay of the Gaussian pdf tails. Also, the proposed models are parameterized in such a way as to include Gaussian pdf as a special case.

16.4.1 Generalized Gaussian Noise

A generalized Gaussian pdf is characterized by two constants, variance σ^2 and an exponential decay-rate parameter $k > 0$. It is symmetric and unimodal, given by [17]

$$f_k(x) = \frac{k}{2A(k)\Gamma(1/k)} e^{-[|x|/A(k)]^k}, \tag{16.39}$$

where

$$A(k) = \left[\sigma^2 \frac{\Gamma(1/k)}{\Gamma(3/k)}\right]^{1/2} \tag{16.40}$$

and Γ is the gamma function:

$$\Gamma(\alpha) := \int_0^\infty x^{\alpha-1} e^{-x} dx. \tag{16.41}$$

When $k = 2$, Equation 16.39 reduces to a Gaussian pdf. For $k < 2$, the tails of f_k decay at a lower rate than for the Gaussian case f_2. The value $k = 1$ leads to the Laplace density (two-sided exponential). It is known that generalized Gaussian density with k around 0.5 can be used to model certain impulsive atmospheric noise [17].

16.4.2 Middleton Class A Noise

Unlike most of the other noise models, the Middleton class A mode is based upon physical modeling considerations rather than an empirical fit to observed data. It is a canonical model based upon the assumption that the noise bandwidth is comparable to, or less than, that of the receiver. The observed noise process is assumed to have two independent components:

$$X(t) = X_G(t) + X_P(t), \tag{16.42}$$

where

- $X_G(t)$ is a stationary background Gaussian noise component
- $X_P(t)$ is the impulsive component

The component $X_P(t)$ is represented by

$$X_P(t) = \sum_i U_i(t, \theta), \tag{16.43}$$

where

- U_i denotes the ith waveform from an interfering source
- θ represents a set of random parameters that describe the scale and structure of the waveform

The arrival time of these independent impulsive events at the receiver is assumed to be Poisson distributed. Under these and some additional assumptions, the class A pdf for the normalized instantaneous amplitude of noise is given by

$$f_A(x) = e^{-A} \sum_{m=0}^{\infty} \frac{A^m}{m!\sqrt{2\pi\sigma_m^2}} e^{-x^2/(2\sigma_m^2)}, \tag{16.44}$$

where

$$\sigma_m^2 = \frac{(m/A) + \Gamma'}{1 + \Gamma'}. \tag{16.45}$$

The parameter A, called the impulsive index, determines how impulsive noise is: a small value of A implies highly impulsive interference (although $A = 0$ degenerates into purely Gaussian $X(t)$). The parameter Γ' is the ratio of power in the Gaussian component of the noise to the power in the Poisson mechanism interference. The term in Equation 16.44 corresponding to $m = 0$ represents the background component of the noise with no impulsive waveform present, whereas the higher order terms represent the occurrence of m impulsive events overlapping simultaneously at the receiver input.

The class A model has been found to provide very good fits to a variety of noise and interference measurements [17].

16.4.3 Stable Noise Distribution

This is another useful noise distribution model which has a drawback that its variance may not be finite. It is most conveniently described by its characteristic function. A stable univariate pdf has characteristic function $\varphi(t)$ of the form [18]

$$\varphi(t) = \exp\{jat - \gamma|t|^{\alpha}\ [1 + j\beta \text{sgn}(t)\omega(t,\alpha)]\}, \tag{16.46}$$

where

$$\omega(t,\alpha) = \begin{cases} \tan(\alpha\pi/2) & \text{for } \alpha \neq 1(47) \\ (2/\pi)\log(|t|) & \text{for } \alpha = 1 \end{cases} \tag{16.47}$$

$$\text{sgn}(t) = \begin{cases} 1 & \text{for } t > 0(49) \\ 0 & \text{for } t = 0(50) \\ -1 & \text{for } t < 0 \end{cases} \tag{16.48}$$

and

$$-\infty < a < \infty, \quad \gamma > 0, \quad 0 < \alpha \leq 2, \quad -1 \leq \beta \leq 1. \tag{16.49}$$

A stable distribution is completely determined by four parameters: location parameter a, the scale parameter γ, the index of skewness β, and the characteristic exponent α. A stable distribution with characteristic exponent α is called alpha-stable.

The characteristic exponent α is a shape parameter and it measures the "thickness" of the tails of the pdf. A small value of α implies longer tails. When $\alpha = 2$, the corresponding stable distribution is Gaussian. When $\alpha = 1$ and $\beta = 0$, then the corresponding stable distribution is Cauchy.

Inverse Fourier transform of $\varphi(t)$ yields the pdf and, therefore, the pdf of noise. No closed-form solution exists in general for the two; however, power series expansion of the pdf is available—details may be found in [18] and references therein.

16.5 Concluding Remarks

In this chapter, several fundamental aspects of fitting linear time-invariant parametric (rational transfer function) models to a given measurement record were considered. Before a linear model is fitted, one needs to test for stationarity, linearity, and Gaussianity of the given data. Statistical test for these properties were discussed in Section 16.2. After a model is fitted, one needs to validate the model and assess the reliability of the fitted model parameters. This aspect was discussed in Section 16.3. A cautionary note is appropriate at this point. All of the tests and procedures discussed in this chapter are based upon asymptotic considerations (as record length tends to ∞). In practice, this implies that sufficiently long record length should be available, particularly when higher order statistics are exploited.

References

1. Brillinger, D.R., An introduction to polyspectra, *Ann. Math. Stat.*, 36: 1351–1374, 1965.
2. Brillinger, D.R., *Time Series, Data Analysis and Theory*, Holt, Rinehart and Winston, New York, 1975.
3. Subba Rao, T. and Gabr, M.M., A test for linearity of stationary time series, *J. Time Ser. Anal.*, 1(2): 145–158, 1980.
4. Hinich, M.J., Testing for Gaussianity and linearity of a stationary time series, *J. Time Ser. Anal.*, 3(3): 169–176, 1982.
5. Tugnait, J.K., Linear model validation and order selection using higher-order statistics, *IEEE Trans. Signal Process.*, SP-42: 1728–1736, July 1994.
6. Tugnait, J.K., Detection of non-Gaussian signals using integrated polyspectrum, *IEEE Trans. Signal Process.*, SP-42: 3137–3149, Nov. 1994. (Corrections in *IEEE Trans. Signal Process.*, SP-43., Nov. 1995.)
7. Tugnait, J.K., Testing for linearity of noisy stationary signals, *IEEE Trans. Signal Process.*, SP-42: 2742–2748, Oct. 1994.
8. Giannakis, G.B. and Tstatsanis, M.K., Time-domain tests for Gaussianity and time-reversibility, *IEEE Trans. Signal Process.*, SP-42: 3460–3472, Dec. 1994.
9. Priestley, M.B., *Nonlinear and Nonstationary Time Series Analysis*, Academic Press, New York, 1988.
10. Jenkins, G.M., General considerations in the estimation of spectra, *Technometrics*, 3: 133–166, 1961.
11. Basseville, M. and Nikiforov, I.V., *Detection of Abrupt Changes*, Prentice-Hall, Englewood Cliffs, NJ, 1993.
12. Nikias, C.L. and Petropulu, A.P., *Higher-Order Spectra Analysis*, Prentice-Hall, Englewood Cliffs, NJ, 1993.
13. Tong, H., *Nonlinear Time Series*, Oxford University Press, New York, 1990.
14. Dalle Molle, J.W. and Hinich, M.J., Tripsectral analysis of stationary time series, *J. Acoust. Soc. Am.*, 97(5), Pt. 1, May 1995.
15. Söderström, T. and Stoica, P., *System Identification*, Prentice Hall International, London, U.K. 1989.
16. Ljung, L., *System Identification: Theory for the User*, Prentice-Hall, Englewood Cliffs, NJ, 1987.
17. Kassam, S.A., *Signal Detection in Non-Gaussian Noise*, Springer-Verlag, New York, 1988.
18. Shao, M. and Nikias, C.L., Signal processing with fractional lower order moments: Stable processes and their applications, *Proc. IEEE*, 81: 986–1010, July 1993.

17

Cyclostationary Signal Analysis

Georgios B. Giannakis
University of Minnesota

17.1 Introduction

Processes encountered in statistical signal processing, communications, and time series analysis applications are often assumed stationary. The plethora of available algorithms testifies to the need for processing and spectral analysis of stationary signals (see, e.g., [42]). Due to the varying nature of physical phenomena and certain man-made operations, however, time-invariance and the related notion of stationarity are often violated in practice. Hence, study of time-varying systems and nonstationary processes is well motivated.

Research in nonstationary signals and time-varying systems has led both to the development of adaptive algorithms and to several elegant tools, including short-time (or running) Fourier transforms, time-frequency representations such as the Wigner–Ville (a member of Cohen's class of distributions), Loeve's and Karhunen's expansions (leading to the notion of evolutionary spectra), and time-scale representations based on wavelet expansions (see [37,45] and references therein). Adaptive algorithms derived from stationary models assume slow variations in the underlying system. On the other hand, time-frequency and time-scale representations promise applicability to general nonstationarities and provide useful visual cues for preprocessing. When it comes to nonstationary signal analysis and estimation in the presence of noise, however, they assume availability of multiple independent realizations.

In fact, it is impossible to perform spectral analysis, detection, and estimation tasks on signals involving generally unknown nonstationarities, when only a single data record is available. For instance, consider extracting a deterministic signal $s(n)$ observed in stationary noise $v(n)$, using regression techniques based on nonstationary data $x(n) = s(n) + v(n)$, $n = 0, 1, \ldots, N-1$. Unless $s(n)$ is finitely parameterized by a $d_{\theta_s} \times 1$ vector θ_s (with $d_{\theta_s} < N$), the problem is ill-posed because adding a new datum, say $x(n_0)$, adds a new unknown, $s(n_0)$, to be determined. Thus, only structured nonstationarities can be handled when rapid variations are present; and only for classes of finitely parameterized nonstationary processes can reliable statistical descriptors be computed using a single time series. One such class is that of (wide-sense) cyclostationary (CS) processes which are characterized by the periodicity they exhibit in their mean, correlation, or spectral descriptors.

An overview of CS signal analysis and applications are the main goals of this section. Periodicity is omnipresent in physical as well as manmade processes, and CS signals occur in various real life problems entailing phenomena and operations of repetitive nature: communications [15], geophysical and atmospheric sciences (hydrology [66], oceanography [14], meteorology [35], and climatology [4]), rotating machinery [43], econometrics [50], and biological systems [48].

In 1961, Gladysev [34] introduced key representations of CS time series, while in 1969, Hurd's thesis [38] offered an excellent introduction to continuous time CS processes. Since 1975 [22], Gardner and coworkers have contributed to the theory of continuous-time CS signals, and especially their applications to communications engineering. Gardner [15] adopts a "non-probabilistic" viewpoint of CS (see [19] for an overview and also [36] and [18] for comments on this approach). Responding to a recent interest in digital periodically varying systems and CS time series, the exposition here is probabilistic and focuses on discrete-time signals and systems, with emphasis on their second-order statistical characterization and their applications to signal processing and communications.

The material in the remaining sections is organized as follows: Section 17.2 provides definitions, properties, and representations of CS processes, along with their relations with stationary and general classes of nonstationary processes. Testing a time series for CS and retrieval of possibly hidden cycles along with single record estimation of cyclic statistics are the subjects of Section 17.3. Typical signal classes and operations inducing CS are delineated in Section 17.4 to motivate the key uses and selected applications described in Section 17.5. Finally, Section 17.6 concludes and presents trade-offs, topics not covered, and future directions.

17.2 Definitions, Properties, Representations

Let $x(n)$ be a discrete-index random process (i.e., a time series) with mean $\mu_x(n) := E\{x(n)\}$ and covariance $c_{xx}(n;\tau) := E\{[x(n)-\mu_x(n)][x(n+\tau)-\mu_x(n+\tau)]\}$. For $x(n)$ complex valued, let also $\bar{c}_{xx}(n;\tau) := c_{xx^*}(n;\tau)$, where $*$ denotes complex conjugation, and n, τ are in the set of integers $\mathcal{Z}$.

Definition 17.1: Process $x(n)$ is (wide-sense) CS iff there exists an integer P such that $\mu_x(n) = \mu_x(n+lP)$, $c_{xx}(n;\tau) = c_{xx}(n+lP;\tau)$, or $\bar{c}_{xx}(n;\tau) = \bar{c}_{xx}(n+lP;\tau)$, $\forall n, l \in \mathcal{Z}$. The smallest of all such P's is called the period. Being periodic, they all accept Fourier series expansions over complex harmonic cycles with the set of cycles defined as $A_{xx}^c := \{\alpha_k = 2\pi k/P, k = 0, \ldots, P-1\}$; e.g., $c_{xx}(n;\tau)$ and its Fourier coefficients called cyclic correlations are related by

$$c_{xx}(n;\tau) = \sum_{k=0}^{P-1} C_{xx}\left(\frac{2\pi}{P}k;\tau\right) e^{j\frac{2\pi}{P}kn} \quad \overset{FS}{\leftrightarrow} \quad C_{xx}\left(\frac{2\pi}{P}k;\tau\right) = \frac{1}{P}\sum_{n=0}^{P-1} c_{xx}(n;\tau) e^{-j\frac{2\pi}{P}kn}. \tag{17.1}$$

Strict sense CS, or periodic (non)stationarity, can also be defined in terms of probability distributions or density functions when these functions vary periodically (in n). But the focus in engineering is on periodically and almost periodically correlated* time series, since real data are often zero-mean, correlated, and with unknown distributions. Almost periodicity is very common in discrete-time

* The term "cyclostationarity" is due to Bennet [3]. CS processes in economics and atmospheric sciences are also referred to as seasonal time series [50].

because sampling a continuous-time periodic process will rarely yield a discrete-time periodic signal; e.g., sampling $\cos(\omega_c t + \theta)$ every T_s seconds results in $\cos(\omega_c n T_s + \theta)$ for which an integer period exists only if $\omega_c T_s = 2\pi/P$. Because $2\pi/(\omega_c T_s)$ is "almost an integer" period, such signals accept generalized (or limiting) Fourier expansions (see also Equation 17.2 and [9] for rigorous definitions of almost periodic functions).

Definition 17.2: Process $x(n)$ is (wide-sense) almost cyclostationary (ACS) iff its mean and correlation(s) are almost periodic sequences. For $x(n)$ zero-mean and real, the time-varying and cyclic correlations are defined as the generalized Fourier series pair:

$$c_{xx}(n;\tau) = \sum_{\alpha_k \in A_{xx}^c} C_{xx}(\alpha_k;\tau) e^{j\alpha_k n} \quad \overset{\text{FS}}{\leftrightarrow} \quad C_{xx}(\alpha_k;\tau) = \lim_{N\to\infty} \frac{1}{N} \sum_{n=0}^{N-1} c_{xx}(n;\tau) e^{-j\alpha_k n}. \tag{17.2}$$

The set of cycles, $A_{xx}^c(\tau) := \{\alpha_k : C_{xx}(\alpha_k;\tau) \neq 0, -\pi < \alpha_k \leq \pi\}$, must be countable and the limit is assumed to exist at least in the mean-square sense [9, Theorem 1.15].

Definition 17.2 and Equation 17.2 for ACS, subsume CS Definition 17.1 and Equation 17.1. Note that the latter require integer period and a finite set of cycles. In the α-domain, ACS signals exhibit lines but not necessarily at harmonically related cycles. The following example will illustrate the cyclic quantities defined thus far:

Example 17.1: Harmonic in Multiplicative and Additive Noise

Let

$$x(n) = s(n)\cos(\omega_0 n) + v(n), \tag{17.3}$$

where $s(n)$ and $v(n)$ are assumed real, stationary, and mutually independent. Such signals appear when communicating through flat-fading channels, and with weather radar or sonar returns when, in addition to sensor noise $v(n)$, backscattering, target scintillation, or fluctuating propagation media give rise to random amplitude variations modeled by $s(n)$ [32]. We will consider two cases:

Case 1: $\mu_s \neq 0$. The mean in Equation 17.3 is $\mu_x(n) = \mu_s \cos(\omega_0 n) + \mu_v$, and the cyclic mean is

$$C_x(\alpha) := \lim_{N\to\infty} \frac{1}{N} \sum_{n=0}^{N-1} \mu_x(n) e^{-j\alpha n} = \frac{\mu_s}{2}[\delta(\alpha - \omega_0) + \delta(\alpha + \omega_0)] + \mu_v \delta(\alpha), \tag{17.4}$$

where in Equation 17.4 we used the definition of Kronecker's delta:

$$\lim_{N\to\infty} \frac{1}{N} \sum_{n=0}^{N-1} e^{j\alpha n} = \delta(\alpha) := \begin{cases} 1 & \alpha = 0 \\ 0 & \text{else} \end{cases}. \tag{17.5}$$

Signal $x(n)$ in Equation 17.3 is thus (first-order) CS with set of cycles $A_x^c = \{\pm\omega_0, 0\}$. If $X_N(\omega) := \sum_{n=0}^{N-1} x(n)\exp(-j\omega n)$, then from Equation 17.4 we find $C_x(\alpha) = \lim_{N\to\infty} N^{-1} E\{X_N(\alpha)\}$; thus, the cyclic mean can be interpreted as an averaged DFT and ω_0 can be retrieved by picking the peak of $|X_N(\omega)|$ for $\omega \neq 0$.

Case 2: $\mu_s = 0$. From Equation 17.3 we find the correlation $c_{xx}(n;\tau) = c_{ss}(\tau)[\cos(2\omega_0 n + \omega_0\tau) + \cos(\omega_0\tau)]/2 + c_{vv}(\tau)$. Because $c_{xx}(n;\tau)$ is periodic in n, $x(n)$ is (second-order) CS with cyclic correlation (cf. Equations 17.2 and 17.5):

$$C_{xx}(\alpha;\tau) = \frac{c_{ss}(\tau)}{4}\left[\delta(\alpha + 2\omega_0)e^{j\omega_0\tau} + \delta(\alpha - 2\omega_0)e^{-j\omega_0\tau}\right] + \left[\frac{c_{ss}(\tau)}{2}\cos(\omega_0\tau) + c_{vv}(\tau)\right]\delta(\alpha). \tag{17.6}$$

The set of cycles is $\mathcal{A}^c_{xx}(\tau) = \{\pm 2\omega_0, 0\}$ provided that $c_{ss}(\tau) \neq 0$ and $c_{vv}(\tau) \neq 0$. The set $\mathcal{A}^c_{xx}(\tau)$ is lag-dependent in the sense that some cycles may disappear while others may appear for different τ's. To illustrate the τ-dependence, let $s(n)$ be an MA process of order q. Clearly, $c_{ss}(\tau) = 0$ for $|\tau| > q$, and thus $\mathcal{A}^c_{xx}(\tau) = \{0\}$ for $|\tau| > q$.

The CS process in Equation 17.3 is just one example of signals involving products and sums of stationary processes such as $s(n)$ with (almost) periodic deterministic sequences $d(n)$, or CS processes $x(n)$. For such signals, the following properties are useful:

Property 17.1: *Finite sums and products of ACS signals are ACS. If $x_i(n)$ is CS with period P_i, then for λ_i constants, $y_1(n) := \sum_{i=1}^{I_1}\lambda_i x_i(n)$ and $y_2(n) := \prod_{i=1}^{I_2}\lambda_i x_i(n)$ are also CS. Unless cycle cancellations occur among $x_i(n)$ components, the period of $y_1(n)$ and $y_2(n)$ equals the least common multiple of the P_i's. Similarly, finite sums and products of stationary processes with deterministic (almost) periodic signals are also ACS processes.*

As examples of random-deterministic mixtures, consider

$$x_1(n) = s(n) + d(n) \quad \text{and} \quad x_2(n) = s(n)d(n), \tag{17.7}$$

where

$s(n)$ is zero-mean stationary

$d(n)$ is deterministic (almost) periodic with Fourier series coefficients $D(\alpha)$

Time-varying correlations are, respectively,

$$c_{x_1x_1}(n;\tau) = c_{ss}(\tau) + d(n)d(n+\tau) \quad \text{and} \quad c_{x_2x_2}(n;\tau) = c_{ss}(\tau)d(n)d(n+\tau). \tag{17.8}$$

Both are (almost) periodic in n, with cyclic correlations

$$C_{x_1x_1}(\alpha;\tau) = c_{ss}(\tau)\delta(\alpha) + D_2(\alpha;\tau) \quad \text{and} \quad C_{x_2x_2}(\alpha;\tau) = c_{ss}(\tau)D_2(\alpha;\tau), \tag{17.9}$$

where $D_2(\alpha;\tau) = \sum_\beta D(\beta)D(\alpha-\beta)\exp[j(\alpha-\beta)\tau]$, since the Fourier series coefficients of the product $d(n)d(n+\tau)$ are given by the convolution of each component's coefficients in the α-domain. To reiterate the dependence on τ, notice that if $d(n)$ is a periodic ± 1 sequence, then $c_{x_2x_2}(n;0) = c_{ss}(0)d^2(n) = c_{ss}(0)$, and hence periodicity disappears at $\tau = 0$.

ACS signals appear often in nature with the underlying periodicity hidden, unknown, or inaccessible. In contrast, CS signals are often man-made and arise as a result of, e.g., oversampling (by a known integer factor P) digital communication signals, or by sampling a spatial waveform with P antennas (see also Section 17.4).

Both CS and ACS definitions could also be given in terms of the Fourier transforms ($\tau \to \omega$) of $c_{xx}(n;\tau)$ and $C_{xx}(\alpha;\tau)$, namely the time-varying and the cyclic spectra which we denote by $S_{xx}(n;\omega)$ and $S_{xx}(\alpha;\omega)$. Suppose $c_{xx}(n;\tau)$ and $C_{xx}(\alpha;\tau)$ are absolutely summable w.r.t. τ for all n in $\mathcal{Z}$ and α_k in $A_{xx}^c(\tau)$. We can then define and relate time-varying and cyclic spectra as follows:

$$S_{xx}(n;\omega) := \sum_{\tau=-\infty}^{\infty} c_{xx}(n;\tau)e^{-j\omega\tau} = \sum_{\alpha_k \in A_{xx}^s} S_{xx}(\alpha_k;\omega)e^{j\alpha_k n} \tag{17.10}$$

$$S_{xx}(\alpha_k;\omega) := \sum_{\tau=-\infty}^{\infty} C_{xx}(\alpha_k;\tau)e^{-j\omega\tau} = \lim_{N\to\infty} \frac{1}{N}\sum_{n=0}^{N-1} S_{xx}(n;\omega)e^{-j\alpha_k n}. \tag{17.11}$$

Absolute summability w.r.t. τ implies vanishing memory as the lag separation increases, and many real-life signals satisfy these so-called mixing conditions [5, Chapter 2]. Power signals are not absolutely summable, but it is possible to define cyclic spectra equivalently (for real-valued $x(n)$) as

$$S_{xx}(\alpha_k;\omega) := \lim_{N\to\infty} \frac{1}{N} E\{X_N(\omega)X_N(\alpha_k - \omega)\}, \quad X_N(\omega) := \sum_{n=0}^{N-1} x(n)e^{-j\omega n}. \tag{17.12}$$

If $x(n)$ is complex ACS, then one also needs $\bar{S}_{xx}(\alpha_k;\omega) := \lim_{N\to\infty} N^{-1}E\{X_N^*(-\omega)\, X_N(\alpha_k - \omega)\}$. Both S_{xx} and $\bar{S}_{xx}$ reveal presence of spectral correlation. This must be contrasted to stationary processes whose spectral components, $X_N(\omega_1)$, $X_N(\omega_2)$ are known to be asymptotically uncorrelated unless $|\omega_1 \pm \omega_2| = 0$ (mod 2π) [5, Chapter 4]. Specifically, we have from Equation 17.12 the following property:

Property 17.2: *If $x(n)$ is ACS or CS, the N-point Fourier transform $X_N(\omega_1)$ is correlated with $X_N(\omega_2)$ for $|\omega_1 \pm \omega_2| = \alpha_k (\text{mod } 2\pi)$, and $\alpha_k \in A_{xx}^s$.*

Before dwelling further on spectral characterization of ACS processes, it is useful to note the diversity of tools available for processing. Stationary signals are analyzed with time-invariant (TI) correlations (lag-domain analysis), or with power spectral densities (frequency-domain analysis). However, CS, ACS, and generally nonstationary signals entail four variables: $(n, \tau, \alpha, \omega) :=$ (time, lag, cycle, frequency). Grouping two variables at a time, four domains of analysis become available and their relationship is summarized in Figure 17.1. Note that pairs $(n;\tau) \leftrightarrow (\alpha;\tau)$, or $(n;\omega) \leftrightarrow (\alpha;\omega)$, have τ or ω fixed and are Fourier series pairs; whereas $(n;\tau) \leftrightarrow (n;\omega)$, or $(\alpha;\tau) \leftrightarrow (\alpha;\omega)$, have n or α fixed and are related by Fourier transforms.

Further insight on the links between stationary and CS processes is gained through the uniform shift (or phase) randomization concept. Let $x(n)$ be CS with period P, and define $y(n) := x(n + \theta)$, where θ is uniformly distributed in $[0, P)$ and independent of $x(n)$. With $c_{yy}(n;\tau) := E_\theta\{E_x[x(n+\theta)x(n+\tau+\theta)]\}$, we find

$$c_{yy}(n;\tau) = \frac{1}{P}\sum_{p=0}^{P-1} c_{xx}(p;\tau) := C_{xx}(0;\tau) := c_{yy}(\tau), \tag{17.13}$$

where the first equality follows because θ is uniform and the second uses the CS definition in Equation 17.1. Noting that c_{yy} is not a function of n, we have established (see also [15,38]).

Property 17.3: *A CS process $x(n)$ can be mapped to a stationary process $y(n)$ using a shift θ, uniformly distributed over its period, and the transformation $y(n) := x(n + \theta)$.*

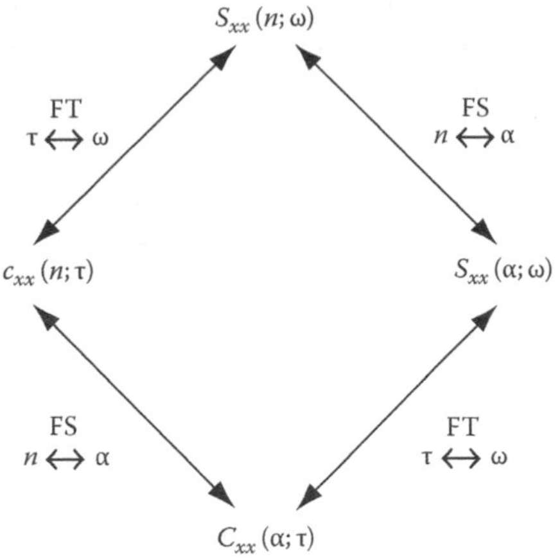

FIGURE 17.1 Four domains for analyzing CS signals.

Such a mapping is often used with harmonic signals; e.g., $x(n) = A\ \exp[j(2\pi n/P + \theta)] + v(n)$ is according to Property 17.2 a CS signal, but can be stationarized by uniform phase randomization. An alternative trick for stationarizing signals which involve complex harmonics is conjugation. Indeed, $c_{xx*}(n;\tau) = A^2 \exp(-j2\pi\tau/P) + c_{vv}(\tau)$ is not a function of n—but why deal with CS or ACS processes if conjugation or phase randomization can render them stationary?

Revisiting Case 2 of Example 17.1 offers a partial answer when the goal is to estimate the frequency ω_0. Phase randomization of $x(n)$ in Equation 17.3 leads to a stationary $y(n)$ with correlation found by substituting $\alpha = 0$ in Equation 17.6. This leads to $c_{yy}(\tau) = (1/2)c_{ss}(\tau)\cos(\omega_0\tau) + c_{vv}(\tau)$, and shows that if $s(n)$ has multiple spectral peaks, or if $s(n)$ is broadband, then multiple peaks or smearing of the spectral peak hamper estimation of ω_0 (in fact, it is impossible to estimate ω_0 from the spectrum of $y(n)$ if $s(n)$ is white). In contrast, picking the peak of $C_{xx}(\alpha;\tau)$ in Equation 17.6 yields ω_0, provided that $\omega_0 \in (0,\pi)$ so that spectral folding is prevented [32]. Equation 17.13 provides a more general answer. Phase randomization restricts a CS process only to one cycle, namely $\alpha = 0$. In other words, the cyclic correlation $C_{xx}(\alpha;\tau)$ contains the "stationarized correlation" $C_{xx}(0;\tau)$ and additional information in cycles $\alpha \neq 0$.

Since CS and ACS processes form a superset of stationary ones, it is useful to know how a stationary process can be viewed as a CS process. Note that if $x(n)$ is stationary, then $c_{xx}(n;\tau) = c_{xx}(\tau)$ and on using Equations 17.2 and 17.5, we find

$$C_{xx}(\alpha;\tau) = c_{xx}(\tau)\left[\lim_{N\to\infty}\frac{1}{N}\sum_{n=0}^{N-1} e^{-j\alpha n}\right] = c_{xx}(\tau)\delta(\alpha). \tag{17.14}$$

Intuitively, Equation 17.14 is justified if we think that stationarity reflects "zero time-variation" in the correlation $c_{xx}(\tau)$. Formally, Equation 17.14 implies

---

Property 17.4: *Stationary processes can be viewed as ACS or CS with cyclic correlation* $C_{xx}(\alpha;\tau) = c_{xx}(\tau)\delta(\alpha)$.

Separation of information bearing ACS signals from stationary ones (e.g., noise) is desired in many applications and can be achieved based on Property 17.4 by excluding the cycle $\alpha = 0$.

Next, it is of interest to view CS signals as special cases of general nonstationary processes with two-dimensional (2D) correlation $r_{xx}(n_1, n_2) := E\{x(n_1)x(n_2)\}$, and 2D spectral densities

$S_{xx}(\omega_1,\omega_2) := \mathrm{FT}[r_{xx}(n_1, n_2)]$ that are assumed to exist.* Two questions arise: What are the implications of periodicity in the (ω_1, ω_2) plane and how does the cyclic spectra in Equations 17.10 through 17.12 relate to $S_{xx}(\omega_1, \omega_2)$? The answers are summarized in Figure 17.2, which illustrates that the support of CS processes in the (ω_1, ω_2) plane consists of $2P-1$ parallel lines (with unity slope) intersecting the axes at equidistant points $2\pi/P$ far apart from each other. More specifically, we have [34]:

Property 17.5: *A CS process with period P is a special case of a nonstationary (harmonizable) process with 2D spectral density given by*

$$S_{xx}(\omega_1, \omega_2) = \sum_{k=-(P-1)}^{P-1} S_{xx}\left(\frac{2\pi}{P}k; \omega_1\right)\delta_D\left(\omega_2 - \omega_1 + \frac{2\pi}{P}k\right), \tag{17.15}$$

where δ_D *denotes the delta of Dirac.*

For stationary processes, only the $k=0$ term survives in Equation 17.15 and we obtain $S_{xx}(\omega_1, \omega_2) = S_{xx}(0; \omega_1)\delta_D(\omega_2 - \omega_1)$; i.e., the spectral mass is concentrated on the diagonal of Figure 17.2. The well-structured spectral support for CS processes will be used to test for presence of CS and estimate the period P. Furthermore, the superposition of lines parallel to the diagonal hints toward representing CS processes as a superposition of stationary processes. Next we will examine two such representations introduced by Gladysev [34] (see also [22,38,49,56]).

We can uniquely write $n_0 = nP + i$ and express $x(n_0) = x(nP + i)$, where the remainder i takes values $0, 1, \ldots, P-1$. For each i, define the subprocess $x_i(n) := x(nP + i)$. In multirate processing, the $P \times 1$ vector $\mathbf{x}(n) := [x_0(n) \ldots x_{P-1}(n)]'$ constitutes the so-called polyphase decomposition of $x(n)$ [51, Chapter 12]. As shown in Figure 17.3, each $x_i(n)$ is formed by downsampling an advanced copy of $x(n)$.

On the other hand, combining upsampled and delayed $x_i(n)$'s, we can synthesize the CS process as

$$x(n) = \sum_{i=0}^{P-1} \sum_{l} x_i(l)\delta(n - i - lP). \tag{17.16}$$

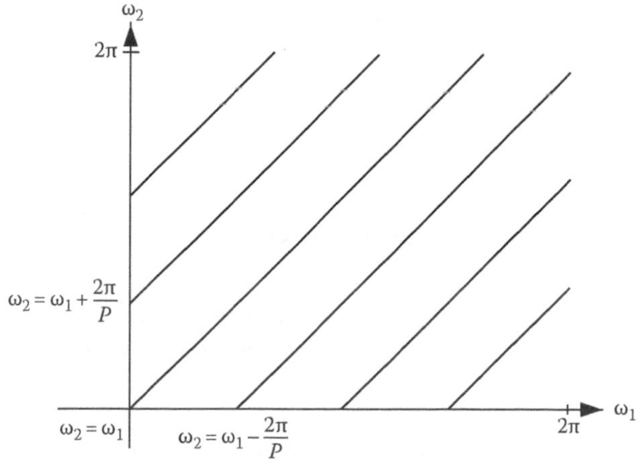

FIGURE 17.2 Support of 2D spectrum $S_{xx}(\omega_1,\omega_2)$ for CS processes.

---

* Nonstationary processes with Fouriers transformable 2D correlations are called harmonizable processes.

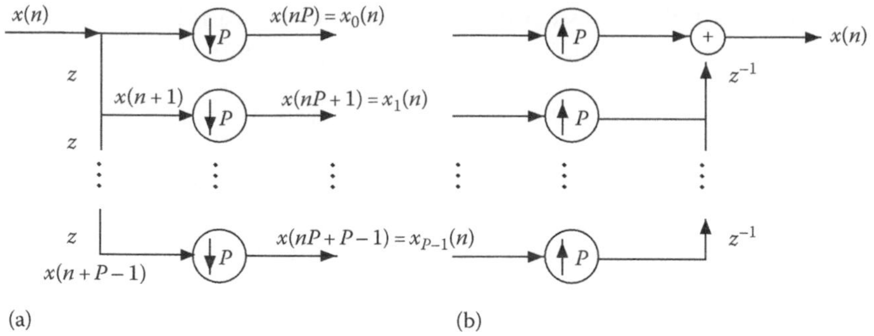

FIGURE 17.3 Representation 17.1: (a) analysis and (b) synthesis.

We maintain that subprocesses $\{x_i(n)\}_{i=0}^{P-1}$ are (jointly) stationary, and thus $\mathbf{x}(n)$ is vector stationary. Suppose for simplicity that $E\{x(n)\} = 0$, and start with $E\{x_{i_1}(n)x_{i_2}(n+\tau)\} = E\{x(nP+i_1)x(nP+\tau P+i_2)\} := c_{xx}(i_1+nP; i_2-i_1+\tau P)$. Because $x(n)$ is CS, we can drop nP and c_{xx} becomes independent of n establishing that $x_{i_1}(n)$, $x_{i_2}(n)$, are (jointly) stationary with correlation:

$$c_{x_{i_1}x_{i_2}}(\tau) = c_{xx}(i_1; i_2 - i_1 + \tau P), \quad i_1, i_2 \in [0, P-1]. \tag{17.17}$$

Using Equation 17.17, it can be shown that auto- and cross-spectra of $x_{i_1}(n)$, $x_{i_2}(n)$, can be expressed in terms of the cyclic spectra of $x(n)$ as [56]

$$S_{x_{i_1}x_{i_2}}(\omega) = \frac{1}{P}\sum_{k_1=0}^{P-1}\sum_{k_2=0}^{P-1} S_{xx}\left(\frac{2\pi}{P}k_1; \frac{\omega - 2\pi k_2}{P}\right) e^{j[(\frac{\omega-2\pi k_2}{P})(i_2-i_1)+\frac{2\pi}{P}k_1 i_1]}. \tag{17.18}$$

To invert Equation 17.18, we Fourier transform Equation 17.16 and use Equation 17.12 to obtain (for $x(n)$ real):

$$S_{xx}\left(\frac{2\pi}{P}k; \omega\right) = \sum_{i_1=0}^{P-1}\sum_{i_2=0}^{P-1} S_{x_{i_1}x_{i_2}}(\omega) e^{j\omega(i_2-i_1)} e^{-j\frac{2\pi}{P}k i_2}. \tag{17.19}$$

Based on Equations 17.16 through 17.19, we infer that CS signals with period P can be analyzed as stationary $P \times 1$ multichannel processes and vice versa. In summary, we have

---

Representation 17.1: (Decimated Components)

CS process $x(n)$ can be represented as a P-variate stationary multichannel process $\mathbf{x}(n)$ with components $x_i(n) = x(nP+i)$, $i = 0, 1, \ldots, P-1$. Cyclic spectra and stationary auto- and cross-spectra are related as in Equations 17.18 and 17.19.

An alternative means of decomposing a CS process into stationary components is by splitting the $(-\pi, \pi]$ spectral support of $X_N(\omega)$ into bands each of width $2\pi/P$ [22]. As shown in Figure 17.4, this can be accomplished by passing modulated copies of $x(n)$ through an ideal low-pass filter $H_0(\omega)$ with spectral support $(-\pi/P, \pi/P]$.The resulting subprocesses $\bar{x}_m(n)$ can be shifted up in frequency and recombined to synthesize the CS process as $x(n) = \sum_{m=0}^{P-1} \bar{x}_m(n) \exp(-j2\pi\, mn/P)$. Within each band, frequencies are

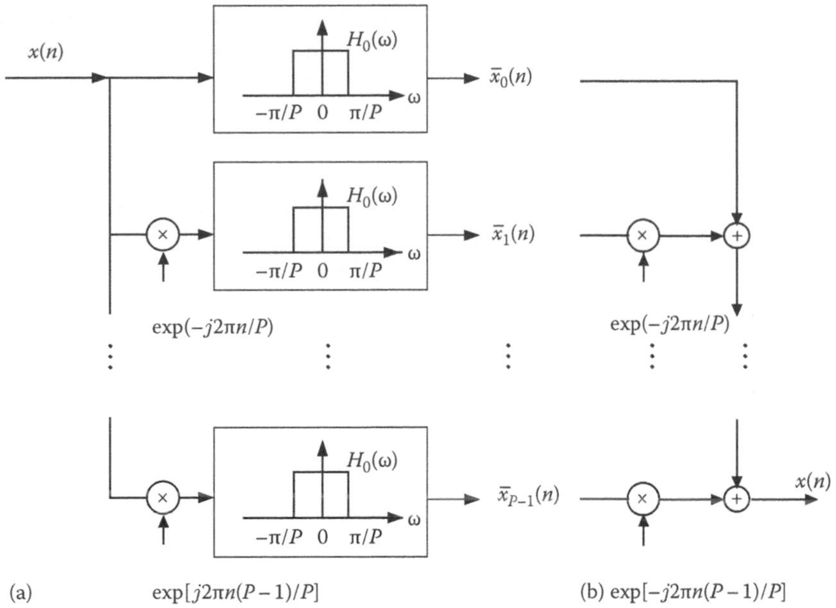

FIGURE 17.4 Representation 17.2: (a) analysis and (b) synthesis.

separated by less than $2\pi/P$ and according to Property 17.2, there is no correlation between spectral components $\bar{X}_{m,N}(\omega_1)$ and $\bar{X}_{m,N}(\omega_2)$; hence, $\bar{x}_m(n)$ components are stationary with auto- and cross-spectra having nonzero support over $-\pi/P < \omega < \pi/P$. They are related with the cyclic spectra as follows:

$$S_{\bar{x}_{m_1}\bar{x}_{m_2}}(\omega) = S_{xx}\left(\frac{2\pi}{P}(m_1 - m_2);\, \omega + \frac{2\pi}{P}m_1\right), |\omega| < \frac{\pi}{P}. \tag{17.20}$$

Equation 17.20 suggests that CS signal analysis is linked with stationary subband processing.

Representation 17.2: (Subband Components)

CS process $x(n)$ can be represented as a superposition of P stationary narrowband subprocesses according to $x(n) = \sum_{m=0}^{P-1} \bar{x}_m(n) \exp(-j2\pi mn/P)$. Auto- and cross-spectra of $\bar{x}_m(n)$ can be found from the cyclic spectra of $x(n)$ as in Equation 17.20.

Because ideal low-pass filters cannot be designed, the subband decomposition seems less practical. However, using Representation 17.1 and exploiting results from uniform DFT filter banks, it is possible using FIR low-pass filters to obtain stationary subband components (see, e.g., [51, Chapter 12]). We will not pursue this approach further, but Representation 17.1 will be used next for estimating time-varying correlations of CS processes based on a single data record.

17.3 Estimation, Time-Frequency Links, and Testing

The time-varying and cyclic quantities introduced in Equations 17.1, 17.2, and 17.10 through 17.12 entail ideal expectations (i.e., ensemble averages) and unless reliable estimators can be devised from finite (and often noisy) data records, their usefulness in practice is questionable. For stationary processes with (at least asymptotically) vanishing memory,* sample correlations and spectral density estimators converge to their ensembles as the record length $N \to \infty$. Constructing reliable (i.e., consistent) estimators for nonstationary processes, however, is challenging and generally impossible. Indeed, capturing time-variations calls for short observation windows, whereas variance reduction demands long records for sample averages to converge to their ensembles.

Fortunately, ACS and CS signals belong to the class of processes with "well-structured" time-variations that under suitable mixing conditions allow consistent single record estimators. The key is to note that although $c_{xx}(n;\tau)$ and $S_{xx}(n;\omega)$ are time-varying, they are expressed in terms of cyclic quantities, $C_{xx}(\alpha_k;\tau)$ and $S_{xx}(\alpha_k;\omega)$, which are TI. Indeed, in Equations 17.2 and 17.10, time-variation is assigned to the Fourier basis.

17.3.1 Estimating Cyclic Statistics

First we will consider ACS processes with known cycles α_k. Simpler estimators for CS processes and cycle estimation methods will be discussed later in this section. If $x(n)$ has nonzero mean, we estimate the cyclic mean as in Example 17.1 using the normalized DFT: $\hat{C}_{xx}(\alpha_k) = N^{-1}\sum_{n=0}^{N-1} x(n)\exp(-j\alpha_k n)$. If the set of cycles is finite, we estimate the time-varying mean as $\hat{c}_{xx}(n) = \sum_{\alpha_k} \hat{C}_{xx}(\alpha_k)\exp(j\alpha_k n)$. Similarly, for zero-mean ACS processes we estimate first cyclic and then time-varying correlations using

$$\hat{C}_{xx}(\alpha_k;\tau) = \frac{1}{N}\sum_{n=0}^{N-1} x(n)x(n+\tau)e^{-j\alpha_k n} \quad \text{and}$$
$$\hat{C}_{xx}(\alpha_k;\tau) = \frac{1}{N}\sum_{n=0}^{N-1} x(n)x(n+\tau)e^{-j\alpha_k n}. \tag{17.21}$$

Note that $\hat{C}_{xx}$ can be computed efficiently using the FFT of the product $x(n)x(n+\tau)$.

For cyclic spectral estimation, two options are available: (1) smoothed cyclic periodograms and (2) smoothed cyclic correlograms. The first is motivated by Equation 17.12 and smoothes the cyclic periodogram, $I_{xx}(\alpha;\omega) := N^{-1} X_N(\omega)X_N(\alpha-\omega)$, using a frequency-domain window $W(\omega)$. The second follows Equation 17.2 and Fourier transforms $\hat{C}_{xx}(\alpha;\tau)$ after smoothing it by a lag-window $w(\tau)$ with support $\tau \in [-M, M]$. Either one of the resulting estimates

$$\hat{S}^{(i)}_{xx}(\alpha;\omega) = \frac{1}{N}\sum_{n=0}^{N-1} W\left(\omega - \frac{2\pi}{N}n\right) I_{xx}\left(\alpha;\frac{2\pi}{N}n\right) \quad \text{or}$$
$$\hat{S}^{(ii)}_{xx}(\alpha;\omega) = \sum_{\tau=-M}^{M} w(\tau)\hat{C}_{xx}(\alpha;\tau)e^{-j\omega\tau} \tag{17.22}$$

can be used to obtain time-varying spectral estimates; e.g., using $\hat{S}^{(i)}_{xx}(\alpha;\omega)$, we estimate $S_{xx}(n;\omega)$ as

$$\hat{S}^{(i)}_{xx}(n;\omega) = \sum_{\alpha_k \in A^s_{xx}} \hat{S}^{(i)}_{xx}(\alpha_k;\omega)e^{j\alpha_k n}. \tag{17.23}$$

Estimates of Equations 17.21 through 17.23 apply to ACS (and hence CS) processes with a finite number of known cycles, and rely on the following steps: (1) estimate the TI (or "stationary") quantities by dropping limits and expectations from the corresponding cyclic definitions, and (2) use the cyclic estimates to obtain time-varying estimates relying on the Fourier synthesis (Equations 17.2 and 17.10). Selection of the windows in Equation 17.22, variance expressions, consistency, and asymptotic normality

* Well-separated samples of such processes are asymptotically independent. Sufficient(so-called mixing) conditions include absolute summability of cumulants and are satisfied by many real-life signals (see [5] and [12, Chapter 2]).

of the estimators in Equations 17.21 through 17.23 under mixing conditions can be found in [11,12,24,39] and references therein.

When $x(n)$ is CS with known integer period P, estimation of time-varying correlations and spectra becomes easier. Recall that thanks to Representations 17.1 and 17.2, not only $c_{xx}(n;\tau)$ and $S_{xx}(n;\omega)$, but the process $x(n)$ itself can be analyzed into P stationary components. Starting with Equation 17.16, it can be shown that $c_{xx}(i;\tau) = c_{x_i x_{i+\tau}}(0)$, where $i = 0, 1, \ldots, P-1$ and subscript $i+\tau$ is understood mod(P). Because the subprocesses $x_i(n)$ and $x_{i+\tau}(n)$ are stationary, their cross-covariances can be estimated consistently using sample averaging; hence, the time-varying correlation can be estimated as

$$\hat{c}_{xx}(i;\tau) = \hat{c}_{x_i x_{i+\tau}}(0) = \frac{1}{[N/P]} \sum_{n=0}^{[N/P]-1} x(nP+i)x(nP+i+\tau), \tag{17.24}$$

where the integer part $[N/P]$ denotes the number of samples per subprocess $x_i(n)$, and the last equality follows from the definition of $x_i(n)$ in Representation 17.1. Similarly, the time-varying periodogram can be estimated using $I_{xx}(n;\omega) = P^{-1}\sum_{k=0}^{P-1} X_P(\omega)X_P(2\pi k/P - \omega)\exp(-j2\pi kn/P)$, and then smoothed to obtain a consistent estimate of $S_{xx}(n;\omega)$.

17.3.2 Links with Time-Frequency Representations

Consistency (and hence reliability) of single record estimates is a notable difference between CS and time-frequency signal analyses. Short-time Fourier transforms, the Wigner–Ville, and derivative representations are valuable exploratory (and especially graphical) tools for analyzing nonstationary signals. They promise applicability on general nonstationarities, but unless slow variations are present and multiple independent data records are available, their usefulness in estimation tasks is rather limited. In contrast, ACS analysis deals with a specific type of structured variation, namely (almost) periodicity, but allows for rapid variations and consistent single record sample estimates. Intuitively speaking, CS provides within a single record, multiple periods that can be viewed as "multiple realizations." Interestingly, for ACS processes there is a close relationship between the normalized asymmetric ambiguity function $A(\alpha;\tau)$ [37], and the sample cyclic correlation in Equation 17.21:

$$N\hat{C}_{xx}(\alpha;\tau) = A(\alpha;\tau) := \sum_{n=0}^{N-1} x(n)x(n+\tau)e^{-j\alpha n}. \tag{17.25}$$

Similarly, one may associate the Wigner–Ville with the time-varying periodogram $I_{xx}(n;\omega) = \sum_{\tau=-(N-1)}^{N-1} x(n)x(n+\tau)\exp(-j\omega\tau)$. In fact, the aforementioned equivalences and the consistency results of [12] establish that ambiguity and Wigner–Ville processing of ACS signals is reliable even when only a single data record is available. The following example uses a chirp signal to stress this point and shows how some of our sample estimates can be extended to complex processes.

Example 17.2: Chirp in Multiplicative and Additive Noise

Consider $x(n) = s(n)\exp(j\omega_0 n^2) + v(n)$, where $s(n)$ and $v(n)$ are zero mean, stationary, and mutually independent; $c_{xx}(n;\tau)$ is nonperiodic for almost every ω_0, and hence $x(n)$ is not (second-order) ACS. Even when $E\{s(n)\} \neq 0$, $E\{x(n)\}$ is also nonperiodic, implying that $x(n)$ is not first-order ACS either. However,

$$\begin{aligned}\tilde{c}_{xx^*}(n;\tau) &:= c_{xx^*}(n+\tau;-2\tau) := E\{x(n+\tau)x^*(n-\tau)\} \\ &= c_{ss}(2\tau)\exp(j4\omega_0\tau n) + c_{vv^*}(2\tau)\end{aligned} \tag{17.26}$$

exhibits (almost) periodicity and its cyclic correlation is given by $\tilde{C}_{xx*}(\alpha;\tau) = c_{ss}(\tau)\delta(\alpha - 4\omega_0\tau) + c_{vv*}(2\tau)\delta(\alpha)$. Assuming $c_{ss}(\tau) \neq 0$, the latter allows evaluation of ω_0 by picking the peak of the sample cyclic correlation magnitude evaluated at, e.g., $\tau = 1$, as follows:

$$\hat{\omega}_0 = -\frac{1}{4}\arg\max_{\alpha\neq 0}|\hat{\tilde{C}}_{xx*}(\alpha;1)|,$$
$$\hat{\tilde{C}}_{xx*}(\alpha;\tau) = \frac{1}{N}\sum_{n=0}^{N-1} x(n+\tau)x^*(n-\tau)e^{-j\alpha n}. \tag{17.27}$$

The $\hat{\tilde{C}}_{xx*}(\alpha;\tau)$ estimate in Equation 17.27 is nothing but the symmetric ambiguity function. Because $x(n)$ is ACS, $\hat{\tilde{C}}_{xx*}$ can be shown to be consistent. This provides yet one more reason for the success of time-frequency representations with chirp signals. Interestingly, Equation 17.27 shows that exploitation of CS allows not only for additive noise tolerance (by avoiding the $\alpha = 0$ cycle in Equation 17.27), but also permits parameter estimation of chirps modulated by stationary multiplicative noise $s(n)$.

17.3.3 Testing for CS

In certain applications involving man-made (e.g., communication) signals, presence of CS and knowledge of the cycles is assured by design (e.g., baud rates or oversampling factors). In other cases, however, only a time series $\{x(n)\}_{n=0}^{N-1}$ is given and two questions arise: How does one detect CS, and if $x(n)$ is confirmed to be CS of a certain order, how does one estimate the cycles present? The former is addressed by testing hypotheses of nonzero $\hat{C}_x(\alpha_k)$, $\hat{C}_{xx}(\alpha_k;\tau)$ or $\hat{S}_{xx}(\alpha_k;\omega)$ over a fine cycle-frequency grid obtained by sufficient zero-padding prior to taking the FFT.

Specifically, to test whether $x(n)$ exhibits CS in $\{\hat{C}_{xx}(\alpha;\tau_l)\}_{l=1}^{L}$ for at least one lag, we form the $(2L+1)\times 1$ vector $\hat{\mathbf{c}}_{xx}(\alpha) := [\hat{C}_{xx}^{R}(\alpha;\tau_1)\cdots\hat{C}_{xx}^{R}(\alpha;\tau_L);\ \hat{C}_{xx}^{I}(\alpha;\tau_1)\cdots\hat{C}_{xx}^{I}(\alpha;\tau_L)]'$ where superscript R(I) denotes real (imaginary) part. Similarly, we define the ensemble vector $\mathbf{c}_{xx}(\alpha)$ and the error $\mathbf{e}_{xx}(\alpha) := \hat{\mathbf{c}}_{xx}(\alpha) - \mathbf{c}_{xx}(\alpha)$. For N large, it is known that $\sqrt{N}\mathbf{e}_{xx}(\alpha)$ is Gaussian with pdf $N(0,\Sigma_c)$. An estimate $\hat{\Sigma}_c$ of the asymptotic covariance can be computed from the data [12]. If α is not a cycle for all $\{\tau_l\}_{l=1}^{L}$, then $\mathbf{c}_{xx}(\alpha) \equiv 0$, $\mathbf{e}_{xx}(\alpha) = \hat{\mathbf{c}}_{xx}(\alpha)$ will have zero mean, and $\hat{D}_{2c}(\alpha) := \hat{\mathbf{c}}_{xx}'(\alpha)\hat{\Sigma}_c^{\dagger}(\alpha)\hat{\mathbf{c}}_{xx}(\alpha)$ will be central chi-square. For a given false-alarm rate, we find from χ^2 tables a threshold Γ and test [10]

$$H_0:\ \hat{D}_{xx}^{c}(\alpha) \geq \Gamma \Rightarrow \alpha \in A_{xx}^{c} \quad \text{vs.} \quad H_1:\ \hat{D}_{xx}^{c}(\alpha) < \Gamma \Rightarrow \alpha \notin A_{xx}^{c}. \tag{17.28}$$

Alternate 2D contour plots revealing presence of spectral correlation rely on Equation 17.15 and more specifically on its normalized version (coherence or correlation coefficient) estimated as [40]

$$\rho_{xx}(\omega_1,\omega_2) := \frac{\frac{1}{M}\sum_{m=0}^{M-1}\left|X_N\left(\omega_1+\frac{2\pi m}{M}\right)X_N^*\left(\omega_2+\frac{2\pi m}{M}\right)\right|^2}{\frac{1}{M}\sum_{m=0}^{M-1}\left|X_N\left(\omega_1+\frac{2\pi m}{M}\right)\right|^2\frac{1}{M}\sum_{m=0}^{M-1}\left|X_N\left(\omega_2+\frac{2\pi m}{M}\right)\right|^2}. \tag{17.29}$$

Plots of $\rho_{xx}(\omega_1,\omega_2)$ with the empirical thresholds discussed in [40] are valuable tools not only for cycle detection and estimation of CS signals but even for general nonstationary processes exhibiting partial (e.g., "transient" lag- or frequency-dependent) CS.

Example 17.3: CS Test

Consider $x(n) = s_1(n)\cos(\pi n/8) + s_2(n)\cos(\pi n/4)$ with $s_1(n)$, $s_2(n)$, and $v(n)$ zero-mean, Gaussian, and mutually independent. To test for CS and retrieve the possible periods present, $N = 2048$ samples were generated; $s_1(n)$ and $s_2(n)$ were simulated as AR(1) with variances $\sigma_{s_1}^2 = \sigma_{s_2}^2 = 2$, while $v(n)$ was white with variance $\sigma_v^2 = 0.1$. Figure 17.5a shows $|\hat{C}_{xx}(\alpha;0)|$ peaking at $\alpha = \pm 2(\pi/8), \pm 2(\pi/4), 0$ as expected, while Figure 17.5b depicts $\rho_{xx}(\omega_1, \omega_2)$ computed as in Equation 17.29 with $M = 64$. The parallel lines in Figure 17.5b are seen at $|\omega_1 - \omega_2| = 0, \pi/8, \pi/4$ revealing the periods present.

One can easily verify from Equation 17.11 that $C_{xx}(\alpha;0) = (2\pi)^{-1}\int_{-\pi}^{\pi} S_{xx}(\alpha;\omega)d\omega$. It also follows from Equation 17.15 that $S_{xx}(\alpha;\omega) = S_{xx}(\omega_1 = \omega, \omega_2 = \omega - \alpha)$; thus, $C_{xx}(\alpha;0) = (2\pi)^{-1}\int_{-\pi}^{\pi} S_{xx}(\omega, \omega - \alpha)d\omega$, and for each α, we can view Figure 17.5a as the (normalized) integral (or projection) of Figure 17.5b along each parallel line [40]. Although $|\hat{C}_{xx}(\alpha;0)|$ is simpler to compute using the FFT of $x^2(n)$, $\rho_{xx}(\omega_1, \omega_2)$ is generally more informative.

Because CS is lag-dependent, as an alternative to $\rho_{xx}(\omega_1, \omega_2)$ one can also plot $|\hat{C}_{xx}(\alpha;\tau)|$ or $|\hat{S}_{xx}(\alpha;\omega)|$ for all τ or ω. Figures 17.6 and 17.7 show perspective and contour plots of $|\hat{C}_{xx}(\alpha;\tau)|$ for $\tau \in [-31, 31]$ and $|\hat{S}_{xx}(\alpha;\omega)|$ for $\omega \in (-\pi, \pi]$, respectively. Both sets exhibit planes (lines) parallel to the τ -axis and ω-axis, respectively, at cycles $\alpha = \pm 2(\pi/8), \pm 2(\pi/4), 0$, as expected.

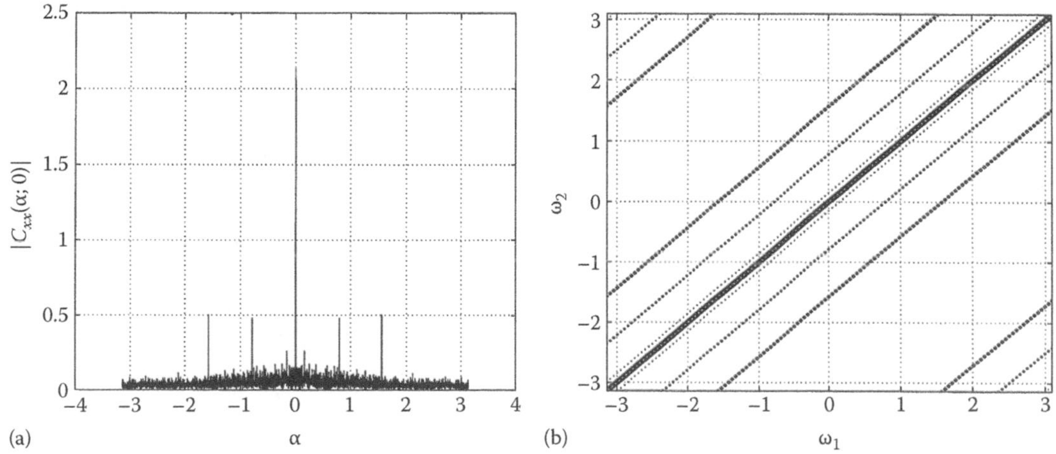

FIGURE 17.5 (a) Cyclic cross-correlation $C_{xx}(\alpha;0)$ and (b) coherence $\rho_{xx}(\omega_1, \omega_2)$ (Example 17.3).

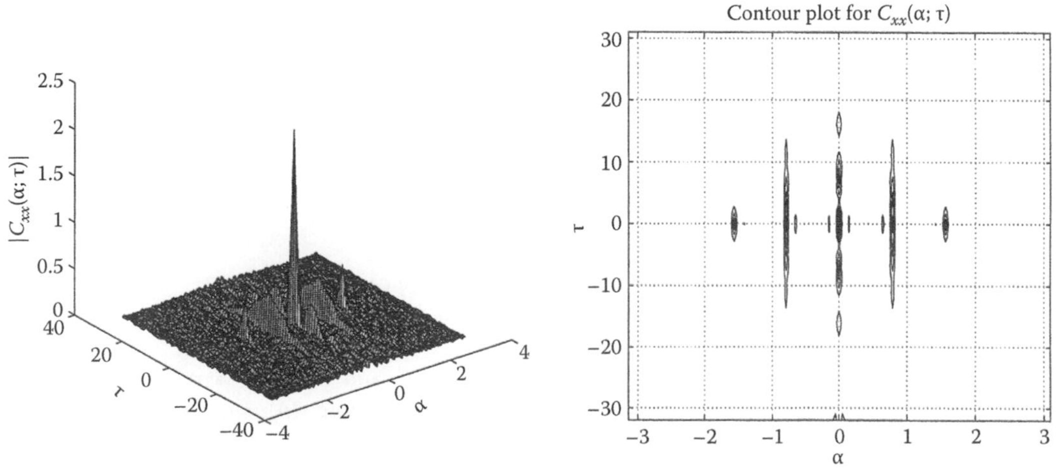

FIGURE 17.6 Cycle detection and estimation (Example 17.3): 3D and contour plots of $\hat{C}_{xx}(\alpha;\tau)$.

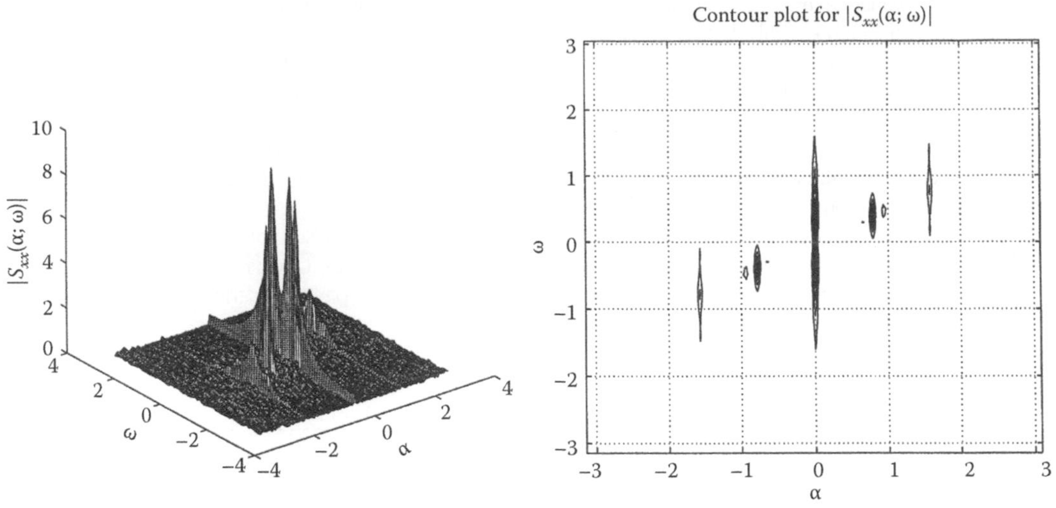

FIGURE 17.7 Cycle detection and estimation (Example 17.3): 3D and contour plots of $\hat{S}_{xx}(\alpha;\omega)$.

17.4 CS Signals and CS-Inducing Operations

We have already seen in Examples 17.1 and 17.2 that amplitude or index transformations of repetitive nature give rise to one class of CS signals. A second category consists of outputs of repetitive (e.g., periodically varying) systems excited by CS or even stationary inputs. Finally, it is possible to have CS emerging in the output due to the data acquisition process (e.g., multiple sensors or fractional sampling).

17.4.1 Amplitude Modulation

General examples in this class include signals $x_1(n)$ and $x_2(n)$ of Equation 17.7 or their combinations as described by Property 17.1. More specifically, we will focus on communication signals where random (often i.i.d.) information data $w(n)$ are D/A converted with symbol period T_0, to obtain the process $w_c(t) = \sum_l w(l)\delta_D(t - lT_0)$, which is CS in the continuous variable t. The continuous-time signal $w_c(t)$ is subsequently pulse shaped by the transmit filter $h_c^{(tr)}(t)$, modulated with the carrier $\exp(j\omega_c t)$, and transmitted over the linear time-invariant (LTI) channel $h_c^{(ch)}(t)$. On reception, the carrier is removed and the data are passed through the receive filter $h_c^{(rec)}(t)$ to suppress stationary additive noise. Defining the composite channel $h_c(t) := h_c^{(tr)} * h_c^{(ch)} * h_c^{(rec)}(t)$, the continuous time received signal at the baseband is

$$r_c(t) = e^{j\omega_{ec}t} \sum_l w(l) h_c(t - lT_0 - \varepsilon) + v_c(t), \qquad (17.30)$$

where

$\varepsilon \in (0, T_0)$ is the propagation delay

ω_{ec} denotes the frequency error between transmit and receive carriers

$v_c(t)$ is AWGN

Signal $r_c(t)$ is CS due to (1) the periodic carrier offset $e^{j\omega_{ec}t}$ and (2) the CS of $w_c(t)$. However, (2) disappears in discrete-time if one samples at the symbol rate because $r(n) := r_c(nT_0)$ becomes

$$r(n) = e^{j\omega_e n} x(n) + v(n), \quad x(n) := \sum_l w(l) h(n - l), \quad n \in [0, N-1], \qquad (17.31)$$

with $\omega_e := \omega_{ec} T_0$, $h(n) := h_c(nT_0 - \varepsilon)$, and $v(n) := v_c(nT_0)$.

If $\omega_e = 0$, $x(n)$ (and thus $v(n)$) is stationary, whereas $\omega_e \neq 0$ renders $r(n)$ similar to the ACS signal in Example 17.1. When $w(n)$ is zero-mean, i.i.d., complex symmetric, we have $E\{w(n)\} \equiv 0$, and $E\{w(n)w(n+\tau)\} \equiv 0$; thus, the cyclic mean and correlations cannot be used to retrieve ω_e. However, peak-picking the cyclic fourth-order correlation (Fourier coefficients of $r^4(n)$) yields $4\omega_e$ uniquely, provided $\omega_e < \pi/4$. If $E\{w^4(n)\} \equiv 0$, higher powers can be used to estimate and recover ω_e.

Having estimated ω_e, we form $\exp(-j\omega_e n)r(n)$ in order to demodulate the signal in Equation 17.31. Traditionally, CS is removed from the discrete-time information signal, although it may be useful for other purposes (e.g., blind channel estimation) to retain CS at the baseband signal $x(n)$. This can be accomplished by multiplying $w(n)$ with a P-periodic sequence $p(n)$ prior to pulse shaping. The noise-free signal in this case is $x(n) = \sum_l p(l)w(l)\, h(n-l)$, and has correlation $\bar{c}_{xx}(n;\tau) = \sigma_w^2 \sum_l |p(n-l)|^2 h(l)h^*(l+\tau)$, which is periodic with period P. Cyclic correlations and spectra are given by [27]

$$\begin{aligned} \bar{C}_{xx}(\alpha;\tau) &= \sigma_w^2 P_2(\alpha) \sum_l h(l)h^*(l+\tau)e^{-j\alpha l}, \\ \bar{S}_{xx}(\alpha;\omega) &= \sigma_w^2 P_2(\alpha) H^*(-\omega)H(\alpha-\omega), \end{aligned} \tag{17.32}$$

where $P_2(\alpha) := P^{-1}\sum_{m=0}^{P-1} |p(m)|^2 \exp(-j\alpha m)$ and $H(\omega) := \sum_{l=0}^{L} h(l)\exp(-j\omega l)$. As we will see later in this section, CS can also be introduced at the transmitter using multirate operations, or at the receiver by fractional sampling. With a CS input, the channel $h(n)$ can be identified using noisy output samples only [27,64,65]—an important step toward blind equalization of (e.g., multipath) communication channels.

If $p(n) = 1$ for $n \in [0, P_1)$ (mod P) and $p(n) = 0$ for $n \in [P_1, P)$, the CS signal $x(n) = p(n)s(n) + v(n)$ can be used to model systematically missing observations. Periodically, the stationary signal $s(n)$ is observed in noise $v(n)$ for P_1 samples and disappears for the next $P - P_1$ data. Using $C_{xx}(\alpha;\tau) = P_2(\alpha;\tau)c_{ss}(\tau)$, the period P (and thus $P_2(\alpha;\tau)$) can be determined. Subsequently, $c_{ss}(\tau)$ can be retrieved and used for parametric or nonparametric spectral analysis of $s(n)$; see [31] and references therein.

17.4.2 Time Index Modulation

Suppose that a random CS signal $s(n)$ is delayed by D samples and received in zero-mean stationary noise $v(n)$ as $x(n) = s(n-D) + v(n)$. With $s(n)$ independent of $v(n)$, the cyclic correlation is $C_{xx}(\alpha;\tau) = C_{ss}(\alpha;\tau)\exp(j\alpha D) + \delta(\alpha)c_{vv}(\tau)$ and the delay manifests itself as a phase of a complex exponential. But even when $s(n)$ models a narrowband deterministic signal, the delay appears in the exponent since $s[n - D(n)] \approx s(n)\exp[jD(n)]$ [53]. Time-delay estimation of CS signals appears frequently in sonar and radar for range estimation where $D(n) = \nu n$ and ν denotes velocity of propagation. $D(n)$ is also used to model Doppler effects that appear when relative motion is present. Note that with time-varying (e.g., accelerating) motion, we have $D(n) = \gamma n^2$ and CS appears in the complex correlation as explained in Example 17.2.

Polynomial delays are one form of time scale transformations. Another one is $d(n) = \lambda n + p(n)$, where λ is a constant and $p(n)$ is periodic with period P (e.g., [38]). For stationary $s(n)$, signal $x(n) = s[d(n)]$ is CS because $c_{xx}(n + lP;\tau) = c_{ss}[d(n + lP + \tau) - d(n + lP)] = c_{ss}[\lambda\tau + p(n) - p(n+\tau)] = c_{xx}(n;\tau)$. A special case is the familiar FM model with $d(n) = \omega_c n + h\sin(\omega_0 n)$ where h here denotes the modulation index. The signal and its periodically varying correlation are given by

$$\begin{aligned} x(n) &= A\cos[\omega_0 n + h\sin(\omega_0 n) + \phi], \\ c_{xx}(n;\tau) &= \frac{A^2}{2}\cos[\omega_0\tau + h\sin(\omega_0(n+\tau)) - h\sin(\omega_0 n)]. \end{aligned} \tag{17.33}$$

In addition to communications, frequency modulated signals appear in sonar and radar when rotating and vibrating objects (e.g., propellers or helicopter blades) induce periodic variations in the phase of incident narrowband waveforms [2,67].

Delays and scale modulations also appear in 2D signals. Consider an image frame at time n with the scene displaced relative to time $n=0$ by $[d_x(n), d_y(n)]$; in spatial and Fourier coordinates, we have [8]

$$\begin{aligned} f(x,y;n) &= f_0(x-d_x(n), y-d_y(n)), \\ F(\omega_x,\omega_y;n) &= F_0(\omega_x,\omega_y)e^{-j\omega_x d_x(n)}e^{-j\omega_y d_y(n)}. \end{aligned} \tag{17.34}$$

Images of moving objects having time-varying velocities can be modeled using polynomial displacements, whereas trigonometric $[d_x(n), d_y(n)]$ can be adopted when the motion is circular, or when the imaging sensor (e.g., camera) is vibrating. In either case, $F(\omega_x, \omega_y; n)$ is CS and thus cyclic statistics can be used for motion estimation and compensation [8].

17.4.3 Fractional Sampling and Multivariate/Multirate Processing

Let $\omega_e = 0$ and suppose we oversample (i.e., fractionally sample) Equation 17.30 by a factor P. With $x(n) := r_c(nT_0/P)$, we obtain (see also Figure 17.8)

$$x(n) = \sum_l w(l)h(n-lP) + v(n), \tag{17.35}$$

where now $h(n) := h_c(nT_0/P - \varepsilon)$ and $v(n) := v_c(nT_0/P)$. Figure 17.8 shows the continuous-time model and the multirate discrete time equivalent of Equation 17.35. With $P=1$, Equation 17.35 reduces to the stationary part of $r(n)$ in Equation 17.31, but with $P>1$, $x(n)$ in Equation 17.35 is CS with correlation $c_{xx}(n;\tau) = \sigma_w^2 \sum_l h(n-lP)h^*(n+\tau-lP) + \sigma_v^2\delta(\tau)$, which can be verified to be periodic with period equal to the oversampling factor P [25,29,61]. Cyclic correlations and cyclic spectra are given, respectively, by

$$\bar{C}_{xx}\left(\frac{2\pi}{P}k;\tau\right) = \frac{\sigma_w^2}{P}\sum_l h(l)h^*(l+\tau)e^{-j\frac{2\pi}{P}kl} + \sigma_v^2\delta(k)\delta(\tau), \tag{17.36}$$

$$\bar{S}_{xx}\left(\frac{2\pi}{P}k;\omega\right) = \frac{\sigma_w^2}{P}H^*(-\omega)H\left(\frac{2\pi}{P}k-\omega\right) + \sigma_v^2\delta(k). \tag{17.37}$$

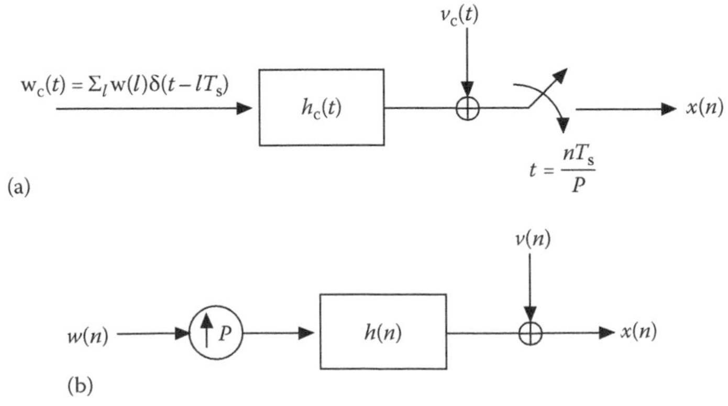

FIGURE 17.8 (a) Fractionally sampled communications model and (b) multirate equivalent.

Although similar, the order of the FIR channel h in Equation 17.35 is, due to oversampling, P times larger than that of Equation 17.31. Cyclic spectra in Equations 17.32 and 17.37 carry phase information about the underlying H, which is not the case with spectra of stationary processes ($P = 1$). Interestingly, Equations 17.35 can be used also to model spread spectrum and direct sequence code-division multiple access data if $h(n)$ includes also the code [63,64]. Relying on $\bar{S}_{xx}$ in Equation 17.37, it is possible to identify $h(n)$ based only on output data—a task traditionally accomplished using higher than second-order statistics (see, e.g., [52]). By avoiding $k = 0$ in Equation 17.36 or 17.37, the resulting cyclic statistics offer a high SNR domain for blind processing in the presence of stationary additive noise of arbitrary color and distribution (cf. Property 17.4).

Oversampling by $P > 1$ also allows for estimating the synchronization parameters ω_l and ε in Equation 17.31 [33,54]. Finally, fractional sampling induces CS in 2D, linear system outputs [28], as well as in outputs of Volterra-type nonlinear systems [30]. In all these cases, relying on Representation 17.1 we can view the CS output $x(n)$ as a $P \times 1$ vector output of a multichannel system. Let us focus on 1D linear channels and evaluate Equation 17.35 at $nP + i$ to obtain the multivariate model

$$x(nP + i) := x_i(n) = \sum_l w(l) h_i(n - l) + v_i(n), \quad i = 0, 1, \ldots, P - 1, \tag{17.38}$$

where $h_i(n) := h(nP + i)$ denotes the polyphase decomposition (decimated components) of the channel $h(n)$. Figure 17.9 shows how the single-input single-output multirate model of Figure 17.8 can be thought of as a single-input P-output multichannel system. The converse interpretation is equally interesting because it illustrates another CS-inducing operation.

Suppose P sensors (e.g., antennas or cameras) are deployed to receive data from a singe source $w(n)$ propagating through P channels $\{h_i(n)\}_{i=0}^{P-1}$. Using Equation 17.16, we can combine the corresponding sensor data $\{x_i(n)\}_{i=0}^{P-1}$ given by Equation 17.38, in order to create a single channel CS process $x(n)$, identical to the one in Equation 17.35. There is a common feature between fractional sampling and multisensor (i.e., spatial) sampling: they both introduce strict CS with known period P.

Strict CS is also induced by multirate operators such as upsamplers in synthesis filterbanks, one branch of which corresponds to the multirate diagram of Figure 17.8b. We infer that outputs of synthesis filter banks are, in general, CS processes (see also [57]). Analysis filter banks, on the other hand, produce CS outputs when their inputs are also CS, but not if their inputs are stationary. Indeed, downsampling does not affect stationarity, and in contrast to upsamplers, downsamplers do not induce CS. Downsamplers can remove CS (as verified by Figure 17.3) and from this point of view, analysis banks can undo CS effects induced by synthesis banks.

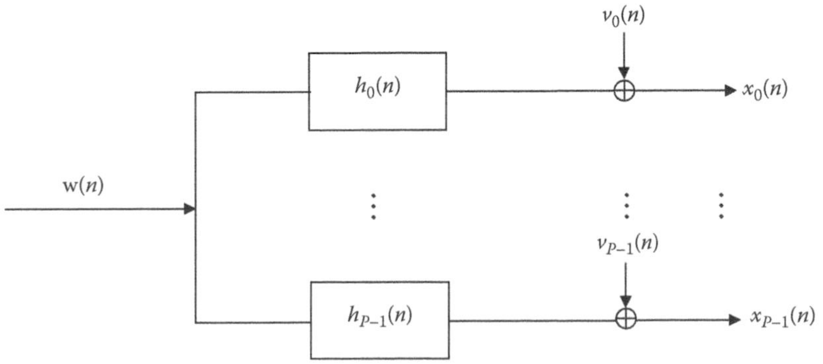

FIGURE 17.9 Multichannel stationary equivalent model of a scalar CS process.

17.4.4 Periodically Varying Systems

Thus far we have dealt with CS signals passing through TI systems. Here we will focus on (almost) periodically time-varying (APTV) systems and input–output relationships such as $x(n) = \Sigma_l h(n; l)w(n-l)$. Because $h(n; l)$ is APTV, following Definition 17.2 it accepts a (generalized) Fourier series expansion $h(n; l) = \Sigma_\beta H(\beta; l)\exp(j\beta n)$. Coefficients $H(\beta; l)$ are TI, and together with their Fourier transform are given by

$$\begin{aligned} H(\beta; l) &:= \text{FS}[h(n; l)] = \lim_{N\to\infty} \frac{1}{N} \sum_{n=0}^{N-1} h(n; l)e^{-j\beta n}, \\ H(\beta; \omega) &:= \text{FT}[H(\beta; l)] = \sum_l H(\beta; l)e^{-j\omega l}. \end{aligned} \tag{17.39}$$

In practice, $h(n; l)$ has finite bandwidth and the set of system cycles is finite; i.e., $\beta \in \{\beta_1, \ldots, \beta_Q\}$. Such a finite parametrization could appear, e.g., with FIR multipath channels entailing path variations due to Doppler effects present with mobile communicators [62]. Note that when the cycles β are available, knowledge of $h(n; l)$ is equivalent to knowing $H(\beta; l)$ or $H(\beta; \omega)$ in Equation 17.39.

The output correlation of an LTI system is given by

$$\bar{c}_{xx}(n; \tau) = \sum_{l_1, l_2} h(n; l_1)\, h^*(n+\tau; l_2)\bar{c}_{ww}(n-l_1; \tau + l_1 - l_2). \tag{17.40}$$

Equation 17.40 shows that if $w(n)$ is ACS, then $x(n)$ is also ACS, regardless of whether h is APTV or TI. More important, if h is APTV, then $x(n)$ is ACS even when $w(n)$ is stationary; i.e., APTV systems are CS inducing operators. Similar observations apply to the input–output cross-correlation $\bar{c}_{xw}(n; \tau) := E\{x(n)w^*(n+\tau)\}$, which is given by

$$\bar{c}_{xw}(n; \tau) = \sum_l h(n; l)\bar{c}_{xw}(n-l; l+\tau). \tag{17.41}$$

If the n-dependence is dropped from Equations 17.40 and 17.41, one recovers the well-known auto- and cross-correlation expressions of stationary processes passing through LTI systems. Relying on definitions of Equations 17.2, 17.11, and 17.37, the auto- and cross-cyclic correlations and cyclic spectra can be found as

$$\begin{aligned} \bar{C}_{xx}(\alpha; \tau) - 3mm = -3mm \sum_{l_1, l_2} \sum_{\beta_1, \beta_2} H(\beta_1; l_1)H^*(\beta_2; l_2)e^{-j(\alpha - \beta_1 + \beta_2)l_1} e^{-j\beta_2 \tau} \\ \times \bar{C}_{ww}(\alpha - \beta_1 + \beta_2; \tau + l_1 - l_2), \end{aligned} \tag{17.42}$$

$$\bar{C}_{xw}(\alpha; \tau) = \sum_\beta \sum_l H(\beta; l)e^{-j(\alpha-\beta)l}\bar{C}_{ww}(\alpha - \beta; l + \tau), \tag{17.43}$$

$$\bar{S}_{xx}(\alpha; \omega) = \sum_{\beta_1, \beta_2} H(\beta_1; \alpha + \beta_2 - \beta_1 - \omega)H^*(\beta_2; -\omega)\bar{S}_{ww}(\alpha - \beta_1 + \beta_2; \omega), \tag{17.44}$$

$$\bar{S}_{xw}(\alpha; \omega) = \sum_\beta H(\beta; \alpha - \beta - \omega)\, \bar{S}_{ww}(\alpha - \beta; \omega). \tag{17.45}$$

Simpler expressions are obtained as special cases of Equations 17.42 through 17.45 when $w(n)$ is stationary; e.g., cyclic auto- and cross-spectra reduce to

$$\begin{aligned} \bar{S}_{xx}(\alpha; \omega) &= \bar{S}_{ww}(\omega) \sum_\beta H(\beta; -\omega)H^*(\alpha - \beta; -\omega), \\ \bar{S}_{xw}(\alpha; \omega) &= \bar{S}_{ww}(\omega)\, H(\alpha; -\omega). \end{aligned} \tag{17.46}$$

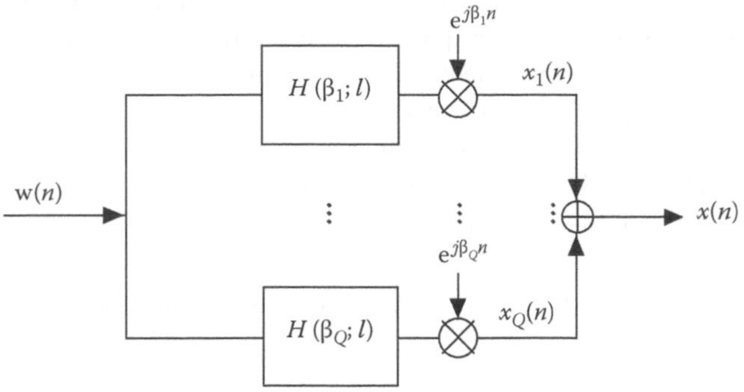

FIGURE 17.10 Multichannel model of a periodically varying system.

If $w(n)$ is i.i.d. with variance σ_w^2, then $H(\alpha;\omega)$ can be easily found from Equation 17.46 as $\bar{S}_{xw}(\alpha;-\omega)/\sigma_w^2$. APTV systems and the four domains of characterizing them, namely $h(n;l)$, $H(\beta;l)$, $H(\beta;\omega)$, and $H(n;\omega)$, offer diversity similar to that exhibited by ACS statistics. Furthermore, with finite cycles $\{\beta_q\}_{q=1}^{Q}$, the input–output relation can be rewritten as

$$x(n) = \sum_{q=1}^{Q} x_q(n) = \sum_{q=1}^{Q}\left[\sum_{l} H(\beta_q; l)\, w(n-l)\right] e^{j\beta_q n}. \tag{17.47}$$

Figure 17.10 depicts Equation 17.47 and illustrates that periodically varying systems can be modeled as a superposition of TI systems weighted by the bases. If separation of the $\{x_q(n)\}_{q=1}^{Q}$ components is possible, identification and equalization of APTV channels can be accomplished using approaches for multichannel TI systems. In [44], separation is achieved based on fractional sampling or multiple antennas.

17.5 Application Areas

CS signals appear in various applications, but here we will deal with problems where CS is exploited for signal extraction, modeling, and system identification. The tools common to all applications are cyclic (cross-)correlations, cyclic (cross-)spectra, or multivariate stationary correlations and spectra which result from the multichannel equivalent stationary processes (recall Representations 17.1 and 17.2, and Section 17.4.3). Because these tools are TI, the resulting approaches follow the lines of similar methods developed for applications involving stationary signals.

As a general rule for problems entailing CS signals, one can either map the scalar CS signal model to a multichannel stationary process, or work in the TI domain of cyclic statistics and follow techniques similar to those developed for stationary signals and TI systems. CS signal analysis exploits two extra features not available with scalar stationary signal processing, namely (1) ability to separate signals on the basis of their cycles and (2) diversity offered by means of cycles. Of course, the cycles must be known or estimated as we discussed in Section 17.3.

Suppose $x(n) = s(n) + v(n)$, where $s(n)$ and $v(n)$ are generally CS, and let α be a cycle which is not in $A_{ss}^{c}(\tau) \cap A_{vv}^{c}(\tau)$. It then follows for their cyclic correlations and spectra that

$$C_{xx}(\alpha;\tau) = \begin{cases} C_{ss}(\alpha;\tau) & \text{if } \alpha \in A_{ss}^{c}(\tau) \\ C_{vv}(\alpha;\tau) & \text{if } \alpha \in A_{vv}^{c}(\tau) \end{cases},$$
$$S_{xx}(\alpha;\omega) = \begin{cases} S_{ss}(\alpha;\omega) & \text{if } \alpha \in A_{ss}^{s}(\omega) \\ S_{vv}(\alpha;\omega) & \text{if } \alpha \in A_{vv}^{s}(\omega) \end{cases}. \tag{17.48}$$

In words, Equation 17.48 says that signals $s(n)$ and $v(n)$ can be separated in the cyclic correlation or the cyclic spectral domains provided that they possess at least one non-common cycle. This important property applies to more than two components and is not available with stationary signals because they all have only one cycle, namely $\alpha = 0$, which they share.

More significantly, if $s(n)$ models a CS information bearing signal and $v(n)$ denotes stationary noise, then working in cyclic domains allows for theoretical elimination of the noise, provided that the $\alpha = 0$ cycle is avoided (see also Property 17.4); i.e.,

$$C_{xx}(\alpha;\tau) = C_{ss}(\alpha;\tau) \quad \text{and} \quad S_{xx}(\alpha;\omega) = S_{ss}(\alpha;\omega), \quad \text{for } \alpha \neq 0. \tag{17.49}$$

In practice, noise affects the estimators' variance so that Equations 17.48 and 17.49 hold approximately for sufficiently long data records. Notwithstanding, Equations 17.48 and 17.49, and SNR improvement in cyclic domains hold true irrespective of the color and distribution of the CS signals or the stationary noise involved.

Example 17.4: Separation Based on Cycles

Consider the mixture of two modulated signals in noise: $x(n) = s_1(n)\exp[j(\omega_1 n + \varphi_1)] + s_2(n)\exp[j(\omega_2 n + \varphi_2)] + v(n)$, where $s_1(n)$, $s_2(n)$, and $v(n)$ are Gaussian zero-mean stationary and mutually uncorrelated. Let $s_1(n)$ be MA (3) with parameters [1, 0.2, 0.3, 0.5] and variance $\sigma_1^2 = 1.38$, $s_2(n)$ be AR (1) with parameters [1, − 0.5] and variance $\sigma_2^2 = 2$, and noise $v(n)$ be MA(1) (i.e., colored) with parameters [1, 0.5] and variance $\sigma_v^2 = 1.25$. Frequencies and phases are $(\omega_1, \varphi_1) = (-0.5, 0.6)$, $(\omega_2, \varphi_2) = (1, 1.8)$, and $N = 2048$ samples are used to compute the correlogram estimates $\hat{S}_{s_1s_1}(\omega)$, $\hat{S}_{s_2s_2}(\omega)$, and $\hat{S}_{vv}(\omega)$ shown in Figure17.11a through c; $\hat{C}_{xx}(\alpha;0)$ is plotted in Figure 17.11d and $\hat{S}_{xx}(\alpha;\omega)$ is depicted in Figure 17.12. The cyclic correlation and cyclic spectrum of $x(n)$ are, respectively

$$\begin{aligned} C_{xx}(\alpha;\tau) = {} & c_{s_1s_1}(\tau)e^{j(\omega_1\tau+\varphi_1)}\delta(\alpha - 2\omega_1) \\ & + c_{s_2s_2}(\tau)e^{j(\omega_2\tau+\varphi_2)}\delta(\alpha - 2\omega_2) + c_{vv}(\tau)\delta(\alpha), \end{aligned} \tag{17.50}$$

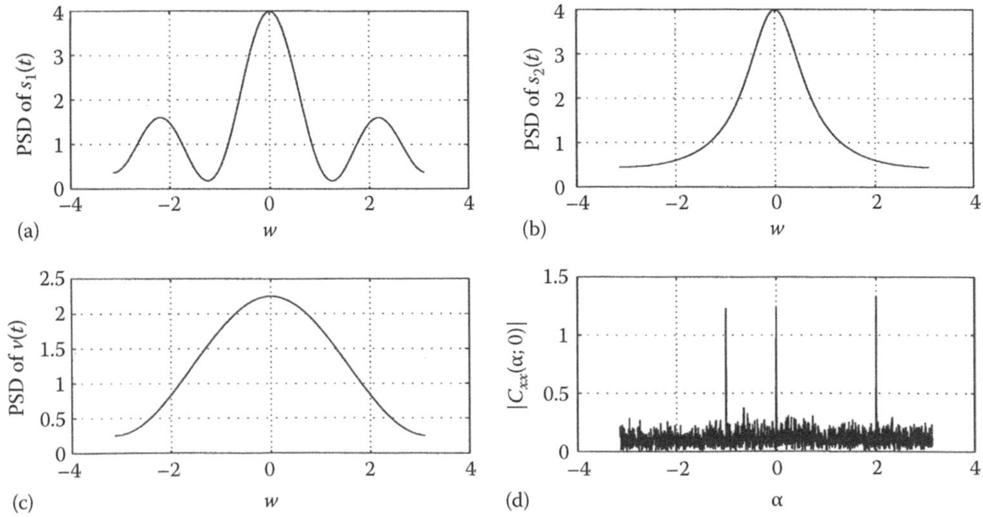

FIGURE 17.11 Spectral densities and cyclic correlation signals in Example 17.4.

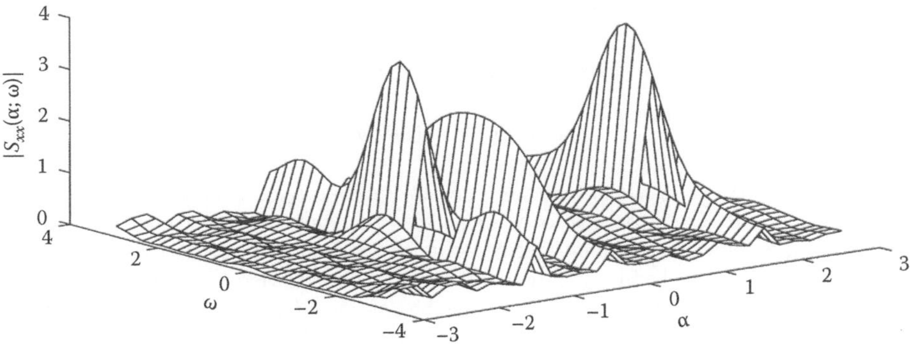

FIGURE 17.12 Cyclic spectrum of $x(n)$ in Example 17.4.

$$S_{xx}(\alpha;\omega) = S_{s_1s_1}(\omega-\omega_1)e^{j2\varphi_1}\delta(\alpha-2\omega_1) + S_{s_2s_2}(\omega-\omega_2)e^{j2\varphi_2}\delta(\alpha-2\omega_2) + S_{vv}(\omega)\delta(\alpha). \quad (17.51)$$

As predicted by Equation 17.50, $|C_{xx}(\alpha;0)| = \sigma_{s_1}^2\delta(\alpha-2\omega_1) + \sigma_{s_2}^2\delta(\alpha-2\omega_2) + \sigma_v^2\delta(\alpha)$, which explains the two peaks emerging in Figure 17.11d at twice the modulating frequencies $(2\omega_1, 2\omega_2) = (-1, 2)$. The third peak at $\alpha = 0$ is due to the stationary noise which can be thought of as being "modulated" by $\exp(j\omega_3 n)$ with $\omega_3 = 0$. Clearly, $2_{\hat{\omega}_1}$, $2_{\hat{\omega}_2}$, $\hat{\sigma}_{s_1}^2$, $\hat{\sigma}_{s_2}^2$, and $\hat{\sigma}_v^2$ can be found from Figure 17.11d, while the phases at the peaks of $\hat{C}_{xx}(\alpha;0)$ will yield $\hat{\varphi}_i = \sigma_{s_i}^{-2}\arg[\hat{C}_{xx}(2_{\hat{\omega}_i};0)]/2$, $i = 1, 2$. In addition, the correlations of $s_i(n)$ can be retrieved as $\hat{c}_{s_is_i}(\tau) = \exp[-j(\hat{\omega}_i\tau + 2\hat{\varphi}_i)]\hat{C}_{xx}(2\hat{\omega}_i;\tau)$, $i = 1, 2$.

Separation based on cycles is illustrated in Figure 17.12, where three distinct slices emerge along the α-axis, each positioned at $\{\alpha_i = 2\omega_i\}_{i=1}^3$, representing the profiles of $\hat{S}_{s_1s_1}(\omega)$, $\hat{S}_{s_2s_2}(\omega)$, and $\hat{S}_{vv}(\omega)$ shown also in Figure 17.11a through c.

In the ensuing example, we will demonstrate how the diversity offered by fractional sampling or by multiple sensors can be exploited for identification of FIR systems when the input is not available. Such a blind scenario appears when estimation and equalization of, e.g., communication channels is to be accomplished without training inputs. Bandwidth efficiency and ability to cope with changing multipath environments provide the motivating reasons for blind processing, while fractional sampling or multiple antennas justify the use of cyclic statistics as discussed in Section 17.4.3.

Example 17.5: Diversity for Channel Estimation

Suppose we sample the output of the receiver's filter every $T_0/2$ seconds, to obtain $x(n)$ samples obeying Equation 17.35 with $P = 2$ (see also Figure 17.8). In the absence of noise, the spectrum of $x(n)$ will be $X_N(\omega) = H(\omega)W_N(2\omega)$. We wish to obtain $H(\omega)$ based only on $X_N(\omega)$ (blind scenario). Note that $W_N(2\omega) = W_N[2(\omega - 2\pi k/2)]$ for any integer k. Considering $k = 1$, we can eliminate the input spectrum $W_N(2\omega)$ from $X_N(\omega)$ and $X_N(\omega - \pi)$, and arrive at [25]

$$H(\omega)X_N(\omega-\pi) = H(\omega-\pi)X_N(\omega). \quad (17.52)$$

With $H(\omega)$ being FIR, the cross-relation (Equation 17.52) has turned the output-only identification problem into an input–output problem. The input is $X_N(\omega - \pi) = \mathrm{FT}[(-1)^n x(n)]$, the output is $X_N(\omega)$, and the pole-zero system is $H(\omega)/H(\omega - \pi)$. If the Z-transform $H(z)$ has no zeros on a circle, separated by π, there is no pole-zero cancellation and $H(\omega)$ can be identified uniquely [61], using standard realization (e.g., Padé) methods [42].

Alternatively, with $P=2$, we can map Equation 17.52 to its one-input two-output TI equivalent model obeying Equation 17.38 with $P=2$. In the absence of noise, the output spectra are $X_i(\omega)=H_i(\omega)W(\omega)$, $i=0,1$, from which $W(\omega)$ can be eliminated to arrive at a similar cross-relation [69]:

$$H_0(\omega)X_1(\omega) = H_1(\omega)X_0(\omega). \tag{17.53}$$

When oversampling by $P=2$, $x_0(n)[h_0(n)]$ correspond to the even samples of $x(n)[h(n)]$, whereas $x_1[n]$ $[h_1(n)]$ to the odd ones. Once again, $H_0(\omega)$ and $H_1(\omega)$ can be uniquely recovered using input–output realization methods, provided that they have no common zeros so that cancellations do not occur in Equation 17.53. The desired channel $h(n)$ can be recovered by interleaving $h_0(n)$ with $h_1(n)$.

As explained in Section 17.4.3, oversampling is not the only means of diversity. Even with symbol rate sampling, if multiple (here two) antennas receive a common source through different channels, then $X_i(\omega)=H_i(\omega)\ W(\omega)$, $i=0,1$, and thus Equation 17.53 is still applicable.

Interestingly, both Equations 17.52 and 17.53 neither restrict the input to be white (or even random) nor do they assume the channel to be minimum phase as univariate stationary spectral factorization approaches require for blind estimation [52]. The diversity (or overdeterminacy) offered by Equation 17.35 or 17.38 guarantees identifiability provided that no cancellations occur in Equation 17.52 or 17.53 and $W(\omega)$ is nonzero for as many frequencies as the number of channel taps to be estimated [69]. Subspace and least-squares methods are also possible for blind channel estimation and useful when noise is present [25,47,60,69].

In the sequel, we will show how cycle-based separation and diversity can be exploited in selected applications.

17.5.1 CS Signal Extraction

In our first application, a mixture of CS sources with distinct cycles will be recovered using samples collected by an array of sensors.

Application 17.1: Array Processing

Suppose N_s CS source signals $\{s_l(n)\}_{l=1}^{N_s}$ are received by N_x sensors $\{x_m(n)\}_{m=1}^{N_x}$ in the presence of undesired sources of interference $\{i_m(n)\}_{m=1}^{N_x}$ and stationary noise $\{v_m(n)\}_{m=1}^{N_x}$. The mth sensor samples are $x_m(n) = \sum_{l=1}^{N_s} \rho_l s_l(n - D_{lm}) + i_m(n) + v_m(n)$, where ρ_l denotes complex gain and D_{lm} the delay experienced by the l th source arriving at the m th sensor relative to the first sensor which is taken as the reference. For uniformly spaced linear arrays $D_{lm}=(m-1)d \sin\theta_l/v$, where d stands for the sensor spacing, v is the propagation velocity, and θ_l denotes the angle of arrival of the lth source. Assuming that the $s_l(n)$ s have a nonzero cycle α not shared by the undesired interferences, we wish to estimate $\theta:=[\theta_1\cdots\theta_{N_s}]$ and subsequently use it to design beamformers that null out the interferences and suppress noise.

For mutually uncorrelated $\{s_l(n), i_m(n), v_m(n)\}$, the time-delay property in Section 17.4.2 yields [68]

$$\bar{C}_{x_m x_m}(\alpha;\tau) = \sum_{l=1}^{N_s} \bar{C}_{s_l s_l}(\alpha;\tau)e^{-j\alpha D_{lm}} + \bar{C}_{i_m i_m}(\alpha;\tau) + \bar{C}_{ww}(\tau)\delta(\alpha). \tag{17.54}$$

Choosing a nonzero α not in the interference set of cycles $A^c_{i_m i_m}(\tau)$ and collecting $\{\bar{C}_{x_m x_m}\}_{m=1}^{N_x}$ in an $N_x\times 1$ vector, we arrive at $\bar{\mathbf{c}}_{x_m}(\alpha;\tau)=\mathbf{A}(\alpha;\theta)\mathbf{c}_{ss}(\alpha;\tau)$, where the $N_x\times N_s$ matrix $\mathbf{A}(\theta)$ is the so-called array manifold containing the propagation parameters. In [68], N_τ lags are used to form the $N_x\times N_\tau$ cyclic correlation matrix

$$\bar{\mathbf{C}}_{xx}(\alpha) := [\bar{\mathbf{c}}_{xx}(\alpha;\tau_1)\cdots\bar{\mathbf{c}}_{xx}(\alpha;\tau_{N_\tau})]' = \mathbf{A}(\alpha;\theta)\bar{\mathbf{C}}_{ss}(\alpha),$$
$$\bar{\mathbf{C}}_{ss}(\alpha) := [\bar{\mathbf{c}}_{ss}(\alpha;\tau_1)\cdots\bar{\mathbf{c}}_{ss}(\alpha;\tau_{N_\tau})]'. \quad (17.55)$$

Standard subspace methods can be employed to recover θ from Equation 17.55. It is worth noting that cycle-based separation of desired from undesired signals and noise is possible for both narrowband and broadband sources [68] (see also [16] for the narrowband case).

With the propagation parameters available, spatiotemporal filtering based on $\bar{\mathbf{C}}_{xx}(\alpha_l;\tau)$ is capable of isolating the source $s_l(n)$ if $\alpha_l \in A^c_{s_l s_l}(\tau)$ and $\alpha_l \notin A^{c}_{s_k s_k}$ for $k \neq l$. Thus, in addition to interference and noise suppression, cyclic beamformers increase resolution by exploiting known separating cycles. In fact, even sources arriving from the same direction can be separated provided that not all of their cycles are common (see [1,6,16,58] for detailed algorithms).

In our next application, the desired CS $d(n)$ we wish to extract from noisy data $x(n)$ is known, or at least its (cross-) correlation with $x(n)$ is available.

Application 17.2: Cyclic Wiener Filtering

In a number of real life problems, CS data $x(n)$ carry information about a desired CS signal $d(n)$ which may not be available, but the cross-correlation $\bar{c}_{dx}(n;\tau)$ is known or can be estimated otherwise. With reference to Figure 17.13, we seek a linear (generally time-varying) filter $f(n;k)$ whose output, $\hat{d}(n) = \sum_k f(n;k)\,x(n-k)$, will come close to the desired $d(n)$ in terms of minimizing $\sigma_e^2(n) = E\{|e(n)|^2\} := E\{|d(n)-\hat{d}(n)|^2\}$. Because both $x(n)$ and $d(n)$ are CS with period P, for $\hat{d}(n)$ to also be CS, filter $f(n;k)$ must be periodically varying with period P; i.e., $f(n;k)$ is equivalent to P TI filters $\{f(n;k)\}_{n=0}^{P-1}$ and accepts a Fourier series expansion with coefficients $F(\alpha;k)$ defined as in Equation 17.39. Note that $e(n)$ is also CS and $E\{|e(n)|^2\}$ should be minimized for $n = 0, 1, \ldots, P-1$.

Solving the minimization problem for each n, we arrive at time-varying normal equations

$$\sum_k f(n;k)\bar{c}_{xx}(n-k;k-\tau) = \bar{c}_{dx}(n;-\tau)\,, \quad n = 0,1,\ldots,P-1, \quad (17.56)$$

where $\bar{c}_{xx}$ can be estimated consistently from the data as discussed in Section 17.3, and similarly for $\bar{c}_{dx}$ if $d(n)$ is available. Note that with sample estimates, Equation 17.56 could have been reached as a result of minimizing the least-squares error (cf. Equation 17.24): $\hat{\sigma}_e^2(n) = [P/N]\sum_{i=0}^{[N/P]-1}|e(iP+n)|^2$. For each $n \in [0, P-1]$, FIR filters of order K_n can be obtained by concatenating equations such as Equation 17.56 for more than K_n lags τ. As with TI Wiener filters, noncausal and IIR designs are possible for each n in the frequency-domain, $F(n;\omega)$, using nonparametric estimates of the time-varying (cross-) spectra. Depending on $d(n)$, APTV (FIR or IIR) filters can thus be constructed for filtering, prediction, and interpolation or smoothing of CS processes.

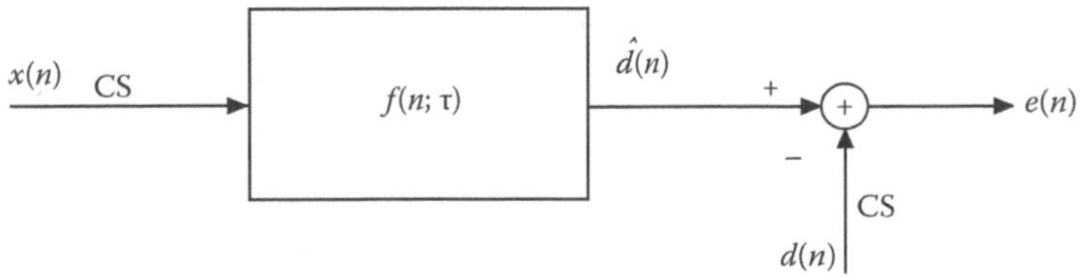

FIGURE 17.13 Cyclic Wiener filtering.

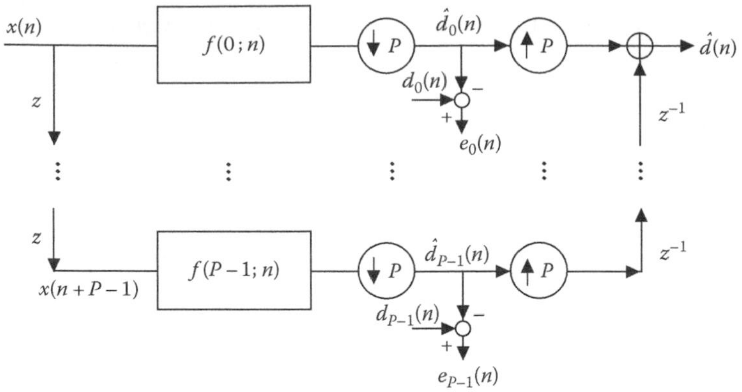

FIGURE 17.14 Multichannel-multirate equivalent of cyclic Wiener filtering.

In Section 17.4.4, we viewed the periodically varying scalar $f(n; k)$ as a TI multichannel filter. Consider the polyphase stationary components $d_i(n)$, $e_i(n)$, and

$$\hat{d}_i(n) := d(nP + i) = \sum_k f(nP + i; k)x(nP + i - k) = \sum_k f(i; k)x(nP + i - k). \qquad (17.57)$$

Equation 17.57 allows us to cast the scalar processing in Figure 17.13 as the filterbank of Figure 17.14. Because $\sigma_{e_i}^2 = E|e(i)|^2$, for $i = 0, 1, \ldots, P-1$, and $d_i(n)$, $\hat{d}_i(n)$, and $e_i(n)$ are stationary, solving for the periodic Wiener filter $f(n; k)$ is equivalent to solving for the P TI Wiener filters $f(i; k)$ in Figure 17.14. Using the multirate (Noble) identity (e.g., [51, Chapter 12]), one can move the downsamplers before the Wiener filters which now have transfer functions $G(i;\omega) = F(i; \omega/P)$. Such an interchange corresponds to feeding a TI $P \times 1$ vector Wiener filter $\mathbf{g}(k) := [g(0; k) \cdots g(P-1; k)]'$, with input the $P \times 1$ polyphase component vector $\mathbf{x}(n) := [x(nP)x(nP+1) \ldots x(nP+P-1)]'$.

An alternative multichannel interpretation is obtained based on the Fourier series expansion $f(n; k) = \sum_\alpha F(\alpha; k) \exp(j\alpha n)$. The resulting Wiener processing allows also for APTV filters, which is particularly useful when $d(n)$, $x(n)$, and thus $\hat{d}(n)$, $e(n)$ are ACS processes. Substituting the expansion in the filter output and multiplying by $\exp(i\alpha k)\exp(-i\alpha k) = 1$, we find [22]

$$\hat{d}(n) = \sum_\alpha \sum_k \left[F(\alpha; k)e^{j\alpha k}\right]\left[x(n-k)e^{j\alpha(n-k)}\right] = \sum_\alpha \left[\sum_k \tilde{F}(\alpha; k)\tilde{x}(n-k)\right], \qquad (17.58)$$

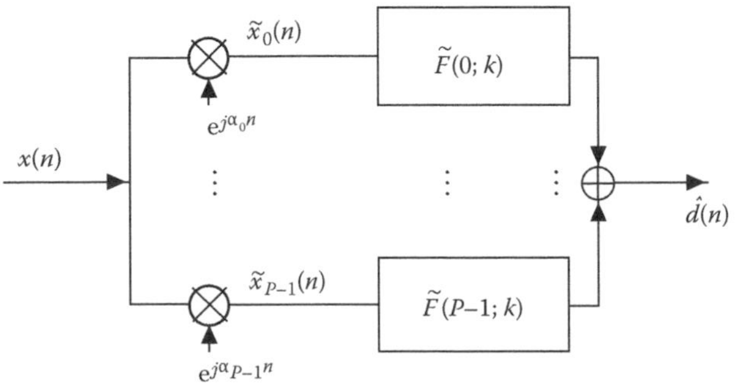

FIGURE 17.15 Multichannel-modulation equivalent of cyclic Wiener filtering.

where $\tilde{F}(\tilde{x})$ are the modulated versions of $F(x)$ shown in the square brackets. For CS processes with period P, the sum over α in Equation 17.58 has finite terms $\{\alpha_i = 2\pi i/P\}_{i=0}^{P-1}$ and shows that scalar cyclic Wiener filtering is equivalent to a superposition of P TI Wiener filters with inputs $\tilde{x}_i(n)$ formed by modulating $x(n)$ with the Fourier bases $\{\exp j(\alpha_i n)\}_{i=1}^{P-1}$ (see also Figure 17.15).

17.5.2 Identification and Modeling

The need to identify TI and APTV systems (or their inverses for equalization) appears in many applications where input–output or output-only CS data are available. Our first problem in this class deals with identifying pure delay TI systems, $h(n) = \delta(n - D)$, given CS input–output signals observed in correlated noise.

Application 17.3: Time-Delay Estimation

We wish to estimate the relative delay D of a CS signal $s(n)$ given data from a pair of sensors

$$x(n) = s(n) + v_x(n) \quad \text{and} \quad y(n) = s(n - D) + v_y(n). \tag{17.59}$$

Signal $s(n)$ is assumed uncorrelated with $v_x(n)$ and $v_y(n)$, but the noises at both sensors are allowed to be colored and correlated with unknown (cross-)spectral characteristics. The time-varying cross-correlation yields the delay (see also [7] and [70] for additional methods relying on cyclic spectra). In addition to suppressing stationary correlated noise, cyclic statistics can also cope with interferences present at both sensors as we show in the following example.

Example 17.6: Time-Delay Estimation

Consider $x(n) = w(n)\exp\{j[-0.5(n) + 0.6]\} + i(n)\exp[j(n + 1.8)] + v_x(n)$, and $y(n) = w(n - D)\exp\{j[-0.5(n - D) + 0.6]\} + i(n - D)\exp[j(n - D + 1.8)] + v_y(n)$, with $D = 20$, $v_x(n)$ white, $v_y(n) = v_x * h(n)$, $h(0) = h(10) = 0.8$ and $h(n) = 0$ for $n \neq 0, 10$.

The magnitude of $\hat{C}_{xy}(\alpha;\tau)$ is computed as in Equation 17.21 with $N = 2048$ samples and is depicted in Figure 17.16 (3D and contour plots). It peaks at the correct delay $D = 20$ at cycles $\alpha = 2(-0.5) = -1$ (due to the signal) and $\alpha = 2(+1) = 2$ (due to the interference). The additional peak at delay 10 occurs at cycle $\alpha = 0$ and reveals the memory introduced in the correlation of $v_y(n)$ due to $h(n)$.

Relying on Equation 17.46, input–output cyclic statistics allow for identification of TI systems, but in certain applications estimation of $h(n)$ or its inverse (call it $g(n)$) is sought based on output data only. In Application 17.2, we outlined two approaches capable of estimating FIR channels blindly in the absence of noise, even when the input $w(n)$ is not white. If $w(n)$ is white, it follows easily from Equation 17.36 that $\bar{C}_{xx}$ for two cycles k_1, k_2 satisfies [25]

$$\sum_{l=0}^{L}\left[\bar{C}_{xx}\left(\frac{2\pi}{P}k_1;\tau + l\right) - e^{j\frac{2\pi}{P}(k_2-k_1)l}\bar{C}_{xx}\left(\frac{2\pi}{P}k_2;\tau + l\right)\right]h(l) = 0, \quad k_1 \neq k_2 \neq 0. \tag{17.60}$$

The matrix equation that results from Equation 17.60 for different τ's can be solved to obtain $\{h(l)\}_{l=0}^{L}$ within a scale (assuming that the matrix involved is full rank), even when stationary colored noise is present. To fix the scale, we either set $h(0) = 1$, or $\sum_{l=0}^{L}|h(l)|^2 = 1$. Having estimated $h(l)$, one could find the cross-correlation $\bar{c}_{xw}(n;\tau)$ via Equation 17.35 and use it in Equation 17.56 to obtain FIR minimum mean-square error (MMSE; i.e., Wiener) equalizers for recovering the desired input $d(n) = w(n)$. However, as we will see next, it is possible to construct blind equalizers directly from the data bypassing the channel estimation step.

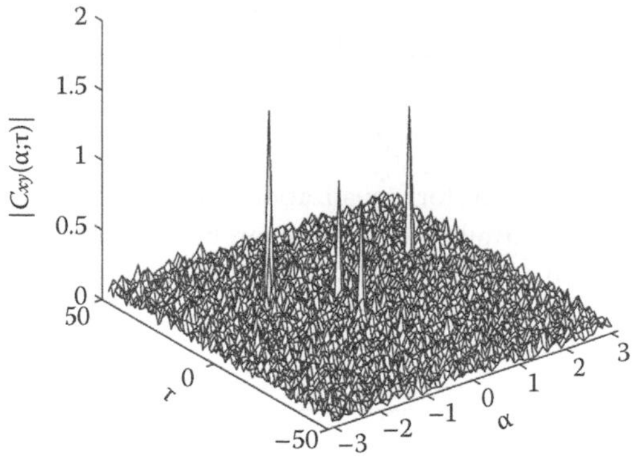

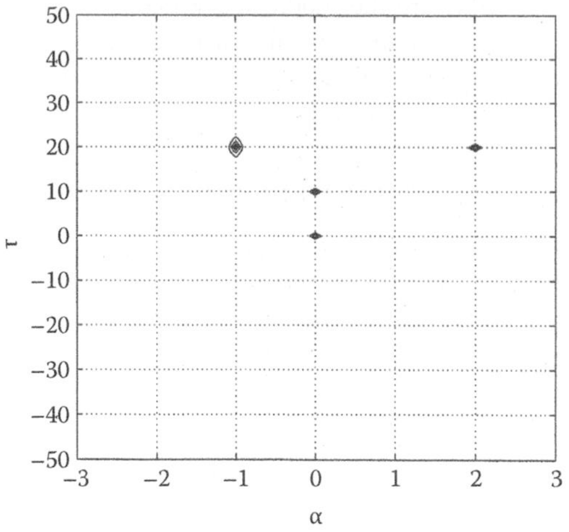

FIGURE 17.16 Cyclic cross-correlation for time-delay estimation.

Application 17.4: Blind Channel Equalization

Our setup is described in Figure 17.8 and the available data satisfy Equation 17.35 with $h(n)$ causal of order L. With reference to Figure 17.17, we seek a Kth order equalizer, $\{g^{(d)}(n)\}_{n=0}^{K}$, parameterized by the delay d, such that $E\{|w(n-d)-\hat{w}(n)|^2\}$ is minimized. Expressing $\hat{w}(n)$ as $\hat{w}(n)=\sum_k g^{(d)}(k)x(nP-k)$, and using the whiteness of $w(n)$ and the independence between $w(n)$ and $v(n)$, we arrive at

$$\sum_{k=0}^{K} g^{(d)}(k)\bar{c}_{xx}(-k;k-m)=\sigma_w^2\, h^*(dP-m)$$
$$=0,\quad \text{for } d=0,\ m>0. \tag{17.61}$$

Equation 17.61 can be solved for the equalizer coefficients in batch or adaptive forms using recursive least-squares or the computationally simpler LMS algorithm suitably modified to compute the cyclic correlation statistics [29]. It turns out that using $\{g^{(0)}(k)\}_{k=0}^{K}$ one can find $\{g^{(d)}(k)\}_{k=0}^{K}$ for $d\in[1,L+K]$, which is important because, in practice, nonzero delay equalizers often achieve lower MSE [29].

Another interesting feature of the overall system in Figure 17.17 is that in the absence of noise ($v(n)\equiv 0$), the FIR equalizer $\{g^{(d)}(n)\}_{k=0}^{K}$ can equalize the FIR channel $h(n)$ perfectly in the zero-forcing (ZF) sense: $\sum_{k=0}^{K} g^{(d)}(k)h(nP-k)=\delta(n-d)$, provided that: (1) the channel $H(z)$ has no equispaced zeros on a circle with each zero separated from the next by $2\pi/P$ and (2) the equalizer has order satisfying $K\geq L/(P-1)-1$. Such a ZF equalizer can be found from the solution of Equation 17.61 provided that conditions (1) and (2) are satisfied. The equalizer obtained is unique when (2) is satisfied as equality, or when the minimum norm solution is adopted [29]. Recall that with symbol rate sampling ($P=1$), FIR-ZF equalizers are impossible because the inverse of an FIR $H(z)$ is always the IIR $G(z):=1/H(z)$. Further with $P=1$, FIR-MMSE (i.e., Wiener) equalizers cannot be ZF. In [29], it is also shown that under conditions (1) and (2), it is possible to have FIR hybrid MMSE-ZF equalizers.

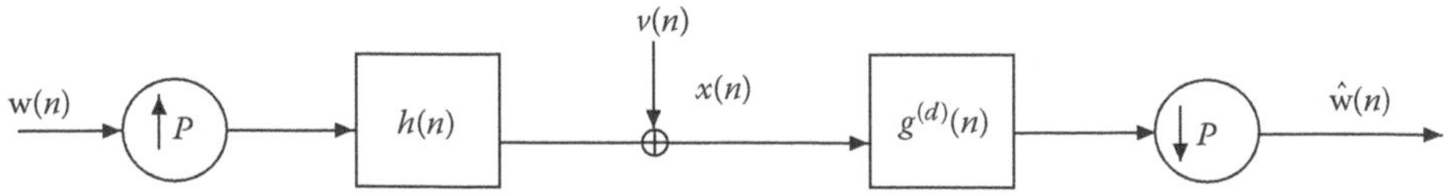

FIGURE 17.17 Cyclic (or multirate) channel-equalizer model.

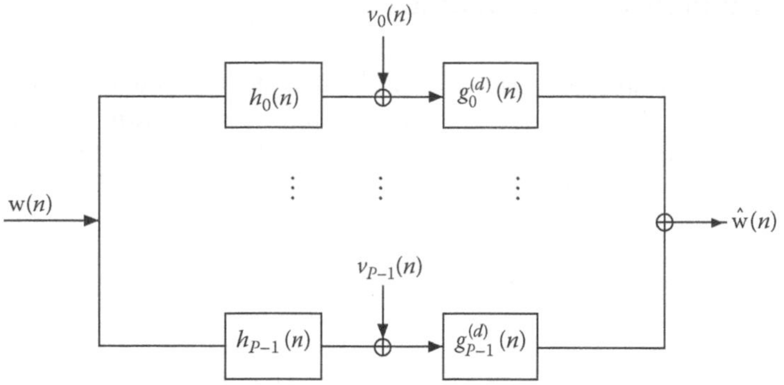

FIGURE 17.18 Multivariate channel-equalizer model.

The FIR channel–FIR equalizer feature can be seen also from the multichannel viewpoint which applies after the CS data $x(n)$ are mapped to the stationary components $\{x_i(n)\}_{i=0}^{P-1}$, or when P sensors collect symbol rate samples as in Equation 17.38. With reference to Figure 17.18, the channel-equalizer transfer functions satisfy, in the absence of noise, the so-called Bezout's identity: $\sum_{i=0}^{P-1} H_i(z)G_i^{(d)}(z) = z^{-d}$, which is analogous to the condition encountered with perfect reconstruction filterbanks. Given the Lth-order FIR analysis bank (H_i), existence and uniqueness of the Kth-order FIR synthesis filters (G_i) is guaranteed when (1) $\{H_i(z)\}_{i=0}^{P-1}$ have no common zeros and (2) $K \geq L/(P-1) - 1$. Next, we illustrate how the blind MMSE equalizer of Equation 17.61 can be used to mitigate inter-symbol interference (ISI) introduced by a two-ray multipath channel.

Example 17.7: Direct Blind Equalization

We generated 16-QAM symbols and passed them through a seventh-order FIR channel obtained by sampling at a rate $T_0/2$ the continuous-time channel $h_c(t) = \exp(-j2\pi 0.15)\rho_c(t - 0.25T_0, 0.35) + 0.8 \exp(-j2\pi 0.6)\rho_c(t - T_0, 0.35)$, where $\rho_c(t, 0.35)$ denotes the raised cosine pulse with roll-off factor 0.35 [53, p. 546]. We estimated the time-varying correlations as in Equation 17.24 and solved Equation 17.61 for the equalizer of order $K = 6$ and $d = 0$. At SNR = 25 dB, Figure 17.19 shows the received and equalized constellations illustrating the ability of the blind equalizer to remove ISI.

In our final application, we will be concerned with parameter estimation of APTV systems.

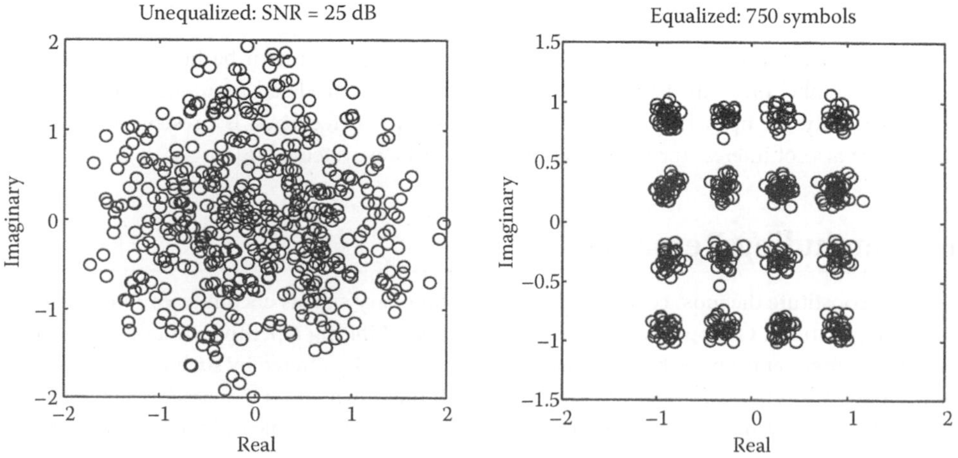

FIGURE 17.19 Before and after equalization (Example 17.7).

Application 5: Parametric APTV Modeling

Seasonal (e.g., atmospheric) time series are often modeled as the CS output of a linear APTV system $h(n;l)$ with i.i.d. input $w(n)$. Suppose that $x(n)$ obeys an autoregressive [AR(p_n)] model with coefficients $a(n;l)$ which are periodic in n with period P_l. The time series $x(n)$ and its correlation $c_{xx}(n;\tau)$ obey the following periodically varying AR recursions:

$$
\begin{aligned}
x(n) + \sum_{l=1}^{p_n} a(n;l)x(n-l) &= w(n), \\
c_{xx}(n;\tau) + \sum_{l=1}^{p_n} a(n;l)c_{xx}(n-l;l-\tau) &= \sigma_w^2(n)\delta(\tau).
\end{aligned}
\tag{17.62}
$$

The "periodic normal equations" in Equation 17.62 can be solved for each n to estimate the $a(n;l)$ parameters. Relying on Representation 17.1, [49] showed how PTV-AR modeling algorithms can be used to estimate multivariate AR coefficient matrices. Usage of single channel cyclic (instead of multivariate) statistics for parametric modeling of multichannel stationary time series was motivated on the basis of potential computational savings; see [49] for details and also [55] for cyclic lattice structures. Maximum likelihood estimation of periodic ARMA (PARMA) models is reported in [66]. PARMA modeling is important for seasonal time series encountered in meteorology, climatology [41], and stratospheric ozone data analysis [4]. Linear methods for estimating periodic MA coefficients along with important TV-MA parameter identifiability issues can be found in [13] using higher than second-order cyclic statistics.

When both input and output CS data are available, it is possible to identify linear PTV systems $h(n;l)$, even in the presence of correlated stationary input and output noise. Taking advantage of nonzero cycles present in the input and/or the system, one employs auto- and cross-cyclic spectra to identify $H(\beta;\omega)$, the cyclic spectrum of $h(n;l)$, relying on Equation 17.45 or 17.46, when $w(n)$ is stationary.

If the underlying system is TI (e.g., a frequency selective communications channel or a dispersive delay medium), a closed-form solution is possible in the frequency domain. With $\beta = 0$, Equation 17.45 yields $H(\omega) = \bar{S}_{xw}(\alpha;\omega)/\bar{S}_{ww}(\alpha;\omega)$, where $\alpha \in A_{ww}^c$ (see also [17]). For Lth-order FIR system identification a parametric approach in the lag-domain may be preferred because it avoids the trade-offs involved in choosing windows for nonparametric cyclic spectral estimates. One simply solves the following system of linear equations formed by cyclic (cross-) correlations [26]

$$
\sum_{l=0}^{L} h(l)\bar{C}_{ww}(\alpha;\tau-l) = \bar{C}_{xw}(\alpha;\tau)
\tag{17.63}
$$

using batch or adaptive algorithms. If desired, pole-zero models can then be fit in the estimated $\hat{h}(n)$ using Padé or Hankel methods. Estimation of TI systems with correlated input–output disturbances is important not only for open-loop identification but also when feedback is present. Therefore, cyclic approaches are also of interest for identification of closed-loop systems [26].

17.6 Concluding Remarks

CS processes constitute the most common class of nonstationary signals encountered in engineering and time series applications. CS appears in signals and systems exhibiting repetitive variations and allows for separation of components on the basis of their cycles. The diversity offered by such a structured variation can be exploited for suppression of stationary noise with unknown spectral characteristics and for blind parameter estimation using a single data record. Variance of finite sample estimates is affected by noise and increases when the cycles are unknown and have to be estimated prior to applying cyclic signal processing algorithms.

Although our discussion focused on linear systems and second-order statistical descriptors, CS appears also with nonlinear systems and certain signals exhibit periodicity in their higher than second-order statistics. The latter are especially useful because in both cases the underlying processes are non-Gaussian and second-order analysis cannot characterize them completely. CS in nonlinear time series of the Volterra-type is exploited in [21,30,46], whereas sample estimation issues and motivating applications of higher order CS can be found in [11,12,23,59] and references therein.

Topics of current interest and future trends include algorithms for nonlinear signal processing, theoretical performance evaluation, and analysis of CS point processes. As far as applications, exploitation of CS is expected to further improve algorithms in manufacturing problems involving vibrating and rotating components, and will continue to contribute in the design of single- and multiuser digital communication systems especially in the presence of fading and time-varying multipath environments.

Acknowledgments

The author wishes to thank his former and current graduate students for shaping up the content and helping with the preparation of this manuscript. This work was supported by ONR Grant N0014-93-1-0485.

References

1. Agee, B.G., Schell, S.V., and Gardner, W.A., Spectral self-coherence restoral: A new approach to blind adaptive signal extraction using antenna arrays, *Proc. IEEE*, 78, 753–767, 1990.
2. Bell, M.R. and Grubbs, R.A., JEM modeling and measurement for radar target identification, *IEEE Trans. Aerosp. Electron. Syst.*, 29, 73–87, 1993.
3. Bennet, W.R., Statistics of regenerative digital transmission, *Bell Syst. Tech. J.*, 37, 1501–1542, 1958.
4. Bloomfield, P., Hurd, H.L., and Lund, R.B., Periodic correlation in stratospheric ozone data, *J. Time Ser. Anal.*, 15, 127–150, 1994.
5. Brillinger, D.R., *Time Series, Data Analysis and Theory*, McGraw-Hill, New York, 1981.
6. Castedo, L., Figueiras, V., and Anibal, R., An adaptive beamforming technique based on cyclostationary signal properties, *IEEE Trans. Signal Process.*, 43, 1637–1650, 1995.
7. Chen, C.-K. and Gardner, W.A., Signal-selective time-difference-of-arrival estimation for passive location of manmade signal sources in highly-corruptive environments: Part II: Algorithms and performance, *IEEE Trans. Signal Process.*, 40, 1185–1197, 1992.
8. Chen, W., Giannakis, G.B., and Nandhakumar, N., Spatio-temporal approach for time-varying image motion estimation, *IEEE Trans. Image Process.*, 10, 1448–1461, 1996.
9. Corduneanu, C., *Almost Periodic Functions*, Interscience Publishers (John Wiley & Sons), New York, 1968.
10. Dandawate, A.V. and Giannakis, G.B., Statistical tests for presence of cyclostationarity, *IEEE Trans. Signal Process.*, 42, 2355–2369, 1994.
11. Dandawate, A.V. and Giannakis, G.B., Nonparametric polyspectral estimators for kth-order (almost) cyclostationary processes, *IEEE Trans. Inf. Theory*, 40, 67–84, 1994.
12. Dandawate, A.V. and Giannakis, G.B., Asymptotic theory of mixed time averages and kth-order cyclic- moment and cumulant statistics, *IEEE Trans. Inf. Theory*, 41, 216–232, 1995.
13. Dandawate, A.V. and Giannakis, G.B., Modeling (almost) periodic moving average processes using cyclic statistics, *IEEE Trans. Signal Process.*, 44, 673–684, 1996.
14. Dragan, Y.P. and Yavorskii, I., The periodic correlation-random field as a model for bidimensional ocean waves, *Peredacha Informatsii*, 51, 15–25, 1982.

15. Gardner, W.A., *Statistical Spectral Analysis: A Nonprobabilistic Theory*, Prentice-Hall, Englewood Cliffs, NJ, 1988.
16. Gardner, W.A., Simplification of MUSIC and ESPRIT by exploitation of cyclostationarity, *Proc. IEEE*, 76, 845–847, 1988.
17. Gardner, W.A., Identification of systems with cyclostationary input and correlated input/output measurement noise, *IEEE Trans. Autom. Control*, 35, 449–452, 1990.
18. Gardner, W.A., Two alternative philosophies for estimation of the parameters of time-series, *IEEE Trans. Inf. Theory*, 37, 216–218, 1991.
19. Gardner, W.A., Exploitation of spectral redundancy in cyclostationary signals, *IEEE Acoust. Speech Signal Process. Mag.*, 8, 14–36, 1991.
20. Garder, W.A., Cyclic Wiener filtering: Theory and method, *IEEE Trans. Commn.*, 41, 151–163, 1993.
21. Gardner, W.A. and Archer, T.L., Exploitation of cyclostationarity for identifying the Volterra kernels of nonlinear systems, *IEEE Trans. Inf. Theory*, 39, 535–542, 1993.
22. Gardner, W.A. and Franks, L.E., Characterization of cyclostationary random processes, *IEEE Trans. Inf. Theory*, 21, 4–14, 1975.
23. Gardner, W.A. and Spooner, C.M., The cumulant theory of cyclostationary time-series, Part I: Foundation, *IEEE Trans. Signal Process.*, 42, 3387–3408, December 1994.
24. Genossar, M.J., Lev-Ari, H., and Kailath, T., Consistent estimation of the cyclic autocorrelation, *IEEE Trans. Signal Process.*, 42, 595–603, 1994.
25. Giannakis, G.B., A linear cyclic correlation approach for blind identification of FIR channels *Proceedings of 28th Asilomar Conference on Signals, Systems, and Computers*, Pacific Grove, CA, October 31–November 2, 1994, pp. 420–424.
26. Giannakis, G.B., Polyspectral and cyclostationary approaches for identification of closed loop systems, *IEEE Trans. Autom. Control*, 40, 882–885, 1995.
27. Giannakis, G.B., Filterbanks for blind channel identification and equalization, *IEEE Signal Process. Lett.*, 4, 184–187, June 1997.
28. Giannakis, G.B. and Chen, W., Blind blur identification and multichannel image restoration using cyclostationarity, *Proceedings of IEEE Workshop on Nonlinear Signal and Image Processing*, Vol. II, Halkidiki, Greece, June 20–22, 1995, pp. 543–546.
29. Giannakis, G.B. and Halford, S., Blind fractionally-spaced equalization of noisy FIR channels: Direct and adaptive solutions, *IEEE Trans. Signal Process.*, 45, 2277–2292, September 1997.
30. Giannakis, G.B. and Serpedin, E., Linear multichannel blind equalizers of nonlinear FIR Volterra channels, *IEEE Trans. Signal Process.*, 45, 67–81, January 1997.
31. Giannakis, G.B. and Zhou, G., Parameter estimation of cyclostationary amplitude modulated time series with application to missing observations, *IEEE Trans. Signal Process.*, 42, 2408–2419, 1994.
32. Giannakis, G.B. and Zhou, G., Harmonics in multiplicative and additive noise: Parameter estimation using cyclic statistics, *IEEE Trans. Signal Process.*, 43, 2217–2221, 1995.
33. Gini, F. and Giannakis, G.B., Frequency offset and timing estimation in slowly-varying fading channels: A cyclostationary approach, *Proceedings of 1st IEEE Signal Processing Workshop on Wireless Communications*, Paris, France, April 16–18, 1997, pp. 393–396.
34. Gladyšev, E.G., Periodically correlated random sequences, *Sov. Math.*, 2, 385–388, 1961.
35. Hasselmann, K. and Barnett, T.P., Techniques of linear prediction of systems with periodic statistics, *J. Atmos. Sci.*, 38, 2275–2283, 1981.
36. Hinich, M.J., *Statistical Spectral Analysis: Nonprobabilistic Theory*, book review in *SIAM Review*, 33, 677–678, 1991.
37. Hlawatsch, F. and Boudreaux-Bartels, G.F., Linear and quadratic time-frequency representations, *IEEE Signal Process. Mag.*, 9(2), 21–67, April 1992.
38. Hurd, H.L., An investigation of periodically correlated stochastic processes, PhD dissertation, Duke University, Durham, NC, 1969.

39. Hurd, H.L., Nonparametric time series analysis of periodically correlated processes, *IEEE Trans. Inf. Theory*, 35(2), 350–359, March 1989.
40. Hurd, H.L. and Gerr, N.L., Graphical methods for determining the presence of periodic correlation, *J. Time Ser. Anal.*, 12, 337–350, 1991.
41. Jones, R.H. and Brelsford, W.M., Time series with periodic structure, *Biometrika*, 54, 403–408, 1967.
42. Kay, S.M., *Modern Spectral Estimation—Theory and Application*, Prentice-Hall, Englewood Cliffs, NJ, 1988.
43. Koenig, D. and Boehme, J., Application of cyclostationarity and time-frequency analysis to engine car diagnostics, *Proceedings of the International Conference on Acoustics, Speech and Signal Processing*, Adelaide, Australia, 1994, pp. 149–152.
44. Liu, H., Giannakis, G.B., and Tsatsanis, M.K., Time-varying system identification: A deterministic blind approach using antenna arrays, *Proceedings of 30th Conference on Information Sciences and Systems*, Princeton University, Princeton, NJ, March 20–22, 1996, pp. 880–884.
45. Longo, G. and Picinbono, B. (Eds.), *Time and Frequency Representation of Signals*, Springer-Verlag, New York, 1989.
46. Marmarelis, V.Z., Practicable identification of nonstationary and nonlinear systems, *IEEE Proc. Part D*, 128, 211–214, 1981.
47. Moulines, E., Duhamel, P., Cardoso, J.-F., and Mayrargue, S., Subspace methods for the blind identification of multichannel FIR filters, *IEEE Trans. Signal Process.*, 43, 516–525, 1995.
48. Newton, H.J., Using periodic autoregressions for multiple spectral estimation, *Technometrics*, 24, 109–116, 1982.
49. Pagano, M., On periodic and multiple autoregressions, *Ann. Stat.*, 6, 1310–1317, 1978.
50. Parzen, E. and Pagano, M., An approach to modeling seasonally stationary time-series, *J. Econometrics*, 9, 137–153, 1979.
51. Porat, B., *A Course in Digital Signal Processing*, John Wiley & Sons, New York, 1997.
52. Porat, B. and Friedlander, B., Blind equalization of digital communication channels using high-order moments, *IEEE Trans. Signal Process.*, 39, 522–526, 1991.
53. Proakis, J., *Digital Communications*, 3rd ed., McGraw-Hill, New York, 1989.
54. Riba, J. and Vazquez, G., Bayesian recursive estimation of frequency and timing exploiting the cyclostationarity property, *Signal Process.*, 40, 21–37, 1994.
55. Sakai, H., Circular lattice filtering using Pagano's method, *IEEE Trans. Acoust. Speech Signal Process.*, 30, 279–287, 1982.
56. Sakai, H., On the spectral density matrix of a periodic ARMA process, *J. Time Ser. Anal.*, 12, 73–82, 1991.
57. Sathe, V.P. and Vaidyanathan, P.P., Effects of multirate systems on the statistical properties of random signals, *IEEE Trans. Signal Process.*, 131–146, 1993.
58. Schell, S.V., An overview of sensor array processing for cyclostationary signals, in *Cyclostationarity in Communications and Signal Processing*, Gardner, W.A. (Ed.), IEEE Press, New York, 1994, pp. 168–239.
59. Spooner, C.M. and Gardner, W.A., The cumulant theory of cyclostationary time-series: Development and applications, *IEEE Trans. Signal Process.*, 42, 3409–3429, 1994.
60. Tong, L., Xu, G., and Kailath, T., Blind identification and equalization based on second-order statistics: A time domain approach, *IEEE Trans. Inf. Theory*, 340–349, 1994.
61. Tong, L., Xu, G., Hassibi, B., and Kailath, T., Blind channel identification based on second-order statistics: A frequency-domain approach, *IEEE Trans. Information Theory*, 41, 329–334, 1995.
62. Tsatsanis, M.K. and Giannakis, G.B., Modeling and equalization of rapidly fading channels, *Int. J. Adaptive Control Signal Process.*, 10, 159–176, 1996.
63. Tsatsanis, M.K. and Giannakis, G.B., Optimal linear receivers for DS-CDMA systems: A signal processing approach, *IEEE Trans. Signal Process.*, 44, 3044–3055, 1996.

64. Tsatsanis, M.K. and Giannakis, G.B., Blind estimation of direct sequence spread spectrum signals in multipath, *IEEE Trans. Signal Process.*, 45, 1241–1252, 1997.
65. Tsatsanis, M.K. and Giannakis, G.B., Transmitter induced cyclostationarity for blind channel equalization, *IEEE Trans. Signal Process.*, 45, 1785–1794, 1997.
66. Vecchia, A.V., Periodic autoregressive-moving average (PARMA) modeling with applications to water resources, *Water Res. Bull.*, 21, 721–730, 1985.
67. Wilbur, J.-E. and McDonald, R.J., Nonlinear analysis of cyclically correlated spectral spreading in modulated signals, *J. Acoust. Soc. Am.*, 92, 219–230, 1992.
68. Xu, G. and Kailath, T., Direction-of-arrival estimation via exploitation of cyclostationarity—A combination of temporal and spatial processing, *IEEE Trans. Signal Process.*, 40, 1775–1786, 1992.
69. Xu, G., Liu, H., Tong, L., and Kailath, T., A least-squares approach to blind channel identification, *IEEE Trans. Signal Process.*, 43, 2982–2993, 1995.
70. Zhou, G. and Giannakis, G.B., Performance analysis of cyclic time-delay estimation algorithms, *Proceedings of 29th Conference. on Information Sciences and Systems*, The Johns Hopkins University, Baltimore, MD, March 22–24, 1995, pp. 780–785.

VI

Adaptive Filtering

Scott C. Douglas
Southern Methodist University

A FILTER IS, IN ITS MOST BASIC SENSE, a device that enhances and/or rejects certain components of a signal. To adapt is to change one's characteristics according to some knowledge about one's environment. Taken together, these two terms suggest the goal of an adaptive filter: to alter its selectivity based on the specific characteristics of the signals that are being processed.

In digital signal processing, the term "adaptive filters" refers to a particular set of computational structures and methods for processing digital signals. While many of the most popular techniques used in adaptive filters have been developed and refined within the past forty years, the field of adaptive filters is part of the larger field of optimization theory that has a history dating back to the scientific work of both Galileo and Gauss in the eighteenth and nineteenth centuries. Modern developments in adaptive filters began in the 1930s and 1940s with the efforts of Kolmogorov, Wiener, and Levinson to formulate and solve linear estimation tasks. For those who desire an overview of many of the structures, algorithms, analyses, and applications of adaptive filters, the seven chapters in this section provide an excellent introduction to several prominent topics in the field.

Chapter 18 presents an overview of adaptive filters, describing many of the applications for which these systems are used today. This chapter considers basic adaptive filtering concepts while providing an introduction to the popular least-mean-square (LMS) adaptive filter that is often used in these applications.

Chapters 19 and 20 focus on the design of the LMS adaptive filter from two different viewpoints. In the former chapter, the behavior of the LMS adaptive filter is analyzed within a statistical framework that has proven to be quite useful for establishing initial choices of the parameter values of this system. The latter chapter studies the behavior of the LMS adaptive filter from a deterministic viewpoint, showing why this system behaves robustly even when modeling errors and finite-precision calculation errors continually perturb the state of this adaptive filter.

Chapter 21 presents the techniques used in another popular class of adaptive systems collectively known as recursive least-squares adaptive filters. Focusing on the numerical methods that are typically employed in the implementations of these systems, the chapter provides a detailed summary of both conventional and "fast" computational methods for these high-performance systems.

Transform domain adaptive filtering is discussed in Chapter 22. Using the frequency-domain and fast convolution techniques described in this chapter, it is possible both to reduce the computational complexity and to increase the performance of LMS adaptive filters when implemented in block form.

The first five chapters of this section focus almost exclusively on adaptive structures of a finite-impulse response form. In Chapter 23, the subtle performance issues surrounding methods for adaptive infinite-impulse-response (IIR) filters are carefully described. The most recent technical results concerning the convergence behavior and stability of each major adaptive IIR algorithm class is provided in an easy-to-follow format.

Finally, Chapter 24 presents an important emerging application area for adaptive filters: blind equalization. This section indicates how an adaptive filter can be adjusted to produce a desirable input/output characteristic without having an example desired output signal on which to be trained.

While adaptive filters have had a long history, new adaptive filter structures and algorithms are continually being developed. In fact, the range of adaptive filtering algorithms and applications is so great that no one paper, chapter, section, or even book can fully cover the field. Those who desire more information on the topics presented in this section should consult works within the extensive reference lists that appear at the end of each chapter.

18

Introduction to Adaptive Filters

Scott C. Douglas
Southern Methodist University

18.1 What Is an Adaptive Filter?

An adaptive filter is a computational device that attempts to model the relationship between two signals in real time in an iterative manner. Adaptive filters are often realized either as a set of program instructions running on an arithmetical processing device such as a microprocessor or DSP chip, or as a set of logic operations implemented in a field-programmable gate array or in a semi-custom or custom VLSI integrated circuit. However, ignoring any errors introduced by numerical precision effects in these implementations, the fundamental operation of an adaptive filter can be characterized independently of the specific physical realization that it takes. For this reason, we shall focus on the mathematical forms of adaptive filters as opposed to their specific realizations in software or hardware. Descriptions of adaptive filters as implemented on DSP chips and on a dedicated integrated circuit can be found in [1–3] and [4], respectively.

An adaptive filter is defined by four aspects:

1. The signals being processed by the filter
2. The structure that defines how the output signal of the filter is computed from its input signal
3. The parameters within this structure that can be iteratively changed to alter the filter's input–output relationship
4. The adaptive algorithm that describes how the parameters are adjusted from one time instant to the next

By choosing a particular adaptive filter structure, one specifies the number and type of parameters that can be adjusted. The adaptive algorithm used to update the parameter values of the system can take on a myriad of forms and is often derived as a form of optimization procedure that minimizes an error criterion that is useful for the task at hand.

In this section, we present the general adaptive filtering problem and introduce the mathematical notation for representing the form and operation of the adaptive filter. We then discuss several different structures that have been proven to be useful in practical applications. We provide an overview of the many and varied applications in which adaptive filters have been successfully used. Finally, we give a simple derivation of the least-mean-square (LMS) algorithm, which is perhaps the most popular method for adjusting the coefficients of an adaptive filter, and we discuss some of this algorithm's properties.

As for the mathematical notation used throughout this section, all quantities are assumed to be real-valued. Scalar and vector quantities shall be indicated by lowercase (e.g., x) and uppercase-bold (e.g., $\mathbf{X}$) letters, respectively. We represent scalar and vector sequences or signals as $x(n)$ and $\mathbf{X}(n)$, respectively, where n denotes the discrete time or discrete spatial index, depending on the application. Matrices and indices of vector and matrix elements shall be understood through the context of the discussion.

18.2 Adaptive Filtering Problem

Figure 18.1 shows a block diagram in which a sample from a digital input signal $x(n)$ is fed into a device, called an adaptive filter, that computes a corresponding output signal sample $y(n)$ at time n. For the moment, the structure of the adaptive filter is not important, except for the fact that it contains adjustable parameters whose values affect how $y(n)$ is computed. The output signal is compared to a second signal $d(n)$, called the desired response signal, by subtracting the two samples at time n. This difference signal, given by

$$e(n) = d(n) - y(n), \tag{18.1}$$

is known as the error signal. The error signal is fed into a procedure which alters or adapts the parameters of the filter from time n to time $(n + 1)$ in a well-defined manner. This process of adaptation is represented by the oblique arrow that pierces the adaptive filter block in the figure. As the time index n is incremented, it is hoped that the output of the adaptive filter becomes a better and better match to the desired response signal through this adaptation process, such that the magnitude of $e(n)$ decreases over time. In this context, what is meant by "better" is specified by the form of the adaptive algorithm used to adjust the parameters of the adaptive filter.

In the adaptive filtering task, adaptation refers to the method by which the parameters of the system are changed from time index n to time index $(n + 1)$. The number and types of parameters within this system depend on the computational structure chosen for the system. We now discuss different filter structures that have been proven useful for adaptive filtering tasks.

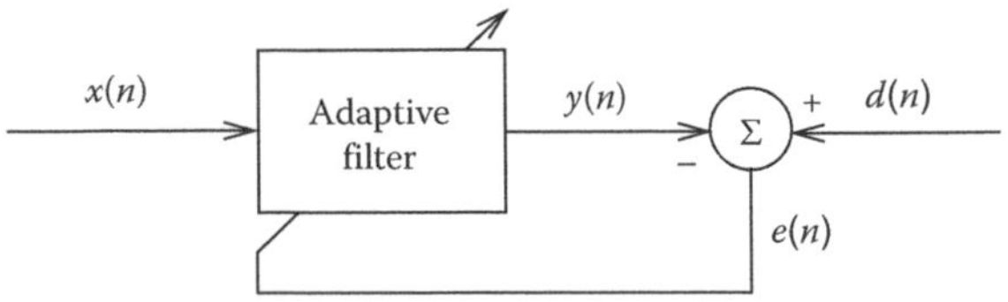

FIGURE 18.1 The general adaptive filtering problem.

18.3 Filter Structures

In general, any system with a finite number of parameters that affect how $y(n)$ is computed from $x(n)$ could be used for the adaptive filter in Figure 18.1. Define the parameter or coefficient vector $\mathbf{W}(n)$ as

$$\mathbf{W}(n) = [w_0(n)\, w_1(n) \cdots w_{L-1}(n)]^T, \tag{18.2}$$

where $\{w_i(n)\}$, $0 \leq i \leq L-1$, are the L parameters of the system at time n. With this definition, we could define a general input-output relationship for the adaptive filter as

$$y(n) = f(\mathbf{W}(n), y(n-1), y(n-2), \&, y(n-N), x(n), x(n-1), \&, x(n-\boldsymbol{M}+1)), \tag{18.3}$$

where

$f(\cdot)$ represents any well-defined linear or nonlinear function
M and N are positive integers

Implicit in this definition is the fact that the filter is causal, such that future values of $x(n)$ are not needed to compute $y(n)$. While noncausal filters can be handled in practice by suitably buffering or storing the input signal samples, we do not consider this possibility.

Although Equation 18.3 is the most general description of an adaptive filter structure, we are interested in determining the best linear relationship between the input and desired response signals for many problems. This relationship typically takes the form of a finite-impulse-response (FIR) or infinite-impulse-response (IIR) filter. Figure 18.2 shows the structure of a direct-form FIR filter, also known as a tapped-delay-line or transversal filter, where z^{-1} denotes the unit delay element and each $w_i(n)$ is a multiplicative gain within the system. In this case, the parameters in $\mathbf{W}(n)$ correspond to the impulse response values of the filter at time n. We can write the output signal $y(n)$ as

$$y(n) = \sum_{i=0}^{L-1} w_i(n)x(n-i) \tag{18.4}$$

$$= \mathbf{W}^T(n)\mathbf{X}(n), \tag{18.5}$$

where

$\mathbf{X}(n) = [x(n)\, x(n-1) \cdots x(n-L+1)]^T$ denotes the input signal vector
superscript T denotes vector transpose

Note that this system requires L multiplies and $L-1$ adds to implement, and these computations are easily performed by a processor or circuit so long as L is not too large and the sampling period for the signals is not too short. It also requires a total of $2L$ memory locations to store the L input signal samples and the L coefficient values, respectively.

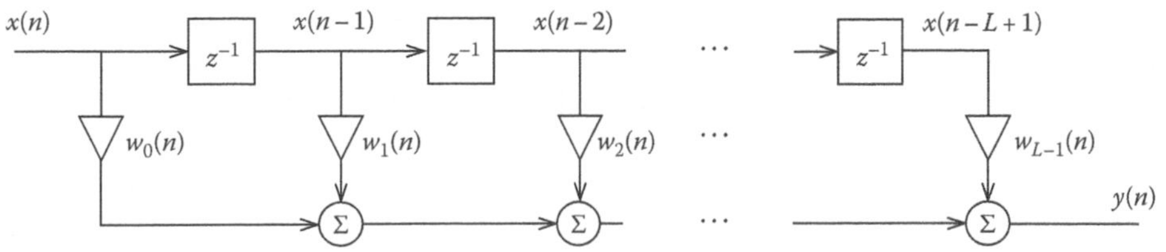

FIGURE 18.2 Structure of an FIR filter.

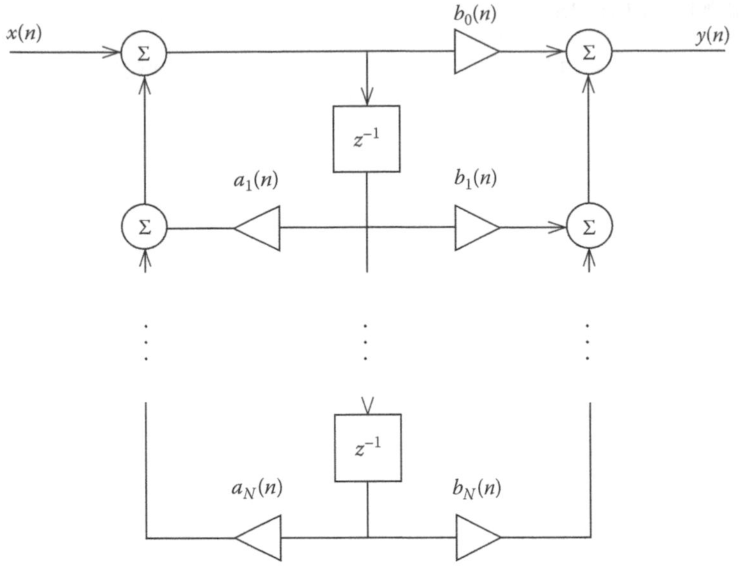

FIGURE 18.3 Structure of an IIR filter.

The structure of a direct-form IIR filter is shown in Figure 18.3. In this case, the output of the system can be represented mathematically as

$$y(n) = \sum_{i=1}^{N} a_i(n)y(n-i) + \sum_{j=0}^{N} b_j(n)x(n-j), \tag{18.6}$$

although the block diagram does not explicitly represent this system in such a fashion.* We could easily write Equation 18.6 using vector notation as

$$y(n) = \mathbf{W}^{\mathrm{T}}(n)\mathbf{U}(n), \tag{18.7}$$

where the $(2N+1)$-dimensional vectors $\mathbf{W}(n)$ and $\mathbf{U}(n)$ are defined as

$$\mathbf{W}(n) = [a_1(n)a_2(n)\cdots a_N(n)b_0(n)b_1(n)\cdots b_N(n)]^{\mathrm{T}} \tag{18.8}$$

$$\mathbf{U}(n) = [y(n-1)y(n-2)\cdots y(n-N)x(n)x(n-1)\cdots x(n-N)]^{\mathrm{T}}, \tag{18.9}$$

respectively. Thus, for purposes of computing the output signal $y(n)$, the IIR structure involves a fixed number of multiplies, adds, and memory locations not unlike the direct-form FIR structure.

A third structure that has proven useful for adaptive filtering tasks is the lattice filter. A lattice filter is an FIR structure that employs $L-1$ stages of preprocessing to compute a set of auxiliary signals $\{b_i(n)\}$, $0 \leq i \leq L-1$ known as backward prediction errors. These signals have the special property that they are uncorrelated, and they represent the elements of $\mathbf{X}(n)$ through a linear transformation. Thus, the backward prediction errors can be used in place of the delayed input signals in a structure similar to that in Figure 18.2, and the uncorrelated nature of the prediction errors can provide improved

* The difference between the *direct form II* or *canonical form structure* shown in Figure 18.3 and the *direct form I* implementation of this system as described by Equation 18.6 is discussed in [5].

convergence performance of the adaptive filter coefficients with the proper choice of algorithm. Details of the lattice structure and its capabilities are discussed in [6].

A critical issue in the choice of an adaptive filter's structure is its computational complexity. Since the operation of the adaptive filter typically occurs in real time, all of the calculations for the system must occur during one sample time. The structures described above are all useful because $y(n)$ can be computed in a finite amount of time using simple arithmetical operations and finite amounts of memory.

In addition to the linear structures above, one could consider nonlinear systems for which the principle of superposition does not hold when the parameter values are fixed. Such systems are useful when the relationship between $d(n)$ and $x(n)$ is not linear in nature. Two such classes of systems are the Volterra and bilinear filter classes that compute $y(n)$ based on polynomial representations of the input and past output signals. Algorithms for adapting the coefficients of these types of filters are discussed in [7]. In addition, many of the nonlinear models developed in the field of neural networks, such as the multilayer perceptron, fit the general form of Equation 18.3, and many of the algorithms used for adjusting the parameters of neural networks are related to the algorithms used for FIR and IIR adaptive filters. For a discussion of neural networks in an engineering context, the reader is referred to [8].

18.4 Task of an Adaptive Filter

When considering the adaptive filter problem as illustrated in Figure 18.1 for the first time, a reader is likely to ask, "If we already have the desired response signal, what is the point of trying to match it using an adaptive filter?" In fact, the concept of "matching" $y(n)$ to $d(n)$ with some system obscures the subtlety of the adaptive filtering task. Consider the following issues that pertain to many adaptive filtering problems:

- In practice, the quantity of interest is not always $d(n)$. Our desire may be to represent in $y(n)$ a certain component of $d(n)$ that is contained in $x(n)$, or it may be to isolate a component of $d(n)$ within the error $e(n)$ that is not contained in $x(n)$. Alternatively, we may be solely interested in the values of the parameters in $\mathbf{W}(n)$ and have no concern about $x(n)$, $y(n)$, or $d(n)$ themselves. Practical examples of each of these scenarios are provided later in this chapter.
- There are situations in which $d(n)$ is not available at all times. In such situations, adaptation typically occurs only when $d(n)$ is available. When $d(n)$ is unavailable, we typically use our most-recent parameter estimates to compute $y(n)$ in an attempt to estimate the desired response signal $d(n)$.
- There are real-world situations in which $d(n)$ is never available. In such cases, one can use additional information about the characteristics of a "hypothetical" $d(n)$, such as its predicted statistical behavior or amplitude characteristics, to form suitable estimates of $d(n)$ from the signals available to the adaptive filter. Such methods are collectively called blind adaptation algorithms. The fact that such schemes even work is a tribute both to the ingenuity of the developers of the algorithms and to the technological maturity of the adaptive filtering field.

It should also be recognized that the relationship between $x(n)$ and $d(n)$ can vary with time. In such situations, the adaptive filter attempts to alter its parameter values to follow the changes in this relationship as "encoded" by the two sequences $x(n)$ and $d(n)$. This behavior is commonly referred to as tracking.

18.5 Applications of Adaptive Filters

Perhaps the most important driving forces behind the developments in adaptive filters throughout their history have been the wide range of applications in which such systems can be used. We now discuss the forms of these applications in terms of more-general problem classes that describe the assumed relationship between $d(n)$ and $x(n)$. Our discussion illustrates the key issues in selecting an adaptive filter for a particular task. Extensive details concerning the specific issues and problems associated with each problem genre can be found in the references at the end of this chapter.

18.5.1 System Identification

Consider Figure 18.4, which shows the general problem of system identification. In this diagram, the system enclosed by dashed lines is a "black box," meaning that the quantities inside are not observable from the outside. Inside this box is (1) an unknown system which represents a general input–output relationship and (2) the signal $\eta(n)$, called the observation noise signal because it corrupts the observations of the signal at the output of the unknown system.

Let $\hat{d}(n)$ represent the output of the unknown system with $x(n)$ as its input. Then, the desired response signal in this model is

$$d(n) = \hat{d}(n) + \eta(n). \tag{18.10}$$

Here, the task of the adaptive filter is to accurately represent the signal $\hat{d}(n)$ at its output. If $y(n) = \hat{d}(n)$, then the adaptive filter has accurately modeled or identified the portion of the unknown system that is driven by $x(n)$.

Since the model typically chosen for the adaptive filter is a linear filter, the practical goal of the adaptive filter is to determine the best linear model that describes the input–output relationship of the unknown system. Such a procedure makes the most sense when the unknown system is also a linear model of the same structure as the adaptive filter, as it is possible that $y(n) = \hat{d}(n)$ for some set of adaptive filter parameters. For ease of discussion, let the unknown system and the adaptive filter both be FIR filters, such that

$$d(n) = \mathbf{W}_{\text{opt}}^{T}(n)\mathbf{X}(n) + \eta(n), \tag{18.11}$$

where $\mathbf{W}_{\text{opt}}(n)$ is an optimum set of filter coefficients for the unknown system at time n. In this problem formulation, the ideal adaptation procedure would adjust $\mathbf{W}(n)$ such that $\mathbf{W}(n) = \mathbf{W}_{\text{opt}}(n)$ as $n \to \infty$. In practice, the adaptive filter can only adjust $\mathbf{W}(n)$ such that $y(n)$ closely approximates $\hat{d}(n)$ over time.

The system identification task is at the heart of numerous adaptive filtering applications. We list several of these applications here.

18.5.1.1 Channel Identification

In communication systems, useful information is transmitted from one point to another across a medium such as an electrical wire, an optical fiber, or a wireless radio link. Nonidealities of the transmission medium or channel distort the fidelity of the transmitted signals, making the deciphering of the received

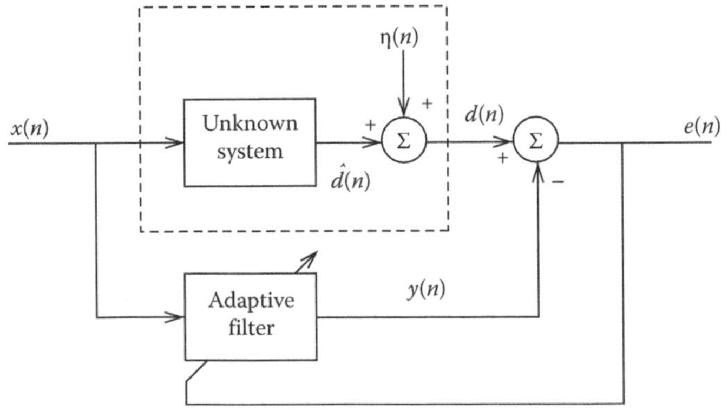

FIGURE 18.4 System identification.

information difficult. In cases where the effects of the distortion can be modeled as a linear filter, the resulting "smearing" of the transmitted symbols is known as inter-symbol interference (ISI). In such cases, an adaptive filter can be used to model the effects of the channel ISI for purposes of deciphering the received information in an optimal manner. In this problem scenario, the transmitter sends to the receiver a sample sequence $x(n)$ that is known to both the transmitter and receiver. The receiver then attempts to model the received signal $d(n)$ using an adaptive filter whose input is the known transmitted sequence $x(n)$. After a suitable period of adaptation, the parameters of the adaptive filter in $\mathbf{W}(n)$ are fixed and then used in a procedure to decode future signals transmitted across the channel.

Channel identification is typically employed when the fidelity of the transmitted channel is severely compromised or when simpler techniques for sequence detection cannot be used. Techniques for detecting digital signals in communication systems can be found in [9].

18.5.1.2 Plant Identification

In many control tasks, knowledge of the transfer function of a linear plant is required by the physical controller so that a suitable control signal can be calculated and applied. In such cases, we can characterize the transfer function of the plant by exciting it with a known signal $x(n)$ and then attempting to match the output of the plant $d(n)$ with a linear adaptive filter. After a suitable period of adaptation, the system has been adequately modeled, and the resulting adaptive filter coefficients in $\mathbf{W}(n)$ can be used in a control scheme to enable the overall closed-loop system to behave in the desired manner.

In certain scenarios, continuous updates of the plant transfer function estimate provided by $\mathbf{W}(n)$ are needed to allow the controller to function properly. A discussion of these adaptive control schemes and the subtle issues in their use is given in [10,11].

18.5.1.3 Echo Cancellation for Long-Distance Transmission

In voice communication across telephone networks, the existence of junction boxes called hybrids near either end of the network link hampers the ability of the system to cleanly transmit voice signals. Each hybrid allows voices that are transmitted via separate lines or channels across a long-distance network to be carried locally on a single telephone line, thus lowering the wiring costs of the local network. However, when small impedance mismatches between the long distance lines and the hybrid junctions occur, these hybrids can reflect the transmitted signals back to their sources, and the long transmission times of the long-distance network—about 0.3 s for a trans-oceanic call via a satellite link—turn these reflections into a noticeable echo that makes the understanding of conversation difficult for both callers. The traditional solution to this problem prior to the advent of the adaptive filtering solution was to introduce significant loss into the long-distance network so that echoes would decay to an acceptable level before they became perceptible to the callers. Unfortunately, this solution also reduces the transmission quality of the telephone link and makes the task of connecting long distance calls more difficult.

An adaptive filter can be used to cancel the echoes caused by the hybrids in this situation. Adaptive filters are employed at each of the two hybrids within the network. The input $x(n)$ to each adaptive filter is the speech signal being received prior to the hybrid junction, and the desired response signal $d(n)$ is the signal being sent out from the hybrid across the long-distance connection. The adaptive filter attempts to model the transmission characteristics of the hybrid junction as well as any echoes that appear across the long-distance portion of the network. When the system is properly designed, the error signal $e(n)$ consists almost totally of the local talker's speech signal, which is then transmitted over the network. Such systems were first proposed in the mid-1960s [12] and are commonly used today. For more details on this application, see [13,14].

18.5.1.4 Acoustic Echo Cancellation

A related problem to echo cancellation for telephone transmission systems is that of acoustic echo cancellation for conference-style speakerphones. When using a speakerphone, a caller would like to turn up the amplifier gains of both the microphone and the audio loudspeaker in order to transmit and hear the voice signals more clearly. However, the feedback path from the device's loudspeaker to its input microphone causes a distinctive howling sound if these gains are too high. In this case, the culprit is the room's response to the voice signal being broadcast by the speaker; in effect, the room acts as an extremely poor hybrid junction, in analogy with the echo cancellation task discussed previously. A simple solution to this problem is to only allow one person to speak at a time, a form of operation called half-duplex transmission. However, studies have indicated that half-duplex transmission causes problems with normal conversations, as people typically overlap their phrases with others when conversing.

To maintain full-duplex transmission, an acoustic echo canceller is employed in the speakerphone to model the acoustic transmission path from the speaker to the microphone. The input signal $x(n)$ to the acoustic echo canceller is the signal being sent to the speaker, and the desired response signal $d(n)$ is measured at the microphone on the device. Adaptation of the system occurs continually throughout a telephone call to model any physical changes in the room acoustics. Such devices are readily available in the marketplace today. In addition, similar technology can and is used to remove the echo that occurs through the combined radio/room/telephone transmission path when one places a call to a radio or television talk show. Details of the acoustic echo cancellation problem can be found in [14].

18.5.1.5 Adaptive Noise Canceling

When collecting measurements of certain signals or processes, physical constraints often limit our ability to cleanly measure the quantities of interest. Typically, a signal of interest is linearly mixed with other extraneous noises in the measurement process, and these extraneous noises introduce unacceptable errors in the measurements. However, if a linearly related reference version of any one of the extraneous noises can be cleanly sensed at some other physical location in the system, an adaptive filter can be used to determine the relationship between the noise reference $x(n)$ and the component of this noise that is contained in the measured signal $d(n)$. After adaptively subtracting out this component, what remains in $e(n)$ is the signal of interest. If several extraneous noises corrupt the measurement of interest, several adaptive filters can be used in parallel as long as suitable noise reference signals are available within the system.

Adaptive noise canceling has been used for several applications. One of the first was a medical application that enabled the electroencephalogram (EEG) of the fetal heartbeat of an unborn child to be cleanly extracted from the much-stronger interfering EEG of the maternal heartbeat signal. Details of this application as well as several others are described in the seminal paper by Widrow and his colleagues [15].

18.5.2 Inverse Modeling

We now consider the general problem of inverse modeling, as shown in Figure 18.5. In this diagram, a source signal $s(n)$ is fed into an unknown system that produces the input signal $x(n)$ for the adaptive filter. The output of the adaptive filter is subtracted from a desired response signal that is a delayed version of the source signal, such that

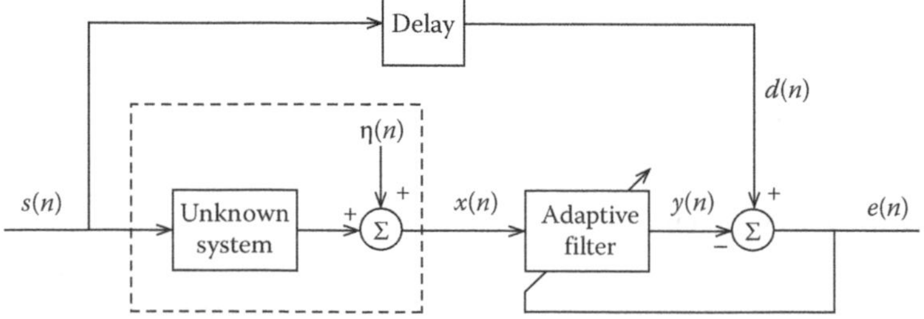

FIGURE 18.5 Inverse modeling.

$$d(n) = s(n - \Delta), \tag{18.12}$$

where Δ is a positive integer value. The goal of the adaptive filter is to adjust its characteristics such that the output signal is an accurate representation of the delayed source signal.

The inverse modeling task characterizes several adaptive filtering applications, two of which are now described.

18.5.2.1 Channel Equalization

Channel equalization is an alternative to the technique of channel identification described previously for the decoding of transmitted signals across nonideal communication channels. In both cases, the transmitter sends a sequence $s(n)$ that is known to both the transmitter and receiver. However, in equalization, the received signal is used as the input signal $x(n)$ to an adaptive filter, which adjusts its characteristics so that its output closely matches a delayed version $s(n - \Delta)$ of the known transmitted signal. After a suitable adaptation period, the coefficients of the system either are fixed and used to decode future transmitted messages or are adapted using a crude estimate of the desired response signal that is computed from $y(n)$. This latter mode of operation is known as decision-directed adaptation.

Channel equalization was one of the first applications of adaptive filters and is described in the pioneering work of Lucky [16]. Today, it remains as one of the most popular uses of an adaptive filter. Practically every computer telephone modem transmitting at rates of 9600 baud (bits per second) or greater contains an adaptive equalizer. Adaptive equalization is also useful for wireless communication systems. Qureshi [17] provides a tutorial on adaptive equalization. A related problem to equalization is deconvolution, a problem that appears in the context of geophysical exploration [18]. Equalization is closely related to linear prediction, a topic that we shall discuss shortly.

18.5.2.2 Inverse Plant Modeling

In many control tasks, the frequency and phase characteristics of the plant hamper the convergence behavior and stability of the control system. We can use a system of the form in Figure 18.5 to compensate for the nonideal characteristics of the plant and as a method for adaptive control. In this case, the signal $s(n)$ is sent at the output of the controller, and the signal $x(n)$ is the signal measured at the output of the plant. The coefficients of the adaptive filter are then adjusted so that the cascade of the plant and adaptive filter can be nearly represented by the pure delay $z^{-\Delta}$. Details of the adaptive algorithms as applied to control tasks in this fashion can be found in [11].

18.5.3 Linear Prediction

A third type of adaptive filtering task is shown in Figure 18.6. In this system, the input signal $x(n)$ is derived from the desired response signal as

$$x(n) = d(n - \Delta), \tag{18.13}$$

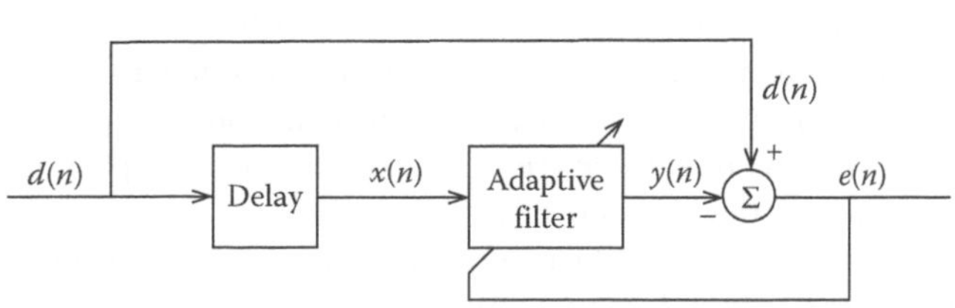

FIGURE 18.6 Linear prediction.

where Δ is an integer value of delay. In effect, the input signal serves as the desired response signal, and for this reason it is always $j \cdot x(n + \Delta)$ at time n is desired, a copy of the adaptive filter whose input is the current sample $x(n)$ can be employed to compute this quantity. However, linear prediction has a number of uses besides the obvious application of forecasting future events, as described in the following two applications.

18.5.3.1 Linear Predictive Coding

When transmitting digitized versions of real-world signals such as speech or images, the temporal correlation of the signals is a form of redundancy that can be exploited to code the waveform in a smaller number of bits than are needed for its original representation. In these cases, a linear predictor can be used to model the signal correlations for a short block of data in such a way as to reduce the number of bits needed to represent the signal waveform. Then, essential information about the signal model is transmitted along with the coefficients of the adaptive filter for the given data block. Once received, the signal is synthesized using the filter coefficients and the additional signal information provided for the given block of data.

When applied to speech signals, this method of signal encoding enables the transmission of understandable speech at only 2.4 kb/s, although the reconstructed speech has a distinctly synthetic quality. Predictive coding can be combined with a quantizer to enable higher quality speech encoding at higher data rates using an adaptive differential pulse-code modulation scheme. In both of these methods, the lattice filter structure plays an important role because of the way in which it parameterizes the physical nature of the vocal tract. Details about the role of the lattice filter in the linear prediction task can be found in [19].

18.5.3.2 Adaptive Line Enhancement

In some situations, the desired response signal $d(n)$ consists of a sum of a broadband signal and a nearly periodic signal, and it is desired to separate these two signals without specific knowledge about the signals (such as the fundamental frequency of the periodic component).

In these situations, an adaptive filter configured as in Figure 18.6 can be used. For this application, the delay Δ is chosen to be large enough such that the broadband component in $x(n)$ is uncorrelated with the broadband component in $x(n - \Delta)$. In this case, the broadband signal cannot be removed by the adaptive filter through its operation, and it remains in the error signal $e(n)$ after a suitable period of adaptation. The adaptive filter's output $y(n)$ converges to the narrowband component, which is easily predicted given past samples. The name line enhancement arises because periodic signals are characterized by lines in their frequency spectra, and these spectral lines are enhanced at the output of the adaptive filter.

For a discussion of the adaptive line enhancement task using LMS adaptive filters, the reader is referred to [20].

18.5.4 Feedforward Control

Another problem area combines elements of both the inverse modeling and system identification tasks and typifies the types of problems encountered in the area of adaptive control known as feedforward control. Figure 18.7 shows the block diagram for this system, in which the output of the adaptive filter passes through a plant before it is subtracted from the desired response to form the error signal. The plant hampers the operation of the adaptive filter by changing the amplitude and phase characteristics of the adaptive filter's output signal as represented in $e(n)$. Thus, knowledge of the plant is generally required in order to adapt the parameters of the filter properly.

An application that fits this particular problem formulation is active noise control, in which unwanted sound energy propagates in air or a fluid into a physical region in space. In such cases, an electroacoustic system employing microphones, speakers, and one or more adaptive filters can be used to create a

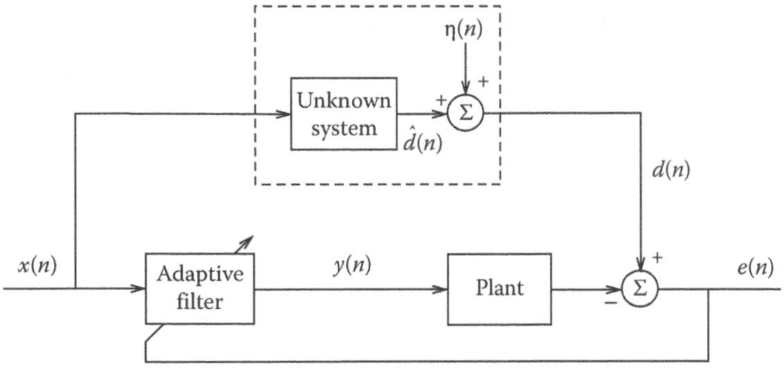

FIGURE 18.7 Feedforward control.

secondary sound field that interferes with the unwanted sound, reducing its level in the region via destructive interference. Similar techniques can be used to reduce vibrations in solid media. Details of useful algorithms for the active noise and vibration control tasks can be found in [21,22].

18.6 Gradient-Based Adaptive Algorithms

An adaptive algorithm is a procedure for adjusting the parameters of an adaptive filter to minimize a cost function chosen for the task at hand. In this section, we describe the general form of many adaptive FIR filtering algorithms and present a simple derivation of the LMS adaptive algorithm. In our discussion, we only consider an adaptive FIR filter structure, such that the output signal $y(n)$ is given by Equation 18.5. Such systems are currently more popular than adaptive IIR filters because (1) the input–output stability of the FIR filter structure is guaranteed for any set of fixed coefficients, and (2) the algorithms for adjusting the coefficients of FIR filters are more simple in general than those for adjusting the coefficients of IIR filters.

18.6.1 General Form of Adaptive FIR Algorithms

The general form of an adaptive FIR filtering algorithm is

$$\mathbf{W}(n+1) = \mathbf{W}(n) + \mu(n)\mathbf{G}(e(n), \mathbf{X}(n), \Phi(n)), \tag{18.14}$$

where

- $\mathbf{G}(\cdot)$ is a particular vector-valued nonlinear function
- $\mu(n)$ is a step size parameter
- $e(n)$ and $\mathbf{X}(n)$ are the error signal and input signal vector, respectively
- $\Phi(n)$ is a vector of states that store pertinent information about the characteristics of the input and error signals and/or the coefficients at previous time instants

In the simplest algorithms, $\Phi(n)$ is not used, and the only information needed to adjust the coefficients at time n are the error signal, input signal vector, and step size.

The step size is so called because it determines the magnitude of the change or "step" that is taken by the algorithm in iteratively determining a useful coefficient vector. Much research effort has been spent characterizing the role that $\mu(n)$ plays in the performance of adaptive filters in terms of the statistical or frequency characteristics of the input and desired response signals. Often, success or failure of an adaptive filtering application depends on how the value of $\mu(n)$ is chosen or calculated to obtain the best performance from the adaptive filter. The issue of choosing $\mu(n)$ for both stable and accurate convergence of the LMS adaptive filter is addressed in Chapter 19.

18.6.2 Mean-Squared Error Cost Function

The form of $\mathbf{G}(\cdot)$ in Equation 18.14 depends on the cost function chosen for the given adaptive filtering task. We now consider one particular cost function that yields a popular adaptive algorithm.

Define the mean-squared error (MSE) cost function as

$$J_{\mathrm{MSE}}(n) = \frac{1}{2}\int_{-\infty}^{\infty} e^2(n)p_n(e(n))de(n) \tag{18.15}$$

$$= \frac{1}{2}E\{e^2(n)\}, \tag{18.16}$$

where

$p_n(e)$ represents the probability density function of the error at time n
$E\{\cdot\}$ is shorthand for the expectation integral on the right-hand side of Equation 18.15

The MSE cost function is useful for adaptive FIR filters because

- $J_{\mathrm{MSE}}(n)$ has a well-defined minimum with respect to the parameters in $\mathbf{W}(n)$.
- The coefficient values obtained at this minimum are the ones that minimize the power in the error signal $e(n)$, indicating that $y(n)$ has approached $d(n)$.
- $J_{\mathrm{MSE}}(n)$ is a smooth function of each of the parameters in $\mathbf{W}(n)$, such that it is differentiable with respect to each of the parameters in $\mathbf{W}(n)$.

The third point is important in that it enables us to determine both the optimum coefficient values given knowledge of the statistics of $d(n)$ and $x(n)$ as well as a simple iterative procedure for adjusting the parameters of an FIR filter.

18.6.3 Wiener Solution

For the FIR filter structure, the coefficient values in $\mathbf{W}(n)$ that minimize $J_{\mathrm{MSE}}(n)$ are well-defined if the statistics of the input and desired response signals are known. The formulation of this problem for continuous-time signals and the resulting solution was first derived by Wiener [23]. Hence, this optimum coefficient vector $\mathbf{W}_{\mathrm{MSE}}(n)$ is often called the Wiener solution to the adaptive filtering problem. The extension of Wiener's analysis to the discrete-time case is attributed to Levinson [24].

To determine $\mathbf{W}_{\mathrm{MSE}}(n)$, we note that the function $J_{\mathrm{MSE}}(n)$ in Equation 18.16 is quadratic in the parameters $\{w_i(n)\}$, and the function is also differentiable. Thus, we can use a result from optimization theory that states that the derivatives of a smooth cost function with respect to each of the parameters is zero at a minimizing point on the cost function error surface. Thus, $\mathbf{W}_{\mathrm{MSE}}(n)$ can be found from the solution to the system of equations

$$\frac{\partial J_{\mathrm{MSE}}(n)}{\partial w_i(n)} = 0, \quad 0 \le i \le L-1. \tag{18.17}$$

Taking derivatives of $J_{\mathrm{MSE}}(n)$ in Equation 18.16 and noting that $e(n)$ and $y(n)$ are given by Equations 18.1 and 18.5, respectively, we obtain

$$\frac{\partial J_{\mathrm{MSE}}(n)}{\partial w_i(n)} = E\left\{e(n)\frac{\partial e(n)}{\partial w_i(n)}\right\} \tag{18.18}$$

$$= -E\left\{e(n)\frac{\partial y(n)}{\partial w_i(n)}\right\} \tag{18.19}$$

$$= -E\{e(n)x(n-i)\} \tag{18.20}$$

$$= -\left(E\{d(n)x(n-i)\} - \sum_{j=0}^{L-1} E\{x(n-i)x(n-j)\}w_j(n)\right), \tag{18.21}$$

where we have used the definitions of $e(n)$ and of $y(n)$ for the FIR filter structure in Equations 18.1 and 18.5, respectively, to expand the last result in Equation 18.21.

By defining the matrix $\mathbf{R}_{XX}(n)$ and vector $\mathbf{P}_{dX}(n)$ as

$$\mathbf{R}_{XX} = E\{\mathbf{X}(n)\mathbf{X}^{\mathrm{T}}(n)\} \quad \text{and} \quad \mathbf{P}_{dX}(n) = E\{d(n)\mathbf{X}(n)\}, \tag{18.22}$$

respectively, we can combine Equations 18.17 and 18.21 to obtain the system of equations in vector form as

$$\mathbf{R}_{XX}(n)\mathbf{W}_{\mathrm{MSE}}(n) - \mathbf{P}_{dX}(n) = 0, \tag{18.23}$$

where **0** is the zero vector. Thus, so long as the matrix $\mathbf{R}_{XX}(n)$ is invertible, the optimum Wiener solution vector for this problem is

$$\mathbf{W}_{\mathrm{MSE}}(n) = \mathbf{R}_{XX}^{-1}(n)\mathbf{P}_{dX}(n). \tag{18.24}$$

18.6.4 The Method of Steepest Descent

The method of steepest descent is a celebrated optimization procedure for minimizing the value of a cost function $J(n)$ with respect to a set of adjustable parameters $\mathbf{W}(n)$. This procedure adjusts each parameter of the system according to

$$w_i(n+1) = w_i(n) - \mu(n)\frac{\partial J(n)}{\partial w_i(n)}. \tag{18.25}$$

In other words, the ith parameter of the system is altered according to the derivative of the cost function with respect to the ith parameter. Collecting these equations in vector form, we have

$$\mathbf{W}(n+1) = \mathbf{W}(n) - \mu(n)\frac{\partial J(n)}{\partial \mathbf{W}(n)}, \tag{18.26}$$

where $\partial J(n)/\partial \mathbf{W}(n)$ is a vector of derivatives $\partial J(n)/\partial w_i(n)$.

For an FIR adaptive filter that minimizes the MSE cost function, we can use the result in Equation 18.21 to explicitly give the form of the steepest descent procedure in this problem. Substituting these results into Equation 18.25 yields the update equation for $\mathbf{W}(n)$ as

$$\mathbf{W}(n+1) = \mathbf{W}(n) + \mu(n)[\mathbf{P}_{dX}(n) - \mathbf{R}_{XX}(n)\mathbf{W}(n)]. \tag{18.27}$$

However, this steepest descent procedure depends on the statistical quantities $E\{d(n)x(n-i)\}$ and $E\{x(n-i)x(n-j)\}$ contained in $\mathbf{P}_{dX}(n)$ and $\mathbf{R}_{XX}(n)$, respectively. In practice, we only have measurements of both $d(n)$ and $x(n)$ to be used within the adaptation procedure. While suitable estimates of the statistical quantities needed for Equation 18.27 could be determined from the signals $x(n)$ and $d(n)$, we instead develop an approximate version of the method of steepest descent that depends on the signal values themselves. This procedure is known as the LMS algorithm.

18.6.5 LMS Algorithm

The cost function $J(n)$ chosen for the steepest descent algorithm of Equation 18.25 determines the coefficient solution obtained by the adaptive filter. If the MSE cost function in Equation 18.16 is chosen, the resulting algorithm depends on the statistics of $x(n)$ and $d(n)$ because of the expectation operation that defines this cost function. Since we typically only have measurements of $d(n)$ and of $x(n)$ available to us, we substitute an alternative cost function that depends only on these measurements. One such cost function is the least-squares cost function given by

$$J_{\mathrm{LS}}(n) = \sum_{k=0}^{n} \alpha(k)(d(k) - \mathbf{W}^{\mathrm{T}}(n)\mathbf{X}(k))^2, \tag{18.28}$$

where $\alpha(n)$ is a suitable weighting sequence for the terms within the summation. This cost function, however, is complicated by the fact that it requires numerous computations to calculate its value as well as its derivatives with respect to each $w_i(n)$, although efficient recursive methods for its minimization can be developed. See Chapter 21 for more details on these methods.

Alternatively, we can propose the simplified cost function $J_{\mathrm{LMS}}(n)$ given by

$$J_{\mathrm{LMS}}(n) = \frac{1}{2}e^2(n). \tag{18.29}$$

This cost function can be thought of as an instantaneous estimate of the MSE cost function, as $J_{\mathrm{MSE}}(n) = E\{J_{\mathrm{LMS}}(n)\}$. Although it might not appear to be useful, the resulting algorithm obtained when $J_{\mathrm{LMS}}(n)$ is used for $J(n)$ in Equation 18.25 is extremely useful for practical applications. Taking derivatives of $J_{\mathrm{LMS}}(n)$ with respect to the elements of $\mathbf{W}(n)$ and substituting the result into Equation 18.25, we obtain the LMS adaptive algorithm given by

$$\mathbf{W}(n+1) = \mathbf{W}(n) + \mu(n)e(n)\mathbf{X}(n). \tag{18.30}$$

Note that this algorithm is of the general form in Equation 18.14. It also requires only multiplications and additions to implement. In fact, the number and type of operations needed for the LMS algorithm is nearly the same as that of the FIR filter structure with fixed coefficient values, which is one of the reasons for the algorithm's popularity.

The behavior of the LMS algorithm has been widely studied, and numerous results concerning its adaptation characteristics under different situations have been developed. For discussions of some of these results, the reader is referred to Chapters 19 and 20. For now, we indicate its useful behavior by noting that the solution obtained by the LMS algorithm near its convergent point is related to the Wiener solution. In fact, analyses of the LMS algorithm under certain statistical assumptions about the input and desired response signals show that

$$\lim_{n\to\infty} E\{\mathbf{W}(n)\} = \mathbf{W}_{\mathrm{MSE}}, \tag{18.31}$$

when the Wiener solution $\mathbf{W}_{\mathrm{MSE}}(n)$ is a fixed vector. Moreover, the average behavior of the LMS algorithm is quite similar to that of the steepest descent algorithm in Equation 18.27 that depends explicitly on the statistics of the input and desired response signals. In effect, the iterative nature of the LMS coefficient updates is a form of time-averaging that smoothes the errors in the instantaneous gradient calculations to obtain a more reasonable estimate of the true gradient.

18.6.6 Other Stochastic Gradient Algorithms

The LMS algorithm is but one of an entire family of algorithms that are based on instantaneous approximations to steepest descent procedures. Such algorithms are known as stochastic gradient algorithms because they use a stochastic version of the gradient of a particular cost function's error surface to adjust the parameters of the filter. As an example, we consider the cost function

$$J_{SA}(n) = |e(n)|, \tag{18.32}$$

where $|\cdot|$ denotes absolute value. Like $J_{LMS}(n)$, this cost function also has a unique minimum at $e(n) = 0$, and it is differentiable everywhere except at $e(n) = 0$. Moreover, it is the instantaneous value of the mean absolute error cost function $J_{MAE}(n) = E\{J_{SA}(n)\}$. Taking derivatives of $J_{SA}(n)$ with respect to the coefficients $\{w_i(n)\}$ and substituting the results into Equation 18.25 yields the sign error algorithm as*

$$\mathbf{W}(n+1) = \mathbf{W}(n) + \mu(n)\text{sgn}[e(n)]\mathbf{X}(n), \tag{18.33}$$

where

$$\text{sgn}(e) = \begin{cases} 1 & \text{if } e > 0 \\ 0 & \text{if } e = 0 \\ -1 & \text{if } e < 0. \end{cases} \tag{18.34}$$

This algorithm is also of the general form in Equation 18.14.

The sign error algorithm is a useful adaptive filtering procedure because the terms $\text{sgn}[e(n)]x(n-i)$ can be computed easily in dedicated digital hardware. Its convergence properties differ from those of the LMS algorithm, however. Discussions of this and other algorithms based on non-MSE criteria can be found in [25].

18.6.7 Finite-Precision Effects and Other Implementation Issues

In all digital hardware and software implementations of the LMS algorithm in Equation 18.30, the quantities $e(n)$, $d(n)$, and $\{x(n-i)\}$ are represented by finite-precision quantities with a certain number of bits. Small numerical errors are introduced in each of the calculations within the coefficient updates in these situations. The effects of these numerical errors are usually less severe in systems that employ floating-point arithmetic, in which all numerical values are represented by both a mantissa and exponent, as compared to systems that employ fixed-point arithmetic, in which a mantissa-only numerical representation is used. The effects of the numerical errors introduced in these cases can be characterized, see [26] for a discussion of these issues.

While knowledge of the numerical effects of finite-precision arithmetic are necessary for obtaining the best performance from the LMS adaptive filter, it can be generally stated that the LMS adaptive filter performs robustly in the presence of these numerical errors. In fact, the apparent robustness of the LMS adaptive filter has led to the development of approximate implementations of Equation 18.30 that are more easily implemented in dedicated hardware. The general form of these implementations is

$$w_i(n+1) = w_i(n) + \mu(n)g_1[e(n)]g_2[x(n-i)], \tag{18.35}$$

where $g_1(\cdot)$ and $g_2(\cdot)$ are odd-symmetric nonlinearities that are chosen to simplify the implementation of the system. Some of the algorithms described by Equation 18.35 include the sign-data $\{g_1(e) = e,$

* Here, we have specified $\partial|e|/\partial e = 0$ for $e = 0$, although the derivative of this function does not exist at this point.

$g_2(x) = \text{sgn}(x)\}$, sign-sign or zero-forcing $\{g_1(e) = \text{sgn}(e), g_2(x) = \text{sgn}(x)\}$, and power-of-two quantized algorithms, as well as the sign error algorithm introduced previously. A presentation and comparative analysis of the performance of many of these algorithms can be found in [27].

18.6.8 System Identification Example

We now illustrate the actual behavior of the LMS adaptive filter through a system identification example in which the impulse response of a small audio loudspeaker in a room is estimated. A Gaussian-distributed signal with a flat frequency spectrum over the usable frequency range of the loudspeaker is generated and sent through an audio amplifier to the loudspeaker. This same Gaussian signal is sent to a 16-bit analog-to-digital (A/D) converter which samples it at an 8 kHz rate. The sound produced by the loudspeaker propagates to a microphone located several feet away from the loudspeaker, where it is collected and digitized by a second A/D converter also sampling at an 8 kHz rate. Both signals are stored to a computer file for subsequent processing and analysis. The goal of the analysis is to determine the combined impulse response of the loudspeaker/room/microphone sound propagation path. Such information is useful if the loudspeaker and microphone are to be used in the active noise control task described previously, and the general task also resembles that of acoustic echo cancellation for speakerphones.

We process these signals using a computer program that implements the LMS adaptive filter within the MATLAB®* signal manipulation environment. In this case, we have normalized the powers of both the Gaussian input signal and desired response signal collected at the microphone to unity, and we have highpass-filtered the microphone signal using a filter with transfer function $H(z) = (1 - z^{-1})/(1 - 0.95z^{-1})$ to remove any DC offset in this signal. For this task, we have chosen an $L = 100$-coefficient FIR filter adapted using the LMS algorithm in Equation 18.30 with a fixed step size of $\mu = 0.0005$ to obtain an accurate estimate of the impulse response of the loudspeaker and room. Figure 18.8 shows the

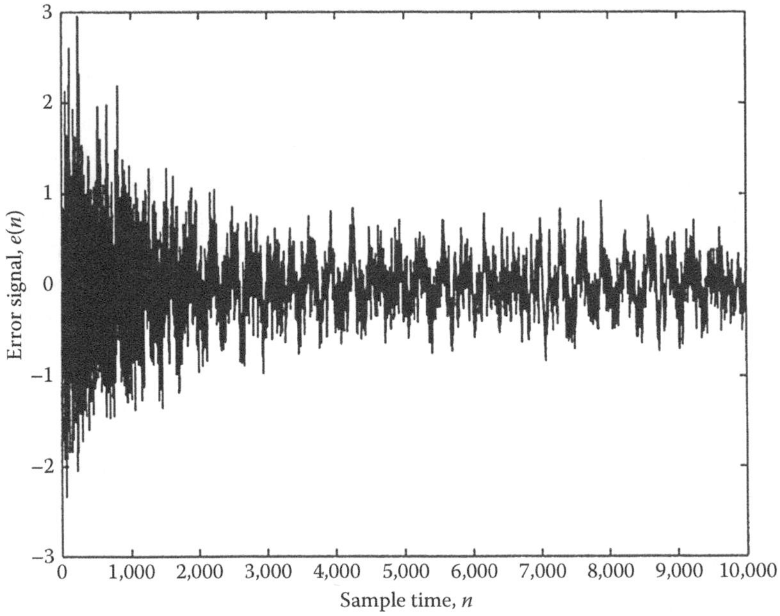

FIGURE 18.8 Convergence of the error signal in the loudspeaker identification experiment.

* MATLAB is a registered trademark of The MathWorks, Inc., Newton, Massachusetts.

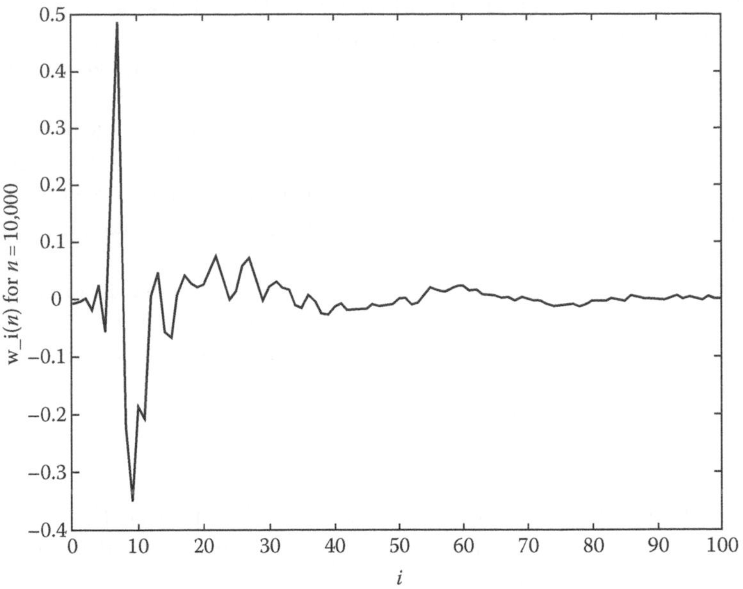

FIGURE 18.9 The adaptive filter coefficients obtained in the loudspeaker identification experiment.

convergence of the error signal in this situation. After about 4000 samples (0.5 s), the error signal has been reduced to a power that is about 1/15 (−12 dB) below that of the microphone signal, indicating that the filter has converged. Figure 18.9 shows the coefficients of the adaptive filter at iteration $n = 10{,}000$. The impulse response of the loudspeaker/room/microphone path consists of a large pulse corresponding to the direct sound propagation path as well as numerous smaller pulses caused by reflections of sounds off walls and other surfaces in the room.

18.7 Conclusions

In this section, we have presented an overview of adaptive filters, emphasizing the applications and basic algorithms that have already proven themselves to be useful in practice. Despite the many contributions in the field, research efforts in adaptive filters continue at a strong pace, and it is likely that new applications for adaptive filters will be developed in the future. To keep abreast of these advances, the reader is urged to consult journals such as the *IEEE Transactions on Signal Processing* as well as the proceedings of yearly conferences and workshops in the signal processing and related fields.

References

1. Kuo, S. and Chen, C., Implementation of adaptive filters with the TMS320C25 or the TMS320C30, in *Digital Signal Processing Applications with the TMS320 Family*, Papamichalis, P. (Ed.), Prentice-Hall, Englewood Cliffs, NJ, 1991, pp. 191–271.
2. Analog Devices, Adaptive filters, in *ADSP-21000 Family Application Handbook*, Vol. 1, Analog Devices, Narwood, MA, 1994, pp. 157–203.
3. El-Sharkawy, M., Designing adaptive FIR filters and implementing them on the DSP56002 processor, in *Digital Signal Processing Applications with Motorola's DSP56002 Processor*, Prentice-Hall, Upper Saddle River, NJ, 1996, pp. 319–342.

4. Borth, D.E., Gerson, I.A., Haug, J.R., and Thompson, C.D., A flexible adaptive FIR filter VLSI IC, *IEEE J. Sel. Areas Commn.*, 6(3): 494–503, April 1988.
5. Oppenheim, A.V. and Schafer, A.W., *Discrete-Time Signal Processing*, Prentice-Hall, Englewood Cliffs, NJ, 1989.
6. Friedlander, B., Lattice filters for adaptive processing, *Proc. IEEE*, 70(8): 829–867, August 1982.
7. Mathews, V.J., Adaptive polynomial filters, *IEEE Signal Process. Mag.*, 8(3): 10–26, July 1991.
8. Haykin, S., *Neural Networks: A Comprehensive Foundation*, Macmillan, New York, 1994.
9. Proakis, J.G. and Salehi, M., *Communication Systems Engineering*, Prentice-Hall, Englewood Cliffs, NJ, 1994.
10. Åström, K.G. and Wittenmark, B., *Adaptive Control*, Addison-Wesley, Reading, MA, 1989.
11. Widrow, B. and Walach, E., *Adaptive Inverse Control*, Prentice-Hall, Upper Saddle River, NJ, 1996.
12. Sondhi, M.M., An adaptive echo canceller, *Bell Sys. Tech. J.*, 46: 497–511, March 1967.
13. Messerschmitt, D.G., Echo cancellation in speech and data transmission, *IEEE J. Sel. Areas Commn.*, SAC-2(2): 283–297, March 1984.
14. Murano, K., Unagami, S., and Amano, F., Echo cancellation and applications, *IEEE Commn. Mag.*, 28(1): 49–55, January 1990.
15. Widrow, B., Glover, J.R., Jr., McCool, J.M., Kaunitz, J., Williams, C.S., Hearn, R.H., Zeidler, J.R., Dong, E., Jr., and Goodlin, R.C., Adaptive noise cancelling: Principles and applications, *Proc. IEEE*, 63(12): 1692–1716, December 1975.
16. Lucky, R.W., Techniques for adaptive equalization of digital communication systems, *Bell Sys. Tech. J.*, 45: 255–286, February 1966.
17. Qureshi, S.U.H., Adaptive equalization, *Proc. IEEE*, 73(9): 1349–1387, September 1985.
18. Robinson, E.A. and Durrani, T., *Geophysical Signal Processing*, Prentice-Hall, Englewood Cliffs, NJ, 1986.
19. Makhoul, J., Linear prediction: A tutorial review, *Proc. IEEE*, 63(4): 561–580, April 1975.
20. Zeidler, J.R., Performance analysis of LMS adaptive prediction filters, *Proc. IEEE*, 78(12): 1781–1806, December 1990.
21. Kuo, S.M. and Morgan, D.R., *Active Noise Control Systems: Algorithms and DSP Implementations*, John Wiley & Sons, New York, 1996.
22. Fuller, C.R., Elliott, S.J., and Nelson, P.A., *Active Control of Vibration*, Academic Press, London, U.K., 1996.
23. Wiener, N., *Extrapolation, Interpolation, and Smoothing of Stationary Time Series, with Engineering Applications*, MIT Press, Cambridge, MA, 1949.
24. Levinson, N., The Wiener RMS (root-mean-square) error criterion in filter design and prediction, *J. Math. Phys.*, 25: 261–278, 1947.
25. Douglas, S.C. and Meng, T.H.-Y., Stochastic gradient adaptation under general error criteria, *IEEE Trans. Signal Process.*, 42(6): 1335–1351, June 1994.
26. Caraiscos, C. and Liu, B., A roundoff error analysis of the LMS adaptive algorithm, *IEEE Trans. Acoust. Speech Signal Process.*, ASSP-32(1): 34–41, February 1984.
27. Duttweiler, D.L., Adaptive filter performance with nonlinearities in the correlation multiplier, *IEEE Trans. Acoust. Speech Signal Process.*, ASSP-30(4): 578–586, August 1982.

19

Convergence Issues in the LMS Adaptive Filter

Scott C. Douglas
Southern Methodist University

Markus Rupp
Technical University of Vienna

19.1 Introduction

In adaptive filtering, the least-mean-square (LMS) adaptive filter [1] is the most popular and widely used adaptive system, appearing in numerous commercial and scientific applications. The LMS adaptive filter is described by the equations

$$\mathbf{W}(n+1) = \mathbf{W}(n) + \mu(n)e(n)\mathbf{X}(n) \tag{19.1}$$

$$e(n) = d(n) - \mathbf{W}^{\mathrm{T}}(n)\mathbf{X}(n), \tag{19.2}$$

where

$\mathbf{W}(n) = [w_0(n)\ w_1(n)\ \cdots\ w_{L-1}(n)]^{\mathrm{T}}$ is the coefficient vector

$\mathbf{X}(n) = [x(n)\ x(n-1)\ \cdots\ x(n-L+1)]^{\mathrm{T}}$ is the input signal vector

$d(n)$ is the desired signal

$e(n)$ is the error signal

$\mu(n)$ is the step size

There are three main reasons why the LMS adaptive filter is so popular. First, it is relatively easy to implement in software and hardware due to its computational simplicity and efficient use of memory. Second, it performs robustly in the presence of numerical errors caused by finite-precision arithmetic. Third, its behavior has been analytically characterized to the point where a user can easily set up the system to obtain adequate performance with only limited knowledge about the input and desired response signals.

Our goal in this chapter is to provide a detailed performance analysis of the LMS adaptive filter so that the user of this system understands how the choice of the step size $\mu(n)$ and filter length L affect the performance of the system through the natures of the input and desired response signals $x(n)$ and $d(n)$, respectively. The organization of this chapter is as follows. We first discuss why analytically characterizing the behavior of the LMS adaptive filter is important from a practical point of view. We then present particular signal models and assumptions that make such analyses tractable. We summarize the analytical results that can be obtained from these models and assumptions, and we discuss the implications of these results for different practical situations. Finally, to overcome some of the limitations of the LMS adaptive filter's behavior, we describe simple extensions of this system that are suggested by the analytical results. In all of our discussions, we assume that the reader is familiar with the adaptive filtering task and the LMS adaptive filter as described in Chapter 18.

19.2 Characterizing the Performance of Adaptive Filters

There are two practical methods for characterizing the behavior of an adaptive filter. The simplest method of all to understand is simulation. In simulation, a set of input and desired response signals are either collected from a physical environment or are generated from a mathematical or statistical model of the physical environment. These signals are then processed by a software program that implements the particular adaptive filter under evaluation. By trial-and-error, important design parameters, such as the step size $\mu(n)$ and filter length L, are selected based on the observed behavior of the system when operating on these example signals. Once these parameters are selected, they are used in an adaptive filter implementation to process additional signals as they are obtained from the physical environment. In the case of a real-time adaptive filter implementation, the design parameters obtained from simulation are encoded within the real-time system to allow it to process signals as they are continuously collected.

While straightforward, simulation has two drawbacks that make it a poor sole choice for characterizing the behavior of an adaptive filter:

- Selecting design parameters via simulation alone is an iterative and time-consuming process. Without any other knowledge of the adaptive filter's behavior, the number of trials needed to select the best combination of design parameters is daunting, even for systems as simple as the LMS adaptive filter.
- The amount of data needed to accurately characterize the behavior of the adaptive filter for all cases of interest may be large. If real-world signal measurements are used, it may be difficult or costly to collect and store the large amounts of data needed for simulation characterizations. Moreover, once this data is collected or generated, it must be processed by the software program that implements the adaptive filter, which can be time-consuming as well.

For these reasons, we are motivated to develop an analysis of the adaptive filter under study. In such an analysis, the input and desired response signals $x(n)$ and $d(n)$ are characterized by certain properties that govern the forms of these signals for the application of interest. Often, these properties are statistical in nature, such as the means of the signals or the correlation between two signals at different time instants. An analytical description of the adaptive filter's behavior is then developed that is based on these signal properties. Once this analytical description is obtained, the design parameters are selected to obtain the best performance of the system as predicted by the analysis. What is considered "best performance" for the adaptive filter can often be specified directly within the analysis, without the need for iterative calculations or extensive simulations.

Usually, both analysis and simulation are employed to select design parameters for adaptive filters, as the simulation results provide a check on the accuracy of the signal models and assumptions that are used within the analysis procedure.

19.3 Analytical Models, Assumptions, and Definitions

The type of analysis that we employ has a long-standing history in the field of adaptive filters [2–6]. Our analysis uses statistical models for the input and desired response signals, such that any collection of samples from the signals $x(n)$ and $d(n)$ have well-defined joint probability density functions (p.d.f.s). With this model, we can study the average behavior of functions of the coefficients $\mathbf{W}(n)$ at each time instant, where "average" implies taking a statistical expectation over the ensemble of possible coefficient values. For example, the mean value of the i th coefficient $w_i(n)$ is defined as

$$E\{w_i(n)\} = \int_{-\infty}^{\infty} w\, p_{w_i}(w, n)\mathrm{d}w, \tag{19.3}$$

where $p_{w_i}(w, n)$ is the probability distribution of the ith coefficient at time n. The mean value of the coefficient vector at time n is defined as

$$E\{\mathbf{W}(n)\} = [E\{w_0(n)\}\; E\{w_1(n)\} \cdots E\{w_{L-1}(n)\}]^{\mathrm{T}}.$$

While it is usually difficult to evaluate expectations such as Equation 19.3 directly, we can employ several simplifying assumptions and approximations that enable the formation of evolution equations that describe the behavior of quantities such as $E\{\mathbf{W}(n)\}$ from one time instant to the next. In this way, we can predict the evolutionary behavior of the LMS adaptive filter on average. More importantly, we can study certain characteristics of this behavior, such as the stability of the coefficient updates, the speed of convergence of the system, and the estimation accuracy of the filter in steady-state. Because of their role in the analyses that follow, we now describe these simplifying assumptions and approximations.

19.3.1 System Identification Model for the Desired Response Signal

For our analysis, we assume that the desired response signal is generated from the input signal as

$$d(n) = \mathbf{W}_{\mathrm{opt}}^{\mathrm{T}}\mathbf{X}(n) + \eta(n), \tag{19.4}$$

where

$\mathbf{W}_{\mathrm{opt}} = [w_{0,\mathrm{opt}}\; w_{1,\mathrm{opt}} \cdots w_{L-1,\mathrm{opt}}]^{\mathrm{T}}$ is a vector of optimum FIR filter coefficients
$\eta(n)$ is a noise signal that is independent of the input signal

Such a model for $d(n)$ is realistic for several important adaptive filtering tasks. For example, in echo cancellation for telephone networks, the optimum coefficient vector $\mathbf{W}_{\mathrm{opt}}$ contains the impulse response of the echo path caused by the impedance mismatches at hybrid junctions within the network, and the noise $\eta(n)$ is the near-end source signal [7]. The model is also appropriate in system identification and modeling tasks such as plant identification for adaptive control [8] and channel modeling for communication systems [9]. Moreover, most of the results obtained from this model are independent of the specific impulse response values within $\mathbf{W}_{\mathrm{opt}}$, so that general conclusions can be readily drawn.

19.3.2 Statistical Models for the Input Signal

Given the desired response signal model in Equation 19.4, we now consider useful and appropriate statistical models for the input signal $x(n)$. Here, we are motivated by two typically conflicting concerns: (1) the need for signal models that are realistic for several practical situations and (2) the tractability of the analyses that the models allow. We consider two input signal models that have proven useful for predicting the behavior of the LMS adaptive filter.

19.3.2.1 Independent and Identically Distributed Random Processes

In digital communication tasks, an adaptive filter can be used to identify the dispersive characteristics of the unknown channel for purposes of decoding future transmitted sequences [9]. In this application, the transmitted signal is a bit sequence that is usually zero mean with a small number of amplitude levels. For example, a nonreturn-to-zero binary signal takes on the values of ± 1 with equal probability at each time instant. Moreover, due to the nature of the encoding of the transmitted signal in many cases, any set of L samples of the signal can be assumed to be independent and identically distributed (i.i.d.). For an i.i.d. random process, the p.d.f. of the samples $\{x(n_1), x(n_2), \ldots, x(n_L)\}$ for any choices of n_i such that $n_i \neq n_j$ is

$$p_X(x(n_1), x(n_2), \ldots, x(n_L)) = p_x[x(n_1)]\, p_x[x(n_2)] \cdots p_x[x(n_L)], \tag{19.5}$$

where $p_x(\cdot)$ and $p_X(\cdot)$ are the univariate and L-variate probability densities of the associated random variables, respectively.

Zero-mean and statistically independent random variables are also uncorrelated, such that

$$E\{x(n_i)x(n_j)\} = 0 \tag{19.6}$$

for $n_i \neq n_j$, although uncorrelated random variables are not necessarily statistically independent. The input signal model in Equation 19.5 is useful for analyzing the behavior of the LMS adaptive filter, as it allows a particularly simple analysis of this system.

19.3.2.2 Spherically Invariant Random Processes

In acoustic echo cancellation for speakerphones, an adaptive filter can be used to electronically isolate the speaker and microphone so that the amplifier gains within the system can be increased [10]. In this application, the input signal to the adaptive filter consists of samples of bandlimited speech. It has been shown in experiments that samples of a bandlimited speech signal taken over a short time period (e.g., 5 ms) have so-called "spherically invariant" statistical properties. Spherically invariant random processes (SIRPs) are characterized by multivariate p.d.f.s that depend on a quadratic form of their arguments, given by $\mathbf{X}^T(n)\mathbf{R}_{XX}^{-1}\mathbf{X}(n)$, where

$$\mathbf{R}_{XX} = E\{\mathbf{X}(n)\mathbf{X}^T(n)\} \tag{19.7}$$

is the L-dimensional input signal autocorrelation matrix of the stationary signal $x(n)$. The best-known representative of this class of stationary stochastic processes is the jointly Gaussian random process for which the joint p.d.f. of the elements of $\mathbf{X}(n)$ is

$$p_X(x(n), \ldots, x(n-L+1)) = [(2\pi)^L \det(\mathbf{R}_{XX})]^{-1/2} \exp\left[-\frac{1}{2}\mathbf{X}^T(n)\mathbf{R}_{XX}^{-1}\mathbf{X}(n)\right], \tag{19.8}$$

where $\det(\mathbf{R}_{XX})$ is the determinant of the matrix $\mathbf{R}_{XX}$. More generally, SIRPs can be described by a weighted mixture of Gaussian processes as

$$p_X(x(n),\ldots,x(n-L+1)) = \int_0^\infty ((2\pi|u|)^L \det(\bar{\mathbf{R}}_{XX}))^{-1/2} \times p_\sigma(u)\exp\left(-\frac{1}{2u^2}\mathbf{X}^T(n)\bar{\mathbf{R}}_{XX}^{-1}\mathbf{X}(n)\right)du, \quad (19.9)$$

where $\bar{\mathbf{R}}_{XX}$ is the autocorrelation matrix of a zero-mean, unit-variance jointly Gaussian random process. In Equation 19.9, the p.d.f. $p_\sigma(u)$ is a weighting function for the value of u that scales the standard deviation of this process. In other words, any single realization of a SIRP is a Gaussian random process with an autocorrelation matrix $u^2\bar{\mathbf{R}}_{XX}$. Each realization, however, will have a different variance u^2.

As described, the above SIRP model does not accurately depict the statistical nature of a speech signal. The variance of a speech signal varies widely from phoneme (vowel) to fricative (consonant) utterances, and this burst-like behavior is uncharacteristic of Gaussian signals. The statistics of such behavior can be accurately modeled if a slowly varying value for the random variable u in Equation 19.9 is allowed. Figure 19.1 depicts the differences between a nearly SIRP and an SIRP. In this system, either the random variable u or a sample from the slowly varying random process $u(n)$ is created and used to scale the magnitude of a sample from an uncorrelated Gaussian random process. Depending on the position of the switch, either an SIRP (upper position) or a nearly SIRP (lower position) is created. The linear filter $F(z)$ is then used to produce the desired autocorrelation function of the SIRP. So long as the value of $u(n)$ changes slowly over time, $\mathbf{R}_{XX}$ for the signal $x(n)$ as produced from this system is approximately the same as would be obtained if the value of $u(n)$ were fixed, except for the amplitude scaling provided by the value of $u(n)$.

The random process $u(n)$ can be generated by filtering a zero-mean uncorrelated Gaussian process with a narrow-bandwidth lowpass filter. With this choice, the system generates samples from the so-called K_0 p.d.f., also known as the MacDonald function or degenerated Bessel function of the second kind [11]. This density is a reasonable match to that of typical speech sequences, although it does not necessarily generate sequences that sound like speech. Given a short-length speech sequence from a particular speaker, one can also determine the proper $p_\sigma(u)$ needed to generate $u(n)$ as well as the form of the filter $F(z)$ from estimates of the amplitude and correlation statistics of the speech sequence, respectively.

In addition to adaptive filtering, SIRPs are also useful for characterizing the performance of vector quantizers for speech coding. Details about the properties of SIRPs can be found in [12].

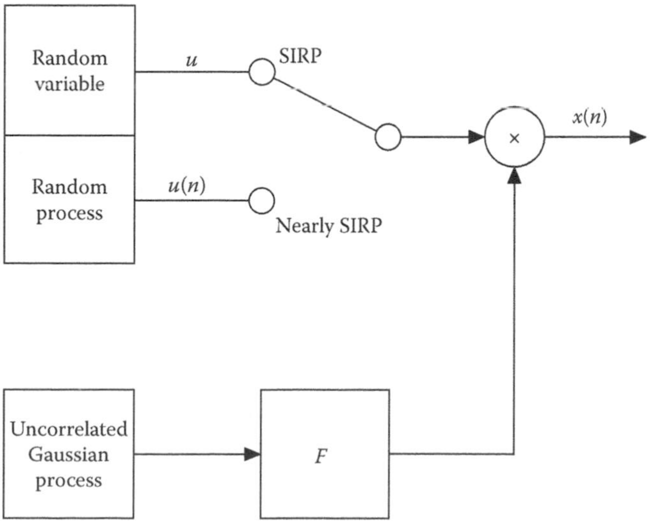

FIGURE 19.1 Generation of SIRPs and nearly SIRPs.

19.3.3 Independence Assumptions

In the LMS adaptive filter, the coefficient vector $\mathbf{W}(n)$ is a complex function of the current and past samples of the input and desired response signals. This fact would appear to foil any attempts to develop equations that describe the evolutionary behavior of the filter coefficients from one time instant to the next. One way to resolve this problem is to make further statistical assumptions about the nature of the input and the desired response signals. We now describe a set of assumptions that have proven to be useful for predicting the behaviors of many types of adaptive filters.

Elements of the vector $\mathbf{X}(n)$ are statistically independent of the elements of the vector $\mathbf{X}(m)$ if $m \neq n$. In addition, samples from the noise signal $\eta(n)$ are i.i.d. and independent of the input vector sequence $\mathbf{X}(k)$ for all k and n.

A careful study of the structure of the input signal vector indicates that the independence assumptions are never true, as the vector $\mathbf{X}(n)$ shares elements with $\mathbf{X}(n-m)$ if $|m| < L$ and thus cannot be independent of $\mathbf{X}(n-m)$ in this case. Moreover, $\eta(n)$ is not guaranteed to be independent from sample to sample. Even so, numerous analyses and simulations have indicated that these assumptions lead to a reasonably accurate characterization of the behavior of the LMS and other adaptive filter algorithms for small step size values, even in situations where the assumptions are grossly violated. In addition, analyses using the independence assumptions enable a simple characterization of the LMS adaptive filter's behavior and provide reasonable guidelines for selecting the filter length L and step size $\mu(n)$ to obtain good performance from the system.

It has been shown that the independence assumptions lead to a first-order-in-$\mu(n)$ approximation to a more accurate description of the LMS adaptive filter's behavior [13]. For this reason, the analytical results obtained from these assumptions are not particularly accurate when the step size is near the stability limits for adaptation. It is possible to derive an exact statistical analysis of the LMS adaptive filter that does not use the independence assumptions [14], although the exact analysis is quite complex for adaptive filters with more than a few coefficients. From the results in [14], it appears that the analysis obtained from the independence assumptions is most inaccurate for large step sizes and for input signals that exhibit a high degree of statistical correlation.

19.3.4 Useful Definitions

In our analysis, we define the minimum mean-squared error (MSE) solution as the coefficient vector $\mathbf{W}(n)$ that minimizes the MSE criterion given by

$$\xi(n) = E\{e^2(n)\}. \tag{19.10}$$

Since $\xi(n)$ is a function of $\mathbf{W}(n)$, it can be viewed as an error surface with a minimum that occurs at the minimum MSE solution. It can be shown for the desired response signal model in Equation 19.4 that the minimum MSE solution is $\mathbf{W}_{\text{opt}}$ and can be equivalently defined as

$$\mathbf{W}_{\text{opt}} = \mathbf{R}_{XX}^{-1}\mathbf{P}_{dX}, \tag{19.11}$$

where

$\mathbf{R}_{XX}$ is as defined in Equation 19.7

$\mathbf{P}_{dX} = E\{d(n)\mathbf{X}(n)\}$ is the cross-correlation of $d(n)$ and $\mathbf{X}(n)$

When $\mathbf{W}(n) = \mathbf{W}_{\text{opt}}$, the value of the minimum MSE is given by

$$\xi_{\min} = \sigma_\eta^2, \tag{19.12}$$

where σ_η^2 is the power of the signal $\eta(n)$.

We define the coefficient error vector $\mathbf{V}(n) = [v_0(n) \cdots v_{L-1}(n)]^T$ as

$$\mathbf{V}(n) = \mathbf{W}(n) - \mathbf{W}_{opt}, \tag{19.13}$$

such that $\mathbf{V}(n)$ represents the errors in the estimates of the optimum coefficients at time n. Our study of the LMS algorithm focuses on the statistical characteristics of the coefficient error vector. In particular, we can characterize the approximate evolution of the coefficient error correlation matrix $\mathbf{K}(n)$, defined as

$$\mathbf{K}(n) = E\{\mathbf{V}(n)\mathbf{V}^T(n)\}. \tag{19.14}$$

Another quantity that characterizes the performance of the LMS adaptive filter is the excess mean-squared error (excess MSE), defined as

$$\begin{aligned}\xi_{ex}(n) &= \xi(n) - \xi_{min} \\ &= \xi(n) - \sigma_\eta^2,\end{aligned} \tag{19.15}$$

where $\xi(n)$ is as defined in Equation 19.10. The excess MSE is the power of the additional error in the filter output due to the errors in the filter coefficients. An equivalent measure of the excess MSE in steady-state is the misadjustment, defined as

$$M = \lim_{n\to\infty} \frac{\xi_{ex}(n)}{\sigma_\eta^2}, \tag{19.16}$$

such that the quantity $(1 + M)\sigma_\eta^2$ denotes the total MSE in steady-state.

Under the independence assumptions, it can be shown that the excess MSE at any time instant is related to $\mathbf{K}(n)$ as

$$\xi_{ex}(n) = \mathrm{tr}[\mathbf{R}_{XX}\mathbf{K}(n)], \tag{19.17}$$

where the trace tr[·] of a matrix is the sum of its diagonal values.

19.4 Analysis of the LMS Adaptive Filter

We now analyze the behavior of the LMS adaptive filter using the assumptions and definitions that we have provided. For the first portion of our analysis, we characterize the mean behavior of the filter coefficients of the LMS algorithm in Equations 19.1 and 19.2. Then, we provide a mean-square analysis of the system that characterizes the natures of $\mathbf{K}(n)$, $\xi_{ex}(n)$, and M in Equations 19.14 through 19.16, respectively.

19.4.1 Mean Analysis

By substituting the definition of $d(n)$ from the desired response signal model in Equation 19.4 into the coefficient updates in Equations 19.1 and 19.2, we can express the LMS algorithm in terms of the coefficient error vector in Equation 19.13 as

$$\mathbf{V}(n+1) = \mathbf{V}(n) - \mu(n)\mathbf{X}(n)\mathbf{X}^T(n)\mathbf{V}(n) + \mu(n)\eta(n)\mathbf{X}(n). \tag{19.18}$$

We take expectations of both sides of Equation 19.18, which yields

$$E\{\mathbf{V}(n+1)\} = E\{\mathbf{V}(n)\} - \mu(n)E\{\mathbf{X}(n)\mathbf{X}^{\mathrm{T}}(n)\mathbf{V}(n)\} + \mu(n)E\{\eta(n)\mathbf{X}(n)\}, \tag{19.19}$$

in which we have assumed that $\mu(n)$ does not depend on $\mathbf{X}(n)$, $d(n)$, or $\mathbf{W}(n)$.

In many practical cases of interest, either the input signal $x(n)$ and/or the noise signal $\eta(n)$ is zero-mean, such that the last term in Equation 19.19 is zero. Moreover, under the independence assumptions, it can be shown that $\mathbf{V}(n)$ is approximately independent of $\mathbf{X}(n)$, and thus the second expectation on the right-hand side of Equation 19.19 is approximately given by

$$\begin{aligned} E\{\mathbf{X}(n)\mathbf{X}^{\mathrm{T}}(n)\mathbf{V}(n)\} &\approx E\{\mathbf{X}(n)\mathbf{X}^{\mathrm{T}}(n)\}E\{\mathbf{V}(n)\} \\ &= \mathbf{R}_{XX}E\{\mathbf{V}(n)\}. \end{aligned} \tag{19.20}$$

Combining these results with Equation 19.19, we obtain

$$E\{\mathbf{V}(n+1)\} = (\mathbf{I} - \mu(n)\mathbf{R}_{XX})E\{\mathbf{V}(n)\}. \tag{19.21}$$

The simple expression in Equation 19.21 describes the evolutionary behavior of the mean values of the errors in the LMS adaptive filter coefficients. Moreover, if the step size $\mu(n)$ is constant, then we can write Equation 19.21 as

$$E\{\mathbf{V}(n)\} = (\mathbf{I} - \mu\mathbf{R}_{XX})^n E\{\mathbf{V}(0)\}. \tag{19.22}$$

To further simplify this matrix equation, note that $\mathbf{R}_{XX}$ can be described by its eigenvalue decomposition as

$$\mathbf{R}_{XX} = \mathbf{Q}\Lambda\mathbf{Q}^{\mathrm{T}}, \tag{19.23}$$

where

$\mathbf{Q}$ is a matrix of the eigenvectors of $\mathbf{R}_{XX}$

Λ is a diagonal matrix of the eigenvalues $\{\lambda_0, \lambda_1, \ldots, \lambda_{L-1}\}$ of $\mathbf{R}_{XX}$, which are all real valued because of the symmetry of $\mathbf{R}_{XX}$

Through some simple manipulations of Equation 19.22, we can express the $(i+1)$th element of $E\{\mathbf{W}(n)\}$ as

$$E\{w_i(n)\} = w_{i,\mathrm{opt}} + \sum_{j=0}^{L-1} q_{ij}(1 - \mu\lambda_j)^n E\{\tilde{v}_j(0)\}, \tag{19.24}$$

where

q_{ij} is the $(i+1, j+1)$th element of the eigenvector matrix $\mathbf{Q}$

$\tilde{v}_j(n)$ is the $(j+1)$th element of the rotated coefficient error vector defined as

$$\tilde{\mathbf{V}}(n) = \mathbf{Q}^{\mathrm{T}}\mathbf{V}(n). \tag{19.25}$$

From Equations 19.21 and 19.24, we can state several results concerning the mean behaviors of the LMS adaptive filter coefficients:

- The mean behavior of the LMS adaptive filter as predicted by Equation 19.21 is identical to that of the method of steepest descent for this adaptive filtering task. Discussed in Chapter 18, the method of steepest descent is an iterative optimization procedure that requires precise knowledge of the statistics of $x(n)$ and $d(n)$ to operate. That the LMS adaptive filter's average behavior is similar to that of steepest descent was recognized in one of the earliest publications of the LMS adaptive filter [1].

- The mean value of any LMS adaptive filter coefficient at any time instant consists of the sum of the optimal coefficient value and a weighted sum of exponentially converging and/or diverging terms. These error terms depend on the elements of the eigenvector matrix $\mathbf{Q}$, the eigenvalues of $\mathbf{R}_{XX}$, and the mean $E\{\mathbf{V}(0)\}$ of the initial coefficient error vector.
- If all of the eigenvalues $\{\lambda_j\}$ of $\mathbf{R}_{XX}$ are strictly positive and

$$0 < \mu < \frac{2}{\lambda_j} \tag{19.26}$$

for all $0 < j < L - 1$, then the means of the filter coefficients converge exponentially to their optimum values. This result can be found directly from Equation 19.24 by noting that the quantity $(1 - \mu\lambda_j)^n \to 0$ as $n \to \infty$ if $|1 - \mu\lambda_j| < 1$.

- The speeds of convergence of the means of the coefficient values depend on the eigenvalues λ_i and the step size μ. In particular, we can define the time constant τ_j of the jth term within the summation on the right-hand side of Equation 19.24 as the approximate number of iterations it takes for this term to reach $(1/e)$th its initial value. For step sizes in the range $0 < \mu \ll 1/\lambda_{max}$ where λ_{max} is the maximum eigenvalue of $\mathbf{R}_{XX}$, this time constant is

$$\tau_j = -\frac{1}{\ln(1 - \mu\lambda_j)} \approx \frac{1}{\mu\lambda_j}. \tag{19.27}$$

Thus, faster convergence is obtained as the step size is increased. However, for step size values greater than $1/\lambda_{max}$, the speeds of convergence can actually decrease. Moreover, the convergence of the system is limited by its mean-squared behavior, as we shall indicate shortly.

19.4.1.1 An Example

Consider the behavior of an $L = 2$-coefficient LMS adaptive filter in which $x(n)$ and $d(n)$ are generated as

$$x(n) = 0.5x(n-1) + \frac{\sqrt{3}}{2} z(n) \tag{19.28}$$

$$d(n) = x(n) + 0.5x(n-1) + \eta(n), \tag{19.29}$$

where $z(n)$ and $\eta(n)$ are zero-mean uncorrelated jointly Gaussian signals with variances of 1 and 0.01, respectively. It is straightforward to show for these signal statistics that

$$\mathbf{W}_{opt} = \begin{bmatrix} 1 \\ 0.5 \end{bmatrix} \quad \text{and} \quad \mathbf{R}_{XX} = \begin{bmatrix} 1 & 0.5 \\ 0.5 & 1 \end{bmatrix}. \tag{19.30}$$

Figure 19.2a depicts the behavior of the mean analysis equation in Equation 19.24 for these signal statistics, where $\mu(n) = 0.08$ and $\mathbf{W}(0) = [4 - 0.5]^T$. Each circle on this plot corresponds to the value of $E\{\mathbf{W}(n)\}$ for a particular time instant. Shown on this $\{w_0, w_1\}$ plot are the coefficient error axes $\{v_0, v_1\}$, the rotated coefficient error axes $\{\tilde{v}_0, \tilde{v}_1\}$, and the contours of the excess MSE error surface ξ_{ex} as a function of w_0 and w_1 for values in the set $\{0.1, 0.2, 0.5, 1, 2, 5, 10, 20\}$. Starting from the initial coefficient vector $\mathbf{W}(0)$, $E\{\mathbf{W}(n)\}$ converge toward $\mathbf{W}_{opt}$ by reducing the components of the mean coefficient error

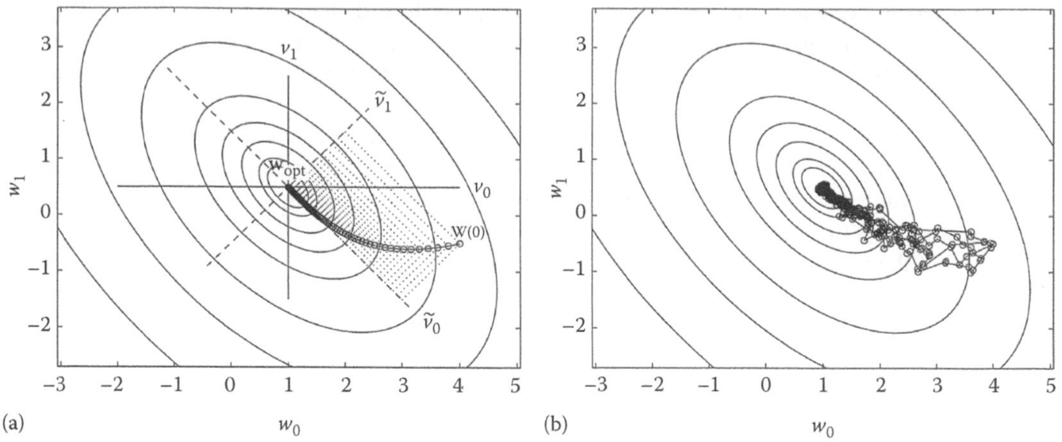

FIGURE 19.2 Comparison of the predicted and actual performances of the LMS adaptive filter in the two-coefficient example: (a) the behavior predicted by the mean analysis and (b) the actual LMS adaptive filter behavior for five different simulation runs.

vector $E\{\mathbf{V}(n)\}$ along the rotated coefficient error axes $\{\tilde{v}_0, \tilde{v}_1\}$ according to the exponential weighting factors $(1 - \mu\lambda_0)^n$ and $(1 - \mu\lambda_1)^n$ in Equation 19.24.

For comparison, Figure 19.2b shows five different simulation runs of an LMS adaptive filter operating on Gaussian signals generated according to Equations 19.28 and 19.29, where $\mu(n) = 0.08$ and $\mathbf{W}(0) = [4 - 0.5]^{\mathrm{T}}$ in each case. Although any single simulation run of the adaptive filter shows a considerably more erratic convergence path than that predicted by Equation 19.24, one observes that the average of these coefficient trajectories roughly follows the same path as that of the analysis.

19.4.2 Mean-Square Analysis

Although Equation 19.24 characterizes the mean behavior of the LMS adaptive filter, it does not indicate the nature of the fluctuations of the filter coefficients about their mean values, as indicated by the actual behavior of the LMS adaptive filter in Figure 19.2b. The magnitudes of these fluctuations can be accurately characterized through a mean-square analysis of the LMS adaptive filter. Because the coefficient error correlation matrix $\mathbf{K}(n)$ as defined in Equation 19.14 is the basis for our mean-square analysis, we outline methods for determining an evolution equation for this matrix. Then, we derive the forms of this evolution equation for both the i.i.d. and SIRP input signal models described previously, and we summarize the resulting expressions for the steady-state values of the misadjustment and excess MSE in Equations 19.16 and 19.17, respectively, for these different signal types. Finally, several conclusions regarding the mean-square behavior of the LMS adaptive filter are drawn.

19.4.2.1 Evolution of the Coefficient Error Correlation Matrix

To derive an evolution equation for $\mathbf{K}(n)$, we post-multiply both sides of Equation 19.18 by their respective transposes, which gives

$$\begin{aligned}\mathbf{V}(n+1)\mathbf{V}^{\mathrm{T}}(n+1) = {} & (\mathbf{I} - \mu(n)\mathbf{X}(n)\mathbf{X}^{\mathrm{T}}(n))\mathbf{V}(n)\mathbf{V}^{\mathrm{T}}(n)(\mathbf{I} - \mu(n)\mathbf{X}(n)\mathbf{X}^{\mathrm{T}}(n)) \\ & + \mu^2(n)\eta^2(n)\mathbf{X}(n)\mathbf{X}^{\mathrm{T}}(n) \\ & + \mu(n)\eta(n)(\mathbf{I} - \mu(n)\mathbf{X}(n)\mathbf{X}^{\mathrm{T}}(n))\mathbf{V}(n)\mathbf{X}^{\mathrm{T}}(n) \\ & + \mu(n)\eta(n)\mathbf{X}(n)\mathbf{V}^{\mathrm{T}}(n)(\mathbf{I} - \mu(n)\mathbf{X}(n)\mathbf{X}^{\mathrm{T}}(n)).\end{aligned} \quad (19.31)$$

Taking expectations of both sides of Equation 19.31, we note that $\eta(n)$ is zero mean and independent of both $\mathbf{X}(n)$ and $\mathbf{V}(n)$ from our models and assumptions, and thus the expectations of the third and fourth terms on the right-hand side of Equation 19.31 are zero. Moreover, by using the independence assumptions, it can be shown that

$$\begin{aligned} E\{\mathbf{X}(n)\mathbf{X}^{\mathrm{T}}(n)\mathbf{V}(n)\mathbf{V}^{\mathrm{T}}(n)\} &\approx E\{\mathbf{X}(n)\mathbf{X}^{\mathrm{T}}(n)\}E\{\mathbf{V}(n)\mathbf{V}^{\mathrm{T}}(n)\} \\ &= \mathbf{R}_{XX}\mathbf{K}(n). \end{aligned} \tag{19.32}$$

Thus, we obtain from Equation 19.31 the expression

$$\begin{aligned} \mathbf{K}(n+1) = \mathbf{K}(n) &- \mu(n)[\mathbf{R}_{XX}\mathbf{K}(n) + \mathbf{K}(n)\mathbf{R}_{XX}] \\ &+ \mu^2(n)E\{\mathbf{X}(n)\mathbf{X}^{\mathrm{T}}(n)\mathbf{K}(n)\mathbf{X}(n)\mathbf{X}^{\mathrm{T}}(n)\} + \mu^2(n)\sigma_\eta^2\mathbf{R}_{XX}, \end{aligned} \tag{19.33}$$

where σ_η^2 is as defined in Equation 19.12.

At this point, the analysis can be simplified depending on how the third term on the right-hand side of Equation 19.33 is evaluated according to the signal models and assumptions.

19.4.2.1.1 Analysis for SIRP Input Signals

In this case, the value of $E\{\mathbf{X}(n)\mathbf{X}^{\mathrm{T}}(n)\mathbf{K}(n)\mathbf{X}(n)\mathbf{X}^{\mathrm{T}}(n)\}$ can be expressed as

$$E\{\mathbf{X}(n)\mathbf{X}^{\mathrm{T}}(n)\mathbf{K}(n)\mathbf{X}(n)\mathbf{X}^{\mathrm{T}}(n)\} = m_z^{(2,2)}[2\mathbf{R}_{XX}\mathbf{K}(n)\mathbf{R}_{XX} + \mathbf{R}_{XX}\mathrm{tr}\{\mathbf{R}_{XX}\mathbf{K}(n)\}], \tag{19.34}$$

where the moment term $m_z^{(2,2)}$ is given by

$$m_z^{(2,2)} = E\{z_i^2(n)z_j^2(n)\} \tag{19.35}$$

for any $0 \leq i \neq j \leq (L-1)$ and

$$z_i(n) = \lambda_i^{1/2}\sum_{l=0}^{L-1} q_{li}x(n-l). \tag{19.36}$$

If $x(n)$ is a Gaussian random process, then $m_z^{(2,2)} = 1$, and it can be shown that $m_z^{(2,2)} \geq 1$ for SIRPs in general. For more details on these results, see [15].

19.4.2.1.2 Analysis for I.I.D. Input Signals

In this case, we can express the (i,j) th element of the matrix $E\{\mathbf{X}(n)\mathbf{X}^{\mathrm{T}}(n)\mathbf{K}(n)\mathbf{X}(n)\mathbf{X}^{\mathrm{T}}(n)\}$ as

$$[E\{\mathbf{X}(n)\mathbf{X}^{\mathrm{T}}(n)\mathbf{K}(n)\mathbf{X}(n)\mathbf{X}^{\mathrm{T}}(n)\}]_{i,j} = \begin{cases} 2\sigma_x^4[\mathbf{K}(n)]_{i,j}, & \text{if } i \neq j \\ \sigma_x^4\left(\gamma[\mathbf{K}(n)]_{i,j} + \sum\limits_{m=1,\, m\neq i}^{L}[\mathbf{K}(n)]_{m,m}\right), & \text{if } i = j, \end{cases} \tag{19.37}$$

where $[\mathbf{K}(n)]_{i,j}$ is the (i,j) th element of $\mathbf{K}(n)$,

$$\sigma_x^2 = E\{x^2(n)\}, \quad \text{and} \quad \gamma = \frac{E\{x^4(n)\}}{\sigma_x^4}, \tag{19.38}$$

respectively. For details, see [5].

19.4.2.1.3 Zeroth-Order Approximation Near Convergence

For small step sizes, it can be shown that the elements of $\mathbf{K}(n)$ are approximately proportional to both the step size and the noise variance σ_η^2 in steady-state. Thus, the magnitudes of the elements in the third term on the right-hand side of Equation 19.33 are about a factor of $\mu(n)$ smaller than those of any other terms in this equation at convergence. Such a result suggests that we could set

$$\mu^2(n)E\{\mathbf{X}(n)\mathbf{X}^{\mathrm{T}}(n)\mathbf{K}(n)\mathbf{X}(n)\mathbf{X}^{\mathrm{T}}(n)\} \approx \mathbf{0} \tag{19.39}$$

in the steady-state analysis of Equation 19.33 without perturbing the analytical results too much. If this approximation is valid, then the form of Equation 19.33 no longer depends on the form of the amplitude statistics of $x(n)$, as in the case of the mean analysis.

19.4.2.2 Excess MSE, Mean-Square Stability, and Misadjustment

Given the results in Equations 19.34 through 19.39, we can use the evolution equation for $\mathbf{K}(n)$ in Equation 19.33 to explore the mean-square behavior of the LMS adaptive filter in several ways:

- By studying the structure of Equation 19.33 for different signal types, we can determine conditions on the step size $\mu(n)$ to guarantee the stability of the mean-square analysis equation.
- By setting $\mathbf{K}(n+1) = \mathbf{K}(n)$ and fixing the value of $\mu(n)$, we can solve for the steady-state value of $\mathbf{K}(n)$ at convergence, thereby obtaining a measure of the fluctuations of the coefficients about their optimum solutions.
- Given a value for $\mathbf{V}(0)$, we can write a computer program to simulate the behavior of this equation for different signal statistics and step size sequences.

Moreover, once the matrix sequence $\mathbf{K}(n)$ is known, we can obtain the values of the excess MSE and misadjustment from $\mathbf{K}(n)$ by employing the relations in Equations 19.16 and 19.17, respectively.

Table 19.1 summarizes many of the analytical results that can be obtained from a careful study of Equation 19.33. Shown in the table are the conditions on the step size $\mu(n) = \mu$ to guarantee stability, sufficient stability conditions on the step size that can be easily calculated, and the misadjustment in steady-state for the three different methods of evaluating $E\{\mathbf{X}(n)\mathbf{X}^{\mathrm{T}}(n)\mathbf{K}(n)\mathbf{X}(n)\mathbf{X}^{\mathrm{T}}(n)\}$ in Equations 19.34 through 19.39. In the table, the quantity C is defined as

$$C = \sum_{i=0}^{L-1} \frac{\lambda_i}{1 - \mu m_z^{(2,2)}\lambda_i}. \tag{19.40}$$

From these results and others that can be obtained from Equation 19.33, we can infer several facts about the mean-square performance of the LMS adaptive filter:

- The value of the excess MSE at time n consists of the sum of the steady-state excess MSE, given by $M\sigma_\eta^2$, and a weighted sum of L exponentially converging and/or diverging terms. Similar to the mean analysis case, these additional terms depend on the elements of the eigenvector matrix $\mathbf{Q}$, the

TABLE 19.1 Summary of MSE Analysis Results

Assumption	MSE Stability Conditions	Sufficient Conditions	Misadjustment
I.I.D. input	$0 < \mu < \frac{2}{(L-1+\gamma)\sigma_x^2}$	$0 < \mu < \frac{2}{(L-1+\gamma)\sigma_x^2}$	$M = \frac{\mu\sigma_x^2 L}{2-\mu\sigma_x^2(L-1+\gamma)}$
SIRP input	$0 < \mu < \frac{2}{m_z^{(2,2)}\lambda_{\max}}$ and $\mu m_z^{(2,2)}C < 2$	$0 < \mu < \frac{2}{3Lm_z^{(2,2)}\sigma_x^2}$	$M = \frac{\mu C}{2-\mu m_z^{(2,2)}C}$
Approximation	$0 < \mu < \frac{1}{\lambda_{\max}}$	$0 < \mu < \frac{1}{L\sigma_x^2}$	$M = \frac{\mu\sigma_x^2 C}{2}$

eigenvalues of $\mathbf{R}_{XX}$, the eigenvalues of $\mathbf{K}(0)$, and the values of $m_z^{(2,2)}$ or γ for the SIRP or i.i.d. input signal models, respectively.

- For all input signal types, approximate conditions on the fixed step size value to guarantee convergence of the evolution equations for $\mathbf{K}(n)$ are of the form

$$0 < \mu < \frac{K}{L\sigma_x^2}, \tag{19.41}$$

where σ_x^2 is the input signal power and where the constant K depends weakly on the nature of the input signal statistics and not on the magnitude of the input signal. All of the sufficient stability bounds on μ as shown in Table 19.1 can be put in the form of Equation 19.26. Because of the inaccuracies within the analysis that are caused by the independence assumptions, however, the actual step size chosen for stability of the LMS adaptive filter should be somewhat smaller than these values, and step sizes in the range $0 < \mu(n) < 0.1/(L\sigma_x^2)$ are often chosen in practice.
- The misadjustment of the LMS adaptive filter increases as the filter length L and step size μ are increased. Thus, a larger step size causes larger fluctuations of the filter coefficients about their optimum solutions in steady-state.

19.5 Performance Issues

When using the LMS adaptive filter, one must select the filter length L and the step size $\mu(n)$ to obtain the desired performance from the system. In this section, we explore the issues affecting the choices of these parameters using the analytical results for LMS adaptive filter's behavior derived in the last section.

19.5.1 Basic Criteria for Performance

The performance of the LMS adaptive filter can be characterized in three important ways: the adequacy of the FIR filter model, the speed of convergence of the system, and the misadjustment in steady-state.

19.5.1.1 Adequacy of the FIR Model

The LMS adaptive filter relies on the linearity of the FIR filter model to accurately characterize the relationship between the input and desired response signals. When the relationship between $x(n)$ and $d(n)$ deviates from the linear one given in Equation 19.4, then the performance of the overall system suffers. In general, it is possible to use a nonlinear model in place of the adaptive FIR filter model considered here. Possible nonlinear models include polynomial-based filters such as Volterra and bilinear filters [16] as well as neural network structures [17].

Another source of model inaccuracy is the finite impulse response length of the adaptive FIR filter. It is typically necessary to tune both the length of the filter L and the relative delay between the input and desired response signals so that the input signal values not contained in $\mathbf{X}(n)$ are largely uncorrelated with the desired response signal sample $d(n)$. However, such a situation may be impossible to achieve when the relationship between $x(n)$ and $d(n)$ is of an infinite-impulse response (IIR) nature. Adaptive IIR filters can be considered for these situations, although the stability and performance behaviors of these systems are much more difficult to characterize. Adaptive IIR filters are discussed in Chapter 23.

19.5.1.2 Speed of Convergence

The rate at which the coefficients approach their optimum values is called the speed of convergence. As the analytical results show, there exists no one quantity that characterizes the speed of convergence, as it depends on the initial coefficient values, the amplitude and correlation statistics of the signals, the filter length L, and the step size $\mu(n)$. However, we can make several qualitative statements relating the speed of convergence to both the step size and the filter length. All of these results assume that the desired response signal model in Equation 19.4 is reasonable and that the errors in the filter coefficients are uniformly distributed across the coefficients on average.

- The speed of convergence increases as the value of the step size is increased, up to step sizes near one-half the maximum value required for stable operation of the system. This result can be obtained from a careful analysis of Equation 19.33 for different input signal types and correlation statistics. Moreover, by simulating the behavior of Equations 19.33 and 19.34 for typical signal scenarios, it is observed that the speed of convergence of the excess MSE actually decreases for large enough step size values. For i.i.d. input signals, the fixed step size providing fastest convergence of the excess MSE is exactly one-half the MSE step size bound as given in Table 19.1 for this type of input signal.
- The speed of convergence decreases as the length of the filter is increased. The reasons for this behavior are twofold. First, if the input signal is correlated, the condition number of $\mathbf{R}_{XX}$, defined as the ratio of the largest and smallest eigenvalues of this matrix, generally increases as L is increased for typical real-world input signals. A larger condition number for $\mathbf{R}_{XX}$ makes it more difficult to choose a good step size to obtain fast convergence of all of the elements of either $E\{\mathbf{V}(n)\}$ or $\mathbf{K}(n)$. Such an effect can be seen in Equation 19.24, as a larger condition number leads to a larger disparity in the values of $(1 - \mu\lambda_j)$ for different j. Second, in the MSE analysis equation of Equation 19.33, the overall magnitude of the expectation term $E\{\mathbf{X}(n)\mathbf{X}^T(n)\mathbf{K}(n)\mathbf{X}(n)\mathbf{X}^T(n)\}$ is larger for larger L, due to the fact that the scalar quantity $\mathbf{X}^T(n)\mathbf{K}(n)\mathbf{X}(n)$ within this expectation increases as the size of the filter is increased. Since this quantity is always positive, it limits the amount that the excess MSE can be decreased at each iteration, and it reduces the maximum step size that is allowed for mean-square convergence as L is increased.
- The maximum possible speed of convergence is limited by the largest step size that can be chosen for stability for moderately correlated input signals. In practice, the actual step size needed for stability of the LMS adaptive filter is smaller than one-half the maximum values given in Table 19.1 when the input signal is moderately correlated. This effect is due to the actual statistical relationships between the current coefficient vector $\mathbf{W}(n)$ and the signals $\mathbf{X}(n)$ and $d(n)$, relationships that are neglected via the independence assumptions. Since the convergence speed increases as μ is increased over this allowable step size range, the maximum stable step size provides a practical limit on the speed of convergence of the system.
- The speed of convergence depends on the desired level of accuracy that is to be obtained by the adaptive filter. Generally speaking, the speed of convergence of the system decreases as the desired level of misadjustment is decreased. This result is due to the fact that the behavior of the system is dominated by the slower converging modes of the system as the length of adaptation time is increased. Thus, if the desired level of misadjustment is low, the speed of convergence is dominated by the slower-converging modes, thus limiting the overall convergence speed of the system.

19.5.1.3 Misadjustment

The misadjustment, defined in Equation 19.16, is the additional fraction of MSE in the filter output above the minimum MSE value σ_η^2 caused by a nonzero adaptation speed. We can draw the following two conclusions regarding this quantity:

- The misadjustment increases as the step size is increased.
- The misadjustment increases as the filter length is increased.

Both results can be proven by direct study of the analytical results for M in Table 19.1.

19.5.2 Identifying Stationary Systems

We now evaluate the basic criteria for performance to provide qualitative guidance as to how to choose μ and L to identify a stationary system.

19.5.2.1 Choice of Filter Length

We have seen that as the filter length L is increased, the speed of convergence of the LMS adaptive filter decreases, and the misadjustment in steady-state increases. Therefore, the filter length should be chosen as short as possible but long enough to adequately model the unknown system, as too short a filter model leads to poor modeling performance. In general, there exists an optimal length L for a given μ that exactly balances the penalty for a finite-length filter model with the increase in misadjustment caused by a longer filter length, although the calculation of such a model order requires more information than is typically available in practice. Modeling criteria, such as Akaike's information criterion [18] and minimum description length [19] could be used in this situation.

19.5.2.2 Choice of Step Size

We have seen that the speed of convergence increases as the step size is increased, up to values that are roughly within a factor of 1/2 of the step size stability limits. Thus, if fast convergence is desired, one should choose a large step size according to the limits in Table 19.1. However, we also observe that the misadjustment increases as the step size is increased. Therefore, if highly accurate estimates of the filter coefficients are desired, a small step size should be chosen. This classical trade-off in convergence speed vs. the level of error in steady state dominates the issue of step size selection in many estimation schemes.

If the user knows that the relationship between $x(n)$ and $d(n)$ is linear and time-invariant, then one possible solution to the above trade-off is to choose a large step size initially to obtain fast convergence, and then switch to a smaller step size to obtain a more accurate estimate of $\mathbf{W}_{opt}$ near convergence. The point to switch to a smaller step size is roughly when the excess MSE becomes a small fraction ($\sim$1/10th) of the minimum MSE of the filter. This method of gearshifting, as it is commonly known, is part of a larger class of time-varying step size methods that we shall explore shortly.

Although we have discussed qualitative criteria by which to choose a fixed step size, it is possible to define specific performance criteria by which to choose μ. For one study of this problem for i.i.d. input signals, see [20].

19.5.3 Tracking Time-Varying Systems

Since the LMS adaptive filter continually adjusts its coefficients to approximately minimize the MSE criterion, it can adjust to changes in the relationship between $x(n)$ and $d(n)$. This behavior is commonly referred to as tracking. In such situations, it clearly is not desirable to reduce the step size to an extremely small value in steady-state, as the LMS adaptive filter would not be able to follow any changes in the relationship between $x(n)$ and $d(n)$.

To illustrate the issues involved, consider the desired response signal model given by

$$d(n) = \mathbf{W}_{opt}^{T}(n)\mathbf{X}(n) + \eta(n), \tag{19.42}$$

where the optimum coefficient vector $\mathbf{W}_{opt}(n)$ varies with time according to

$$\mathbf{W}_{opt}(n+1) = \mathbf{W}_{opt}(n) + \mathbf{M}(n), \tag{19.43}$$

and $\{\mathbf{M}(n)\}$ is a sequence of vectors whose elements are all i.i.d. This nonstationary model is similar to others used in other tracking analyses of the LMS adaptive filter [4,21]; it also enables a simple analysis that is similar to the stationary system identification model discussed earlier.

Applying the independence assumptions, we can analyze the behavior of the LMS adaptive filter for this desired response signal model. For brevity, we only summarize the results of an approximate analysis in which terms of the form $\mu^2 E\{\mathbf{X}(n)\mathbf{X}^T(n)\mathbf{K}(n)\mathbf{X}(n)\mathbf{X}^T(n)\}$ are neglected in the MSE behavioral equations [4]. The misadjustment of the system in steady-state is

$$M_{\text{non}} = \frac{L}{2}\left(\mu\sigma_x^2 + \frac{\sigma_m^2}{\mu\sigma_\eta^2}\right), \tag{19.44}$$

where σ_m^2 is the power in any one element of $\mathbf{M}(n)$. Details of a more-accurate tracking analysis can be found in [21].

In this case, the misadjustment is the sum of two terms. The first term is the same as that for the stationary case and is proportional to μ. The second term is the lag error and is due to the fact that the LMS coefficients follow or "lag" behind the optimum coefficient values. The lag error is proportional to the speed of variation of the unknown system through σ_m^2 and is inversely proportional to the step size, such that its value increases as the step size is decreased.

In general, there exists an optimum fixed step size that minimizes the misadjustment in steady-state for an LMS adaptive filter that is tracking changes in $\mathbf{W}_{\text{opt}}(n)$. For the approximate analysis used to derive Equation 19.44, the resulting step size is

$$\mu_{\text{opt}} = \frac{\sigma_m}{\sigma_\eta\sigma_x}. \tag{19.45}$$

As the value of σ_m^2 increases, the level of nonstationarity increases such that a larger step size is required to accurately track changes in the unknown system. Similar conclusions can be drawn from other analyses of the LMS adaptive filter in tracking situations [22].

19.6 Selecting Time-Varying Step Sizes

The analyses of the previous sections enable one to choose a fixed step size μ for the LMS adaptive filter to meet the system's performance requirements when the general characteristics of the input and desired response signals are known. In practice, the exact statistics of $x(n)$ and $d(n)$ are unknown or vary with time. A time-varying step size $\mu(n)$, if properly computed, can provide stable, robust, and accurate convergence behavior for the LMS adaptive filter in these situations. In this section, we consider useful online procedures for computing $\mu(n)$ in the LMS adaptive filter to meet these performance requirements.

19.6.1 Normalized Step Sizes

For the LMS adaptive filter to be useful, it must operate in a stable manner so that its coefficient values do not diverge. From the stability results in Table 19.1 and the generalized expression for these stability bounds in Equation 19.41, the upper bound for the step size is inversely proportional to the input signal power σ_x^2 in general. In practice, the input signal power is unknown or varies with time. Moreover, if one were to choose a small fixed step size value to satisfy these stability bounds for the largest anticipated input signal power value, then the convergence speed of the system would be unnecessarily slow during periods when the input signal power is small.

These concerns can be addressed by calculating a normalized step size $\mu(n)$ as

$$\mu(n) = \frac{\bar{\mu}}{\delta + L\hat{\sigma}_x^2(n)}, \tag{19.46}$$

where
$\widehat{\sigma_x^2}(n)$ is an estimate of the input signal power
$\bar{\mu}$ is a constant somewhat smaller than the value of K required for system stability in Equation 19.41
δ is a small constant to avoid a divide-by-zero should $\widehat{\sigma_x^2}(n)$ approach zero

To estimate the input signal power, a lowpass filter can be applied to the sequence $x^2(n)$ to track its changing envelope. Typical estimators include

- Exponentially weighted estimate:

$$\widehat{\sigma_x^2}(n) = (1 - c)\widehat{\sigma_x^2}(n - 1) + cx^2(n) \tag{19.47}$$

- Sliding-window estimate:

$$\widehat{\sigma_x^2}(n) = \frac{1}{N}\sum_{i=0}^{N-1} x^2(n - i) \tag{19.48}$$

where the parameters c, $0 < c \ll 1$, and $N, N \geq L$, control the effective memories of the two estimators, respectively.

19.6.1.1 Normalized LMS Adaptive Filter

By choosing a sliding window estimate of length $N = L$, the LMS adaptive filter with $\mu(n)$ in Equation 19.46 becomes

$$\mathbf{W}(n + 1) = \mathbf{W}(n) + \frac{\bar{\mu}e(n)}{p(n)}\mathbf{X}(n) \tag{19.49}$$

$$p(n) = \delta + \|\mathbf{X}(n)\|^2, \tag{19.50}$$

where $\|\mathbf{X}(n)\|^2$ is the L_2-norm of the input signal vector. The value of $p(n)$ can be updated recursively as

$$p(n) = p(n - 1) + x^2(n) - x^2(n - L), \tag{19.51}$$

where $p(0) = \delta$ and $x(n) = 0$ for $n \leq 0$. The adaptive filter in Equation 19.49 is known as the normalized LMS (NLMS) adaptive filter. It has two special properties that make it useful for adaptive filtering tasks:

- The NLMS adaptive filter is guaranteed to converge for any value of $\bar{\mu}$ in the range

$$0 < \bar{\mu} < 2, \tag{19.52}$$

  regardless of the statistics of the input signal. Thus, selecting the value of $\bar{\mu}$ for stable behavior of this system is much easier than selecting μ for the LMS adaptive filter.
- With the proper choice of $\bar{\mu}$, the NLMS adaptive filter can often converge faster than the LMS adaptive filter. In fact, for noiseless system identification tasks in which $\eta(n)$ in Equation 19.4 is zero, one can obtain $\mathbf{W}_{opt}(n) = \mathbf{W}_{opt}$ after L iterations of Equation 19.49 for $\bar{\mu} = 1$. Moreover, for SIRP input signals, the NLMS adaptive filter provides more uniform convergence of the filter coefficients, making the selection of $\bar{\mu}$ an easier proposition than the selection of μ for the LMS adaptive filter.

A discussion of these and other results on the NLMS adaptive filter can be found in [15,23–25].

19.6.2 Adaptive and Matrix Step Sizes

In addition to stability, the step size controls both the speed of convergence and the misadjustment of the LMS adaptive filter through the statistics of the input and desired response signals. In situations where the statistics of $x(n)$ and/or $d(n)$ are changing, the value of $\mu(n)$ that provides the best performance from the system can change as well. In these situations, it is natural to consider $\mu(n)$ as an adaptive parameter to be optimized along with the coefficient vector $\mathbf{W}(n)$ within the system. While it may seem novel, the idea of computing an adaptive step size has a long history in the field of adaptive filters [2]. Numerous such techniques have been proposed in the scientific literature. One such method uses a stochastic gradient procedure to adjust the value of $\mu(n)$ to iteratively minimize the MSE within the LMS adaptive filter. A derivation and performance analysis of this algorithm is given in [26].

In some applications, the task at hand suggests a particular strategy for adjusting the step size $\mu(n)$ to obtain the best performance from an LMS adaptive filter. For example, in echo cancellation for telephone networks, the signal-to-noise ratio of $d(n)$ falls to extremely low values when the near-end talker signal is present, making accurate adaptation during these periods difficult. Such systems typically employ double-talk detectors, in which estimates of the statistical characteristics of $x(n), d(n)$, and/or $e(n)$ are used to raise and lower the value of $\mu(n)$ in an appropriate manner. A discussion of this problem and a method for its solution are given in [27].

While our discussion of the LMS adaptive filter has assumed a single step size value for each of the filter coefficients, it is possible to select L different step sizes $\mu_i(n)$ for each of the L coefficient updates within the LMS adaptive filter. To select fixed values for each $\mu_i(n) = \mu_i$, these matrix step size methods require prior knowledge about the statistics of $x(n)$ and $d(n)$ and/or the approximate values of $\mathbf{W}_{\text{opt}}$. It is possible, however, to adapt each $\mu_i(n)$ according to a suitable performance criterion to obtain improved convergence behavior from the overall system. A particularly simple adaptive method for calculating matrix step sizes is provided in [28].

19.6.3 Other Time-Varying Step Size Methods

In situations where the statistics of $x(n)$ and $d(n)$ do not change with time, choosing a variable step size sequence $\mu(n)$ is still desirable, as one can decrease the misadjustment over time to obtain an accurate estimate of the optimum coefficient vector $\mathbf{W}_{\text{opt}}$. Such methods have been derived and characterized in a branch of statistical analysis known as stochastic approximation [29]. Using this formalism, it is possible to prove under certain assumptions on $x(n)$ and $d(n)$ that the value of $\mathbf{W}(n)$ for the LMS adaptive filter converges to $\mathbf{W}_{\text{opt}}$ as $n \to \infty$ if $\mu(n)$ satisfies

$$\sum_{n=0}^{\infty} |\mu(n)| \to \infty \quad \text{and} \quad \sum_{n=0}^{\infty} \mu^2(n) < \infty, \tag{19.53}$$

respectively. One step size function satisfying these constraints is

$$\mu(n) = \frac{\mu(0)}{n+1}, \tag{19.54}$$

where $\mu(0)$ is an initial step size parameter. The gearshifting method described in Section 19.5.2 can be seen as a simple heuristic approximation to Equation 19.54. Moreover, one can derive an optimum step size sequence $\mu_{\text{opt}}(n)$ that minimizes the excess MSE at each iteration under certain situations, and the limiting form of the resulting step size values for stationary signals are directly related to Equation 19.54 as well [24].

19.7 Other Analyses of the LMS Adaptive Filter

While the analytical techniques employed in this section are useful for selecting design parameters for the LMS adaptive filter, they are but one method for characterizing the behavior of this system. Other forms of analyses can be used to determine other characteristics of this system, such as the p.d.f.s of the adaptive filter coefficients [30] and the probability of large excursions in the adaptive filter coefficients for different types of input signals [31]. In addition, much research effort has focused on characterizing the stability of the system without extensive assumptions about the signals being processed. One example of such an analysis is given in [32]. Other methods for analyzing the LMS adaptive filter include the method of ordinary differential equations [33], the stochastic approximation methods described previously [29], computer-assisted symbolic derivation methods [14], and averaging techniques that are particularly useful for deterministic signals [34].

19.8 Analysis of Other Adaptive Filters

Because of the difficulties in performing multiplications in the first digital hardware implementations of adaptive filters, many of these systems employed nonlinearities in the coefficient update terms to simplify their hardware requirements. An example of one such algorithm is the sign-error adaptive filter, in which the coefficient update is

$$\mathbf{W}(n+1) = \mathbf{W}(n) + \mu(n)\mathrm{sgn}(e(n))\mathbf{X}(n), \tag{19.55}$$

where the value of $\mathrm{sgn}[e(n)]$ is either 1 or -1 depending on whether $e(n)$ is positive or negative, respectively. If $\mu(n)$ is chosen as a power of 2, this algorithm only requires a comparison and bit shift per coefficient to implement in hardware. Other algorithms employing nonlinearities of the input signal vector $\mathbf{X}(n)$ in the updates are also useful [35].

Many of the analysis techniques developed for the LMS adaptive filter can be applied to algorithms with nonlinearities in the coefficient updates, although such methods require additional assumptions to obtain accurate results. For presentations of two such analyses, see [36,37]. It should be noted that the performance characteristics and stability properties of these nonlinearly modified versions of the LMS adaptive filter can be quite different from those of the LMS adaptive filter. For example, the sign-error adaptive filter in Equation 19.55 is guaranteed to converge for any fixed positive step size value under fairly loose assumptions on $x(n)$ and $d(n)$ [38].

19.9 Conclusions

In summary, we have described a statistical analysis of the LMS adaptive filter, and through this analysis, suggestions for selecting the design parameters for this system have been provided. While useful, analytical studies of the LMS adaptive filter are but one part of the system design process. As in all design problems, sound engineering judgment, careful analytical studies, computer simulations, and extensive real-world evaluations and testing should be combined when developing an adaptive filtering solution to any particular task.

References

1. Widrow, B. and Hoff, M.E., Adaptive switching circuits, *IRE WESCON Conv. Rec.*, 4: 96–104, Aug. 1960.
2. Widrow, B., Adaptive sampled-data systems—A statistical theory of adaptation, *IRE WESCON Conv. Rec.*, 4: 74–85, Aug. 1959.

3. Senne, K.D., Adaptive linear discrete-time estimation, PhD thesis, Stanford University, Stanford, CA, June 1968.
4. Widrow, B., McCool, J., Larimore, M.G., and Johnson, C.R., Jr., Stationary and nonstationary learning characteristics of the LMS adaptive filter, *Proc. IEEE*, 64(8): 1151–1162, Aug. 1976.
5. Gardner, W.A., Learning characteristics of stochastic-gradient-descent algorithms: A general study, analysis, and critique, *Signal Process.*, 6(2): 113–133, Apr. 1984.
6. Feuer, A. and Weinstein, E., Convergence analysis of LMS filters with uncorrelated data, *IEEE Trans. Acoust. Speech Signal Process.*, ASSP-331: 222–230, Feb. 1985.
7. Messerschmitt, D.G., Echo cancellation in speech and data transmission, *IEEE J. Sel. Areas Commn.*, 2(2): 283–301, Mar. 1984.
8. Widrow, B. and Walach, E., *Adaptive Inverse Control*, Prentice-Hall, Upper Saddle River, NJ, 1996.
9. Proakis, J.G., *Digital Communications*, 3rd ed., McGraw-Hill, New York, 1995.
10. Murano, K., Unagami, S., and Amano, F., Echo cancellation and applications, *IEEE Commn. Mag.*, 28(1): 49–55, Jan. 1990.
11. Gradsteyn, I.S. and Ryzhik, I.M., *Table of Integrals, Series and Products*, Academic Press, New York, 1980.
12. Brehm, H. and Stammler, W., Description and generation of spherically invariant speech-model signals, *Signal Process.*, 12(2): 119–141, Mar. 1987.
13. Mazo, J.E., On the independence theory of equalizer convergence, *Bell Sys. Tech. J.*, 58(5): 963–993, May–June 1979.
14. Douglas, S.C. and Pan, W., Exact expectation analysis of the LMS adaptive filter, *IEEE Trans. Signal Process.*, 43(12): 2863–2871, Dec. 1995.
15. Rupp, M., The behavior of LMS and NLMS algorithms in the presence of spherically invariant processes, *IEEE Trans. Signal Process.*, 41(3): 1149–1160, Mar. 1993.
16. Mathews, V.J., Adaptive polynomial filters, *IEEE Signal Proc. Mag.*, 8(3): 10–26, July 1991.
17. Haykin, S., *Neural Networks*, Prentice-Hall, Englewood Cliffs, NJ, 1995.
18. Akaike, H., A new look at the statistical model identification, *IEEE Trans. Autom. Control*, AC-19(6): 716–723, Dec. 1974.
19. Rissanen, J., Modelling by shortest data description, *Automatica*, 14(5): 465–471, Sept. 1978.
20. Bershad, N.J., On the optimum gain parameter in LMS adaptation, *IEEE Trans. Acoust. Speech Signal Process.*, ASSP-35(7): 1065–1068, July 1987.
21. Gardner, W.A., Nonstationary learning characteristics of the LMS algorithm, *IEEE Trans. Circuit Syst.*, 34(10): 1199–1207, Oct. 1987.
22. Farden, D.C., Tracking properties of adaptive signal processing algorithms, *IEEE Trans. Acoust. Speech Signal Process.*, ASSP-29(3): 439–446, June 1981.
23. Bitmead, R.R. and Anderson, B.D.O., Performance of adaptive estimation algorithms in dependent random environments, *IEEE Trans. Autom. Control*, AC-25(4): 788–794, Aug. 1980.
24. Slock, D.T.M., On the convergence behavior of the LMS and the normalized LMS algorithms, *IEEE Trans. Signal Process.*, 41(9): 2811–2825, Sept. 1993.
25. Douglas, S.C. and Meng, T.H.-Y., Normalized data nonlinearities for LMS adaptation, *IEEE Trans. Signal Process.*, 42(6): 1352–1365, June 1994.
26. Mathews, V.J. and Xie, Z., A stochastic gradient adaptive filter with gradient adaptive step size, *IEEE Trans. Signal Process.*, 41(6): 2075–2087, June 1993.
27. Ding, Z., Johnson, C.R. Jr., and Sethares, W.A., Frequency-dependent bursting in adaptive echo cancellation and its prevention using double-talk detectors, *Int. J. Adaptive Control Signal Process.*, 4(3): 219–236, May–June 1990.
28. Harris, R.W., Chabries, D.M., and Bishop, F.A., A variable step (VS) adaptive filter algorithm, *IEEE Trans. Acoust. Speech Signal Process.*, ASSP-34(2): 309–316, Apr. 1986.
29. Kushner, H.J. and Clark, D.S., *Stochastic Approximation Methods for Constrained and Unconstrained Systems*, Springer-Verlag, New York, 1978.

30. Bershad, N.J. and Qu, L.Z., On the probability density function of the LMS adaptive filter weights, *IEEE Trans. Acoust. Speech Signal Process.*, ASSP-37(1): 43–56, Jan. 1989.
31. Rupp, M., Bursting in the LMS algorithm, *IEEE Trans. on Signal Process.*, 43(10): 2414–2417, Oct. 1995.
32. Macchi, O. and Eweda, E., Second-order convergence analysis of stochastic adaptive linear filtering, *IEEE Trans. Autom. Control*, AC-28(1): 76–85, Jan. 1983.
33. Benveniste, A., Métivier, M., and Priouret, P., *Adaptive Algorithms and Stochastic Approximations*, Springer-Verlag, New York, 1990.
34. Solo, V. and Kong, X., *Adaptive Signal Processing Algorithms: Stability and Performance*, Prentice-Hall, Englewood Cliffs, NJ, 1995.
35. Duttweiler, D.L., Adaptive filter performance with nonlinearities in the correlation multiplier, *IEEE Trans. Acoust. Speech Signal Process.*, ASSP-30(4): 578–586, Aug. 1982.
36. Bucklew, J.A., Kurtz, T.J., and Sethares, W.A., Weak convergence and local stability properties of fixed step size recursive algorithms, *IEEE Trans. Info. Theory*, 39(3): 966–978, May 1993.
37. Douglas, S.C. and Meng, T.H.-Y., Stochastic gradient adaptation under general error criteria, *IEEE Trans. Signal Process.*, 42(6): 1335–1351, June 1994.
38. Cho, S.H. and Mathews, V.J., Tracking analysis of the sign algorithm in nonstationary environments, *IEEE Trans. Acoust. Speech Signal Process.*, ASSP-38(12): 2046–2057, Dec. 1990.

20

Robustness Issues in Adaptive Filtering

Ali H. Sayed
University of California at Los Angeles

Markus Rupp
Technical University of Vienna

Adaptive filters are systems that adjust themselves to a changing environment. They are designed to meet certain performance specifications and are expected to perform reasonably well under the operating conditions for which they have been designed. In practice, however, factors that may have been ignored or overlooked in the design phase of the system can affect the performance of the adaptive scheme that has been chosen for the system. Such factors include unmodeled dynamics, modeling errors, measurement noise, and quantization errors, among others, and their effect on the performance of an adaptive filter could be critical to the proposed application. Moreover, technological advancements in digital circuit and VLSI design have spurred an increase in the range of new adaptive filtering applications in fields ranging from biomedical engineering to wireless communications. For these new areas, it is increasingly important to design adaptive schemes that are tolerant to unknown or nontraditional factors and effects. The aim of this chapter is to explore and determine the robustness properties of some classical adaptive schemes. Our presentation is meant as an introduction to these issues, and many of the relevant details of specific topics discussed in this section, and alternative points of view, can be found in the references at the end of the chapter.

20.1 Motivation and Example

A classical application of adaptive filtering is that of system identification. The basic problem formulation is depicted in Figure 20.1, where z^{-1} denotes the unit-time delay operator. The diagram contains two system blocks: one representing the unknown plant or system and the other containing a time-variant

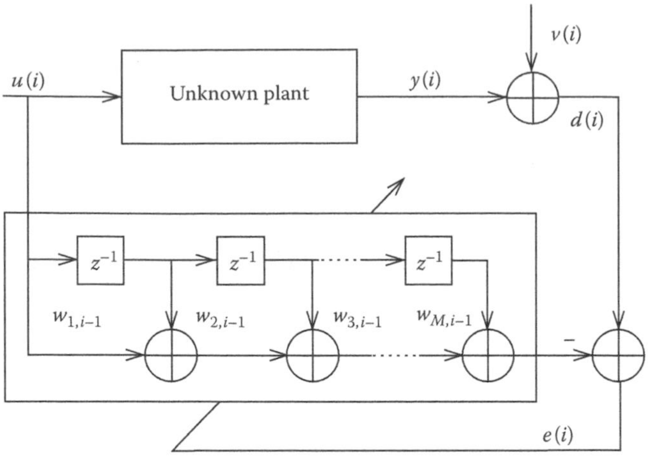

FIGURE 20.1 A system identification example.

tapped-delay-line or finite-impulse-response (FIR) filter structure. The unknown plant represents an arbitrary relationship between its input and output. This block might implement a pole-zero transfer function, an all-pole or autoregressive transfer function, a fixed or time-varying FIR system, a nonlinear mapping, or some other complex system. In any case, it is desired to determine an FIR model for the unknown system of a predetermined impulse response length M, and whose coefficients at time $i-1$ are denoted by $\{w_{1,i-1}, w_{2,i-1}, \ldots, w_{M,i-1}\}$. The unknown system and the FIR filter are excited by the same input sequence $\{u(i)\}$, where the time origin is at $i=0$.

If we collect the FIR coefficients into a column vector, say $\mathbf{w}_{i-1} = \text{col}\{w_{1,i-1}, w_{2,i-1}, \ldots, w_{M,i-1}\}$, and define the state vector of the FIR model at time i as $\mathbf{u}_i = \text{col}\{u(i), u(i-1), \ldots, u(i-M+1)\}$, then the output of the FIR filter at time i is the inner product $\mathbf{u}_i^T\mathbf{w}_{i-1}$. In principle, this inner product should be compared with the output $y(i)$ of the unknown plant in order to determine whether or not the FIR output is a good enough approximation for the output of the plant and, therefore, whether or not the current coefficient vector $\mathbf{w}_{i-1}$ should be updated.

In general, however, we do not have direct access to the uncorrupted output $y(i)$ of the plant but rather to a noisy measurement of it, say $d(i) = y(i) + v(i)$. The purpose of an adaptive scheme is to employ the output error sequence $\{e(i) = d(i) - \mathbf{u}_i^T\mathbf{w}_{i-1}\}$, which measures how far $d(i)$ is from $\mathbf{u}_i^T\mathbf{w}_{i-1}$, in order to update the entries of $\mathbf{w}_{i-1}$ and provide a better model, say $\mathbf{w}_i$, for the unknown system. That is, the purpose of the adaptive filter is to employ the available data at time i, $\{d(i), \mathbf{w}_{i-1}, \mathbf{u}_i\}$, in order to update the coefficient vector $\mathbf{w}_{i-1}$ into a presumably better estimate vector $\mathbf{w}_i$.

In this sense, we may regard the adaptive filter as a recursive estimator that tries to come up with a coefficient vector $\mathbf{w}$ that "best" matches the observed data $\{d(i)\}$ in the sense that, for all i, $d(i) \approx \mathbf{u}_i^T\mathbf{w} + v(i)$ to good accuracy. The successive $\mathbf{w}_i$ provide estimates for the unknown and desired $\mathbf{w}$.

20.2 Adaptive Filter Structure

We may reformulate the above adaptive problem in mathematical terms as follows. Let $\{\mathbf{u}_i\}$ be a sequence of regression vectors and let $\mathbf{w}$ be an unknown column vector to be estimated or identified. Given noisy measurements $\{d(i)\}$ that are assumed to be related to $\mathbf{u}_i^T\mathbf{w}$ via an additive noise model of the form

$$d(i) = \mathbf{u}_i^T\mathbf{w} + v(i), \tag{20.1}$$

we wish to employ the given data $\{d(i), \mathbf{u}_i\}$ in order to provide recursive estimates for $\mathbf{w}$ at successive time instants, say $\{\mathbf{w}_0, \mathbf{w}_1, \mathbf{w}_2, \ldots\}$. We refer to these estimates as weight estimates since they provide estimates for the coefficients or weights of the tapped-delay model.

Most adaptive schemes perform this task in a recursive manner that fits into the following general description: starting with an initial guess for $\mathbf{w}$, say $\mathbf{w}_{-1}$, iterate according to the learning rule

$$\begin{pmatrix} \text{new weight} \\ \text{estimate} \end{pmatrix} = \begin{pmatrix} \text{old weight} \\ \text{estimate} \end{pmatrix} + \begin{pmatrix} \text{correction} \\ \text{term} \end{pmatrix},$$

where the correction term is usually a function of $\{d(i), \mathbf{u}_i, \text{old weight estimate}\}$. More compactly, we may write $\mathbf{w}_i = \mathbf{w}_{i-1} + f[d(i), \mathbf{u}_i, \mathbf{w}_{i-1}]$, where $\mathbf{w}_i$ denotes an estimate for $\mathbf{w}$ at time i and f denotes a function of the data $\{d(i), \mathbf{u}_i, \mathbf{w}_{i-1}\}$ or of previous values of the data, as in the case where only a filtered version of the error signal $d(i) - \mathbf{u}_i^{\mathrm{T}}\mathbf{w}_{i-1}$ is available. In this context, the well-known least-mean-square (LMS) algorithm has the form

$$\mathbf{w}_i = \mathbf{w}_{i-1} + \mu \cdot \mathbf{u}_i \cdot [d(i) - \mathbf{u}_i^{\mathrm{T}} \cdot \mathbf{w}_{i-1}], \tag{20.2}$$

where μ is known as the step-size parameter.

20.3 Performance and Robustness Issues

The performance of an adaptive scheme can be studied from many different points of view. One distinctive methodology that has attracted considerable attention in the adaptive filtering literature is based on stochastic considerations that have become known as the independence assumptions. In this context, certain statistical assumptions are made on the natures of the noise signal $\{v(i)\}$ and of the regression vectors $\{\mathbf{u}_i\}$, and conclusions are derived regarding the steady-state behavior of the adaptive filter.

The discussion in this chapter avoids statistical considerations and develops the analysis in a purely deterministic framework that is convenient when prior statistical information is unavailable or when the independence assumptions are unreasonable. The conclusions discussed herein highlight certain features of the adaptive algorithms that hold regardless of any statistical considerations in an adaptive filtering task.

Returning to the data model in Equation 20.1, we see that it assumes the existence of an unknown weight vector $\mathbf{w}$ that describes, along with the regression vectors $\{\mathbf{u}_i\}$, the uncorrupted data $\{y(i)\}$. This assumption may or may not hold.

For example, if the unknown plant in the system identification scenario of Figure 20.1 is itself an FIR system of length M, then there exists an unknown weight vector $\mathbf{w}$ that satisfies Equation 20.1. In this case, the successive estimates provided by the adaptive filter attempt to identify the unknown weight vector of the plant.

If, on the other hand, the unknown plant of Figure 20.1 is an autoregressive model of the simple form

$$\frac{1}{1 - cz^{-1}} = 1 + cz^{-1} + c^2z^{-2} + c^3z^{-3} + \cdots,$$

where $|c| < 1$, then an infinitely long tapped-delay line is necessary to justify a model of the form Equation 20.1. In this case, the first term in the linear regression model Equation 20.1 for a finite order M cannot describe the uncorrupted data $\{y(i)\}$ exactly, and thus modeling errors are inevitable. Such modeling errors can naturally be included in the noise term $v(i)$. Thus, we shall use the term $v(i)$ in Equation 20.1 to account not only for measurement noise but also for modeling errors, unmodeled dynamics, quantization effects, and other kind of disturbances within the system. In many cases, the performance of the adaptive filter depends on how these unknown disturbances affect the weight estimates.

A second source of error in the adaptive system is due to the initial guess $\mathbf{w}_{-1}$ for the weight vector. Due to the iterative nature of our chosen adaptive scheme, it is expected that this initial weight vector plays less of a role in the steady-state performance of the adaptive filter. However, for a finite number of iterations of the adaptive algorithm, both the noise term $v(i)$ and the initial weight error vector $(\mathbf{w} - \mathbf{w}_{-1})$ are disturbances that affect the performance of the adaptive scheme, particularly since the system designer often has little control over them.

The purpose of a robust adaptive filter design, then, is to develop a recursive estimator that minimizes in some well-defined sense the effect of any unknown disturbances on the performance of the filter. For this purpose, we first need to quantify or measure the effect of the disturbances. We address this concern in the following sections.

20.4 Error and Energy Measures

Assuming that the model Equation 20.1 is reasonable, two error quantities come to mind. The first one measures how far the weight estimate $\mathbf{w}_{i-1}$ provided by the adaptive filter is from the true weight vector $\mathbf{w}$ that we are trying to identify. We refer to this quantity as the weight error at time $(i-1)$, and we denote it by $\tilde{\mathbf{w}}_{i-1} = \mathbf{w} - \mathbf{w}_{i-1}$. The second type of error measures how far the estimate $\mathbf{u}_i^T\mathbf{w}_{i-1}$ is from the uncorrupted output term $\mathbf{u}_i^T\mathbf{w}$. We shall call this the *a priori* estimation error, and we denote it by $e_a(i) = \mathbf{u}_i^T\tilde{\mathbf{w}}_{i-1}$. Similarly, we define an *a posteriori* estimation error as $e_p(i) = \mathbf{u}_i^T\tilde{\mathbf{w}}_i$. Comparing with the definition of the *a priori* error, the *a posteriori* error employs the most recent weight error vector.

Ideally, one would like to make the estimation errors $\{\tilde{\mathbf{w}}_i, e_a(i)\}$ or $\{\tilde{\mathbf{w}}_i, e_p(i)\}$ as small as possible. This objective is hindered by the presence of the disturbances $\{\tilde{\mathbf{w}}_{-1}, v(i)\}$. For this reason, an adaptive filter is said to be robust if the effects of the disturbances $\{\tilde{\mathbf{w}}_{-1}, v(i)\}$ on the resulting estimation errors $\{\tilde{\mathbf{w}}_i, e_a(i)\}$ or $\{\tilde{\mathbf{w}}_i, e_p(i)\}$ is small in a well-defined sense. To this end, we can employ one of several measures to denote how "small" these effects are. For our discussion, a quantity known as the energy of a signal will be used to quantify these effects. The energy of a sequence $x(i)$ of length N is measured by $E_x = \sum_{i=0}^{N-1}|x(i)|^2$. A finite energy sequence is one for which $E_x < \infty$ as $N \to \infty$. Likewise, a finite power sequence is one for which

$$P_x = \lim_{N\to\infty}\left(\frac{1}{N}\sum_{i=0}^{N-1}|x(i)|^2\right) < \infty.$$

20.5 Robust Adaptive Filtering

We can now quantify what we mean by robustness in the adaptive filtering context. Let A denote any adaptive filter that operates causally on the input data $\{d(i), \mathbf{u}_i\}$. A causal adaptive scheme produces a weight vector estimate at time i that depends only on the data available up to and including time i. This adaptive scheme receives as input the data $\{d(i), \mathbf{u}_i\}$ and provides as output the weight vector estimates $\{\mathbf{w}_i\}$. Based on these estimates, we introduce one or more estimation error quantities such as the pair $\{\tilde{\mathbf{w}}_{i-1}, e_a(i)\}$ defined above. Even though these quantities are not explicitly available because $\mathbf{w}$ is unknown, they are of interest to us as their magnitudes determine how well or how poorly a candidate adaptive filtering scheme might perform.

Figure 20.2 indicates the relationship between $\{d(i), \mathbf{u}_i\}$ to $\{\tilde{\mathbf{w}}_{i-1}, e_a(i)\}$ in block diagram form. This schematic representation indicates that an adaptive filter A operates on $\{d(i), \mathbf{u}_i\}$ and that its performance relies on the sizes of the error quantities $\{\tilde{\mathbf{w}}_{i-1}, e_a(i)\}$, which could be replaced by the error quantities $\{\tilde{\mathbf{w}}_i, e_p(i)\}$ if desired. This representation explicitly denotes the quantities $\{\tilde{\mathbf{w}}_{-1}, v(i)\}$ as disturbances to the adaptive scheme.

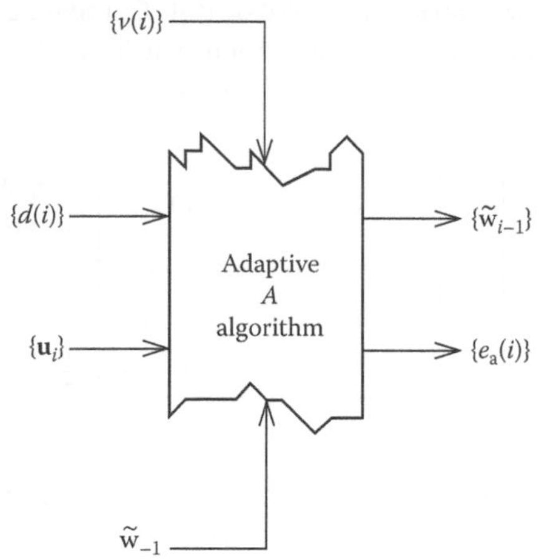

FIGURE 20.2 Input–output map of a generic adaptive scheme.

In order to measure the effect of the disturbances on the performance of an adaptive scheme, it will be helpful to determine the explicit relationship between the disturbances and the estimation errors that is provided by the adaptive filter. For example, we would like to know what effect the noise terms and the initial weight error guess $\{\tilde{\mathbf{w}}_{-1}, v(i)\}$ would have on the resulting *a priori* estimation errors and the final weight error, $\{e_a(i), \tilde{\mathbf{w}}_N\}$, for a given adaptive scheme. Knowing such a relationship, we can then quantify the robustness of the adaptive scheme by determining the degree to which disturbances affect the size of the estimation errors.

We now illustrate how this disturbances-to-estimation-errors relationship can be determined by considering the LMS algorithm in Equation 20.2. Since $d(i) - \mathbf{u}_i^T\mathbf{w}_{i-1} = e_a(i) + v(i)$, we can subtract $\mathbf{w}$ from both sides of Equation 20.2 to obtain the weight-error update equation

$$\tilde{\mathbf{w}}_i = \tilde{\mathbf{w}}_{i-1} - \mu \cdot \mathbf{u}_i \cdot [e_a(i) + v(i)]. \tag{20.3}$$

Assume that we run N steps of the LMS recursion starting with an initial guess $\tilde{\mathbf{w}}_{-1}$. This operation generates the weight error estimates $\{\tilde{\mathbf{w}}_0, \tilde{\mathbf{w}}_1, \ldots, \tilde{\mathbf{w}}_N\}$ and the *a priori* estimation errors $\{e_a(0), \ldots, e_a(N)\}$. Define the following two column vectors:

$$\underline{\text{dist}} = \text{col}\left\{\frac{1}{\sqrt{\mu}}\tilde{\mathbf{w}}_{-1}, v(0), v(1), \ldots, v(N)\right\}, \quad \underline{\text{error}} = \text{col}\left\{e_a(0), e_a(1), \ldots, e_a(N), \frac{1}{\sqrt{\mu}}\tilde{\mathbf{w}}_N\right\}.$$

The vector dist contains the disturbances that affect the performance of the adaptive filter. The initial weight error vector is scaled by $\mu^{-1/2}$ for convenience. Likewise, the vector error contains the *a priori* estimation errors and the final weight error vector which has also been scaled by $\mu^{-1/2}$. The weight error update relation in Equation 20.3 allows us to relate the entries of both vectors in a straightforward manner. For example,

$$e_a(0) = \mathbf{u}_0^T\tilde{\mathbf{w}}_{-1} = \left(\sqrt{\mu}\mathbf{u}_0^T\right)\left(\frac{1}{\sqrt{\mu}}\tilde{\mathbf{w}}_{-1}\right),$$

which shows how the first entry of error relates to the first entry of dist. Similarly, for $e_a(1) = \mathbf{u}_1^T\tilde{\mathbf{w}}_0$, we obtain

$$e_a(1) = \left(\sqrt{\mu}\mathbf{u}_1^T\left[I - \mu\mathbf{u}_0\mathbf{u}_0^T\right]\right)\frac{1}{\sqrt{\mu}}\tilde{\mathbf{w}}_{-1} - \left(\mu\mathbf{u}_1^T\mathbf{u}_0\right)v(0),$$

which relates $e_a(1)$ to the first two entries of the vector $\underline{\text{dist}}$. Continuing in this manner, we can relate $e_a(2)$ to the first three entries of $\underline{\text{dist}}$, $e_a(3)$ to the first four entries of $\underline{\text{dist}}$, and so on.

In general, we can compactly express this relationship as

$$\underbrace{\begin{bmatrix} e_a(0) \\ e_a(1) \\ \vdots \\ e_a(N) \\ \frac{1}{\sqrt{\mu}}\tilde{\mathbf{w}}_N \end{bmatrix}}_{\underline{\text{error}}} = \underbrace{\begin{bmatrix} \times & & & & & \\ \times & \times & & O & & \\ \vdots & & & \ddots & & \\ \times & \times & \times & \times & \times & \times \end{bmatrix}}_{\mathcal{T}} \underbrace{\begin{bmatrix} \frac{1}{\sqrt{\mu}}\tilde{\mathbf{w}}_{-1} \\ v(0) \\ v(1) \\ \vdots \\ v(N) \end{bmatrix}}_{\underline{\text{dist}}},$$

where the symbol $\times$ is used to denote the entries of the lower triangular mapping $\mathcal{T}$ relating $\underline{\text{dist}}$ to $\underline{\text{error}}$. The specific values of the entries of $\mathcal{T}$ are not of interest for now, although we have indicated how the expressions for these $\times$ terms can be found. However, the causal nature of the adaptive algorithm requires that $\mathcal{T}$ be of lower triangular form.

Given the above relationship, our objective is to quantify the effect of the disturbances on the estimation errors. Let E_d and E_e denote the energies of the vectors $\underline{\text{dist}}$ and $\underline{\text{error}}$, respectively, such that

$$E_e = \frac{1}{\mu}\|\tilde{\mathbf{w}}_N\|^2 + \sum_{i=0}^{N}|e_a(i)|^2 \quad \text{and} \quad E_d = \frac{1}{\mu}\|\tilde{\mathbf{w}}_{-1}\|^2 + \sum_{i=0}^{N}|v(i)|^2,$$

where $\|\cdot\|$ denotes the Euclidean norm of a vector. We shall say that the LMS adaptive algorithm is robust with level γ if a relation of the form

$$\frac{E_e}{E_d} \leq \gamma^2, \tag{20.4}$$

holds for some positive γ and for any nonzero, finite-energy disturbance vector $\underline{\text{dist}}$. In other words, no matter what the disturbances $\{\tilde{\mathbf{w}}_{-1}, v(i)\}$ are, the energy of the resulting estimation errors will never exceed γ^2 times the energy of the associated disturbances.

The form of the mapping $\mathcal{T}$ affects the value of γ in Equation 20.4 for any particular algorithm. To see this result, recall that for any finite-dimensional matrix A, its maximum singular value, denoted by $\bar{\sigma}(A)$, is defined by $\bar{\sigma}(A) = \max_{x\neq 0}\frac{\|Ax\|}{\|x\|}$. Hence, the square of the maximum singular value, $\bar{\sigma}^2(A)$, measures the maximum energy gain from the vector x to the resulting vector Ax. Therefore, if a relation of the form Equation 20.4 should hold for any nonzero disturbance vector $\underline{\text{dist}}$, then it means that

$$\max_{\underline{\text{dist}}\neq 0} \frac{\|\mathcal{T}\underline{\text{dist}}\|}{\|\underline{\text{dist}}\|} \leq \gamma.$$

Consequently, the maximum singular value of $\mathcal{T}$ must be bounded by γ. This imposes a condition on the allowable values for γ; its smallest value cannot be smaller than the maximum singular value of the resulting $\mathcal{T}$.

Ideally, we would like the value of γ in Equation 20.4 to be as small as possible. In particular, an algorithm for which the value of γ is 1 would guarantee that the estimation error energy will never exceed the disturbance energy, no matter what the natures of the disturbances are! Such an algorithm would possess a good degree of robustness since it would guarantee that the disturbance energy will never be unnecessarily magnified.

Before continuing our study, we ask and answer the obvious questions that arise at this point:

- What is the smallest possible value for γ for the LMS algorithm? It turns out for the LMS algorithm that, under certain conditions on the step-size parameter, the smallest possible value for γ is 1. Thus, $\varepsilon_e \leq \varepsilon_d$ for the LMS algorithm.
- Does there exist any other causal adaptive algorithm that would result in a value for γ in Equation 20.4 that is smaller than 1? It can be argued that no such algorithm exists for the model Equation 20.1 and criterion Equation 20.4.

In other words, the LMS algorithm is in fact the most robust adaptive algorithm in the sense defined by Equation 20.4. This result provides a rigorous basis for the excellent robustness properties that the LMS algorithm, and several of its variants, have shown in practical situations. The references at the end of the chapter provide an overview of the published works that have established these conclusions. Here, we only motivate them from first principles. In so doing, we shall also discuss other results (and tools) that can be used in order to impose certain robustness and convergence properties on other classes of adaptive schemes.

20.6 Energy Bounds and Passivity Relations

Consider the LMS recursion in Equation 20.2, with a time-varying step-size $\mu(i)$ for purposes of generality, as given by

$$\mathbf{w}_i = \mathbf{w}_{i-1} + \mu(i) \cdot \mathbf{u}_i \cdot [d(i) - \mathbf{u}_i^T \cdot \mathbf{w}_{i-1}] \,. \tag{20.5}$$

Subtracting the optimal coefficient vector $\mathbf{w}$ from both sides and squaring the resulting expressions, we obtain

$$||\tilde{\mathbf{w}}_i||^2 = ||\tilde{\mathbf{w}}_{i-1} - \mu(i) \cdot \mathbf{u}_i \cdot [e_a(i) + v(i)]||^2 \,.$$

Expanding the right-hand side of this relationship and rearranging terms leads to the equality

$$||\tilde{\mathbf{w}}_i||^2 - ||\tilde{\mathbf{w}}_{i-1}||^2 + \mu(i) \cdot |e_a(i)|^2 - \mu(i) \cdot |v(i)|^2 = \mu(i) \cdot |e_a(i) + v(i)|^2 \cdot [\mu(i) \cdot ||\mathbf{u}_i||^2 - 1].$$

The right-hand side in the above equality is the product of three terms. Two of these terms, $\mu(i)$ and $|e_a(i) + v(i)|^2$, are nonnegative, whereas the term $(\mu(i) \cdot ||\mathbf{u}_i||^2 - 1)$ can be positive, negative, or zero depending on the relative magnitudes of $\mu(i)$ and $||\mathbf{u}_i||^2$. If we define $\bar{\mu}(i)$ as (assuming nonzero regression vectors):

$$\bar{\mu}(i) = ||\mathbf{u}_i||^{-2} \,, \tag{20.6}$$

then the following relations hold:

$$\frac{||\tilde{\mathbf{w}}_i||^2 + \mu(i)|e_a(i)|^2}{||\tilde{\mathbf{w}}_{i-1}||^2 + \mu(i)|v(i)|^2} \begin{cases} \leq 1 & \text{for } 0 < \mu(i) < \bar{\mu}(i) \\ = 1 & \text{for } \quad \mu(i) = \bar{\mu}(i) \\ \geq 1 & \text{for } \quad \mu(i) > \bar{\mu}(i). \end{cases}$$

The result for $0 < \mu(i) \leq \bar{\mu}(i)$ has a nice interpretation. It states that, no matter what the value of $v(i)$ is and no matter how far $\mathbf{w}_{i-1}$ is from $\mathbf{w}$, the sum of the two energies $||\tilde{\mathbf{w}}_i||^2 + \mu(i) \cdot |e_a(i)|^2$ will always be smaller than or equal to the sum of the two disturbance energies $||\tilde{\mathbf{w}}_{i-1}||^2 + \mu(i) \cdot |v(i)|^2$. This relationship is a statement of the passivity of the algorithm locally in time, as it holds for every time instant. Similar relationships can be developed in terms of the *a posteriori* estimation error.

Since this relationship holds for each time instant i, it also holds over an interval of time such that

$$\frac{\|\tilde{\mathbf{w}}_N\|^2 + \sum_{i=0}^{N} |\bar{e}_a(i)|^2}{\|\tilde{\mathbf{w}}_{-1}\|^2 + \sum_{i=0}^{N} |\bar{v}(i)|^2} \leq 1, \tag{20.7}$$

where we have introduced the normalized *a priori* residuals and noise signals

$$\bar{e}_a(i) = \sqrt{\mu(i)}e_a(i) \quad \text{and} \quad \bar{v}(i) = \sqrt{\mu(i)}v(i),$$

respectively. Equation 20.7 states that the lower-triangular matrix that maps the normalized noise signals $\{\bar{v}(i)\}_{i=0}^{N}$ and the initial uncertainty $\tilde{\mathbf{w}}_{-1}$ to the normalized *a priori* residuals $\{\bar{e}_a(i)\}_{i=0}^{N}$ and the final weight error $\tilde{\mathbf{w}}_N$ has a maximum singular value that is less than one. Thus, it is a contraction mapping for $0 < \mu(i) \leq \bar{\mu}(i)$. For the special case of a constant step-size μ, this is the same mapping $\mathcal{T}$ that we introduced earlier Equation 20.4.

In the above derivation, we have assumed for simplicity of presentation that the denominators of all expressions are nonzero. We can avoid this restriction by working with differences rather than ratios. Let $\Delta_N(\mathbf{w}_{-1}, v(\cdot))$ denote the difference between the numerator and the denominator of Equation 20.7, such that

$$\Delta_N(\mathbf{w}_{-1}, v(\cdot)) = \left\{ \|\tilde{\mathbf{w}}_N\|^2 + \sum_{i=0}^{N} |\bar{e}_a(i)|^2 \right\} - \left\{ \|\tilde{\mathbf{w}}_{-1}\|^2 + \sum_{i=0}^{N} |\bar{v}(i)|^2 \right\}. \tag{20.8}$$

Then, a similar argument that produced Equation 20.7 can be used to show that for any $\{\mathbf{w}_{-1}, v(\cdot)\}$,

$$\Delta_N(\mathbf{w}_{-1}, v(\cdot)) \leq 0. \tag{20.9}$$

20.7 Min–Max Optimality of Adaptive Gradient Algorithms

The property in Equation 20.7 or 20.9 is valid for any initial guess $\mathbf{w}_{-1}$ and for any noise sequence $v(\cdot)$, so long as the $\mu(i)$ are properly bounded by $\bar{\mu}(i)$. One might then wonder whether the bound in Equation 20.7 is tight or not. In other words, are there choices $\{\mathbf{w}_{-1}, v(\cdot)\}$ for which the ratio in Equation 20.7 can be made arbitrarily close to one or Δ_N in Equation 20.9 arbitrarily close to zero? We now show that there are. We can rewrite the gradient recursion of Equation 20.5 in the equivalent form

$$\mathbf{w}_i = \mathbf{w}_{i-1} + \mu(i) \cdot \mathbf{u}_i \cdot [e_a(i) + v(i)]. \tag{20.10}$$

Envision a noise sequence $v(i)$ that satisfies $v(i) = -e_a(i)$ at each time instant i. Such a sequence may seem unrealistic but is entirely within the realm of our unrestricted model of the unknown disturbances. In this case, the above gradient recursion trivializes to $\mathbf{w}_i = \mathbf{w}_{i-1}$ for all i, thus leading to $\mathbf{w}_N = \mathbf{w}_{-1}$. Thus, Δ_N in Equation 20.8 will be zero for this particular experiment. Therefore,

$$\max_{\{\mathbf{w}_{-1}, v(\cdot)\}} \{\Delta_N(\mathbf{w}_{-1}, v(\cdot))\} = 0 .$$

We now consider the following question: How does the gradient recursion in Equation 20.5 compare with other possible causal recursive algorithms for the update of the weight estimate? Let A denote any given causal algorithm. Suppose that we initialize algorithm A with $\mathbf{w}_{-1} = \mathbf{w}$, and suppose the noise sequence is given by $v(i) = -e_a(i)$ for $0 \leq i \leq N$. Then, we have

$$\sum_{i=0}^{N} |\bar{v}(i)|^2 = \sum_{i=0}^{N} |\bar{e}_a(i)|^2 \leq \|\tilde{\mathbf{w}}_N\|^2 + \sum_{i=0}^{N} |\bar{e}_a(i)|^2,$$

no matter what the value of $\tilde{\mathbf{w}}_N$ is. This particular choice of initial guess ($\mathbf{w}_{-1} = \mathbf{w}$) and noise sequence $\{v(\cdot)\}$ will always result in a nonnegative value of Δ_N in Equation 20.8, implying for any causal algorithm A that

$$\max_{\{\mathbf{w}_{-1}, v()\}} \{\Delta_N(\mathbf{w}_{-1}, v(\cdot))\} \geq 0.$$

For the gradient recursion in Equation 20.5, the maximum has to be exactly zero because the global property Equation 20.9 provided us with an inequality in the other direction. Therefore, the algorithm in Equation 20.5 solves the following optimization problem:

$$\min_{\text{Algorithm}} \left\{ \max_{\{\mathbf{w}_{-1}, v()\}} \Delta_N(\mathbf{w}_{-1}, v(\cdot)) \right\},$$

and the optimal value is equal to zero. More details and justification can be found in the references at the end of this chapter, especially connections with so-called H_∞ estimation theory.

As explained before, Δ_N measures the difference between the output energy and the input energy of the algorithm mapping $\mathcal{T}$. The gradient algorithm in Equation 20.5 minimizes the maximum possible difference between these two energies over all disturbances with finite energy. In other words, it minimizes the effect that the worst-possible input disturbances can have on the resulting estimation-error energy.

20.8 Comparison of LMS and RLS Algorithms

To illustrate the ideas in our discussion, we compare the robustness performance of two classical algorithms: the LMS algorithm Equation 20.2 and the recursive least-squares (RLS) algorithm. More details on the example given below can be found in the reference section at the end of the chapter.

Consider the data model in Equation 20.1 where $\mathbf{u}_i$ is a scalar that randomly assumes the values $+1$ and -1 with equal probability. Let $\mathbf{w} = 0.25$, and let $v(i)$ be an uncorrelated Gaussian noise sequence with unit variance. We first employ the LMS recursion in Equation 20.2 and compute the initial 150 estimates $\mathbf{w}_i$, starting with $\mathbf{w}_{-1} = 0$ and using $\mu = 0.97$. Note that μ satisfies the requirement $\mu \leq 1/\|\mathbf{u}_i\|^2 = 1$ for all i. We then evaluate the entries of the resulting mapping $\mathcal{T}$, now denoted by $\mathcal{T}_{\text{lms}}$, that we defined in Equation 20.4. We then compute the corresponding $\mathcal{T}_{\text{rls}}$ for the RLS algorithm for these signals, which for this special data model can be expressed as

$$\mathbf{w}_{i+1} = \mathbf{w}_i + \frac{p_i \mathbf{u}_i}{1 + p_i}[d(i) - \mathbf{u}_i^{\text{T}} \mathbf{w}_{i-1}], \quad p_{i+1} = \frac{p_i}{1 + p_i}.$$

The initial condition chosen for p_i is $p_0 = \mu = 0.97$.

Figure 20.3 shows a plot of the 150 singular values of the resulting mappings $\mathcal{T}_{\text{lms}}$ and $\mathcal{T}_{\text{rls}}$. As predicted from our analysis, the singular values of $\mathcal{T}_{\text{lms}}$, indicated by an almost horizontal line at unity, are all bounded by one, whereas the maximum singular value of $\mathcal{T}_{\text{rls}}$ is approximately 1.65. This result indicates that the LMS algorithm is indeed more robust than the RLS algorithm, as is predicted by the earlier analysis.

Observe, however, that most of the singular values of $\mathcal{T}_{\text{rls}}$ are considerably smaller than one, whereas the singular values of $\mathcal{T}_{\text{lms}}$ are clustered around one. This has an interesting interpretation that we explain as follows. An $N \times N$-dimensional matrix A has N singular values $\{\sigma_i\}$ that are equal to the positive square-roots of the eigenvalues of AA^{T}. For each σ_i, there exists a unit-norm vector x_i such that the energy gain from x_i to Ax_i is equal to σ_i^2, i.e., $\sigma_i = \|Ax_i\|/\|x_i\|$. The vector x_i can be chosen as the ith right singular

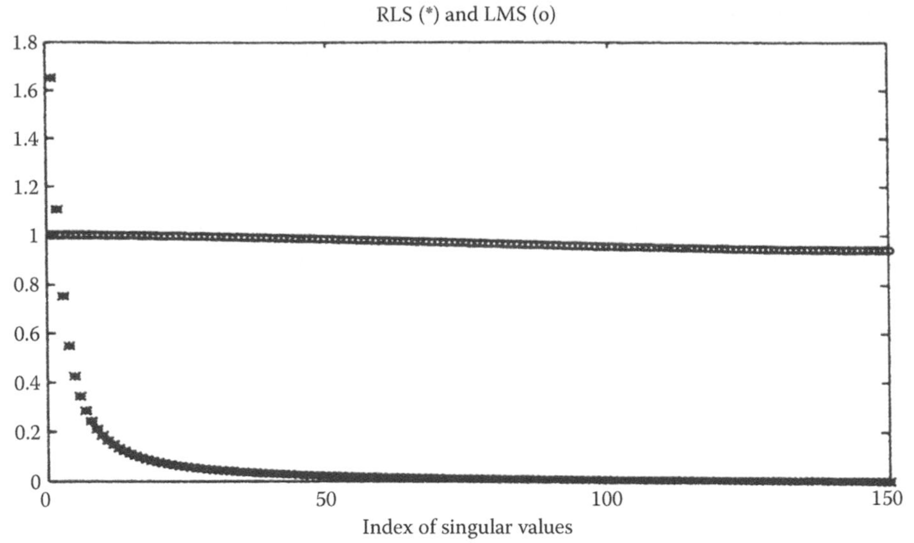

FIGURE 20.3 Singular value plot.

vector of A. Now, recall that $\mathcal{T}_{\text{lms}}$ and $\mathcal{T}_{\text{rls}}$ are finite-dimensional matrices that map a disturbance vector $\underline{\text{dist}}$ to the estimation-errors vector $\underline{\text{error}}$. Considering the plot of the singular values of $\mathcal{T}_{\text{rls}}$, we see that if the disturbance vector $\underline{\text{dist}}$ happens to lie in the range space of the right singular vectors associated with the smaller singular values in this plot, then its effect will be significantly attenuated. This fact indicates that while the performance of the LMS algorithm guards against worst-case disturbances, the RLS algorithm is likely to have a better performance than the LMS algorithm on average, as is well known.

20.9 Time-Domain Feedback Analysis

Robust adaptive filters are designed to induce contractive mappings between sequences of numbers. This fact also has important implications on the convergence performance of a robust adaptive scheme. In the remaining sections of this chapter, we discuss the combined issues of robustness and convergence from a deterministic standpoint. In particular, the following issues are discussed:

- We show that each step of the update equation of the gradient algorithm in Equation 20.5 can be described in terms of an elementary section that possesses a useful feedback structure.
- The feedback structure provides insights into the robust and convergence performance of the adaptive scheme. This is achieved by studying the energy flow through a cascade of elementary sections and by invoking a useful tool from system theory known as the small gain theorem.
- The feedback analysis extends to more general update relations. The example considered here is filtered-error LMS algorithm, although the methodology can be extended to other structures such as perceptrons. Details can be found in the references at the end of this chapter.

20.9.1 Time-Domain Analysis

From the update equation in Equation 20.5, $\tilde{\mathbf{w}}_i$ satisfies

$$\tilde{\mathbf{w}}_i = \tilde{\mathbf{w}}_{i-1} - \mu(i) \cdot \mathbf{u}_i \cdot [e_a(i) + v(i)]. \qquad (20.11)$$

If we multiply both sides of Equation 20.11 by $\mathbf{u}_i^T$ from the left, we obtain the following relation among $\{e_p(i), e_a(i), v(i)\}$:

$$e_p(i) = \left(1 - \frac{\mu(i)}{\bar{\mu}(i)}\right) e_a(i) - \frac{\mu(i)}{\bar{\mu}(i)} v(i), \tag{20.12}$$

where $\bar{\mu}(i)$ is given by Equation 20.6. Using Equations 20.12 and 20.5 can be rewritten in the equivalent form

$$\begin{aligned} \mathbf{w}_i &= \mathbf{w}_{i-1} + \bar{\mu}(i) \cdot \mathbf{u}_i \cdot [e_a(i) - e_p(i)], \\ &= \mathbf{w}_{i-1} + \bar{\mu}(i) \cdot \mathbf{u}_i \cdot [e_a(i) + r(i)], \end{aligned} \tag{20.13}$$

where we have defined the signal $r(i) = -e_p(i)$ for convenience. The expression Equation 20.13 shows that Equation 20.5 can be rewritten in terms of a new step-size $\bar{\mu}(i)$ and a modified "noise" term $r(i)$.

Therefore, if we follow arguments similar to those prior to Equation 20.6, we readily conclude that for algorithm Equation 20.5 the following equality holds for all $\{\mu(i), v(i)\}$:

$$\frac{\|\tilde{\mathbf{w}}_i\|^2 + \bar{\mu}(i)|e_a(i)|^2}{\|\tilde{\mathbf{w}}_{i-1}\|^2 + \bar{\mu}(i)|r(i)|^2} = 1. \tag{20.14}$$

This relation establishes a lossless map (denoted by $\overline{\mathcal{T}}_i$) from the signals $\{\tilde{\mathbf{w}}_{i-1}, \sqrt{\bar{\mu}(i)}r(i)\}$ to the signals $\{\tilde{\mathbf{w}}_i, \sqrt{\bar{\mu}(i)}e_a(i)\}$. Correspondingly, using Equation 20.12, the map from the original weighted disturbance $\sqrt{\bar{\mu}(i)}v(i)$ to the weighted estimation error signal $\sqrt{\bar{\mu}(i)}e_a(i)$ can be expressed in terms of the feedback structure shown in Figure 20.4.

The feedback description provides useful insights into the behavior of the adaptive scheme. Because the map $\overline{\mathcal{T}}_i$ in the feedforward path is lossless or energy-preserving, the design and analysis effort can be concentrated on the terms contained in the feedback path. This feedback block controls

- How much energy is fed back into the input of each section and whether energy magnification or demagnification may occur (i.e., stability)
- How sensitive the estimation error is to noise and disturbances (i.e., robustness)
- How fast the estimation error energy decays (i.e., convergence rate)

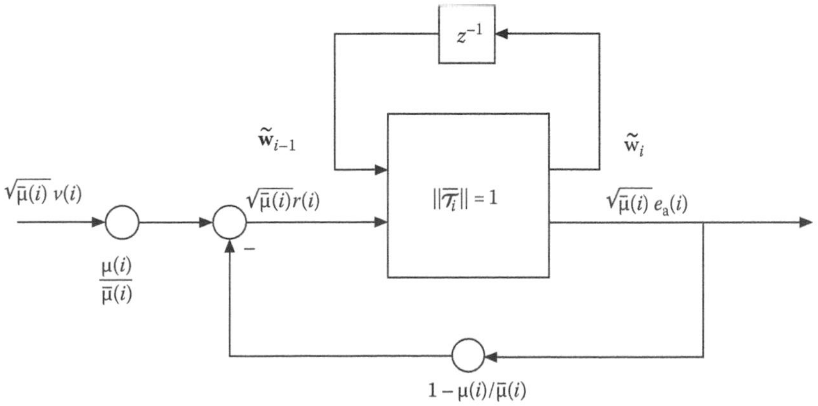

FIGURE 20.4 A time-variant lossless mapping with gain feedback for gradient algorithms. (From Rupp, M. and Sayed, A.H., *IEEE Trans. Signal Process.*, 44(6), 1428, June 1996.)

20.9.2 l_2–Stability and the Small Gain Condition

We start by reconsidering the robustness issue. Recall that if the step-sizes $\mu(i)$ are chosen such that $\mu(i) \leq \bar{\mu}(i)$, then robustness is guaranteed in that the ratio of the energies of the signals in Equation 20.7 will be bounded by one.

The condition on $\mu(i)$ can be relaxed at the expense of guaranteeing energy ratios that are bounded by some other positive number, say

$$\frac{\text{Weighted estimation error energy}}{\text{Weighted disturbance energy}} \leq \gamma^2, \tag{20.15}$$

for some constant γ to be determined. This is still a desirable property because it means that the disturbance energy will be, at most, scaled by a factor of γ. This fact can in turn lead to useful convergence conclusions, as argued later.

In order to guarantee robustness conditions according to Equation 20.15, for some γ, we rely on the observation that feedback configurations of the form shown in Figure 20.4 can be analyzed using a tool known in system theory as the small gain theorem. In loose terms, this theorem states that the stability of a feedback configuration such as that in Figure 20.4 is guaranteed if the product of the norms of the feedforward and the feedback mappings are strictly bounded by one. Since the feedforward mapping $\overline{\mathcal{T}}_i$ has a norm (or maximum singular value) of one, the norm of the feedback map needs to be strictly bounded by one for stability of this system.

To illustrate these concepts more fully, consider the feedback structure in Figure 20.5 that has a lossless mapping $\mathcal{T}$ in its feedforward path and an arbitrary mapping $\mathcal{F}$ in its feedback path. The input/output signals of interest are denoted by $\{x, y, r, v, e\}$. In this system, the signals x, v play the role of the disturbances.

The losslessness of the feedforward path implies conservation of energy such that

$$\|y\|^2 + \|e\|^2 = \|x\|^2 + \|r\|^2.$$

Consequently, $\|e\| \leq \|x\| + \|r\|$. On the other hand, the triangle inequality of norms implies that

$$\|r\| \leq \|v\| + \|\mathcal{F}\| \cdot \|e\|,$$

where the notation $\|\mathcal{F}\|$ denotes the maximum singular value of the mapping $\mathcal{F}$. Provided that the small gain condition $\|\mathcal{F}\| < 1$ is satisfied, we have

$$\|e\| \leq \frac{1}{1 - \|\mathcal{F}\|} \cdot [\|x\| + \|v\|]. \tag{20.16}$$

Thus, a contractive $\mathcal{F}$ guarantees a robust map from $\{x, v\}$ to $\{e\}$ with a robustness level that is determined by the factor $1/(1 - \|\mathcal{F}\|)$. In this case, we shall say that the map from $\{x, v\}$ to $\{e\}$ is l_2–stable.

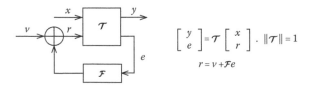

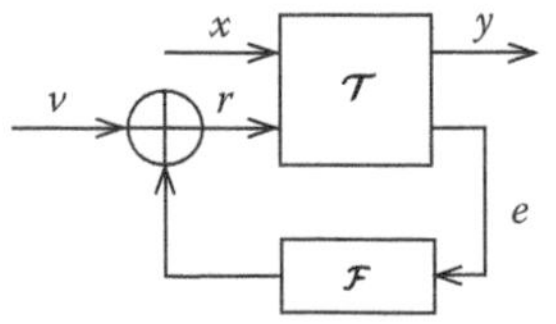

$$\begin{bmatrix} y \\ e \end{bmatrix} = \mathcal{T} \begin{bmatrix} x \\ r \end{bmatrix}, \quad \|\mathcal{T}\| = 1$$

$$r = v + \mathcal{F}e$$

FIGURE 20.5 A feedback structure.

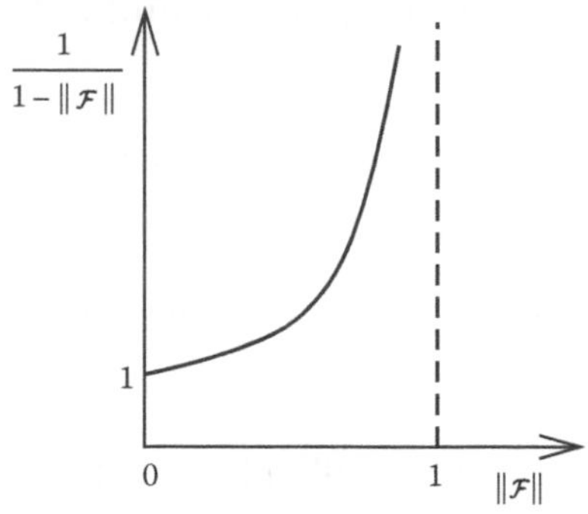

FIGURE 20.6 Plot of the l_2-gain.

A plot of the factor $1/(1 - \|\mathcal{F}\|)$, as a function of $\|\mathcal{F}\|$, is shown in Figure 20.6. It can be seen that the smaller the value of $\|\mathcal{F}\|$:

- The smaller the effect of $\{x, v\}$ on $\{e\}$
- The smaller the upper bound on $\|e\|$

Moreover, we shall argue that smaller values of $\|\mathcal{F}\|$ are associated with faster convergence. Therefore, controlling the norm of $\mathcal{F}$, is important for both the robustness and convergence performance of an adaptive algorithm. In most cases, the feedback filter $\mathcal{F}$ will depend on several quantities, such as the step-sizes $\{\mu(i)\}$ and the data vectors $\{\mathbf{u}_i\}$ (as in Figure 20.4). It may also depend on error filters and on regression filters that appear in more general adaptive schemes.

Referring to Figure 20.4, define

$$\eta(N) = \max_{0 \le i \le N} \left|1 - \frac{\mu(i)}{\bar{\mu}(i)}\right| \quad \text{and} \quad \xi(N) = \max_{0 \le i \le N} \frac{\mu(i)}{\bar{\mu}(i)}.$$

where

$\eta(N)$ is the maximum absolute value of the gain of the feedback loop over the interval of time $0 \le i \le N$

$\xi(N)$ is the maximum value of the scaling factor $\mu(i)/\bar{\mu}(i)$ at the input of the feedback interconnection

In this context, the small gain condition requires that $\eta(N) < 1$. This condition is equivalent to choosing the step-size parameter $\mu(i)$ such that $0 < \mu(i) < 2\bar{\mu}(i)$. Under this condition, the general relation Equation 20.16 can be used to deduce either of the following two relationships:

$$\sqrt{\sum_{i=0}^{N} \bar{\mu}(i)|e_a(i)|^2} \le \frac{1}{1 - \eta(N)} \left[\|\tilde{\mathbf{w}}_{-1}\| + \xi(N) \sqrt{\sum_{i=0}^{N} \bar{\mu}(i)|v(i)|^2} \right] \tag{20.17}$$

or

$$\sqrt{\sum_{i=0}^{N} \mu(i)|e_a(i)|^2} \le \frac{\xi^{1/2}(N)}{1 - \eta(N)} \left[\|\tilde{\mathbf{w}}_{-1}\| + \xi^{1/2}(N) \sqrt{\sum_{i=0}^{N} \mu(i)|v(i)|^2} \right]. \tag{20.18}$$

Note that in either case the upper bound on $\mu(i)$ is now $2\bar{\mu}(i)$ and the robustness level is essentially determined by

$$\frac{1}{1-\eta(N)} \quad \text{or} \quad \frac{\xi^{1/2}(N)}{1-\eta(N)},$$

depending on how the estimation errors $\{e_a(i)\}$ and the noise terms $\{v(i)\}$ are normalized (by $\mu(\cdot)$ or $\bar{\mu}(\cdot)$).

20.9.3 Energy Propagation in the Feedback Cascade

By studying the energy flow in the feedback interconnection of Figure 20.4, we can also obtain some physical insights into the convergence behavior of the gradient recursion Equation 20.5.

Assume that $\mu(i) = \bar{\mu}(i)$, such that the feedback loop of Figure 20.4 is disconnected. In this situation, there is no energy flowing back into the lower input of the lossless section from its lower output $e_a(\cdot)$. The losslessness of the feedforward path then implies that

$$E_w(i) = E_w(i-1) + E_v(i) - E_e(i),$$

where we have defined the energy terms

$$E_e(i) = \bar{\mu}(i)|e_a(i)|^2, \quad E_v(i) = \bar{\mu}(i)|v(i)|^2, \quad \text{and} \quad E_w(i) = \|\tilde{\mathbf{w}}_i\|^2 .$$

In the noiseless case where $v(i) = 0$, the above expression implies that the weight-error energy is a nonincreasing function of time, i.e., $E_w(i) \leq E_w(i-1)$.

However, what happens if $\mu(i) \neq \bar{\mu}(i)$? In this case, the feedback path is active and the convergence speed will be affected because the rate of decrease in the energy of the estimation error will change. Indeed, for $\mu(i) \neq \bar{\mu}(i)$ we obtain for $E_v(i) = 0$

$$E_w(i) = E_w(i-1) - \left(1 - \left|1 - \frac{\mu(i)}{\bar{\mu}(i)}\right|^2\right) E_e(i),$$

where, due to the small gain condition, the coefficient multiplying $E_e(i)$ can be seen to be smaller than 1.

Loosely speaking, this energy argument indicates for $v(i) = 0$ that the smaller the maximum singular value of feedback block F, for a generic feedback interconnection of the form shown in Figure 20.5, the faster the convergence of the algorithm will be, since less energy is fed back to the input of each section.

20.9.4 A Deterministic Convergence Analysis

The energy argument can be pursued in order to provide sufficient deterministic conditions for the convergence of the weight estimates $\mathbf{w}_i$ to the true weight vector $\mathbf{w}$. The argument follows as a consequence of the energy relations (or robustness bounds) Equations 20.17 and 20.18, which essentially establishes that the adaptive gradient algorithm Equation 20.5 maps a finite-energy sequence to another finite-energy sequence.

To clarify this point, we define the quantities

$$\eta = \sup_i \left|1 - \frac{\mu(i)}{\bar{\mu}(i)}\right| \quad \text{and} \quad \xi = \sup_i \left[\frac{\mu(i)}{\bar{\mu}(i)}\right],$$

and note that if the step-size parameter $\mu(i)$ is chosen such that $\mu(i)\|\mathbf{u}_i\|^2$ is uniformly bounded by 2, then we guarantee $\xi < 2$ and $\eta < 1$. We further note that it follows from the weight-error update relation Equation 20.11 that $\tilde{\mathbf{w}}_i$ satisfies

$$\tilde{\mathbf{w}}_i = \tilde{\mathbf{w}}_{i-1} - \bar{\mathbf{u}}_i^{\mathrm{T}}[\bar{e}_a(i) + \bar{v}(i)], \tag{20.19}$$

where we have defined $\bar{\mathbf{u}}_i = \sqrt{\mu(i)}\mathbf{u}_i$ (likewise for $\bar{e}_a(i), \bar{v}(i)$). The following conclusions can now be established under the stated conditions:

- Finite noise energy condition. We assume that the normalized sequence $\{\bar{v}(\cdot) = \sqrt{\mu(i)}v(i)\}$ has finite energy, i.e.,

$$\sum_{i=0}^{\infty} \mu(i)|v(i)|^2 < \infty. \tag{20.20}$$

  This in turn implies that $\bar{v}(i) \to 0$ as $i \to \infty$ (but not necessarily $v(i) \to 0$). If the initial weight-error vector is finite, $\|\tilde{\mathbf{w}}_{-1}\| < \infty$, then condition Equation 20.20 along with the energy bound Equation 20.18 (as $N \to \infty$) allows us to conclude that $\sum_{i=0}^{\infty} \mu(i)|e_a(i)|^2 < \infty$. Consequently, $\lim_{i\to\infty} \bar{e}_a(i) \to 0$ (but not necessarily $e_a(i) \to 0$).
- Persistent excitation condition. We also assume that the normalized vectors $\{\bar{\mathbf{u}}_i\}$ are persistently exciting. By this we mean that there exists a finite integer $L \geq M$ such that the smallest singular value of

$$\mathrm{col}\{\bar{\mathbf{u}}_i^{\mathrm{T}}, \bar{\mathbf{u}}_{i+1}^{\mathrm{T}}, \ldots, \bar{\mathbf{u}}_{i+L}^{\mathrm{T}}\}$$

  is uniformly bounded from below by a positive quantity for sufficiently large i. The persistence of excitation condition can be used to further conclude from $\bar{e}_a(i) \to 0$ that $\lim_{i\to\infty}\mathbf{w}_i = \mathbf{w}$.

The above statements can also be used to clarify the behavior of the adaptive algorithm Equation 20.5 in the presence of finite-power (rather than finite-energy) normalized noise sequences $\{\bar{v}(\cdot)\}$, i.e., for $v(\cdot)$ satisfying

$$\lim_{N\to\infty} \frac{1}{N} \sum_{i=0}^{N-1} \mu(i)|v(i)|^2 = P_v < \infty.$$

For this purpose, we divide both sides of Equation 20.18 by $\sqrt{N}$ and take the limit as $N \to \infty$ to conclude that

$$\lim_{N\to\infty} \frac{1}{N} \sum_{i=0}^{N-1} \mu(i)|e_a(i)|^2 \leq \frac{\xi^2 P_v}{(1-\eta)^2}.$$

In other words, a bounded noise power leads to a bounded estimation error power.

20.10 Filtered-Error Gradient Algorithms

The feedback analysis of the former sections can be extended to gradient algorithms that employ filtered versions of the error signal $d(i) - \mathbf{u}_i^{\mathrm{T}}\mathbf{w}_{i-1}$. Such algorithms are useful in applications such as active noise and vibration control and in adaptive IIR filters, where a filtered error signal is more easily observed or measured. Figure 20.7 depicts the context of this problem. The symbol F denotes the filter that operates

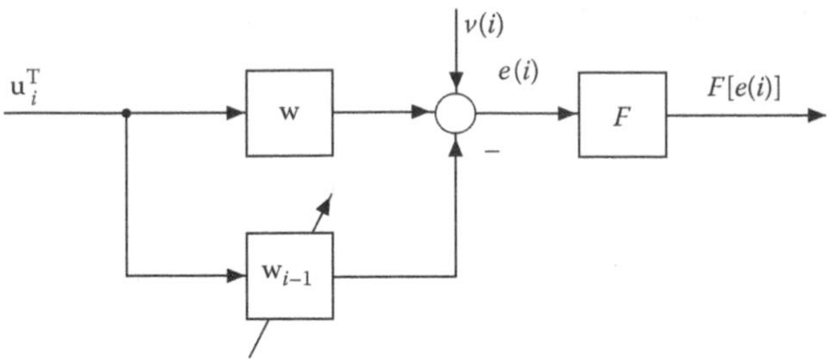

FIGURE 20.7 Structure of filtered-error gradient algorithms. (From Rupp, M. and Sayed, A.H., *IEEE Trans. Signal Process.*, 44(6), 1428, June 1996.)

on $e(i)$. For our discussion, we assume that F is a finite-impulse response filter of order M_F, such that the z-transform of its impulse response is $F(z) = \sum_{j=0}^{M_F-1} f_j z^{-j}$.

For purposes of discussion, we focus on one particular form of adaptive update known as the filtered-error LMS algorithm:

$$\mathbf{w}_i = \mathbf{w}_{i-1} + \mu(i) \cdot \mathbf{u}_i \cdot F[d(i) - \mathbf{u}_i^T \mathbf{w}_{i-1}]. \quad (20.21)$$

Comparing Equation 20.21 with Equation 20.5, the only difference between the two updates is the filter F that acts on the error $d(i) - \mathbf{u}_i^T \mathbf{w}_{i-1}$.

Following the discussion that led to Equation 20.13, it can be verified that Equation 20.21 is equivalent to the following update:

$$\mathbf{w}_i = \mathbf{w}_{i-1} + \bar{\mu}(i) \cdot \mathbf{u}_i \cdot [e_a(i) + r(i)], \quad (20.22)$$

where $\bar{\mu}(i) = 1/\|\mathbf{u}_i\|^2$, $e_a(i) = \mathbf{u}_i^T \tilde{\mathbf{w}}_{i-1}$, and $r(i)$ is defined as

$$\bar{\mu}(i) r(i) = \mu(i) F[v(i)] - \bar{\mu}(i) e_a(i) + \mu(i) F[e_a(i)]. \quad (20.23)$$

Expression 20.22 is of the same form as Equation 20.13, which implies that the following relation also holds:

$$\frac{\|\tilde{\mathbf{w}}_i\|^2 + \bar{\mu}(i)|e_a(i)|^2}{\|\tilde{\mathbf{w}}_{i-1}\|^2 + \bar{\mu}(i)|r(i)|^2} = 1. \quad (20.24)$$

This establishes that the map $\overline{\mathcal{T}}_i$ from the signals $\{\tilde{\mathbf{w}}_{i-1}, \sqrt{\bar{\mu}(i)} r(i)\}$ to the signals $\{\tilde{\mathbf{w}}_i, \sqrt{\bar{\mu}(i)} e_a(i)\}$ is lossless. Moreover, the map from the original disturbance $\sqrt{\bar{\mu}(\cdot)} v(\cdot)$ to the signal $\sqrt{\bar{\mu}(\cdot)} e_a(\cdot)$ can be expressed in terms of a feedback structure, as shown in Figure 20.8. We remark that the notation $1 - \frac{\mu(i)}{\sqrt{\bar{\mu}(i)}} F[\cdot] \frac{1}{\sqrt{\bar{\mu}(i)}}$ should be interpreted as follows. We first divide $\sqrt{\bar{\mu}(i)} e_a(i)$ by $\sqrt{\bar{\mu}(i)}$ before filtering it by the filter F and then scaling the result by $\mu(i)/\sqrt{\bar{\mu}(i)}$. Similarly, the term $\sqrt{\bar{\mu}(i)} v(i)$ is first divided by $\sqrt{\bar{\mu}(i)}$, then filtered by F, and is finally scaled by $\mu(i)/\sqrt{\bar{\mu}(i)}$.

The feedback path now contains a dynamic system. The small gain theorem dictates that this system will be robust if the feedback path is a contractive system. For the special case of the projection filtered-error LMS algorithm that employs the step-size $\mu(i) = \alpha \bar{\mu}(i), \alpha > 0$,

$$\mathbf{w}_i = \mathbf{w}_{i-1} + \alpha \frac{\mathbf{u}_i}{\|\mathbf{u}_i\|^2} F[d(i) - \mathbf{u}_i^T \mathbf{w}_{i-1}], \quad (20.25)$$

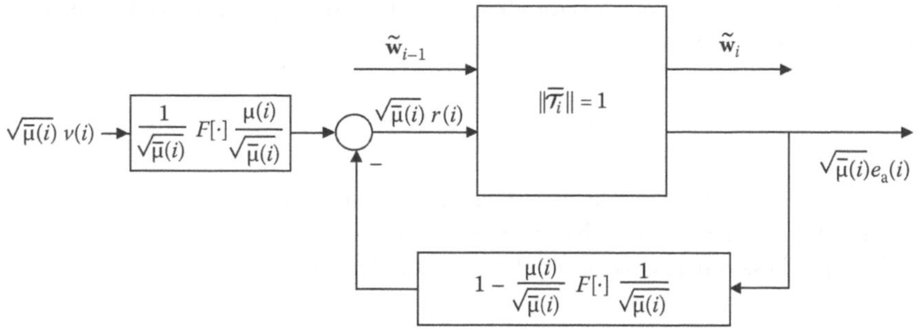

FIGURE 20.8 Filtered-error LMS algorithm as a time-variant lossless mapping with dynamic feedback. (From Rupp, M. and Sayed, A.H., *IEEE Trans. Signal Process.*, 44(6), 1428, June 1996.)

the small-gain condition implies that the following matrix should be strictly contractive:

$$\mathbf{P}_N = \begin{pmatrix} 1-\alpha f_0 & & & \mathbf{O} \\ -\alpha\frac{\sqrt{\bar{\mu}(1)}}{\sqrt{\bar{\mu}(0)}} f_1 & 1-\alpha f_0 & & \\ -\alpha\frac{\sqrt{\bar{\mu}(2)}}{\sqrt{\bar{\mu}(0)}} f_2 & -\alpha\frac{\sqrt{\bar{\mu}(2)}}{\sqrt{\bar{\mu}(1)}} f_1 & 1-\alpha f_0 & \\ \vdots & & & \ddots \end{pmatrix}.$$

Here, the $\{f_i\}$ are the coefficients of the FIR filter F. Since, in practice, the length M_F of this filter is usually much smaller than the length of the regression vector $\mathbf{u}_i$, the energy of the input sequence $\mathbf{u}_i$ does not change very rapidly over the filter length M_F, such that

$$\bar{\mu}(i) \approx \cdots \approx \bar{\mu}(i - M_F).$$

In this case, $\mathbf{P}_N$ becomes

$$\mathbf{P}_N \approx \mathbf{I} - \alpha\mathbf{F}_N, \tag{20.26}$$

where $\mathbf{F}_N$ is the lower triangular Toeplitz matrix that describes the convolution of the filter F on an input sequence. This is generally a banded matrix since $M_F \ll M$, as shown below for the special case of $M_F = 3$,

$$\mathbf{F}_N = \begin{bmatrix} f_0 & & & & \\ f_1 & f_0 & & & \\ f_2 & f_1 & f_0 & & \\ & f_2 & f_1 & f_0 & \\ & & \ddots & \ddots & \ddots \end{bmatrix}.$$

In this case, the strict contractivity of $(\mathbf{I} - \alpha\mathbf{F}_N)$ can be guaranteed by choosing the step-size parameter α such that

$$\max_{\Omega} \left|1 - \alpha F(e^{j\Omega})\right| < 1, \tag{20.27}$$

where $F(z)$ is the transfer function of the error filter. For better convergence performance, we may choose α by solving the min–max problem

$$\min_{\alpha} \max_{\Omega} \left|1 - \alpha F(e^{j\Omega})\right|. \tag{20.28}$$

If the resulting minimum is less than one, then the corresponding optimum value of α will result in faster convergence, and it will also guarantee the robustness of the scheme.

We now illustrate these concepts via a simulation example. The error-path filter for this example is

$$F(z) = 1 - 1.2z^{-1} + 0.72z^{-2}.$$

We use an FIR filter adapted by the algorithm in Equation 20.25, where the input signal to the adaptive filter consists of a single sinusoid of frequency $\Omega_0 = 1.2/\pi$. In this case, if we assume that the *a priori* error signal is dominated by the frequency component Ω_0, we can solve for the optimum α via the simpler expression (cf. Equation 20.28) $\min_\alpha \left|1 - \alpha F\left(e^{j\Omega_0}\right)\right|$. The resulting optimum value of α is

$$\alpha_{\text{opt}} = \text{Real}\left\{\frac{1}{F(e^{-j\Omega_0})}\right\}.$$

This step size provides the fastest convergence speed. In addition, the stability limits for α can be shown to be $0 < \alpha < 2\alpha_{\text{opt}}$ using a similar procedure.

Figure 20.9 shows three convergence curves of the average squared error $\text{Av}[|e(i)|^2] = \frac{1}{50}\sum_{j=1}^{50}\left|e_j(i)\right|^2$, as determined from 50 simulation runs of the projection filtered-error LMS algorithm for the choices $\alpha = 0.085, \alpha = 0.15$, and $\alpha = 0.18$, respectively. In this case, we have generated an input sequence of the form $u(i) = \sin(1.2i + \varphi)$ for each simulation run, where φ is uniformly chosen from the interval $[-\pi, \pi]$ to obtain smoother learning curves after averaging, and the $M = 10$ coefficients of the unknown system were all set to unity. Moreover, the additive noise $v(i)$ corrupting the signal $d(i)$ is uncorrelated Gaussian-distributed with a level that is -40 dB below that of the input signal power. The optimal step-size

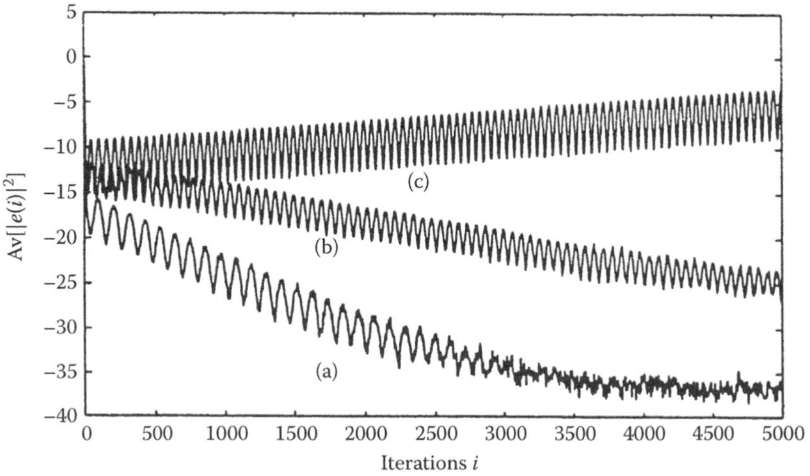

FIGURE 20.9 Convergence behavior for FELMS algorithm with sinusoidal input sequence and various step-sizes: (a) $\alpha = 0.085$, (b) $\alpha = 0.15$, and (c) $\alpha = 0.18$. (From Rupp, M. and Sayed, A.H., *IEEE Trans. Signal Process.*, 44(6), 1428, June 1996.)

α_{opt} in this case can be calculated to be $\alpha_{opt} = 0.085$ and the stability bounds for the system are $0 < \alpha < 0.17$. As expected, choosing $\alpha = 0.085$ provides the fastest convergence speed for this situation. We also see that for values of α greater than $2\alpha_{opt}$, the error of the system diverges.

20.11 Concluding Remarks

The intent of this chapter was to highlight certain robustness and convergence issues that arise in the study of adaptive algorithms in the presence of uncertain data. More details, extensions, and related discussions can be found in several of the references indicated in the reference section. The references are not intended to be complete but rather indicative of the work in the different areas. More complete lists can be found in several of the textbooks mentioned herein.

Detailed discussions on the different forms of adaptive algorithms and their potential applications can be found in [1–4]. The fundamentals of robust or H^∞ design, both in filtering and control applications, can be found in [5–10]. The small gain analysis is a standard tool in linear and nonlinear system theory. More advanced and detailed treatments can be found in [11,12].

The LMS algorithm usually has been presented in the literature as an instantaneous-gradient-based approximation for the steepest descent algorithm. Its robustness properties, and the interesting observation that it is in fact the exact solution of a min-max (or H^∞) optimization problem, have been first noted in [13].

Also, more details on the example comparing the performance of LMS and RLS can be found in the above reference. Extensions of the discussion to the backpropagation algorithm for neural network training, and other related results in adaptive filtering and H_∞ estimation and control can be found in [14–16].

Extensions of the feedback analysis to perceptron training in neural network can be found in [17]. A Cauchy-Schwarz argument that further highlights the robustness property of adaptive gradient algorithms, along with other local energy bounds, are given in [4,18]. The time-domain feedback and small gain analyses of adaptive filters, along with extensions to nonlinear settings and connections with Gauss-Newton updates and H^∞ filters, are discussed in [19–21]. Discussions of the singular value decomposition and its properties can be found in [22].

References

1. Haykin, S., *Adaptive Filter Theory*, 3rd ed., Prentice-Hall, Englewood Cliffs, NJ, 1996.
2. Proakis, J.G., Rader, C.M., Ling, F., and Nikias, C.L., *Advanced Digital Signal Processing*, Macmillan Publishing, New York, 1992.
3. Widrow, B. and Stearns, S.D., *Adaptive Signal Processing*, Prentice-Hall, Englewood Cliffs, NJ, 1985.
4. Sayed, A.H. and Kailath, T., A state-space approach to adaptive RLS filtering, *IEEE Signal Processing Magazine*, 11(3): 18–60, July 1994.
5. Green, M. and Limebeer, D.J.N., *Linear Robust Control*, Prentice-Hall, Englewood Cliffs, NJ, 1995.
6. Zhou, K., Doyle, J.C., and Glover, K., *Robust and Optimal Control*, Prentice Hall, Englewood Cliffs, NJ, 1996.
7. Khargonekar, P.P. and Nagpal, K.M., Filtering and smoothing in an H^∞ -setting, *IEEE Transactions on Automatic Control*, 36: 151–166, 1991.
8. Shaked, U. and Theodor, Y., H^∞-optimal estimation: A tutorial, *Proceedings of the IEEE Conference on Decision and Control*, Tucson, AZ, Dec. 1992, pp. 2278–2286.
9. Hassibi, B., Sayed, A.H., and Kailath, T., Linear estimation in Krein spaces—Part I: Theory, *IEEE Transactions on Automatic Control*, 41(1): 18–33, Jan. 1996.
10. Hassibi, B., Sayed, A.H., and Kailath, T., Linear estimation in Krein spaces—Part II: Applications, *IEEE Transactions on Automatic Control*, 41(1): 34–49, Jan. 1996.
11. Khalil, H.K., *Nonlinear Systems*, 2nd ed., Macmillan, New York, 1996.

12. Vidyasagar, M., *Nonlinear Systems Analysis*, 2nd ed., Prentice Hall, Englewood Cliffs, NJ, 1993.
13. Hassibi, B., Sayed, A.H., and Kailath, T., H^{∞} optimality of the LMS algorithm, *IEEE Transactions on Signal Processing*, 44(2): 267–280, Feb. 1996. See also *Proceedings of the Conference on Decision and Control*, San Antonio, TX, Dec. 1993, Vol. 1, pp. 74–79.
14. Hassibi, B., Sayed, A.H., and Kailath, T., LMS and backpropagation are minimax filters, in *Neural Computation and Learning*, Roychowdhurys, V., Siu, K.Y., and Orlitsky, A. (Eds.), Kluwer Academic Publishers, Norwell, MA, 1994, pp. 425–447.
15. Hassibi, B., Indefinite metric spaces in estimation, control, and adaptive filtering, Ph.D. dissertation, Stanford University, Stanford, CA, Aug. 1996.
16. Hassibi, B., Sayed, A.H., and Kailath, T., *Indefinite Quadratic Estimation and Control: A Unified Approach to H_2 and H_{∞} Theories*, Studies in Applied Mathematics Series, SIAM, Philadelphia, PA, 1997.
17. Rupp, M. and Sayed, A.H., Supervised learning of perceptron and output feedback dynamic networks: A feedback analysis via the small gain theorem, *IEEE Transactions on Neural Networks*, 8(3): 612–622, May 1997.
18. Sayed, A.H. and Rupp, M., Error energy bounds for adaptive gradient algorithms, *IEEE Transactions on Signal Processing*, 44(8): 1982–1989, Aug. 1996.
19. Rupp, M. and Sayed, A.H., A time-domain feedback analysis of filtered-error adaptive gradient algorithms, *IEEE Transactions on Signal Processing*, 44(6): 1428–1439, June 1996.
20. Rupp, M. and Sayed, A.H., Robustness of Gauss-Newton recursive methods: A deterministic feedback analysis, *Signal Processing*, 50(3): 165–188, June 1996.
21. Sayed, A.H. and Rupp, M., An l_2−stable feedback structure for nonlinear adaptive filtering and identification, *Automatica*, 33(1): 13–30, 1997.
22. Golub, G.H. and Van Loan, C.F., *Matrix Computations*, 3rd ed., The Johns Hopkins University Press, Baltimore, MD, 1996.

21

Recursive Least-Squares Adaptive Filters

Ali H. Sayed
University of California at Los Angeles

Thomas Kailath
Stanford University

The central problem in estimation is to recover, to good accuracy, a set of unobservable parameters from corrupted data. Several optimization criteria have been used for estimation purposes over the years, but the most important, at least in the sense of having had the most applications, are criteria that are based on quadratic cost functions. The most striking among these is the linear least-squares criterion, which was perhaps first developed by Gauss (ca. 1795) in his work on celestial mechanics. Since then, it has enjoyed widespread popularity in many diverse areas as a result of its attractive computational and statistical properties. Among these attractive properties, the most notable are the facts that least-squares solutions:

- Can be explicitly evaluated in closed forms
- Can be recursively updated as more input data are made available
- Are maximum likelihood (ML) estimators in the presence of Gaussian measurement noise

TABLE 21.1 Most Common RLS Adaptive Schemes

Adaptive Algorithm	Order Recursive	Fixed Order	Cost per Iteration
RLS		x	$O(M^2)$
QR and Inverse QR		x	$O(M^2)$
FTF and FAEST		x	$O(M)$
LSL	x		$O(M)$
QRD-LSL	x		$O(M)$

The aim of this chapter is to provide an overview of adaptive filtering algorithms that result when the least-squares criterion is adopted. Over the last several years, a wide variety of algorithms in this class has been derived. They all basically fall into the following main groups (or variations thereof): recursive least-squares (RLS) algorithms and the corresponding fast versions (known as FTF [fast transversal filter] and FAEST [fast a posteriori error sequential technique]), QR and inverse QR algorithms, least-squares lattice (LSL), and QR decomposition-based least-squares lattice (QRD-LSL) algorithms.

Table 21.1 lists these different variants and classifies them into order-recursive and fixed-order algorithms. The acronyms and terminology are not important at this stage and will be explained as the discussion proceeds. Also, the notation $O(M)$ is used to indicate that each iteration of an algorithm requires of the order of M floating point operations (additions and multiplications). In this sense, some algorithms are fast (requiring only $O(M)$), while others are slow (requiring $O(M^2)$). The value of M is the filter order that will be introduced in due time.

It is practically impossible to list here all the relevant references and all the major contributors to the rich field of adaptive RLS filtering. The reader is referred to some of the textbooks listed at the end of this chapter for more comprehensive treatments and bibliographies.

Here we wish to stress that, apart from introducing the reader to the fundamentals of RLS filtering, one of our goals in this exposition is to present the different versions of the RLS algorithm in computationally convenient so-called array forms. In these forms, an algorithm is described as a sequence of elementary operations on arrays of numbers. Usually, a prearray of numbers has to be triangularized by a rotation, or a sequence of elementary rotations, in order to yield a postarray of numbers. The quantities needed to form the next prearray can then be read off from the entries of the postarray, and the procedure can be repeated. The explicit forms of the rotation matrices are not needed in most cases.

Such array descriptions are more truly algorithms in the sense that they operate on sets of numbers and provide other sets of numbers, with no explicit equations involved. The rotations themselves can be implemented in a variety of well-known ways: as a sequence of elementary circular or hyperbolic rotations, in square-root- and/or division-free forms, as Householder transformations, etc. These may differ in computational complexity, numerical behavior, and ease of hardware (VLSI) implementation. But, if preferred, explicit expressions for the rotation matrices can also be written down, thus leading to explicit sets of equations in contrast to the array forms.

For this reason, and although the different RLS algorithms that we consider here have already been derived in many different ways in earlier places in the literature, the derivation and presentation in this chapter are intended to provide an alternative unifying exposition that we hope will help a reader get a deeper appreciation of this class of adaptive algorithms.

Notation

We use small boldface letters to denote column vectors (e.g., $\mathbf{w}$) and capital boldface letters to denote matrices (e.g., $\mathbf{A}$). The symbol $\mathbf{I}_n$ denotes the identity matrix of size $n \times n$, while 0 denotes a zero column. The superscript T denotes transposition. This chapter deals with real-valued data. The case of complex-valued data is essentially identical and is treated in many of the references at the end of this chapter.

Square-Root Factors

A symmetric positive-definite matrix $\mathbf{A}$ is one that satisfies $\mathbf{A} = \mathbf{A}^{\mathrm{T}}$ and $\mathbf{x}^{\mathrm{T}}\mathbf{A}\mathbf{x} > 0$ for all nonzero column vectors $\mathbf{x}$. Any such matrix admits a factorization (also known as eigen-decomposition) of the form $\mathbf{A} = \mathbf{U}\Sigma\mathbf{U}^{\mathrm{T}}$, where $\mathbf{U}$ is an orthogonal matrix, namely a square matrix that satisfies $\mathbf{U}\mathbf{U}^{\mathrm{T}} = \mathbf{U}^{\mathrm{T}}\mathbf{U} = \mathbf{I}$, and Σ is a diagonal matrix with real positive entries. In particular, note that $\mathbf{A}\mathbf{U} = \mathbf{U}\Sigma$, which shows that the columns of $\mathbf{U}$ are the right eigenvectors of $\mathbf{A}$ and the entries of Σ are the corresponding eigenvalues.

Note also that we can write $\mathbf{A} = \mathbf{U}\Sigma^{1/2}(\Sigma^{1/2})^{\mathrm{T}}\mathbf{U}^{\mathrm{T}}$, where $\Sigma^{1/2}$ is a diagonal matrix whose entries are (positive) square-roots of the diagonal entries of Σ. Since $\Sigma^{1/2}$ is diagonal, $(\Sigma^{1/2})^{\mathrm{T}} = \Sigma^{1/2}$. If we introduce the matrix notation $\mathbf{A}^{1/2} = \mathbf{U}\Sigma^{1/2}$, then we can alternatively write $\mathbf{A} = (\mathbf{A}^{1/2})(\mathbf{A}^{1/2})^{\mathrm{T}}$. This can be regarded as a square-root factorization of the positive-definite matrix $\mathbf{A}$. Here, the notation $\mathbf{A}^{1/2}$ is used to denote one such square-root factor, namely the one constructed from the eigen-decomposition of $\mathbf{A}$.

Note, however, that square-root factors are not unique. For example, we may multiply the diagonal entries of $\Sigma^{1/2}$ by ± 1s and obtain a new square-root factor for Σ and, consequently, a new square-root factor for $\mathbf{A}$.

Also, given any square-root factor $\mathbf{A}^{1/2}$, and any orthogonal matrix Θ (satisfying $\Theta\Theta^{\mathrm{T}} = \mathbf{I}$) we can define a new square-root factor for $\mathbf{A}$ as $\mathbf{A}^{1/2}\Theta$ since

$$(\mathbf{A}^{1/2}\Theta)(\mathbf{A}^{1/2}\Theta)^{\mathrm{T}} = \mathbf{A}^{1/2}(\Theta\Theta^{\mathrm{T}})(\mathbf{A}^{1/2})^{\mathrm{T}} = \mathbf{A}.$$

Hence, square factors are highly nonunique. We shall employ the notation $\mathbf{A}^{1/2}$ to denote any such square-root factor. They can be made unique, for example, by insisting that the factors be symmetric or that they be triangular (with positive diagonal elements). In most applications, the triangular form is preferred. For convenience, we also write

$$(\mathbf{A}^{1/2})^{\mathrm{T}} = \mathbf{A}^{\mathrm{T}/2}, \ (\mathbf{A}^{1/2})^{-1} = \mathbf{A}^{-1/2}, \quad \text{and} \quad (\mathbf{A}^{-1/2})^{\mathrm{T}} = \mathbf{A}^{-\mathrm{T}/2}.$$

Thus, note the expressions $\mathbf{A} = \mathbf{A}^{1/2}\mathbf{A}^{\mathrm{T}/2}$ and $\mathbf{A}^{-1} = \mathbf{A}^{-\mathrm{T}/2}\mathbf{A}^{-1/2}$.

21.1 Array Algorithms

The array form is so important that it will be worthwhile to explain its generic form here.

An array algorithm is described via rotation operations on a prearray of numbers, chosen to obtain a certain zero pattern in a postarray. Schematically, we write

$$\begin{bmatrix} x & x & x & x \\ x & x & x & x \\ x & x & x & x \\ x & x & x & x \end{bmatrix} \Theta = \begin{bmatrix} x & 0 & 0 & 0 \\ x & x & 0 & 0 \\ x & x & x & 0 \\ x & x & x & x \end{bmatrix},$$

where Θ is any rotation matrix that triangularizes the prearray. In general, Θ is required to be a $\mathbf{J}$-orthogonal matrix in the sense that it should satisfy the normalization $\Theta\mathbf{J}\Theta^{\mathrm{T}} = \mathbf{J}$, where $\mathbf{J}$ is a given signature matrix with ± 1s on the diagonal and zeros elsewhere. The orthogonal case corresponds to $\mathbf{J} = \mathbf{I}$ since then $\Theta\Theta^{\mathrm{T}} = \mathbf{I}$.

A rotation Θ that transforms a prearray to triangular form can be achieved in a variety of ways: by using a sequence of elementary Givens and hyperbolic rotations, Householder transformations, or square-root-free versions of such rotations. Here we only explain the elementary forms. The other choices are discussed in some of the references at the end of this chapter.

21.1.1 Elementary Circular Rotations

An elementary 2×2 orthogonal rotation Θ (also known as Givens or circular rotation) takes a row vector $[a \quad b]$ and rotates it to lie along the basis vector $[1 \quad 0]$. More precisely, it performs the transformation

$$[a \quad b]\Theta = \left[\pm\sqrt{|a|^2 + |b|^2} \quad 0\right]. \tag{21.1}$$

The quantity $\pm\sqrt{|a|^2 + |b|^2}$ that appears on the right-hand side is consistent with the fact that the prearray, $[a \quad b]$, and the postarray, $\left[\pm\sqrt{|a|^2 + |b|^2} \quad 0\right]$, must have equal Euclidean norms (since an orthogonal transformation preserves the Euclidean norm of a vector).

An expression for Θ is given by

$$\Theta = \frac{1}{\sqrt{1+\rho^2}}\begin{bmatrix} 1 & -\rho \\ \rho & 1 \end{bmatrix}, \quad \text{where } \rho = \frac{b}{a}, \ a \neq 0. \tag{21.2}$$

In the trivial case $a = 0$ we simply choose Θ as the permutation matrix,

$$\Theta = \begin{bmatrix} 0 & 1 \\ 1 & 0 \end{bmatrix}.$$

The orthogonal rotation (Equation 21.2) can also be expressed in the alternative form:

$$\Theta = \begin{bmatrix} c & -s \\ s & c \end{bmatrix},$$

where the so-called cosine and sine parameters, c and s, respectively, are defined by

$$c = \frac{1}{\sqrt{1+\rho^2}} \quad \text{and} \quad s = \frac{\rho}{\sqrt{1+\rho^2}}.$$

The name circular rotation for Θ is justified by its effect on a vector; it rotates the vector along the circle of equation $x^2 + y^2 = |a|^2 + |b|^2$, by an angle θ that is determined by the inverse of the above cosine and/or sine parameters, $\theta = \tan^{-1}\rho$, in order to align it with the basis vector $[1 \quad 0]$. The trivial case $a = 0$ corresponds to a 90° rotation in an appropriate clockwise (if $b \geq 0$) or counterclockwise (if $b < 0$) direction.

21.1.2 Elementary Hyperbolic Rotations

An elementary 2×2 hyperbolic rotation Θ takes a row vector $[a \quad b]$ and rotates it to lie either along the basis vector $[1 \quad 0]$ (if $|a| > |b|$) or along the basis vector $[0 \quad 1]$ (if $|a| < |b|$). More precisely, it performs either of the transformations:

$$[a \quad b]\Theta = \left[\pm\sqrt{|a|^2 - |b|^2} \quad 0\right] \quad \text{if } |a| > |b|, \tag{21.3}$$

$$[a \quad b]\Theta = \left[0 \quad \pm\sqrt{|b|^2 - |a|^2}\right] \quad \text{if } |a| < |b|. \tag{21.4}$$

The quantity $\sqrt{\pm(|a|^2 - |b|^2)}$ that appears on the right-hand side of the above expressions is consistent with the fact that the prearray, $[a \quad b]$, and the postarrays must have equal hyperbolic "norms." By the hyperbolic "norm" of a row vector $\mathbf{x}^T$ we mean the indefinite quantity $\mathbf{x}^T\mathbf{J}\mathbf{x}$, which can be positive or negative. Here,

$$\mathbf{J} = \begin{bmatrix} 1 & 0 \\ 0 & -1 \end{bmatrix} = (1 \oplus -1) .$$

An expression for a hyperbolic rotation Θ that achieves Equation 21.3 or 21.4 is given by

$$\Theta = \frac{1}{\sqrt{1-\rho^2}} \begin{bmatrix} 1 & -\rho \\ -\rho & 1 \end{bmatrix}, \tag{21.5}$$

where

$$\rho = \begin{cases} \dfrac{b}{a} & \text{when } a \neq 0 \text{ and } |a| > |b| \\[2ex] \dfrac{a}{b} & \text{when } b \neq 0 \text{ and } |b| > |a| . \end{cases}$$

The hyperbolic rotation (Equation 21.5) can also be expressed in the alternative form:

$$\Theta = \begin{bmatrix} ch & -sh \\ -sh & ch \end{bmatrix},$$

where the so-called hyperbolic cosine and sine parameters, *ch* and *sh*, respectively, are defined by

$$ch = \frac{1}{\sqrt{1-\rho^2}} \quad \text{and} \quad sh = \frac{\rho}{\sqrt{1-\rho^2}} .$$

The name hyperbolic rotation for Θ is again justified by its effect on a vector; it rotates the original vector along the hyperbola of equation $x^2 - y^2 = |a|^2 - |b|^2$, by an angle θ determined by the inverse of the above hyperbolic cosine and/or sine parameters, $\theta = \tanh^{-1}[\rho]$, in order to align it with the appropriate basis vector. Note also that the special case $|a| = |b|$ corresponds to a row vector $[a \quad b]$ with zero hyperbolic norm since $|a|^2 - |b|^2 = 0$. It is then easy to see that there does not exist a hyperbolic rotation that will rotate the vector to lie along the direction of one basis vector or the other.

21.1.3 Square-Root-Free and Householder Transformations

We remark that the above expressions for the circular and hyperbolic rotations involve square-root operations. In many situations, it may be desirable to avoid the computation of square-roots because it is usually expensive. For this and other reasons, square-root- and division-free versions of the above elementary rotations have been developed and constitute an attractive alternative.

Therefore one could use orthogonal or J-orthogonal Householder reflections (for given $\mathbf{J}$) to simultaneously annihilate several entries in a row, for example, to transform $[x \quad x \quad x \quad x]$ directly to the form $[x' \quad 0 \quad 0 \quad 0]$. Combinations of rotations and reflections can also be used.

We omit the details here but the idea is clear. There are many different ways in which a prearray of numbers can be rotated into a postarray of numbers.

21.1.4 A Numerical Example

Assume we are given a 2×3 prearray **A**,

$$\mathbf{A} = \begin{bmatrix} 0.875 & 0.15 & 1.0 \\ 0.675 & 0.35 & 0.5 \end{bmatrix}, \tag{21.6}$$

and wish to triangularize it via a sequence of elementary circular rotations, i.e., reduce **A** to the form

$$\mathbf{A}\Theta = \begin{bmatrix} x & 0 & 0 \\ x & x & 0 \end{bmatrix}. \tag{21.7}$$

This can be obtained, among several different possibilities, as follows. We start by annihilating the (1, 3) entry of the prearray in Equation 21.6 by pivoting with its (1, 1) entry. According to Equation 21.2, the orthogonal transformation Θ_1 that achieves this result is given by

$$\Theta_1 = \frac{1}{\sqrt{1+\rho_1^2}} \begin{bmatrix} 1 & -\rho_1 \\ \rho_1 & 1 \end{bmatrix} = \begin{bmatrix} 0.6585 & -0.7526(9) \\ 0.7526 & 0.6585 \end{bmatrix}, \quad \text{where } \rho_1 = \frac{1}{0.875}.$$

Applying Θ_1 to the prearray in Equation 21.6 leads to (recall that we are only operating on the first and third columns, leaving the second column unchanged)

$$\begin{bmatrix} 0.875 & 0.15 & 1 \\ 0.675 & 0.35 & 0.5 \end{bmatrix} \begin{bmatrix} 0.6585 & 0 & -0.7526 \\ 0 & 1 & 0 \\ 0.7526 & 0 & 0.6585 \end{bmatrix} = \begin{bmatrix} 1.3288 & 0.1500 & 0.0000 \\ 0.8208 & 0.3500 & -0.1788 \end{bmatrix}. \tag{21.8}$$

We now annihilate the (1, 2) entry of the resulting matrix in the above equation by pivoting with its (1, 1) entry. This requires that we choose

$$\Theta_2 = \frac{1}{\sqrt{1+\rho_2^2}} \begin{bmatrix} 1 & -\rho_2 \\ \rho_2 & 1 \end{bmatrix} = \begin{bmatrix} 0.9937 & -0.1122(12) \\ 0.1122 & 0.9937 \end{bmatrix}, \quad \text{where } \rho_2 = \frac{0.1500}{1.3288}. \tag{21.9}$$

Applying Θ_2 to the matrix on the right-hand side of Equation 21.8 leads to (now we leave the third column unchanged)

$$\begin{bmatrix} 1.3288 & 0.1500 & 0.0000 \\ 0.8208 & 0.3500 & 0.1788 \end{bmatrix} \begin{bmatrix} 0.9937 & -0.1122 & 0 \\ 0.1122 & 0.9937 & 0 \\ 0 & 0 & 1 \end{bmatrix} = \begin{bmatrix} 1.3373 & 0.0000 & 0.0000 \\ 0.8549 & 0.2557 & 0.1788 \end{bmatrix}. \tag{21.10}$$

We finally annihilate the (2, 3) entry of the resulting matrix in Equation 21.10 by pivoting with its (2, 2) entry. In principle this requires that we choose

$$\Theta_3 = \frac{1}{\sqrt{1+\rho_3^2}} \begin{bmatrix} 1 & -\rho_3 \\ \rho_3 & 1 \end{bmatrix} = \begin{bmatrix} 0.8195 & 0.5731(16) \\ -0.5731 & 0.8195 \end{bmatrix}, \quad \text{where } \rho_3 = \frac{0.1788}{-0.2557}, \tag{21.11}$$

and apply it to the matrix on the right-hand side of Equation 21.10, which would then lead to

$$\begin{bmatrix} 1.3373 & 0.0000 & 0.0000 \\ 0.8549 & -0.2557 & 0.1788 \end{bmatrix} \begin{bmatrix} 1 & 0 & 0 \\ 0 & 0.8195 & 0.5731 \\ 0 & -0.5731 & 0.8195 \end{bmatrix} = \begin{bmatrix} 1.3373 & 0.0000 & 0.0000 \\ 0.8549 & -0.3120 & 0.0000 \end{bmatrix}. \tag{21.12}$$

Alternatively, this last step could have been implemented without explicitly forming Θ_3. We simply replace the row vector $[-0.2557 \quad 0.1788]$, which contains the (2, 2) and (2, 3) entries of the prearray in Equation 21.12, by the row vector $[\pm\sqrt{(-0.2557)^2 + (0.1788)^2} \quad 0.0000]$, which is equal to $[\pm 0.3120 \quad 0.0000]$. We choose the positive sign in order to conform with our earlier convention that the diagonal entries of triangular square-root factors are taken to be positive. The resulting postarray is therefore

$$\begin{bmatrix} 1.3373 & 0.0000 & 0.0000 \\ 0.8549 & 0.3120 & 0.0000 \end{bmatrix}. \tag{21.13}$$

We have exhibited a sequence of elementary orthogonal transformations that triangularizes the prearray of numbers in Equation 21.6. The combined effect of the sequence of transformations $\{\Theta_1, \Theta_2, \Theta_3\}$ corresponds to the orthogonal rotation Θ required in Equation 21.7. However, note that we do not need to know or to form $\Theta = \Theta_1\Theta_2\Theta_3$.

It will become clear throughout our discussion that the different adaptive RLS schemes can be described in array forms, where the necessary operations are elementary rotations as described above. Such array descriptions lend themselves rather directly to parallelizable and modular implementations. Indeed, once a rotation matrix is chosen, then all the rows of the prearray undergo the same rotation transformation and can thus be processed in parallel. Returning to the above example, where we started with the prearray **A**, we see that once the first rotation is determined, both rows of **A** are then transformed by it, and can thus be processed in parallel, and by the same functional (rotation) block, to obtain the desired postarray. The same remark holds for prearrays with multiple rows.

21.2 Least-Squares Problem

Now that we have explained the generic form of an array algorithm, we return to the main topic of this chapter and formulate the least-squares problem and its regularized version. Once this is done, we shall then proceed to describe the different variants of the RLS solution in compact array forms.

Let $\mathbf{w}$ denote a column vector of n unknown parameters that we wish to estimate, and consider a set of $(N+1)$ noisy measurements $\{d(i)\}$ that are assumed to be linearly related to $\mathbf{w}$ via the additive noise model

$$d(j) = \mathbf{u}_j^{\mathrm{T}}\mathbf{w} + v(j),$$

where the $\{\mathbf{u}_j\}$ are given column vectors. The $(N+1)$ measurements can be grouped together into a single matrix expression:

$$\underbrace{\begin{bmatrix} d(0) \\ d(1) \\ \vdots \\ d(N) \end{bmatrix}}_{\mathbf{d}} = \underbrace{\begin{bmatrix} \mathbf{u}_0^{\mathrm{T}} \\ \mathbf{u}_1^{\mathrm{T}} \\ \vdots \\ \mathbf{u}_N^{\mathrm{T}} \end{bmatrix}}_{\mathbf{A}} \mathbf{w} + \underbrace{\begin{bmatrix} v(0) \\ v(1) \\ \vdots \\ v(N) \end{bmatrix}}_{\mathbf{v}},$$

or, more compactly, $\mathbf{d} = \mathbf{A}\mathbf{w} + \mathbf{v}$. Because of the noise component $\mathbf{v}$, the observed vector $\mathbf{d}$ does not lie in the column space of the matrix $\mathbf{A}$. The objective of the least-squares problem is to determine the vector in the column space of $\mathbf{A}$ that is closest to $\mathbf{d}$ in the least-squares sense.

More specifically, any vector in the range space of **A** can be expressed as a linear combination of its columns, say $\mathbf{A}\hat{\mathbf{w}}$ for some $\hat{\mathbf{w}}$. It is therefore desired to determine the particular $\hat{\mathbf{w}}$ that minimizes the distance between **d** and $\mathbf{A}\hat{\mathbf{w}}$,

$$\min_{\mathbf{w}} \|\mathbf{d} - \mathbf{A}\mathbf{w}\|^2. \tag{21.14}$$

The resulting $\hat{\mathbf{w}}$ is called the least-squares solution and it provides an estimate for the unknown **w**. The term $\mathbf{A}\hat{\mathbf{w}}$ is called the linear least-squares estimate of **d**.

The solution to Equation 21.14 always exists and it follows from a simple geometric argument. The orthogonal projection of **d** onto the column span of **A** yields a vector $\hat{\mathbf{d}}$ that is the closest to **d** in the least-squares sense. This is because the resulting error vector $(\mathbf{d} - \hat{\mathbf{d}})$ will be orthogonal to the column span of **A**.

In other words, the closest element $\hat{\mathbf{d}}$ to **d** must satisfy the orthogonality condition:

$$\mathbf{A}^{\mathrm{T}}(\mathbf{d} - \hat{\mathbf{d}}) = 0.$$

That is, and replacing $\hat{\mathbf{d}}$ by $\mathbf{A}\hat{\mathbf{w}}$, the corresponding $\hat{\mathbf{w}}$ must satisfy

$$\mathbf{A}^{\mathrm{T}}\mathbf{A}\hat{\mathbf{w}} = \mathbf{A}^{\mathrm{T}}\mathbf{d}.$$

These equations always have a solution $\hat{\mathbf{w}}$. But while a solution $\hat{\mathbf{w}}$ may or may not be unique (depending on whether **A** is or is not full rank), the resulting estimate $\hat{\mathbf{d}} = \mathbf{A}\hat{\mathbf{w}}$ is always unique no matter which solution $\hat{\mathbf{w}}$ we pick. This is obvious from the geometric argument because the orthogonal projection of **d** onto the span of **A** is unique.

If **A** is assumed to be a tall full rank matrix then $\mathbf{A}^{\mathrm{T}}\mathbf{A}$ is invertible and we can write

$$\hat{\mathbf{w}} = (\mathbf{A}^{\mathrm{T}}\mathbf{A})^{-1}\mathbf{A}^{\mathrm{T}}\mathbf{d}. \tag{21.15}$$

21.2.1 Geometric Interpretation

The quantity $\mathbf{A}\hat{\mathbf{w}}$ provides an estimate for **d**; it corresponds to the vector in the column span of **A** that is closest in Euclidean norm to the given **d**. In other words,

$$\hat{\mathbf{d}} = \mathbf{A}(\mathbf{A}^{\mathrm{T}}\mathbf{A})^{-1}\mathbf{A}^{\mathrm{T}} \cdot \mathbf{d} \overset{\Delta}{=} \mathcal{P}_A \cdot \mathbf{d},$$

where $\mathcal{P}_A$ denotes the projector onto the range space of **A**. Figure 21.1 is a schematic representation of this geometric construction, where $\mathcal{R}(\mathbf{A})$ denotes the column span of **A**.

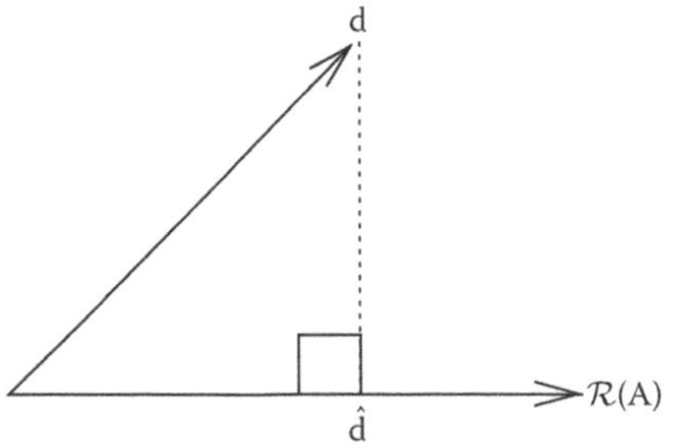

FIGURE 21.1 Geometric interpretation of the least-squares solution.

21.2.2 Statistical Interpretation

The least-squares solution also admits an important statistical interpretation. For this purpose, assume that the noise vector $\mathbf{v}$ is a realization of a vector-valued random variable that is normally distributed with zero mean and identity covariance matrix, written $\mathbf{v} \sim N[0, \mathbf{I}]$. In this case, the observation vector $\mathbf{d}$ will be a realization of a vector-valued random variable that is also normally distributed with mean $\mathbf{Aw}$ and covariance matrix equal to the identity $\mathbf{I}$. This is because the random vectors are related via the additive model $\mathbf{d} = \mathbf{Aw} + \mathbf{v}$. The probability density function of the observation process $\mathbf{d}$ is then given by

$$\frac{1}{\sqrt{(2\pi)^{(N+1)}}} \cdot \exp\left[-\frac{1}{2}(\mathbf{d} - \mathbf{Aw})^{\mathrm{T}}(\mathbf{d} - \mathbf{Aw})\right]. \tag{21.16}$$

It follows, in this case, that the least-squares estimator $\hat{\mathbf{w}}$ is also the ML estimator because it maximizes the probability density function over $\mathbf{w}$, given an observation vector $\mathbf{d}$.

21.3 Regularized Least-Squares Problem

A more general optimization criterion that is often used instead of Equation 21.14 is the following:

$$\min_{\mathbf{w}} \left[(\mathbf{w} - \bar{\mathbf{w}})^{\mathrm{T}}\Pi_0^{-1}(\mathbf{w} - \bar{\mathbf{w}}) + \|\mathbf{d} - \mathbf{Aw}\|^2\right]. \tag{21.17}$$

This is still a quadratic cost function in the unknown vector $\mathbf{w}$, but it includes the additional term

$$(\mathbf{w} - \bar{\mathbf{w}})^{\mathrm{T}}\Pi_0^{-1}(\mathbf{w} - \bar{\mathbf{w}}),$$

where

Π_0 is a given positive-definite (weighting) matrix
$\bar{\mathbf{w}}$ is also a given vector

Choosing $\Pi_0 = \infty \cdot \mathbf{I}$ leads us back to the original expression (Equation 21.14).

A motivation for Equation 21.17 is that the freedom in choosing Π_0 allows us to incorporate additional *a priori* knowledge into the statement of the problem. Indeed, different choices for Π_0 would indicate how confident we are about the closeness of the unknown $\mathbf{w}$ to the given vector $\bar{\mathbf{w}}$.

Assume, for example, that we set $\Pi_0 = \varepsilon \cdot \mathbf{I}$, where ε is a very small positive number. Then the first term in the cost function (Equation 21.17) becomes dominant. It is then not hard to see that, in this case, the cost will be minimized if we choose the estimate $\hat{\mathbf{w}}$ close enough to $\bar{\mathbf{w}}$ in order to annihilate the effect of the first term. In simple words, a "small" Π_0 reflects a high confidence that $\bar{\mathbf{w}}$ is a good and close enough guess for $\mathbf{w}$. On the other hand, a "large" Π_0 indicates a high degree of uncertainty in the initial guess $\bar{\mathbf{w}}$.

One way of solving the regularized optimization problem (Equation 21.17) is to reduce it to the standard least-squares problem (Equation 21.20). This can be achieved by introducing the change of variables $\mathbf{w}' = \mathbf{w} - \bar{\mathbf{w}}$ and $\mathbf{d}' = \mathbf{d} - \mathbf{A}\bar{\mathbf{w}}$. Then Equation 21.14 becomes

$$\min_{\mathbf{w}'} \left[(\mathbf{w}')^{\mathrm{T}}\Pi_0^{-1}\mathbf{w}' + \|\mathbf{d}' - \mathbf{Aw}'\|^2\right],$$

which can be further rewritten in the equivalent form

$$\min_{\mathbf{w}'} \left\| \begin{bmatrix} 0 \\ \mathbf{d}' \end{bmatrix} - \begin{bmatrix} \Pi_0^{-1/2} \\ \mathbf{A} \end{bmatrix} \mathbf{w}' \right\|^2.$$

This is now of the same form as our earlier minimization problem (Equation 21.14), with the observation vector **d** in Equation 21.14 replaced by

$$\begin{bmatrix} 0 \\ \mathbf{d}' \end{bmatrix},$$

and the matrix **A** in Equation 21.14 replaced by

$$\begin{bmatrix} \Pi_0^{-1/2} \\ \mathbf{A} \end{bmatrix}.$$

21.3.1 Geometric Interpretation

The orthogonality condition can now be used, leading to the equation

$$\begin{bmatrix} \Pi_0^{-1/2} \\ \mathbf{A} \end{bmatrix}^{\mathrm{T}} \left(\begin{bmatrix} 0 \\ \mathbf{d}' \end{bmatrix} - \begin{bmatrix} \Pi_0^{-1/2} \\ \mathbf{A} \end{bmatrix} \mathbf{w}' \right) = 0,$$

which can be solved for the optimal estimate $\hat{\mathbf{w}}$,

$$\hat{\mathbf{w}} = \bar{\mathbf{w}} + \left[\Pi_0^{-1} + \mathbf{A}^{\mathrm{T}}\mathbf{A}\right]^{-1}\mathbf{A}^{\mathrm{T}}[\mathbf{d} - \mathbf{A}\bar{\mathbf{w}}]. \tag{21.18}$$

Comparing with Equation 21.15, we see that instead of requiring the invertibility of $\mathbf{A}^{\mathrm{T}}\mathbf{A}$, we now require the invertibility of the matrix $\left[\Pi_0^{-1} + \mathbf{A}^{\mathrm{T}}\mathbf{A}\right]$. This is yet another reason in favor of the modified criterion in Equation 21.17 because it allows us to relax the full rank condition on **A**.

The solution of Equation 21.18 can also be reexpressed as the solution of the following linear system of equations:

$$\underbrace{\left[\Pi_0^{-1} + \mathbf{A}^{\mathrm{T}}\mathbf{A}\right]}_{\Phi}(\hat{\mathbf{w}} - \bar{\mathbf{w}}) = \underbrace{\mathbf{A}^{\mathrm{T}}[\mathbf{d} - \mathbf{A}\bar{\mathbf{w}}]}_{\mathbf{s}}, \tag{21.19}$$

where we have denoted, for convenience, the coefficient matrix by Φ and the right-hand side by **s**.

Moreover, it further follows that the value of Equation 21.17 at the minimizing solution in Equation 21.18, denoted by $E_{\min}$, is given by either of the following two expressions:

$$\begin{aligned} E_{\min} &= \|\mathbf{d} - \mathbf{A}\bar{\mathbf{w}}\|^2 - \mathbf{s}^{\mathrm{T}}(\hat{\mathbf{w}} - \bar{\mathbf{w}}) \\ &= (\mathbf{d} - \mathbf{A}\bar{\mathbf{w}})^{\mathrm{T}}[\mathbf{I} + \mathbf{A}\Pi_0\mathbf{A}^{\mathrm{T}}]^{-1}(\mathbf{d} - \mathbf{A}\bar{\mathbf{w}}). \end{aligned} \tag{21.20}$$

Expressions in Equations 21.19 and 21.20 are often rewritten into the so-called normal equations:

$$\begin{bmatrix} \|\mathbf{d} - \mathbf{A}\bar{\mathbf{w}}\|^2 & \mathbf{s}^{\mathrm{T}} \\ \mathbf{s} & \Phi \end{bmatrix} \begin{bmatrix} 1 \\ -(\hat{\mathbf{w}} - \bar{\mathbf{w}}) \end{bmatrix} = \begin{bmatrix} E_{\min} \\ 0 \end{bmatrix}. \tag{21.21}$$

The results of this section are summarized in Table 21.2.

TABLE 21.2 Linear Least-Squares Estimation

Optimization/Problem	Solution
$\{\mathbf{w}, \mathbf{d}\}$	
$\min_{\mathbf{w}} \|\mathbf{d} - \mathbf{A}\mathbf{w}\|^2$	$\hat{\mathbf{w}} = (\mathbf{A}^T\mathbf{A})^{-1}\mathbf{A}^T\mathbf{d}$
A full rank	
$\{\mathbf{w}, \mathbf{d}, \mathbf{w}, \Pi_0\}$	
$\min_{\mathbf{w}}\left[(\mathbf{w} - \hat{\mathbf{w}})^T\Pi_0^{-1}(\mathbf{w} - \hat{\mathbf{w}}) + \|\mathbf{d} - \mathbf{A}\mathbf{w}\|^2\right]$	$\hat{\mathbf{w}} = \bar{\mathbf{w}} + \left[\Pi_0^{-1} + \mathbf{A}^T\mathbf{A}\right]^{-1}\mathbf{A}^T[\mathbf{d} - \mathbf{A}\bar{\mathbf{w}}]$
Π_0 positive-definite	Minimum value $= (\mathbf{d} - \mathbf{A}\bar{\mathbf{w}})^T\left(\mathbf{I} + \mathbf{A}\Pi_0\mathbf{A}^T\right)^{-1}(\mathbf{d} - \mathbf{A}\bar{\mathbf{w}})$

21.3.2 Statistical Interpretation

A statistical interpretation for the regularized problem can be obtained as follows. Given two vector-valued zero-mean random variables $\mathbf{w}$ and $\mathbf{d}$, the minimum-variance unbiased (MVU) estimator of $\mathbf{w}$ given an observation of $\mathbf{d}$ is $\hat{\mathbf{w}} = E(\mathbf{w}|\mathbf{d})$, the conditional expectation of $\mathbf{w}$ given $\mathbf{d}$. If the random variables $(\mathbf{w}, \mathbf{d})$ are jointly Gaussian, then the MVU estimator for $\mathbf{w}$ given $\mathbf{d}$ can be shown to collapse to

$$\hat{\mathbf{w}} = (E\mathbf{w}\mathbf{d}^T)(E\mathbf{d}\mathbf{d}^T)^{-1}\mathbf{d}. \tag{21.22}$$

Therefore, if $(\mathbf{w}, \mathbf{d})$ are further linearly related, say

$$\mathbf{d} = \mathbf{A}\mathbf{w} + \mathbf{v}, \quad \text{where } \mathbf{v} \sim N(0, \mathbf{I}) \quad \text{and} \quad \mathbf{w} \sim N(0, \Pi_0) \tag{21.23}$$

with a zero-mean noise vector $\mathbf{v}$ that is uncorrelated with $\mathbf{w}$ ($E\mathbf{w}\mathbf{v}^T = 0$), then the expressions for $(E\mathbf{w}\mathbf{d}^T)$ and $(E\mathbf{d}\mathbf{d}^T)$ can be evaluated as

$$E\mathbf{w}\mathbf{d}^T = E\mathbf{w}(\mathbf{A}\mathbf{w} + \mathbf{v})^T = \Pi_0\mathbf{A}^T \quad \text{and} \quad E\mathbf{d}\mathbf{d}^T = \mathbf{A}\Pi_0\mathbf{A}^T + \mathbf{I}.$$

This shows that Equation 21.22 evaluates to

$$\hat{\mathbf{w}} = \Pi_0\mathbf{A}^T\left(\mathbf{I} + \mathbf{A}\Pi_0\mathbf{A}^T\right)^{-1}\mathbf{d}. \tag{21.24}$$

By invoking the useful matrix inversion formula (for arbitrary matrices of appropriate dimensions and invertible $\mathbf{E}$ and $\mathbf{C}$):

$$(\mathbf{E} + \mathbf{B}\mathbf{C}\mathbf{D})^{-1} = \mathbf{E}^{-1} - \mathbf{E}^{-1}\mathbf{B}(\mathbf{D}\mathbf{E}^{-1}\mathbf{B} + \mathbf{C}^{-1})^{-1}\mathbf{D}\mathbf{E}^{-1},$$

we can rewrite Equation 21.24 in the equivalent form

$$\hat{\mathbf{w}} = \left(\Pi_0^{-1} + \mathbf{A}^T\mathbf{A}\right)^{-1}\mathbf{A}^T\mathbf{d}. \tag{21.25}$$

This expression coincides with the regularized solution (Equation 21.24) for $\bar{\mathbf{w}} = 0$ (the case $\bar{\mathbf{w}} \neq 0$ follows from similar arguments by assuming a nonzero mean random variable $\mathbf{w}$).

Therefore, the regularized least-squares solution is the MVU estimate of $\mathbf{w}$ given observations $\mathbf{d}$ that are corrupted by additive Gaussian noise as in Equation 21.23.

21.4 Recursive Least-Squares Problem

The RLS formulation deals with the problem of updating the solution $\hat{\mathbf{w}}$ of a least-squares problem (regularized or not) when new data are added to the matrix **A** and to the vector **d**. This is in contrast to determining afresh the least-squares solution of the new problem. The distinction will become clear as we proceed in our discussions. In this section, we formulate the RLS problem as it arises in the context of adaptive filtering.

Consider a sequence of $(N+1)$ scalar data points, $\{d(j)\}_{j=0}^{N}$, also known as reference or desired signals, and a sequence of $(N+1)$ row vectors $\{\mathbf{u}_j^{\mathrm{T}}\}_{j=0}^{N}$, also known as input signals. Each input vector $\mathbf{u}_j^{\mathrm{T}}$ is a $1 \times M$ row vector whose individual entries we denote by $\{u_k(j)\}_{k=1}^{M}$, viz.,

$$\mathbf{u}_j^{\mathrm{T}} = \begin{bmatrix} u_1(j) & u_2(j) & \cdots & u_M(j) \end{bmatrix}. \tag{21.26}$$

The entries of $\mathbf{u}_j$ can be regarded as the values of M input channels at time j: channels 1 through M.

Consider also a known column vector $\bar{\mathbf{w}}$ and a positive-definite weighting matrix Π_0. The objective is to determine an $M \times 1$ column vector $\mathbf{w}$, also known as the weight vector, so as to minimize the weighted error sum:

$$E(N) = (\mathbf{w} - \bar{\mathbf{w}})^{\mathrm{T}} \left[\lambda^{-(N+1)}\Pi_0\right]^{-1} (\mathbf{w} - \bar{\mathbf{w}}) + \sum_{j=0}^{N} \lambda^{N-j} \left| d(j) - \mathbf{u}_j^{\mathrm{T}}\mathbf{w} \right|^2, \tag{21.27}$$

where λ is a positive scalar that is less than or equal to one (usually $0 \ll \lambda \leq 1$). It is often called the forgetting factor since past data is exponentially weighted less than the more recent data. The special case $\lambda = 1$ is known as the growing memory case, since, as the length N of the data grows, the effect of past data is not attenuated. In contrast, the exponentially decaying memory case ($\lambda < 1$) is more suitable for time-variant environments.

Also, and in principle, the factor $\lambda^{-(N+1)}$ that multiplies Π_0 in the error-sum expression (Equation 21.27) can be incorporated into the weighting matrix Π_0. But it is left explicit for convenience of exposition.

We further denote the individual entries of the column vector $\mathbf{w}$ by $\{w(j)\}_{j=1}^{M}$,

$$\mathbf{w} = \mathrm{col}\{w(1), w(2), \ldots, w(M)\}.$$

A schematic description of the problem is shown in Figure 21.2. At each time instant j, the inputs of the M channels are linearly combined via the coefficients of the weight vector and the resulting signal is compared with the desired signal $d(j)$. This results in a residual error $e(j) = d(j) - \mathbf{u}_j^{\mathrm{T}}\mathbf{w}$, for every j, and the objective is to find a weight vector $\mathbf{w}$ in order to minimize the (exponentially weighted and regularized) squared-sum of the residual errors over an interval of time, say from $j = 0$ up to $j = N$.

The linear combiner is said to be of order M since it is determined by M coefficients $\{w(j)\}_{j=1}^{M}$.

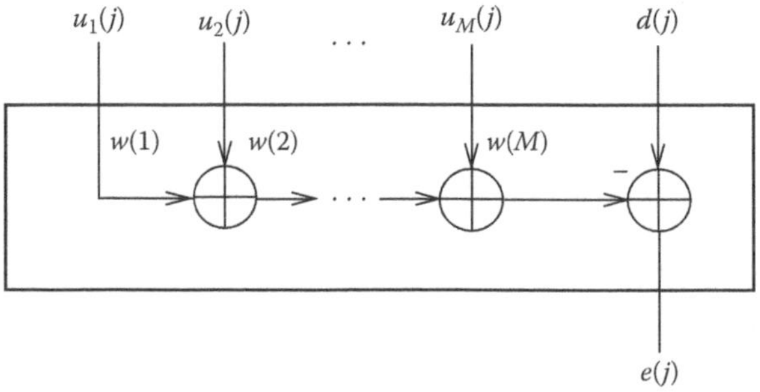

FIGURE 21.2 A linear combiner.

21.4.1 Reducing to the Regularized Form

The expression for the weighted error-sum (Equation 21.27) is a special case of the regularized cost function (Equation 21.17). To clarify this, we introduce the residual vector $\mathbf{e}_N$, the reference vector $\mathbf{d}_N$, the data matrix $\mathbf{A}_N$, and a diagonal weighting matrix Λ_N:

$$\mathbf{e}_N = \underbrace{\begin{bmatrix} d(0) \\ d(1) \\ d(2) \\ \vdots \\ d(N) \end{bmatrix}}_{\mathbf{d}_N} - \underbrace{\begin{bmatrix} u_1(0) & u_2(0) & \dots & u_M(0) \\ u_1(1) & u_2(1) & \dots & u_M(1) \\ u_1(2) & u_2(2) & \dots & u_M(2) \\ \vdots & \vdots & \vdots & \vdots \\ u_1(N) & u_2(N) & \dots & u_M(N) \end{bmatrix}}_{\mathbf{A}_N} \mathbf{w},$$

$$\Lambda_N^{1/2} = \begin{bmatrix} \left[\lambda^{\frac{1}{2}}\right]^N & & & & \\ & \left[\lambda^{\frac{1}{2}}\right]^{N-1} & & & \\ & & \ddots & & \\ & & & \left[\lambda^{\frac{1}{2}}\right]^2 & \\ & & & & 1 \end{bmatrix}.$$

We now use a subscript N to indicate that the above quantities are determined by data that is available up to time N.

With these definitions, we can write $E(N)$ in the equivalent form:

$$E(N) = (\mathbf{w} - \bar{\mathbf{w}})^{\mathrm{T}} \left[\lambda^{-(N+1)}\Pi_0\right]^{-1} (\mathbf{w} - \bar{\mathbf{w}}) + \left\|\Lambda_N^{1/2}\mathbf{e}_N\right\|^2,$$

which is a special case of Equation 21.17 with

$$\Lambda_N^{1/2}\mathbf{d}_N \quad \text{and} \quad \Lambda_N^{1/2}\mathbf{A}_N \tag{21.28}$$

replacing

$$\mathbf{d}_N \quad \text{and} \quad \mathbf{A}_N, \tag{21.29}$$

respectively, and with $\lambda^{-(N+1)}\Pi_0$ replacing Π_0.

We therefore conclude from Equation 21.19 that the optimal solution $\hat{\mathbf{w}}$ of Equation 21.27 is given by

$$(\hat{\mathbf{w}} - \bar{\mathbf{w}}) = \Phi_N^{-1}\mathbf{s}_N, \tag{21.30}$$

where we have introduced

$$\Phi_N = \left[\lambda^{(N+1)}\Pi_0^{-1} + \mathbf{A}_N^{\mathrm{T}}\Lambda_N\mathbf{A}_N\right], \tag{21.31}$$

$$\mathbf{s}_N = \mathbf{A}_N^{\mathrm{T}}\Lambda_N[\mathbf{d}_N - \mathbf{A}_N\bar{\mathbf{w}}]. \tag{21.32}$$

The coefficient matrix Φ_N is clearly symmetric and positive-definite.

21.4.2 Time Updates

It is straightforward to verify that Φ_N and $\mathbf{s}_N$ so defined satisfy simple time-update relations, viz.,

$$\Phi_{N+1} = \lambda\Phi_N + \mathbf{u}_{N+1}\mathbf{u}_{N+1}^{\mathrm{T}}, \tag{21.33}$$

$$\mathbf{s}_{N+1} = \lambda\mathbf{s}_N + \mathbf{u}_{N+1}\left[d(N+1) - \mathbf{u}_{N+1}^{\mathrm{T}}\bar{\mathbf{w}}\right], \tag{21.34}$$

with initial conditions $\Phi_{-1} = \Pi_0^{-1}$ and $\mathbf{s}_{-1} = 0$. Note that Φ_{N+1} and $\lambda\Phi_N$ differ only by a rank-one matrix.

The solution $\hat{\mathbf{w}}$ obtained by solving Equation 21.30 is the optimal weight estimate based on the available data from time $i = 0$ up to time $i = N$. We shall denote it from now on by $\mathbf{w}_N$,

$$\Phi_N(\mathbf{w}_N - \bar{\mathbf{w}}) = \mathbf{s}_N.$$

The subscript N in $\mathbf{w}_N$ indicates that the data up to, and including, time N were used. This is to differentiate it from the estimate obtained by using a different number of data points.

This notational change is necessary because the main objective of the RLS problem is to show how to update the estimate $\mathbf{w}_N$, which is based on the data up to time N, to the estimate $\mathbf{w}_{N+1}$, which is based on the data up to time $(N+1)$, without the need to solve afresh a new set of linear equations of the form

$$\Phi_{N+1}(\mathbf{w}_{N+1} - \bar{\mathbf{w}}) = \mathbf{s}_{N+1}.$$

Such a recursive update of the weight estimate should be possible since the coefficient matrices $\lambda\Phi_N$ and Φ_{N+1} of the associated linear systems differ only by a rank-one matrix. In fact, a wide variety of algorithms has been devised for this end and our purpose in this chapter is to provide an overview of the different schemes.

Before describing these different variants, we note in passing that it follows from Equation 21.20 that we can express the minimum value of $E(N)$ in the form

$$E_{\min}(N) = \left\|\Lambda_N^{1/2}(\mathbf{d}_N - \mathbf{A}_N\bar{\mathbf{w}})\right\|^2 - \mathbf{s}_N^{\mathrm{T}}(\mathbf{w}_N - \bar{\mathbf{w}}). \tag{21.35}$$

21.5 RLS Algorithm

The first recursive solution that we consider is the famed RLS algorithm, usually referred to as the RLS algorithm. It can be derived as follows.

Let $\mathbf{w}_{i-1}$ be the solution of an optimization problem of the form as Equation 21.27 that uses input data up to time $(i-1)$ (i.e., for $N = (i-1)$). Likewise, let $\mathbf{w}_i$ be the solution of the same optimization problem but with input data up to time i $[N = i]$.

The RLS algorithm provides a recursive procedure that computes $\mathbf{w}_i$ from $\mathbf{w}_{i-1}$. A classical derivation follows by noting from Equation 21.30 that the new solution $\mathbf{w}_i$ should satisfy

$$\mathbf{w}_i - \bar{\mathbf{w}} = \Phi_i^{-1}\mathbf{s}_i = \left[\lambda\Phi_{i-1} + \mathbf{u}_i\mathbf{u}_i^{\mathrm{T}}\right]^{-1}\left(\lambda\mathbf{s}_{i-1} + \mathbf{u}_i\left[d(i) - \mathbf{u}_i^{\mathrm{T}}\bar{\mathbf{w}}\right]\right),$$

where we have also used the time-updates for $\{\Phi_i, \mathbf{s}_i\}$.

Introduce the quantities

$$\mathbf{P}_i = \Phi_i^{-1} \quad \text{and} \quad \mathbf{g}_i = \Phi_i^{-1}\mathbf{u}_i. \tag{21.36}$$

Expanding the inverse of $[\lambda\Phi_{i-1} + \mathbf{u}_i\mathbf{u}_i^T]$ by using the matrix inversion formula (stated after Equation 21.24), and grouping terms, leads after some straightforward algebra to the RLS procedure:

- Initial conditions: $\mathbf{w}_{-1} = \bar{\mathbf{w}}$ and $\mathbf{P}_{-1} = \Pi_0$.
- Repeat for $i \geq 0$:

$$\mathbf{w}_i = \mathbf{w}_{i-1} + \mathbf{g}_i\left[d(i) - \mathbf{u}_i^T\mathbf{w}_{i-1}\right], \tag{21.37}$$

$$\mathbf{g}_i = \frac{\lambda^{-1}\mathbf{P}_{i-1}\mathbf{u}_i}{1 + \lambda^{-1}\mathbf{u}_i^T\mathbf{P}_{i-1}\mathbf{u}_i}, \tag{21.38}$$

$$\mathbf{P}_i = \lambda^{-1}\left[\mathbf{P}_{i-1} - \mathbf{g}_i\mathbf{u}_i^T\mathbf{P}_{i-1}\right]. \tag{21.39}$$

- The computational complexity of the algorithm is $O(M^2)$ per iteration.

21.5.1 Estimation Errors and the Conversion Factor

With the RLS problem we associate two residuals at each time instant i: the *a priori* estimation error $e_a(i)$, defined by

$$e_a(i) = d(i) - \mathbf{u}_i^T\mathbf{w}_{i-1},$$

and the *a posteriori* estimation error $e_p(i)$, defined by

$$e_p(i) = d(i) - \mathbf{u}_i^T\mathbf{w}_i.$$

Comparing the expressions for $e_a(i)$ and $e_p(i)$, we see that the latter employs the most recent weight-vector estimate.

If we replace $\mathbf{w}_i$ in the definition for $e_p(i)$ by its update expression (Equation 21.37), say

$$e_p(i) = d(i) - \mathbf{u}_i^T\left\{\mathbf{w}_{i-1} + \mathbf{g}_i\left[d(i) - \mathbf{u}_i^T\mathbf{w}_{i-1}\right]\right\},$$

some straightforward algebra will show that we can relate $e_p(i)$ and $e_a(i)$ via a factor $\gamma(i)$ known as the conversion factor:

$$e_p(i) = \gamma(i)e_a(i),$$

where $\gamma(i)$ is equal to

$$\gamma(i) = \frac{1}{1 + \lambda^{-1}\mathbf{u}_i^T\mathbf{P}_{i-1}\mathbf{u}_i} = 1 - \mathbf{u}_i^T\mathbf{P}_i\mathbf{u}_i. \tag{21.40}$$

That is, the *a posteriori* error is a scaled version of the *a priori* error. The scaling factor $\gamma(i)$ is defined in terms of $\{\mathbf{u}_i, \mathbf{P}_{i-1}\}$ or $\{\mathbf{u}_i, \mathbf{P}_i\}$. Note that $0 \leq \gamma(i) \leq 1$.

Note further that the expression for $\gamma(i)$ appears in the definition of the so-called gain vector $\mathbf{g}_i$ in Equation 21.38 and, hence, we can alternatively rewrite Equations 21.38 and 21.39 in the forms

$$\mathbf{g}_i = \lambda^{-1}\gamma(i)\mathbf{P}_{i-1}\mathbf{u}_i, \tag{21.41}$$

$$\mathbf{P}_i = \lambda^{-1}\mathbf{P}_{i-1} - \gamma^{-1}(i)\mathbf{g}_i\mathbf{g}_i^T. \tag{21.42}$$

21.5.2 Update of the Minimum Cost

Let $E_{\min}(i)$ denote the value of the minimum cost of the optimization problem (Equation 21.27) with data up to time i. It is given by an expression of the form as Equation 21.35 with N replaced by i,

$$E_{\min}(i) = \left[\sum_{j=0}^{i} \lambda^{i-j} d(j) - \mathbf{u}_j^{\mathrm{T}} \bar{\mathbf{w}}^2 \right] - \mathbf{s}_i^{\mathrm{T}} (\mathbf{w}_i - \bar{\mathbf{w}}).$$

Using the RLS update (Equation 21.37) for $\mathbf{w}_i$ in terms of $\mathbf{w}_{i-1}$, as well as the time-update (Equation 21.34) for $\mathbf{s}_i$ in terms of $\mathbf{s}_{i-1}$, we can derive the following time-update for the minimum cost:

$$E_{\min}(i) = \lambda E_{\min}(i-1) + e_{\mathrm{p}}(i) e_{\mathrm{a}}(i), \tag{21.43}$$

where $E_{\min}(i-1)$ denotes the value of the minimum cost of the same optimization problem (Equation 21.27) but with data up to time $(i-1)$.

21.6 RLS Algorithms in Array Forms

As mentioned in the introduction, we intend to stress the array formulations of the RLS solution due to their intrinsic advantages:

- They are easy to implement as a sequence of elementary rotations on arrays of numbers.
- They are modular and parallelizable.
- They have better numerical properties than the classical RLS description.

21.6.1 Motivation

Note from Equation 21.39 that the RLS solution propagates the variable $\mathbf{P}_i$ as the difference of two quantities. This variable should be positive-definite. But due to roundoff errors, however, the update (Equation 21.39) may not guarantee the positive-definiteness of $\mathbf{P}_i$ at all times i. This problem can be ameliorated by using the so-called array formulations. These alternative forms propagate square-root factors of either $\mathbf{P}_i$ or $\mathbf{P}_i^{-1}$, namely, $\mathbf{P}_i^{1/2}$ or $\mathbf{P}_i^{-1/2}$, rather than $\mathbf{P}_i$ itself. By squaring $\mathbf{P}_i^{1/2}$, for example, we can always recover a matrix $\mathbf{P}_i$ that is more likely to be positive-definite than the matrix obtained via Equation 21.39,

$$\mathbf{P}_i = \mathbf{P}_i^{1/2} \mathbf{P}_i^{\mathrm{T}/2}.$$

21.6.2 A Very Useful Lemma

The derivation of the array variants of the RLS algorithm relies on a very useful matrix result that encounters applications in many other scenarios as well. For this reason, we not only state the result but also provide one simple proof.

LEMMA 21.1

Given two $n \times m (n \leq m)$ matrices $\mathbf{A}$ and $\mathbf{B}$, then $\mathbf{A}\mathbf{A}^{\mathrm{T}} = \mathbf{B}\mathbf{B}^{\mathrm{T}}$ if, and only if, there exists an $m \times m$ orthogonal matrix Θ ($\Theta\Theta^{\mathrm{T}} = \mathbf{I}_m$) such that $\mathbf{A} = \mathbf{B}\Theta$.

Proof 21.1 One implication is immediate. If there exists an orthogonal matrix Θ such that $\mathbf{A} = \mathbf{B}\Theta$ then

$$\mathbf{A}\mathbf{A}^{\mathrm{T}} = (\mathbf{B}\Theta)(\mathbf{B}\Theta)^{\mathrm{T}} = \mathbf{B}(\Theta\Theta^{\mathrm{T}})\mathbf{B}^{\mathrm{T}} = \mathbf{B}\mathbf{B}^{\mathrm{T}}.$$

One proof for the converse implication follows by invoking the singular value decompositions of the matrices $\mathbf{A}$ and $\mathbf{B}$,

$$\mathbf{A} = \mathbf{U}_A[\,\Sigma_A \quad 0\,]\mathbf{V}_A^{\mathrm{T}},$$
$$\mathbf{B} = \mathbf{U}_B[\,\Sigma_B \quad 0\,]\mathbf{V}_B^{\mathrm{T}},$$

where

$\mathbf{U}_A$ and $\mathbf{U}_B$ are $n \times n$ orthogonal matrices
$\mathbf{V}_A$ and $\mathbf{V}_B$ are $m \times m$ orthogonal matrices
Σ_A and Σ_B are $n \times n$ diagonal matrices with nonnegative (ordered) entries.

The squares of the diagonal entries of $\Sigma_A(\Sigma_B)$ are the eigenvalues of $\mathbf{A}\mathbf{A}^{\mathrm{T}}(\mathbf{B}\mathbf{B}^{\mathrm{T}})$. Moreover, $\mathbf{U}_A(\mathbf{U}_B)$ are constructed from an orthonormal basis for the right eigenvectors of $\mathbf{A}\mathbf{A}^{\mathrm{T}}(\mathbf{B}\mathbf{B}^{\mathrm{T}})$.

Hence, it follows from the identity $\mathbf{A}\mathbf{A}^{\mathrm{T}} = \mathbf{B}\mathbf{B}^{\mathrm{T}}$ that we have $\Sigma_A = \Sigma_B$ and we can choose $\mathbf{U}_A = \mathbf{U}_B$. Let $\Theta = \mathbf{V}_B\mathbf{V}_A^{\mathrm{T}}$. We then obtain $\Theta\Theta^{\mathrm{T}} = \mathbf{I}_m$ and $\mathbf{B}\Theta = \mathbf{A}$.

21.6.3 Inverse QR Algorithm

We now employ the above result to derive an array form of the RLS algorithm that is known as the inverse QR algorithm.

Let $\mathbf{P}_{i-1}^{1/2}$ denote a (preferably lower triangular) square-root factor of $\mathbf{P}_{i-1}$, i.e., any matrix that satisfies

$$\mathbf{P}_{i-1} = \mathbf{P}_{i-1}^{1/2}\,\mathbf{P}_{i-1}^{\mathrm{T}/2}.$$

(The triangular square-root factor of a symmetric positive-definite matrix is also known as the Cholesky factor.)

Now note that the RLS recursions (Equations 21.38 and 21.39) can be expressed in factored form as follows:

$$\begin{bmatrix} 1 & \frac{1}{\sqrt{\lambda}}\mathbf{u}_i^{\mathrm{T}}\mathbf{P}_{i-1}^{1/2} \\ 0 & \frac{1}{\sqrt{\lambda}}\mathbf{P}_{i-1}^{1/2} \end{bmatrix} \begin{bmatrix} 1 & 0^{\mathrm{T}} \\ \frac{1}{\sqrt{\lambda}}\mathbf{P}_{i-1}^{\mathrm{T}/2}\mathbf{u}_i & \frac{1}{\sqrt{\lambda}}\mathbf{P}_{i-1}^{\mathrm{T}/2} \end{bmatrix} = \begin{bmatrix} \gamma^{-1/2}(i) & 0^{\mathrm{T}} \\ \mathbf{g}_i\gamma^{-1/2}(i) & \mathbf{P}_i^{1/2} \end{bmatrix} \begin{bmatrix} \gamma^{-1/2}(i) & \mathbf{g}_i^{\mathrm{T}}\gamma^{-1/2}(i) \\ 0 & \mathbf{P}_i^{\mathrm{T}/2} \end{bmatrix}.$$

To verify that this is indeed the case, we simply multiply the factors and compare terms on both sides of the equality.

The point to note is that the above equality fits nicely into the statement of the previous lemma by taking

$$\mathbf{A} = \begin{bmatrix} 1 & \frac{1}{\sqrt{\lambda}}\mathbf{u}_i^{\mathrm{T}}\mathbf{P}_{i-1}^{1/2} \\ 0 & \frac{1}{\sqrt{\lambda}}\mathbf{P}_{i-1}^{1/2} \end{bmatrix} \tag{21.44}$$

and

$$\mathbf{B} = \begin{bmatrix} \gamma^{-1/2}(i) & \mathbf{0}^{\mathrm{T}} \\ \mathbf{g}_i\gamma^{-1/2}(i) & \mathbf{P}_i^{1/2} \end{bmatrix}. \tag{21.45}$$

We therefore conclude that there should exist an orthogonal matrix Θ_i that relates the arrays **A** and **B** in the form

$$\begin{bmatrix} 1 & \frac{1}{\sqrt{\lambda}}\mathbf{u}_i^{\mathrm{T}}\mathbf{P}_{i-1}^{1/2} \\ 0 & \frac{1}{\sqrt{\lambda}}\mathbf{P}_{i-1}^{1/2} \end{bmatrix}\Theta_i = \begin{bmatrix} \gamma^{-1/2}(i) & \mathbf{0}^{\mathrm{T}} \\ \mathbf{g}_i\gamma^{-1/2}(i) & \mathbf{P}_i^{1/2} \end{bmatrix}.$$

That is, there should exist an orthogonal Θ_i that transforms the prearray **A** into the postarray **B**.

Note that the prearray contains quantities that are available at step i, namely $\{\mathbf{u}_i, \mathbf{P}_{i-1}^{1/2}\}$, while the postarray provides the (normalized) gain vector $\mathbf{g}_i\gamma^{-1/2}(i)$, which is needed to update the weight-vector estimate $\mathbf{w}_{i-1}$ into $\mathbf{w}_i$, as well as the square-root factor of the variable $\mathbf{P}_i$, which is needed to form the prearray for the next iteration.

But how do we determine Θ_i? The answer highlights a remarkable property of array algorithms. We do not really need to know or determine Θ_i explicitly!

To clarify this point, we first remark from the expressions in Equations 21.44 and 21.45 for the pre- and postarrays that Θ_i is an orthogonal matrix that takes an array of numbers of the form (assuming a vector $\mathbf{u}_i$ of dimension $M = 3$)

$$\begin{bmatrix} 1 & x & x & x \\ 0 & x & 0 & 0 \\ 0 & x & x & 0 \\ 0 & x & x & x \end{bmatrix} \tag{21.46}$$

and transforms it to the form

$$\begin{bmatrix} x & 0 & 0 & 0 \\ x & x & 0 & 0 \\ x & x & x & 0 \\ x & x & x & x \end{bmatrix}. \tag{21.47}$$

That is, Θ_i annihilates all the entries of the top row of the prearray (except for the left-most entry).

Now assume we form the prearray **A** in Equation 21.44 and choose any Θ_i (say as a sequence of elementary rotations) so as to reduce **A** to the triangular form (Equation 21.47), i.e., in order to annihilate the desired entries in the top row.

Let us denote the resulting entries of the postarray arbitrarily as

$$\begin{bmatrix} 1 & \frac{1}{\sqrt{\lambda}}\mathbf{u}_i^{\mathrm{T}}\mathbf{P}_{i-1}^{1/2} \\ 0 & \frac{1}{\sqrt{\lambda}}\mathbf{P}_{i-1}^{1/2} \end{bmatrix}\Theta_i = \begin{bmatrix} a & \mathbf{0}^{\mathrm{T}} \\ \mathbf{b} & \mathbf{C} \end{bmatrix}, \tag{21.48}$$

where $\{a, \mathbf{b}, \mathbf{C}\}$ are quantities that we wish to identify (a is a scalar, **b** is a column vector, and **C** is a lower triangular matrix). The claim is that by constructing Θ_i in this way (i.e., by simply requiring that it achieves the desired zero pattern in the postarray), the resulting quantities $\{a, \mathbf{b}, \mathbf{C}\}$ will be meaningful and can in fact be identified with the quantities in the postarray **B**.

To verify that the quantities $\{a, \mathbf{b}, \mathbf{C}\}$ can indeed be identified with $\{\gamma^{-1/2}(i), \mathbf{g}_i\gamma^{-1/2}(i), \mathbf{P}_i^{1/2}\}$, we proceed by squaring both sides of Equation 21.48,

$$\begin{bmatrix} 1 & \frac{1}{\sqrt{\lambda}}\mathbf{u}_i^T\mathbf{P}_{i-1}^{1/2} \\ 0 & \frac{1}{\sqrt{\lambda}}\mathbf{P}_{i-1}^{1/2} \end{bmatrix} \underbrace{\Theta_i\Theta_i^T}_{I} \begin{bmatrix} 1 & 0 \\ \frac{1}{\sqrt{\lambda}}\mathbf{P}_{i-1}^{T/2}\mathbf{u}_i & \frac{1}{\sqrt{\lambda}}\mathbf{P}_{i-1}^{T/2} \end{bmatrix} = \begin{bmatrix} a & \mathbf{0}^T \\ \mathbf{b} & \mathbf{C} \end{bmatrix} \begin{bmatrix} a & \mathbf{b}^T \\ 0 & \mathbf{C}^T \end{bmatrix},$$

and comparing terms on both sides of the equality to get the identities:

$$\begin{aligned} a^2 &= 1 + \lambda^{-1}\mathbf{u}_i^T\mathbf{P}_{i-1}\mathbf{u}_i = \gamma^{-1}(i), \\ \mathbf{b}a &= \lambda^{-1}\mathbf{P}_{i-1}\mathbf{u}_i = \mathbf{g}_i\gamma^{-1}(i), \\ \mathbf{C}\mathbf{C}^T &= \lambda^{-1}\mathbf{P}_{i-1} - \mathbf{b}\mathbf{b}^T = \lambda^{-1}\mathbf{P}_{i-1} - \gamma^{-1}(i)\mathbf{g}_i\mathbf{g}_i^T. \end{aligned}$$

Hence, as desired, we can make the identifications

$$a = \gamma^{-1/2}(i), \quad \mathbf{b} = \mathbf{g}_i\gamma^{-1/2}(i), \quad \text{and} \quad \mathbf{C} = \mathbf{P}_i^{1/2}.$$

In summary, we have established the validity of an array alternative to the RLS algorithm, known as the inverse QR algorithm (also as square-root RLS). It is listed in Table 21.3. The recursions are known as inverse QR since they propagate $\mathbf{P}_i^{1/2}$, which is a square-root factor of the inverse of the coefficient matrix Φ_i.

21.6.4 QR Algorithm

The RLS recursion (Equation 21.39) and the inverse QR recursion of Table 21.3 propagate the variable $\mathbf{P}_i$ or a square-root factor of it. The starting condition for both algorithms is therefore dependent on the weighting matrix Π_0 or its square-root factor $\Pi_0^{1/2}$.

This situation becomes inconvenient when the initial condition Π_0 assumes relatively large values, say $\Pi_0 = \sigma\mathbf{I}$ with $\sigma \gg 1$. A particular instance arises, for example, when we take $\sigma \to \infty$ in which case the regularized least-squares problem (Equation 21.33) reduces to a standard least-squares problem of the form

$$\min_{\mathbf{w}} \left[E(N) = \sum_{j=0}^{N} \lambda^{N-j} \left| d(j) - \mathbf{u}_j^T\mathbf{w} \right|^2 \right]. \tag{21.49}$$

TABLE 21.3 Inverse QR Algorithm

Initialization: Start with $\mathbf{w}_{-1} = \bar{\mathbf{w}}$ and $\mathbf{P}_{-1}^{1/2} = \Pi_0^{1/2}$.

- Repeat for each time instant $i \geq 0$:

$$\begin{bmatrix} 1 & \frac{1}{\sqrt{\lambda}}\mathbf{u}_i^T\mathbf{P}_{i-1}^{1/2} \\ 0 & \frac{1}{\sqrt{\lambda}}\mathbf{P}_{i-1}^{1/2} \end{bmatrix} \Theta_i = \begin{bmatrix} \gamma^{-1/2}(i) & \mathbf{0}^T \\ \mathbf{g}_i\gamma^{-1/2}(i) & \mathbf{P}_i^{1/2} \end{bmatrix},$$

where Θ_i is any orthogonal rotation that produces the zero pattern in the postarray.

The weight-vector estimate is updated via

$$\mathbf{w}_i = \mathbf{w}_{i-1} + \left[\frac{\mathbf{g}_i}{\gamma^{-1/2}(i)}\right]\left[\frac{1}{\gamma^{-1/2}(i)}\right]^{-1}[d(i) - \mathbf{u}_i^T\mathbf{w}_{i-1}],$$

where the quantities $\{\gamma^{-1/2}(i), \mathbf{g}_i\gamma^{-1/2}(i)\}$ are read from the entries of the postarray.

The computational cost is $O(M^2)$ per iteration.

For such problems, it is preferable to propagate the inverse of the variable $\mathbf{P}_i$ rather than $\mathbf{P}_i$ itself. Recall that the inverse of $\mathbf{P}_i$ is Φ_i since we have defined earlier $\mathbf{P}_i = \Phi_i^{-1}$.

The QR algorithm is a recursive procedure that propagates a square-root factor of Φ_i. Its validity can be verified in much the same way as we did for the inverse QR algorithm. We form a prearray of numbers and then choose a sequence of rotations that induces a desired zero pattern in the postarray. Then by squaring and comparing terms on both sides of an equality we can identify the resulting entries of the postarray as meaningful quantities in the RLS context. For this reason, we shall be brief and only highlight the main points.

Let $\Phi_{i-1}^{1/2}$ denote a square-root factor (preferably lower triangular) of Φ_{i-1}, $\Phi_{i-1} = \Phi_{i-1}^{1/2}\,\Phi_{i-1}^{T/2}$, and define, for notational convenience, the quantity

$$\mathbf{q}_{i-1} = \Phi_{i-1}^{T/2}\,\mathbf{w}_{i-1}. \tag{21.50}$$

At time $(i-1)$, we form the prearray of numbers

$$\mathbf{A} = \begin{bmatrix} \sqrt{\lambda}\Phi_{i-1}^{1/2} & \mathbf{u}_i \\ \sqrt{\lambda}\mathbf{q}_{i-1}^{T} & d(i) \\ 0^{T} & 1 \end{bmatrix},$$

whose entries have the following pattern (shown for $M = 3$):

$$\mathbf{A} = \begin{bmatrix} x & 0 & 0 & x \\ x & x & 0 & x \\ x & x & x & x \\ x & x & x & x \\ 0 & 0 & 0 & 1 \end{bmatrix}.$$

Now implement an orthogonal transformation Θ_i that reduces $\mathbf{A}$ to the form

$$\mathbf{B} = \begin{bmatrix} x & 0 & 0 & 0 \\ x & x & 0 & 0 \\ x & x & x & 0 \\ x & x & x & x \\ x & x & x & x \end{bmatrix} = \begin{bmatrix} \mathbf{C} & 0 \\ \mathbf{b}^{T} & a \\ \mathbf{h}^{T} & f \end{bmatrix},$$

where the quantities $\{\mathbf{C}, \mathbf{b}, \mathbf{h}, a, f\}$ need to be identified. By comparing terms on both sides of the equality

$$\begin{bmatrix} \sqrt{\lambda}\Phi_{i-1}^{1/2} & \mathbf{u}_i \\ \sqrt{\lambda}\mathbf{q}_{i-1}^{T} & d(i) \\ 0^{T} & 1 \end{bmatrix} \underbrace{\Theta_i\Theta_i^{T}}_{I} \begin{bmatrix} \sqrt{\lambda}\Phi_{i-1}^{1/2} & \mathbf{u}_i \\ \sqrt{\lambda}\mathbf{q}_{i-1}^{T} & d(i) \\ 0^{T} & 1 \end{bmatrix}^{T} = \begin{bmatrix} \mathbf{C} & 0 \\ \mathbf{b}^{T} & a \\ \mathbf{h}^{T} & f \end{bmatrix} \begin{bmatrix} \mathbf{C} & 0 \\ \mathbf{b}^{T} & a \\ \mathbf{h}^{T} & f \end{bmatrix}^{T},$$

we can make the identifications:

$$\mathbf{C} = \Phi_i^{1/2}, \quad \mathbf{b}^{T} = \mathbf{q}_i^{T}, \quad \mathbf{h}^{T} = \mathbf{u}_i^{T}\Phi_i^{-T/2},$$
$$a = e_a(i)\gamma^{1/2}(i), \quad \text{and} \quad f = \gamma^{1/2}(i),$$

where $e_a(i) = d(i) - \mathbf{u}_i^{T}\mathbf{w}_{i-1}$ is the *a priori* estimation error. This derivation establishes the so-called QR algorithm (listed in Table 21.4).

TABLE 21.4 QR Algorithm

Initialization: Start with $\mathbf{w}_{-1} = \bar{\mathbf{w}}$, $\Phi_{-1}^{1/2} = \Pi_0^{-T/2}$, $\mathbf{q}_{-1} = \Pi_0^{-1/2}\bar{\mathbf{w}}$.

- Repeat for each time instant $i \geq 0$:

$$\begin{bmatrix} \sqrt{\lambda}\Phi_{i-1}^{1/2} & \mathbf{u}_i^T \\ \sqrt{\lambda}\mathbf{q}_{i-1}^T & d(i) \\ 0^T & 1 \end{bmatrix} \Theta_i = \begin{bmatrix} \Phi_i^{1/2} & 0 \\ \mathbf{q}_i^T & e_a(i)\gamma^{1/2}(i) \\ \mathbf{u}_i^T\Phi_i^{-T/2} & \gamma^{1/2}(i) \end{bmatrix},$$

where Θ_i is any orthogonal rotation that produces the zero pattern in the postarray.

The weight-vector estimate can be obtained by solving the triangular linear systems of equations:

$$\Phi_i^{T/2}\mathbf{w}_i = \mathbf{q}_i,$$

where the quantities $\{\Phi_i^{1/2}, \mathbf{q}_i\}$ are available from the entries of the postarray.

The computational complexity is still $O(M^2)$ per iteration.

The QR solution determines the weight-vector estimate $\mathbf{w}_i$ by solving a triangular linear system of equations, for example, via back-substitution. A major drawback of a back-substitution step is that it involves serial operations and, therefore, does not lend itself to a fully parallelizable implementation.

An alternative procedure for computing the estimate $\mathbf{w}_i$ can be obtained by appending one more block row to the arrays of the QR algorithm, leading to the equations:

$$\begin{bmatrix} \sqrt{\lambda}\Phi_{i-1}^{1/2} & \mathbf{u}_i \\ \sqrt{\lambda}\mathbf{q}_{i-1}^T & d(i) \\ 0^T & 1 \\ \frac{1}{\sqrt{\lambda}}\Phi_{i-1}^{-T/2} & 0 \end{bmatrix} \Theta_i = \begin{bmatrix} \Phi_i^{1/2} & 0 \\ \mathbf{q}_i^T & e_a(i)\gamma^{1/2}(i) \\ \mathbf{u}_i^T\Phi_i^{-T/2} & \gamma^{1/2}(i) \\ \Phi_i^{\ T/2} & -\mathbf{g}_i\gamma^{\ 1/2}(i) \end{bmatrix}. \tag{21.51}$$

In this case, the last row of the postarray provides the gain vector $\mathbf{g}_i$ that can be used to update the weight-vector estimate as follows:

$$\mathbf{w}_i = \mathbf{w}_{i-1} + \left[\frac{\mathbf{g}_i}{\gamma^{1/2}(i)}\right]\left[e_a(i)\gamma^{1/2}(i)\right].$$

Note, however, that the pre- and postarrays now propagate both $\Phi_i^{1/2}$ and its inverse, which may lead to numerical difficulties.

21.7 Fast Transversal Algorithms

The earlier RLS solutions require $O(M^2)$ floating point operations per iteration, where M is the size of the input vector $\mathbf{u}_i$:

$$\mathbf{u}_i^T = [\, u_1(i) \quad u_2(i) \quad \cdots \quad u_M(i)\,].$$

It often happens in practice that the entries of $\mathbf{u}_i$ are time-shifted versions of each other. More explicitly, if we denote the value of the first entry of $\mathbf{u}_i$ by $u(i)$ (instead of $u_1(i)$), then $\mathbf{u}_i$ will have the form

$$\mathbf{u}_i^T = [\, u(i) \quad u(i-1) \quad \cdots \quad u(i-M+1)\,]. \tag{21.52}$$

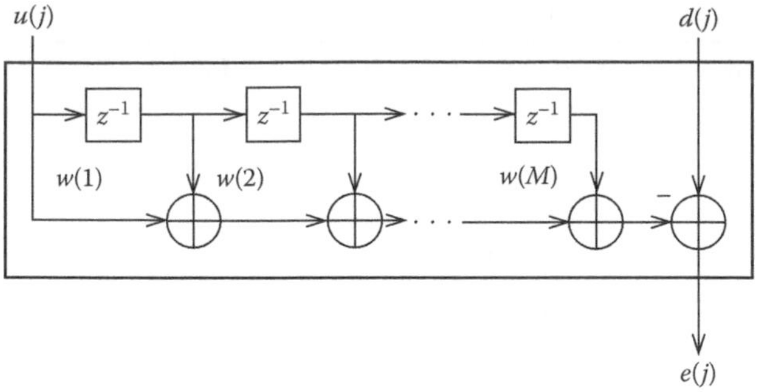

FIGURE 21.3 A linear combiner with shift structure in the input channels.

This has the pictorial representation shown in Figure 21.3. The term z^{-1} represents a unit-time delay. The structure that takes $u(j)$ as an input and provides the inner product $\sum_{k=1}^{M} u(j+1-k)w(k)$ as an output is known as a transversal or FIR (finite-impulse response) filter.

The shift structure in $\mathbf{u}_i$ can be exploited in order to derive fast variants to the RLS solution that would require $O(M)$ operations per iteration rather than $O(M^2)$. This can be achieved by showing that, in this case, the $M \times M$ variables $\mathbf{P}_i$ that are needed in the RLS recursion (Equation 21.39) exhibit certain matrix structure that allows us to replace the RLS recursions by an alternative set of recursions that we now motivate.

21.7.1 Prewindowed Case

We first assume that no input data are available prior to and including time $i = 0$. That is, $u(i) = 0$ for $i \leq 0$. In this case, the values at time 0 of the variables $\{\mathbf{u}_i, \mathbf{g}_i, \gamma(i), \mathbf{P}_i\}$ become

$$\mathbf{u}_0 = 0, \quad \mathbf{g}_0 = 0, \quad \gamma(0) = 1, \quad \text{and} \quad \mathbf{P}_0 = \lambda^{-1}\mathbf{P}_{-1} = \lambda^{-1}\Pi_0.$$

It then follows that the following equality holds:

$$\begin{bmatrix} \mathbf{P}_0 & 0 \\ 0^T & 0 \end{bmatrix} - \begin{bmatrix} 0 & 0^T \\ 0 & \mathbf{P}_{-1} \end{bmatrix} = \begin{bmatrix} \lambda^{-1}\Pi_0 & 0 \\ 0^T & 0 \end{bmatrix} - \begin{bmatrix} 0 & 0^T \\ 0 & \Pi_0 \end{bmatrix}$$

Note that we have embedded $\mathbf{P}_0$ and $\mathbf{P}_{-1}$ into larger matrices (of size $(M+1) \times (M+1)$ each) by adding one zero row and one zero column. This embedding will allow us to suggest a suitable choice for the initial weighting matrix Π_0 in order to enforce a low-rank difference matrix on the right-hand side of the above expression. In so doing, we guarantee that $(\mathbf{P}_0 \oplus 0)$ can be obtained from $(0 \oplus \mathbf{P}_{-1})$ via a low rank update.

Strikingly enough, the argument will further show that because of the shift structure in the input vectors $\mathbf{u}_i$, if this low-rank property holds for the initial time instant then it also holds for the successive time instants! Consequently, the successive matrices $(\mathbf{P}_i \oplus 0)$ will also be low-rank modifications of earlier matrices $(0 \oplus \mathbf{P}_{i-1})$.

In this way, a fast procedure for updating the $\mathbf{P}_i$ can be developed by replacing the propagation of $\mathbf{P}_i$ via Equation 21.39 by a recursion that instead propagates the low-rank factors that generate the $\mathbf{P}_i$. We will verify that this procedure also allows us to update the weight-vector estimates rapidly (in $O(M)$ operations).

21.7.2 Low-Rank Property

Assume we choose Π_0 in the special diagonal form

$$\Pi_0 = \delta \cdot \text{diag}\{\lambda^2, \lambda^3, \ldots, \lambda^{M+1}\}, \tag{21.53}$$

where δ is a positive quantity (usually much larger than one, $\delta \gg 1$). In this case, we are led to a rank-two difference of the form

$$\begin{bmatrix} \lambda^{-1}\Pi_0 & 0 \\ 0^T & 0 \end{bmatrix} - \begin{bmatrix} 0 & 0^T \\ 0 & \Pi_0 \end{bmatrix} = \delta \cdot \lambda \cdot \begin{bmatrix} 1 & & \\ & 0 & \\ & & -\lambda^M \end{bmatrix},$$

which can be factored as

$$\begin{bmatrix} \mathbf{P}_0 & 0 \\ 0^T & 0 \end{bmatrix} - \begin{bmatrix} 0 & 0^T \\ 0 & \mathbf{P}_{-1} \end{bmatrix} = \lambda \cdot \bar{L}_0 \mathbf{S}_0 \bar{L}_0^T, \tag{21.54}$$

where $\bar{L}_0$ is $(M+1) \times 2$ and $\mathbf{S}_0$ is a 2×2 signature matrix that are given by

$$\bar{L}_0 = \sqrt{\delta} \cdot \begin{bmatrix} 1 & 0 \\ 0 & 0 \\ 0 & \lambda^{\frac{M}{2}} \end{bmatrix} \quad \text{and} \quad \mathbf{S}_0 = \begin{bmatrix} 1 & 0 \\ 0 & -1 \end{bmatrix}.$$

21.7.3 Fast Array Algorithm

We now argue by induction, and by using the shift property of the input vectors $\mathbf{u}_i$, that if the low-rank property holds at a certain time instant i, say

$$\begin{bmatrix} \mathbf{P}_i & 0 \\ 0^T & 0 \end{bmatrix} - \begin{bmatrix} 0 & 0^T \\ 0 & \mathbf{P}_{i-1} \end{bmatrix} = \lambda \cdot \bar{L}_i \mathbf{S}_i \bar{L}_i^T, \tag{21.55}$$

then three important facts hold:

- The low-rank property also holds at time $i+1$, say

$$\begin{bmatrix} \mathbf{P}_{i+1} & 0 \\ 0^T & 0 \end{bmatrix} - \begin{bmatrix} 0 & 0^T \\ 0 & \mathbf{P}_i \end{bmatrix} = \lambda \cdot \bar{L}_{i+1} \mathbf{S}_{i+1} \bar{L}_{i+1}^T.$$

- There exists an array algorithm that updates $\bar{L}_i$ to $\bar{L}_{i+1}$. Moreover, the algorithm also provides the gain vector $\mathbf{g}_i$ that is needed to update the weight-vector estimate in the RLS solution.
- The signature matrices $\{\mathbf{S}_i, \mathbf{S}_{i+1}\}$ are equal! That is, all successive low-rank differences have the same signature matrix as the initial difference and, hence,

$$\mathbf{S}_i = \mathbf{S}_0 = \begin{bmatrix} 1 & 0 \\ 0 & -1 \end{bmatrix} \quad \text{for all } i.$$

To verify these claims, consider Equation 21.55 and form the prearray

$$\mathbf{A} = \begin{bmatrix} \gamma^{-1/2}(i) & [u(i+1) \;\; \mathbf{u}_i^T]\bar{L}_i \\ \begin{bmatrix} 0 \\ \mathbf{g}_i \gamma^{-1/2}(i) \end{bmatrix} & \bar{L}_i \end{bmatrix}.$$

For $M = 3$, the prearray has the following generic form (recall that $\bar{L}_i$ is $(M+1) \times 2$):

$$\mathbf{A} = \begin{bmatrix} x & x & x \\ 0 & x & x \\ x & x & x \\ x & x & x \\ x & x & x \end{bmatrix}.$$

Now let Θ_i be a matrix that satisfies

$$\Theta_i \begin{bmatrix} 1 & & \\ & 1 & \\ & & -1 \end{bmatrix} \Theta_i^{\mathrm{T}} = \begin{bmatrix} 1 & & \\ & 1 & \\ & & -1 \end{bmatrix} = \begin{bmatrix} 1 & \\ & \mathbf{S}_i \end{bmatrix},$$

and such that it transforms $\mathbf{A}$ into the form

$$\mathbf{B} = \begin{bmatrix} x & 0 & 0 \\ x & x & x \\ x & x & x \\ x & x & x \\ x & x & x \end{bmatrix} = \begin{bmatrix} a & \mathbf{0}^{\mathrm{T}} \\ \mathbf{b} & \mathbf{C} \end{bmatrix}.$$

That is, Θ_i annihilates two entries in the top row of the prearray. This can be achieved by employing a circular rotation that pivots with the left-most entry of the first row and annihilates its second entry. We then employ a hyperbolic rotation that pivots again with the left-most entry and annihilates the last entry of the top row.

The unknown entries $\{a, \mathbf{b}, \mathbf{C}\}$ can be identified by resorting to the same technique that we employed earlier during the derivation of the QR and inverse QR algorithms. By comparing entries on both sides of the equality

$$\mathbf{A}\begin{bmatrix} 1 & \\ & \mathbf{S}_i \end{bmatrix}\mathbf{A}^{\mathrm{T}} = \begin{bmatrix} a & \mathbf{0}^{\mathrm{T}} \\ \mathbf{b} & \mathbf{C} \end{bmatrix}\begin{bmatrix} 1 & \\ & \mathbf{S}_i \end{bmatrix}\begin{bmatrix} a & \mathbf{0}^{\mathrm{T}} \\ \mathbf{b} & \mathbf{C} \end{bmatrix}^{\mathrm{T}},$$

we obtain several equalities. For example, by equating the (1, 1) entries, we obtain the following relation:

$$\gamma^{-1}(i) + \begin{bmatrix} u(i+1) & \mathbf{u}_i^{\mathrm{T}} \end{bmatrix} \bar{L}_i \mathbf{S}_i \bar{L}_i^{\mathrm{T}} \begin{bmatrix} u(i+1) \\ \mathbf{u}_i \end{bmatrix} = a^2. \tag{21.56}$$

By using Equation 21.55 for $\bar{L}_i\mathbf{S}_i\bar{L}_i$ and by noting that we can rewrite the vector $\begin{bmatrix} u(i+1) & \mathbf{u}_i^{\mathrm{T}} \end{bmatrix}$ in two equivalent forms (due to its shift structure):

$$\begin{bmatrix} u(i+1) & \mathbf{u}_i^{\mathrm{T}} \end{bmatrix} = \begin{bmatrix} \mathbf{u}_{i+1}^{\mathrm{T}} & u(i-M+1) \end{bmatrix}, \tag{21.57}$$

we readily conclude that Equation 21.56 collapses to

$$\gamma^{-1}(i) + \lambda^{-1}\mathbf{u}_{i+1}^{\mathrm{T}}\mathbf{P}_i\mathbf{u}_{i+1} - \lambda^{-1}\mathbf{u}_i^{\mathrm{T}}\mathbf{P}_{i-1}\mathbf{u}_i = a^2.$$

But $\gamma^{-1}(i) = 1 + \lambda^{-1}\mathbf{u}_i^{\mathrm{T}}\mathbf{P}_{i-1}\mathbf{u}_i$. Therefore,

$$a^2 = 1 + \lambda^{-1}\mathbf{u}_{i+1}^{\mathrm{T}}\mathbf{P}_i\mathbf{u}_{i+1} = \gamma^{-1}(i+1),$$

which shows that we can identify a as

$$a = \gamma^{-1/2}(i+1).$$

A similar argument allows us to identify **b**. By comparing the (2, 1) entries we obtain

$$\mathbf{a}\mathbf{b} = \begin{bmatrix} 0 \\ \mathbf{g}_i\gamma^{-1}(i) \end{bmatrix} + \bar{L}_i\mathbf{S}_i\bar{L}_i^{\mathrm{T}} \begin{bmatrix} u(i+1) \\ \mathbf{u}_i \end{bmatrix}. \tag{21.58}$$

Again, by Equation 21.55 for $\bar{L}_i\mathbf{S}_i\bar{L}_i^{\mathrm{T}}$, Equation 21.57 for the vector $\begin{bmatrix} u(i+1) & \mathbf{u}_i^{\mathrm{T}} \end{bmatrix}$, and by noting from the definition of $\mathbf{g}_i$ that

$$\begin{bmatrix} 0 \\ \mathbf{g}_i\gamma^{-1}(i) \end{bmatrix} = \begin{bmatrix} 0 \\ \lambda^{-1}\mathbf{P}_{i-1}\mathbf{u}_i \end{bmatrix},$$

we obtain

$$\mathbf{b} = \begin{bmatrix} \mathbf{g}_{i+1}\gamma^{-1/2}(i+1) \\ 0 \end{bmatrix}.$$

Finally, for the last term **C** we compare the (2, 2) entries to obtain

$$\mathbf{C}\mathbf{S}_i\mathbf{C}^{\mathrm{T}} = \begin{bmatrix} \mathbf{P}_{i+1} & 0 \\ 0^{\mathrm{T}} & 0 \end{bmatrix} - \begin{bmatrix} 0 & 0^{\mathrm{T}} \\ 0 & \mathbf{P}_i \end{bmatrix}.$$

The difference on the right-hand side is by definition $\lambda\bar{L}_{i+1}\mathbf{S}_{i+1}\bar{L}_{i+1}^{\mathrm{T}}$. This shows that we can make the identifications

$$\mathbf{C} = \sqrt{\lambda}\cdot\bar{L}_{i+1} \quad \text{and} \quad \mathbf{S}_{i+1} = \mathbf{S}_i.$$

In summary, we have established the validity of the array algorithm shown in Table 21.5, which minimizes the cost function (Equation 21.27) in the prewindowed case and for the special choice of Π_0 in Equation 21.53.

Note that this fast procedure computes the required gain vectors $\mathbf{g}_i$ without explicitly evaluating the matrices $\mathbf{P}_i$. Instead, the low-rank factors $\bar{L}_i$ are propagated, which explains the lower computational requirements.

TABLE 21.5 Fast Array Algorithm

Input: Prewindowed data $\{d(j), u(j)\}$ for $j \geq 1$ and Π_0 as in Equation 21.53 in the cost (Equation 21.27).

Initialization: Set

$\mathbf{w}_{-1} = \bar{\mathbf{w}}, \quad \gamma^{-1/2}(0) = 1,$

$$\mathrm{L}_o = \sqrt{\delta}.\begin{bmatrix} 1 & 0 \\ 0 & 0 \\ 0 & \lambda^{\frac{M}{2}} \end{bmatrix}, \quad \text{and} \quad \mathrm{S}_0 = \begin{bmatrix} 1 & 0 \\ 0 & -1 \end{bmatrix}.$$

Repeat for each time instant $i \geq 0$:

$$\begin{bmatrix} \gamma^{-1/2}(i) & \begin{bmatrix} u(i+1) & \mathbf{u}_i^{\mathrm{T}} \end{bmatrix}\bar{\mathrm{L}}_i \\ \begin{bmatrix} 0 \\ \mathbf{g}_i\gamma^{-1/2}(i) \end{bmatrix} & \bar{\mathrm{L}}_i \end{bmatrix}\Theta_i = \begin{bmatrix} \gamma^{-1/2}(i+1) & 0^{\mathrm{T}} \\ \begin{bmatrix} \mathbf{g}_{i+1}\gamma^{-1/2}(i+1) \\ 0 \end{bmatrix} & \sqrt{\lambda}\bar{\mathrm{L}}_{i-1} \end{bmatrix},$$

where Θ_i is any $(1 \oplus \mathrm{S}_0)$-orthogonal matrix that produces the zero pattern in the postarray, and $\bar{\mathrm{L}}_i$ is a two-column matrix.

The weight-vector estimate is updated via

$$\mathbf{w}_i = \mathbf{w}_{i-1} + \begin{bmatrix} \dfrac{\mathbf{g}_i}{\gamma^{1/2}(i)} \end{bmatrix}\begin{bmatrix} \gamma^{-1/2}(i) \end{bmatrix}^{-1}\begin{bmatrix} d(i) - \mathbf{u}_i^{\mathrm{T}}\mathbf{w}_{i-1} \end{bmatrix}.$$

The computational cost is $O(M)$ per iteration.

21.7.4 Fast Transversal Filter

The fast algorithm of the last section is an array version of fast RLS algorithms known as FTF and FAEST. In contrast to the above array description, where the transformation Θ_i that updates the data from time i to time $(i+1)$ is left implicit, the FTF and FAEST algorithms involve explicit sets of equations.

The derivation of these explicit sets of equations can be motivated as follows. Note that the factorization in Equation 21.54 is highly nonunique. What is special about Equation 21.54 (and also Equation 21.55) is that we have forced $\mathbf{S}_0$ to be a signature matrix, i.e., a matrix with ± 1s on its diagonal. More generally, we can allow for different factorizations with an $\mathbf{S}_0$ that is not restricted to be a signature matrix. Different choices lead to different sets of equations.

More explicitly, assume we factor the difference matrix in Equation 21.55 as

$$\begin{bmatrix} \mathbf{P}_i & 0 \\ 0^T & 0 \end{bmatrix} - \begin{bmatrix} 0 & 0^T \\ 0 & \mathbf{P}_{i-1} \end{bmatrix} = \lambda \cdot L_i \mathbf{M}_i L_i^T, \tag{21.59}$$

where

L_i is an $(M+1) \times 2$ matrix

$\mathbf{M}_i$ is a 2×2 matrix that is not restricted to be a signature matrix

(We already know from the earlier array-based argument that this difference is always low-rank.)

Given the factorization (Equation 21.59), it is easy to verify that two successive gain vectors satisfy the following relation:

$$\begin{bmatrix} \mathbf{g}_{i+1}\gamma^{-1}(i+1) \\ 0 \end{bmatrix} = \begin{bmatrix} 0 \\ \mathbf{g}_i \gamma^{-1}(i) \end{bmatrix} + L_i \mathbf{M}_i L_i^T \begin{bmatrix} u(i+1) \\ \mathbf{u}_i \end{bmatrix}.$$

This is identical to Equation 21.58 except that $\mathbf{S}_i$ is replaced by $\mathbf{M}_i$ and $\bar{L}_i$ is replaced by L_i. The fast array algorithm of the previous section provides one possibility for enforcing this relation and, hence, of updating $\mathbf{g}_i$ to $\mathbf{g}_{i+1}$ via updates of $\bar{L}_i$.

The FTF and FAEST algorithms follow by employing one such alternative factorization, where the two columns of the factor L_i turn out to be related to the solution of two fundamental problems in adaptive filter theory: the so-called forward and backward prediction problems. Moreover, the $\mathbf{M}_i$ factor turns out to be diagonal with entries equal to the so-called forward and backward minimum prediction energies. An explicit derivation of the FTF equations can be pursued along these lines. We omit the details and continue to focus on the square-root formulation. We now proceed to discuss order-recursive adaptive filters within this framework.

21.8 Order-Recursive Filters

The RLS algorithms that were derived in the previous sections are all fixed-order solutions of Equation 21.27 in the sense that they recursively evaluate successive weight estimates $\mathbf{w}_i$ that correspond to a fixed-order combiner of order M. This form of computing the minimizing solution $\mathbf{w}_N$ is not convenient from an order-recursive point of view. In other words, assume we pose a new optimization problem of the same form as Equation 21.27 but where the vectors $\{\mathbf{w},\mathbf{u}_j\}$ are now of order $(M+1)$ rather than M. How do the weight estimates of this new higher-dimensional problem relate to the weight estimates of the lower dimensional problem?

Before addressing this issue any further, it is apparent at this stage that we need to introduce a notational modification in order to keep track of the proper sizes of the variables. Indeed, from now on, we shall explicitly indicate the size of a variable by employing an additional subscript. For example, we shall write $\{\mathbf{w}_M, \mathbf{u}_{M,j}\}$ instead of $\{\mathbf{w}, \mathbf{u}_j\}$ to denote vectors of size M.

Returning to the point raised in the previous paragraph, let $\mathbf{w}_{M+1,N}$ denote the optimal solution of the new optimization problem (with $(M+1)$ – dimensional vectors $\{\mathbf{w}_{M+1}, \mathbf{u}_{M+1,j}\}$. The adaptive algorithms of the previous sections give an explicit recursive (time-update) relation between $\mathbf{w}_{M,N}$ and $\mathbf{w}_{M,N-1}$. But they do not provide a recursive (order-update) relation between $\mathbf{w}_{M,N}$ and $\mathbf{w}_{M+1,N}$.

There is an alternative to the FIR implementation of Figure 21.3 that allows us to easily carry over the information from previous computations for the order M filter. This is the so-called lattice filter.

From now on we assume, for simplicity of presentation, that the weighting matrix Π_0 in Equation 21.27 is very large, i.e., $\Pi_0 \rightarrow \infty \mathbf{I}$. This assumption reduces Equation 21.27 to a standard least-squares formulation:

$$\min_{\mathbf{w}_M} \left[\sum_{j=0}^{N} \lambda^{N-i} \left| d(j) - \mathbf{u}_{M,j}^{\mathrm{T}} \mathbf{w}_M \right|^2 \right]. \tag{21.60}$$

The order-recursive filters of this section deal with this kind of minimization.

Now suppose that our interest in solving Equation 21.60 is not to explicitly determine the weight estimate $\mathbf{w}_{M,N}$, but rather to determine estimates for the reference signals $\{d(\cdot)\}$, say

$$d_M(N) = \mathbf{u}_{M,N}^{\mathrm{T}} \mathbf{w}_{M,N} = \text{estimate of } d(N) \text{ of order } M.$$

Likewise, for the higher order problem,

$$d_{M+1}(N) - \mathbf{u}_{M+1,N}^{\mathrm{T}} \mathbf{w}_{M+1,N} = \text{estimate of } d(N) \text{ of order } M+1.$$

The resulting estimation errors will be denoted by

$$e_M(N) = d(N) - d_M(N) \quad \text{and} \quad e_{M+1}(N) = d(N) - d_{M+1}(N).$$

The lattice solution allows us to update $e_M(N)$ to $e_{M+1}(N)$ without explicitly computing the weight estimates $\mathbf{w}_{M,N}$ and $\mathbf{w}_{M+1,N}$.

The discussion that follows relies heavily on the orthogonality property of least-squares solutions and, therefore, serves as a good illustration of the power and significance of this property. It will further motivate the introduction of the forward and backward prediction problems.

21.8.1 Joint Process Estimation

For the sake of illustration, and without loss of generality, the discussion in this section assumes particular values for M and λ, say $M = 3$ and $\lambda = 1$. These assumptions simplify the exposition without affecting the general conclusions. In particular, a nonunity λ can always be incorporated into the discussion by properly normalizing the vectors involved in the derivation (cf. Equations 21.28 and 21.29) and we will do so later. We continue to assume prewindowed data (i.e., the data are zero for time instants $i \leq 0$).

To begin with, assume we solve the following problem (as suggested by Equation 21.60): minimize over $\mathbf{w}_3$ the cost function

$$\left\| \underbrace{\begin{bmatrix} 0 \\ d(1) \\ d(2) \\ \vdots \\ d(N) \end{bmatrix}}_{\mathbf{d}_N} - \underbrace{\begin{bmatrix} 0 & 0 & 0 \\ u(1) & 0 & 0 \\ u(2) & u(1) & 0 \\ \vdots & \vdots & \vdots \\ u(N) & u(N-1) & u(N-2) \end{bmatrix}}_{\mathbf{A}_{3,N}} \underbrace{\begin{bmatrix} w_3(1) \\ w_3(2) \\ w_3(3) \end{bmatrix}}_{\mathbf{w}_3} \right\|^2 \tag{21.61}$$

where

$\mathbf{d}_N$ denotes the vector of desired signals up to time N

$\mathbf{A}_{3,N}$ denotes a three-column matrix of input data $\{u(\cdot)\}$, also up to time N

The optimal solution is denoted by $\mathbf{w}_{3,N}$. The subscript N indicates that it is an estimate based on the data $u(\cdot)$ up to time N. Determining $\mathbf{w}_{3,N}$ corresponds to determining the entries of a three-dimensional weight vector so as to approximate the column vector $\mathbf{d}_N$ by the linear combination $\mathbf{A}_{3,N}\mathbf{w}_{3,N}$ in the least-squares sense (Equation 21.61). We thus say that expression in Equation 21.61 defines a third-order estimator for the reference sequence $\{d(\cdot)\}$. The resulting *a posteriori* estimation error vector is denoted by

$$\mathbf{e}_{3,N} = \mathbf{d}_N - \mathbf{A}_{3,N}\mathbf{w}_{3,N},$$

where, for example, the last entry of $\mathbf{e}_{3,N}$ is given by

$$e_3(N) = d(N) - \mathbf{u}_{3,N}^{\mathrm{T}}\mathbf{w}_{3,N},$$

and it denotes the *a posteriori* estimation error in estimating $d(N)$ from a linear combination of the three most recent inputs.

We already know from the orthogonality property of least-squares solutions that the *a posteriori* residual vector $\mathbf{e}_{3,N}$ has to be orthogonal to the data matrix $\mathbf{A}_{3,N}$, viz.,

$$\mathbf{A}_{3,N}^{\mathrm{T}}\mathbf{e}_{3,N} = 0.$$

We also know that the optimal solution $\mathbf{w}_{3,N}$ provides an estimate vector $\mathbf{A}_{3,N}\mathbf{w}_{3,N}$ that is the closest element in the column space of $\mathbf{A}_{3,N}$ to the column vector $\mathbf{d}_N$.

Now assume that we wish to solve the next higher order problem, viz., of order $M = 4$: minimize over $\mathbf{w}_4$ the cost function

$$\|\mathbf{d}_N - \mathbf{A}_{4,N}\mathbf{w}_4\|^2, \tag{21.62}$$

where

$$\mathbf{A}_{4,N} = \begin{bmatrix} 0 & 0 & 0 & 0 \\ u(1) & 0 & 0 & 0 \\ u(2) & u(1) & 0 & 0 \\ \vdots & \vdots & \vdots & \vdots \\ u(N-1) & u(N-2) & u(N-3) & u(N-4) \\ u(N) & u(N-1) & u(N-2) & u(N-3) \end{bmatrix} \quad \text{and} \quad \mathbf{w}_4 = \begin{bmatrix} w_4(1) \\ w_4(2) \\ w_4(3) \\ w_4(4) \end{bmatrix}.$$

This statement is very close to Equation 21.61 except for an extra column in the data matrix $\mathbf{A}_{4,N}$: the first three columns of $\mathbf{A}_{4,N}$ coincide with those of $\mathbf{A}_{3,N}$, while the last column of $\mathbf{A}_{4,N}$ contains the extra new data that are needed for a fourth-order estimator. More specifically, $\mathbf{A}_{3,N}$ and $\mathbf{A}_{4,N}$ are related as follows:

$$\mathbf{A}_{4,N} = \begin{bmatrix} & 0 \\ & 0 \\ & 0 \\ \mathbf{A}_{3,N} & \vdots \\ & u(N-4) \\ & u(N-3) \end{bmatrix}. \tag{21.63}$$

The problem in Equation 21.62 requires us to linearly combine the four columns of $\mathbf{A}_{4,N}$ in order to compute the fourth-order estimates of $\{0, d(1), d(2), \ldots, d(N)\}$. In other words, it requires us to determine the closest element in the column space of $\mathbf{A}_{4,N}$ to the same column vector $\mathbf{d}_N$.

We already know what is the closest element to $\mathbf{d}_N$ in the column space of $\mathbf{A}_{3,N}$, which is a submatrix of $\mathbf{A}_{4,N}$. This suggests that we should try to decompose the column space of $\mathbf{A}_{4,N}$ into two orthogonal subspaces, viz.,

$$\text{Range}(\mathbf{A}_{4,N}) = \text{Range}(\mathbf{A}_{3,N}) \oplus \text{Range}(\mathbf{m}), \tag{21.64}$$

where $\mathbf{m}$ is a column vector that is orthogonal to $\mathbf{A}_{3,N}$, $\mathbf{A}_{3,N}^{T}\mathbf{m} = 0$. The notation $\text{Range}(\mathbf{A}_{3,N}) \oplus \text{Range}(\mathbf{m})$ also means that every element in the column space of $\mathbf{A}_{4,N}$ can be expressed as a linear combination of the columns of $\mathbf{A}_{3,N}$ and of $\mathbf{m}$.

The desired decomposition motivates the backward prediction problem.

21.8.2 Backward Prediction Error Vectors

We continue to assume $\lambda = 1$ and $M = 3$, and we note that the required decomposition can be accomplished by projecting the last column of $\mathbf{A}_{4,N}$ onto the column space of its first three columns (i.e., onto the column space of $\mathbf{A}_{3,N}$) and keeping the residual vector as the desired vector $\mathbf{m}$. This is nothing but a Gram–Schmidt orthogonalization step and it is equivalent to the following minimization problem: minimize over $\mathbf{w}_3^b$

$$\left\| \underbrace{\begin{bmatrix} 0 \\ 0 \\ 0 \\ \vdots \\ u(N-4) \\ u(N-3) \end{bmatrix}}_{\substack{\text{Last column} \\ \text{of } \mathbf{A}_{4,N}}} - \mathbf{A}_{3,N} \underbrace{\begin{bmatrix} w_3^b(1) \\ w_3^b(2) \\ w_3^b(3) \end{bmatrix}}_{\mathbf{w}_3^b} \right\|^2. \tag{21.65}$$

This is also a special case of Equation 21.60 where we have replaced the sequence

$$\{0, d(1), \ldots, d(N)\}$$

by the sequence

$$\{0, 0, 0, \ldots, u(N-4), u(N-3)\}.$$

We denote the optimal solution by $\mathbf{w}^{b}_{3,N}$. The subscript N indicates that it is an estimate based on the data $u(\cdot)$ up to time N. Determining $\mathbf{w}^{b}_{3,N}$ corresponds to determining the entries of a three-dimensional weight vector so as to approximate the last column of $\mathbf{A}_{4,N}$ by a linear combination of the columns of $\mathbf{A}_{3,N}$, viz., $\mathbf{A}_{3,N}\mathbf{w}^{b}_{3,N}$, in the least-squares sense.

Note that the entries in every row of the data matrix $\mathbf{A}_{3,N}$ are the three "future" values corresponding to the entry in the last column of $\mathbf{A}_{4,N}$. Hence, the last element of the above linear combination serves as a backward prediction of $u(N-3)$ in terms of $\{u(N), u(N-1), u(N-2)\}$. A similar remark holds for the other entries. The superscript b stands for backward.

We thus say that the expression of Equation 21.65 defines a third-order backward prediction problem. The resulting *a posteriori* backward prediction error vector is denoted by

$$\mathbf{b}_{3,N} = \begin{bmatrix} 0 \\ 0 \\ 0 \\ \vdots \\ u(N-4) \\ u(N-3) \end{bmatrix} - \mathbf{A}_{3,N}\mathbf{w}^{b}_{3,N}.$$

In particular, the last entry of $\mathbf{b}_{3,N}$ is defined as the *a posteriori* backward prediction error in estimating $u(N-3)$ from a linear combination of the future three inputs. It is denoted by $b_3(N)$ and is given by

$$b_3(N) = u(N-3) - \mathbf{u}^{T}_{3,N}\mathbf{w}^{b}_{3,N}. \tag{21.66}$$

We further know, from the orthogonality property of least-squares solutions, that the *a posteriori* backward residual vector $\mathbf{b}_{3,N}$ has to be orthogonal to the data matrix $\mathbf{A}_{3,N}$, $\mathbf{A}^{T}_{3,N}\mathbf{b}_{3,N} = 0$, which therefore implies that it can be taken as the **m** column that we mentioned earlier, viz., we can write

$$\text{Range}(\mathbf{A}_{4,N}) = \text{Range}(\mathbf{A}_{3,N}) \oplus \text{Range}(\mathbf{b}_{3,N}). \tag{21.67}$$

Our original motivation for introducing the *a posteriori* backward residual vector $\mathbf{b}_{3,N}$ was the desire to solve the fourth-order problem in Equation 21.62, not afresh, but in a way so as to exploit the solution of lower order, thus leading to an order-recursive algorithm.

Assume now that we have available the estimation error vectors $\mathbf{e}_{3,N}$ and $\mathbf{b}_{3,N}$, which are both orthogonal to $\mathbf{A}_{3,N}$. Knowing that $\mathbf{b}_{3,N}$ leads to an orthogonal decomposition of the column space of $\mathbf{A}_{4,N}$ as in Equation 21.67, then updating $\mathbf{e}_{3,N}$ into a fourth-order *a posteriori* residual vector $\mathbf{e}_{4,N}$, which has to be orthogonal to $\mathbf{A}_{4,N}$, simply corresponds to projecting the vector $\mathbf{e}_{3,N}$ onto the vector $\mathbf{b}_{3,N}$. More explicitly, it corresponds to determining a scalar coefficient k_3 that solves the optimization problem

$$\min_{k_3} \|\mathbf{e}_{3,N} - k_3\mathbf{b}_{3,N}\|^2. \tag{21.68}$$

This is a standard least-squares problem and its optimal solution is denoted by

$$k_3(N) = \frac{1}{\mathbf{b}^{T}_{3,N}\mathbf{b}_{3,N}}\mathbf{b}^{T}_{3,N}\mathbf{e}_{3,N}. \tag{21.69}$$

We now know how to update $\mathbf{e}_{3,N}$ into $\mathbf{e}_{4,N}$ by projecting $\mathbf{e}_{3,N}$ onto $\mathbf{b}_{3,N}$. In order to be able to proceed with this order update procedure, we still need to know how to order-update the backward residual vector. That is, we need to know how to go from $\mathbf{b}_{3,N}$ to $\mathbf{b}_{4,N}$.

21.8.3 Forward Prediction Error Vectors

We continue to assume $\lambda = 1$ and $M = 3$. The order-update of the backward residual vector motivates us to introduce the forward prediction problem: minimize over $\mathbf{w}_3^f$ the cost function

$$\left\| \begin{bmatrix} u(1) \\ u(2) \\ u(3) \\ \vdots \\ u(N+1) \end{bmatrix} - \mathbf{A}_{3,N} \underbrace{\begin{bmatrix} w_3^f(1) \\ w_3^f(2) \\ w_3^f(3) \end{bmatrix}}_{\mathbf{w}_3^f} \right\|^2. \tag{21.70}$$

We denote the optimal solution by $\mathbf{w}_{3,N+1}^f$. The subscript indicates that it is an estimate based on the data $u(\cdot)$ up to time $N+1$. Determining $\mathbf{w}_{3,N+1}^f$ corresponds to determining the entries of a three-dimensional weight vector so as to approximate the column vector

$$\begin{bmatrix} u(1) \\ u(2) \\ u(3) \\ \vdots \\ u(N+1) \end{bmatrix}$$

by a linear combination of the columns of $\mathbf{A}_{3,N}$, viz., $\mathbf{A}_{3,N}\mathbf{w}_{3,N+1}^f$.

Note that the entries of the successive rows of the data matrix $\mathbf{A}_{3,N}$ are the past three inputs relative to the corresponding entries of the column vector. Hence, the last element of the linear combination $\mathbf{A}_{3,N}\mathbf{w}_{3,N+1}^f$ serves as a forward prediction of $u(N+1)$ in terms of $\{u(N), u(N-1), u(N-2)\}$. A similar remark holds for the other entries. The superscript f stands for forward.

We thus say that the expression of Equation 21.70 defines a third-order forward prediction problem. The resulting *a posteriori* forward prediction error vector is denoted by

$$\mathbf{f}_{3,N+1} = \begin{bmatrix} u(1) \\ u(2) \\ u(3) \\ \vdots \\ u(N+1) \end{bmatrix} - \mathbf{A}_{3,N}\mathbf{w}_{3,N+1}^f.$$

In particular, the last entry of $\mathbf{f}_{3,N+1}$ is defined as the *a posteriori* forward prediction error in estimating $u(N+1)$ from a linear combination of the past three inputs. It is denoted by $f_3(N+1)$ and is given by

$$f_3(N+1) = u(N+1) - \mathbf{u}_{3,N}\mathbf{w}_{3,N+1}^f. \tag{21.71}$$

Now assume that we wish to solve the next-higher order problem, viz., of order $M = 4$: minimize over $\mathbf{w}_4^{\mathrm{f}}$ the cost function

$$\left\| \begin{bmatrix} u(1) \\ u(2) \\ u(3) \\ \vdots \\ u(N+1) \end{bmatrix} - \mathbf{A}_{4,N} \begin{bmatrix} w_4^{\mathrm{f}}(1) \\ w_4^{\mathrm{f}}(2) \\ w_4^{\mathrm{f}}(3) \\ w_4^{\mathrm{f}}(4) \end{bmatrix} \right\|^2 . \tag{21.72}$$

We again observe that this statement is very close to Equation 21.70 except for an extra column in the data matrix $\mathbf{A}_{4,N}$, in precisely the same way as happened with $\mathbf{e}_{4,N}$ and $\mathbf{b}_{3,N}$. We can therefore obtain $\mathbf{f}_{4,N+1}$ by projecting $\mathbf{f}_{3,N+1}$ onto $\mathbf{b}_{3,N}$ and taking the residual vector as $\mathbf{f}_{4,N+1}$,

$$\min_{k_3^{\mathrm{f}}} \left\| \mathbf{f}_{3,N+1} - k_3^{\mathrm{f}} \mathbf{b}_{3,N} \right\|^2 . \tag{21.73}$$

This is also a standard least-squares problem and we denote its optimal solution by $k_3^{\mathrm{f}}(N+1)$,

$$k_3^{\mathrm{f}}(N+1) = \frac{\mathbf{b}_{3,N}^{\mathrm{T}} \mathbf{f}_{3,N+1}}{\mathbf{b}_{3,N}^{\mathrm{T}} \mathbf{b}_{3,N}}, \tag{21.74}$$

with

$$\mathbf{f}_{4,N+1} = \mathbf{f}_{3,N+1} - k_3^{\mathrm{f}}(N+1) \mathbf{b}_{3,N}. \tag{21.75}$$

Similarly, the backward residual vector $\mathbf{b}_{3,N}$ can be updated to $\mathbf{b}_{4,N+1}$ by projecting $\mathbf{b}_{3,N}$ onto $\mathbf{f}_{3,N+1}$,

$$\min_{k_3^{\mathrm{b}}} \left\| \mathbf{b}_{3,N} - k_3^{\mathrm{b}} \mathbf{f}_{3,N+1} \right\|^2 , \tag{21.76}$$

and we get, after denoting the optimal solution by $k_3^{\mathrm{b}}(N+1)$,

$$\mathbf{b}_{4,N+1} = \mathbf{b}_{3,N} - k_3^{\mathrm{b}}(N+1) \mathbf{f}_{3,N+1}, \tag{21.77}$$

where

$$k_3^{\mathrm{b}}(N+1) = \frac{\mathbf{f}_{3,N+1}^{\mathrm{T}} \mathbf{b}_{3,N}}{\mathbf{f}_{3,N+1}^{\mathrm{T}} \mathbf{f}_{3,N+1}}. \tag{21.78}$$

Note the change in the time index as we move from $\mathbf{b}_{3,N}$ to $\mathbf{b}_{4,N+1}$. This is because $\mathbf{b}_{4,N+1}$ is obtained by projecting $\mathbf{b}_{3,N}$ onto $\mathbf{f}_{3,N+1}$, which corresponds to the following definition for $\mathbf{b}_{4,N+1}$,

$$\mathbf{b}_{4,N+1} = \begin{bmatrix} 0 \\ 0 \\ 0 \\ \vdots \\ u(N-4) \\ u(N-3) \end{bmatrix} - \mathbf{A}_{4,N+1} \underbrace{\begin{bmatrix} w_{4,N+1}^{\mathrm{b}}(1) \\ w_{4,N+1}^{\mathrm{b}}(2) \\ w_{4,N+1}^{\mathrm{b}}(3) \\ w_{4,N+1}^{\mathrm{b}}(4) \end{bmatrix}}_{\mathbf{w}_{4,N+1}^{\mathrm{b}}} .$$

Finally, in view of Equation 21.69, the joint process estimation problem involves a recursion of the form

$$\mathbf{e}_{4,N} = \mathbf{e}_{3,N} - k_3(N)\mathbf{b}_{3,N}, \tag{21.79}$$

where

$$k_3(N) = \frac{\mathbf{b}_{3,N}^{\mathrm{T}}\mathbf{e}_{3,N}}{\mathbf{b}_{3,N}^{\mathrm{T}}\mathbf{b}_{3,N}}. \tag{21.80}$$

21.8.4 A Nonunity Forgetting Factor

For a general filter order M and for a nonunity λ, an extension of the above arguments would show that the prediction vectors can be updated as follows:

$$\begin{aligned}
\mathbf{f}_{M+1,N+1} &= \mathbf{f}_{M,N+1} - k_M^{\mathrm{f}}(N+1)\mathbf{b}_{M,N}, \\
\mathbf{b}_{M+1,N+1} &= \mathbf{b}_{M,N} - k_M^{\mathrm{b}}(N+1)\mathbf{f}_{M,N+1}, \\
\mathbf{e}_{M+1,N} &= \mathbf{e}_{M,N} - k_M(N)\mathbf{b}_{M,N}, \\
k_M^{\mathrm{f}}(N+1) &= \frac{\mathbf{b}_{M,N}^{\mathrm{T}}\Lambda_N\mathbf{f}_{M,N+1}}{\mathbf{b}_{M,N}^{\mathrm{T}}\Lambda_N\mathbf{b}_{M,N}}, \\
k_M^{\mathrm{b}}(N+1) &= \frac{\mathbf{f}_{M,N+1}^{\mathrm{T}}\Lambda_N\mathbf{b}_{M,N}}{\mathbf{f}_{M,N+1}^{\mathrm{T}}\Lambda_N\mathbf{f}_{M,N+1}}, \\
k_M(N) &= \frac{\mathbf{b}_{M,N}^{\mathrm{T}}\Lambda_N\mathbf{e}_{M,N}}{\mathbf{b}_{M,N}^{\mathrm{T}}\Lambda_N\mathbf{b}_{M,N}},
\end{aligned}$$

where $\Lambda_N = \mathrm{diag}\{\lambda^N, \lambda^{N-1}, \ldots, \lambda, 1\}$.

For completeness, we also include the defining relations for the *a priori* and *a posteriori* prediction errors:

$$\begin{aligned}
\beta_M(N) &= u(N-M) - \mathbf{u}_{M,N}^{\mathrm{T}}\mathbf{w}_{M,N-1}^{\mathrm{b}}, \\
b_M(N) &= u(N-M) - \mathbf{u}_{M,N}^{\mathrm{T}}\mathbf{w}_{M,N}^{\mathrm{b}}, \\
\alpha_M(N+1) &= u(N+1) - \mathbf{u}_{M,N}^{\mathrm{T}}\mathbf{w}_{M,N}^{\mathrm{f}}, \\
f_M(N+1) &= u(N+1) - \mathbf{u}_{M,N}^{\mathrm{T}}\mathbf{w}_{M,N+1}^{\mathrm{f}}.
\end{aligned}$$

Using the definition of Equation 21.40 for a conversion factor in a least-squares formulation, it is easy to see that the same factor converts the *a priori* prediction errors to the corresponding *a posteriori* prediction errors. This factor will be denoted by $\gamma_M(N)$.

Table 21.6 summarizes, for ease of reference, the definitions and relations that have been introduced thus far. In particular, the last two lines of the table also provide time-update relations for the minimum costs of the forward and backward prediction problems. These costs are denoted by $\xi_M^{\mathrm{f}}(N+1)$ and $\xi_M^{\mathrm{b}}(N)$ and they are equal to the quantities $\mathbf{f}_{M,N+1}^{\mathrm{T}}\Lambda_N\mathbf{f}_{M,N+1}$ and $\mathbf{b}_{M,N}^{T}\Lambda_N\mathbf{b}_{M,N}$ that appear in the denominators of some of the earlier expressions. The last two relations of Table 21.6 use the result in

TABLE 21.6 Useful Relations for the Prediction Problems

Variable	Definition or Relation
A priori forward error	$\alpha_M(N+1) = u(N+1) - \mathbf{u}_{M,N}^{\mathrm{T}}\mathbf{w}_{M,N-1}^{\mathrm{f}}$
A priori backward error	$\beta_M(N) = u(N-M) - \mathbf{u}_{M,N}^{\mathrm{T}}\mathbf{w}_{M,N-1}^{\mathrm{b}}$
A posteriori forward error	$f_M(N+1) = u(N+1) - \mathbf{u}_{M,N}^{\mathrm{T}}\mathbf{w}_{M,N}^{\mathrm{f}}$
A posteriori backward error	$b_M(N) = u(N-M) - \mathbf{u}_{M,N}^{\mathrm{T}}\mathbf{w}_{M,N}^{\mathrm{b}}$
Forward error by conversion	$f_M(N+1) = \alpha_M(N+1)\gamma_M(N)$
Backward error by conversion	$b_M(N) = \beta_M(N)\gamma_M(N)$
Gain vector	$g_{M,N} = \Phi_{M,N}^{-1}\mathbf{u}_{M,N}$
Conversion factor	$\gamma_M(N) = 1 - \mathbf{u}_{M,N}^{\mathrm{T}}\Phi_{M,N}^{-1}\mathbf{u}_{M,N}$
Minimum forward-prediction error energy	$\xi_M^{\mathrm{f}}(N+1) = \lambda\xi_M^{\mathrm{f}}(N) + \lvert\bar{f}_M(N+1)\rvert^2$
Minimum backward-prediction error energy	$\xi_M^{\mathrm{b}}(N+1) = \lambda\xi_M^{\mathrm{b}}(N) + \lvert\bar{b}_M(N+1)\rvert^2$

Equation 21.43 to express the minimum costs in terms of the so-called angle-normalized prediction errors:

$$\bar{f}_M(N+1) = \alpha_M(N+1)\gamma_M^{1/2}(N), \tag{21.81}$$

$$\bar{b}_M(N) = \beta_M(N)\gamma_M^{1/2}(N). \tag{21.82}$$

We can derive, in different ways, similar update relations for the inner product terms

$$\Delta_M(N+1) = \mathbf{f}_{M,N+1}^{\mathrm{T}}\Lambda_N\mathbf{b}_{M,N},$$
$$\rho_M(N) = \mathbf{b}_{M,N}^{\mathrm{T}}\Lambda_N\mathbf{e}_{M,N}.$$

One possibility is to note, after some algebra and using the orthogonality principle, that the following relation holds:

$$\Delta_M(N+1) = \begin{bmatrix} 1 & -(\mathbf{w}_{M,N}^{\mathrm{f}})^{\mathrm{T}} & 0 \end{bmatrix}\Phi_{M+2,N+1}\begin{bmatrix} 0 \\ -\mathbf{w}_{M,N}^{\mathrm{b}} \\ 1 \end{bmatrix},$$

where

$$\Phi_{M+2,N+1} = \sum_{j=0}^{N+1}\lambda^{N+1-j}\mathbf{u}_{M+2,j}\mathbf{u}_{M+2,j}^{\mathrm{T}}$$

If we now invoke the time-update expression

$$\Phi_{M+2,N+1} = \lambda\Phi_{M+2,N} + \mathbf{u}_{M+2,N+1}^{\mathrm{T}}\mathbf{u}_{M+2,N+1},$$

we conclude that $\Delta_M(N+1)$ satisfies the time-update formula:

$$\begin{aligned}\Delta_M(N+1) &= \lambda\Delta_M(N) + \alpha_M(N+1)b_M(N)\\ &= \lambda\Delta_M(N) + \frac{f_M(N+1)b_M(N)}{\gamma_M(N)}.\end{aligned}$$

A similar argument for $\rho_M(N)$ shows that it satisfies the time-update relation:

$$\rho_M(N) = \lambda\rho_M(N-1) + \frac{e_M(N)b_M(N)}{\gamma_M(N)}.$$

Finally, the orthogonality principle can again be invoked to derive order-update (rather than time-update) relations for $\xi^f_M(N+1)$ and $\xi^b_M(N)$. Indeed, using $\mathbf{f}^T_{M+1,N+1}\Lambda_N\mathbf{b}_{M,N} = 0$, we obtain

$$\begin{aligned}\xi^f_{M+1}(N+1) &= \mathbf{f}^T_{M+1,N+1}\Lambda_N\mathbf{f}_{M+1,N+1} = \mathbf{f}^T_{M+1,N+1}\Lambda_N\mathbf{f}_{M,N+1}\\ &= \xi^f_M(N+1) - \frac{\|\Delta_M(N+1)\|^2}{\xi^b_M(N)}.\end{aligned}$$

Likewise,

$$\xi^b_{M+1}(N+1) = \xi^b_M(N) - \frac{\|\Delta_M(N+1)\|^2}{\xi^f_M(N+1)}.$$

Table 21.7 summarizes the order-update relations derived thus far.

21.8.5 QRD-LSL Filter

There are many variants of adaptive lattice algorithms. In this section we present one such variant in square-root form. Most, if not all, other alternatives can be obtained as special cases. Some alternatives propagate the *a posteriori* prediction errors $\{f_M(N+1), b_M(N)\}$, while others employ the *a priori* prediction errors $\{\alpha_M(N+1), \beta_M(N)\}$. The QRD-LSL algorithm we present here is invariant to the particular choice of *a posteriori* or *a priori* errors because it propagates the angle-normalized prediction errors that we introduced earlier in Equations 21.81 and 21.82, viz.,

$$\begin{aligned}\bar{f}_M(i+1) &= \alpha_M(i+1)\gamma_M^{1/2}(i) = [u(i+1) - \mathbf{u}^T_{M,i}\mathbf{w}^f_{M,i}]\gamma_M^{1/2}(i),\\ \bar{b}_M(i) &= \beta_M(i)\gamma_M^{1/2}(i) = [u(i-M) - \mathbf{u}^T_{M,i}\mathbf{w}^b_{M,i-1}]\gamma_M^{1/2}(i).\end{aligned}$$

TABLE 21.7 Order-Update Relations

$\Delta_M(N+1) = \lambda\Delta_M(N) + \frac{f_M(N+1)b_M(N)}{\gamma_M(N)}$
$\rho_M(N) = \lambda\rho_M(N-1) + \frac{e_M(N)b_M(N)}{\gamma_M(N)}$
$\xi^f_M(N+1) = \lambda\xi^f_M(N) + \frac{\lvert f_M(N+1)\rvert^2}{\gamma_M(N)}$
$\xi^b_M(N) = \lambda\xi^b_M(N-1) + \frac{\lvert b_M(N)\rvert^2}{\gamma_M(N)}$
$k^f_M(N+1) = \Delta_M(N+1)/\xi^b_M(N)$
$k^b_M(N+1) = \Delta_M(N+1)/\xi^f_M(N+1)$
$k_M(N) = \rho_M(N)/\xi^b_M(N)$
$f_{M+1}(N+1) = f_M(N+1) - k^f_M(N+1)b_M(N)$
$b_{M+1}(N+1) = b_M(N) - k^b_M(N+1)f_M(N+1)$
$e_{M+1}(N) = e_M(N) - k_M(N)b_M(N)$
$\xi^f_{M+1}(N+1) = \xi^f_M(N+1) - \frac{\lvert\Delta_M(N+1)\rvert^2}{\xi^b_M(N)}$
$\xi^b_{M+1}(N+1) = \xi^b_M(N) - \frac{\lvert\Delta_M(N+1)\rvert^2}{\xi^f_M(N+1)}$

The QRD-LSL algorithm can be motivated as follows. Assume we form the following two vectors of angle-normalized prediction errors:

$$\bar{\mathbf{f}}_{M,N+1} = \begin{bmatrix} \bar{f}_M(1) \\ \bar{f}_M(2) \\ \vdots \\ \bar{f}_M(N+1) \end{bmatrix} \quad \text{and} \quad \bar{\mathbf{b}}_{M,N} = \begin{bmatrix} \bar{b}_M(0) \\ \bar{b}_M(1) \\ \vdots \\ \bar{b}_M(N) \end{bmatrix}. \tag{21.83}$$

We then conclude from the time-updates in Table 21.6 for $\xi^{f}_{M}(N+1)$ and $\xi^{b}_{M}(N)$ that $\xi^{f}_{M}(N+1)$ and $\xi^{b}_{M}(N)$ are the (weighted) squared Euclidean norms of the angle normalized vectors $\bar{\mathbf{f}}_M(N+1)$ and $\bar{\mathbf{b}}_M(N)$, respectively. That is, $\xi^{f}_{M}(N+1) = \bar{\mathbf{f}}^{T}_{M,N+1}\Lambda_N\bar{\mathbf{f}}_{M,N+1}$ and $\xi^{b}_{M}(N) = \bar{\mathbf{b}}^{T}_{M,N}\Lambda_N\bar{\mathbf{b}}_{M,N}$. Likewise, it follows from the time-update for $\Delta_M(N+1)$ that it is equal to the inner product of the angle normalized vectors:

$$\Delta_M(N+1) = \bar{\mathbf{b}}^{T}_{M,N}\Lambda_N\bar{\mathbf{f}}_{M,N+1}. \tag{21.84}$$

Consequently, the coefficients $k^{f}_{M}(N+1)$ and $k^{b}_{M}(N+1)$ are also equal to the ratios of the inner product of the angle normalized vectors to their energies. But recall that $k^{f}_{M}(N+1)$ is the coefficient we need in order to project $\mathbf{f}_{M,N+1}$ onto $\mathbf{b}_{M,N}$. This means that we can alternatively evaluate the same coefficient by posing the problem of projecting $\bar{\mathbf{f}}_{M,N+1}$ onto $\bar{\mathbf{b}}_{M,N}$. In a similar fashion, $k^{b}_{M}(N+1)$ can be evaluated alternatively by projecting $\bar{\mathbf{b}}_{M,N}$ onto $\bar{\mathbf{f}}_{M,N+1}$. (The inner products and projections are to be understood here to include the additional weighting by Λ_N.)

We are therefore reduced to two simple projection problems that involve projecting a vector onto another vector (with exponential weighting). But these are special cases of standard least-squares problems. In particular, recall that the QR solution of Table 21.4 solves the problem of projecting a given vector $\mathbf{d}_N$ onto the range space of a data matrix $\mathbf{A}_N$ (whose rows are $\mathbf{u}^{T}_{j}$).

In a similar fashion, we can write down the QR solution that would solve the problem of projecting $\bar{\mathbf{f}}_{M,N+1}$ onto $\bar{\mathbf{b}}_{M,N}$. For this purpose, we introduce the scalar variables $q^{f}_{M}(N+1)$ and $q^{b}_{M}(N+1)$ (recall the earlier notation in Equation 21.50):

$$q^{b}_{M}(N+1) = \frac{\Delta_M(N+1)}{\xi^{b/2}_{M}(N)} \quad \text{and} \quad q^{f}_{M}(N+1) = \frac{\Delta_M(N+1)}{\xi^{f/2}_{M}(N+1)}. \tag{21.85}$$

The QR array that updates the forward prediction errors can now be obtained as follows. Form the 3×2 prearray (this is a special case of the QR array of Table 21.4):

$$\mathbf{A} = \begin{bmatrix} \sqrt{\lambda}\xi^{b/2}_{M}(N-1) & \bar{b}_M(N) \\ \sqrt{\lambda}q^{b}_{M}(N) & \bar{f}_M(N+1) \\ 0 & 1 \end{bmatrix}$$

and choose an orthogonal rotation $\Theta^{b}_{M,N}$ that reduces it to the form

$$\mathbf{A}\Theta^{b}_{M,N} = \begin{bmatrix} x & 0 \\ a & b \\ y & c \end{bmatrix}.$$

That is, it annihilates the second entry in the top row of the prearray. The scalar quantities $\{a, b, c, x, y\}$ can be identified, as before, by squaring and comparing entries of the resulting equality. This step allows us to make the following identifications very immediately:

$$\begin{aligned} x &= \xi_M^{b/2}(N), \\ a &= q_M^b(N+1), \\ y &= \bar{b}_M(N)\xi_M^{-b/2}(N), \\ bc &= \gamma_M^{-1/2}(N)f_{M+1}(N+1), \\ b^2 &= |\bar{f}_{M+1}(N+1)|^2, \end{aligned}$$

where for the last equality we used the following relation that follows immediately from the last two lines of Table 21.7:

$$q_M^b(N+1)^2 + \bar{f}_{M+1}(N+1)^2 = \lambda q_M^b(N)^2 + \bar{f}_M(N+1)^2.$$

Therefore, $b^2c^2 = \frac{\gamma_{M+1}(N)}{\gamma_M(N)}|\bar{f}_{M+1}(N+1)|^2$ and we can make the identifications:

$$c = \frac{\gamma_{M+1}^{1/2}(N)}{\gamma_M^{1/2}(N)} \quad \text{and} \quad b = \bar{f}_{M+1}(N+1).$$

A similar argument leads to an array equation for the update of the backward errors. In summary, we obtain the QRD-LSL algorithm (listed in Table 21.8) for the update of the angle-normalized forward and backward prediction errors with prewindowed data that correspond to the minimization problem:

$$\min_{\mathbf{w}_M} \sum_{j=0}^{N} \lambda^{N-j}|d(j) - \mathbf{u}_{M,j}^T \mathbf{w}_M|^2.$$

TABLE 21.8 QRD-LSL Algorithm

Input: Prewindowed data $\{d(j), u(j)\}$ for $j \geq 1$.

Initialization: For each $M = 0, 1, 2, \ldots, M_{\max}$ set

$$\xi_M^{f/2}(0) = 0, \ \xi_M^{b/2}(-1) = 0, \quad \text{and} \quad q_M^b(0) = 0 = q_M^f(0).$$

- For each time instant $N \geq 0$ do

$$\gamma_0(N) = 1, \bar{f}_0(N) = u(N), \quad \text{and} \quad \bar{b}_0(N) = u(N).$$

- For each $M = 0, 1, 2, \ldots, M_{\max} - 1$ do

$$\begin{bmatrix} \sqrt{\lambda \xi^{b/2}(N-1)} & \bar{b}_M(N) \\ \sqrt{\lambda q_M^b(N)} & \bar{f}_M(N+1) \\ 0 & \gamma_M^{1/2}(N) \end{bmatrix} \Theta_{M,N}^b = \begin{bmatrix} \xi_M^{b/2}(N) & 0 \\ q_M^b(N+1) & \bar{f}_{M+1}(N+1) \\ \bar{b}_M(N)\xi_M^{-b/2}(N) & \gamma_{M+1}^{1/2}(N) \end{bmatrix},$$

$$\begin{bmatrix} \sqrt{\lambda \xi_M^{f/2}(N)} & \bar{f}_M(N+1) \\ \sqrt{\lambda q_M(N)} & \bar{b}_M(N) \end{bmatrix} \Theta_{M,N+1}^f = \begin{bmatrix} \xi_M^{f/2}(N+1) & 0 \\ q_M^f(N+1) & \bar{b}_{M+1}(N+1) \end{bmatrix}.$$

The orthogonal matrices $\Theta_{M,N}^b$ and $\Theta_{M,N+1}^f$ are chosen so as to annihilate the (1, 2) entries in the corresponding postarrays.

TABLE 21.9 Array for Joint Process Estimation

Input: Prewindowed data $\{d(j), u(j)\}$ for $j \geq 1$.

Initialization: For each $M = 0, 1, 2, \ldots, M_{max}$ set

$$\xi_M^{b/2}(-1) = 0, \quad q_M^{d}(-1) = 0, \quad \text{and} \quad q_M^{b}(0) = 0$$

- For each time instant $N \geq 0$ do

$$\gamma_0(N) = 1, \quad \bar{e}_0(N) = d(N), \quad \text{and} \quad \bar{b}_0(N) = u(N).$$

- For each $M = 0, 1, 2, \ldots, M_{max} - 1$ do

$$\begin{bmatrix} \sqrt{\lambda \xi_M^{b/2}(N-1)} & \bar{b}_M(N) \\ \sqrt{\lambda q_M^{d}(N-1)} & \bar{e}_M(N) \end{bmatrix} \Theta_{M,N}^{b} = \begin{bmatrix} \xi_M^{b/2}(N) & 0 \\ q_M^{d}(N) & \bar{e}_{M+1}(N) \end{bmatrix},$$

where the orthogonal matrix $\Theta_{M,N}^{b}$ is the same as in the QRD-LSL algorithm.

The recursions of the table can be shown to collapse, by squaring and comparing terms on both sides of the resulting equality, to several lattice forms that are available in the literature. We forgo the details here.

21.8.6 Filtering or Joint Process Array

We now return to the estimation of the sequence $\{d(\cdot)\}$. We argued earlier that if we are given the backward residual vector $\mathbf{b}_{M,N}$ and the estimation residual vector $\mathbf{e}_{M,N}$, then the higher order estimation residual vector $\mathbf{e}_{M+1,N}$ can be obtained by projecting $\mathbf{e}_{M,N}$ onto $\mathbf{b}_{M,N}$ and using the corresponding residual vector as $\mathbf{e}_{M+1,N}$.

Arguments similar to what we have done in the previous section will readily show that the array for the joint process estimation problem is the following: define the angle-normalized residual

$$\bar{e}_M(i) = e_M(i)\gamma_M^{-1/2}(i) = [d(i) - \mathbf{u}_{M,i}^{T}\mathbf{w}_{M,i}]\gamma_M^{-1/2}(i),$$

as well as the scalar quantity

$$q_M^{d}(N) = \frac{\rho_M(N)}{\xi_M^{b/2}(N)}.$$

Then the array for the filtering process is what is shown in Table 21.9. Note that it uses precisely the same rotation as the first array in the QRD-LSL algorithm. Hence, the second line in the above array can be included as one more line in the first array of QRD-LSL, thus completing the algorithm to also include the joint-process estimation part.

21.9 Concluding Remarks

The intent of this chapter was to provide an overview of the fundamentals of RLS estimation, with emphasis on array formulations of the varied algorithms (slow or fast) that are available for this purpose. More details and related discussion can be found in several of the references indicated in this section. The references are not intended to be complete but rather indicative of the work in the different areas. More complete lists can be found in several of the textbooks mentioned herein.

Detailed discussions on the different forms of RLS adaptive algorithms and their potential applications can be found in [1–7]. The array formulation that we emphasized in this chapter is motivated by the state-space approach developed in [8]. This reference also clarifies the connections between adaptive RLS filtering and Kalman filter theory and treats other forms of lattice filters.

A detailed discussion of the square-root formulation in the context of Kalman filtering can be found in [9]. Further motivation, and earlier discussion, on lattice algorithms can be found in several places in the literature [10–12]. The fast fixed-order RLS algorithms (FTF and FAEST) were independently derived in [13,14]. These algorithms, however, suffer from numerical instability problems. Some variables that are supposed to remain positive or bounded by one may lose this property due to roundoff errors. A treatment of these issues appears in [15].

More discussion on the QRD-LSL filter, including alternative derivations that are based on the QR decomposition of certain data matrices, can be found in [16–19]. More discussion and examples of elementary and square-root-free rotations and householder transformations can be found in [20–23]. Fast fixed-order adaptive algorithms that consider different choices of the initial weighting matrix Π_0, and also the case of data that is not necessarily prewindowed, can be found in [24]. Gauss' original exposition of the least-squares criterion can be found in [25].

References

1. Sayed, A.H., *Adaptive Filters*, Wiley, NJ, 2008.
2. Sayed, A.H., *Fundamentals of Adaptive Filtering*, Willey, NJ, 2003.
3. Haykin, S., *Adaptive Filter Theory*, 3rd ed., Prentice-Hall, Englewood Cliffs, NJ, 1996.
4. Proakis, J.G., Rader, C.M., Ling, F., and Nikias, C.L., *Advanced Digital Signal Processing*, Macmillan, New York, 1992.
5. Honig, M.L. and Messerschmitt, D.G., *Adaptive Filters—Structures, Algorithms and Applications*, Kluwer Academic Publishers, Boston, MA, 1984.
6. Orfanidis, S.J., *Optimum Signal Processing*, 2nd ed., McGraw-Hill, New York, 1988.
7. Kalouptsidis, N. and Theodoridis, S., *Adaptive System Identification and Signal Processing Algorithms*, Prentice-Hall, Englewood Cliffs, NJ, 1993.
8. Sayed, A.H. and Kailath, T., A state-space approach to adaptive RLS filtering, *IEEE Signal Processing Magazine*, 11(3): 18–60, July 1994.
9. Morf, M. and Kailath, T. Square root algorithms for least squares estimation, *IEEE Transactions on Automatic Control*, AC-20(4): 487–497, Aug. 1975.
10. Lee, D.T.L., Morf, M., and Friedlander, B., Recursive least-squares ladder estimation algorithms, *IEEE Transactions on Circuits and Systems*, CAS-28(6): 467–481, June 1981.
11. Friedlander, B., Lattice filters for adaptive processing, *Proceedings of the IEEE*, 70(8): 829–867, Aug. 1982.
12. Lev-Ari, H., Kailath, T., and Cioffi, J., Least squares adaptive lattice and transversal filters: A unified geometrical theory, *IEEE Transactions on Information Theory*, IT-30(2): 222–236, Mar. 1984.
13. Carayannis, G., Manolakis, D., and Kalouptsidis, N., A fast sequential algorithm for least squares filtering and prediction, *IEEE Transactions on Acoustics, Speech and Signal Processing*, ASSP-31(6): 1394–1402, Dec. 1983.
14. Cioffi, J. and Kailath, T., Fast recursive-least-squares transversal filters for adaptive filtering, *IEEE Transactions on Acoustics, Speech and Signal Processing*, ASSP-32: 304–337, Apr. 1984.
15. Slock, D.T.M. and Kailath, T., Numerically stable fast transversal filters for recursive least squares adaptive filtering, *IEEE Transactions on Signal Processing*, SP-39(1): 92–114, Jan. 1991.
16. Cioffi, J., The fast adaptive rotor's RLS algorithm, *IEEE Trans. Acoustics, Speech and Signal Processing*, ASSP-38: 631–653, 1990.
17. Proudler, I.K., McWhirter, J.G., and Shepherd, T.J., Computationally efficient QR decomposition approach to least squares adaptive filtering, *IEEE Proceedings*, 138(4): 341–353, Aug. 1991.
18. Regalia, P.A. and Bellanger, M.G., On the duality between fast QR methods and lattice methods in least squares adaptive filtering, *IEEE Transactions on Signal Processing*, 39(4): 879–891, Apr. 1991.

19. Yang, B. and Böhme, J.F., Rotation-based RLS algorithms: Unified derivations, numerical properties, and parallel implementations, *IEEE Transactions on Signal Processing*, SP-40(5): 1151–1167, May 1992.
20. Golub, G.B. and Van Loan, C.F., *Matrix Computations*, 2nd ed., The Johns Hopkins University Press, Baltimore, MD, 1989.
21. Rader, C.M. and Steinhardt, A.O., Hyperbolic householder transformations, *IEEE Transactions on Acoustics, Speech and Signal Processing*, ASSP-34(6): 1589–1602, Dec. 1986.
22. Bojanczyk, A.W. and Steinhardt, A.O., Stabilized hyperbolic householder transformations, *IEEE Transactions on Acoustics, Speech and Signal Processing*, ASSP-37(8): 1286–1288, Aug. 1989.
23. Hsieh, S.F., Liu, K.J.R., and Yao, K., A unified square-root-free approach for QRD-based recursive least-squares estimation, *IEEE Transactions on Signal Processing*, SP-41(3): 1405–1409, Mar. 1993.
24. Houacine, A., Regularized fast recursive least squares algorithms for adaptive filtering, *IEEE Transactions on Signal Processing*, SP-39(4): 860–870, Apr. 1991.
25. Gauss, C.F., *Theory of the Motion of Heavenly Bodies*, Dover, New York, 1963 (English translation of *Theoria Motus Corporum Coelestium*, 1809).

22

Transform Domain Adaptive Filtering

W. Kenneth Jenkins
The Pennsylvania State University

C. Radhakrishnan
The Pennsylvania State University

Daniel F. Marshall
Raytheon Company

One of the earliest works on transform domain adaptive filtering (TDAF) was published in 1978 by Dentino et al. [1], in which the concept of adaptive filtering in the frequency domain was proposed. Many publications have since appeared that further develop the theory and expand the current understanding of the performance characteristics for this class of adaptive filters. In addition to the discrete Fourier transform (DFT), other orthogonal transforms such as the discrete cosine transform (DCT) and the Walsh Hadamard transform (WHT) can also be used effectively as a means to improve the LMS algorithm without adding too much computational complexity. For this reason, the general term transform domain adaptive filtering is used in the following discussion to mean that the input signal is preprocessed by decomposing the input vector into orthogonal components, which are in turn used as inputs to a parallel bank of simpler adaptive subfilters. With an orthogonal transformation, the adaptation takes place in the transform domain, as it is possible to show that the adjustable parameters are indeed related to an equivalent set of time domain filter coefficients by means of the same transformation that is used for the real-time processing [2–5].

A direct form finite impulse response (FIR) digital filter structure is shown in Figure 22.1. The direct form requires $N-1$ delays, N multiplications, and $N-1$ additions for each output sample that is produced. The amount of hardware (as well as power) required to implement the direct form structure depends on the degree of hardware multiplexing that can be utilized within the speed demands of the application. A fully parallel implementation consisting of N delay registers, N multipliers, and a tree of two-input adders would be needed for very high-frequency applications. At the opposite end of the performance spectrum, a sequential implementation consisting of a length N delay line and a single time multiplexed multiplier and accumulation adder would provide the cheapest (and slowest) implementation. This latter structure would be characteristic of a filter that is implemented in software on one of the many commercially available DSP chips [6–9].

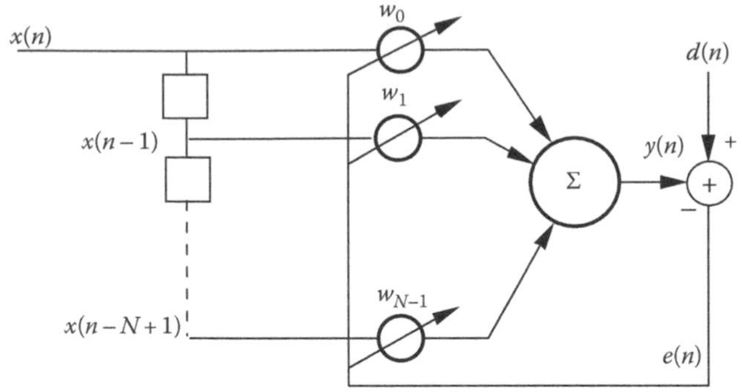

FIGURE 22.1 Direct-form adaptive filter structure.

Regardless of the hardware complexity that results from a particular implementation, the computational complexity of the filter is determined by the requirements of the algorithm and, as such, remains invariant with respect to different hardware structures. In particular, the computational complexity of the direct form FIR filter is $O[N]$, since N multiplications and $(N-1)$ additions must be performed at each iteration. When designing an adaptive filter it is reasonable to seek an adaptive algorithm whose order of complexity is no greater than the order of complexity of the basic filter structure itself. This goal is achieved by the LMS algorithm, which is the major contributing factor to the enormous success of that algorithm. Extending this principle for two-dimensional (2-D) adaptive filters implies that desirable 2-D adaptive algorithms have an order of complexity of $O[N^2]$, since a 2-D FIR direct form filter has $O[N^2]$ complexity inherent in its basic structure [10,11].

The transform domain adaptive filter (TDAF) is a generalization of the LMS FIR structure, in which a linear transformation is performed on the input signal and each transformed "channel" is power normalized to improve the convergence rate of the adaptation process. The linear transform is characterized throughout the following discussions as a sliding window operator that consists of a transformation matrix multiplying an input vector [4]. At each iteration the input vector includes one new input sample $x(n)$, and $N-1$ past input samples $x(n-k), k=1,\ldots,N-1$. As the window slides forward sample by sample, filtered outputs are produced continuously at each value of the index n.

Since the input transformation is represented by a matrix-vector product, it might appear that the computational complexity of the transform domain filter is at least $O[N^2]$. However, many transformations can be implemented with fast algorithms that have complexities less than $O[N^2]$. For example, the DFT can be implemented by the FFT algorithm, resulting in a complexity of $O[N \log_2 N]$ per iteration. Some transformations can be implemented recursively in a bank of parallel filters, resulting in a net complexity of $O[N]$ per iteration. The main point to be made here is that the complexity of the transform domain filter typically falls between $O[N]$ and $O[N^2]$, with the actual complexity depending on the specific algorithm that is used to compute the sliding window transform operator [3].

22.1 LMS Adaptive Filter Theory

The LMS algorithm is derived as an approximation to the steepest descent optimization strategy. The fact that the field of adaptive signal processing is based on elementary principles from optimization theory suggests that more advanced adaptive algorithms can be developed by incorporating other results from the field of optimization. This point of view recurs throughout this discussion, as concepts are borrowed from the field of optimization and modified for adaptive filtering as appropriate. In particular, one of the borrowed ideas that appears later is the quasi-Newton optimization strategy. It will be shown that TDAF algorithms are closely related to quasi-Newton algorithms, but have computational complexity that is closer to the simple requirements of the LMS algorithm.

For a length N FIR filter with the input expressed as a column vector

$$\mathbf{x}(n) = [x(n), x(n-1), \ldots, x(n-N+1)]^{\mathrm{T}},$$

the filter output $y(n)$ is easily expressed as

$$y(n) = \mathbf{w}^{\mathrm{T}}(n)\mathbf{x}(n), \tag{22.1}$$

where $\mathbf{w}(n) = [w_0(n), w_1(n), \ldots, w_{N-1}(n)]^{\mathrm{T}}$ is the time-varying vector of filter coefficients (tap weights) the superscript T denotes vector transpose.

The output error is formed as the difference between the filter output and a training signal $d(n)$, i.e., $e(n) = d(n) - y(n)$. Strategies for obtaining an appropriate $d(n)$ vary from one application to another. In many cases, the availability of a suitable training signal determines whether an adaptive filtering solution will be successful in a particular application. The ideal cost function is defined by the mean-squared error (MSE) criterion, $E\left[|e(n)|^2\right]$. The LMS algorithm is derived by approximating the ideal cost function by the instantaneous squared error, resulting in $J_{\mathrm{LMS}}(n) = |e(n)|^2$. While the LMS seems to make a rather crude approximation at the very beginning, the approximation results in an unbiased estimator. In many applications, the LMS algorithm is quite robust and is able to converge rapidly to a small neighborhood of the optimum Wiener solution [6].

The steepest descent optimization strategy is given by

$$\mathbf{w}(n+1) = \mathbf{w}(n) - \mu\nabla_{\mathrm{E}[|e^2|]}(n), \tag{22.2}$$

where $\nabla_{\mathrm{E}[|e^2|]}(n)$ is the gradient of the cost function with respect to the coefficient vector $\mathbf{w}(n)$. When the gradient is formed using the LMS cost function $J_{\mathrm{LMS}}(n) = |e(n)|^2$, the conventional LMS results:

$$\begin{aligned} \mathbf{w}(n+1) &= \mathbf{w}(n) + \mu e(n)\mathbf{x}(n), \\ e(n) &= d(n) - y(n), \quad \text{and} \\ y(n) &= \mathbf{x}(n)^{\mathrm{T}}\mathbf{w}(n). \end{aligned} \tag{22.3}$$

(Note: Many sources include a "2" before the μ factor in Equation 22.3 because this factor arises during the derivation of Equation 22.3 from Equation 22.2. In this discussion, we assume this factor is absorbed into μ, and so it will not appear explicitly.) Since the LMS algorithm is treated in considerable detail in other sections of this book, we will not present any further derivation or analysis of it here. However, the following observations will be useful when other algorithms are compared to the LMS as a baseline design [6–9].

1. Assume that all of the signals and filter variables are real-valued. The filter itself requires N multiplications and $N-1$ additions to produce $y(n)$ at each value of n. The coefficient update algorithm requires $2N$ multiplications and N additions, resulting in a total computational burden of $3N$ multiplications and $2N-1$ additions per iteration. Since N is generally much larger than the factor of three, the order of complexity of the LMS algorithm is $O[N]$.
2. The cost function given for the LMS algorithm is a simplified form of the one used for the RLS algorithm. This implies that the LMS algorithm is a simplified version of the RLS algorithm, where averages are replaced by single instantaneous terms.
3. The (power normalized) LMS algorithm is also a simplified form of the TDAF which results by setting the transform matrix equal to the identity matrix.

4. The LMS algorithm is also a simplified form of the Gauss–Newton optimization strategy which introduces second-order statistics (the input autocorrelation function) to accelerate the rate of convergence. In order to obtain the LMS algorithm from the Gauss–Newton algorithm, two approximations must be made: (i) The gradient must be approximated by the instantaneous error squared, and (ii) the inverse of the input autocorrelation matrix must be crudely approximated by the identity matrix.

These observations suggest that many of the seemingly distinct adaptive filtering algorithms that appear scattered about in the literature are indeed closely related, and can be considered to be members of a family whose hereditary characteristics have their origins in Gauss–Newton optimization theory [12,13]. The different members of this family inherit their individual characteristics from approximations that are made on the pure Gauss–Newton algorithm at various stages of their derivations. However, after the individual derivations are complete and each algorithm is packaged in its own algorithmic form, the algorithms look considerably different from one another. Unless a conscious effort is made to reveal their commonality, the fact that they have evolved from common roots may be entirely obscured.

The convergence behavior of the LMS algorithm, as applied to a direct form FIR filter structure, is controlled by the autocorrelation matrix $\mathbf{R}_x$ of the input process, where

$$\mathbf{R}_x \equiv E[\mathbf{x}^*(n)\mathbf{x}^T(n)]. \tag{22.4}$$

(The * in Equation 22.4 denotes complex conjugate to account for the general case of complex input signals, although throughout most of the following discussions it will be assumed that $x(n)$ and $d(n)$ are both real-valued signals.) The autocorrelation matrix $\mathbf{R}_x$ is usually positive definite, which is one of the conditions necessary to guarantee convergence to the Wiener solution. Another necessary condition for convergence is $0 < \mu < 1/\lambda_{max}$, where λ_{max} is the largest eigenvalue of $\mathbf{R}_x$. It is also well established that the convergence of this algorithm is directly related to the eigenvalue spread of $\mathbf{R}_x$. The eigenvalue spread is measured by the condition number of $\mathbf{R}_x$, defined as $\kappa = \lambda_{max}/\lambda_{min}$, where λ_{min} is the minimum eigenvalue of $\mathbf{R}_x$. Ideal conditioning occurs when $\kappa = 1$ (white noise); as this ratio increases, slower convergence results. The eigenvalue spread (condition number) depends on the spectral distribution of the input signal and can be shown to be related to the maximum and minimum values of the input power spectrum. From this line of reasoning it becomes clear that white noise is the ideal input signal for rapidly training an LMS adaptive filter. The adaptive process becomes slower and requires more computation for input signals that are more severely colored [7].

Convergence properties are reflected in the geometry of the MSE surface, which is simply the mean-squared output error $E[|e(n)|^2]$ expressed as a function of the N adaptive filter coefficients in $(N+1)$-space. An expression for the error surface of the direct form filter is

$$J(\mathbf{z}) \equiv E\left[|e(n)|^2\right] = J_{min} + \mathbf{z}^{*T}\mathbf{R}_x\mathbf{z}, \tag{22.5}$$

with $\mathbf{R}_x$ defined in Equation 22.4 and $\mathbf{z} \equiv \mathbf{w} - \mathbf{w}_{opt}$, where $\mathbf{w}_{opt}$ is the vector of optimum filter coefficients in the sense of minimizing the MSE ($\mathbf{w}_{opt}$ is known as the Wiener solution). An example of an error surface for a simple 2-tap filter is shown in Figure 22.2. In this example, $x(n)$ was specified to be a colored noise input signal with an autocorrelation matrix

$$\mathbf{R}_x = \begin{bmatrix} 1.0 & 0.9 \\ 0.9 & 1.0 \end{bmatrix}.$$

Figure 22.2 shows three equal-error contours on the 3-D surface. The term $\mathbf{z}^{*T}\mathbf{R}_x\mathbf{z}$ in Equation 22.2 is a quadratic form that describes the bowl shape of the FIR error surface. When $\mathbf{R}_x$ is positive definite, the equal-error contours of the surface are hyperellipses (N-dimensional ellipses) centered at the origin of

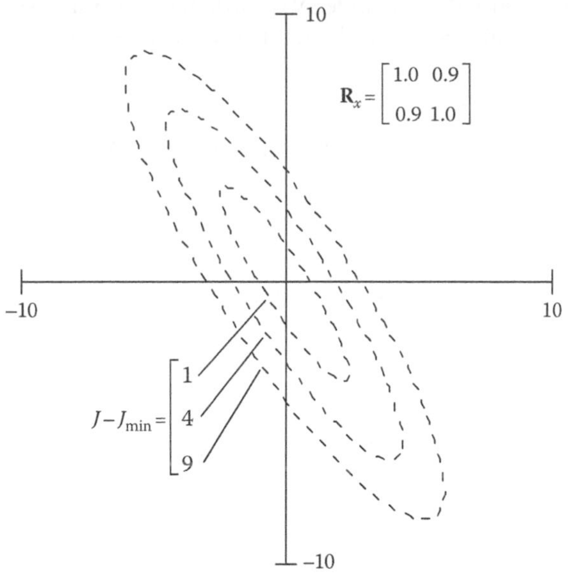

FIGURE 22.2 Example of an error surface for a simple 2-tap filter.

the coefficient parameter space. Furthermore, the principle axes of these hyperellipses are the eigenvectors of $\mathbf{R}_x$, and their lengths are proportional to the eigenvalues of $\mathbf{R}_x$. Since the convergence rate of the LMS algorithm is inversely related to the ratio of the maximum to the minimum eigenvalues of $\mathbf{R}_x$, large eccentricity of the equal-error contours implies slow convergence of the adaptive system. In the case of an ideal white noise input, $\mathbf{R}_x$ has a single eigenvalue of multiplicity N, so that the equal-error contours are hyperspheres [14].

22.2 Orthogonalization and Power Normalization

The TDAF structure is shown in Figure 22.3. The input $x(n)$ and desired signal $d(n)$ are assumed to be zero mean and jointly stationary. The input to the filter is a vector of N current and past input samples, defined in the previous section and denoted as $\mathbf{x}(n)$. This vector is processed by a unitary

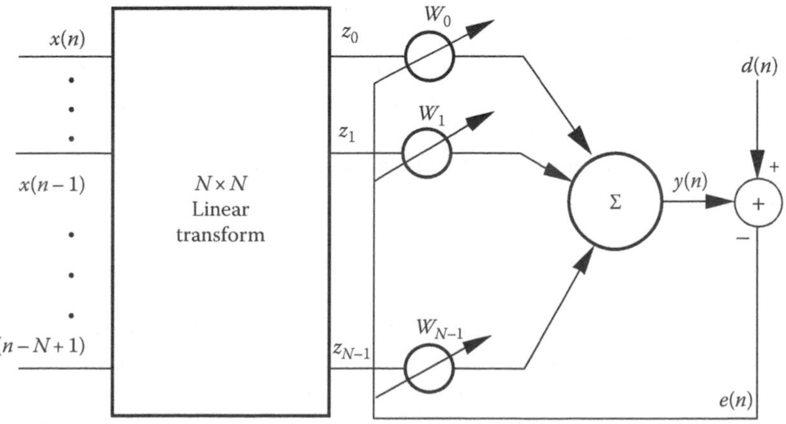

FIGURE 22.3 TDAF structure.

transform, such as the DFT. Once the filter order N is fixed, the transform is simply an $N \times N$ matrix $\mathbf{T}$, which is in general complex, with orthonormal rows. The transformed outputs form a vector $\mathbf{v}(n)$ that is given by

$$\mathbf{v}(n) = [v_0(n), v_1(n), \ldots, v_{N-1}(n)]^{\mathrm{T}} = \mathbf{T}\mathbf{x}(n). \tag{22.6}$$

With an adaptive tap vector defined as

$$\mathbf{W}(n) = [W_0(n), W_1(n), \ldots, W_{N-1}(n)]^{*\mathrm{T}}, \tag{22.7}$$

the filter output is given by

$$y(n) = \mathbf{W}^{\mathrm{T}}(n)\mathbf{v}(n) = \mathbf{W}^{\mathrm{T}}(n)\mathbf{T}\mathbf{x}(n). \tag{22.8}$$

The instantaneous output error

$$e(n) = d(n) - y(n) \tag{22.9}$$

is then used to update the adaptive filter taps using a modified form of the LMS algorithm [11]:

$$\begin{aligned} \mathbf{W}(n+1) &= \mathbf{W}(n) + \mu e(n)\Lambda^{-2}\mathbf{v}^*(n) \\ \Lambda^2 &\equiv \operatorname{diag}\left[\sigma_0^2, \sigma_1^2, \ldots, \sigma_{N-1}^2\right], \end{aligned} \tag{22.10}$$

where $\sigma_i^2 = E\left[|v_i(n)|^2\right]$.

As before, the superscript asterisk in Equation 22.10 indicates complex conjugation to account for the most general case in which the transform is complex. Also, the use of the uppercase coefficient vector in Equation 22.10 denotes that $\mathbf{W}(n)$ is a transform domain variable. The power estimates σ_i^2 can be developed online by computing an exponentially weighted average of past samples according to

$$\sigma_i^2(n) = \alpha\sigma_i^2(n-1) + |v_i(n)|^2, \quad 0 < \alpha < 1. \tag{22.11}$$

If σ_i^2 becomes too small due to an insufficient amount of energy in the ith channel, the update mechanism becomes ill-conditioned due to a very large effective step size. In some cases the process will become unstable and register overflow will cause the adaptation to catastrophically fail. So the algorithm given by Equation 22.10 should have the update mechanism disabled for the ith orthogonal channel, if σ_i^2 falls below a critical threshold.

Alternatively, the transform domain algorithm may be stabilized by adding small positive constants ε to the diagonal elements of Λ^2, resulting in

$$\hat{\Lambda}^2 = \Lambda^2 + \varepsilon\mathbf{I}. \tag{22.12}$$

Then $\hat{\Lambda}^2$ is used in the place of Λ^2 in Equation 22.10. For most input signals $\sigma_i^2 \gg \varepsilon$, the inclusion of the stabilization factors is transparent to the performance of the algorithm. However, whenever $\sigma_i^2 \approx \varepsilon$, the stabilization terms begin to have a significant effect. Within this operating region, the power in the channels will not be uniformly normalized and the convergence rate of the filter will begin to degrade but catastrophic failure will be avoided.

The motivation for using the TDAF adaptive system instead of a simpler LMS- based system is to achieve rapid convergence of the filter's coefficients when the input signal is not white, while maintaining a reasonably low computational complexity requirement. In the following section, this convergence rate improvement of the TDAF will be explained geometrically.

22.3 Convergence of the Transform Domain Adaptive Filter

In this section, the convergence rate improvement of the TDAF is described in terms of the MSE surface. From Equations 22.4 and 22.6, it is found that $\mathbf{R}_v = \mathbf{T}^*\mathbf{R}_x\mathbf{T}^{\mathrm{T}}$, so that for the transform structure without power normalization Equation 22.5 becomes

$$J(\mathbf{z}) = E\left[|e(n)|^2\right] = J_{\min} + \mathbf{z}^{*\mathrm{T}}[\mathbf{T}^*\mathbf{R}_x\mathbf{T}^{\mathrm{T}}]\mathbf{z}. \tag{22.13}$$

The difference between Equations 22.5 and 22.13 is the presence of $\mathbf{T}$ in the quadratic term of Equation 22.13. When $\mathbf{T}$ is a unitary matrix, its presence in Equation 22.13 gives a rotation and/or a reflection of the surface. The eccentricity of the surface is unaffected by the transform, so the convergence rate of the system is unchanged by the transformation alone.

However, the signal power levels at the adaptive coefficients are changed by the transformation. Consider the intersection of the equal-error contours with the rotated axes: letting $\mathbf{z} = [0 \cdots z_i \cdots 0]^{\mathrm{T}}$, with z_i in the ith position, Equation 22.13 becomes

$$J(\mathbf{z}) - J_{\min} = \left[\mathbf{T}^*\mathbf{R}_x\mathbf{T}^{\mathrm{T}}\right]_i z_i^2 \approx \sigma_i^2 z_i^2. \tag{22.14}$$

If the equal-error contours are hyperspheres (the ideal case), then for a fixed value of the error $J(n)$, Equation 22.14 must give $|z_i| = |z_j|$ for all i and j, since all points on a hypersphere are equidistant from the origin. When the filter input is not white, this will not hold in general. But, since the power levels σ_i^2 are easily estimated, the rotated axes can be scaled to have this property. Let $\Lambda^{-1}\hat{\mathbf{z}} = \mathbf{z}$, where Λ is defined in Equation 22.10. Then the error surface of the TDAF, with transform $\mathbf{T}$ and including power normalization, is given by

$$J(\hat{\mathbf{z}}) = J_{\min} + \hat{\mathbf{z}}^{*\mathrm{T}}\left[\Lambda^{-1}\mathbf{T}^*\mathbf{R}_x\mathbf{T}^{\mathrm{T}}\Lambda^{-1}\right]\hat{\mathbf{z}}. \tag{22.15}$$

The main diagonal entries of $\Lambda^{-1}\mathbf{T}^*\mathbf{R}_x\mathbf{T}^{\mathrm{T}}\Lambda^{-1}$ are all equal to one, so Equation 22.14 becomes $J(\mathbf{z}) - J_{\min} = \hat{z}_i^2$, which has the property described above.

Thus, the action of the TDAF system is to rotate the axes of the filter coefficient space using a unitary rotation matrix $\mathbf{T}$, and then to scale these axes so that the error surface contours become approximately hyperspherical at the points where they can be easily observed, i.e., the points of intersection with the new (rotated) axes. Usually this scaling reduces the eccentricity of the error surface contours and results in faster convergence.

Transform domain processing can now be added to the previous example, as illustrated in Figures 22.4 and 22.5. The error surface shown in Figure 22.4 was created by using the following (arbitrary) transform:

$$\mathbf{T} = \begin{bmatrix} 0.866 & 0.500 \\ 0.500 & 0.866 \end{bmatrix},$$

on the error surface shown in Figure 22.2, which produces clockwise rotation of the ellipsoidal contours so that the major and minor axes more closely align with the coordinate axes than they did without

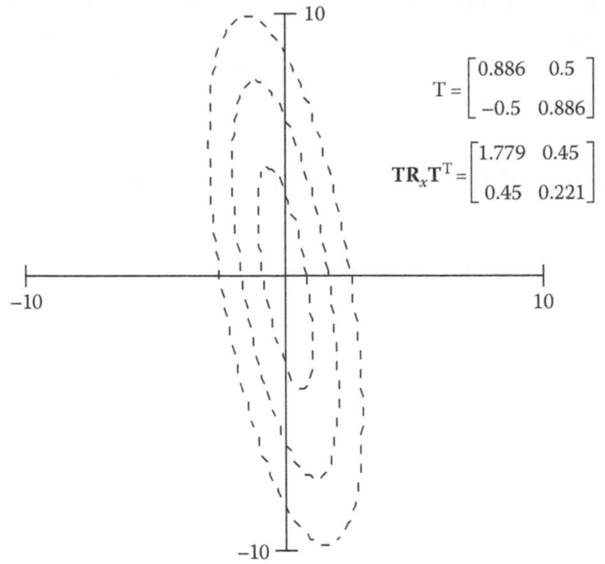

FIGURE 22.4 Error surface for the TDAF with transform **T**.

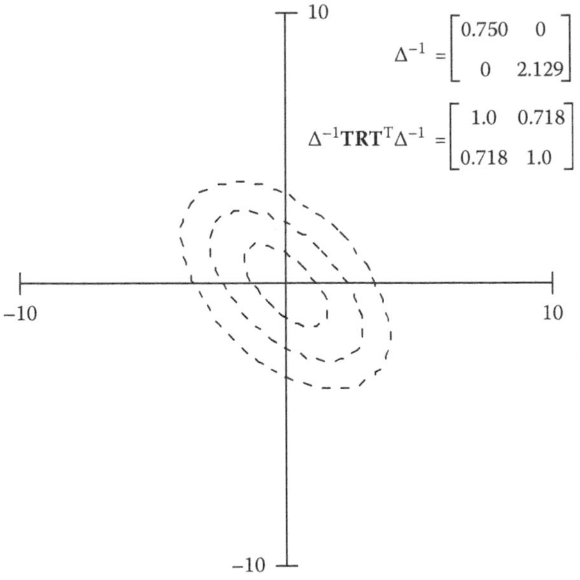

FIGURE 22.5 Error surface with transform and power normalization.

the transform. Power normalization was then applied using the normalization matrix Λ^{-1} as shown in Figure 22.5, which represents the transformed and power-normalized error surface. Note that the elliptical contours after transform domain processing are nearly circular in shape, and in fact they would have been perfectly circular if the rotation of Figure 22.4 had brought the contours into precise alignment with the coordinate axes. Perfect alignment did not occur in this example because **T** was not able to perfectly diagonalize the input autocorrelation matrix for this particular $x(n)$. Since **T** is a fixed transform in the TDAF structure, it clearly cannot properly diagonalize $\mathbf{R}_x$ for an arbitrary $x(n)$; hence, the surface rotation (orthogonalization) will be less than perfect for most input signals. It should be noted here that a well-known conventional algorithm called recursive least squares (RLS) is known to achieve near optimum convergence rates by forming an estimate of $\mathbf{R}_x^{-1}$, the inverse of the autocorrelation matrix. This type of algorithm automatically adjusts to whiten any input signal, and it also varies over time if the input signal is a nonstationary process. Unfortunately, the computation required for the RLS algorithm is large and not easily implemented in real time within the resource limitations of many practical applications. The RLS algorithm falls into the general class of quasi-Newton optimization techniques, which are thoroughly treated in numerous places throughout the literature.

There are two different ways to interpret the mechanism that brings about improved convergence rates achieved through transform domain processing [13]. The first point of view considers the combined operations of orthogonalization and power normalization to be the effective transformation $\Lambda^{-1}\boldsymbol{T}$, an interpretation that is implied by Equation 22.15. This line of thinking leads to an understanding of the transformed error surfaces as illustrated by the example in Figures 22.4 and 22.5 and leads to the logical conclusion that the faster convergence rate is due to the conventional LMS algorithm operating on an improved error surface that has been rendered more properly oriented and more symmetrical via the transformation. While this point of view is useful in understanding the principles of transform domain processing, it is not generally implementable from a practical point of view. This is because for an arbitrary input signal, the power normalization factors that constitute the Λ^{-1} part of the input transformation are not known *a priori*, and must be estimated after $\mathbf{T}$ is used to decompose the input signal into orthogonal channels.

The second point of view interprets the transform domain equations as operating on the transformed error surface (without power normalization) with a modified LMS algorithm where the step sizes are adjusted differently in the various channels according to $\mu(n) = \mu\Lambda^{-1}$, where $\mu(n) = \text{diag}[\mu_i(n)]$ is a diagonal matrix that contains the step size for the ith channel at location (i, i). The dependence of the $\mu_i(n)$s on the iteration (time) index n acknowledges that the steps sizes are a function of the power normalization factors, which are updated in real time as part of the online algorithm. This suggests that the TDAF should be able to track nonstationary input statistics within the limited abilities of the transformation $\mathbf{T}$ to orthogonalize the input and within the accuracy limits of the power normalization factors. Furthermore, when the input signal is white, all of the σ_i^2s are identical and each is equal to the power in the input signal. In this case, the TDAF with power normalization becomes the conventional-normalized LMS algorithm.

It is straightforward to show mathematically that the above two points of view are indeed compatible [11]. Let $\hat{\mathbf{v}}(n) = \Lambda^{-1}\mathbf{T}\mathbf{x}(n) = \Lambda^{-1}\mathbf{v}(n)$ and let the filter tap vector be denoted $\hat{\mathbf{W}}(n)$ when the matrix $\Lambda^{-1}\boldsymbol{T}$ is treated as the effective transformation. For the resulting filter to have the same response as the filter in Figure 22.3, we must have

$$\mathbf{v}^{*\mathrm{T}}(n)\mathbf{W} = y(n) = \hat{\mathbf{v}}^{*\mathrm{T}}\hat{\mathbf{W}} = \mathbf{v}^{*\mathrm{T}}(n)\Lambda^{-1}\hat{\mathbf{W}}, \quad \forall\mathbf{v}(n), \tag{22.16}$$

which implies that $\mathbf{W} = \Lambda^{-1}\hat{\mathbf{W}}$. If the tap vector $\hat{\mathbf{w}}$ is updated using the LMS algorithm, then

$$\begin{aligned}\mathbf{W}(n+1) &= \Lambda^{-1}\hat{\mathbf{W}}(n+1) = \Lambda^{-1}[\hat{\mathbf{W}}(n) + \mu e(n)\hat{\mathbf{v}}^*(n)] \\ &= \Lambda^{-1}\hat{\mathbf{W}}(n) + \mu e(n)\Lambda^{-1}\hat{\mathbf{v}}^*(n) \\ &= \mathbf{W}(n) + \mu e(n)\Lambda^{-2}\mathbf{v}^*(n),\end{aligned} \tag{22.17}$$

which is precisely the algorithm in Equation 22.10. This analysis demonstrates that the two interpretations are consistent, and they are, in fact, alternate ways to explain the fundamentals of transform domain processing.

22.4 Discussion and Examples

It is clear from the above development that the power estimates σ_i^2 are the optimum scale factors, as opposed to $|\sigma_i|$ or some other statistic. Also, it is significant to note that no convergence rate improvement can be realized without power normalization. This is the same conclusion that was reached in [7] where the frequency domain LMS algorithm was analyzed with a constant convergence factor. From the error surface description of the TDAF's operation, it is seen that an optimal transform rotates the axes of the hyperellipsoidal equal-error contours into alignment with the coordinate axes. The prescribed power normalization scheme then gives the ideal hyperspherical contours, and the convergence rate becomes the same as if the input were white. The optimal transform is composed of the orthonormal eigenvectors of the input autocorrelation matrix and is known in the literature as the Karhunen–Loe've transform (KLT). The KLT is signal dependent and usually cannot be easily computed in real time. Note that real-valued signals have real-valued KLTs, suggesting the use of real transforms in the TDAF (in contrast to complex transforms such as the DFT).

Since the optimal transform for the TDAF is signal dependent, a universally optimal fixed parameter transform can never be found. It is also clear that once the filter order has been chosen, any unitary matrix of correct dimensions is a possible choice for the transform; there is no need to restrict attention to classes of known transforms. In fact, if a prototype input power spectrum is available, its KLT can be constructed and used. One factor that must be considered in choosing a transform for real-time applications is computational complexity. In this respect, real transforms are superior to complex ones, transforms with fast algorithms are superior to those without, and transforms whose elements are all powers-of-2 are attractive since only additions and shifts are needed to compute them. Throughout the literature the DFT, the DCT, and the WHT have received considerable attention as possible candidates for use in the TDAF [14]. In spite of the fact that the DFT is a complex transform and not computationally optimal from that point of view, it is often used in practice because of the availability of efficient FFT algorithms.

Figure 22.6 shows learning characteristics for computer-generated TDAF examples using six different orthogonal transforms to decorrelate the input signal. The examples presented are for system identification experiments, where the desired signal was derived by passing the input through an 8-tap

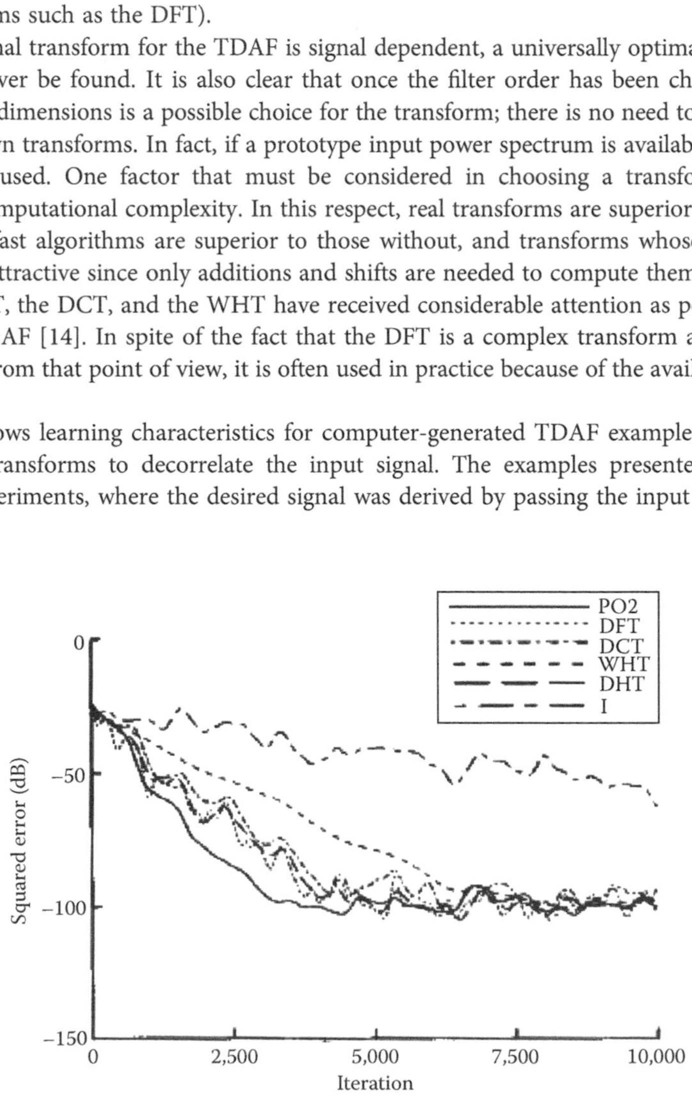

FIGURE 22.6 Comparison of (smoothed) learning curves for five different transforms operating on a colored noise input signal with condition number 681.

FIR filter, which serves as the model system to be identified. Computer-generated white pseudo-noise, uncorrelated with the input signal, was added to the output of the model system, creating a -100 dB noise floor. The filter inputs were generated by filtering white pseudo-noise with a 32-tap linear phase FIR noise-coloring filter to produce an input autocorrelation eigenvalue ratio of 681. Experiments were then performed using the DFT, the DCT, the WHT, discrete Hartley transform, and a specially designed computationally efficient "power-of-2" (PO2) transform, as listed in Figure 22.6. The eigenvalue ratios that result from transform processing with each of these transforms are reduced relative to the identity-transform case; the PO2 transform with power normalization reduces the input condition number from 681 to 128, resulting in the most effective transform for this particular input coloring. All of the transforms used in this experiment are able to reduce the input condition number and greatly improve convergence rates, although some transforms are seen to be more effective than others for the coloring chosen for these examples.

22.5 Quasi-Newton Adaptive Algorithms

The dependence of the adaptive system's convergence rate on the input power spectrum can be reduced by using second-order statistics via the Gauss–Newton method [12,15]. The Gauss–Newton algorithm is well known in the field of optimization as one of the basic accelerated search techniques. In recent years it has also appeared in various forms in publications on adaptive filtering. In this section a brief introduction to quasi-Newton adaptive filtering methods is presented. When the quasi-Newton concept is integrated into the LMS algorithm, the resulting adaptive strategy is closely related to the TDAF, but where the transform is computed online as an approximation to the Hessian acceleration matrix. For FIR structures it turns out that the Hessian is equivalent to the input autocorrelation matrix inverse, and therefore the quasi-Newton LMS algorithm effectively implements a transform that adjusts to the statistics of the input signal and is capable of tracking slowly varying nonstationary input signals.

The basic Gauss–Newton coefficient update algorithm for an FIR adaptive filter is given by

$$\mathbf{w}(n+1) = \mathbf{w}(n) - \mu\mathbf{H}(n)\nabla_{E[e^2]}(n), \tag{22.18}$$

where

$\mathbf{H}(n)$ is the Hessian matrix

$\nabla_{E[e^2]}(n)$ is the gradient of the cost function at iteration n

For an FIR adaptive filter with a stationary input the Hessian is equal to $\mathbf{R}_x^{-1}$. If the gradient is estimated with the instantaneous error squared, as in the LMS algorithm, the result is

$$\mathbf{w}(n+1) = \mathbf{w}(n) + \mu e(n)\hat{\mathbf{R}}_x^{-1}(n)\mathbf{x}(n), \tag{22.19}$$

where $\hat{\mathbf{R}}_x^{-1}(n)$ is an estimate of $\mathbf{R}_x^{-1}$ that varies as a function of the index n. Equation 22.19 characterizes the quasi-Newton LMS algorithm. Note that Equation 22.18 is the starting point for the development of many practical adaptive algorithms that can be obtained by making approximations to one or both of the Hessian and the gradient. Therefore, we typically refer to all such algorithms derived from Equation 22.18 as the family of quasi-Newton algorithms.

The autocorrelation estimate $\hat{\mathbf{R}}_x(n)$ is constructed from data received up to time step n. It must then be inverted for use in Equation 22.19. This is in general an $O[N^3]$ operation, which must be performed for every iteration of the algorithm. However, the use of certain autocorrelation estimators allows more economical matrix inversion techniques to be applied. Using this approach, the conventional sequential regression algorithm [6] and the RLS algorithm [16] achieve quasi-Newton implementations with a computational requirement of only $O[N^2]$.

The RLS algorithm is probably the best-known member of the class of quasi-Newton algorithms [7]. The drawback that has prevented its widespread use in real-time signal processing is its $O[N^2]$ computational requirement, which is still too high for many applications (and is an order of magnitude higher than the order of complexity of the FIR filter itself). This problem appeared to have been solved by the formulation of $O[N]$ versions of the RLS algorithm. Unfortunately, many of these more efficient forms of the RLS tend to be numerically ill-conditioned. They are often unstable in finite precision implementations, especially in low signal-to-noise applications or where the input signal is highly colored. This behavior is caused by the accumulation of finite precision errors in internal variables of the algorithm and is essentially the same source of numerical instability that occurs in the standard $O[N^2]$ RLS algorithm, although the problem is greater in the $O[N]$ case since these algorithms typically have a larger number of coupled internal recursions. Considerable work has been reported in the literature to stabilize $O[N^2]$ RLS algorithm and produce a numerically robust $O[N]$ RLS algorithm.

22.5.1 Fast Quasi-Newton Algorithm

The quasi-Newton algorithms discussed above achieve reduced computation through the use of particular autocorrelation estimators which lend themselves to efficient matrix inversion techniques. This section reviews a particular quasi-Newton algorithm that was developed to provide a numerically robust $O[N]$ algorithm [12]. This particular 1-D algorithm is discussed here simply as a representative algorithm from the quasi-Newton class; numerous variations of the Newton optimization strategy are reported in various places throughout the adaptive filtering literature. The fast quasi-Newton (FQN) algorithm described below has also been extended successfully to 2-D FIR adaptive filters [11].

To derive the $O[N]$ FQN algorithm, an autocorrelation matrix estimate is used which permits the use of more robust and efficient computation techniques. Assuming stationarity, the autocorrelation matrix $\mathbf{R}_x$ has a high degree of structure; it is symmetric and Toeplitz, and thus has only N free parameters, the elements of the first row. This structure can be imposed on the autocorrelation estimate, since this incorporates prior knowledge of the autocorrelation into the estimation process. The estimation problem then becomes that of estimating the N autocorrelation lags $r_i, i = 0, \ldots, N-1$, which comprise the first row of $\mathbf{R}_x$. The autocorrelation estimate is also required to be positive definite to ensure the stability of the adaptive update process.

A standard positive semidefinite autocorrelation lag estimator for a block of data is given by

$$\hat{r}_i = \frac{1}{M+1} \sum_{k=i}^{M} x(k-i)x(k), \tag{22.20}$$

where

$x(k), k = 0, \ldots, M$, is a block of real data samples

i ranges from 0 to M

However, the preferred form of the estimation equation for use in an adaptive system, from an implementation standpoint, is an exponentially weighted recursion. Thus, Equation 22.20 must be expressed in an exponentially weighted recursive form, without destroying its positive semidefiniteness property. Consider the form of the sum in Equation 22.20: it is the (deterministic) correlation of the data sequence $x(k), k = 0, \ldots, M$, with itself. Thus, $\hat{r}_i, i = 0, \ldots, M$ is the deterministic autocorrelation sequence of the sequence $x(k)$. (Note that $\hat{r}_i$ must also be defined for $i = -M, \ldots, -1$, according to the requirement that $\hat{r}_i = \hat{r}_{-i}$.) In fact, the deterministic autocorrelation for any sequence is positive semidefinite. The goal of exponential weighting, in a general sense, is to weight recent data most heavily and forget old data by using progressively smaller weighting factors. To construct an exponentially weighted, positive definite autocorrelation estimate, we must weight the data first, then form its deterministic autocorrelation to guarantee positive semidefiniteness. At time step n, the available data are $x(k), k = 0, \ldots, n$. If these samples are exponentially weighted using $\sqrt{\alpha}$, the result is $\alpha^{(n-k)/2}x(k), k = 0, \ldots, n$. Using Equation 22.20 and assuming $n > N - 1$, the result becomes

$$\begin{aligned}\hat{r}_i(n) &= \sum_{k=i}^{n} [\alpha^{(n-k+i)/2}x(k-i)][\alpha^{(n-k)/2}x(k)] \\ &= \alpha \sum_{k=i}^{n-1} \alpha^{(n-1-k)}\alpha^{i/2}x(k-i)x(k) + \alpha^{i/2}x(n-i)x(n) \\ &= \alpha\hat{r}_i(n-1) + \alpha^{i/2}x(n-i)x(n) \quad \text{for } i = 0, \ldots, N-1. \end{aligned} \tag{22.21}$$

A normalization term is omitted in Equation 22.21, and initialization is ignored. With regard to the latter point, the simplest way to consistently generate $\hat{r}_i(n)$ for $0 \leq n \leq N - 1$ is to assume that $x(n) = 0$ for $n < 0$, set $\hat{r}_i(-1) = 0$ for all i, and then use the above recursion. A small positive constant δ may be added to $\hat{r}_0(n)$ to ensure positive definiteness of the estimated autocorrelation matrix.

With this choice of an autocorrelation matrix estimate, a quasi-Newton algorithm is determined. Thus, the FQN algorithm is given by Equations 22.19 and 22.21, where $\hat{\mathbf{R}}_x(n) \approx \mathbf{R}_x$ is understood to be the Toeplitz symmetric matrix whose first row consists of the autocorrelation lag estimates $\hat{r}_i(n), i = 0, \ldots, N - 1$, generated by Equation 22.21. Because $\hat{\mathbf{R}}_x(n)$ is Toeplitz, its inverse can be obtained using the Levinson recursion, leading to an $O[N]$ implementation of this algorithm. The step size μ for the FQN algorithm is given by

$$\frac{1}{2}\mu^{-1} = \varepsilon + \mathbf{x}^{\mathrm{T}}(n)\hat{\mathbf{R}}_x^{-1}(n-1)\mathbf{x}(n). \tag{22.22}$$

This step size is used in other quasi-Newton algorithms [2], and seems nearly optimal. The parameter ε is intended to be small relative to the average value of $x^{\mathrm{T}}(n)\hat{\mathbf{R}}_x^{-1}(n-1)x(n)$. Then the normalization term omitted from Equation 22.21, which is a function of α but not of i, cancels out of the coefficient update, since $\mathbf{R}_x^{-1}(n)$ appears in both the numerator and the denominator. Thus, the normalization can be safely ignored.

22.5.2 Examples

The previous examples are used again here to compare the performance of the FQN algorithm with the RLS, which provides a baseline for performance comparisons. The RLS examples are shown in Figure 22.7a for different values of the exponential forgetting factor α, and the FQN examples are shown in Figure 22.7b. Note that the FQN algorithm is somewhat slower to converge due to the fact that the autocorrelation inverse matrix is updated only once every eight samples. In comparison the RLS algorithm converges more quickly, but has a computational complexity of $O[N^2]$ as compared to a complexity of $O[N]$ for the FQN algorithm. But, more important, note that the convergence rate of the FQN algorithm is much faster than any of the transform domain examples shown previously.

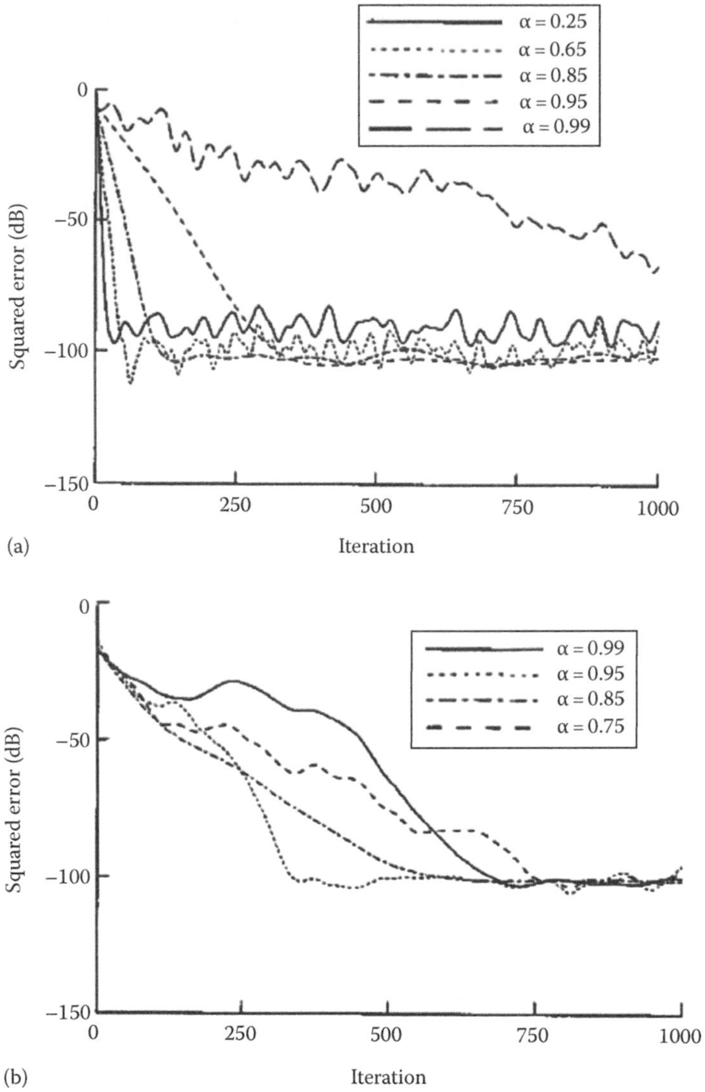

FIGURE 22.7 Comparison of the RLS and FQN performance. Simulated learning curves for (a) the RLS algorithm and (b) the FQN algorithm.

22.6 2-D Transform Domain Adaptive Filter

Many successful 1-D FIR algorithms have been extended to 2-D filters [10,11,17]. Transform domain adaptive algorithms are also well suited to 2-D signal processing. Orthogonal transforms with power normalization can be used to accelerate the convergence of an adaptive filter in the presence of a colored input signal.

The 2-D TDAF structure is shown in Figure 22.8 with the corresponding (possibly complex) LMS algorithm given as

$$\mathbf{w}_{k+1}(m_1, m_2) = \mathbf{w}_k(m_1, m_2) + \mu e(n_1, n_2)\Lambda_u^{-2}\mathbf{u}_k^*(n_1, n_2), \tag{22.23}$$

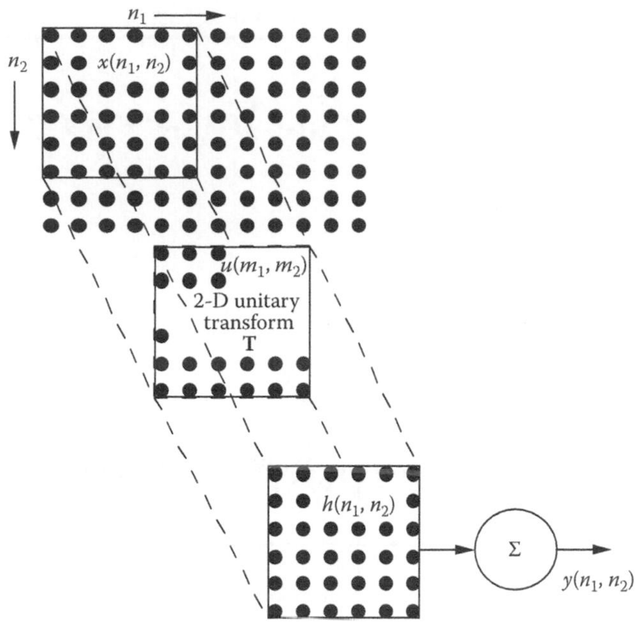

FIGURE 22.8 2-D TDAF structure.

where $\mathbf{u}_k(n_1, n_2)$ is the column-ordered vector formed by premultiplying the input column-ordered vector $\mathbf{x}_k(n_1, n_2)$ by the 2-D unitary transform **T**, i.e.,

$$\mathbf{u}_k(n_1, n_2) = \mathbf{T}\mathbf{x}_k(n_1, n_2). \tag{22.24}$$

Channel normalization results from including $\Lambda_u^2 = \text{diag}[\,\sigma_u^2(0,0) \quad \sigma_u^2(1,1) \quad \cdots \quad \sigma_u^2(N,N)\,]$ in Equation 22.23 where $\sigma_u^2(n_1, n_2) \approx E[|\mathbf{u}(n_1, n_2)|^2]$.

Ideally, the KLT is used to achieve optimal convergence, but this requires *a priori* knowledge of the input statistical properties. The KLT corresponding to the input autocorrelation matrix $\mathbf{R}_x$ is constructed using as rows of **T** the orthonormal eigenvectors of $\mathbf{R}_x$. Therefore, with unitary $\mathbf{Q}_x^{\text{H}} = [\mathbf{q}_1, \ldots, \mathbf{q}_{M^2}]$ and $\Lambda_x = \text{diag}[\lambda_1, \ldots, \lambda_{M^2}]$ $(M = N + 1$ for convenience), the unitary similarity transformation is $\mathbf{R}_x = \mathbf{Q}_x^{-1}\Lambda_x\mathbf{Q}_x$, and the KLT is given by $\mathbf{T} = \mathbf{Q}_x$. However, since the statistical properties of the input process are usually unknown and time varying, the KLT cannot be implemented in practice. Researchers have found that many fixed transforms do provide good orthogonalization for a wide class of input signals. Those include the DFT or FFT, the DCT, and the WHT. For example, the DFT provides only approximate channel decorrelation since it is well known that a "sliding" DFT implements a parallel bank of overlapping band-pass filters with center frequencies evenly distributed over the interval $[0, 2\pi]$. Furthermore, the DFT (or FFT) is hampered by the fact that it requires complex arithmetic. It is still a very effective method of orthogonalization which we compare here to the 2-D FQN algorithm.

The convergence plots in Figure 22.9 show the comparison between the 2-D FQN, the 2-D TDAF (with the DFT), and the simple 2-D LMS with the same fourth-order low-pass coloring filter. The adaptive filter is second order, and the 2-D FQN algorithm, as expected, outperforms the 2-D TDAF. The 2-D FQN algorithm is effectively attempting to estimate the KLT online so that, while not able to perfectly orthogonalize the training signal, it does offer improved convergence over that of the fixed transform algorithm. Similar results appear in Figure 22.10 with the same coloring filter and a fourth-order adaptive filter.

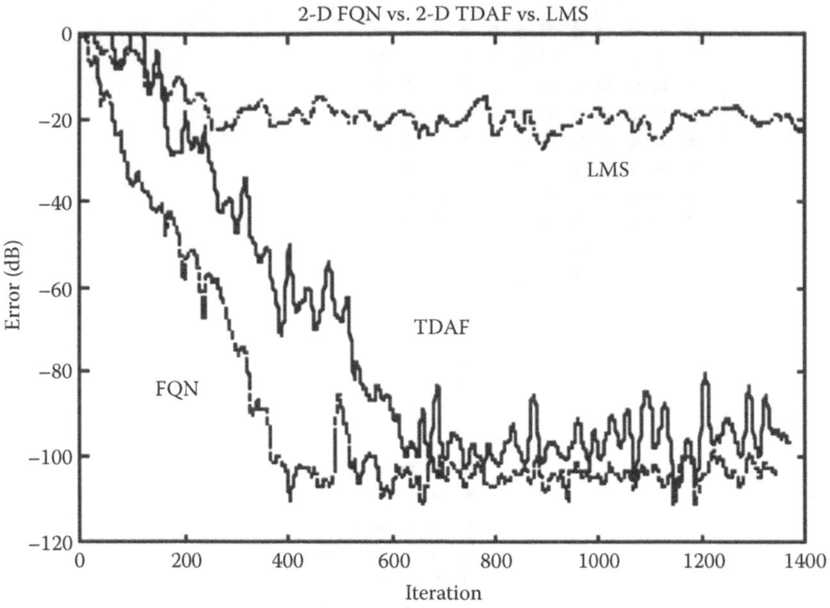

FIGURE 22.9 Convergence plot for 3 × 3 FIR 2-D LMS, 2-D TDAF, and 2-D FQN adaptive filters in the system identification configuration with low-pass colored inputs.

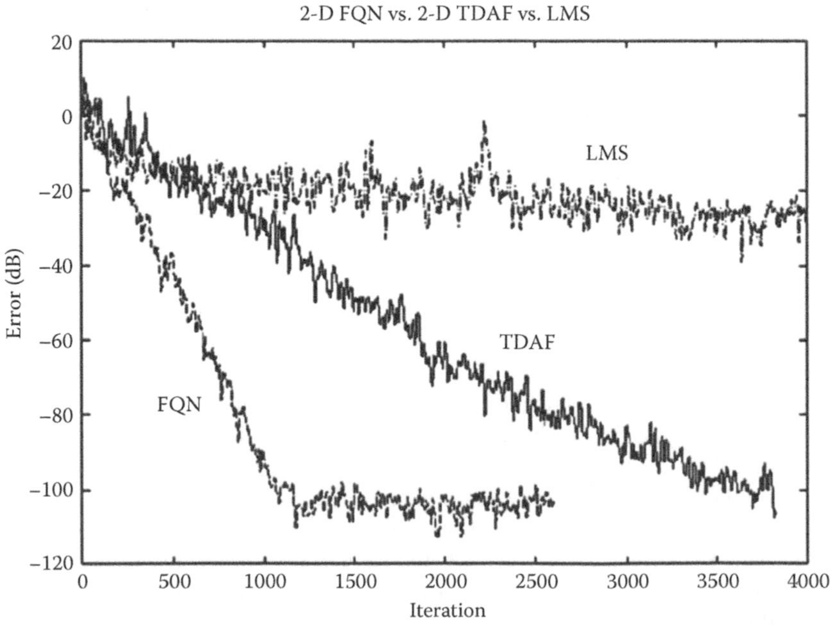

FIGURE 22.10 Convergence plot for 5 × 5 FIR 2-D LMS, 2-D TDAF, and 2-D FQN adaptive filters in the system identification configuration with low-pass colored inputs.

22.7 Fault-Tolerant Transform Domain Adaptive Filters

Reducing feature sizes and lowering supply voltages have been popular approaches to enable high-speed low-power complex systems to be built on a single chip. But technology scaling has also led to more problems with increased vulnerability to particle hits and noise effects, high operating temperatures, and process variations making these circuits more susceptible to transient errors and fixed hardware ("stuck-at") faults [18,19]. When adaptive systems such as echo cancellers, channel equalizers, noise cancellers, and LPC data compressors are implemented in nanoscale VLSI circuits, there is a concern about how such systems will perform in the presence of both transient and permanent errors.

Adaptive systems adjust their parameters to minimize a specified error criterion under normal operating conditions. Fixed errors or hardware faults will prevent the system from minimizing the error criterion, but at the same time the system will adapt the parameters such that the best possible solution is reached under constraints imposed by the fault conditions. In adaptive fault-tolerant (AFT) filter structures the inherent adaptive property is used to compensate for failures in correctly adjusting the adaptive coefficients. This mechanism can be used with specially designed structures whose redundant coefficients have the ability to compensate for failures of other coefficients.

AFT concepts were originally developed for FIR adaptive filters using vector space concepts [20]. Consider a 3-tap direct form FIR adaptive filter that has the following tap weight vector:

$$\mathbf{W}(n) = w_0(n)\begin{bmatrix}1\\0\\0\end{bmatrix} + w_1(n)\begin{bmatrix}0\\1\\0\end{bmatrix} + w_2(n)\begin{bmatrix}0\\0\\1\end{bmatrix}. \tag{22.25}$$

If one of the adaptive coefficients incurs a "stuck-at" fault that prevents it from updating, the other taps cannot compensate for the failure. However, by adding a fourth adaptive tap, whose input is the sum of the signals driving the original taps, an effective length $N = 3$ adaptive filter is achieved with an impulse response given by

$$\mathbf{W}(n) = w_0(n)\begin{bmatrix}1\\0\\0\end{bmatrix} + w_1(n)\begin{bmatrix}0\\1\\0\end{bmatrix} + w_2(n)\begin{bmatrix}0\\0\\1\end{bmatrix} + w_3(n)\begin{bmatrix}1\\1\\1\end{bmatrix}, \tag{22.26}$$

where $\tilde{w}_j(n) = w_j(n) + w_3(n)$, for $j = 0, 1, 2$. The presence of the additional adaptively weighted column makes it possible for the remaining three adaptively weighted columns to match any Wiener solution when any one of the four coefficients incurs a stuck-at fault condition.

Most AFT filter structures are based on adding extra coefficients and using the adaptive algorithm to automatically compensate for failures in the adaptive coefficients to adjust properly. An AFT adaptive filter architecture with R redundant coefficients, where R is greater than or equal to one, is able to achieve the fault-free minimum MSE despite the occurrence of R coefficient failures. All the coefficients work together to match the Wiener solution, and when a subset of the coefficients fails, the remaining functional coefficients continue to operate normally until the Wiener solution is again achieved.

A more general form of a fault-tolerant adaptive filter is based on the TDAF structure [20]. For the transform domain fault-tolerant adaptive filter (TDFTAF) structure, the transformed data vector $\mathbf{V}(n)$ is

$$\mathbf{V}(n) = \mathbf{T_M}\mathbf{X_e}(n) \tag{22.27}$$

where

$\mathbf{X_e}(n)$ is $\mathbf{X}(n)$ zero-padded with R zeros
$\mathbf{T_M}$ is a $M \times M$ unitary transform matrix, where $M = L + R$

The structure is characterized by the following relationships:

$$\mathbf{X_e}(n) = [\mathbf{X}(n) \quad 0 \quad 0 \quad \cdots \quad 0]^{\mathrm{T}} \tag{22.28}$$

$$\mathbf{V}(n) = \mathbf{T_M}\mathbf{X_e}(n) \tag{22.29}$$

$$y(n) = \mathbf{W}(n)^{\mathrm{T}}\mathbf{V}(n), \tag{22.30}$$

where

$\mathbf{X_e}(n)$ is a length M vector

$\mathbf{W}(n)$ is the vector of M adaptive coefficients

If the power-normalized LMS algorithm is used to update the coefficients of the TDAF, then the relevant equations are

$$e(n) = d(n) - y(n) \tag{22.31}$$

$$\mathbf{W}(n+1) = \mathbf{W}(n) + e(n)\tilde{\mu}\mathbf{V}(n), \tag{22.32}$$

where $\tilde{\mu}$ is a diagonal matrix of time-varying step size parameters that results from online power normalization.

As an example of the performance of the FFT-based TDFTAF, a 10th-order TDFTAF with two redundant taps was implemented in system identification mode with a white noise input signal. Since the unknown system was a 10th-order FIR filter, the noise floor of the adaptive system is a result of finite machine precision. In this example, a fault occurs at iteration 750, where tap weight 5 becomes "stuck" at zero. The resulting error curve is generated by averaging over 100 independent runs and is presented in Figure 22.11. From this figure it is clearly seen that before the fault occurs, the error converges rapidly to the noise floor in about 650 iterations. The occurrence of the fault causes the error to jump to a large value after which it converges back to the same noise floor at a somewhat reduced rate.

The capability of adaptive fault tolerance also applies to cases of single and multiple stuck-at bit errors occurring in one or more of the adaptive filter coefficients for fixed-point number representation [21]. In an over-parameterized adaptive filter with a large number of equivalent MSE solutions, the occurrence of a single stuck-at bit error reduces the number of available solutions, although multiple minimum MSE

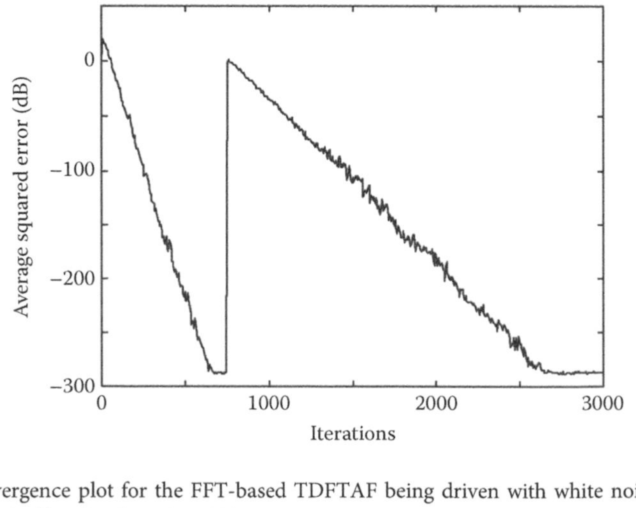

FIGURE 22.11 Convergence plot for the FFT-based TDFTAF being driven with white noise having $N = 10$ and $R = 2$. The fault occurs in Tap 5 at Iteration 750.

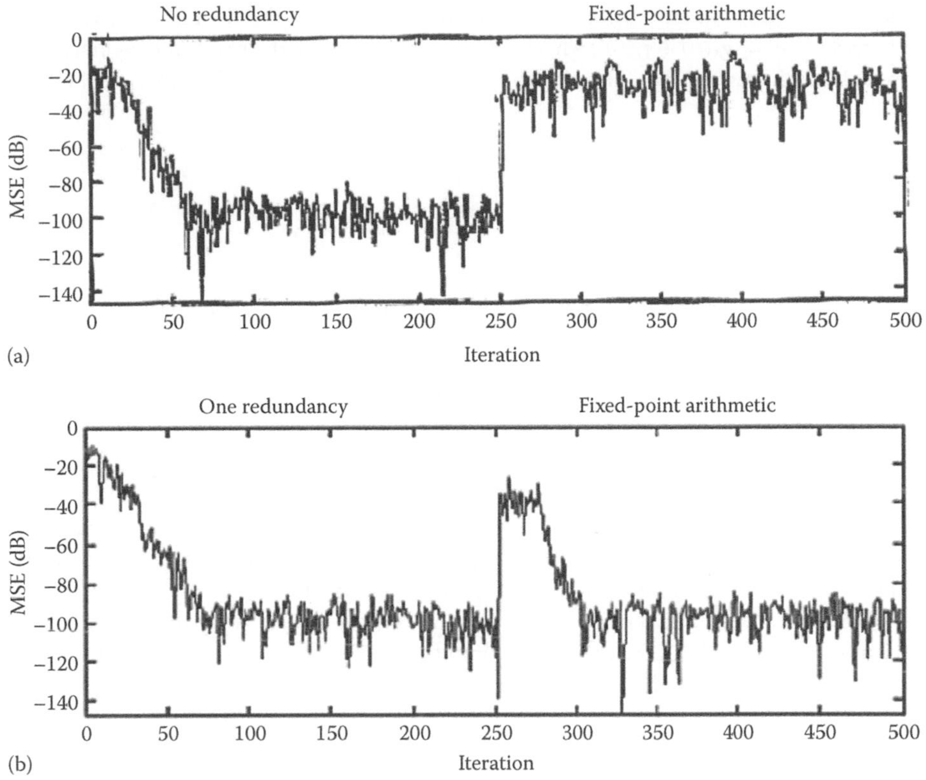

FIGURE 22.12 Learning curves for a fault-tolerant DFT transform domain LMS adaptive filter with one stuck-at-1 fault in the most significant bit of the real part of the first coefficient $w_1(n)$ (fixed point).

solutions still exist. Examples of fixed single bit faults (fixed-point number representation) in a DFT-based TDAF are shown in Figure 22.12. Figure 22.12a shows the case where no hardware redundancy is included, so the single stuck-at bit error in a transform domain coefficient results in catastrophic failure of the adaptive algorithm (no re-convergence occurs after the fault occurs). Figure 22.12b shows a similar experiment where a single coefficient redundancy was included in the design based on the TDFTAF structure described previously. In the single bit error case the filter relearns through proper coefficient adjustments and the adaptive filter regains its pre-failure performance after experiencing a period of transient response. It has been shown that the principles of AFT can also overcome arbitrary patterns of stuck-at bit errors in adaptive filters implemented with floating-point binary codes [22].

When a transform domain FTAF operates on real-valued input and desired signals, the complex arithmetic required by the DFT matrix is generally considered a disadvantage in this application. However, the introduction of complex tap weights leads to additional free parameters if the coefficient updates are based only on the real part of the output error $e[n]$. In [23] it was shown that an FFT-based transform domain FTAF (FFT-TDAF) algorithm operating on real-valued signals does not provide full fault tolerance, although it can provide a high degree of fault coverage without introducing extra redundant hardware through zero padding. In particular it has been shown that the fault conditions not covered are when multiple errors occur in either the real parts or imaginary parts of the transform domain coefficients that are conjugate pairs. However, when the filter operates on real-valued signals the complex arithmetic provides extensive fault-tolerant capabilities that may be useful for achieving fault-tolerant performance in highly scaled low-power VLSI realizations where permanent faults are becoming an increasing concern.

Consider the situation where $x(n)$ and $d(n)$ are real-valued, but the filter is treated as producing complex outputs, $y(n) = y_R(n) + jy_I(n)$, where $y_R(n)$ and $y_I(n)$ denote the real and imaginary parts of the filter output. Then the minimum MSE using the complex output error can be expressed as follows:

$$|e(n)|^2 = [d(n) - y_R(n)]^2 + y_I(n)^2$$

and

$$\min_{w_T} \{|e(n)|^2\} = \min_{w_T} \{[d(n) - y_R(n)]^2\} + \min_{w_T} \{y_I^2(n)\}. \tag{22.33}$$

Since the desired signal is real, the minimization operation in Equation 22.23 will result in

$$\min_{w} \{y_I^2(n)\} = 0, \tag{22.34}$$

thereby imposing additional constraints on $W(n)$. This implies that there are N (even) adjustable parameters ($\frac{N}{2}$ real parameters and $\frac{N}{2}$ imaginary parameters) in the frequency domain that uniquely specify the N real coefficients in the time domain. However, if only the real part of $e(n)$ is used in the minimization of Equation 22.33, then the constraint of Equation 22.34 is relaxed. In this case there are more than N parameters in the frequency domain to define the N real-valued tap weights in the time domain, and hence there is an inherent over parameterization introduced by minimizing only the real part of $e(n)$.

To demonstrate this concept, two examples are presented below to demonstrate the fault-tolerant behavior of the FFT-FTAF when used without zero padding to minimize the real part of the output error in a system identification application. The unknown system that is identified in this example is a 64-tap FIR low-pass filter with real-valued coefficients. The training signal used was a Gaussian white noise with

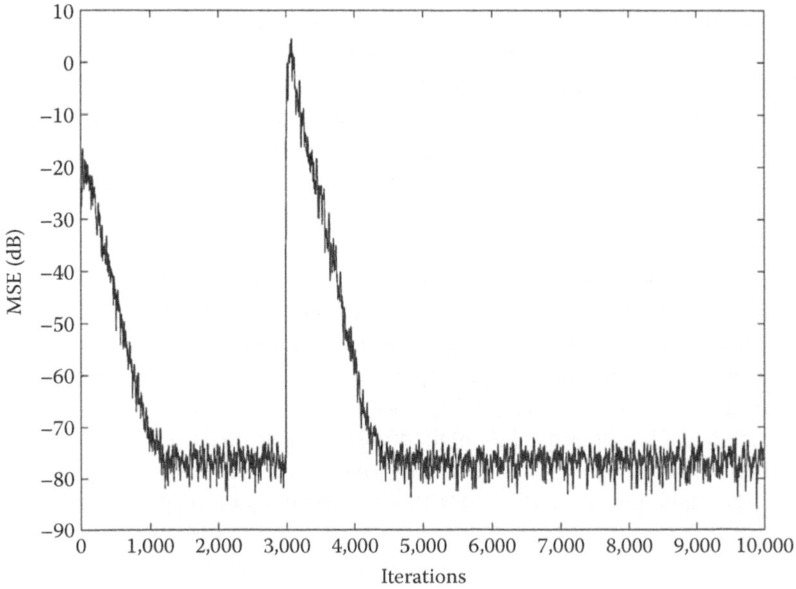

FIGURE 22.13 Fault in tap $(W_{1R}[n] + jW_{1I}[n])$.

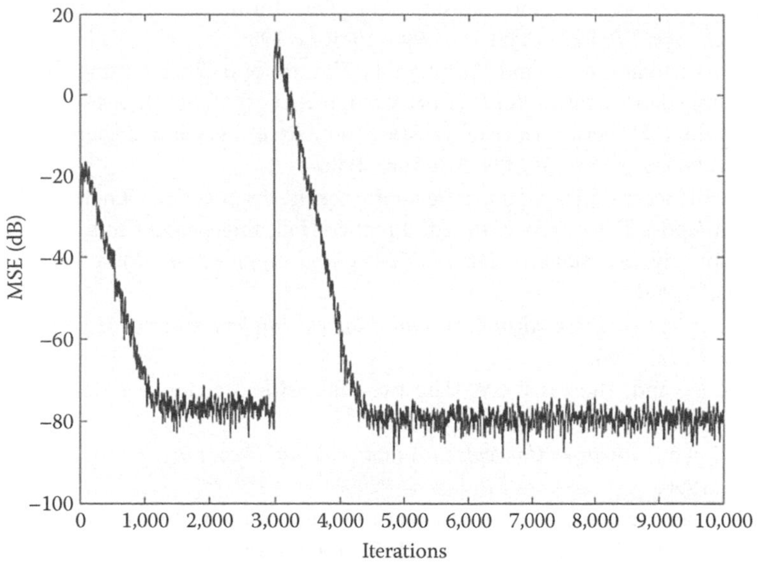

FIGURE 22.14 Faults in taps $(W_{1R}[n] + jW_{1I}[n], \ldots, W_{31R}[n] + jW_{31I}[n])$.

unit variance and the noise floor was set at −80 dB relative to the training signal. An error in the weight update computation for the kth tap weight, $W_k(n)$, results in an incorrect value for that coefficient. Hence, the fault condition is simulated by setting the erroneous filter coefficient to an arbitrary random value at the 3000th iteration, chosen so that the filter has reached the Wiener solution prior to the occurrence of the error. The final mean-square error curve was obtained by computing an ensemble average over 50 trials and time averaging the results over a window of length 10.

The results are shown in Figures 22.13 and 22.14. Figure 22.13 shows the case when a fixed fault occurs in the second transform domain tap $(W_{1R}(n) + jW_{1I}(n))$ of the filter. The filter is abruptly reinitialized by the fault but then converges to the correct solution. An extreme case of fixed faults occurring in 31 of the 64 total coefficients is shown in Figure 22.14. In this case, the faults were introduced in the 2nd to the 32nd taps $(W_{1R}(n) + jW_{1I}(n), \ldots, W_{31R}(n) + jW_{31I}(n))$. Since none of these taps are conjugates of each other, this case falls within the error coverage of the algorithm FFT-TDAF and the filter reconverges to the correct solution.

The analysis presented in [22] has determined that an N-tap filter with real coefficients (N even) can recover from up to $N/2$ hard faults as long as the faults do not occur in transform domain tap weights that are real-valued or simultaneously in both the real or the imaginary parts of conjugate symmetric positions in the transform domain. Furthermore, for each redundant tap that is added in the transform domain via zero padding in the time domain, additional fault tolerance is achieved for either the real-valued tap weights or one of the conjugate pairs in the complex frequency domain. The result is that full fault tolerance can be achieved by zero padding the input vector with $N/2$ zeros in the time domain, resulting in the addition of $N/2$ redundant coefficients in the transform domain.

References

1. Dentino, M., McCool, J., and Widrow, B., Adaptive filtering in the frequency domain, *Proc. IEEE*, 66, 1658–1659, Dec. 1978.
2. Gitlin, R.D. and Magee, F.R., Jr., Self-orthogonalizing adaptive equalization algorithms, *IEEE Trans. Commn.*, COM-25(7), 666–672, July 1977.

3. Narayan, S.S., Peterson, A.M., and Narasima, M.J., Transform domain LMS algorithm, *IEEE Trans. Acoust. Speech Signal Process.*, ASSP-34, 499–510, June 1986.
4. Marshall, D.F., Jenkins, W.K., and Murphy, J.J., The use of orthogonal transforms for improving performance of adaptive filters, *IEEE Trans. Circuits Syst.*, CAS-36(4), 474–484, Apr. 1989.
5. Lee, J.C. and Un, C.K., Performance of transform domain LMS adaptive filters, *IEEE Trans. Acoust. Speech Signal Process.*, ASSP-34, 499–510, June 1986.
6. Widrow, B. and Stearns, S.D., *Adaptive Signal Processing*, Prentice-Hall, Englewood Cliffs, NJ, 1985.
7. Haykin, S., *Adaptive Filter Theory*, 4th ed., Prentice-Hall, Englewood Cliffs, NJ, 2001.
8. Farhang-Boroujeny, B., *Adaptive Filters: Theory and Applications*, John Wiley and Sons, Ltd, Southgate, UK, 1999.
9. Diniz, P.S.R., *Adaptive filters: Algorithms and Practical Implementation*, 3rd ed., Springer Publishing Co., New York, 2008.
10. Hadhoud, M.M. and Thomas, D.W., The two-dimensional adaptive LMS (TDLMS) algorithm, *IEEE Trans. Circuits Syst.*, 35, 485–494, 1988.
11. Jenkins, W.K. et al., *Advanced Concepts in Adaptive Signal Processing*, Kluwer Academic Publishers, Boston, MA, 1996.
12. Marshall, D.F. and Jenkins, W.K., A fast quasi-Newton adaptive filtering algorithm, *IEEE Trans. Acoust. Speech Signal Process.*, ASSP-40(7), 1652–1662, July 1992.
13. Marshall, D.F., Computationally efficient techniques for rapid convergence of adaptive digital filters, PhD dissertation, University of Illinois, Urbana-Champaign, IL, 1988.
14. Honig, M.L. and Messerschmidt, D.G., *Adaptive Filters: Structures, Algorithms, and Applications*, Kluwer Academic Press, Boston, MA, 1984.
15. Hull, A.W. and Jenkins, W.K., A preconditioned conjugate gradient method for block adaptive filtering, *Proceedings of the IEEE International Symposium on Circuits and Systems*, Singapore, June 1991, pp. 540–543.
16. Goodwin, G.C. and Sin, K.S., *Adaptive Filtering Prediction and Control*, Prentice Hall, Englewood Cliffs, NJ, 1984.
17. Shapiro, J.M., Algorithms and systolic architectures for real-time multidimensional adaptive filtering of frequency domain multiplexed video signals, PhD dissertation, M.I.T., Cambridge, MA, 1990.
18. Srinivasan, J., Adve, S.V., Bose, P., and Rivers, J.A., The impact of technology scaling on lifetime reliability, *Proceedings of International Conference on Dependable Systems and Networks*, Florence, Italy, June 28–July 1, 2004, pp. 177–186.
19. Reviriego, P., Maestro, J.A., and Ruano, O., Efficient protection techniques against SEU's for adaptive filters: An echo canceller case study, *IEEE Trans. Nucl. Sci.*, 55(3), 1700–1707, June 2008.
20. Schnaufer, B.A. and Jenkins, W.K., Adaptive fault tolerance for reliable LMS adaptive filtering, *IEEE Trans. Circuits Systems—II: Analog Digital Signal Process.*, 44(12), pp. 1001–1014, Dec. 1997.
21. Leon, G and Jenkins, W.K., Adaptive fault tolerant digital filters with single and multiple bit errors in fixed-point arithmetic, *Proceedings of the 33rd Annual Asilomar Conference on Signals, Systems, and Computers*, Pacific Grove, CA, Oct. 1999.
22. Leon, G. and Jenkins, W.K., Adaptive fault tolerant digital filters with single and multiple bit errors in floating-point arithmetic, *Proceedings of International Symposium on Circuits and Systems*, Geneva, Switzerland, May 2000, Vol. 3, pp. III.630–III.633.
23. Radhakrishnan, C. and Jenkins, W.K., Fault tolerance in transform domain adaptive filters operating with real-valued signals, *IEEE Transactions on Circuits and Systems I*, to appear.

23

Adaptive IIR Filters

Geoffrey A. Williamson
Illinois Institute of Technology

23.1 Introduction

In comparison with adaptive finite impulse response (FIR) filters, adaptive infinite impulse response (IIR) filters offer the potential to implement an adaptive filter meeting desired performance levels, as measured by mean-square error, for example, with much less computational complexity. This advantage stems from the enhanced modeling capabilities provided by the pole/zero transfer function of the IIR structure, compared to the "all-zero" form of the FIR structure.

However, adapting an IIR filter brings with it a number of challenges in obtaining stable and optimal behavior of the algorithms used to adjust the filter parameters. Since the 1970s, there has been much active research focused on adaptive IIR filters, but many of these challenges to date have not been completely resolved. As a consequence, adaptive IIR filters are not found in commercial practice in anywhere near the frequency that adaptive FIR filters are. Nonetheless, recent advances in adaptive IIR filter research have provided new results and insights into the behavior of several methods for adapting the filter parameters, and new algorithms have been proposed that address some of the problems and open issues in these systems. Hence, this class of adaptive filter continues to maintain promise as a potentially effective and efficient adaptive filtering option.

In this section, we provide an up-to-date overview of the different approaches to the adaptive IIR filtering problem. Due to the extensive literature on the subject, many readers may wish to peruse several earlier general treatments of the topic. Johnson's 1984 [11] and Shynk's 1989 papers [23] are still current in the sense that a number of open issues cited therein remain open today. More recently, Regalia's 1995 book [19] provides a comprehensive view of the subject.

23.1.1 System Identification Framework for Adaptive IIR Filtering

The spread of issues associated with adaptive IIR filters is most easily understood if one adopts a system identification perspective to the filtering problem. To this end, consider the diagram presented in Figure 23.1. Available to the adaptive filter are two external signals: the input signal $x(n)$ and the desired output signal $d(n)$. The adaptive filtering problem is to adjust the parameters of the filter acting on $x(n)$ so that its output $y(n)$ approximates $d(n)$. From the system identification perspective, the task at hand is to adjust the parameters of the filter generating $y(n)$ from $x(n)$ in Figure 23.1 so that the filtering operation itself matches in some sense the system generating $d(n)$ from $x(n)$. These two viewpoints are closely related because if the systems are the same, then their outputs will be close. However, by adopting the convention that there is a system generating $d(n)$ from $x(n)$, clearer insights into the behavior and design of adaptive algorithms are obtained. This insight is useful even if the "system" generating $d(n)$ from $x(n)$ has only a statistical and not a physical basis in reality.

The standard adaptive IIR filter is described by

$$y(n)+a_1(n)y(n-1)+\cdots+a_N(n)y(n-N)=b_0(n)x(n)+b_1(n)x(n-1)+\cdots+b_M(n)x(n-M), \quad (23.1)$$

or equivalently

$$(1+a_1(n)q^{-1}+\cdots+a_N(n)q^{-N})y(n)=(b_0(n)+b_1(n)q^{-1}+\cdots+b_M(n)q^{-M})x(n)\,. \quad (23.2)$$

As is shown in Figure 23.1, Equation 23.2 may be written in shorthand as

$$y(n)=\frac{B(q^{-1},n)}{A(q^{-1},n)}x(n), \quad (23.3)$$

where $B(q^{-1},n)$ and $A(q^{-1},n)$ are the time-dependent polynomials in the delay operator q^{-1} appearing in Equation 23.2. The parameters that are updated by the adaptive algorithm are the coefficients of these polynomials. Note that the polynomial $A(q^{-1},n)$ is constrained to be monic, such that $a_0(n)=1$.

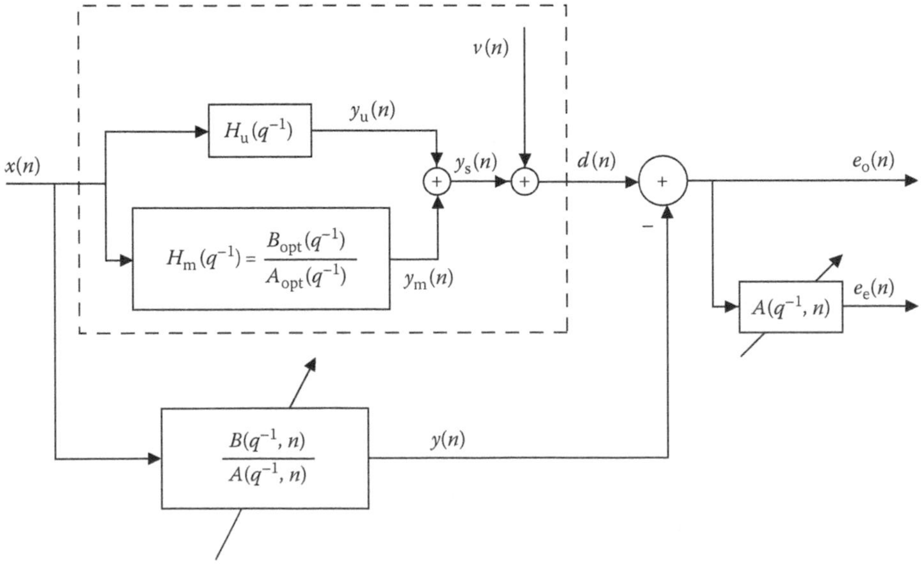

FIGURE 23.1 System identification configuration of the adaptive IIR filter.

We adopt a rather more general description for the unknown system, assuming that $d(n)$ is generated from the input signal $x(n)$ via some linear time-invariant system $H(q^{-1})$, with the addition of a noise signal $v(n)$ to reflect components in $d(n)$ that are independent of $x(n)$. We further break down $H(q^{-1})$ into a transfer function $H_{\mathrm{m}}(q^{-1})$ that is explicitly modeled by the adaptive filter, and a transfer function $H_{\mathrm{u}}(q^{-1})$ that is unmodeled. In this way, we view $d(n)$ as a sum of three components: the signal $y_{\mathrm{m}}(n)$ that is modeled by the adaptive filter, the signal $y_{\mathrm{u}}(n)$ that is unmodeled but that depends on the input signal, and the signal $v(n)$ that is independent of the input. Hence,

$$d(n) = y_{\mathrm{m}}(n) + y_{\mathrm{u}}(n) + v(n) \tag{23.4}$$

$$= y_{\mathrm{s}}(n) + v(n), \tag{23.5}$$

where $y_{\mathrm{s}}(n) = y_{\mathrm{m}}(n) + y_{\mathrm{u}}(n)$. The modeled component of the system output is viewed as

$$y_{\mathrm{m}}(n) = \frac{B_{\mathrm{opt}}(q^{-1})}{A_{\mathrm{opt}}(q^{-1})} x(n), \tag{23.6}$$

with $B_{\mathrm{opt}}(q^{-1}) = \sum_{i=0}^{M} b_{i,\mathrm{opt}} q^{-i}$ and $A_{\mathrm{opt}}(q^{-1}) = 1 + \sum_{i=i}^{N} a_{i,\mathrm{opt}} q^{-i}$. Note that Equation 23.6 has the same form as Equation 23.3. The parameters $\{a_{i,\mathrm{opt}}\}$ and $\{b_{i,\mathrm{opt}}\}$ are considered to be the optimal values for the adaptive filter parameters, in a manner that we describe shortly.

Figure 23.1 shows two error signals: $e_{\mathrm{e}}(n)$, termed the equation error, and $e_{\mathrm{o}}(n)$, termed the output error. The parameters of the adaptive filter are usually adjusted so as to minimize some positive function of one or the other of these error signals. However, the figure of merit for judging adaptive filter performance that we will apply throughout this section is the mean-square output error $E\{e_{\mathrm{o}}^2(n)\}$. In most adaptive filtering applications, the desired signal, $d(n)$, is available only during a "training phase" in which the filter parameters are adapted. At the conclusion of the training phase, the filter will be operated to produce the output signal $y(n)$ as shown in the figure, with the difference between the filter output $y(n)$ and the (now unmeasurable) system output $d(n)$ the error. Thus, we adopt the convention that $\{a_{i,\mathrm{opt}}\}$ and $\{b_{i,\mathrm{opt}}\}$ are defined such that when $a_i(n) \equiv a_{i,\mathrm{opt}}$ and $b_i(n) \equiv b_{i,\mathrm{opt}}$, $E\{e_{\mathrm{o}}^2(n)\}$ is minimized, with $A_{\mathrm{opt}}(q^{-1})$ constrained to be stable.

At this point it is convenient to set down some notation and terminology. Define the regressor vectors

$$\mathbf{U}_{\mathrm{e}}(n) = [x(n) \cdots x(n-M) - d(n-1) - \cdots - d(n-N)]^{\mathrm{T}}, \tag{23.7}$$

$$\mathbf{U}_{\mathrm{o}}(n) = [x(n) \cdots x(n-M) - y(n-1) - \cdots - y(n-N)]^{\mathrm{T}}, \tag{23.8}$$

$$\mathbf{U}_{\mathrm{m}}(n) = [x(n) \cdots x(n-M) - y_{\mathrm{m}}(n-1) - \cdots - y_{\mathrm{m}}(n-N)]^{\mathrm{T}}. \tag{23.9}$$

These vectors are the equation error regressor, output error regressor, and modeled system regressor vectors, respectively. Define a noise regressor vector

$$\mathbf{V}(n) = [0 \cdots 0 - v(n-1) - \cdots - v(n-N)]^{\mathrm{T}} \tag{23.10}$$

with $M+1$ leading zeros corresponding to the $x(n-i)$ values in the preceding regressors. Furthermore, define the parameter vectors

$$\mathbf{W}(n) = [b_0(n) b_1(n) \cdots b_M(n) a_1(n) \cdots a_N(n)]^{\mathrm{T}}, \tag{23.11}$$

$$\mathbf{W}_{\mathrm{opt}} = [b_{0,\mathrm{opt}} b_{1,\mathrm{opt}} \cdots b_{M,\mathrm{opt}} a_{1,\mathrm{opt}} \cdots a_{N,\mathrm{opt}}]^{\mathrm{T}}, \tag{23.12}$$

$$\tilde{\mathbf{W}}(n) = \mathbf{W}_{\text{opt}} - \mathbf{W}(n), \tag{23.13}$$

$$\mathbf{W}_{\infty} = \lim_{n\to\infty} E\{\mathbf{W}(n)\}. \tag{23.14}$$

We will have occasion to use $\mathbf{W}$ to refer to the adaptive filter parameter vector when the parameters are considered to be held at fixed values. With this notation, we may for instance write $y_{\text{m}}(n) = \mathbf{U}_{\text{m}}^{\text{T}}(n)\mathbf{W}_{\text{opt}}$ and $y(n) = \mathbf{U}_{\text{o}}^{\text{T}}(n)\mathbf{W}(n)$.

The situation in which $y_{\text{u}}(n) \equiv 0$ is referred to as the sufficient order case. The situation in which $y_{\text{u}}(n) \not\equiv 0$ is termed the undermodeled case.

23.1.2 Algorithms and Performance Issues

A number of different algorithms for the adaptation of the parameter vector $W(n)$ in Equation 23.11 have been suggested. These may be characterized with respect to the form of the error criterion employed by the algorithm. Each algorithm attempts to drive to zero either the equation error, the output error, or some combination or hybrid of these two error criteria. Major algorithm classes that we consider for the equation error approach include the standard least-squares (LS) and least mean-square (LMS) algorithms, which parallel the algorithms used in adaptive FIR filtering. For equation error methods, we also examine the instrumental variables (IV) algorithm, as well as algorithms that constrain the parameters in the denominator of the adaptive filter's transfer function to improve estimation properties. In the output error class, we examine gradient algorithms and hyperstability-based algorithms. Within the equation and output error hybrid algorithm class, we focus predominantly on the Steiglitz–McBride (SM) algorithm, though there are several algorithms that are more straightforward combinations of equation and output error approaches.

In general, we desire that the adaptive filtering algorithm adjusts the parameter vector $\mathbf{W}_n$ so that it converges to $\mathbf{W}_{\text{opt}}$, the parameters that minimize the mean-square output error. The major issues for adaptive IIR filtering on which we will focus herein are

1. Conditions for the stability and convergence of the algorithm used to adapt $\mathbf{W}(n)$
2. Asymptotic value of the adapted parameter vector $\mathbf{W}_{\infty}$, and its relationship to $\mathbf{W}_{\text{opt}}$.

This latter issue relates to the minimum mean-square error achievable by the algorithm, as noted above. Other issues of importance include the convergence speed of the algorithm, its ability to track time variations of the "true" parameter values, and numerical properties, but these will receive less attention here. Of these, convergence speed is of particular concern to practitioners, especially as adaptive IIR filters tend to converge at a far slower rate than their FIR counterparts. However, we emphasize the stability and nature of convergence over the speed because if the algorithm fails to converge or converges to an undesirable solution, the rate at which it does so is of less concern. Furthermore, convergence speed is difficult to characterize for adaptive IIR filters due to a number of factors, including complicated dependencies on algorithm initializations, input signal characteristics, and the relationship between $x(n)$ and $d(n)$.

23.1.3 Some Preliminaries

Unless otherwise indicated, we assume in our discussion that all signals in Figure 23.1 are stationary, zero mean, random signals with finite variance. In particular, the properties we ascribe to the various algorithms are stated with this assumption and are presumed to be valid. Results that are based on a deterministic framework are similar to those developed here; see [1] for an example.

We shall also make use of the following definitions.

Definition 23.1: A (scalar) signal is persistently exciting (PE) of order L if, with

$$\mathbf{X}(n) = [x(n) \cdots x(n-L+1)]^{\mathrm{T}}, \tag{23.15}$$

there exist α and β satisfying $0 < \alpha < \beta < \infty$ such that $\alpha I < E\{\mathbf{X}(n)\mathbf{X}^{\mathrm{T}}(n)\} < \beta I$. The (vector) signal $\mathbf{X}(n)$ is then also said to be PE.

If $x(n)$ contains at least $L/2$ distinct sinusoidal components, then $x(n)$ is PE of order L. Any random signal $x(n)$ whose power spectrum is nonzero over a interval of nonzero width will be PE for any value of L in Equation 23.15. Such is the case, for example, if $x(n)$ is uncorrelated or if $x(n)$ is modeled as an AR, MA, or ARMA process driven by uncorrelated noise. PE conditions are required of all adaptive algorithms to ensure good behavior because if there is inadequate excitation to provide information to the algorithm, convergence of the adapted parameters estimates will not necessary follow [22].

Definition 23.2: A transfer function $H(q^{-1})$ is said to be strictly positive real (SPR) if $H(q^{-1})$ is stable and the real part of its frequency response is positive at all frequencies.

An SPR condition will be required to ensure convergence for a few of the algorithms that we discuss. Note that such a condition cannot be guaranteed in practice when $H(q^{-1})$ is an unknown transfer function, or when $H(q^{-1})$ depends on an unknown transfer function.

23.2 Equation Error Approach

To motivate the equation error approach, consider again Figure 23.1. Suppose that $y(n)$ in the figure were actually equal to $d(n)$. Then the system relationship $A(q^{-1}, n)y(n) = B(q^{-1}, n)x(n)$ would imply that $A(q^{-1}, n)d(n) = B(q^{-1}, n)x(n)$. But of course this last equation does not hold exactly, and we term its error the "equation error" $e_e(n)$. Hence, we define

$$e_e(n) = A(q^{-1}, n)d(n) - B(q^{-1}, n)x(n). \tag{23.16}$$

Using the notation developed in Equations 23.7 through 23.14, we find that

$$e_e(n) = d(n) - \mathbf{U}_e^{\mathrm{T}}(n)\mathbf{W}(n). \tag{23.17}$$

Equation error methods for adaptive IIR filtering typically adjust $\mathbf{W}(n)$ so as to minimize the mean-squared error (MSE) $J_{\mathrm{MSE}}(n) = E\{e_e^2(n)\}$, where $E\{\cdot\}$ denotes statistical expectation, or the exponentially weighted LS error $J_{\mathrm{LS}}(n) = \sum_{k=0}^{n} \lambda^{n-k} e_e^2(k)$.

23.2.1 LMS and LS Equation Error Algorithms

The equation error $e_e(n)$ of Equation 23.17 is the difference between $d(n)$ and a prediction of $d(n)$ given by $\mathbf{U}_e^{\mathrm{T}}(n)\mathbf{W}(n)$. Noting that $\mathbf{U}_e^{\mathrm{T}}(n)$ does not depend on $\mathbf{W}(n)$, we see that equation error adaptive IIR filtering is a type of linear prediction, and in particular the form of the prediction is identical to that arising in adaptive FIR filtering. One would suspect that many adaptive FIR filter algorithms would then apply directly to adaptive IIR filters with an equation error criterion, and this is in fact the case.

Two adaptive algorithms applicable to equation error adaptive IIR filtering are the LMS algorithm given by

$$\mathbf{W}(n+1) = \mathbf{W}(n) + \mu(n)\mathbf{U}_e(n)e_e(n). \tag{23.18}$$

and the recursive least-squares (RLS) algorithm given by

$$\mathbf{W}(n+1) = \mathbf{W}(n) + P(n)\mathbf{U}_e(n)e_e(n), \tag{23.19}$$

$$P(n) = \frac{1}{\lambda}\left(P(n-1) - \frac{P(n-1)\mathbf{U}_e(n)U_e^T(n)P(n-1)}{\lambda + \mathbf{U}_e^T(n)P(n-1)\mathbf{U}_e(n)}\right), \tag{23.20}$$

where the above expression for $P(n)$ is a recursive implementation of

$$P(n) = \left(\sum_{k=0}^{n} \lambda^{n-k}\mathbf{U}_e(k)\mathbf{U}_e^T(k)\right)^{-1}. \tag{23.21}$$

Some typical choices for $\mu(n)$ in Equation 23.18 are $\mu(n) \equiv \mu_0$, a constant, or $\mu(n) = \bar{\mu}/\left[\varepsilon + \mathbf{U}_e^T(n)\mathbf{U}_e(n)\right]$, a normalized step size. For convergence of the gradient algorithm in Equation 23.18, μ_0 is chosen in the range $0 < \mu_0 < 1/\left[(M+1)\sigma_x^2 + N\sigma_d^2\right]$, where $\sigma_x^2 = E\{x^2(n)\}$ and $\sigma_d^2 = E\{d^2(n)\}$. Typically, values of μ_0 in the range $0 < \mu_0 < 0.1/\left[(M+1)\sigma_x^2 + N\sigma_d^2\right]$ are chosen. With the normalized step size, we require $0 < \bar{\mu} < 2$ and $\varepsilon > 0$ for stability, with typical choices of $\bar{\mu} = 0.1$ and $\varepsilon = 0.001$. In Equation 23.20, we require that λ satisfy $0 < \lambda \leq 1$, with λ typically close to or equal to one, and we initialize $P(0) = \gamma I$ with γ a large, positive number. These results are analogous to the FIR filter cases considered in the earlier sections of this chapter.

These algorithms possess nice convergence properties, as we now discuss.

Property 23.1: *Given that x is PE of order $N + M + 1$, under Equation 23.18 and under Equations 23.19 and 23.20, with algorithm parameters chosen to satisfy the conditions noted above, then $E\{W(n)\}$ converges to a value $\mathbf{W}_\infty$ minimizing $J_{\mathrm{MSE}}(n)$ and $J_{\mathrm{LS}}(n)$, respectively, as $n \to \infty$.*

This property is desirable in that global convergence to parameter values optimal for the equation error cost function is guaranteed, just as with adaptive FIR filters. The convergence result holds whether the filter is operating in the sufficient order case or the undermodeled case. This is an important advantage of the equation error approach over other approaches. The reader is referred to Chapters 19 through 21 for further details on the convergence behaviors of these algorithms and their variations. As in the FIR case, the eigenvalues of the matrix $R = E\{\mathbf{U}_e(n)\mathbf{U}_e^T(n)\}$ determine the rates of convergence for the LMS algorithm. A large eigenvalue disparity in R engenders slow convergence in the LMS algorithm and ill-conditioning, with the attendant numerical instabilities, in the RLS algorithm. For adaptive IIR filters, compared to the FIR case, the presence of $d(n)$ in $\mathbf{U}_e(n)$ tends to increase the eigenvalue disparity, so that slower convergence is typically observed for these algorithms.

Of importance is the value of the convergence points for the LMS and RLS algorithms with respect to the modeling assumptions of the system identification configuration of Figure 23.1. For simplicity, let us first assume that the adaptive filter is capable of modeling the unknown system exactly; that is, $H_u(q^{-1}) = 0$. One may readily show that the parameter vector $\mathbf{W}$ that minimizes the mean-square equation error (or equivalently the asymptotic LS equation error, given ergodic stationary signals) is

$$\begin{aligned}\mathbf{W} &= E\{\mathbf{U}_e(n)\mathbf{U}_e^T(n)\}^{-1}E\{\mathbf{U}_e(n)d(n)\} \\ &= \left(E\{\mathbf{U}_m(n)\mathbf{U}_m^T(n)\} + E\{\mathbf{V}(n)\mathbf{V}^T(n)\}\right)^{-1} \qquad (23.22)\\ &= \left(E\{\mathbf{U}_m(n)y_m(n)\} + E\{\mathbf{V}(n)v(n)\}\right). \qquad (23.23)\end{aligned}$$

Clearly, if $v(n) \equiv 0$, the $\mathbf{W}$ so obtained must equal $\mathbf{W}_{opt}$, so that we have

$$\mathbf{W}_{opt} = E\{\mathbf{U}_m(n)\mathbf{U}_m^T(n)\}^{-1}E\{\mathbf{U}_m(n)y_m(n)\}. \tag{23.24}$$

By comparing Equations 23.23 and 23.24, we can easily see that when $v(n) \not\equiv 0, \mathbf{W} \neq \mathbf{W}_{opt}$. That is, the parameter estimates provided by Equations 23.18 through 23.20 are, in general, biased from the desired values, even when the noise term $v(n)$ is uncorrelated.

What effect on adaptive filter performance does this bias impose? Since the parameters that minimize the mean-square equation error are not the same as $\mathbf{W}_{opt}$, the values that minimize the mean-square output error, the adaptive filter performance will not be optimal. Situations can arise in which this bias is severe, with correspondingly significant degradation of performance.

Furthermore, a critical issue with regard to the parameter bias is the input–output stability of the resulting IIR filter. Because the equation error is formed as $A(q^{-1})d(n) - B(q^{-1})x(n)$, a difference of two FIR filtered signals, there are no built in constraints to keep the roots of $A(q^{-1})$ within the unit circle in the complex plane. Clearly, if an unstable polynomial results from the adaptation, then the filter output $y(n)$ can grow unboundedly in operational mode, so that the adaptive filter fails. An example of such a situation is given in [25]. An important feature of this example is that the adaptive filter is capable of precisely modeling the unknown system, and that interactions of the noise process within the algorithm are all that is needed to destabilize the resulting model.

Nonetheless, under certain operating conditions, this kind of instability can be shown not to occur, as described in the following.

Property 23.2: *[18] Consider the adaptive filter depicted in Figure 23.1, where $y(n)$ is given by Equation 23.2. If $x(n)$ is an autoregressive process of order no more than N, and $v(n)$ is independent of $x(n)$ and of finite variance, then the adaptive filter parameters minimizing the mean-square equation error $E\{e_e^2(n)\}$ are such that $A(q^{-1})$ is stable.*

For instance, if $x(n)$ is an uncorrelated signal, then the convergence point of the equation error algorithms corresponds to a stable filter.

To summarize, for LMS and RLS adaptation in an equation error setting, we have guaranteed global convergence, but bias in the presence of additive noise even in the exact modeling case, and an estimated model guaranteed to be stable only under a limited set of conditions.

23.2.2 Instrumental Variable Algorithms

A number of different approaches to adaptive IIR filtering have been proposed with the intention of mitigating the undesirable biased properties of the LMS- and RLS-based equation error adaptive IIR filters. One such approach, still within the equation error context, is the IV method. Observe that the bias problem illustrated above stems from the presence of $v(n)$ in both $\mathbf{U}_e(n)$ and in $e_e(n)$ in the update terms in Equations 23.18 and 23.19, so that second-order terms in $v(n)$ then appear in Equation 23.23. This simultaneous presence creates, in expectation, a nonzero, noise-dependent driving term to the adaptation. The IV algorithm approach addresses this by replacing $\mathbf{U}_e(n)$ in these algorithms with a vector $\mathbf{U}_{iv}(n)$ of IV that are independent of $v(n)$. If $\mathbf{U}_{iv}(n)$ remains correlated with $\mathbf{U}_m(n)$, the noiseless regressor, convergence to unbiased filter parameters is possible.

The IV algorithm is given by

$$\mathbf{W}(n+1) = \mathbf{W}(n) + \mu(n)P_{iv}(n)\mathbf{U}_{iv}(n)e_e(n). \tag{23.25}$$

$$P_{iv}(n) = \frac{1}{\lambda(n)}\left(P_{iv}(n-1) - \frac{P_{iv}(n-1)\mathbf{U}_{iv}(n)\mathbf{U}_e^T(n)P_{iv}(n-1)}{[\lambda(n)/\mu(n)] + \mathbf{U}_e^T(n)P_{iv}(n-1)\mathbf{U}_{iv}(n)}\right). \quad (23.26)$$

with $\lambda(n) = 1 - \mu(n)$. Common choices for $\lambda(n)$ are to set $\lambda(n) \equiv \lambda_0$, a fixed constant in the range $0 < \lambda < 1$ and usually chosen in the range between 0.9 and 0.99, or to choose $\mu(n) = 1/n$ and $\lambda(n) = 1 - \mu(n)$. As with RLS methods, $P(0) = \gamma I$ with γ a large, positive number. The vector $\mathbf{U}_{iv}(n)$ is typically chosen as

$$\mathbf{U}_{iv}(n) = [x(n) \cdots x(n-M) - z(n-1) - \cdots - z(n-N)]^T \quad (23.27)$$

with either

$$z(n) = -x(n-M) \quad \text{or} \quad z(n) = \frac{\bar{B}(q^{-1})}{\bar{A}(q^{-1})}x(n)\,. \quad (23.28)$$

In the first case, $\mathbf{U}_{iv}(n)$ is then simply an extended regressor in the input $x(n)$, while the second choice may be viewed as a regressor parallel to $\mathbf{U}_m(n)$, with $z(n)$ playing the role of $y_m(n)$. For this choice, one may think of $\bar{A}(q^{-1})$ and $\bar{B}(q^{-1})$ as fixed filters chosen to approximate $A_{opt}(q^{-1})$ and $B_{opt}(q^{-1})$, but the exact choice of $\bar{A}(q^{-1})$ and $\bar{B}(q^{-1})$ is not critical to the qualitative behavior of the algorithm. In both cases, note that $\mathbf{U}_{iv}(n)$ is independent of $v(n)$, since $d(n)$ is not employed in its construction.

The convergence of this algorithm is described by the following property, derived in [15].

Property 23.3: *In the sufficient order case with $x(n)$ PE of order at least $N + M + 1$, the IV algorithm in Equations 23.25 and 23.26 with $\mathbf{U}_{iv}(n)$ chosen according to Equations 23.27 or 23.28 causes $E\{\mathbf{W}(n)\}$ to converge to $\mathbf{W}_\infty = \mathbf{W}_{opt}$.*

There are a few additional technical conditions an $A_{opt}(q^{-1}), B_{opt}(q^{-1}), \bar{A}(q^{-1})$, and $\bar{B}(q^{-1})$ that are required for the property to hold. These conditions will be satisfied in almost all circumstances; for details, the reader is referred to [15]. This convergence property demonstrates that the IV algorithm does in fact achieve unbiased parameter estimates in the sufficient order case.

In the undermodeled case, little has been said regarding the behavior and performance of the IV algorithm. A convergence point $\mathbf{W}_\infty$ must satisfy $E\{\mathbf{U}_{iv}(n) - [d(n)\mathbf{U}_e^T(n)\mathbf{W}_\infty]\} = 0$, but no characterization of such points exists if N and M are not of sufficient order. Furthermore, it is possible for the IV algorithm to converge to a point such that $1/A(q^{-1})$ is unstable [9].

Notice that Equations 23.25 and 23.26 are similar in form to the RLS algorithm. One may postulate an "LMS-style" IV algorithm as

$$\mathbf{W}(n+1) = \mathbf{W}(n) + \mu(n)\mathbf{U}_{iv}(n)e_e(n), \quad (23.29)$$

which is computationally much simpler than the "RLS-style" IV algorithm of Equations 23.25 and 23.26. However, the guarantee of convergence of the algorithm to $\mathbf{W}_{opt}$ in the sufficient order case for the RLS-style algorithm is now complicated by an additional requirement on $\mathbf{U}_{iv}(n)$ for convergence of the algorithm in Equation 23.29. In particular, all eigenvalues of

$$R_{iv} = E\{\mathbf{U}_{iv}(n)\mathbf{U}_e^T(n)\} \quad (23.30)$$

must lie strictly in the right half of the complex plane. Since the properties of $\mathbf{U}_e(n)$ depend on the unknown relationship between $x(n)$ and $d(n)$, one is generally unable to guarantee *a priori* satisfaction of such conditions. This situation has parallels with the stability-theory approach to output error algorithms, as discussed later in this section.

Summarizing the IV algorithm properties, we have that in the sufficient order case, the RLS-style IV algorithm is guaranteed to converge to unbiased parameter values. However, an understanding and characterization of its behavior in the undermodeled case is yet incomplete, and the IV algorithm may produce unstable filters.

23.2.3 Equation Error Algorithms with Unit Norm Constraints

A different approach to mitigating the parameter bias in equation error methods arises as follows. Consider modifying the equation error of Equation 23.17 to

$$e_e(n) = a_0(n)d(n) - \mathbf{U}_e^T(n)\mathbf{W}(n). \tag{23.31}$$

In terms of the expression in Equation 23.16, this change corresponds to redefining the adaptive filter's denominator polynomial to be

$$A(q^{-1}, n) = a_0(n) + a_1(n)q^{-1} + \cdots + a_N(n)q^{-N}, \tag{23.32}$$

and allowing for adaptation of the new parameter $a_0(n)$. One can view the equation error algorithms that we have already discussed as adapting the coefficients of this version of $A(q^{-1}, n)$, but with a monic constraint that imposes $a_0(n) = 1$. Recently, several algorithms have been proposed that consider instead equation error methods with a unit norm constraint. In these schemes, one adapts $\mathbf{W}(n)$ and $a_0(n)$ subject to the constraint

$$\sum_{i=0}^{N} a_i^2(n) = 1. \tag{23.33}$$

Note that if $A(q^{-1}, n)$ is defined as in Equation 23.32, then $e_e(n)$ as constructed in Figure 23.1 is in fact the error $e_e(n)$ given in Equation 23.31.

The effect on the parameter bias stemming from this change from a monic to a unit norm constraint is as follows.

Property 23.4: *[18] Consider the adaptive filter in Figure 23.1 with $A(q^{-1}, n)$ given by Equation 23.32, with $v(n)$ an uncorrelated signal and with $H_u(q^{-1}) = 0$ (the sufficient order case). Then the parameter values* $\mathbf{W}$ *and a_0 that minimize $E\{e_e^2(n)\}$ subject to the unit norm constraint (Equation 23.33) satisfy* $\mathbf{W}/a_0 = \mathbf{W}_{opt}$.

That is, the parameter estimates are unbiased in the sufficient order case with uncorrelated output noise. Note that normalizing the coefficients in $\mathbf{W}$ by a_0 recovers the monic character of the denominator for $\mathbf{W}_{opt}$:

$$\frac{B(q^{-1})}{A(q^{-1})} = \frac{b_0 + b_1 q^{-1} + \cdots + b_{Mq-M}}{a_0 + a_1 q^{-1} + \cdots + a_{Nq-N}} \tag{23.34}$$

$$= \frac{(b_0/a_0) + (b_1/a_0)q^{-1} + \cdots + (b_M/a_0)q^{-M}}{1 + (a_1/a_0)q^{-1} + \cdots + (a_N/a_0)q^{-N}}. \tag{23.35}$$

In the undermodeled case, we have the following.

Property 23.5: *[18] Consider the adaptive filter in Figure 23.1 with $A(q^{-1}, n)$ given by Equation 23.32. If $x(n)$ is an autoregressive process of order no more than N, and $v(n)$ is independent of $x(n)$ and of finite variance, then the parameter values* **W** *and a_0 that minimize $E\{e_e^2(n)\}$ subject to the unit norm constraint (Equation 23.33) are such that $A(q^{-1})$ is stable. Furthermore, at those minimizing parameter values, if $x(n)$ is an uncorrelated input, then*

$$E\{e_e^2(n)\} \leq \sigma_{N+1}^2 + \sigma_v^2, \tag{23.36}$$

where σ_{N+1} is the $(N+1)$th *Hankel singular value of $H(z)$.*

Notice that Property 23.5 is similar to Property 23.2, except that we have the added bonus of a bound on the mean-square equation error in terms of the Hankel singular values of $H(q^{-1})$. Note that the $(N+1)$th Hankel singular value of $H(q^{-1})$ is related to the achievable modeling error in an Nth order, reduced order approximation to $H(q^{-1})$ (see [19] for details). This bound thus indicates that the optimal unit norm constrained equation error filter will in fact do about as well as can be expected with an Nth order filter. However, this adaptive filter will suffer, just as with the equation error approaches with the monic constraint on the denominator, from a possibly unstable denominator if the input $x(n)$ is not an autoregressive process.

An adaptive algorithm for minimizing the mean-square equation error subject to the unit norm constraint can be found in [4]. The algorithm of [4] is formulated as a recursive total LS algorithm using a two-channel, fast transversal filter implementation. The connection between total LS and the unit norm constrained equation error adaptive filter implies that the correlation matrices that are embedded within the adaptive algorithm will be more poorly conditioned than the correlation matrices arising in the RLS algorithm. Consequently, convergence will be slower for the unit norm constrained approach than in the standard, monic constraint approach.

More recently, several new algorithms that generalize the above approach to confer unbiasedness in the presence of correlated output noises $v(n)$ have been proposed [5]. These algorithms require knowledge of the statistics of $v(n)$, though versions of the algorithms in which these statistics are estimated online are also presented in [5]. However, little is known about the transient behaviors or the local stabilities of these adaptive algorithms, particularly in the undermodeled case.

In conclusion, minimizing the equation error cost function with a unit norm constraint on the autoregressive parameter vector provides bias-free estimates in the sufficient order case and a bias level similar to the standard equation error methods in the undermodeled case. Adaptive algorithms for constrained equation error minimization are under development, and their convergence properties are largely unknown.

23.3 Output Error Approach

We have already noted that the error of merit for adaptive IIR filters is the output error $e_o(n)$. We now describe a class of algorithms that explicitly uses the output error in the parameter updates. We distinguish between two categories within this class: those algorithms that directly attempt to minimize the LS or mean-square output error, and those formulated using stability theory to enforce convergence to the "true" system parameters. This class of algorithms has the advantage of eliminating the parameter bias that occurs in the equation error approach. However, as we will see, the price paid is that convergence of the algorithms becomes more complicated, and unlike in the equation error methods, global convergence to the desired parameter values is no longer guaranteed.

Critical to the formulation of these output error algorithms is an understanding of the relationship of $\mathbf{W}(n)$ to $e_o(n)$. With reference to Figure 23.1, we have

$$e_o(n) = d(n) - \frac{B(q^{-1}, n)}{A(q^{-1}, n)} x(n). \tag{23.37}$$

Using the notation in Equations 23.7 through 23.14 and following a standard derivation, [19] shows that

$$y_{\mathrm{m}}(n) - y(n) = \frac{1}{A_{\mathrm{opt}}(q^{-1})}\left[\mathbf{U}_{\mathrm{o}}^{\mathrm{T}}(n)\tilde{\mathbf{W}}(n)\right], \tag{23.38}$$

so that

$$e_{\mathrm{o}}(n) = \frac{1}{A_{\mathrm{opt}}(q^{-1})}\left[\mathbf{U}_{\mathrm{o}}^{\mathrm{T}}(n)\tilde{\mathbf{W}}(n)\right] + y_{\mathrm{u}}(n) + v(n). \tag{23.39}$$

The expression in Equation 23.39 makes clear two characteristics of $e_{\mathrm{o}}(n)$. First, $e_{\mathrm{o}}(n)$ separates the error due to the modeled component, which is the term based on $\tilde{\mathbf{W}}(n)$, from the error due to the unmodeled effects in $d(n)$, that is $y_{\mathrm{u}}(n) + v(n)$. Neither $y_{\mathrm{u}}(n)$ nor $v(n)$ appear in the term based on $\tilde{\mathbf{W}}(n)$. Second, $e_{\mathrm{o}}(n)$ is nonlinear in $\mathbf{W}(n)$, since $\mathbf{U}_{\mathrm{o}}(n)$ depends on $\mathbf{W}(n)$. The first feature leads to the desirable unbiasedness characteristic of output error methods, while the second is a source of difficulty for defining globally convergent algorithms.

23.3.1 Gradient-Descent Algorithms

An output error-based gradient-descent algorithm may be defined as follows. Set

$$x_{\mathrm{f}}(n) = \frac{1}{A(q^{-1}, n)}x(n), \quad y_{\mathrm{f}}(n) = \frac{1}{A(q^{-1}, n)}y(n), \tag{23.40}$$

and define

$$\mathbf{U}_{\mathrm{of}}(n) = [x_{\mathrm{f}}(n)\cdots x_{\mathrm{f}}(n-M) - y_{\mathrm{f}}(n-1) - \cdots - y_{\mathrm{f}}(n-N)]^{\mathrm{T}}. \tag{23.41}$$

Then

$$\mathbf{W}(n+1) = \mathbf{W}(n) + \mu(n)\mathbf{U}_{\mathrm{of}}(n)e_{\mathrm{o}}(n) \tag{23.42}$$

defines an approximate stochastic-gradient (SG) algorithm for adapting the parameter vector $\mathbf{W}(n)$. The direction of the update term in Equation 23.42 is opposite to the gradient of $e_{\mathrm{o}}(n)$ with respect to $\mathbf{W}(n)$, assuming that the parameter vector $\mathbf{W}(n)$ varies slowly in time. To see how a gradient descent results in this algorithm, note that the output error may be written as

$$e_{\mathrm{o}}(n) = d(n) - y(n) \tag{23.43}$$

$$= d(n) - \sum_{i=0}^{M} b_i(n)x(n-i) + \sum_{i=1}^{N} a_i(n)y(n-i) \tag{23.44}$$

so that

$$\frac{\partial e_{\mathrm{o}}(n)}{\partial b_i(n)} = -x(n-i) + \sum_{i=1}^{N} a_i(n)\frac{\partial y(n-1)}{\partial b_i(n)}. \tag{23.45}$$

Noting that $\partial e_{\mathrm{o}}(n)/\partial b_i(n) = -\partial y(n)/\partial b_i(n)$, and assuming that the parameter $b_i(n)$ varies slowly enough so that

$$\frac{\partial e(n-i)}{\partial b_i(n)} \approx \frac{\partial e(n-i)}{\partial b_i(n-i)}, \tag{23.46}$$

Equation 23.45 becomes

$$\frac{\partial e(n)}{\partial b_i(n)} \approx -x(n-i) - \sum_{i=1}^{N} a_i(n)\frac{\partial e(n-i)}{\partial b_i(n-i)}. \tag{23.47}$$

This equation can be rearranged to

$$\frac{\partial e(n-i)}{\partial b_i(n)} \approx -A(q^{-1},n)x(n-i) = -x_f(n-i). \tag{23.48}$$

The relation

$$\frac{\partial e(n-i)}{\partial a_i(n)} \approx A(q^{-1},n)y(n-i) = y_f(n-i) \tag{23.49}$$

may be found in a similar fashion. Since the gradient descent algorithm is

$$b_i(n+1) = b_i(n) - \frac{\mu}{2}\frac{\partial e_o^2(n)}{\partial b_i(n)} \tag{23.50}$$

$$\approx b_i(n) + \mu x_f(n-i)e_o(n) \tag{23.51}$$

$$a_i(n+1) = a_i(n) - \frac{\mu}{2}\frac{\partial e_o^2(n)}{\partial a_i(n)} \tag{23.52}$$

$$\approx b_i(n) - \mu y_f(n-i)e_o(n), \tag{23.53}$$

Equation 23.42 follows.

The step size $\mu(n)$ is typically chosen either as a constant μ_0, or normalized by $\mathbf{U}_{of}(n)$ as

$$\mu(n) = \frac{\bar{\mu}}{1 + \bar{\mu}\mathbf{U}_{of}^T(n)\mathbf{U}_{of}(n)}. \tag{23.54}$$

Due to the nonlinear relationship between the parameters and the output error, selection of values for $\mu(n)$ is less straightforward than in the equation error case. Roughly speaking, one would like that $0 < \mu(n) \leq 1/\mathbf{U}_{of}^T(n)\mathbf{U}_{of}(n)$, or more conservatively, $\mu(n) = 0.1/\mathbf{U}_{of}^T(n)\mathbf{U}_{of}(n)$. This suggests setting $\mu_0 = 0.1/E\{\mathbf{U}_{of}^T(n)\mathbf{U}_{of}(n)\}$, given an estimate of the expected value, or $\bar{\mu}$ at about the same value. The behavior of the algorithm using the normalized step size of Equation 23.54 is in general less sensitive to variations in $\bar{\mu}$ than is the unnormalized version with respect to choice of μ_0.

Another alternative to Equation 23.42 is the Gauss–Newton (GN) algorithm given by

$$\mathbf{W}(n+1) = \mathbf{W}(n) + \mu(n)P(n)\mathbf{U}_{of}(n)e_o(n), \tag{23.55}$$

$$P(n) = \frac{1}{\lambda(n)}\left(P(n-1) - \frac{P(n-1)\mathbf{U}_{of}(n)\mathbf{U}_{of}^T(n)P(n-1)}{[\lambda(n)/\mu(n)] + \mathbf{U}_{of}^T(n)P(n-1)\mathbf{U}_{of}(n)}\right), \tag{23.56}$$

while setting $\lambda(n) = 1 - \mu(n)$, and $P(0) = \gamma I$ just as for the IV algorithm. Most frequently $\lambda(n)$ is chosen as a constant in the range between 0.9 and 0.99. Another choice is to set $\mu(n) = 1/n$, a decreasing adaptation gain, but when $\mu(n)$ tends to zero, one loses adaptability. The GN algorithm is a descent strategy utilizing approximate second order information, with the matrix $P(n)$ being an approximation of the inverse of the Hessian of $e_o(n)$ with respect to $\mathbf{W}(n)$. Note the similarity of Equations 23.55 and 23.56 to Equations 23.19 and 23.20. In fact, replacing $\mathbf{U}_{of}(n)$ in the GN algorithm with $\mathbf{U}_e(n)$, replacing $e_o(n)$ with $e_e(n)$, and setting $\mu(n) = (1-\lambda)/(1-\lambda^n)$, one recovers the RLS algorithm, though the interpretation of $P(n)$ in this form is slightly different. As n gets large, the choice of constant λ and $\mu = 1-\lambda$ approximates RLS (with $0 < \lambda < 1$).

Precise convergence analyses of these two algorithms are quite involved and rely on a number of technical assumptions. Analyses fall into two categories. One approach treats the step size $\mu(n)$ in Equation 23.42 and in Equations 23.55 and 23.56 as a quantity that tends to zero, satisfying the following properties:

$$(1)\ \mu(n) \to 0, \quad (2)\ \lim_{L\to\infty} \sum_{n=0}^{L} \mu(n) \to \infty, \quad \text{and} \quad (3)\ \lim_{L\to\infty} \sum_{n=0}^{L} \mu^2(n) < \infty; \tag{23.57}$$

for instance $\mu(n) = 1/n$, as noted above. The ODE analysis of [15] applies in this situation. Assuming a decreasing step size is a necessary technicality to enable convergence of the adapted parameters to their optimum values in a random environment. The second approach allows μ to remain as a fixed, but small, step size, as in [3]. The results describe the probabilistic behavior of $\mathbf{W}(n)$ over finite intervals of time, with the extent of the interval increasing and the degree of variability of $\mathbf{W}(n)$ decreasing as the fixed value of μ becomes smaller.

However, in both cases, the conclusions are essentially the same. The behavior of these algorithms with small enough step size μ is to follow gradient descent of the mean-square output error $E\{e_o^2(n)\}$. We should note that a technical requirement for the analyses to remain valid is that signals within the adaptive filter remain bounded, and to insure this the stability for the polynomial $A(q^{-1}, n)$ must be maintained to ensure this requirement. Therefore, at each iteration of the gradient descent algorithm, one must check the stability of $A(q^{-1}, n)$ and, if it is unstable, prevent the update of the $a_i(n+1)$ values in $\mathbf{W}(n+1)$, or project $\mathbf{W}(n+1)$ back into the set of parameter vector values whose corresponding $A(q^{-1}, n)$ polynomial is stable [11,15]. For direct-form adaptive filters, this stability check can be computationally burdensome, but it is necessary as the algorithm often fails to converge without its implementation, especially if $A_{opt}(q^{-1})$ has roots near to the unit circle. Imposing a stability check at each iteration of the algorithm guarantees the following result.

Property 23.6: *For the SG or GN algorithm with decreasing $\mu(n)$ satisfying Equation 23.57,* $\mathbf{W}(n)$ *converges to a value locally minimizing the mean-square output error or locks up on a point on the stability boundary where* $\mathbf{W}$ *represents a marginally stable filter. For the SG or GN algorithm with constant μ that is small enough,* $\mathbf{W}(n)$ *remains close in probability to a trajectory approaching a value locally minimizing the mean-square output error.*

This property indicates that the value of $\mathbf{W}(n)$ found by these algorithms does in practice approach a local minimum of the mean-square output error surface. A stronger analytic statement of this expected convergence is unfortunately not possible, and in fact, the probability of large deviations of the algorithms from a minimum point becomes large with time with constant μ. As a practical matter, however, one can expect the parameter vector to approach and stay near a minimizing parameter value using these methods.

More problematic, however, is whether effective convergence to a global minimum is achieved. A thorough treatment of this issue appears in [16] and [26], with conclusions as follows:

Property 23.7: *In the sufficient order case ($y_u \equiv 0$) with an uncorrelated input $x(n)$, all minima of $E\{e_o^2(n)\}$ are global minima.*

The same conclusion holds if $x(n)$ is generated as an ARMA process, given satisfaction of certain conditions on the orders of the adaptive filter, the unknown system, and the system generating $x(n)$; see [26] for details. However, in the undermodeled case ($y_u(n) \not\equiv 0$), it is possible for the system to converge to a local but not global minimum. Several examples of this are presented in [16]. Since the insufficient order case will likely be the one encountered in practice, the possibility of convergence to a local but not global minimum will always exist with these gradient descent output error algorithms. It is possible that these local minima will provide a level of mean-square output error much greater than that obtained at the global minimum, so serious performance degradation may result. However, any such minimum must correspond to a stable parametrization of the adaptive filter, in contrast with the equation error methods for which there is no such guarantee in the most general of circumstances.

We have the following summary. The output error gradient descent algorithms converge to a stable filter parametrization that locally minimizes the mean-square output error. These algorithms are unbiased, and reach a global minimum, when $y_u(n) \equiv 0$, but when the true system has been undermodeled, convergence to a local but not global minimum is likely.

23.3.2 Output Error Algorithms Based on Stability Theory

One of the first adaptive IIR filters to be proposed employs the parameter update

$$\mathbf{W}(n+1) = \mathbf{W}(n) + \mu(n)\mathbf{U}_o(n)e_o(n). \tag{23.58}$$

This algorithm, often referred to as a pseudolinear regression, Landau's algorithm, or Feintuch's algorithm, was proposed as an alternative to the gradient descent algorithm of Equation 23.42. This algorithm is similar in form to Equation 23.18, save that the regressor vector and error signal of the output error formulation appear in place of their equation error counterparts. In essence, the update in Equation 23.58 ignores the nonlinear dependence of $e_o(n)$ on $\mathbf{W}(n)$, and takes the form of an algorithm for a linear regression, hence the label "pseudolinear" regression. One advantage of Equation 23.58 over the algorithm in Equation 23.42 is that the former is computationally simpler as it avoids the filtering operations necessary to generate Equation 23.40. An additional requirement is needed for stability of the algorithm, however, as we now discuss.

The algorithm of Equation 23.58 is one possibility among a broad range of algorithms studied in [21], given by

$$\mathbf{W}(n+1) = \mathbf{W}(n) + \mu(n)F(q^{-1},n)[\mathbf{U}_o(n)]G(q^{-1},n)[e_o(n)], \tag{23.59}$$

where $F(q^{-1},n)$ and $G(q^{-1},n)$ are possibly time-varying filters, and where $F(q^{-1},n)$ acting on the vector $\mathbf{U}_o(n)$ denotes an element-by-element filtering operation. We can see that setting $F(q^{-1}) = G(q^{-1}) = 1$ yields Equation 23.58. The convergence of these algorithms has been studied using the theory of hyperstability and the theory of averaging [1]. For this reason, we classify this family as "stability theory based" approaches to adaptive IIR filtering. The method behind Equation 23.59 can be understood by considering the algorithm subclass represented by

$$\mathbf{W}(n+1) = \mathbf{W}(n) + \mu(n)\mathbf{U}_o(n)G(q^{-1})[e_o(n)], \tag{23.60}$$

which is known as the simplified hyperstable adaptive recursive filter or SHARF algorithm [12]. A GN-like alternative is

$$\mathbf{W}(n+1) = \mathbf{W}(n) + \mu(n)P(n)\mathbf{U}_o(n)G(q^{-1})[e_o(n)]\ , \tag{23.61}$$

$$P(n) = \frac{1}{\lambda(n)}\left(P(n-1) - \frac{P(n-1)\mathbf{U}_o(n)\mathbf{U}_o^T(n)P(n-1)}{[\lambda(n)/\mu(n)] + \mathbf{U}_o^T(n)P(n-1)\mathbf{U}_o(n)}\right). \tag{23.62}$$

with again $\lambda(n) = 1 - \mu(n)$. Choice of $\mu(n), \lambda(n)$, and $P(0)$ are similar to those for other algorithms.

The averaging analyses applied in [21] to Equations 23.60 through 23.62 obtain the following convergence results, with reference to Definitions 23.3 and 23.15 in the Introduction.

Property 23.8: *If $G(q^{-1})/A_{opt}(q^{-1})$ is SPR and $\mathbf{U}_m(n)$ is a bounded, PE vector sequence,* then when $y_u(n) = v(n) = 0$, there exists a μ_0 such that $0 < \mu < \mu_0$ implies that Equation 23.60 is locally exponentially stable about $\mathbf{W} = \mathbf{W}_{opt}$. It also follows that nonzero, but small, $y_u(n)$ and $v(n)$ result in a bounded perturbation of $\mathbf{W}$ from $\mathbf{W}_{opt}$. If the SPR condition is strengthened to $[G(q^{-1})/A_{opt}(q^{-1})] - (1/2)$ being SPR, then the results apply to Equations 23.61 and 23.62 with μ constant and small enough.*

The essence of the analysis is to describe the average behavior of the parameter error $\tilde{\mathbf{W}}(n)$ under Equation 23.60 by

$$\tilde{\mathbf{W}}_{avg}(n+1) = (I - \mu R)\tilde{\mathbf{W}}_{avg}(n) + \mu\xi_1(n) + \mu^2\xi_2(n). \tag{23.63}$$

In Equation 23.63, the signal $\xi_1(n)$ is a term dependent on $y_u(n)$ and $v(n)$, the signal $\xi_2(n)$ represents the approximation error made in linearizing and averaging the update, and the matrix R is given by

$$R = \text{avg}\left\{\mathbf{U}_m(n)\left(\frac{G(q^{-1})}{A_{opt}(q^{-1})}[\mathbf{U}_m(n)]\right)^T\right\}. \tag{23.64}$$

The SPR and PE conditions imply that the eigenvalues of R all have positive real part, so that μ may then be chosen small enough so that the eigenvalues of $I - \mu R$ are all less than one in magnitude. Then $\tilde{\mathbf{W}}_{avg}(n+1) = (I - \mu R)\tilde{\mathbf{W}}_{avg}(n)$ is exponentially stable, and Property 23.8 follows. The exponential stability of Equation 23.63 is the property that allows the algorithm to behave robustly in the presence of a number of effects, including that the undermodeling [1].

The above convergence result is local in nature and is the best that can be stated for this class of algorithms in the general case. In the variation proposed for system identification by Landau and interpreted for adaptive filters by Johnson [10], a stronger statement of convergence can be made in the exact modeling case when $y_u(n)$ is zero and assuming that $v(n) = 0$. In that situation, given satisfaction of the SPR and PE conditions, $\mathbf{W}(n)$ can be shown to converge to $\mathbf{W}_{opt}$. For nonzero $v(n)$, analyses with a vanishing step size $\mu(n) \to \infty$ have established this convergence, again assuming exact modeling, even in the presence of a correlated noise term $v(n)$ [20]. One advantage on this convergence result in comparison to the exact modeling convergence result for the gradient algorithm (Equation 23.42) is that the PE condition on the input is less restrictive than the conditions that enable global convergence of the gradient algorithm. Nonetheless, in the undermodeled case, convergence is local in nature, and although the robustness conferred by the local, exponential stability to some extent mitigates this problem, it represents a drawback to the practical application of these techniques.

A further drawback of this technique is the SPR condition that $G(q^{-1})/A_{opt}(q^{-1})$ must satisfy. The polynomial $A_{opt}(q^{-1})$ is of course unknown to the adaptive filter designer, presenting difficulties in the selection of $G(q^{-1})$ to ensure that $G(q^{-1})/A_{opt}(q^{-1})$ is SPR. Recent research into choices of filters $G(q^{-1})$ that render $G(q^{-1})/A_{opt}(q^{-1})$ SPR for all $A_{opt}(q^{-1})$ within a set of filters, a form of "robust SPR"

* The PE condition applies to $\mathbf{U}_m(n)$, rather than $\mathbf{U}_o(n)$, since this an analysis local to $\mathbf{W} = \mathbf{W}_{opt}$, where $\mathbf{U}_o(n) = \mathbf{U}_m(n)$.

result, has begun to address this issue [2], but the problem of selecting $G(q^{-1})$ has not yet been completely resolved.

To summarize, for the SHARF algorithm and its cousins, we have convergence to unbiased parameter values guaranteed in the sufficient order case when there is adequate excitation and an SPR condition is satisfied. Satisfaction of the SPR condition cannot be guaranteed without *a priori* knowledge of the optimal filter, however. In the undermodeled case, no general results can be stated, but as long as the unmodeled component of the optimal filter is small in some sense, the exponential convergence in the sufficient order case implies stable behavior in this situation. Filter order selection to make the unmodeled component small again requires *a priori* knowledge about the optimal filter.

23.4 Equation-Error/Output-Error Hybrids

We have seen that equation error methods enjoy global convergence of their parameters, but suffer from parameter estimation bias, while output error methods enjoy unbiased parameter estimates while suffering from difficulties in their convergence properties. A number of algorithms have been proposed that in a sense strive to combine the best of both of these approaches. The most important of these is the SM algorithm, which we consider in detail below. Several other algorithms in this class work by using convex combinations of terms in the equation error and output error parameter updates. Two such algorithms include the bias remedy LMS algorithm of [14] and the composite regressor algorithm of [13]. We will not consider these algorithms here; for details, see [13] and [14].

23.4.1 Steiglitz–McBride Family of Algorithms

The SM algorithm is adapted from an off-line system identification method that iteratively minimizes the squared equation error criterion using prefiltered data. The prefiltering operations are based on the results of the previous iteration in such a way that the algorithm bears a close relationship to an output error approach.

A clear understanding of the algorithm in an adaptive filtering context is best obtained by first considering the original off-line method. Given a finite record of input and output sequences $x(n)$ and $d(n)$, one first forms the equation error according to Equation 23.16. The parameters of $A(q^{-1})$ and $B(q^{-1})$ minimizing the LS criterion for this error are then found, and the minimizing polynomials are labeled as $A^{(0)}(q^{-1})$ and $B^{(0)}(q^{-1})$. The SM method then proceeds iteratively by minimizing the LS criterion for

$$e_{\mathrm{e}}^{(i)}(n) = A(q^{-1})d_{\mathrm{f}}^{(i)}(n) - B(q^{-1})x_{\mathrm{f}}^{(i)}(n) \tag{23.65}$$

to find $A^{(i)}(q^{-1})$ and $B^{(i)}(q^{-1})$, where

$$d_{\mathrm{f}}^{(i)}(n) = \frac{1}{A^{(i-1)}(q^{-1})}d(n) \quad \text{and} \quad x_{\mathrm{f}}^{(i)}(n) = \frac{1}{A^{(i-1)}(q^{-1})}x(n). \tag{23.66}$$

Notice that at each iteration, we find $A^{(i)}(q^{-1})$ and $B^{(i)}(q^{-1})$ through equation error minimization, for which we have globally convergent methods as discussed previously.

Let $A^{(\infty)}(q^{-1})$ and $B^{(\infty)}(q^{-1})$ denote the polynomials obtained at a convergence point of this algorithm. Then minimizing the LS criterion applied to

$$e_{\mathrm{e}}^{(\infty)}(n) = A(q^{-1})\frac{1}{A^{(\infty)}(q^{-1})}d(n) - B(q^{-1})\frac{1}{A^{(\infty)}(q^{-1})}x(n) \tag{23.67}$$

results again in $A(q^{-1}) = A^{(\infty)}(q^{-1})$ and $B(q^{-1}) = B^{(\infty)}(q^{-1})$ by virtue of this solution being a convergence point, and the error signal at this minimizing choice of parameters is

$$e_{\mathrm{e}}^{(\infty)}(n) = d(n) - \frac{B^{(\infty)}(q^{-1})}{A^{(\infty)}(q^{-1})}x(n). \tag{23.68}$$

Comparing Equation 23.68 to Equation 23.37, we see that at a convergence point of the SM algorithm, $e_{\mathrm{e}}^{(\infty)}(n) = e_{\mathrm{o}}(n)$, thereby drawing the connection between equation error and output error approaches in the SM approach.* Because of this connection, one expects that the parameter bias problem is mitigated, and in fact this is the case, as demonstrated by the following property.

Property 23.9: *[27] If $y_u(n) \equiv 0$ and $v(n)$ is white noise, then with $x(n)$ PE of order at least $N + M + 1, B(q^{-1}) = B_{\mathrm{opt}}(q^{-1})$ and $A(q^{-1}) = A_{\mathrm{opt}}(q^{-1})$ is the only convergence point of the SM algorithm, and this point is locally stable.*

The local stability implies that if the initial denominator estimate $A^{(0)}(q^{-1})$ is close enough to $A_{\mathrm{opt}}(q^{-1})$, then the algorithm converges to the unbiased solution in the uncorrelated noise case.

The on-line variation of the SM algorithm useful for adaptive filtering applications is given as follows. Set $x_{\mathrm{f}}(n)$ as in Equation 23.40 and set

$$d_{\mathrm{f}}(n) = \frac{1}{A(q^{-1}, n+1)}d(n). \tag{23.69}$$

The $(n+1)$ index in the above filter is reasonable as only past $d_{\mathrm{f}}(n)$ samples shall appear in the parameter updates at time n. Then by defining the SM regressor vector as

$$\mathbf{U}_{\mathrm{ef}}(n) = [x_{\mathrm{f}}(n) \cdots x_{\mathrm{f}}(n-M) - d_{\mathrm{f}}(n-1) - \cdots - d_{\mathrm{f}}(n-N)]^{\mathrm{T}}, \tag{23.70}$$

the algorithm is

$$\mathbf{W}(n+1) = \mathbf{W}(n) + \mu(n)\mathbf{U}_{\mathrm{ef}}(n)e_{\mathrm{o}}(n)\ . \tag{23.71}$$

Alternatively, we may employ the GN-style version given by

$$\mathbf{W}(n+1) = \mathbf{W}(n) + \mu(n)P_{\mathrm{ef}}(n)\mathbf{U}_{\mathrm{ef}}(n)e_{\mathrm{o}}(n), \tag{23.72}$$

$$P_{\mathrm{ef}}(n) = \frac{1}{\lambda(n)}\left(P_{\mathrm{ef}}(n-1) - \frac{P_{\mathrm{ef}}(n-1)\mathbf{U}_{\mathrm{ef}}(n)\mathbf{U}_{\mathrm{ef}}^{\mathrm{T}}(n)P_{\mathrm{ef}}(n-1)}{[\lambda(n)/\mu(n)] + \mathbf{U}_{\mathrm{ef}}^{\mathrm{T}}(n)P_{\mathrm{ef}}(n-1)\mathbf{U}_{\mathrm{ef}}(n)}\right), \tag{23.73}$$

with $\lambda(n), \mu(n)$, and $P(0)$ chosen in the same fashion as with the IV and GN algorithms. For these algorithms, the signal $e_{\mathrm{o}}(n)$ is the output error, constructed as shown in Figure 23.1, a reflection of the connection of the SM and output error approaches noted above. Also, note that $\mathbf{U}_{\mathrm{ef}}(n)$ is a filtered version of the equation error regressor $\mathbf{U}_{\mathrm{e}}(n)$, but with the time index of the filtering operation of Equation 23.69 set to $n+1$ rather than n, reflecting the derivation of the algorithm from the iterative off-line procedure. This form of the algorithm is only one of several variations; see [8] or [19] for others.

---

* Realize, however, that minimizing the square of $e_{\mathrm{o}}^{(\infty)}(n)$ in Equation 23.67 is *not* equivalent to minimizing the squared output error, and in general these two approaches can result in different values for $A(q^{-1})$ and $B(q^{-1})$.

Assuming that one monitors and maintains the stability of the adapted polynomial $A(q^{-1}, n)$, in order that the signals $x_f(n)$ and $d_f(n)$ remain bounded, this algorithm has the following properties [19].

Property 23.10: *[6] In the sufficient order case where $y_u(n) \equiv 0$ and with $v(n)$ an uncorrelated noise sequence and $x(n)$ PE of order at least $N + M + 1$, the online SM algorithm converges to $\mathbf{W}_\infty = \mathbf{W}_{opt}$ or locks up on the stability boundary.*

Property 23.11: *[19] In the sufficient order case where $y_u(n) \equiv 0$ with $v(n)$ a correlated sequence, and in the undermodeled case where $y_u(n) \not\equiv 0$, the existence of convergence points $\mathbf{W}_\infty$ of the online SM algorithm is not guaranteed, and if these convergence points exist, they are generally biased away from $\mathbf{W}_{opt}$.*

Property 23.12: *[19] In the undermodeled case with the order of the adaptive filter numerator and denominator both equal to N, and with $x(n)$ an uncorrelated sequence, then at the convergence points of the online SM algorithm, if they exist,*

$$E\{e_o^2(n)\} \leq \sigma_{N+1}^2 + \frac{\max_\omega S_v(e^{j\omega}) - \sigma_v^2}{\sigma_u^2} + \sigma_v^2, \tag{23.74}$$

where

σ_{N+1} is the $(N + 1)$th Hankel singular value of $H(z) = H_m(z) + H_u(z)$
$S_v(e^{j\omega})$ is the power spectral density function of $v(n)$

Note that in the off-line version for either the sufficient order case with correlated $v(n)$ or the undermodeled case, the SM algorithm can possibly converge to a set of parameters yielding an unstable filter. The stability check and projection steps noted above will prevent such convergence in the online version, contributing in part to the possibility of non-convergence.

The pessimistic nature of Properties 23.11 and 23.12 with regard to existence of convergence points is somewhat unfair in the following sense. In practice, one finds that in most circumstances the SM algorithm does converge, and furthermore that the convergence point is close to the minimum of the mean-square output error surface [7]. Property 23.12 quantifies this closeness. The $(N + 1)$th Hankel singular value of a transfer function is an upper bound for the minimum mean-square output error of an Nth order transfer function approximation [19]. Hence, one sees that $\mathbf{W}_\infty$ under the SM algorithm is guaranteed to remain close in this sense to $\mathbf{W}_{opt}$, and this fact remains true regardless of the existence or relative values of local minima on the mean-square output error surface.

The second term in Equation 23.74 describes the effect of the noise term. The fact that $\max_\omega S_v(e^{j\omega}) = \sigma_v^2$ for uncorrelated $v(n)$ shows the disappearance of noise effects in that case. For strongly correlated $v(n)$, the effect of this noise term will increase, as the adaptive filter attempts to model the noise as well as the unknown system, and of course this effect is reduced as the signal-to-noise ratio of $d(n)$ is increased. One sees, then, that with strongly correlated noise, the SM algorithm may produce a significantly biased solution $\mathbf{W}_\infty$.

To summarize, given adequate excitation, the SM algorithm in the sufficient order case converges to unbiased parameter values when the noise $v(n)$ is uncorrelated, and generally converges to biased parameter values when $v(n)$ is correlated. The SM algorithm is not guaranteed to converge in the undermodeled case and, furthermore, if it converges, there is no general guarantee of stability of the resulting filter. However, a bound of the modeling error in these instances quantifies what can be considered as the good performance of the algorithm when it converges.

23.5 Alternate Parametrizations

Thus far we have couched our discussion of adaptive IIR filters in terms of a direct-form implementation of the system. Direct-form implementations suffer from poor finite precision effects both in terms of coefficient quantization and round-off effects in their computations. Furthermore, in output error adaptive IIR filtering, one must check the stability of the adaptive filter's denominator polynomial at each iteration. In direct-form implementations, this stability check is cumbersome and computationally expensive to implement. For these reasons, adaptive IIR filters implemented in alternative realizations such as parallel-form, cascade-form, and lattice-form have been proposed. For these structures, a stability check is easily implemented.

The SG and GN algorithms of Equations 23.42, 23.55, and 23.56, respectively, are easily adapted for many of the alternate parametrizations. The resulting updates are

$$\mathbf{W}(n+1) = \mathbf{W}(n) + \mu(n)\mathbf{U}_{\text{alt}}(n)e_{\text{o}}(n) \tag{23.75}$$

and

$$\mathbf{W}(n+1) = \mathbf{W}(n) + \mu(n)P(n)\mathbf{U}_{\text{alt}}(n)e_{\text{o}}(n) \tag{23.76}$$

$$P(n) = \frac{1}{\lambda(n)}\left(P(n-1) - \frac{P(n-1)\mathbf{U}_{\text{alt}}(n)\mathbf{U}_{\text{alt}}^{\text{T}}(n)P(n-1)}{[\lambda(n)/\mu(n)] + \mathbf{U}_{\text{alt}}^{\text{T}}(n)P(n-1)\mathbf{U}_{\text{alt}}(n)}\right), \tag{23.77}$$

respectively, where all signal definitions parallel those for the direct-form algorithms, save that $\mathbf{U}_{\text{alt}}(n)$ equals the gradient of the filter output with respect to the adapted parameters. Note that $\mathbf{U}_{\text{of}}(n)$ in Equations 23.42, 23.55, and 23.56 is such a gradient for the direct-form implementation. The output gradient $\mathbf{U}_{\text{of}}(n)$ for the direct-form was constructed as shown in Equations 23.40 and 23.41. For alternate parametrizations, these output gradients may be constructed as described in [29]. In [29], the implementation for a two-multiplier lattice adaptive IIR filter is shown, but the methodology is applicable to cascade and parallel implementations, resulting for instance in the same algorithm for the parallel-form filter that appears in [24]. We should note that the complexity of the output gradient generation may be an issue; however, implementations for parallel and cascade realizations exist where this complexity is equivalent to that of the direct form. The lattice implementation of [29] presents a sizable computational burden in gradient generation, but the normalized lattice of [17] (see below) can be implemented with the same complexity as the direct form.

The convergence results we have noted for previous output error approaches for the most part apply as well for these alternate realizations. Differences in these results for cascade- and parallel-form filters stem from the fact that permutations of some of the filter parameters yield equivalent filter transfer functions, but these differences do not affect the convergence results. In general, gradient algorithms for alternate implementations appear to converge more slowly than their direct-form counterparts; however, the reasons for this difference in convergence speed are poorly understood.

Algorithms other than the gradient approach have not been extensively explored for the alternate parametrizations. There may be fundamental limitations in this regard. For direct-form implementations, the signals of the unknown system corresponding to internal states of the adaptive filter are available through the delayed outputs $d(n-i)$, and it is these signals that are used in the equation error-based algorithms. However, the analogous signals for alternate implementations are unavailable, and so equation error methods, as well as the SM approach, are a challenge to even devise, let alone implement. Stability theory-based approaches are difficult to formulate, and the results of [28] indicate that simple algorithms of the form of Equation 23.60 would not be stable in a wide set of operating conditions.

One promising alternate structure is the normalized lattice of [17]. The normalized structure is by nature stable, and hence no stability check is necessary in the adaptive algorithm. Furthermore, a clever implementation of the output gradient calculation keeps the computational burden of the SG and GN algorithms for this normalized lattice comparable to direct-form implementations.

Convergence rates for this structure appear to be comparable to the direct-form structure as well [19]. While we have noted that SM approaches are, in general, infeasible for alternate parametrizations, it is in fact possible to implement an SM algorithm for the normalized lattice through use of an invertibility property held by stages of the lattice [17]. The convergence results we have noted for SM apply as well to the normalized lattice implementation.

We summarize as follows. Alternate parametrizations of adaptive IIR filters enable the stability of the adaptive system to be easily checked. Convergence results for gradient-based algorithms typically apply to these alternate structures. However, the complexities of the gradient calculations can be large for certain systems, and GN approaches appear to be difficult to implement and stabilize for these systems.

23.6 Conclusions

Adaptive IIR filtering remains an open area of research. The preceding survey has examined a number of different approaches to algorithm design within this field. We have considered equation error algorithm designs, including the well-known LMS and RLS algorithms, but also the IV approach and the more recent equation error algorithms with a unit-norm constraint. Output error algorithm designs that we have treated are gradient descent methods and methods based on stability theory. Somewhere in between these two categories is the SM approach to adaptive IIR filtering.

Each of these approaches has certain advantages but also disadvantages. We have evaluated each approach in terms of convergence conditions and also with regard to the nature of the filter parameters to which the algorithm converges. We have taken special interest in whether the algorithm converges to or is biased away from the optimal filter parameters, both in the presence of undermodeling and also measurement noise effects, and a further concern has been the stability of the resulting filter. We have placed less emphasis on convergence speed, as this issue is highly dependent on the particular environment in which the filter is to operate.

Unfortunately, no one algorithm possesses satisfactory properties in all of these regards. Therefore, the choice of algorithm in a given application will depend on which property is most critical in the application setting. Meanwhile, research seeking improvement in adaptive IIR filtering algorithms continues.

References

1. Anderson, B.D.O. et al., *Stability of Adaptive Systems: Passivity and Averaging Analysis*, MIT Press, Cambridge, MA, 1987.
2. Anderson, B.D.O. et al., Robust strict positive realness: Characterization and construction, *IEEE Trans. Circuits Syst.*, 37(7), 869–876, 1990.
3. Benveniste, A., Metivier, M., and Priouret, P., *Adaptive Algorithms and Stochastic Approximations*, Springer-Verlag, New York, 1990.
4. Davila, C.E., An algorithm for efficient, unbiased, equation-error infinite impulse response adaptive filtering, *IEEE Trans. Signal Process.*, 42(5), 1221–1226, 1994.
5. Douglas, S.C. and Rupp, M., On bias removal and unit-norm constraints in equation-error adaptive IIR filters, *Proceedings of the 30th Annual Asilomar Conference on Signals, Systems, and Computers*, Pacific Grove, CA, 1996.
6. Fan, H., Application of Benveniste's convergence results in the study of adaptive IIR filtering algorithms, *IEEE Trans. Inf. Theory*, 34(7), 692–709, 1988.
7. Fan, H. and Doroslovacki, D. On "global convergence" of Steiglitz-McBride adaptive algorithm, *IEEE Trans. Circuits Syst. II*, 40(2), 73–87, 1993.
8. Fan, H. and Jenkins, W.K., Jr., A new adaptive IIR filter, *IEEE Trans. Circuits Syst.*, 33(10), 939–947, 1986.

9. Fan, H. and Nayeri, M., On reduced order identification: Revisiting on some system identification techniques for adaptive filtering, *IEEE Trans. Circuits Syst.*, 37(9), 1144–1151, 1990.
10. Johnson, C.R., Jr., A convergence proof for a hyperstable adaptive recursive filter, *IEEE Trans. Inf. Theory*, 25(6), 745–749, 1979.
11. Johnson, C.R., Jr., Adaptive IIR filtering: Current results and open issues, *IEEE Trans. Inf. Theory*, 30(2), 237–250, 1984.
12. Johnson, C.R., Jr., Larimore, M.G., Treichler, J.R., and Anderson, B.D.O., SHARF convergence properties, *IEEE Trans. Circuits Syst.*, 28(6), 499–510, 1984.
13. Kenney, J.B. and Rohrs, C.E., The composite regressor algorithm for IIR adaptive systems, *IEEE Trans. Signal Process.*, 41(2), 617–628, 1993.
14. Lin, J.-N. and Unbehauen, R., Bias-remedy least mean square equation error algorithm for IIR parameter recursive estimation, *IEEE Trans. Signal Process.*, 40(1), 62–69, 1992.
15. Ljung, L. and Söderström, T., *Theory and Practice of Recursive System Identification*, MIT Press, Cambridge, MA, 1983.
16. Nayeri, M., Fan, H., and Jenkins, W.K., Jr., Some characteristics of error surfaces for insufficient order adaptive IIR filters, *IEEE Trans. Acoust. Speech Signal Process.*, 38(7), 1222–1227, 1990.
17. Regalia, P.A., Stable and efficient lattice algorithms for adaptive IIR filtering, *IEEE Trans. Signal Process.*, 40(2), 375–388, 1992.
18. Regalia, P.A., An unbiased equation error identifier and reduced-order approximations, *IEEE Trans. Signal Process.*, 42(6), 1397–1412, 1994.
19. Regalia, P.A., *Adaptive IIR Filtering in Signal Processing and Control*, Marcel-Dekker, New York, 1995.
20. Ren, W. and Kumar, P.R., Stochastic parallel model adaptation: Theory and applications to active noise canceling, feedforward control, IIR filtering, and identification, *IEEE Trans. Autom. Control*, 37(5), 566–578, 1992.
21. Sethares, W.A., Anderson, B.D.O., and Johnson, C.R., Jr., Adaptive algorithms with filtered regressor and filtered error, *Math. Control Signals Syst.*, 2, 381–403, 1988.
22. Sethares, W.A., Lawrence, D.A., Johnson, Jr., C.R., and Bitmead, R.R., Parameter drift in LMS adaptive filters, *IEEE Trans. Acoust. Speech Signal Process.*, 34(8), 868–879, 1986.
23. Shynk, J.J., Adaptive IIR filtering, *IEEE Acoust. Speech Signal Process. Mag.*, 6(2), 4–21, 1989.
24. Shynk, J.J., Adaptive IIR filtering using parallel-form realizations, *IEEE Trans. Acoust. Speech Signal Process.*, 37(4), 519–533, 1989.
25. Söderström, T. and Stoica, P., On the stability of dynamic models obtained by least-squares identification, *IEEE Trans. Autom. Control*, 26(2), 575–577, 1981.
26. Söderström, S. and Stoica, P., Some properties of the output error method, *Automatica*, 18(1), 93–99, 1982.
27. Stoica, P. and Söderström, S., The Steiglitz-McBride identification algorithm revisited—Convergence analysis and accuracy aspects, *IEEE Trans. Autom. Control*, 26(3), 712–717, 1981.
28. Williamson, G.A., Anderson, B.D.O., and Johnson, C.R., Jr., On the local stability properties of adaptive parameter estimators with composite errors and split algorithms, *IEEE Trans. Autom. Control*, 36(4), 463–473, 1991.
29. Williamson, G.A., Johnson, C.R., Jr., and Anderson, B.D.O., Locally robust identification of linear systems containing unknown gain elements with application to adapted IIR lattice models, *Automatica*, 27(5), 783–798, 1991.

24

Adaptive Filters for Blind Equalization

Zhi Ding
University of California at Davis

24.1 Introduction

One of the earliest and most successful applications of adaptive filters is adaptive channel equalization in digital communication systems. Using the standard least mean square (LMS) algorithm, an adaptive equalizer is a finite impulse response (FIR) filter whose desired reference signal is a known training sequence sent by the transmitter over the unknown channel. The reliance of an adaptive channel equalizer on a training sequence requires that the transmitter cooperates by (often periodically) resending the training sequence, lowering the effective data rate of the communication link.

In many high-data-rate bandlimited digital communication systems, the transmission of a training sequence is either impractical or very costly in terms of data throughput. Conventional LMS adaptive filters depending on the use of training sequences cannot be used. For this reason, blind adaptive channel equalization algorithms that do not rely on training signals have been developed. Using these "blind" algorithms, individual receivers can begin self-adaptation without transmitter assistance. This ability of blind startup also enables a blind equalizer to self-recover from system breakdowns. This self-recovery ability is critical in broadcast and multicast systems where channel variation often occurs.

In this chapter, we provide an introduction to the basics of blind adaptive equalization. We describe commonly used blind algorithms, highlight important issues regarding convergence properties of various blind equalizers, outline common initialization tactics, present several open problems, and discuss recent advances in this field.

24.2 Channel Equalization in QAM Data Communication Systems

In data communication, digital signals are transmitted by the sender through an analog channel to the receiver. Nonideal analog media such as telephone cables and radio channels typically distort the transmitted signal.

The problem of blind channel equalization can be described using the simple system diagram shown in Figure 24.1. The complex baseband model for a typical QAM (quadrature amplitude modulated) data communication system consists of an unknown linear time-invariant (LTI) channel $h(t)$ which represents all the interconnections between the transmitter and the receiver at baseband. The matched filter is also included in the LTI channel model. The baseband-equivalent transmitter generates a sequence of complex-valued random input data $\{a(n)\}$, each element of which belongs to a complex alphabet $\mathcal{A}$ (or constellation) of QAM symbols. The data sequence $\{a(n)\}$ is sent through a baseband-equivalent complex LTI channel whose output $x(t)$ is observed by the receiver. The function of the receiver is to estimate the original data $\{a(n)\}$ from the received signal $x(t)$.

For a causal and complex-valued LTI communication channel with impulse response $h(t)$, the input/output relationship of the QAM system can be written as

$$x(t) = \sum_{n=-\infty}^{\infty} a(n)h(t - nT + t_0) + w(t), \quad a(n) \in \mathcal{A}, \tag{24.1}$$

where T is the symbol (or baud) period. Typically the channel noise $w(t)$ is assumed to be stationary, Gaussian, and independent of the channel input $a(n)$.

In typical communication systems, the matched filter output of the channel is sampled at the known symbol rate $1/T$ assuming perfect timing recovery. For our model, the sampled channel output

$$x(nT) = \sum_{k=-\infty}^{\infty} a(k)h(nT - kT + t_0) + w(nT) \tag{24.2}$$

is a discrete time stationary process. Equation 24.2 relates the channel input to the sampled matched filter output. Using the notations

$$x(n) \stackrel{\Delta}{=} x(nT), \quad w(n) \stackrel{\Delta}{=} w(nT), \quad \text{and} \quad h(n) \stackrel{\Delta}{=} h(nT + t_0), \tag{24.3}$$

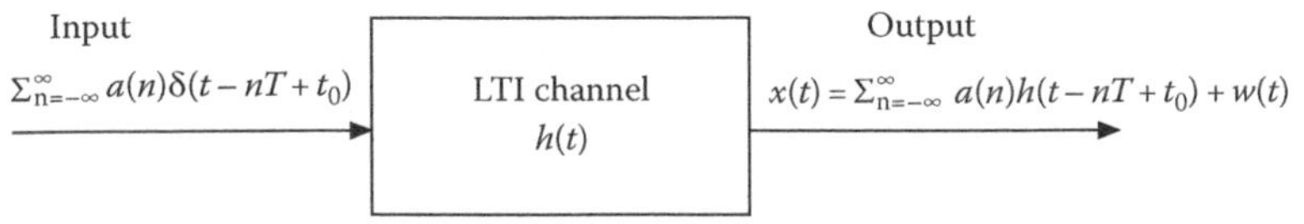

FIGURE 24.1 Baseband representation of a QAM data communication system.

the relationship in Equation 24.2 can be written as

$$x(n) = \sum_{k=-\infty}^{\infty} a(k)h(n-k) + w(n). \tag{24.4}$$

When the channel is nonideal, its impulse response $h(n)$ is nonzero for $n \neq 0$. Consequently, undesirable signal distortion is introduced as the channel output $x(n)$ depends on multiple symbols in $\{a(n)\}$. This phenomenon, known as intersymbol interference (ISI), can severely corrupt the transmitted signal. ISI is usually caused by limited channel bandwidth, multipath, and channel fading in digital communication systems. A simple memoryless decision device acting on $x(n)$ may not be able to recover the original data sequence under strong ISI. Channel equalization has proven to be an effective means of significant ISI removal. A comprehensive tutorial on nonblind adaptive channel equalization by Qureshi [1] contains detailed discussions on various aspects of channel equalization.

Figure 24.2 shows the combined communication system with adaptive equalization. In this system, the equalizer $G(z, \mathbf{W})$ is a linear FIR filter with parameter vector $\mathbf{W}$ designed to remove the distortion caused by channel ISI. The goal of the equalizer is to generate an output signal $y(n)$ that can be quantized to yield a reliable estimate of the channel input data as

$$\hat{a}(n) = Q(y(n)) = a(n-\delta), \tag{24.5}$$

where δ is a constant integer delay. Typically any constant but finite amount of delay introduced by the combined channel and equalizer is acceptable in communication systems.

The basic task of equalizing a linear channel can be translated to that task of identifying the equivalent discrete channel, defined in z-transform notation as

$$H(z) = \sum_{k=0}^{\infty} h(k)z^{-k}. \tag{24.6}$$

With this notation, the channel output becomes

$$x(n) = H(z)a(n) + w(n), \tag{24.7}$$

where

$H(z)a(n)$ denotes linear filtering of the sequence $a(n)$ by the channel and

$w(n)$ is a white (for a root-raised-cosine matched filter [1]) stationary noise with constant power spectrum N_0.

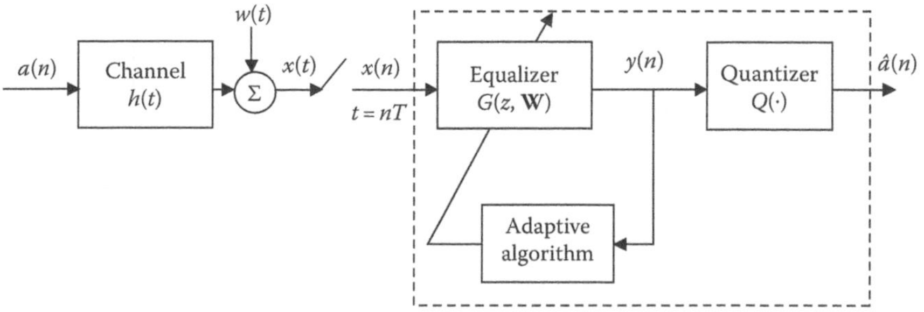

FIGURE 24.2 Adaptive blind equalization system.

Once the channel has been identified, the equalizer can be constructed according to the minimum mean square error (MMSE) criterion between the desired signal $a(n-\delta)$ and the output $y(n)$ as

$$G_{\mathrm{MMSE}}(z, \mathbf{W}) = \frac{H^*(z^{-1})z^{-\delta}}{H(z)H^*(z^{-1}) + N_0}, \tag{24.8}$$

where * denotes complex conjugate. Alternatively, if the zero-forcing (ZF) criterion is employed, then the optimum ZF equalizer is

$$G_{\mathrm{ZF}}(z, \mathbf{W}) = \frac{z^{-\delta}}{H(z)}, \tag{24.9}$$

which causes the combined channel-equalizer response to become a purely δ-sample delay with zero ISI. ZF equalizers tend to perform poorly when the channel noise is significant and when the channels $H(z)$ have zeros near the unit circle.

Both the MMSE equalizer (Equation 24.8) and the ZF equalizer (Equation 24.9) are of a general infinite impulse response (IIR) form. However, adaptive linear equalizers are usually implemented as FIR filters due to the difficulties inherent in adapting IIR filters. Adaptation is then based on a well-defined criterion such as the MMSE between the ideal IIR and truncated FIR impulse responses or the MMSE between the training signal and the equalizer output.

24.3 Decision-Directed Adaptive Channel Equalizer

Adaptive channel equalization was first developed by Lucky [2] for telephone channels. Figure 24.3 depicts the traditional adaptive equalizer. The equalizer begins adaptation with the assistance of a known training sequence initially transmitted over the channel. Since the training signal is known, standard gradient-based adaptive algorithms such as the LMS algorithm can be used to adjust the equalizer coefficients to minimize the mean square error (MSE) between the equalizer output and the training sequence. It is assumed that the equalizer coefficients are sufficiently close to their optimum values and that much of the ISI is removed by the end of the training period. Once the channel input sequence $\{a(n)\}$ can be accurately recovered from the equalizer output through a memoryless decision device such as a quantizer, the system is switched to the decision-directed mode whereby the adaptive equalizer obtains its reference signal from the decision output.

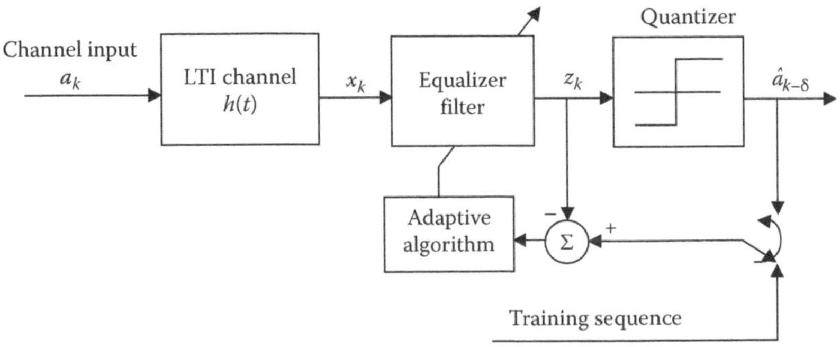

FIGURE 24.3 Decision-directed channel equalization algorithm.

One can construct a blind equalizer by employing decision-directed adaptation without a training sequence. The algorithm minimizes the MSE between the quantizer output

$$\hat{a}(n-\delta) = Q(y(n)) \tag{24.10}$$

and the equalizer output $y(n)$. Naturally, the performance of the decision-directed algorithm depends on the accuracy of the estimate $Q[y(n)]$ for the true symbol $a(n-\delta)$. Undesirable convergence to a local minimum with severe residual ISI can occur in this situation such that $Q[y(n)]$ and $a(n-\delta)$ differ sufficiently often. Thus, the challenge of blind equalization lies in the design of special adaptive algorithms that eliminate the need for training without compromising the desired convergence to near the optimum MMSE or ZF equalizer coefficients.

24.4 Basic Facts on Blind Adaptive Equalization

In blind equalization, the desired signal or input to the channel is unknown to the receiver, except for its probabilistic or statistical properties over some known alphabet $\mathcal{A}$. As both the channel $h(n)$ and its input $a(n)$ are unknown, the objective of blind equalization is to recover the unknown input sequence based solely on its probabilistic and statistical properties.

The first comprehensive analytical study of the blind equalization problem was presented by Benveniste, Goursat, and Ruget in 1980 [3]. In fact, the very term "blind equalization" can be attributed to Benveniste and Goursat from the title of their 1984 paper [4]. The seminal paper of Benveniste et al. [3] established the connection between the task of blind equalization and the use of higher order statistics (HOS) of the channel output. Through rigorous analysis, they generalized the original Sato algorithm [5] into a class of algorithms based on non-MSE cost functions. More importantly, the convergence properties of the proposed algorithms were carefully investigated.

Based on the work of [3], the following facts about blind equalization are generally noted:

1. Second order statistics of $x(n)$ alone only provide the magnitude information of the linear channel and are insufficient for blind equalization of a mixed phase channel $H(z)$ containing zeros inside and outside the unit circle in the z-plane.
2. A mixed phase linear channel $H(z)$ cannot be identified from its outputs when the input signal is i.i.d. Gaussian, since only second order statistical information is available.
3. Although the exact inverse of a nonminimum phase channel is unstable, a truncated anticausal expansion can be delayed by δ to allow a causal approximation to a ZF equalizer.
4. ZF equalizers cannot be implemented for channels $H(z)$ with zeros on the unit circle.
5. The symmetry of QAM constellations $\mathcal{A} \subset \mathbb{C}$ causes an inherent phase ambiguity in the estimate of the channel input sequence or the unknown channel when input to the channel is uniformly distributed over $\mathcal{A}$. This phase ambiguity can be overcome by differential encoding of the channel input.

Due to the absence of a training signal, it is important to exploit various available information about the input symbol and the channel output to improve the quality of blind equalization. Usually, the following information is available to the receiver for blind equalization:

- The power spectral density of the channel output signal $x(t)$, which contains information on the magnitude of the channel transfer function
- The HOS of the T-sampled channel output $\{x(kT)\}$, which contains information on the phase of the channel transfer function
- Cyclostationary second order statistics and HOS of the channel output signal $x(t)$, which contain additional phase information of the channel
- The finite channel input alphabet, which can be used to design quantizers or decision devices with memory to improve the reliability of the channel input estimate

Naturally in some cases, these information sources are not necessarily independent as they contain overlapping information. Efficient and effective blind equalization schemes are more likely to be designed when all useful information is exploited at the receiver. We now describe various algorithms for blind channel identification and equalization.

24.5 Adaptive Algorithms and Notations

There are basically two different approaches to the problem of blind equalization. The stochastic gradient descent (SGD) approach iteratively minimizes a chosen cost function over all possible choices of the equalizer coefficients, while the statistical approach uses sufficient stationary statistics collected over a block of received data for channel identification or equalization. The latter approach often exploits HOS or cyclostationary statistical information directly. In this discussion, we focus on the adaptive online equalization methods employing the gradient descent approach, as these methods are most closely related to other topics in this chapter. Consequently, the design of special, non-MSE cost functions that implicitly exploits the HOS of the channel output is the key issue in our methods and discussions.

For reasons of practicality and ease of adaptation, a linear channel equalizer is typically implemented as an FIR filter $G(z, \mathbf{W})$. Denote the equalizer parameter vector as

$$\mathbf{W} \triangleq [\, w_0 \quad w_1 \quad \cdots \quad w_m \,]^{\mathrm{T}}, \quad m < \infty.$$

In addition, define the received signal vector as

$$\mathbf{X}(n) \triangleq [x(n)\; x(n-1)\; \cdots\; x(n-m)\,]^{\mathrm{T}}. \tag{24.11}$$

The output signal of the linear equalizer is thus

$$\begin{aligned} y(n) &= \mathbf{W}^{\mathrm{T}}\mathbf{X}(n) \\ &= G(z,\mathbf{W})\{x(n)\}, \end{aligned} \tag{24.12}$$

where we have defined the equalizer transfer function as

$$G(z, \mathbf{W}) = \sum_{i=0}^{m} w_i z^{-i}. \tag{24.13}$$

All the ISI is removed by a ZF equalizer if

$$H(z)G(z, \mathbf{W}) = gz^{-\delta}, \quad g \neq 0 \tag{24.14}$$

such that the noiseless equalizer output becomes $y(n) = ga(n-\delta)$, where g is a complex-valued scaling factor. Hence, a ZF equalizer attempts to achieve the inverse of the channel transfer function with a possible gain difference g and/or a constant time delay δ.

Denoting the parameter vector of the equalizer at sample instant n as $\mathbf{W}(n)$, the conventional LMS adaptive equalizer employing a training sequence is given by

$$\mathbf{W}(n+1) = \mathbf{W}(n) + \mu[a(n-\delta) - y(n)]\mathbf{X}^*(n), \tag{24.15}$$

where

* denotes complex conjugates

μ is a small positive stepsize

Naturally, this algorithm requires that the channel input $a(n-\delta)$ be available. The equalizer iteratively minimizes the MSE cost function

$$E\{|e_n|^2\} = E\{|a(n-\delta) - y(n)|^2\}.$$

If the MSE is so small after training that the equalizer output $y(n)$ is a close estimate of the true channel input $a(n-\delta)$, then $Q[y(n)]$ can replace $a(n-\delta)$ in a decision-directed algorithm that continues to track modest time-variations in the channel dynamics [1].

In blind equalization, the channel input $a(n-\delta)$ is unavailable, and thus different minimization criteria are explored. The crudest blind equalization algorithm is the decision-directed scheme that updates the adaptive equalizer coefficients as

$$\mathbf{W}(n+1) = \mathbf{W}(n) + \mu[Q[y(n)] - y(n)]\mathbf{X}^*(n). \tag{24.16}$$

The performance of the decision-directed algorithm depends on how close $\mathbf{W}(n)$ is to its optimum setting $\mathbf{W}_{opt}$ under the MMSE or the ZF criterion. The closer $\mathbf{W}(n)$ is to $\mathbf{W}_{opt}$, the smaller the ISI is and the more accurate the estimate $Q[y(n)]$ is to $a(n-\delta)$. Consequently, the algorithm in Equation 24.16 is likely to converge to $\mathbf{W}_{opt}$ if $\mathbf{W}(n)$ is initially close to $\mathbf{W}_{opt}$. The validity of this intuitive argument is shown analytically in [6,7]. On the other hand, $\mathbf{W}(n)$ can also converge to parameter values that do not remove sufficient ISI from certain initial parameter values $\mathbf{W}(0)$, as $Q[y(n)] \neq a(n-\delta)$ sufficiently often in some cases [6,7].

The ability of the equalizer to achieve the desired convergence result when it is initialized with sufficiently small ISI accounts for the key role that the decision-directed algorithm plays in channel equalization. In the system of Figure 24.3, the training session is designed to help $\mathbf{W}(n)$ converge to a parameter vector such that most of the ISI has been removed, from which adaptation can be switched to the decision-directed mode. Without direct training, a blind equalization algorithm is therefore used to provide a good initialization for the decision-directed equalizer because of the decision-directed equalizer's poor convergence behavior under high ISI.

24.6 Mean Cost Functions and Associated Algorithms

Under the ZF criterion, the objective of the blind equalizer is to adjust $\mathbf{W}(n)$ such that Equation 24.14 can be achieved using a suitable rule of self-adaptation. We now describe the general methodology of blind adaptation and introduce several popular algorithms.

Unless otherwise stated, we focus on the blind equalization of pulse-amplitude modulation (PAM) signals, in which the input symbol is uniformly distributed over the following M levels,

$$\{\pm(M-1)d, \pm(M-3)d, \ldots, \pm 3d, \pm d\}, \quad M \text{ even}. \tag{24.17}$$

We study this particular case because (1) algorithms are often defined only for real signals when first developed [3,5], and (2) the extension to complex (QAM) systems is generally straightforward [4].

Blind adaptive equalization algorithms are often designed by minimizing special non-MSE cost functions that do not involve the use of the original input $a(n)$ but still reflect the current level of ISI in the equalizer output. Define the mean cost function as

$$J(\mathbf{W}) \stackrel{\Delta}{=} E\{\Psi(y(n))\}, \tag{24.18}$$

where $\Psi(\cdot)$ is a scalar function of its argument. The mean cost function $J(\mathbf{W})$ should be specified such that its minimum point $\mathbf{W}$ corresponds to a minimum ISI or MSE condition. Because of the symmetric distribution of $a(n)$ over $\mathcal{A}$ in Equation 24.17, the function Ψ should be even ($\Psi(-x)=\Psi(x)$), so that both $y(n)=a(n-\delta)$ and $y(n)=-a(n-\delta)$ are desired objectives or global minima of the mean cost function.

Using Equation 24.18, the SGD minimization algorithm is easily derived as [3]

$$\begin{aligned}\mathbf{W}(n+1) &= \mathbf{W}(n) - \mu\frac{\partial}{\partial \mathbf{W}(n)}\Psi[y(n)] \\ &= \mathbf{W}(n) - \mu\Psi'(\mathbf{X}^{\mathrm{T}}(n)\mathbf{W}(n))\mathbf{X}^{*}(n).\end{aligned} \tag{24.19}$$

Define the first derivative of Ψ as

$$\psi(x) \triangleq \Psi'(x) = \frac{\partial}{\partial x}\Psi(x).$$

The resulting blind equalization algorithm can then be written as

$$\mathbf{W}(n+1) = \mathbf{W}(n) - \mu\psi(\mathbf{X}^{\mathrm{T}}(n)\mathbf{W}(n))\mathbf{X}^{*}(n). \tag{24.20}$$

Hence, a blind equalizer can either be defined by its cost function $\Psi(x)$, or equivalently, by the derivative $\psi(x)$ of its cost function, which is also called the error function since it replaces the prediction error in the LMS algorithm. Correspondingly, we have the following relationship:

Minima of the mean cost $J(\mathbf{W}) \Leftrightarrow$ Stable equilibria of the algorithm in Equation 24.20.

The design of the blind equalizer thus translates into the selection of the function Ψ (or ψ) such that local minima of $J(\mathbf{W})$, or equivalently, the locally stable equilibria of the algorithm (Equation 24.20) correspond to a significant removal of ISI in the equalizer output.

24.6.1 Sato Algorithm

The first blind equalizer for multilevel PAM signals was introduced by Sato [5] and is defined by the error function

$$\psi_1[y(n)] = y(n) - R_1\operatorname{sgn}(y(n)), \tag{24.21}$$

where

$$R_1 \triangleq \frac{E|a(n)|^2}{E|a(n)|}.$$

Clearly, the Sato algorithm effectively replaces $a(n-\delta)$ with $R_1\operatorname{sgn}[y(n)]$, known as the slicer output. The multilevel PAM signal is viewed as an equivalent binary input signal in this case, so that the error function often has the same sign for adaptation as the LMS error $y(n)-a(n-\delta)$.

24.6.2 BGR Extensions of Sato Algorithm

The Sato algorithm was extended by Benveniste, Goursat, and Ruget [3] who introduced a class of error functions given by

$$\psi_b[y(n)] = \tilde{\psi}(y(n)) - R_b\operatorname{sgn}(y(n)), \tag{24.22}$$

where

$$R_{\rm b} \stackrel{\Delta}{=} \frac{E\{\tilde{\psi}[a(n)]a(n)\}}{E|a(n)|}. \tag{24.23}$$

Here, $\tilde{\psi}(x)$ is an odd and twice differentiable function satisfying

$$\tilde{\psi}''(x) \geq 0, \quad \forall x \geq 0. \tag{24.24}$$

The use of the function $\tilde{\psi}$ generalizes the linear function $\tilde{\psi}(x) = x$ in the Sato algorithm. The class of algorithms satisfying Equations 24.22 and 24.24 are called BGR algorithms. They are individually represented by the explicit specification of the $\tilde{\psi}$ function, as with the Sato algorithm.

The generalization of these algorithms to complex signals (QAM) and complex equalizer parameters is straightforward by separating signals into their real and the imaginary as

$$\psi_{\rm b}[y(n)] = \tilde{\psi}\{{\rm Re}[y(n)]\} - R_{\rm b}{\rm sgn}\{{\rm Re}[y(n)]\} + j(\tilde{\psi}\{Im[y(n)]\} - R_{\rm b}{\rm sgn}\{{\rm Im}[y(n)]\}). \tag{24.25}$$

24.6.3 Constant Modulus or Godard Algorithms

Integrating the Sato error function $\psi_1(x)$ shows that the Sato algorithm has an equivalent cost function

$$\Psi_1[y(n)] = \frac{1}{2}(|y(n)| - R_1)^2.$$

This cost function was generalized by Godard into another class of algorithms that are specified by the cost functions [8]

$$\Psi_{\rm q}[y(n)] = \frac{1}{2q}\left[|y(n)|^q - R_q\right]^2, \quad q = 1, 2, \ldots, \tag{24.26}$$

where $R_q \stackrel{\Delta}{=} \dfrac{E|a(n)|^{2q}}{E|a(n)|^q}$.

This class of Godard algorithms is indexed by the positive integer q. Using the SGD approach, the Godard algorithms given by

$$\mathbf{W}(n+1) = \mathbf{W}(n) - \mu\left(|\mathbf{X}(n)^{\rm H}\mathbf{W}(n)|^q - R_q\right)|\mathbf{X}(n)^{\rm T}\mathbf{W}(n)|^{q-2}\mathbf{X}(n)^{\rm T}\mathbf{W}(n)\mathbf{X}^*(n). \tag{24.27}$$

The Godard algorithm for the case $q = 2$ was independently developed as the "constant modulus algorithm" (CMA) by Treichler and co-workers [9] using the philosophy of property restoral. For channel input signal that has a constant modulus $|a(n)|^2 = R_2$, the CMA equalizer penalizes output samples $y(n)$ that do not have the desired constant modulus characteristics. The modulus error is simply

$$e(n) = |y(n)|^2 - R_2,$$

and the squaring of this error yields the constant modulus cost function that is the identical to the Godard cost function.

This modulus restoral concept has a particular advantage in that it allows the equalizer to be adapted independent of carrier recovery. A carrier frequency offset of Δ_f causes a possible phase rotation of the equalizer output so that

$$y(n) = |y(n)| \exp[j(2\pi\Delta_f n + \varphi(n))].$$

Because the CMA cost function is insensitive to the phase of $y(n)$, the equalizer parameter adaptation can occur independently and simultaneously with the operation of the carrier recovery system. This property also allows CMA to be applied to analog modulation signals with constant amplitude such as those using frequency or phase modulation [9].

24.6.4 Stop-and-Go Algorithms

Given the standard form of the blind equalization algorithm in Equation 24.20, it is apparent that the convergence characteristics of these algorithms are largely determined by the sign of the error signal $\psi[y(n)]$. In order for the coefficients of a blind equalizer to converge to the vicinity of the optimum MMSE solution as observed through LMS adaptation, the sign of its error signal should agree with the sign of the LMS prediction error $y(n) - a(n - \delta)$ most of the time. Slow convergence or convergence of the parameters to local minima of the cost function $J(\mathbf{W})$ that do not provide proper equalization can occur if the signs of these two errors differ sufficiently often. In order to improve the convergence properties of blind equalizers, the so-called stop-and-go methodology was proposed by Picchi et al. [10]. We now describe its simple concept.

The idea behind the stop-and-go algorithms is to allow adaptation "to go" only when the error function is more likely to have the correct sign for the gradient descent direction. Since there are several criteria for blind equalization, one can expect a more accurate descent direction when more than one of the existing algorithms provide the same sign of the error function. When the error signs differ for a particular output sample, parameter adaptation is "stopped." Consider two algorithms with error functions $\psi_1(y)$ and $\psi_2(y)$. We can devise the following stop-and-go algorithm:

$$\mathbf{W}(k+1) = \begin{cases} \mathbf{W}(k) - \mu\psi_1[y(n)]\mathbf{X}^*(n), & \text{if } \operatorname{sgn}[\psi_1(\mathrm{y(n)})] = \operatorname{sgn}[\psi_2(\mathrm{y(n)})]; \\ \mathbf{W}(k), & \text{if } \operatorname{sgn}[\psi_1(\mathrm{y(n)})] \neq \operatorname{sgn}[\psi_2(\mathrm{y(n)})]. \end{cases} \tag{24.28}$$

In their work, Picchi and Prati combined only the Sato and the decision-directed algorithms with faster convergence results through the corresponding error function

$$\psi[y(n)] = \frac{1}{2}\{y(n) - Q[y(n)]\} + \frac{1}{2}|y(n) - Q[y(n)]|\operatorname{sgn}\{y(n) - R_1\operatorname{sgn}[y(n)]\}.$$

However, given the number of existing algorithms, the stop-and-go methodology can include many different combinations of error functions. One that combines Sato and Godard algorithms was tested by Hatzinakos [11].

24.6.5 Shalvi and Weinstein Algorithms

Unlike previously introduced algorithms, the methods of Shalvi–Weinstein [12] are based on HOS of the equalizer output. Define the kurtosis of the equalizer output signal $y(n)$ as

$$K_y \triangleq E|y(n)^4| - 2E^2|y(n)^2| - |E[y(n)^2]|^2. \tag{24.29}$$

The Shalvi–Weinstein algorithm maximizes $|K_y|$ subject to the constant power constraint $E|y(n)|^2 = E|a(n)^2|$. Define c_n as the combined channel-equalizer impulse response given by

$$c_n = \sum_{k=0}^{m} h_k w_{n-k}, \qquad -\infty < n < \infty. \tag{24.30}$$

Using the fact that $a(n)$ is i.i.d., it can be shown [13] that

$$E|y(n)^2| = E|a(n)^2| \sum_{i=-\infty}^{\infty} |c_i|^2 \tag{24.31}$$

$$K_y = K_a \sum |c_n|^4, \tag{24.32}$$

where K_a is the kurtosis of the channel input, a quantity that is nonzero for most QAM and PAM signals. Hence, the Shalvi–Weinstein equalizer is equivalent to the following criterion:

$$\text{maximize} \quad \sum_{n=-\infty}^{\infty} |c_n|^4 \quad \text{subject to} \quad \sum_{n=-\infty}^{\infty} |c_n|^2 = 1. \tag{24.33}$$

It can be shown [14] that there is a one-to-one correspondence between the minima of the cost function surface searched by this algorithm and those of the Godard algorithm with $q = 2$. However, the methods of adaptation given in [12] can exhibit convergence characteristics different from those of the CMA.

24.6.6 Summary

Over the years, there have been many attempts to derive new algorithms and equalization methods that are more reliable and faster than the existing methods. Nonetheless, the algorithms presented above are still the most commonly used methods in blind equalization due to their computational simplicity and practical effectiveness. In particular, CMA has proven to be useful not only in blind equalization but also in blind array signal processing systems. Because it does not rely on the accuracy of the decision device output nor the knowledge of the channel input signal constellation, CMA is a versatile algorithm that can be used not only for digital communication signals but also for analog signals that do not conform to a finite constellation alphabet.

As a practical example, consider a QAM system in which the channel impulse response is shown in Figure 24.4a. This sampled composite channel response results from a continuous time system in which the transmitter and receiver filters both have identical root-raised cosine frequency response with the roll-off factor of 0.13, while the channel between the two filters is nonideal with several nondominant multipaths. The channel input signal is generated from a rectangular 64-QAM constellation as shown in Figure 24.4b. The channel output points are shown in Figure 24.4c. The channel output signal clearly has significant ISI such that a simple quantizer based on the nearest neighbor principle is likely to make many decision errors. We use a CMA equalizer with 25 parameter taps. The equalizer input is normalized by its power and a stepsize $\mu = 10^{-3}$ is used in the CMA adaptation. After 20,000 iterations, the final equalizer output after parameter convergence is shown in Figure 24.4d. The tighter clustering of the equalizer output shows that the decision error rate will be very low so that the equalizer can be switched to the decision-directed or decision-feedback algorithm mode at this point.

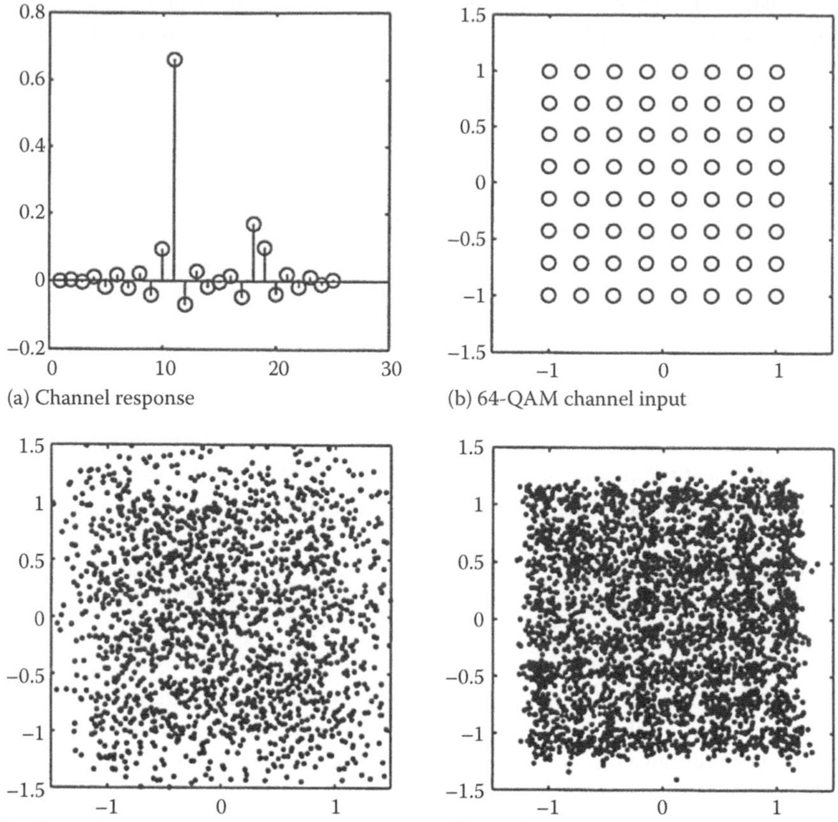

FIGURE 24.4 An example of a CMA equalizer in a 64-QAM communication system.

24.7 Initialization and Convergence of Blind Equalizers

The success and effectiveness of a QAM blind equalization algorithm clearly hinges on its convergence behavior in practical QAM systems with distortive channels. A desired globally convergent algorithm should only produce stable equilibria that are close to the optimum MMSE or ZF equalizer coefficients. If an equalization algorithm has local equilibria, then the initial equalizer parameter values are critical in determining the final values of parameters at convergence. Due to the analytical difficulty in locating and characterizing these local minima, most analytical studies of blind equalizers focus on the noiseless environment. For noiseless channels, the optimum MMSE and ZF equalizers are identical. The goal in the noiseless system is to remove sufficient ISI so that the open eye condition or errorless decision output, given by

$$Q[y(n)] = a(n - \delta),$$

holds.

Although the problem of blind equalization has been studied for over two decades, useful convergence analyses of most blind adaptive algorithms have proven to be difficult to perform. While some recent analytical results have helped to characterize the behavior of several popular algorithms, the overall knowledge of the behaviors of most known effective algorithms is still quite limited. Consequently, practical implementations of blind equalizers still employ heuristic measures to improve their convergence characteristics. We summarize several issues regarding the convergence and initialization of blind equalizers in this subsection.

24.7.1 A Common Analysis Approach

Although many readers may be surprised by the apparent lack of convergence proofs for most blind equalization algorithms, a closer look at the cost functions for these algorithms shows the analytical difficulty of the problem. Specifically, the stable stationary points of the blind algorithm in Equation 24.20 correspond to the local minima of the mean cost function

$$J(\mathbf{W}) = E\left\{\Psi\left(\sum_{i=0}^{m} w_i x(n-i)\right)\right\}. \tag{24.34}$$

The convergence of the adaptive algorithm is thus determined by the geometry of the error function $J(\mathbf{W})$ over the equalizer parameters $\{w_i\}$. An analysis of the convergence of the algorithm in terms of its parameters $\{w_i\}$ is difficult because the statistical characterization of the channel output signal $x(n)$ is highly dependent on the channel impulse response. For this reason, most blind equalization algorithms have initially been presented with only simulation results and without a rigorous convergence analysis.

Faced with this difficulty, several researchers have studied the global behavior of the equalizer in the combined parameter space c_i of Equation 24.30 since

$$J(\mathbf{W}) = E\left\{\Psi\left(\sum_{i=-\infty}^{\infty} w_i x(n-i)\right)\right\} = E\left\{\Psi\left(\sum_{i=-\infty}^{\infty} c_i a(n-i)\right)\right\}. \tag{24.35}$$

Because the probabilistic information of the signal $a(n)$ is completely known, the convergence analysis of the c_i parameters tends to be much simpler than that of the equalizer parameters w_i. The following convergence results are known from these analyses:

- For channel input signals with uniform or sub-Gaussian probability distributions, Sato and BGR algorithms are globally convergent under zero channel noise. The corresponding cost functions only have global minima at parameter settings that result in zero ISI [3].
- For uniform and discrete PAM channel input distributions, undesirable local minima of the Sato and the BGR algorithms exist that do not satisfy the open-eye condition [6,7,15].
- For uniform and discrete PAM (or QAM) channel input distributions, the Godard algorithm with $q = 2$ (CMA) and the Shalvi–Weinstein algorithm have no local minima under zero channel noise. Only global minima exist at parameter settings that result in zero ISI [12,16]. In other words, all minima satisfy the ZF condition

$$c_n^2 = \begin{cases} 1, & n = \delta \\ 0, & n \neq \delta. \end{cases} \tag{24.36}$$

24.7.2 Local Convergence of Blind Equalizers

In order for the convergence analysis of the c_i parameters to be valid for the w_i parameters, a one-to-one linear mapping must exist between the two parameter spaces. A cost function of two variables will still have the same number of minima, maxima, and saddle points after a linear one-to-one coordinate change. On the other hand, a mapping that is not one-to-one can turn a nonstationary or saddle point into a local minimum.

If a one-to-one linear mapping exists between the two parameter spaces $\{w_i\}$ and $\{c_i\}$, then a stationary point for the equalizer coefficients w_i must correspond to a stationary point in the c_i parameters. Consequently, the convergence properties in the c_i parameter space will be equivalent to those in the w_i parameter space. However, $c_i = \sum_{k=0}^{m} h_k w_{i-k}$ does not provide a one-to-one mapping. The linear mapping is one-to-one if and only if $c_i = \sum_{k=-\infty}^{\infty} h_k w_{i-k}$, i.e., the equalizer coefficients w_i must exist for $-\infty < i < \infty$.

In this case, the equalizer parameter vector $\mathbf{W}$ needs to be doubly infinite. Hence, unless the equalizer has an infinite number of parameters and is infinitely noncausal, the convergence behavior of the c_i parameters do not completely characterize the behavior of the finite-length equalizer [17].

Undesirable local convergence of the Godard algorithm to a high ISI equalizer was initially thought to be impossible due to some overzealous interpretations of the global convergence results in the combined c_i space [16]. The local convergence of the Godard ($q = 2$) algorithm or CMA is accurately analyzed by Ding et al. [18], where it is shown that even for noiseless channels whose ISI can be completely eliminated by an FIR equalizer, there can be local convergence of this equalizer to undesirable minima of the cost surface. Furthermore, these equilibria still remain under moderate channel noise. Based on the convergence similarity between the Godard algorithm and the Shalvi–Weinstein algorithm, the local convergence of the Shalvi–Weinstein algorithm to undesirable minima is established in [14]. Using a similar method, Sato and BGR algorithms have also been seen to have additional local minima previously undetected in the combined parameter space [15].

The proof that existing blind equalization algorithms previously thought to be robust can converge to poor solutions demonstrates that rigorous convergence analyses of blind equalizers must be based on the actual equalizer coefficients. Moreover, the undesirable local convergence behavior of existing algorithms indicates the importance of algorithm parameter initialization, which can avoid these local convergent points.

24.7.3 Initialization Issues

In [19], it is shown that local minima of a CMA equalizer cost surface tend to exist near MMSE parameter settings if the delay δ is chosen to be too short or too long. In other words, convergence to local minima is more likely to occur when the equalizer has large tap weights concentrated near either end of the finite equalizer coefficient vector. This type of lopsided parameter weight distribution was also suggested in [16] as being indicative of a local convergence phenomenon. To avoid local convergence to a lopsided tap weight vector, Foschini [16] introduced a tap-centering initialization strategy that requires the gravity center of the equalizer coefficient vector be centered through periodic tap-shifting. A more recent result [14] shows that, by over-parameterization and tap-centering, the Godard algorithm or CMA can effectively reduce the probability of local convergence. This tap-centering method has also been proposed for the Shalvi–Weinstein algorithm [20].

In practice, the tap-centering initialization approach has become an integral part of most blind equalization algorithms. Although a thorough analysis of its effect has not been shown, most reported successful uses of blind equalization algorithms typically rely on tap-centering or center-spike initialization scheme [17]. Although special channels exist that can foil the successful convergence of Sato and BGR algorithms using tap-centering, such channels are atypical [15]. Hence, unless global convergence of the equalizer can be proven, tap-centering is commonly recommended for most blind equalizers.

24.8 Globally Convergent Equalizers

24.8.1 Linearly Constrained Equalizer with Convex Cost

Without a proof of global convergence and a thorough analysis on initialization of existing equalization methods, one can design new and possibly better blind algorithms that can proven to always result in the global minimization of ISI. Here we present one strategy based on highly specialized convex cost functions coupled with a constrained equalizer parameterization designed to avoid ill-convergence.

Recall that the goal of blind equalization is to remove ISI so that the equalizer output is

$$y(n) = ga(n - \delta), \quad g \neq 0. \tag{24.37}$$

Blind equalization of PAM systems without gain recovery has been proposed in [21]. The idea is to fix the center tap w_0 as a nonzero constant in order to prevent equalizer to the trivial minimum with all zero coefficient values in a convex cost function. For QAM input, a nontrivial extension is shown here.

For the particular equalizer design, assume that the input QAM constellation is square, which resembles the constellation in Figure 24.4b. The cost function to be minimized is

$$J(\mathbf{W}) \triangleq \max|\mathrm{Re}(y(n))| = \max|\mathrm{Im}(y(n))|. \tag{24.38}$$

The convexity of $J(\mathbf{W})$ with respect to the equalizer coefficient vector $\mathbf{W}$ follows from the triangle inequality under the assumption that all input sequences are possible. We constrain the equalizer coefficients $\mathbf{W}(n)$ with the following linear constraint:

$$\mathrm{Re}(w_0) + \mathrm{Im}(w_0) = 1, \tag{24.39}$$

where w_0 is the center tap. Due to the linearity of this constraint, the convexity of the cost function (Equation 24.38) with respect to both the real and imaginary parts of the equalizer coefficients is maintained, and global convergence is therefore assured.

Because of its convexity, this cost function is unimodal with a unique global minimum for almost all channels. It can then be shown [22] that a doubly infinite noncausal equalizer under the linear constraint is globally convergent to the condition in [37].

The linear constraint in Equation 24.39 can be changed to any weighted linear combination of the two terms in Equation 24.39. More general linear constraints on the equalizer coefficients can also be employed [23]. This fact is particularly important for preserving the global convergence property when causal finite-length equalizers are used. This behavior is a direct consequence of convexity, since restricting most of the equalizer taps to zero values as in FIR is a form of linear constraint. Convexity also ensures that one can approximate arbitrarily closely the performance of the ideal nonimplementable double infinite noncausal equalizer with a finite length FIR equalizer. These facts are important since many of the limitations illustrated earlier for convergence analyses of other equalizers can be overcome in this case.

For an actual implementation of this algorithm, a gradient descent method can be derived by using an l_p-norm cost function to approximate Equation 24.38 as

$$J(\mathbf{W}) \approx E|\mathrm{Re}(z_k)|^p, \tag{24.40}$$

where p is a large integer. As the cost function in Equation 24.40 is strictly convex, linear constraints such as truncation preserve convexity. Simulation examples of this algorithm can be found in [22,24].

24.8.2 Fractionally Spaced Blind Equalizers

A so-called fractionally spaced equalizer (FSE) is obtained from the system in Figure 24.2 if the channel output is sampled at a rate faster than the baud or symbol rate $1/T$. Recent work on the blind FSE has been motivated by several new results on nonadaptive blind equalization based on second order cyclostationary statistics. In addition to the first noted work by Tong et al. [25], new nonadaptive algorithms are also presented in [26–29]. Here we only focus on the adaptive framework.

Let p be an integer such that the sampling interval be $\Delta = T/p$. As long as the channel bandwidth is greater than the minimum $1/(2T)$, sampling at higher than $1/T$ can retain channel diversity as shown here.

Let the sequence of sampled channel output be

$$x(k\Delta) = \sum_{n=0}^{\infty} a(n)h(k\Delta - np\Delta + t_0) + w(k\Delta). \tag{24.41}$$

For notational simplicity, the oversampled channel output $x(k\Delta)$ can be divided into p linearly independent subsequences:

$$x^{(i)}(n) \stackrel{\Delta}{=} x[(np + i)\Delta] = x(nT + i\Delta), \quad i = 1, \ldots, p. \tag{24.42}$$

Define K as the effective channel length based on

$$\begin{aligned} h_0^{(i)} &\neq 0, \quad \text{for some } 1 \leq i \leq p \\ h_K^{(i)} &\neq 0, \quad \text{for some } 1 \leq i \leq p. \end{aligned} \tag{24.43}$$

By denoting the sub-channel transfer function as

$$H_i(z) = \sum_{k=0}^{K} h_k^{(i)} z^{-k} \quad \text{where} \quad h_k^{(i)} \stackrel{\Delta}{=} h(kT + i\Delta + t_0), \tag{24.44}$$

the p subsequences can be written as

$$x^{(i)}(n) = H_i(z)a(n) + w(nT + i\Delta), \quad i = 1, \ldots, p. \tag{24.45}$$

Thus, these p subsequences can be viewed as stationary outputs of p discrete FIR channels with a common input sequence $a(n)$ as shown in Figure 24.5. Naturally, they can also represent physical sub-channels in multisensor receivers.

The vector representation of the FSE is shown in Figure 24.5. One equalizer filter is provided for each subsequence $x^{(i)}(n)$. In fact, the actual equalizer is a vector of filters

$$G_i(z) = \sum_{k=0}^{m} w_{i,k} z^{-k}, \quad i = 1, \ldots, p. \tag{24.46}$$

The p filter outputs $\{y(n)^{(i)}\}$ are summed to form the stationary equalizer output

$$y(n) = \mathbf{W}^T \mathbf{X}(n), \tag{24.47}$$

where

$$\mathbf{W} \stackrel{\Delta}{=} \begin{bmatrix} w_{1,0} & \cdots & w_{1,m} & \cdots & w_{p,0} & \cdots & w_{p,m} \end{bmatrix}^T.$$

$$\mathbf{X}(n) \stackrel{\Delta}{=} \begin{bmatrix} x(n)^{(1)} & \cdots & x(n-m)^{(1)} & \cdots & x(n)^{(p)} & \cdots & x(n-m)^{(p)} \end{bmatrix}^T.$$

Given the equalizer output and parameter vector, any T-sampled blind equalization adaptive algorithm can be applied to the FSE via SGD techniques.

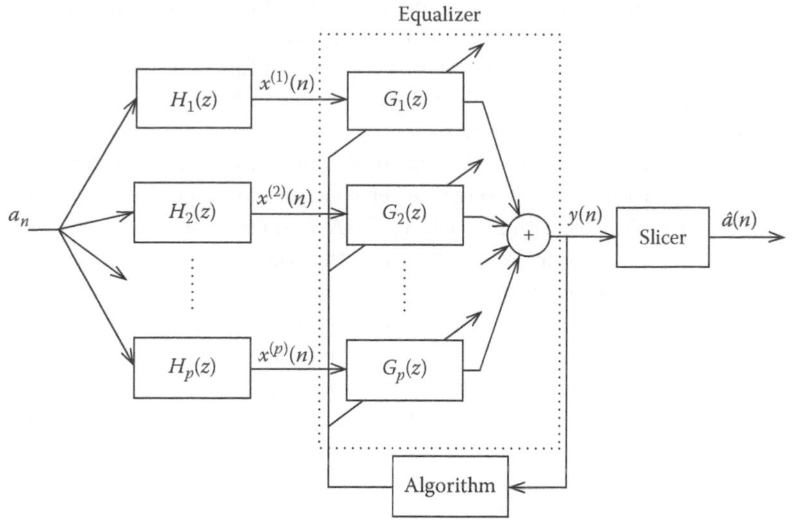

FIGURE 24.5 Vector representation for an FSE.

Since their first use, adaptive blind equalizers have often been implemented as FSEs. When training data are available, FSEs have the known advantage of suppressing timing phase sensitivity [30]. In fact, a blind FSE has another important advantage: there exists a one-to-one mapping between the combined parameter space and the equalizer parameter space, as shown in [31], under the following length and zero conditions:

- The equalizer length satisfies $(m + 1) \geq K$.
- The p discrete sub-channels $\{H_i(z)\}$ do not share any common zeros.

Note that for T-sampled equalizers, only one ($p = 1$) sub-channel exists and all zeros are common zeros, and, thus, the length and zero conditions cannot be satisfied. In most practical implementations, p is either 2 or 3. So long as the above conditions hold, the convergence behaviors of blind adaptive FSEs can be characterized completely in the combined parameter space. Based on the work of [12,16], for QAM channel inputs, there do not exist any algorithm-dependent stable equilibria other than the desired global minima [36] for FSEs driven by the Godard ($q = 2$) algorithm (CMA) and the Shalvi–Weinstein algorithms. Thus, the Godard and the Shalvi–Weinstein algorithms are globally convergent for FSEs satisfying these conditions [31].

Notice that global convergence of the Godard FSE is only proven for noiseless channels under the no-common zero condition. There have been recent advances in analyzing the performance of blind equalizers in the presence of Gaussian noise and the existence of common sub-channel zeros. While all possible delays of [36] are global minima for noiseless channels, the locations and effects of minima vary when channel noises are present. An analysis by Zeng and Tong shows that for noisy channels, CMA equalizer parameters have minima near the MMSE equilibria [32].

The effects of noise and common zeros was also studied by Fijalkow et al. [33,34], providing further indications of the robustness of CMA when implemented as an FSE.

24.9 Concluding Remarks

Adaptive channel equalization and blind equalization are among the most successful applications of adaptive filtering. We have introduced the basic concept of blind equalization along with some of the most commonly used blind equalization algorithms. Without the aid of training signals, the key challenge of blind adaptive equalizers lies in the design of special cost functions whose minimization is consistent with the goal of ISI removal. We have also summarized key results on the convergence of blind equalizers. The idea of constrained minimization of a convex cost function to assure global convergence of the blind equalizer was described. Finally, the blind adaptation in FSEs and multichannel receivers was shown to possess useful convergence properties.

It is important to note that the problem of blind equalization has not been completely solved by any means. In addition to the fact that the convergence behaviors of most algorithms are still unknown, the rates of convergence of typical algorithms such as CMA is quite slow, often needing thousands of iterations to achieve acceptable output. The difficulty of the convergence analysis and the slow rate of convergence of these algorithms have prompted many efforts to modify blind error functions to obtain faster and better algorithms. Furthermore, nonadaptive algorithms that explicitly exploit HOS [35–38] and second order cyclostationary statistics [25–29] appear to be quite efficient in exploiting small amount of channel output data. A detailed discussion of these methods is beyond the scope of this chapter. Interested readers may refer to two collected works edited by Haykin [24] and Gardner [39] and the references therein.

References

1. Qureshi, S.U.H., Adaptive equalization, *Proc. IEEE*, 73:1349–1387, Sept. 1985.
2. Lucky, R.W., Techniques for adaptive equalization of digital communication systems, *Bell Syst. Tech. J.*, 45:255–286, Feb. 1966.
3. Benveniste, A., Goursat, M., and Ruget, G., Robust identification of a nonminimum phase system, *IEEE Trans. Autom. Control*, AC-25:385–399, June 1980.
4. Benveniste, A. and Goursat, M., Blind equalizers, *IEEE Trans. Commn.*, 32:871–882, Aug. 1982.
5. Sato, Y., A method of self-recovering equalization for multi-level amplitude modulation, *IEEE Trans. Commn.*, COM-23:679–682, June 1975.
6. Macchi, O. and Eweda, E., Convergence analysis of self-adaptive equalizers, *IEEE Trans. Inf. Theory*, IT-30:162–176, Mar. 1984.
7. Mazo, J.E., Analysis of decision-directed equalizer convergence, *Bell Syst. Tech. J.*, 59:1857–1876, Dec. 1980.
8. Godard, D.N., Self-recovering equalization and carrier tracking in two-dimensional data communication systems, *IEEE Trans. Commn.*, COM-28:1867–1875, 1980.
9. Treichler, J.R. and Agee, B.G., A new approach to multipath correction of constant modulus signals, *IEEE Trans. Acoust. Speech Signal Process.*, ASSP-31:349–372, 1983.
10. Picchi, G. and Prati, G., Blind equalization and carrier recovery using a "stop-and-go" decision-directed algorithm, *IEEE Trans. Commn.*, COM-35: 877–887, Sept. 1987.
11. Hatzinakos, D., Blind equalization using stop-and-go criterion adaptation rules, *Opt. Eng.*, 31:1181–1198, June 1992.
12. Shalvi, O. and Weinstein, E., New criteria for blind deconvolution of nonminimum phase systems (channels), *IEEE Trans. Inf. Theory*, IT-36:312–321, Mar. 1990.
13. Brillinger, D.R. and Rosenblatt, M., Computation and interpretation of k-th order spectra, in *Spectral Analysis of Time Series*, B. Harris (Ed.), Wiley, New York, 1967.
14. Li, Y. and Ding, Z., Convergence analysis of finite length blind adaptive equalizers, *IEEE Trans. Signal Process.*, 43:2120–2129, Sept. 1995.
15. Ding, Z., Kennedy, R.A., Anderson, B.D.O., and Johnson, C.R., Jr., Local convergence of the Sato blind equalizer and generalizations under practical constraints, *IEEE Trans. Inf. Theory*, IT-39:129–144, Jan. 1993.
16. Foschini, G.J., Equalization without altering or detect data, *AT&T Tech. J.*, 64:1885–1911, Oct. 1985.
17. Ding, Z., Kennedy, R.A., and Johnson, C.R., Jr., On the (non)existence of undesirable equilibria of Godard blind equalizer, *IEEE Trans. Signal Process.*, 40:2425–2432, Oct. 1992.

18. Ding, Z., Kennedy, R.A., Anderson, B.D.O., and Johnson, C.R., Jr., Ill-convergence of Godard blind equalizers in data communications, *IEEE Trans. Commn.*, 39:1313–1328, Sept. 1991.
19. Minardi, M.J. and Ingram, M.A., Finding misconvergence in blind equalizers and new variance constraint cost functions to mitigate the problem, *Proceedings of 1996 International Conference on Acoustics, Speech, and Signal Processing*, Atlanta, GA, May 7–10, 1996, Vol. 3, pp. 1723–1726.
20. Tugnait, J.K., Shalvi, O., and Weinstein, E., Comments on new criteria for blind deconvolution of nonminimum phase systems (channels), *IEEE Trans. Inf. Theory*, IT-38:210–213, Jan. 1992.
21. Rupprecht, W.T., Adaptive equalization of binary NRZ-signals by means of peak value minimization, in *Proceedings of 7th Europe Conference on Circuit Theory and Design*, Prague, Czech Republic, 1985, pp. 352–355.
22. Kennedy, R.A. and Ding, Z., Blind adaptive equalizers for QAM communication systems based on convex cost functions, *Opt. Eng.*, 31:1189–1199, June 1992.
23. Yamazaki, K. and Kennedy, R.A., Reformulation of linearly constrained adaptation and its application to blind equalization, *IEEE Trans. Signal Process.*, SP-42:1837–1841, 1994.
24. Haykin, S. (Ed.), *Blind Deconvolution*, Prentice-Hall, Englewood Cliffs, NJ, 1994.
25. Tong, L., Xu, G., and Kailath, T., Blind channel identification and equalization based on second-order statistics: A time-domain approach, *IEEE Trans. Inf. Theory*, IT-40:340–349, Mar. 1994.
26. Moulines, E. et al., Subspace methods for the blind identification of multichannel FIR filters, *Proceedings of IEEE International Conference on Acoustics, Speech, and Signal Processing*, Adelaide, Australia, 1994, Vol. 4, pp. 573–576.
27. Li, Y. and Ding, Z., ARMA system identification based on second order cyclostationarity, *IEEE Trans. Signal Process.*, 42(12):3483–3493, Dec. 1994.
28. Meriam, K.A., Duhamel, P., Gesbert, D., Loubaton, P., Mayrarague, S., Moulines, E., and Slock, D., Prediction error methods for time-domain blind identification of multichannel FIR filters, *Proceedings of IEEE International Conference on Acoustics, Speech, and Signal Processing*, Detroit, MI, May 9–12, 1995, Vol. 3, pp. 1968–1971.
29. Hua, Y., Fast maximum likelihood for blind identification of multiple FIR channels, *IEEE Trans. Signal Process.*, SP-44:661–672, Mar. 1996.
30. Gitlin, R.D. and Weinstein, S.B., Fractionally spaced equalization: An improved digital transversal equalizer, *Bell Syst. Tech. J.*, 60:275–296, 1981.
31. Li, Y. and Ding, Z., Global convergence of fractionally spaced Godard adaptive equalizers, *IEEE Trans. Signal Process.*, SP-44:818–826, Apr. 1996.
32. Zeng, H. and Tong, L., On the performance of CMA in the presence of noise, *Proceedings of the Conference on Information Sciences and Systems*, Princeton, NJ, Mar. 1996.
33. Fijalkow, I., Treichler, J.R., and Johnson, C.R., Jr., Fractionally spaced blind equalization: Loss of channel diversity, *Proceedings of IEEE International Conference on Acoustics, Speech, and Signal Processing*, Detroit. MI, May 9–12, 1995, Vol. 3, pp. 1988–1991.
34. Touzni, A., Fijalkow, I., and Treichler, J.R., Fractionally spaced CMA under channel noise, *Proceedings of IEEE International Conference on Acoustics, Speech, and Signal Processing*, Atlanta, GA, May 7–10, 1996, Vol. 5, pp. 2674–2677.
35. Tugnait, J.K., Identification of linear stochastic systems via second and fourth-order cumulant matching, *IEEE Trans. Inf. Theory*, IT-33:393–407, May 1987.
36. Giannakis, G.B. and Mendel, J.M., Identification of nonminimum phase systems using via higher order statistics, *IEEE Trans. Acoust. Speech Signal Process.*, ASSP-37:360–377, 1989.
37. Hatzinakos, D. and Nikias, C.L., Blind equalization using a tricepstrum based algorithm, *IEEE Trans. Commn.*, 39:669–682, May 1991.
38. Shalvi, O. and Weinstein, E., Super-exponential methods for blind deconvolution, *IEEE Trans. Inf. Theory*, IT-39:504–519, Mar. 1993.
39. Gardner, W.A. (Ed.), *Cyclostationarity in Communications and Signal Processing*, IEEE Press, New York, 1994.

VII

Inverse Problems and Signal Reconstruction

Richard J. Mammone
Rutgers University

THERE ARE MANY SITUATIONS WHERE A DESIRED SIGNAL cannot be measured directly. The measurement might be degraded by physical limitations of the signal source and/or by the measurement device itself. The acquired signal is thus a transformation of the desired signal. The inversion of such transformations is the subject of this section. In the following chapters we will review several inverse problems and various methods of implementation of the inversion or recovery process. The methods differ in the ability to deal with the specific limitations present in each application. For example, the *a priori* constraint of nonnegativity is important for image recovery, but not so for adaptive array processing. The goal of the following chapters is to present the basic approaches of inversion and signal recovery. Each chapter focuses on a particular application area and describes the appropriate methods for that area.

Chapter 25 reviews the basic problem of signal recovery. The idea of projection onto convex sets (POCs) is introduced as an elegant solution to the signal recovery problem. The inclusion of linear and nonlinear constraints is addressed. The POCs method is shown to be a subset of the set theoretic approach to signal estimation. The application of image of restoration is described in detail.

Chapter 26 presents methods to reconstruct the interiors of objects from data collected based on transmitted or emitted radiation. The problem occurs in a wide range of application areas. The computer algorithms used for achieving the reconstructions are discussed. The basic techniques of image reconstruction from projections are classified into "Transform Methods" (including "Filtered Backprojection" and the "Linogram Methods") and "Series Expansion Methods" (including, in particular, the "Algebraic Reconstruction Techniques" and the method of "Expectation Maximization"). In addition, a performance comparison of the various algorithms for computed tomography is given.

The performance of speech and speaker recognition systems is significantly affected by the acoustic environment. The background noise level, the filtering effects introduced by the microphone and the communication channel dramatically affect the performance of recognition systems. It is therefore critical that these speech recognition systems capable of detecting the ambient acoustic environment continue and inverse their effects from the speech signal. This is the inverse problem in robust speech processing that will be addressed in Chapter 27. A general approach to solving this inverse problem is presented based on an affine transform model in the cepstrum domain.

In Chapter 28, a computational approach to 3-D (three-dimensional) coordinate restoration is presented. The problem is to obtain high-resolution coordinates of 3-D volume-elements (voxels) from observations of their corresponding 2-D picture-elements (pixels). The problem is posed as a combinatorial optimization problem and borrowing from our understanding of statistical mechanics, we show

how to adapt the tool of simulated annealing to solve this problem. This method is highly amenable to parallel and distributed processing.

In Chapter 29, the image recovery/reconstruction problem is formulated as a maximum-likelihood problem in which the image is recovered by maximizing an appropriately defined likelihood function. These likelihood functions are often highly nonlinear and when some of the variables involved are not directly observable, they can only be specified in integral form (i.e., averaging over the "hidden variables"). The expectation-maximization algorithm is revised and applied to some typical image recovery problems. Examples include image restoration using the Markov random field model and single and multiple channel image restoration with blur identification.

Array processing uses multiple sensors to improve signal reception by reducing the effects of interfering signals that originate from different spatial locations. Array processing algorithms are generally implemented via narrowband and broadband arrays, both of which are discussed in Chapter 30. Two classical approaches, namely sidelobe canceler and Frost beam formers, are reviewed. These algorithms are formulated as an inverse problem and an iterative approach for solving the resulting inverse problem is provided.

In Chapter 31 the relationship between communication channel equalization and the inversion of a linear system of equations is examined. A regularized method of inversion is an inversion process in which the noise dominated modes of the restored signal are attenuated. Channel equalization is the process that reduces the effects of a band-limited channel at the receiver of a communication system. A regularized method of channel equalization is presented in this section. Although there are many ways to accomplish this, the method presented uses linear and adaptive filters, which makes the transition to matrix inversion possible.

The response of an acoustic enclosure is, in general, a non-minimum phase function and hence not invertible. In Chapter 32, we discuss techniques using microphone arrays that attempt to recover speech signals degraded by the filtering effect of acoustic enclosures by either approximately or exactly "inverting" the room response. The aim of such systems is to force the impulse response of the overall system, after de-reverberation, to be an impulse function. Beamforming and matched-filtering techniques (that approximate this ideal case) and the Diophantine inverse filtering method (a technique that provides an exact inverse) are discussed in detail.

A synthetic aperture radar (SAR) is a radar sensor that provides azimuth resolution superior to that achievable with its real beam by synthesizing a long aperture by platform motion. Chapter 33 presents an overview of the basics of SAR phenomenology and the associated algorithms that are used to form the radar image and to enhance it. The chapter begins with an overview of SAR applications, historical development, fundamental phenomenology, and a survey of modern SAR systems. It also presents examples of SAR imagery. This is followed by a discussion of the basic principles of SAR image formation that begins with side looking radar, progresses to unfocused SAR, and finishes with focused SAR. A discussion of SAR image enhancement techniques, such as the polarimetric whitening filters, follows. Finally, a brief discussion of automatic target detection and classification techniques is offered.

In Chapter 34, a class of iterative restoration algorithms is presented. Such algorithms provide solutions to the problem of recovering an original signal or image from a noisy and blurred observation of it. This situation is encountered in a number of important applications, ranging from the restoration of images obtained by the Hubble space telescope to the restoration of compressed images. The successive approximation methods form the basis of the material presented in this section.

The sample of applications and methods described in this chapter are meant to be representative of the large volume of work performed in this field. There is no claim of completeness, any omissions of significant contributors or other errors are solely the responsibility of the section editor, and all praiseworthy contributions are due solely to the chapter authors.

25

Signal Recovery from Partial Information

Christine Podilchuk
Rutgers University

25.1 Introduction

Signal recovery has been an active area of research for applications in many different scientific disciplines. A central reason for exploring the feasibility of signal recovery is due to the limitations imposed by a physical device on the amount of data one can record. For example, for diffraction-limited systems, the finite aperture size of the lens constrains the amount of frequency information that can be captured. The image degradation is due to attenuation of high frequency components resulting in a loss of details and other high frequency information. In other words, the finite aperture size of the lens acts like a lowpass filter on the input data. In some cases, the quality of the recorded image data can be improved by building a more costly recording device but many times the required condition for acceptable data quality is physically unrealizable or too costly. Other times signal recovery may be necessary is for the recording of a unique event that cannot be reproduced under more ideal recording conditions.

Some of the earliest work on signal recovery includes the work by Sondhi [1] and Slepian [2] on recovering images from motion blur and Helstrom [3] on least squares restoration. A sampling of some of the signal recovery algorithms applied to different types of problems can be found in [4–21]. Further reading includes the other sections in this book, Chapter 15 of *Digital Signal Processing: Video, Speech, Audio, and Associated Standards,* and the extended list of references provided by all the authors.

The simple signal degradation model described in the next section turns out to be a useful representation for many different problems encountered in practice. Some examples that can be formulated using the general signal recovery paradigm include image restoration, image reconstruction, spectral estimation, and filter design. We distinguish between image restoration, which pertains to image recovery based on a measured distorted version of the original image, and image reconstruction, which refers most commonly to medical imaging where the image is reconstructed from a set of indirect measurements, usually projections. For many of the signal recovery applications, it is desirable to extrapolate a signal outside of a known interval. Extrapolating a signal in the spatial or temporal domain could result in improved spectral resolution and applies to such problems as power spectrum estimation, radio astronomy, radar target detection, and geophysical exploration. The dual problem, extrapolating the signal in the frequency domain, also known as superresolution, results in improved spatial or temporal resolution and is desirable in many image restoration problems. As will be shown later, the standard inverse filtering techniques are not able to resolve the signal estimate beyond the diffraction limit imposed by the physical measuring device.

The observed signal is degraded from the original signal by both the measuring device as well as external conditions. Besides the measured, distorted output signal we may have some additional information about the following: the measuring system and external conditions, such as noise, as well as some *a priori* knowledge about the desired signal to be restored or reconstructed. In order to produce a good estimate of the original signal, we should take advantage of all the available information.

Although the data recovery algorithms described here apply in general to any data type, we derive most of the techniques based on two-dimensional input data for image processing applications. For most cases, it is straightforward to adapt the algorithms to other data types. Examples of data recovery techniques for different inputs are illustrated in the other sections in this book as well as Chapter 15 of *Digital Signal Processing: Video, Speech, Audio, and Associated Standards* for image restoration. The material in this section requires some basic knowledge of linear algebra as found in [22].

Section 25.2 presents the signal degradation model and formulates the signal recovery problem. The early attempts of signal recovery based on inverse filtering are presented in Section 25.3. The concept of projection onto convex sets (POCS) described in Section 25.4 allows us to introduce *a priori* knowledge about the original signal in the form of linear as well as nonlinear constraints into the recovery algorithm. Convex set theoretic formulations allow us to design recovery algorithms that are extremely flexible and powerful. Sections 25.5 and 25.6 present some basic POCS-based algorithms and Section 25.7 presents a POCS-based algorithm for image restoration as well as some results. The sample algorithms presented here are not meant to be exhaustive and the reader is encouraged to read the other sections in this chapter as well as the references for more details.

25.2 Formulation of the Signal Recovery Problem

Signal recovery can be viewed as an estimation process in which operations are performed on an observed signal in order to estimate the ideal signal that would be observed if no degradation was present. In order to design a signal recovery system effectively, it is necessary to characterize the degradation effects of the physical measuring system. The basic idea is to model the signal degradation effects as accurately as possible and perform operations to undo the degradations and obtain a restored signal. When the degradation cannot be modeled sufficiently, even the best recovery algorithms will not yield satisfactory results. For many applications, the degradation system is assumed to be linear and can be modeled as a Fredholm integral equation of the first kind expressed as

$$g(x) = \int_{-\infty}^{\infty} h(x; a) f(a) \mathrm{d}a + n(x). \tag{25.1}$$

This is the general case for a one-dimensional signal where f and g are the original and measured signals, respectively, n represents noise, and $h(x; a)$ is the impulse response or the response of the measuring system to an impulse at coordinate a.* A block diagram illustrating the general one-dimensional signal

* This corresponds to the case of a shift-varying impulse response.

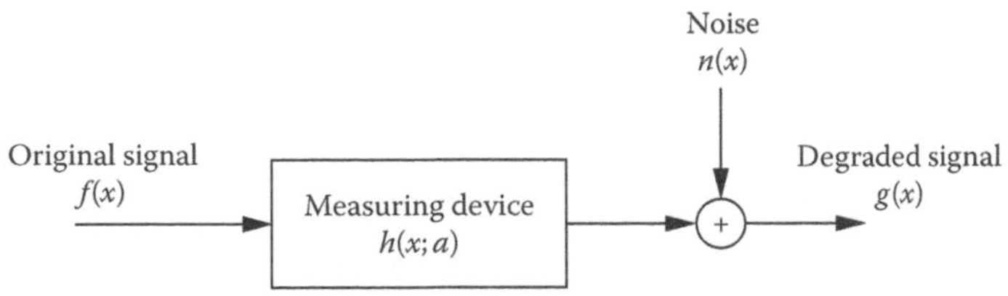

FIGURE 25.1 Block diagram of the signal recovery problem.

degradation system is shown in Figure 25.1. For image processing applications, we modify this equation to the two-dimensional case, that is,

$$g(x,y) = \int_{-\infty}^{+\infty}\int_{-\infty}^{+\infty} h(x,y;a,b)f(a,b)\mathrm{d}a\mathrm{d}b + n(x,y). \tag{25.2}$$

The degradation operator h is commonly referred to as a point spread function (PSF) in imaging applications because in optics, h is the measured response of an imaging system to a point of light.

The Fourier transform of the PSF $h(x,y)$ denoted as $\mathcal{H}(w_x, w_y)$ is known as the optical transfer function (OTF) and can be expressed as

$$\mathcal{H}(w_x, w_y) = \frac{\iint_{-\infty}^{\infty} h(x,y)\exp\left(-i(w_x x + w_y y)\right)\mathrm{d}x\mathrm{d}y}{\iint_{-\infty}^{\infty} h(x,y)\mathrm{d}x\mathrm{d}y}. \tag{25.3}$$

The absolute value of the OTF is known as the modulation transfer function. A commonly used optical image formation system is a circular thin lens. The recovery problem is considered ill-posed when a small change in the observed image, g, results in a large change in the solution, f. Most signal recovery problems in practice are ill-posed.

The continuous version of the degradation system for two-dimensional signals formulated in Equation 25.2 can be expressed in discrete form by replacing the continuous arguments with arrays of samples in two dimensions, that is,

$$g(i,j) = \sum_m \sum_n h(i,j;m,n)f(m,n) + n(i,j). \tag{25.4}$$

It is convenient for image recovery purposes to represent the discrete formulation given in Equation 25.4 as a system of linear equations expressed as

$$\mathbf{g} = \mathbf{Hf} + \mathbf{n}, \tag{25.5}$$

where

- $\mathbf{g}$, $\mathbf{f}$, and $\mathbf{n}$ are the lexicographic row-stacked versions of the discretized versions of g, f, and n in Equation 25.4
- $\mathbf{H}$ is the degradation matrix composed of the PSF

This section presents an overview of some of the techniques proposed to estimate $\mathbf{f}$ when the recovery problem can be modeled by Equation 25.5. If there is no external noise or measurement error and the set of equations is consistent, Equation 25.5 reduces to

$$\mathbf{g} = \mathbf{Hf}. \tag{25.6}$$

It is usually not the case that a practical system can be described by Equation 25.6. In this section, we will focus on recovery algorithms where an estimate of the distortion operation represented by the matrix **H** is known. For recovery problems where both the desired signal, **f**, and the degradation operator, **H**, are unknown, refer to other chapters in this book.

For most systems, the degradation matrix **H** is highly structured and quite sparse. The additive noise term due to measurement errors and external and internal noise sources is represented by the vector **n**. At first glance, the solution to the signal recovery problem seems to be straightforward—find the inverse of the matrix **H** to solve for the unknown vector **f**. It turns out that the solution is not so simple because in practice the degradation operator is usually ill-conditioned or rank-deficient and the problem of inconsistencies or noise must be addressed. Other problems that may arise include computational complexity due to extremely large problem dimensions especially for image processing applications. The algorithms described here try to address these issues for the general signal recovery problem described by Equation 25.5.

25.2.1 Prolate Spheroidal Wavefunctions

We introduce the problem of signal recovery by examining a one-dimensional, linear, time-invariant system that can be expressed as

$$g(x) = \int_{-T}^{+T} f(\alpha)h(x-\alpha)d\alpha, \tag{25.7}$$

where

$g(x)$ is the observed signal

$f(\alpha)$ is the desired signal of finite support on the interval $(-T, +T)$

$h(x)$ denotes the degradation operator

Assuming that the degradation operator in this case is an ideal lowpass filter, h can be described mathematically as

$$h(x) = \frac{\sin(x)}{x}. \tag{25.8}$$

For this particular case, it is possible to solve for the exact signal $f(x)$ with prolate spheroidal wavefunctions [23]. The key to successfully solving for f lies in the fact that prolate spheroidal wavefunctions are the eigenfunctions of the integral equation expressed by Equation 25.7 with Equation 25.8 as the degradation operator. This relationship is expressed as

$$\int_{-T}^{+T} \psi_n(\alpha)\frac{\sin(x-\alpha)}{x-\alpha}d\alpha = \lambda_n\psi_n(x), \quad n = 0, 1, 2, \ldots, \tag{25.9}$$

where

$\psi_n(x)$ are the prolate spheroidal wavefunctions

λ_n are the corresponding eigenvalues

A critical feature of prolate spheroidal wavefunctions is that they are complete orthogonal bases in the interval $(-\infty, +\infty)$ as well as the interval $(-T, +T)$, that is,

$$\int_{-\infty}^{+\infty} \psi_n(x)\psi_m(x)\mathrm{d}x = \begin{cases} 1, & \text{if } n = m, \\ 0, & \text{if } n \neq m, \end{cases} \tag{25.10}$$

and

$$\int_{-T}^{+T} \psi_n(x)\psi_m(x)\mathrm{d}x = \begin{cases} \lambda_n, & \text{if } n = m, \\ 0, & \text{if } n \neq m. \end{cases} \tag{25.11}$$

This allows the functions $g(x)$ and $f(x)$ to be expressed as the series expansion:

$$g(x) = \sum_{n=0}^{\infty} c_n \psi_n(x), \tag{25.12}$$

$$f(x) = \sum_{n=0}^{\infty} d_n \psi_{Ln}(x), \tag{25.13}$$

where $\psi_{Ln}(x)$ are the prolate spheroidal functions truncated to the interval $(-T, T)$. The coefficients c_n and d_n are given by

$$c_n = \int_{-\infty}^{\infty} g(x)\psi_n(x)\mathrm{d}x \tag{25.14}$$

and

$$d_n = \frac{1}{\lambda_n} \int_{-T}^{T} f(x)\psi_n(x)\mathrm{d}x. \tag{25.15}$$

If we substitute the series expansions given by Equations 25.12 and 25.13 into Equation 25.7, we get

$$\begin{aligned} g(x) &= \sum_{n=0}^{\infty} c_n \psi_n(x) \\ &= \int_{-T}^{+T} \left[\sum_{n=0}^{\infty} d_n \psi_{Ln}(\alpha) \right] h(x-\alpha)\mathrm{d}\alpha \qquad (25.16) \\ &= \sum_{n=0}^{\infty} d_n \left[\int_{-T}^{+T} \psi_n(\alpha) h(x-\alpha)\mathrm{d}\alpha \right]. \qquad (25.17) \end{aligned}$$

Combining this result with Equation 25.9,

$$\sum_{n=0}^{\infty} c_n \psi_n(x) = \sum_{n=0}^{\infty} \lambda_n d_n \psi_n(x), \tag{25.18}$$

where

$$c_n = \lambda_n d_n, \tag{25.19}$$

and

$$d_n = \frac{c_n}{\lambda_n}. \tag{25.20}$$

We get an exact solution for the unknown signal $f(x)$ by substituting Equation 25.20 into Equation 25.13, that is,

$$f(x) = \sum_{n=0}^{\infty} \frac{c_n}{\lambda_n} \psi_{Ln}(x). \tag{25.21}$$

Therefore, in theory, it is possible to obtain the exact image $f(x)$ from the diffraction-limited image, $g(x)$, using prolate spheroidal wavefunctions. The difficulties of signal recovery become more apparent when we examine the simple diffraction-limited case in relation to prolate spheroidal wavefunctions as described in Equation 25.21. The finite aperture size of a diffraction-limited system translates to eigenvalues λ_n which exhibit a unit-step response; that is, the several largest eigenvalues are approximately one followed by a succession of eigenvalues that rapidly fall off to zero. The solution given by Equation 25.21 will be extremely sensitive to noise for small eigenvalues λ_n. Therefore, for the general problem represented in vector space by Equation 25.5, the degradation operator H is ill-conditioned or rank-deficient due to the small or zero-valued eigenvalues, and a simple inverse operation will not yield satisfactory results. Many algorithms have been proposed to find a compromise between exact deblurring and noise amplification. These techniques include Wiener filtering and pseudoinverse filtering. We begin our overview of signal recovery techniques by examining some of the methods that fall under the category of optimization-based approaches.

25.3 Least Squares Solutions

The earliest attempts toward signal recovery are based on the concept of inverting the degradation operator to restore the desired signal. Because in practical applications the system will often be ill-conditioned, several problems can arise. Specifically, high detail signal information may be masked by observation noise, or a small amount of observation noise may lead to an estimate that contains very large false high frequency components. Another potential problem with such an approach is that for a rank-deficient degradation operator, the zero-valued eigenvalues cannot be inverted. Therefore, the general inverse filtering approach will not be able to resolve the desired signal beyond the diffraction limit imposed by the measuring device. In other words, referring to the vector–space description, the data that has been nulled out by the zero-valued eigenvalues cannot be recovered.

25.3.1 Wiener Filtering

Wiener filtering combines inverse filtering with *a priori* statistical knowledge about the noise and unknown signal [24] in order to deal with the problems associated with an ill-conditioned system.

The impulse response of the restoration filter is chosen to minimize the mean square error (mse) as defined by

$$\varepsilon_f = E\{(\mathbf{f} - \hat{\mathbf{f}})^2\} \tag{25.22}$$

where

$\hat{\mathbf{f}}$ denotes the estimate of the ideal signal $\mathbf{f}$

$E\{\cdot\}$ denotes the expected value

The Wiener filter estimate is expressed as

$$\mathbf{H}_{\mathrm{W}}^{-1} = \frac{\mathbf{R}_{ff}\mathbf{H}^{\mathrm{T}}}{\mathbf{H}\mathbf{R}_{ff}\mathbf{H}^{\mathrm{T}} + \mathbf{R}_{nn}} \tag{25.23}$$

where $\mathbf{R}_{ff}$ and $\mathbf{R}_{nn}$ are the covariance matrices of $\mathbf{f}$ and $\mathbf{n}$, respectively, and $\mathbf{f}$ and $\mathbf{n}$ are assumed to be uncorrelated; that is,

$$\mathbf{R}_{ff} = E\{\mathbf{f}\mathbf{f}^{\mathrm{T}}\}, \tag{25.24}$$

$$\mathbf{R}_{nn} = E\{\mathbf{n}\mathbf{n}^{\mathrm{T}}\}, \tag{25.25}$$

and

$$\mathbf{R}_{fn} = 0. \tag{25.26}$$

The superscript T in the above equations denotes transpose. The Wiener filter can also be expressed in the Fourier domain as

$$\mathcal{H}_{\mathrm{W}}^{-1} = \frac{\mathcal{H}^{*}S_{ff}}{|\mathcal{H}|^{2}S_{ff} + S_{nn}} \tag{25.27}$$

where

S denotes the power spectral density

the superscript * denotes the complex conjugate

$\mathcal{H}$ denotes the Fourier transform of $\mathbf{H}$

Note that when the noise power is zero, the Wiener filter reduces to the inverse filter; that is,

$$\mathcal{H}_{\mathrm{W}}^{-1} = \mathcal{H}^{-1}. \tag{25.28}$$

The Wiener filter approach for signal recovery assumes that the power spectra are known for the input signal and the noise. Also, this approach assumes that finding a least squares solution that optimizes Equation 25.22 is meaningful. For the case of image processing, it has been shown, specifically in the context of image compression, that the mse does not predict subjective image quality [25]. Many signal processing algorithms are based on the least squares paradigm because the solutions are tractable and, in practice, such approaches have produced some useful results. However, in order to define a more meaningful optimization metric in the design of image processing algorithms, we need to incorporate a human visual model into the algorithm design. In the area of image coding, several coding schemes based on perceptual criteria have been shown to produce improved results over schemes based on maximizing signal-to-noise ratio (SNR) or minimizing mse [25]. Likewise, the Wiener filtering approach will not necessarily produce an estimate that maximizes perceived image or signal quality. Another limitation of the Wiener filter approach is that the solution will not necessarily be consistent with any *a priori* knowledge about the desired signal characteristics. In addition, the Wiener filter approach does not resolve the desired signal beyond the diffraction limit imposed by the measuring system. For more details on Wiener filtering and the various applications, see other chapters in this book.

25.3.2 Pseudoinverse Solution

The Wiener filters attempt to minimize the noise amplification obtained in a direct inverse by providing a taper determined by the statistics of the signal and noise process under consideration. In practice, the power spectra of the noise and desired signal might not be known. Here we present what is commonly referred to as the generalized inverse solution. This will be the framework for some of the signal recovery algorithms described later.

The pseudoinverse solution is an optimization approach that seeks to minimize the least squares error as given by

$$\varepsilon_n = \mathbf{n}^{\mathrm{T}}\mathbf{n} = (\mathbf{g} - \mathbf{H}\mathbf{f})^{\mathrm{T}}(\mathbf{g} - \mathbf{H}\mathbf{f}). \tag{25.29}$$

The least squares solution is not unique when the rank of the $M \times N$ matrix $\mathbf{H}$ is $r < N \leq M$. In other words, there are many solutions that satisfy Equation 25.29. However, the Moore–Penrose generalized inverse or pseudoinverse [26] does provide a unique least squares solution based on determining the least squares solution with minimum norm. For a consistent set of equations as described in Equation 25.6, a solution is sought that minimizes the least squares estimation error; that is,

$$\begin{aligned} \varepsilon_f &= (\mathbf{f} - \hat{\mathbf{f}})^{\mathrm{T}}(\mathbf{f} - \hat{\mathbf{f}}) \\ &= \mathrm{tr}\{(\mathbf{f} - \hat{\mathbf{f}})(\mathbf{f} - \hat{\mathbf{f}})^{\mathrm{T}}\}, \end{aligned} \tag{25.30}$$

where
- $\mathbf{f}$ is the desired signal vector
- $\hat{\mathbf{f}}$ is the estimate
- $\mathrm{tr}\{\cdot\}$ denotes the trace [22]

The generalized inverse provides an optimum solution that minimizes the estimation error for a consistent set of equations. Thus, the generalized inverse provides an optimum solution for both the consistent and inconsistent set of equations as defined by the performance functions ε_f and ε_n, respectively. The generalized inverse solution satisfies the normal equations

$$\mathbf{H}^{\mathrm{T}}\mathbf{g} = \mathbf{H}^{\mathrm{T}}\mathbf{H}\mathbf{f}. \tag{25.31}$$

The generalized inverse solution, also known as the Moore–Penrose generalized inverse, pseudoinverse, or least squares solution with minimum norm is defined as

$$\mathbf{f}^{\dagger} = (\mathbf{H}^{\mathrm{T}}\mathbf{H})^{-1}\mathbf{H}^{\mathrm{T}}\mathbf{g} = \mathbf{H}^{\dagger}\mathbf{g}, \tag{25.32}$$

where
- the dagger $\dagger$ denotes the pseudoinverse
- the rank of $\mathbf{H}$ is $r = N \leq M$

For the case of an inconsistent set of equations as described in Equation 25.5, the pseudoinverse solution becomes

$$\mathbf{f}^{\dagger} = \mathbf{H}^{\dagger}\mathbf{g} = \mathbf{H}^{\dagger}\mathbf{H}\mathbf{f} + \mathbf{H}^{\dagger}\mathbf{n}, \tag{25.33}$$

where $\mathbf{f}^{\dagger}$ is the minimum norm, least squares solution. If the set of equations are overdetermined with rank $r = N < M$, $\mathbf{H}^{\dagger}\mathbf{H}$ becomes an identity matrix of size N denoted as $\mathbf{I}_N$ and the pseudoinverse solution reduces to

$$\begin{aligned}\mathbf{f}^{\dagger} &= \mathbf{f} + \mathbf{H}^{\dagger}\mathbf{n} \\ &= \mathbf{f} + \Delta\mathbf{f}.\end{aligned} \tag{25.34}$$

A straightforward result from linear algebra is the bound on the relative error:

$$\frac{\|\Delta\mathbf{f}\|}{\|\mathbf{f}\|}\,\|\mathbf{H}^{\dagger}\|\,\|\mathbf{H}\|\,\frac{\|\mathbf{n}\|}{\|\mathbf{g}\|}, \tag{25.35}$$

where the product $\|\mathbf{H}^{\dagger}\|\,\|\mathbf{H}\|$ is the condition number of **H**. This quantity determines the relative error in the estimate in terms of the ratio of the vector norm of the noise to the vector norm of the observed image. The condition number of H is defined as

$$C_H = \|\mathbf{H}^{\dagger}\|\,\|\mathbf{H}\| = \frac{\sigma_1}{\sigma_N} \tag{25.36}$$

where σ_1 and σ_N denote the largest and smallest singular values of the matrix H, respectively. The larger the condition number, the greater the sensitivity to noise perturbations. A matrix with a large condition number, typically greater than 100, results in an ill-conditioned system.

The pseudoinverse solution is best described by diagonalizing the degradation matrix **H** using singular value decomposition (SVD) [22]. SVD provides a way to diagonalize any arbitrary $M \times N$ matrix. In this case, we wish to diagonalize **H**; that is,

$$\mathbf{H} = \mathbf{U}\boldsymbol{\Sigma}\mathbf{V}^{\mathrm{T}} \tag{25.37}$$

where

- **U** is a unitary matrix composed of the orthonormal eigenvectors of $\mathbf{H}^{\mathrm{T}}\mathbf{H}$
- **V** is a unitary matrix composed of the orthonormal eigenvectors of $\mathbf{H}\mathbf{H}^{\mathrm{T}}$
- $\boldsymbol{\Sigma}$ is a diagonal matrix composed of the singular values of **H**

The number of nonzero diagonal terms denotes the rank of **H**. The degradation matrix can be expressed in series form as

$$\mathbf{H} = \sum_{i=1}^{r} \sigma_i \mathbf{u}_i \mathbf{v}_i^{\mathrm{T}} \tag{25.38}$$

where

- $\mathbf{u}_i$ and $\mathbf{v}_i$ are the ith columns of **U** and **V**, respectively
- r is the rank of **H**

From Equations 25.37 and 25.38, the pseudoinverse of **H** becomes as

$$\mathbf{H}^{\dagger} = \mathbf{V}\boldsymbol{\Sigma}^{\dagger}\mathbf{U}^{\mathrm{T}} = \sum_{i=1}^{r} \sigma_i^{-1} \mathbf{v}_i \mathbf{u}_i^{\mathrm{T}}. \tag{25.39}$$

Therefore, from Equation 25.39, the pseudoinverse solution can be expressed as

$$\mathbf{f}^{\dagger} = \mathbf{H}^{\dagger}\mathbf{g} = \mathbf{V}\boldsymbol{\Sigma}^{\dagger}\mathbf{U}^{\mathrm{T}}\mathbf{g} \tag{25.40}$$

or

$$\mathbf{f}^{\dagger} = \sum_{i=1}^{r} \sigma_i^{-1} \mathbf{v}_i \mathbf{u}_i^{\mathrm{T}} \mathbf{g} = \sum_{i=1}^{r} \sigma_i^{-1} \left(\mathbf{u}_i^{\mathrm{T}} \mathbf{g} \right) \mathbf{v}_i. \tag{25.41}$$

The series form of the pseudoinverse solution using SVD allows us to solve for the pseudoinverse solution using a sequential restoration algorithm expressed as

$$\mathbf{f}^{\dagger(k+1)} = \mathbf{f}^{\dagger(k)} + \sigma_k^{-1} \left(\mathbf{u}_k^{\mathrm{T}} \mathbf{g} \right) \mathbf{v}_k. \tag{25.42}$$

The iterative approach for finding the pseudoinverse solution is advantageous when dealing with ill-conditioned systems and noise corrupted data. The iterative form can be terminated before the inversion of small singular values resulting in an unstable estimate. This technique becomes quite easy to implement for the case of a circulant degradation matrix **H**, where the unitary matrices in Equation 25.37 reduce to the discrete Fourier transform.

25.3.3 Regularization Techniques

Smoothing and regularization techniques [27–29] have been proposed in an attempt to overcome the problems associated with inverting ill-conditioned degradation operators for signal recovery. These methods attempt to force smoothness on the solution of a least squares error problem. The problem can be formulated in two different ways. One way of formulating the problem is

minimize:

$$\hat{\mathbf{f}}^{\mathrm{T}} \mathbf{S} \hat{\mathbf{f}} \tag{25.43}$$

subject to:

$$(\mathbf{g} - \mathbf{H}\hat{\mathbf{f}})^{\mathrm{T}} \mathbf{W} (\mathbf{g} - \mathbf{H}\hat{\mathbf{f}}) = e \tag{25.44}$$

where

S represents a smoothing matrix
W is an error weighting matrix
e is a residual scalar estimation error

The error weighting matrix can be chosen as $\mathbf{W} = \mathbf{R}_{nn}^{-1}$. The smoothing matrix is typically composed of the first or second order difference. For this case, we wish to find the stationary point of the Lagrangian expression:

$$F(\hat{\mathbf{f}}, \lambda) = \hat{\mathbf{f}}^{\mathrm{T}} \mathbf{S} \hat{\mathbf{f}} + \lambda [(\mathbf{g} - \mathbf{H}\hat{\mathbf{f}})^{\mathrm{T}} \mathbf{W} (\mathbf{g} - \mathbf{H}\hat{\mathbf{f}}) - e]. \tag{25.45}$$

The solution is found by taking derivatives with respect to **f** and λ and setting them equal to zero. The solution for a nonsingular overdetermined set of equations becomes

$$\hat{\mathbf{f}} = \left(\mathbf{H}^{\mathrm{T}} \mathbf{W} \mathbf{H} + \frac{1}{\lambda} \mathbf{S} \right)^{-1} \mathbf{H}^{\mathrm{T}} \mathbf{W} \mathbf{g}, \tag{25.46}$$

where λ is chosen to satisfy the compromise between residual error and smoothness in the estimate.

Alternately, this problem can be formulated as

minimize:

$$(\mathbf{g}-\mathbf{H}\hat{\mathbf{f}})^{\mathrm{T}}\mathbf{W}(\mathbf{g}-\mathbf{H}\hat{\mathbf{f}}) \tag{25.47}$$

subject to:

$$\hat{\mathbf{f}}^{\mathrm{T}}\mathbf{S}\hat{\mathbf{f}} = d \tag{25.48}$$

where d represents a fixed degree of smoothness. The Lagrangean expression for this formulation becomes

$$G(\hat{\mathbf{f}},\gamma) = (\mathbf{g}-\mathbf{H}\hat{\mathbf{f}})^{\mathrm{T}}\mathbf{W}(\mathbf{g}-\mathbf{H}\hat{\mathbf{f}})^{\mathrm{T}} + \gamma(\hat{\mathbf{f}}^{\mathrm{T}}\mathbf{S}\hat{\mathbf{f}} - \mathbf{d}) \tag{25.49}$$

and the solution for a nonsingular overdetermined set of equations becomes

$$\hat{\mathbf{f}} = (\mathbf{H}^{\mathrm{T}}\mathbf{W}\mathbf{H} + \gamma\mathbf{S})^{-1}\mathbf{H}^{\mathrm{T}}\mathbf{W}\mathbf{g}. \tag{25.50}$$

Note that for the two problem formulations, the results as given by Equations 25.46 and 25.50 are identical if $\gamma = 1/\lambda$. The shortcomings of such a regularization technique is that the smoothing function **S** must be estimated and either the degree of smoothness, d, or the degree of error, e, must be known to determine γ or λ.

Constrained restoration techniques have also been developed [30] to overcome the problem of an ill-conditioned system. Linear equality constraints and linear inequality constraints have been enforced to yield one-step solutions similar to those described in this section. All the techniques described thus far attempt to overcome the problem of noise corrupted data and ill-conditioned systems by forcing some sort of taper on the inverse of the degradation operator. The sampling of algorithms discussed thus far fall under the category of optimization techniques where the objective function to be minimized is the least squares error. Recovery algorithms that fall under the category of optimization-based algorithms include maximum likelihood (ML), maximum *a posteriori* (MAP), and maximum entropy methods [17].

We now introduce the concept of POCS, which will be the framework for a much broader and more powerful class of signal recovery algorithms.

25.4 Signal Recovery using Projection onto Convex Sets

A broad set of recovery algorithms has been proposed to conform to the general framework introduced by the theory of POCS [31]. The POCS framework enables one to define an iterative recovery algorithm that can incorporate a number of linear as well as nonlinear constraints that satisfy certain properties. The more *a priori* information about the desired signal that one can incorporate into the algorithm, the more effective the algorithm becomes. In [21], POCS is presented as a particular example of a much broader class of algorithms described as set theoretic estimation. The author distinguishes between two basic approaches to a signal estimation or recovery problem: optimization-based approaches and set theoretic approaches. The effectiveness of optimization-based approaches is highly dependent on defining a valid optimization criterion that, in practice, is usually determined by computational tractability rather than how well it models the problem. The optimization-based approaches seek a unique solution based on some predefined optimization criterion. The optimization-based approaches include the least squares techniques of the previous section as well as ML, MAP, and maximum entropy techniques. Set theoretic estimation is based on the concept of finding a feasible solution, that is, a solution that is consistent with all the available *a priori* information. Unlike the optimization-based approaches which seek to find one optimum solution, the set theoretic approaches usually determine one of many possible feasible solutions. Many problems in signal recovery can be approached using the set theoretic paradigm. POCS has been one of the most extensively studied set theoretic approaches in the literature due to its convergence properties and flexibility to handle a wide range of signal characteristics. We limit our discussion here to POCS-based algorithms. The more general case of signal estimation using nonconvex as well as convex sets is covered in [21]. The rest of this section will focus on defining the POCS framework and describing several useful algorithms that fall into this general category.

25.4.1 POCS Framework

A projection operator onto a closed convex set is an example of a nonlinear mapping that is easily analyzed and contains some very useful properties. Such projection operators minimize error distance and are non-expansive. These are two very important properties of ordinary linear orthogonal projections onto closed linear manifolds (CLMs). The benefit of using POCS for signal restoration is that one can incorporate nonlinear constraints of a certain type into the POCS framework. Linear image restoration algorithms cannot take advantage of *a priori* information based on nonlinear constraints.

The method of POCS depends on the set of solutions that satisfies *a priori* characteristics of the desired signal to lie in a well-defined closed convex set. For such properties, **f** is restricted to lie in the region defined by the intersection of all the convex sets, that is,

$$\mathbf{f} \in C_0 = \cap_{i=1}^{l} C_i. \tag{25.51}$$

Here C_i denotes the ith closed convex set corresponding to the ith property of **f**, $C_i \in S$, and $i \in \mathcal{I}$. The unknown signal **f** can be restored by using the corresponding projection operators P_i onto each convex set C_i. A property of closed convex sets is that a projection of a point onto the convex set is unique. This is known as the unique-nearest-neighbor property. The general form of the POCS-based recovery algorithm is expressed as

$$\mathbf{f}^{(k+1)} = P_{i_k}\mathbf{f}^{(k)} \tag{25.52}$$

where
 k denotes the iteration
 i_k denotes a sequence of indices in $\mathcal{I}$

A common technique for iterating through the projections is referred to as cyclic control where the projections are applied in a cyclic manner, that is, $i_k = k(\text{modulo } l) + 1$. A geometric interpretation of the POCS algorithm for the simple case of two convex sets is illustrated in Figure 25.2. The original POCS formulation is further generalized by introducing a relaxation parameter expressed as

$$\mathbf{f}^{(k+1)} = \mathbf{f}^{(k)} + \lambda_k\left(\mathbf{P}_{i_k}(\mathbf{f}^{(k)}) - \mathbf{f}^{(k)}\right), \quad 0 < \lambda_k < 2 \tag{25.53}$$

where λ_k denotes the relaxation parameter. If $\lambda_k < 1$, the algorithm is said to be *underrelaxed* and if $\lambda_k > 1$, the algorithm is *overrelaxed*. Refer to [31] for further details on the convergence properties of POCS.

Common constraints that apply to many different signals in practice and whose solution space obeys the properties of convex sets are described in [10]. Some examples from [10] include frequency limits, spatial/temporal bounds, nonnegativity, sparseness, intensity or energy bounds, and partial knowledge of the spectral or spatial/temporal components. For further details on commonly used convex sets, see [10]. Most of the commonly used constraints for different signal processing applications fall under the

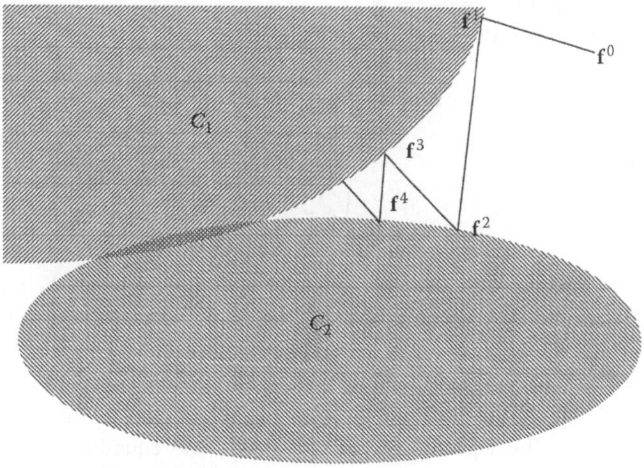

FIGURE 25.2 Geometric interpretation of POCS.

category of convex sets which provide weak convergence. However, in practice, most of the POCS algorithms provide strong convergence.

Many of the commonly used iterative signal restoration techniques are specific examples of the POCS algorithm. The Kaczmarz algorithm [32], Landweber's iteration [33], and the method of alternating projections [9] are all POCS-based algorithms. It is worth noting that the image restoration technique developed independently by Gerchberg and Saxton [4] and Papoulis [5] are also versions of POCS. The algorithm developed by Gerchberg addressed phase retrieval from two images and Papoulis addressed superresolution by iterative methods. The Gerchberg–Papoulis (GP) algorithm is based on applying constraints on the estimate in the signal space and the Fourier space in an iterative fashion until the estimate converges to a solution. For the image restoration problem, the high frequency components of the image are extrapolated by imposing the finite extent of the object in the spatial domain and by imposing the known low frequency components in the frequency domain. The dual problem involves spectral estimation where the signal is extrapolated in the time or spatial domain. The algorithm consists of imposing the known part of the signal in the time domain and imposing a finite bandwidth constraint in the frequency domain. The GP algorithm assumes a space-invariant (or time-invariant) degradation operator.

We now present several signal recovery algorithms that conform to the POCS paradigm which are broadly classified under two categories: row-based and block-based algorithms.

25.5 Row-Based Methods

As early as 1937, Kaczmarz [32] developed an iterative projection technique to solve the inverse problem for a linear set of equations as given by Equation 25.5. The algorithm takes the following form:

$$\mathbf{f}^{(k+1)} = \mathbf{f}^{(k)} + \lambda_k \frac{\mathbf{g}_{i_k} - (\mathbf{h}_{i_k}, \mathbf{f}^{(k)})}{\|\mathbf{h}_{i_k}\|^2} \mathbf{h}_{i_k}, \tag{25.54}$$

where

the relaxation parameter λ_k is bound by $0 \leq \lambda_k \leq 2$

$\mathbf{h}$ represents a row of the matrix $\mathbf{H}$

i_k denotes a sequence of indices corresponding to a row in $\mathbf{H}$

$\mathbf{g}_i$ represents the *ith* element of the vector $\mathbf{g}$
$(\cdot,\cdot)$ is the standard inner product between two vectors
k denotes the iteration
$\|\cdot\|$ denotes the Euclidean or $\mathcal{L}_2$ norm of a vector defined as

$$\|g\| = \left(\sum_{i=1}^{N} g_i^2\right)^{1/2}. \tag{25.55}$$

Kaczmarz proved that Equation 25.54 converges to the unique solution when the relaxation parameter is unity and H represents a square, nonsingular matrix, that is, H possesses an inverse and under certain conditions, the solution will converge to the minimum norm least squares or pseudoinverse solution. For further reading on the Kaczmarz algorithm and conditions for convergence, see [7,8,34,35].

In general, the order in which one performs the Kaczmarz algorithm on the M existing equations can differ. Cyclic control, where the algorithm iterates through the equations in a periodic fashion is described as $i_k = k(\text{modulo}\, M) + 1$ where M is the number of rows in $\mathbf{H}$. Almost cyclic control exists when M sequential iterations of the Kaczmarz algorithm yield exactly one operation per equation in any order. Remotest set control exists when one performs the operations on the most distant equation first; most distant in the sense that the projection onto the hyperplane represented by the equation is the furthest away. The measure of distance is determined by the norm. This type of control is seldomly used since it requires a measurement dependent on all the equations.

The method of Kaczmarz for $\lambda = 1.0$, can be expressed geometrically as follows. Given $\mathbf{f}^{(k)}$ and the hyperplane $H_{i_k} = \{\mathbf{f} \in R^n | (\mathbf{h}_{i_k}, \mathbf{f}) = \mathbf{g}_{i_k}\}, \mathbf{f}^{(k+1)}$ is the orthogonal projection of $\mathbf{f}^{(k)}$ onto H_{i_k}. This is illustrated in Figure 25.3. Note that by changing the relaxation parameter, the next iterate can be a point anywhere along the line segment connecting the previous iterate and its orthogonal reflection with respect to the hyperplane.

The technique of Kaczmarz to solve for a set of linear equations has been rediscovered over the years for many different applications where the general problem formulation can be expressed as Equation 25.5. For this reason, the Kaczmarz algorithm appears as the algebraic reconstruction technique in the field of medical imaging for computerized tomography [7], as well as the Widrow–Hoff least mean squares algorithm [36] for channel equalization, echo cancellation, system identification, and adaptive array processing.

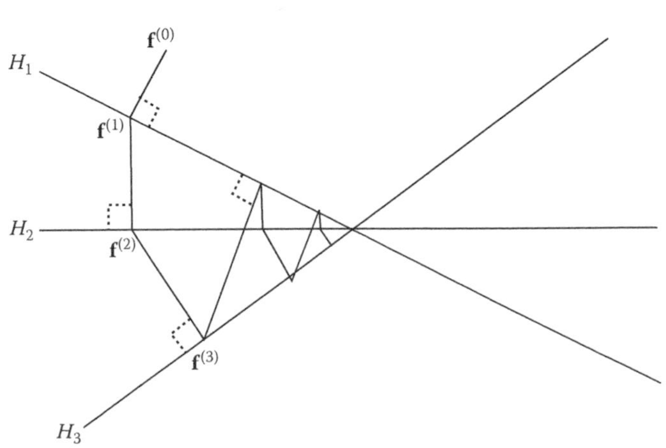

FIGURE 25.3 Geometric interpretation of the Kaczmarz algorithm.

For the case of solving linear inequalities where Equation 25.5 is replaced with

$$\mathbf{Hf} \leq \mathbf{g}, \tag{25.56}$$

a method very similar to Kaczmarz's algorithm is developed by Agmon [37] and Motzkin and Schoenberg [38],

$$\begin{aligned} \mathbf{f}^{(k+1)} &= \mathbf{f}^{(k)} + c^{(k)}\mathbf{h}_{i_k} \\ c^{(k)} &= \min\left(0,\ \lambda_k \frac{g_{i_k} - (\mathbf{h}_{i_k}, \mathbf{f}^{(k)})}{\|\mathbf{h}_{i_k}\|^2}\right). \end{aligned} \tag{25.57}$$

Once again, the relaxation parameter is defined on the interval $0 \leq \lambda_k \leq 2$. The method of solving linear inequalities by Agmon and Motzkin and Schoenberg is mathematically identical to the perceptron convergence theorem from the theory of learning machines (see [39]).

25.6 Block-Based Methods

A generalization of the Kaczmarz algorithm introduced in the previous section has been suggested by Eggermont [35] which can be described as a block, iterative algorithm. Recall the set of linear equations given by Equation 25.5 where the dimensions of the problem are redefined so that $H \in \mathcal{R}^{LM \times N}$, $\mathbf{f} \in \mathcal{R}^N$, and $\mathbf{g} \in \mathcal{R}^{LM}$. In order to describe the generalization of the Kaczmarz algorithm, the matrix $\mathbf{H}$ is partitioned into M blocks of length L:

$$\mathbf{H} = \begin{pmatrix} \mathbf{h}_1^{\mathrm{T}} \\ \mathbf{h}_2^{\mathrm{T}} \\ \vdots \\ \mathbf{h}_{LM}^{\mathrm{T}} \end{pmatrix} = \begin{pmatrix} \mathbf{H}_1 \\ \mathbf{H}_2 \\ \vdots \\ \mathbf{H}_M \end{pmatrix} \tag{25.58}$$

and $\mathbf{g}$ is partitioned as

$$\mathbf{g} = \begin{pmatrix} g_1 \\ g_2 \\ \vdots \\ g_{LM} \end{pmatrix} = \begin{pmatrix} \mathbf{G}_1 \\ \mathbf{G}_2 \\ \vdots \\ \mathbf{G}_M \end{pmatrix}, \tag{25.59}$$

where

$\mathbf{G}_i, i = 1, 2, \ldots, M$, is a vector of length L
the subblocks $\mathbf{H}_i$ are of dimension $L \times N$

The generalized group-iterative variation of the Kaczmarz algorithm is expressed as

$$\mathbf{f}^{(k+1)} = \mathbf{f}^{(k)} + \mathbf{H}_{i_k}^{\mathrm{T}} \Sigma_k \left[\mathbf{G}_{i_k} - \mathbf{H}_{i_k}\mathbf{f}^{(k)}\right] \tag{25.60}$$

where $\mathbf{f}^{(0)} \in \mathcal{R}^N$. Eggermont gives details of convergence as well as conditions for convergence to the pseudoinverse solution [35].

A further generalization of Kaczmarz's algorithm led Eggermont [35] to the following form of the general block Kaczmarz algorithm:

$$\mathbf{f}^{(k+1)} = \mathbf{f}^{(k)} + \mathbf{H}_{i_k}^{\dagger} \Lambda_k \left[\mathbf{G}_{i_k} - \mathbf{H}_{i_k} x^{(k)}\right], \tag{25.61}$$

where once again $\mathbf{H}_{i_k}^{\dagger}$ denotes the Moore–Penrose inverse of $\mathbf{H}_{i_k}$, Λ_k is the $L \times L$ relaxation matrix, and cyclic control is defined as $i_k = k(\text{modulo}\, M) + 1$.

When the block size L given in Equation 25.60 is equal to the number of equations M, the algorithm becomes identical to Landweber's iteration [33] for solving Fredholm equations of the first kind; that is,

$$\mathbf{f}^{(k+1)} = \mathbf{f}^{(k)} + \mathbf{H}^{\mathrm{T}}\Sigma_k(\mathbf{g} - \mathbf{H}\mathbf{f}^{(k)}). \tag{25.62}$$

The resulting Landweber iteration becomes

$$\mathbf{f}^{(k+1)} = \mathbf{H}^{\mathrm{T}}\mathbf{g} + (\mathbf{I} - \mathbf{H}^{\mathrm{T}}\mathbf{H})\mathbf{f}^{(k)}. \tag{25.63}$$

Another interesting approach that is similar to the generalized block-Kaczmarz algorithm, with the block size L equal to the number of equations M, is the method of alternating orthogonal projections described by Youla [9] where alternating orthogonal projections are made onto CLMs.

The row-based and block-based algorithms described here correspond to a POCS framework where the only *a priori* information incorporated into the algorithm is the original problem formulation as described by Equation 25.5. At times, the only information we may have is the original measurement $\mathbf{g}$ and an estimate of the degradation operator $\mathbf{H}$ and these algorithms are suited for such applications. However, for most applications, other *a priori* information is known about the desired signal and an effective algorithm should utilize this information.

We now describe a POCS-based algorithm suited for the problem of image restoration where additional *a priori* signal information is incorporated into the algorithm.

25.7 Image Restoration Using POCS

Here we describe an image recovery algorithm [18,40] that is based on the POCS framework and show some image restoration results [19,20]. The list of references includes other examples of POCS-based recovery algorithms.

The least squares minimum norm or pseudoinverse solution can be formulated as

$$\mathbf{f}^{\dagger} = \mathbf{H}^{\dagger}\mathbf{H}\mathbf{f} = \mathbf{V}\Lambda\mathbf{V}^{\mathrm{T}}\mathbf{f}, \tag{25.64}$$

where

the dagger † denotes the pseudoinverse

$\mathbf{V}$ is the unitary matrix found in the diagonalization of $\mathbf{H}$

Λ is the following diagonal matrix whose first r diagonal terms are equal to one

$$\Lambda = \begin{pmatrix} 1_1 & 0 & \cdots & & & \\ 0 & 1_2 & 0 & & & \\ & & & \ddots & & \\ & & & & 1_r & \\ & & & & & 0 \end{pmatrix} \tag{25.65}$$

By defining

$$\mathbf{P} = \mathbf{V}\Lambda\mathbf{V}^{\mathrm{T}}, \tag{25.66}$$

the orthogonal complement to the operator **P** is given by the projection operator

$$\mathbf{Q} = \mathbf{I} - \mathbf{P} = \mathbf{V}\Lambda^{C}\mathbf{V}^{T}, \tag{25.67}$$

where

$$\Lambda^{C} = \begin{pmatrix} 0 & \cdots & & & \\ \vdots & \ddots & & & \\ & & 1 & & \\ & & & 1_{r+1} & \\ & & & & \ddots \end{pmatrix}. \tag{25.68}$$

The diagonal matrix Λ^{C} contains ones in the last $N - r$ diagonal positions and zeros elsewhere. The superscript C denotes the complement.

Any arbitrary vector **f** can be decomposed as follows:

$$\mathbf{f} = \mathbf{Pf} + \mathbf{Qf}, \tag{25.69}$$

where the projection operator **P** projects **f** onto the range space of the degradation matrix $\mathbf{H}^{T}\mathbf{H}$ and the orthogonal projection operator **Q** projects **f** onto the null-space of the degradation matrix $\mathbf{H}^{T}\mathbf{H}$. The component **Pf** will be referred to as the "in-band" term and the component **Qf** will be referred to as the "out-of-band" term.

In general, the least squares family of solutions to the image restoration problem can be stated as

$$\begin{aligned} \mathbf{f} &= \mathbf{f}_{\text{in-band}} + \mathbf{f}_{\text{out-of-band}} \\ &= \mathbf{f}^{\dagger} + K_{r+1}\mathbf{v}_{r+1} + K_{r+2}\mathbf{v}_{r+2} + \cdots + K_{N}\mathbf{v}_{N}. \end{aligned} \tag{25.70}$$

The vectors $\mathbf{v}_i$ correspond to the eigenvectors of $\{\sigma_{r+1}^2, \sigma_{r+2}^2, \ldots, \sigma_N^2\}$ for $\mathbf{H}^{T}\mathbf{H}$; they are the eigenvectors associated with zero valued eigenvalues. The out-of-band solution $K_{r+1}\mathbf{v}_{r+1} + \cdots + K_N\mathbf{v}_N$ must satisfy

$$\mathbf{H}\mathbf{f}_{\text{out-of-band}} = 0. \tag{25.71}$$

Adding the terms $\{K_{r+1}\mathbf{v}_{r+1}, K_{r+2}\mathbf{v}_{r+2}, \ldots, K_N\mathbf{v}_N\}$ to the pseudoinverse solution $\mathbf{f}^{\dagger}$ does not change the L_2 norm of the error since

$$\begin{aligned} \|n\| &= \|\mathbf{g} - \mathbf{Hf}\| \\ &= \|\mathbf{g} - \mathbf{H}(\mathbf{f}^{\dagger} + K_{r+1}\mathbf{v}_{r+1} + \cdots + K_N\mathbf{v}_N)\| \\ &= \|\mathbf{g} - \mathbf{Hf}^{\dagger} - \mathbf{H}K_{r+1}\mathbf{v}_{r+1} - \cdots - \mathbf{H}K_N\mathbf{v}_N\| \\ &= \|\mathbf{g} - \mathbf{Hf}^{\dagger}\| \end{aligned} \tag{25.72}$$

which is the least squares error. The terms $\mathbf{H}K_{r+1}\mathbf{v}_{r+1}, \ldots, \mathbf{H}K_N\mathbf{v}_N$ are all equal to zero because the vectors $\mathbf{v}_{r+1}, \ldots, \mathbf{v}_N$ are in the null-space of **H**. Therefore, any linear combination of $\mathbf{v}_i$ in the null-space of **H** can be added to the pseudoinverse solution without affecting the least squares cost function. The pseudoinverse solution, $\mathbf{f}^{\dagger}$, provides the unique least squares estimate with minimum norm:

$$\min\|\mathbf{f}_{\text{LS}}\| = \|\mathbf{f}^{\dagger}\|, \tag{25.73}$$

where $\mathbf{f}_{\text{LS}}$ denotes the least squares solution. In practice, it is unlikely that the desired solution is required to possess the minimum norm out of all feasible solutions so that $\mathbf{f}^{\dagger}$ is not necessarily the optimum solution. The image restoration algorithm described here provides a framework that allows *a priori* information in the form of signal constraints to be incorporated into the algorithm in order to obtain a better estimate than the least squares minimum norm solution $\mathbf{f}^{\dagger}$. The constraint operator will be represented by **C** and can incorporate a variety of linear and nonlinear *a priori* signal characteristics as long as they obey the properties of convex set theory. In the case of image restoration, the constraint operator **C** includes nonnegativity which can be described by

$$(\mathbf{C}_{+}\mathbf{f})_i = \begin{cases} f_i & f_i \geq 0 \\ 0 & f_i < 0. \end{cases} \tag{25.74}$$

Concatenating the vectors $\mathbf{v}_i$ in Equation 25.70 yields

$$\mathbf{f} = \mathbf{f}^{\dagger} + \mathbf{V}\Lambda^{\text{C}}\mathbf{K}, \tag{25.75}$$

where

$$\mathbf{K} = \begin{pmatrix} K_1 \\ K_2 \\ \vdots \\ K_N \end{pmatrix} \tag{25.76}$$

and

$$\mathbf{V}\Lambda^{\text{C}} = (\mathbf{v}_1 \quad \mathbf{v}_2 \quad \cdots \quad \mathbf{v}_N) \begin{pmatrix} 0 & & & & \\ & \ddots & & & \\ & & 1_{r+1} & & \\ & & & \ddots & \\ & & & & 1_N \end{pmatrix} \tag{25.77}$$

We would like to find the solution to the unknown vector K in Equation 25.75. A reasonable approach is to start with the constrained pseudoinverse solution and solve for K in a least squares manner; that is,

minimize:

$$\|\mathbf{C}_{+}\mathbf{f}^{\dagger} - \{\mathbf{f}^{\dagger} + \mathbf{V}\Lambda^{\text{C}}\mathbf{K}\}\| \tag{25.78}$$

subject to:

$$\mathbf{C}_{+}\mathbf{f}^{\dagger} = \mathbf{f}^{\dagger} + \mathbf{V}\Lambda^{\text{C}}\mathbf{K}. \tag{25.79}$$

The least squares solution becomes

$$\begin{aligned} \mathbf{C}_{+}\mathbf{f}^{\dagger} - \mathbf{f}^{\dagger} &= \mathbf{V}\Lambda^{\mathrm{C}}\mathbf{K} \\ \mathbf{K} &= \Lambda^{\mathrm{C}}\mathbf{V}^{\mathrm{T}}\left(\mathbf{C}_{+}\mathbf{f}^{\dagger} - \mathbf{f}^{\dagger}\right). \end{aligned} \tag{25.80}$$

Since $\Lambda^{\mathrm{C}}\mathbf{V}^{\mathrm{T}}\mathbf{f}^{\dagger} = 0$, we get

$$\mathbf{K} = \Lambda^{\mathrm{C}}\mathbf{V}^{\mathrm{T}}\mathbf{C}_{+}\mathbf{f}^{\dagger}. \tag{25.81}$$

Substituting Equation 25.81 into Equation 25.79 yields

$$\mathbf{C}_{+}\mathbf{f}^{\dagger} = \mathbf{f}^{\dagger} + \mathbf{Q}\mathbf{C}_{+}\mathbf{f}^{\dagger} + e, \tag{25.82}$$

where e denotes a residual vector. The process of enforcing the overall least squares solution and solving for the out-of-band component to fit the constraints can be implemented in an iterative fashion. The resulting recursion is

$$\mathbf{C}_{+}\mathbf{f}^{(k)} = \mathbf{f}^{\dagger} + \mathbf{Q}\mathbf{C}_{+}\mathbf{f}^{(k)} + e^{(k)}. \tag{25.83}$$

By defining

$$\mathbf{f}^{(k+1)} \equiv \mathbf{C}_{+}\mathbf{f}^{(k)} - e^{(k)}, \tag{25.84}$$

the final iterative algorithm becomes

$$\begin{aligned} \mathbf{f}^{(0)} &= \mathbf{f}^{\dagger} \\ \mathbf{f}^{(k+1)} &= \mathbf{f}^{\dagger} + \mathbf{Q}\mathbf{C}_{+}\mathbf{f}^{(k)},\ k = 0, 1, 2, \ldots. \end{aligned} \tag{25.85}$$

Note that the recursion yields the least squares solution while enforcing the *a priori* constraints through the out-of-band signal component. It is apparent that such an approach will yield a better estimate for the unknown signal $\mathbf{f}$ than the minimum norm least squares solution $\mathbf{f}^{\dagger}$. Note that this algorithm can easily be generalized to other problems by replacing the non-negativity constraint $\mathbf{C}_{+}$ with the signal appropriate constraints. In the case when $\mathbf{f}^{\dagger}$ satisfies all the constraints exactly, the solution to iterative algorithm reduces to the pseudoinverse solution. For more details on this algorithm, convergence issues, and stopping criterion, refer to [18,20,40]. By looking at this algorithmic framework from the set theoretic viewpoint described in [21], the original set of solutions is given by all the solutions that satisfy the least squares error criterion. The addition of *a priori* signal constraints attempts to reduce the feasible set of solutions and to provide a better estimate than the pseudoinverse solution.

Finally, we would like to show some image restoration results based on the method described in [19,20]. The technique is a modification of the Kaczmarz method described here using the theory of POCS. Original, degraded, restored images using the original Kaczmarz algorithm and the restored images using the modified algorithm based on the POCS framework are shown in Figure 25.4. Similarly, we show the original, degraded, and restored images in the frequency domain in Figure 25.5. The details of the algorithm are found in [19].

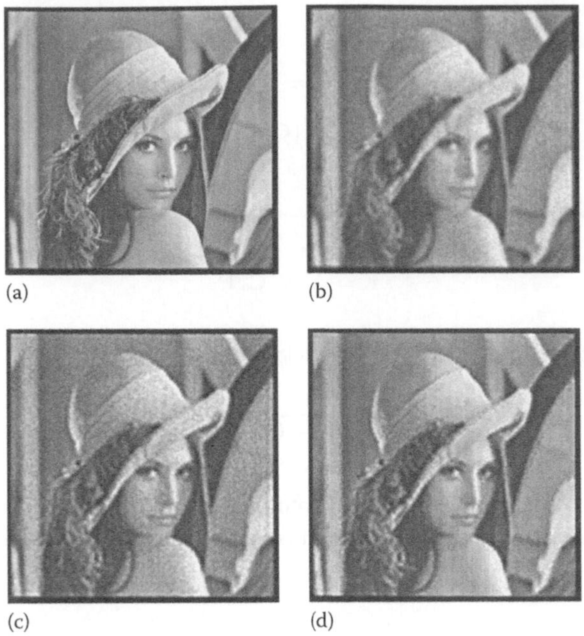

FIGURE 25.4 (a) Original image, (b) degraded image at 25 dB SNR, (c) restored image using Kaczmarz iterations, and (d) restored image using the modified Kaczmarz algorithm in a POCS framework. (Courtesy of IEEE: Kuo, S.S. and Mammone, R.J., *IEEE Trans. Signal Process.*, 40, 159, 1992.)

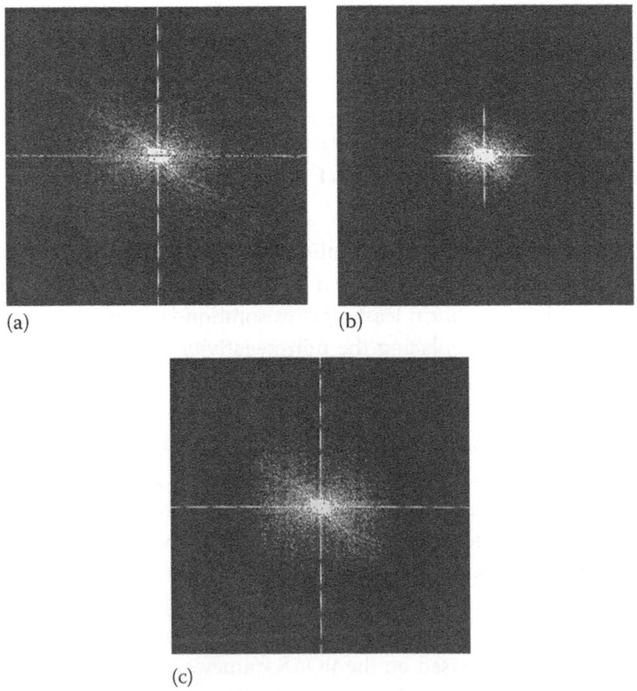

FIGURE 25.5 Spatial frequency response of the (a) original image, (b) degraded image, and (c) restored image using the new algorithm. (Courtesy of IEEE: Kuo, S.S. and Mammone, R.J., *IEEE Trans. Signal Process.*, 40, 159, 1992.)

References

1. Sondhi, M.M., Image restoration: The removal of spatially invariant degradations, *Proc. IEEE*, 60(7), 842–853, July 1972.
2. Slepian, D., Restoration of photographs blurred by image motion, *Bell Syst. Tech. J.*, XLVI, 2353–2362, 1967.
3. Helstrom, C.W., Image restoration by the method of least squares, *J. Opt. Soc. Am.*, 57, 297–303, 1967.
4. Gerchberg, R.W. and Saxton, W.O., A practical algorithm for the determination of phase from image and diffraction plane pictures, *Optik*, 35, 237–246, 1972.
5. Papoulis, A., A new algorithm in spectral analysis and band-limited extrapolation, *IEEE Trans. Circuits Syst.*, 22, 735–742, 1975.
6. Hayes, M.H., Lim, J.S., and Oppenheim, A.V., Signal reconstruction from phase or magnitude, *IEEE Trans. Acoust. Speech Signal Process.*, ASSP-28, 672–680, 1980.
7. Lent, A., Herman G.T., and Rowland, S.W., Art: Mathematics and applications, *J. Theor. Biol.*, 42, 1–32, 1973.
8. Lent, A., Herman, G.T., and Lutz, P.H., Relaxation methods for image reconstruction, *Commn. Assoc. Comput. Mach.*, 21, 152–158, 1978.
9. Youla, D.C., Generalized image restoration by the method of alternating projections, *IEEE Trans. Circuits Syst.*, CAS-25, 694–702, September 1978.
10. Youla, D.C. and Webb, H., Image restoration by the method of convex projections: Part I—Theory, *IEEE Trans. Med. Imaging*, 1, 81–94, 1982.
11. Sezan, M.I. and Stark, H., Image restoration by the method of convex projections: Part II—Applications and numerical results, *IEEE Trans. Med. Imaging*, 1, 95–101, 1982.
12. Schafer, R.W., Mersereau, R.M., and Richards, M.A., Constrained iterative restoration algorithms, *Proc. IEEE*, 69(4), 432–449, April 1981.
13. Civanlar, M.R. and Trussell, H.J., Digital signal restoration using fuzzy sets, *IEEE Trans. Acoust. Speech Signal Process.*, 34, 919–936, 1986.
14. Trussell, H.J. and Civanlar, M.R., The feasible solution in signal restoration, *IEEE Trans. Acoust. Speech Signal Process.*, 32, 201–212, 1984.
15. Sezan, M.I. and Trussell, H.J., Prototype image constraints for set-theoretic image restoration, *IEEE Trans. Signal Process.*, 39, 2275–2285, 1991.
16. Sezan, M.I. and Tekalp, A.M., Adaptive image restoration with artifact suppression using the theory of convex projections, *IEEE Trans. Acoust. Speech Signal Process.*, 38, 181–185, January 1990.
17. Stark, H., Ed., *Image Recovery Theory and Applications*, Academic Press, New York, 1987.
18. Podilchuk, C.I. and Mammone, R.J., Image recovery by convex projections using a least-squares constraint, *J. Opt. Soc. Am. A*, 7, 517–521, March 1990.
19. Kuo, S.S. and Mammone, R.J., Image restoration by convex projections using adaptive constraints and the l_1 norm, *IEEE Trans. Signal Process.*, 40, 159–168, 1992.
20. Mammone, R.J., Ed., *Computational Methods of Signal Recovery and Recognition*, John Wiley & Sons, New York, 1992.
21. Combettes, P.L., The foundations of set theoretic estimation, *Proc. IEEE*, 81, 182–208, 1993.
22. Noble, B. and Daniel, J.W., *Applied Linear Algebra*, Prentice-Hall, Englewood Cliffs, NJ, 1977.
23. Landau, H.J. and Miranker, W.L., The recovery of distorted bandlimited signals, *J. Math. Anal. Appl.*, 2, 97–104, 1961.
24. Wiener, N., On the factorization of matrices, *Commentarii Mathematici Helvetici*, 29, 97–111, 1955.
25. Jayant, N.S., Johnston, J.D., and Safranek, R.J., Signal compression based on models of human perception, *Proc. IEEE*, 81(10), 1385–1422, October 1993.
26. Pratt, W.K. and Davarian, F., Fast computational techniques for pseudoinverse and Wiener image restoration, *IEEE Trans. Comput.*, 26, 571–580, 1977.

27. Twomey, S., On the numerical solution of fredholm integral equations of the first kind by the inversion of the linear system produced by quadrature, *J. Assoc. Comput. Mach.*, 10, 97–101, 1963.
28. Tikonov, A.N., Regularization of incorrectly posed problems, *Sov. Math.*, 4, 1624–1627, 1963.
29. Phillips, D.L., A technique for the numerical solution of certain integral equations of the first kind, *J. Assoc. Comput. Mach.*, 9, 84–97, 1964.
30. Mascarenhas, N.D.A. and Pratt, W.K., Digital image restoration under a regression model, *IEEE Trans. Circuits Syst.*, 22, 252–266, 1975.
31. Polyak, B.T., Gubin, L.G., and Raik, E.V., The method of projections for finding the common point of convex sets, *U.S.S.R. Comput. Math. Phys.*, 7, 1–24, 1967.
32. Kaczmarz, S., Angenaherte au flosung von systemen linearer gleichungen, *Bull. Acad. Pol. Sci. Lett. A*, 6(8A), 355–357, 1937.
33. Strand, O.N., Theory and methods related to the singular-function expansion and Landweber's iteration for integral equations of the first kind, *SIAM J. Numerical Anal.*, 11, 798–825, 1974.
34. Tanabe, K., Projection method for solving a singular system of linear equations and its applications, *Numerical Math.*, 17, 203–214, 1971.
35. Eggermont, P.P.B., Iterative algorithms for large partitioned linear systems with applications to image reconstruction, *Linear Algebra Appl.*, 40, 37–67, 1981.
36. Widrow, B. and McCool, J.M., A comparison of adaptive algorithms based on the methods of steepest descent and random search, *IEEE Trans. Antennas Propagation*, 24, 615–637, 1976.
37. Agmon, S., The relaxation method for linear inequalities, *Can. J. Math.*, 6, 382–392, 1954.
38. Motzkin, T.S. and Schoenberg, I.J., The relaxation method for linear inequalities, *Can. J. Math.*, 6, 393–404, 1954.
39. Minsky, M. and Papert, S., *Perceptrons: An Introduction to Computational Geometry*, MIT Press, Cambridge, MA, 1969.
40. Podilchuk, C.I. and Mammone, R.J., Step size for the general iterative image recovery algorithm, *Opt. Eng.*, 27, 806–811, 1988.

26

Algorithms for Computed Tomography

Gabor T. Herman
City University of New York

26.1 Introduction

Computed tomography (CT) is the process of reconstructing the interiors of objects from data collected based on transmitted or emitted radiation. The problem occurs in a wide range of application areas. Here, we discuss the computer algorithms used for achieving the reconstructions.

26.2 Reconstruction Problem

We want to solve the following general problem. There is a three-dimensional structure whose internal composition is unknown to us. We subject this structure to some kind of radiation, either by transmitting the radiation through the structure or by introducing the emitter of the radiation into the structure. We measure the radiation transmitted through, or emitted from, the structure at a number of points. CT is the process of obtaining from these measurements the distribution of the physical parameter(s) inside the structure that have an effect on the measurements. The problem occurs in a wide range of areas, such as x-ray CT, emission tomography, photon migration imaging, and electron microscopic reconstruction (see, e.g., [1,2]). All of these are inverse problems of various sorts (see, e.g., [3]).

Where it is not otherwise stated, we will be discussing the special reconstruction problem of estimating a function of two variables from estimates of its line integrals. As it is quite reasonable for any application, we will assume that the domain of the function is contained in a finite region of the plane. In what follows, we will introduce all the needed notation and terminology; in most cases, these agree with those used in [1].

Suppose f is a function of the two polar variables r and φ. Let $[\mathcal{R}f](\ell, \theta)$ denote the line integral of f along the line that is at a distance ℓ from the origin and makes an angle θ with the vertical axis.

We refer to this operator $\mathcal{R}$ as the Radon transform (it has also been referred to in the literature as the x-ray transform).

The input data to a reconstruction algorithm are estimates (based on physical measurements) of the values of $[\mathcal{R}f](\ell, \theta)$ for a finite number of pairs (ℓ, θ); its output is an estimate, in some sense, of f. More precisely, suppose that the estimates of $[\mathcal{R}f](\ell, \theta)$ are known for I pairs: (ℓ_i, θ_i), $1 \leq i \leq I$. We use y to denote the I-dimensional column vector (called the measurement vector) whose ith component, y_i, is the available estimate of $[\mathcal{R}f](\ell_i, \theta_i)$. The task of a reconstruction algorithm is

given the data y, **estimate** the function f.

Following [1], reconstruction algorithms are characterized either as transform methods or as series expansion methods. In the following subsections, we discuss the underlying ideas of these two approaches and give detailed descriptions of two algorithms from each category.

26.3 Transform Methods

The Radon transform has an inverse, R^{-1}, defined as follows. For a function p of ℓ and θ,

$$\left[R^{-1}p\right](r, \varphi) = \frac{1}{2\pi^2} \int_0^{\pi} \int_{-\infty}^{\infty} \frac{1}{r\cos(\theta - \varphi) - \ell} p_1(\ell, \theta) d\ell d\theta, \tag{26.1}$$

where $p_1(\ell, \theta)$ denotes the partial derivative of p with respect to its first variable ℓ. (Note that it is intrinsically assumed in this definition that p is sufficiently smooth for the existence of the integral in Equation 26.1). It is known [1] that for any function f that satisfies some physically reasonable conditions (such as continuity and boundedness) we have, for all points (r, φ),

$$\left[R^{-1}Rf\right](r, \varphi) = f(r, \varphi). \tag{26.2}$$

Transform methods are numerical procedures that estimate values of the double integral on the right-hand side of Equation 26.1 from given values of $p(\ell_i, \theta_i)$, for $1 \leq i \leq I$. We now discuss two such methods: the widely adopted filtered backprojection (FBP) algorithm and the more recently developed linogram method.

26.4 Filtered Backprojection

In this algorithm, the right-hand side of Equation 26.1 is approximated by a two-step process (for derivational details see [1] or, in a more general context, [3]). First, for fixed values of θ, convolutions defined by

$$[p *_Y q](\ell', \theta) = \int_{-\infty}^{\infty} p(\ell, \theta) q(\ell' - \ell, \theta) d\ell \tag{26.3}$$

are carried out, using a convolving function q (of one variable) whose exact choice will have an important influence on the appearance of the final image. Second, our estimate f^* of f is obtained by backprojection as follows:

$$f^*(r, \varphi) = \int_0^{\pi} [p *_Y q](r\cos(\theta - \varphi), \theta) d\theta. \tag{26.4}$$

To make explicit the implementation of this for a given measurement vector, let us assume that the data function p is known at points $(nd, m\Delta)$, $-N \leq n \leq N$, $0 \leq m \leq M-1$, and $M\Delta = \pi$. Let us further assume that the function f is to be estimated at points (r_j, φ_j), $1 \leq j \leq J$. The computer algorithm operates as follows.

A sequence $f_0, \ldots, f_{M-1}, f_M$ of estimates is produced; the last of these is the output of the algorithm. First we define

$$f_0(r_j, \varphi_j) = 0, \tag{26.5}$$

for $1 \leq j \leq J$. Then, for each value of $m, 0 \leq m \leq M-1$, we produce the $(m+1)$th estimate from the mth estimate by a two-step process:

1. For $-N \leq n' \leq N$, calculate

$$p_c(n'd, m\Delta) = d \sum_{n=-N}^{N} p(nd, m\Delta) q[(n'-n)d], \tag{26.6}$$

   using the measured values of $p(nd, m\Delta)$ and precalculated values (same for all m) of $q[(n'-n)d]$. This is a discretization of Equation 26.3.
2. For $1 \leq j \leq J$, we set

$$f_{m+1}(r_j, \varphi_j) = f_m(r_j, \varphi_j) + \Delta p_c(r_j \cos(m\Delta - \varphi_j), m\Delta). \tag{26.7}$$

This is a discretization of Equation 26.4. To do it, we need to interpolate the first variable of p_c from the values calculated in Equation 26.6 to obtain the values needed in Equation 26.7. In practice, once $f_{m+1}(r_j, \varphi_j)$ has been calculated, $f_m(r_j, \varphi_j)$ is no longer needed and the computer can reuse the same memory location for $f_0(r_j, \varphi_j), \ldots, f_{M-1}(r_j, \varphi_j), f_M(r_j, \varphi_j)$.

In a complete execution of the algorithm, the uses of Equation 26.6 require $M(2N+1)$ multiplications and additions, while all the uses of Equation 26.7 require MJ interpolations and additions. Since J is typically of the order of N^2 and N itself in typical applications is between 100 and 1000, we see that the cost of backprojection is likely to be much more computationally demanding than the cost of convolution. In any case, reconstruction of a typical 512×512 cross-section from data collected by a typical x-ray CT device is not a challenge to the state-of-the art computational capabilities; it is routinely done in the order of a second or so and can be done, using a pipeline architecture, in a fraction of a second [4].

26.5 Linogram Method

The basic result that justifies this method is the well-known projection theorem that says that "taking the two-dimensional Fourier transform is the same as taking the Radon transform and then applying the Fourier transform with respect to the first variable" [1]. The method was first proposed in [5] and the reason for the name of the method can be found there. The basic reason for proposing this method is its speed of execution and we return to this below. In the description that follows, we use the approach of [6]. That paper deals with the fully three-dimensional problem; here, we simplify it to the two-dimensional case.

For the linogram approach, we assume that the data were collected in a special way (i.e., at points whose locations will be precisely specified below); if they were collected otherwise, we need to interpolate prior to reconstruction. If the function is to be estimated at an array of points with rectangular coordinates $\{(id, jd)|, -N \leq i \leq N, -N \leq j \leq N\}$ (this array is assumed to cover the object to be reconstructed), then the data function p needs to be known at points

$$(nd_m, \theta_m), \ -2N-1 \leq n \leq 2N+1, \ -2N-1 \leq m \leq 2N+1 \tag{26.8}$$

and at points

$$\left(nd_m, \frac{\pi}{2}+\theta_m\right), \ -2N-1 \leq n \leq 2N+1, \ -2N-1 \leq m \leq 2N+1, \tag{26.9}$$

where

$$\theta_m = \tan^{-1}\frac{2m}{4N+3} \quad \text{and} \quad d_m = d\cos\theta_m. \tag{26.10}$$

The linogram method produces from such data estimates of the function values at the desired points using a multistage procedure. We now list these stages, but first point out two facts. One is that the most expensive computation that needs to be used in any of the stages is the taking of discrete Fourier transforms (DFTs), which can always be implemented (possibly after some padding by zeros) very efficiently by the use of the fast Fourier transform (FFT). The other is that the output of any stage produces estimates of function values at exactly those points where they are needed for the discrete computations of the next stage; there is never any need to interpolate between stages. It is these two facts that indicate why the linogram method is both computationally efficient and accurate. (From the point of view of this book, these facts justify the choice of sampling points in Equations 26.8 through 26.10; a geometrical interpretation is given in [7].)

1. Fourier transforming of the data—For each value of the second variable, we take the DFT of the data with respect to the first variable in Equations 26.8 and 26.9. By the projection theorem, this provides us with estimates of the two-dimensional Fourier transform F of the object at points (in a rectangular coordinate system)

$$\left(\frac{k}{(4N+3)d'}, \frac{k}{(4N+3)d}\tan\theta_m\right), \ -2N-1 \leq k \leq 2N+1, \ -2N-1 \leq m \leq 2N+1 \tag{26.11}$$

and at points (also in a rectangular coordinate system)

$$\left(\frac{k}{(4N+3)d}\tan\left(\frac{\pi}{2}+\theta_m\right), \frac{k}{(4N+3)d}\right), \ -2N-1 \leq k \leq 2N+1, \ -2N-1 \leq m \leq 2N+1. \tag{26.12}$$

2. Windowing—At this point we may suppress those frequencies that we suspect to be noise-dominated by multiplying with a window function (corresponding to the convolving function in FBP).
3. Separating into two functions—The sampled Fourier transform F of the object to be reconstructed is written as the sum of two functions, G and H. G has the same values as F at all the points specified in Equation 26.11 except at the origin and is zero-valued at all other points. H has the same values as F at all the points specified in Equation 26.12 except at the origin and is zero-valued at all other points. Clearly, except at the origin, $F = G + H$. The idea is that by first taking the two-dimensional inverse Fourier transforms of G and H separately and then adding the results, we get an estimate (except for a DC term that has to be estimated separately, see [6]) of f. We only follow what needs to be done with G; the situation with H is analogous.

4. Chirp z-transforming in the second variable—Note that the way the θ_m was selected implies that if we fix k, then the sampling in the second variable of Equation 26.11 is uniform. Furthermore, we know that the value of G is zero outside the sampled region. Hence, for each fixed k, $0 < |k| \leq 2N + 1$, we can use the chirp z-transform to estimate the inverse DFT in the second variable at points

$$\left(\frac{k}{(4N+3)d}, jd\right), \quad -2N - 1 \leq k \leq 2N + 1, \; -N \leq j \leq N. \tag{26.13}$$

The chirp z-transform can be implemented using three FFTs, see [7].

5. Inverse transforming in the first variable—The inverse Fourier transform of G can now be estimated at the required points by taking, for every fixed j, the inverse DFT in the first variable of the values at the points of Equation 26.13.

26.6 Series Expansion Methods

This approach assumes that the function, f, to be reconstructed can be approximated by a linear combination of a finite set of known and fixed basis functions,

$$f(r,\varphi) \approx \sum_{j=1}^{J} x_j b_j(r,\varphi), \tag{26.14}$$

and that our task is to estimate the unknowns, x_j. If we assume that the measurements depend linearly on the object to be reconstructed (certainly true in the special case of line integrals) and that we know (at least approximately) what the measurements would be if the object to be reconstructed was one of the basis functions (we use $r_{i,j}$ to denote the value of the ith measurement of the jth basis function), then we can conclude [1] that the ith of our measurements of f is approximately

$$\sum_{j=1}^{J} r_{i,j} x_j. \tag{26.15}$$

Our problem is then to estimate x_j from the measured approximations (for $1 \leq i \leq I$) to Equation 26.15. The estimate can often be selected as one that satisfies some optimization criterion.

To simplify the notation, the image is represented by a J-dimensional image vector x (with components x_j) and the data form an I-dimensional measurement vector y. There is an assumed projection matrix R (with entries $r_{i,j}$). We let r_i denote the transpose of the ith row of R ($1 \leq i \leq I$), and so the inner product $\langle r_i, x\rangle$ is the same as the expression in Equation 26.15. Then y is approximately Rx and there may be further information that x belongs to a subset C of $\mathbf{R}^J$, the space of J-dimensional real-valued vectors. In this formulation R, C, and y are known and x is to be estimated. Substituting the estimated values of x_j into Equation 26.14 will then provide us with an estimate of the function f.

The simplest way of selecting the basis functions is by subdividing the plane into pixels (or space into voxels) and choosing basis functions whose value is 1 inside a specific pixel (or voxel) and is 0 everywhere else. However, there are other choices that may be preferable; for example, [8] uses spherically symmetric basis functions that are not only spatially limited, but also can be chosen to be very smooth. The smoothness of the basis functions then results in smoothness of the reconstructions, while the spherical symmetry allows easy calculation of the $r_{i,j}$. It has been demonstrated [9] that, for the case of fully three-dimensional positron emission tomography (PET) reconstruction, such basis functions indeed lead to statistically significant improvements in the task-oriented performance of series expansion reconstruction methods.

In many situations only a small proportion of $r_{i,j}$ is nonzero. (For example, if the basis functions are based on voxels in a $200 \times 200 \times 100$ array and the measurements are approximate line integrals, then the percent of nonzero $r_{i,j}$ is less than 0.01, since a typical line will intersect fewer than 400 voxels.) This makes certain types of iterative methods for estimating the x_j surprisingly efficient. This is because one can make use of a subroutine that, for any i, returns a list of those js for which $r_{i,j}$ is not zero, together with the values of the $r_{i,j}$ [1,10]. We now discuss two such iterative approaches: the so-called algebraic reconstruction techniques (ART) and the use of expectation maximization (EM).

26.7 Algebraic Reconstruction Techniques

The basic version of ART operates as follows [1]. The method cycles through the measurements repeatedly, considering only one measurement at a time. Only those x_j are updated for which the corresponding $r_{i,j}$ for the currently considered measurement i is nonzero and the change made to x_j is proportional to $r_{i,j}$. The factor of proportionality is adjusted so that if Equation 26.15 is evaluated for the resulting x_j, then it will match exactly the ith measurement. Other variants will use a block of measurements in one iterative step and will update the x_j in different ways to ensure that the iterative process converges according to a chosen estimation criterion.

Here we discuss only one specific optimization criterion and the associated algorithm. (Others can be found, for example, in [1]). Our task is to find the x in $\mathbf{R}^J$ that **minimizes**

$$r^2\|y - R_x\|^2 + \|x - \mu_x\|^2 \tag{26.16}$$

($\|\cdot\|$ indicates the usual Euclidean norm), for a given constant scalar r (called the regularization parameter) and a given constant vector μ_x.

The algorithm makes use of an I-dimensional vector u of additional variables, one for each measurement. First we define $u^{(0)}$ to be the I-dimensional zero vector and $x^{(0)}$ to be the J-dimensional zero vector. Then, for $k \geq 0$, we set

$$\begin{aligned} u^{(k+1)} &= u^{(k)} + c^{(k)} e_{i_k}, \\ x^{(k+1)} &= x^{(k)} + rc^{(k)} r_{i_k}, \end{aligned} \tag{26.17}$$

where e_i is an I-dimensional vector whose ith component is 1 with all other components being 0 and

$$c^{(k)} = \lambda^{(k)} \frac{r\left(y_{i_k} - \left\langle r_{i_k}, x^{(k)} \right\rangle\right) - u_{i_k}^{(k)}}{1 + r^2\|r_{i_k}\|^2}, \tag{26.18}$$

with $i_k = [k(\text{mod}I) + 1]$.

THEOREM 26.1

(see [1] for a proof). Let y be any measurement vector, $\mathbf{r}$ be any real number, and μ_x be any element of $\mathbf{R}^J$. Then for any real numbers $\lambda^{(k)}$ satisfying

$$0 < \varepsilon_1 \leq \lambda^{(k)} \leq \varepsilon_2 < 2, \tag{26.19}$$

the sequence $x^{(0)}, x^{(1)}, x^{(2)}, \ldots$ *determined by the algorithm given above converges to the unique vector* x *that minimizes Equation 26.16.*

The implementation of this algorithm is hardly more complicated than that of basic ART, which is described at the beginning of this section. We need an additional sequence of I-dimensional vectors $u^{(k)}$, but in the kth iterative step only one component of $u^{(k)}$ is needed or altered. Since the i_ks are defined in a cyclic order, the components of the vector $u^{(k)}$ (just as the components of the measurement vector y) can be sequentially accessed. (The exact choice of this—often referred to as the data access ordering—is very important for fast initial convergence; it is described in [11]. The underlying principle is that in any subsequence of steps, we wish to have the individual actions to be as independent as possible.) We also use, for every integer $k \geq 0$, a positive real number $\lambda^{(k)}$. (These are the so-called relaxation parameters. They are free parameters of the algorithm and in practice need to be optimized [11].) The r_is are usually not stored at all, but the location and size of their nonzero elements are calculated as and when needed. Hence, the algorithm described by Equations 26.17 and 26.18 shares the storage-efficient nature of basic ART and its computational requirements are essentially the same. Assuming, as is reasonable, that the number of nonzero $r_{i,j}$ is of the same order as N, we see that the cost of cycling through the data once using ART is of the order NJ, which is approximately the same as the cost of reconstructing using FBP. (That this is indeed so is confirmed by the timings reported in [12].) An important thing to note about Theorem 26.1 is that there are no restrictions of consistency in its statement. Hence, the algorithm of Equations 26.17 and 26.18 will converge to the minimizer of Equation 26.16—the so-called regularized least-squares solution—using the real data collected in any application.

26.8 Expectation Maximization

We may wish to find x such that it maximizes the likelihood of observing the actual measurements, based on the assumption that the ith measurement comes from a Poisson distribution whose mean is given by Equation 26.15. An iterative method to do exactly that, based on the so-called EM approach, was proposed in [13]. Here, we discuss a variant of this approach that was designed for a somewhat more complicated optimization criterion [14], which enforces smoothness of the results where the original maximum likelihood criterion may result in noisy images.

Let $\mathbf{R}_+^J$ denote those elements of $\mathbf{R}^J$ in which all components are non-negative. Our task is to find the x in $\mathbf{R}_+^J$ that **minimizes**

$$\sum_{i=1}^{I} [\langle r_i, x\rangle - y_i \ln\langle r_i, x\rangle] + \frac{\gamma}{2} x^{\mathrm{T}} S x, \tag{26.20}$$

where the $J \times J$ matrix S (with entries denoted by $s_{j,u}$) is a modified smoothing matrix [1], which has the following property. (This definition is only applicable if we use pixels to define the basis functions.) Let N denote the set of indexes corresponding to pixels that are not on the border of the digitization. Each such pixel has eight neighbors, let N_j denote the indexes of the pixels associated with the neighbors of the pixel indexed by j. Then,

$$x^{\mathrm{T}} S x = \sum_{j \in N} \left(x_j - \frac{1}{8} \sum_{k \in N_j} x_k \right)^2. \tag{26.21}$$

Consider the following rules for obtaining $x^{(k+1)}$ from $x^{(k)}$:

$$p_j^{(k)} = \frac{\sum_{i=1}^{I} r_{i,j}}{9\gamma s_{j,j}} - x_j^{(k)} + \frac{1}{9s_{j,j}} \sum_{u=1}^{J} s_{j,u} x_u^{(k)}, \tag{26.22}$$

$$q_j^{(k)} = \frac{x_j^{(k)}}{9\gamma s_{j,j}} \sum_{i=1}^{I} \frac{r_{i,j} y_i}{\langle r_i, x^{(k)} \rangle}, \tag{26.23}$$

$$x_j^{(k+1)} = \frac{1}{2}\left(-p_j^{(k)} + \sqrt{\left(p_j^{(k)}\right)^2 + 4q_j^{(k)}}\right). \tag{26.24}$$

Since the first term of Equation 26.22 can be precalculated, the execution of Equation 26.22 requires essentially no more effort than multiplying $x^{(k)}$ with the modified smoothing matrix. As explained in [1], there is a very efficient way of doing this. The execution of Equation 26.23 requires approximately the same effort as cycling once through the data set using ART (see Equation 26.18). Algorithmic details of efficient computations of Equation 26.23 appeared in [15]. Clearly, the execution of Equation 26.24 requires a trivial amount of computing. Thus, we see that one iterative step of the EM algorithm of Equations 26.22 through 26.24 requires, in total, approximately the same computing effort as cycling through the data set once with ART, which costs about the same as a complete reconstruction by FBP. A basic difference between the ART method and the EM method is that the former updates its estimate based on one measurement at a time, while the latter deals with all measurements simultaneously.

THEOREM 26.2

(see [14] for a proof). For any $x^{(0)}$ with only positive components, the sequence $x^{(0)}, x^{(1)}, x^{(2)}, \ldots$ generated by the algorithm of Equations 26.22 through 26.24 converges to the minimizer of Equation 26.20 in $\mathbf{R}_+^J$.

26.9 Comparison of the Performance of Algorithms

We have discussed four very different-looking algorithms and the literature is full of many others, only some of which are surveyed in books such as [1]. Many of the algorithms are available in general purpose image reconstruction software packages, such as SNARK09 [10]. The novice faced with a problem of reconstruction is justified in being puzzled as to which algorithm to use. Unfortunately, there is no generally valid answer: the right choice may very well be dependent on the area of application and the instrument used for gathering the data. Here, we make only some general comments regarding the four approaches discussed above, followed by some discussion of the methodologies that are available for comparative evaluation of reconstruction algorithms for a particular application.

Concerning the two transform methods we have discussed, the linogram method is faster than FBP (essentially an $N^2 \log N$ method, rather than an N^3 method as is the FBP) and, when the data are collected according to the geometry expressed by Equations 26.8 and 26.9, the linogram method is likely to be more accurate because it requires no interpolations. However, data are not normally collected this way and the need for an initial interpolation together with the more complicated-looking expressions that need to be implemented for the linogram method may indeed steer some users toward FBP, in spite of its extra computational requirements.

Advantages of series expansion methods over transform methods are their flexibility (no special relationship needs to be assumed between the object to be reconstructed and the measurements taken, such as that the latter are uniform samples of the Radon transform of the former) and the ability to control the type of solution we want by specifying the exact sense in which the image vector is to be estimated from the measurement vector (see Equations 26.16 and 26.20). The major disadvantage is that it is computationally more intensive to find these precise estimators than to numerically evaluate Equation 26.1. Also, if the model (the basis functions, the projection matrix, and the estimation criterion) is not well chosen, then the resulting estimate may be inferior to that provided by a transform method. The recent literature has demonstrated that usually there are models that make the efficacy of a reconstruction provided by a series expansion method at least as good as that provided by a transform method. To avoid the problem of computational expense, one usually stops the iterative process involved in the optimization long before the method has converged to the mathematically specified estimator. Practical experience indicates that this can be done very efficaciously. For example, as reported in [12], in the area of fully three-dimensional PET, the reconstruction times for FBP are slightly longer than for cycling through the data just once with a version of ART using spherically symmetric basis functions and the accuracy of FBP is significantly worse than what is obtained by this very early iterate produced by ART.

Since the iterative process is, in practice, stopped early, in evaluating the efficacy of the result of a series expansion method one should look at the actual outputs rather than the ideal mathematical optimizer. Reported experiences comparing an optimized version of ART with an optimized version of EM [9,11] indicate that the former can obtain as good or better reconstructions as the latter, but at a fraction of the computational cost. This computational advantage appears to be due to not trying to make use of all the measurements in each iterative step.

The proliferation of image reconstruction algorithms imposes a need to evaluate the relative performance of these algorithms and understand the relationship between their attributes (free parameters) and their performance. In a specific application of an algorithm, choices have to be made regarding its parameters (such as the basis functions, the optimization criterion, constraints, relaxation, etc.). Such choices affect the performance of the algorithm and there is a need for an efficient and objective evaluation procedure that enables us to select the best variant of an algorithm for a particular task and compare the efficacy of different algorithms for that task.

An approach to evaluating an algorithm is to first start with a specification of the task for which the image is to be used and then define a figure of merit (FOM) that determines quantitatively how helpful the image is, and hence the reconstruction algorithm, for performing that task. In the numerical observer approach [1,11,16,17], a task-specific FOM is computed for each image. Based on the FOMs for all the images produced by two different techniques, we can calculate the statistical significance at which we can reject the null hypothesis that the methods are equally helpful for solving a particular task in favor of the alternative hypothesis that the method with the higher average FOM is more helpful for solving that task. Different imaging techniques can then be rank-ordered on the basis of their average FOMs. It is strongly advised that a reconstruction algorithm should not be selected based on the appearance of a few sample reconstructions, but rather on a study carried out along the lines indicated above.

In addition to the efficacy of images produced by the various algorithms, one should also be aware of the computational possibilities that exist for executing them. A survey from this point of view can be found in [2].

26.10 Further Reading

A good understanding of some of the important recent developments in the field of algorithms for CT can be obtained by studying the books [1,18–21] that were published between 2001 and 2009.

References

1. Herman, G.T., *Fundamentals of Computerized Tomography: Image Reconstruction from Projections*, 2nd edition, Springer, London, UK, 2009.
2. Herman, G.T., Image reconstruction from projections, *J. Real-Time Imaging*, 1, 3–18, 1995.

3. Herman, G.T., Tuy, H.K., Langenberg, K.J., and Sabatier, P.C., *Basic Methods of Tomography and Inverse Problems*, Adam Hilger, Bristol, UK, 1987.
4. Sanz, J.L.C., Hinkle, E.B., and Jain, A.K., *Radon and Projection Transform-Based Computer Vision*, Springer-Verlag, Berlin, Germany, 1988.
5. Edholm, P. and Herman, G.T., Linograms in image reconstruction from projections, *IEEE Trans. Med. Imaging*, 6, 301–307, 1987.
6. Herman, G.T., Roberts, D., and Axel, L., Fully three-dimensional reconstruction from data collected on concentric cubes in Fourier space: Implementation and a sample application to MRI, *Phys. Med. Biol.*, 37, 673–687, 1992.
7. Edholm, P., Herman, G.T., and Roberts, D.A., Image reconstruction from linograms: Implementation and evaluation, *IEEE Trans. Med. Imaging*, 7, 239–246, 1988.
8. Lewitt, R.M., Alternatives to voxels for image representation in iterative reconstruction algorithms, *Phys. Med. Biol.*, 37, 705–716, 1992.
9. Matej, S., Herman, G.T., Narayan, T.K., Furuie, S.S., Lewitt, R.M., and Kinahan, P., Evaluation of task-oriented performance of several fully 3-D PET reconstruction algorithms, *Phys. Med. Biol.*, 39, 355–367, 1994.
10. Davidi, R., Herman, G.T., and Klukowska, J., SNARK09: A programming system for the reconstruction of 2D images from 1D projections, http://www.snark09.com/.
11. Herman, G.T. and Meyer, L.B., Algebraic reconstruction techniques can be made computationally efficient, *IEEE Trans. Med. Imaging*, 12, 600–609, 1993.
12. Matej, S. and Lewitt, R.M., Efficient 3D grids for image reconstruction using spherically symmetric volume elements, *IEEE Trans. Nucl. Sci.*, 42, 1361–1370, 1995.
13. Shepp, L.A. and Vardi, Y., Maximum likelihood reconstruction in positron emission tomography, *IEEE Trans. Med. Imaging*, 1, 113–122, 1982.
14. Herman, G.T., De Pierro, A.R., and Gai, N., On methods for maximum *a posteriori* image reconstruction with a normal prior, *J. Vis. Commn. Image Representation*, 3, 316–324, 1992.
15. Herman, G.T., Odhner, D., Toennies, K.D., and Zenios, S.A., A parallelized algorithm for image reconstruction from noisy projections, in Coleman, T.F. and Li, Y. (Eds.), *Large-Scale Numerical Optimization*, SIAM, Philadelphia, PA, 1990, pp. 3–21.
16. Hanson, K.M., Method of evaluating image-recovery algorithms based on task performance, *J. Opt. Soc. Am. A*, 7, 1294–1304, 1990.
17. Furuie, S.S., Herman, G.T., Narayan, T.K., Kinahan, P., Karp, J.S., Lewitt, R.M., and Matej, S., A methodology for testing for statistically significant differences between fully 3-D PET reconstruction algorithms, *Phys. Med. Biol.*, 39, 341–354, 1994.
18. Natterer, F. and Wübbeling, F., *Mathematical Methods in Image Reconstruction*, SIAM, Philadelphia, PA, 2001.
19. Kalender, W.A., *Computed Tomography: Fundamentals, System Technology, Image Quality, Applications*, 2nd edition, Wiley-VCH, Berlin, Germany, 2006.
20. Herman, G.T. and Kuba, A., *Advances in Discrete Tomography and Its Applications*, Birkhauser, Boston, MA, 2007.
21. Banhart, J., *Advanced Tomographic Methods in Materials Research and Engineering*, Oxford University Press, Oxford, UK, 2008.

27

Robust Speech Processing as an Inverse Problem

Richard J. Mammone
Rutgers University

Xiaoyu Zhang
Rutgers University

27.1 Introduction

This section addresses the inverse problem in robust speech processing. A problem that speaker and speech recognition systems regularly encounter in the commercialized applications is the dramatic degradation of performance due to the mismatch of the training and operating environments. The mismatch generally results from the diversity of the operating environments. For applications over the telephone network, the operating environments may vary from offices and laboratories to household places and airports. The problem becomes worse when speech is transmitted over the wireless network. Here the system experiences cross-channel interferences in addition to the channel and noise degradations that exist in the regular telephone network. The key issue in robust speech processing is to obtain good performance regardless of the mismatch in the environmental conditions. The inverse problem in this sense refers to the process of modeling the mismatch in the form of a transformation and resolving it via an inverse transformation. In this section, we introduce the method of modeling the mismatch as an affine transformation.

Before getting into the details of the inverse problem in robust speech processing, we would like to give a brief review of the mechanism of speech production, as well as the retrieval of useful information from the speech for the recognition purposes.

27.2 Speech Production and Spectrum-Related Parameterization

The speech signal consists of time-varying acoustic waveforms produced as a result of acoustical excitation of the vocal tract. It is nonstationary in that the vocal tract configuration changes over time. A time-varying digital filter is generally used to describe the vocal tract characteristics. The steady-state system function of the filter is of the form [1,2]:

$$S(z) = \frac{G}{1 - \sum_{i=1}^{p} a_i z^{-i}} = \frac{G}{\prod_{i=1}^{p} (1 - z_i z^{-1})}, \tag{27.1}$$

where

p is the order of the system

z_i denote the poles of the transfer function

The time domain representation of this filter is

$$s(n) = \sum_{i=1}^{p} a_i s(n-i) + Gu(n). \tag{27.2}$$

The speech sample $s(n)$ is predicted as a linear combination of previous p samples plus the excitation $Gu(n)$, where G is the gain factor. The factor G is generally ignored in the recognition-type tasks to allow for robustness to variations in the energy of speech signals. This speech production model is often referred to as the linear prediction (LP) model, or the autoregressive model, and the coefficients a_i are called the predictor coefficients.

The cepstrum of the speech signal $s(n)$ is defined as

$$c(n) = \int_{-\pi}^{\pi} \log\left|S(e^{j\omega})\right| e^{j\omega n} \frac{d\omega}{2\pi}. \tag{27.3}$$

It is simply the inverse Fourier transform of the logarithm of the magnitude of the Fourier transform $S(e^{j\omega})$ of the signal $s(n)$.

From the definition of cepstrum in Equation 27.3, we have

$$\sum_{n=-\infty}^{n=\infty} c(n) e^{-j\omega n} = \log\left|S(e^{j\omega})\right| = \left|\log \frac{1}{1 - \sum_{n=1}^{p} a_n e^{-j\omega n}}\right|. \tag{27.4}$$

If we differentiate both sides of the equation with respect to ω and equate the coefficients of like powers of $e^{j\omega}$, the following recursion is obtained:

$$c(n) = \begin{cases} \log G & n = 0 \\ a(n) + \frac{1}{n} \sum_{i=1}^{n-1} i c(i) a(n-i) & n > 0 \end{cases}. \tag{27.5}$$

The cepstral coefficients can be calculated using the recursion once the predictor coefficients are solved. The zeroth order cepstral coefficient is generally ignored in speech and speaker recognition due to its sensitivity to the gain factor, G.

An alternative solution for the cepstral coefficients is given by

$$c(n) = \frac{1}{q}\sum_{i=1}^{p} z_i^n. \tag{27.6}$$

It is obtained by equating the terms of like powers of z^{-1} in the following equation:

$$\sum_{n=-\infty}^{n=\infty} c(n)z^{-n} = \log\frac{1}{\prod_{n=1}^{p}(1 - z_n z^{-1})} = -\sum_{i=1}^{p} \log\left[1 - z_n z^{-1}\right], \tag{27.7}$$

where the logarithm terms can be written as a power series expansion given as

$$log\left[1 - z_n z^{-1}\right] = \sum_{k=1}^{\infty} \frac{1}{k} z_n^k z^{-k}. \tag{27.8}$$

There are two standard methods of solving for the predictor coefficients, a_i, namely, the autocorrelation method and the covariance method [3–6]. Both approaches are based on minimizing the mean square value of the estimation error $e(n)$ as given by

$$e(n) = s(n) - \sum_{i=1}^{p} a_i s(n - i). \tag{27.9}$$

The two methods differ with respect to the details of numerical implementation. The autocorrelation method assumes that the speech samples are zero outside the processing interval of N samples. This results in a nonzero prediction error, $e(n)$, outside the interval. The covariance method fixes the interval over which the prediction error is computed and has no constraints on the sample values outside the interval. The autocorrelation method is computationally simpler than the covariance approach and assures a stable system where all poles of the transfer function lie within the unit circle. A brief description of the autocorrelation method is given as follows.

The autocorrelation of the signal $s(n)$ is defined as

$$r_s(k) = \sum_{n=0}^{N-1-k} s(n)s(n + k) = s(n) \otimes s(-n), \tag{27.10}$$

where

N is the number of samples in the sequence $s(n)$

the sign $\otimes$ denotes the convolution operation

The definition of autocorrelation implies that $r_s(k)$ is an even function. The predictor coefficients a_i can therefore be obtained by solving the following set of equations:

$$\begin{pmatrix} r_s(0) & r_s(1) & \cdots & r_s(p-1) \\ r_s(1) & r_s(0) & \cdots & r_s(p-2) \\ \vdots & \vdots & \ddots & \vdots \\ r_s(p-1) & r_s(p-2) & \cdots & r_s(0) \end{pmatrix} \begin{pmatrix} a_1 \\ \vdots \\ a_p \end{pmatrix} = \begin{pmatrix} r_s(1) \\ \vdots \\ r_s(p) \end{pmatrix}.$$

Denoting the $p \times p$ Toeplitz autocorrelation matrix on the left-hand side by $\mathbf{R}_s$, the predictor coefficient vector by $\mathbf{a}$, and the autocorrelation coefficients by $\mathbf{r}_s$, we have

$$\mathbf{R}_s\mathbf{a} = \mathbf{r}_s. \tag{27.11}$$

The solution for the predictor coefficient vector $\mathbf{a}$ can be solved by the inverse relation

$$\mathbf{a} = \mathbf{R}_s^{-1}\mathbf{r}_s.$$

This equation will be used throughout the analysis in the rest of this article. Since the matrix $\mathbf{R}_s$ is Toeplitz, a computationally efficient algorithm known as Levinson–Durbin recursion can be used to solve for $\mathbf{a}$ [3].

27.3 Template-Based Speech Processing

The template-based matching algorithms for speech processing are generally conducted using the similarity of the vocal tract characteristics inhabited in the spectrum of a particular speech sound. There are two types of speech sounds, namely, voiced and unvoiced sounds. Figure 27.1 shows the speech waveforms, the spectra, and the spectral envelopes of the voiced and the unvoiced sounds. Voiced sounds such as the vowel /*a*/ and the nasal sound /*n*/ are produced by the passage of a quasi-periodic air wave through the vocal tract that creates resonances in the speech waveforms known as formants. The quasi-periodic air wave is generated as a result of the vibration of the vocal cord. The fundamental frequency of the vibration is known as the pitch. In the case of generating fricative sounds such as /*sh*/, the vocal tract is excited by random noise, resulting in speech waveforms exhibiting no periodicity, as can be seen in Figure 27.1. Therefore, the spectral envelopes of voiced sounds constantly exhibit the pitch as well as three to five formants when the sampling rate is 8 kHz, whereas the spectral envelopes of the unvoiced sounds reveal no pitch and formant characteristics. In addition, the formants of different voiced sounds differ with respect to the shape and the location of the center frequencies of the formants. This is due to the unique shape of the vocal tract formed to produce a particular sound. Thus, different sounds can be distinguished based on attributes of the spectral envelope.

The cepstral distance given by

$$d = \sum_{n=-\infty}^{\infty} [c(n) - c'(n)]^2 \tag{27.12}$$

is one of the metrics for measuring the similarity of two spectra envelopes. The reason is as follows. From the definition of cepstrum, we have

$$\begin{aligned} \sum_{n=\infty}^{\infty} [c(n) - c'(n)]e^{j\omega n} &= \log|S(e^{j\omega})| - \log|S'(e^{j\omega})| \\ &= \log\frac{|S(e^{j\omega})|}{|S'(e^{j\omega})|}. \end{aligned} \tag{27.13}$$

The Fourier transform of the difference between a pair of cepstra is equal to the difference between the corresponding spectra pair. By applying the Parseval's theorem, the cepstral distance can be related to the log spectral distance as

$$d = \sum_{n=\infty}^{\infty} [c(n) - c'(n)]^2 = \int_{-\pi}^{\pi} [\log|S(e^{j\omega})| - \log|S'(e^{j\omega})|]^2 \frac{d\omega}{2\pi}. \tag{27.14}$$

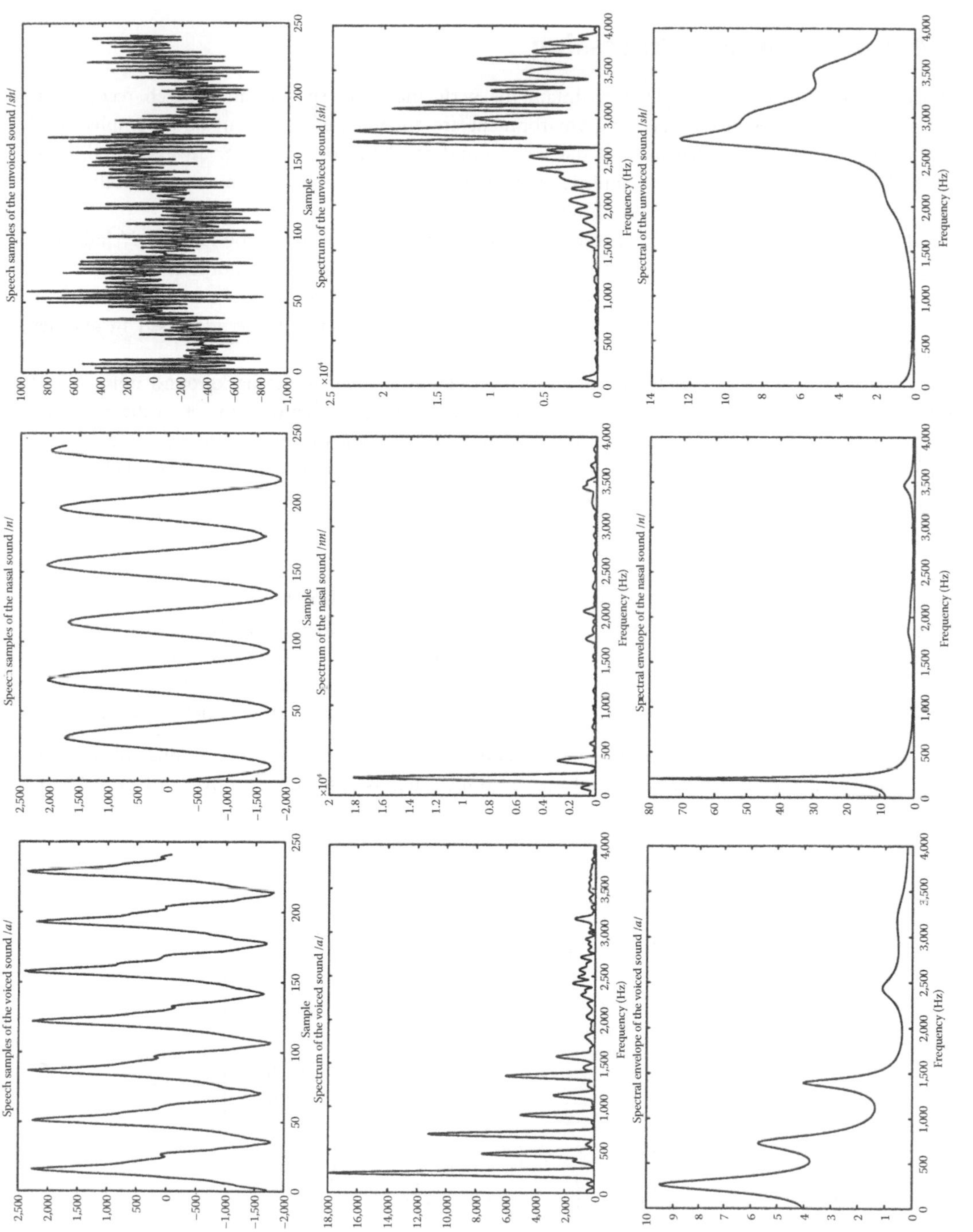

FIGURE 27.1 Illustration of voiced/unvoiced speech.

The cepstral distance is usually approximated by the distance between the first few lower order cepstral coefficients, the reason being that the magnitude of the high order cepstral coefficients is small and has a negligible contribution to the cepstral distance.

27.4 Robust Speech Processing

Robust speech processing attempts to maintain the performance of speaker and speech recognition system when variations in the operating environment are encountered. This can be accomplished if the similarity in vocal tract structures of the same sound can be recovered under adverse conditions. Figure 27.2 illustrates how the deterministic channel and random noise contaminate a speech signal during the recording and transmission of the signal.

First of all, at the front end of the speech acquisition system, additive background noise $N_1(\omega)$ from the speaking environment distorts the speech waveform. Adverse background conditions are also found to put stress on the speech production system and change the characteristics of the vocal tract. It is equivalent to performing a linear filtering of the speech. This problem will be addressed in another chapter and will not be discussed here.

After being sampled and quantized, the speech samples corrupted by the background noise $N_1(\omega)$ are then passed through the transmission channel such as a telephone network to get to the receiver's site. The transmission channel generally involves two types of degradation sources: the deterministic and convolutional filter with the transfer function $H(\omega)$, and the additive noise denoted by $N_2(\omega)$ in Figure 27.2.

The signal observed at the output of the system is, therefore,

$$Y(\omega) = H(\omega)[X(\omega) + N_1(\omega)] + N_2(\omega). \tag{27.15}$$

The spectrum of the output signal is corrupted by both additive and multiplicative interferences. The multiplicative interference due to the linear channel $H(\omega)$ is sometimes referred to as the multiplicative noise.

The various sources of degradation cause distortions of the predictor coefficients and the cepstral coefficients. Figure 27.4 shows the change of spatial clustering of the cepstral coefficients due to interferences of the linear channel, white noise, and the composite effect of both linear channel and white noise.

- When the speech is interfered by a linear bandpass channel, the frequency response of which is shown in Figure 27.3, a translation of the cepstral clusters is observed, as shown in Figure 27.4b.

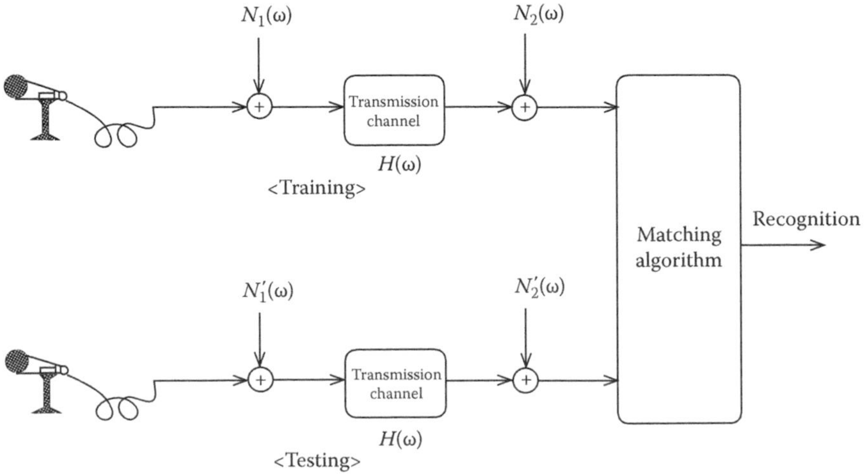

FIGURE 27.2 Speech acquisition system.

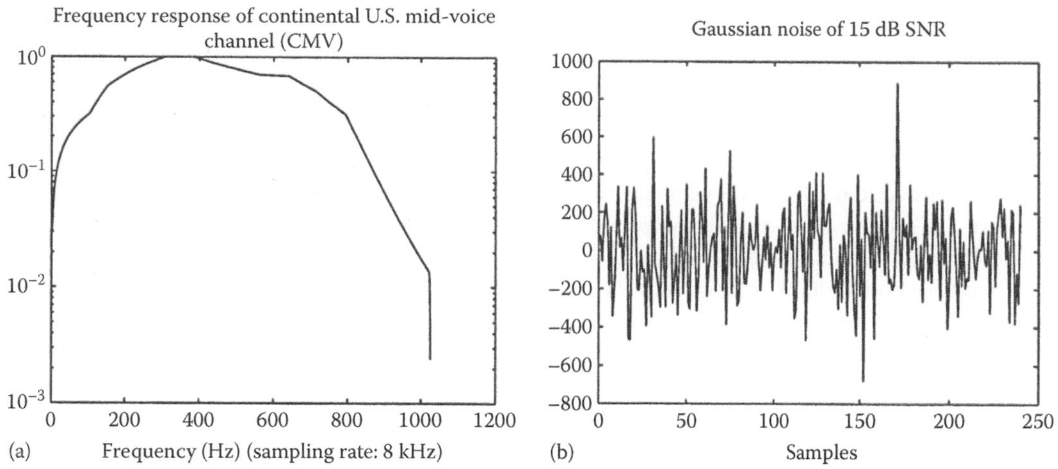

FIGURE 27.3 Simulated environmental interference: (a) medium voiced channel and (b) Gaussian white noise.

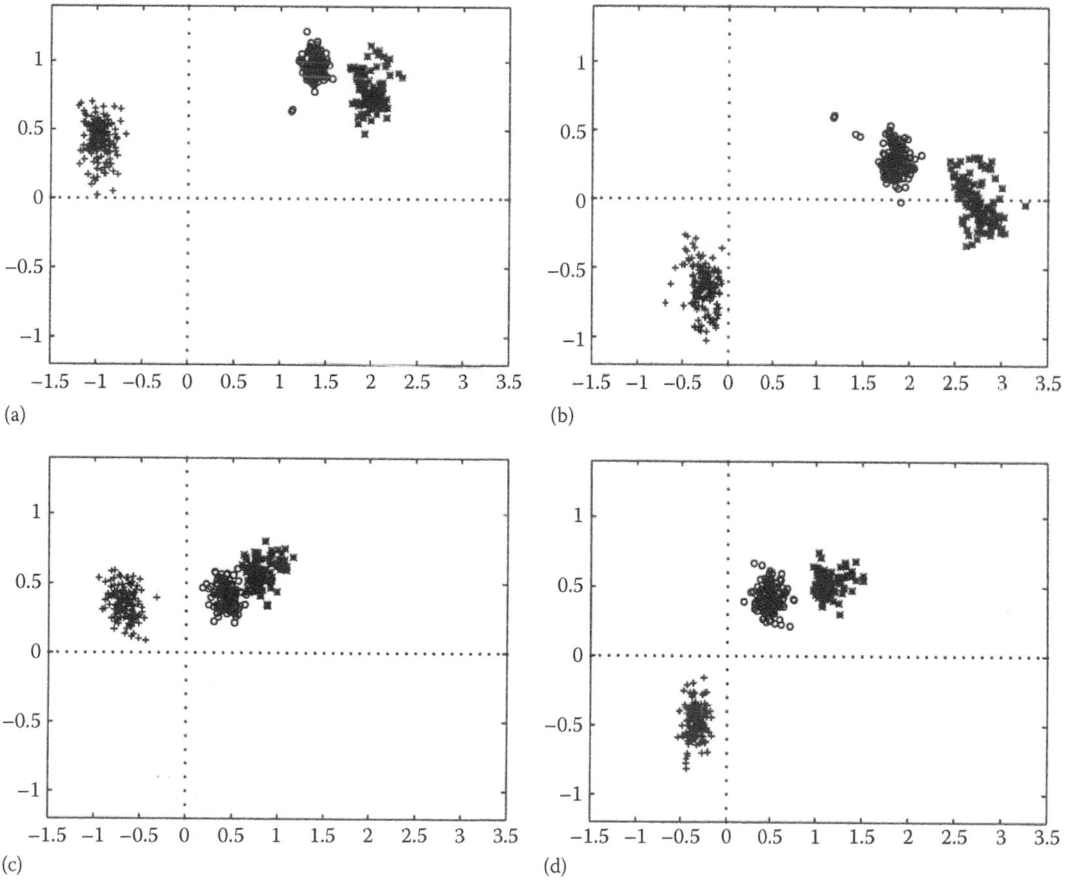

FIGURE 27.4 Spatial distribution of cepstral coefficients under various conditions, "*" for the sound /*a*/, "o" for the sound /*n*/, and "+" for the sound /*sh*/: (a) cepstrum of the clean speech; (b) cepstrum of signals filtered by continental U.S. mid-voice channel (CMV); (c) cepstrum of signals with 15 dB SNR, the noise type is additive white Gaussian (AWG); and (d) cepstrum of speech corrupted by both CMV channel and AWG noise of 15 dB SNR.

- When the speech is corrupted by Gaussian white noise of 15 dB SNR, a shrinkage of the cepstral vectors results. This is shown in Figure 27.4c, where it can be seen that the cepstral clusters move toward the origin.
- When the speech is degraded by both the linear channel and Gaussian white noise, the cepstral vectors are translated and scaled simultaneously.

There are three underlying thoughts behind the various solutions to robust speech processing. The first is to recover the speech signal from the noisy observation by removing an estimate of the noise from the signal. This is also known as the speech enhancement approach. Methods that are executed in the speech sample domain include noise suppression [7] and noise masking [8]. Other speech enhancement methods are carried out in the feature domain, for example, cepstral mean subtraction (CMS) and pole-filtered CMS. In this category, the key to the problem is to find feature sets that are invariant* to the changes of transmission channel and environmental noise. Liftered cepstrum [9] and the adaptive component weighted cepstrum [10] are examples of the feature enhancement approach. A third category consists of methods for matching the testing features with the models after adaptation of environmental conditions [11–14]. In this case, the presence of noise in the training and testing environments are tolerable as long as an adaptation algorithm can be found to match the conditions. The adaptations can be performed in either of the following two directions, i.e., adapt the training data to the testing environment, or adapt the testing data to the environment.

The focus of the following discussion will be on viewing the robust speech processing as an inverse problem. We utilize the fact that both deterministic and nondeterministic noise introduce a sound-dependent linear transformation of the predictor coefficients of speech. This can be approximated by an affine transformation in the cepstrum domain. The mismatch can, therefore, be resolved by solving for the inverse affine transformation of the cepstral coefficients.

27.5 Affine Transform

An affine transform $\mathbf{y}$ of a vector $\mathbf{x}$ is defined as

$$\mathbf{y} = \mathbf{A}\mathbf{x} + \mathbf{b}, \quad \text{for } \mathbf{b} \neq 0. \tag{27.16}$$

The matrix, $\mathbf{A}$, represents the linear transformation of the vector, $\mathbf{x}$, and $\mathbf{b}$ is a nonzero vector representing the translation of the vector. Note that the addition of the vector $\mathbf{b}$ to the equation causes the transform to become nonlinear.

The singular value decomposition (SVD) of the matrix, $\mathbf{A}$, can be used to gain some insight into the geometry of an affine transform, i.e.,

$$\mathbf{y} = \mathbf{U}\boldsymbol{\Sigma}\mathbf{V}^{\mathrm{T}}\mathbf{x} + \mathbf{b}, \tag{27.17}$$

where

$\mathbf{U}$ and $\mathbf{V}^{\mathrm{T}}$ are unitary matrices

$\boldsymbol{\Sigma}$ is a diagonal matrix

The geometric interpretation is thus seen to be that x is rotated by unitary matrix $\mathbf{V}^{\mathrm{T}}$, rescaled by the diagonal matrix $\boldsymbol{\Sigma}$, rotated again by the unitary matrix $\mathbf{U}$, and finally translated by the vector $\mathbf{b}$.

* In practice, it is difficult to find a set of features invariant to the environmental changes. The robust features currently used are mostly less sensitive to environmental changes.

27.6 Transformation of Predictor Coefficients

It will be proved in this section that the contamination of a speech signal by a stationary convolutional channel and random white noise is equivalent to a signal dependent linear transformation of the predictor coefficients. The conclusion drawn here will be used in the next section to show that the effect of environmental interference is equivalent to an affine transform in the cepstrum domain.

27.6.1 Deterministic Convolutional Channel as a Linear Transform

When a sample sequence is passed through a convolutional channel of impulse response $h(n)$, the filtered signal $s'(n)$ obtained at the output of the channel is

$$s'(n) = h(n) \otimes s(n). \tag{27.18}$$

If the power spectra of the signals $s(n)$ and $s'(n)$ are denoted $S_s(\omega)$, and $S_{s'}(\omega)$, respectively, then

$$S_{s'}(\omega) = |H(\omega)|^2 S_s(\omega). \tag{27.19}$$

Therefore, in the time domain,

$$r_{s'}(k) = [h(n) \otimes h(-n)] \otimes r_s(k) = r_h(k) \otimes r_s(k), \tag{27.20}$$

where $r_s(k)$ and $r_{s'}(k)$ are the autocorrelation of the input and output signals. The autocorrelation of the impulse response $h(n)$ is denoted $r_h(k)$ and by definition,

$$r_h(k) = h(n) \otimes h(-n). \tag{27.21}$$

If the impulse response $h(n)$ is assumed to be zero outside the interval $[0, p-1]$, then

$$r_h(k) = 0 \quad \text{for } |k| > p-1. \tag{27.22}$$

Equation 27.20 can therefore be rewritten in matrix form as

$$\begin{pmatrix} r_{s\prime}(0)(23) \\ r_{s\prime}(1)(24) \\ \vdots(25) \\ r_{s\prime}(p-1) \end{pmatrix} = \begin{pmatrix} r_h(0) & r_h(1) & r_h(2) & \cdots & r_h(p-1)(26) \\ r_h(1) & r_h(0) & r_h(1) & \cdots & r_h(p-2)(27) \\ \vdots & \vdots & \ddots & & \vdots(28) \\ r_h(p-1) & r_h(p-2) & r_h(p-3) & \cdots & r_h(0) \end{pmatrix} \begin{pmatrix} r_s(0)(29) \\ r_s(1)(30) \\ \vdots(31) \\ r_s(p-1) \end{pmatrix}$$

$$= \mathbf{R}_{h1}\mathbf{r}_s. \tag{27.23}$$

$\mathbf{R}_{h1}$ refers to the autocorrelation matrix of the impulse response of the channel on the right-hand side of the above equation.

The autocorrelation matrix $\mathbf{R}_{s'}$ of the filtered signal $s'(n)$ is then

$$\mathbf{R}_{s'} = \begin{pmatrix} r_{s'}(0) & r_{s'}(1) & r_{s'}(2) & \cdots & r_{s'}(p-1)(33) \\ r_{s'}(1) & r_{s'}(0) & r_{s'}(1) & \cdots & r_{s'}(p-2)(34) \\ \vdots & \vdots & \ddots & & \vdots(35) \\ r_{s'}(p-1) & r_{s'}(p-2) & r_{s'}(p-3) & \cdots & r_{s'}(0) \end{pmatrix}$$

$$= \begin{pmatrix} r_h(0) & r_h(1) & r_h(2) & \cdots & r_h(p-1)(36) \\ r_h(1) & r_h(0) & r_h(1) & \cdots & r_h(p-2)(37) \\ \vdots & \vdots & \ddots & & \vdots(38) \\ r_h(p-1) & r_h(p-2) & r_h(p-3) & \cdots & r_h(0) \end{pmatrix}$$

$$\times \begin{pmatrix} r_s(0) & r_s(1) & r_s(2) & \cdots & r_s(p-1)(39) \\ r_s(1) & r_s(0) & r_s(1) & \cdots & r_s(p-2)(40) \\ \vdots & \vdots & \ddots & & \vdots(41) \\ r_s(p-1) & r_s(p-2) & r_s(p-3) & \cdots & r_s(0) \end{pmatrix}$$

$$= \mathbf{R}_{h1}\mathbf{R}_s. \tag{27.24}$$

Also, the autocorrelation vector $\mathbf{r}_{s'}$ of the filtered signal $s'(n)$ is

$$\mathbf{r}_{s'} = \begin{pmatrix} r_{s'}(1)(43) \\ r_{s'}(2)(44) \\ \vdots(45) \\ r_{s'}(p)(46) \end{pmatrix}$$

$$= \begin{pmatrix} r_h(1) & r_h(0) & r_h(1) & \cdots & r_h(p-2)(47) \\ r_h(2) & r_h(1) & r_h(0) & \cdots & r_h(p-3)(48) \\ \vdots & \vdots & \ddots & & \vdots(49) \\ r_h(p) & r_h(p-1) & r_h(p-2) & \cdots & r_h(1) \end{pmatrix} \begin{pmatrix} r_s(1)(50) \\ r_s(2)(51) \\ \vdots(52) \\ r_s(p) \end{pmatrix}$$

$$= \mathbf{R}_{h2}\mathbf{r}_s, \tag{27.25}$$

where $\mathbf{R}_{h2}$ denotes the matrix on the right-hand side.

The predictor coefficients of the output signal $s'(n)$ is thus given by

$$\mathbf{a}_{s'} = \mathbf{R}_{s'}^{-1}\mathbf{r}_{s'} = (\mathbf{R}_{h1}\mathbf{R}_s)^{-1} \times (\mathbf{R}_{h2}\mathbf{r}_s) = \mathbf{R}_s^{-1}\left(\mathbf{R}_{h1}^{-1}\mathbf{R}_{h2}\right)\mathbf{R}_s\mathbf{a}. \tag{27.26}$$

Therefore, the predictor coefficients of a speech signal filtered by a convolutional channel can be obtained via taking a linear transformation of the predictor coefficients of the input speech signal. Note that the transformation in Equation 27.26 is sound dependent, as the estimates of the autocorrelation matrices assume stationary.

27.6.2 Additive Noise as a Linear Transform

The random noise arising from the background and the fluctuation of the transmission channel is generally assumed to be additive white noise (AWN). The resulted noisy observation of the original speech signal is given by

$$s'(n) = s(n) + e(n), \tag{27.27}$$

where

$$E[e(n)] = 0 \quad \text{and} \quad E[e^2(n)] = \sigma^2 \tag{27.28}$$

and $s'(n)$ results from the original speech signal $s(n)$ being corrupted by the noise $e(n)$.

The autocorrelation of the corrupted speech signal $s'(n)$ is

$$r_{s'}(k) = [s(n) + e(n)] \otimes [s(-n) + e(-n)] = r_s(k) + r_{se}(k) + r_{es}(k) + r_e(k), \tag{27.29}$$

where

$r_s(k)$ and $r_e(k)$ denote the autocorrelation of the signal $s(n)$ and the noise $e(n)$, respectively
$r_{se}(k)$ and $r_{es}(k)$ represent the cross-correlation of $s(n)$ and $e(n)$

Since

$$\begin{aligned}
r_e(k) &= E\left[\sum_{m=0}^{N-1-k} e(m)e(m+k)\right] = \begin{cases} \sigma^2 & k = 0(58) \\ 0 & \text{otherwise,} \end{cases} \\
r_{se}(k) &= E\left[\sum_{m=0}^{N-1-k} s(m)e(m+k)\right] = \sum_{m=0}^{N-1-k} s(m)E[e(m+k)] = 0, \quad \text{and} \\
r_{es}(k) &= E\left[\sum_{m=0}^{N-1-k} e(m)s(m+k)\right] = \sum_{m=0}^{N-1-k} s(m+k)E[e(m)] = 0,
\end{aligned} \tag{27.30}$$

the autocorrelation of the signal $s'(n)$ presented in Equation 27.29 becomes

$$r_{s'}(k) = \begin{cases} r_s(k) + \sigma^2 & k = 0 \\ r_s(k) & \text{otherwise.} \end{cases} \tag{27.31}$$

Hence, the predictor coefficients as given by Equation 27.11 are

$$\begin{aligned}
\mathbf{a}' &= \mathbf{R}_{s'}^{-1}\mathbf{r}_{s'} \\
&= \begin{pmatrix} r_s(0)+\sigma^2 & r_s(1) & r_s(2) & \cdots & r_s(p-1)(61) \\ r_s(1) & r_s(0)+\sigma^2 & r_s(1) & \cdots & r_s(p-2)(62) \\ \vdots & \vdots & \ddots & & \vdots(63) \\ r_s(p-1) & r_s(p-2) & r_s(p-3) & \cdots & r_s(0)+\sigma \end{pmatrix}^{-1} \begin{pmatrix} r_s(1)(64) \\ r_s(2)(65) \\ \vdots(66) \\ r_s(p) \end{pmatrix} \\
&= \left(\mathbf{R}_s + \sigma^2\mathbf{I}\right)^{-1}\mathbf{r}_s = \left(\mathbf{R}_s + \sigma^2\mathbf{I}\right)^{-1}\mathbf{R}_s\mathbf{a}.
\end{aligned} \tag{27.32}$$

It can be seen from Equation 27.32 that the addition of AWN to the speech is also equivalent to taking a linear transformation of the predictor coefficients. The linear transformation depends on the autocorrelation of the speech and thus in a spectrum-based model, all the spectrally similar predictors will be mapped by a similar linear transform.

The SVD will gain us some insight into what the transformation in Equation 27.32 actually does. Assume that the Toeplitz autocorrelation matrix of the original speech signal $\mathbf{R}_s$ is decomposed as

$$\mathbf{R}_s = \mathbf{U}\Lambda\mathbf{U}^{\mathrm{T}}, \tag{27.33}$$

where

$\mathbf{U}$ is a unitary matrix

Λ is a diagonal matrix whose diagonal elements are the eigenvalues of the matrix $\mathbf{R}_s$

Then the autocorrelation matrix of the noise-corrupted signal is

$$\mathbf{R}_s' = \mathbf{R}_s + \sigma^2\mathbf{I} = \mathbf{U}(\Lambda + \sigma^2 I)\mathbf{U}^{\mathrm{T}}. \tag{27.34}$$

Therefore, Equation 27.32 can be rewritten as

$$\begin{aligned}\mathbf{a}' &= \left[\mathbf{U}(\Lambda + \sigma^2 I)\mathbf{U}^{\mathrm{T}}\right]^{-1}(\mathbf{U}\Lambda\mathbf{U}^{\mathrm{T}})\mathbf{a} = \mathbf{U}\left[(\Lambda + \sigma^2 I)^{-1}\Lambda\right]\mathbf{U}^{\mathrm{T}} \\ &= \mathbf{U}\begin{pmatrix} \frac{\lambda_1^2}{\lambda_1^2+\sigma^2} & & & (70) \\ & \frac{\lambda_2^2}{\lambda_2^2+\sigma^2} & & (71) \\ & & \ddots & (72) \\ & & & \frac{\lambda_n^2}{\lambda_n^2+\sigma^2} \end{pmatrix}\mathbf{U}^{\mathrm{T}}.\end{aligned} \tag{27.35}$$

From the above equation we can see that the norm of the predictor coefficients is reduced when the speech is perturbed by white noise.

27.7 Affine Transform of Cepstral Coefficients

Most speaker and speech recognition systems use a spectrum-based similarity measure to group the vectors, which are normally the LP cepstrum vectors. Thus, we shall investigate the spectrum as to whether or not the cepstral vectors are affinely mapped.

Consider the cepstrum of a speech signal as defined by

$$c_n = Z^{-1}\left[\log\frac{1}{A(z)}\right], \tag{27.36}$$

where $\frac{1}{A(z)} = \frac{1}{1-\sum_{i=1}^{p} a_i z^{-i}}$ is the transfer function of the linear predictive system. Taking the first-order partial derivative of c_n with respect to a_i yields

$$\frac{\partial c_n}{\partial a_i} = \frac{\partial Z^{-1}\left[\log\left(\frac{1}{1-\sum_{i=1}^{p} a_i z^{-i}}\right)\right]}{\partial a_i} \tag{27.37}$$

$$= Z^{-1}\left[\frac{\partial \log\left(\frac{1}{1-\sum_{i=1}^{p} a_i z^{-i}}\right)}{\partial a_i}\right] \tag{27.38}$$

$$= -h(n-i), \tag{27.39}$$

where $h(n-i)$ is the nth impulse response delayed by i taps. Therefore, if $\mathbf{c}$ is the vector of the first p cepstral coefficients of the clean speech $s(n)$, then

$$\mathbf{dc} = \mathbf{Hda}, \tag{27.40}$$

where

$$\mathbf{H} = -\begin{pmatrix} h(0) & 0 & \cdots & 0 \\ h(1) & h(0) & \cdots & 0 \\ \vdots & \vdots & \ddots & \vdots \\ h(p-1) & h(p-2) & \cdots & h(0) \end{pmatrix}. \tag{27.41}$$

Note that the impulse response matrix $\mathbf{H}$ would be the same for a group of spectrally similar cepstral vectors. The relationship between a degradation in the predictor coefficients and the corresponding degradation in the cepstral coefficients is given by Equation 27.40. A degraded set of spectrally similar vectors would undergo the transformation

$$\mathbf{dc}' = \mathbf{H}'\mathbf{da}', \tag{27.42}$$

where

$\mathbf{c}'$ and $\mathbf{a}'$ are the degraded cepstrum and predictor coefficients, respectively
$\mathbf{H}'$ is a lower triangular matrix corresponding to the impulse response of the test signal

Since the predictor coefficients satisfy the linear relation $\mathbf{a}' = \mathbf{Aa}$, as shown in Equations 27.26 and 27.32, differentiating both sides of the equation yields

$$\mathbf{da}' = \mathbf{Ada}. \tag{27.43}$$

If we integrate the above three equations, we have

$$\frac{\mathbf{dc}'}{\mathbf{dc}} = \left(\frac{\mathbf{dc}'}{\mathbf{da}'}\right)\left(\frac{\mathbf{da}'}{\mathbf{da}}\right)\left(\frac{\mathbf{da}}{\mathbf{dc}}\right) = \mathbf{H}'\mathbf{AH}^{-1}. \tag{27.44}$$

The degraded cepstrum is then given by

$$\mathbf{c}' = \mathbf{H}'\mathbf{AH}^{-1}\mathbf{c} + \mathbf{b}_c \tag{27.45}$$

In order to draw the conclusion that there exists an affine transform for the cepstral coefficients, all the variables on the right-hand side of the above equation must be expressed as an explicit function of the training data. However, this is not the case for the matrix $\mathbf{H}'$ in the equation. Since $\mathbf{H}'$ consists of the impulse response of the prediction model of the test data $h'(n)$, we need to represent the impulse response as a function of the training data.

Consider the cases of channel interferences and noise corruption, respectively.

- Assume the training data is of the form

$$s(n) = h_{\text{ch1}}(n) \otimes h_{\text{sig}}(n) \otimes e(n) = h_{\text{ch1}}(n) \otimes s_0(n), \tag{27.46}$$

where
$e(n)$ represents the innovation sequence
$h_{\text{sig}}(n)$ is the impulse response of the all-pole model of the vocal tract
$h_{\text{ch1}}(n)$ is the impulse response of the transmission channel

The convolution of the innovation sequence with the impulse response of the vocal tract yields the clean speech signal $s_0(n)$, the convolution of which with the transmission channel generates the observed sequence $s(n)$.

Similarly, the test data is

$$s'(n) = h_{\text{ch2}}(n) \otimes h_{\text{sig}}(n) \otimes e(n) = h_{\text{ch2}}(n) \otimes s_0(n), \tag{27.47}$$

where $h_{\text{ch2}}(n)$ is the impulse response of the transmission channel in the operating environment. In practice, the all-pole model is applied to the observation sequence involving channel interference rather than the clean speech signal. The estimated impulse response of the observed signal $h'(n)$ is actually given by

$$h'(n) = h_{\text{ch2}}(n) \otimes h_{\text{sig}}(n). \tag{27.48}$$

The matrix $\mathbf{H}'$ can therefore be written as

$$\begin{aligned}\mathbf{H}' &= \begin{pmatrix} h'(0) & 0 & \cdots & 0(87) \\ h'(1) & h'(0) & \cdots & 0(88) \\ \vdots & \vdots & \ddots & \vdots(89) \\ h'(p-1) & h'(p-2) & \cdots & h'(0) \end{pmatrix} \\ &= \begin{pmatrix} h_{\text{ch2}}(0) & 0 & \cdots & 0(90) \\ h_{\text{ch2}}(1) & h_{\text{ch2}}(0) & \cdots & 0(91) \\ \vdots & \vdots & \ddots & \vdots(92) \\ h_{\text{ch2}}(p-1) & h_{\text{ch2}}(p-2) & \cdots & h_{\text{ch2}}(0) \end{pmatrix} \begin{pmatrix} h_{\text{sig}}(0) & 0 & \cdots & 0(93) \\ h_{\text{sig}}(1) & h_{\text{sig}}(0) & \cdots & 0(94) \\ \vdots & \vdots & \ddots & \vdots(95) \\ h_{\text{sig}}(p-1) & h_{\text{sig}}(p-2) & \cdots & h_{\text{sig}}(0) \end{pmatrix}\end{aligned} \tag{27.49}$$

- When the speech is corrupted by additive noise, the autocorrelation matrix $\mathbf{R}_s'$ can also be written as

$$\mathbf{R}_s' = \mathbf{H}'\mathbf{H}'^{\text{T}}. \tag{27.50}$$

Equating the right-hand side of Equations 27.34 and 27.50 yields

$$\mathbf{H}' = \mathbf{U}\left(\Lambda + \sigma^2\mathbf{I}\right)^{1/2}. \tag{27.51}$$

where $\mathbf{H}'$ is an explicit function of the training data and the noise.

At this point, we can conclude that the cepstrum coefficients are affinely mapped by mismatches in the noise and channel conditions. Note again that the parameters of the affine mapping are spectrally dependent.

27.8 Parameters of Affine Transform

Assume the knowledge of the correspondence between the set of training cepstral vectors $\{\mathbf{c}_i = (c_{i1}, c_{i2}, \ldots, c_{iq})^{\mathrm{T}} | i = 1, 2, \ldots, N\}$ and the set of testing cepstral vectors $\{\mathbf{c}_i' = (c_{i1}', c_{i2}', \ldots, c_{iq}')^{\mathrm{T}} | i = 1, 2, \ldots, N\}$. Here N is the number of vectors in the vector set and q is the order of the cepstral coefficients. The affine transform holds for the corresponding vectors $\mathbf{c}_i$ and $\mathbf{c}_i'$ in the following way:

$$\mathbf{c}_i'^{\mathrm{T}} = \mathbf{A}\mathbf{c}_i^{\mathrm{T}} + \mathbf{b}$$

$$\Downarrow$$

$$\begin{pmatrix} c_{i1}'(99) \\ \vdots(100) \\ c_{iq}' \end{pmatrix} = \begin{pmatrix} \alpha_{11} & \cdots & \alpha_{1q}(101) \\ \vdots & \ddots & \vdots(102) \\ \alpha_{q1} & \cdots & \alpha_{qq} \end{pmatrix} \begin{pmatrix} c_{i1}(103) \\ \vdots(104) \\ c_{iq} \end{pmatrix} + \begin{pmatrix} b_1(105) \\ \vdots(106) \\ b_q \end{pmatrix}, \quad \text{for } i = 1, 2, \ldots, N. \tag{27.52}$$

The entries $\{\alpha_{ij}\}$ and $\{b_j\}$ can be solved in the row by row order since for the jth row of the matrix, i.e., $(\alpha_{j1}, \alpha_{j2}, \ldots, \alpha_{jq})$, there exists a set of equations given by

$$\begin{pmatrix} c_{1j}'(108) \\ \vdots(109) \\ c_{Nj}' \end{pmatrix} = \begin{pmatrix} c_{11} & \cdots & c_{1q} & 1(110) \\ \vdots & \ddots & \vdots & \vdots(111) \\ c_{N1} & \cdots & c_{Nq} & 1 \end{pmatrix} \begin{pmatrix} \alpha_{j1}(112) \\ \vdots(113) \\ \alpha_{jq}(114) \\ b_j \end{pmatrix}, \quad \text{for } j = 1, 2, \ldots, q. \tag{27.53}$$

Denoting the vector on the left-hand side of the above equation by γ_j', the matrix and the vector on the right-hand side by Γ and α_j, respectively, we have

$$\gamma_j' = \Gamma \alpha_j. \tag{27.54}$$

The least squares solution to the above systems of equation is

$$\alpha_j = \begin{pmatrix} \sum_{i=1}^{N} \mathbf{c}_i \mathbf{c}_i^{\mathrm{T}} & \sum_{i=1}^{N} \mathbf{c}_i(117) \\ \left(\sum_{i=1}^{N} \mathbf{c}_i\right)^{\mathrm{T}} & N \end{pmatrix}^{-1} \times \begin{pmatrix} \mathbf{c}_1 & \cdots & \mathbf{c}_N(118) \\ 1 & \cdots & 1 \end{pmatrix} \gamma_j' \quad \text{for } j = 1, \ldots, q, \tag{27.55}$$

where

$$\sum_{i=1}^{N} \mathbf{c}_i \mathbf{c}_i^{\mathrm{T}} = \sum_{i=1}^{N} \begin{pmatrix} c_{i1}(120) \\ c_{i2}(121) \\ \vdots(122) \\ c_{iq} \end{pmatrix} (c_{i1}, c_{i2}, \ldots, c_{iq}), \quad \text{for } i = 1, \ldots, N, \tag{27.56}$$

is the summation of a series of matrices.

The testing vectors can then be adapted to the model by an inverse affine transformation of the form

$$\hat{\mathbf{c}} = \mathbf{A}^{-1}(\mathbf{c}' - \mathbf{b}) \tag{27.57}$$

or vice versa. The adaptation removes the mismatch of environmental conditions due to channel and noise variability.

In the case that the matrix A is diagonal, i.e.,

$$A = \begin{pmatrix} \alpha_{11} & & \\ & \ddots & \\ & & \alpha_{qq} \end{pmatrix}, \tag{27.58}$$

the solutions of α_{ij} in Equation 27.55 can be simplified as

$$\begin{aligned} a_{jj} &= \frac{\sum_{i=1}^{N} \gamma'_{ij}\gamma_{ij} - \left(\sum_{i=1}^{N} \gamma'_{ij} \sum_{i=1}^{N} \gamma_{ij}\right)/N}{\sum_{i=1}^{N} \gamma_{ij}^2 - \left(\sum_{i=1}^{N} \gamma_{ij}\right)^2/N} \\ &= \frac{E[\gamma'_j,\gamma_j] - E[\gamma'_j]E[\gamma_j]}{E[\gamma_j,\gamma_j] - E^2[\gamma_j] = \frac{\mathrm{Cov}[\gamma'_j,\gamma_j]}{\mathrm{Var}[\gamma_j,\gamma_j]}} \end{aligned} \tag{27.59}$$

and

$$b_j = \frac{1}{N}\left(\sum_{i=1}^{N} \gamma'_{ij} - \alpha_{jj}\sum_{i=1}^{N} \gamma_{ij}\right) = E[\gamma_j'] - \alpha_{jj}E[\gamma_j]. \tag{27.60}$$

where

$E[\cdot]$ is the expected value operator

$\mathrm{Var}[\cdot]$ and $\mathrm{Cov}[\cdot]$ represent the variance and covariance operators, respectively

As can be seen from Equation 27.60, the diagonal entries α_{jj} are the weighted covariance of the model and the testing vector, and the value of b_j is equal to the weighted difference between the mean of the training vectors and that of the testing vectors. There are three cases of interest:

1. If the training and operating conditions are matched, then

$$E[\gamma_j'] = E[\gamma_j] \quad \text{and} \quad \mathrm{Cov}[\gamma_j', \gamma_j] = \mathrm{Var}[\gamma_j, \gamma_j]. \tag{27.61}$$

Therefore,

$$\alpha_{jj} = 1 \quad \text{and} \quad b_j = 0, \quad \text{for } j = 1, 2, \ldots, q \quad \Rightarrow \quad \hat{\mathbf{c}} = \mathbf{c}'. \tag{27.62}$$

No adaptation is necessary in this case.

2. If the operating environment differs from the training environment due to convolutional distortions, then all the testing vectors are translated by a constant amount as given by

$$\mathbf{c}_i' = \mathbf{c}_i + \mathbf{c}^0, \tag{27.63}$$

and

$$E[\gamma_j'] = E[\gamma_j] + \mathbf{c}^0 \quad \text{and} \quad \mathrm{Cov}[\gamma_j', \gamma_j] = \mathrm{Var}[\gamma_j, \gamma_j]. \tag{27.64}$$

Therefore,

$$\alpha_{jj} = 1 \quad \text{and} \quad b_j = c_j^0, \quad \text{for } j = 1, 2, \ldots, q \quad \Rightarrow \quad \hat{\mathbf{c}} = \mathbf{c}' - \mathbf{b}_c. \tag{27.65}$$

This is equivalent to the method of CMS [6].

3. If the mismatch is caused by both channel and random noise, the testing vector is translated as well as shrunk. The shrinkage is measured by α_{jj} and the translation by b_j. The smaller the covariance of the model and the testing data, the greater the scaling of the testing vectors by noise.

   The affine matching is similar to matching the z scores of the training and testing cepstral vectors. The z score of a set of vectors, c_i, is defined as

$$\mathbf{z}_{c_i} = \frac{\mathbf{c}_i - \mu_c}{\sigma_c}, \tag{27.66}$$

where

μ_c is the mean of the vectors, $\mathbf{c}_i$

σ_c is the variance

Thus, we could form

$$\mathbf{z}_{c_i'} = \sigma_{c'} \left[\frac{\mathbf{c}_i - \mu_c}{\sigma_c} \right] + \mu_{c'}. \tag{27.67}$$

In the above analysis, we show that the cepstrum domain distortions due to channel and noise interference can be modeled as an affine transformation. The parameters of the affine transformation can be optimally estimated using the least squares method which yields the general result given by Equation 27.55. In the special case of a similarity transform, we get the result given by Equation 27.60.

27.9 Correspondence of Cepstral Vectors

While solving for the affine transform parameters, **A** and **b**, we assume to have *a priori* knowledge of the correspondence between the cepstral vectors. A straightforward solution to finding the correspondence is to align the sound units in a speech utterance in terms of the time stamp of the sounds. However, this is generally not realizable in practice due to variations in the content of speech, the identity of a speaker, as well as the rate of speaking. For example, in a speaker recognition system, the text of the testing speech may not be the same as that of the training speech, resulting in a sequence of sounds in a completely different order than the training sequence. Furthermore, even if the text of the speech is the same, the speaking rate may change over time as well as speakers. The time stamp of a particular sound is still not sufficient for lining up corresponding sounds. A valuable solution to the correspondence problem [11] is to use the expectation-maximization (EM) algorithm, also known as the Baum–Welch algorithm [15].

The EM algorithm approaches the optimal solution to a system by repeating the procedure of (1) estimating a set of prespecified system parameters and (2) optimizing the system solution based on these parameters. The step of estimating the parameters is known as the expectation step, and the step of optimizing the solution is the maximization step. The second step is usually realized via the maximum-likelihood method.

With the EM algorithm, the parameters of the affine transform in Equation 27.52 can be solved at the same time as the correspondence of the cepstral vectors are found. The method can be stated as follows.

- **Expectation**
  Solve for the parameters $\{\mathbf{A}, \mathbf{b}\}$ using Equation 27.55. The vector correspondence is found based on the optimization results obtained in the maximization step.

Maximization

Compute the *a posteriori* probability $P(\mathbf{c}_j^{\mathrm{ATC}}|\mathbf{c}_i)$ and find the optimal matching by maximizing the *a posteriori* probability. This can be formulated as

$$k = \mathrm{argmax}_i P\left(\mathbf{c}_j^{\mathrm{ATC}}|\mathbf{c}_i\right), \quad \text{for all } j. \tag{27.68}$$

Here, $\mathbf{c}_i^{\mathrm{ATC}}$ represents the affine-transformed cepstrum that can be obtained by

$$\mathbf{c}_i^{\mathrm{ATC}} = \mathbf{A}^{-1}(\mathbf{c}_i{}' - \mathbf{b}_c) \tag{27.69}$$

Therefore, we have a set of vector pairs denoted by $(\mathbf{c}_k, \mathbf{c}_j{}')$.

The definition of the *a posteriori* probability is dependent on the models employed by the classifier. In general, for the VQ-based classifiers, the *a posteriori* likelihood probability is defined as a Gaussian given by

$$P\left(\mathbf{c}_j^{\mathrm{ATC}}|\mathbf{c}_i\right) = \frac{1}{\sqrt{2\pi}\Sigma^{1/2}} \exp\left[\frac{1}{2}\left(\mathbf{c}_j^{\mathrm{ATC}} - \mathbf{c}_i\right)^{\mathrm{T}} \Sigma^{-1} \left(\mathbf{c}_j^{\mathrm{ATC}} - \mathbf{c}_i\right)\right], \tag{27.70}$$

where Σ is the variance matrix. If we assume that every cepstral coefficient has a unit variance, namely, $\Sigma = \mathbf{I}$, where $\mathbf{I}$ is the identity matrix, then the maximization of the likelihood probability is equivalent to finding the cepstral vector in the VQ codebook that has minimum Euclidean distance to the affine-transformed vector $\mathbf{c}_j^{\mathrm{ATC}}$.

References

1. Flanagan, J.L., *Speech Analysis, Synthesis, and Perception*, Springer-Verlag, Berlin, Germany, 1983.
2. Fant, G., *Acoustic Theory of Speech Production*, Mouton and Co., Gravenhage, the Netherlands, 1960.
3. Rabiner, L.R. and Schafer, R.W., *Digital Processing of Speech Signals*, Prentice-Hall, Englewood Cliffs, NJ, 1978.
4. Atal, B.S., Effectiveness of linear prediction characteristics of the speech wave for automatic speaker identification and verification, J. *Acoust. Soc. Am.*, 55, 1304–1312, 1974.
5. Atal, B.S., Automatic recognition of speakers from their voices, *Proc. IEEE*, 64, 460–475, April 1976.
6. Furui, S., Cepstral analysis techniques for automatic speaker verification, *IEEE Trans. Acoust. Speech Signal Process.*, 29, 254–272, April 1981.
7. Boll, S.F., Suppression of acoustic noise in speech using spectral subtraction, *IEEE Trans. Acoust. Speech Signal Process.*, 27, 113–120, April 1979.
8. Klatt, D.H., A digital filter bank for spectral matching, *International Conference on Acoustics, Speech, and Signal Processing*, Philadelphia, PA, 1976, pp. 573–576.
9. Juang, B.H., Rabiner, L.R., and Wilpon, J.G., On the use of bandpass liftering in speech recognition, *IEEE Trans. Acoust. Speech Signal Process.*, 35, 947–954, July 1987.
10. Assaleh, K.T. and Mammone, R.J., New 1p-derived features for speaker identification, *IEEE Trans. Speech Audio Process.*, 2, 630–638, October 1994.
11. Sankar, A. and Lee, C.H., Robust speech recognition based on stochastic matching, *International Conference on Acoustics, Speech, and Signal Processing*, Detroit, MI, May 9–12, 1995, Vol. 1, pp. 121–124.
12. Neumeyer, L. and Weintraub, M., Probabilistic optimum filtering for robust speech recognition, *Proceedings of IEEE International Conference on Acoustics, Speech, and Signal Processing*, Adelaide, Australia, April 19–22, 1994, Vol. 1, pp. 417–420.

13. Nadas, A., Nahamoo D., and Picheny, M.A., Adaptive labeling: Normalization of speech by adaptive transformation based on vector quantization, *Proceedings of IEEE International Conference on Acoustics, Speech, and Signal Processing*, New York, April 11–14, 1988, Vol. 1, pp. 521–524.
14. Gish, H., Ng, K., and Rohlicek, J.R., Robust mapping of noisy speech parameters for HMM word spotting, *International Conference on Acoustics, Speech, and Signal Processing*, San Francisco, CA, March 23–26, 1992, Vol. 2, pp. 109–112.
15. Baum, L.E., An inequality and associated maximization technique in statical estimation for probabilistic functions of Markov processes, *Inequalities*, 3, 1–8, 1972.

28

Inverse Problems, Statistical Mechanics, and Simulated Annealing

K. Venkatesh Prasad
Ford Motor Company

28.1 Background

The focus of this chapter is on inverse problems—what they are, where they manifest themselves in the realm of digital signal processing (DSP), and how they might be "solved."* Inverse problems deal with estimating hidden causes, such as a set of transmitted symbols $\{\mathbf{t}\}$, given observable effects such as a set of received symbols $\{\mathbf{r}\}$and a system ($\mathbf{H}$) responsible for mapping $\{\mathbf{t}\}$ into $\{\mathbf{r}\}$. Inverse problems are succinctly stated using vector-space notation and take the form of estimating $\mathbf{t} \in \mathcal{R}^M$, given

$$\mathbf{r} = \mathbf{H}\mathbf{t}, \tag{28.1}$$

where $\mathbf{r} \in \mathcal{R}^N$ and $\mathbf{H} \in \mathcal{R}^{M \times N}$ and $\mathcal{R}$ denotes the space of real numbers whose dimensions are specified in the superscript(s). Such problems call for the inversion of $\mathbf{H}$, an operation which may or may not be numerically possible. We will shortly address these issues, but we should note here for completeness that these problems contrast with direct problems—where $\mathbf{r}$ is to be directly (without matrix inversion) estimated, given $\mathbf{H}$ and $\mathbf{t}$.

28.2 Inverse Problems in DSP

Inverse problems manifest themselves in a broad range of DSP applications in fields as diverse as digital astronomy, electronic communications, geophysics [2], medicine [3], and oceanography. The core of all these problems takes the form shown in Equation 28.1. This, in fact, is the discrete version of the

* The quotes are used to stress that unique deterministic solutions might not exist for such problems and the observed effects might not continuously track the underlying causes. Formally speaking, this is a result of such problems of being ill-posed in the sense of Hadmard [1]. What is typically sought is an optimal solution, such as a minimum norm/minimum energy solution.

Fredholm integral equation of the first kind for which, by definition,* the limits of integration are fixed and the unknown function **f** appears only inside the integral. To motivate our discussion, we will describe an application-specific problem, and in the process introduce some of the notations and concepts to be used in the later sections. The inverse problem in the field of electronic communications has to do with estimating **t**, given **r** which is often received with noise, commonly modeled to be additive white Gaussian (AWG) in nature. The communication system and the transmission channel are typically stochastically characterizable and are represented by a linear system matrix (**H**). The problem, therefore, is to solve for **t** in the system of linear equations:

$$\mathbf{r} = \mathbf{H}\mathbf{t} + \mathbf{n}, \tag{28.2}$$

where vector **n** denotes AWG noise. Two tempting solutions might come to mind: if matrix **H** is invertible, i.e., $\mathbf{H}^{-1}$ exists, then why not solve for **t** as

$$\mathbf{t} = \mathbf{H}^{-1}(\mathbf{r} - \mathbf{n}), \tag{28.3}$$

or else why not compute a minimum-norm solution such as the pseudoinverse solution:

$$\mathbf{t} = \mathbf{H}^{\dagger}(\mathbf{r} - \mathbf{n}), \tag{28.4}$$

where $\mathbf{H}^{\dagger}$ is referred to as the pseudoinverse [5] of **H** and is defined to be $[\mathbf{H}'\mathbf{H}]^{-1}\mathbf{H}'$, where $\mathbf{H}'$ denotes the transpose of **H**. There are several reasons why neither solution (Equation 28.3 or 28.4) might be viable. One reason is that the dimensions of the system might be extremely large, placing a greater computational load than might be affordable. Another reason is that **H** is often numerically ill-conditioned, implying that inversions or pseudoinversions might not be reliable even if otherwise reliable numerical inversion procedures, such as Gaussian elimination or singular value decomposition [6,19], were to be employed. Furthermore, even if preconditioning [6] were possible on the system of linear equations $\mathbf{r} = \mathbf{H}\mathbf{t} + \mathbf{n}$, resulting in a numerical improvement of the coefficients of **H**, there is one even more overbearing hurdle that has often to be dealt with, and this has to do with the fact that such problems are frequently ill-posed. In practical terms,† this means that small changes in the inputs might result in arbitrarily large changes in outputs. For all these reasons the most tempting solution-approaches are often ruled out. As we describe in the next section, inverse problems may be recast as combinatorial optimization problems. We will then show how combinatorial optimization problems may be solved using a powerful tool called simulated annealing [7] that has evolved from our understanding of statistical mechanics [8] and the simulation of the annealing (cooling) behavior of physical matter [9].

28.3 Analogies with Statistical Mechanics

Understanding the analogies of inverse problems in DSP to problems in statistical mechanics is valuable to us because we can then draw upon the analytical and computational tools developed over the past century to solve inverse problems in the field of statistical mechanics [8]. The broad analogy is that just as the received symbols **r** in Equation 28.1 are the observed effects of hidden underlying causes (the transmitted symbols **t**)—the measured temperature and state (solid, liquid, or gaseous) of physical matter are the effects of underlying causes such as the momenta and velocities of the particles that compose the matter. A more specific analogy comes from the reasoning that if the inverse problem

* There exist two classes of integral equations ([4], p. 865): (1) if the limits of integration are fixed, the equations are referred to as Fredholm integral equations and (2) if one of the limits is a variable, the equations are referred to as Volterra integral equations. Further, if the unknown function appears only inside the integral, the equation is called "first kind," but if it appears both inside and outside the integral, the equation is called "second kind."

† For a more complete description see [1].

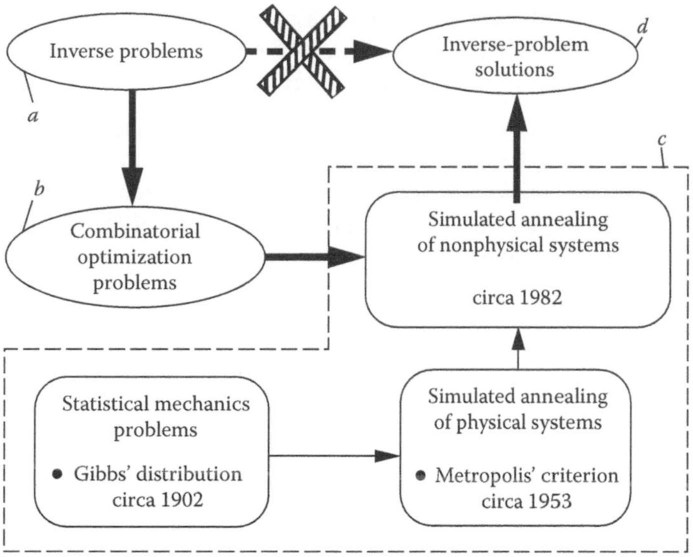

FIGURE 28.1 The direct path ($a \rightarrow d$) to solving the inverse problem is often not viable since it relies on the inversion of a system matrix. An optimal solution, however, may be obtained by an indirect path ($a \rightarrow b \rightarrow c \rightarrow d$) which involves recasting the inverse problem as an equivalent combinatorial optimization problem and then solving this problem using simulated annealing.

were to be treated as a combinatorial optimization problem, where each candidate solution is one possible configuration (or combination of the scalar elements of **t**), then we could use the criterion developed by Metropolis et al. [9] for physical systems to select the optimal configuration. The Metropolis criterion is based on the assumption that candidate configurations have probabilistic distributions of the form originally described by Gibbs [8] to guarantee statistical equilibrium of ensembles of systems. In order to apply Metropolis' selection criterion, we must make one final analogy: we need to treat the combinatorial optimization problem as if it were the outcome of an imaginary physical system in which matter has been brought to boil. When such a physical system is gradually cooled (a process referred to as annealing) then, provided the cooling rate is neither too fast nor too slow, the system will eventually solidify into a minimum energy configuration. As depicted in Figure 28.1 to solve inverse problems we first recast the problem as a combinatorial optimization problem and then solve this recasted problem using simulated annealing—a procedure that numerically mimics the annealing of physical systems. In this section we will describe the basic principles of combinatorial optimization, Metropolis' criterion to select or discard potential configurations, and the origins of Gibbs' distribution. We will outline the simulated annealing algorithm in the following section and will follow that with examples of implementation and applications.

28.3.1 Combinatorial Optimization

The optimal solution to the inverse problem (Equation 28.1), as explained above, amounts to estimating vector **t**. Under the assumptions enumerated below, the inverse problem can be recast as a combinatorial problem whose solution then yields the desired optimal solution to the inverse problem. The assumptions required are

1. Each (scalar) element $t(i)$, $1 \leq i \leq M$, of $\mathbf{t} \in \mathcal{R}^M$ can take on only a finite set of finite values. That is $-\infty < t^j(i) < \infty$; $\forall i$ and j, where $t^j(i)$ denotes the jth possible value that the ith element of **t** can take, and j is a finite valued index $j \leq J^i < \infty$; $\forall i$. J^i denotes the number of possible values the ith element of **t** can take.

2. Let each combination of M scalar values $t(i)$ of $\mathbf{t}$ be referred to as a candidate vector or a feasible configuration $\mathbf{t}_k$, where the index $k \leq K < \infty$. Associated with each candidate vector $\mathbf{t}_k$ we must have a quantifiable measure of error, cost, or energy (E_k).

Given the above assumptions, the combinatorial form of the inverse problem may be stated as: out of K possible candidate vectors $\mathbf{t}_k$, $1 \leq k \leq K$, search for the vector $\mathbf{t}_{k_{\text{opt}}}$ with the lowest error $E_{k_{\text{opt}}}$. Although easily stated, the time and computational efficiency with which the solution is obtained hinges on at least two significant factors—the design of the error-function and the choice of the search strategy. The error-function (E_k) must provide a quantifiable measure of dissimilarity or distance, between a feasible configuration ($\mathbf{t}_k$) and the true (but unknown) configuration ($\mathbf{t}_{\text{true}}$), i.e.,

$$E(\mathbf{t}_k) \stackrel{\Delta}{=} d(\mathbf{t}_k - \mathbf{t}_{\text{true}}), \tag{28.5}$$

where d denotes a distance function. The goal of the combinatorial optimization problem is to efficiently search through the combinatorial space and stop at the optimal, minimum-error (E_{opt}), configuration—$\mathbf{t}_{k_{\text{opt}}}$:

$$E_{\text{opt}} = E(\mathbf{t}_{k_{\text{opt}}}) = \delta \leq E(\mathbf{t}_k), \quad \forall k \neq k_{\text{opt}}, \tag{28.6}$$

where k_{opt} denotes the value of index k associated with the optimal configuration. In the ideal case, when $\delta = 0$, from Equation 28.5, we have that $\mathbf{t}_{k_{\text{opt}}} = \mathbf{t}_{\text{true}}$. In practice, however, owing to a combination of factors such as noise (Equation 28.2), or the system (Equation 28.1) being underdetermined, $E_{\text{opt}} = \delta > 0$, implying that $\mathbf{t}_{k_{\text{opt}}} \neq \mathbf{t}_{\text{true}}$, but that $\mathbf{t}_{k_{\text{opt}}}$ is the best possible solution given what is known about the problem and its solutions. In general the error-function must satisfy the requirements of a distance function or metric (adapted from [10], p. 237):

$$E(\mathbf{t}_k) = 0 <=> \mathbf{t}_k = \mathbf{t}_{\text{true}}, \tag{28.7a}$$

$$E(\mathbf{t}_k) = E(-\mathbf{t}_k) \stackrel{\Delta}{=} d(\mathbf{t}_{\text{true}} - \mathbf{t}_k), \tag{28.7b}$$

$$E(\mathbf{t}_k) \leq E(-\mathbf{t}_j) + d(\mathbf{t}_k - \mathbf{t}_j), \tag{28.7c}$$

where Equation 28.7a follows from Equation 28.5, and where, like k, index j is defined in the range $(1, K)$ and $K < \infty$. Equation 28.7a stated that if the error is zero, t_k is the true configuration. The implication of Equation 28.7b is that error is a function of the absolute value of the distance of a configuration from the true configuration. Equation 28.7c implies that the triangle inequality law holds.

In designing the error-function, one can classify the sources of error into two distinct categories: The first category of error, denoted by E_k^{signal}, provides a measure of error (or distance) between the observed signal ($\mathbf{r}_k$) and the estimated signal ($\hat{\mathbf{r}}_k$)—computed for the current configuration $\mathbf{t}_k$ using Equation 28.1. The second category, denoted by $E_k^{\text{constraints}}$, accounts for the price to be "paid" when an estimated solution deviates from the constraints we would want to impose on them based on our understanding of the physical world. The physical world, for instance, might suggest that each element of the signal is very probably positive valued. In this case, a negative valued estimate of a signal element will result in an error-value that is proportionate to the magnitude of the signal negativity. This constraint is popularly known as the nonnegativity constraint. Another constraint might arise from the assumption that the solution is expected to be smooth [11]:

$$\hat{t}' S \hat{t} = \delta_{\text{smooth}}, \tag{28.8}$$

where

S is a smoothing matrix

δ_{smooth} is the degree of smoothness of the signal

The error-function, therefore, takes the following form:

$$
\begin{aligned}
E_k &\triangleq E_k^{\text{signal}} + E_k^{\text{constraints}} \quad \text{where,} \\
E_{\text{signal}}^k &\triangleq \| \mathbf{r}_k - \hat{\mathbf{r}}_k \|_2 \quad \text{where} \\
\hat{\mathbf{r}}_k &= \mathbf{H} \cdot \mathbf{t_k}, \quad \text{and} \\
E_k^{\text{constraints}} &\triangleq \sum_{c \in C} (\alpha_c \cdot E_c),
\end{aligned} \tag{28.9}
$$

where

$E_{constraints}$ represents the total error from all other factors or constraints that might be imposed on the solution

$\{C\}$ represents the set of constraint indices

α_c and E_c represent the weight and the error-function, respectively, associated with cth constraint

28.3.2 Metropolis Criterion

The core task in solving the combinatorial optimization described above is to search for a configuration $\mathbf{t}_k$ for which the error-function E_K is a minimum. Standard gradient descent methods [6,12,13] would have been the natural choice had the E_k been a function with just one minimum (or maximum) value, but this function typically has multiple minimas (or maximas)—gradient descent methods would tend to get locked into a local minimum. The simulated annealing procedure (Figure 28.2—discussed in the next

```
/* SIMULATED ANNEALING */
/* Set initial conditions: */
/* Temperature: T_initial = T_0 */
/* Configuration t_0 = t_initial */
/* Minimum-cost configuration t_opt = t_0 */
while(stopping criterion is not satisfied){
   while(configuration is not in equilibrium){
      Perturb(t_k → t_{k+1});
      ComputeErrorDifference(ΔE_{k+1} = E_{k+1} − E_k);
      if ΔE_{k+1} ≤ 0 then accept else
      if exp(−ΔE_{k+1}/T) > random (0, 1] then accept;
      if (accept) then {
      Update(E_opt ← E_{k+1}); /* remember the lowest error value */
      Update(t_opt ← t_{k+1})/* remember the lowest error config. */
      } end /* when equilibrium is reached */;
   Cool(T ← T_k)
   k = k + 1
}end /* when stopping criterion is satisfied */
return(); /* the global minimum-error configuration */
```

FIGURE 28.2 The outline of the annealing algorithm.

section), suggested by Metropolis et al. [9] for the problem of finding stable configurations of interacting atoms and adapted for combinatorial optimization by Kirkpatrick [7], provides a scheme to traverse the surface of the E_k, get out of local minimas, and eventually cool into a global minimum. The contribution of Metropolis et al., commonly referred to in the literature as Metropolis' criterion, is based on the assumption that the difference in the error of two consecutive feasible configurations (denoted as $\Delta E \triangleq E_{k+1} - E_k$) takes the form of Gibbs' distribution (Equation 28.11). The criterion states that even if a configuration were to result in increased error, i.e., $\Delta E > 0$, one can select the new configuration if

$$\text{random} \leq e^{\frac{-\Delta E}{kT}}, \tag{28.10}$$

where
random denotes a random number drawn from a uniform distribution in the range (0, 1)
T denotes the temperature of the physical system

28.3.3 Gibbs' Distribution

At the turn of the twentieth century, Gibbs [8], building upon the work of Clausius, Maxwell, and Boltzmann in statistical mechanics, proposed the probability distribution P:

$$P = e^{\frac{\psi - \varepsilon}{\Theta}}, \tag{28.11}$$

where
ψ and Θ were constants
ε denoted the free energy in a system

This distribution was crafted to satisfy the condition of statistical equilibrium ([8], p. 32) for ensembles of (thermodynamical) systems:

$$\sum \left(\frac{dP}{dP_1} \dot{p}_i + \frac{dP}{dq_1} \dot{q}_i \right) = 0, \tag{28.12}$$

where p_i and q_i represented the generalized momentum and velocity, respectively, of the ith degree of freedom. The negative sign on ε in Equation 28.11 was required to satisfy the condition

$$\underbrace{\int \cdots \int}_{\text{all phases}} P dp_1 \cdots dq_n = 1. \tag{28.13}$$

28.4 Simulated Annealing Procedure

The simulated annealing algorithm as outlined in Figure 28.2 mimics the annealing (or controlled cooling) of an imaginary physical system. The unknown parameters are treated like particles in a physical system. An initial configuration t_{initial} is chosen along with an initial ("boiling") temperature value (T_{initial}). The choice of T_{initial} is made so as to ensure that a vast majority, say 90%, of configurations are acceptable even if they result in a negative ΔE_k. The initial configuration is perturbed, either by using a random number generator or by sequential selection, to create a second configuration, and ΔE_2 is computed. The Metropolis criterion is applied to decide whether or not to accept the new configuration. After equilibrium is reached, i.e., after $|\Delta E_2| \leq \delta_{\text{equilib}}$, where δ_{equilib} is a small heuristically chosen threshold, the temperature is lowered according to a cooling schedule and the process is repeated until

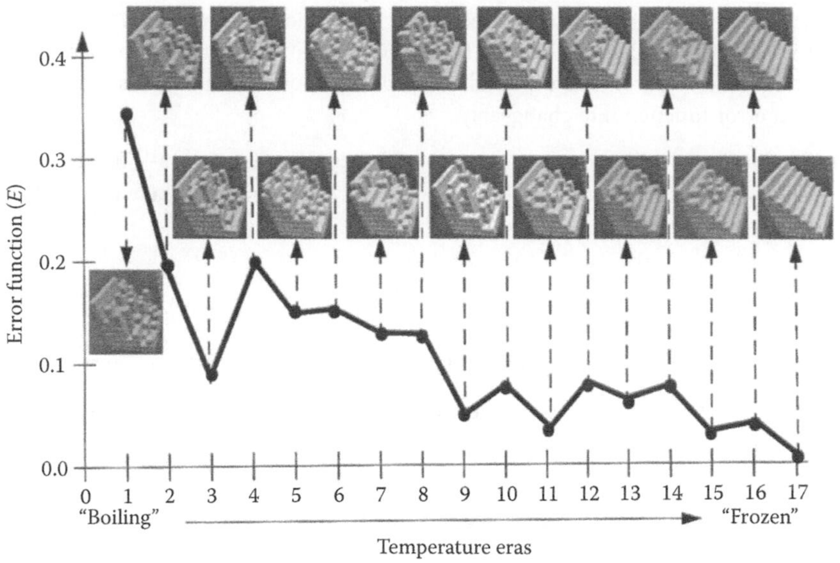

FIGURE 28.3 Three-dimensional signal recovery using simulated annealing. The staircase object shown corresponding to era 17 is recovered from a defocused image by testing a number of feasible configurations and applying the Metropolis criterion to a simulated annealing procedure.

a preselected frozen temperature is reached. Several different cooling schedules have been proposed in the literature ([18], p. 59). In one popular schedule [18,19] each subsequent temperature T_{k+1} is less than the current temperature T_k, by a fixed percentage of T_k, i.e., $T_{k+1} = \beta_k T_k$, where β_k is typically in the range of 0.8 to unity. Based on the behavior of physical systems which attain minimum (free) energy (or global minimum) states when they freeze at the end of an annealing process, the assumption underlying the simulated annealing procedure is that the $\mathbf{t}_{opt}$ that is finally attained is also globally minimum.

The results of applying the simulated annealing procedure to the problems of three-dimensional signal restoration [14] is shown in Figure 28.3. In this problem, a defocused image, vector **r**, of an opaque eight-step staircase object was provided along with the space-varying point-spread-function matrix (**H**), and a well-focused image. The unknown vector **t** represented the intensities of the volume elements (voxels) with the visible voxels taking on positive values and hidden voxels having a value of zero. The vector **t** was lexicographically indexed so that by knowing which elements of **t** were positive, one could reconstruct the three-dimensional structure. Using simulated annealing, and constraints (opacity, non-negativity of intensity, smoothness of intensity and depth, and tight bounds on the voxel intensity values obtained from the well-focused image), the original object was reconstructed.

Defining Terms

In the following definitions, as in the preceding discussion, $\mathbf{t} \in \mathcal{R}^M$, $\mathbf{r} \in \mathcal{R}^N$, and $\mathbf{H} \in R^{M \times N}$.

Combinatorial Optimization: The process of selecting the optimal (lowest cost) configuration from a large space of candidate or feasible configurations.

Configuration: Any vector **t** is a configuration. The term is used in the combinatorial optimization literature.

Cost/energy/error function: The terms cost, energy, or error function are frequently used interchangeably in the literature. Cost function is often used in the optimization literature to represent the mapping of a candidate vector into a (scalar) functional whose value is indicative of the optimality of the candidate vector. Energy function is frequently used in electronic communication theory as a pseudonym for the L_2 norm or root-mean-square value of a vector. Error function is typically used to measure a mismatch between an estimated (vector) and its expected value. For purposes of this discussion we use the terms cost, energy, and error function interchangeably.

Gibbs' distribution: The distribution (in reality a probability density function [pdf]) in which the η the index of probability (P) is a linear function of energy, i.e., $\eta = \log P = \frac{\psi - \varepsilon}{\Theta}$, where ψ and Θ are constants and ε represents energy, giving the familiar pdf:

$$P = e^{\frac{\psi - \varepsilon}{\Theta}}, \tag{28.14}$$

Inverse problem: Given matrix **H** and vector **r**, find **t** that satisfies $\mathbf{r} = \mathbf{Ht}$.

Metropolis' criterion: The criterion first suggested by Metropolis et al. [9] to decide whether or not to accept a configuration that results in an increased error, when trying to search for minimum error configurations in a combinatorial optimization problem.

Minimum-norm: The norm between two vectors is a (scalar) measure of distance (such as the L_1, L_2) (or Euclidean), L_∞ norms or the Mahalanobis distance ([10], p. 24), or the Manhattan metric [7]) between them. Minimum-norm, unless otherwise noted, implies minimum Euclidean (L_2) norm (denoted by $\| \cdot \|$):

$$\min_{\text{among all } \mathbf{t}} \| \mathbf{Ht} - \mathbf{r} \| . \tag{28.15}$$

Pseudoinverse: Let $\mathbf{t}_{\text{opt}}$ be the unique minimum norm vector, therefore,

$$\| \mathbf{Ht}_{\text{opt}} - \mathbf{r} \| = \min_{\text{among all } \mathbf{t}} \| \mathbf{Ht} - \mathbf{r} \| . \tag{28.16}$$

The pseudoinverse of matrix **H** denoted by $\mathbf{H}^{\dagger} \in \mathcal{R}^{N \times M}$ is the matrix mapping all **r** into its corresponding $\mathbf{t}_{\text{opt}}$.

Statistical mechanics: That branch of mechanics in which the problem is to find the statistical distribution of the parameters of ensembles (large numbers) of systems (each differing not just infinitesimally, but embracing every possible combination of the parameters) at a desired instant in time, given those distributions at the present time. Maxwell, according to Gibbs [8], coined the term "statistical mechanics." This field owes its origin to the desire to explain the laws of thermodynamics as stated by Gibbs ([8], p. viii): "The laws of thermodynamics, as empirically determined, express the approximate and probable behavior of systems of a great number of particles, or, more precisely, they express the laws of mechanics for such systems as they appear to beings who have not the fineness of perception to enable them to appreciate quantities of the order of magnitude of those which relate to single particles, and who cannot repeat their experiments often enough to obtain any but the most probable results."

Further Reading

- Inverse problems: The classic by Tikhonov [15] provides a good introduction to the subject matter. For a description of inverse problems related to synthetic aperture radar application see [16].
- Statistical mechanics: Gibbs' [8] work is historical treasure.
- Vector spaces and optimization: The books by Leunberger [12] and Gill and Murray [13] provide a broad introductory foundation.

- Simulated annealing: Two recent books by van Laarhoven and Aarts [17] and Aarts and Korst [18] contain a comprehensive coverage of the theory and application of simulated annealing. A useful simulated annealing algorithm, along with tips for numerical implementation and random number generation, can be found in *Numerical Recipes in C* [19]. An alternative simulated annealing procedure (in which the temperature T is kept constant) can be found in the widely cited work of Geman and Geman [20], applied to image restoration.

References

1. Hadamard, J., Sur les problèmes aux dérivés partilles et leur signification physique, Princeton Univ. Bull., 13: 49–52, 1902.
2. Frolik, J.L. and Yagle A.E., Reconstruction of multilayered lossy dielectrics from plane-wave impulse responses at 2 angles of incidence, *IEEE Trans. Geosci. Remote Sens.*, 33: 268–279, March 1995.
3. Greensite, F., Well-posed formulation of the inverse problem of electrocardiography, *Ann. Biomed. Eng.*, 22 (2): 172–183, 1994.
4. Arfken, G., *Mathematical Methods for Physicists*, Academic Press, Orlando, FL, 1985.
5. Greville, T.N.E., The pseudoinverse of a rectangular or singular matrix and its application to the solution of systems of linear equations, *SIAM Rev.*, 1: 38–43, 1959.
6. Golub, G.H. and Van Loan, C.F., *Matrix Computations*, 2nd ed., The Johns Hopkins University Press, Baltimore, MD, 1989.
7. Kirkpatrick, S., Optimization by simulated annealing: Quantitative studies, *J. Stat. Phys.*, 34(5 and 6): 975–986, 1984.
8. Gibbs, J.W., *Elementary Particles in Statistical Mechanics*, Yale University Press, New Haven, CT, 1902.
9. Metropolis, N., Rosenbluth, A., Rosenbluth, M., Teller, A., and Teller, E., Equation of state calculations by fast computing machines, *J. Chem. Phys.*, 21: 1087–1092, June 1953.
10. Duda, R.O. and Hart, P.E., *Pattern Classification and Scene Analysis*, John Wiley & Sons, New York, 1973.
11. Pratt, W.K., *Digital Image Processing*, John Wiley, New York, 1978.
12. Luenberger, D.G., *Optimization by Vector Space Methods*, John Wiley & Sons, New York, 1969.
13. Gill, P.E. and Murray, W., Quasi-Newton methods for linearly constrained optimization, in *Numerical Methods for Constrained Optimization*, Gill, P.E. and Murray, W. (Eds.), Academic Press, London, U.K., 1974.
14. Prasad, K.V., Mammone, R.J., and Yogeshwar, J., 3-D image restoration using constrained optimization techniques, *Opt. Eng.*, 29: 279–288, April 1990.
15. Tikhonov, A.N. and Arsenin, V.Y., *Solutions of Ill-Posed Problems*, V.H. Winston & Sons, Washington D.C., 1977.
16. Soumekh, M., Reconnaissance with ultra wideband UHF synthetic aperture radar, *IEEE Acoust. Speech Signal Process.*, 12: 21–40, July 1995.
17. van Laarhoven, P.J.M. and Aarts, E.H.L., *Simulated Annealing: Theory and Applications*, D. Riedel, Dordrecht, the Netherlands, 1987.
18. Aarts, E. and Korst, J., *Simulated Annealing and Boltzmann Machines*, John Wiley, New York, 1989.
19. Press, W.H., Flannery, B.D., Teukolsky, S.A., and Vetterling, W.T., *Numerical Recipes in C*, Cambridge University Press, Cambridge, UK, 1988.
20. Geman, S. and Geman, D., Stochastic relaxation, Gibbs distributions and the Bayesian restorations of images, *IEEE Trans. Pattern Recognit. Mach. Intell.*, PAMI-6: 721–741, November 1984.

29

Image Recovery Using the EM Algorithm

Jun Zhang
University of Milwaukee

Aggelos K. Katsaggelos
Northwestern University

29.1 Introduction

Image recovery constitutes a significant portion of the inverse problems in image processing. Here, by image recovery we refer to two classes of problems, image restoration and image reconstruction. In image restoration, an estimate of the original image is obtained from a blurred and noise-corrupted image. In image reconstruction, an image is generated from measurements of various physical quantities, such as x-ray energy in CT and photon counts in single photon emission tomography and positron emission tomography. Image restoration has been used to restore pictures in remote sensing, astronomy, medical imaging, art history studies, e.g., see [1], and more recently, it has been used to remove picture artifacts due to image compression, e.g., see [2] and [3]. While primarily used in biomedical imaging [4], image reconstruction has also found applications in materials studies [5].

Due to the inherent randomness in the scene and imaging process, images and noise are often best modeled as multidimensional random processes called random fields. Consequently, image recovery becomes the problem of statistical inference. This amounts to estimating certain unknown parameters of a probability density function (pdf) or calculating the expectations of certain random fields from the observed image or data. Recently, the maximum-likelihood estimate (MLE) has begun to play a central role in image recovery and led to a number of advances [6,8]. The most significant advantage of the MLE over traditional techniques, such as the Wiener filtering, is perhaps that it can work more autonomously. For example, it can be used to restore an image with unknown blur and noise level by estimating them and the original image simultaneously [8,9]. The traditional Wiener filter and other least mean square error techniques, on the other hand, would require the knowledge of the blur and noise level.

In the MLE, the likelihood function is the pdf evaluated at an observed data sample conditioned on the parameters of interest, e.g., blur filter coefficients and noise level, and the MLE seeks the parameters that maximize the likelihood function, i.e., best explain the observed data. Besides being intuitively appealing, the MLE also has several good asymptotic (large sample) properties [10] such as consistency (the estimate converges to the true parameters as the sample size increases). However, for many nontrivial image recovery problems, the direct evaluation of the MLE can be difficult, if not impossible. This difficulty is due to the fact that likelihood functions are usually highly nonlinear and often cannot be written in closed forms (e.g., they are often integrals of some other pdfs). While the former case would prevent analytic solutions, the latter case could make any numerical procedure impractical.

The EM algorithm, proposed by Dempster, Laird, and Rubin in 1977 [11], is a powerful iterative technique for overcoming these difficulties. Here, EM stands for expectation maximization. The basic idea behind this approach is to introduce an auxiliary function (along with some auxiliary variables) such that it has similar behavior to the likelihood function but is much easier to maximize. By similar behavior, we mean that when the auxiliary function increases, the likelihood function also increases. Intuitively, this is somewhat similar to the use of auxiliary lines for the proofs in elementary geometry.

The EM algorithm was first used by Shepp and Verdi [7] in 1982 in emission tomography (medical imaging). It was first used by Katsaggelos and Lay [8] and Lagendijk et al. [9] for simultaneous image restoration and blur identification around 1989. The work of using the EM algorithm in image recovery has since flourished with impressive results. A recent search on the Compendex database with key words "EM" and "image" turned up more than 60 journal and conference papers, published over the two and a half year period from January 1993 to June 1995.

Despite these successes, however, some fundamental problems in the application of the EM algorithm to image recovery remain. One is convergence. It has been noted that the estimates often do not converge, converge rather slowly, or converge to unsatisfactory solutions (e.g., spiky images) [12,13]. Another problem is that, for some popular image models such as Markov random fields (MRFs), the conditional expectation in the E-step of the EM algorithm can often be difficult to calculate [14]. Finally, the EM algorithm is rather general in that the choice of auxiliary variables and the auxiliary function is not unique. Is it possible that one choice is better than another with respect to convergence and expectation calculations [17]?

The purpose of this chapter is to demonstrate the application of the EM algorithm in some typical image recovery problems and survey the latest research work that addresses some of the fundamental problems described above. The chapter is organized as follows. In Section 29.2, the EM algorithm is reviewed and demonstrated through a simple example. In Section 29.3, recent work in convergence, expectation calculation, and the selection of auxiliary functions is discussed. In Section 29.4, more complicated applications are demonstrated, followed by a summary in Section 29.5. Most of the examples in this chapter are related to image restoration. This choice is motivated by two considerations—the mathematical formulations for image reconstruction are often similar to that of image restoration and a good account on image reconstruction is available in Snyder and Miller [6].

29.2 EM Algorithm

Let the observed image or data in an image recovery problem be denoted by $\mathbf{y}$. Suppose that $\mathbf{y}$ can be modeled as a collection of random variables defined over a lattice $\mathbf{S}$ with $\mathbf{y} = \{y_i, i \in \mathbf{S}\}$. For example, $\mathbf{S}$ could be a square lattice of N^2 sites. Suppose that the pdf of $\mathbf{y}$ is $p_{\mathbf{y}}(\mathbf{y}|\theta)$, where θ is a set of parameters. In this chapter, $p(\cdot)$ is a general symbol for pdf and the subscript will be omitted whenever there is no confusion. For example, when $\mathbf{y}$ and $\mathbf{x}$ are two different random fields, their pdfs are represented as $p(\mathbf{y})$ and $p(\mathbf{x})$, respectively.

29.2.1 The Algorithm

Under statistical formulations, image recovery often amounts to seeking an estimate of θ, denoted by $\hat{\theta}$, from an observed $\mathbf{y}$. The MLE approach is to find $\hat{\theta}_{\mathrm{ML}}$ such that

$$\hat{\theta}_{\mathrm{ML}} = \arg\max_{\theta} p(\mathbf{y}|\theta) = \arg\max_{\theta} \log p(\mathbf{y}|\theta), \tag{29.1}$$

where $p(\mathbf{y}|\theta)$, as a function of θ, is called the likelihood. As described previously, a direct solution of Equation 29.1 can be difficult to obtain for many applications. The EM algorithm attempts to overcome this problem by introducing an auxiliary random field $\mathbf{x}$ with pdf $p(\mathbf{x}|\theta)$. Here, $\mathbf{x}$ is somewhat "more informative" [17] than $\mathbf{y}$ in that it is related to $\mathbf{y}$ by a many-to-one mapping:

$$\mathbf{y} = \mathbf{H}(\mathbf{x}). \tag{29.2}$$

That is, $\mathbf{y}$ can be regarded as a partial observation of $\mathbf{x}$, or incomplete data, with $\mathbf{x}$ being the complete data.

The EM algorithm attempts to obtain the incomplete data MLE of Equation 29.1 through an iterative procedure. Starting with an initial estimate θ^0, each iteration k consists of two steps:

- E-step: Compute the conditional expectation* $\langle \log p(\mathbf{x}|\theta)|\mathbf{y}, \theta^k\rangle$. This leads to a function of θ, denoted by $Q(\theta|\theta^k)$, which is the auxiliary function mentioned previously.
- M-step: Find θ^{k+1} from

$$\theta^{k+1} = \arg\max_{\theta} Q(\theta|\theta^k). \tag{29.3}$$

It has been shown that the EM algorithm is monotonic [11], i.e., $\log p(\mathbf{y}|\theta^k) \geq \log p(\mathbf{y}|\theta^{k+1})$. It has also been shown that under mild regularity conditions, such as that the true θ must lie in the interior of a compact set and that the likelihood functions involved must have continuous derivatives, the estimate of θ from the EM algorithm converges, at least to a local maxima of $p(\mathbf{y}|\theta)$ [20,21]. Finally, the EM algorithm extends easily to the case in which the MLE is used along with a penalty or a prior on θ. For example, suppose that $q(\theta)$ is a penalty to be minimized. Then, the M-step is modified to maximizing $Q(\theta|\theta^k) - q(\theta)$ with respect to θ.

29.2.2 Example: A Simple MRF

As an illustration of the EM algorithm, we consider a simple image restoration example. Let $\mathbf{S}$ be a two-dimensional (2D) square lattice. Suppose that the observed image $\mathbf{y}$ and the original image $\mathbf{u} = \{u_i, i \in \mathbf{S}\}$ are related through

$$\mathbf{y} = \mathbf{u} + \mathbf{w}, \tag{29.4}$$

where $\mathbf{w} = \{u_i, i \in \mathbf{S}\}$ is an i.i.d. additive zero-mean white Gaussian noise with variance σ^2. Suppose that $\mathbf{u}$ is modeled as a random field with an exponential or Gibbs pdf

$$p(\mathbf{u}) = Z^{-1} e^{-\beta E(\mathbf{u})}, \tag{29.5}$$

where $E(\mathbf{u})$ is an energy function with

$$E(\mathbf{u}) = \frac{1}{2} \sum_{i} \sum_{j \in N_i} \phi(u_i, u_j) \tag{29.6}$$

* In this chapter, we use $\langle \cdot \rangle$ rather than $E[\cdot]$ to represent expectations since E is used to denote energy functions of the MRF.

and Z is a normalization factor

$$Z = \sum_{\mathbf{u}} e^{-\beta E(\mathbf{u})} \tag{29.7}$$

called the partition function whose evaluation generally involves all possible realizations of $\mathbf{u}$. In the energy function, N_i is a set of neighbors of i (e.g., the nearest four neighbors) and $\phi(\cdot,\cdot)$ is a nonlinear function called the clique function. The model for $\mathbf{u}$ is a simple but nontrivial case of the MRF [22,23] which, due to its versatility in modeling spatial interactions, has emerged as a powerful model for various image processing and computer vision applications [24].

A restoration that is optimal in the sense of minimum mean square error is

$$\hat{\mathbf{u}} = \langle \mathbf{u}|\mathbf{y}\rangle = \int \mathbf{u}p(\mathbf{u}|\mathbf{y})d\mathbf{u}. \tag{29.8}$$

If parameters β and σ^2 are known, the above expectation can be computed, at least approximately (see Section 29.3.1 for details). To estimate the parameters, now denoted by $\theta = (\beta, \sigma^2)$, one could use the MLE. Since $\mathbf{u}$ and $\mathbf{w}$ are independent,

$$p(\mathbf{y}|\theta) = \int p_{\mathbf{u}}(\mathbf{v}|\theta)p_{\mathbf{w}}(\mathbf{y}-\mathbf{v}|\theta)d\mathbf{v} = (p_{\mathbf{u}} {*} p_{\mathbf{w}})(\mathbf{y}|\theta), \tag{29.9}$$

where $*$ denotes convolution, and we have used some subscripts to avoid ambiguity. Notice that the integration involved in the convolution generally does not have a closed-form expression. Furthermore, for most types of clique functions, Z is a function of β and its evaluation is exponentially complex. Hence, direct MLE does not seem possible.

To try with the EM algorithm, we first need to select the complete data. A natural choice here, e.g., is to let

$$\mathbf{x} = (\mathbf{u}, \mathbf{w}) \tag{29.10}$$

$$\mathbf{y} = \mathbf{H}(\mathbf{x}) = \mathbf{H}(\mathbf{u}, \mathbf{w}) = \mathbf{u} + \mathbf{w}. \tag{29.11}$$

Clearly, many different $\mathbf{x}$ can lead to the same $\mathbf{y}$. Since $\mathbf{u}$ and $\mathbf{w}$ are independent, $p(\mathbf{x}|\theta)$ can be found easily as

$$p(\mathbf{x}|\theta) = p(\mathbf{u})p(\mathbf{w}). \tag{29.12}$$

However, as the reader can verify, one encounters difficulty in the derivation of $p(\mathbf{x}|\mathbf{y}, \theta^k)$ which is needed for the conditional expectation of the E-step. Another choice is to let

$$\mathbf{x} = (\mathbf{u}, \mathbf{y}) \tag{29.13}$$

$$\mathbf{y} = H(\mathbf{u}, \mathbf{y}) = \mathbf{y}. \tag{29.14}$$

The log likelihood of the complete data is

$$\begin{aligned} \log p(\mathbf{x}|\theta) &= \log p(\mathbf{y}, \mathbf{u}|\theta) \\ &= \log p(\mathbf{y}|\mathbf{u}, \theta)p(\mathbf{u}|\theta) \\ &= c - \sum_i \frac{(y_i - u_i)^2}{2\sigma^2} - \log Z(\beta) - \frac{\beta}{2}\sum_i \sum_{j \in N_i} \phi(u_i, u_j), \end{aligned} \tag{29.15}$$

where c is a constant. From this we see that in the E-step, we only need to calculate three types of terms, $\langle u_i \rangle$, $\langle u_i^2 \rangle$, and $\langle \phi(u_i, u_j) \rangle$. Here, the expectations are all conditioned on $\mathbf{y}$ and θ^k. To compute these expectations, one needs the conditional pdf $p(\mathbf{u}|\mathbf{y}, \theta^k)$ which is, from Bayes' formula,

$$\begin{aligned} p(\mathbf{u}|\mathbf{y}, \theta^k) &= \frac{p(\mathbf{y}|\mathbf{u}, \theta^k)p(\mathbf{u}|\theta^k)}{p(\mathbf{y}|\theta^k)} \\ &= [2\pi\sigma^2]^{-\|\mathbf{S}\|/2} e^{-\sum_i (y_i - u_i)^2/2(\sigma^2)^k} Z^{-1} e^{-\beta^k E(\mathbf{u})} [p(\mathbf{y}|\theta^k)]^{-1}. \end{aligned} \tag{29.16}$$

Here, the superscript k denotes the kth iteration rather than the kth power. Combining all the constants and terms in the exponentials, the above equation becomes that of a Gibbs distribution:

$$p(\mathbf{u}|\mathbf{y}, \theta^k) = Z_1^{-1}(\theta^k) e^{-E_1(\mathbf{u}|\mathbf{y}, \theta^k)}, \tag{29.17}$$

where the energy function is

$$E_1(\mathbf{u}|\mathbf{y}, \theta^k) = \sum_i \left[\frac{(y_i - u_i)^2}{2(\sigma^2)^k} + \frac{\beta^k}{2} \sum_{j \in N_i} \phi(u_i, u_j) \right]. \tag{29.18}$$

Even with this, the computation of the conditional expectation in the E-step can still be a difficult problem due to the coupling of the u_i and u_j in E_1. This is one of the fundamental problems of the EM algorithm that will be addressed in Section 29.3. For the moment, we assume that the E-step can be performed successfully with

$$\begin{aligned} Q(\theta|\theta^k) &= \langle \log p(\mathbf{x}|\theta)|\mathbf{y}, \theta^k \rangle \\ &= c - \sum_i \frac{\langle (y_i - x_i)^2 \rangle^k}{2\sigma^2} - \log Z(\beta) - \frac{\beta}{2} \sum_i \sum_{j \in N_i} \langle \phi(u_i, u_j) \rangle^k, \end{aligned} \tag{29.19}$$

where $\langle \cdot \rangle^k$ is an abbreviation for $\langle \cdot|\mathbf{y}, \theta^k \rangle$. In the M-step, the update for θ can be found easily by setting

$$\frac{\partial}{\partial \sigma^2} Q(\theta|\theta^k) = 0, \quad \frac{\partial}{\partial \beta} Q(\theta|\theta^k) = 0. \tag{29.20}$$

From the first of these,

$$(\sigma^2)^{k+1} = \|\mathbf{S}\|^{-1} \sum_i \langle (y_i - u_i)^2 \rangle^k. \tag{29.21}$$

The solution of the second equation, on the other hand, is generally difficult due to the well-known difficulties of evaluating the partition function $Z(\beta)$ (see also Equation 29.7) which needs to be dealt with via specialized approximations [22,25]. However, as demonstrated by Bouman and Sauer [26], some simple yet important cases exist in which the solution is straightforward. For example, when $\phi(u_i, u_j) = (u_i - u_j)^2$, $Z(\beta)$ can be written as

$$\begin{aligned} Z(\beta) &= \int e^{-\frac{\beta}{2} \sum_i \sum_{j \in N_i} (u_i - u_j)^2} d\mathbf{u} \\ &= \beta^{-\|\mathbf{S}\|/2} \int e^{-\frac{1}{2} \sum_i \sum_{j \in N_i} (v_i - v_j)^2} d\mathbf{v} = \beta^{-\|\mathbf{S}\|/2} Z(1). \end{aligned} \tag{29.22}$$

Here, we have used a change of variable, $v_i = \sqrt{\beta u_i}$. Now, the update of β can be found easily as

$$\beta^{k+1} = \|\mathbf{S}\|^{-1} \sum_i \sum_{j \in N_i} \langle (u_i - u_j)^2 \rangle^k. \tag{29.23}$$

This simple technique applies to a wider class of clique functions characterized by $\phi(u_i, u_j) = |u_i - u_j|^r$ with any $r > 0$ [26].

29.3 Some Fundamental Problems

As is in many other areas of signal processing, the power and versatility of the EM algorithm has been demonstrated in a large number of diverse image recovery applications. Previous work, however, has also revealed some of its weaknesses. For example, the conditional expectation of the E-step can be difficult to calculate analytically and too time-consuming to compute numerically, as is in the MRF example in the previous section. To a lesser extent, similar remarks can be made to the M-step. Since the EM algorithm is iterative, convergence can often be a problem. For example, it can be very slow. In some applications, e.g., emission tomography, it could converge to the wrong result—the reconstructed image gets spikier as the number of iterations increases [12,13]. While some of these problems, such as slow convergence, are common to many numerical algorithms, most of their causes are inherent to the EM algorithm [17,19].

In previous work, the EM algorithm has mostly been applied in a "natural fashion" (e.g., in terms of selecting incomplete and complete data sets) and the problems mentioned above were dealt with on an ad hoc basis with mixed results. Recently, however, there has been interest in seeking more fundamental solutions [14,19]. In this section, we briefly describe the solutions to two major problems related to the EM algorithm, namely, the conditional expectation computation in the E-step when the data is modeled as MRFs and fundamental ways of improving convergence.

29.3.1 Conditional Expectation Calculations

When the complete data is an MRF, the conditional expectation of the E-step of the EM algorithm can be difficult to perform. For instance, consider the simple MRF in Section 29.2, where it amounts to calculating $\langle u_i \rangle$, $\langle u_i^2 \rangle$, and $\langle \phi(u_i, u_j) \rangle$ and the expectations are taken with respect to $p(\mathbf{u}|\mathbf{y}, \theta^k)$ of Equation 29.17. For example, we have

$$\langle u_i \rangle = Z_1^{-1} \int u_i e^{-E_1(\mathbf{u})} d\mathbf{u}. \tag{29.24}$$

Here, for the sake of simplicity, we have omitted the superscript k and the parameters, and this is done in the rest of this section whenever there is no confusion. Since the variables u_i and u_j are coupled in the energy function for all i and j that are neighbors, the pdf and Z_1 cannot be factored into simpler terms, and the integration is exponentially complex, i.e., it involves all possible realizations of $\mathbf{u}$. Hence, some approximation scheme has to be used. One of these is the Monte Carlo simulation. For example, Gibbs samplers [23] and Metropolis techniques [27] have been used to generate samples according to $p(\mathbf{u}|\mathbf{y}, \theta^k)$ [26,28]. A disadvantage of these is that, generally, hundreds of samples of $\mathbf{u}$ are needed and if the image size is large, this can be computation intensive. Another technique is based on the mean field theory (MFT) of statistical mechanics [25]. This has the advantage of being computationally inexpensive while providing satisfactory results in many practical applications. In this section, we will outline the essentials of this technique.

Let $\mathbf{u}$ be an MRF with pdf

$$p(\mathbf{u}) = Z^{-1} e^{-\beta E(\mathbf{u})}. \tag{29.25}$$

For the sake of simplicity, we assume that the energy function is of the form

$$E(\mathbf{u}) = \sum_i \left[h_i(u_i) + \frac{1}{2} \sum_{j \in N_i} \phi(u_i, u_j) \right], \tag{29.26}$$

where $h_i(\cdot)$ and $\phi(\cdot, \cdot)$ are some suitable, and possibly nonlinear, functions. The MFT attempts to derive a pdf $p_{\text{MF}}(\mathbf{u})$ that is an approximation to $p(\mathbf{u})$ and can be factored like an independent pdf.

The MFT used previously can be divided into two classes: the local mean field energy (LMFE) and the ones based on the Gibbs–Bogoliubov–Feynman (GBF) inequality. The LMFE scheme is based on the idea that when calculating the mean of the MRF at a given site, the influence of the random variables at other sites can be approximated by the influence of their means. Hence, if we want to calculate the mean of u_i, a local energy function can be constructed by collecting all the terms in Equation 29.26 that are related to u_i and replacing the u_j's by their mean. Hence, for this energy function we have

$$E_i^{\text{MF}}(u_i) = h_i(u_i) + \sum_{i \in N_i} \phi(u_i, \langle u_j \rangle) \tag{29.27}$$

$$p_i^{\text{MF}}(u_i) = Z_i^{-1} e^{-\beta E_i^{\text{MF}}(u_i)} \tag{29.28}$$

$$p_{\text{MF}}(\mathbf{u}) = \prod_i p_i^{\text{MF}}(u_i). \tag{29.29}$$

Using this mean field pdf, the expectation of u_i and its functions can be found easily.

Again we use the MRF example from Section 29.2.2 as an illustration. Its energy function is Equation 29.18 and for the sake of simplicity, we assume that $\phi(u_i, u_j) = |u_i - u_j|^2$. By the LMFE scheme,

$$E_i^{\text{MF}} = \frac{(y_i - u_i)^2}{2\sigma^2} + \sum_{j \in N_i} \beta(u_i - \langle u_j \rangle)^2 \tag{29.30}$$

which is the energy of a Gaussian. Hence, the mean can be found easily by completing the square in Equation 29.30 with

$$\langle u_i \rangle = \frac{y_i/\sigma^2 + 2\beta \sum_{j \in N_i} \langle u_j \rangle}{1/\sigma^2 + 2\beta \|N_i\|}. \tag{29.31}$$

When $\phi(\cdot, \cdot)$ is some general nonlinear function, numerical integration might be needed. However, compared to Equation 29.24 such integrals are all with respect to one or two variables and are easy to compute.

Compared to the physically motivated scheme above, the GBF is an optimization approach. Suppose that $p_0(\mathbf{u})$ is a pdf which we want to use to approximate another pdf, $p(\mathbf{u})$. According to information theory, e.g., see [29], the directed-divergence between p_0 and p is defined as

$$D(p_0 \| p) = \langle \log p_0(\mathbf{u}) - \log p(\mathbf{u}) \rangle_0, \tag{29.32}$$

where the subscript 0 indicates that the expectation is taken with respect to p_0, and it satisfies

$$D(p_0 \| p) \geq 0 \tag{29.33}$$

with equality holds if and only if $p_0 = p$. When the pdfs are Gibbs distributions, with energy functions E_0 and E and partition functions Z_0 and Z, respectively, the inequality becomes

$$\log Z \geq \log Z_0 - \beta\langle E - E_0\rangle_0 = \log Z_0 - \beta\langle \Delta E\rangle_0, \tag{29.34}$$

which is known as the GBF inequality.

Let p_0 be a parametric Gibbs pdf with a set of parameters ω to be determined. Then, one can obtain an optimal p_0 by maximizing the right-hand side of Equation 29.34. As an illustration, consider again the MRF example in Section 29.2 with the energy function (Equation 29.18) and a quadratic clique function, as we did for the LMFE scheme. To use the GBF, let the energy function of p_0 be defined as

$$E_0(\mathbf{u}) = \sum_i \frac{(u_i - m_i)^2}{2v_i^2}, \tag{29.35}$$

where $\{m_i, v_i^2, i \in \mathbf{S}\} = \omega$ is the set of parameters to be determined in the maximization of the GBF. Since this is the energy for an independent Gaussian, Z_0 is just

$$Z_0 = \prod_i \sqrt{2\pi v_i^2}. \tag{29.36}$$

The parameters of p_0 can be obtained by finding an expression for the right-hand side of the GBF inequality, letting its partial derivatives (with respect to the parameters m_i and v_i^2) be zero, and solving for the parameters. Through a somewhat lengthy but straightforward derivation, one can find that [30]

$$m_i = \frac{y_i/\sigma^2 + 2\beta\sum_{j\in N_i}\langle u_j\rangle}{1/\sigma^2 + 2\beta\|N_i\|}. \tag{29.37}$$

Since $m_i = \langle u_i\rangle$, the GBF produces the same result as the LMEF. This, however, is an exception rather than the rule [30] and it is due to the quadratic structures of both energy functions.

We end this section with several remarks. First, compared to the LMFE, the GBF scheme is an optimization scheme, hence more desirable. However, if the energy function of the original pdf is highly nonlinear, the GBF could require the solution of a difficult nonlinear equation in many variables (see, e.g., [30]). The LMFE, though not optimal, can always be implemented relatively easily. Secondly, while the MFT techniques are significantly more computation efficient than the Monte Carlo techniques and provide good results in many applications, no proof exists as yet that the conditional mean computed by the MFT will converge to the true conditional mean. Finally, the performance of the mean field approximations may be improved by using "high-order" models. For example, one simple scheme is to consider LMFEs with a pair of neighboring variables [25,31]. For the energy function in Equation 29.26, e.g., the "second-order" LMFE is

$$E_{i,j}^{\mathrm{MF}}(u_i, u_j) = h_i(u_i) + h_i(u_j) + \beta\sum_{i'\in N_i}\phi(u_i,\langle u_{i'}\rangle) + \beta\sum_{j'\in N_j}\phi(u_j,\langle u_{j'}\rangle) \tag{29.38}$$

and

$$p_{\mathrm{MF}}(u_i, u_j) = Z_{\mathrm{MF}}^{-1} e^{-\beta E_{i,j}^{\mathrm{MF}}(u_i,u_j)}, \tag{29.39}$$

$$p_{\mathrm{MF}}(u_i) = \int p_{\mathrm{MF}}(u_i, u_j)du_j. \tag{29.40}$$

Notice that Equation 29.40 is not the same as Equation 29.28 in that the fluctuation of u_j is taken into consideration.

29.3.2 Convergence Problem

Research on the EM algorithm-based image recovery has so far suggested two causes for the convergence problems mentioned previously. The first is whether the random field models used adequately capture the characteristics and constraints of the underlying physical phenomenon. For example, in emission tomography the original EM procedure of Shepp and Verdi tends to produce spikier and spikier images as the number of iteration increases [13]. It was found later that this is due to the assumption that the densities of the radioactive material at different spatial locations are independent. Consequently, various smoothness constraints (density dependence between neighboring locations) have been introduced as penalty functions or priors and the problem has been greatly reduced. Another example is in blind image restoration. It has been found that in order for the EM algorithm to produce reasonable estimate of the blur, various constraints need to be imposed. For instance, symmetry conditions and good initial guesses (e.g., a lowpass filter) are used in [8,9]. Since the blur tends to have a smooth impulse response, orthonormal expansion (e.g., the DCT) has also been used to reduce (compress) the number of parameters in its representation [15].

The second factor that can be quite influential to the convergence of the EM algorithm, noticed earlier by Feder and Weinstein [16], is how the complete data is selected. In their work [18], Fessler and Hero found that for some EM procedures, it is possible to significantly increase the convergence rate by properly defining the complete data. Their idea is based on the observation that the EM algorithm, which is essentially a MLE procedure, often converges faster if the parameters are estimated sequentially in small groups rather than simultaneously. Suppose, e.g., that 100 parameters are to be estimated. It is much better to estimate, in each EM cycle, the first 10 while holding the next 90 constant, then estimate the next 10 holding the remaining 80 and the newly updated 10 parameters constant, and so on. This type of algorithm is called the SAGE (space alternating generalized EM) algorithm.

We illustrate this idea through a simple example used by Fessler and Hero [18]. Consider a simple image recovery problem, modeled as

$$\mathbf{y} = \mathbf{A}_1\theta_1 + \mathbf{A}_2\theta_2 + \mathbf{n}, \tag{29.41}$$

where

column θ_1 and θ_2 represent two original images or two data sources
$\mathbf{A}_1$ and $\mathbf{A}_2$ are two blur functions represented as matrices
$\mathbf{n}$ is an additive white Gaussian noise source

In this model, the observed image $\mathbf{y}$ is the noise-corrupted combination of two blurred images (or data sources). A natural choice for the complete data is to view $\mathbf{n}$ as the combination of two smaller noise sources, each associated with one original image, i.e.,

$$\mathbf{x} = [\mathbf{A}_1\theta_1 + \mathbf{n}_1, \mathbf{A}_2\theta_2 + \mathbf{n}_2]', \tag{29.42}$$

where

$\mathbf{n}_1$ and $\mathbf{n}_2$ are i.i.d additive white Gaussian noise vectors with covariance matrix $\frac{\sigma^2}{2}\mathbf{I}$
$'$ denotes transpose

The incomplete data $\mathbf{y}$ can be obtained from $\mathbf{x}$ by

$$\mathbf{y} = [\mathbf{I}, \mathbf{I}]\mathbf{x}. \tag{29.43}$$

Notice that this is a Gaussian problem in that both **x** and **y** are Gaussian and they are jointly Gaussian as well. From the properties of jointly Gaussian random variables [32], the EM cycle can be found relatively straightforwardly as

$$\theta_1^{k+1} = \theta_1^k + (\mathbf{A}_1'\mathbf{A}_1)^{-1}\mathbf{A}_1'\hat{\varepsilon}/2\sigma^2 \tag{29.44}$$

$$\theta_2^{k+1} = \theta_2^k + (\mathbf{A}_2'\mathbf{A}_2)^{-1}\mathbf{A}_2'\hat{\varepsilon}/2\sigma^2, \tag{29.45}$$

where

$$\hat{\varepsilon} = \left(\mathbf{y} - \mathbf{A}_1\theta_1^k - \mathbf{A}_2\theta_2^k\right)/\sigma^2. \tag{29.46}$$

The SAGE algorithm for this simple problem is obtained by defining two smaller "complete data sets":

$$\mathbf{x}_1 = \mathbf{A}_1\theta_1 + \mathbf{n} \quad \text{and} \quad \mathbf{x}_2 = \mathbf{A}_2\theta_2 + \mathbf{n}. \tag{29.47}$$

Notice that now the noise **n** is associated "totally" with each smaller complete data set. The incomplete data **y** can be obtained from both $\mathbf{x}_1$ and $\mathbf{x}_2$, e.g.,

$$\mathbf{y} = \mathbf{x}_1 + \mathbf{A}_2\theta_2. \tag{29.48}$$

The SAGE algorithm amounts to two sequential and "smaller" EM algorithms. Specifically, corresponding to each classical EM cycle (Equations 29.44 through 29.46), the first SAGE cycle is a classical EM cycle with $\mathbf{x}_1$ $P_{\text{MF}}(u_i, u_j) = Z_{\text{MF}}^{-1} e^{-\beta E_{i,j}^{\text{MF}}(u_i,u_j)}$, as the complete data and θ_1 as the parameter set to be updated. The second SAGE cycle is a classical EM cycle with $\mathbf{x}_2$ as the complete data and θ_2 as the parameter set to be updated. The new update of θ_1 is also used. The specific algorithm is

$$\theta_1^{k+1} = \theta_1^k + (\mathbf{A}_1'\mathbf{A}_1)^{-1}\mathbf{A}_1'\hat{\varepsilon}_1/2\sigma^2 \tag{29.49}$$

$$\theta_2^{k+1} = \theta_2^k + (\mathbf{A}_2'\mathbf{A}_2)^{-1}\mathbf{A}_2'\hat{\varepsilon}_2/2\sigma^2, \tag{29.50}$$

where

$$\hat{\varepsilon}_1 = \left(\mathbf{y} - \mathbf{A}_1\theta_1^k - \mathbf{A}_2\theta_2^k\right)/\sigma^2 \tag{29.51}$$

$$\hat{\varepsilon}_2 = \left(\mathbf{y} - \mathbf{A}_1\theta_1^{k+1} - \mathbf{A}_2\theta_2^k\right)/\sigma^2. \tag{29.52}$$

We end this subsection with several remarks. First, for a wide class of random field models including the simple one above, Fessler and Hero have shown that the SAGE converges significantly faster than the classical EM [17]. In some applications, e.g., tomography, an acceleration of 5–10 times may be achieved. Secondly, just as for the EM algorithm, various constraints on the parameters are often needed and can be imposed easily as penalty functions in the SAGE algorithm. Finally, notice that in Equation 29.41, the original images are treated as parameters (with constraints) rather than as random variables with their own pdfs. It would be of interest to investigate a Bayesian counterpart of the SAGE algorithm.

29.4 Applications

In this section, we describe the application of the EM algorithm to the simultaneous identification of the blur and image model and the restoration of single and multichannel images.

29.4.1 Single Channel Blur Identification and Image Restoration

Most of the work on restoration in the literature was done under the assumption that the blurring process (usually modeled as a linear space-invariant [LSI] system and specified by its point spread function [PSF]) is exactly known (for recent reviews of the restoration work in the literature see [8,33]). However, this may not be the case in practice since usually we do not have enough knowledge about the mechanism of the degradation process. Therefore, the estimation of the parameters that characterize the degradation operator needs to be based on the available noisy and blurred data.

29.4.1.1 Problem Formulation

The observed image $y(i,j)$ is modeled as the output of a 2D LSI system with PSF $\{d(p,q)\}$. In the following we will use (i,j) to denote a location on the lattice **S**, instead of a single subscript. The output of the LSI system is corrupted by additive zero-mean Gaussian noise $v(i,j)$ with covariance matrix Λ_v, which is uncorrelated with the original image $u(i,j)$. That is, the observed image $y(i,j)$ is expressed as

$$y(i,j) = \sum_{(p,q)\in \mathbf{S}_\mathrm{D}} d(p,q)u(i-p,j-q) + v(i,j), \tag{29.53}$$

where $\mathbf{S}_\mathrm{D}$ is the finite support region of the distortion filter. We assume that the arrays $y(i,j)$, $u(i,j)$, and $v(i,j)$ are of size $N \times N$. By stacking them into $N^2 \times 1$ vectors, Equation 29.53 can be rewritten in matrix/vector form as [35]

$$\mathbf{y} = \mathbf{D}\mathbf{u} + \mathbf{v}, \tag{29.54}$$

where $\mathbf{D}$ is an $N^2 \times N^2$ matrix.

The vector $\mathbf{u}$ is modeled as a zero-mean Gaussian random field. Its pdf is equal to

$$p(\mathbf{u}) = |2\pi\Lambda_\mathrm{U}|^{-1/2} \exp\left\{\frac{-1}{2}\mathbf{u}^\mathrm{H}\Lambda_\mathrm{U}{}^{-1}\mathbf{u}\right\}, \tag{29.55}$$

where

Λ_U is the covariance matrix of $\mathbf{u}$

superscript H denotes the Hermitian (i.e., conjugate transpose) of a matrix and a vector

$|\cdot|$ denotes the determinant of a matrix

A special case of this representation is when $u(i,j)$ is described by an autoregressive (AR) model. Then Λ_U can be parameterized in terms of the AR coefficients and the covariance of the driving noise [38,57].

Equation 29.53 can be written in the continuous frequency domain according to the convolution theorem. Since the discrete Fourier transform (DFT) will be used in implementing convolution, we assume that Equation 29.53 represents circular convolution (2D sequences can be padded with zeros in such a way that the result of the linear convolution equals that of the circular convolution, or the observed image can be preprocessed around its boundaries so that Equation 29.53 is consistent with the circular convolution of $\{d(p,q)\}$ with $\{u(p,q)\}$ [36]). Matrix $\mathbf{D}$ then becomes block circulant [35].

29.4.1.2 Maximum Likelihood Parameter Identification

The assumed image and blur models are specified in terms of the deterministic parameters $\theta = \{\Lambda_\mathrm{U}, \Lambda_\mathrm{V}, \mathbf{D}\}$. Since $\mathbf{u}$ and $\mathbf{v}$ are uncorrelated, the observed image $\mathbf{y}$ is also Gaussian with pdf equal to

$$\begin{aligned} p(\mathbf{y}/\theta) &= |2\pi(\mathbf{D}\Lambda_\mathrm{U}\mathbf{D}^\mathrm{H} + \Lambda_\mathrm{V})|^{-1/2} \\ &\quad\times \exp\left\{\frac{-1}{2}\mathbf{y}^\mathrm{T}(\mathbf{D}\Lambda_\mathrm{U}\mathbf{D}^\mathrm{H} + \Lambda_\mathrm{V})^{-1}\mathbf{y}\right\}, \end{aligned} \tag{29.56}$$

where the inverse of the matrix $(\mathbf{D}\Lambda_\mathrm{U}\mathbf{D}^\mathrm{H} + \Lambda_\mathrm{V})$ is assumed to be defined since covariance matrices are symmetric positive definite.

Taking the logarithm of Equation 29.56 and disregarding constant additive and multiplicative terms, the maximization of the log-likelihood function becomes the minimization of the function $L(\theta)$, given by

$$L(\theta) = \log\left|\mathbf{D}\Lambda_\mathrm{U}\mathbf{D}^\mathrm{H} + \Lambda_\mathrm{V}\right| + \left[\mathbf{y}^\mathrm{T}\left(\mathbf{D}\Lambda_\mathrm{U}\mathbf{D}^\mathrm{H} + \Lambda_\mathrm{V}\right)^{-1}\mathbf{y}\right]. \tag{29.57}$$

By studying the function $L(\theta)$ it is clear that if no structure is imposed on the matrices $\mathbf{D}$, Λ_U, and Λ_V, the number of unknowns involved is very large. With so many unknowns and only one observation (i.e., $\mathbf{y}$), the ML identification problem becomes unmanageable. Furthermore, the estimate of $\{d(p,q)\}$ is not unique, because the ML approach to image and blur identification uses only second-order statistics of the blurred image, since all pdfs are assumed to be Gaussian. More specifically, the second-order statistics of the blurred image do not contain information about the phase of the blur, which, therefore, is in general undetermined. In order to restrict the set of solutions and hopefully obtain a unique solution, additional information about the unknown parameters needs to be incorporated into the solution process.

The structure we are imposing on Λ_U and Λ_V results from the commonly used assumptions in the field of image restoration [35]. First we assume that the additive noise $\mathbf{v}$ is white, with variance σ_V^2, i.e.,

$$\Lambda_\mathrm{V} = \sigma_\mathrm{V}^2\mathbf{I}. \tag{29.58}$$

Further we assume that the random process $\mathbf{u}$ is stationary which results in Λ_U being a block Toeplitz matrix [35]. A block Toeplitz matrix is asymptotically equivalent to a block circulant matrix as the dimension of the matrix becomes large [37]. For average size images, the dimensions of Λ_U are large indeed; therefore, the block circulant approximation is a valid one. Associated with Λ_U are the 2D sequences $\{l_\mathrm{U}(p,q)\}$. The matrix $\mathbf{D}$ in Equation 29.54 was also assumed to be block circulant. Block circulant matrices can be diagonalized with a transformation matrix constructed from discrete Fourier kernels [35]. The diagonal matrices corresponding to Λ_U and $\mathbf{D}$ are denoted respectively by $\mathbf{Q}_\mathrm{U}$ and $\mathbf{Q}_\mathrm{D}$. They have as elements the raster scanned 2D DFT values of the 2D sequences $\{l_\mathrm{U}(p,q)\}$ and $\{d(p,q)\}$, denoted respectively by $S_\mathrm{U}(m,n)$ and $\Delta(m,n)$.

Due to the above assumptions Equation 29.57 can be written in the frequency domain as

$$L(\theta) = \sum_{m=0}^{N-1}\sum_{n=0}^{N-1}\left\{\log\left[|\Delta(m,n)|^2 S_\mathrm{U}(m,n) + \sigma_\mathrm{V}^2\right] + \frac{|Y(m,n)|^2}{|\Delta(m,n)|^2 S_\mathrm{U}(m,n) + \sigma_\mathrm{V}^2}\right\}, \tag{29.59}$$

where $Y(m,n)$ is the 2D DFT of $y(i,j)$. Equation 29.59 more clearly demonstrates the already mentioned nonuniqueness of the ML blur solution, since only the magnitude of $\Delta(m,n)$ appears in $L(\theta)$. If the blur is zero-phase, as is the case with $\mathbf{D}$ modeling atmospheric turbulence with long exposure times and mild defocussing ($\{d(p,q)\}$ is 2D Gaussian in this case), then a unique solution may be obtained. Nonuniqueness of the estimation of $\{d(p,q)\}$ can in general be avoided by enforcing the solution to satisfy a set of constraints. Most PSFs of practical interest can be assumed to be symmetric, i.e., $d(p,q) = d(-p,-q)$. In this case the phase of the DFT of $\{d(p,q)\}$ is zero or $\pm\pi$. Unfortunately, uniqueness of the ML solution is not always established by the symmetry assumption, due primarily to the phase ambiguity. Therefore, additional constraints may alleviate this ambiguity. Such additional constraints are the following: (1) The PSF coefficients are nonnegative, (2) the support $\mathbf{S_D}$ is finite, and (3) the blurring mechanism preserves energy [35], which results in

$$\sum_{(i,j)\in \mathbf{S_D}} d(i,j) = 1. \tag{29.60}$$

29.4.1.3 EM Iterations for the ML Estimation of θ

The next step to be taken in implementing the EM algorithm is the determination of the mapping **H** in Equation 29.2. Clearly Equation 29.54 can be rewritten as

$$\mathbf{y} = [0 \quad \mathbf{I}]\begin{bmatrix}\mathbf{u}\\ \mathbf{y}\end{bmatrix} = [\mathbf{D} \quad \mathbf{I}]\begin{bmatrix}\mathbf{u}\\ \mathbf{v}\end{bmatrix} = [\mathbf{I} \quad \mathbf{I}]\begin{bmatrix}\mathbf{Du}\\ \mathbf{v}\end{bmatrix}, \tag{29.61}$$

where 0 and $\mathbf{I}$ represent the $N^2 \times N^2$ zero and identity matrices, respectively. Therefore, according to Equation 29.61, there are three candidates for representing the complete data **x**, namely, $\{\mathbf{u},\mathbf{y}\}$, $\{\mathbf{u},\mathbf{v}\}$, and $\{\mathbf{Du},\mathbf{v}\}$. All three cases are analyzed in the following. However, as it will be shown, only the choice of $\{\mathbf{u},\mathbf{y}\}$ as the complete data fully justifies the term "complete data", since it results in the simultaneous identification of all unknown parameters and the restoration of the image.

For the case when **H** in Equation 29.2 is linear, as are the cases represented by Equation 29.61, and the data **y** is modeled as a zero-mean Gaussian process, as is the case under consideration expressed by Equation 29.56, the following general result holds for all three choices of the complete data [38,39,57].

The E-step of the algorithm results in the computation of $Q(\theta/\theta^k) = constant - \mathrm{F}(\theta/\theta^k)$ where

$$\begin{aligned} F(\theta/\theta^k) &= \log|\Lambda_\mathbf{X}| + \mathrm{tr}\left(\Lambda_\mathbf{X}^{-1}\mathbf{C}_{\mathbf{X}|\mathbf{y}}^k\right) \\ &= \log|\Lambda_\mathbf{X}| + \mathrm{tr}\left(\Lambda_\mathbf{X}^{-1}\Lambda_{\mathbf{X}|\mathbf{y}}^k\right) + \mu_{\mathbf{X}|\mathbf{y}}^{(k)\mathrm{H}}\Lambda_\mathbf{X}^{-1}\mu_{\mathbf{X}|\mathbf{y}}^k, \end{aligned} \tag{29.62}$$

where $\Lambda_\mathbf{X}$ is the covariance of the complete data **x** which is also a zero-mean Gaussian process,

$$\begin{aligned} \mathbf{C}_{\mathbf{X}|\mathbf{y}}^k &= \langle \mathbf{x}\mathbf{x}^\mathrm{H}|\mathbf{y};\theta^k\rangle = \Lambda_{\mathbf{X}|\mathbf{y}}^k + \mu_{\mathbf{X}|\mathbf{y}}^k\mu_{\mathbf{X}|\mathbf{y}}^{(k)\mathrm{H}}, \\ \mu_{\mathbf{X}|\mathbf{y}}^k &= \langle \mathbf{x}|\mathbf{y};\theta^k\rangle = \Lambda_{\mathbf{XY}}\Lambda_\mathbf{Y}^{-1}\mathbf{y} = \Lambda_\mathbf{X}\mathbf{H}^\mathrm{H}\left(\mathbf{H}\Lambda_{\mathbf{XH}}^{\mathrm{H}}\right)^{-1}\mathbf{y}, \end{aligned} \tag{29.63}$$

and

$$\begin{aligned} \Lambda_{\mathbf{X}|\mathbf{y}} &= \left\langle (\mathbf{x}-\mu_{\mathbf{X}|\mathbf{y}})(\mathbf{x}-\mu_{\mathbf{X}|\mathbf{y}})^\mathrm{H}|\mathbf{y};\theta^k\right\rangle = \Lambda_\mathbf{X} - \Lambda_{\mathbf{XY}}\Lambda_\mathbf{Y}^{-1}\Lambda_{\mathbf{YX}} \\ &= \Lambda_\mathbf{X} - \Lambda_\mathbf{X}\mathbf{H}^\mathrm{H}\left(\mathbf{H}\Lambda_\mathbf{X}\mathbf{H}^\mathrm{H}\right)^{-1}\mathbf{H}\Lambda_\mathbf{X}. \end{aligned} \tag{29.64}$$

The M-step of the algorithm is described by the following equation

$$\theta^{(k+1)} = \arg\left\{\min_{\{\theta\}} F(\theta/\theta^k)\right\}. \tag{29.65}$$

In our formulation of the identification/restoration problem the original image is not one of the unknown parameters in the set θ. However, as it will be shown in the next section, the restored image will be obtained in the E-step of the iterative algorithm.

29.4.1.3.1 $\{\mathbf{u}, \mathbf{y}\}$ as the Complete Data (CD_uy Algorithm)

Choosing the original and observed images as the complete data, we obtain $\mathbf{H} = [0\mathbf{I}]$ and $\mathbf{x} = [\mathbf{u}^H\mathbf{y}^H]^H$. The covariance matrix of $\mathbf{x}$ takes the form

$$\Lambda_\mathbf{x} = \langle \mathbf{x}\mathbf{x}^H \rangle = \begin{bmatrix} \Lambda_U & \Lambda_U \mathbf{D}^H \\ \mathbf{D}\Lambda_U & \mathbf{D}\Lambda_U\mathbf{D}^H + \Lambda_V \end{bmatrix}, \tag{29.66}$$

and its inverse is equal to [40]

$$\Lambda_\mathbf{X}^{-1} = \begin{bmatrix} \Lambda_U^{-1} + \mathbf{D}^H\Lambda_V^{-1}\mathbf{D} & -\mathbf{D}^H\Lambda_V^{-1} \\ -\Lambda_V^{-1}\mathbf{D} & \Lambda_V^{-1} \end{bmatrix}. \tag{29.67}$$

Substituting Equations 29.66 and 29.67 into Equations 29.62 through 29.64, we obtain

$$\begin{aligned} F(\theta/\theta^k) = {} & \log|\Lambda_U| + \log|\Lambda_V| + \operatorname{tr}\left\{ \left(\Lambda_U^{-1} + \mathbf{D}^H\Lambda_V^{-1}\,\mathbf{D}\right)\Lambda_{U|\mathbf{y}}^k \right\} \\ & + \mu_{U|\mathbf{y}}^{(k)H}\left(\Lambda_U^{-1} + \mathbf{D}^H\Lambda_V^{-1}\mathbf{D}\right)\mu_{U|\mathbf{y}}^k \\ & - 2\mathbf{y}^H\Lambda_V^{-1}\mathbf{D}\mu_{U|\mathbf{y}}^k + \mathbf{y}^H\Lambda_V^{-1}\mathbf{y}, \end{aligned} \tag{29.68}$$

where

$$\mu_{U|\mathbf{y}}^k = \Lambda_U^k\,\mathbf{D}^{(k)H}\left(\mathbf{D}^k\Lambda_U^k\,\mathbf{D}^{(k)H} + \Lambda_V^k\right)^{-1}\mathbf{y} \tag{29.69}$$

and

$$\Lambda_{U|\mathbf{y}}^k = \Lambda_U^k - \Lambda_U^k\mathbf{D}^{(k)H}\left(\mathbf{D}^k\Lambda_U^k\mathbf{D}^{(k)H} + \Lambda_V^k\right)^{-1}\mathbf{D}^k\Lambda_U^k. \tag{29.70}$$

Due to the constraints on the unknown parameters described in the subsection Equation 29.62 can be written in the discrete frequency domain as follows:

$$\begin{aligned} F(\theta/\theta^k) = {} & N^2\log\sigma_V^2 \\ & + \frac{1}{\sigma_V^2}\sum_{m=0}^{N-1}\sum_{n=0}^{N-1}\left\{ |\Delta(m,n)|^2\left(S_{U|\mathbf{y}}^k(m,n) + \frac{1}{N^2}\left|M_{U|\mathbf{y}}^k(m,n)\right|^2 \right)\right. \\ & \left. + \frac{1}{N^2}\left(|Y(m,n)|^2 - 2\operatorname{Re}\left[Y^*(m,n)\Delta(m,n)M_{U|\mathbf{y}}^k(m,n) \right] \right) \right\} \\ & + \sum_{m=0}^{N-1}\sum_{n=0}^{N-1}\left\{ \log S_U(m,n) + \frac{1}{S_U(m,n)}\left(S_{U|\mathbf{y}}^k(m,n) + \frac{1}{N^2}\left|M_{U|\mathbf{y}}^k(m,n)\right|^2 \right) \right\}, \end{aligned} \tag{29.71}$$

where

$$M_{U|\mathbf{y}}^k(m,n) = \frac{\Delta^{(k)^*}(m,n)S_U^k(m,n)}{|\Delta^k(m,n)|^2 S_U^k(m,n) + \sigma_V^{2(p)}}\,Y(m,n), \tag{29.72}$$

$$S^k_{\mathbf{U}|\mathbf{y}}(m,n) = \frac{S^k_{\mathbf{U}}(m,n)\sigma^{2(k)}_{\mathbf{V}}}{|\Delta^k(m,n)|^2 S^k_{\mathbf{U}}(m,n) + \sigma^{2(k)}_{\mathbf{V}}}. \quad (29.73)$$

In Equation 29.71, $Y(m,n)$ is the 2D DFT of the observed image $y(i,j)$ and $M^k_{\mathbf{U}|\mathbf{y}}(m,n)$ is the 2D DFT of the unstacked vector $\mu^k_{\mathbf{U}|\mathbf{y}}$ into an $N \times N$ array. Taking the partial derivatives of $F(\theta/\theta^k)$ with respect to $S_{\mathbf{U}}(m,n)$ and $\Delta(m,n)$ and setting them equal to zero, we obtain the solutions that minimize $F(\theta/\theta^k)$, which represent $S^{(k+1)}_{\mathbf{U}}(m,n)$ and $\Delta^{(k+1)}(m,n)$. They are equal to

$$S^{(k+1)}_{\mathbf{U}}(m,n) = S^k_{\mathbf{U}|\mathbf{y}}(m,n) + \frac{1}{N^2}\left|M^k_{\mathbf{U}|\mathbf{y}}(m,n)\right|^2, \quad (29.74)$$

$$\Delta^{(k+1)}(m,n) = \frac{1}{N^2}\frac{Y(m,n)M^{(k)^*}_{\mathbf{U}|\mathbf{y}}(m,n)}{S^k_{\mathbf{U}|\mathbf{y}}(m,n) + \frac{1}{N^2}\left|M^k_{\mathbf{U}|\mathbf{y}}(m,n)\right|^2}, \quad (29.75)$$

where $M^k_{\mathbf{U}|\mathbf{y}}(m,n)$ and $S^k_{\mathbf{U}|\mathbf{y}}(m,n)$ are computed by Equations 29.72 and 29.73. Substituting Equation 29.75 into Equation 29.71 and then minimizing $F(\theta/\theta^k)$ with respect to $\sigma^2_{\mathbf{V}}$, we obtain

$$\sigma^{2(k+1)}_{\mathbf{V}} = \frac{1}{N^2}\sum_{m=0}^{N-1}\sum_{n=0}^{N-1}\left\{|\Delta^{(k+1)}(m,n)|^2\left(S^k_{\mathbf{U}|\mathbf{y}}(m,n) + \frac{1}{N^2}\left|M^k_{\mathbf{U}|\mathbf{y}}(m,n)\right|^2\right)\right.$$
$$\left. + \frac{1}{N^2}\left(|Y(m,n)|^2 - 2\mathrm{Re}\left[Y^*(m,n)\Delta^{(k+1)}(m,n)M^k_{\mathbf{U}|\mathbf{y}}(m,n)\right]\right)\right\}. \quad (29.76)$$

According to Equation 29.72 the restored image (i.e., $M^k_{\mathbf{U}|\mathbf{y}}(m,n)$) is the output of a Wiener filter, based on the available estimate of θ, with the observed image as input.

29.4.1.3.2 $\{\mathbf{u},\mathbf{v}\}$ as the Complete Data (CD_uv Algorithm)

The second choice of the complete data is $\mathbf{x} = [\mathbf{u}^H\mathbf{v}^H]^H$, therefore, $\mathbf{H} = [\mathbf{DI}]$. Following similar steps as in the previous case it has been shown that the equations for evaluating the spectrum of the original image are the same as in the previous case, i.e., Equations 29.72 through 29.74 hold true. The other two unknowns, i.e., the variance of the additive noise and the DFT of the PSF are given by

$$\sigma^{2(k+1)}_{\mathbf{V}} = \frac{1}{N^2}\sum_{m=0}^{N-1}\sum_{n=0}^{N-1}\left(S^k_{\mathbf{V}|\mathbf{y}}(m,n) + \frac{1}{N}\left|M^k_{\mathbf{V}|\mathbf{y}}(m,n)\right|^2\right), \quad (29.77)$$

where

$$M^k_{\mathbf{V}|\mathbf{y}}(m,n) = \frac{\sigma^{2(k)}_{\mathbf{V}}}{|\Delta^k(m,n)|^2 S^k_{\mathbf{U}}(m,n) + \sigma^{2(k)}_{\mathbf{V}}}Y(m,n), \quad (29.78)$$

$$S^k_{\mathbf{V}|\mathbf{y}}(m,n) = \frac{|\Delta^k(m,n)|^2 S^k_{\mathbf{U}}(m,n)\sigma^{2(k)}_{\mathbf{V}}}{|\Delta^k(m,n)|^2 S^k_{\mathbf{U}}(m,n) + \sigma^{2(k)}_{\mathbf{V}}}, \quad (29.79)$$

and

$$|\Delta^k(m,n)|^2 = \begin{cases} \dfrac{\frac{1}{N^2}|Y(m,n)|^2 - \sigma^{2(k)}_{\mathbf{V}}}{S^k_{\mathbf{U}}(m,n)}, & \text{if } \frac{1}{N^2}|Y(m,n)|^2 > \sigma^{2(k)}_{\mathbf{V}} \\ 0, & \text{otherwise.} \end{cases} \quad (29.80)$$

From Equation 29.80 we observe that only the magnitude of $\Delta^k(m,n)$ is available, as was mentioned earlier. A similar observation can be made for Equation 29.75, according to which the phase of $\Delta(m,n)$ is equal to the phase of $\Delta^0(m,n)$.

In deriving the above expressions the set of unknown parameters θ was divided into two sets $\theta_1 = \{\Lambda_U, \Lambda_V\}$ and $\theta_2 = \{\mathbf{D}\}$. $F(\theta_1/\theta^k)$ was then minimized with respect to θ_1, resulting in Equations 29.74 and 29.77. The likelihood function in Equation 29.59 was then minimized directly with respect to $\Delta(m,n)$ assuming knowledge of θ_1^k, resulting in Equation 29.80. The effect of mixing the optimization procedure into the EM algorithm has not been completely analyzed theoretically. That is, the convergence properties of the EM algorithm do not necessarily hold, although the application of the resulting equations increases the likelihood function. Based on the experimental results, the algorithm derived in this section always converges to a stationary point. Furthermore, the results are comparable to the ones obtained with the CD_{uy} algorithm.

29.4.1.3.3 $\{\mathbf{Dx}, \mathbf{v}\}$ *as the Complete Data (CD_Dx,v Algorithm)*

The third choice of the complete data is $\mathbf{x} = [(\mathbf{Du})^H, \mathbf{v}^H]^H$. In this case, $\mathbf{D}$ and $\mathbf{x}$ cannot be estimated separately, since various combinations of $\mathbf{D}$ and $\mathbf{u}$ can result in the same $\mathbf{Du}$. The two quantities $\mathbf{D}$ and $\mathbf{u}$ are lumped into one quantity $\mathbf{t} = \mathbf{Du}$.

Following similar steps as in the two previous cases it has been shown [38,39,57] that the variance of the additive noise is computed according to Equation 29.77, while the spectrum of the noise-free but blurred image $\mathbf{t}$ by the iterations

$$S_{\mathrm{T}}^{(k+1)}(m,n) = S_{\mathrm{T|y}}^{k}(m,n) + \frac{1}{N^2}\left|M_{\mathrm{T|y}}^{k}(m,n)\right|^2, \tag{29.81}$$

where

$$M_{\mathrm{T|y}}^{k}(m,n) = \frac{S_{\mathrm{T}}^{k}(m,n)}{S_{\mathrm{T}}^{k}(m,n) + \sigma_{\mathrm{V}}^{2(k)}} Y(m,n) \tag{29.82}$$

and

$$S_{\mathrm{T|y}}^{k}(m,n) = S_{\mathrm{T}}^{k}(m,n) - \frac{S_{\mathrm{T}}^{(k)2}(m,n)}{S_{\mathrm{T}}^{k}(m,n) + \sigma_{\mathrm{V}}^{2(k)}} Y(m,n). \tag{29.83}$$

29.4.1.4 Iterative Wiener Filtering

In this subsection, we deviate somewhat from the original formulation of the identification problem by assuming that the blur function is known. The problem at hand then is the restoration of the noisy-blurred image. Although there are a great number of approaches that can be followed in this case, the Wiener filtering approach represents a commonly used choice. However, in Wiener filtering knowledge of the power spectrum of the original image ($\mathbf{S}_\mathrm{U}$) and the additive noise ($\mathbf{S}_\mathrm{V}$) is required. A standard assumption is that of ergodicity, i.e., ensemble averages are equal to spatial averages. Even in this case, the estimation of the power spectrum of the original image has to be based on the observed noisy-blurred image, since the original image is not available. Assuming that the noise is white, its variance σ_v^2 needs also to be estimated from the observed image. Approaches, according to which the power spectrum of the original image is computed from images with similar statistical properties, have been suggested in the literature [35]. However, a reasonable idea is to successively use the Wiener-restored image as an improved prototype for updating the unknown S_U and σ_V^2. This idea is precisely implemented by the CD_uy algorithm.

More specifically, now that the blur function is known, Equation 29.75 is removed from the EM iterations. Thus, Equations 29.74 and 29.76 are used to estimate $\mathbf{S_U}$ and σ_V^2, respectively, while Equation 29.72 is used to compute the Wiener-filtered image. The starting point $\mathbf{S_U}^0$ for the Wiener iteration can be chosen to be equal to

$$S_U^0(m,n) = \hat{S}_Y(m,n), \tag{29.84}$$

where $\hat{S}_Y(m,n)$ is an estimate of the power spectral density of the observed image. The value of $\sigma_V^{2(0)}$ can be determined from flat regions in the observed image, since this represents a commonly used approach for estimating the noise variance.

29.4.2 Multichannel Image Identification and Restoration

29.4.2.1 Introduction

We use the term multichannel images to define the multiple image planes (channels) which are typically obtained by an imaging system that measures the same scene using multiple sensors. Multichannel images exhibit strong between-channel correlations. Representative examples are multispectral images [41], microwave radiometric images [42], and image sequences [43]. In the first case such images are acquired for remote sensing and facilities/military surveillance applications. The channels are the different frequency bands (color images represent a special case of great interest). In the last case the channels are the different time frames after motion compensation. More recent applications of multichannel filtering theory include the processing of the wavelet decomposed single-channel image [44] and the reconstruction of a high resolution image from multiple low-resolution images [45–48].

Although the problem of single channel image restoration has been thoroughly researched, significantly less work has been done on the problem of multichannel restoration. The multichannel formulation of the restoration problem is necessary when cross-channel degradations exist. It can be useful, however, in the case when only within-channel degradations exist, since cross-correlation terms are exploited to achieve better restoration results [49,50]. The cross-channel degradations may come in the form of channel crosstalks, leakage in detectors, and spectral blurs [51]. Work on restoring multichannel images is reported in [42,49–55], when the within- and cross-channel (where applicable) blurs are known.

29.4.3 Problem Formulation

The degradation process is modeled again as [35]

$$\mathbf{y} = \mathbf{D}\mathbf{u} + \mathbf{v}, \tag{29.85}$$

where $\mathbf{y}, \mathbf{u}$, and $\mathbf{v}$ are the observed (noisy and degraded) image, the original undistorted image, and the noise process, respectively, all of which have been lexicographically ordered, and $\mathbf{D}$ the resulting degradation matrix. The noise process is assumed to be white Gaussian, independent of $\mathbf{u}$.

Let P be the number of channels, each of size $N \times N$. If $\mathbf{u}_i, i = 0, 1, \ldots, P-1$, represents the ith channel. Then using the ordering of [56], the multichannel image $\mathbf{u}$ can be represented in vector form as

$$\mathbf{u} = \left[u_1(0)u_2(0)\cdots u_P(0)u_1(1)\cdots u_P(1)\cdots u_1(N^2-1)\cdots u_P(N^2-1)\right]^{\mathrm{T}}. \tag{29.86}$$

Defining $\mathbf{y}$ and $\mathbf{v}$ similar to that of Equation 29.86, we can now use the degradation model of Equation 29.85, recognizing that $\mathbf{y}$, $\mathbf{u}$, and $\mathbf{v}$ are of size $PN^2 \times 1$, and $\mathbf{D}$ is of size $PN^2 \times PN^2$.

Assuming that the distortion system is linear shift invariant, $\mathbf{D}$ is a $PN^2 \times PN^2$ matrix of the form

$$\mathbf{D} = \begin{bmatrix} \mathbf{D}(0) & \mathbf{D}(1) & .. & \mathbf{D}(N^2-1) \\ \mathbf{D}(N^2-1) & \mathbf{D}(0) & .. & \mathbf{D}(N^2-2) \\ \vdots & \vdots & .. & \vdots \\ \mathbf{D}(1) & \mathbf{D}(2) & .. & \mathbf{D}(0) \end{bmatrix}, \quad (29.87)$$

where the $P \times P$ sub-matrices (subblocks) have the form

$$\mathbf{D}(m) = \begin{bmatrix} D_{11}(m) & D_{12}(m) & .. & D_{1P}(m) \\ D_{21}(m) & D_{22}(m) & .. & D_{2P}(m) \\ \vdots & \vdots & .. & \vdots \\ D_{P1}(m) & D_{P2}(m) & .. & D_{PP}(m) \end{bmatrix}, \quad 0 \le m \le N^2 - 1. \quad (29.88)$$

Note that $D_{ii}(m)$ represents the intrachannel blur, while $D_{ij}(m), i \neq j$ represents the interchannel blur. The matrix $\mathbf{D}$ in Equation 29.87 is circulant at the block level. However, for $\mathbf{D}$ to be block-circulant, each of its subblocks $\mathbf{D}(m)$ also needs to be circulant, which, in general, is not the case. Matrices of this form are called semiblock circulant (SBC) matrices [56]. The singular values of such matrices can be found with the use of the DFT kernels. Equation 29.85 can therefore be written in the vector DFT domain [56].

Similarly, the covariance matrix of the original signal, Λ_{U}, and the covariance matrix of the noise process, Λ_{V}, are also SBC (assuming $\mathbf{u}$ and $\mathbf{v}$ are stationary). Note that Λ_{U} is not block-circulant because there is no justification to assume stationarity between channels (i.e., $\Lambda_{\mathrm{U}_i\mathrm{U}_j}(m) = E[\mathbf{u}_i(m)\mathbf{u}_j(m)^*]$ is not equal to $\Lambda_{\mathrm{U}_{i+p}\mathrm{U}_{j+p}}(m) = E[\mathbf{u}_{i+p}(m)\mathbf{u}_{j+p}(m)^*]$ [50], where $\Lambda_{\mathrm{U}_i\mathrm{U}_j}(m)$ is the (i,j)th submatrix of Λ_{U}). However, Λ_{U} and Λ_{V} are SBC because $\mathbf{u}_i$ and $\mathbf{v}_i$ are assumed to be stationary within each channel.

29.4.4 E-Step

We follow here similar steps to the ones presented in the previous section. We choose $[\mathbf{u}^{\mathrm{H}}\mathbf{y}^{\mathrm{H}}]^{\mathrm{H}}$ as the complete data. Since the matrices Λ_{U}, Λ_{V}, and D, are assumed to be SBC, the E-step requires the evaluation of

$$F(\theta;\theta^k) = \sum_{m=0}^{N-1}\sum_{n=0}^{N-1} J(m,n), \quad (29.89)$$

where

$$\begin{aligned} J(m,n) &= \log|\Theta_{\mathrm{U}}(m,n)| + \log|\Theta_{\mathrm{V}}(m,n)| + \mathrm{tr}\Big\{\Big[\Theta_{\mathbf{U}}^{-1}(m,n) \\ &\quad + \Theta_{\mathbf{D}}^{\mathrm{H}}(m,n)\Theta_{\mathbf{V}}^{-1}(m,n)\Theta_{\mathrm{D}}(m,n)\Big]\Theta_{\mathbf{U}|\mathbf{y}}^{k}(m,n)\Big\} \\ &\quad + \frac{1}{N^2}\mathrm{tr}\Big\{\Big[\Theta_{\mathbf{U}}^{-1}(m,n) + \Theta_{\mathbf{D}}^{\mathrm{H}}(m,n)\Theta_{\mathbf{V}}^{-1}(m,n)\Theta_{\mathrm{D}}(m,n)\Big]\mathbf{M}_{\mathbf{U}|\mathbf{y}}^{k}(m,n)\mathbf{M}_{\mathbf{U}|\mathbf{y}}^{(k)\mathrm{H}}(m,n)\Big\} \\ &\quad - \frac{1}{N^2}\Big(\mathbf{Y}^{\mathrm{H}}(m,n)\Theta_{\mathbf{V}}^{-1}(m,n)\Theta_{\mathrm{D}}(m,n)\mathbf{M}_{\mathbf{U}|\mathbf{y}}^{k}(m,n) \\ &\quad + \mathbf{M}_{\mathbf{U}|\mathbf{y}}^{(k)\mathrm{H}}(m,n)\Theta_{\mathbf{D}}^{\mathrm{H}}(m,n)\Theta_{\mathbf{V}}^{-1}(m,n)\mathbf{Y}(m,n)\Big) \\ &\quad + \frac{1}{N^2}\mathbf{Y}^{\mathrm{H}}(m,n)\Theta_{\mathbf{V}}^{-1}(m,n)\mathbf{Y}(m,n). \end{aligned} \quad (29.90)$$

The derivation of Equation 29.90 is presented in detail in [48,57,58]. Equation 29.89 is the corresponding equation to Equation 29.71 for the multichannel case.

In Equation 29.90, $\Theta_U(m,n)$ is the (m,n)th component matrix of Θ_U, which is related to Λ_U by a similarity transformation using 2D discrete Fourier kernels [56,57]. To be more specific, for $P=3$, the matrix,

$$\Theta_U(m,n) = \begin{bmatrix} S_{11}(m,n) & S_{12}(m,n) & S_{13}(m,n) \\ S_{21}(m,n) & S_{22}(m,n) & S_{23}(m,n) \\ S_{31}(m,n) & S_{32}(m,n) & S_{33}(m,n) \end{bmatrix}, \tag{29.91}$$

consists of all the (m,n)th component of the power and cross-power spectra of the original color image (without loss of generality in the subsequent discussion three-channel examples will be used). It is worthwhile noting here that the power spectra $S_{ii}(m,n), i=1,2,$ and 3, which are the diagonal entries of $\Theta_U(m,n)$, are real-valued, while the cross power spectra (the off-diagonal entries) are complex. This illustrates one of the main differences between working with multichannel images as opposed to single-channel images. In addition to each frequency component being a $P \times P$ matrix versus a scalar quantity for the single-channel case, the cross power spectra is complex versus being real for the single-channel case. Similarly, the (m,n)th component of the inverse of the noise spectrum matrix is given by

$$\Theta_V{}^{-1}(m,n) = \begin{bmatrix} z_{11}(m,n) & z_{12}(m,n) & z_{13}(m,n) \\ z_{21}(m,n) & z_{22}(m,n) & z_{23}(m,n) \\ z_{31}(m,n) & z_{32}(m,n) & z_{33}(m,n) \end{bmatrix}. \tag{29.92}$$

One simplifying assumption that we can make about Equation 29.92 is that the noise is white within channels and zero across channels. This results in $\Theta_V(m,n)$ being the same diagonal matrix for all (m,n).

$\Theta_D(m,n)$ in Equation 29.90 is equal to

$$\Theta_D(m,n) = \begin{bmatrix} \Delta_{11}(m,n) & \Delta_{12}(m,n) & \Delta_{13}(m,n) \\ \Delta_{21}(m,n) & \Delta_{22}(m,n) & \Delta_{23}(m,n) \\ \Delta_{31}(m,n) & \Delta_{32}(m,n) & \Delta_{33}(m,n) \end{bmatrix}, \tag{29.93}$$

where

$\Delta_{ij}(m,n)$ is the within-channel $(i=j)$ or cross-channel $(i \neq j)$ frequency response of the blur system

$\mathbf{Y}(m,n)$ is the (m,n)th component of the DFT of the observed image

$\Theta^k_{U|y}(m,n)$ and $\mathbf{M}^k_{U|y}(m,n)$ are the (m,n)th frequency component matrix and vector of the multichannel counterparts of $\Lambda_{U|y}$ and $\mu_{U|y}$, respectively, computed by

$$\begin{aligned} \Theta^k_{U|y}(m,n) = \Theta^k_U(m,n) - \Theta^k_U(m,n)\Theta^{(k)H}_D(m,n)\Big[\Theta^k_V(m,n) \\ + \Theta^k_D(m,n)\Theta^k_U(m,n)\Theta^{(k)H}_D(m,n)\Big]^{-1}\Theta^k_D(m,n)\Theta^k_U(m,n) \end{aligned} \tag{29.94}$$

and

$$\begin{aligned} \mathbf{M}^k_{U|y}(m,n) = \Theta^k_U(m,n)\Theta^{(k)H}_D(m,n)\Big[\Theta^k_V(m,n) \\ + \Theta^k_D(m,n)\Theta^k_U(m,n)\Theta^{(k)H}_D(m,n)\Big]^{-1}\mathbf{Y}(m,n). \end{aligned} \tag{29.95}$$

29.4.5 M-Step

The M-step requires the minimization of $J(m, n)$ with respect to $\Theta_U(m, n)$, $\Theta_V(m, n)$, and $\Theta_D(m, n)$. The resulting solutions become $\Theta_U^{(k+1)}(m, n)$, $\Theta_V^{(k+1)}(m, n)$ and $\Theta_D^{(k+1)}(m, n)$, respectively.

The minimization of $J(m, n)$ with respect to Θ_U is straightforward, since Θ_U is decoupled from $\Theta_V(\mathbf{m}, \mathbf{n})$ and Θ_D. An equation similar to Equation 29.74 results. The minimization of $J(m, n)$ with respect to Θ_D is not as straightforward; Θ_D is coupled with Θ_V. Therefore, in order to minimize $J(m, n)$ with respect to Θ_D, Θ_V must be solved first in terms of Θ_D, substituted back into Equation 29.90, and then minimized with respect to Θ_D.

It is shown in [48,58] that two conditions must be met in order to obtain explicit equations for the blur. First, the noise spectrum matrix, $\Theta_V(m, n)$, must be a diagonal matrix, which is frequently encountered in practice. Second, all of the blurs must be symmetric, so that there is no phase when working in the discrete frequency domain. The first condition arises from the fact that $\Theta_V(m, n)$ and $\Theta_D(m, n)$ are coupled. The second condition arises from the Cauchy–Riemann theorem, and must be satisfied in order to guarantee the existence of a derivative at every point.

With these conditions, the iterations for $\Delta(m, n)$ and $\sigma_V(m, n)$ are derived in [48,58], which are similar respectively to Equations 29.75 and 29.76. Special cases are also analyzed in [48,58], when the number of unknowns is reduced. For example, if Θ_D is known, the multichannel Wiener filter results.

29.5 Experimental Results

The effectiveness of both the single channel and multi-channel restoration and identification algorithms is demonstrated experimentally. The red, green, and blue (RGB) channels of the original Lena image used for this experiment are shown in Figure 29.1. A 5 × 5 truncated Gaussian blur is used for each channel and Gaussian white noise is added resulting in a blurred signal-to-noise ratio (SNR) of 20 dB. The degraded channels are shown in Figure 29.2. Three different experiments were performed with the available

FIGURE 29.1 Original RGB Lena.

FIGURE 29.2 Degraded RGB Lena, intrachannel blurs only, 20 dB SNR.

FIGURE 29.3 Restored RGB by the decoupled single channel EM algorithm.

FIGURE 29.4 Restored RGB Lena by the multichannel EM algorithm.

FIGURE 29.5 Restored RGB Lena by the iterative multichannel Wiener algorithm.

degraded data. The single-channel algorithm of Equations 29.74 through 29.76 was first run for each of the RGB channels independently. The restored images are shown in Figure 29.3. The corresponding multichannel algorithm was then run, resulting in the restored channels shown in Figure 29.4. Finally the multichannel Wiener filter was also run, in demonstrating the upper bound of the algorithm's performance, since the blurs are now exactly known. The resulting restored images are shown in Figure 29.5. The improvement in SNR for the three experiments and for each channel is shown in Table 29.1. According to this table, the performance of the algorithm increases from the first to the last experiment. This is to be expected, since in considering the multichannel algorithm over the single channel algorithm the correlation between channels is taken into account, which brings additional information into the problem.

A photographically blurred image is shown next in Figure 29.6. The restorations of it by the CD_uy and CD_uv algorithms are shown, respectively, in Figures 29.7 and 29.8.

TABLE 29.1 Improvement in SNR (in dB)

η	Decoupled EM	Multichannel EM	Wiener
Red	1.5573	2.1020	2.3420
Green	1.3814	2.0086	2.3181
Blue	1.1520	1.5148	1.8337

FIGURE 29.6 Photographically blurred image.

FIGURE 29.7 Restored image by the CD_uy algorithm.

FIGURE 29.8 Restored image by the CD_uv algorithm.

29.5.1 Comments on the Choice of Initial Conditions

The likelihood function which is optimized is highly nonlinear and a number of local minima exist. Although the incorporation of the various constraints, discussed earlier, restricts the set of possible solutions, a number of local minima still exist. Therefore, the final result depends on the initial conditions. Based on our experience in implementing the EM iterations of the previous sections for the single-channel and the multi-channel image restoration cases, the following comments and observations are in order.

It was observed experimentally that the final results are quite insensitive to variations in the values of the noise variance(s) and the original image power spectra. An estimate of the noise variances from flat regions of the noisy and blurred images were used as initial condition. It was observed that using initial estimates of the noise variances larger than the actual ones produced good final results.

The final results are quite sensitive, however, to variations in the values of the PSF. Knowledge of the support of the PSF is quite important. In [38] after convergence of the EM algorithm the estimate of the PSF was truncated, normalized, and used as an initial condition in restarting another iteration cycle.

29.6 Summary and Conclusion

In this chapter, we have described and illustrated how the EM algorithm can be used in image recovery problems. The basic approach can be summarized by the following steps:

1. Select a statistical model for the observed data and formulate the image recovery problem as an MLE problem.
2. If the likelihood function is difficult to optimize directly, the EM algorithm can be used by properly selecting the complete data.
3. Constraints on the parameters or image to be estimated, proper initial conditions, and multiple complete data spaces can be considered to improve the uniqueness and convergence of the estimates.
4. Derive the equations for the E-step and M-step.

We end this chapter with several remarks. We want to emphasize again that the EM algorithm only guarantees convergence to a local optimum. Therefore, the initial conditions are quite critical, as is also discussed in the previous section. Depending on the number of the unknown parameters, one could consider evaluating in a systematic fashion the likelihood function directly at a number of points and use as initial condition the point which results in the largest value of the likelihood function. Improved results can be obtained potentially if the number of the unknown parameters is reduced by parameterizing the unknown functions. For example, separable and nonseparable exponential covariance models are used in [46–48], and an AR model in [38,57] to model the original image, and parameterized blur models are discussed in [38]. We want to mention also that the EM algorithm can be implemented in different domains. For example, it is implemented in both spatial and frequency domains, respectively, in Sections 29.3 and 29.4. Other domains are also possible by applying proper transforms, e.g., the wavelet transform [59].

References

1. Jain, A.K., *Fundamentals of Digital Image Processing*, Prentice Hall, Englewood Cliffs, NJ, 1989.
2. Yang, Y., Galatsanos, N.P., and Katsaggelos, A.K., Regularized image reconstruction to remove blocking artifacts from block discrete cosine transform compressed images, *IEEE Trans. Circuits Syst. Video Technol.*, 3(6): 421–432, December 1993.
3. Yang, Y., Galatsanos, N.P., and Katsaggelos, A.K., Projection-based spatially-adaptive reconstruction of block transform compressed images, *IEEE Trans. Image Process.*, 4(7): 896–908, July 1995.
4. Parker, A.J., *Image Reconstruction in Radiology*, CRC Press, Boca Raton, FL, 1990.
5. Russ, J.C., *The Image Processing Handbook*, CRC Press, Boca Raton, FL, 1992.
6. Snyder, D.L. and Miller, M.I., *Random Processes in Time and Space*, 2nd ed., Springer-Verlag, New York, 1991.
7. Shepp, L. and Vardi, Y., Maximum-likelihood reconstruction for emission tomography, *IEEE Trans. Med. Imaging*, 1: 113–122, October 1982.
8. Katsaggelos, A.K. (Ed.), *Digital Image Restoration*, Springer-Verlag, New York, 1991.
9. Lagendijk, R.L. and Biemond, J., *Iterative Identification and Restoration of Images*, Kluwer Academic Publishers, Boston, MA, 1991.
10. Cox, D.R and Hinkley, D.V., *Theoretical Statistics*, Chapman and Hall, London, UK, 1974.
11. Dempster, A.P., Laird, N.M., and Rubin, D.B., Maximum likelihood from incomplete data via the EM algorithm, *J. R. Soc. Stat., Ser. B*, 39: 1–38, 1977.
12. Hebert, T. and Leahy, R., A generalized EM algorithm for 3-D Bayesian reconstruction from Poisson data using Gibbs priors, *IEEE Trans. Med. Imaging*, 8: 194–202, June 1989.
13. Green, P.J., On use of the EM algorithm for penalized likelihood estimation, *J. R. Soc. Stat., Ser. B*, 52: 443–452, 1990.
14. Zhang, J., The mean field theory in EM procedures for Markov random fields, *IEEE Trans. Acoust. Speech Signal Process.*, 40: 2570–2583, October 1992.
15. Zhang, J., The mean field theory in EM procedures for blind Markov random field image restoration, *IEEE Trans. Image Process.*, 2: 27–40, January 1993.
16. Feder, M. and Weinstein, E., Parameter estimation of superimposed signals using the EM algorithm, *IEEE Trans. Acoust Speech Signal Process*, 36: 477–489, April 1988.
17. Fessler, J.A. and Hero, A.O., Space alternating generalized expectation-maximization algorithm, *IEEE Trans. Signal Process.*, 42: 2664–2678, October 1994.
18. Fessler, J.A. and Hero, A.O., Complete data space and generalized EM algorithm, *Proceedings of International Conference on Acoustics, Speech, and Signal Processing*, Vol. IV, pp. 1–4, Mineappolis, MN, April 27–30, 1993.
19. Hero, A.O and Fessler, J.A., Convergence in norm for alternating expectation-maximization (EM) type algorithms, *Statistica Sin.*, 5: 41–54, January 1995.
20. Wu, J., On the convergence properties of the EM algorithm, *Ann. Stat.*, 11: 95–103, 1983.
21. Redner, R.A. and Walker, H.F., Mixture densities, maximum likelihood and the EM algorithm, *SIAM Rev.*, 26(2): 195–239, 1984.

22. Besag, J., Spatial interaction and the statistical analysis of lattice systems, *J. R. Stat. Soc., Ser. B*, 36: 192–226, 1974.
23. Geman, S. and Geman, D., Stochastic relaxation, Gibbs distribution, and the Bayesian restoration of images, *IEEE Trans. Pattern Anal. Mach. Intell.*, 6: 721–741, November 1984.
24. Chellappa, R. and Jain, A. (Eds.), *Markov Random Fields—Theory and Applications*, Academic Press, New York, 1993.
25. Chandler, D., *Introduction to Modern Statistical Mechanics*, Oxford University Press, New York, 1987.
26. Bouman, C. and Sauer, K., Maximum likelihood scale estimation for a class of Markov random fields, *Proceedings of International Conference on Acoustics, Speech, and Signal Processing*, Vol. 5, pp. 537–540, Adelaide, Australia, April 19-22, 1994.
27. Metropolis, N., Rosenbluth, A.W., Rosenbluth, M.N., Teller, A.H., and Teller, E., Equation of state calculation by fast computing machines, *J. Chem. Phys.*, 21(6): 1087–1092, 1953.
28. Konrad, J. and Dubois, E., Comparison of stochastic and deterministic solution methods in Bayesian estimation of 2D motion, *Image Vis. Comput.*, 8(4): 304–317, November 1990.
29. Cover, T. and Thomas, J., *Elements of Information Theory*, John Wiley & Sons, New York, 1992.
30. Zhang, J., The application of the Gibbs-Bogoliubov-Feynmann inequality in the mean field theory for Markov random fields, Preprint, 1995.
31. Wu, C.-H. and Doerschuk, P.C., Cluster expansions for the deterministic computation of Bayesian estimators based on Markov random fields, *IEEE Trans. Pattern Anal. Mach. Intell.*, 17: 275–293, March 1995.
32. Anderson, B.D.O. and Moore, J. B., *Optimal Filtering*, Prentice-Hall, Englewood Cliffs, NJ, 1979.
33. Banham, M.R. and Katsaggelos, A.K., Digital restoration of images, *IEEE Signal Process. Mag.*, 14(2): 24–41, March 1997.
34. Tekalp, A.M., Kaufman, H., and Woods, J.W., Identification of image and blur parameters for the restoration of non-causal blurs, *IEEE Trans. Acoust. Speech Signal Process*, 34: 963–972, 1986.
35. Andrews, H.C. and Hunt, B.R., *Digital Image Restoration*, Prentice-Hall, Englewood Cliffs, NJ, 1977.
36. Dudgeon, D.E. and Mersereau, R.M., *Multidimensional Digital Signal Processing*, Prentice-Hall, Englewood Cliffs, NJ, 1984.
37. Gray, R.M., On unbounded toeplitz matrices and nonstationary time series with an application to information theory, *Information and Control*, 24: 181–196, 1974.
38. Katsaggelos, A.K. and Lay, K.T., Identification and restoration of images using the expectation maximization algorithm, in *Digital Image Restoration*, Katsaggelos, A.K. (Ed.), Springer Series in Information Sciences No. 23, Chapter 6, Springer-Verlag, Heidelberg, Berlin, Germany, 1991.
39. Lay, K.T. and Katsaggelos, A.K., Image identification and restoration based on the expectation-maximization algorithm, *Opt. Eng.*, 29: 436–445, May 1990.
40. Kailath, T., *Linear Systems*, Prentice-Hall, Englewood Cliffs, NJ, 1980.
41. Lee, J.B., Woodyatt, A.S., and Berman, M., Enhancement of high spectral resolution remote-sensing data by a noise adjusted principle component transform, *IEEE Trans. Geosci. Remote Sens.*, 28(3): 295–304, 1990.
42. Chin, R.T., Yeh, C.L., and Olson, W.S., Restoration of multichannel microwave radiometric images, *IEEE Trans. Pattern Anal. Mach. Intell.*, PAMI-7(4): 475–484, July 1985.
43. Choi, M.G., Galatsanos, N.P., and Katsaggelos, A.K., Multichannel regularized iterative restoration of image sequences, *J. Vis. Commn. Image Representation*, 7(3): 244–258, September 1996.
44. Banham, M.R., Galatsanos, N.P., Gonzalez, H., and Katsaggelos, A.K., Multichannel restoration of single channel images using a wavelet-based subband decomposition, *IEEE Trans. Image Process.*, 3(6): 821–833, November 1994.
45. Tsai, R.Y. and Huang, T.S., Multiframe image restoration and registration, in *Advances in Computer Vision and Registration*, Vol. 1, *Image Reconstruction from Incomplete Observations*, Huang, T.S. (Ed.), Chapter 7, pp. 317–339, JAI Press, Greenwich, CT, 1984.

46. Tom, B.C. and Katsaggelos, A.K., Reconstruction of a high resolution image from multiple degraded mis-registered low resolution images, *Proc. SPIE, Visual Communications and Image Processing*, Chicago, IL, Vol. 2308, Part. 2, pp. 971–981, September 1994.
47. Tom, B.C., Katsaggelos, A.K., and Galatsanos, N.P., Reconstruction of a high resolution from registration and restoration of low resolution images, *IEEE Proceedings of International Conference on Image Processing*, Austin, TX, Vol. 3, pp. 553–557, November 1994.
48. Tom, B.C., Reconstruction of a high resolution image from multiple degraded mis-registered low resolution images, PhD thesis, Department of Electrical Engineering and Computer Science, Northwestern University, Evanston, IL, June 1995.
49. Hunt, B.R. and Kübler, O., Karhunen-Loeve multispectral image restoration, Part I : Theory, *IEEE Trans. Acoust. Speech Signal Process.*, ASSP-32(3): 592–600, June 1984.
50. Galatsanos, N.P. and Chin, R.T., Digital restoration of multichannel images, *IEEE Trans. Acoust. Speech Signal Process.*, ASSP-37(3): 415–421, March 1989.
51. Galatsanos, N.P. and Chin, R.T., Restoration of color images by multichannel Kalman filtering, *IEEE Trans. Signal Process.*, 39(10): 2237–2252, October 1991.
52. Galatsanos, N.P., Katsaggelos, A.K., Chin, R.T., and Hillery, A.D., Least squares restoration of multichannel images, *IEEE Trans. Signal Process.*, 39: 2222–2236, October 1991.
53. Tekalp, A.M. and Pavlovic, G., Multichannel image modeling and Kalman filtering for multispectral image restoration, *IEEE Trans. Signal Process.*, 19(3): 221–232, March 1990.
54. Kang, M.G. and Katsaggelos, A.K., Simultaneous multichannel image restoration and estimation of the regularization parameters, *IEEE Trans. Image Process.*, 6(5) 774–778, May 1997.
55. Zhu, W., Galatsanos, N.P., and Katsaggelos, A.K., Regularized multichannel restoration using cross-validation, *Graphical Models Image Process.*, 57(1): 38–54, January 1995.
56. Katsaggelos, A.K., Lay, K.T., and Galatsanos, N.P., A general framework for frequency domain multichannel signal processing, *IEEE Trans. Image Process.*, 2(3): 417–420, July 1993.
57. Lay, K.T., Blur identification and image restoration using the EM algorithm, PhD thesis, Department of Electrical Engineering and Computer Science, Northwestern University, Evanston, IL, December 1991.
58. Tom, B.C.S., Lay, K.T., and Katsaggelos, A.K., Multi-channel image identification and restoration using the expectation-maximization algorithm, *Opt. Eng.*, Special Issue on Visual Communications and Image Processing, 35(1): 241–254, January 1996.
59. Banham, M.R., Wavelet based image restoration techniques, PhD thesis, Department of Electrical Engineering and Computer Science, Northwestern University, Evanston, IL, June 1994.

30

Inverse Problems in Array Processing

Kevin R. Farrell
T-NETIX, Inc.

30.1 Introduction

Signal reception has numerous applications in communications, radar, sonar, and geoscience among others. However, the adverse effects of noise in these applications limit their utility. Hence, the quest for new and improved noise removal techniques is an ongoing research topic of great importance in a vast number of applications of signal reception.

When certain characteristics of noise are known, their effects can be compensated. For example, if the noise is known to have certain spectral characteristics, then a finite impulse response (FIR) or infinite impulse response filter can be designed to suppress the noise frequencies. Similarly, if the statistics of the noise are known, then a Weiner filter can be used to alleviate its effects. Finally, if the noise is spatially separated from the desired signal, then multisensor arrays can be used for noise suppression. This last case is discussed in this article.

A multisensor array consists of a set of transducers, i.e., antennas, microphones, hydrophones, seismometers, geophones, etc. that are arranged in a pattern which can take advantage of the spatial location of signals. A two-element television antenna provides a good example. To improve signal reception and/or mitigate the effects of a noise source, the antenna pattern is manually adjusted to steer a low gain component of the antenna pattern toward the noise source. Multisensor arrays typically achieve this adjustment through the use of an array processing algorithm. Most applications of multisensor arrays involve a fixed pattern of transducers, such as a linear array. Antenna pattern adjustments are made by applying weights to the outputs of each transducer. If the noise arrives from a specific non-changing spatial location, then the weights will be fixed. Otherwise, if the noise arrives from random, changing locations then the weights must be adaptive. So, in a military communications application where a communications channel is subject to jamming from random spatial locations, an adaptive array processing algorithm would be the appropriate solution. Commercial applications of microphone arrays include teleconferencing [6] and hearing aids [9].

There are several methods for obtaining the weight update equations in array processing. Most of these are derived from statistically based formulations. The resulting optimal weight vector is then generally expressed in terms of the input autocorrelation matrix. An alternative formulation is to express the array processing problem as a linear system of equations to which iterative matrix inversion techniques can be applied. The matrix inverse formulation will be the focus of this article.

The following section provides a background overview of wave propagation, spatial sampling, and spatial filtering. Next, narrowband and broadband beamforming arrays are described along with the standard algorithms used for these implementations. The narrowband and broadband algorithms are then reformulated in terms of an inverse problem and an iterative technique for solving this system of equations is provided. Finally, several examples are given along with a summary.

30.2 Background Theory

Array processing uses information regarding the spatial locations of signals to aid in interference suppression and signal enhancement. The spatial locations of signals may be determined by the wavefronts that are emanated by the signal sources. Some background theory regarding wave propagation and spatial frequency is necessary to fully understand the interference suppression techniques used within array processing. The following subsections provide this background material.

30.2.1 Wave Propagation

An adaptive array consists of a number of sensors typically configured in a linear pattern that utilizes the spatial characteristics of signals to improve the reception of a desired signal and/or cancellation of undesired signals. The analysis used in this chapter assumes that a linear array is being used, which corresponds to the sensors being configured along a line. Signals may be spatially characterized by their angle of arrival with respect to the array. The angle of arrival of a signal is defined as the angle between the propagation path of the signal and the perpendicular of the array. Consider the wavefront emanating from a point source as is illustrated in Figure 30.1. Here, the angle of arrival is shown as θ.

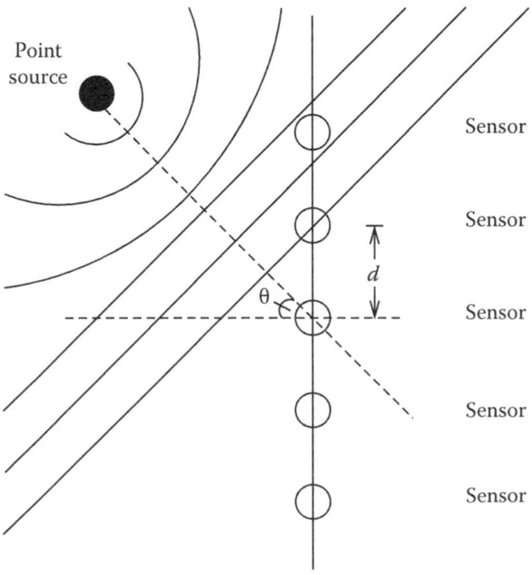

FIGURE 30.1 Propagating wavefront.

Note in Figure 30.1 that wavefronts emanating from a point source may be characterized by plane waves (i.e., the locus of constant phase form straight lines) when originating from the far field or Fraunhofer, region. The far field approximation is valid for signals that satisfy the following condition:

$$s \geq \frac{D^2}{\lambda}, \tag{30.1}$$

where s is the distance between the signal and the array, λ is the wavelength of the signal, and D is the length of the array. Wavefronts that originate closer than D^2/λ are considered to be from the near field or Fresnel, region. Wavefronts originating from the near field exhibit a convex shape when striking the array sensors. These wavefronts do not create linear phase shifts between consecutive sensors. However, the curvature of the wavefront allows algorithms to determine point source location in addition to direction of arrival [1]. The remainder of this article assumes that all wavefronts arrive from the far field region.

30.2.2 Spatial Sampling

In Figure 30.1 it can be seen that the signal waveform experiences a time delay between crossing each sensor, assuming that it does not arrive perpendicular to the array. The time delay, τ, of the waveform striking the first and then second sensors in Figure 30.1 may be calculated as

$$\tau = \frac{d}{c}\sin\theta, \tag{30.2}$$

where

d is the sensor spacing

c is the speed of propagation of the given waveform for a particular medium (i.e., 3×10^8 m/s for electromagnetic waves through air, 1.5×10^3 m/s for sound waves through water, etc.)

θ is the angle of arrival of the wavefront

This time delay corresponds to a shift in phase of the signal as observed by each sensor. The phase shift, ϕ, or electrical angle observed at each sensor due to the angle of arrival of the wavefront may be found as

$$\phi = \frac{2\pi d}{\lambda_o}\sin\theta = \frac{\omega_o d}{c}\sin\theta. \tag{30.3}$$

Here, λ_o is the wavelength of the signal at frequency f_o as defined by

$$\lambda_o = \frac{c}{f_o}. \tag{30.4}$$

Hence, a signal $x(k)$ that crosses the sensor array and exhibits a phase shift ϕ between uniformly spaced, consecutive sensors can be characterized by the vector $\mathbf{x}(k)$, where

$$\mathbf{x}(k) = x(k)\begin{bmatrix} 1 \\ e^{-j\phi} \\ e^{-2j\phi} \\ \vdots \\ e^{-j(K-1)\phi} \end{bmatrix}. \tag{30.5}$$

Uniform sensor spacing is assumed throughout the remainder of this article.

30.2.3 Spatial Frequency

The angle of arrival of a wavefront defines a quantity known as the spatial frequency. Adaptive arrays use information regarding the spatial frequency to suppress undesired signals that originate from different locations than that of the target signal. The spatial frequency is determined from the periodicity that is observed across an array of sensors due to the phase shift of a signal arriving at some angle of arrival.

Signals that arrive perpendicular to the array (known as boresight) create identical waveforms at each sensor. The spatial frequency of such signals is zero. Signals that do not arrive perpendicular to the array will not create waveforms that are identical at each sensor assuming that there is no spatial aliasing due to insufficiently spaced sensors. In general, as the angle increases, so does the spatial frequency. It can also be deduced that retaining signals having an angle of arrival equal to zero degrees while suppressing signals from other directions is equivalent to low pass filtering the spatial frequency. This provides the motivation for conventional or fixed-weight beamforming techniques. Here, the sensor values can be computed via a windowing technique, such as a rectangular, Hamming, etc. to yield a fixed suppression of non-boresight signals. However, adaptive techniques can locate the specific spatial frequency of an interfering signal and position a null in that exact location to achieve greater suppression.

There are two types of beamforming, namely conventional, or "fixed weight," beamforming and adaptive beamforming. A conventional beamformer can be designed using windowing and FIR filter theory. They utilize fixed weights and are appropriate in applications where the spatial locations of noise sources are known and are not changing. Adaptive beamformers make no such assumptions regarding the locations of the signal sources. The weights are adapted to accommodate the changing signal environment.

Arrays that have a visible region of $-90°$ to $+90°$ (i.e., the azimuth range for signal reception) require that the sensor spacing satisfy the relation

$$d \leq \frac{\lambda}{2}. \tag{30.6}$$

The above relation for sensor spacing is analogous to the Nyquist sampling rate for frequency domain analysis. For example, consider a signal that exhibits exactly one period between consecutive sensors. In this case, the output of each sensor would be equivalent, giving the false impression that the signal arrives normal to the array. In terms of the antenna pattern, insufficient sensor spacing results in grating lobes. Grating lobes are lobes other than the main lobe that appear in the visible region and can amplify undesired directional signals.

The spatial frequency characteristics of signals enable numerous enhancement opportunities via array processing algorithms. Array processing algorithms are typically realized through the implementation of narrowband or broadband arrays. These two arrays are discussed in the following sections.

30.3 Narrowband Arrays

Narrowband adaptive arrays are used in applications where signals can be characterized by a single frequency and thus occupy a relatively narrow bandwidth. A signal whose envelope does not change during the time their wavefront is incident on the transducers is considered to be narrowband. A narrowband adaptive array consists of an array of sensors followed by a set of adjustable gains, or weights. The outputs of the weighted sensors are summed to produce the array output. A narrowband array is shown in Figure 30.2.

The input vector $\mathbf{x}(k)$ consists of the sum of the desired signal $\mathbf{s}(k)$ and noise $\mathbf{n}(k)$ vectors and is defined as

$$\mathbf{x}(k) = \mathbf{s}(k) + \mathbf{n}(k), \tag{30.7}$$

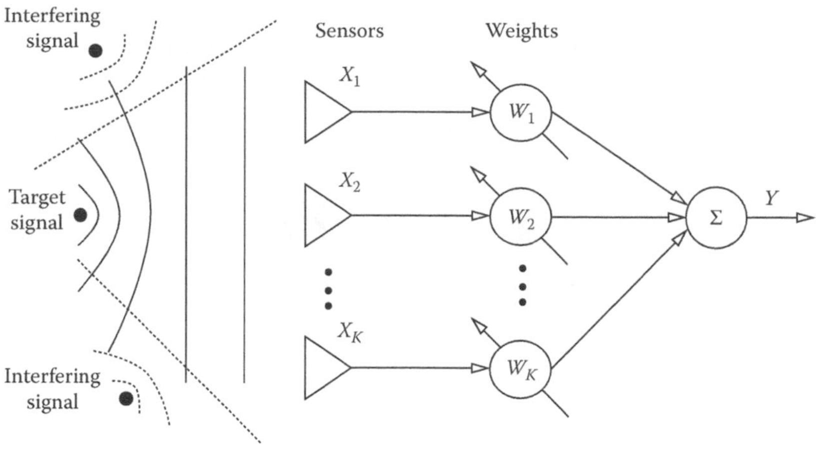

FIGURE 30.2 Narrowband array.

where k denotes the time instant of the input vector. The noise vector $\mathbf{n}(k)$ will generally consist of thermal noise and directional interference. At each time instant, the input vector is multiplied with the weight vector to obtain the array output, which is given as

$$y(k) = \mathbf{x}^{\mathrm{T}}(k)\mathbf{w}, \ \mathbf{x}, \mathbf{w} \in C^K, \tag{30.8}$$

where C^K is the complex space of dimension K. The array output is then passed to the signal processor which uses the previous value of the output and current values of the inputs to determine the adjustment to make to the weights. The weights are then adjusted and multiplied with the new input vector to obtain the next output. The output feedback loop allows the weights to be adjusted adaptively, thus accommodating nonstationary environments.

In Equation 30.8, it is desired to find a weight vector that will allow the output y to approximately equal the true target signal. For the derivation of the weight update equations, it is necessary to know what *a priori* information is being assumed. One form of *a priori* information could be the spatial location of the target signal, also known as the "look-direction." For example, many array processing algorithms assume that the target signal arrives normal to the array, or else a steering vector is used to make it appear as such. Another form of *a priori* information is to use a signal at the receiving end that is correlated with the input signal, i.e., a pilot signal. Each of these criteria will be considered in the following subsections.

30.3.1 Look-Direction Constraint

One of the first narrowband array algorithms was proposed by Applebaum [2]. This algorithm is known as the sidelobe canceler and assumes that the direction of the target signal is known. The algorithm does not attempt to maximize the signal gain, but instead adjusts the sidelobes so that interfering signals coincide with the nulls of the antenna pattern. This concept is illustrated in Figure 30.3.

Applebaum derived the weight update equation via maximization of the signal to interference plus thermal noise ratio (SINR). As derived in [2], this optimization results in the optimal weight vector as given by Equation 30.9:

$$\mathbf{w}_{\mathrm{opt}} = \mu R_{xx}^{-1}\mathbf{t}, \tag{30.9}$$

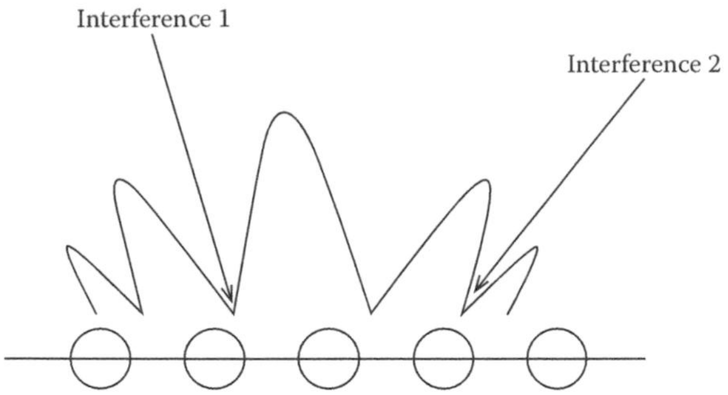

FIGURE 30.3 Sidelobe canceling.

where

R_{xx} is the covariance matrix of the input
μ is a constant related to the signal gain
t is a steering vector that corresponds to the angle of arrival of the desired signal

This steering vector is equivalent to the phase shift vector of Equation 30.5. Note that if the angle of arrival of the desired signal is zero, then the **t** vector will simply contain ones.

A discretized implementation of the Applebaum algorithm appears as follows:

$$\mathbf{w}^{(j+1)} = \mathbf{w}^{(j)} + \alpha\left(\mathbf{w}_{\mathrm{q}} - \mathbf{w}^{(j)}\right) - \beta\mathbf{x}(k)y(k), \tag{30.10}$$

where

$\mathbf{w}_{\mathrm{q}}$ represents the quiescent weight vector (i.e., when no interference is present)
the superscript j refers to the iteration
α is a gain parameter for the steering vector
β is a gain parameter controlling the adaptation rate and variance about the steady state solution

30.3.2 Pilot Signal Constraint

Another form of *a priori* information is to use a pilot signal that is correlated with the target signal. This results in a beamforming algorithm that will concentrate on maintaining a beam directed toward the target signal, as opposed to, or in addition to, positioning the nulls as in the case of the sidelobe canceler. One such adaptive beamforming algorithm was proposed by Widrow [20,21]. The resulting weight update equation is based on minimizing the quantity $[y(k) - p(k)]^2$ where $p(k)$ is the pilot signal. The resulting weight update equation is

$$\mathbf{w}^{(j+1)} = \mathbf{w}^{(j)} + \mu\varepsilon(k)\mathbf{x}(k). \tag{30.11}$$

This corresponds to the least means square (LMS) algorithm, where ε is the current error, namely $[y(k) - p(k)]$, and μ is a scaling factor.

30.4 Broadband Arrays

Narrowband arrays rely on the assumption that wavefronts normal to the array will create identical waveforms at each sensor and wavefronts arriving at angles not normal to the array will create a linear phase shift at each sensor. Signals that occupy a large bandwidth and do not arrive normal to the array violate this assumption since the phase shift is a function of f_o and varying frequency will cause a varying phase shift. Broadband signals that arrive normal to the array will not be subject to frequency dependent phase shifts at each sensor as will broadband signals that do not arrive normal to the array. This is attributed to the coherent summation of the target signal at each sensor where the phase shift will be a uniform random variable with zero mean. A modified array structure, however, is necessary to compensate the interference waveform inconsistencies that are caused by variations about the center frequency. This can be achieved by having the weight for a sensor being a function of frequency, i.e., a FIR filter, instead of just being a scalar constant as in the narrowband case. Broadband adaptive arrays consist of an array of sensors followed by tapped delay lines, which is the major implementation difference between a broadband and narrowband array. A broadband array is shown in Figure 30.4.

Consider the transfer functions for a given sensor of the narrowband and broadband arrays, shown by

$$H_{\text{narrow}}(w) = w_1 \tag{30.12}$$

and

$$H_{\text{broad}}(w) = w_1 + w_2 e^{-jwT} + w_3 e^{-2jwT} + \cdots + w_J e^{-j(J-1)wT}. \tag{30.13}$$

The narrowband transfer function has only a single weight that is constant with frequency. However, the broadband transfer function, which is actually a Fourier series expansion, is frequency dependent and allows for choosing a weight vector that may compensate phase variations due to signal bandwidth. This property of tapped delay lines provides the necessary flexibility for processing broadband signals. Note that typically four or five taps will be sufficient to compensate most bandwidth variances [14].

The broadband array shown in Figure 30.4 obtains values at each sensor and then propagates these values through the array at each time interval. Therefore, if the values x_1 through x_K are input at time instant one, then at time instant two, x_{K+1} through x_{2K} will have the values previously held by x_1 through x_K, x_{2K+1} through x_{3K} will have the values previously held by x_{K+1} through x_{2K}, etc. Also, at each time

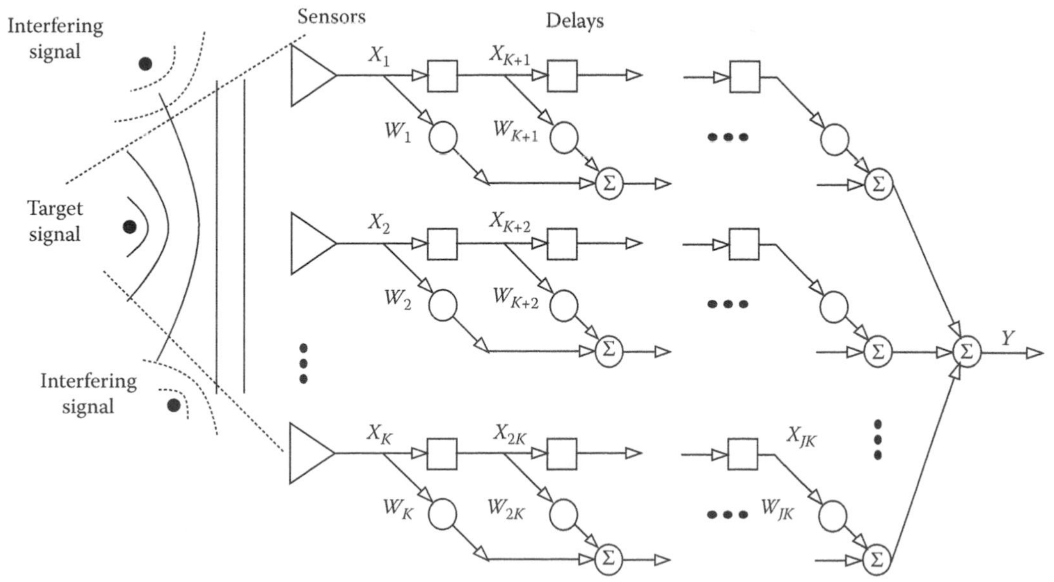

FIGURE 30.4 Broadband array.

instant, a scalar value y will be calculated as the inner product of the input vector $\mathbf{x}$ and the weight vector $\mathbf{w}$. This array output is calculated as

$$y(k) = \mathbf{x}^{\mathrm{T}}(k)\mathbf{w}, \mathbf{x}, \mathbf{w} \in C^{JK}, \tag{30.14}$$

where C^{JK} is the complex space of dimension JK.

Although not shown in Figure 30.4, a signal processor exists as in the narrowband array, which uses the previous output and current inputs to determine the adjustments to make to the weight vector $\mathbf{w}$. The output signal y will approach the value of the desired signal as the interfering signals are canceled until it converges to the desired signal in the least squares sense.

Broadband arrays have been analyzed by Widrow [21], Griffiths [10,12], and Frost [7]. Widrow [21] proposed a LMS algorithm that minimizes the square of the difference between the observed output and the expected output, which was estimated with a pilot signal. This approach assumes that the angle of arrival and a pilot signal are available *a priori*. Griffiths [10] proposed a LMS algorithm that assumes knowledge of the cross-correlation matrix between the input and output data instead of the pilot signal. This method assumes that the angle of arrival and second order signal statistics are known *a priori*. The methods proposed by Widrow and Griffiths are forms of unconstrained optimization. Frost [7] proposed a LMS algorithm that assumes *a priori* knowledge of the angle of arrival and the frequency band of interest. The Frost algorithm utilizes a constrained optimization technique, which Griffiths later derived an unconstrained formulation that utilizes the same constraints [12]. The Frost algorithm will be the focus of this section.

The Frost algorithm implements the look-direction and frequency response constraints as follows. For the broadband array shown in Figure 30.4, a target signal waveform propagating normal to the array, or steered to appear as such, will create identical waveforms at each sensor. Since the taps in each column, i.e., w_1 through w_K, see the same signal, this array may be collapsed to a single sensor FIR filter. Hence, to constrain the frequency range of the target signal, one just has to constrain the sum of the taps for each column to be equal to the corresponding tap in a FIR filter having J taps and the desired frequency response for the target signal.

These look-direction and frequency response constraints can be implemented by the following optimization problem:

$$\text{minimize:} \quad \mathbf{w}^{\mathrm{T}}\mathbf{R}_{xx}\mathbf{w} \tag{30.15}$$

$$\text{subject to:} \quad \mathbf{C}^{\mathrm{T}}\mathbf{w} = \mathbf{h}, \tag{30.16}$$

where

$\mathbf{R}_{xx}$ is the covariance matrix of the received signals
$\mathbf{h}$ is the vector of FIR filter coefficients defining the desired frequency response
$\mathbf{C}^{\mathrm{T}}$ is the constraint matrix given by

$$\mathbf{C}^{\mathrm{T}} = \begin{bmatrix} 11 & \cdots & 1 & 00 & \cdots & 0 & \cdots & 00 & \cdots & 0 \\ 00 & \cdots & 0 & 11 & \cdots & 1 & \cdots & 00 & \cdots & 0 \\ \vdots & & & & & & & & & \\ 00 & \cdots & 0 & 00 & \cdots & 0 & \cdots & 11 & \cdots & 1 \end{bmatrix}.$$

The number of rows in $\mathbf{C}^{\mathrm{T}}$ is equal to the number of taps of the array and the number of ones in each row is equal to the number of sensors. The optimal weight vector $\mathbf{w}_{\mathrm{opt}}$ will minimize the output power of the noise sources subject to the constraint that the sum of each column vector of weights is equal to a coefficient of a FIR filter defining the desired impulse response of the array.

The Frost algorithm [7] is a constrained LMS method derived by solving Equations 30.15 and 30.16 via Lagrange Multipliers to obtain an expression for the optimum weight vector, Frost [7] derived the constrained LMS algorithm for broadband array processing using Lagrange multipliers. The function to be minimized may be defined as

$$H(\mathbf{w}) = \frac{1}{2}\mathbf{w}^T\mathbf{R}_{xx}\mathbf{w} + \lambda^T(\mathbf{C}^T\mathbf{w} - \mathbf{h}), \tag{30.17}$$

where

λ is a Lagrange multiplier

F is a vector representative of the desired frequency response

Minimizing the function $H(\mathbf{w})$ with respect to $\mathbf{w}$ will obtain the following optimal weight vector:

$$\mathbf{w}_{\text{opt}} = \mathbf{R}_{xx}^{-1}\mathbf{C}(\mathbf{C}^T\mathbf{R}_{xx}^{-1}\mathbf{C})^{-1}\mathbf{h}. \tag{30.18}$$

An iterative implementation of this algorithm was implemented via the following equations:

$$\mathbf{w}^{(j+1)} - \mathbf{P}\left[\mathbf{w}^{(j)} - \mu\mathbf{R}_{xx}\mathbf{w}^{(j)}\right] + \mathbf{C}(\mathbf{C}^T\mathbf{C})^{-1}\mathbf{h}, \tag{30.19}$$

where μ is a step size parameter and

$$\begin{aligned} \mathbf{P} &= \mathbf{I} - \mathbf{C}(\mathbf{C}^T\mathbf{C})^{-1}\mathbf{C}^T \\ \mathbf{w}(0) &= \mathbf{C}(\mathbf{C}^T\mathbf{C})^{-1}\mathbf{h}, \end{aligned}$$

where $\mathbf{I}$ is the identity matrix and

$$\mathbf{h} = [\,h_1 \quad h_2 \quad \cdots \quad h_j\,].$$

30.5 Inverse Formulations for Array Processing

The array processing algorithms discussed thus far have all been derived through statistical analysis and/or adaptive filtering techniques. An alternative approach is to view the constraints as equations that can be expressed in a matrix-vector format. This allows for a simple formulation of array processing algorithms to which additional constraints can be easily incorporated. Additionally, this formulation allows for efficient iterative matrix inversion techniques that can be used to adapt the weights in real time.

30.5.1 Narrowband Arrays

Two algorithms were discussed for narrowband arrays, namely, the sidelobe canceler and pilot signal algorithms. We will consider the sidelobe canceler algorithm here. The derivation of the sidelobe canceler is based on the optimization of the SINR and yields an expression for the optimum weight vector as a function of the input autocorrelation matrix. We will use the same constraints as the sidelobe canceler to yield a set of linear equations that can be put in a matrix vector format.

Consider the narrowband array description provided in Section 30.3. In Equation 30.7, $\mathbf{s}(k)$ is the vector representing the desired signal whose wavefront is normal to the array and $\mathbf{n}(k)$ is the sum of the interfering signals arriving from different directions. A weight vector is desired that will allow the signal vector $\mathbf{s}(k)$ to pass through the array undistorted while nulling any contribution of the noise vector $\mathbf{n}(k)$. An optimal weight vector $\mathbf{w}_{\text{opt}}$ that satisfies these conditions is represented by

$$\mathbf{w}_{\text{opt}}^{\text{T}}\mathbf{s}(k) = s(k) \tag{30.20}$$

and

$$\mathbf{w}_{\text{opt}}^{\text{T}}\mathbf{n}(k) = 0, \tag{30.21}$$

where $s(k)$ is the scalar value of the desired signal. Since the sidelobe canceler does not have access to $s(k)$, an alternative approach must be taken to implement the condition of Equation 30.20. One method for finding this constraint is to minimize the expectation of the output power [7]. This expectation can be approximated by the quantity y^2, where $y = \mathbf{x}^{\text{T}}(k)\mathbf{w}$. Minimizing y^2 subject to the look-direction constraint will tend to cancel the noise vector while maintaining the signal vector. This criteria can be represented by the linear equation:

$$\mathbf{x}^{\text{T}}(k)\mathbf{w} = 0. \tag{30.22}$$

Note that Equation 30.22 implies that the weight vector be orthogonal to the composite input vector as opposed to just the noise component. However, the look-direction constraint imposed by the following equation will maintain the desired signal

$$[1 \quad 1 \quad \cdots \quad 1]\mathbf{w} = 1. \tag{30.23}$$

This equation satisfies the look-direction constraint that a signal arriving perpendicular to the array will have unity gain in the output.

The constraints imposed by Equations 30.22 and 30.23 can be expressed in a matrix-vector form as follows:

$$\begin{bmatrix} x_1(k) & x_2(k) & \cdots & x_K(k) \\ 1 & 1 & \cdots & 1 \end{bmatrix}\mathbf{w} = \begin{bmatrix} 0 \\ 1 \end{bmatrix} \tag{30.24}$$

or, equivalently,

$$\mathbf{A}\mathbf{w} = \mathbf{b}.$$

30.5.2 Broadband Arrays

The broadband array considered in this section will utilize the constraints considered by Frost [7], namely the look-direction and frequency range of the target signal. The linear equations that represent the Frost algorithm are similar to those used for the narrowband formulation derived in the previous section. Once again, the minimization of the cost function in Equation 30.15 can be achieved by Equation 30.22, assuming that the target signal arrives normal to the array. The constraint for the desired frequency response in the look direction can also be implemented in a similar fashion to that of the narrowband array in Equation 30.23. Instead of constraining the sum of the weights to be one, as in the narrowband array, the broadband array implementation will constrain the sum of each column of weights to be equal to a corresponding tap value in a FIR filter with the desired frequency response for the target signal.

Hence, the broadband array problem represented by Equations 30.15 and 30.16 can be expressed as a linear system of equations by creating a matrix that has the cost function given by Equation 30.15 augmented with the linear constraint equations given by Equation 30.16. The problem can now be expressed as

$$\begin{bmatrix} x_1 & \cdots & x_K & \cdots & x_{(J-1)K+1} & \cdots & x_{JK} \\ 1 & \cdots & 1 & \cdots & 0 & \cdots & 0 \\ 0 & \cdots & 0 & \cdots & 0 & \cdots & 0 \\ \vdots & & & & & & \\ 0 & \cdots & 0 & \cdots & 1 & \cdots & 1 \end{bmatrix} \begin{bmatrix} w_1 \\ w_2 \\ \vdots \\ w_{JK} \end{bmatrix} = \begin{bmatrix} 0 \\ h_1 \\ \vdots \\ h_J \end{bmatrix},$$

or

$$\mathbf{Aw} = \mathbf{h}', \tag{30.25}$$

where $\mathbf{h}'$ is the vector of FIR filter coefficients augmented with a zero.

30.5.3 Row-Action Projection Method

The matrix-vector formulation for the narrowband beamforming problem, as represented in Equation 30.24 or the broadband array problem formulated in Equation 30.25 can now be expressed as an inverse problem. For example, if $\mathbf{A}$ is $\mathbf{n} \times \mathbf{n}$ and $\text{rank}[\mathbf{A}|\mathbf{b}] = \text{rank}[\mathbf{A}]$, then a unique solution for $\mathbf{w}$ can be found as

$$\mathbf{w} = \mathbf{A}^{-1}\mathbf{b}. \tag{30.26}$$

If instead, $\mathbf{A}$ is $\mathbf{m} \times \mathbf{n}$, then a least squares solution can be implemented as

$$\mathbf{w} = (\mathbf{A}^{\mathrm{T}}\mathbf{A})^{-1}\mathbf{A}^{\mathrm{T}}\mathbf{b}. \tag{30.27}$$

Another solution can be obtained by using the Moore–Penrose generalized inverse, or pseudo-inverse, of $\mathbf{A}$ via

$$\mathbf{w}^{\dagger} = \mathbf{A}^{\dagger}\mathbf{b}, \tag{30.28}$$

where $\mathbf{A}^{\dagger}$ and $\mathbf{w}^{\dagger}$ represent the pseudo-inverse of $\mathbf{A}$ and the pseudo-inverse solution for $\mathbf{w}$, respectively.

These methods all provide an immediate solution for the weight vector, $\mathbf{w}$, however, at the expense of requiring a matrix inversion along with any instabilities that may be apparent if the matrix is ill-conditioned. A more convenient approach to solve for the weights is to use an iterative approach. The method that we shall use here is known as the row-action projection (RAP) algorithm. The RAP algorithm is an iterative technique for solving a system of linear equations. The RAP method has found numerous applications in digital signal processing [16] and is applied here to adaptive beamforming.

The RAP method for iteratively solving the system in Equation 30.24 is given by the update equation:

$$\mathbf{w}^{(j+1)} = \mathbf{w}^{(j)} + \mu \frac{\varepsilon_i}{\|\mathbf{a}_i\|} \frac{\mathbf{a}_i^{\mathrm{T}}}{\|\mathbf{a}_i\|}, \tag{30.29}$$

where ε_i is the error term for the ith row defined as:

$$\varepsilon_i = b_i - \mathbf{a}_i\mathbf{w}^{(k)}. \tag{30.30}$$

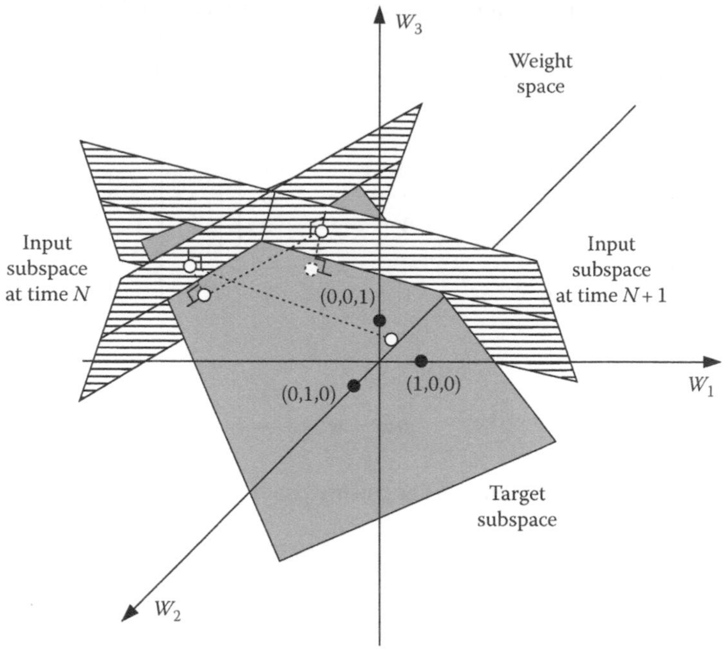

FIGURE 30.5 Orthogonal projections in weight space.

In Equations 30.29 and 30.30, the superscript j denotes the iteration, the subscript i refers to the row number of the matrix or vector, and μ is a gain parameter, which is known to be stable for values between zero and two. The choice of μ is important for performance characteristics and has the tradeoff that a large μ will provide faster convergence, while a small μ will provide greater accuracy. Also, note that choosing μ between one and two may, in some instances, prevent convergence to the LMS solution.

The RAP method operates by creating orthogonal projections in the space defined by the data matrix **A** in Equation 30.24. A graphical representation of the RAP algorithm, as applied to a three sensor beamforming array, is illustrated in Figure 30.5.

In Figure 30.5, the target signal subspace consists of the plane represented by the look-direction constraint, namely $w_1 + w_2 + w_3 = 1$. The input signal subspace, given by $w_1x_1(k) + w_2x_2(k) + w_3x_3(k) = 0$, will consist of a different plane for each discrete time index k. The RAP method first creates an orthogonal projection to the input subspace (i.e., satisfying $\mathbf{w}^T\mathbf{x}(k) = 0$). A projection is then made to the target signal subspace. This procedure will be repeated for the next input subspace, etc. Intuitively, this procedure will find a solution as "orthogonal as possible" to the different input subspaces, which lies in the target signal subspace. Since the RAP method consists of only row operations, it is convenient for parallel implementations. This technique, described by Equations 30.24, 30.29, and 30.30, comprises the RAP method for array processing.

30.6 Simulation Results

Several simulations were performed to compare the inverse formulation of the array processing problem to the more traditional adaptive filtering approaches. These simulations compare the inverse formulation to the sidelobe canceler implementation of the narrowband array and to the Frost implementation of the broadband array.

TABLE 30.1 Input Scenario for Narrowband Experiment

Signal	Angle (degree)	Frequency (KHz)
Target signal	0	2.0
Interference 1	28	3.0
Interference 2	41	1.0
Interference 3	72	4.0

30.6.1 Narrowband Results

The sidelobe canceler application is evaluated with both the Applebaum algorithm and the inverse formulation. Both arrays are simulated for a nine-sensor narrowband array. The RAP algorithm for the inverse formulation uses a gain value of $\mu = 0.001$ and the Applebaum array uses values of $\alpha = 0.25$ and $\beta = 0.01$. The signal environment for the scenario consists of unit amplitude tones whose spectral and spatial characteristics are summarized by Table 30.1. The input spectrum of the narrowband scenario is shown in Figure 30.6. The input and output spectrums for the inverse formulation and Applebaum algorithm are shown in Figures 30.6 through 30.8. The inverse formulation and Applebaum algorithms demonstrate similar performance for this example.

30.6.2 Broadband Results

The broadband array application is also evaluated with both the inverse formulation and Frost algorithm. The algorithms are both evaluated for a broadband array that consists of nine sensors, each followed by five taps. The signal environment used for the scenario consists of several signals of varying spectral and spatial characteristics as summarized by Table 30.2.

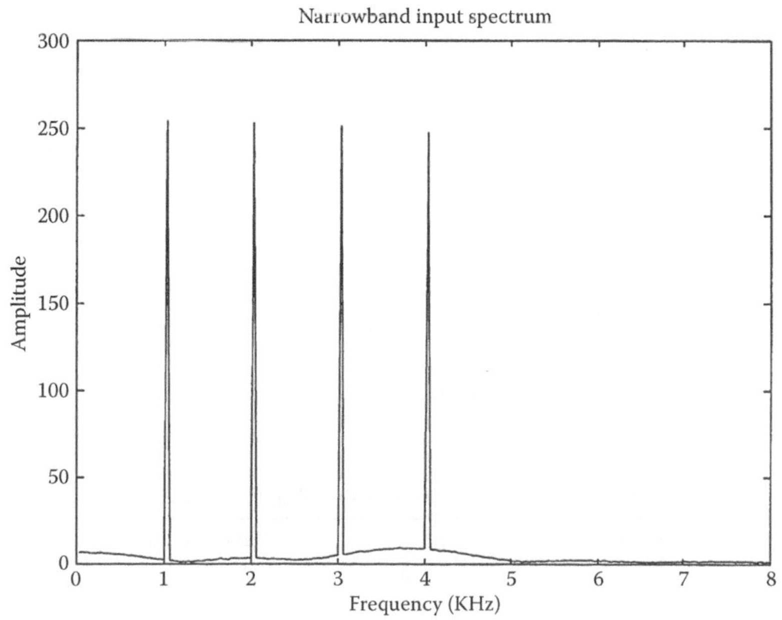

FIGURE 30.6 Narrowband input spectrum.

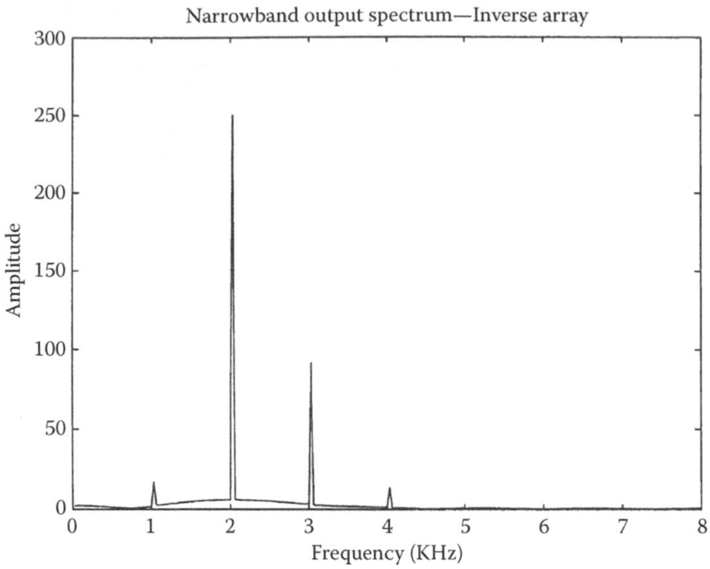

FIGURE 30.7 Output spectrum for inverse formulation.

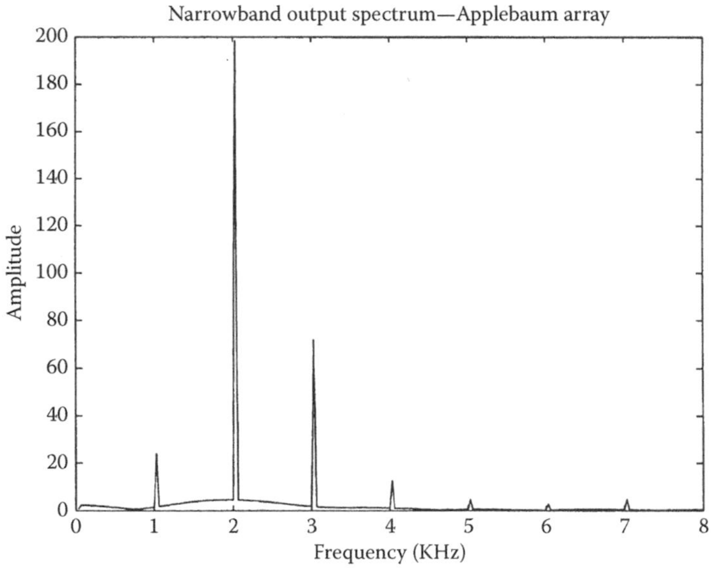

FIGURE 30.8 Output spectrum for Applebaum array.

TABLE 30.2 Input Scenario for Broadband Experiment

Signal	Angle (degree)	Frequency (KHz)
Target signal	0	3.0
Interference 1	27	1.5
Interference 2	41	4.0

The RAP algorithm used for the inverse has a gain value $\mu = 0.5$ and the Frost algorithm uses the gain value $\mu = 0.05$. The **h** vector specifies a low pass frequency response with a passband up to 4 KHz. The input and output signal spectrums are shown in Figures 30.9 through 30.11. The inverse formulation and Frost algorithms again demonstrate similar performance.

The broadband array processing algorithms are also evaluated for a microphone array application [5]. The simulation uses a microphone array with nine equispaced transducers each followed by 13 taps. The microphone spacing is chosen as 4.3 cm and the sampling rate for the speech signals is 16 KHz.

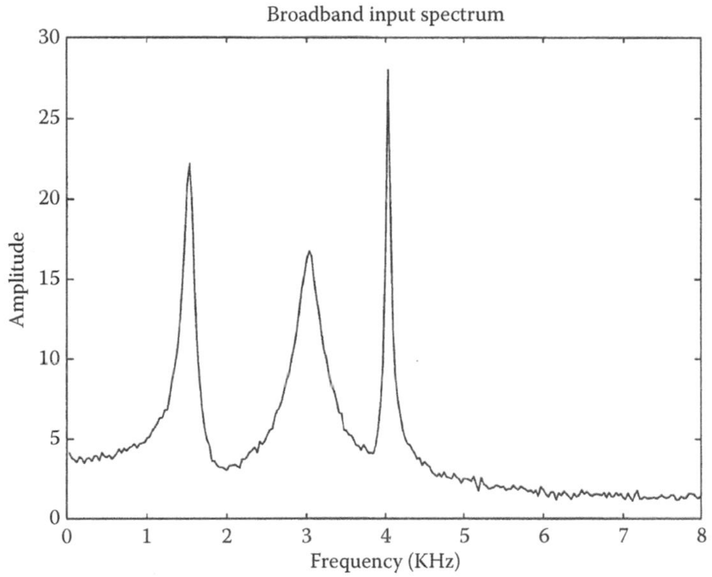

FIGURE 30.9 Broadband input spectrum.

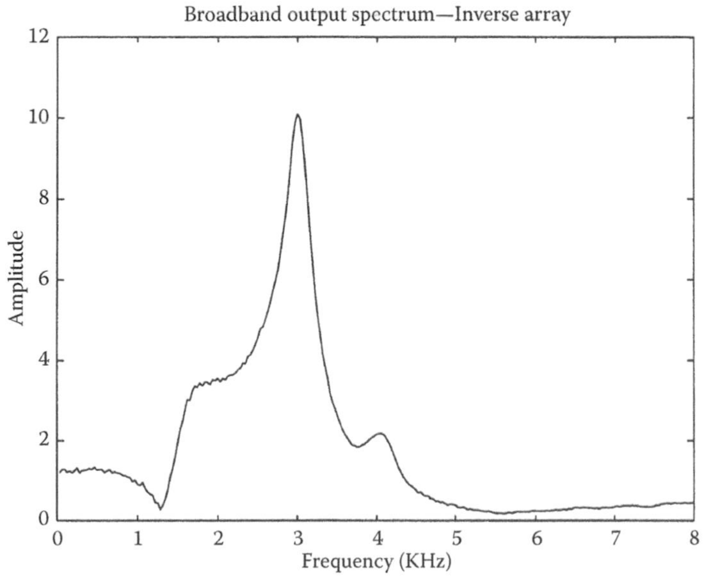

FIGURE 30.10 Output spectrum for inverse array.

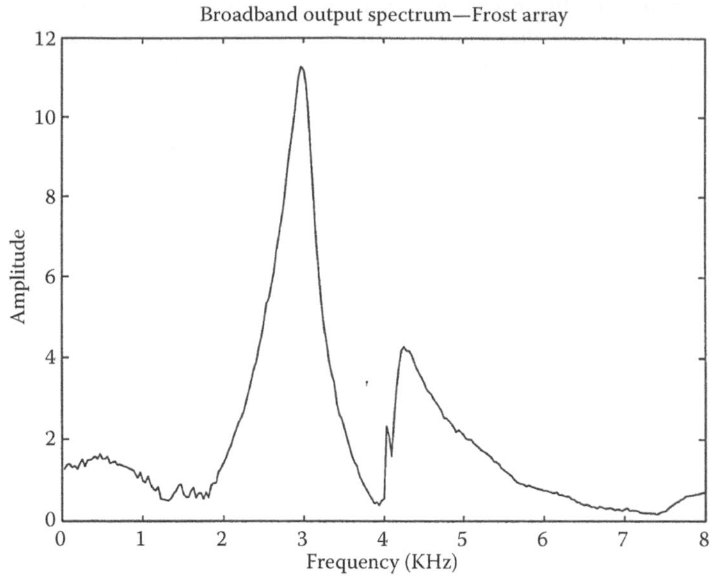

FIGURE 30.11 Output spectrum for Frost array.

The **h** vector contains coefficients for a low pass FIR filter designed with a Hamming window for a passband of 0–4 KHz. The signal environment consists of two speech signals. The target signal arrives normal to the array. The interfering signal is applied to the array at uniformly spaced angles ranging from $-90°$ to $+90°$ in unit increments. The interference power is 2.6 dB greater than the desired signal. The resulting interference suppression observed in the array output is illustrated in Figure 30.12. The maximum interference suppression (i.e., for interference arriving at $\pm 90°$) is 11.0 dB for the RAP method and 11.2 dB for the Frost method.

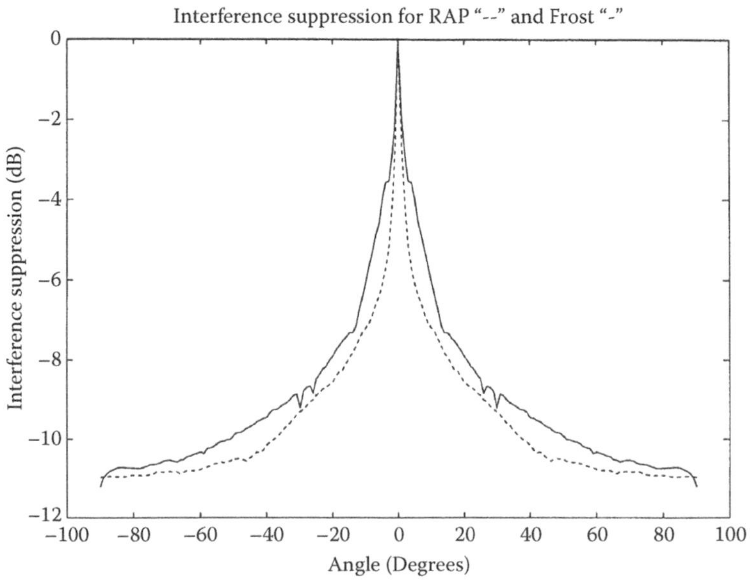

FIGURE 30.12 Interference suppression.

30.7 Summary

This chapter has formulated the array processing problem as an inverse problem. Inverse formulations for both narrowband and broadband arrays were discussed. Specifically, the sidelobe canceler algorithm for narrowband array processing and Frost algorithm for broadband array processing were analyzed. The inverse formulations provide a flexible, intuitive implementation of the constraints that are used by each algorithm. The inverse formulations were then solved through use of the RAP method. The RAP method is a simple technique for creating orthogonal projections within a space defined by a set of hyperplanes. The RAP method can easily be applied to unconstrained and constrained optimization problems whose solution lies in a convex set (i.e., no local maxima or minima). Many array processing algorithms fall into this category and it has been shown that the RAP method is a viable solution for this application. Since the RAP method only involves row operations, it is also more convenient for parallel processing implementations such as systolic arrays [15].

These algorithms have also been simulated for both narrowband and broadband implementations. The narrowband simulation consisted of a set of tones arriving at different spatial locations. The broadband array was evaluated for a simulation of several signals with differing spatial locations and bandwidths, in addition to a speech enhancement application. For all scenarios, the inverse formulations were found to perform comparable to the traditional approaches.

References

1. Adugna, E., Speech enhancement using microphone arrays, PhD thesis, CAIP Center, Rutgers University, Piscataway, NJ, June 1994.
2. Applebaum, S.P., Adaptive arrays, *IEEE Trans. Antennas Propagation*, AP-24, 585–598, 1976.
3. Censor, Y., Row-action techniques for huge and sparse systems and their applications, *SIAM Rev.*, 23(4), 444–466, Oct. 1981.
4. DeFatta, D., Lucas, J., and Hodgkiss, W., *Digital Signal Processing: A System Design Approach*, John Wiley & Sons, New York, 1988.
5. Farrell, K.R., Mammone, R.J., and Flanagan, J.L., Beamforming microphone arrays for speech enhancement, in *Proceedings of International Conference on Acoustics, Speech, and Signal Processing*, San Francisco, CA, Mar. 23–26, 1992, Vol. 1, pp. 285–288.
6. Flanagan, J.L., Johnston, J.D., Zahn, R., and Elko, G.W., Computer-steered microphone arrays for sound transduction in large rooms, *J. Acoust. Soc. Am.*, 78(11), 1508–1518, Nov. 1985.
7. Frost, O.L., III, An algorithm for linearly constrained adaptive array processing, *Proc. IEEE*, 60(8), 926–935, Aug. 1972.
8. Giordano, A. and Hsu, F., *Least Square Estimation with Applications to Digital Signal Processing*, John Wiley & Sons, New York, 1985.
9. Greenberg, J.E. and Zurek, P.M., Evaluation of an adaptive beamforming method for hearing aids, *J. Acoust. Soc. Am.*, 91(3), 1662–1676, Mar. 1992.
10. Griffiths, L.J., A simple adaptive algorithm for real-time processing in antenna arrays, *Proc. IEEE*, 57(10), 1696–1704, Oct. 1969.
11. Griffiths, L.J., Linearly-constrained adaptive signal processing methods, in *Advanced Algorithms and Architectures for Signal Processing II*, SPIE, Bellingham, WA, 1987, pp. 96–100.
12. Griffiths, L.J. and Jim, C.W., An alternative approach to linearly constrained adaptive beamforming, *IEEE Trans. Antennas Propagation*, AP-30(1), 27–34, Jan. 1982.
13. Haykin, W., *Adaptive Filter Theory*, Prentice-Hall, Englewood Cliffs, NJ, 1991.
14. Hudson, J.E., *Adaptive Array Principles*, Institute of Electrical Engineers, Peregrinus, New York; Stevenage, UK, 1981.
15. Kung, S.Y., *VLSI Array Processors*, Prentice-Hall, Englewood Cliffs, NJ, 1988.

16. Mammone, R.J., *Computational Methods of Signal Recovery and Recognition*, John Wiley & Sons, New York, 1992.
17. Noble, B. and Daniel, J.W., *Applied Linear Algebra*, Prentice-Hall, Englewood Cliffs, NJ, 1988.
18. Papoulis, A., *Probability, Random Variables, and Stochastic Process*, McGraw-Hill, New York, 1984.
19. Takao, K., Fujita, M., and Nishi, T., An adaptive antenna array under directional constraint, *IEEE Trans. Antennas Propagation*, AP-24(9), 662–669, Sept. 1976.
20. Widrow, B. and Stearns, S.D., *Adaptive Signal Processing*, Prentice-Hall, Englewood Cliffs, NJ, 1985.
21. Widrow, B., Mantey, P.E., and Goode, B.B., Adaptive antenna systems, *Proc. IEEE*, 55(12), 2143–2158, Dec. 1967.

31

Channel Equalization as a Regularized Inverse Problem

John F. Doherty
The Pennsylvania State University

31.1 Introduction

In this chapter we examine the problem of communication channel equalization and how it relates to the inversion of a linear system of equations. Channel equalization is the process by which the effect of a band-limited channel may be diminished, that is, equalized, at the sink of a communication system. Although there are many ways to accomplish this, we will concentrate on linear filters and adaptive filters. It is through the linear filter approach that the analogy to matrix inversion is possible. Regularized inversion refers to a process in which noise dominated modes of the observed signal are attenuated.

31.2 Discrete-Time Intersymbol Interference Channel Model

Intersymbol interference (ISI) is a phenomenon observed by the equalizer caused by frequency distortion of the transmitted signal. This distortion is usually caused by the frequency selective characteristics of the transmission medium. However, it can also be due to deliberate time dispersion of the transmitted pulse to affect realizable implementations of the transmit filter. In any case, the purpose of the equalizer is to remove deleterious effects of the ISI on symbol detection. The ISI generation mechanism is described next with a description of equalization techniques to follow. The information transmitted by a digital communication system is comprised of a set of discrete symbols. Likewise, the ultimate form of the received information is cast into a discrete form. However, the intermediate components of the digital communications system operate with continuous waveforms which carry the information. The major

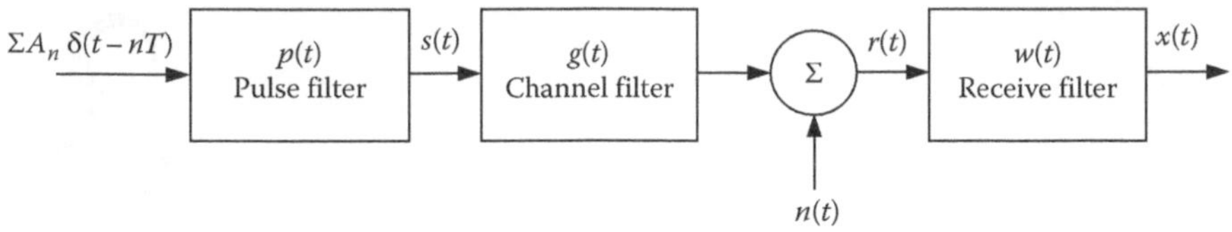

FIGURE 31.1 The signal flow block diagram for the equivalent channel description. The equalizer observes $x(nT)$, a sampled version of the receive filter output $x(t)$.

portions of the communications link are the transmitter pulse shaping filter, the modulator, the channel, the demodulator, and the receiver filter. It will be advantageous to transform the continuous part of the communication system into an equivalent discrete time channel description for simulation purposes. The discrete formulation should be transparent to both the information source and the equalizer when evaluating performance. The equivalent discrete time channel model is attained by combining the transmit filter, $p(t)$, the channel filter, $g(t)$, and the receive filter, $w(t)$, into a single continuous filter, that is,

$$h(t) = w(t) * g(t) * p(t) \tag{31.1}$$

Refer to Figure 31.1. The effect of the sampler preceding the decision device is to discretize the aggregate filter.

The equivalent discrete time channel as a means to simulate the performance of digital communications systems was advanced by Proakis [1] and has found subsequent use throughout the communications literature [2,3].

It has been shown that a bandpass transmitted pulse train has an equivalent low-pass representation [1]:

$$s(t) = \sum_{n=0}^{\infty} A_n p(t - nT) \tag{31.2}$$

where

$\{A_n\}$ is the information bearing symbol set
$p(t)$ is the equivalent low-pass transmit pulse waveform
T is the symbol rate

The observed signal at the input of the receiver is

$$r(t) = \sum_{n=0}^{\infty} A_n \int_{-\infty}^{+\infty} p(t - nT)g(t - nT - \tau)d\tau + n(t) \tag{31.3}$$

where $g(t)$ is the equivalent low-pass bandlimited impulse response of the channel and the channel noise, $n(t)$, is modeled as white Gaussian noise. The optimum receiver filter, $w(t)$, is the matched filter which is designed to give maximum correlation with the received pulse [4]. The output of the receiver filter, that is, the signal seen by the sampler, can be written as

$$x(t) = \sum_{n=0}^{\infty} A_n h(t - nT) + \nu(t) \tag{31.4}$$

$$h(t) = \int_{-\infty}^{+\infty} \left[\int_{-\infty}^{+\infty} p(t - nT)g(t - nT - \lambda)d\lambda \right] w(t - \tau)d\tau \tag{31.5}$$

$$v(t) = \int_{-\infty}^{+\infty} n(t)w(t-\tau)\mathrm{d}\tau \tag{31.6}$$

where

$h(t)$ is the response of the receiver filter to the received pulse, representing the overall impulse response between the transmitter and the sampler

$v(t) = \int_{-\infty}^{+\infty} n(t)w(t-\tau)\mathrm{d}\tau$ is a filtered version of the channel noise

The input to the equalizer is a sampled version of Equation 31.4, that is, sampling at times $t = kT$ produces

$$x(kT) = \sum_{n=0}^{\infty} A_n h(kt - nT) + v(kT) \tag{31.7}$$

as the input to the discrete time equalizer. By normalizing with respect to the sampling interval and rearranging terms, Equation 31.7 becomes

$$x_k = \underbrace{h_0 A_k}_{\text{desired symbol}} + \underbrace{\sum_{\substack{n=0\\n\neq k}}^{\infty} A_n h_{k-n}}_{\text{ISI}} + v_k \tag{31.8}$$

31.3 Channel Equalization Filtering

31.3.1 Matrix Formulation of the Equalization Problem

The task of finding the optimum linear equalizer coefficients can be described by casting the problem into a system of linear equations,

$$\begin{bmatrix} \tilde{d}_1 \\ \tilde{d}_2 \\ \vdots \\ \tilde{d}_L \end{bmatrix} = \begin{bmatrix} \boldsymbol{x}_1^{\mathrm{T}} \\ \boldsymbol{x}_2^{\mathrm{T}} \\ \vdots \\ \boldsymbol{x}_L^{\mathrm{T}} \end{bmatrix} c + \begin{bmatrix} e_1 \\ e_2 \\ \vdots \\ e_L \end{bmatrix} \tag{31.9}$$

$$\boldsymbol{x}_k = [x_{k+N-1}, \ldots, x_{k-1}]^{\mathrm{T}} \tag{31.10}$$

where the superscript T denotes the transpose operation. The received sample at time k is x_k, which consists of the channel output corrupted by additive noise. The elements of the $N \times 1$ vector c_k are the coefficients of the equalizer filter at time k. The equalizer is said to be in decision directed mode when $\tilde{d}_k$ is taken as the output of the nonlinear decision device. The equalizer is in training, or reference directed, mode when $\tilde{d}_k$ is explicitly made identical to the transmitted sequence A_k. In either case, e_k is the error between the desired equalizer output, $\tilde{d}_k$, and the actual equalizer output, $\boldsymbol{x}_k^{\mathrm{T}}\boldsymbol{c}$. We will assume that $\tilde{d}_k = A_{k+N}$, then the notation in Equation 31.9 can be written in the compact form,

$$\boldsymbol{d} = \boldsymbol{X}\boldsymbol{c} + \boldsymbol{e} \tag{31.11}$$

by defining $\boldsymbol{d} = \left[\tilde{d}_1, \ldots, \tilde{d}_L\right]^{\mathrm{T}}$ and by making the obvious associations with Equation 31.9. Note that the parameter L determines the number of rows of the time varying matrix $\boldsymbol{X}$. Therefore, choosing L is analogous to choosing an observation interval for the estimation of the filter coefficients.

31.4 Regularization

We seek a solution for the filter coefficients of the form $\boldsymbol{c} = \boldsymbol{Yd}$, where $\boldsymbol{Y}$ is in some sense an inverse of the data matrix $\boldsymbol{X}$. The least squares solution requires that

$$\boldsymbol{Y} = [\boldsymbol{X}^{\mathrm{T}}\boldsymbol{X}]^{-1}\boldsymbol{X}^{\mathrm{T}} \tag{31.12}$$

where $\boldsymbol{X}^{\#} \triangleq [\boldsymbol{X}^{\mathrm{T}}\boldsymbol{X}]^{-1}\boldsymbol{X}^{\mathrm{T}}$ represents the Moore–Penrose (M–P) inverse of $\boldsymbol{X}$. If one or more of the eigenvalues of the matrix $\boldsymbol{X}^{\mathrm{T}}\boldsymbol{X}$ is zero, then the M–P inverse does not exist.

To investigate the behavior of the inverse, we will decompose the data matrix into the form $\boldsymbol{X} = \boldsymbol{X}_{\mathrm{S}} + \boldsymbol{X}_{\mathrm{N}}$, where $\boldsymbol{X}_{\mathrm{S}}$ is the signal component and $\boldsymbol{X}_{\mathrm{N}}$ is the noise component. Generally, the noise data matrix is full rank and the signal data matrix may be nearly rank deficient from the spectral nulls in the transmission channel. This is illustrated by examining the smallest eigenvalue of $\boldsymbol{X}_{\mathrm{S}}^{\mathrm{T}}\boldsymbol{X}_{\mathrm{S}}$

$$\lambda_{\min} = S_{\mathrm{R}\,\min} + O(N^{-k}) \tag{31.13}$$

where

S_{R} is the continuous power spectral density (PSD) of the received data x_k
$S_{\mathrm{R}\,\min}$ is the minimum value of the PSD
k is the number of nonvanishing derivatives of S_{R} at $S_{\mathrm{R}\,\min}$
N is the equalizer filter length

Any spectral loss in the signal caused by the channel is directly translated into a corresponding decrease in the minimum eigenvalue of the received signal. If $\lambda_{\min}$ becomes small, but nonzero, the data correlation matrix $\boldsymbol{X}^{\mathrm{T}}\boldsymbol{X}$ becomes ill-conditioned and its inversion becomes sensitive to the noise. The sensitivity is expressed in the quantity

$$\delta \triangleq \frac{\|\tilde{\boldsymbol{c}} - \boldsymbol{c}\|}{\|\boldsymbol{c}\|} \leq \frac{\sigma_n^2}{\lambda_{\min}} + O\left(\sigma_n^4\right) \tag{31.14}$$

where the noiseless least squares filter coefficient vector solution, $\boldsymbol{c}$, has been perturbed by adding a white noise to the data with variance $\sigma_n^2 \ll 1$, to produce the least squares solution c. Substituting Equation 31.13 into Equation 31.14 yields

$$\delta \leq \frac{\sigma_n^2}{S_{\mathrm{R}\,\min} + O(N^{-k})} + O\left(\sigma_n^4\right) \approx \frac{\sigma_n^2}{S_{\mathrm{R}\,\min}} \tag{31.15}$$

The relation in Equation 31.15 is an indicator of the potential numerical problems in solving for the equalizer filter coefficients when the data is spectrally deficient.

We see that direct inversion of the data matrix is not recommendable when the channel has severe spectral nulls. This situation is equivalent to stating that the original estimation problem $\boldsymbol{d} = \boldsymbol{Xc}$ is ill-posed. That is, the equalizer is asked to reproduce components of the channel input that are unobservable at the channel output or are obscured by noise. Thus, it is reasonable to ascertain the modes of the input dominated by noise and give them little weight, relative to the signal dominated components, when solving for the equalizer filter coefficients. This process of weighting is called regularization.

Regularization can be described by relying on a generalization of the M–P inverse that depends on the singular value decomposition (SVD) of the data matrix

$$\boldsymbol{X} = \boldsymbol{U\Sigma V}^{\mathrm{T}} \tag{31.16}$$

where

$\boldsymbol{U}$ is an $L \times N$ unitary matrix
$\boldsymbol{V}$ is an $N \times N$ unitary matrix
$\boldsymbol{\Sigma} = \text{diag}(\sigma_1, \sigma_2, \ldots, \sigma_N)$ is a diagonal matrix of singular values where $\sigma_i \geq 0$, $\sigma_1 > \sigma_2 > \cdots > \sigma_N$

It is assumed in Equation 31.16 that $L > N$, which is typical in the equalization problem.

We define the generalized pseudo-inverse of X as

$$\boldsymbol{X}^{\dagger} = \boldsymbol{V}\boldsymbol{\Sigma}^{\dagger}\boldsymbol{U}^{\mathrm{T}} \tag{31.17}$$

where $\boldsymbol{\Sigma}^{\dagger} = \text{diag}\left(\sigma_1^{\dagger}, \sigma_2^{\dagger}, \ldots, \sigma_N^{\dagger}\right)$ and

$$\sigma_i^{\dagger} = \begin{cases} \sigma_i^{-1} & \sigma_i \neq 0 \\ 0 & \sigma_i = 0 \end{cases} \tag{31.18}$$

The M–P inverse can be reformulated using the SVD as follows:

$$\boldsymbol{X}^{\#} = [\boldsymbol{V}\boldsymbol{\Sigma}^2\boldsymbol{V}^{\mathrm{T}}]^{-1}\boldsymbol{V}\boldsymbol{\Sigma}\boldsymbol{U}^{\mathrm{T}} = \boldsymbol{V}\boldsymbol{\Sigma}^{-1}\boldsymbol{U}^{\mathrm{T}} \tag{31.19}$$

Upon examination of Equations 31.17 and 31.19, we note that $\boldsymbol{X}^{\#} = \boldsymbol{X}^{\dagger}$ only if all the singular values of $\boldsymbol{X}$ are nonzero, $\sigma_i \neq 0$. Another item to note is that $\boldsymbol{V}\boldsymbol{\Sigma}^2\boldsymbol{V}^{\mathrm{T}}$ is the eigenvalue decomposition of $\boldsymbol{X}^{\mathrm{T}}\boldsymbol{X}$, which implies that the eigenvalues of $\boldsymbol{X}^{\mathrm{T}}\boldsymbol{X}$ are the squares of the singular values of $\boldsymbol{X}$.

The generalized pseudo-inverse in Equation 31.17 provides an eigenvalue spectral weighting given by Equation 31.18, which differs from the M–P inverse only when one or more of the eigenvalues of $\boldsymbol{X}^{\mathrm{T}}\boldsymbol{X}$ are identically zero. However, this form of regularization is rather restrictive since complete annihilation of the spectral components is rarely encountered in practice. A more likely condition for the eigenvalues of $\boldsymbol{X}^{\mathrm{T}}\boldsymbol{X}$ is that a small band of signal eigenmodes are much smaller in magnitude than the corresponding noise modes. Direct inversion of these eigenmodes, although well-defined mathematically, leads to noise enhancement at the equalizer output and to noise sensitivity in the filter coefficient solution. An alternative to the generalized pseudo-inverse is to use a regularized inverse wherein the eigenmodes are weighted prior to inversion [5]. This approach leads to a trade-off between the noise immunity of the equalizer filter weights and the signal fidelity at the equalizer filter output. To demonstrate this trade-off, let

$$\boldsymbol{c}\mathcal{D}\boldsymbol{X}^{\dagger}\boldsymbol{d} \tag{31.20}$$

be the least squares solution. Let the regularized inverse be $\boldsymbol{Y}_n$ such that $\lim_{n\to\infty}\boldsymbol{Y}_n = \boldsymbol{X}^{\dagger}$. The regularized estimate for an observation perturbed by a random noise vector, $\boldsymbol{n}$, is

$$\boldsymbol{c}_n = \boldsymbol{Y}_n(\boldsymbol{d}\mathcal{C}\boldsymbol{n}) \tag{31.21}$$

The effects of the regularized inverse and the noise vector are indicated by

$$\|\boldsymbol{c}_n - \boldsymbol{c}\| = \|\boldsymbol{Y}_n\boldsymbol{n} + (\boldsymbol{Y}_n - \boldsymbol{X}^{\dagger})\boldsymbol{d}\| \leq \|\boldsymbol{Y}_n\boldsymbol{n}\| + \|\boldsymbol{Y}_n - \boldsymbol{X}^{\dagger}\|\|\boldsymbol{d}\| \tag{31.22}$$

The term $\|\boldsymbol{Y}_n\boldsymbol{n}\|$ is the part of the coefficient error due to the noise and is likely to increase as $n \to \infty$. The term $\|\boldsymbol{Y}_n - \boldsymbol{X}^{\dagger}\|$ represents the contribution due to the regularization error in approximating the pseudo-inverse. This error tends to zero as $n \to \infty$. The trade-off between noise attenuation and regularization error is evident upon inspection of Equation 31.22, which also points out an idiosyncratic property of the regularization process. At first, the equalizer output error tends to decrease, due to decreasing regularization error, $\| \boldsymbol{Y}_n - \boldsymbol{X}^{\dagger} \|$. Then, as n increases further, the output error is likely to increase due to the noise amplification component, $\| \boldsymbol{Y}_n \boldsymbol{n} \|$. This behavior leads to the question regarding the best choice for the parameter n. A widely accepted procedure is to use the discrepancy principle, which states that n should satisfy

$$\| \boldsymbol{X} \boldsymbol{c}_{n'} - (\boldsymbol{d} \boldsymbol{C} \boldsymbol{n}) \| = \| \boldsymbol{n} \| \tag{31.23}$$

Letting $n > n'$ usually results in noise amplification at the equalizer output.

31.5 Discrete-Time Adaptive Filtering

We will next examine three adaptive algorithms in terms of their regularization properties in deriving the equalizer filter. These algorithms are the normalized least mean squares (NLMS) algorithm, the recursive least squares (RLS) algorithm, and the block-iterative NLMS (BINLMS) algorithm. These algorithms are representative of the wider class of adaptive algorithms of which they belong.

31.5.1 Adaptive Algorithm Recapitulation

31.5.1.1 NLMS

The NLMS algorithm update is given by

$$\boldsymbol{c}_n = \boldsymbol{c}_{n-1} + \mu \left(d_n - \boldsymbol{x}_n^{\mathrm{T}} \boldsymbol{c}_{n-1} \right) \frac{\boldsymbol{x}_n}{\| \boldsymbol{x}_n \|^2} \tag{31.24}$$

for $n = 1, \ldots, L$. This is rewritten as

$$\boldsymbol{c}_n = \left(\boldsymbol{I} - \mu \frac{\boldsymbol{x}_n \boldsymbol{x}_n^{\mathrm{T}}}{\| \boldsymbol{x}_n \|^2} \right) \boldsymbol{c}_{n-1} + \mu \frac{d_n \boldsymbol{x}_n}{\| \boldsymbol{x}_n \|^2} \tag{31.25}$$

Define $\boldsymbol{P}_n \triangleq \left(\boldsymbol{I} - \mu \boldsymbol{x}_n \boldsymbol{x}_n^{\mathrm{T}} / \| \boldsymbol{x}_n \|^2 \right)$ and $\boldsymbol{p}_n \triangleq \mu d_n \boldsymbol{x}_n / \| \boldsymbol{x}_n \|^2$, then Equation 31.25 becomes

$$\boldsymbol{c}_L = \boldsymbol{Q} \boldsymbol{c}_0 + \boldsymbol{q} \tag{31.26}$$

where

$$\boldsymbol{Q} \mathcal{D} \boldsymbol{P}_L \boldsymbol{P}_{L-1} \cdots \boldsymbol{P}_1 \tag{31.27}$$

and

$$\boldsymbol{q} = [\boldsymbol{P}_L \cdots \boldsymbol{P}_2] \boldsymbol{p}_1 + [\boldsymbol{P}_L \cdots \boldsymbol{P}_3] \boldsymbol{p}_2 + \cdots + \boldsymbol{P}_L \boldsymbol{p}_{L-1} + \boldsymbol{p}_L \tag{31.28}$$

31.5.1.2 BINLMS

The BINLMS algorithm relies on observing the entire block of filter vectors $\boldsymbol{x}_n$, $1 \leq n \leq L$, in Equation 31.9. The BINLMS update procedure is

$$\boldsymbol{c}_{n+1} = \boldsymbol{c}_n + \mu \left(d_j - \boldsymbol{x}_j^{\mathrm{T}} \boldsymbol{c}_n \right) \frac{\boldsymbol{x}_j}{\| \boldsymbol{x}_j \|^2} \tag{31.29}$$

where $j = n \bmod L$. The update in Equation 31.29 is related to the NLMS update by considering Equation 31.26. That is, Equation 31.29 is equivalent to

$$\boldsymbol{c}_{nL} = \boldsymbol{Q}\boldsymbol{c}_{(n-1)L} + \boldsymbol{q} \tag{31.30}$$

where L updates of Equation 31.29 are compacted into a single update in Equation 31.30. Note that only L updates are possible using Equation 31.24 compared to an arbitrary number of updates in Equation 31.29.

31.5.1.3 RLS

The update procedure for the RLS algorithm is

$$\boldsymbol{g}_n = \frac{\lambda^{-1}\boldsymbol{Y}_{n-1}\boldsymbol{x}_n}{1 + \lambda^{-1}\boldsymbol{x}_n^{\mathrm{T}}\boldsymbol{Y}_{n-1}\boldsymbol{x}_n} \tag{31.31}$$

$$e_n = d_n - \boldsymbol{c}_{n-1}^{\mathrm{T}}\boldsymbol{x}_n \tag{31.32}$$

$$\boldsymbol{c}_n = \boldsymbol{c}_{n-1} + e_n\boldsymbol{g}_n \tag{31.33}$$

$$\boldsymbol{Y}_n = \lambda^{-1}\left[\boldsymbol{Y}_{n-1} - \boldsymbol{g}_n\boldsymbol{x}_n^{\mathrm{T}}\boldsymbol{Y}_{n-1}\right] \tag{31.34}$$

where

$\boldsymbol{g}_n$ is called the gain vector
$\boldsymbol{Y}_n$ is the estimate of $\left[\boldsymbol{X}_n^{\mathrm{T}}\boldsymbol{X}_n\right]^{-1}$ using the matrix inversion lemma
$\boldsymbol{x}_n$ represents the first n rows of $\boldsymbol{X}$ in Equation 31.9

The forgetting factor $0 < \lambda \ll 1$ allows the RLS algorithm to weight more recent samples providing a tracking capability for time-varying channels. The matrix inversion recursion is initialized with $\boldsymbol{Y}_0 = \delta^{-1}\boldsymbol{I}$, where $0 < \delta \ll 1$. The initialization constant transforms the data correlation matrix into

$$\boldsymbol{X}_n^{\mathrm{T}}\boldsymbol{\Lambda}_n\boldsymbol{X}_n + \lambda^n\delta\boldsymbol{I} \tag{31.35}$$

where $\boldsymbol{\Lambda}_n = \mathrm{diag}(1, \lambda, \ldots, \lambda^{n-1})$.

31.5.2 Regularization Properties of Adaptive Algorithms

In this section we examine how each of the adaptive algorithms achieve regularization of the equalizer filter solution. We begin with the BINLMS and will subsequently take the NLMS as a special case. The BINLMS update of Equation 31.30 is equivalent to

$$\boldsymbol{c}_l = \boldsymbol{Q}\boldsymbol{c}_{l-1} + \boldsymbol{q} \tag{31.36}$$

where an increment in l is equivalent to L increments of n in Equation 31.29. The recursion in Equation 31.36 is also equivalent to

$$\boldsymbol{c}_l = \boldsymbol{B}_l\boldsymbol{d} \tag{31.37}$$

where $\lim_{l\to\infty}\boldsymbol{B}_l = \boldsymbol{X}^{\dagger}$. Let $\hat{\sigma}_{k,l}$ represent the singular values of $\boldsymbol{B}_l$, then the relationship among the singular values of $\boldsymbol{B}_l$ and the singular values of $\boldsymbol{X}$ is [6]

$$\hat{\sigma}_{k,l} = \begin{cases} \frac{1}{\sigma_k}\left[1 - \left(1 - \frac{\mu}{N}\sigma_k^2\right)^{l+1}\right], & \sigma_k \neq 0 \\ 0, & \sigma_k = 0 \end{cases} \tag{31.38}$$

The regularization property of the BINLMS depends on both μ and l. Since the step size parameter μ is chosen to guarantee convergence, that is, $0 < \left(1 - \frac{\mu}{N}\sigma_1^2\right) < 1$, the regularization is primarily controlled by the iteration index l. The regularization behavior of the BINLMS given by Equation 31.38 is that the signal dominant modes are inverted first, followed by the weaker noise dominant modes, as the index l increases.

The regularization behavior of the NLMS algorithm is directly derived from the BINLMS by setting $l = 1$ in Equation 31.38. We see that the only control over the regularization for the NLMS algorithm is to decrease the step size μ. However, this leads to a potentially undesirable reduction in the convergence rate of the adaptive equalizer filter.

The RLS algorithm weighting of the singular values is derived upon inspection of Equation 31.35. The RLS equalizer filter coefficient estimate is

$$\boldsymbol{c}_{\text{LS}} = \left[\boldsymbol{X}^{\text{T}}\boldsymbol{\Lambda}_L\boldsymbol{X} + \lambda^L\delta\boldsymbol{I}\right]^{-1}\boldsymbol{X}^{\text{T}}\left(\boldsymbol{\Lambda}_L^{1/2}\right)^{\text{T}}\boldsymbol{d} \tag{31.39}$$

Let $\hat{\sigma}_{\text{LS},k}$ represent the singular values of the effective inverse used in the RLS algorithm, then

$$\hat{\sigma}_{\text{LS},k} = \frac{\sqrt{\lambda_k}\,\sigma_k}{\lambda_k\sigma_k^2 + \lambda^L\delta} \tag{31.40}$$

There are several points to note about Equation 31.40. In the absence of the forgetting factor, $\lambda = 1$, and the initialization constant, $\delta = 0$, the RLS algorithm provides the exact inverse of the singular values, as expected. The constant δ prevents the dominator of Equation 31.40 from getting too small. However, this regularization is lost if $\lambda^L \to 0$, which is the case when the observation interval L becomes large.

The behavior of the regularization functions (Equations 31.38 and 31.40) is illustrated in Figure 31.2.

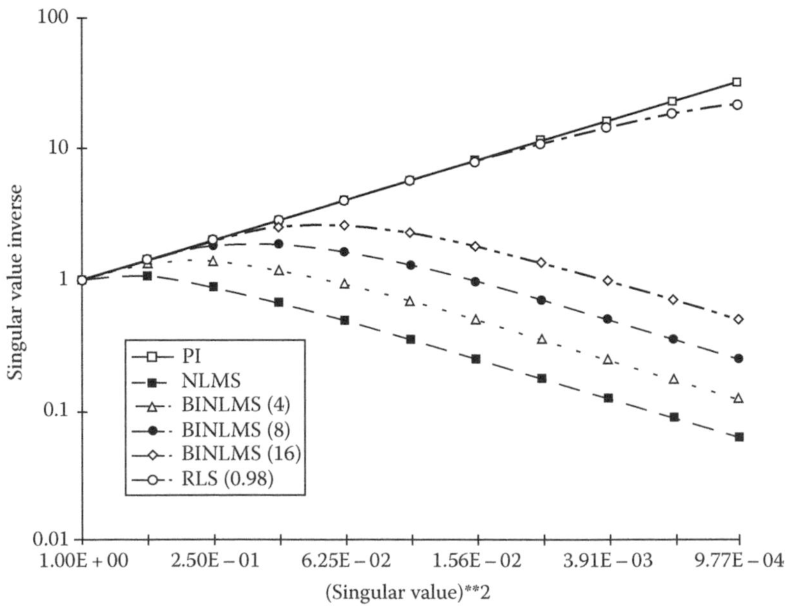

FIGURE 31.2 The regularization functions of the NLMS, BINLMS, and RLS algorithms.

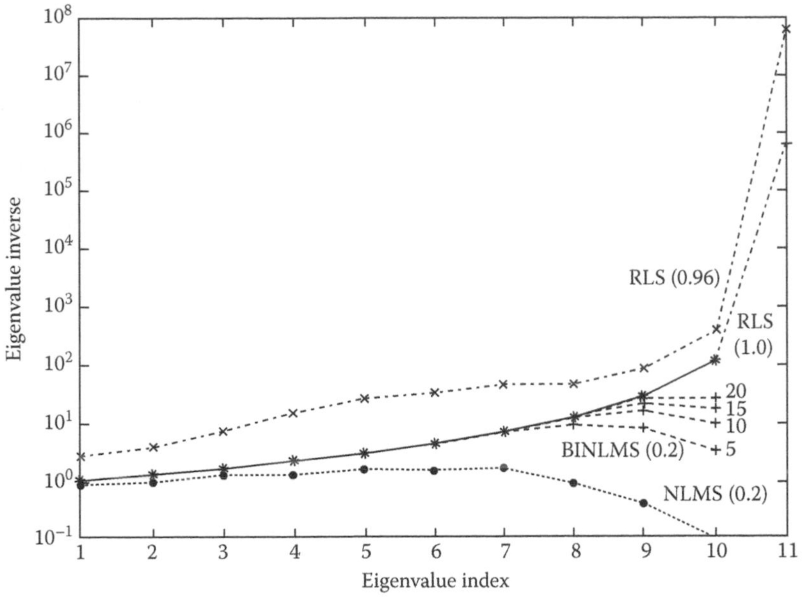

FIGURE 31.3 The regularization behavior of the NLMS, BINLMS, and the RLS adaptive algorithms is shown. The BINLMS curves represent block iterations of 5, 10, 15, and 20. The RLS algorithm uses $\lambda = 1.0$ and $\lambda = 0.96$.

31.6 Numerical Results

A numerical example of the regularization characteristics of the adaptive equalization algorithms discussed is now presented. A data matrix $\boldsymbol{X}$ $\boldsymbol{X}$ is constructed with dimensions $L = 50$ and $N = 11$, which has the singular value matrix $\Sigma = \text{diag}(1.0, 0.9, \ldots, 0.1, 0.0)$. The step size $\mu = 0.2$ is chosen. Since the RLS algorithm computes an estimate of $[\boldsymbol{X}^{\mathrm{T}}\boldsymbol{X}]^{-1}$, it is sensitive to the eigenvalues of $\boldsymbol{X}^{\mathrm{T}}\boldsymbol{X}$. A graph similar to Figure 31.2 is produced with the exception that the eigenvalue inverses of $\boldsymbol{X}^{\mathrm{T}}\boldsymbol{X}$ are plotted for the RLS algorithm. These results are shown in Figure 31.3 using the eigenvalues of $\boldsymbol{X}$ given by $\sigma_i^2 = [1 - (i-1)/10]^2$ for $1 \leq i \leq 10$ and $\sigma_{11}^2 = 0$. The RLS algorithm exhibits large dynamic range in the eigenvalue inverse using the matrix inversion lemma, which may lead to unstable operation of the adaptive equalizer filter.

31.7 Conclusion

A short introduction to the basic concepts of regularization analysis are presented in this chapter. Some further development in the application of this analysis to decision-feedback equalization may be found in [6]. The choice of which adaptive algorithm to use is application-dependent and each one comes with its associated advantages and disadvantages. The LMS-type algorithms are low-complexity solutions that have relatively slow convergence. The RLS-type algorithms have much faster convergence but are typically plagued by stability problems associated with error propagation and unregularized matrix inversion. Circumventing these stability problems tends to lead to more complex algorithm implementation. The BINLMS algorithm is a trade-off between the convergence speed of the RLS-type algorithms and the stability of the LMS-type algorithms. A disadvantage of the BINLMS algorithm is that instantaneous throughput may be high due to the block-processing required.

References

1. Proakis, J., *Digital Communications*, 2nd ed., McGraw-Hill, New York, 1989.
2. Hatzinakos, D. and Nikias, C., Estimation of multipath channel response in frequency selective channels, *IEEE J. Sel. Areas Commn.*, SAC-7, 12–19, Jan. 1989.
3. Eleftheriou, E. and Falconer, D., Adaptive equalization techniques for HF channels, *IEEE J. Sel. Areas Commn.*, SAC-5, 238–247, Feb. 1987.
4. Wozencraft, J. and Jacobs, I., *Principles of Communication Engineering*, John Wiley & Sons, New York, 1965.
5. Tikhonov, A. and Arsenin, V., *Solutions to Ill-Posed Problems*, V.H. Winston and Sons, Washington DC, 1977.
6. Doherty, J. and Mammone, R., An adpative algorithm for stable decision-feedback filtering, *IEEE Trans. Circuits Syst. II: Analog Digital Signal Process.*, 40(1), 1–9, Jan. 1993.

32

Inverse Problems in Microphone Arrays

A.C. Surendran
Lucent Technologies, Bell Laboratories

32.1 Introduction: Dereverberation Using Microphone Arrays

An acoustic enclosure usually reduces the intelligibility of the speech transmitted through it because the transmission path is not ideal. Apart from the direct signal from the source, the sound is also reflected off one or more surfaces (usually walls) before reaching the receiver. The resulting signal can be viewed as the output of a convolution in the time domain of the speech signal and the room impulse response. This phenomenon affects the quality of the transmitted sound in important applications such as teleconferencing, cellular telephony, and automatic voice activated systems (speaker and speech recognizers). Room reverberation can be perceptually separated into two broad classes. Early room echoes are manifested as irregularities or "ripples" in the amplitude spectrum. This effect dominates in small rooms, typically offices. Long-term reverberation is typically exhibited as an echo "tail" following the direct sound [1].

If the transfer function $G(z)$ of the system is known, it might be possible to remove the deleterious multi-path effects by inverse filtering the output using a filter $H(z)$ where

$$H(z) = \frac{1}{G(z)}. \tag{32.1}$$

Typically $G(z)$ is the transform of the impulse response of the room $g(n)$. In general, the transfer function of a reverberant environment is a non-minimum phase function, i.e., all the zeros of the function do not necessarily lie inside $|z| = 1$. A minimum phase function has a stable causal inverse, while the inverse of a non-minimum phase function is acausal and, in general, infinite in length.

In general, $G(z)$ can be expressed as a product of a minimum-phase function and a non-minimum phase function:

$$G(z) = G_{\min}(z) \cdot G_{\max}(z). \tag{32.2}$$

Many approaches have been proposed for dereverberating signals. The aim of all the compensation schemes is to bring the impulse response of the system after dereverberation as close as possible to an impulse function. Homomorphic filtering techniques were used to estimate the minimum phase part of $G(z)$ [2,3]. In [2], the minimum phase component was estimated by zeroing out the cepstrum for negative frequencies. Then the output signal was filtered by the inverse of the minimum phase transfer function. But this technique still did not remove the reverberation contributed by the maximum-phase part of the room response. In [3], the inverse of the maximum-phase part was also estimated from the delayed and truncated version of the acausal inverse. But, the delay can be inordinate and care must be taken to avoid temporal aliasing.

An alternate approach to dereverberation is to calculate, in some form, the least squares estimate of the inverse of the transmission path, i.e., calculate the least squares solution of the equation

$$h(n)^{*}g(n) = d(n), \tag{32.3}$$

where

$d(n)$ is the impulse function

$*$ denotes convolution

Assuming that the system can be modeled by an FIR filter, Equation 32.3 can be expressed in matrix form as

$$\begin{pmatrix} g(0) & & & \\ g(1) & g(0) & & \\ \vdots & g(1) & \cdots & 0 \\ g(m) & \vdots & \cdots & g(0) \\ 0 & g(m) & \cdots & g(1) \\ 0 & 0 & \cdots & \vdots \\ & & & g(m) \end{pmatrix} \begin{pmatrix} h(0) \\ h(1) \\ \vdots \\ h(i) \end{pmatrix} = \begin{pmatrix} 1 \\ 0 \\ \vdots \\ 0 \end{pmatrix}, \tag{32.4}$$

or

$$GH = D, \tag{32.5}$$

where D is the unity matrix and G, H, and D are matrices of appropriate dimensions as shown in Equation 32.4. The least squares method finds an approximate solution given by

$$\hat{H}(z) = (G^{T}G)^{-1}G^{T}D. \tag{32.6}$$

Thus, the error vector can be written as

$$\begin{aligned} \varepsilon &= [\mathbf{D} - \mathbf{G}\hat{\mathbf{H}}] \\ &= [\mathbf{I} - \mathbf{G}(\mathbf{G}^{T}\mathbf{G})^{-1}\mathbf{G}^{T}]\mathbf{D} \\ &= \mathbf{E}\mathbf{D}, \end{aligned}$$

where $E = [I - G\,(G^T\,G)^{-1}G^T]$. The mean square error or the energy in the error vector is

$$\|\varepsilon\|_2 = \|ED\|_2 \leq |E|\|D\|_2 \leq \frac{\lambda_{max}}{\lambda_{min}}\|D\|_2, \tag{32.7}$$

where
- $|E|$ is the norm of E
- λ_{max} and λ_{min} are the maximum and minimum eigenvalues of E

The ratio between the maximum and minimum eigenvalues is called the condition number of a matrix and it specifies the noise amplification of the inversion process [4].

Typically, the operation is done on the full-band signal. Sub-band approaches have been proposed in [5–8]. All these approaches use a single microphone.

The amplitude spectrum of the room response has "ripples" which produce pronounced notches in the signal output spectrum. As the location of the microphone in the room changes, the room response for the same source changes and, as a result, the position of the notches in the amplitude spectrum varies. This property was used to advantage in [1]. In this method, multiple microphones were located in the room. Then, the output of each microphone was divided into multiple bands of equal bandwidth. For each band, by choosing the microphone whose output has the maximum energy, the ripples were reduced. In [9], the signals from all the microphones in each band were first co-phased, and then weighted by a gain calculated from a normalized cross-correlation function calculated based on the outputs of different microphones. Since the reverberation tails are uncorrelated, the cross-correlation-based gain turned off the tail of the signal. These techniques have had modest success in combating reverberation.

In recent years, great progress has been made in the quality, availability, and cost of high performance microphones. Fast digital signal processors that permit complex algorithms to operate in real time have been developed. These advances have enabled the use of large microphone arrays that deploy more sophisticated algorithms for dereverberation. Figure 32.1 shows a generic microphone array system which can "invert" the room acoustics. Different choices of $H_i(z)$ lead to different algorithms, each with their own advantages and disadvantages. In this report, we shall discuss single and multiple beamforming, matched filtering, and Diophantine inverse filtering through multiple input–output (MINT) modeling. In all cases we assume that the source location and the room configuration or, alternatively, the $G_i(z)$'s are known.

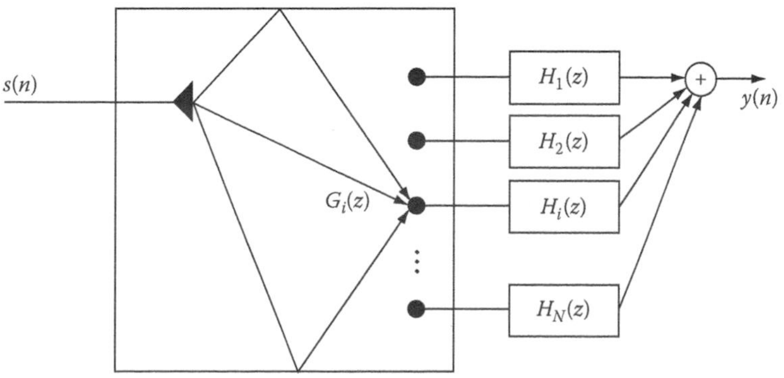

FIGURE 32.1 Modeling a room with a microphone array as a multiple output FIR system.

32.2 Simple Delay-and-Sum Beamformers

Arrays that form a single beam directed toward the source of the sound have been designed and built [11]. In these simple delay-and-sum beamformers, the processing filter has the impulse response

$$h_i(n) = \delta(n - n_i), \tag{32.8}$$

where $n_i = d_i/c$, d_i is the distance of the ith microphone from the source and c is the speed of sound in air. Sound propagation in the room can be modeled by a set of successive reflections off the surfaces (typically the walls) [10]. Figure 32.2 illustrates the impulse response of a single beamformer. The delay at the output of each microphone coheres the sound that arrives at the microphone directly from the source. It can be seen from Figure 32.2 that in the resulting response, the strength of the coherent pulse is N and there are $N(K-1)$ distributed pulses. So, ideally, the signal-to-reverberant noise ratio (SRNR; measured as the ratio of undistorted signal power to reverberant noise power) is $N^2/N(K-1)$ [13]. In a highly reverberant room, as the number of images K increases toward infinity, the signal-to-noise ratio (SNR) improvement, $N/K-1$, falls to zero.

The single-beamforming system reported in [11] can automatically determine the direction of the source and rapidly steer the array. But, as the beam is steered away from the broadside, the system exhibits a reduction in spatial discrimination because the beam pattern broadens [12]. Further,

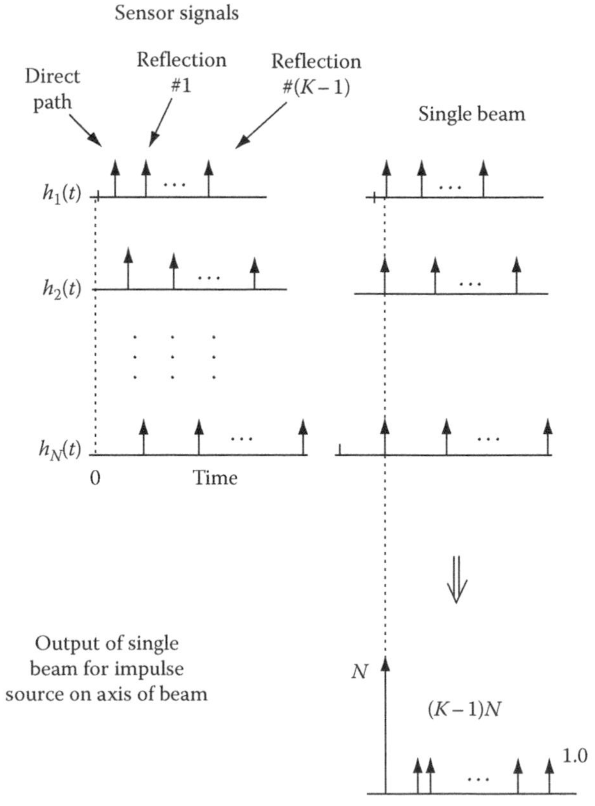

FIGURE 32.2 A single beamformer. (From Flanagan, J.L., Surendran, A.C., and Jan, E.-E., *Speech Commn.*, 13, 207, 1993. With permission.)

beamwidth varies with frequency, so an array has an approximate "useful bandwidth" given by the upper and lower frequencies [12]:

$$f_{\text{upper}} = \frac{c}{d|\cos\phi - \cos\phi'|_{\max}}, \tag{32.9}$$

and

$$f_{\text{lower}} = \frac{f_{\text{upper}}}{N}, \tag{32.10}$$

where
- c is the speed of sound in air
- N is the number of sensors in the array
- d is the sensor spacing
- ϕ' is the steering angle measured with respect to the axis of the array
- ϕ is the direction of the source

For example, consider an array with seven microphones and a sensor spacing of 6.5 cm. Further, suppose the desired range of steering is $\pm 30°$ from broadside. Then, $|\cos\phi - \cos\phi'|_{\max} = 1.5$ and hence $f_{\text{upper}} \approx 3500$ Hz and $f_{\text{lower}} \approx 500$ Hz. So, to cover the bandwidth of speech, say from 250 Hz to 7 kHz, three harmonically nested arrays of spacing 3.25, 6.5, and 13 cm can be used. Further, the beamwidth also depends on the frequency of the signal as well as the steering direction. If the beam is steered to an angle ϕ', then the direction of the source for which the beam response falls to half its power is [12]

$$\phi_{3\text{dB}} = \cos^{-1}\left\{\cos\phi' \pm \frac{2.8}{N\omega d}\right\}, \tag{32.11}$$

where $\omega = 2\pi f$ and f is the frequency of the signal.

Equation 32.11 shows that the smaller the array, the wider the beam. Since most of the energy of a typical room interfering noise lies at lower frequencies, it would be advantageous to build arrays that have higher directivity (smaller beamwidth) at lower frequencies. This, combined with the fact that the array spacing is larger for lower frequency bands, gives yet another reason to harmonically nest arrays (see Figure 32.3).

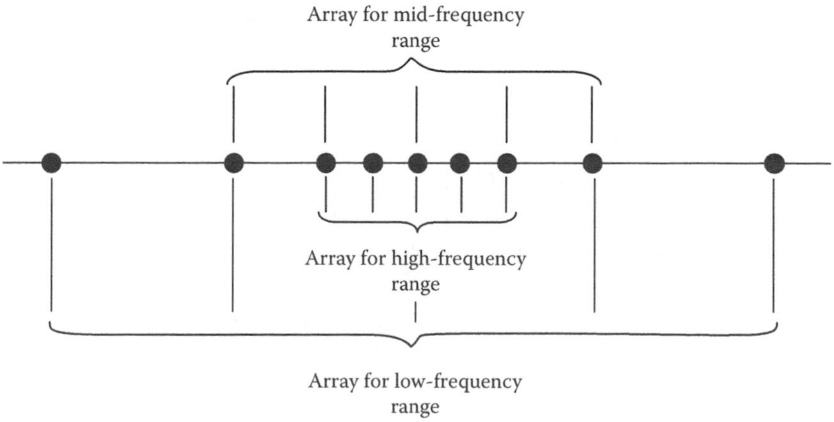

FIGURE 32.3 Harmonically nested array that covers three frequency ranges.

Just as linear one-dimensional arrays display significant fattening of the beams when steered toward the axis of the array, two-dimensional arrays exhibit widening of the beams when steered at angles acute to the plane of the array. Three-dimensional microphone arrays can be constructed [13] that have essentially a constant beamwidth over 4π steradians. Multiple beamforming using three-dimensional arrays of sensors not only provides selectivity in azimuth and elevation but also selectivity in the direction of the beam, i.e., it provides range selectivity.

The performance of single beamformers can degrade severely in the presence of other interfering noise sources, especially if they fall in the direction of the sidelobes. This problem can be mitigated using adaptive arrays. Adaptive arrays are briefly discussed in the next section.

32.2.1 A Brief Look at Adaptive Arrays

Adaptive signal processing techniques can be used to form a beam at the desired source while simultaneously forming a null in the direction of the interfering noise source. Such arrays are called "adaptive arrays." Though adaptive arrays are not effective under conditions of severe reverberation, they are included here because problems in adaptive arrays can be formulated as inverse problems. Hence, we shall discuss adaptive arrays briefly without providing a quantitative analysis of them. Broadband arrays have been analyzed in [14–19]. In all these methods, the direction of arrival of the signal is assumed to be known.

Let the array have N sensors and M delay taps per sensor. If $X(k) = [x_1(k) \cdots x_i(k) \cdots x_{NM}(k)]^{\mathrm{T}}$ (see Figure 32.4) is the set of signals observed at the tap points, then $X(k) = S(k) + N(k)$, where $S(k)$ is the contribution of the desired signal at the tap points and $N(k)$ is the contribution of the unknown interfering noise. The inputs to the sensors, $x_{(jM+1)}(k)$, $j = 0, \ldots, (N-1)$, are the noisy versions of $g(k)$, the actual signal at the source. Now, the filter output $y(k) = W^{\mathrm{T}}X(k)$, where $W^{\mathrm{T}} = [w_{11}, \ldots, w_{1M}, w_{21}, \ldots, w_{2M}, \ldots, w_{N1}, \ldots, w_{NM}]$ is the set of weights at the tap points. The goal of the system is to make the output $y(k)$ as close as possible to the source $g(k)$. One way of doing this is to minimize the error $E\{[g(k) - y(k)]^2\}$. The weight W^* that achieves this least mean square (LMS) error is also called the Weiner filter, and is given by

$$W^* = R_{XX}^{-1} C_{gX}, \tag{32.12}$$

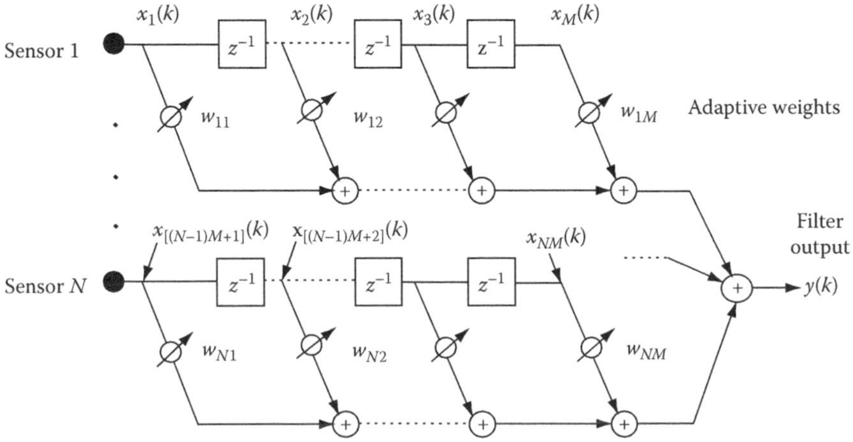

FIGURE 32.4 General form of an adaptive filter.

where

R_{XX} is the autocorrelation of $X(k)$

C_{gX} is the set of cross-correlations between $g(k)$ and each element of $X(k)$

If $g(k)$ and $N(k)$ are uncorrelated, then

$$\begin{aligned} C_{gX} &= E\{g(k)X(k)\} = E\{g(k)S(k)\} + E\{g(k)N(k)\} \\ &= E\{g(k)S(k)\} \end{aligned}$$

and

$$\begin{aligned} R_{XX} &= E\{X(k)X^{\mathrm{T}}(k)\} = E\{[S(k)+N(k)]\,[S(k)+N(k)]^{\mathrm{T}}\} \\ &= R_{SS} + R_{NN}, \end{aligned}$$

where R_{SS} and R_{NN} are the autocorrelation matrices for the signal and noise.

Usually R_{NN} is not known. In such cases, the exact inverse cannot be calculated and an iterative approach to update the weights is needed. In Widrow's approach [15], a known pilot-signal $g(k)$ is injected into the array. Then, the weights are updated using the Widrow–Hopf algorithm that increments the weight vector in the direction of the negative gradient of the error:

$$W^{k+1} = W^k + \mu[g(k) - y(k)]X(k),$$

where

W^{k+1} is the weight vector after the kth update

μ is the step size

Griffiths' method also uses the LMS approach, but minimizes the mean square error based on the autocorrelation and the cross-correlation values between the input and the output, rather than the signals themselves. Since the mean square error can be written as

$$E\{[g(k) - y(k)]^2\} = R_{gg} - 2C_{gS}^{\mathrm{T}}W + W^{\mathrm{T}}R_{XX}W,$$

where

R_{gg} is the auto-correlation matrix of $g(k)$

C_{gS} is the set of cross-correlation matrix between $g(k)$ and each element of $S(k)$

the weight update can also be done by

$$W^{k+1} = W^k + \mu[C_{gS} - R_{XX}W^k] \tag{32.13}$$

$$= W^k + \mu[C_{gS} - X(k)X^{\mathrm{T}}(k)W^k] \tag{32.14}$$

$$= W^k + \mu[C_{gS} - y(k)X(k)]. \tag{32.15}$$

In the above methods, significant distortion is observed in the primary beam due to null-steering. Constrained LMS techniques which place constraints on the performance of the main lobe can be used to reduce distortion [18,19]. By specifying the broadband response and the array beam characteristics as constraints, more robust beams can be formed. The problem now can be formulated as an optimization technique that minimizes the output power of the system. Given that the output power is

$$\begin{aligned} E\{y^2(k)\} &= E\{W^{\mathrm{T}}X(k)X^{\mathrm{T}}(k)W\} = W^{\mathrm{T}}R_{XX}W \\ &= W^{\mathrm{T}}R_{SS}W + W^{\mathrm{T}}R_{NN}W, \end{aligned}$$

if W can be chosen such that $W^{\mathrm{T}}\ R_{NN}W = 0$, the noise can be eliminated. It was proposed [18] that once the array is steered toward the source with appropriate delays, minimizing the output power is equivalent to removing directional interference, since in-phase signals add coherently. In an accurately steered array, the wavefronts arriving from the direction of steering generate identical signals at each sensor. Hence, the array may be collapsed to a single sensor implementation which is equivalent to an FIR filter [18], i.e., the columns of the broadband array sum to an FIR filter. Additional constraints can be placed on this FIR filter. If the weights of the filters can be written as a matrix:

$$\hat{W} = \begin{pmatrix} w_{11} & w_{12} & \dots & w_{1M} \\ \vdots & \vdots & \vdots & \vdots \\ w_{N1} & w_{N2} & \dots & w_{NM} \end{pmatrix},$$

then it can be specified that $\sum_{i=1}^{N} w_{ij} = f_j,\ j = 1, \dots, M$, where $f_j, j = 1, \dots, M$ are the taps of an FIR filter that provides the desired filter response. Hence, using this method, directional interference can be suppressed by minimizing the output power and spectral interference can be suppressed by constraining the columns of the weight coefficients.

Thus, the problem can be formulated as

$$\text{minimize:} \quad W^{\mathrm{T}}R_{XX}W \tag{32.16}$$

$$\text{subject to:} \quad C^{\mathrm{T}}W = F, \tag{32.17}$$

where F is the desired FIR filter and

$$C = \begin{pmatrix} 1 & 0 & 0 & \dots & 0 & 1 & 0 & 0 & \dots & 0 & \dots & 1 & 0 & 0 & \dots & 0 \\ 0 & 1 & 0 & \dots & 0 & 0 & 1 & 0 & \dots & 0 & \dots & 0 & 1 & 0 & \dots & 0 \\ & & & \vdots & & & & & \vdots & & \vdots & & & & \vdots & \\ 0 & 0 & 0 & \dots & 1 & 0 & 0 & 0 & \dots & 1 & \dots & 0 & 0 & 0 & \dots & 1 \end{pmatrix}. \tag{32.18}$$

C has M rows with NM entries on each row. The first row of C in Equation 32.18 has ones in positions $1, (M+1), \dots, (N-1)*M+1$; the second row has ones in positions $2, (M+2), \dots, (N-1)*M+2$, etc. Equation 32.17 can be solved using Lagrange multipliers [18]. This optimization problem can alternatively be posed as an inverse problem.

32.2.2 Constrained Adaptive Beamforming Formulated as an Inverse Problem

Using a similar cost function and the same constraint, the system can be formulated as an inverse problem [19]. The function to be optimized, $W^{\mathrm{T}}R_{XX}W = 0$, can be approximated by $X^{\mathrm{T}}W = 0$. This, combined with the constraint in Equation 32.17, is written as

$$\begin{pmatrix} x_1 & \dots & x_M & \dots & x_{(N-1)*M+1} & \dots & x_{N*M} \\ 1 & \dots & 0 & \dots & 1 & \dots & 0 \\ & \vdots & & \vdots & & \vdots & \\ 0 & \dots & 1 & \dots & 0 & \dots & 1 \end{pmatrix} * \begin{pmatrix} w_{11} \\ \vdots \\ w_{1M} \\ \vdots \\ w_{N1} \\ \vdots \\ w_{NM} \end{pmatrix} = \begin{pmatrix} 0 \\ f_1 \\ \vdots \\ f_M \end{pmatrix}, \tag{32.19}$$

$$AW = F. \tag{32.20}$$

This equation can be solved with any technique that can invert a matrix. There are several problems in solving Equation 32.20. In general, the equation can be inconsistent. In addition, the system is rank deficient. Further, traditional methods used to solve Equation 32.20 are not robust to errors such as round-off errors in digital computers, measurement inaccuracies, and noise corruption. In the least squares solution (Equation 32.6), the noise amplification is dictated by the condition number of the error matrix, i.e., the ratio of the highest and the lowest eigenvalues of E. In the extreme case when $\lambda_{\min} = 0$, the system is rank-deficient. In such cases, the pseudo-inverse solution can be used.

Any matrix A can be written using the singular value decomposition as

$$A = UDV^{\mathrm{T}},$$

where

$$\boldsymbol{D} = \begin{pmatrix} \sigma_1 & 0 & \dots & 0 \\ 0 & \sigma_2 & \dots & 0 \\ \vdots & \vdots & \ddots & \vdots \\ 0 & 0 & \dots & \sigma_N \end{pmatrix},$$

then,

$$A^{-1} = VD^{-1}U^{\mathrm{T}},$$

where

$$\boldsymbol{D}^{-1} = \begin{pmatrix} \frac{1}{\sigma_1} & 0 & \dots & 0 \\ 0 & \frac{1}{\sigma_2} & \dots & 0 \\ \vdots & \vdots & \ddots & \vdots \\ 0 & 0 & \dots & \frac{1}{\sigma_N} \end{pmatrix}.$$

σ_i^2, $i = 1, \dots N$ are the eigenvalues of AA^{T}. The matrices U and V are made up of the eigenvectors of AA^{T} and $A^{\mathrm{T}}A$, respectively.

Extending this definition to rank-deficient matrices, the pseudo-inverse can be written as

$$A^{\dagger} = VD^{\dagger}U^{\mathrm{T}},$$

where

$$D^{\dagger} = \begin{pmatrix} \frac{1}{\sigma_1} & 0 & & \cdots & & 0 \\ 0 & \frac{1}{\sigma_2} & & \cdots & & 0 \\ 0 & 0 & \cdots & \frac{1}{\sigma_r} & & \cdots \\ & & & & 0 & \\ & & & & & 0 \end{pmatrix},$$

where r is the rank of the matrix $\boldsymbol{A}$.

The rank-deficient system has infinite number of solutions. The pseudo-inverse solution can be shown to be the least squares solution with minimum energy. It can also be viewed as the projection of the least squares solution in the range space of A. An iterative technique called the row action projection (RAP) algorithm [4,19] can be used to solve Equation 32.20.

32.2.2.1 Row Action Projection

An effective way to find a solution for Equation 32.20 is to use the RAP method [4], which has been shown to be effective in providing a fast and stable solution to a system of simultaneous equations. Traditional least squares methods need a block of data to calculate the estimate. Most of these methods demand a lot of memory and processing power. RAP operates on only one row at a time, which makes it a useful sample-by-sample method in adaptive signal processing. Further, the matrix A in Equation 32.20 is a sparse matrix. RAP has been shown to be effective in solving systems with sparse matrices [4].

For a given system of equations,

$$\begin{aligned} a_{01}w_1 + a_{02}w_2 + \cdots + a_{0,NM}w_{NM} &= f_0 \\ a_{11}w_1 + a_{12}w_2 + \cdots + a_{1,NM}w_{NM} &= f_1 \\ &\vdots \\ a_{M1}w_1 + a_{M2}w_2 + \cdots + a_{M,NM}w_{NM} &= f_M \,, \end{aligned}$$

each equation can be viewed as a "hyperplane" in NM dimensional space. If a unique solution exists, then it is at the point of intersection of all the hyperplanes. If the equations are inconsistent or ill-defined, then the solution set is a region in space.

The RAP method defines an iterative method to arrive at a point in the solution set and is as follows: Starting from an initial guess W^0, the algorithm iterates over all the equations by repeatedly projecting the solution on the hyperplanes represented by the equations. At step $i+1$ the weight vector is updated as

$$W^{i+1} = W^i + \lambda \frac{e_i}{\|\mathbf{a}_p\|^2}\mathbf{a}_p, \tag{32.21}$$

where $\mathbf{a}_\mathrm{p}$ is the pth row of A, λ is the step size, and

$$e_i = f_p - \mathbf{a}_\mathrm{p}^\mathrm{T}\mathbf{W}^\mathrm{i} \tag{32.22}$$

is the error at the ith iteration. At the ith iteration, we use the pth row, where $p = i \bmod (M+1)$, i.e., we cycle over all the equations.

The RAP method is a special case of the projection onto convex sets algorithm. The geometrical interpretation of the above algorithm is given in Figure 32.5. Each equation is modeled as a hyperplane in

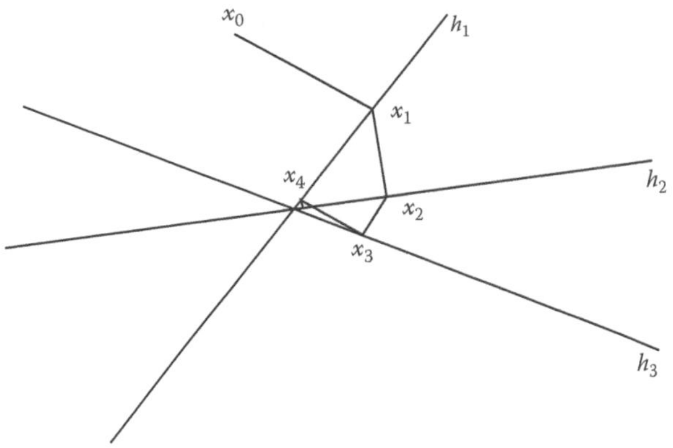

FIGURE 32.5 Geometrical interpretation of RAP.

the solution space. Here, in the figure, it is shown as a line. The initial guess is projected onto the first hyperplane to obtain the second guess. This point is again projected onto the next hyperplane to get the third guess. It can be shown that by repeated projection on to the hyperplanes, the point converges to the solution [4]. λ $(0 \leq \lambda \leq 1)$ is called the relaxation parameter. It dictates how far we should proceed along the direction of the estimate. It is also a measure of confidence in the estimate, i.e., if the measurements are noisy, then usually λ is given a small value; if the values are relatively less noisy, then a larger value of λ can be used to speed up convergence. The algorithm is guaranteed to converge to the actual solution (if it exists). If a solution does not exist, then the "guess" is guaranteed to converge to the pseudo-inverse solution. The pseudo-inverse solution is the least squares solution which minimizes the energy in the solution vector. The RAP method provides stable estimates at each iteration. Since the method uses only one row at a time, the system can be made adaptive, i.e., as the source moves around in the room, the system response can be varied. For a detailed discussion of adaptive arrays, the reader is referred to [20].

32.2.3 Multiple Beamforming

In a highly reverberant environment, many images of the sound source fall along the bore of the beam of a single beamformer. Hence, delay-and-sum single beamformers have limited success in combating reverberation [13]. As shown earlier, the SNR improvement is poor under severe reverberation. Instead of forming a single beam on the source, many beams can be formed, each directed toward the source and its major images [13]. This is called multiple beamforming. In a multiple beamformer (Figure 32.6), the SRNR is $\frac{(BN)^2}{BN(K-1)} = \frac{BN}{(K-1)}$. As B, the number of beams, approaches K, the number of images, the SNR approaches N, or the number of microphones. Multiple beamforming, when $B = K$, can be shown to be equivalent to matched filtering.

32.3 Matched Filtering

Matched filtering techniques can be applied to microphone arrays for dereverberation. In this technique, each microphone output is filtered by a causal approximation of the time reverse of the impulse response to that microphone [13]. Thus, if $g_i(n)$ is the impulse response to microphone i, then

$$h_i(n) = g_i(n_0 - n) \tag{32.23}$$

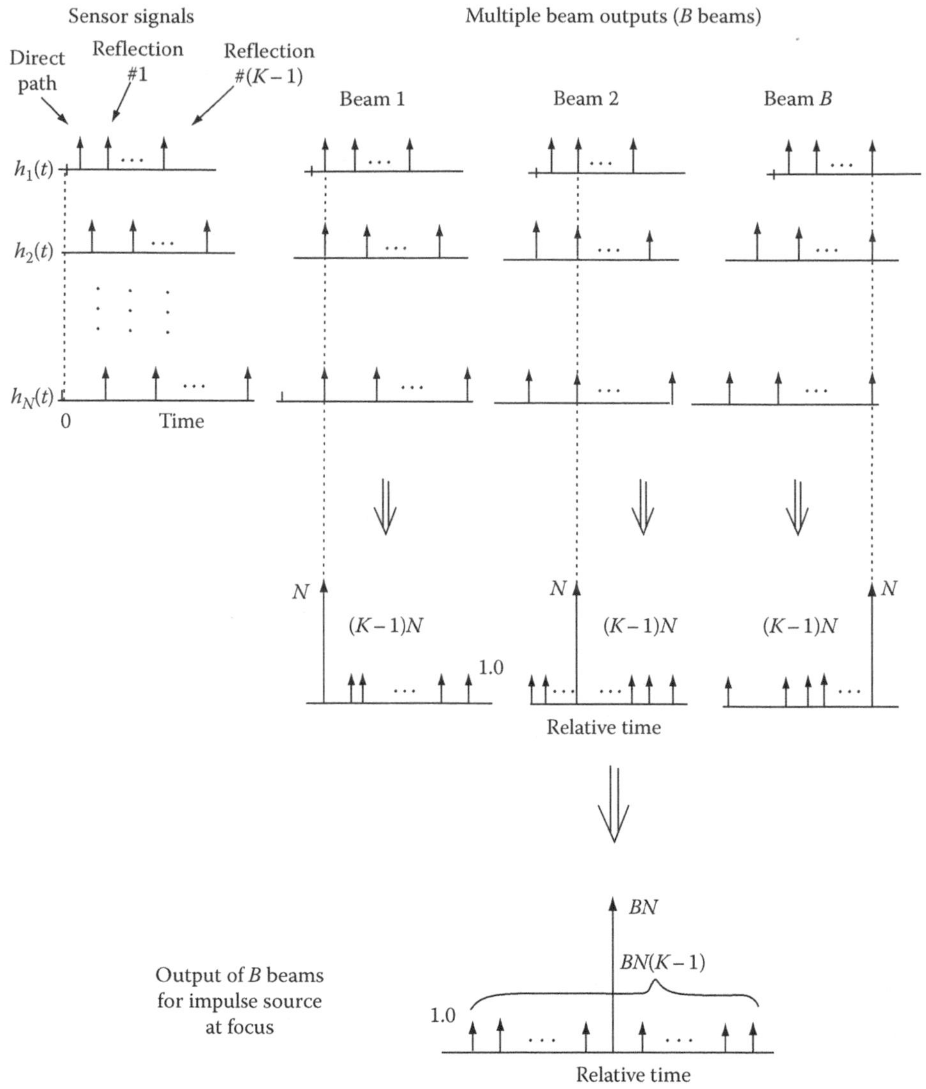

FIGURE 32.6 A multiple beamformer. (From Flanagan, J.L., Surendran, A.C., and Jan, E.-E., *Speech Commn.*, 13, 207, 1993. With permission.)

and

$$H_i(z) = z^{-n_0} G_i\left(\frac{1}{z}\right). \tag{32.24}$$

Since it is desirable for the delay n_0 to be suitably small, the time-reversed response is typically truncated. But careful choice of n_0 leads to a good compromise between delay of the system and high SNR. The matched filter can also be viewed as a special case of a multiple beamformer, when a beam is directed at every image, and when the output of the ith microphone contributing to the beam directed to the jth image is weighted by $\frac{1}{d_{ij}}$, where d_{ij} is the distance of the ith microphone from the jth image. Figure 32.7

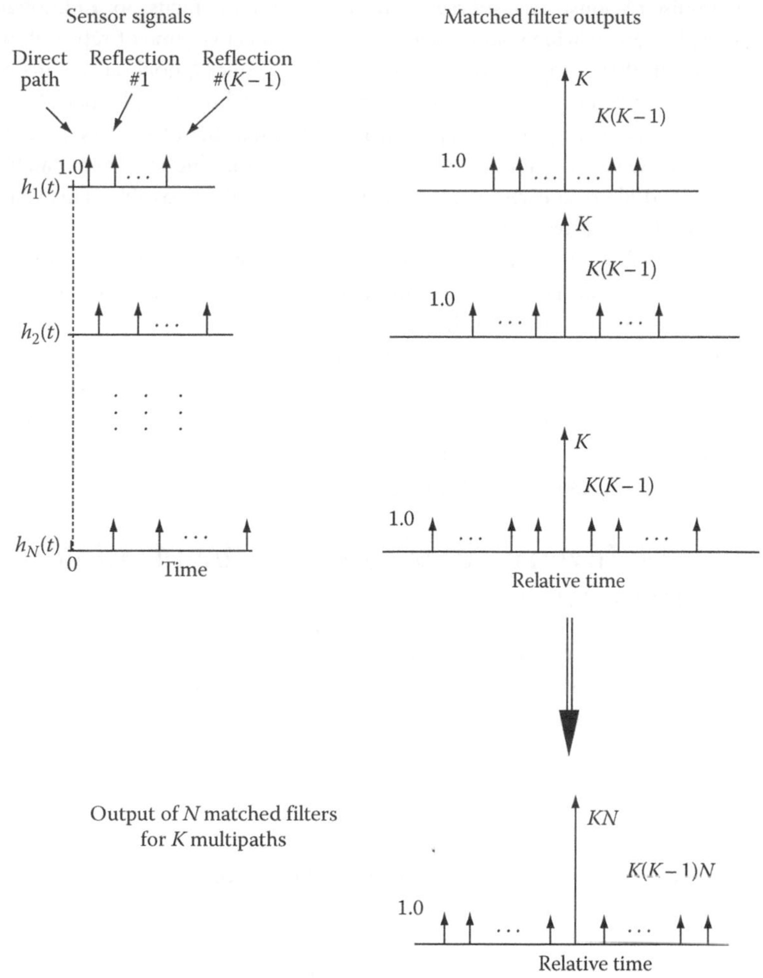

FIGURE 32.7 Principle of a matched filter. (From Flanagan, J.L., Surendran, A.C., and Jan, E.-E., *Speech Commn.*, 13, 207, 1993. With permission.)

shows the principle of a matched filter. The SNR analysis of a matched filter is similar to the multiple beamformer when $B = K$.

Thus, for a source $s(n)$ located at the focal point, the output of the system is

$$o(n) = s(n)^{*}\left\{\sum_{i=1}^{N} g_i(n)^{*}g_i(n_0 - n)\right\}, \tag{32.25}$$

and the output for a source away from the focus is

$$o(t) = s(t)^{*}\left\{\sum_{i=1}^{N} g_i'(n)^{*}g_i(n_0 - n)\right\}, \tag{32.26}$$

where $g_i'(n)$ is the impulse response for a source located away from the focus. So, additional to mitigating reverberation, matched filters provide volume selectivity, i.e., a focal volume of retrieval, which depends on the spatial correlation of the impulse responses $g_i(n)$. Using microphone arrays instead of a single microphone provides not only a smoother frequency response [22], but also a higher SNR improvement, which, even in the worst case, asymptotically approaches N, the number of sensors used [13]. Since each individual matched filter seeks to smooth out the spectral minima due to other matched filters, it is desirable that the matched filters at each microphone be as different as possible. This is a motivation to use a random distribution of sensors [22].

The aim of the matched filter is to maximize the power of the output of the array for a source located at the focus and minimize the power of off-focus sources. This is an important property, which we shall contrast with the exact inverse discussed in the next section.

The power of matched filtering in mitigating reverberation and suppressing interfering noise is demonstrated through examples in Section 32.5. Figure 32.11 shows the response of a matched filter system. It is clear that the matched filter response is similar to, but cannot be exactly, an ideal impulse, i.e., it cannot provide an exact inverse to the room transfer function. Next, we discuss a method that can provide an exact inverse to the room transfer function.

32.4 Diophantine Inverse Filtering Using the Multiple Input–Output Model

Miyoshi and Kaneda [23] proposed a novel method to find the exact inverse of a point in a room by using multiple inputs and outputs, each input–output pair modeled by an FIR system. For example, a two-input single-output system is described by the two speaker-to-single-microphone responses, $G_1(z)$ and $G_2(z)$. The inputs need to be pre-processed by the two FIR filters, $H_1(z)$ and $H_2(z)$, such that

$$H_1(z)G_1(z) + H_2(Z)G_2(Z) = 1. \tag{32.27}$$

This is a Diophantine equation which has an infinite number of solutions. That is, if $H_1(z)$ and $H_2(z)$ satisfy Equation 32.27, then

$$H_1' = H_1(z) + G_2(z)K(z) \tag{32.28}$$

$$H_2' = H_2(z) - G_1(z)K(z), \tag{32.29}$$

where $K(z)$ is an arbitrary polynomial, is also a solution for Equation 32.27. But, if $G_1(z)$ and $G_2(z)$ do not have common zeros in the z-plane, and if the orders of $H_1(z)$ and $H_2(z)$ are less than that of $G_2(z)$ and $G_1(z)$, respectively, by Euclid's theorem, a unique solution is guaranteed to exist [23,24].

The above system can be used with a microphone array for dereverberation (Figure 32.1). The problem is to find $H_i(z)$, $i = 1, 2, \ldots, N$ such that

$$G_1(z)H_1(z) + G_2(z)H_2(z) + \cdots + G_N(z)H_N(z) = 1. \tag{32.30}$$

As the number of microphones in the array increases, the chances that all the $G_i(z)$'s share a common zero in the z-plane diminishes. This assures that the multiple microphone system yields a unique and exact solution.

In time domain, the previous expression can be written as

$$d(k) = g_1(k)^* h_1(k) + \cdots + g_N(k)^* h_N(k), \tag{32.31}$$

where N is the number of microphones. Now,

$$\begin{pmatrix} g_1(0) & & & & g_N(0) & & \\ g_1(1) & & & & g_N(1) & & \\ \vdots & \cdots & 0 & \cdots & \vdots & \cdots & 0 \\ g_1(m) & \cdots & g_1(0) & \cdots & g_N(l) & \cdots & g_N(0) \\ 0 & \cdots & g_1(1) & \cdots & 0 & \cdots & g_N(1) \\ 0 & \cdots & \vdots & \cdots & 0 & \cdots & \vdots \\ & & g_1(m) & & & & g_N(l) \end{pmatrix} \begin{pmatrix} h_1(0) \\ \vdots \\ h_1(i) \\ \vdots \\ h_N(0) \\ \vdots \\ h_N(k) \end{pmatrix} = \begin{pmatrix} 1 \\ 0 \\ \vdots \\ 0 \end{pmatrix}, \tag{32.32}$$

$$(G_1 \cdots G_N) \begin{pmatrix} H_1 \\ \vdots \\ H_N \end{pmatrix} = D. \tag{32.33}$$

Thus,

$$\begin{pmatrix} H_1 \\ \vdots \\ H_N \end{pmatrix} = (G_1 \cdots G_N)^{-1} D. \tag{32.34}$$

The RAP algorithm described in Section 32.2.2.1 is an effective method to solve Equation 32.34. In the MINT modeling, even if the different $G_i(z)$'s share a common zero, RAP can provide a stable inverse. Even if the data are "noisy," or if the system is ill-conditioned, the algorithm is guaranteed to converge. From computer simulations, it can be shown that the solution converges very fast (see Figure 32.8). Hence, the system can adapt to the varying conditions without having to recalculate the FIR filters.

Figure 32.8 shows the rate of convergence of the RAP algorithm when the number of microphones in the array is varied. The results suggest that increasing the number of microphones used in the array increases the speed of convergence and also provides more accurate results.

32.5 Results

In this section, computer simulations are presented to demonstrate the effect of matched filtering and the Diophantine inverse filtering method. A room ($20 \times 16 \times 5$ m in size) was simulated using the image model [10]. The source was located at (14, 9.5, 1.7) m. Fifth order images were assumed and wall reflectivity was assumed to be $\alpha = 0.1$. Sensor spacing was considered to be 40 cm. A large spacing between sensors was chosen to make the impulse responses as dissimilar as possible.

The SNR of the output was calculated using the formula:

$$\text{SNR(dB)} = 10 \log_{10} \frac{\sum s(n)^2}{\sum (y(n) - s(n))^2}, \tag{32.35}$$

where

$s(n)$ is the input speech signal
$y(n)$ is the output speech signal

The two signals are sufficiently staggered to account for the delay in the processing.

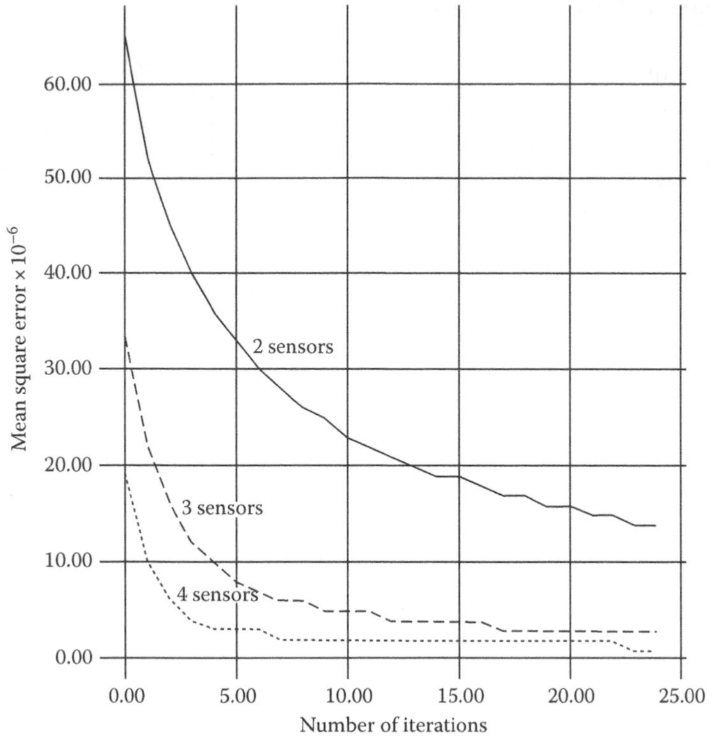

FIGURE 32.8 Rate of convergence of RAP for calculating the exact inverse filters.

The SNRs were calculated as follows:

No. of mics	SNR(dB)
2	15
3	27
4	37

For comparison, the SNR gains of a single beamforming, multiple beamforming, and matched filter linear arrays using five microphones are presented below. The multiple beamformer has one beam directed at each image of the source.

Method	SNR(dB)
Single beamformer	−1
Multiple beamformer	11
Matched filter	13

Figure 32.9 shows the impulse response of the room using an unsteered array system consisting of four microphones. Figures 32.10 and 32.11 are the system responses of a single beamformer and the matched filter. The matched filter system response is a much better approximation of an ideal impulse than the single beamformer. But the tail of the response is still significant compared to the exact inverse system (Figure 32.12) whose final response is very close to an ideal impulse.

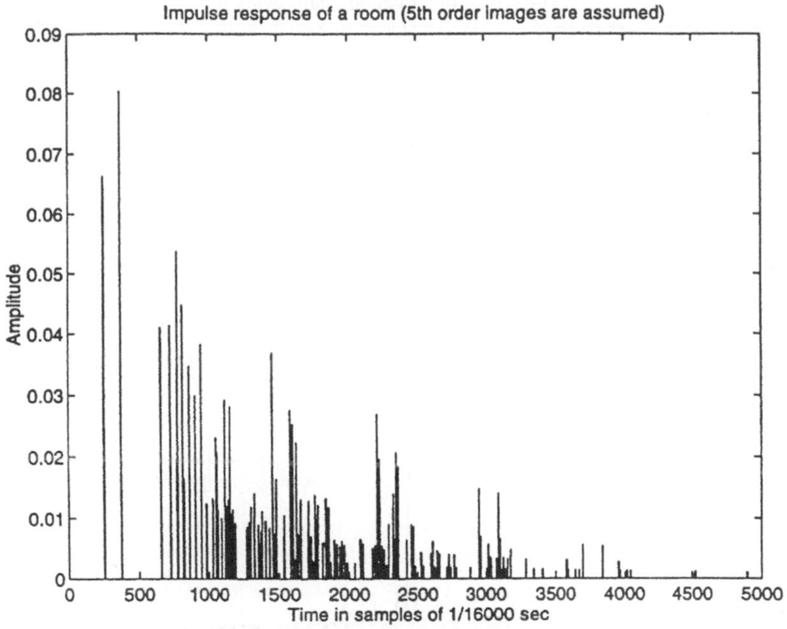

FIGURE 32.9 Impulse response of a room (images up to fifth order are used).

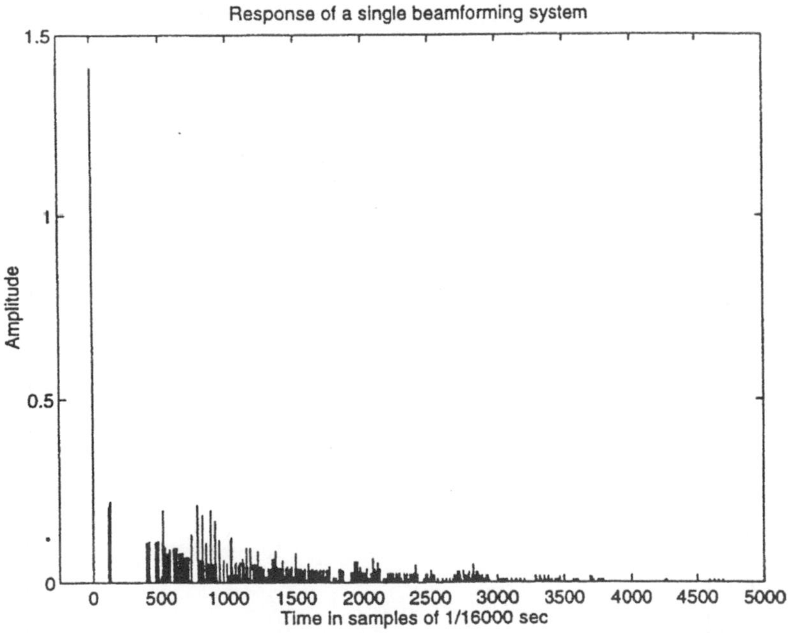

FIGURE 32.10 Response of a single beamformer for a source located on the axis.

For obtaining the same SNR gain, the exact inverse requires a lesser number of microphones than either the matched filter or the multiple beamformer. The Diophantine inverse filtering method does not suffer from the effects of spatial aliasing that may affect traditional beamformers using periodically spaced microphones. Finding the exact inverse is also more computationally intensive than matched filtering or multiple beamforming.

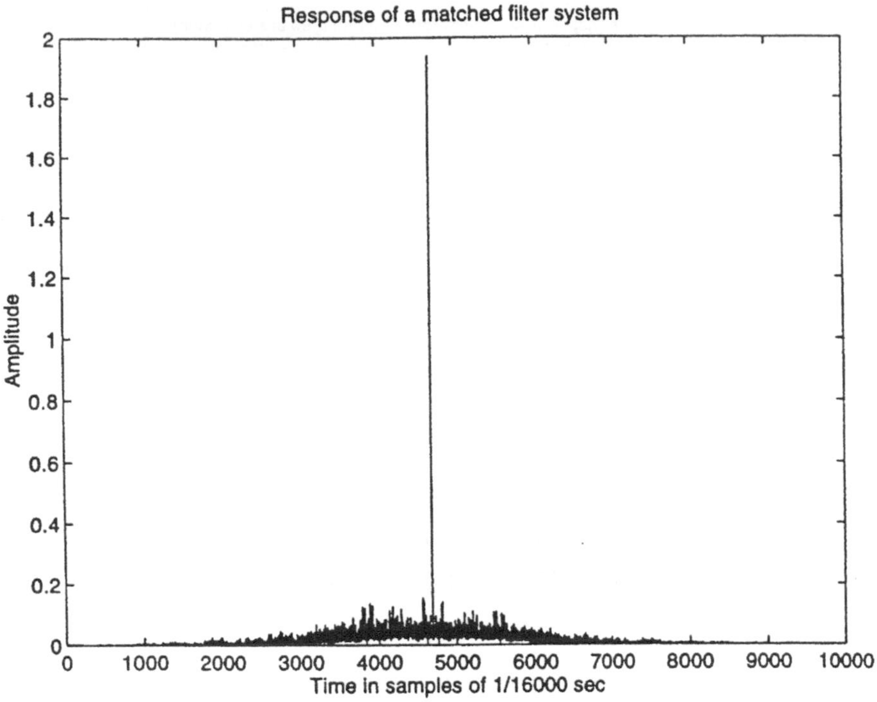

FIGURE 32.11 Response of a matched filtering system for a source located at the focus.

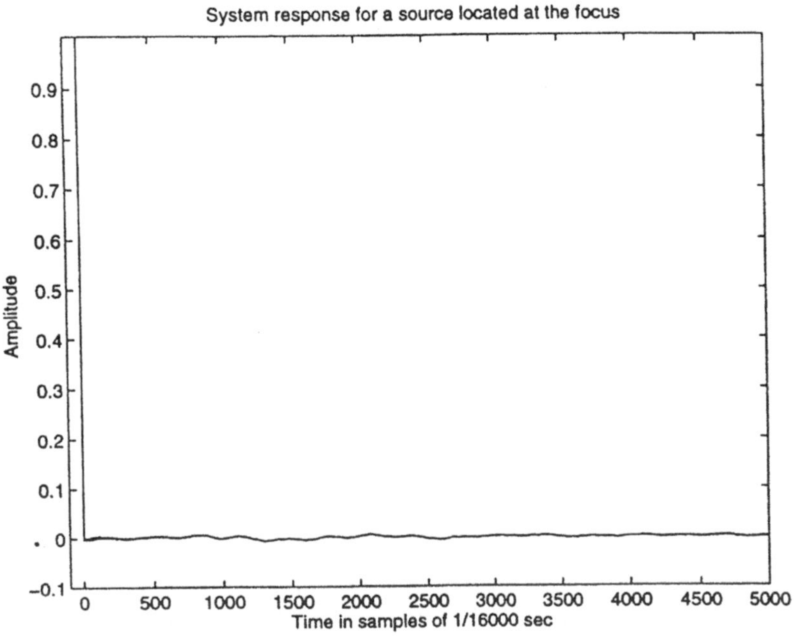

FIGURE 32.12 Response of the Diophantine inverse filtering system (the delay involved is not shown).

32.5.1 Speaker Identification

A simple speaker identification experiment was done to test the acoustic fidelity of the exact inverse system. The dimensions of the simulated room, the location of the source and the other conditions was assumed to be identical to the experiment reported in the previous section. A part of the TIMIT database with 38 speakers, all from the New England area, was used. Five sentences were used for training and five were used for testing. Twelve cepstral vectors were used and a learning vector quantizer was used for identification [25].

	Testing data		
Training data	CLS (%)	One mic (%)	Array output (%)
Speaker identification accuracy for the exact inverse system:			
CLS	91.6	36.3	90
Array output	—	—	92.6
Speaker identification accuracy for the exact inverse system when an interfering Gaussian noise source at 15 dB signal-to-competing-noise ratio is present:			
CLS	91.6	14.2	9.5
Array output	—	—	49

The identification accuracy when trained and tested on clean speech recorded through a close talking microphone (CLS) was 91.6%. The performance dropped to 36.3% when the same system was tested on a single microphone located at the center of the array. Once the Diophantine inverse filtering was used to clean up the speech, the performance jumped back to 90%. The identification accuracy when the system was trained and tested on the Diophantine inverse filtered output was 92.6%.

But the performance was poor even in the presence of modest interference. When a Gaussian noise source at 15 dB signal-to-competing-noise ratio levels was introduced at (3.0, 5.0, 1.0) m, the performance on the output of the exact inverse filtering system (9.5%) was worse than the single microphone (14.2%). Under matched training and testing conditions, the performance of the exact inverse system was significantly lower (49%).

Recently, speaker identification results were reported on the output of a matched-filtered system [26]. The room dimensions and conditions were similar to the ones in this report and the data sets used for training and testing were the same. The performance under matched conditions for close talking microphone was 94.7% and for the matched filtered output was 88.4%. In the presence of an interfering source producing Gaussian noise at 15 dB signal-to-competing-noise ratio levels, the performance when trained on close talking microphone and tested on the matched filtered output was 80%; the performance when trained and tested on the matched filtered output in the presence of noise was approximately 88% [26].

From these results, it is clear that though the exact inverse filtering outperforms the matched filter under clean conditions, it performs significantly poorer when there are interfering noise sources. This can be attributed to the fact that the exact inverse system attempts to maximize the SRNR for a source at the focus. Though it maximizes the SRNR for a source at the focus and lowers the SRNR for any source located away from the focus, it does not guarantee that the contribution of interfering source to the output power will also be lowered. Figure 32.13 shows the impulse response of the exact inverse system for the location of the interfering noise source. It is clear that the SNR of the source at this location would be poor (the effective response does not look like an ideal impulse). But the signal is effectively amplified. On the other hand, the matched filter maximizes the output power for a source located at the focus and minimizes the output power for all other sources thus providing lower SNR ratio improvement, but higher levels of spatial discrimination.

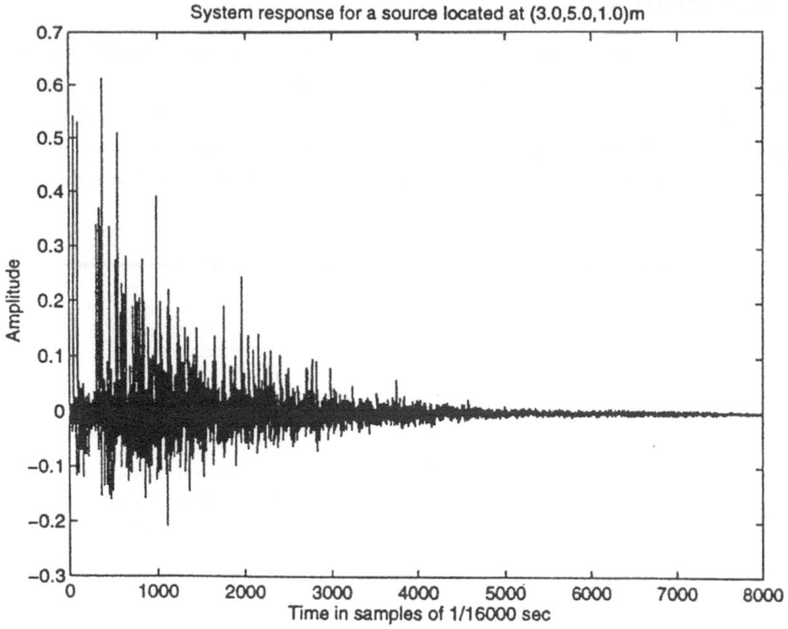

FIGURE 32.13 Response of the Diophantine inverse filtering system for a source located away from the focus.

32.6 Summary

Microphone arrays can be successfully used in "inverting" room acoustics. A simple single beamformer is not effective in combating room reverberation, especially in the presence of interfering noise sources. Adaptive algorithms that project a null in the direction of the interferer can be used, but they introduce significant distortion in the main signal. Constrained adaptive arrays mitigate this problem but they are of limited capability in severely reverberant environments. Processing algorithms such as multiple beamforming and matched filtering, combined with three-dimensional array of sensors, though only providing an approximation to the inverse, give robust dereverberant systems that provide selectivity in a spatial volume and thus immunity from interfering noise sources. An exact inverse using Diophantine inverse filtering using the MINT model can be found. Though this method provides a higher SNR for a source at the focus, it does not provide immunity from noise interference that the matched filtering can offer. Speaker identification results are provided that substantiate the performance analysis of these systems.

References

1. Flanagan, J.L. and Lummis, R.C., Signal processing to reduce multipath distortions in small rooms, *J. Acoust. Soc. Am.*, 47, 1475–1481, Feb. 1970.
2. Neely, S. and Allen, J., Invertibility of a room response, *J. Acoust. Soc. Am.*, 66, 165–169, 1979.
3. Mourjopoulos, J., Clarkson, P.M., and Hammond, J.K., A comparative study of least-squares and homomorphic techniques for the inversion of mixed phase signals, *Proceedings of IEEE Conference on Acoustics, Speech, and Signal Processing*, Paris, France, May 1982, Vol. 7, pp. 1858–1861.
4. Mammone, R.J., *Computational Methods of Signal Recognition and Recovery*, John Wiley & Sons, New York, 1992.
5. Mourjopoulos, J. and Hammond, J.K., Modelling and enhancement of reverberant speech using an envelope convolution method, *Proceedings of IEEE Conference on Acoustics, Speech, and Signal Processing*, Boston, MA, Apr. 1983, Vol. 8, pp. 1144–1147.

6. Stockham, T.G., Cannon, T.M., and Ingebresten, B.R., Blind deconvolution through digital signal processing, *Proc. IEEE*, 63(4), 678–692, 1975.
7. Langhans, T. and Strube, H.W., Speech enhancement by nonlinear multiband envelope filtering, *Proceedings of IEEE Conference on Acoustics, Speech, and Signal Processing*, New York, May 1982, Vol. 7, pp. 156–159.
8. Wang, H. and Itakura, F., Dereverberation of speech signals based on sub-band envelope estimation, *ICIE Trans.*, E 74(11), 3576–3583, Nov. 1991.
9. Allen, J.B., Berkeley, D.A., and Blauert, J., Multimicrophone signal processing technique to remove room reverberation from speech signals, *J. Acoust. Soc. Am.*, 62, 912–915, Oct. 1977.
10. Allen, J.B. and Berkeley, D.A., Image method for efficiently simulating small-room acoustics, *J. Acoust. Soc. Am.*, 65(4), 943–950, Apr. 1979.
11. Flanagan, J.L., Berkeley, D.A., Elko, G.W., and Sondhi, M.M., Autodirective microphone systems, *Acustica*, 73, 58–71, 1991.
12. Flanagan, J.L., Beamwidth and usable bandwidth of delay-steered microphone arrays, *AT&T Tech. J.*, 64(4), 983–995, Apr. 1985.
13. Flanagan, J.L., Surendran, A.C., and Jan, E.-E., Spatially selective sound capture for speech and audio processing, *Speech Commn.*, 13, 207–222, 1993.
14. Widrow, B. and Stearns, S.T., *Adaptive Signal Processing*, Prentice-Hall, Englewood Cliffs, NJ, 1985.
15. Widrow, B., Mantey, P.E., Griffiths, L.J., and Goode, B.B., Adaptive antenna systems, *Proc. IEEE*, 55, 2143–2159, Dec. 1967.
16. Griffiths, L.J., A simple adaptive algorithm for real-time processing in antenna arrays, *Proc. IEEE*, 57(10), 1696–1704, Oct. 1969.
17. Griffiths, L.J. and Jim, C.W., An alternative approach to linearly constrained adaptive beamforming, *IEEE Trans. Antennas Propagation*, AP-30(1), 27–34, Jan. 1982.
18. Frost III, O.L., An algorithm for linearly constrained adaptive array processing, *Proc. IEEE*, 60(8), 926–935, 1972.
19. Farrell, K., Mammone, R.J., and Flanagan, J.L., Beamforming microphone arrays for speech enhancement, *Proceedings of IEEE Conference on Acoustics, Speech, and Signal Processing*, San Francisco, CA, Mar. 23–26, 1992, Vol. 1, pp. 285–288.
20. *IEEE Trans. Antennas Propagation: Special Issues on Adaptive Arrays*, 34(3), Mar. 1986.
21. Applebaum, S.P., Adaptive arrays, *IEEE Trans. Antennas Propagation*, AP-24(5), 585–599, Sept. 1976.
22. Jan, E.-E. and Flanagan, J.L., Microphone arrays for speech processing, *International Symposium on Signals, Systems, and Electronics*, San Francisco, CA, Oct. 1995, pp. 373–376.
23. Miyoshi, M. and Kaneda, Y., Inverse filtering of room acoustics, *IEEE Trans. Acoust. Speech Signal Process.*, 36(2), 145–152, Feb., 1988.
24. Sondhi, M.M., Personal communication.
25. Surendran, A.C. and Flanagan, J.L., Stable dereverberation using microphone arrays for speaker identification, *J. Acoust. Soc. Am.*, 96(5), 3261, Nov. 1994.
26. Lin, Q., Jan, E.-E., and Flanagan, J.L., Microphone arrays and speaker identification, *IEEE Trans. Speech Audio Process.*, 2(4), 622–629, Oct., 1994.

33

Synthetic Aperture Radar Algorithms

Clay Stewart
Science Applications International Corporation

Vic Larson
Science Applications International Corporation

33.1 Introduction

A synthetic aperture radar (SAR) is a radar sensor that provides azimuth resolution superior to that achievable with its real beam by synthesizing a long aperture using platform motion. The geometry for the production of the SAR image is shown in Figure 33.1. The SAR is used to generate an electromagnetic map of the surface of the earth from an airborne or spaceborne platform. This electromagnetic map of the surface contains information that can be used to distinguish different types of objects that make up the surface. The sensor is called an SAR because a synthetic aperture is used to achieve the narrow beamwidth necessary to get a high cross-range resolution. In SAR imagery the two dimensions are range (perpendicular to the sensor) and cross-range (parallel to the sensor). The range resolution is achieved using a high bandwidth pulsed waveform. The cross-range resolution is achieved by making use of the forward motion of the radar platform to synthesize a long aperture giving a narrow beamwidth and high cross-range resolution. The pulse returns collected along this synthetic aperture are coherently combined to create the high cross-range resolution image. A SAR sensor is advantageous compared to an optical sensor because it can operate day and night through clouds, fog, and rain, as well as at very long ranges. At very low nominal operating frequencies, less than 1 GHz, the radar even penetrates foliage and can image objects below the tree canopy. The resolution of a SAR ground map is also not fundamentally limited by the range from the sensor to the ground. If a given resolution is desired at a longer range, the synthetic aperture can simply be made longer to achieve the desired cross-range resolution.

A SAR image may contain "speckle" or coherent noise because it results from coherent processing of the data. This speckle noise is a common characteristic of high frequency SAR imagery and reducing speckle, or building algorithms that minimize speckle, is a major part of processing SAR imagery beyond the image formation stage. Traditional techniques averaged the intensity of adjacent pixels, resulting in a smoother but lower resolution image. Advanced SAR sensors can collect multiple polarimetric and/or frequency channels where each channel contains unique information about the surface. Recent systems have also used elevation angle diversity to produce three-dimensional (3-D) SAR images using interferometric techniques. In all of these techniques, some sort of averaging is employed to reduce the speckle.

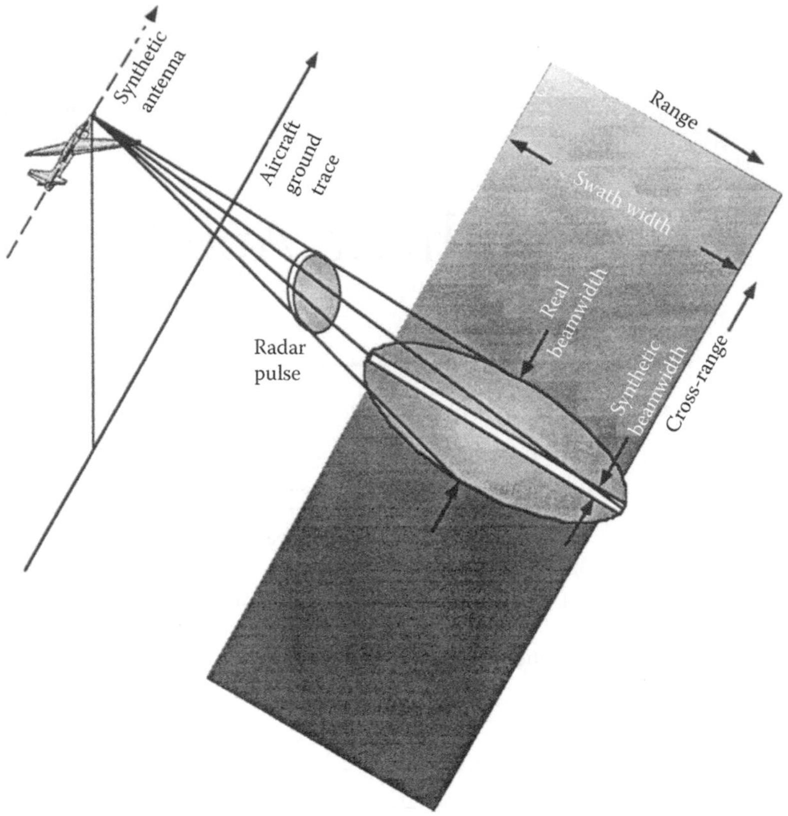

FIGURE 33.1 SAR imaging geometry.

The largest consumers of SAR sensors and products are the defense and intelligence communities. These communities use SAR to locate and target relocatable and fixed objects. Manmade objects, especially ones with sharp corners, have very bright signals in SAR imagery, making these objects particularly easy to locate with a SAR sensor. A technology similar to SAR is inverse synthetic aperture radar (ISAR) which employs motion of the platform to image the target in cross-range. The ISAR data can be collected from a fixed radar platform since the target motion creates the viewing angle diversity necessary to achieve a given cross-range resolution. ISAR systems have been used to image ships, aircraft, and ground vehicles.

In addition to the defense and intelligence applications of SAR, there are several commercial remote sensing applications. Because a SAR sensor can operate day and night and in all weather, it provides the ability to collect data at regular intervals uninterrupted by natural influences. This stable source of ground mapping information is invaluable in tracking agriculture and other natural resources. SAR sensors have also been used to track oil spills (oil-coated water has a different backscatter than natural water), image underground rock formations (at some frequencies the radar will penetrate some soils), track ice conditions in the Arctic, and collect digital terrain elevation data.

Radar is an abbreviation for RAdio Detection And Ranging. Radar was developed in the 1930s and 1940s to detect and track ships and aircraft. These surveillance and tracking radars were designed so that a target was contained in a single resolution cell. The size of the resolution cell was a critical design parameter. Smaller resolution cells allowed one to determine the location of a target more accurately and increased the target-to-clutter ratio, improving the ability to detect a target. In the 1950s it was observed that one could map the ground (an extended target that takes up more than one resolution cell) by mounting the radar on the side of an aircraft and building a surface map from the radar returns. High range resolution was achieved by using a short pulse or high bandwidth waveform. The cross-range resolution was limited by the size of the antenna, with the cross-range resolution roughly proportional to R/L_a where R is the range from the sensor to the ground and L_a is the length of the antenna. The physical length of the antenna was constrained, limiting the resolution. In 1951, Carl Wiley of the Goodyear Aircraft Corporation noted that the reflections from two fixed targets in the antenna beam, but at different angular positions relative to the velocity vector of the platform, could be resolved by frequency analysis of the along track (or cross-range) signal spectrum. Wiley simply observed that each target had different Doppler characteristics because of its relative position to the radar platform and that one could exploit the Doppler to separate the targets. The Doppler effect is, of course, the change in frequency of a signal transmitted or received from a moving platform discovered by Christian J. Doppler in 1853:

$$f_d = \nu/\lambda$$

where
- f_d is the Doppler shift
- ν is the radial velocity between the radar and target
- λ is the radar wavelength

While the Doppler effect had been used in radar processing before the 1950s to separate moving targets from stationary ground clutter, Wiley's contribution was to discover that with a side looking airborne radar (SLAR), Doppler could be used to improve the cross-range spatial resolution of the radar. Other early work on SAR was done independently of Wiley at the University of Illinois and the University of Michigan during the 1950s. The first demonstration of SAR mapping was done in 1953 by the University of Illinois by performing frequency analysis of data collected by a radar operating at a 3 cm wavelength from a C-46 aircraft. Much work has been accomplished perfecting SAR hardware and processing algorithms since the first demonstration. For a much more detailed description of the history of SAR including the development of focused SAR, phase compensation techniques, calibration techniques, and autofocus, see the recent book by Curlander and McDonough [1].

Before offering a brief description of some processing approaches for forming, enhancing, and interpreting SAR imagery, we give two examples of existing SAR systems and their applications. The first system is the Shuttle Imaging Radar (SIR) developed by the NASA Jet Propulsion Laboratory (JPL) and flown on several space shuttle missions. This system was designed for nonmilitary collection of geographic data. The second example is the Advanced Detection Technology Sensor (ADTS) built by the Loral Corporation for the MIT Lincoln Laboratory. The ADTS sensor was designed to demonstrate the capability of a SAR to detect and classify military targets. Table 33.1 contains the basic parameters for the ADTS and SIR SAR systems along with details on several other SAR systems.

Figure 33.2 shows an example image formed from data collected by the SIR SAR. The JPL engineers describe this image as follows:

> This is a radar image of Mount Rainier in Washington state... This image was acquired by the Spaceborne Imaging Radar-C and X-band Synthetic Aperture Radar (SIR-C/X-SAR) aboard the space shuttle Endeavor on its 20th orbit on October 1, 1994. The area shown in the image is approximately 59 kilometers by 60 kilometers (36.5 miles by 37 miles). North is toward the top left of the image, which was composed by assigning red and green colors to the L-band, horizontally transmitted and vertically received, and the L-band, horizontally transmitted and vertically received. Blue indicates the C-band, horizontally transmitted and vertically received. In addition to highlighting topographic slopes facing the space shuttle, SIR-C records rugged areas as brighter and smooth areas as darker. The scene was illuminated by the shuttle's radar from the northwest so that northwest-facing slopes are brighter and southeast-facing slopes are dark. Forested regions are

TABLE 33.1 Example SAR Systems

Platform	Bands Polarization	Resolution (m)	Swath Width	Interferometry
JPL AIRSAR	C, L, P-Full	4	10–18 km	Cross track L, C Along track L, C
SIR-C/X-SAR	C, L-Full, X-VV	30 × 30	15–90	Multi-pass
ERIM IFSARE	X-HH	2.5 × 0.8	10 km	Cross track
ERIM DCS	X-Full	<1	1 km	Cross track
MIT LL ADTS	Ka (33 GHz)-Full	0.33	400 m	Multi-pass
NORDEN G11	Ku-VV	1, 3	5 km	3 Along track 3 Cross track Phase centers
SRI UWB	100–300 MHz,	1 × 1	400–600 m	None
FOLPEN 2	200–400 MHz, 300–500 MHz, HH			
LORAL UHF	500–800 MHz	0.6 × 0.6	280 m	None
MSAR	Full			
NAWC P-3	C, L, X-Full	1.5 × 0.7	5 km	Along track X, C
NAWC P-3	600 MHz-Full tunable over 200–900 MHz	0.33 × 0.66	930 km	None
Tier II+ UAV SAR	X	1 and 0.3	10 km	None

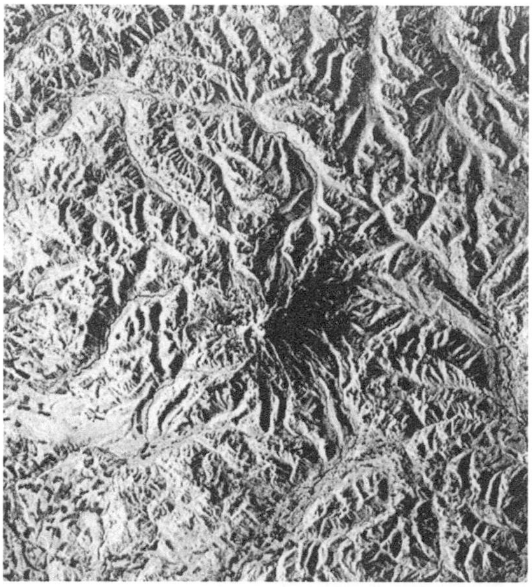

FIGURE 33.2 SAR image of Mt. Rainier in Washington State taken from shuttle imaging radar.

pale green in color; clear cuts and bare ground are bluish or purple; ice is dark green and white. The round cone at the center of the image is the 14,435-foot (4,399-meter) active volcano, Mount Rainier. On the lower slopes is a zone of rock ridges and rubble (purple to reddish) above coniferous forests (in yellow/green). The western boundary of Mount Rainier National Park is seen as a transition from protected, old-growth forest to heavily logged private land, a mosaic of recent clear cuts (bright purple/blue) and partially regrown timber plantations (pale blue).

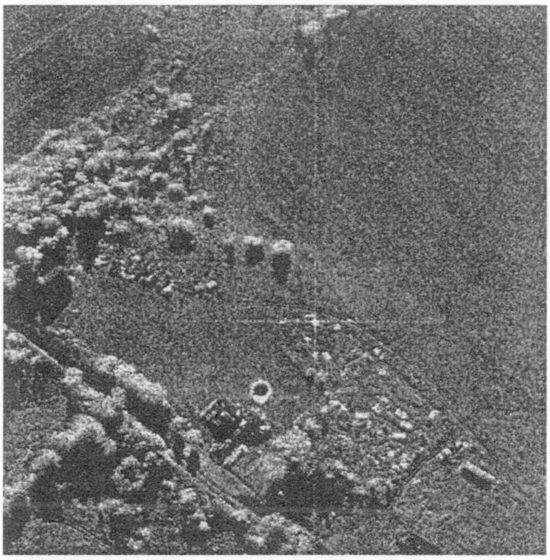

FIGURE 33.3 SAR image near Stockbridge, New York, collected by the ADTS.

Figure 33.3 is an example image collected by the ADTS system. The ADTS system operates at a nominal frequency of 33 GHz and collects fully polarimetric, 1 ft resolution data. This image was formed using the polarimetric whitening filter (PWF) combination of three polarimetric channels to reduce the speckle noise. The output of the PWF is an estimate of radar backscatter intensity. The image displayed in Figure 33.3 is based on a false color map which maps low intensity to black followed by green, yellow, and finally white. The color map simply gives the noncolor radar sensor output false colors that make the low intensity shadows look black, the grass look green, the trees look yellow, and bright objects look white. This sample image was collected near Stockbridge, New York, and is of a house with an above ground swimming pool and several junked cars in the backyard. The radar is at the top of the image looking down at a 20° depression angle. The scene contains large areas of grass or crops and some foliage. Note the bright returns from the manmade objects, including the circular aboveground swimming pool, and strong corner reflector scattering from some of the cars in the backyard. Also note the relatively strong return from the foliage canopy. At this frequency the radar does not penetrate the foliage canopy. Note the shadows behind the trees where there is no radar illumination.

In this chapter on SAR algorithms, we give a brief introduction to the image formation process in Section 33.2. We review a few simple algorithms for reducing speckle noise in SAR imagery and automatic detection of manmade objects in Section 33.3. We review a few simple automatic object classification algorithms for SAR imagery in Section 33.4. This brief introduction to SAR only contains a few example algorithms. In the section "Further Reading," we recommend some starting points for further reading on SAR algorithms, and discuss several open issues under current research in the SAR community.

33.2 Image Formation

In this section, we discuss some basic principles of SAR image formation. For more detailed information about SAR image formation, the reader is directed to the references given at the end of this chapter. One fundamental scenario under which SAR data is collected is shown in Figure 33.1. An aircraft flies in a straight path at a constant velocity and collects radar data at a boresight of 90°. In practice it is impossible for an aircraft to fly in a perfectly straight line at a constant velocity (at least within a wavelength), so motion (phase) compensation of the received radar signal is needed to account for aircraft perturbations.

The radar on the aircraft transmits a short pulsed waveform or uses frequency modulation to achieve high range resolution imaging of the surface. The pulses collected from several positions along the trajectory of the aircraft are coherently combined to synthesize a long synthetic aperture in order to achieve a high cross-range resolution on the surface. In this section, we first discuss SLAR where only range processing is performed. Next, we discuss unfocused SAR where both range and cross-range processing are executed. Finally, we discuss focused SAR where "focusing" is performed in addition to range and cross-range processing to achieve the highest resolution and best image quality. At the end of this section we briefly mention several other important SAR image formation topics such as phase compensation, clutter-lock, autofocus, spotlight SAR, and ISAR. The details of these topics can be found in [1–3].

33.2.1 Side-Looking Airborne Radar

SLAR is the earliest radar system for remote surveillance of a surface. These radar systems could only perform range processing to form the 2-D reflectivity map of the surface, so the cross-range resolution is limited by the real antenna beamwidth. These SLAR systems typically operated at high frequencies (microwave or millimeter-wave) to maximize the cross-range resolution. We cover SLAR systems because SLAR performs the same range processing as SAR, and the limitations of a SLAR motivate the need for SAR processing.

The resolution of a SLAR system is limited by the radar pulse width in the range dimension, and the beamwidth and slant range in the cross-range dimension:

$$\delta_r = cT/2\cos\eta$$
$$\delta_{cr} = R\lambda/L_a$$

where

λ/L_a represents the approximate 3 dB beamwidth of the antenna
δ_r is the range resolution
δ_{cr} is the cross-range resolution
c is the speed of wave propagation
T is the compressed pulse width
η is the angle between the radar beam and the surface
R is the slant range to the surface
λ is the wavelength
L_a is the length of the antenna

The goal is to design the SLAR with a narrow beamwidth, short slant range, and a short pulsewidth to achieve high resolution. In practice, the pulsewidth of the radar is limited by hardware constraints and the amount of "energy on target" required to get sufficient signal-to-noise ratio to obtain a good image. To achieve a high range resolution without a short pulse, frequency modulation can be used to synthesize an effectively short pulse. This process of generating a narrow synthetic pulsewidth is called pulse compression. The approach is to introduce a modulation on the transmitted pulse, and then pass the received signal through a filter matched to the transmit signal modulation. The most common transmit waveforms used for pulse compression are linear FM (or chirp) and phase coded. Some radars use a digital version of linear FM called a stepped frequency waveform.

We illustrate pulse compression with the ideal application of the linear FM waveform. The square pulse is modulated by a linear FM signal, and the resulting transmit signal is

$$s(t) = \begin{cases} \cos\left(\omega_0^{\dagger} - \frac{1}{2}\mu^{\dagger 2}\right) & |\dagger| \leq T/2 \\ 0 & |\dagger| > T/2 \end{cases}$$

where the bandwidth (frequency deviation) introduced by the linear FM is

$$\Delta f = T\mu/2\pi$$

If this transmit pulse is perfectly reflected from a stationary point target, range losses are ignored, and we shift in time to remove the two-way delay; the received signal is exactly the same as the transmitted signal. The matched filter response for the transmitted signal is

$$h(t) = \left(\frac{2\mu}{\pi}\right)^{1/2} \cos\left(\omega_0^\dagger + \frac{1}{2}\mu^{\dagger 2}\right)$$

The output of the received signal applied to the matched filter is

$$\Psi(^\dagger) = \left(\frac{\mu T^2}{2\pi}\right)^{1/2} \frac{\sin(\mu T^\dagger/2)}{(\mu T^\dagger/2)} \mathrm{Re}\left[e^{j\left(\omega_0^\dagger + \frac{1}{2}\mu^{\dagger}2 + \pi/4\right)}\right]$$

This output has a mainlobe that has a 4 dB beamwidth of $1/\Delta f$. The resulting compressed pulse can be significantly narrower than the width of the transmitted pulse with a pulse compression ratio of $T\Delta f$. The range resolution of the radar has been increased by this pulse compression factor and is now given by

$$\delta_r \approx c/2\Delta f \cos\eta$$

Note that the range resolution in the ideal case is now completely independent of the physical width of the transmitted pulse. Performing range compression against real radar targets that Doppler shift the frequency of the receive signal introduces ambiguities resulting in additional signal processing issues that must be addressed. There is a trade-off between the ability of a radar waveform to resolve a target in range and frequency. The performance of a waveform in range-frequency space is given by its ambiguity. The ambiguity function is the output of the matched filter for the signal for which it is matched and for frequency shifted versions of that signal. The references contain a much more detailed description of ambiguity functions and radar waveform design.

Using pulse compression, a SLAR system can achieve a very high range resolution on the order of 1 ft or less, but the cross-range resolution of the SLAR is limited by the physical beamwidth of the antenna, the operating frequency, and the slant range. This cross-range resolution limitation of SLAR motivates the use of a synthetic array antenna to increase the cross-range resolution.

33.2.2 Unfocused Synthetic Aperture Radar

Figure 33.1 provides a good geometric description of SAR. As with SLAR, the radar platform moves along a straight line collecting radar data from the surface. The SAR system goes one step further than SLAR by coherently combining pulses collected along the flight path to synthesize a long synthetic array. The beamwidth of this synthetic aperture is significantly narrower than the physical beamwidth (real beam) of the real antenna. The ideal synthetic beamwidth of this synthetic aperture is

$$\theta_B = \lambda/2L_\theta$$

The factor of two results from the two-way propagation from the moving platform. The unfocused SAR can be implemented by performing FFT processing in the cross-range dimension for the samples in each range bin. This is simply the conventional beamformer for an array antenna. The difference between SAR and real beam radar is that the aperture samples that comprise the SAR are collected at different times by a moving platform. There are several design constraints on a SAR system, including

- The speed of the platform and pulse repetition rate (PRF) of the radar must be mutually selected so that the sample points of the synthetic array are separated by less than $\lambda/2$ to avoid grating lobes.
- The PRF must be selected so that the swath width is unambiguously sampled.
- A point on the ground must be visible to the radar real beam across the entire length of the synthetic array. This limits the size of the real beam antenna. This constraint leads to the observation that with SAR, the smaller the real-beam antenna, the better the resolution, whereas with SLAR the larger the real-beam antenna, the better the resolution.
- The SAR assumes that a ground target has an isotropic signal across the collection angle of the radar platform as it flies along the synthetic array.

The resolution of the unfocused SAR is limited because the slant range to a scatterer at a fixed location on the surface changes along the synthetic aperture. If we limit the synthetic aperture to a length so that the range from every array point in the aperture to a fixed surface location differs by less than $\lambda/8$, then the cross-range resolution of the unfocused SAR is limited to

$$\delta_{\mathrm{cr}} = \sqrt{R\lambda/2}$$

33.2.3 Focused Synthetic Aperture Radar

The cross-range limitation of an unfocused SAR can be removed by focusing the data, as in optics. The focusing procedure for the SAR involves adjusting the phase of the received signal for every range sample in the image so that all of the points processed in cross-range through the synthetic beamformer appear to be at the same range. The phase error at each range sample used to form the SAR image is

$$\Delta\phi = \frac{2\pi}{\lambda}\left(\frac{d_n^2}{R}\right) \quad \text{[radiar]}$$

where
d_n is the cross-range distance from the beam center
R is the slant range to the point on the ground from the beam center
λ is the wavelength

The range samples can be focused before cross-range processing by removing this phase error from the phase history data. Note that each data point has a different phase correction based on the along-track position of the sensor and the point's range from the sensor.

When focusing is performed, the resulting SAR image resolution is independent of the slant range between the sensor and ground. This can be shown as follows:

$$\delta_{\mathrm{cr}} = R\theta_{\mathrm{s}}$$

where $\theta_{\mathrm{s}} \approx \dfrac{\lambda}{2L_{\mathrm{e}}}$ and $L_{\mathrm{e}} \approx \dfrac{R\lambda}{L_{\mathrm{a}}}$
therefore, $\delta_{\mathrm{cr}} \approx L_{\mathrm{a}}/2$

The effective beamwidth of the synthetic aperture is approximately $\lambda/2L_{\mathrm{e}}$ where the factor of two comes from the two-way propagation of the energy (the exact effective beamwidth depends on the synthetic array taper used to control sidelobes). The length of the effective aperture (L_{e}) is limited by the fact that a given scatterer on the surface must be in the mainbeam of the real radar beam for every position along the synthetic aperture. The result is that the resolution of the SAR when the data is focused is approximately $L_a/2$.

SAR processing can also be developed by considering the Doppler of the radar signal from the surface as first done by Wiley in 1951. When the real beamwidth of the SAR is small, a point on the surface has an approximately linearly decreasing Doppler frequency as it passes through the main beam of the real SAR beamwidth. This time varying Doppler frequency has been shown to be approximately

$$f_d(t) = \frac{2v^2|t - t_0|}{\lambda R}$$

where

v is the velocity of the platform

t_0 is the time that the point scatterer is in the center of the main beam

The change in Doppler frequency as the point passes through the main beam is $2v^2T_d/\lambda R$, and T_d is the time that the point is in the main beam. As with linear FM pulse compression, covered in Section 33.2.1, this Doppler signal can be processed through a filter to produce a higher cross-range resolution signal which is limited by the size of the real aperture just as with the synthetic antenna interpretation ($\delta_{cr} = L_a/2$). In a modern SAR system, typically both pulse compression (synthetic range processing) and a synthetic aperture (synthetic cross-range processing) are employed. In most cases, these transformations are separable where the range processing is referred to as "fast time" processing and the cross-range processing is referred to as "slow-time" processing.

A modern SAR system requires several additional signal processing algorithms to achieve high resolution imagery. In practice, the platform does not fly a straight and level path, so the phase of the raw receive signal must be adjusted to account for aircraft perturbations, a procedure called motion compensation. In addition, since it is difficult to exactly estimate the platform parameters necessary to focus the SAR image, an autofocus algorithm is used. This algorithm derives the platform parameters from the raw SAR data to focus the imagery. There is also an interpolation algorithm that converts from polar to rectangular formats for the imagery display. Most modern SAR systems form imagery digitally using either an FFT or a bank of matched filters. Typically, a SAR will operate in either a stripmap or spotlight mode. In the stripmap mode, the SAR antenna is typically pointed perpendicular to the flight path (although it may be squinted slightly to one side). A stripmap SAR keeps its antenna position fixed and collects SAR imagery along a swath to one side of the platform. A spotlight SAR can move its antenna to point at a position on the ground for a longer period of time (thus actually achieving cross-range resolutions even greater than the aperture length over two). Many SAR systems support both stripmap and spotlight modes, using the stripmap mode to cover large areas of the surface in a slightly lower resolution mode, and spotlight modes to perform very high resolution imaging of areas of high interest.

33.3 SAR Image Enhancement

In this section we review a few techniques for removing speckle noise from SAR imagery. Removing the speckle can make it easier to extract information from SAR imagery and improves the visual quality.

Coherent noise or speckle can be a major distortion in high resolution, high frequency SAR imagery. The speckle is caused when the intensity of a resolution cell results from the coherent combination of many wavefronts resulting from randomly oriented clutter surfaces within a resolution cell. These wavefronts can combine constructively or destructively resulting in intensity variations across the image. When the number of wavefronts approaches infinity (i.e., large resolution cell collected by a high frequency radar) the Rayleigh clutter model can be used to represent the speckle under the right statistical assumptions. When the number of wavefronts is less than infinity, the K-distribution and other product models do a better job of theoretically and empirically modeling the clutter.

When the combination of the radar system design and clutter properties results in images that contain large amounts of speckle, it is desirable to perform additional processing to reduce the speckle. One approach for speckle reduction is to noncoherently spatially average adjacent resolution cells, sacrificing resolution for the speckle reduction. This spatial averaging can be performed as a part of the image formation analogous to the Bartlett method of spectral estimation. Another approach for reducing speckle is to average across polarimetric channels if multiple polarimetric channels are available.

The PWF reduces the speckle content while preserving the image resolution. The PWF was derived by Novak et al. [5] as a quadratic filter that minimizes a specific speckle metric (defined as the ratio of the clutter standard deviation to its mean). The PWF first whitens the polarimetric data with respect to the clutter's polarimetric covariance, and then noncoherently averages across the polarimetric channels. This whitening filter essentially diagonalizes the covariance matrix of the complex backscatter vector $[HH, HV, VV]^T$, such that the resulting new linear polarization basis $[HH', HV', VV']^T$ has equal power in each component, where

$$\begin{bmatrix} HH' \\ HV' \\ VV' \end{bmatrix} = \begin{bmatrix} HH \\ \dfrac{HV}{\sqrt{\varepsilon}} \\ \dfrac{VV - \rho^* \sqrt{\gamma} HH}{\sqrt{\gamma(1 - |\rho|^2)}} \end{bmatrix} \tag{33.1}$$

where

$$\varepsilon = \frac{E(|HV|^2)}{E(|HH|^2)}, \quad \gamma = \frac{E(|W|^2)}{E(|HH|^2)}, \quad \rho = \frac{E(HH \cdot W^*)}{\sqrt{E(|HH|^2) \cdot E(|W|^2)}} \tag{33.2}$$

The polarization scattering matrix (using a linear-polarization basis) can then be expressed as

$$\sum = \sigma_{HH} \begin{bmatrix} 1 & 0 & \rho\sqrt{\gamma} \\ 0 & \varepsilon & 0 \\ \rho^*\sqrt{\gamma} & 0 & \gamma \end{bmatrix} \tag{33.3}$$

The pixel intensity (power) is then derived through noncoherent averaging of the power in each of the new polarization components:

$$Y = |HH|^2 + \left|\frac{HV}{\sqrt{\varepsilon}}\right|^2 + \left|\frac{W - \rho^* \sqrt{\gamma} HH}{\sqrt{\gamma(1 - |p|^2)}}\right|^2 \tag{33.4}$$

yielding a minimal speckle image at the original image resolution. Novak et al. [5] have shown that on the ADTS SAR data, the PWF reduces the clutter standard deviation by 2.0–2.7 dB compared with the standard deviation of single-polarimetric-channel data. The PWF has a dramatic effect on the visual quality of the SAR imagery and the performance of automatic detection and classification algorithms applied to SAR images. The PWF does not take into account the effect of the speckle reduction operation on target signals. It only minimizes the clutter. There has been recent work on polarimetric speckle reduction

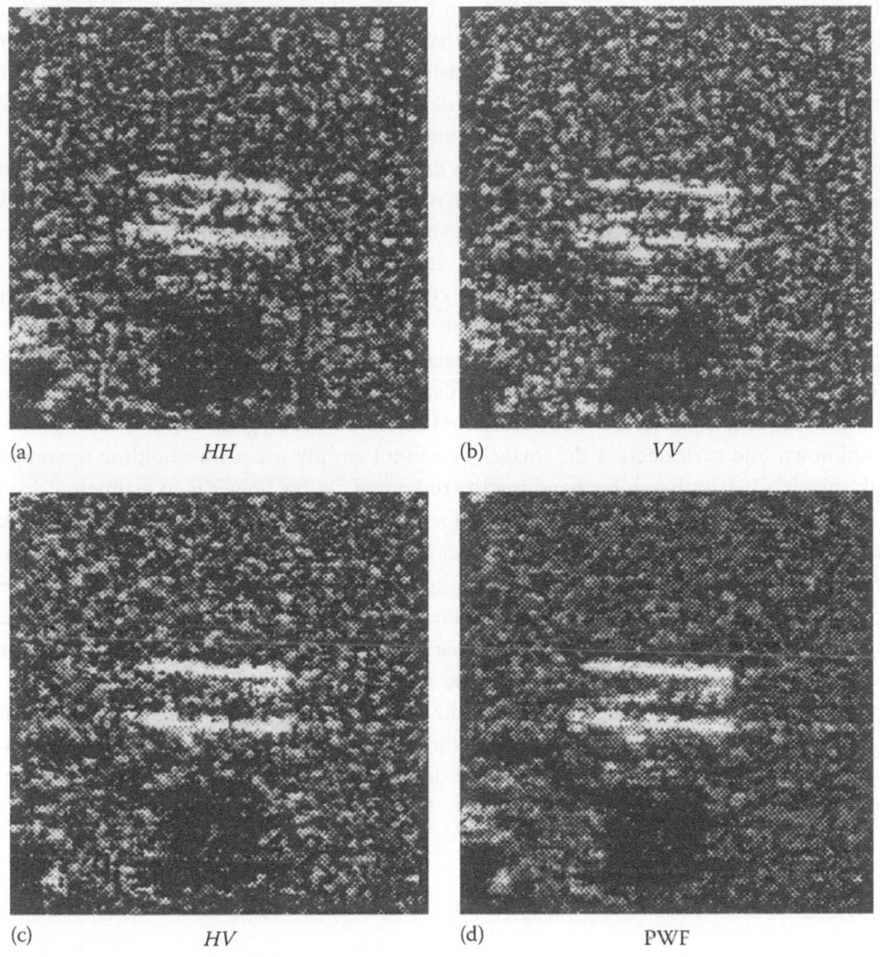

FIGURE 33.4 Polarimetric processing of SAR data to reduce speckle.

filters that both reduce the clutter speckle while preserving the target signal. Figure 33.4 shows the three polarimetric channels and the resulting PWF image for an ADTS SAR chip of a target-like object.

33.4 Automatic Object Detection and Classification in SAR Imagery

SAR algorithmic tasks of high interest to the defense and intelligence communities include automatic target detection and recognition (ATD/R). Since SAR imagery has very different target and clutter characteristics as compared with visual and infrared imagery, uniquely designed ATD/R algorithms are required for SAR data. In this section, we describe a few basic ATD/R algorithms that have been developed for high resolution, high frequency SAR imagery (10 GHz or above) [6–8].

Performing target detection and classification against remote sensing imagery and, in particular, SAR imagery is very different from the classical pattern recognition problem. In the classical pattern recognition problem, we have models defining *N* classes, and the goal is to design a classifier to separate sensor data into one of the *N* classes. In SAR target classification, the imagery contains regions of diffuse clutter which can be represented to some degree by models, but the imagery also contains a possibly uncountable set of target-like discrete unknown and unmodelable objects. The goal is to reject both the diffuse clutter and the unknown discrete objects and to classify the target objects. This need to handle the unknown object means that the classifier must have the unknown class as a possible outcome of the classifier. Since the unknown class cannot be modeled, most SAR ATR systems solve the problem by employing a distance metric to compare the sensor data with models for each target of interest, and if the distance is too great, the data is classified as an unknown object.

Another design issue for a SAR ATD/R system is the need to process hundreds of square kilometers of data in near real-time to be of practical benefit. One widely used approach for solving this computational problem is to use a simple focus-of-attention or pre-detection algorithm to reject most of the diffuse clutter and pass only regions of interest (ROIs), including all of the targets. These ROIs are then processed through a set of computationally more complicated classifiers which classify objects in the ROIs as one of the targets or as an unknown object.

In high frequency SAR imagery most target signatures have extremely bright peaks caused by physical corners on the target. One effective pre-detection technique involves applying a single pixel detector to find the bright pixels caused by corner reflectors on the targets. Since the background clutter power is unknown and varies across the image, we cannot simply use a thresholding operation to find these bright pixels. One approach for handling the unknown clutter power is to estimate it from clutter samples surrounding a test pixel. This approach for target detection is referred to as a constant false alarm rate (CFAR) detector because with the proper clutter and target models, it can be shown that the output of the detector has a CFAR in the presence of unknown clutter parameters. Figure 33.5 depicts one design for a CFAR template. The clutter parameters are estimated using the auxiliary samples along a box with a test sample in the center. This test sample may or may not be on a target. The size of the box containing the auxiliary samples is sized so that the auxiliary samples do not overlap a target when the test sample is on the target. We also need to keep the size of the box containing the auxiliary samples as small as possible, so that we get a good local estimate of the clutter parameters. With these design constraints, a good choice for the CFAR template is just over twice the maximum dimension of the targets of interest.

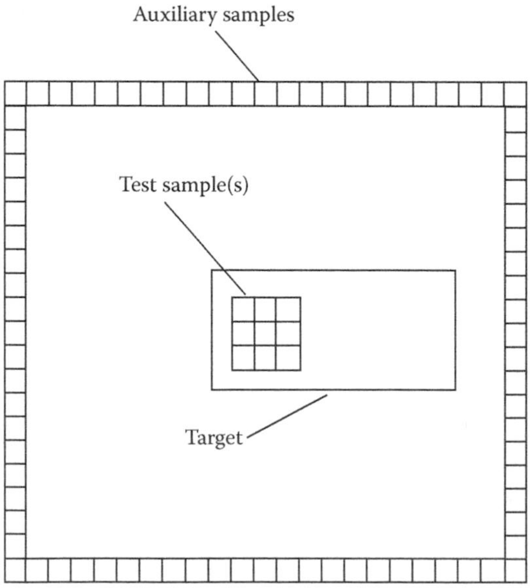

FIGURE 33.5 CFAR template.

One of these CFAR algorithms, first developed by Goldstein [9], is referred to as the two-parameter CFAR or the log-t test:

$$\frac{\log x - \frac{1}{N}\sum_{i=1}^{N}\log y_i}{\sqrt{\frac{1}{N-1}\sum_{i=1}^{N}\left(\log y_i - \frac{1}{N}\sum_{i=1}^{N}\log y_i\right)^2}} \mathop{\gtrless}_{H_0}^{H_1} t$$

where
  x is the test sample
  $y_1, \ldots, y_N$ are the auxiliary samples

This test is performed for every pixel in the SAR scene and the output is thresholded with the threshold t. When N is large, the test statistic is approximately Gaussian if the SAR data is log normally distributed. In this case, Gaussian statistics can be used to determine the threshold for a given probability of false alarm. In practice, it is much more accurate to determine the threshold with a set of training data. This is primarily a corner reflector detector, and the output will almost always get more than one detection per target. In practice, a simple clustering algorithm can be used based on the size of the targets and the expected spacing of targets to get one detection per target and reduce the number of false alarms which are usually also clustered. The two-parameter CFAR test is one example of a simple SAR target detector. Researchers have also developed more sophisticated ordered statistic detectors, multi-polarimetric channel detectors, and feature-based discriminators to get improved SAR target detector performance [6–8].

This simple pre-detector gets a large number of false alarms (hundreds per square kilometer in single polarimetric channel, 1 ft resolution imagery) [5]. In order to further reduce the false alarm rate and classify the targets, further processing is necessary on the output of the pre-detector. One widely used approach for performing this classification operation is to apply a linear filter bank classifier to the ROIs identified by the pre-detector. Researchers have developed a large number of approaches for designing these linear filter bank classifiers including spatial matched filters [7], synthetic discriminant functions [7], and vector quantization/learning vector quantization [8]. The simplest approach is to build the spatial matched filters by breaking the target into angle subclasses, and averaging the training signatures in a given angle subclass to represent that subclass. In practice, the templates must be normalized because the absolute energy of a given target signature is unknown. The exact location of a target in the ROI is also unknown, so the matched filter must be applied for every possible spatial position of the target. This is performed more efficiently in the frequency domain as follows:

$$\rho_{ij} = \max\left\{FFT^{-1}\left[FFT(\mathbf{t}_{ij}) \cdot FFT(\mathbf{x})^*\right]\right\}$$

where
  $\mathbf{x}$ is an ROI
  $\mathbf{t}_{ij}$ is the spatial matched filter representing the ith target and the jth angle subclass of that target.

The ρ_{ij} is computed for every angle subclass of every target, and the maximum represents the estimate of the correct target and angle subclass. The output can be thresholded to reject false alarms. In practice the level of the threshold is determined by testing on both target and false alarm data.

In this section, we have reviewed a few basic concepts in SAR ATD/R. For a much more detailed treatment of this topic, consult the references and the recommended further reading given below.

Further Reading and Open Research Issues

A very brief overview of SAR with a few example algorithms is given here. The items in the reference list give a more detailed treatment of the topics covered in this chapter. SAR is a very active research topic. Articles on SAR algorithms are regularly published in many journals and conferences, including

- Journals: *IEEE Transactions on Aerospace and Electronic Systems, IEEE Transactions on Geoscience and Remote Sensing, IEEE Transactions on Antennas and Propagation, IEEE Transactions on Signal Processing,* and *IEEE Transactions on Image Processing.*
- Conferences: *IEEE National Radar Conference, IEEE International Radar Conference,* and the International Society for Optical Engineering (SPIE) has held several SAR conferences.

There are numerous open areas of research on SAR signal processing algorithms including

- Still developing an understanding of the utility and applications of multi-polarimetric, multi-frequency, and 3-D SAR
- Performance/robustness of model-based image formation not completely understood
- Performance/robustness of different detection, discrimination, and classification algorithms given radar, clutter, and target parameters not completely understood
- No fundamental theoretical understanding of performance limitations given radar, clutter, and target parameters (i.e., no Shannon theory)

References

1. Curlander, J.C. and McDonough, R.N., *Synthetic Aperture Radar: Systems and Signal Processing*, John Wiley & Sons, New York, 1991.
2. Wehner, D.R., *High Resolution Radar*, 2nd ed., Artech House, Boston, MA, 1995.
3. Stimson, G.W., *Introduction to Airborne Radar*, Hughes Aircraft Company, El Segundo, CA, 1983.
4. Skolnik, M., *Introduction to Radar Systems*, 2nd ed., McGraw-Hill, New York, 1980.
5. Novak, L., Burl, M., and Irving, B., Optimal polarimetric processing for enhanced target detection, *IEEE Trans. Aerosp. Electron. Syst.*, 29(1), 234–244, Jan. 1993.
6. Stewart, C., Moghaddam, B., Hintz, K., and Novak, L., Fractional Brownian motion for synthetic aperture radar imagery scene segmentation, *Proc. IEEE*, 81(10), 1511–1522, Oct. 1993.
7. Novak, L., Owirka, G., and Netishen, C., Radar target identification using spatial matched filters, *Pattern Recognit.*, 27(4), 607–617, Apr. 1994.
8. Stewart, C., Lu, Y.-C., and Larson, V., A neural clustering approach for high resolution radar target classification, *Pattern Recognit.*, 27(4), 503–513, Apr. 1994.
9. Goldstein, G., False-alarm regulation in log-normal and Weibull clutter, *IEEE Trans. Aerosp. Electron. Syst.*, 9, 84–92, 1972.

34

Iterative Image Restoration Algorithms

Aggelos K. Katsaggelos
Northwestern University

34.1 Introduction

In this chapter we consider a class of iterative restoration algorithms. If $\boldsymbol{y}$ is the observed noisy and blurred signal, $\boldsymbol{D}$ the operator describing the degradation system, $\boldsymbol{x}$ the input to the system, and $\boldsymbol{n}$ the noise added to the output signal, the input–output relation is described by [3,51]

$$\boldsymbol{y} = \boldsymbol{D}\boldsymbol{x} + \boldsymbol{n}. \tag{34.1}$$

Henceforth, boldface lowercase letters represent vectors and boldface uppercase letters represent a general operator or a matrix. The problem, therefore, to be solved is the inverse problem of recovering $\boldsymbol{x}$ from knowledge of $\boldsymbol{y}$, $\boldsymbol{D}$, and $\boldsymbol{n}$. Although the presentation will refer to and apply to signals of any dimensionality, the restoration of grayscale images is the main application of interest.

There are numerous imaging applications which are described by Equation 34.1 [3,5,23,36,52]. $\boldsymbol{D}$, for example, might represent a model of the turbulent atmosphere in astronomical observations with ground-based telescopes, or a model of the degradation introduced by an out-of-focus imaging device.

$\boldsymbol{D}$ might also represent the quantization performed on a signal, or a transformation of it, for reducing the number of bits required to represent the signal (compression application).

The success in solving any recovery problem depends on the amount of the available prior information. This information refers to properties of the original signal, the degradation system (which is in general only partially known), and the noise process. Such prior information can, for example, be represented by the fact that the original signal is a sample of a stochastic field, or that the signal is "smooth," or that the signal takes only nonnegative values. Besides defining the amount of prior information, the ease of incorporating it into the recovery algorithm is equally critical.

After the degradation model is established, the next step is the formulation of a solution approach. This might involve the stochastic modeling of the input signal (and the noise), the determination of the model parameters, and the formulation of a criterion to be optimized. Alternatively it might involve the formulation of a functional to be optimized subject to constraints imposed by the prior information. In the simplest possible case, the degradation equation defines directly the solution approach. For example, if $\boldsymbol{D}$ is a square invertible matrix, and the noise is ignored in Equation 34.1, $\boldsymbol{x}=\boldsymbol{D}^{-1}\boldsymbol{y}$ is the desired unique solution. In most cases, however, the solution of Equation 34.1 represents an ill-posed problem [56]. Application of regularization theory transforms it to a well-posed problem which provides meaningful solutions to the original problem.

There are a large number of approaches providing solutions to the image restoration problem. For recent reviews of such approaches refer, for example, to [5,23]. The intention of this chapter is to concentrate only on a specific type of iterative algorithm, the successive approximation algorithm, and its application to the signal and image restoration problem. The basic form of such an algorithm is presented and analyzed first in detail to introduce the reader to the topic and address the issues involved. More advanced forms of the algorithm are presented in subsequent sections.

34.2 Iterative Recovery Algorithms

Iterative algorithms form an important part of optimization theory and numerical analysis. They date back at least to the Gauss years, but they also represent a topic of active research. A large part of any textbook on optimization theory or numerical analysis deals with iterative optimization techniques or algorithms [43,44]. In this chapter we review certain iterative algorithms which have been applied to solving specific signal recovery problems in the last 15–20 years. We will briefly present some of the more basic algorithms and also review some of the recent advances.

A very comprehensive paper describing the various signal processing inverse problems which can be solved by the successive approximations iterative algorithm is the paper by Schafer et al. [49]. The basic idea behind such an algorithm is that the solution to the problem of recovering a signal which satisfies certain constraints from its degraded observation can be found by the alternate implementation of the degradation and the constraint operator. Problems reported in [49] which can be solved with such an iterative algorithm are the phase-only recovery problem, the magnitude-only recovery problem, the bandlimited extrapolation problem, the image restoration problem, and the filter design problem [10]. Reviews of iterative restoration algorithms are also presented in [7,22]. There are certain advantages associated with iterative restoration techniques, such as [22,49] (1) there is no need to determine or implement the inverse of an operator, (2) knowledge about the solution can be incorporated into the restoration process in a relatively straightforward manner, (3) the solution process can be monitored as it progresses, and (4) the partially restored signal can be utilized in determining unknown parameters pertaining to the solution.

In the following we first present the development and analysis of two simple iterative restoration algorithms. Such algorithms are based on a simpler degradation model, when the degradation is linear and spatially invariant, and the noise is ignored. The description of such algorithms is intended to provide a good understanding of the various issues involved in dealing with iterative algorithms. We then proceed to work with the matrix-vector representation of the degradation model and the iterative algorithms. The degradation systems described now are linear but not necessarily spatially invariant. The relation between the matrix-vector and scalar representation of the degradation equation and the iterative solution is also presented. Various forms of regularized solutions and the resulting iterations are briefly presented. As it will become clear, the basic iteration is the basis for any of the iterations to be presented.

34.3 Spatially Invariant Degradation

34.3.1 Degradation Model

Let us consider the following degradation model

$$y(i,j) = d(i,j)^{*}x(i,j), \tag{34.2}$$

where

$y(i,j)$ and $x(i,j)$ represent, respectively, the observed degraded and original image, $d(i,j)$ the impulse response of the degradation system

* denotes two-dimensional (2D) convolution

We rewrite Equation 34.2 as follows:

$$\Phi[x(i,j)] = y(i,j) - d(i,j)^{*}x(i,j) = 0. \tag{34.3}$$

The restoration problem, therefore, of finding an estimate of $x(i,j)$ given $y(i,j)$ and $d(i,j)$ becomes the problem of finding a root of $\Phi[x(i,j)] = 0$.

34.3.2 Basic Iterative Restoration Algorithm

The following identity holds for any value of the parameter β:

$$x(i,j) = x(i,j) + \beta\Phi[x(i,j)]. \tag{34.4}$$

Equation 34.4 forms the basis of the successive approximation iteration by interpreting $x(i,j)$ on the left-hand side as the solution at the current iteration step and $x(i,j)$ on the right-hand side as the solution at the previous iteration step. That is,

$$\begin{aligned} x_0(i,j) &= 0 \\ x_{k+1}(i,j) &= x_k(i,j) + \beta\Phi[x_k(i,j)] \\ &= \beta y(i,j) + [\delta(i,j) - \beta d(i,j)]^{*}x_k(i,j), \end{aligned} \tag{34.5}$$

where

$\delta(i,j)$ denotes the discrete delta function

β is the relaxation parameter which controls the convergence as well as the rate of convergence of the iteration

Iteration in Equation 34.5 is the basis of a large number of iterative recovery algorithms, some of which will be presented in the subsequent sections [1,14,17,31,33,38]. This is the reason it will be analyzed in quite some detail. What differentiates the various iterative algorithms is the form of the function $\Phi[x(i,j)]$. Perhaps the earliest reference to iteration in Equation 34.5 was by Van Cittert [61] in the 1930s. In this case the gain β was equal to one. Jansson et al. [17] modified the Van Cittert algorithm by replacing β with a relaxation parameter that depends on the signal. Also Kawata et al. [31,33] used Equation 34.5 for image restoration with a fixed or a varying parameter β.

34.3.3 Convergence

Clearly if a root of $\Phi[x(i,j)]$ exists, this root is a fixed point of iteration in Equation 34.5, that is $x_{k+1}(i,j) = x_k(i,j)$. It is not guaranteed, however, that iteration in Equation 34.5 will converge even if Equation 34.3 has one or more solutions. Let us, therefore, examine under what conditions (sufficient conditions) iteration in Equation 34.5 converges. Let us first rewrite it in the discrete frequency domain, by taking the 2D discrete Fourier transform (DFT) of both sides. It should be mentioned here that the arrays involved in iteration in Equation 34.5 are appropriately padded with zeros so that the result of 2D circular convolution equals the result of 2D linear convolution in Equation 34.2. The required padding by zeros determines the size of the 2D DFT. Iteration in Equation 34.5 then becomes

$$\begin{aligned} X_0(u,v) &= 0 \\ X_{k+1}(u,v) &= \beta Y(u,v) + [1 - \beta D(u,v)]X_k(u,v), \end{aligned} \tag{34.6}$$

where $X_k(u,v)$, $Y(u,v)$, and $D(u,v)$ represent the 2D DFT of $x_k(i,j)$, $y(i,j)$, and $d(i,j)$, respectively, and (u,v) the discrete 2D frequency lattice. We express next $X_k(u,v)$ in terms of $X_0(u,v)$. Clearly,

$$\begin{aligned} X_1(u,v) &= \beta Y(u,v) \\ X_2(u,v) &= \beta Y(u,v) + [1 - \beta D(u,v)]\beta Y(u,v) \\ &= \sum_{\ell=0}^{1} [1 - \beta D(u,v)]^{\ell} \beta Y(u,v) \\ &\vdots \\ X_k(u,v) &= \sum_{\ell=0}^{k-1} [1 - \beta D(u,u)]^{\ell} \beta Y(u,v) \\ &= \frac{1 - [1 - \beta D(u,v)]^k}{1 - [1 - \beta D(u,v)]} \beta Y(u,v) \\ &= \{1 - [1 - \beta D(u,v)]^k\} X(u,v) \end{aligned} \tag{34.7}$$

if $D(u,v) \neq 0$. For $D(u,v) = 0$,

$$X_k(u,v) = k \cdot \beta Y(u,v) = 0, \tag{34.8}$$

since $Y(u,v) = 0$ at the discrete frequencies (u,v) for which $D(u,v) = 0$. Clearly, from Equation 34.7 if

$$|1 - \beta D(u,v)| < 1, \tag{34.9}$$

then

$$\lim_{k \to \infty} X_k(u,v) = X(u,v). \tag{34.10}$$

Having a closer look at the sufficient condition for convergence, Equation 34.9, it can be rewritten as

$$\begin{aligned} &|1 - \beta \mathrm{Re}\{D(u,v)\} - \beta \mathrm{Im}\{D(u,v)\}|^2 < 1 \\ &\Rightarrow [1 - \beta \mathrm{Re}\{D(u,v)\}]^2 < 1. \end{aligned} \tag{34.11}$$

Inequality Equation 34.11 defines the region inside a circle of radius $1/\beta$ centered at $c = (1/\beta, 0)$ in the $(\mathrm{Re}\{D(u,v)\}, \mathrm{Im}\{D(u,v)\})$ domain, as shown in Figure 34.1. From this figure it is clear that the left

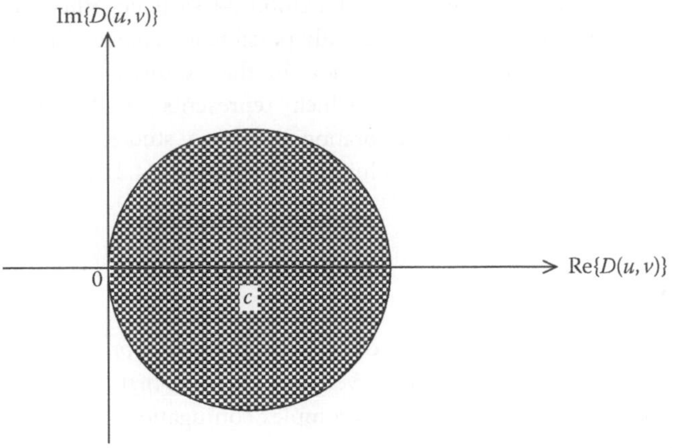

FIGURE 34.1 Geometric interpretation of the sufficient condition for convergence of the basic iteration, where $c = (1/\beta, 0)$.

half-plane is not included in the region of convergence. That is, even though by decreasing β the size of the region of convergence increases, if the real part of $D(u, v)$ is negative, the sufficient condition for convergence cannot be satisfied. Therefore, for the class of degradations that this is the case, such as the degradation due to motion, iteration in Equation 34.5 is not guaranteed to converge.

The following form of Equation 34.11 results when $\text{Im}\{D(u, v)\} = 0$, which means that $d(i, j)$ is symmetric

$$0 < \beta < \frac{2}{D_{\max}(u, v)}, \tag{34.12}$$

where $D_{\max}(u, v)$ denotes the maximum value of $D(u, v)$ over all frequencies (u, v). If we now also take into account that $d(i, j)$ is typically normalized, that is, $\sum_{i,j} d(i, j) = 1$, and represents a low pass degradation, then $D(0, 0) = D_{\max}(u, v) = 1$. In this case Equation 34.11 becomes

$$0 < \beta < 2. \tag{34.13}$$

From the above analysis, when the sufficient condition for convergence is satisfied, the iteration converges to the original signal. This is also the inverse solution obtained directly from the degradation equation. That is, by rewriting Equation 34.2 in the discrete frequency domain

$$Y(u, v) = D(u, v) \cdot X(u, v), \tag{34.14}$$

we obtain, for $D(u, v) \neq 0$,

$$X(u, v) = \frac{Y(u, v)}{D(u, v)}. \tag{34.15}$$

An important point to be made here is that, unlike the iterative solution, the inverse solution Equation 34.15 can be obtained without imposing any requirements on $D(u, v)$. That is, even if Equation 34.2 or 34.14 has a unique solution, that is, $D(u, v) \neq 0$ for all (u, v), iteration in Equation 34.5 may not converge if the sufficient condition for convergence is not satisfied. It is not, therefore, the appropriate iteration to solve the problem. Actually iteration in Equation 34.5 may not offer any advantages over the direct implementation of the inverse filter of Equation 34.15 if no other features of the iterative algorithms are used, as will be explained later. The only possible advantage of iteration in Equation 34.5 over Equation 34.15 is that the noise amplification in the restored image can be controlled by terminating the iteration before convergence, which represents another form of regularization. The effect of noise on the quality of the restoration has been studied experimentally in [47]. An iteration which will converge to the inverse solution of Equation 34.2 for any $d(i,j)$ is described in the next section.

34.3.4 Reblurring

The degradation Equation 34.2 can be modified so that the successive approximations iteration converges for a larger class of degradations. That is, the observed data $y(i,j)$ are first filtered (reblurred) by a system with impulse response $d^*(-i,-j)$, where * denotes complex conjugation [32]. The degradation Equation 34.2, therefore, becomes

$$\begin{aligned}\tilde{y}(i,j) = y(i,j) * d^*(-i,-j) &= d^*(-i,-j) * d(i,j) * x(i,j)\\ &= \tilde{d}(i,j) * x(i,j).\end{aligned} \tag{34.16}$$

If we follow the same steps as in the previous section substituting $y(i,j)$ by $\tilde{y}(i,j)$ and $d(i,j)$ by $\tilde{d}(i,j)$ the iteration providing a solution to Equation 34.16 becomes

$$\begin{aligned}x_0(i,j) &= 0\\ x_{k+1}(i,j) &= x_k(i,j) + \beta d^*(-i,-j) * [y(i,j) - d(i,j) * x_k(i,j)]\\ &= \beta d^*(-i,-j) * y(i,j) + [\delta(i,j) - \beta d^*(-i,-j) * d(i,j)] * x_k(i,j).\end{aligned} \tag{34.17}$$

Now, the sufficient condition for convergence, corresponding to condition in Equation 34.9, becomes

$$|1 - \beta|D(u,v)|^2| < 1 \tag{34.18}$$

which can be always satisfied for

$$0 < \beta < \frac{2}{\max_{u,v}|D(u,v)|^2}. \tag{34.19}$$

The presentation so far has followed a rather simple and intuitive path, hopefully demonstrating some of the issues involved in developing and implementing an iterative algorithm. We move next to the matrix-vector formulation of the degradation process and the restoration iteration. We borrow results from numerical analysis in obtaining the convergence results of the previous section but also more general results.

34.4 Matrix-Vector Formulation

What became clear from the previous sections is that in applying the successive approximations iteration the restoration problem to be solved is brought first into the form of finding the root of a function (see Equation 34.3). In other words, a solution to the restoration problem is sought which satisfies

$$\Phi(\boldsymbol{x}) = 0, \tag{34.20}$$

where

$\boldsymbol{x} \in \mathcal{R}^N$ is the vector representation of the signal resulting from the stacking or ordering of the original signal

$\Phi(\boldsymbol{x})$ represents a nonlinear in general function

The row-by-row from left-to-right stacking of an image $x(i, j)$ is typically referred to as lexicographic ordering.

Then the successive approximations iteration which might provide us with a solution to Equation 34.20 is given by

$$\begin{aligned} \boldsymbol{x}_0 &= 0 \\ \boldsymbol{x}_{k+1} &= \boldsymbol{x}_k + \beta\Phi(\boldsymbol{x}_k) \\ &= \Psi(\boldsymbol{x}_k). \end{aligned} \tag{34.21}$$

Clearly if $\boldsymbol{x}^*$ is a solution to $\Phi(\boldsymbol{x}) = 0$, that is, $\Phi(\boldsymbol{x}^*) = 0$, then $\boldsymbol{x}^*$ is also a fixed point to the above iteration since $\boldsymbol{x}_{k+1} = \boldsymbol{x}_k = \boldsymbol{x}^*$. However, as was discussed in the previous section, even if $\boldsymbol{x}^*$ is the unique solution to Equation 34.20, this does not imply that iteration in Equation 34.21 will converge. This again underlines the importance of convergence when dealing with iterative algorithms. The form iteration in Equation 34.21 takes for various forms of the function $\Phi(\boldsymbol{x})$ will be examined in the following sections.

34.4.1 Basic Iteration

From the degradation Equation 34.1, the simplest possible form $\Phi(\boldsymbol{x})$ can take, when the noise is ignored, is

$$\Phi(\boldsymbol{x}) = \boldsymbol{y} - \boldsymbol{D}\boldsymbol{x}. \tag{34.22}$$

Then Equation 34.21 becomes

$$\begin{aligned} \boldsymbol{x}_0 &= 0 \\ \boldsymbol{x}_{k+1} &= \boldsymbol{x}_k + \beta(y - \boldsymbol{D}\boldsymbol{x}_k) \\ &= \beta y + (\mathbf{I} - \beta \boldsymbol{D})\boldsymbol{x}_k \\ &= \beta \boldsymbol{y} + \mathbf{G}_1\boldsymbol{x}_k, \end{aligned} \tag{34.23}$$

where $\mathbf{I}$ is the identity operator.

34.4.2 Least-Squares Iteration

A least-squares approach can be followed in solving Equation 34.1. That is, a solution is sought which minimizes

$$M(\mathbf{x}) = \|y - \boldsymbol{D}x\|^2. \tag{34.24}$$

A necessary condition for $M(\boldsymbol{x})$ to have a minimum is that its gradient with respect to $\boldsymbol{x}$ is equal to zero, which results in the normal equations

$$\boldsymbol{D}^{\mathrm{T}}\boldsymbol{D}x = \boldsymbol{D}^{\mathrm{T}}y \tag{34.25}$$

or

$$\Phi(\boldsymbol{x}) = \boldsymbol{D}^{\mathrm{T}}(\boldsymbol{y} - \boldsymbol{D}\boldsymbol{x}) = 0, \tag{34.26}$$

where superscript T denotes the transpose of a matrix or vector. Application of iteration in Equation 34.21 then results in

$$\begin{aligned} \boldsymbol{x}_0 &= 0 \\ \boldsymbol{x}_{k+1} &= \boldsymbol{x}_k + \beta \boldsymbol{D}^{\mathrm{T}}(\boldsymbol{y} - \boldsymbol{D}\boldsymbol{x}_k) \\ &= \beta \boldsymbol{D}^{\mathrm{T}}\boldsymbol{y} + (\mathbf{I} - \beta \boldsymbol{D}^{\mathrm{T}}\boldsymbol{D})\boldsymbol{x}_k \\ &= \beta \boldsymbol{D}^{\mathrm{T}}\boldsymbol{y} + \mathbf{G}_2\boldsymbol{x}_k. \end{aligned} \tag{34.27}$$

It is mentioned here that the matrix-vector representation of an iteration does not necessarily determine the way the iteration is implemented. In other words, the pointwise version of the iteration may be more efficient from the implementation point of view than the matrix-vector form of the iteration.

34.5 Matrix-Vector and Discrete Frequency Representations

When Equations 34.22 and 34.26 are obtained from Equation 34.2, the resulting iterations in Equations 34.23 and 34.27 should be identical to iterations in Equations 34.5 and 34.17, respectively, and their frequency domain counterparts. This issue, of representing a matrix-vector equation in the discrete frequency domain, is addressed next.

Any matrix can be diagonalized using its singular value decomposition. Finding, in general, the singular values of a matrix with no special structure is a formidable task, given also the size of the matrices involved in image restoration. For example, for a 256×256 image, $\boldsymbol{D}$ is of size 64 K $\times$ 64 K. The situation is simplified, however, if the degradation model of Equation 34.2, which represents a special case of the degradation model of Equation 34.1, is applicable. In this case, the degradation matrix $\boldsymbol{D}$ is block circulant [3]. This implies that the singular values of $\mathbf{D}$ are the DFT values of $d(i,j)$, and the eigenvectors are the complex exponential basis functions of the DFT. In matrix form, this relationship can be expressed by

$$\boldsymbol{D} = \boldsymbol{W}\tilde{\boldsymbol{D}}\boldsymbol{W}^{-1}, \tag{34.28}$$

where

$\tilde{\boldsymbol{D}}$ is a diagonal matrix with entries the DFT values of $d(i,j)$

$\boldsymbol{W}$ is the matrix formed by the eigenvectors of $\boldsymbol{D}$

The product $\boldsymbol{W}^{-1}z$, where z is any vector, provides us with a vector which is formed by lexicographically ordering the DFT values of $z(i,j)$, the unstacked version of z. Substituting $\boldsymbol{D}$ from Equation 34.28 into iteration in Equation 34.23 and premultiplying both sides by $\boldsymbol{W}^{-1}$, iteration in Equation 34.5 results. The same way iteration in Equation 34.17 results from iteration in Equation 34.27. In this case, reblurring, as was named when initially proposed, is nothing else than the least squares solution to the inverse problem. In general, if in a matrix-vector equation all matrices involved are block circulant, a 2D discrete frequency domain equivalent expression can be obtained. Clearly, a matrix-vector representation encompasses a considerably larger class of degradations than the linear spatially invariant degradation.

34.6 Convergence

In dealing with iterative algorithms, their convergence as well as their rate of convergence are very important issues. Some general convergence results will be presented in this section. These results will be presented for general operators, but also equivalent representations in the discrete frequency domain can be obtained if all matrices involved are block circulant.

The contraction mapping theorem usually serves as a basis for establishing convergence of iterative algorithms. According to it, iteration in Equation 34.21 converges to a unique fixed point $\boldsymbol{x}^*$, that is, a point such that $\Psi(\boldsymbol{x}^*) = \boldsymbol{x}^*$ for any initial vector if the operator or transformation $\Psi(\boldsymbol{x})$ is a contraction. This means that for any two vectors z_1 and z_2 in the domain of $\Psi(\boldsymbol{x})$ the following relation holds:

$$\| \Psi(z_1) - \Psi(z_2) \| \leq \eta \, \| z_1 - z_2 \| \,, \tag{34.29}$$

where
 η is strictly less than one
 $\|\cdot\|$ denotes any norm

It is mentioned here that condition in Equation 34.29 is norm dependent, that is, a mapping may be contractive according to one norm, but not according to another.

34.6.1 Basic Iteration

For iteration in Equation 34.23 the sufficient condition for convergence in Equation 34.29 results in

$$\|\mathbf{I} - \beta \boldsymbol{D}\| < 1, \quad \text{or} \quad \|\mathbf{G}_1\| < 1. \tag{34.30}$$

If the l_2 norm is used, then condition in Equation 34.30 is equivalent to the requirement that

$$\max_i |\sigma_i(\mathbf{G}_1)| < 1, \tag{34.31}$$

where $|\sigma_i(\mathbf{G}_1)|$ is the absolute value of the ith singular value of $\mathbf{G}_1$ [54].

The necessary and sufficient condition for iteration in Equation 34.23 to converge to a unique fixed point is that

$$\max_i |\lambda_i(\mathbf{G}_1)| < 1, \quad \text{or} \quad \max_i |1 - \beta\lambda_i(\boldsymbol{D})| < 1, \tag{34.32}$$

where $|\lambda_i(\boldsymbol{A})|$ represents the magnitude of the ith eigenvalue of the matrix $\boldsymbol{A}$. Clearly for a symmetric matrix $\boldsymbol{D}$ conditions in Equations 34.30 and 34.32 are equivalent. Conditions in Equations 34.29 through 34.32 are used in defining the range of values of β for which convergence of iteration in Equation 34.23 is guaranteed.

Of special interest is the case when matrix $\boldsymbol{D}$ is singular ($\boldsymbol{D}$ has at least one zero eigenvalue), since it represents a number of typical distortions of interest (e.g., distortions due to motion, defocusing, etc.). Then there is no value of β for which conditions in Equation 34.31 or 34.32 are satisfied. In this case $\mathbf{G}_1$ is a nonexpansive mapping (η in Equation 34.29 is equal to one). Such a mapping may have any number of fixed points (zero to infinitely many). However, a very useful result is obtained if we further restrict the properties of $\boldsymbol{D}$ (this results in no loss of generality, as it will become clear in the following sections). That is, if $\boldsymbol{D}$ is a symmetric, semi-positive definite matrix (all its eigenvalues are nonnegative), then according to Bialy's theorem [6], iteration in Equation 34.23 will converge to the minimum norm solution of Equation 34.1, if this solution exists, plus the projection of x_0 onto the null space of

D for $0 < \beta < 2 \cdot \|D\|^{-1}$. The theorem provides us with the means of incorporating information about the original signal into the final solution with the use of the initial condition.

Clearly, when D is block circulant the conditions for convergence shown above can be written in the discrete frequency domain. More specifically, conditions in Equations 34.31 and 34.9 are identical in this case.

34.6.2 Iteration with Reblurring

The convergence results presented above also holds for iteration in Equation 34.27, by replacing $\mathbf{G}_1$ by $\mathbf{G}_2$ in Equations 34.30 through 34.32. If $D^T D$ is singular, according to Bialy's theorem, iteration in Equation 34.27 will converge to the minimum norm least-squares solution of Equation 34.1, denoted by x^+, for $0 < \beta < 2 \cdot \|D\|^{-2}$, since $D^T y$ is in the range of $D^T D$.

The rate of convergence of iteration in Equation 34.27 is linear. If we denote by D^+ the generalized inverse of D, that is, $x^+ = D^+ y$, then the rate of convergence of Equation 34.27 is described by the relation [24]

$$\frac{\|x_k - x^+\|}{\|x^+\|} \leq c^{k+1}, \tag{34.33}$$

where

$$c = \max\{|1 - \beta\|D\|^2|, |1 - \beta\|D^+\|^{-2}|\}. \tag{34.34}$$

The expression for c in Equation 34.34 will also be used in Section 34.8, where higher order iterative algorithms are presented.

34.7 Use of Constraints

Iterative signal restoration algorithms regained popularity in the 1970s due to the realization that improved solutions can be obtained by incorporating prior knowledge about the solution into the restoration process. For example, we may know in advance that x is bandlimited or space-limited, or we may know on physical grounds that x can only have nonnegative values. A convenient way of expressing such prior knowledge is to define a constraint operator C, such that

$$x = Cx, \tag{34.35}$$

if and only if x satisfies the constraint. In general, C represents the concatenation of constraint operators. With the use of constraints, iteration in Equation 34.21 becomes [49]

$$\begin{aligned} x_0 &= 0, \\ x_k &= Cx_k, \\ x_{k+1} &= \Psi(\tilde{x}_k). \end{aligned} \tag{34.36}$$

The already mentioned recent popularity of constrained iterative restoration algorithms is also due to the fact that solutions to a number of recovery problems, such as the bandlimited extrapolation problem [48,49] and the reconstruction from phase or magnitude problem [49,57], were provided with the use of algorithms of the form of Equation 34.36 by appropriately describing the distortion and constraint operators.

These operators are defined in the discrete spatial or frequency domains. A review of the problems which can be solved by an algorithm of the form of Equation 34.36 is presented by Schafer et al. [49].

The contraction mapping theorem can again be used as a basis for establishing convergence of constrained iterative algorithms. The resulting sufficient condition for convergence is that at least one of the operators C and Ψ is contractive while the other is nonexpansive. Usually it is harder to prove convergence and determine the convergence rate of the constrained iterative algorithm, taking also into account that some of the constraint operators are nonlinear, such as the positivity constraint operator.

34.7.1 Method of Projecting onto Convex Sets

The method of projecting onto convex sets (POCS) describes an alternative approach in incorporating prior knowledge about the solution into the restoration process. It reappears in the engineering literature in the early 1980s [64], and since then it has been successfully applied to the solution of different restoration problems (e.g., from the reconstruction from phase or magnitude [52] to the removal of blocking artifacts [62,63]). According to the method of POCS the incorporation of prior knowledge into the solution can be interpreted as the restriction of the solution to be a member of a closed convex set that is defined as the set of vectors which satisfy a particular property. If the constraint sets have a nonempty intersection, then a solution that belongs to the intersection set can be found by the method of POCS. Indeed, any solution in the intersection set is consistent with the *a priori* constraints and, therefore, it is a feasible solution.

More specifically, let $Q_1, Q_2, \ldots, Q_m$ be closed convex sets in a finite dimensional vector space, with $\boldsymbol{P}_1, \boldsymbol{P}_2, \ldots, \boldsymbol{P}_m$ their respective projectors. Then, the iterative procedure,

$$\boldsymbol{x}_{k+1} = \boldsymbol{P}_1\boldsymbol{P}_2\cdots\boldsymbol{P}_m\boldsymbol{x}_k, \tag{34.37}$$

converges to a vector which belongs to the intersection of the sets Q_i, $i = 1, 2, \ldots, m$, for any starting vector x_0. It is interesting to note that the resulting set intersection is also a closed convex set.

Clearly, the application of a projection operator $\boldsymbol{P}$ and the constraint C, discussed in the previous section, express the same idea. Projection operators represent nonexpansive mappings.

34.8 Class of Higher Order Iterative Algorithms

One of the drawbacks of the iterative algorithms presented in the previous sections is their linear rate of convergence. In [24] a unified approach is presented in obtaining a class of iterative algorithms with different rates of convergence, based on a representation of the generalized inverse of a matrix. That is, the algorithm,

$$\begin{aligned}
\boldsymbol{x}_0 &= \beta\boldsymbol{D}^{\mathrm{T}}\boldsymbol{y} \\
\boldsymbol{D}_0 &= \beta\boldsymbol{D}^{\mathrm{T}}\boldsymbol{D} \\
\boldsymbol{\Omega}_{k+1} &= \sum_{i=0}^{p-1}(\mathbf{I} - \boldsymbol{D}_k)^i \\
\boldsymbol{D}_{k+1} &= \boldsymbol{\Omega}_k\boldsymbol{D}_k \\
\boldsymbol{x}_{k+1} &= \boldsymbol{\Omega}_k\boldsymbol{x}_k,
\end{aligned} \tag{34.38}$$

converges to the minimum norm least squares solution of Equation 34.1, with $\boldsymbol{n} = 0$. If iteration in Equation 34.38 is thought of as corresponding to iteration in Equation 34.27, then an iteration similar to Equation 34.38 which corresponds to iteration in Equation 34.23 has also been derived [24, 41].

Algorithm in Equation 34.38 exhibits a pth order of convergence. That is, the following relation holds [24]:

$$\frac{\|\boldsymbol{x}_k - \boldsymbol{x}^+\|}{\|\boldsymbol{x}^+\|} \leq c^{p^k}, \tag{34.39}$$

where the convergence factor c is described by Equation 34.34.

It is observed that the matrix sequences $\{\Omega_k\}$ and $\boldsymbol{D}_k$ can be computed in advance or off-line. When $\boldsymbol{D}$ is block circulant, substantial computational savings result with the use of iteration in Equation 34.38 over the linear algorithms. Questions dealing with the best order p of algorithm in Equation 34.38 to be used in a given application, as well as comparisons of the trade-off between speed of computation and computational load, are addressed in [24]. One of the drawbacks of the higher order algorithms is that the application of constraints may lead to erroneous results. Combined adaptive or nonadaptive linear and higher order algorithms have been proposed in overcoming this difficulty [11,24].

34.9 Other Forms of $\Phi(x)$

34.9.1 Ill-Posed Problems and Regularization Theory

The two most basic forms of the function $\Phi(\boldsymbol{x})$ have only been considered so far. These two forms are represented by Equations 34.22 and 34.26, and are meaningful when the noise in Equation 34.1 is not taken into account. Without ignoring the noise, however, the solution of Equation 34.1 represents an ill-posed problem. If the image formation process is modeled in a continuous infinite dimensional space, $\boldsymbol{D}$ becomes an integral operator and Equation 34.1 becomes a Fredholm integral equation of the first kind. Then the solution of Equation 34.1 is almost always an ill-posed problem [42,45,59,60]. This means that the unique least-squares solution of minimal norm of Equation 34.1 does not depend continuously on the data, or that a bounded perturbation (noise) in the data results in an unbounded perturbation in the solution, or that the generalized inverse of D is unbounded [42]. The integral operator $\boldsymbol{D}$ has a countably infinite number of singular values that can be ordered with their limit approaching zero [42]. Since the finite dimensional discrete problem of image restoration results from the discretization of an ill-posed continuous problem, the matrix $\boldsymbol{D}$ has (in addition to possibly a number of zero singular values) a cluster of very small singular values. Clearly, the finer the discretization (the larger the size of $\boldsymbol{D}$) the closer the limit of the singular values is approximated. Therefore, although the finite dimensional inverse problem is well posed in the least-squares sense [42], the ill-posedness of the continuous problem translates into an ill-conditioned matrix $\boldsymbol{D}$.

A regularization method replaces an ill-posed problem by a well-posed problem, whose solution is an acceptable approximation to the solution of the given ill-posed problem [39,56]. In general, regularization methods aim at providing solutions which preserve the fidelity to the data but also satisfy our prior knowledge about certain properties of the solution. A class of regularization methods associates both the class of admissible solutions and the observation noise with random processes [12]. Another class of regularization methods regards the solution as a deterministic quantity. We give examples of this second class of regularization methods in the following.

34.9.2 Constrained Minimization Regularization Approaches

Most regularization approaches transform the original inverse problem into a constrained optimization problem. That is, a functional needs to be optimized with respect to the original image and possibly other parameters. By using the necessary condition for optimality, the gradient of the functional with respect to the original image is set equal to zero, therefore determining the mathematical form of $\Phi(\boldsymbol{x})$.

The successive approximations iteration becomes in this case a gradient method with a fixed step (determined by β). We briefly mention next the general form of some of the commonly used functionals.

34.9.2.1 Set Theoretic Formulation

With this approach the problem of solving Equation 34.1 is replaced by the problem of searching for vectors x which belong to both sets [22,28,30]

$$\|\boldsymbol{Dx} - \boldsymbol{y}\| \leq \varepsilon \tag{34.40}$$

and

$$\|\boldsymbol{Cx}\| \leq E, \tag{34.41}$$

where

ε is an estimate on the data accuracy (noise norm)
E is a prescribed constant
C is a high-pass operator

Inequality in Equation 34.41 constrains the energy of the signal at high frequencies, therefore requiring that the restored signal is smooth. On the other hand, inequality in Equation 34.40 requires that the fidelity to the available data is preserved.

Inequalities in Equations 34.40 and 34.41 can be respectively rewritten as [22,30]

$$(\boldsymbol{x} - \boldsymbol{x}^+)^{\mathrm{T}} \frac{\boldsymbol{D}^{\mathrm{T}}\boldsymbol{D}}{\varepsilon^2} (\boldsymbol{x} - \boldsymbol{x}^+) \leq 1 \tag{34.42}$$

and

$$\boldsymbol{x}^{\mathrm{T}} \frac{\boldsymbol{C}^{\mathrm{T}}\boldsymbol{C}}{E_2} \boldsymbol{x} \leq 1, \tag{34.43}$$

where $\boldsymbol{x}^+ = \boldsymbol{D}^+\boldsymbol{y}$. That is, each of them represents an N-dimensional ellipsoid, where N is the dimensionality of the vectors involved. The intersection of the two ellipsoids (assuming it is not empty) is also a convex set but not an ellipsoid. The center of one of the ellipsoids which bounds the intersection can be chosen as the solution to the problem [50]. Clearly, even if the intersection is not empty, the center of the bounding ellipsoid may not belong to the intersection, and, therefore, a posterior test is required. The equation the center of one of the bounding ellipsoids is satisfying is given by [22,30]

$$\Phi(\boldsymbol{x}) = (\boldsymbol{D}^{\mathrm{T}}\boldsymbol{D} + \alpha \boldsymbol{C}^{\mathrm{T}}\boldsymbol{C})\boldsymbol{x} - \boldsymbol{D}^{\mathrm{T}}\boldsymbol{y} = 0, \tag{34.44}$$

where α, the regularization parameter, is equal to $(\varepsilon/E)^2$.

34.9.2.2 Projection onto Convex Sets Approach

Iteration in Equation 34.37 can also be applied in finding a solution which belongs to both ellipsoids in Equations 34.42 and 34.43. The respective projections $P_1\boldsymbol{x}$ and $P_2\boldsymbol{x}$ are defined by [22]

$$\mathbf{P}_1\boldsymbol{x} = \boldsymbol{x} + \lambda_1(\mathbf{I} + \lambda_1\boldsymbol{D}^{\mathrm{T}}\boldsymbol{D})^{-1}\boldsymbol{D}^{\mathrm{T}}(y - \boldsymbol{Dx}) \tag{34.45}$$

$$\mathbf{P}_2\boldsymbol{x} = [\mathbf{I} - \lambda_2(\mathbf{I} + \lambda_2\boldsymbol{C}^{\mathrm{T}}\boldsymbol{C})^{-1}\boldsymbol{C}^{\mathrm{T}}\boldsymbol{C}]\boldsymbol{x}, \tag{34.46}$$

where λ_1 and λ_2 need to be chosen so that conditions in Equations 34.42 and 34.43 are satisfied, respectively. Clearly, a number of other projection operators can be used in Equation 34.37 which force the signal to exhibit certain known *a priori* properties expressed by convex sets.

34.9.2.3 Functional Minimization Approach

The determination of the value of the regularization parameter is a critical issue in regularized restoration. A number of approaches for determining its value are presented in [13]. If only one of the parameters ε or E in Equations 34.40 and 34.41 is known, a constrained least-squares formulation can be followed [9,15]. With it, the size of one of the ellipsoids is minimized, subject to the constraint that the solution belongs to the surface of the other ellipsoid (the one defined by the known parameter). Following the Lagrangian approach, which transforms the constrained optimization problem into an unconstrained one, the following functional is minimized

$$M(\alpha, \boldsymbol{x}) = \|\boldsymbol{D}\boldsymbol{x} - y\|^2 + \alpha\|\boldsymbol{C}\boldsymbol{x}\|^2. \tag{34.47}$$

The necessary condition for a minimum is that the gradient of $M(\alpha, x)$ is equal to zero. That is, in this case

$$\Phi(\boldsymbol{x}) = \nabla_{\boldsymbol{x}} M(\alpha, \boldsymbol{x}) = (\boldsymbol{D}^{\mathrm{T}}\boldsymbol{D} + \alpha\boldsymbol{C}^{\mathrm{T}}\boldsymbol{C})\boldsymbol{x} - \boldsymbol{D}^{\mathrm{T}}y, \tag{34.48}$$

which is identical to Equation 34.44, with the only difference that α now is not known, but needs to be determined.

34.9.2.4 Spatially Adaptive Iteration

Spatially adaptive image restoration is the next natural step in improving the quality of the restored images. There are various ways to argue the introduction of spatial adaptivity, the most commonly used ones being the nonhomogeneity or nonstationarity of the image field and the properties of the human visual system. In either case, the functional to be minimized takes the form [21,29,34]

$$M(\alpha, \boldsymbol{x}) = \|\boldsymbol{D}\boldsymbol{x} - \boldsymbol{y}\|^2_{W_1} + \alpha\|C\boldsymbol{x}\|^2_{W_2}, \tag{34.49}$$

in which case

$$\Phi(\boldsymbol{x}) = \nabla_{\boldsymbol{x}} M(\alpha, \boldsymbol{x}) = (\boldsymbol{D}^{\mathrm{T}}\boldsymbol{W}_1^{\mathrm{T}}\boldsymbol{W}_1\boldsymbol{D} + \alpha C^{\mathrm{T}}\boldsymbol{W}_2^{\mathrm{T}}\boldsymbol{W}_2 C)\boldsymbol{x} - \boldsymbol{D}^{\mathrm{T}}\boldsymbol{W}_1 y. \tag{34.50}$$

The choice of the diagonal weighting matrices $\boldsymbol{W}_1$ and $\boldsymbol{W}_2$ can be justified in various ways. In [16, 21, 22, 29] both matrices are determined by the noise visibility matrix $\boldsymbol{V}$ [2,46]. That is, $\boldsymbol{W}_1 = \boldsymbol{V}^{\mathrm{T}}\boldsymbol{V}$ and $\boldsymbol{W}_2 = \mathbf{I} - \boldsymbol{V}^{\mathrm{T}}\boldsymbol{V}$. The entries of $\mathbf{V}$ take values between 0 and 1. They are equal to 0 at the edges (noise is not visible), equal to 1 at the flat regions (noise is visible), and take values in between at the regions with moderate spatial activity. A study of the mapping between the level of spatial activity and the values of the visibility function appears in [11]. The weighting matrices can also be defined by considering the relationship of the restoration approach presented here to the MAP restoration approach [26]. Then, the weighting matrices $\boldsymbol{W}_1$ and $\boldsymbol{W}_2$ contain information about the nonstationarity and/or the nonwhiteness of the high-pass filtered image and noise, respectively.

34.9.2.5 Robust Functionals

Robust functionals can be employed for the representation of both the noise and the signal statistics. They allow for the efficient suppression of a wide variety of noise processes and permit the reconstruction of sharper edges than their quadratic counterparts. In a robust set-theoretic setup a solution is sought by minimizing [65]

$$M(\alpha, \boldsymbol{x}) = R_{\mathrm{n}}(\boldsymbol{y} - \boldsymbol{D}\boldsymbol{x}) + \alpha R_{\mathrm{x}}(C\boldsymbol{x}). \tag{34.51}$$

where $R_n()$ and $R_x()$ are referred to as the residual and stabilizing functionals, respectively, and they are defined in terms of their kernel functions. The derivative of the kernel function is called the influence function.

$\Phi(\boldsymbol{x})$ in this case equals the gradient of $M(\alpha, x)$ in Equation 34.51. A large number of robust functionals have been proposed in the literature. The properties of potential functions to be used in robust Bayesian estimation are listed in [35]. A robust maximum absolute entropy and a robust minimum absolute-information functionals are introduced in [65]. Clearly since the functionals $R_n()$ and $R_x()$ are typically nonlinear and may not be convex, the convergence analysis of iteration in Equation 34.21 or 34.36 is considerably more complicated.

34.9.3 Iteration Adaptive Image Restoration Algorithms

As it has become clear by now there are various pieces of information needed by any regularization algorithms in determining the unknown parameters. In the context of deterministic regularization, the most commonly needed parameter is the regularization parameter. Its determination depends on the noise statistics and the properties of the image. With the set theoretic regularization approach, it is required that the original image is smooth, in which case a bound on the energy of the high-pass filtered image is needed. This bound is proportional to the variance of the image in a stochastic context. In addition, knowledge of the noise variance is also required. In a MAP framework such parameters are called hyperparameters [8,40]. Clearly, such parameters are not typically available and need to be estimated from the available noisy and blurred data. Various techniques for estimating the regularization parameter are discussed, for example, in [13].

In the following we briefly describe a new paradigm we have introduced in the context of iterative image restoration algorithms [18–20,25,26]. According to it, the required information by the deterministic regularization approach is updated at each restoration step, based on the partially restored image.

34.9.3.1 Spatially Adaptive Algorithm

For the spatially adaptive algorithm we mentioned above, the proposed general form of the weighted smoothing functional whose minimization will result in a restored image is written as

$$\begin{aligned} M_w(\lambda_w(\boldsymbol{x}), \boldsymbol{x}) &= \|\boldsymbol{y} - \boldsymbol{D}\boldsymbol{x}\|^2_{\boldsymbol{A}(\boldsymbol{x})} + \lambda_w(\boldsymbol{x})\|\boldsymbol{C}\boldsymbol{x}\|^2_{\boldsymbol{B}(\boldsymbol{x})} \\ &= \|\boldsymbol{n}\|^2_{\boldsymbol{A}(\boldsymbol{x})} + \lambda_w(\boldsymbol{x})\|\boldsymbol{C}\boldsymbol{x}\|^2_{\boldsymbol{B}(\boldsymbol{x})}, \end{aligned} \tag{34.52}$$

where the weighting matrices $\boldsymbol{A}(\boldsymbol{x})$ and $\boldsymbol{B}(\boldsymbol{x})$, both functions of the original image, are used to incorporate noise and image characteristics into the restoration process, respectively. The regularization parameter, also a function of $\boldsymbol{x}$, is defined in such a way as to make the smoothing functional in Equation 34.52 convex with a unique global minimizer.

One of the $\lambda_w(\boldsymbol{x})$ we have proposed is given by

$$\lambda_w(x) = \frac{\|\boldsymbol{y} - \boldsymbol{D}\boldsymbol{x}\|^2_{\boldsymbol{A}(\boldsymbol{x})}}{(1/\gamma) - \|\boldsymbol{C}\boldsymbol{x}\|^2_{\boldsymbol{B}(\boldsymbol{x})}}, \tag{34.53}$$

where the parameter γ is determined from the convergence and convexity analyses.

The main objective with this approach is to employ an iterative algorithm to estimate the regularization parameter and the proper weighting matrices at the same time with the restored image. The available estimate of the restored image at each iteration step will be used for determining the value of the regularization parameter. That is, the regularization parameter is defined as a function of the original image (and eventually in practice of an estimate of it). Of great importance is the form of this functional, so that the smoothing functional to be minimized preserves its convexity and exhibits a global minimizer. $\lambda_w(x)$ maps a vector x onto the positive real line. Its purpose is as before to control the relative contribution of the error term $\|y - \boldsymbol{D}\boldsymbol{x}\|^2_{A(\boldsymbol{x})}$, which enforces "faithfulness" to the data, and the stabilizing functional $\|Cx\|^2_{B(\boldsymbol{x})}$, which enforces smoothness on the solution. Its dependency, however, on the original image, as well as the available data, is explicitly utilized. This dependency on the other hand is implicitly utilized in the constrained least-squares approach, according to which the minimization of $M_w(\lambda_w(\boldsymbol{x}),\boldsymbol{x})$ and the determination of the regularization parameter $\lambda_w(\boldsymbol{x})$ are completely separate steps. The desired properties of $\lambda_w(\boldsymbol{x})$ and $M_w(\lambda_w(\boldsymbol{x}),\boldsymbol{x})$ are analyzed in [20]. The relationship of the resulting forms to the hierarchical Bayesian approach toward image restoration and estimation of the regularization parameters is explored in [40].

In this case, therefore, $\Phi(\boldsymbol{x}) = \nabla_x M_w(\lambda_w(\boldsymbol{x}),\boldsymbol{x})$. The successive approximations iteration after some simplifications takes the form [20,30]

$$\boldsymbol{x}_{k+1} = \boldsymbol{x}_k + \{\boldsymbol{D}^{\mathrm{T}}A(\boldsymbol{x}_k)y - [\boldsymbol{D}^{\mathrm{T}}\boldsymbol{A}(\boldsymbol{x}_k)\boldsymbol{D} + \lambda_w(\boldsymbol{x}_k)C^{\mathrm{T}}B(\boldsymbol{x}_k)C]\boldsymbol{x}_k\}. \tag{34.54}$$

The information required in defining the regularization parameter and the weights for introducing the spatial adaptivity are defined based on the available information about the restored image at the kth iteration step. Clearly for all this to make sense the convergence of iteration in Equation 34.54 has to be guaranteed. Furthermore, convergence to a unique fixed point, which removes the dependency of the final result on the initial conditions, is also desired. These issues are addressed in detail in [20,30]. A major advantage of the proposed algorithm is that the convexity of the smoothing functional and the convergence of the resulting algorithm are guaranteed regardless of the choice of the weighting matrices. Another advantage of this algorithm is that the proposed adaptive algorithm simultaneously determines the regularization parameter and the desirable weighting matrices based on the restored image at each iteration step and restores the image, without any prior knowledge.

34.9.3.2 Frequency Adaptive Algorithm

Adaptivity is now introduced into the restoration process by using a constant smoothness constraint, but by assigning a different regularization parameter at each discrete frequency location. We can now "fine-tune" the regularization of each frequency component, thereby achieving improved results and at the same time speeding up the convergence of the iterative algorithm. The regularization parameters are evaluated simultaneously with the restored image based on the partially restored image.

In this algorithm, the following two ellipsoids QE_x and $QE_{x/y}$ are used

$$QE_x = \{\boldsymbol{x} \,|\, \|C\boldsymbol{x}\|_{\boldsymbol{R}} \leq E_{\boldsymbol{R}}\} \tag{34.55}$$

and

$$QE_{x/y} = \{\boldsymbol{x} \,|\, \|y - \boldsymbol{D}\boldsymbol{x}\|_{\boldsymbol{P}} \leq \varepsilon_{\boldsymbol{P}}\}, \tag{34.56}$$

where $\boldsymbol{P}$ and $\boldsymbol{R}$ are both block-circulant weighting matrices. Then a solution which belongs to the intersection of QE_x and $QE_{x/y}$ is given by

$$(\boldsymbol{D}^{\mathrm{T}}\boldsymbol{P}^{\mathrm{T}}\boldsymbol{P}\boldsymbol{D} + \lambda \boldsymbol{C}^{\mathrm{T}}R^{\mathrm{T}}RC)\boldsymbol{x} = \boldsymbol{D}^{\mathrm{T}}\boldsymbol{P}^{\mathrm{T}}\boldsymbol{P}y, \tag{34.57}$$

where $\lambda = (\varepsilon_{\mathbf{P}}/E_{\mathbf{R}})^2$. Let us define $\boldsymbol{P}^{\mathrm{T}}\boldsymbol{P} = \boldsymbol{B}$, $\boldsymbol{R} = \boldsymbol{P}\boldsymbol{C}$, and $\lambda \boldsymbol{C}^{\mathrm{T}}\boldsymbol{C} = \boldsymbol{A}$. Then Equation 34.57 can be written as

$$\boldsymbol{B}(\boldsymbol{D}^{\mathrm{T}}\boldsymbol{D} + \boldsymbol{A}\boldsymbol{C}^{\mathrm{T}}\boldsymbol{C})x = \boldsymbol{B}\boldsymbol{D}^{\mathrm{T}}y, \tag{34.58}$$

since all matrices are block circulant and they therefore commute. The regularization matrix $\boldsymbol{A}$ is defined based on the set theoretic regularization as

$$\boldsymbol{A} = \|\boldsymbol{y} - \boldsymbol{Dx}\|^2(\|\boldsymbol{Cx}\|^2\mathbf{I} + \Delta)^{-1}, \tag{34.59}$$

where Δ is a block-circulant matrix used to ensure convergence. $\boldsymbol{B}$ plays the role of the "shaping" matrix [53] for maximizing the speed of convergence at every frequency component as well as for compensating for the near-singular frequency components [19].

With the above formulation, therefore,

$$\Phi(\boldsymbol{x}) = B[(\boldsymbol{D}^{\mathrm{T}}\boldsymbol{D} + AC^{\mathrm{T}}\boldsymbol{C})\boldsymbol{x} - \boldsymbol{D}^{\mathrm{T}}y], \tag{34.60}$$

and the successive approximations iteration in Equation 34.21 becomes

$$\boldsymbol{x}_{k+1} = \boldsymbol{x}_k + \boldsymbol{B}[\boldsymbol{D}^{\mathrm{T}}y - (\boldsymbol{D}^{\mathrm{T}}\boldsymbol{D} + A_k\boldsymbol{C}^{\mathrm{T}}\boldsymbol{C})\boldsymbol{x}_k], \tag{34.61}$$

where $\boldsymbol{A}_k = \|\boldsymbol{y} - \boldsymbol{D}_{x_k}\|^2(\|\boldsymbol{C}_{x_k}\|^2\mathbf{I} + \Delta_k)^{-1}$. It is mentioned here that iteration in Equation 34.61 can also be derived from the regularized equation

$$(\boldsymbol{D}^{\mathrm{T}}\boldsymbol{D} + \boldsymbol{AC}^{\mathrm{T}}\boldsymbol{C})\boldsymbol{x} = \boldsymbol{D}^{\mathrm{T}}\boldsymbol{y}, \tag{34.62}$$

using the generalized Landweber's iteration [53]. Since all matrices in iteration in Equation 34.61 are block-circulant, the iteration can be written in the discrete frequency domain as

$$X_{k+1}(\underline{p}) = X_k(\underline{p}) + \beta(\underline{p})\left[D^*(\underline{p})Y(\underline{p}) - \left(|D(\underline{p})|^2 + \lambda_k(\underline{p})|C(\underline{p})|^2\right)X_k(\underline{p})\right], \tag{34.63}$$

where $\underline{p} = (p_1, p_2)$, $0 \leq p_1 \leq N-1$, $0 \leq p_2 \leq N-1, X_{k+1}(\underline{p})$ and $Y(\underline{p})$ represent the 2D DFT of the unstacked image estimate x_{k+1}, and the noisy-blurred image $\boldsymbol{y}$ and $D(\underline{p})$, $C(\underline{p})$, $\beta(\underline{p})$, and $\lambda_k(\underline{p})$ represent 2D DFTs of the 2D sequences which form the block-circulant matrices $\boldsymbol{D}$, $\boldsymbol{C}$, $\boldsymbol{B}$, and $\boldsymbol{A}_k$, respectively. Since Δ_k is block-circulant $\lambda_k(\underline{p})$ is given by

$$\lambda_k(\underline{p}) = \frac{\sum_m |Y(\underline{m}) - D(\underline{m})X_k(\underline{m})|^2}{\sum_n |C(\underline{n})X_k(\underline{n})|^2 + \delta_k(\underline{p})}, \tag{34.64}$$

where $\delta_k(\underline{p})$ is the 2D DFT of the sequence which forms Δ_k.

The allowable range of each regularization and control parameter and the convergence analysis of the iterative algorithm are developed in detail in [19]. It is shown that the algorithm has more than two fixed points. The first fixed point is the inverse or generalized inverse solution of Equation 34.58. The second type of fixed points are regularized approximations to the original image. Since there is more than one solution to iteration in Equation 34.63, the determination of the initial condition becomes important. It has been verified experimentally [19] that if a "smooth" image is used for $X_0(\underline{p})$ almost identical fixed points result independently of X_0. The use of spectral filtering functions [53] is also incorporated into the iteration, as shown in [19].

34.10 Discussion

In this chapter we briefly described the application of the successive approximations-based class of iterative algorithms to the problem of restoring a noisy and blurred signal. We analyzed in some detail the simpler forms of the algorithm, while making reference to work which deals with more complicated forms of the algorithms. There are obviously a number of algorithms and issues pertaining to such algorithms which have not been addressed at all. For example, iterative algorithms with a varying relaxation parameter β, such as the steepest descent and conjugate gradient methods, can be applied to the image restoration problem [4,37]. The number of iterations also represents a means for regularizing the restoration problem [55,58]. Iterative algorithms which depend on more than one previous restoration steps (multistep algorithms) have also been considered, primarily for implementation reasons [27].

It is the hope and the expectation of the author that the material presented will form a good introduction to the topic for engineers or graduate students who would like to work in this area.

References

1. Abbiss, J.B., DeMol, C., and Dhadwal, H.S., Regularized iterative and noniterative procedures for object restoration from experimental data, *Opt. Acta*, 30: 107–124, 1983.
2. Anderson, G.L. and Netravali, A.N., Image restoration based on a subjective criterion, *IEEE Trans. Syst. Man Cybern.*, SMC-6: 845–853, Dec. 1976.
3. Andrews, H.C. and Hunt, B.R., *Digital Image Restoration*, Prentice-Hall, Englewood Cliffs, NJ, 1977.
4. Angel, E.S. and Jain, A.K., Restoration of images degraded by spatially varying point spread functions by a conjugate gradient method, *Appl. Opt.*, 17: 2186–2190, July 1978.
5. Banham, M. and Katsaggelos, A.K., Digital image restoration, *Signal Process. Mag.*, 14(2): 24–41, Mar. 1997.
6. Bialy, H., Iterative Behandlung Linearen Funktionalgleichungen, *Arch. Rational Mech. Anal.*, 4: 166–176, July 1959.
7. Biemond, J., Lagendijk, R.L., and Mersereau, R.M., Iterative methods for image deblurring, *Proc. IEEE*, 78(5): 856–883, May 1990.
8. Demoment, G., Image reconstruction and restoration: Overview of common estimation structures and problems, *IEEE Trans. Acoust. Speech Signal Process.*, 37(12): 2024–2036, Dec. 1989.
9. Dines, K.A. and Kak, A.C., Constrained least squares filtering, *IEEE Trans. Acoust. Speech Signal Process.*, ASSP-25: 346–350, 1977.
10. Dudgeon, D.E. and Mersereau, R.M., *Multidimensional Digital Signal Processing*, Prentice-Hall, Englewood Cliffs, NJ, 1984.
11. Efstratiadis, S.N. and Katsaggelos, A.K., Adaptive iterative image restoration with reduced computational load, *Opt. Eng.*, 29: 1458–1468, Dec. 1990.
12. Franklin, J.N., Well-posed stochastic extensions of ill-posed linear problems, *J. Math. Anal.*, 31: 682–716, 1970.
13. Galatsanos, N.P. and Katsaggelos, A.K., Methods for choosing the regularization parameter and estimating the noise variance in image restoration and their relation, *IEEE Trans. Image Process.*, 1: 322–336, July 1992.
14. Huang, T.S., Barker, D.A., and Berger, S.P., Iterative image restoration, *Appl. Opt.*, 14: 1165–1168, May 1975.
15. Hunt, B.R., The application of constrained least squares estimation to image restoration by digital computers, *IEEE Trans. Comput.*, C-22: 805–812, Sept. 1973.
16. Ichioka, Y. and Nakajima, N., Iterative image restoration considering visibility, *J. Opt. Soc. Am.*, 71: 983–988, Aug. 1981.
17. Jansson, P.A., Hunt, R.H., and Pyler, E.K., Resolution enhancement of spectra, *J. Opt. Soc. Am.*, 60: 596–599, May 1970.
18. Kang, M.G. and Katsaggelos, A.K., Iterative image restoration with simultaneous estimation of the regularization parameter, *IEEE Trans. Signal Process.*, 40(9): 2329–2334, Sept. 1992.
19. Kang, M.G. and Katsaggelos, A.K., Frequency domain adaptive iterative image restoration and evaluation of the regularization parameter, *Opt. Eng.*, 33(10): 3222–3232, Oct. 1994.

20. Kang, M.G. and Katsaggelos, A.K., General choice of the regularization functional in regularized image restoration, *IEEE Trans. Image Process.*, 4(5): 594–602, May 1995.
21. Katsaggelos, A.K., A general formulation of adaptive iterative image restoration algorithms, *Proceedings of 1986 Conference on Information Sciences and Systems*, Princeton, NJ, Mar. 1986, pp. 42–47.
22. Katsaggelos, A.K., Iterative image restoration algorithm, *Opt. Eng.*, 28(7): 735–748, July 1989.
23. Katsaggelos, A.K. (Ed.), *Digital Image Restoration*, Springer Series in Information Sciences, Vol. 23, Springer-Verlag, Heidelberg, Germany, 1991.
24. Katsaggelos, A.K. and Efstratiadis, S.N., A class of iterative signal restoration algorithms, *IEEE Trans. Acoust. Speech Signal Process.*, 38: 778–786, May 1990 (reprinted in *Digital Image Processing*, R. Chellappa (Ed.), IEEE Computer Society Press).
25. Katsaggelos, A.K. and Kang, M.G., Iterative evaluation of the regularization parameter in regularized image restoration, *J. Vis. Commn. Image Rep.*, special issue on *Image Restoration*, 3(6): 446–455, Dec. 1992.
26. Katsaggelos, A.K. and Kang, M.G., A spatially adaptive iterative algorithm for the restoration of astronomical images, *Int. J. Image Syst. Technol.*, special issue on *Image Reconstruction and Restoration in Astronomy*, 6(4): 305–313, Winter, 1995.
27. Katsaggelos, A.K. and Kumar, S.P.R., Single and multistep iterative image restoration and VLSI implementation, *Signal Process.*, 16(1): 29–40, Jan. 1989.
28. Katsaggelos, A.K., Biemond, J., Mersereau, R.M., and Schafer, R.W., A general formulation of constrained iterative restoration algorithms, *Proceedings of 1985 International Conference on Acoustics, Speech and Signal Processing*, Tampa, FL, Mar. 1985, pp. 700–703.
29. Katsaggelos, A.K., Biemond, J., Mersereau, R.M., and Schafer, R.W., Nonstationary iterative image restoration, *Proceedings of 1985 International Conference on Acoustics, Speech and Signal Processing*, Tampa, FL, Mar. 1985, pp. 696–699.
30. Katsaggelos, A.K., Biemond, J., Mersereau, R.M., and Schafer, R.W., A regularized iterative image restoration algorithm, *IEEE Trans. Signal Process.*, 39(4): 914–929, Apr. 1991.
31. Kawata, S. and Ichioka, Y., Iterative image restoration for linearly degraded images, I. Basis, *J. Opt. Soc. Am.*, 70: 762–768, July 1980.
32. Kawata, S. and Ichioka, Y., Iterative image restoration for linearly degraded images, II. Reblurring procedure, *J. Opt. Soc. Am.*, 70: 768–772, July 1980.
33. Kawata, S., Ichioka, Y., and Suzuki, T., Application of man-machine interactive image processing system to iterative image restoration, *Proceedings of the 4th International Conference on Pattern Recognition*, Kyoto, Japan, 1978, pp. 525–529.
34. Lagendijk, R.L., Biemond, J., and Boekee, D.E., Regularized iterative image restoration with ringing reduction, *IEEE Trans. Acoust. Speech Signal Process.*, 36: 1804–1887, Dec. 1988.
35. Lange, K., Convergence of EM image reconstruction algorithms with Gibbs smoothing, *IEEE Trans. Med. Imaging*, 9(4): 439–446, Dec. 1990.
36. Mammone, R.J., *Computational Methods of Signal Recovery and Recognition*, Wiley, New York, 1992.
37. Marucci, R., Mersereau, R.M., and Schafer, R.W., Constrained iterative deconvolution using a conjugate gradient algorithm, *Proceedings of 1982 IEEE International Conference on Acoustics, Speech and Signal Processing*, Paris, France, May 1982, pp. 1845–1848.
38. Mersereau, R.M. and Schafer, R.W., Comparative study of iterative deconvolution algorithms, *Proceedings of 1978 IEEE International Conference on Acoustics, Speech and Signal Processing*, Atlanta, GA, Apr. 1978, pp. 192–195.
39. Miller, K., Least-squares method for ill-posed problems with a prescribed bound, *SIAM J. Math. Anal.*, 1: 52–74, Feb. 1970.
40. Molina, R. and Katsaggelos, A.K., The hierarchical approach to image restoration and the iterative evaluation of the regularization parameter, *Proceedings of 1994 SPIE Conference on Visual Communication and Image Processing*, Chicago, IL, Sept. 1994, pp. 244–251.

41. Morris, C.E., Richards M.A., and Hayes, M.H., Fast reconstruction of linearly distorted signals, *IEEE Trans. Acoust. Speech Signal Process.*, 36: 1017–1025, July 1988.
42. Nashed, M.Z., Operator theoretic and computational approaches to ill-posed problems with application to antenna theory, *IEEE Trans. Antennas Propagation*, AP-29: 220–231, Mar. 1981.
43. Ortega, J.M., *Numerical Analysis: A Second Course*, Academic Press, New York, 1972.
44. Ortega, J.M. and Rheinboldt, W.C., *Iterative Solution of Nonlinear Equations in Several Variables*, Academic Press, New York, 1970.
45. Phillips, D.L., A technique for the numerical solution of certain integral equations of the first kind, *Assoc. Comput. Mach.*, 9: 84–97, 1962.
46. Rajala, S.S. and DeFigueiredo, R.J.P., Adaptive nonlinear image restoration by a modified Kalman filtering approach, *IEEE Trans. Acoust. Speech Signal Process.*, ASSP-29: 1033–1042, Oct. 1981.
47. Richards, M.A., Schafer, R.W., and Mersereau, R.M., An experimental study of the effects of noise on a class of iterative deconvolution algorithms, *Proceedings of 1979 International Conference on Acoustics, Speech and Signal Processing*, Atlanta, GA, Apr. 1979, pp. 401–404.
48. Sanz, J.L.C. and Huang, T.S., Iterative time-limited signal restoration, *IEEE Trans. Acoust. Speech Signal Process.*, ASSP-31: 643–649, June 1983.
49. Schafer, R.W., Mersereau, R.M., and Richards, M.A., Constrained iterative restoration algorithms, *Proc. IEEE*, 69: 432–450, Apr. 1981.
50. Schweppe, F.C., *Uncertain Dynamic Systems*, Prentice-Hall, Englewood Cliffs, NJ, 1973.
51. Sondhi, M.M., Image restoration: The removal of spatially invariant degradations, *Proc. IEEE*, 60: 842–853, July 1972.
52. Stark, H., *Image Recovery: Theory and Applications*, Academic Press, New York, 1987.
53. Strand, O.N., Theory and methods related to the singular-function expansion and Landweber's iteration for integral equations of the first kind, *SIAM J. Numerical Anal.*, 11: 798–825, Sept. 1974.
54. Strang, G., *Linear Algebra and Its Applications*, 2nd ed., Academic Press, New York, 1980.
55. Sullivan, B.J. and Katsaggelos, A.K., A new termination rule for linear iterative image restoration algorithms, *Opt. Eng.*, 29: 471–477, May 1990.
56. Tikhonov, A.N. and Arsenin, V.Y., *Solution of Ill-Posed Problems*, Winston-Wiley, New York, 1977.
57. Tom, V.T., Quatieri, T.F., Hayes, M.H., and McClellan, J.M., Convergence of iterative nonexpansive signal reconstruction algorithms, *IEEE Trans. Acoust. Speech Signal Process.*, ASSP-29: 1052–1058, Oct. 1981.
58. Trussell, H.J., Convergence criteria for iterative restoration methods, *IEEE Trans. Acoust. Speech Signal Process.*, ASSP-31: 129–136, Feb. 1983.
59. Twomey, S., On the numerical solution of Fredholm integral equations of the first kind by the inversion of the linear system produced by quadrature, *Assoc. Comput. Mach.*, 10: 97–101, 1963.
60. Twomey, S., The application of numerical filtering of the solution of integral equations encountered in indirect sensing measurements, *J. Franklin Inst.*, 279: 95–109, Feb. 1965.
61. Van Citttert, P.H., Zum Einfluss der Spaltbreite auf die Intensitatswerteilung in Spektrallinien II, *Z. Physik*, 69: 298–308, 1931.
62. Yang, Y., Galatsanos N.P., and Katsaggelos, A.K., Regularized image reconstruction from incomplete block discrete cosine transform data, *IEEE Trans. Circuits Syst. Video Technol.*, 3(6): 421–432, Dec. 1993.
63. Yang, Y., Galatsanos N.P., and Katsaggelos, A.K., Set theoretic spatially-adaptive reconstruction of block transform compressed images, *IEEE Trans. Image Process.*, 4(7): 896–908, July 1995.
64. Youla, D.C. and Webb, H., Image reconstruction by the method of convex projections, Part 1—Theory, *IEEE Trans. Med. Imaging*, MI-1(2): 81–94, Oct. 1982.
65. Zervakis, M.E., Katsaggelos, A.K., and Kwon, T.M., A class of robust entropic functionals for image restoration, *IEEE Trans. Image Process.*, 4(6): 752–773, June 1995.

VIII

Time–Frequency and Multirate Signal Processing

Cormac Herley
Microsoft Research

Kambiz Nayebi
Beena Vision Systems Inc.

AN IMPORTANT PROBLEM IN SIGNAL PROCESSING is the choice of how to represent a signal. It is for this reason that importance is attached to the choice of bases for the linear expansion of signals. That is, given a discrete-time signal $x(n)$ how to find $a_i(n)$ and $b_i(n)$ such that we can write

$$x(n) = \sum_i < x(n),\, a_i(n) > b_i(n). \tag{VIII.1}$$

If $b_i(n) = a_i(n)$, then Equation VIII.1 is the familiar orthonormal basis expansion formula [1]. Otherwise, the $b_i(n)$ are a set of biorthogonal functions with the property

$$< b_j(n),\ a_i(n) > = \delta_{i-j}.$$

The function δ is defined such that $\delta_{i-j} = 0$, unless $i = j$, in which case $\delta_0 = 1$. We shall consider cases where the summation in Equation VIII.1 is infinite, but restrict our attention to the case where it is finite for the moment; that is, where we have a finite number N of data samples, and so the space is finite dimensional.

We next set up the basic notation used throughout the chapter. Assume that we are operating in C^N, and that we have N basis vectors, the minimum number to span the space. Since the transform is linear, it can be written as a matrix. That is, if the $\mathbf{a}_i^*$ are the rows of a matrix $\mathbf{A}$, then

$$\mathbf{A} \cdot \mathbf{x} = \begin{bmatrix} < x(n),\ a_0(n) > \\ < x(n),\ a_1(n) > \\ \vdots \\ < x(n),\ a_{N-2}(n) > \\ < x(n),\ a_{N-1}(n) > \end{bmatrix} \qquad \text{(VIII.2)}$$

and if $\mathbf{b}_i$ are the columns of $\mathbf{B}$ then

$$\mathbf{x} = \mathbf{B} \cdot \mathbf{A} \cdot \mathbf{x}. \qquad \text{(VIII.3)}$$

Clearly $\mathbf{B} = \mathbf{A}^{-1}$; if $\mathbf{B} = \mathbf{A}^*$ then $\mathbf{A}$ is unitary, $b_i(n) = a_i(n)$ and we have that Equation VIII.1 is the orthonormal basis expansion.

Clearly the construction of bases is not difficult: any nonsingular $N \times N$ matrix will do for this space. Similarly, to get an orthonormal basis we need merely take the rows of any unitary $N \times N$ matrix, for example the identity $\mathbf{I}_N$. There are many reasons for desiring to carry out such an expansion. Much as Taylor or Fourier series are used in mathematics to simplify solutions to certain problems, the underlying goal is that a cleverly chosen expansion may make a given signal processing task simpler.

A major application is signal compression, where we wish to quantize the input signal in order to transmit it with as few bits as possible, while minimizing the distortion introduced. If the input vector comprises samples of a real signal, then the samples are probably highly correlated, and the identity basis (where the ith vector contains 1 in the ith position and is zero elsewhere) with scalar quantization will end up using many of its bits to transmit information which does not vary much from sample to sample. If we can choose a matrix $\mathbf{A}$ such that the elements of $\mathbf{A} \cdot \mathbf{x}$ are much less correlated than those of $\mathbf{x}$, then the job of efficient quantization becomes a great deal simpler [2]. In fact, the Karhunen–Loève transform, which produces uncorrelated coefficients, is known to be optimal in a mean squared error sense [2].

Since in Equation VIII.1 the signal is written as a superposition of the basis sequences $b_i(n)$, we can say that if $b_i(n)$ has most of its energy concentrated around time $n = n_0$, then the coefficient $< x(n),\ a_i(n) >$ measures to some degree the concentration of $x(n)$ at time $n = n_0$. Equally, taking the discrete Fourier transform of Equation VIII.1,

$$X(k) = \sum_i < x(n),\ a_i(n) > B_i(k).$$

Thus, if $B_i(k)$ has most of its energy concentrated about frequency $k = k_0$, then $< x(n),\ a_i(n) >$ measures to some degree the concentration of $X(k)$ at $k = k_0$. This basis function is mostly localized about the point $(n_0,\ k_0)$ in the discrete-time discrete-frequency plane. Similarly, for each of the basis functions $b_i(n)$ we can find the area of the discrete-time discrete-frequency plane where most of their energy lies. All of the basis functions together will effectively cover the plane, because if any part were not covered there would be a "hole" in the basis, and we would not be able to completely represent all sequences in the space. Similarly the localization areas, or tiles, corresponding to distinct basis functions should not overlap by too much, since this would represent a redundancy in the system.

Choosing a basis can then be loosely thought of as choosing some tiling of the discrete-time discrete-frequency plane. For example, Figure VIII.1 shows the tiling corresponding to various orthonormal bases in C^{64}. The horizontal axis represents discrete-time, and the vertical axis discrete-frequency. Naturally, each of the diagrams contains 64 tiles, since this is the number of vectors required for a basis, and each tile can be thought of as containing 64 points out of the total of 64^2 in this discrete-time discrete-frequency plane. The first is the identity basis, which has narrow vertical strips as tiles, since the basis sequences $\delta(n+k)$ are perfectly localized in time, but have energy spread equally at all discrete frequencies. That is, the tile is one discrete-time point wide and 64 discrete-frequency points long. The second, shown in Figure VIII.1b, corresponds to the discrete Fourier transform basis vectors $e^{j2\pi in/N}$; these of course are perfectly localized at the frequencies $i = 0, 1, \ldots, N-1$, but have equal energy at all times (i.e., 64 points wide, one point long). Figure VIII.1c shows the tiling corresponding to a discrete

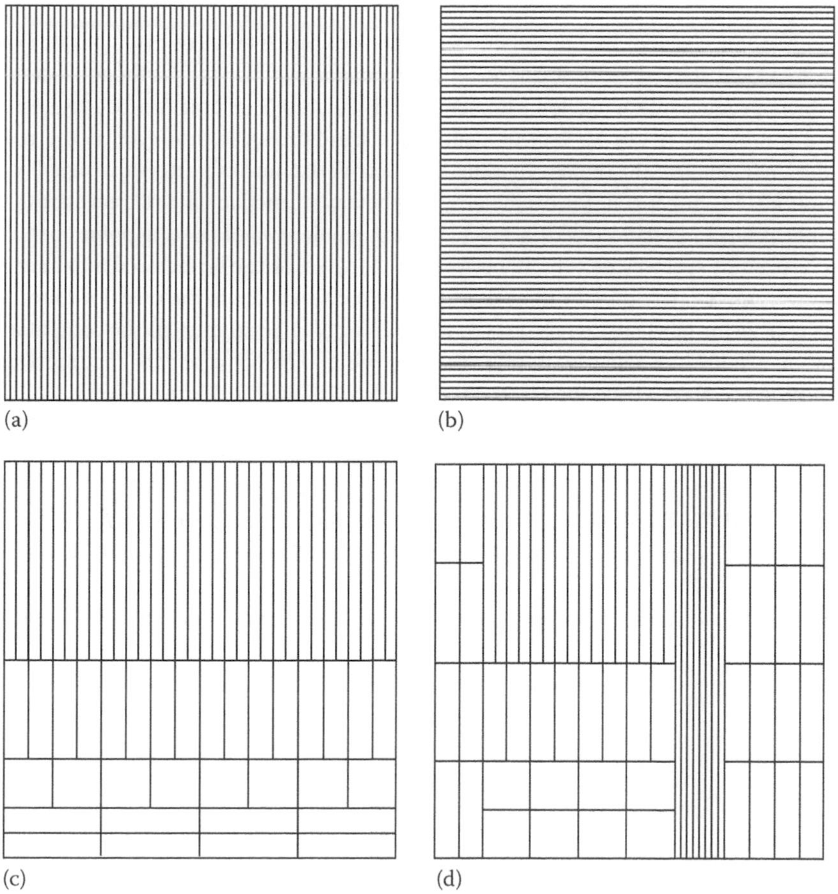

FIGURE VIII.1 Examples of tilings of the discrete-time discrete-frequency plane; time is the horizontal axis, frequency the vertical: (a) identity transform, (b) discrete Fourier transform, (c) finite length discrete wavelet transform, and (d) arbitrary finite length transform.

orthogonal wavelet transform (or logarithmic subband coder) operating over a finite length signal. Figure VIII.1d shows the tiling corresponding to a discrete orthogonal wavelet packet transform operating over a finite length signal, with arbitrary splits in time and frequency; construction of such schemes is discussed in Section 35.1 of Chapter 35. In Figure VIII.1c and d, the tiles have varying shapes but still contain 64 points each.

It should be emphasized that the localization of the energy of a basis function to the area covered by one of the tiles is only approximate. In practice, of course, we will always deal with real signals, and in general we will restrict the basis functions to be real also. When this is so, $\mathbf{B}^* = \mathbf{B}^T$ and the basis is orthonormal provided $\mathbf{A}^T\mathbf{A} = \mathbf{I} = \mathbf{A}\mathbf{A}^T$. Of the bases shown in Figure VIII.1 only the discrete Fourier transform will be excluded with this restriction. One can, however, consider a real transform which has many properties in common with the DFT, for example the discrete Hartley transform [3].

While the above description was given in terms of finite-dimensional signal spaces, the interpretation of the linear transform as a matrix operation, and the tiling approach remains essentially unchanged in the case of infinite length discrete-time signals. In fact, for bases with the structure we desire, construction in the infinite-dimensional case is easier than in the finite-dimensional case. The modifications necessary for the transition from R^N to $l^2(R)$ are that an infinite number of basis functions is required instead of N, the matrices $\mathbf{A}$ and $\mathbf{B}$ become doubly infinite, and the tilings are in the discrete-time continuous-frequency plane (the time axis ranges over Z, the frequency axis goes from 0 to π, assuming real signals).

Good decorrelation is one of the important factors in the construction of bases. If this were the only requirement, we would always use the Karhunen–Loève transform, which is an orthogonal data-dependent transform which produces uncorrelated samples. This is not used in practice, because estimating the coefficients of the matrix $\mathbf{A}$ can be very difficult. Very significant also, however, is the complexity of calculating the coefficients of the transform using Equation VIII.2, and of putting the signal back together using Equation VIII.3. In general, for example, using the basis functions for R^N, evaluating each of the matrix multiplications in Equations VIII.2 and VIII.3 will require $O(N^2)$ floating point operations, unless the matrices have some special structure. If, however, $\mathbf{A}$ is sparse, or can be factored into matrices that are sparse, then the complexity required can be dramatically reduced. This is the case, for example, with the discrete Fourier transform, where there is an efficient $O(N \log N)$ algorithm to do the computations, which has been responsible for its popularity in practice. This will also be the case with the transforms that we consider, $\mathbf{A}$ and $\mathbf{B}$ will always have special structure to allow efficient implementation.

References

1. Gohberg, I. and Goldberg, S., *Basic Operator Theory*, Birkhäuser, Boston, MA, 1981.
2. Gersho, A. and Gray, R.M., *Vector Quantization and Signal Compression*, Kluwer Academic, Norwell, MA, 1992.
3. Bracewell, R., *The Fourier Transform and Its Applications*, 2nd ed., McGraw-Hill, New York, 1986.

35

Wavelets and Filter Banks

Cormac Herley
Microsoft Research

35.1 Filter Banks and Wavelets

The methods of designing bases that we will employ draw on ideas first used in the construction of multirate filter banks. The idea of such systems is to take an input system and split it into subsequences using banks of filters. This simplest case involves splitting into just two parts using a structure such as that shown in Figure 35.1. This technique has a long history of use in the area of subband coding: first of speech [1,2] and more recently of images [3,4]. In fact, the most successful image coding schemes are based on filter bank expansions [5–7]. Recent texts on the subject are [8–10]. We will consider only the two-channel case in this section. If $\hat{X}(z) = X(z)$, then the filter bank has the perfect reconstruction property.

It is easily shown that the output $\hat{X}(z)$ of the overall analysis/synthesis system is given by

$$\begin{aligned}\hat{X}(z) &= \frac{1}{2}[G_0(z)\ G_1(z)]\begin{bmatrix} H_0(z) & H_0(-z)(1) \\ H_1(z) & H_1(-z) \end{bmatrix}\begin{bmatrix} X(z)(2) \\ X(-z) \end{bmatrix} \\ &= \frac{1}{2}[H_0(z)G_0(z) + H_1(z)G_1(z)]X(z) \\ &\quad + \frac{1}{2}[H_0(-z)G_0(z) + H_1(-z)G_1(z)]X(-z).\end{aligned} \tag{35.1}$$

Call the above 2×2 matrix $\mathbf{H}_m(z)$. This gives that the unique choice for the synthesis filters is

$$\begin{aligned}\begin{bmatrix} G_0(z) \\ G_1(z) \end{bmatrix} &= \begin{bmatrix} H_0(z) & H_0(-z) \\ H_1(z) & H_1(-z) \end{bmatrix}^{-1} \cdot \begin{bmatrix} 2 \\ 0 \end{bmatrix} \\ &= \frac{2}{\Delta_m(z)}\begin{bmatrix} H_1(-z) \\ -H_0(-z) \end{bmatrix},\end{aligned} \tag{35.2}$$

where $\Delta_m(z) = \det \mathbf{H}_m(z)$.

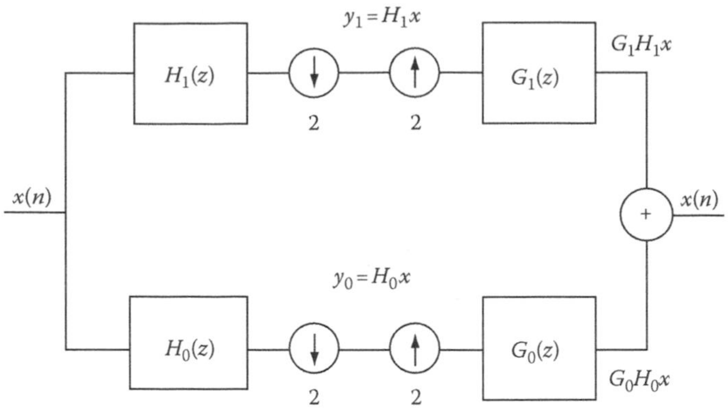

FIGURE 35.1 Maximally decimated two-channel multirate filter bank.

If we observe that $\Delta_m(z) = -\Delta_m(-z)$ and define $P(z) = 2 \cdot H_0(z)H_1(-z)/\Delta_m(z) = H_0(z)G_0(z)$, it follows from Equation 35.2 that $G_1(z)H_1(z) = 2 \cdot H_1(z)H_0(-z)/\Delta_m(-z) = P(-z)$. We can then write that the necessary and sufficient condition for perfect reconstruction of Equation 35.1 is

$$P(z) + P(-z) = 2. \tag{35.3}$$

Since this condition plays an important role in what follows, we will refer to any function having this property as valid. The implication of this property is that all but one of the even-indexed coefficients of $P(z)$ are zero. That is

$$\begin{aligned} P(z) + P(-z) &= \sum_n [p(n)z^{-n} + p(n)(-z)^{-n}] \\ &= \sum_n 2 \cdot p(2n)z^{-(2n+1)}. \end{aligned}$$

For this to satisfy Equation 35.3 requires $p(2n) = \delta_n$; thus, one of the polyphase components of $P(z)$ must be the unit sample. By polyphase components we mean the set of even-indexed samples, and the set of the odd-indexed samples. Such a function is illustrated in Figure 35.2a.

Constructing such a function is not difficult. In general, however, we will wish to impose additional constraints on the filter banks. So, $P(z)$ will have to satisfy other constraints in addition to Equation 35.3.

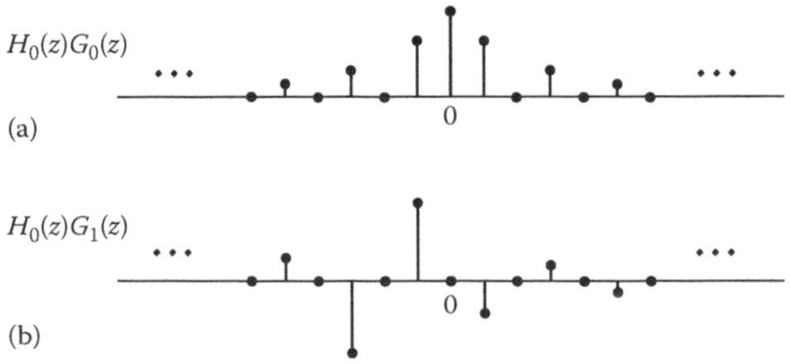

FIGURE 35.2 Zeros of the correlation functions: (a) autocorrelation $H_0(-z)H_0(z^{-1})$ and (b) cross-correlation $H_0(-z)H_1(z^{-1})$.

Observe that as a consequence of Equation 35.2 $G_0(z)H_1(z)$, that is, the cross-correlation of $g_1(n)$ and the time-reversed filter $h_0(-n)$, and $G_1(z)H_0(z)$, the cross-correlation of $g_1(n)$ and $h_0(-n)$, have only odd-indexed coefficients, just as for the function in Figure 35.2b, that is,

$$< g_0(n),\ h_1(2k-n) > = 0, \tag{35.4}$$

$$< g_1(n),\ h_0(2k-n) > = 0 \tag{35.5}$$

(note the time reversal in the inner product). Define now the matrix $\mathbf{H}_0$ as

$$\mathbf{H}_0 = \begin{bmatrix} & \vdots & \vdots & \vdots & \vdots & \vdots & \vdots & \vdots & \\ \ddots & h_0(L-1) & h_0(L-2) & \cdots & \cdots & h_0(0) & 0 & 0 & \ddots \\ & 0 & 0 & h_0(L-1) & \cdots & h_0(2) & h_0(1) & h_0(0) & \\ & \vdots & \vdots & \vdots & \vdots & \vdots & \vdots & \vdots & \end{bmatrix}, \tag{35.6}$$

which has as its kth row the elements of the sequence $h_0(2k-n)$. Pre-multiplying by $\mathbf{H}_0$ corresponds to filtering by $H_0(z)$ followed by subsampling by a factor of 2. Also define

$$\mathbf{G}_0^{\mathrm{T}} = \begin{bmatrix} & \vdots & \vdots & \vdots & \vdots & \vdots & \vdots & \vdots & \\ \ddots & g_0(0) & g_0(1) & \cdots & \cdots & g_0(L-1) & 0 & 0 & \ddots \\ & 0 & 0 & g_0(0) & \cdots & g_0(L-3) & g_0(L-2) & g_0(L-1) & \\ & \vdots & \vdots & \vdots & \vdots & \vdots & \vdots & \vdots & \end{bmatrix}, \tag{35.7}$$

so $\mathbf{G}_0$ has as its kth column the elements of the sequence $g_0(n-2k)$. Define $\mathbf{H}_1$ by replacing the coefficients of $h_0(n)$ with those of $h_1(n)$ in Equation 35.6 and $\mathbf{G}_1$ by replacing the coefficients of $g_0(n)$ with those of $g_1(n)$ in Equation 35.7.

We find that Equation 35.4 gives that all rows of $\mathbf{H}_1$ are orthogonal to all columns of $\mathbf{G}_0$. Similarly we find, from Equation 35.5, that all of the columns of $\mathbf{G}_1$ are orthogonal to the rows of $\mathbf{H}_0$. So, in matrix notation,

$$\mathbf{H}_0\mathbf{G}_1 = 0 = \mathbf{H}_1\mathbf{G}_0. \tag{35.8}$$

Now $P(z) = G_0(z)H_0(z) = z^{-1}H_0(z)H_1(-z)$ and $P(-z) = G_1(z)H_1(z)$ are both valid and have the form given in Figure 35.2a. Hence, the impulse responses of $g_i(n)$ and $h_i(n)$ are orthogonal with respect to even shifts

$$< g_i(n),\ h_i(2l-n) > = \delta_l. \tag{35.9}$$

In operator notation,

$$\mathbf{H}_0\mathbf{G}_0 = \mathbf{I} = \mathbf{H}_1\mathbf{G}_1. \tag{35.10}$$

Since we have a perfect reconstruction system we get

$$\mathbf{G}_0\mathbf{H}_0 + \mathbf{G}_1\mathbf{H}_1 = \mathbf{I}. \tag{35.11}$$

Of course Equation 35.11 indicates that no nonzero vector can lie in the column null-spaces of both $\mathbf{G}_0$ and $\mathbf{G}_1$. Note that Equation 35.10 implies that $\mathbf{G}_0\mathbf{H}_0$ and $\mathbf{G}_1\mathbf{H}_1$ are each projections (since $\mathbf{G}_i\mathbf{H}_i\mathbf{G}_i\mathbf{H}_i = \mathbf{G}_i\mathbf{H}_i$). They project onto subspaces which are not, in general, orthogonal (since the operators are not self-adjoint). Because of Equations 35.4, 35.5, and 35.9 the analysis/synthesis system is termed biorthogonal. If we interleave the rows of $\mathbf{H}_0$ and $\mathbf{H}_1$, much as was done in the orthogonal case, and form again a block Toeplitz matrix:

$$\mathbf{A} = \begin{bmatrix} & \vdots & \vdots & \vdots & \vdots & \vdots & \vdots & \vdots & \\ & h_0(L-1) & h_0(L-2) & \cdots & \cdots & h_0(0) & 0 & 0 & \\ \ddots & h_1(L-1) & h_1(L-2) & \cdots & \cdots & h_1(0) & 0 & 0 & \ddots \\ & 0 & 0 & h_0(L-1) & \cdots & h_0(2) & h_0(1) & h_0(0) & \\ & 0 & 0 & h_1(L-1) & \cdots & h_1(2) & h_1(1) & h_1(0) & \\ & \vdots & \vdots & \vdots & \vdots & \vdots & \vdots & \vdots & \end{bmatrix}, \tag{35.12}$$

we find that the rows of $\mathbf{A}$ form a basis for $l^2(Z)$. If we form $\mathbf{B}$ by interleaving the columns of $\mathbf{G}_0$ and $\mathbf{G}_1$, we find

$$\mathbf{B} \cdot \mathbf{A} = \mathbf{I}.$$

In the special case where we have a unitary solution, one finds $\mathbf{G}_0 = \mathbf{H}_0^{\mathrm{T}}$ and $\mathbf{G}_1 = \mathbf{H}_1^{\mathrm{T}}$, and Equation 35.8 gives that we have projections onto subspaces which are mutually orthogonal. The system then simplifies to the orthogonal case, where $\mathbf{B} = \mathbf{A}^{-1} = \mathbf{A}^{\mathrm{T}}$.

A point that we wish to emphasize is that in the conditions for perfect reconstruction, Equations 35.2 and 35.3, the filters $H_0(z)$ and $G_0(z)$ are related via their product $P(z)$. It is the choice of the function $P(z)$ and the factorization taken that determines the properties of the filter bank. We conclude the introduction with a proposition that sums up the foregoing.

PROPOSITION 35.1

To design a two-channel perfect reconstruction filter bank, it is necessary and sufficient to find a $P(z)$ satisfying Equation 35.3, factor it $P(z) = G_0(z)H_0(z)$ and assign the filters as given in Equation 35.2.

35.1.1 Deriving Continuous-Time Bases from Discrete-Time Ones

We have seen that the construction of bases from discrete-time signals can be accomplished easily by using a perfect reconstruction filter bank as the basic building block. This gives us bases that have a certain structure, and for which the analysis and synthesis can be efficiently performed. The design of bases for continuous-time signals appears more difficult. However, it works out that we can mimic many of the ideas used in the discrete-time case, when we go about the construction of continuous-time bases.

In fact, there is a very close correspondence between the discrete-time bases generated by two-channel filter banks, and dyadic wavelet bases. These are continuous-time bases formed by the stretches and translates of a single function, where the stretches are integer powers of 2:

$$\left\{\psi_{jk}(x) = 2^{-j/2}\psi(2^{-j}x - k), \quad j, k, \in Z\right\} \tag{35.13}$$

This relation has been thoroughly explored in [11,12].

To be precise, a basis of the form in Equation 35.13 necessarily implies the existence of an underlying two-channel filter bank. Conversely, a two-channel filter bank can be used to generate a basis as in

Equation 35.13 provided that the lowpass filter $H_0(z)$ is regular. It is not our intention to go into the details of this connection, but the generation of wavelets from filter banks goes briefly as follows.

Considering the logarithmic tree of discrete-time filters in Figure 35.3, one notices that the lower branch is a cascade of filters $H_0(z)$ followed by subsampling by 2. It is easily shown [12] that the cascade of i blocks of filtering operations, followed by subsampling by 2, is equivalent to a filter $H_0^{(i)}(z)$ with z-transform

$$H_0^{(i)}(z) = \prod_{l=0}^{i-1} H_0(z^{2^l}), \quad i = 1, 2 \ldots, \tag{35.14}$$

followed by subsampling by 2^i. We define $H_0^{(0)}(z) = 1$ to initialize the recursion. Now, in addition to the discrete-time filter, consider the function $f^{(i)}(x)$ which is piecewise constant on intervals of length $1/2^i$, and equal to

$$f^{(i)}(x) = 2^{i/2} h_0^{(i)}(n), \quad n/2^i \leq x < (n+1)/2^i. \tag{35.15}$$

Note that the normalization by $2^{i/2}$ ensures that if $\sum [h_0^{(i)}(n)]^2 = 1$ then $\int [f^{(i)}(x)]^2 \mathrm{d}x = 1$ as well. Also, it can be checked that $\|h_0^{(i)}\|_2 = 1$ when $\|h_0^{(i-1)}\|_2 = 1$. The relation between the sequence $H_0^{(i)}(z)$ and the function $f^{(i)}(x)$ is clarified in Figure 35.3, where the first three iterations of each is shown for the simple case of a filter of length 4.

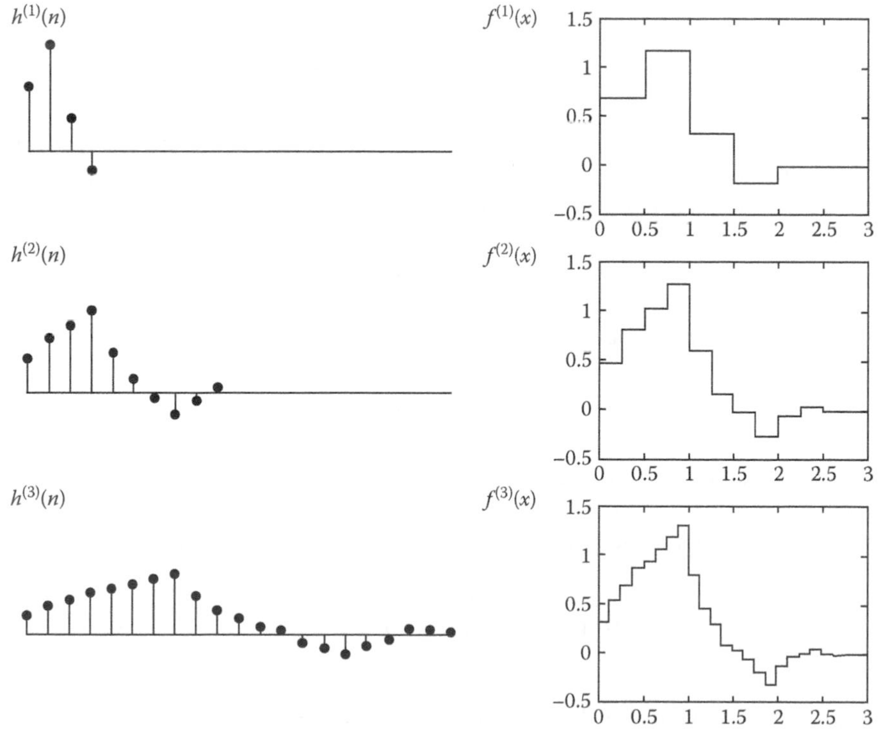

FIGURE 35.3 Iterations of the discrete-time filter (Equation 35.14) and the continuous-time function (Equation 35.15) for the case of a length 4 filter $H_0(z)$. The length of the filter $H_0^{(i)}(z)$ increases without bound, while the function $f^{(i)}(x)$ actually has bounded support.

We are going to use the sequence of functions $f^{(i)}(x)$ to converge to the scaling function $\phi(x)$ of a wavelet basis. Hence, a fundamental question is to find out whether and to what the function $f^{(i)}(x)$ converges as $i \rightarrow \infty$. First assume that the filter $H_0(z)$ has a zero at the half sampling frequency, or $H_0(e^{j\pi}) = 0$. This together with the fact that the filter impulse response is orthogonal to its even translates is equivalent to $\sum h_0(n) = H_0(1) = \sqrt{2}$. Define $M_0(z) = 1/\sqrt{2} \cdot H_0(z)$, that is $M_0(1) = 1$. Now factor $M_0(z)$ into its roots at π (there is at least one by assumption) and a remainder polynomial $K(z)$, in the following way:

$$M_0(z) = [(1 + z^{-1})/2]^N K(z).$$

Note that $K(1) = 1$ from the definitions. Now call B the supremum of $|K(z)|$ on the unit circle:

$$B = \sup_{\omega \in [0, 2\pi]} |K(e^{j\omega})|.$$

Then the following result from [11] holds:

PROPOSITION 35.2

If $B < 2^{N-1}$ and

$$\sum_{n=-\infty}^{\infty} |k(n)|^2 |n|^{\varepsilon} < \infty, \quad \text{for some } \varepsilon > 0, \tag{35.16}$$

then the piecewise constant function $f^{(i)}(x)$ defined in Equation 35.15 converges pointwise to a continuous function $f^{(\infty)}(x)$.

This is a sufficient condition to ensure pointwise convergence to a continuous function, and can be used as a simple test. We shall refer to any filter for which the infinite product converges as regular.

If we indeed have convergence, then we define

$$f^{(\infty)}(x) = \phi(x)$$

as the analysis scaling function, and

$$\psi(x) = 2^{-1/2} \sum h_1(n) \phi(2x - n), \tag{35.17}$$

as the analysis wavelet. It can be shown that if the filters $h_0(n)$ and $h_1(n)$ are from a perfect reconstruction filter bank, then Equation 35.13 indeed forms a continuous-time basis.

In a similar way we examine the cascade of i blocks of the synthesis filter $g_0(n)$:

$$G_0^{(i)}(z) = \prod_{l=0}^{i-1} G_0(z^{2^l}), \quad i = 1, 2, \ldots. \tag{35.18}$$

Again, define $G_0^{(0)}(z) = 1$ to initialize the recursion, and normalize $G_0(1) = 1$. From this define a function which is piecewise constant on intervals of length $1/2^i$:

$$\check{f}^{(i)}(x) = 2^{i/2} \cdot g_0^{(i)}(-n), \quad n/2^i \leq x < (n+1)/2^i. \tag{35.19}$$

We call the limit $\check{f}^{(\infty)}(x)$, if it exists, $\check{\phi}(x)$ the synthesis scaling function, and we find

$$\check{\phi}(x) = 2^{1/2} \sum_{n=0}^{L-1} g_0(-n)\check{\phi}(2x - n) \tag{35.20}$$

$$\check{\psi}(x) = 2^{1/2} \sum_{n=0}^{L-1} g_1(-n)\check{\phi}(2x - n). \tag{35.21}$$

The biorthogonality properties of the analysis and synthesis continuous-time functions follow from the corresponding properties of the discrete-time ones. That is, Equation 35.9 leads to

$$< \check{\phi}(x), \phi(x - k) > = \delta_k \tag{35.22}$$

and

$$< \check{\psi}(x), \psi(x - k) > = \delta_k. \tag{35.23}$$

Similarly

$$< \check{\phi}(x), \psi(x - k) > = 0 \quad \text{and} \tag{35.24}$$

$$< \check{\psi}(x), \phi(x - k) > = 0 \tag{35.25}$$

come from Equations 35.4 and 35.5, respectively.

We have shown that the conditions for perfect reconstruction on the filter coefficients lead to functions that have the biorthogonality properties as shown above. Orthogonality across scales is also easily verified:

$$< \check{\psi}(2^j x), \psi(2^i x - k) > = \delta_{i-j}\delta_k.$$

Thus, the set $\{\psi(2^j x), \check{\psi}(2^i x - k), i, j, \text{ and } k \in Z\}$ is biorthogonal. That it is complete can be verified as in the orthogonal case [13]. Hence, any function from $L^2(R)$ can be written as

$$f(x) = \sum_j \sum_l < f(x), 2^{-j/2}\psi(2^j x - l) > 2^{-j/2}\check{\psi}(2^j x - l).$$

Note that $\psi(x)$ and $\check{\psi}(x)$ play interchangeable roles.

35.1.2 Two-Channel Filter Banks and Wavelets

We have seen that the design of discrete-time bases is not difficult: using two-channel filter banks as the basic building block they can be easily derived. We also know that, using Equations 35.15 and 35.19, we can generate continuous-time bases quite easily as well. If we were just interested in the construction of bases, with no further requirements, we could stop here. However, for applications such as compression, we will often be interested in other properties of the basis functions, for example, whether or not they have any symmetry or finite support, and whether or not the basis is an orthonormal one. We examine these three structural properties for the remainder of this section. Chapter 37 deals with the design of the filters. Chapter 37 deals with time-varying filter banks, where the filters used, or the tree structure employing them, varies over time. Chapter 38 deals with the case of lapped transforms, a very important class of multirate filter banks that have achieved considerable success.

From the filter bank point of view, the properties we are most interested in are the following:

- **Orthogonality:**

$$< h_0(n), h_0(n+2k) > = \delta_k = < h_1(n), h_1(n+2k) > , \tag{35.26}$$

$$< h_0(n), h_1(n+2k) > = 0. \tag{35.27}$$

- **Linear phase:** $H_0(z)$, $H_1(z)$, $G_0(z)$, and $G_1(z)$ are all linear phase filters.
- **Finite support:** $H_0(z)$, $H_1(z)$, $G_0(z)$, and $G_1(z)$ are all FIR filters.

The reason for our interest is twofold. First, these properties are possibly of value in perfect reconstruction filter banks used in subband coding schemes. For example, orthogonality implies that the quantization noise in the two channels will be independent; linear phase is possibly of interest in very low bit-rate coding of images, and FIR filters have the advantage of having very simple low-complexity implementations. Second, these properties are carried over to the wavelets that are generated. So, if we design a filter bank with a certain set of properties, then the continuous-time basis that it generates will also have these properties.

PROPOSITION 35.3

If the filters belong to an orthogonal filter bank, we shall have

$$< \phi(x), \phi(x+k) >= \delta_k = < \psi(x), \psi(x+k) > ,$$
$$< \phi(x), \psi(x+k) > = 0.$$

Proof 35.1 From the definition in Equation 35.15, $f^{(0)}(x)$ is just the indicator function on the interval $[0, 1)$; so we immediately get orthogonality at the 0th level, that is, $< f^{(0)}(x-l), f^{(0)}(x-k) > = \delta_{kl}$. Now we assume orthogonality at the ith level:

$$< f^{(i)}(x-l), f^{(i)}(x-k) > = \delta_{kl}, \tag{35.28}$$

and prove that this implies orthogonality at the $(i+1)$th level:

$$< f^{(i+1)}(x-l), f^{(i+1)}(x-k) > = 2\sum_n \sum_m h_0(n)h_0(m)$$

$$< f^{(i)}(2x-2l-n), f^{(i)}(2x-2k-m) > \frac{\delta_{n+2l-2k-m}}{2} = \sum_n h_0(n)h_0(n+2l-2k)$$
$$= \delta_{kl}.$$

Hence, by induction Equation 35.28 holds for all i. So in the limit $i \to \infty$,

$$< \phi(x-l), \phi(x-k) > = \delta_{kl}. \tag{35.29}$$

The orthogonal case gives considerable simplification, both in the discrete-time and continuous-time cases.

PROPOSITION 35.4

If the filters belong to an FIR filter bank, then $\phi(x)$, $\psi(x)$, $\check{\phi}(x)$, and $\check{\psi}(x)$ will have support on some finite interval.

Proof 35.2 The filters $H_0^{(i)}(z)$ and $G_0^{(i)}(z)$ defined in Equation 35.14 have respective lengths $(2^i - 1)(L_a - 1) + 1$ and $(2^i - 1)(L_s - 1) + 1$ where L_a and L_s are the lengths of $H_0(z)$ and $G_0(z)$. Hence, $f^{(i)}(x)$ in Equation 35.15 is supported on the interval $[0, L_a - 1)$ and $f^{(i)}(x)$ on the interval $[0, L_s - 1)$. This holds $\forall\ i$; hence, in the limit $i \to \infty$ this gives the support of the scaling functions $\phi(x)$ and $\check{\phi}(x)$. That $\psi(x)$ and $\check{\psi}(x)$ have bounded support follow from Equation 35.20 and 35.21.

PROPOSITION 35.5

If the filters belong to a linear phase filter bank, then $\phi(x)$, $\psi(x)$, $\check{\phi}(x)$, and $\check{\psi}(x)$ will be symmetric or antisymmetric.

Proof 35.3 The filter $H_0^{(i)}(z)$ will have linear phase if $H_0(z)$ does. If $H_0^{(i)}(z)$ has length $(2^i - 1)(L_a - 1) + 1$, the point of symmetry is $(2^i - 1)(L_a - 1)/2$ which need not be an integer. The point of symmetry for $f^{(i)}(x)$ will then be $[(2^i - 1)(L_a - 1) + 1]/2^{i+1}$ or $[(2^i - 1)(L_a - 1) + 2]/2^{i+1}$. In either case, by taking the limit $i \to \infty$ we find that $\phi(x)$ is symmetric about the point $(L_a - 1)/2$ and similarly for the other cases.

Thus having established the relation between wavelets and filter banks we can examine the structure of filter banks in detail, and afterward use them to generate wavelets as described above. It should be emphasized that we are speaking of the two-channel, one-dimensional case. Multidimensional filter banks are a large subject in their own right [8,10].

35.1.3 Structure of Two-Channel Filter Banks

We saw already that it is the choice of the function $P(z)$ and the factorization taken that determines the properties of the filter bank. In terms of $P(z)$, we give necessary and sufficient conditions for the three properties mentioned above:

- **Orthogonality:** $P(z)$ is an autocorrelation, and $H_0(z)$ and $G_0(z)$ are its spectral factors.
- **Linear phase:** $P(z)$ is linear phase, and $H_0(z)$ and $G_0(z)$ are its linear phase factors.
- **Finite support:** $P(z)$ is FIR, and $H_0(z)$ and $G_0(z)$ are its FIR factors.

Obviously the factorization is not unique in any of the cases above. The FIR case has been examined in detail in [11,12,14–16] and the linear phase case in [12,15,17]. In the rest of this chapter we will present new results on the orthogonal case, but we shall also review the solutions that explicitly satisfy simultaneous constraints.

PROPOSITION 35.6

To have an orthogonal filter bank it is necessary and sufficient that $P(z)$ be an autocorrelation, and that $H_0(z)$ and $G_0(z)$ be its spectral factors.

PROPOSITION 35.7

To have a linear phase filter bank it is necessary and sufficient that $P(z)$ be a linear phase, and that $H_0(z)$ and $G_0(z)$ be its linear phase factors.

PROPOSITION 35.8

To have an FIR filter bank it is necessary and sufficient that $P(z)$ be FIR, and that $H_0(z)$ and $G_0(z)$ be its FIR factors.

Proofs can be found in [18]. Having seen that the design problem can be considered in terms of $P(z)$ and its factorizations, we consider the three conditions of interest from this point of view.

35.1.3.1 Orthogonality

In the case where the filter bank is to be orthogonal, we can obtain a complete constructive characterization of the solutions, as given by the following theorem, taken from [18].

THEOREM 35.1

All orthogonal rational two channel filter banks can be formed as follows:

1. *Choosing an arbitrary polynomial $R(z)$, form*

$$P(z) = \frac{2R(z)R(z^{-1})}{R(z)R(z^{-1}) + R(-z)R(-z^{-1})}$$

2. *Factor as $P(z) = H(z)H(z^{-1})$*
3. *Form the filter $H_0(z) = A_0(z)H(z)$, where $A_0(z)$ is an arbitrary allpass*
4. *Choose $H_1(z) = z^{2k-1}H_0(-z^{-1})A_1(z^2)$, where $A_1(z)$ is again an arbitrary allpass*
5. *Choose $G_0(z) = H_0(z^{-1})$ and $G_1(z) = -H_1(z^{-1})$*

For a proof, see [18,19].

Example 35.1

Take $R(z) = (1 + z^{-1})^N$ as above and $N = 7$. It works out that in this case there is a closed-form factorization for the filters:

$$\begin{aligned} P(z) &= \frac{(1, 14, 91, 364, 1001, 2002, 3003, 3432, 3003, 2002, 1001, 364, 91, 14, 1)z^7}{14z^6 + 364z^4 + 2002z^2 + 3432 + 2002z^{-2} + 364z^{-4} + 14z^{-6}} \\ &= \frac{E(z)E(z^{-1})}{K(z)K(z^{-1})}, \end{aligned}$$

where

$$\frac{E(z)}{K(z)} = \frac{(1 + 7z^{-1} + 21z^{-2} + 35z^{-3} + 35z^{-4} + 21z^{-5} + 7z^{-6} + z^{-7})}{\sqrt{2}(1 + 21z^{-2} + 35z^{-4} + 7z^{-6})}.$$

Note that we have used the following shorthand notation to list the coefficients of a causal FIR sequence:

$$\sum_{n=0}^{N-1} a_n z^{-n} = (a_0, a_1, a_2, \ldots, a_{N-1}).$$

So, using the description of the filters in Theorem 35.1, with the simplest case $A_0(z) = A_1(z) = 1$ and $k = 0$ we find

$$H_0(z) = \frac{(1 + 7z^{-1} + 21z^{-2} + 35z^{-3} + 35z^{-4} + 21z^{-5} + 7z^{-6} + z^{-7})}{\sqrt{2}(1 + 21z^{-2} + 35z^{-4} + 7z^{-6})}$$

$$H_1(z) = z^{-1}\frac{(1 - 7z^{1} + 21z^{2} - 35z^{3} + 35z^{4} - 21z^{5} + 7z^{6} - z^{7})}{\sqrt{2}(1 + 21z^{2} + 35z^{4} + 7z^{6})}$$

$$G_0(z) = H_0(z^{-1})$$

$$G_1(z) = H_1(z^{-1})$$

In the notation of Proposition 35.2, $B = 8 < 2^6$ so that for this choice of $H_0(z)$ the left-hand side of Equation 35.15 converges to a continuous function. The wavelet, scaling function, and their spectra are shown in Figure 35.4.

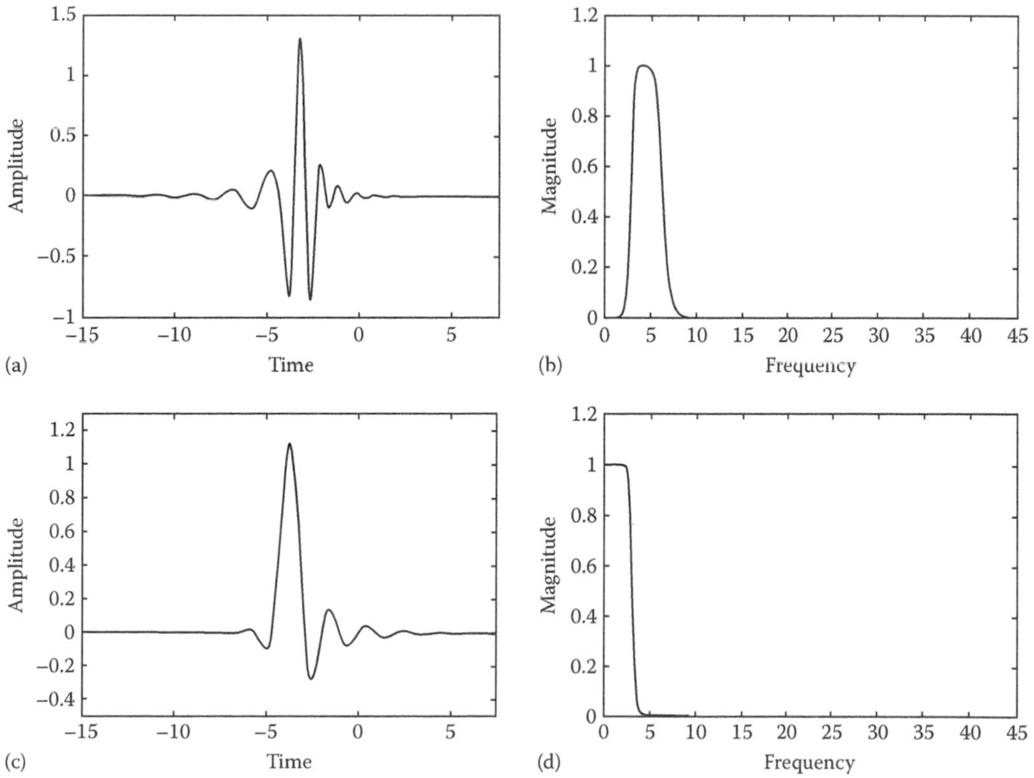

FIGURE 35.4 Example of Butterworth orthogonal wavelet, here $N = 7$, and the closed-form factorization has been used: (a) wavelet, (b) spectrum of the wavelet, (c) scaling function, and (d) spectrum of the scaling function.

35.1.3.2 Finite Impulse Response and Symmetric Solutions

In the case where the filters are to be FIR, we merely require that $P(z)$ be FIR; it is trivially easy to design one. Similarly to have symmetric filters, we merely force $P(z)$ to be symmetric. Obviously any symmetric $P(z)$ which is FIR and satisfies Equation 35.3 can be used to give symmetric FIR filters. We would like, in addition, that the lowpass filters are regular, so that we get symmetric bounded support continuous-time basis functions.

One strategy would be to design a $P(z)$ with the desired properties and then factor to find the filters. Alternatively, we can choose one of the factors, and then find the other necessary to make the product $P(z)$ satisfy Equation 35.3. We will use this approach and, to ensure regularity, choose one factor to be $(1+z^{-1})^{2N}$. This can be done by solving a linear system of equations (see Figure 35.5) [12].

Example 35.2

If we choose $N=3$ we must find the complement to $(1+z^{-1})^6$; so we solve the 3×3 system found by imposing the constraints on the coefficients of the odd powers of z^{-1} of

$$P(z) = \left(k_0 + k_1 z^{-1} + k_2 z^{-2} + k_1 z^{-3} + k_0 z^{-4}\right) \cdot \left(1 + 6z^{-1} + 15z^{-2} + 20z^{-3} + 15z^{-4} + 6z^{-5} + z^{-6}\right) \cdot z^5.$$

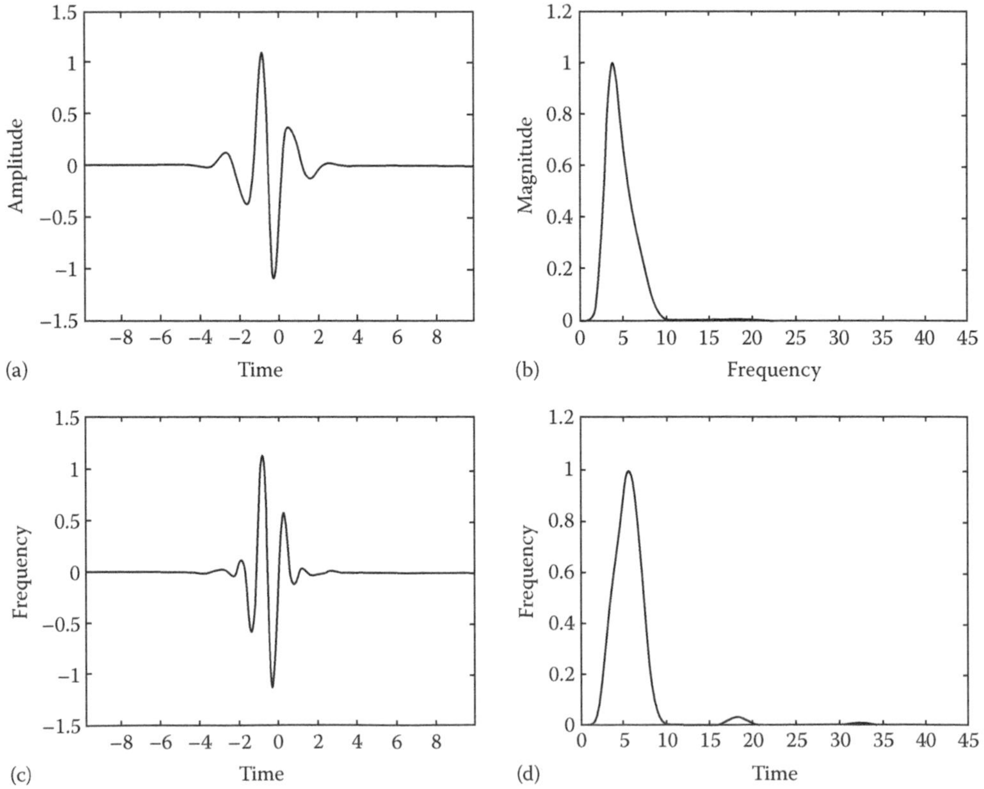

FIGURE 35.5 Biorthogonal wavelets generated by filters of length 18 given in [12]: (a) analysis wavelet function $\psi(x)$, (b) spectrum of analysis wavelet, (c) synthesis wavelet function $\psi(x)$, and (d) spectrum of synthesis wavelet.

So we solve

$$\begin{pmatrix} 6 & 1 & 0 \\ 20 & 16 & 6 \\ 12 & 30 & 20 \end{pmatrix} \begin{pmatrix} k_0 \\ k_1 \\ k_2 \end{pmatrix} = \begin{pmatrix} 0 \\ 0 \\ 1 \end{pmatrix},$$

giving $\mathbf{k}_6 = (3/2, -9, 19)/128$.

In general, therefore, we solve the system:

$$\mathbf{F}_{2N} \cdot \mathbf{k}_{2N} = \mathbf{e}_{2N}, \tag{35.30}$$

where

$\mathbf{F}_{2N}$ is the $N \times N$ matrix
$\mathbf{k}_{2N} = (k_0, \ldots, k_{(k-1)})$
$\mathbf{e}_{2N}$ is the length k vector $(0, 0, \ldots, 1)$

Having found the coefficients of $K_{2N}(z)$, we factor it into linear phase components and then regroup these factors of $K_{2N}(z)$ and the $2N$ zeros at $z = -1$ to form two filters: $H_0(z)$ and $H_1(-z)$, both of which are to be regular.

35.1.4 Putting the Pieces Together

An important consideration that is often encountered in the design of wavelets, or of the filter banks that generate them, is the necessity of satisfying competing design constraints. This makes it necessary to clearly understand whether desired properties are mutually exclusive.

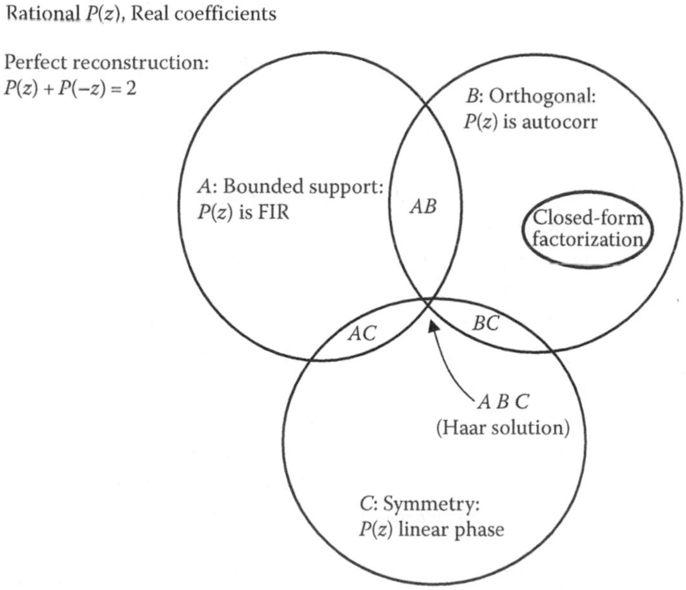

FIGURE 35.6 Two-channel perfect reconstruction filter banks. The Venn diagram illustrates which competing constraints can be simultaneously satisfied. The sets A, B, and C contain FIR, orthogonal, and linear phase solutions, respectively. Solutions in the intersection $A \cap B$ are examined in [11,14,23,24], those in the intersection $A \cap C$ are detailed in [12,13,15,17,25], and solutions in $B \cap C$ are constructed in [18]. The intersection $A \cap B \cap C$ contains only trivial solutions.

Perfect reconstruction solutions, with the constraint that $P(z)$ be rational with real coefficients, must satisfy Equation 35.3. Such general solutions, which do not necessarily have additional properties, were given in [14].

The solutions of set A, where all of the filters involved are FIR, were studied in [14,15]. Set B contains all orthogonal solutions, and has been the main focus of this chapter. A complete characterization of this set was given in Theorem 35.1. A very different characterization, based on lattice structures, is given in [20]. Particular cases of orthogonal solutions were also given in [21]. Set C contains the solutions where all filters are linear phase, first examined in [15].

The earliest examples of perfect reconstruction solutions [22,23] were orthogonal and FIR; that is, they were in $A \cap B$. A constructive parametrization of $A \cap B$ was given in [24]. The construction and characterization of examples which converge to wavelets was first done in [11]. Filter banks with FIR linear phase filters (i.e., $A \cap C$) were first given in [15], and also studied in terms of lattices in [17,25]. The construction of wavelet examples is given in [12,13]. Filter banks, which are linear phase and orthogonal, were constructed in [18].

That there exist only trivial solutions which are linear phase, orthogonal and FIR is indicated by the intersection $A \cap B \cap C$; the only solutions are two tap filters [11,12,26].

It warrants emphasis that Figure 35.6 illustrates the filter bank solutions; if the filters are regular, then they will lead to wavelets. Of the dyadic wavelet bases known to the authors, the only ones based on filters where $P(z)$ is not rational are those of Meyer [27], and the only ones where the filter coefficients are complex are those of Lawton [28]. For the case of the Battle–Lemarié wavelets, while the filters themselves are not rational, the $P(z)$ function is; hence, the filters would belong to $B \cap C$ in the figure.

References

1. Croisier, A., Esteban, D., and Galand, C., Perfect channel splitting by use of interpolation, decimation, tree decomposition techniques, in *International Conference on Information Sciences and Systems*, Patras, Greece, Aug. 1976, pp. 443–446.
2. Crochiere, R.E., Weber, S.A., and Flanagan, J.L., Digital coding of speech in subbands, *Bell Syst. Tech. J.*, 55, 1069–1085, Oct. 1976.
3. Vetterli, M., Multidimensional subband coding: Some theory and algorithms, *Signal Process.*, 6, 97–112, Feb. 1984.
4. Woods, J.W. and O'Neil S.D., Subband coding of images, *IEEE Trans. Acoust. Speech Signal Process.*, 34(5), 1278–1288, 1986.
5. Shapiro, J.M., Embedded image coding using zerotrees of wavelet coefficients, *IEEE Trans. Signal Process.*, 41, 3445–3462, Dec. 1993.
6. Said, A. and Pearlman, W.A., An image multiresolution representation for lossless and lossy compression, *IEEE Trans. Image Process.*, 5(9), 1303–1310, 1996.
7. Xiong, Z., Ramchandran, K., and Orchard, M.T., Wavelet packet image coding using space-frequency quantization, *IEEE Trans. Image Process.*, submitted, 1996.
8. Vaidyanathan, P.P., *Multirate Systems and Filter Banks*, Prentice-Hall, Englewood Cliffs, NJ, 1992.
9. Malvar, H.S., *Signal Processing with Lapped Transforms*, Artech House, Norwood, MA, 1992.
10. Vetterli, M. and Kovacevic, J., *Wavelet and Subband Coding*, Prentice-Hall, Englewood Cliffs, NJ, 1995.
11. Daubechies, I., Orthonormal bases of compactly supported wavelets, *Commn. Pure Appl. Math.*, XLI, 909–996, 1988.
12. Vetterli, M. and Herley, C., Wavelets and filter banks: Theory and design, *IEEE Trans. Signal Process.*, 40, 2207–2232, Sept. 1992.
13. Cohen, A., Daubechies, I., and Feauveau, J.-C., Biorthogonal bases of compactly supported wavelets, *Commn. Pure Appl. Math.*, 45, 485–560, 1992.

14. Smith, M.J.T. and Barnwell, T.P., III, Exact reconstruction for tree-structured subband coders, *IEEE Trans. Acoust. Speech Signal Process.*, 34, 434–441, June 1986.
15. Vetterli, M., Filter banks allowing perfect reconstruction, *Signal Process.*, 10(3), 219–244, 1986.
16. Vaidyanathan, P.P., Multirate digital filters, filter banks, polyphase networks, and applications: A tutorial, *Proc. IEEE*, 78, 56–93, Jan. 1990.
17. Nguyen, T.Q. and Vaidyanathan, P.P., Two-channel perfect-reconstruction FIR QMF structures which yield linear-phase analysis and synthesis filters, *IEEE Trans. Acoust. Speech Signal Process.*, 37, 676–690, May 1989.
18. Herley, C. and Vetterli, M., Wavelets and recursive filter banks, *IEEE Trans. Signal Process.*, 41, 2536–2556, Aug. 1993.
19. Herley, C., Wavelets and filter banks, PhD thesis, Columbia University, New York, Apr. 1993. Available by anonymous ftp at ftp.ctr.columbia.edu directory: CTR-Research/advent/public/papers/PhD-theses/Herley.
20. Doğanata, Z. and Vaidyanathan, P.P., Minimal structures for the implementation of digital rational lossless systems, *IEEE Trans. Acoust. Speech Signal Process.*, 38, 2058–2074, Dec. 1990.
21. Smith, M.J.T., IIR analysis/synthesis systems, in *Subband Coding of Images*, Woods, J.W. (Ed.), Kluwer Academic, Norwell, MA, 1991.
22. Smith, M.J.T. and Barnwell, T.P., III, A procedure for designing exact reconstruction filter banks for tree structured subband coders, in *Proceedings of the IEEE International Conference on Acoustics, Speech and Signal Processing*, San Diego, CA, Mar. 1984, pp. 27.1.1–27.1.4.
23. Mintzer, F., Filters for distortion-free two-band multirate filter banks, *IEEE Trans. Acoust. Speech Signal Process.*, 33, 626–630, June 1985.
24. Vaidyanathan, P.P. and Hoang, P.-Q., Lattice structures for optimal design and robust implementation of two-band perfect reconstruction QMF banks, *IEEE Trans. Acoust. Speech Signal Process.*, 36, 81–94, Jan. 1988.
25. Vetterli, M. and Le Gall, D., Perfect reconstruction FIR filter banks: Some properties and factorizations, *IEEE Trans. Acoust. Speech Signal Process.*, 37, 1057–1071, July 1989.
26. Vaidyanathan, P.P. and Doğanata, Z., The role of lossless systems in modern digital signal processing, *IEEE Trans. Educ.*, 32, 181–197, Aug. 1989. Special issue on Circuits and Systems.
27. Meyer, Y., *Ondelettes*, Vol. 1 of *Ondelettes et Opérateurs*, Hermann, Paris, France, 1990.
28. Lawton, W., Application of complex-valued wavelet transforms to subband decomposition, *IEEE Trans. Signal Process.*, submitted, 1992.

36

Filter Bank Design

Joseph Arrowood
IvySys Technologies, LLC

Tami Randolph
Georgia Institute of Technology

Mark J.T. Smith
Purdue University

The interest in digital filter banks has grown dramatically over the last few years. Owing to the trend toward lower cost, higher speed microprocessors, digital solutions are becoming attractive for a wide variety of applications. Filter banks allow signals to be decomposed into subbands, often facilitating more efficient and effective processing. They are particularly visible in the areas of image compression, speech coding, and image analysis.

The desired characteristics of a subband decomposition will naturally vary from application to application. Moreover, within any given application, there are a myriad of issues to consider. First, one might consider whether to use FIR or IIR filters. IIR designs can offer computational advantages, while FIR designs can offer greater flexibility in filter characteristics. In this chapter we focus exclusively on FIR design. Second, one might identify the time-frequency or space-frequency representation that is most appropriate. Uniform decompositions and octave-band decompositions are particularly popular at present. At the next level, characteristics of the analysis filters should be defined. This involves imposing specifications on the analysis filter passband deviations, transition bands, and stopband deviations. Alternately or in addition, time domain characteristics may be imposed, such as limits on the step response ripples, and degree of regularity.

One can consider similar constraints for the synthesis filters. For coding applications, the characteristics of the synthesis filters often have a dominant effect on the subjective quality of the output. Finally, one should consider analysis-synthesis characteristics. That is, one has flexibility to specify the overall behavior of the system. In most cases, one views having exact reconstruction as being ideal. Occasionally, however, it may be possible to trade some small loss in reconstruction quality for significant gains in computation, speed, or cost. In addition to specifying the quality of reconstruction, it is generally possible to control the overall delay of the system from end to end. In some applications, such as two-way speech and video coding, latency represents a source of quality degradation. Thus, having explicit control over the analysis-synthesis delay can lead to improvement in quality.

The intelligent design of applications-specific filter banks involves first identifying the relevant parameters and optimizing the system with respect to them. As is typical, the filter bank analysis and reconstruction equations lead to complex tradeoffs among complexity, system delay, filter quality, filter length, and quality of performance. This chapter is devoted to presenting an introduction to filter bank design. Filter bank design has reached a state of maturity in many regards. To cover all of the important contributions in any level of detail would be impossible in a single chapter. However, it is possible to gain some insight and appreciation for general design strategies germane to this topic. In addition to discussing design methodologies for linear analysis-synthesis systems, we also consider the design of a couple of new nonlinear classes of filter banks that are currently receiving attention in the literature. This discussion along with the referenced articles should provide a convenient introduction to the design of many useful filter banks.

36.1 Filter Bank Equations

A broad class of linear filter banks can be represented by the block diagram shown in Figure 36.1. This is a linear time-varying system that decomposes the input into $\boldsymbol{M}$-subbands, each one of which is decimated by a factor of R. When $R = \boldsymbol{M}$, the system is said to be critically sampled or maximally decimated. Maximally decimated systems are generally the ones of choice because they can be information preserving, and are not data expansive.

The simplest filter bank of this class is the two-band system, an example of which is shown in Figure 36.2. Here, there are only two analysis filters: $\boldsymbol{H}_0(z)$, a lowpass filter, and $\boldsymbol{H}_1(z)$, a highpass filter. Similarly, there are two synthesis filters: a lowpass $G_0(z)$, and a highpass $G_1(z)$. Let us consider this two-band filter bank first. In the process, we will develop a design methodology that can be extended to the more complex problem of M-band systems.

Examining the two-band filter bank in Figure 36.2, we see that the input $x[n]$ is lowpass and highpass filtered, resulting in $v_0[n]$ and $v_1[n]$. These signals are then downsampled by a factor of two, leading to the analysis section outputs, $y_0[n]$ and $y_1[n]$. The downsampling operation is time varying, which implies a non-trivial relationship between $v_k[n]$ and $y_k[n]$ (where $k = 0, 1$). In general, downsampling a signal $v_k[n]$ by an integer factor R is described in the time domain by the equation

$$y_k[n] = v_k[Rn].$$

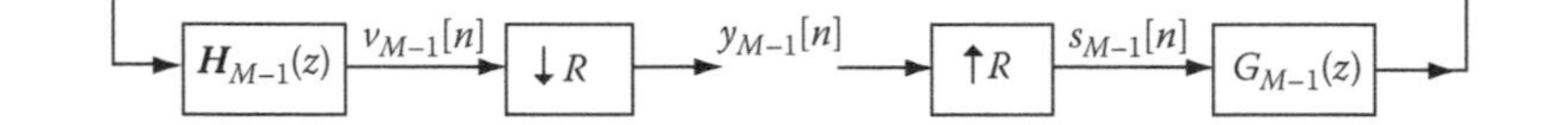

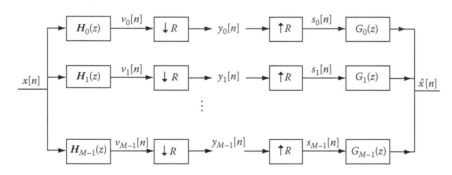

FIGURE 36.1 Multi-band analysis-synthesis filter bank.

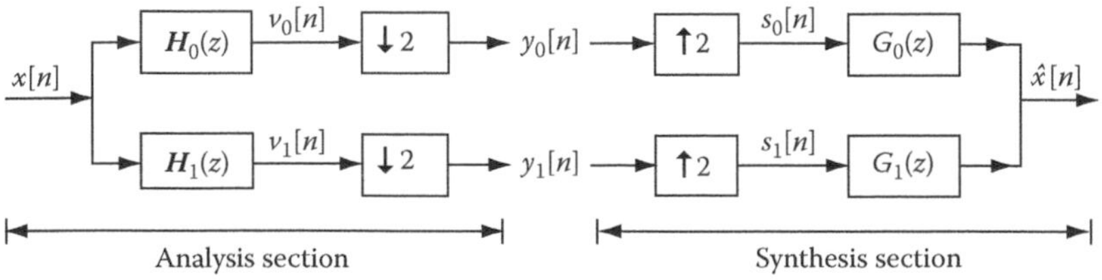

FIGURE 36.2 Two-band analysis-synthesis filter bank.

In the frequency domain, this relationship is given by

$$Y_k(e^{j\omega}) = \frac{1}{R}\sum_{r=0}^{R-1} V_k\left(e^{j\left(\frac{\omega}{R}+\frac{2\pi r}{R}\right)}\right).$$

The equivalent equation in the z-domain is

$$Y_k(z) = \frac{1}{R}\sum_{r=0}^{R-1} V_k\left(W_R^r z^{\frac{1}{R}}\right),$$

where $W_R^r = e^{-j\frac{2\pi r}{R}}$.

In the synthesis section, the subband signals $y_0[n]$ and $y_1[n]$ are upsampled to give $s_0[n]$ and $s_1[n]$. They are then filtered by the lowpass and highpass filters, $G_0(z)$ and $G_1(z)$, respectively, before being summed together. The upsampling operation (for an arbitrary positive integer R) can be defined by

$$s_k[n] = \begin{cases} y_k[n/R] & \text{for } n = 0, \pm R, \pm 2R, \pm 3R, \ldots \\ 0 & \text{otherwise} \end{cases}$$

in the time domain, and

$$S_k(e^{j\omega}) = Y_k(e^{jR\omega}) \quad \text{and} \quad S_k(z) = Y_k(z^R)$$

in the frequency and z domains, respectively.

Using the expressions for the downsampling and upsampling operations, we can describe the two-band filter bank in terms of z-domain equations. The outputs after analysis filtering are

$$V_k(z) = H_k(z)X(z), \quad k = 0, 1.$$

After decimation and recognizing that $W_2^1 = -1$, we obtain

$$Y_k(z) = \frac{1}{2}\left[H_k\left(z^{\frac{1}{2}}\right)X\left(z^{\frac{1}{2}}\right) + H_k\left(-z^{\frac{1}{2}}\right)X\left(-z^{\frac{1}{2}}\right)\right], \quad k = 0, 1. \tag{36.1}$$

Thus, Equation 36.1 defines completely the input–output relationship for the analysis section in the z-domain.

In the synthesis section, the subbands are upsampled giving

$$S_k(z) = Y_k(z^2), \quad k = 0,1.$$

This implies that

$$S_k(z) = \frac{1}{2}(H_k(z)X(z) + H_k(-z)X(-z)), \quad k = 0,1.$$

Passing $S_k(z)$ through the synthesis filters and then summing yields the reconstructed output

$$\begin{aligned}\hat{X}(z) = &\frac{1}{2}G_0(z)[H_0(z)X(z) + H_0(-z)X(-z)] \\ &+\frac{1}{2}G_1(z)[H_1(z)X(z) + H_1(-z)X(-z)].\end{aligned} \tag{36.2}$$

For virtually any application for which one can conceive, the synthesis filters should allow the input to be reconstructed exactly or with a minimal amount of distortion. In other words, ideally we want

$$\hat{X}(z) = z^{-n_0}X(z),$$

where n_0 is the integer system delay. An intuitive approach to handing this problem is to use the AC-matrix formulation, which we introduce next.

36.1.1 AC Matrix

The aliasing component matrix (or AC matrix) represents a simple and intuitive idea originally introduced in [6] for handling analysis and reconstruction. The analysis-synthesis (Equation 36.2) for the two-band case can be expressed as

$$\begin{aligned}\hat{X}(z) &= \frac{1}{2}[H_0(z)G_0(z) + H_1(z)G_1(z)]X(z)\\ &\quad + \frac{1}{2}[H_0(-z)G_0(z) + H_1(-z)G_1(z)]X(-z).\end{aligned}$$

The idea of the AC matrix is to represent the equations in matrix form. For the two-band system, this results in

$$\hat{X}(z) = \frac{1}{2}[X(z), X(-z)]\underbrace{\begin{bmatrix} H_0(z) & H_1(z) \\ H_0(-z) & H_1(-z)\end{bmatrix}}_{\text{AC matrix}}\begin{bmatrix} G_0(z) \\ G_1(z)\end{bmatrix},$$

where the AC matrix is as shown above. The AC matrix is so designated because it contains the analysis filters and all the associated aliasing components. Exact reconstruction is then obtained when

$$\begin{bmatrix} H_0(z) & H_1(z) \\ H_0(-z) & H_1(-z)\end{bmatrix}\begin{bmatrix} G_0(z) \\ G_1(z)\end{bmatrix} = \begin{bmatrix} T(z) \\ 0\end{bmatrix},$$

where $T(z)$ is required to be the scaled integer delay $2z^{-n_0}$. The term $T(z)$ is the transfer function of the overall system. The zero term below $T(z)$ determines the amount of aliasing present in the reconstructed signal. Because this term is zero, all aliasing is explicitly removed.

With the equations expressed in matrix form, we can solve for the synthesis filters, which yields

$$\begin{bmatrix} G_0(z) \\ G_1(z)\end{bmatrix} = \frac{1}{H_0(z)H_1(-z) - H_0(-z)H_1(z)}\begin{bmatrix} H_1(-z) & -H_1(z) \\ -H_0(-z) & H_0(z)\end{bmatrix}\begin{bmatrix} T(z) \\ 0\end{bmatrix}. \tag{36.3}$$

Often for a variety of reasons, we would like both the analysis and synthesis filters to be FIR. This means the determinant of the AC matrix should be a constant delay. The earliest solution to the FIR filter bank problem was presented by Croisier et al. in 1976 [18]. Their solution was to let

$$H_1(z) = H_0(-z)$$

and

$$\begin{aligned}G_0(z) &= H_0(z)\\ G_1(z) &= -H_0(-z).\end{aligned}$$

This is the quadrature mirror filter (QMF) solution. From Equation 36.3, it can be seen that this solution cancels all the aliasing and results in a system transfer function

$$T(z) = H_0(z)H_1(-z) - H_0(-z)H_1(z).$$

As it turns out, with careful design $T(z)$ can be made to be close to a constant delay. However, some amount of distortion will always be present. In 1980, Johnston designed a set of optimized QMFs which are now widely used. The coefficient values may be found in several sources [16,17,19].

Interestingly, Equation 36.3 implies that exact reconstruction is possible by forcing the AC-matrix determinant to be a constant delay. The design of such exact reconstruction filters is discussed in the next section.

36.1.2 Spectral Factorization

The question at hand is how do we determine $H_0(z)$ and $H_1(z)$ such that $T(z)$ is an integer delay z^{-n_0}. A solution to this problem was introduced in 1984 [7], based on the observation that $H_0(z)H_1(-z)$ is a lowpass filter (which we denote $F_0(z)$) and $H_0(-z)H_1(z)$ is its corresponding frequency shifted highpass filter. A unity transfer function can be constructed by forcing $F_0(z)$ and $F_0(-z)$ to be complementary half-band lowpass and highpass filters. Many fine techniques are available for the design of half-band lowpass filters, such as the Parks–McClellan algorithm, Kaiser window design, Hamming window design, the eigenfilter method, and others. Zero-phase half-band filters have the property that zeros occur in the impulse response at $n = \pm 2, \pm 4, \pm 6$, etc. An illustration is shown in Figure 36.3. Once designed, $F_0(z)$ can be factored into two lowpass filters, $H_0(z)$ and $H_1(-z)$. The design procedure can be summarized as follows:

1. First design a $(2N - 1)$-tap half-band lowpass filter, using the Parks-McClellan algorithm, for example. This can be done by constraining the passband and stopband cutoff frequencies to be $\omega_p = \pi - \omega_s$, and using equal passband and stopband error weightings. The resulting filter will have equal passband and stopband ripples, that is, $\delta_p = \delta_s = \delta$.
2. Add the value δ to the $f[0]$ (center) tap value. This forces $F(e^{j\omega}) \geq 0$ for all ω.
3. Spectrally factor $F(z)$ into two lowpass filters, $H_0(z)$ and $H_1(-z)$. Generally the best way to factor $F(z)$ is such that $H_1(-z) = H_0(z^{-1})$. Note that the factorization will not be unique and the roots should be split so that if a particular root is assigned to $H_0(z)$, its reciprocal should be given to $H_0(z^{-1})$.

The result of the above procedure is that $H_0(z)$ will be a power complementary, even length, FIR filter that will form the basis for a perfect reconstruction filter bank. Note that since $H_1(z)$ is just a time-reversed, spectrally shifted version of $H_0(z)$,

$$|H_0(e^{j\omega})| = |H_1(-e^{j\omega})|.$$

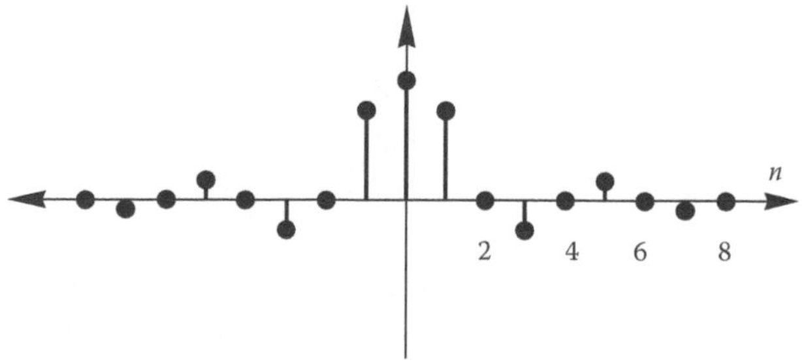

FIGURE 36.3 Example of a zero-phase half-band lowpass filter.

TABLE 36.1 CQF (Smith–Barnwell) Filter Bank Coefficients with 40 dB Attenuation

32-Tap Filter	16-Tap Filter	8-Tap Filter
8.494372478233170D−03	2.193598203004352D−02	3.489755821785150D−02
−9.617816873474045D−05	1.578616497663704D−03	−1.098301946252854D−02
−8.795047132402801D−03	−6.025449102875281D−02	−6.286453934951963D−02
7.087795490845020D−04	−1.189065962053910D−02	0.223907720892568D+00
1.220420156035413D−02	0.137537915636625D+00	0.556856993531445D+00
−1.762639314795336D−03	5.745450056390939D−02	0.357976304997285D+00
−1.558455903573829D−02	−0.321670296165893D+00	−2.390027056113145D−02
4.082855675060479D−03	−0.528720271545339D+00	−7.594096379188282D−02
1.765222024089335D−02	−0.295779674500919D+00	
−8.385219782884901D−03	2.043110845170894D−04	
−1.674761388473688D−02	2.906699709446796D−02	
1.823906210869841D−02	−3.533486088708146D−02	
5.781735813341397D−03	−6.821045322743358D−03	
−4.692674090907675D−02	2.606678468264118D−02	
5.725005445073179D−02	1.033363491944126D−03	
0.354522945953839D+00	−1.435930957477529D−02	
0.504811839124518D+00		
0.264955363281817D+00		
−8.329095161140063D−02		
−0.139108747584926D+00		
3.314036080659188D−02		
9.035938422033127D−02		
−1.468791729134721D−02		
−6.103335886707139D−02		
6.606122638753900D−03		
4.051555088035685D−02		
−2.631418173168537D−03		
−2.592580476149722D−02		
9.319532350192227D−04		
1.535638959916169D−02		
−1.196832693326184D−04		
−1.057032258472372D−02		

Smith and Barnwell designed and published a set of optimal exact reconstruction filters [1]. The filter coefficients for $H_0(z)$ are given in Table 36.1. The analysis and synthesis filters are obtained from $H_0(z)$ by

$$
\begin{aligned}
G_0(z) &= H_0(z^{-1}) \\
G_1(z) &= H_0(-z) \\
H_1(z) &= H_0(-z^{-1}).
\end{aligned}
$$

A complete discussion of this approach can be found in many references [1,6,7,25,27,28].

For the M-channel case shown in Figure 36.1, where the bands are assumed to be maximally decimated, the same AC-matrix approach can be employed, leading to the equations

$$\hat{X}(z) = \frac{1}{M}\underbrace{\left[X(z),\ldots,X\left(zW_M^{M-1}\right)\right]}_{\mathbf{x}^T} \times \underbrace{\begin{bmatrix} H_0(z) & \cdots & H_{M-1}(z) \\ H_0\left(zW_M^1\right) & \cdots & H_{M-1}\left(zW_M^1\right) \\ \vdots & & \vdots \\ H_0\left(zW_M^{M-1}\right) & \cdots & H_{M-1}\left(zW_M^{M-1}\right) \end{bmatrix}}_{\mathbf{H}} \underbrace{\begin{bmatrix} G_0(z) \\ G_1(z) \\ \vdots \\ G_{M-1}(z) \end{bmatrix}}_{\mathbf{g}},$$

where $W_M = e^{-j\frac{2\pi}{M}}$. This can be rewritten compactly as

$$\hat{X}(z)\frac{1}{M}\mathbf{x}^{\mathrm{T}}(z)\mathbf{H}(z)\mathbf{g}(z),$$

where
- $\mathbf{x}$ is the input vector
- $\mathbf{g}$ is the synthesis filter vector
- $\mathbf{H}$ is the AC matrix

However, the AC-matrix determinant for systems with $M > 2$ is typically too intricate for the spectral factorization approach outlined above. An effective approach for handling the design of M-band systems was introduced by Vaidyanathan in [30]. It is based on a lattice implementation structure and is discussed next.

36.1.3 Lattice Implementations

In addition to the direct form structures shown in Figures 36.1 and 36.2, filter banks can be implemented using lattice structures. For simplicity, consider the two-band case first. An example of a lattice structure for a two-band analysis system is shown in Figure 36.4. It is composed of a cascade of crisscross elements, each of which has a set of coefficients associated with it. Conveniently, each section, which we denote $\mathbf{R}_m$, can be described by a matrix. For the two-band lattice, these matrices have the form

$$\mathbf{R}_m = \begin{bmatrix} 1 & r_m \\ -r_m & 1 \end{bmatrix}.$$

Interspersed between the coefficient matrices are delay matrices, $\Lambda(z)$, having the form

$$\Lambda(z) = \begin{bmatrix} 1 & 0 \\ 0 & z^{-1} \end{bmatrix}.$$

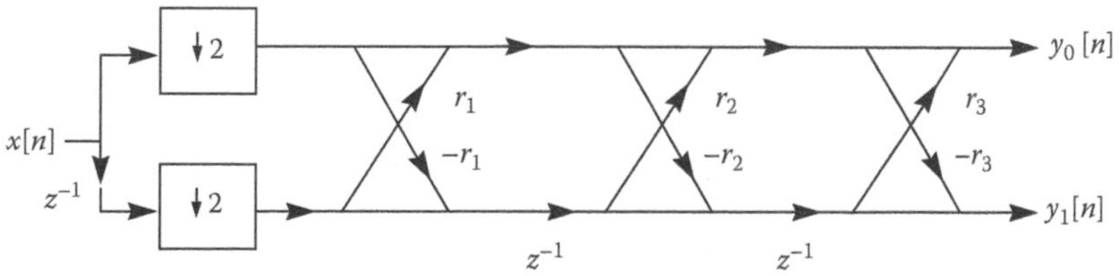

FIGURE 36.4 Flow graph of a two-band lattice structure with three stages.

It can be shown [27] that lattice filters can represent a wide class of exact reconstruction filter banks. Two points regarding lattice filter banks are particularly noteworthy. First, the lattice structure provides an efficient form of implementation. Moreover, the synthesis filter bank is directly related to the analysis bank, since each matrix in the analysis cascade is invertible. Consequently, the synthesis bank consists of the cascade of inverse section matrices. Second, the structure also provides a convenient way to design the filter bank. Each lattice coefficient can be optimized using standard minimization routines to minimize a passband–stopband error cost function for the filters. This approach to design can be used for two-band as well as M-band filter banks [5,27,28].

36.1.4 Time-Domain Design

One of the most flexible design approaches is the time domain formulation proposed by Nayebi et al. [3,8]. This formulation has enabled the discovery of previously unknown classes of filter banks, such as low and variable delay systems [12], time-varying filter banks [4], and block decimation systems [9]. It is attractive because it enables the design of virtually all linear filter banks. The idea underlying this approach is that the conditions for exact reconstruction can be expressed in the time domain in a convenient matrix form. Let us explore this approach in the context of an M-band filter bank. Because of the decimation operations, the overall M-band analysis-synthesis system is periodically time-varying. Thus, we can view an arbitrary maximally decimated M-band system as having M linear time invariant transfer functions associated with it. One can think of the problem as trying to devise M subsampled systems, each one of which exactly reconstructs. This is equivalent to saying that for each impulse input, $\delta[n-i]$, to the analysis-synthesis system, that impulse should appear at the system output at time $n = i + n_0$, where $i = 0, 1, 2, \ldots, M-1$ and n_0 is the system delay.

This amounts to setting up an overconstrained linear system $\mathbf{AS} = \mathbf{B}$, where the matrix $\mathbf{A}$ is created using the analysis filter coefficients, the matrix $\mathbf{B}$ is the desired response of zeros except at the appropriate delay points (i.e., $\delta[n-n_0]$), and $\mathbf{S}$ is a matrix containing synthesis filter coefficients. Particular linear combinations of analysis and synthesis filter coefficients occur at different points in time for different input impulses. The idea is to make $\mathbf{A}$, $\mathbf{S}$, and $\mathbf{B}$ such that they describe completely all M transfer functions that comprise the periodically time-varying system.

The matrix $\mathbf{A}$ is a matrix of filter coefficients and zeros that effectively describe the decimated convolution operations inherent in the filter bank. For convenience, we express the analysis coefficients as a matrix $\mathbf{h}$, where

$$\mathbf{h} = \begin{bmatrix} h_0[0] & h_1[0] & \cdots & h_{M-1}[0] \\ h_0[1] & h_1[1] & \cdots & h_{M-1}[1] \\ \vdots & \vdots & \vdots & \vdots \\ h_0[N-1] & h_1[N-1] & \cdots & h_{M-1}[N-1] \end{bmatrix}.$$

The zeros are represented by an $M \times M$ matrix of zeros, denoted $\mathbf{O_M}$. With these terms, we can write the $(2N-M) \times N$ matrix $\mathbf{A}$:

$$\mathbf{A} = \begin{bmatrix} [\mathbf{h}[n]] & \mathbf{O_M} & \cdots & \mathbf{O_M} \\ \mathbf{O_M} & [\mathbf{h}[n]] & \cdots & \vdots \\ \vdots & \vdots & \vdots & \mathbf{O_M} \\ \mathbf{O_M} & \mathbf{O_M} & \cdots & [\mathbf{h}[n]] \end{bmatrix}.$$

The synthesis filters **S** can be expressed most conveniently in terms of the $M \times M$ matrix

$$\mathbf{Q}_i = \begin{bmatrix} g_0[i] & g_0[i+1] & \cdots & g_0[i+M-1] \\ g_1[i] & g_1[i+1] & \cdots & g_1[i+M-1] \\ \vdots & \vdots & \vdots & \vdots \\ g_{M-1}[i] & g_{M-1}[i+1] & \cdots & g_{M-1}[i+M-1] \end{bmatrix},$$

where $i = 0, 1, \ldots, L-1$ and N is assumed to be equal to LM. The synthesis matrix **S** is then given by

$$\mathbf{S} = \begin{bmatrix} \mathbf{Q}_0 \\ \mathbf{Q}_M \\ \vdots \\ \mathbf{Q}_{iM} \\ \vdots \\ \mathbf{Q}_{(L-1)M} \end{bmatrix}.$$

Finally, to achieve exact reconstruction we want the impulse responses associated with each of the M constituent transfer functions in the periodically time-varying system to be an impulse. Therefore, **B** is a matrix of zero-element column vectors, each with a single "one" at the location of the particular transfer function group delay. More specifically, the matrix has the form

$$\mathbf{B} = \begin{bmatrix} \mathbf{O_M} \\ \mathbf{O_M} \\ \vdots \\ \mathbf{J_M} \\ \vdots \\ \mathbf{O_M} \\ \mathbf{O_M} \end{bmatrix},$$

where $\mathbf{J_M}$ is the $M \times M$ antidiagonal identity matrix

$$\mathbf{J_M} = \begin{bmatrix} 0 & \cdots & 0 & 1 \\ 0 & \cdots & 1 & 0 \\ \vdots & \vdots & \vdots & \vdots \\ 1 & \cdots & 0 & 0 \end{bmatrix}.$$

It is important to mention here that the location of $\mathbf{J_M}$ within the matrix **B** is a system design issue. The case shown here, where it is centered within **B**, corresponds to an overall system delay of $N-1$. This is the natural case for systems with N-tap filters. There are many fine points associated with these time domain conditions. For a complete discussion, the reader is referred to [3].

With the reconstruction equations in place, we now turn our attention to the design of the filters. The problem here is that this is an over-constrained system. The matrix **A** is of size $(2N-M) \times N$. If we think of the synthesis filter coefficients as the parameters to be solved for, we find $M(2N-M)$ equations and MN unknowns. Clearly, the best we can hope for is to determine $\mathbf{B}$ in an approximate sense. Using least-squares approximation, we let

$$\mathbf{S} = (\mathbf{A}^{\mathrm{T}}\mathbf{A})^{-1}\mathbf{B}.$$

Here, it is assumed that $(\mathbf{A}^{\mathrm{T}}\mathbf{A})^{-1}$ exists. This is not automatically the case. However, if reasonable lowpass and highpass filters are used as an initial starting point, there is rarely a problem.

This solution gives the best synthesis filter set for a particular analysis set and system delay $N-1$. The resulting matrix $\mathbf{AS} = \hat{\mathbf{B}}$ will be close to $\mathbf{B}$ but not equal to it in general. The next step in the design is to allow the analysis filter coefficients to vary in an optimization routine to reduce the Frobenius matrix norm, $\|\hat{\mathbf{B}} - \mathbf{B}\|_{\mathrm{F}}^2$. The locally optimal solution will be

$$\mathbf{S} = (\mathbf{A}^{\mathrm{T}}\mathbf{A})^{-1}\mathbf{B}, \quad \text{such that } \|\hat{\mathbf{B}} - \mathbf{B}\|_{\mathrm{F}}^2 \text{ is minimized.}$$

Any number of routines may be used to find this minimum. A simple gradient search that updates the analysis filter coefficients will suffice in most cases. Note that, as written, there are no constraints on the analysis filters other than that they provide an invertible $\mathbf{A}^{\mathrm{T}}\mathbf{A}$ matrix. One can easily start imposing constraints relevant to system quality. Most often we find it appropriate to include constraints on the frequency domain characteristics of the individual analysis filters. This can be done conveniently by creating a cost function comprised of the passband and stopband filter errors. For example, in the two-band case, inclusion of such filter frequency constraints gives rise to the overall error function:

$$\varepsilon = \|\hat{\mathbf{B}} - \mathbf{B}\|_{\mathrm{F}}^2 + \int_0^{\pi_{\mathrm{p}}} \left|1 - H_1(\mathrm{e}^{j\omega})\right|^2 \mathrm{d}\omega + \int_{\pi_{\mathrm{s}}}^{\pi} \left|H_0(\mathrm{e}^{j\omega})\right|^2 \mathrm{d}\omega.$$

This reduces the overall system error of the filter bank while at the same time reducing the stopband errors in analysis filters. Other options in constructing the error function can address control over the step response of the filters, the width of the transition bands, and whether an l_2 norm or an l_∞ norm is used as an optimality criterion.

By properly weighting the reconstruction and frequency response terms in the error function, exact reconstruction can be obtained, if such a solution exists. If an exact reconstruction solution does not exist, the design algorithm will find the locally optimal solution subject to the specified constraints.

36.1.4.1 Functionality of the Design Formulation

One of the distinct advantages of the time-domain design method is its flexibility. The discussion above assumed that the system delay was $N-1$ where N is the filter length. For the time-domain formulation, the amount of overall system delay can be thought of as an input to the design algorithm. In other words, one can pre-specify the desired system delay and then find the locally optimal set of analysis and synthesis filters that reduce the cost function while maintaining the specified delay. Control over the system delay is given by the position of $\mathbf{J}_{\mathbf{M}}$ in the matrix $\mathbf{B}$. Placing $\mathbf{J}_{\mathbf{M}}$ at or near the top of $\mathbf{B}$ lowers the system delay while positioning it at or near the bottom increases the system delay. One consideration here is the effect on filter bank quality. Experiments have shown that as the delay moves toward the extremes, the impact of the overconstrained equations is more severe. One is forced to either tolerate poorer frequency response characteristics or perhaps allow a little distortion in the reconstruction.

The cost function allows for an infinite variety of systems to be designed. The algorithm will converge to a filter set that optimizes the cost function as it is given. This provides the freedom to tradeoff among reconstruction error, frequency domain characteristics, and time domain characteristics. To aid in finding a particular locally optimal solution, the cost function can be allowed to be "adaptive." If exact reconstruction is desired, a heavy weighting may be placed on the reconstruction term in the cost function initially, until that term goes to zero. Then the cost function can be adjusted with new weightings that address reducing the error associated with the remaining distortion components.

This time domain formulation has been used to design an unprecedented variety of filter banks, including the first block decimation systems, the first time-varying systems, the first low delay systems, cosine modulated filter banks, nonuniform band filter banks, and many others [3,4,9–11]. One of the most important in this list is cosine modulated filter banks because they can be implemented very efficiently by using FFT-class algorithms. Cosine modulated filter banks may be designed in a variety of ways. Excellent discussions on this topic are given by Malvar [20,24], Vaidyanathan [21,27], Vetterli [23], and many others.

Linear filter banks have proven to be effective in many applications. Perhaps their most widespread use is in the area of coding. Subband coders for speech audio, image, and video signals tend to work very well. However, at low bit rates, distortions can be detected. Thus, there is interest in designing filter banks that are less prone to producing annoying distortions in these cases. Other nonlinear classes of filter banks can be considered that display different forms of distortion at low bit rates. In the remainder of this chapter, we discuss the design of two nonlinear filter banks that are presently being studied.

36.2 Finite Field Filter Banks

A new and interesting variant of the classical analysis-synthesis system can be achieved by imposing the explicit constraint that the discrete amplitude range of the subbands is confined. For conventional filter banks, we assume the input signal has a finite number of amplitude values. For instance, in the case of subband image coding, the input will typically contain 8 bits or 256 amplitude levels. However, the subband outputs may contain millions of possible amplitude values. For a coding application, we can think of the input as having a small alphabet (e.g., 256 unique values), and the analysis filter output as having a large alphabet (millions). Conceivably, one might be able to improve coding performance in some situations by designing a filter bank that constrains the output alphabet to be small. With this as motivation, we consider the problem of designing exact reconstruction filter banks with this constraint, an idea originally introduced by Vaidyanathan [37].

To begin our discussion, consider an input image with an alphabet size N (e.g., 256 gray levels). The output is expanded to an alphabet size of $M \times N$ after subband filtering. The value of M is governed by the length and coefficient values of the filter. M can be very large. The design task of interest here is to construct a filter bank where M is very small, ideally unity. In other words, we are constraining the system to operate in a finite field of our choosing, for example, GF(N). In order to meet this finite field condition, an operational change is needed. Specifically, the finite field filter bank should operate in an integer field. Consequently, the filters used should be perfect reconstruction filters with integer coefficients. This modification makes it possible to perform wrap-around arithmetic. Wrap-around arithmetic restricts outputs to a finite field by performing all operation modulo N.

The design of a finite field filter bank is relatively simple. The image is passed through analysis filters using wrap-around arithmetic. This means that every operation is either modulo-N addition or modulo-N multiplication. Hence, the subband outputs will have an integer alphabet of size N. To reconstruct, the image is passed through the synthesis filters using the same wrap-around arithmetic within the same finite integer field. The bands are then combined using modulo-N addition. As it turns out, the resulting signal will not match the original. However, the signal can be corrected by applying a mapping based on the gain of the filter banks, M, and the dynamic range, N. Let us assume that the input is an image with N' discrete levels, and that all operations have been performed modulo N. Each value of the output image is found in set **B** and can be mapped into set **A**, where

$$\mathbf{A} = \{0, 1, 2, \ldots, N' - 1\} \quad \text{and} \quad \mathbf{B} = [(M \times \mathbf{A})]_N.$$

The resulting output image $\hat{x}$ will be, under certain conditions, an exact reconstruction of the input image x.

There are two conditions that must be satisfied in order to obtain exact reconstruction. First, the subband output alphabet size N must be equal to or greater than the input alphabet size N'. This is a necessary condition in order to unambiguously resolve all values of the input. Second, the system gain M is constrained in relation to the subband output size N. The system gain is governed by the analysis and synthesis filters in the following way:

$$M = \left(\sum_n |h_0[n]| \times \sum_n |g_0[n]|\right) + \left(\sum_n |h_1[n]| \times \sum_n |g_1[n]|\right),$$

where
$h_0[n]$ and $h_1[n]$ are the analysis filters
$g_0[n]$ and $g_1[n]$ are the synthesis filters

The relation between M and N is crucial in obtaining perfect reconstruction. These two numbers must be relatively prime. That is, M and N can have no common factors. For example, if M is two, any odd value of N would be valid. Ideally, we might want $N = N'$. However, to satisfy the last condition M is determined by the system and N is adjusted slightly up from N'. It is typically easier to adjust N.

To illustrate the differences in outputs obtained from conventional and finite field filter banks, consider the following comparison. For a conventional two-band system with two-tap Haar analysis filters, an input of

$$x = 0, 0, 0, 4, 2, 3, 0, 1, 2, 0, 0, \ldots$$

will yield the outputs

$$y_0 = 0, 4, 5, 1, 2, \ldots$$
$$y_1 = 0, 4, 1, 1, -2, \ldots.$$

However, for the equivalent finite field system (like the one shown in Figure 36.5), the outputs are noticeably different. For the finite field case, all operations are performed modulo N. Thus, for the same input the outputs produced are

$$y_0 = 0, 4, 0, 1, 2, \ldots$$
$$y_1 = 0, 4, 1, 1, 3, \ldots.$$

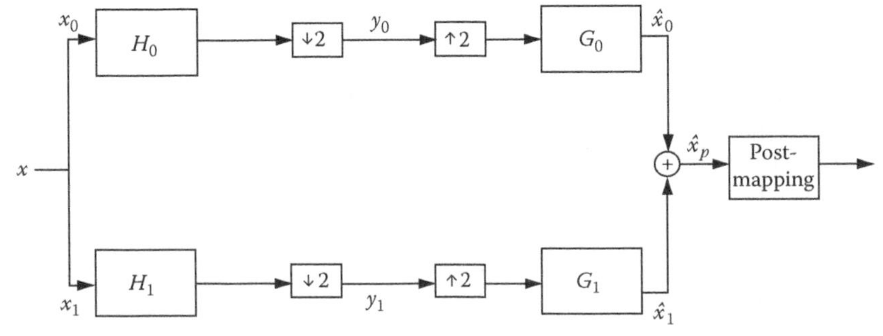

FIGURE 36.5 Block diagram of a two-band finite field filter bank.

Notice that the alphabet here is confined to the integers 0, 1, 2, 3, and 4 because we have set $N = 5$. For the reconstruction, the outputs shown in the figure will be

$$\hat{x}_0 = 0, 4, 4, 0, 0, 1, 1, 2, 2, \ldots$$
$$\hat{x}_1 = 0, 1, 4, 4, 1, 4, 1, 2, 3, \ldots.$$

Adding these together, modulo N gives

$$\hat{x}_p = 0, 0, 3, 4, 1, 0, 2, 4, 0, \ldots.$$

Now unscrambling them in the post-mapping step shown in the figure gives

$$\hat{x} = 0, 0, 4, 2, 3, 0, 1, 2, 0, \ldots = x.$$

It is interesting to compare the analysis section outputs of finite field and conventional filter banks for the two-band case. The lower band output of a conventional filter bank has a dynamic range that is usually much greater than the dynamic range of the input. The values in the lower band tend to have a Gaussian distribution over the range. By constraining the alphabet size, the first-order entropy can be reduced. The amount of the reduction depends on the size of M. The higher band in the conventional filter bank has a dynamic range that might be larger than N; however, the values are clustered around zero. When modulo operations are performed, the negative values go to a high value so not much overlap is obtained. Therefore, the alphabet constraint has little or no affect on the higher bands. The finite field filter bank reduces the overall first-order entropy because the entropy is reduced in the lower band. The degree by which the entropy is reduced is greatly dependent on the image and the filter gains.

How do finite field filter banks affect input images with different dynamic ranges? This effect is dependent on the same two components that have previously been discussed, the system gain M, and the subband output size N. Let us assume the subband output range N is set equal to the input image range N'. Now we can examine the effects of different system gains given N. For example, if the image is binary ($N = 2$), the system gain must be odd. Examining the decomposition of such an image, we can see that it appears very noisy. This is because the dynamic range of the system is small and the gain is large. The image is essentially wrapping around on itself so many times it is difficult to observe the original image in the bands. In a case where $N > 2$, a filter with a smaller gain is more realizable. For example, if $N = 255$, we can choose a system gain of 2. In this decomposition (Figure 36.6), the lower band image is not what we are accustomed to observing in a conventional decomposition. This case does have a lower first-order entropy than its conventional counterpart.

Finite field filter banks are still in their early phases of study. As a result of the constraints, filter quality is limited. Thus, the net gains achievable in an application could be favorable or unfavorable. One must pay careful attention to the subband output size, filter length, and coefficient values during the design of the filter bank. Nonetheless, it seems that finite field filter banks are potentially attractive in some applications.

36.3 Nonlinear Filter Banks

One of the driving forces for research in filter banks is image coding for low bit rate applications. Presently, subband image coders represent the best approach known for image compression. As with any coder, at low rates distortions occur. Subband coders based on conventional linear filter banks suffer from ringing effects due to the Gibbs phenomenon. These ringing effects occur around edges or high

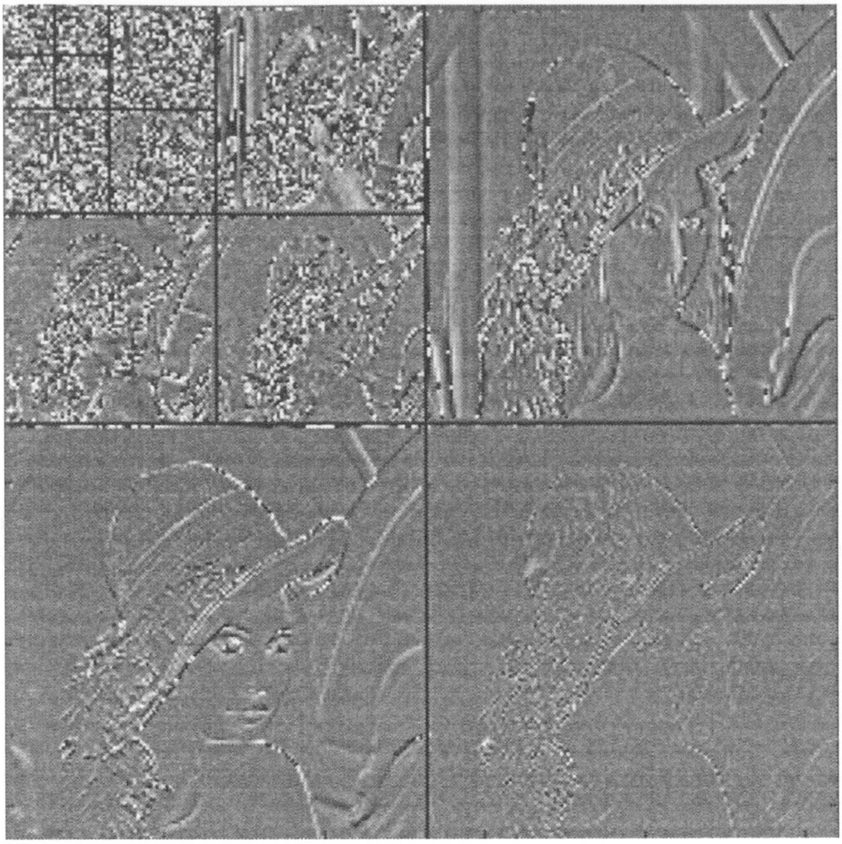

FIGURE 36.6 A four-level octave band decomposition using finite field filter banks.

contrast regions. One way to eliminate ringing is to use nonlinear filter banks. There are pros and cons regarding the utility of nonlinear filter banks. However, the design of the systems is rather new and interesting.

Nonlinear filter banks can be constructed within a general two-band framework. A nonlinear filter may be placed in the highpass analysis and in the lowpass synthesis block of the systems. The condition for exact reconstruction will be discussed later. What type of nonlinear filter is an open question. While there are many candidates, the constraints of the overall system restrict the design of filters in terms of type and degrees of freedom in optimization. The most widely used nonlinear filter is the rank-order filter. In this discussion, we consider rank-order filters, more specifically, median filters. The performance of such filters is determined by the rank used and the region of support. The popular N-point median filter has a rank of $(N+1)/2$, where N is assumed to be odd. Egger et al. [31] suggested a simple two-band nonlinear filter bank that upholds the exact reconstruction property. The lowpass channel consists of direct downsampling, while the highpass channel involves a median filter (differencing) operation to achieve a highpass representation for the other channel. Because straight downsampling and median filtering are involve, there is an inherent finite field constraining property built in to the system. Although these features seem attractive, the system is severely limited by its lack of filtering power. Most notably, the lowpass channel has massive aliasing since no filtering is performed. For many applications, aliasing of this type is not desirable. This problem can be addressed somewhat by using the modified filter bank introduced by Florencio and Schafer [32]. In the two-band system of Florencio

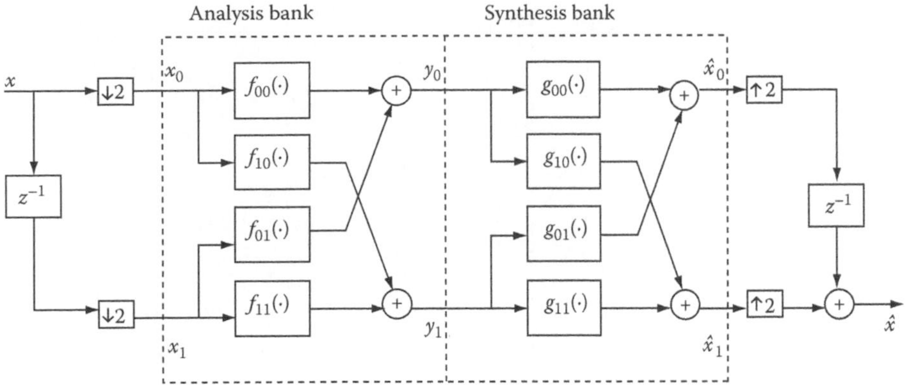

FIGURE 36.7 A two-band polyphase nonlinear filter bank.

and Schafer shown in Figure 36.7, each channel can be expressed as a filtered combination of the input. This structure can be recognized as a classical polyphase implementation for a two-band filter bank. Here, however, we allow the polyphase filters f_{ij} and g_{ij} to be nonlinear filters. Thus,

$$y_0[n] = f_{00}(x_0[n]) + f_{01}(x_1[n])$$
$$y_1[n] = f_{10}(x_0[n]) + f_{11}(x_1[n]),$$

where $f_{ij}(\cdot)$ are the linear or nonlinear polyphase analysis filters. To reconstruct the signal, the output can be expressed as a filtered combination of the channels,

$$\hat{x}_0 = g_{00}(y_0) + g_{01}(y_1)$$
$$\hat{x}_1 = g_{10}(y_0) + g_{11}(y_1),$$

where $g_{ij}(\cdot)$ are the linear or nonlinear polyphase synthesis filters. The perfect reconstruction conditions are based on these different classes or structures. The Type I structure consists of $f_{00}(\cdot) = f_{11}(\cdot) = I$ (identity), and either $f_{10}(\cdot) = 0$ or $f_{01}(\cdot) = 0$. The other is any causal transformation. To obtain perfect reconstruction, $g_{00}(\cdot) = g_{11}(\cdot) = I$, $g_{10}(\cdot) = f_{10}(\cdot)$, and $g_{01}(\cdot) = -f_{01}(\cdot)$. The Type II structure consists of $f_{10}(\cdot) = f_{01}(\cdot) = 0$ and both $f_{00}(\cdot)$ and $f_{11}(\cdot)$ being invertible functions. To obtain perfect reconstruction $g_{01}(\cdot) = g_{10}(\cdot) = 0$, $g_{00}(\cdot) = f_{00}^{-1}(\cdot)$, and $g_{11}(\cdot) = f_{11}^{-1}(\cdot)$. The Type III structure consists of $f_{10}(\cdot) = f_{01}(\cdot) = I$ and $f_{00}(\cdot) = f_{11}(\cdot) = 0$. To obtain perfect reconstruction, $g_{01}(\cdot) = g_{10}(\cdot) = I$ and $g_{00}(\cdot) = g_{11}(\cdot) = 0$.

Similar to linear filter banks, this nonlinear filter bank achieves an overall reduction in first-order entropy. Since perfect reconstruction is achieved in the two-band decomposition, perfect reconstruction can be maintained when used in tree structured systems for compression applications. After quantization, coding, and reconstruction, different features will be affected in different ways. The main advantage of nonlinear filtering is that the edges associated with high contrast features are preserved well, and no "ringing" occurs. However, because of the nature of the sampling in the lower band, texture regions are distorted. Using cascaded sections is a way to help preserve the texture. As it turns out, sections can be cascaded in a way that preserves exact reconstruction. For example, let the first stage of the filter contain $f_{01}(\cdot) = 0$, with $f_{10}(\cdot)$ being a four-point median $f_{01}(\cdot)$ being a four-point median filter with a 0.5 gain (to maintain the dynamic range of the input). The resulting two bands are similar to the bands of the comparable linear case but have the advantages of a nonlinear system.

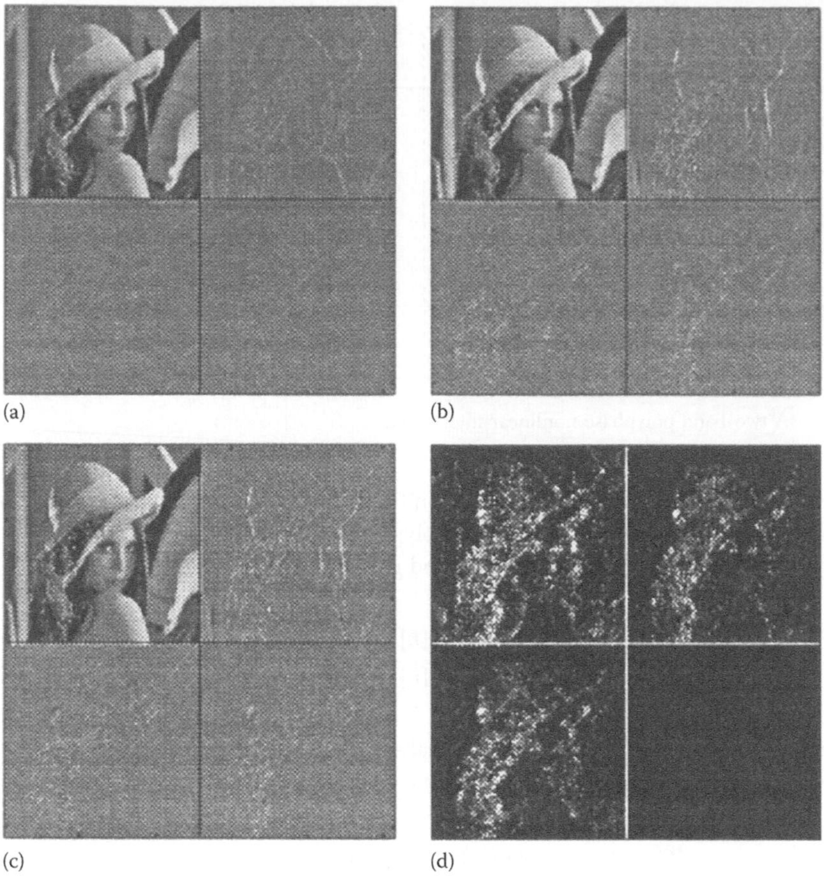

FIGURE 36.8 Comparison of outputs from one linear and three nonlinear filter banks: (a) a four-band linear decomposition using four-tap QMFs, (b) a four-band nonlinear decomposition using the method of Egger and Li, (c) a four-band nonlinear decomposition using the two-stage method of Florencio and Schafer, and (d) the residual image obtained from subtracting the nonlinear decomposition result in (b) from the result in (c).

Most notably, the lower band of the nonlinear case has a reduction in higher frequencies, very similar to the linear case. These differences are illustrated in Figure 36.8 for a four-band decomposition of an image. A conventional QMF decomposition is shown in Figure 36.8a. Next to it in Figure 36.8b and c are the nonlinear decompositions obtained using the Egger and Li approach, and the two-stage approach of Florencio and Schafer, respectively. All show similarities. However, more energy is contained in the high frequency subbands of the nonlinear results. In comparing carefully the two nonlinear results, we can observe that the two-stage approach of Florencio and Schafer has less aliasing in the lowest band and more closely follows the linear result. The difference image between the two nonlinear results is given in Figure 36.8d.

It is clear that there are many possibilities for constructing nonlinear filter banks. What is less obvious at this point is the impact of these systems in practical situations. Given that development related to these filter banks is only in the formative stages, only time will tell. Regardless of whether conventional or nonlinear filter banks are ultimately employed, the variety of design options and design techniques offer many useful solutions to engineering problems. More in-depth discussions on applications can be found in the references.

References

1. Smith, M. and Barnwell, T., The design of digital filters for exact reconstruction in subband coding, *Trans. Acoust. Speech Signal Process.*, ASSP-34(3), 434–441, June 1986.
2. Smith, M. and Barnwell, T., A new filter bank theory for time-frequency representation, *Trans. Acoust. Speech Signal Process.*, ASSP-35(3), 314–327, March 1987.
3. Nayebi, K., Barnwell, T., and Smith, M., Time domain filter bank analysis: A new design theory, *IEEE Trans. Signal Process.*, 40(6), 1412–1429, June 1992.
4. Nayebi, K., Barnwell, T. and Smith, M., Analysis-synthesis systems based on time varying filter banks, *International Conference on Acoustics, Speech, and Signal Processing*, San Francisco, CA, March 1992, Vol. 4, pp. 617–620.
5. Schuller, G.D.T. and Smith, M.J.T., A new framework for modulated perfect reconstruction filter banks, *IEEE Trans. Signal Process.*, 44(8), 1941–1954, August 1996.
6. Smith, M. and Barnwell, T., A unifying framework for analysis/synthesis based on maximally decimated analysis/synthesis systems, *Proceedings of the International Conference on Acoustics, Speech, and Signal Processing*, Tampa, FL, March 1985, pp. 521–524.
7. Smith, M. and Barnwell, T., A procedure for designing exact reconstruction filter banks for tree-structured subband coders, *Proceedings of the International Conference on Acoustics, Speech, and Signal Processing*, San Diego, CA, March 1984, pp. 27.1.1–27.1.4.
8. Nayebi, K., Barnwell, T., and Smith, M., Time domain conditions for exact reconstruction in analysis/synthesis systems based on maximally decimated filter banks, *19th Southeastern Symposium on System Theory*, Clemson, SC, March 1987, pp. 498–502.
9. Nayebi, K., Barnwell, T., and Smith, M., Block decimated analysis-synthesis filter banks, *IEEE International Symposium on Circuits and Systems*, San Diego, CA, May 1992, pp. 947–950.
10. Nayebi, K., Barnwell, T., and Smith, M., Design and implementation of computationally efficient modulated filter banks, *Proceedings of the International Symposium on Circuits and Systems*, Singapore, June 12–14, 1991, pp. 650–653.
11. Nayebi, K., Barnwell, T.P., and Smith, M.J.T., Design of perfect reconstruction nonuniform band filter banks, *Proceedings of the International Conference on Acoustics, Speech, and Signal Processing*, Toronto, Canada, May 14–17, 1991, pp. 1781–1784.
12. Nayebi, K., Barnwell, T.P., and Smith, M.J.T., Design of low delay FIR analysis-synthesis filter bank systems, *Proceedings of the Conference on Information Sciences and Systems*, Baltimore, MD, March 1991.
13. Nayebi, K., Barnwell, T.P., and Smith, M.J.T., Time-domain view of filter banks and wavelets, *25th Asilomar Conference on Signals, Systems and Computers*, Pacific Grove, CA, November 4–6, 1991, Vol. 2, pp. 736–740.
14. Mersereau, R.M. and Smith, M.J.T., *Digital Filtering: A Computer Laboratory Textbook*, John Wiley & Sons, New York, 1993.
15. Akansu, A. and Smith, M. (Eds.), *Subband and Wavelet Transforms: Design and Applications*, Kluwer Academic Publishers, Dordrecht, the Netherlands, 1995.
16. Smith, M. and Docef, A., *A Study Guide to Digital Image Processing*, Scientific Publishers, Riverdale, GA, 1997.
17. Johnston, J., A filter family designed for use in quadrature mirror filter banks, *Proceedings of the IEEE International Conference on Acoustics, Speech, and Signal Processing*, Denver, CO, April 1980, Vol. 5, pp. 291–294.
18. Croisier, A., Esteban, D., and Galand, C., Perfect channel splitting by use of interpolation/decimation/tree decomposition techniques, *Proceedings of International Conference on Information Sciences and Systems*, Patras, Greece, August 1976, pp. 443–446.
19. Crochiere, R.E. and Rabiner, L.R., *Multirate Digital Signal Processing*, Prentice-Hall, Englewood Cliffs, NJ, 1983.

20. Malvar, H.S., *Signal Processing with Lapped Transforms*, Artech House, Norwood, MA, 1991.
21. Koilpillai, R.D. and Vaidyanathan, P.P., New results on cosine modulated FIR filter banks satisfying perfect reconstruction, *Proceedings of IEEE International Conference on Acoustics, Speech, and Signal Processing*, Toronto, ON, April 14–17, 1991, Vol. 3, pp. 1793–1796.
22. Rothweiler, J., Polyphase quadrature mirror filters—A new sub-band coding technique, *Proceedings of IEEE International Conference on Acoustics, Speech, and Signal Processing*, Boston, MA, 1983, pp. 1280–1283.
23. Nussbaumer, H.J. and Vetterli, M., Computationally efficient QMF filter banks, *Proceedings of IEEE International Conference on Acoustics, Speech, and Signal Processing*, San Diego, CA, March 1984, Vol. 9, pp. 437–440.
24. Malvar, H., Modulated QMF filter banks with perfect reconstruction, *Electron. Lett.*, 26(13), 906–907, June 1990.
25. Mintzer, F., Filters for distortion-free two-band multirate filter banks, *IEEE Trans. Acoustics Speech Signal Process.*, ASSP-33, 626–630, June 1985.
26. Akansu, A.N. and Haddad, R.A., *Multiresolution Signal Decomposition*, Academic Press, San Diego, CA, 1992.
27. Vaidyanathan, P.P., *Multirate Systems and Filterbanks*, Prentice-Hall, Englewood Cliffs, NJ, 1993.
28. Vetterli, M. and Kovacevic, J., *Wavelets and Subband Coding*, Prentice-Hall, Englewood Cliffs, NJ, 1995.
29. Fleige, N.J., *Multirate Digital Signal Processing*, John Wiley & Sons, New York, 1993.
30. Vaidyanathan, P.P., Quadrature mirror filter banks, M-band extensions and perfect reconstruction techniques, *IEEE Trans. Acoust. Speech Signal Process.*, 4(3), 4–20, July 1987.
31. Egger, O. and Li, W., Very low bit rate image coding using morphological operators and adaptive decompositions, *IEEE International Conference on Image Processing (ICIP'94)*, Austin, TX, Nov. 13–16, 1994, Vol. 2, pp. 326–330.
32. Florencio, D.A.F. and Schafer, R.W., Perfect reconstructing nonlinear filter banks, *IEEE International Conference on Acoustics, Speech, and Signal Processing (ICASSP'96)*, Atlanta, GA, 1996, Vol. 3, pp. 1814–1817.
33. Florencio, D.A.F. and Schafer, R.W., A non-expansive pyramidal morphological image coder, *IEEE International Conference on Image Processing (ICIP'94)*, Austin, TX, Nov. 13–16, 1994, Vol. 2, pp. 331–334.
34. Sun, F.-K. and Maragos, P., Experiments on image compression using morphological pyramids, *Visual Communication and Image Processing IV (VCIP'89)*, Philadelphia, PA, Nov. 1989, Vol. 1141, pp. 1303–1312.
35. Toet, A., A morphological pyramidal image decomposition, *Pattern Recognit. Lett.*, 9, 255–261, May 1989.
36. Bruekers, F.A.M.L. and van den Enden, A.W.M., New networks for perfect inversion and perfect reconstruction, *IEEE J. Sel. Areas Commn.*, 10, 130–137, Jan. 1992.
37. Vaidyanathan, P.P., Unitary and paraunitary systems in finite fields, *Proceedings of 1990 IEEE International Symposium on Circuits and Systems*, New Orleans, LA, 1990, pp. 1189–1192.
38. Tewfik, A.H., Hosur, S., and Sowelam, S., Recent progress in the application of wavelet in surveillance systems, *Opt. Eng.*, 33, 2509–2519, Aug. 1994.
39. Swanson, M. and Tewfik, A.H., A binary wavelet decomposition of binary images, *IEEE Trans. Image Process.*, 5, 1637–1650, Dec. 1996.
40. Flornes, K., Grossman, A., Hoschneider, M., and Torresani, B., Wavelets on finite fields, preprint, Nov. 1993.

37

Time-Varying Analysis-Synthesis Filter Banks

Iraj Sodagar
PacketVideo

37.1 Introduction

Time-frequency representations (TFR) combine the time-domain and frequency-domain representations into a single framework to obtain the notion of time-frequency. TFR offer the time localization vs. frequency localization trade-off between two extreme cases of time-domain and frequency-domain representations. The short-time Fourier transform (STFT) [1–5] and the Gabor transform [6] are the classical examples of linear time-frequency transforms which use time-shifted and frequency-shifted basis functions.

In conventional time-frequency transforms, the underlying basis functions are fixed in time and define a specific tiling of the time-frequency plane. The term time-frequency tile of a particular basis function is meant to designate the region in the plane that contains most of that function's energy. The STFT and the wavelet transform are just two of many possible tilings of the time-frequency plane. These two are illustrated in Figure 37.1a and b, respectively. In these figures, the rectangular representation for a tile is purely symbolic, since no function can have compact support in both time and frequency. Other arbitrary tilings of the time-frequency plane are possible such as the example shown in Figure 37.1c. In the discrete domain, linear time-frequency transforms can be implemented in the form of filter bank structures.

It is well known that the time-frequency energy distribution of signals often changes with time. Thus, in this sense, the conventional linear time-frequency transform paradigm is fundamentally mismatched to many signals of interest. A more flexible and accurate approach is obtained if the basis functions of the transform are allowed to adapt to the signal properties. An example of such a time-varying tiling is shown in Figure 37.1d. In this scenario, the time-frequency tiling of the transform can be changed from good frequency localization to good time localization and vice versa. Time-varying filter banks provide such flexible and adaptive time-frequency tilings.

The concept of time varying (or adaptive) filter banks was originally introduced in [7] by Nayebi et al. The ideas underlying their method were later developed and extended to a more general case in which it was also shown that the number of frequency bands could also be made adaptive [8–11]. De Queiroz and Rao [12] reported time-varying extended lapped transforms and Herley et al. [13–15] introduced another

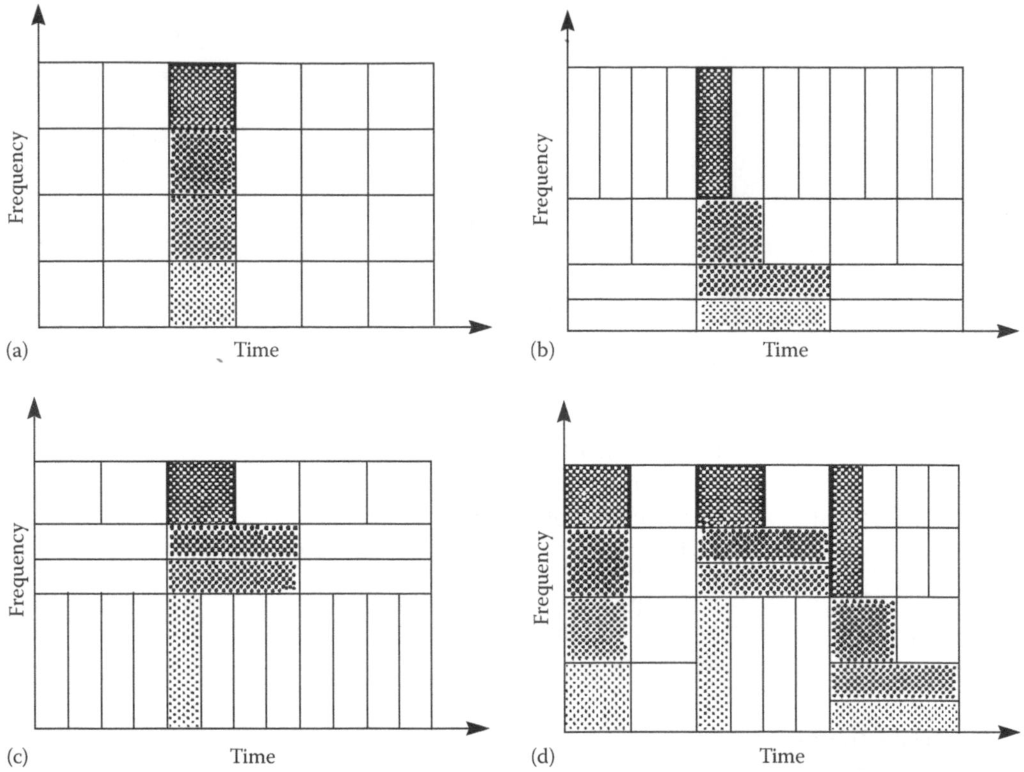

FIGURE 37.1 The time-frequency tiling for different time-frequency transforms: (a) the STFT, (b) the wavelet transform, (c) an example of general tiling, and (d) an example of the time-varying tiling.

time-domain approach for designing time-varying lossless filter banks. Arrowood and Smith [16] demonstrated a method for switching between filter banks using lattice structures. In [17], the authors presented yet another formulation for designing time-varying filter banks using a different factorization of the paraunitary transform. Chen and Vaidyanathan [18] reported a noncausal approach to time-varying filter banks by using time-reversed filters. Phoong and Vaidyanathan [19] studied time-varying paraunitary filter banks using polyphase approach. In [11,20–22], the post filtering technique for designing time-varying filter bank was reported. The design of multidimensional time-varying filter bank was addressed in [23,24]. In this chapter, we introduce the notion of the time-varying filter banks and briefly discuss some design methods.

37.2 Analysis of Time-Varying Filter Banks

Time-varying filter banks are analysis-synthesis systems in which the analysis filters, the synthesis filters, the number of bands, the decimation rates, and the frequency coverage of the bands are changed (in part or in total) in time, as is shown in Figure 37.2. By carefully adapting the analysis section to the temporal properties of the input signal, better performance can be achieved in processing the signal. In the absence of processing errors, the reconstructed output $\hat{x}(n)$ should closely approximate a delayed version of the original signal $x(n)$. When $\hat{x}(n-\Delta)=x(n)$ for some integer constant, Δ, then we say that the filter bank is perfectly reconstructing (PR). The intent of the design is to choose the time-varying analysis and synthesis filters along with the time-varying down/up samplers so that the system requirements are met subject to the constraint that the analysis-synthesis filter bank be PR at all times.

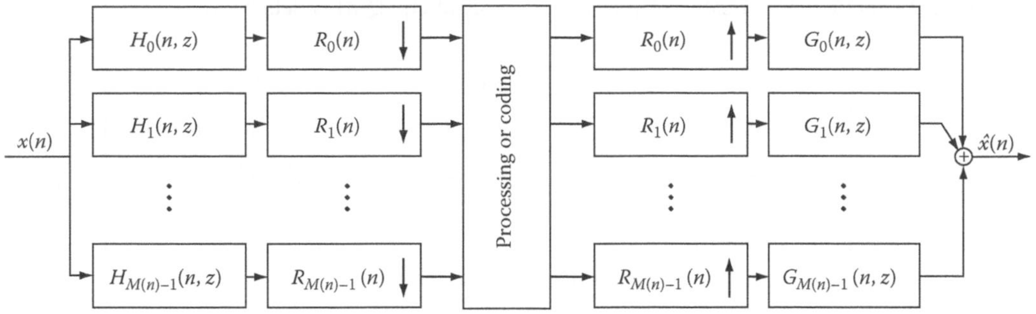

FIGURE 37.2 The time-varying filter bank structure with time-varying filters and time-dependent down/up samplers.

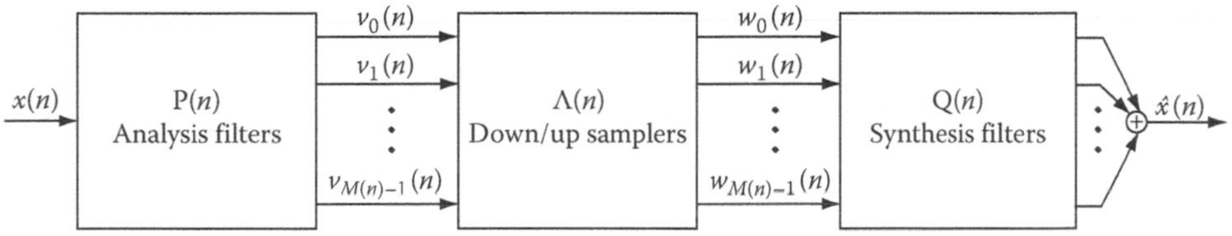

FIGURE 37.3 Time-varying filter bank as a cascade of analysis filters, down/up samplers, and synthesis filters.

One general method for analysis of time-varying filter banks is the time-domain formulation reported in [10,22]. In this method, the time-varying impulse response of the entire filter bank is derived in terms of the analysis and synthesis filter coefficients.

Figure 37.3 shows the diagram of a time-varying filter bank. In this figure, the filter bank is divided into three stages: the analysis filters, the down/up samplers, and the synthesis filters. The signals $x(n)$ and $\hat{x}(n)$ are the filter bank input and output at time n, respectively. The outputs of the analysis filters are shown by $\mathbf{v}(n) = [v_0(n), v_1(n), \ldots, v_{M(n)-1}(n)]^{\mathrm{T}}$, where $v_i(n)$ is the output of the ith analysis filter at time n. The outputs of the down/up samplers at time n are called $\mathbf{w}(n) = [w_0(n), w_1(n), \ldots, w_{M(n)-1}(n)]^{\mathrm{T}}$.

The input/output relation of the analysis filters can be expressed by

$$\mathbf{v}(n) = \mathbf{P}(n)\mathbf{x}_N(n). \tag{37.1}$$

where

$\mathbf{P}(n)$ is an $M(n) \times N(n)$ matrix whose mth row is comprised of the coefficients of the mth analysis filter at time n

$\mathbf{x}_N(n)$ is the input vector of length $N(n)$ at time n:

$$\mathbf{x}_N(n) = [x(n), x(n-1), x(n-2), \ldots, x(n - N(n) + 1)]^{\mathrm{T}}. \tag{37.2}$$

The input/output function of down/up samplers can be expressed in the form

$$\mathbf{w}(n) = \Lambda(n)\mathbf{v}(n), \tag{37.3}$$

where $\Lambda(n)$ is a diagonal matrix of size $M(n) \times M(n)$. The mth diagonal element of $\Lambda(n)$, at time n, is 1 if the input and output of the mth down/up sampler are identical, otherwise it is zero.

To write the input/output relationship of the synthesis filters, $\mathbf{Q}(n)$ is defined as

$$\mathbf{Q}(n) = \begin{bmatrix} g_0(n,0) & g_0(n,1) & g_0(n,2) & \dots & g_0(n,N(n)-1) \\ g_1(n,0) & g_1(n,1) & g_1(n,2) & \dots & g_1(n,N(n)-1) \\ g_2(n,0) & g_2(n,1) & g_2(n,2) & \dots & g_2(n,N(n)-1) \\ \vdots & \vdots & \vdots & \vdots & \vdots \\ g_{M(n)-1}(n,0) & g_{M(n)-1}(n,1) & g_{M(n)-1}(n,2) & \dots & g_{M(n)-1}(n,N(n)-1) \end{bmatrix}$$
$$= \left[\mathbf{q}_0(n) \quad \mathbf{q}_1(n) \quad \mathbf{q}_2(n) \quad \dots \quad \mathbf{q}_{N(n)-1}(n) \right], \tag{37.4}$$

where
$\mathbf{q}_i(n) = [g_0(n,i), g_1(n,i), g_2(n,i), \dots, g_{M(n)-1}(n,i)]^{\mathrm{T}}$ is a vector of length $M(n)$
$g_i(n,j)$ denotes the jth coefficient of the ith synthesis filter

At time n, the mth synthesis filter is convolved with vector $[w_m(n), w_m(n-1), \dots, w_m[n-N(n)+1]]^{\mathrm{T}}$ and all outputs are added together. Using Equation 37.4, the output of the filter bank at time n can be written as

$$\hat{x}(n) = \sum_{i=0}^{N(n)-1} \mathbf{q}_i^{\mathrm{T}}(n)\mathbf{w}(n-i). \tag{37.5}$$

If $\mathbf{s}(n)$ and $\hat{\mathbf{w}}(n)$ are defined as

$$\mathbf{s}(n) = \left[\mathbf{q}_0^{\mathrm{T}}(n), \mathbf{q}_1^{\mathrm{T}}(n), \mathbf{q}_2^{\mathrm{T}}, \dots, \mathbf{q}_{N(n)-1}^{\mathrm{T}}(n) \right]^{\mathrm{T}} \tag{37.6}$$

$$\hat{\mathbf{w}}(n) = [\mathbf{w}^{\mathrm{T}}(n), \mathbf{w}^{\mathrm{T}}(n-1), \mathbf{w}^{\mathrm{T}}(n-2), \dots, \mathbf{w}^{\mathrm{T}}[n-N(n)+1]]^{\mathrm{T}}, \tag{37.7}$$

then Equation 37.5 can be written in the form of one inner product:

$$\hat{x}(n) = \mathbf{s}^{\mathrm{T}}(n)\hat{\mathbf{w}}(n), \tag{37.8}$$

where $\mathbf{s}(n)$ and $\hat{\mathbf{w}}(n)$ are vectors of length $N(n)M(n)$. Using Equations 37.1, 37.3, 37.7, and 37.8, the input/output function of the filter bank can be written as

$$\hat{x}(n) = \mathbf{s}^{\mathrm{T}}(n) \begin{bmatrix} \Lambda(n)\mathbf{P}(n)\mathbf{x}_N(n) \\ \Lambda(n-1)\mathbf{P}(n-1)\mathbf{x}_N(n-1) \\ \Lambda(n-2)\mathbf{P}(n-2)\mathbf{x}_N(n-2) \\ \vdots \\ \Lambda(n-N(n)+1)\mathbf{P}(n-N(n)+1)\mathbf{x}_N(n-N(n)+1) \end{bmatrix}. \tag{37.9}$$

As the last $N(n)-1$ elements of vector $\mathbf{x}_N(n-i)$ are identical to the first $N(n)-1$ elements of vector $\mathbf{x}_N(n-i-1)$, the latter equation can be expressed by

$$\hat{x}(n) = \mathbf{s}^{\mathrm{T}}(n) \begin{bmatrix} [& \Lambda(n)\mathbf{P}(n) &]\mathbf{O}\ldots\ldots\ldots\ldots\mathbf{O} & & \\ \mathbf{O}[& \Lambda(n-1)\mathbf{P}(n-1) &]\mathbf{O}\ldots\ldots\ldots\ldots\mathbf{O} & & \\ \mathbf{O}\mathbf{O}[& \Lambda(n-2)\mathbf{P}(n-2) &]\mathbf{O}\ldots\ldots\ldots\ldots\mathbf{O} & & \\ & \ddots & & & \\ \mathbf{O}\ldots\ldots\ldots\ldots & & \mathbf{O}\{\Lambda[n-N(n)+1]\mathbf{P}[n-N(n)+1]\} & & \end{bmatrix} \times \begin{bmatrix} x(n) \\ x(n-1) \\ x(n-2) \\ \vdots \\ x[n-2N(n)+1] \end{bmatrix}, \qquad (37.10)$$

where $\mathbf{O}$ is the zero column vector with length $M(n)$. Thus, the input/output function of a time-varying filter bank can be expressed in the form of

$$\hat{x}(n) = \mathbf{z}^{\mathrm{T}}(n)\mathbf{x}_I(n), \qquad (37.11)$$

where

$\mathbf{x}_I(n) = [x(n), x(n-1), \ldots, x(n-I+1)]^{\mathrm{T}}$

$I(n) = 2N(n) - 1$

$\mathbf{z}(n)$ is the time-varying impulse response vector of the filter bank at time n:

$$\mathbf{z}(n) = \mathbf{A}(n)\mathbf{s}(n). \qquad (37.12)$$

The matrix $\mathbf{A}(n)$ is the $[2N(n)-1] \times [N(n)M(n)]$ matrix

$$\mathbf{A}(n) = \begin{bmatrix} \left[\mathbf{P}(n)^{\mathrm{T}}\ \Lambda(n)\right] & \mathbf{O}^{\mathrm{T}} & & \mathbf{O}^{\mathrm{T}} \\ \mathbf{O}^{\mathrm{T}} & \left[\mathbf{P}(n-1)^{\mathrm{T}}\ \Lambda(n-1)\right] & & \vdots \\ \mathbf{O}^{\mathrm{T}} & \mathbf{O}^{\mathrm{T}} & \ddots & \mathbf{O}^{\mathrm{T}} \\ \vdots & \vdots & & \left\{\mathbf{P}[n-N(n)+1]^{\mathrm{T}}\ \Lambda[n-N(n)+1]\right\} \\ \mathbf{O}^{\mathrm{T}} & \mathbf{O}^{\mathrm{T}} & & \end{bmatrix}. \qquad (37.13)$$

For a perfect reconstruction filter bank with a delay of Δ, it is necessary and sufficient that all elements but the $(\Delta+1)$th in $\mathbf{z}(n)$ be equal to zero at all times. The $(\Delta+1)$th entry of $\mathbf{z}(n)$ must be equal to one. If the ideal impulse response is $\mathbf{b}(n)$, the filter bank is PR if and only if

$$\mathbf{A}(n)\mathbf{s}(n) = \mathbf{b}(n) \quad \text{for all } n. \qquad (37.14)$$

37.3 Direct Switching of Filter Banks

Changing from one arbitrary filter bank to another independently designed filter bank without using any intermediate filters is called direct switching. Direct switching is the simplest switching scheme and does not require additional steps in switching between two filter banks. But such switching will result in a substantial amount of reconstruction distortion during the transition period. This is because during the transition, none of the synthesis filters satisfies the exact reconstruction conditions. Figure 37.4 shows an example of a direct switching filter bank. Figure 37.5 shows the time-varying impulse response of the above system around the transition periods. In this figure, $z(n, m)$ is the response of the system at time n to the unit input at time m. For a PR system, $z(n, m)$ has a height of 1 along the diagonal and 0 everywhere else in the (m, n)-plane. As is shown, the time-varying filter bank is PR before and after but not during the transition periods. In this case, each switching operation generates a distortion with an eight-sample duration. One way to reduce the distortion is to switch the synthesis filters with an appropriate delay with respect to the analysis switching time. This delay may reduce the output distortion, but it can not eliminate it.

37.4 Time-Varying Filter Bank Design Techniques

The basic time-varying filter bank design methods are summarized in Table 37.1. These techniques can be divided into two major approaches which are briefly described in the following sections.

37.4.1 Approach I: Intermediate Analysis-Synthesis

In the first approach, both analysis and synthesis filters are allowed to change during the transition period to maintain perfect reconstruction. We refer to this approach as the intermediate analysis-synthesis (IAS) approach.

In [16], the authors have chosen to start with the lattice implementation of time-invariant two-band filter banks, originally proposed by Vaidyanathan [25] for time-invariant case. Consider the lattice structure shown in Figure 37.6. Figure 37.6a represents a lossless two-band analysis filter bank, consisting of $J+1$ lattice stages. The corresponding synthesis filter bank is shown in Figure 37.6b. As is shown,

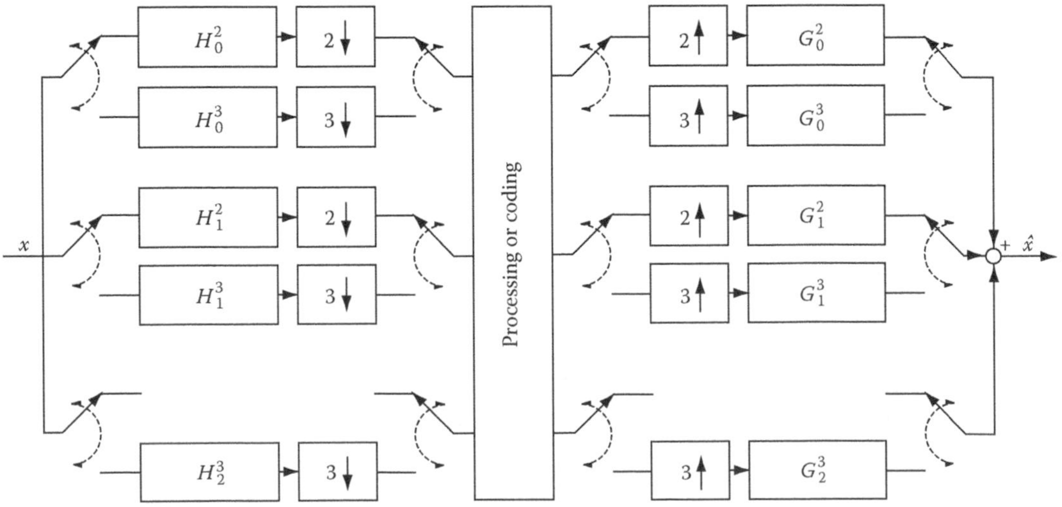

FIGURE 37.4 Block diagram of a time-varying analysis/synthesis filter bank that switches between a two- and three-band decomposition.

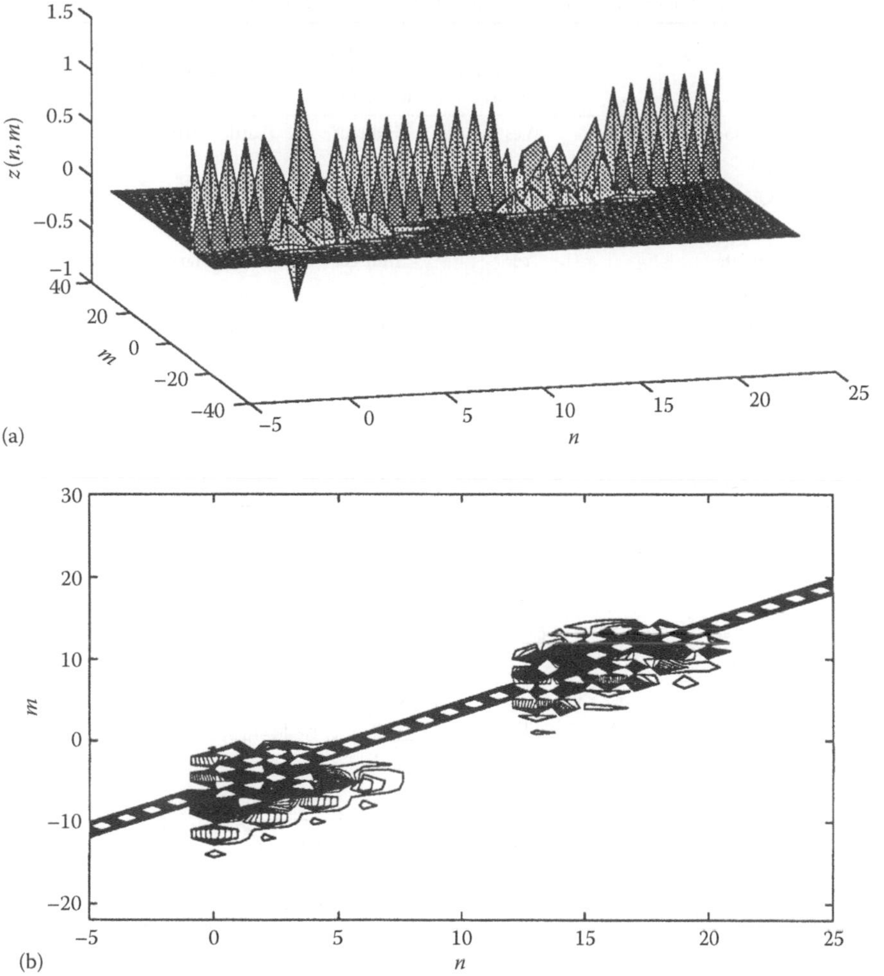

FIGURE 37.5 The time-varying impulse response for direct switching between the two- and the three-band system. The filter bank is switched from the two-band to the three-band at time $n = 0$ and switched back at time $n = 13$: (a) surface plot and (b) contour plot.

for each stage in the analysis filter bank, there exists a corresponding stage in the synthesis filter bank with similar, but inverse functionality. As long as each two corresponding lattice stages in the analysis and synthesis sections are PR, the overall system is PR. To switch one filter bank to another, the lattice stages of the analysis section are changed from one set to another. If the corresponding lattice stages of the synthesis section are also changed according to the changes of the analysis section, the PR property will hold during transition. Due to the existence of delay elements, any change in the analysis section must be followed with the corresponding change in the synthesis section, but with an appropriate delay. For example, the parameter α_j of the analysis and synthesis filter banks can be changed instantaneously. But any change in parameter α_{j-1} in the analysis filter bank must be followed with the similar change in the synthesis filter bank after one sample delay. Because of such delays, switching between two PR filter banks can occur only by going through a transition period in which both analysis and synthesis filter banks are changing in time.

In [12,26], the design of time-varying extended lapped transform (ELT) [27,28] was reported. The ELT is a cosine-modulated filter bank with an additional constraint on the filter lengths. Here, the design

TABLE 37.1 Comparison of Time-Varying Filter Bank Different Designing Methods

		Intermediate Analysis	Changing Frequency Resolution	Filter Bank Requirement	Computational Complexity
Intermediate analysis	Arrowood Smith	Yes	Indirect	Lattice structures	Low
	de Queiroz Rao	Yes	Indirect	ELT	Low
	Gopinath Burrus	Yes	Indirect	Paraunitary	Low
Synthesis	Herley et al.	Yes	Direct	Paraunitary	Low
IAS	Chen Vaidyanathan	Yes	Direct	Noncausal synthesis	Low
ITS	LS synthesis	No	Direct	General (not PR)	Low
	Redesigning analysis	No	Direct	General	High
	Post filtering	No	Direct	General	Low

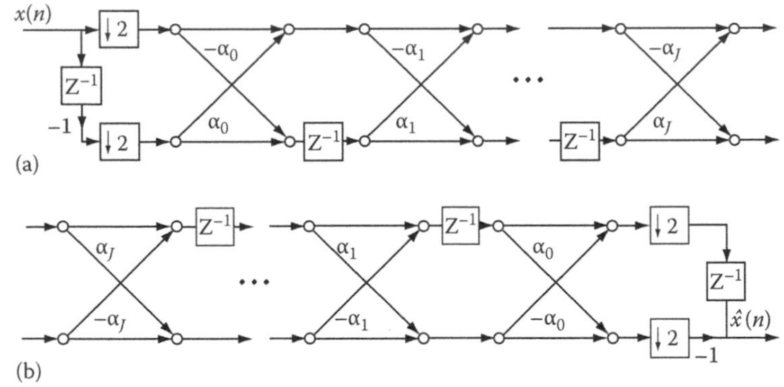

FIGURE 37.6 The block diagram of a two-band paraunitary filter bank in lattice form: (a) analysis lattice and (b) synthesis lattice.

procedure is based on factorization of the time-domain transform matrix into permutation and rotation matrices. As the ELT is paraunitary, the inverse transform can be obtained by reversing the order of the matrix multiplication. Since any orthogonal transform is a succession of plane rotations, any changes in these rotation angles result in changing the filter bank without losing the orthogonality property. The authors derived a general frame work for M-band ELT transforms compared to the two-band case approach in [16]. This method parallels the lattice technique [16] except with the mild modification of imposing the additional ELT constraints. In [17], the authors presented yet another formulation for designing time-varying filter banks. In this chapter, a different factorization of the paraunitary transform has been shown which is not based on plane rotations unlike the ones in [12,26]. Using this factorization, a paraunitary filter bank can be implemented in the form of some cascade structures. Again, to switch one filter bank to another, the corresponding structures in the analysis and synthesis filter bank are changed similarly but with an appropriate delay. If the orthogonality property in each cascade structure is maintained, the time-varying filter bank remains PR. This formulation is very similar to the ones in [12,16,26], but represent a more general form of factorization. In fact, all above procedures consider similar frameworks of structures that inherently guarantee the exact reconstruction.

Herley et al. [13–15,29] introduced a time-domain method for designing time-varying paraunitary filter banks. In this approach, the time-invariant analysis transforms do not overlap. As a simple example, consider the case of switching between two paraunitary time-invariant filter banks. The analysis transform around the transition period can be written as

$$\mathbf{T} = \begin{bmatrix} [\quad \mathbf{P}_1 \quad] & & \\ & [\quad \mathbf{P}_T \quad] & \\ & & [\quad \mathbf{P}_2 \quad] \end{bmatrix}. \tag{37.15}$$

The matrices $\mathbf{P}_1$ and $\mathbf{P}_2$ represent paraunitray transforms and therefore are unitary matrices. Their nonzero columns also do not overlap with each other. The matrix $\mathbf{P}_T$ represents the analysis filter bank during the transition period. In order to find this filter bank, the matrix $\mathbf{P}_T$ is initially replaced with a zero matrix. Then, the null space of the transform $\mathbf{T}$ is found. Any matrix that spans this subspace can be a candidate vector for $\mathbf{P}_T$. By choosing enough independent vectors of this null space and applying the Gram–Schimidt procedure to them, an orthogonal transform can be selected for $\mathbf{P}_T$. This method has also been applied to time-varying modulated lapped transforms [24] and two-dimensional time-varying paraunitary filter banks [30].

The basic property of all above procedures is the use of intermediate analysis transforms in the transition period. The characteristics of these analysis transforms are not easy to control and typically the intermediate filters are not well behaved.

37.4.2 Approach II: Instantaneous Transform Switching

In the second approach, the analysis filters are switched instantaneously and time-varying synthesis filters are used in the transition period. We refer to this approach as the instantaneous transform switching (ITS) approach. In the ITS approach, the analysis filter bank may be switched to another set of analysis filters arbitrarily. This means that the basis vectors and the tiling of the time-frequency plane can be changed instantaneously. To achieve PR at each time in the transition period, a new synthesis section is designed to ensure proper reconstruction.

In the least squares (LS) method [10], for any given set of analysis filters, an LS solution of Equation 37.14 can be used to obtain the "best" synthesis filters of the corresponding system (in $L2$ norm):

$$\mathbf{s}(n)^{\mathrm{LS}} = (\mathbf{A}(n)^{\mathrm{T}}\mathbf{A}(n))^{-1}\mathbf{A}(n)^{\mathrm{T}}\mathbf{b}(n). \tag{37.16}$$

The advantage of the LS approach is that there is no limitation on the number of analysis filter banks that can be used in the system. The disadvantage of the LS method is that it does not achieve PR. However, experiments have shown that the reconstruction is significantly improved in this method compared to direct switching [10].

In the LS solution, $\mathbf{b}(n)$ is projected onto the column space of $\mathbf{A}(n)$. For PR, the projection error should be zero. Thus, to obtain time-varying PR filter banks, the reconstruction error, $\|\mathbf{A}(n)\mathbf{s}(n) - \mathbf{b}(n)\|^2$, can be brought to zero with an optimization procedure. The optimization operates on the analysis filter coefficients and modifies the range space of $\mathbf{A}(n)$ until $\mathbf{b}(n) \in \text{range}[\mathbf{A}(n)]$. Although the $\mathbf{s}(n)$'s at different states are independent of each other, since the $\mathbf{A}(n)$'s have some common elements, optimization procedures should be applied to all analysis sections at the same time. This method is referred to as "redesigning analysis" [10].

The last ITS method, post filtering, uses conventional filter banks with time-varying coefficients followed by a time-varying post filter. The post filter provides exact reconstruction during transition periods, while it operates as a constant delay elsewhere. Assume at time n_0 the time-varying filter bank is switched from the first filter bank to the second. If the length of the transition period is L samples,

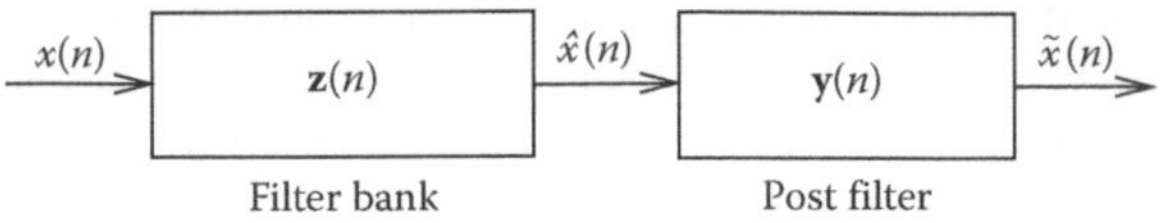

FIGURE 37.7 The block diagram of time-varying filter bank and post filter.

the output of the filter bank in the interval $[n_0, n_0+L-1]$ is distorted because of switching. The post filter removes this distortion. The block diagram of such a system is shown in Figure 37.7. In this figure, $\mathbf{z}(n)$ and $\mathbf{y}(n)$ are the analysis/synthesis filter bank and post filter impulse responses, respectively. If the delays of the filter bank and the post filter are denoted Δ and Θ, respectively, we can write

$$\hat{x}(n) = \begin{cases} \text{Distorted} & \text{if } n_0 \leq n < n_0 + L \\ x(n-\Delta) & \text{otherwise.} \end{cases} \tag{37.17}$$

The desired output of the post filter is

$$\tilde{x}(n) = x(n - \Theta - \Delta). \tag{37.18}$$

The input/output relation of the time-varying filter bank during the transition period can be written as

$$\hat{x}(n) = \mathbf{z}^{\mathrm{T}}(n)\mathbf{x}_I(n), \tag{37.19}$$

where

$\mathbf{x}_I(n)$ is the input vector at time n:

$$\mathbf{x}_I(n) = [x(n), x(n-1), x(n-2), \ldots, x(n-I+1)]^{\mathrm{T}} \text{ and}$$

$\mathbf{z}(n)$ is a vector of length I and represents the time-varying impulse response of the filter bank at time n.

If the transition impulse response matrix is defined to be

$$\mathbf{Z} = \begin{bmatrix} [\mathbf{z}(n_0+L-1)] & O & & O \\ O & [\mathbf{z}(n_0+L-2)] & & O \\ O & O & \ddots & \vdots \\ \vdots & \vdots & & O \\ O & O & & [\mathbf{z}(n_0)] \end{bmatrix}, \tag{37.20}$$

then the input/output relation of the filter bank in the transition period can be described as

$$\hat{\mathbf{x}}_L(n_0+L-1) = \mathbf{Z}^{\mathrm{T}}\mathbf{x}_K(n_0+L-1) \tag{37.21}$$

where $\mathbf{Z}$ is a $K \times L$ matrix and $K = I + L - 1$. In Equation 37.21, the $I - \Delta - 1$ samples before and Δ samples after the transition period are used to evaluate the output. The above intervals are called the tail and head of the transition period, respectively. Since the first and second filter banks are PR, the tail and head samples are exactly reconstructed. We write $\mathbf{x}_K(n_0+L-1)$ as the concatenation of three vectors:

$$\mathbf{x}_K(n_0+L-1) = \begin{bmatrix} \mathbf{x}_\mathrm{a} \\ \mathbf{x}_\mathrm{t} \\ \mathbf{x}_\mathrm{b} \end{bmatrix}, \tag{37.22}$$

where $\mathbf{x}_a$ and $\mathbf{x}_b$ are the input signals in the head and tail regions while $\mathbf{x}_t$ represents the input samples which are distorted during the transition period. Using this notation, Equation 37.21 can be written as

$$\hat{\mathbf{x}}_L(n_0 + L - 1) = \mathbf{Z}_a^T\mathbf{x}_a + \mathbf{Z}_t^T\mathbf{x}_t + \mathbf{Z}_b^T\mathbf{x}_b, \tag{37.23}$$

where

$$\mathbf{Z} = \begin{bmatrix} \mathbf{Z}_a \\ \mathbf{Z}_t \\ \mathbf{Z}_b \end{bmatrix}. \tag{37.24}$$

By replacing vectors $\mathbf{x}_b$ and $\mathbf{x}_a$ with their corresponding output vectors $\hat{\mathbf{x}}_a$ and $\hat{\mathbf{x}}_b$, $\mathbf{x}_t$ of Equation 37.23 can be written as

$$\begin{aligned}\mathbf{x}_t &= \left(\mathbf{Z}_t^T\right)^{-1}\left(\hat{\mathbf{x}}_t - \mathbf{Z}_a^T\hat{\mathbf{x}}_a - \mathbf{Z}_b^T\hat{\mathbf{x}}_b\right) \\ &= \mathbf{Y}^T\hat{\mathbf{x}}_K.\end{aligned} \tag{37.25}$$

Equation 37.25 describes the post filter input/output relationship during the transition region. In this equation, $\mathbf{Y}$ is the time-varying post filter impulse response which is defined as

$$\mathbf{Y} = \begin{bmatrix} -\mathbf{Z}_a\mathbf{Z}_t^{-1} \\ \mathbf{Z}_t^{-1} \\ -\mathbf{Z}_b\mathbf{Z}_t^{-1} \end{bmatrix}. \tag{37.26}$$

From Equation 37.25, it is obvious that the condition for causal post filtering is

$$\Theta \geq L + \Delta - 1. \tag{37.27}$$

The post filter exists if $\mathbf{Z}_t$ has an inverse. It can be shown that the transition response matrix $\mathbf{Z}_t$, can be described by a matrix, product of the form

$$\mathbf{Z}_t = \Psi_L\mathbf{S}, \tag{37.28}$$

where

Ψ_L is the analysis transform applied to those input samples that are distorted during the transition period

$\mathbf{S}$ contains the synthesis filters during the transition period

In order for $\mathbf{Z}_t$ to be invertible, it is necessary (but not sufficient) that Ψ_L and $\mathbf{S}$ be full rank matrices. The analysis sections are defined by the required properties of the first and second filter banks and Ψ_L is fixed. Therefore, a filter bank is switchable to another filter bank if the corresponding Ψ_L is a full rank matrix. In this case, by proper design of the synthesis section, both $\mathbf{S}$ and $\mathbf{Z}_t$ will be full rank. Two methods to obtain proper synthesis filters are shown in [20,22].

37.5 Conclusion

In this chapter, we briefly review some analysis and design methods of time-varying filter banks. Time-varying filter banks can provide a more flexible and accurate approach in which the basis functions of the time-frequency transform are allowed to adapt to the signal properties.

A simple form of time-varying filter bank is achieved by changing the filters of an analysis-synthesis system among a number of choices. Even if all the analysis and synthesis filters are PR sets, exact reconstruction will not normally be achieved during the transition periods. To eliminate all distortion during a transition period, new time-varying analysis and/or synthesis sections are required for the transition periods.

Two different approaches for the design were discussed here. In the first approach, both analysis and synthesis filters are allowed to change during the transition period to maintain PR and so it is called the intermediate analysis-synthesis approach. In the second approach, the analysis filters are switched instantaneously and time-varying synthesis filters are used in the transition period. This approach is known as the instantaneous transform switching approach.

In the IAS approach, both analysis and synthesis filters can change during the transitions rather than only the synthesis filters in ITS approach. That implies that maintaining PR conditions is easier in the IAS approach. Note that the analysis filters in the transition periods are designed only to satisfy PR conditions and they do not usually meet the desired time and frequency characteristics.

In the ITS approach, only synthesis filters are allowed to be time-varying in the transition periods. These methods have the advantage of providing instantaneous switching between the analysis transforms compared to IAS methods. But they have different drawbacks: the LS method does not satisfy PR conditions at all times, the redesigning analysis method requires jointly optimization of the time-invariant analysis section, and finally the post filtering method has the drawback of additional computational complexity required for post filtering.

The analysis and design methods of the time-varying filter bank have been developed to design adaptive time-frequency transforms. These adaptive transforms have many potential applications in areas such as TFR, subband image and video coding, and speech and audio coding. But since the developments of the time-varying filter bank theory is very new, its applications have not been investigated yet.

References

1. Allen, J.B., Short-term spectral analysis, synthesis, and modification by discrete Fourier transform, *IEEE Trans. Acoust. Speech Signal Process.*, 25, 235–238, June 1977.
2. Allen, J.B. and Rabiner, L.R., A unified approach to STFT analysis and synthesis, *Proc. IEEE*, 65, 1558–1564, Nov. 1977.
3. Rabiner, L.R. and Schafer, R.W., *Digital Processing of Speech Signals*, Prentice-Hall, Englewood Cliffs, NJ, 1978.
4. Portnoff, M.R., Time-frequency representation of digital signals and systems based on short-time Fourier analysis, *IEEE Trans. Acoust. Speech Signal Process.*, 55–69, Feb. 1980.
5. Nawab, S.N. and Quatieri, T.F., *Short-Time Fourier Transform, Chapter in Advanced Topics in Signal Processing*, Prentice-Hall, Englewood Cliffs, NJ, 1988.
6. Gabor, D., Theory of communication, *J. IEE (London)*, 93(III), 429–457, Nov. 1946.
7. Nayebi, K., Barnwell, T.P., and Smith, M.J.T., Analysis-synthesis systems with time-varying filter bank structures, *Proceedings of the International Conference on Acoustics, Speech, and Signal Processing*, Toronto, ON, Canada, Mar. 1991.
8. Nayebi, K., Sodagar, I., and Barnwell, T.P., III, The wavelet transform and time-varying tiling of the time-frequency plane, *IEEE-SP International Symposium on Time-Frequency and Time-Scale Analysis*, Victoria, BC, Canada, Oct. 4–6, 1992, pp. 147–150.
9. Sodagar, I., Nayebi, K., and Barnwell, T.P., III, A class of time-varying wavelet transforms, *Proceedings of the International Conference on Acoustics, Speech, and Signal Processing*, Minneapolis, MN, Apr. 27–30, 1993, Vol. 3, pp. 201–204.
10. Sodagar, I., Nayebi, K., Barnwell, T.P., and Smith, M.J.T., Time-varying filter banks and wavelets, *IEEE Trans. Signal Process.*, 42(11): 2983–2996, Nov. 1994.
11. Sodagar, I., Analysis and design of time-varying filter banks, PhD thesis, Georgia Institute of Technology, Atlanta, GA, Dec. 1994.

12. de Queiroz, R.L. and Rao, K.R., Adaptive extended lapped transforms, *Proceedings of the International Conference on Acoustics, Speech, and Signal Processing*, Minneapolis, MN, Apr. 27–30, 1993, Vol. 3, pp. 217–220.
13. Herley, C., Kovacevic, J., Ramchandran, K., and Vetterli, M., Arbitrary orthogonal tilings of the time-frequency plane, *IEEE-SP International Symposium on Time-Frequency and Time-Scale Analysis*, Victoria, BC, Canada, Oct. 1992, pp. 11–14.
14. Herley, C. and Vetterli, M., Orthogonal time-varying filter banks and wavelets, *Proceedings of the International Symposium on Circuits and Systems*, Chicago, IL, May 3–6, 1993, Vol. 1, pp. 391–394.
15. Herley, C., Wavelets and filter banks, PhD thesis, Columbia University, New York, 1993.
16. Arrowood, J.L. and Smith, M.J.T., Exact reconstruction analysis/synthesis filter banks with time-varying filters, *Proceedings of the International Conference on Acoustics, Speech, and Signal Processing*, Minneapolis, MN, Apr. 27–30, 1993, Vol. 3, pp. 233–236.
17. Gopinath, R.A., Factorization approach to time-varying filter banks and wavelets, *Proceedings of the International Conference on Acoustics, Speech, and Signal Processing*, Adelaide, Australia, Apr. 19–22, 1994, Vol. 3, pp. III/109–III/112.
18. Chen, T. and Vaidyanathan, P.P., Time-reversed inversion for time-varying filter banks, *Proceedings of the 27th Asilomar Conference on Signals, Systems, and Computers*, Pacific Grove, CA, Nov. 1–3, 1993, Vol. 1, pp. 55–59.
19. Phoong, S. and Vaidyanathan, P.P., On the study of lossless time-varying filter banks, *Proceedings of the 29th Asilomar Conference on Signals, Systems, and Computers*, Pacific Grove, CA, Oct. 30–Nov. 1, 1995, Vol. 1, pp. 51–55.
20. Sodagar, I., Nayebi, K., Barnwell, T.P., III, and Smith, M.J.T., A new approach to time-varying FIR filter banks, *Proceedings of the 27th Asilomar Conference on Signals, Systems, and Computers*, Pacific Grove, CA, Nov. 1–3, 1993, Vol. 2, pp. 1271–1275.
21. Sodagar, I., Nayebi, K., Barnwell, T.P., and Smith, M.J.T., A novel structure for time-varying FIR filter banks, *Proceedings of the International Conference on Acoustics, Speech, and Signal Processing*, Adelaide, Australia, Apr. 19–22, 1994, Vol. 3, pp. 157–160.
22. Sodagar, I., Nayebi, K., and Barnwell, T.P., and Smith, M.J.T., Time-varying analysis-synthesis systems based on filter banks and post filtering, *IEEE Trans. Signal Processing*, 43(11), 2512–2524, Nov. 1995.
23. Sodagar, I., Nayebi, K., Barnwell, T.P., and Smith, M.J.T., Perfect reconstruction multidimensional filter banks with time-varying basis functions, *Proceedings of the 27th Asilomar Conference on Signals, Systems, and Computers*, Pacific Grove, CA, Nov. 1–3, 1993, Vol. 1, pp. 50–54.
24. Kovacevic, J. and Vetterli, M., Time-varying modulated lapped transforms, *Proceedings of the 27th Asilomar Conference on Signals, Systems, and Computers*, Pacific Grove, CA, Nov. 1–3, 1993, Vol. 1, pp. 481–485.
25. Vaidyanathan, P.P., Theory and design of M channel maximally decimated QMF with arbitrary M, having perfect reconstruction property, *IEEE Trans. Acoust. Speech Signal Process.*, 35(4), 476–492, Apr. 1987.
26. de Queiroz, R.L. and Rao, K.R., Time-varying lapped transforms and wavelet packets, *IEEE Trans. Signal Process.*, 41(12), 3293–3305, Dec. 1993.
27. Malvar, H.S. and Staelin, D.H., The LOT: Transform coding without blocking effects, *IEEE Trans. Acoust. Speech Signal Process.*, 37(4), 553–559, Apr. 1989.
28. Malvar, H.S., Lapped transforms for efficient transform/subband coding, *IEEE Trans. Acoust. Speech Signal Process.*, 38(6), 969–978, June 1990.
29. Herley, C. and Vetterli, M., Orthogonal time-varying filter banks and wavelet packets, *IEEE Trans. Signal Process.*, 42(10), 2650–2663, Oct. 1994.
30. Herley, C. and Kovacevic, J., Spatially varying two-dimensional filter banks, *Proceedings of the 27th Asilomar Conference on Signals, Systems, and Computers*, Pacific Grove, CA, Nov. 1–3, 1993, Vol. 1, pp. 60–64.

38

Lapped Transforms

Ricardo L. de Queiroz
Universidade de Brasilia

38.1 Introduction

The idea of a lapped transform (LT) maintaining orthogonality and non-expansion of the samples was developed in the early 1980s at MIT by a group of researchers unhappy with the blocking artifacts so common in traditional block transform coding of images. The idea was to extend the basis function beyond the block boundaries, creating an overlap, in order to eliminate the blocking effect. This idea was not new, but the new ingredient to overlapping blocks would be the fact that the number of transform coefficients would be the same as if there was no overlap, and that the transform would maintain orthogonality. Cassereau [1] introduced the lapped orthogonal transform (LOT), and Malvarn [5,6,13] gave the LOT its design strategy and a fast algorithm. The equivalence between an LOT and a multirate filter bank was later pointed out by Malvar [7]. Based on cosine modulated filter banks [15], modulated lapped transforms (MLTs) were designed [8,25]. Modulated transforms were generalized for an arbitrary overlap later creating the class of extended lapped transforms (ELTs) [9–12]. Recently a new class of LTs with symmetric bases was developed yielding the class of generalized LOTs (GenLOTs) [16,20,21]. As we mentioned, filter banks and LTs are the same, although studied independently in the past. We, however, refer to LTs for paraunitary uniform FIR filter banks with fast implementation algorithms based on special factorizations of the basis functions.

We assume a one-dimensional input sequence $x(n)$ which is transformed into several coefficients $y_i(n)$, where $y_i(n)$ would belong to the ith subband. We also will use the discrete cosine transform [24] and another cosine transform variation, which we abbreviate as DCT and DCT-IV (DCT type 4), respectively [24].

38.2 Orthogonal Block Transforms

In traditional block-transform processing, such as in image and audio coding, the signal is divided into blocks of M samples, and each block is processed independently [2,3,11,14,22–24]. Let the samples in the mth block be denoted as

$$\mathbf{x}_m^{\mathrm{T}} = [x_0(m), x_1(m), \ldots, x_{M-1}(m)], \tag{38.1}$$

for $x_k(m) = x(mM + k)$ and let the corresponding transform vector be

$$\mathbf{y}_m^{\mathrm{T}} = [y_0(m), y_1(m), \ldots, y_{M-1}(m)]. \tag{38.2}$$

For a real unitary transform **A**, $\mathbf{A}^{\mathrm{T}} = \mathbf{A}^{-1}$. The forward and inverse transforms for the mth block are

$$\mathbf{y}_m = \mathbf{A}\mathbf{x}_m, \tag{38.3}$$

and

$$\mathbf{x}_m = \mathbf{A}^{\mathrm{T}}\mathbf{y}_m. \tag{38.4}$$

The rows of **A**, denoted $\mathbf{a}_n^{\mathrm{T}} (0 \leq n \leq M-1)$, are called the basis vectors because they form an orthogonal basis for the M-tuples over the real field [23]. The transform vector coefficients $[y_0(m), y_1(m), \ldots, y_{M-1}(m)]$ represent the corresponding weights of vector $\mathbf{x}_m$ with respect to this basis.

If the input signal is represented by vector **x** while the subbands are grouped into blocks in vector **y**, we can represent the transform **T** which operates over the entire signal as a block diagonal matrix:

$$\mathbf{T} = \mathrm{diag}\{\ldots, \mathbf{A}, \mathbf{A}, \mathbf{A}, \ldots\}, \tag{38.5}$$

where, of course, **T** is an orthogonal matrix.

38.2.1 Orthogonal Lapped Transforms

For LTs [11], the basis vectors can have length L, such that $L > M$, extending across traditional block boundaries. Thus, the transform matrix is no longer square and most of the equations valid for block transforms do not apply to an LT. We will concentrate our efforts on orthogonal LTs [11] and consider $L = NM$, where N is the overlap factor. Note that N, M, and hence L are all integers. As in the case of block transforms, we define the transform matrix as containing the orthonormal basis vectors as its rows. An LT matrix **P** of dimensions $M \times L$ can be divided into square $M \times M$ submatrices $\mathbf{P}_i$ $(i = 0, 1, \ldots, N-1)$ as

$$\mathbf{P} = [\mathbf{P}_0\ \mathbf{P}_1 \cdots \mathbf{P}_{N-1}]. \tag{38.6}$$

The orthogonality property does not hold because **P** is no longer a square matrix and it is replaced by other properties which we will discuss later.

If we divide the signal into blocks, each of size M, we would have vectors $\mathbf{x}_m$ and $\mathbf{y}_m$ such as in Equations 38.1 and 38.2. These blocks are not used by LTs in a straightforward manner. The actual vector which is transformed by the matrix **P** has to have L samples and, at block number m, it is composed of the samples of $\mathbf{x}_m$ plus $L - M$ samples. These samples are chosen by picking $(L - M)/2$ samples at each side of the block $\mathbf{x}_m$, as shown in Figure 38.1, for $N = 2$. However, the number of transform

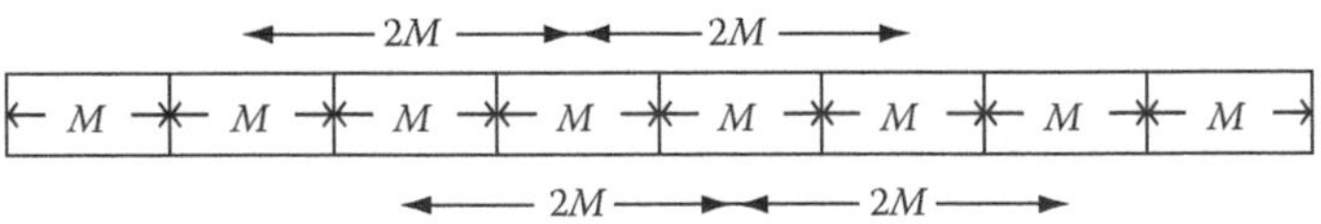

FIGURE 38.1 The signal samples are divided into blocks of M samples. The LT uses neighboring block samples, as in this example for $N = 2$, i.e., $L = 2M$, yielding an overlap of $(L - M)/2 = M/2$ samples on either side of a block.

coefficients at each step is M, and, in this respect, there is no change in the way we represent the transform-domain blocks $\mathbf{y}_m$.

The input vector of length L is denoted as $\mathbf{v}_m$, which is centered around the block $\mathbf{x}_m$, and is defined as

$$\mathbf{v}_m^{\mathrm{T}} = \left[x\left(mM - (N-1)\frac{M}{2}\right) \cdots x\left(mM + (N+1)\frac{M}{2} - 1\right)\right]. \tag{38.7}$$

Then, we have

$$\mathbf{y}_m = \mathbf{P}\mathbf{v}_m. \tag{38.8}$$

The inverse transform is not direct as in the case of block transforms, i.e., with the knowledge of $\mathbf{y}_m$ we do not know the samples in the support region of $\mathbf{v}_m$, and neither in the support region of $\mathbf{x}_m$. We can reconstruct a vector $\hat{\mathbf{v}}_m$ from $\mathbf{y}_m$, as

$$\hat{\mathbf{v}}_m = \mathbf{P}^{\mathrm{T}}\mathbf{y}_m, \tag{38.9}$$

where $\hat{\mathbf{v}}_m \neq \mathbf{v}_m$. To reconstruct the original sequence, it is necessary to accumulate the results of the vectors $\hat{\mathbf{v}}_m$, in a sense that a particular sample $x(n)$ will be reconstructed from the sum of the contributions it receives from all $\hat{\mathbf{v}}_m$, such that $x(n)$ was included in the region of support of the corresponding $\mathbf{v}_m$. This additional complication comes from the fact that $\mathbf{P}$ is not a square matrix [11]. However, the whole analysis-synthesis system (applied to the entire input vector) is orthogonal, assuring the PR property using Equation 38.9.

We can also describe the process using a sliding rectangular window applied over the samples of $x(n)$. As an M-sample, block $\mathbf{y}_m$ is computed using $\mathbf{v}_m$, $\mathbf{y}_{m+1}$ is computed from $\mathbf{v}_{m+1}$ which is obtained by shifting the window to the right by M samples, as shown in Figure 38.2.

As the reader may have noticed, the region of support of all vectors $\mathbf{v}_m$ is greater than the region of support of the input vector. Hence, a special treatment has to be given to the transform at the borders. We will discuss this fact later and assume infinite-length signals until then, or assume the length is very large and the borders of the signal are far enough from the region to which we are focusing our attention.

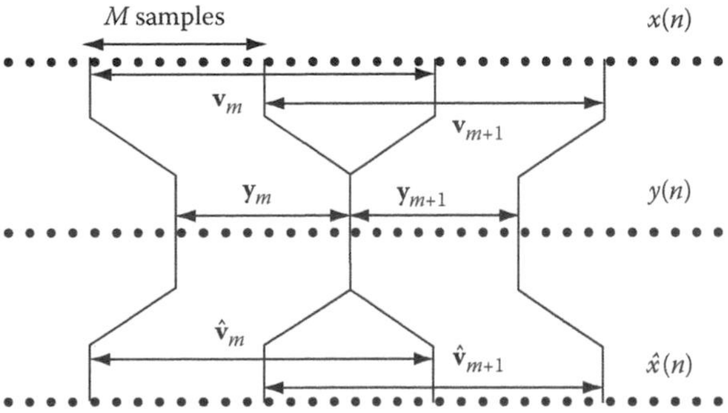

FIGURE 38.2 Illustration of an LT with $N=2$ applied to signal $x(n)$, yielding transform domain signal $y(n)$. The input L-tuple as vector $\mathbf{v}_m$ is obtained by a sliding window advancing M samples, generating $\mathbf{y}_m$. This sliding is also valid for the synthesis side.

If we denote by $\mathbf{x}$ the input vector and by $\mathbf{y}$ the transform-domain vector, we can be consistent with our notation of transform matrices by defining a matrix $\mathbf{T}$ such that $\mathbf{y} = \mathbf{T}\mathbf{x}$ and $\hat{\mathbf{x}} = \mathbf{T}^{\mathrm{T}}\mathbf{y}$. In this case, we have

$$\mathbf{T} = \begin{bmatrix} \ddots & & & \\ & \mathbf{P} & & \\ & & \mathbf{P} & \\ & & & \mathbf{P} \\ & & & & \ddots \end{bmatrix}, \tag{38.10}$$

where the displacement of the matrices $\mathbf{P}$ obeys the following:

$$\mathbf{T} = \begin{bmatrix} \ddots & \ddots & & \ddots & & \\ & \mathbf{P}_0 & \mathbf{P}_1 & \cdots & \mathbf{P}_{N-1} & \\ & & \mathbf{P}_0 & \mathbf{P}_1 & \cdots & \mathbf{P}_{N-1} \\ & & & \ddots & \ddots & & \ddots \end{bmatrix}. \tag{38.11}$$

$\mathbf{T}$ has as many block-rows as transform operations over each vector $\mathbf{v}_m$.

Let the rows of $\mathbf{P}$ be denoted by $1 \times L$ vectors $\mathbf{p}_i^T (0 \le i \le M-1)$, so that $\mathbf{P}^{\mathrm{T}} = [\mathbf{p}_0, \ldots, \mathbf{p}_{M-1}]$. In an analogy to the block transform case, we have

$$y_i(m) = \mathbf{p}_i^T \mathbf{v}_m. \tag{38.12}$$

The vectors $\mathbf{p}_i$ are the basis vectors of the LT. They form an orthogonal basis for an M-dimensional subspace (there are only M vectors) of the L-tuples over the real field.

Assuming that the entire input and output signals are represented by the vectors $\mathbf{x}$ and $\mathbf{y}$, respectively, and that the signals have infinite length, then, from Equation 38.10, we have

$$\mathbf{y} = \mathbf{T}\mathbf{x} \tag{38.13}$$

and, if $\mathbf{T}$ is orthogonal,

$$\mathbf{x} = \mathbf{T}^T\mathbf{y}. \tag{38.14}$$

The conditions for orthogonality of the LT are expressed as the orthogonality of $\mathbf{T}$. Therefore, the following equations are equivalent in a sense that they state the PR property along with the orthogonality of the LT:

$$\sum_{i=0}^{N-1-l} \mathbf{P}_i \mathbf{P}_{i+l}^T = \sum_{i=0}^{N-1-l} \mathbf{P}_i^T \mathbf{P}_{i+l} = \delta(l)\mathbf{I}_M \tag{38.15}$$

$$\mathbf{T}\mathbf{T}^T = \mathbf{T}^T\mathbf{T} = \mathbf{I}_\infty. \tag{38.16}$$

It is worthwhile to reaffirm that orthogonal LTs are a uniform maximally decimated FIR filter bank. Assume the filters in such a filter bank have L-tap impulse responses $f_i(n)$ and $g_i(n)$ $(0 \le i \le M-1, 0 \le n \le L-1)$, for the analysis and synthesis filters, respectively. If the filters originally have a length smaller than L, one can pad the impulse response with 0s until $L = NM$. In other words, we force the basis vectors to have a common length which is an integer multiple of the block size. Assume the entries of **P** are denoted by $\{p_{ij}\}$. One can translate the notation from LTs to filter banks by using

$$p_{kn} = f_k(L-1-n) = g_k(n). \tag{38.17}$$

38.3 Useful Transforms

38.3.1 Extended Lapped Transform

Cosine modulated filter banks are filter banks based on a low-pass prototype filter modulating a cosine sequence. By a proper choice of the phase of the cosine sequence, Malvar developed the MLT [8], which led to the so-called ELT [9–12]. The ELT allows several overlapping factors N, generating a family of LTs with good filter frequency response and fast implementation algorithm.

In the ELTs, the filter length L is basically an even multiple of the block size M, as $L = NM = 2kM$. The MLT-ELT class is defined by

$$p_{k,n} = h(n)\cos\left[\left(k+\frac{1}{2}\right)\left(\left(n-\frac{L-1}{2}\right)\frac{\pi}{M}+(N+1)\frac{\pi}{2}\right)\right] \tag{38.18}$$

for $k = 0, 1, \ldots, M-1$ and $n = 0, 1, \ldots, L-1$. $h(n)$ is a symmetric window modulating the cosine sequence and the impulse response of a low-pass prototype (with cutoff frequency at $\pi/2M$) which is translated in frequency to M different frequency slots in order to construct the uniform filter bank. The ELTs have as their major plus a fast implementation algorithm, which is depicted in Figure 38.3 in an example for $M = 8$. The free parameters in the design of an ELT are the coefficients of the prototype filter. Such degrees of freedom are translated in the fast algorithm as rotation angles.

For the case $N = 4$ there is a useful parameterized design [10–12]. In this design, we have

$$\theta_{k0} = -\frac{\pi}{2} + \mu_{M/2+k} \tag{38.19}$$

$$\theta_{k1} = -\frac{\pi}{2} + \mu_{M/2-1-k}, \tag{38.20}$$

where

$$\mu_i = \left[\left(\frac{1-\gamma}{2M}\right)(2k+1)+\gamma\right] \tag{38.21}$$

and γ is a control parameter, for $0 \leq k \leq (M/2)-1$. γ controls the trade-off between the attenuation and transition region of the prototype filter. For $N = 4$, the relation between angles and $h(n)$ is

$$h(k) = \cos(\theta_{k0})\cos(\theta_{k1}) \tag{38.22}$$

$$h(M-1-k) = \cos(\theta_{k0})\sin(\theta_{k1}) \tag{38.23}$$

$$h(M+k) = \sin(\theta_{k0})\cos(\theta_{k1}) \tag{38.24}$$

$$h(2M-1-k) = -\sin(\theta_{k0})\sin(\theta_{k1}) \tag{38.25}$$

for $k = 0, 1, \ldots, M/2-1$. See [11] for optimized angles for ELTs. Further details on ELTs can be found in [9–12,16].

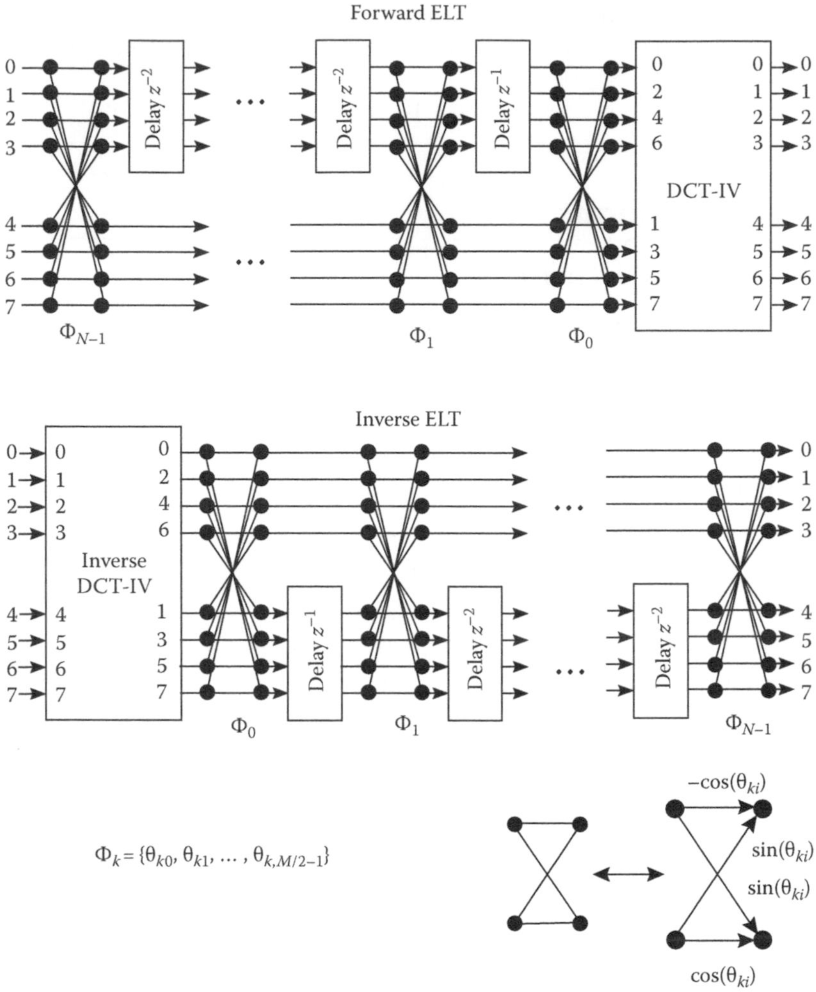

FIGURE 38.3 Implementation flow graph for the ELT with $M = 8$.

38.3.2 Generalized Linear-Phase Lapped Orthogonal Transform

The generalized linear-phase lapped orthogonal transform (GenLOT) is also a useful family of LTs possessing symmetric bases (linear-phase filters). The use of linear-phase filters is a popular requirement in image processing applications. Let

$$\mathbf{W} = \frac{1}{\sqrt{2}}\begin{bmatrix} \mathbf{I}_{M/2} & \mathbf{I}_{M/2} \\ \mathbf{I}_{M/2} & -\mathbf{I}_{M/2} \end{bmatrix} \quad \text{and} \quad \Psi_i = \begin{bmatrix} \mathbf{U}_i & 0_{M/2} \\ 0_{M/2} & \mathbf{V}_i \end{bmatrix}, \tag{38.26}$$

where $\mathbf{U}_i$ and $\mathbf{V}_i$ can be any $M/2 \times M/2$ orthogonal matrices. Let the transform matrix $\mathbf{P}$ for the GenLOT be constructed interactively. Let $\mathbf{P}^{(i)}$ be the partial reconstruction of $\mathbf{P}$ after including up to the ith stage. We start by setting $\mathbf{P}^{(0)} = \mathbf{E}_0$ where $\mathbf{E}_0$ is an orthogonal matrix with symmetric rows. The recursion is given by

$$\mathbf{P}^{(i)} = \Psi_i \mathbf{W}\mathbf{Z}\begin{bmatrix} \mathbf{W}\mathbf{P}^{(i-1)} & 0_M \\ 0_M & \mathbf{W}\mathbf{P}^{(i-1)} \end{bmatrix}, \tag{38.27}$$

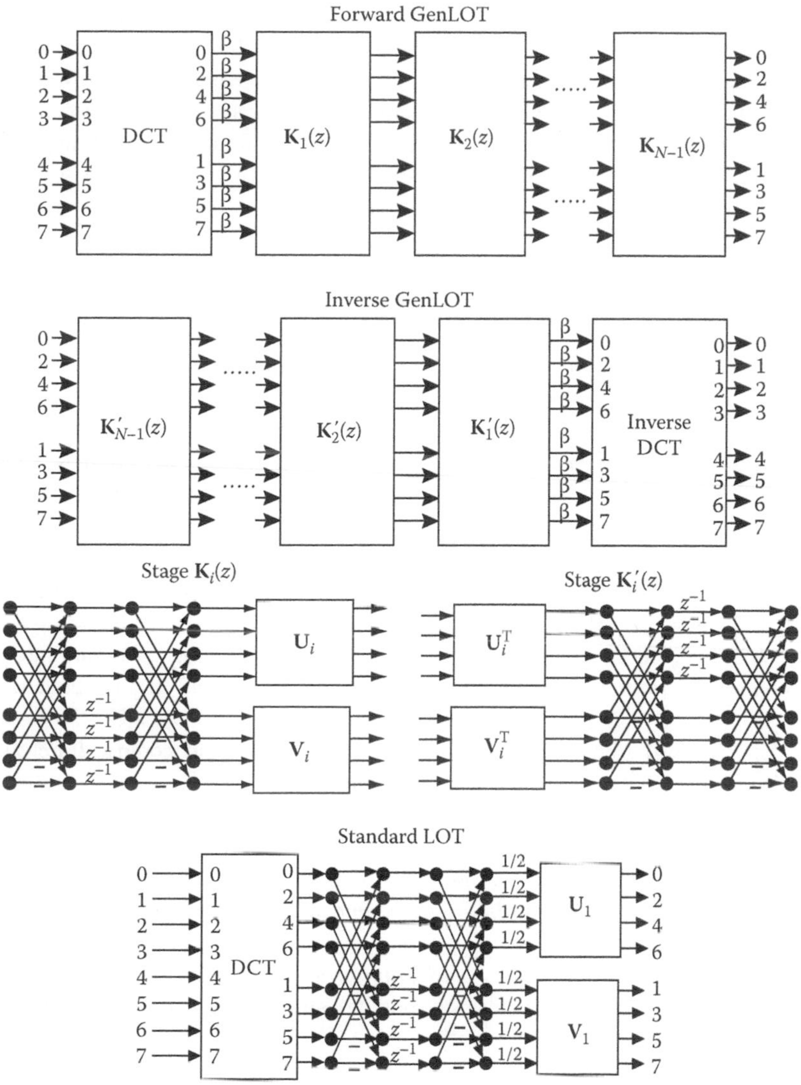

FIGURE 38.4 Implementation flow graph for the GenLOT with $M = 8$, where $\beta = 2^{N-1}$.

where

$$\mathbf{Z} = \begin{bmatrix} 0_{M/2} & 0_{M/2} & \mathbf{I}_{M/2} & 0_{M/2} \\ 0_{M/2} & \mathbf{I}_{M/2} & 0_{M/2} & 0_{M/2} \end{bmatrix}. \tag{38.28}$$

At the final stage we set $\mathbf{P} = \mathbf{P}^{(N-1)}$. $\mathbf{E}_0$ is usually the DCT while the other factors ($\mathbf{U}_i$ and $\mathbf{V}_i$) are found through optimization routines. More details on GenLOTs and their design can be found in [16,20,21]. The implementation flow-graph of a GenLOT with $M = 8$ is shown in Figure 38.4.

38.4 Remarks

We hope this chapter is helpful in understanding the basic concepts of LTs. Filter banks are covered in other parts of this book. An excellent book by Vaidyanathan [28] has a thorough coverage of such subject. The interrelations of filter banks and LTs are well covered by Malvar [11] and Queiroz [16]. For image processing and coding, it is necessary to process finite-length signals. As we discussed, such an issue is not so straightforward in a general case. Algorithms to implement LTs over finite-length signals are discussed in [11,13,16–19]. These algorithms can be general or specific. The specific algorithms are generally targeted to a particular LT invariantly seeking a very fast implementation. In general, Malvar's book [11] is an excellent reference for LTs and their related topics.

References

1. Cassereau, P., A new class of optimal unitary transforms for image processing, Master's thesis, MIT, Cambridge, MA, May 1985.
2. Clarke, R.J., *Transform Coding of Images*, Academic Press, Orlando, FL, 1985.
3. Jayant, N.S. and Noll, P., *Digital Coding of Waveforms*, Prentice-Hall, Englewood Cliffs, NJ, 1984.
4. Jozawa, H. and Watanabe, H., Intrafield/interfield adaptive lapped transform for compatible HDTV coding, *4th International Workshop on HDTV and Beyond*, Torino, Italy, Sept. 4–6, 1991.
5. Malvar, H.S., Optimal pre- and post-filtering in noisy sampled-data systems, PhD dissertation, MIT, Cambridge, MA, Aug. 1986.
6. Malvar, H.S., Reduction of blocking effects in image coding with a lapped orthogonal transform, *Proceeding of International Conference on Acoustics, Speech, and Signal Processing*, Glasgow, Scotland, U.K., Apr. 1988, pp. 781–784.
7. Malvar, H.S., The LOT: A link between block transform coding and multirate filter banks, *Proceedings of International Symposium on Circuits and Systems*, Espoo, Finland, June 1988, pp. 835–838.
8. Malvar, H.S., Lapped transforms for efficient transform/subband coding, *IEEE Trans. Acoust. Speech Signal Process.*, ASSP-38, 969–978, June 1990.
9. Malvar, H.S., Modulated QMF filter banks with perfect reconstruction, *Electron. Lett.*, 26, 906–907, June 1990.
10. Malvar, H.S., Extended lapped transform: Fast algorithms and applications, *Proceedings of International Conference on Acoustics, Speech, and Signal Processing*, Toronto, Canada, 1991, pp. 1797–1800.
11. Malvar, H.S., *Signal Processing with Lapped Transforms*, Artech House, Norwood, MA, 1992.
12. Malvar, H.S., Extended lapped transforms: Properties, applications and fast algorithms, *IEEE Trans. Signal Process.*, 40, 2703–2714, Nov. 1992.
13. Malvar, H.S. and Staelin, D.H., The LOT: Transform coding without blocking effects, *IEEE Trans. Acoust. Speech Signal Process.*, ASSP-37, 553–559, Apr. 1989.
14. Pennebaker, W.B. and Mitchell, J.L., *JPEG: Still Image Compression Standard*, Van Nostrand Reinhold, New York, 1993.
15. Princen, J.P. and Bradley, A.B., Analysis/synthesis filter bank design based on time domain aliasing cancellation, *IEEE Trans. Acoust. Speech Signal Process.*, ASSP-34, 1153–1161, Oct. 1986.
16. de Queiroz, R.L., On lapped transforms, PhD dissertation, University of Texas, Arlington, TX, Aug. 1994.
17. de Queiroz, R.L. and Rao, K.R., Time-varying lapped transforms and wavelet packets, *IEEE Trans. Signal Process.*, 41, 3293–3305, Dec. 1993.
18. de Queiroz, R.L. and Rao, K.R., The extended lapped transform for image coding, *IEEE Trans. Image Process.*, 4, 828–832, June 1995.
19. de Queiroz, R.L. and Rao, K.R., On orthogonal transforms of images using paraunitary filter banks, *J. Vis. Commn. Image Representation*, 6(2), 142–153, June 1995.
20. de Queiroz, R.L., Nguyen, T.Q., and Rao, K.R., The generalized lapped orthogonal transforms, *Electron. Lett.*, 30(2), 107–108, Jan. 1994.

21. de Queiroz, R.L., Nguyen, T.Q., and Rao, K.R., GENLOT: Generalized linear-phase lapped orthogonal transforms, *IEEE Trans. Signal Process.*, 44, 497–507, Apr. 1996.
22. Rabbani, M. and Jones, P.W., *Digital Image Compression Techniques*, SPIE Optical Engineering Press, Bellingham, WA, 1991.
23. Rao, K.R. (Ed.), *Discrete Transforms and Their Applications*, Van Nostrand Reinhold, New York, 1985.
24. Rao, K.R. and Yip, P., *Discrete Cosine Transform: Algorithms, Advantages, Applications*, Academic Press, San Diego, CA, 1990.
25. Schiller, H., Overlapping block transform for image coding preserving equal number of samples and coefficients, *Proc. SPIE, Vis. Commn. Image Process.*, 1001, 834–839, 1988.
26. Soman, A.K., Vaidyanathan, P.P. and Nguyen, T.Q., Linear-phase paraunitary filter banks: Theory, factorizations and applications, *IEEE Trans. Signal Process.*, 41, 3480–3496, Dec. 1993.
27. Temerinac, M. and Edler, B., A unified approach to lapped orthogonal transforms, *IEEE Trans. Image Process.*, 1, 111–116, Jan. 1992.
28. Vaidyanathan, P.P., *Multirate Systems and Filter Banks*, Prentice-Hall, Englewood Cliffs, NJ, 1993.
29. Young, R.W. and Kingsbury, N.G., Frequency domain estimation using a complex lapped transform, *IEEE Trans. Image Process.*, 2, 2–17, Jan. 1993.

Index

A

B

D

E

F

G

H

I

J

K

L

M

N

O

P

S

T

U

V

W

Y

Z